W9-AFA-027

Tab it. Do it. Ace it.

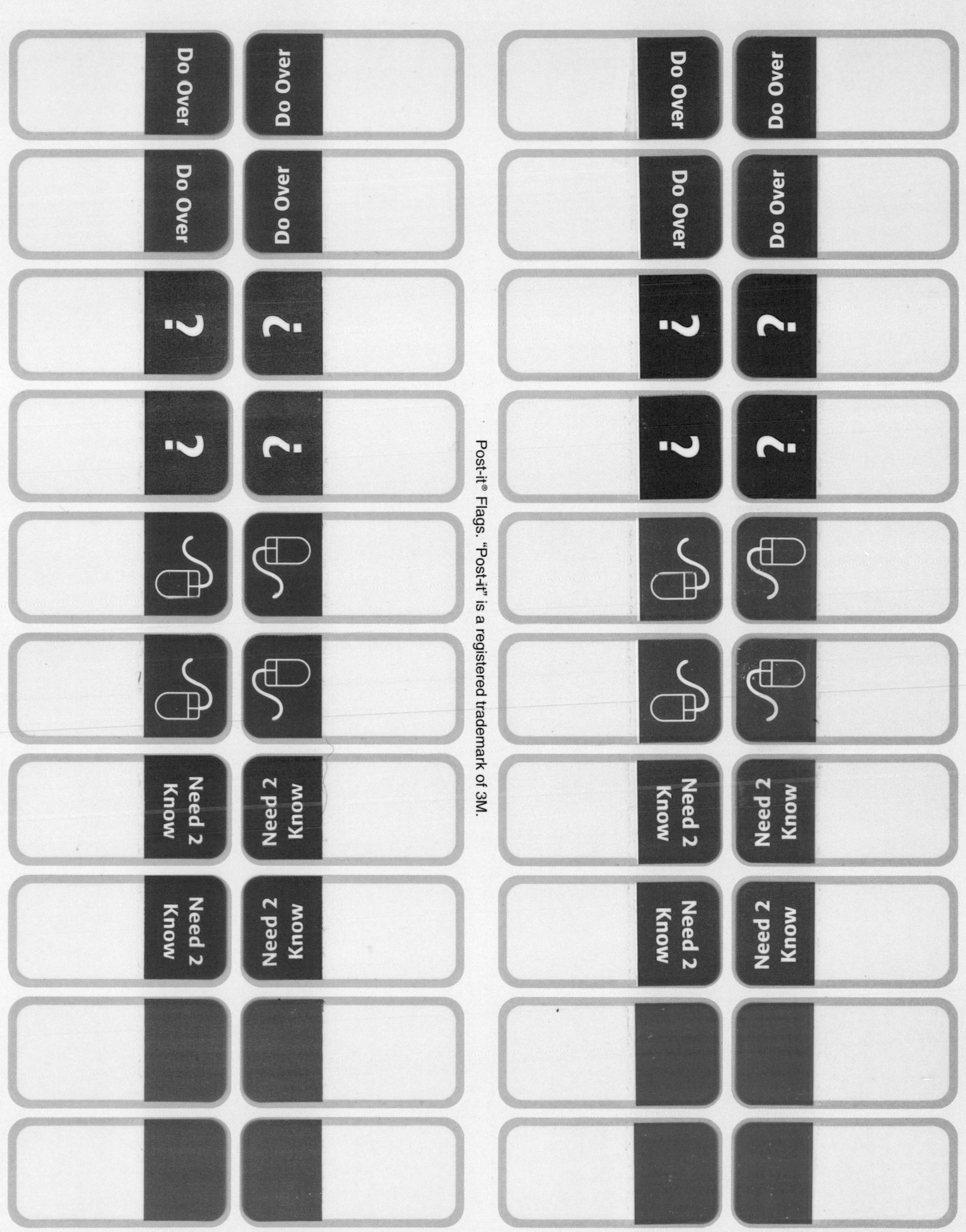

Tab it. Do it. Ace it.

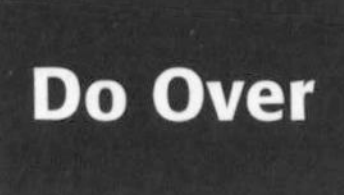

Do you need to review something? Try again? Work it out on your own after class? Tab it.

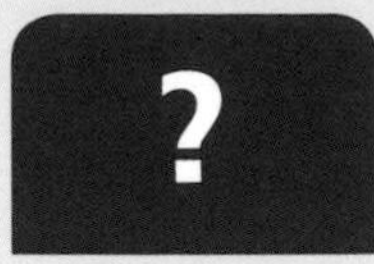

Got a question for office hours? Do you need to review an example on your own to get a full understanding? Do you need to look something up before moving on? Tab it.

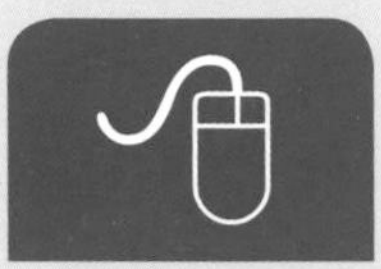

Do you need to see the video? Check out an online source? Complete your online homework? Tab it.

Need 2 Know

Is this going to be on the test? Need to mark a key formula? Do you need to memorize these steps? Tab it.

Do you have your own study system? Do you need to make a note? Do you want to express yourself? Tab it.

Tab it. Do it. Ace it.

ISBN 13: 978-0-495-55855-2
ISBN 10: 0-4955-5855-9

Digital Vision

EIGHTH EDITION

Intermediate Algebra

Richard N. Aufmann
Palomar College

Joanne S. Lockwood
Nashua Community College

BROOKS/COLE
CENGAGE Learning™

Australia • Brazil • Japan • Korea • Mexico • Singapore • Spain • United Kingdom • United States

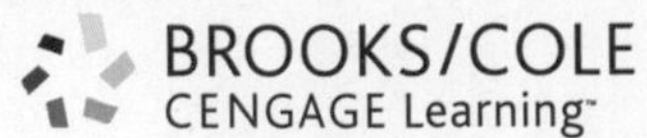

Intermediate Algebra, **Eighth Edition**
Richard N. Aufmann, Joanne S. Lockwood

Developmental Math Editor: Marc Bove

Development Editors: Erin Brown, Stefanie Beeck

Assistant Editor: Shaun Williams

Editorial Assistant: Zack Crockett

Media Editor: Heleny Wong

Marketing Manager: Gordon Lee

Marketing Coordinator: Shannon Maier

Marketing Communications Manager: Mary Anne Payumo

Sr. Content Project Manager: Tanya Nigh

Design Director: Rob Hugel

Sr. Art Director: Vernon Boes

Print Buyer: Karen Hunt

Rights Acquisitions Specialist: Tom McDonough

Production Service: Graphic World Inc.

Text Designer: Geri Davis

Photo Researcher: Chris Althof, Bill Smith Group

Text Researcher: Pablo d'Stair

Copy Editor: Jean Bermingham

Cover Designer: Lisa Henry

Cover Image: Kevin Twomey

Compositor: Graphic World Inc.

Library of Congress Control Number: 2011934064

Student Edition:

ISBN-13: 978-1-111-57949-4

ISBN-10: 1-111-57949-0

Loose-leaf Edition:

ISBN-13: 978-1-133-10373-8

ISBN-10: 1-133-10373-1

Brooks/Cole
20 Davis Drive
Belmont, CA 94002-3098
USA

Cengage Learning is a leading provider of customized learning solutions with office locations around the globe, including Singapore, the United Kingdom, Australia, Mexico, Brazil, and Japan. Locate your local office at **www.cengage.com/global.**

Cengage Learning products are represented in Canada by Nelson Education, Ltd.

To learn more about Brooks/Cole, visit **www.cengage.com/brooks/cole.**

Purchase any of our products at your local college store or at our preferred online store **www.CengageBrain.com.**

Printed in the United States of America
1 2 3 4 5 6 7 15 14 13 12 11

Contents

Chapter 10 Exponential and Logarithmic Functions 529

Chapter 11 Sequences and Series 581

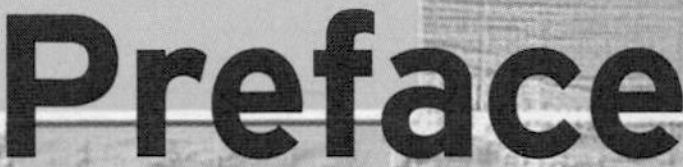

Preface

Digital Vision

Among the many questions we ask when we begin the process of revising a textbook, the most important is, "How can we improve the learning experience for the student?" We find answers to this question in a variety of ways but most commonly by talking to students and instructors and evaluating the written feedback we receive from our customers. As we set out to create the eighth edition of *Intermediate Algebra*, bearing in mind the feedback we received, our ultimate goal was to increase our *focus on the student.*

In the eighth edition, as in previous editions, popular features such as "Take Note" and "Point of Interest" have been retained. We have also retained the worked Examples and accompanying Problems, with complete worked-out solutions to the Problems given at the back of the textbook. New to this edition is the "Focus on Success" feature that appears at the beginning of each chapter. Focus on Success offers practical tips for improving study habits and performance on tests and exams.

Also new to the eighth edition are "How It's Used" boxes. These boxes present real-world scenarios that demonstrate the utility of selected concepts from the text. New "Focus On" examples offer detailed instruction on solving a variety of problems. "In the News" exercises are new application exercises appearing in many of the exercise sets. These exercises are based on newsworthy data and facts and are drawn from current events. The definition/key concept boxes have been enhanced in this edition; they now include examples to show how the general case translates to specific cases.

We trust that the new and enhanced features of the eighth edition will help students engage more successfully with the content. By narrowing the gap between the concrete and the abstract, between the real world and the theoretical, students should more plainly see that mastering the skills and topics presented is well within their reach and well worth the effort.

Updates to This Edition

- NEW! Chapter Openers have been revised and now include "Prep Tests" and "Focus on Success" vignettes.
- NEW! "Try Exercise" prompts are included at the end of each Example/Problem pair.
- NEW! "How It's Used" boxes are featured in each chapter.
- NEW! "Focus On" examples provide detailed instructions for solving problems.
- NEW! "Concept Check" exercises have been added to the beginning of each exercise set.
- NEW! "In the News" applications appear in many of the end-of-section exercise sets.
- NEW! "Projects or Group Activities" exercises are included at the end of each exercise set.
- Definition/key concept boxes have been enhanced with examples.
- Revised exercise sets include new applications.
- Improved Chapter Summaries now include a separate column containing an objective and page number for quick reference.

Organizational Changes

We have made the following changes, based on the feedback we received, in order to improve the effectiveness of the textbook and enhance the student's learning experience.

- Chapter 1 has been reorganized. Section 1.2 of the previous edition, *Operations on Rational Numbers*, is now separated into two sections. Section 1.2 focuses on operations with integers, and Section 1.3 focuses on operations with rational numbers.

  1.1 *Introduction to Real Numbers*
  1.2 *Operations on Integers*
  1.3 *Operations on Rational Numbers*
  1.4 *Variable Expressions*
  1.5 *Verbal Expressions and Variable Expressions*

  Objective 1.1.2, *Interval notation and set operations*, has been reorganized to create a better link between interval notation and set-builder notation. New examples have been added.

- Chapter 2 has been reorganized. Section 2.2 of the previous edition, *Coin, Stamp, and Integer Problems,* has been deleted, and Sections 2.3 through 2.6 have been renumbered.

  2.1 *Equations in One Variable*
  2.2 *Value Mixture and Motion Problems*
  2.3 *Applications: Problems Involving Percent*
  2.4 *Inequalities in One Variable*
  2.5 *Absolute Value Equations and Inequalities*

  In Section 2.4, the material on compound inequalities has been reorganized. A new example and new exercises covering the applications of inequalities have been added.

- In Chapter 3, two objectives from the previous edition—Objective 3.1.1, *Points on a rectangular coordinate system*, and Objective 3.1.2, *Find the length and midpoint of a line segment*—have been combined to form a new Objective 3.1.1, *Distance and midpoint formulas*. Objective 3.1.2 now covers graphing an equation in two variables. Objective 3.1.3 from the previous edition, *Graph a scatter diagram*, has been deleted. Objective 3.1.4, *Average rate of change*, has been moved to Section 3.4 as an application of the concept of slope.

- The material on the y-intercept in Section 3.3 of the previous edition has been moved to Section 3.4, *Slope of a Straight Line*. This change keeps all discussion of the equation $y = mx + b$ in one section. New exercises have been added.

- In Section 3.4, the approach to graphing equations using the slope and y-intercept has changed so that students are instructed first to move up or down from the y-intercept and then to move right or left to plot a second point.

- As suggested by reviewers, Section 5.5, previously titled *Factoring Polynomials*, has been separated into two sections. The revised Section 5.5, now called *Introduction to Factoring*, deals with factoring a monomial from a polynomial and factoring by grouping. The newly written Section 5.6, *Factoring Trinomials*, teaches students how to factor trinomials of the form $x^2 + bx + c$ and $ax^2 + bx + c$. The exercise sets have been modified accordingly.

- Chapter 8 has been reorganized. Nonlinear inequalities now appear in the last section.

  8.1 *Solving Quadratic Equations by Factoring or by Taking Square Roots*
  8.2 *Solving Quadratic Equations by Completing the Square and by Using the Quadratic Formula*
  8.3 *Equations That Are Reducible to Quadratic Equations*
  8.4 *Applications of Quadratic Equations*
  8.5 *Properties of Quadratic Functions*
  8.6 *Applications of Quadratic Functions*
  8.7 *Nonlinear Inequalities*

  Section 8.1 has been extensively revised. Objectives 8.1.1 (*Solve quadratic equations by factoring*) and 8.1.2 (*Write a quadratic equation given its solutions*) of the previous edition have been combined. New examples have been added, and the exercises have been revised.

  In Section 8.7, *Nonlinear Inequalities*, the answers to examples and problems are now given using both interval notation and set-builder notation.

Intermediate Algebra is organized around a carefully constructed hierarchy of OBJECTIVES. This "objective-based" approach provides an integrated learning environment that allows both the student and the instructor to easily find resources such as assessment tools (both within the text and online), videos, tutorials, and additional exercises.

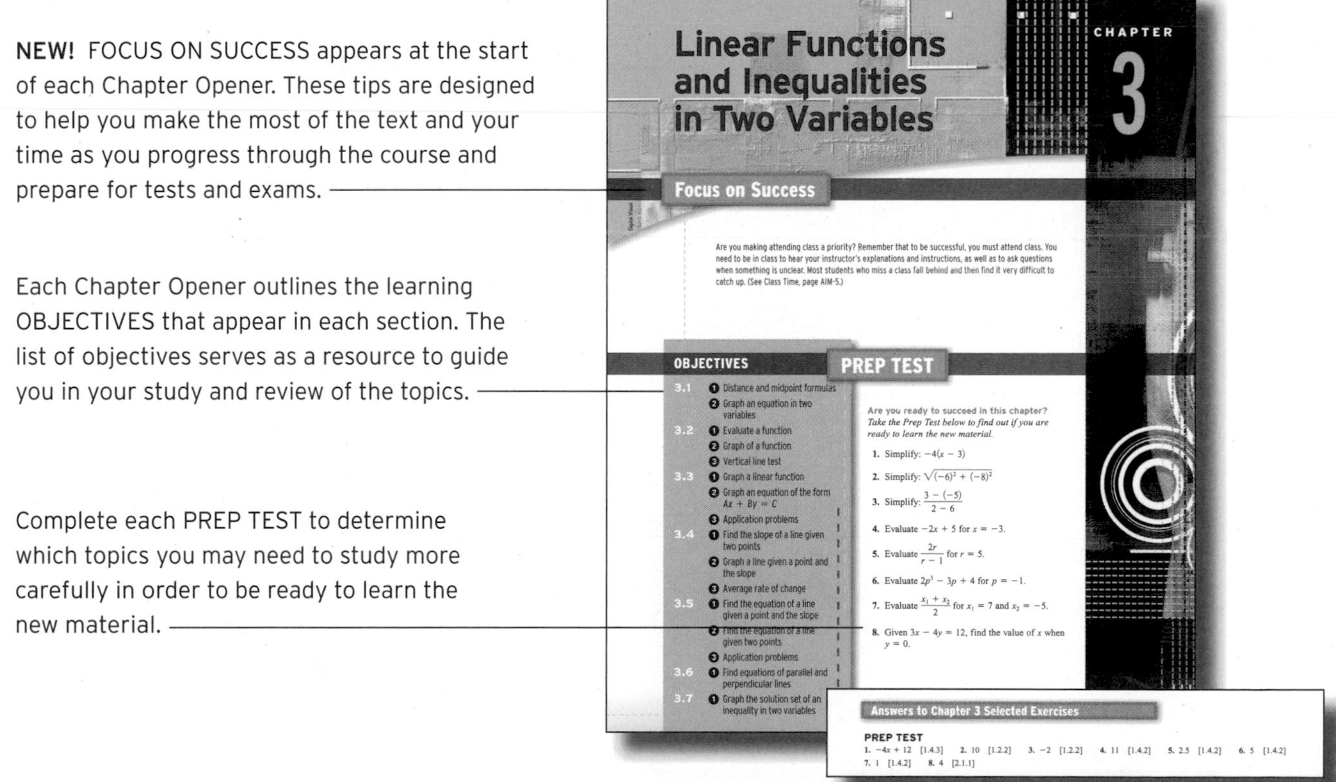

NEW! FOCUS ON SUCCESS appears at the start of each Chapter Opener. These tips are designed to help you make the most of the text and your time as you progress through the course and prepare for tests and exams.
Each Chapter Opener outlines the learning OBJECTIVES that appear in each section. The list of objectives serves as a resource to guide you in your study and review of the topics.
Complete each PREP TEST to determine which topics you may need to study more carefully in order to be ready to learn the new material.
Linear Functions and Inequalities in Two Variables
CHAPTER 3
Focus on Success
OBJECTIVES
PREP TEST
Answers to Chapter 3 Selected Exercises

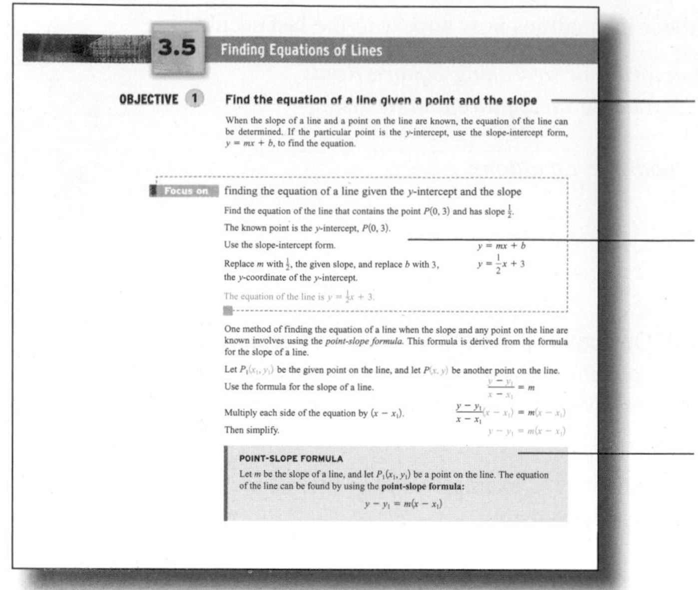

3.5 Finding Equations of Lines

OBJECTIVE 1 **Find the equation of a line given a point and the slope**

When the slope of a line and a point on the line are known, the equation of the line can be determined. If the particular point is the y-intercept, use the slope-intercept form, $y = mx + b$, to find the equation.

Focus on finding the equation of a line given the y-intercept and the slope

Find the equation of the line that contains the point $P(0, 3)$ and has slope $\frac{1}{2}$.

The known point is the y-intercept, $P(0, 3)$.

Use the slope-intercept form. $y = mx + b$

Replace m with $\frac{1}{2}$, the given slope, and replace b with 3, the y-coordinate of the y-intercept. $y = \frac{1}{2}x + 3$

The equation of the line is $y = \frac{1}{2}x + 3$.

One method of finding the equation of a line when the slope and any point on the line are known involves using the *point-slope formula.* This formula is derived from the formula for the slope of a line.

Let $P_1(x_1, y_1)$ be the given point on the line, and let $P(x, y)$ be another point on the line.

Use the formula for the slope of a line. $\frac{y - y_1}{x - x_1} = m$

Multiply each side of the equation by $(x - x_1)$. $\frac{y - y_1}{x - x_1}(x - x_1) = m(x - x_1)$

Then simplify. $y - y_1 = m(x - x_1)$

POINT-SLOPE FORMULA

Let m be the slope of a line, and let $P_1(x_1, y_1)$ be a point on the line. The equation of the line can be found by using the **point-slope formula:**

$$y - y_1 = m(x - x_1)$$

In every section, OBJECTIVE STATEMENTS introduce each new topic of discussion.

NEW! FOCUS ON boxes alert you to the specific type of problem you must master in order to succeed with the homework exercises or on a test. Each FOCUS ON problem is accompanied by detailed explanations for each step of the solution.

NEW! Many of the DEFINITION/KEY CONCEPTS boxes now contain examples to illustrate how each definition or key concept applies in practice.

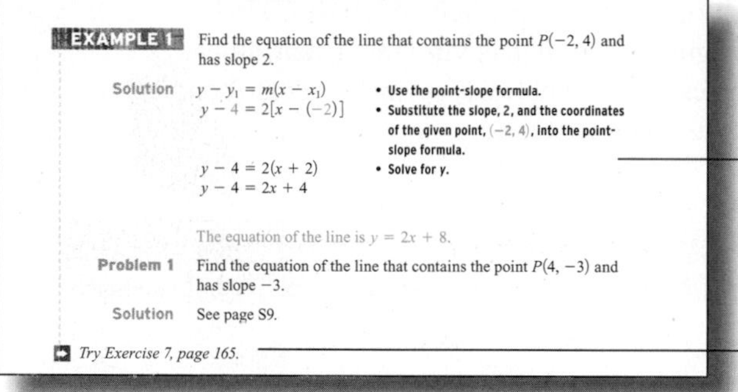

EXAMPLE 1 Find the equation of the line that contains the point $P(-2, 4)$ and has slope 2.

Solution
$y - y_1 = m(x - x_1)$ • Use the point-slope formula.
$y - 4 = 2[x - (-2)]$ • Substitute the slope, 2, and the coordinates of the given point, $(-2, 4)$, into the point-slope formula.
$y - 4 = 2(x + 2)$ • Solve for y.
$y - 4 = 2x + 4$

The equation of the line is $y = 2x + 8$.

Problem 1 Find the equation of the line that contains the point $P(4, -3)$ and has slope -3.

Solution See page S9.

Try Exercise 7, page 165.

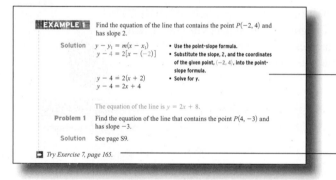

The EXAMPLE/PROBLEM matched pairs are designed to actively involve you in the learning process. The Problems are based on the Examples. They are paired so that you can easily refer to the steps in the Example as you work through the accompanying Problem.

NEW! TRY EXERCISE prompts are given at the end of each Example/Problem pair. They point you to a similar exercise at the end of the section. By following the prompts, you can immediately apply the techniques presented in worked Examples to homework exercises.

SECTION 3.5

Problem 1

$y - y_1 = m(x - x_1)$ • Use the point-slope formula.
$y - (-3) = -3(x - 4)$ • $m = -3$ and $(x_1, y_1) = (4, -3)$.
$y + 3 = -3x + 12$ • Solve for y.
$y = -3x + 9$

The equation of the line is $y = -3x + 9$.

Problem 2

$(x_1, y_1) = (2, 0)$ and $(x_2, y_2) = (5, 3)$.

$m = \frac{y_2 - y_1}{x_2 - x_1} = \frac{3 - 0}{5 - 2} = \frac{3}{3} = 1$ • Find the slope.

$y - y_1 = m(x - x_1)$ • Point-slope formula
$y - 0 = 1(x - 2)$ • Substitute the slope and the coordinates of P_1.
$y = x - 2$ • Solve for y.

The equation of the line is $y = x - 2$.

Complete WORKED-OUT SOLUTIONS to the Problems are found in an appendix at the back of the text. Compare your solution to the solution given in the appendix to obtain immediate feedback and reinforcement of the concept(s) you are studying.

Intermediate Algebra contains a WIDE VARIETY OF EXERCISES that promote skill building, skill maintenance, concept development, critical thinking, and problem solving.

NEW! CONCEPT CHECK exercises promote conceptual understanding. Completing these exercises will deepen your understanding of the topics in the section.

NEW! IN THE NEWS application exercises help you see the usefulness of mathematics in our everyday world. They are based on information culled from popular media sources, including newspapers, magazines, and the Internet.

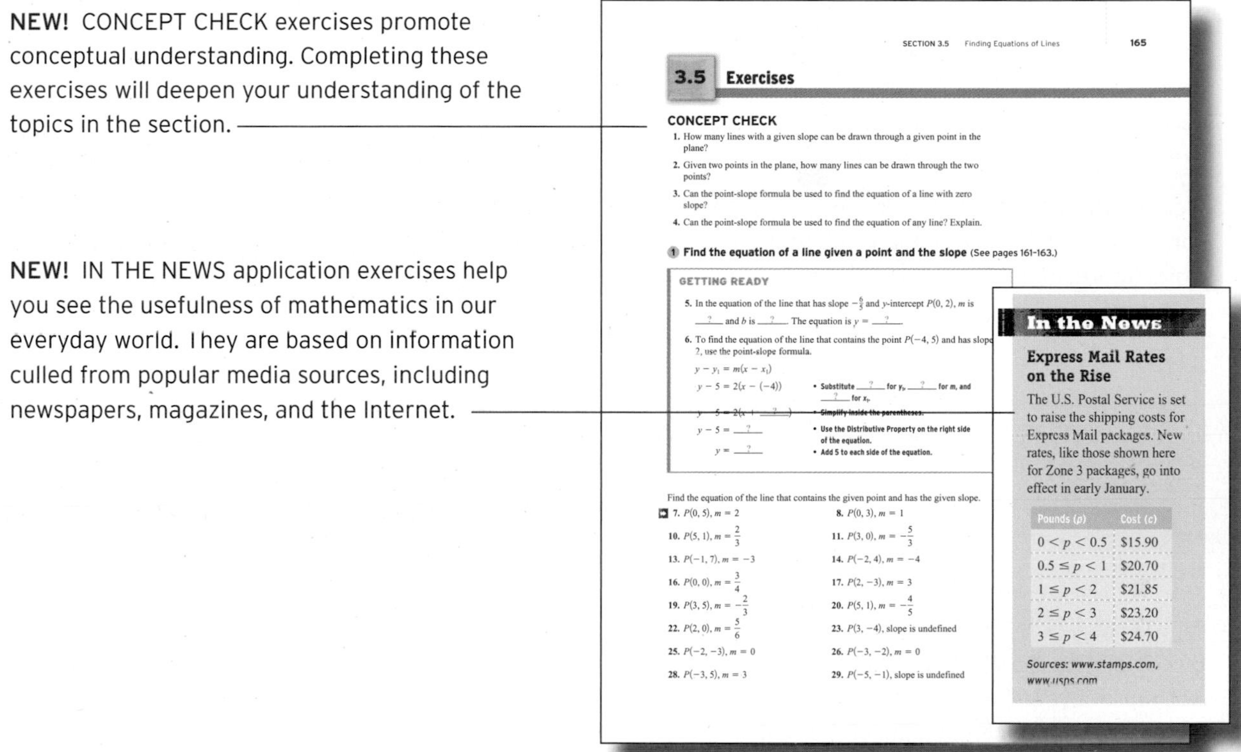
SECTION 3.5 Finding Equations of Lines 165

3.5 Exercises

CONCEPT CHECK

1. How many lines with a given slope can be drawn through a given point in the plane?

2. Given two points in the plane, how many lines can be drawn through the two points?

3. Can the point-slope formula be used to find the equation of a line with zero slope?

4. Can the point-slope formula be used to find the equation of any line? Explain.

1 Find the equation of a line given a point and the slope (See pages 161–163.)

GETTING READY

5. In the equation of the line that has slope $-\frac{6}{5}$ and y-intercept $P(0, 2)$, m is __?__ and b is __?__. The equation is $y =$ __?__.

6. To find the equation of the line that contains the point $P(-4, 5)$ and has slope 2, use the point-slope formula.

$y - y_1 = m(x - x_1)$

$y - 5 = 2(x - (-4))$ • Substitute __?__ for y_1, __?__ for m, and __?__ for x_1.

$y - 5 = 2(x +$ __?__$)$ • Simplify inside the parentheses.

$y - 5 =$ __?__ • Use the Distributive Property on the right side of the equation.

$y =$ __?__ • Add 5 to each side of the equation.

Find the equation of the line that contains the given point and has the given slope.

7. $P(0, 5)$, $m = 2$ 8. $P(0, 3)$, $m = 1$

10. $P(5, 1)$, $m = \frac{2}{3}$ 11. $P(3, 0)$, $m = -\frac{5}{3}$

13. $P(-1, 7)$, $m = -3$ 14. $P(-2, 4)$, $m = -4$

16. $P(0, 0)$, $m = \frac{3}{4}$ 17. $P(2, -3)$, $m = 3$

19. $P(3, 5)$, $m = -\frac{2}{3}$ 20. $P(5, 1)$, $m = -\frac{4}{5}$

22. $P(2, 0)$, $m = \frac{5}{6}$ 23. $P(3, -4)$, slope is undefined

25. $P(-2, -3)$, $m = 0$ 26. $P(-3, -2)$, $m = 0$

28. $P(-3, 5)$, $m = 3$ 29. $P(-5, -1)$, slope is undefined

In the News

Express Mail Rates on the Rise

The U.S. Postal Service is set to raise the shipping costs for Express Mail packages. New rates, like those shown here for Zone 3 packages, go into effect in early January.

Pounds (p)	Cost (c)
$0 < p < 0.5$	\$15.90
$0.5 \le p < 1$	\$20.70
$1 \le p < 2$	\$21.85
$2 \le p < 3$	\$23.20
$3 \le p < 4$	\$24.70

Sources: www.stamps.com, www.usps.com

GETTING READY exercises appear in most end-of-section exercise sets. These exercises provide guided practice and test your understanding of the underlying concepts in a lesson. They act as stepping stones to the remaining exercises for the objective.

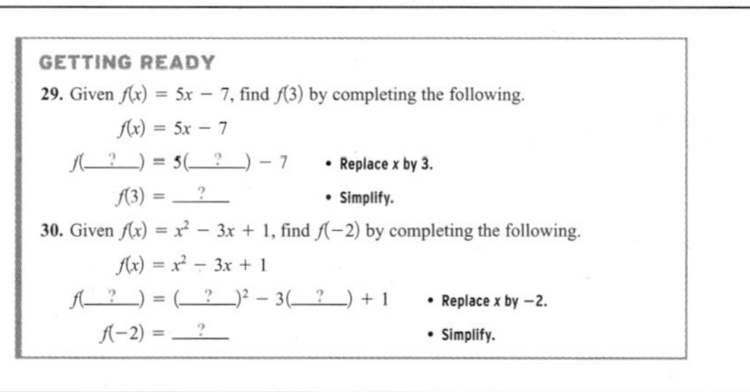
GETTING READY

29. Given $f(x) = 5x - 7$, find $f(3)$ by completing the following.

$f(x) = 5x - 7$

$f($__?__$) = 5($__?__$) - 7$ • Replace x by 3.

$f(3) =$ __?__ • Simplify.

30. Given $f(x) = x^2 - 3x + 1$, find $f(-2)$ by completing the following.

$f(x) = x^2 - 3x + 1$

$f($__?__$) = ($__?__$)^2 - 3($__?__$) + 1$ • Replace x by -2.

$f(-2) =$ __?__ • Simplify.

NEW! TRY EXERCISE icons are used to link exercises back to Examples from the section.

55. Evaluate $s(t) = -16t^2 + 48t$ when $t = 3$.

56. Evaluate $T(s) = s^2 - 4s + 1$ when $s = \frac{3}{4}$.

57. Evaluate $P(x) = 3x^3 - 4x^2 + 6x - 7$ when $x = 2$.

58. Evaluate $R(s) = s^3 - 2s^2 - 5s + 2$ when $s = -3$.

59. Evaluate $R(p) = \frac{3p}{2p - 3}$ when $p = -3$.

60. Evaluate $f(x) = \frac{x + 1}{3x - 1}$ when $x = 3$.

THINK ABOUT IT exercises promote conceptual understanding. Completing these exercises will deepen your understanding of the concept being addressed.

APPLYING CONCEPTS exercises may involve further exploration and analysis of topics, or they may integrate concepts introduced earlier in the text. **Optional** graphing calculator exercises are included, denoted by .

APPLYING CONCEPTS

69. For what values of k will the following system of equations be independent?

$$2x + 3y = 6$$
$$2x + ky = 9$$

70. If the following system of equations is inconsistent, how are the values of C and D related?

$$3x - 4y = C$$
$$3x - 4y = D$$

71. Suppose the following system of equations is an independent system of equations. What is the relationship between $\frac{a_1}{b_1}$ and $\frac{a_2}{b_2}$?

$$a_1x + b_1y = c_1$$
$$a_2x + b_2y = c_2$$

72. Suppose the following system of equations is a dependent or inconsistent system of equations. What is the relationship between $\frac{a_1}{b_1}$ and $\frac{a_2}{b_2}$?

$$a_1x + b_1y = c_1$$
$$a_2x + b_2y = c_2$$

Use a graphing utility to solve each of the following systems of equations. Round answers to the nearest hundredth.

73. $y = -\frac{1}{2}x + 2$, $y = 2x - 1$

74. $y = 1.2x + 2$, $y = -1.3x - 3$

75. $y = \sqrt{2}x - 1$, $y = -\sqrt{3}x + 1$

76. $y = \pi x - \frac{2}{3}$, $y = -x + \frac{\pi}{2}$

Working through the application exercises that contain REAL DATA will prepare you to use real-world information to answer questions and solve problems.

Fuel Economy Use the information in the article at the right for Exercises 29 and 30.

29. One week, the owner of a hybrid car drove 394 mi and spent \$34.74 on gasoline. How many miles did the owner drive in the city? On the highway?

30. Gasoline for one week of driving cost the owner of a hybrid car \$26.50. The owner would have spent \$51.50 for gasoline to drive the same number of miles in a traditional car. How many miles did the owner drive in the city? On the highway?

31. **Health Science** A pharmacist has two vitamin-supplement powders. The first powder is 25% vitamin B_1 and 15% vitamin B_2. The second is 15% vitamin B_1 and 20% vitamin B_2. How many milligrams of each of the two powders should the pharmacist use to make a mixture that contains 117.5 mg of vitamin B_1 and 120 mg of vitamin B_2?

32. **Chemistry** A chemist has two alloys, one of which is 10% gold and 15% lead and the other of which is 30% gold and 40% lead. How many grams of each of the two alloys should be used to make an alloy that contains 60 g of gold and 88 g of lead?

33. **Business** On Monday, a computer manufacturing company sent out three shipments. The first order, which contained a bill for \$114,000, was for 4 Model II, 6 Model VI, and 10 Model IX computers. The second shipment, which contained a bill for \$72,000, was for 8 Model II, 3 Model VI, and 5 Model IX computers. The third shipment, which contained a bill for \$81,000, was for 2 Model II, 9 Model VI, and 5 Model IX computers. What does the manufacturer charge for a Model VI computer?

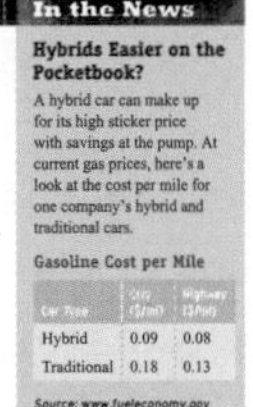
In the News

Hybrids Easier on the Pocketbook?

A hybrid car can make up for its high sticker price with savings at the pump. At current gas prices, here's a look at the cost per mile for one company's hybrid and traditional cars.

Gasoline Cost per Mile

Car Type	City (\$/mi)	Highway (\$/mi)
Hybrid	0.09	0.08
Traditional	0.18	0.13

Source: www.fueleconomy.gov

By completing the WRITING EXERCISES, you will improve your communication skills while increasing your understanding of mathematical concepts.

3.6 Exercises

CONCEPT CHECK

1. Given the slopes of two lines, explain how to determine whether the two lines are parallel.
2. Given the slopes of two lines, how can you determine whether the two lines are perpendicular?
3. Complete the following sentence. Parallel lines have the same __?__.
4. What is the negative reciprocal of $-\frac{3}{4}$?

NEW! PROJECTS OR GROUP ACTIVITIES appear at the end of each set of exercises. Your instructor may assign these individually, or you may be asked to work through the activities in groups.

PROJECTS OR GROUP ACTIVITIES

A set of points in a plane is a **convex set** if each line segment connecting a pair of points in the set is contained completely within the set.

35. Which of the following are convex sets?

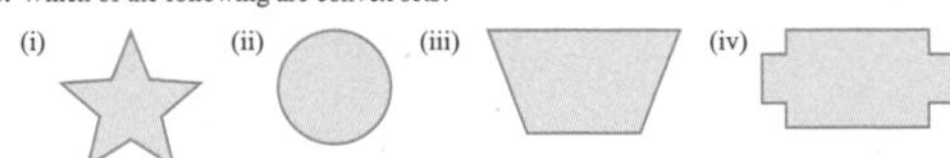

36. Graph the system of inequalities given below. Is the solution set a convex set?

$$x + y \le 10$$
$$2x + y \le 15$$
$$x \ge 0, y \ge 0$$

Intermediate Algebra addresses a broad range of study styles by offering a WIDE VARIETY OF TOOLS FOR REVIEW.

At the end of each chapter, you will find a SUMMARY outlining KEY WORDS and ESSENTIAL RULES AND PROCEDURES presented in the chapter. Each entry includes an objective-level reference and a page reference to show you where in the chapter the concept was introduced. An example demonstrating the concept is also included.

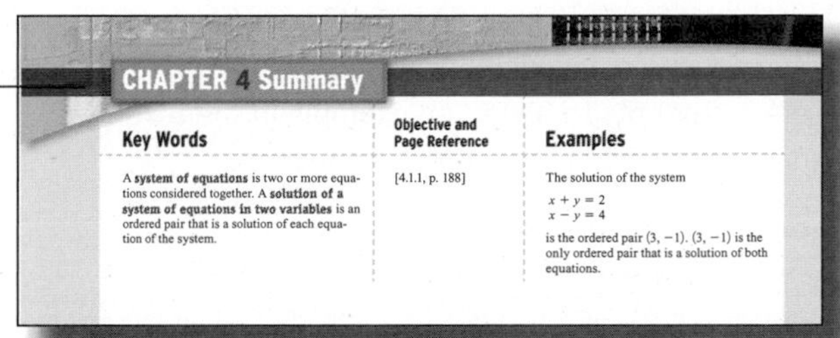

CHAPTER 4 Summary

Key Words	Objective and Page Reference	Examples
A **system of equations** is two or more equations considered together. A **solution of a system of equations in two variables** is an ordered pair that is a solution of each equation of the system.	[4.1.1, p. 188]	The solution of the system $x + y = 2$, $x - y = 4$ is the ordered pair $(3, -1)$. $(3, -1)$ is the only ordered pair that is a solution of both equations.

In the CHAPTER REVIEW EXERCISES, the order in which different types of problems appear is different from the order in which the topics were presented in the chapter. The ANSWERS to these exercises include references to the section objectives upon which the exercises are based. This will help you to quickly identify where to go to review a concept if you need more practice.

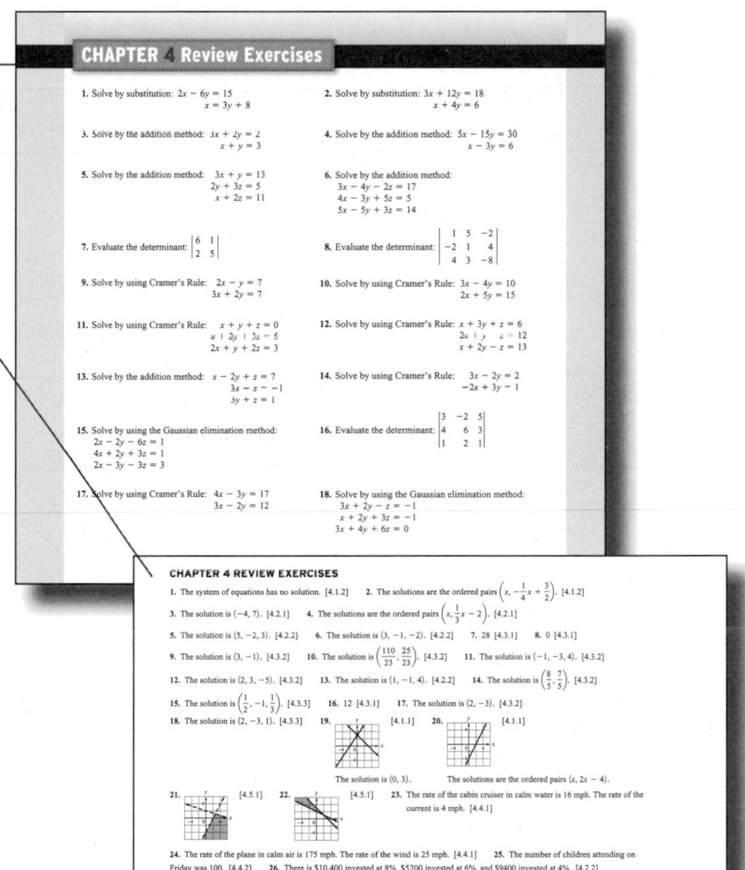

CHAPTER 4 Review Exercises

1. Solve by substitution: $2x - 6y = 15$, $x = 3y + 8$
2. Solve by substitution: $3x + 12y = 18$, $x + 4y = 6$
3. Solve by the addition method: $3x + 2y = 2$, $x + y = 3$
4. Solve by the addition method: $5x - 15y = 30$, $x - 3y = 6$
5. Solve by the addition method: $3x + y = 13$, $2y + 3z = 5$, $x + 2z = 11$
6. Solve by the addition method: $3x - 4y - 2z = 17$, $4x - 3y + 5z = 5$, $5x - 5y + 3z = 14$
7. Evaluate the determinant: $\begin{vmatrix} 6 & 1 \\ 2 & 5 \end{vmatrix}$
8. Evaluate the determinant: $\begin{vmatrix} 1 & 5 & -2 \\ -2 & 1 & 4 \\ 4 & 3 & -8 \end{vmatrix}$
9. Solve by using Cramer's Rule: $2x - y = 7$, $3x + 2y = 7$
10. Solve by using Cramer's Rule: $3x - 4y = 10$, $2x + 5y = 15$
11. Solve by using Cramer's Rule: $x + y + z = 0$, $x + 2y + 3z = 5$, $2x + y + 2z = 3$
12. Solve by using Cramer's Rule: $x + 3y + z = 6$, $2x + y - z = 12$, $x + 2y - z = 13$
13. Solve by the addition method: $x - 2y + z = 7$, $3x - z = -1$, $3y + z = 1$
14. Solve by using Cramer's Rule: $3x - 2y = 2$, $-2x + 3y = 1$
15. Solve by using the Gaussian elimination method: $2x - 2y - 6z = 1$, $4x + 2y + 3z = 1$, $2x - 3y - 3z = 3$
16. Evaluate the determinant: $\begin{vmatrix} 3 & -2 & 5 \\ 4 & 6 & 3 \\ 1 & 2 & 1 \end{vmatrix}$
17. Solve by using Cramer's Rule: $4x - 3y = 17$, $3x - 2y = 12$
18. Solve by using the Gaussian elimination method: $3x + 2y - z = -1$, $x + 2y + 3z = -1$, $3x + 4y + 6z = 0$

CHAPTER 4 REVIEW EXERCISES

1. The system of equations has no solution. [4.1.2] 2. The solutions are the ordered pairs $\left(x, -\frac{1}{4}x + \frac{3}{2}\right)$. [4.1.2] 3. The solution is $(-4, 7)$. [4.2.1] 4. The solutions are the ordered pairs $\left(x, \frac{1}{3}x - 2\right)$. [4.2.1] 5. The solution is $(5, -2, 3)$. [4.2.2] 6. The solution is $(3, -1, -2)$. [4.2.2] 7. 28 [4.3.1] 8. 0 [4.3.1] 9. The solution is $(3, -1)$. [4.3.2] 10. The solution is $\left(\frac{110}{23}, \frac{25}{23}\right)$. [4.3.2] 11. The solution is $(-1, -3, 4)$. [4.3.2] 12. The solution is $(2, 3, -5)$. [4.3.2] 13. The solution is $(1, -1, 4)$. [4.2.2] 14. The solution is $\left(\frac{8}{5}, \frac{7}{5}\right)$. [4.3.2] 15. The solution is $\left(\frac{1}{2}, -1, \frac{1}{3}\right)$. [4.3.3] 16. 12 [4.3.1] 17. The solution is $(2, -3)$. [4.3.2] 18. The solution is $(2, -3, 1)$. [4.3.3] 19. [4.1.1] The solution is $(0, 3)$. 20. [4.1.1] The solutions are the ordered pairs $(x, 2x - 4)$. 21. [4.5.1] 22. [4.5.1] 23. The rate of the cabin cruiser in calm water is 16 mph. The rate of the current is 4 mph. [4.4.1] 24. The rate of the plane in calm air is 175 mph. The rate of the wind is 25 mph. [4.4.1] 25. The number of children attending on Friday was 100. [4.4.2] 26. There is $10,400 invested at 8%, $5200 invested at 6%, and $9400 invested at 4%. [4.2.2]

Each CHAPTER TEST is designed to simulate a typical test of the concepts covered in the chapter. The ANSWERS include references to section objectives. Also provided is a reference to an Example, Problem, or Focus On, which refers students to a worked example in the text that is similar to the given test question.

CHAPTER 4 Test

1. Solve by substitution: $3x + 2y = 4$, $x = 2y - 1$
2. Solve by substitution: $5x + 2y = -23$, $2x + y = -10$
3. Solve by substitution: $y = 3x - 7$, $y = -2x + 3$
4. Solve by using the Gaussian elimination method: $3x + 4y = -2$, $2x + 5y = 1$
5. Solve by the addition method: $4x - 6y = 5$, $6x - 9y = 4$
6. Solve by the addition method: $3x - y = 2x + y - 1$, $5x + 2y = y + 6$
7. Solve by the addition method: $2x + 4y - z = 3$, $x + 2y + z = 5$, $4x + 8y - 2z = 7$
8. Solve by using the Gaussian elimination method: $x - y - z = 5$, $2x + z = 2$, $3y - 2z = 1$

CHAPTER 4 TEST

1. The solution is $\left(\frac{3}{4}, \frac{7}{8}\right)$. [4.1.2, Example 3A] 2. The solution is $(-3, -4)$. [4.1.2, Focus On, page 191]
3. The solution is $(2, -1)$. [4.1.2, Example 3A] 4. The solution is $(-2, 1)$. [4.3.3, Example 7]
5. The system of equations has no solution. [4.2.1, Problem 1B] 6. The solution is $(1, 1)$. [4.2.1, Examples 1 and 2]
7. The system of equations has no solution. [4.2.2, Focus On, part B, page 200] 8. The solution is $(2, -1, -2)$. [4.3.3, Example 8]

CUMULATIVE REVIEW EXERCISES, which appear at the end of each chapter (beginning with Chapter 2), help you maintain the skills you learned previously. The ANSWERS include references to the section objectives upon which the exercises are based.

Cumulative Review Exercises

1. Solve: $\frac{3}{2}x - \frac{3}{8} + \frac{1}{4}x = \frac{7}{12}x - \frac{5}{6}$
2. Find the equation of the line that contains the points $P_1(2, -1)$ and $P_2(3, 4)$.
3. Simplify: $3[x - 2(5 - 2x) - 4x] + 6$
4. Evaluate $a + bc \div 2$ when $a = 4$, $b = 8$, and $c = -2$.
5. Solve: $2x - 3 < 9$ or $5x - 1 < 4$
6. Solve: $|x - 2| - 4 < 2$
7. Solve: $|2x - 3| > 5$
8. Given $f(x) = 3x^3 - 2x^2 + 1$, evaluate $f(-3)$.
9. Find the range of $f(x) = 3x^2 - 2x$ if the domain is $\{-2, -1, 0, 1, 2\}$.
10. Given $F(x) = x^2 - 3$, find $F(2)$.
11. Given $f(x) = 3x - 4$, write $f(2 + h) - f(2)$ in simplest form.
12. Graph: $\{x|x \le 2\} \cap \{x|x > -3\}$
13. Find the equation of the line that contains the point $P(-2, 3)$ and has slope $-\frac{2}{3}$.
14. Find the equation of the line that contains the point $P(-1, 2)$ and is perpendicular to the graph of $2x - 3y = 7$.

CUMULATIVE REVIEW EXERCISES

1. The solution is $-\frac{11}{28}$. [2.1.1] 2. The equation is $y = 5x - 11$. [3.5.2] 3. $3x - 24$ [1.4.3] 4. -4 [1.4.2]
5. The solution set is $\{x|x < 6\}$. [2.4.2] 6. The solution set is $\{x| -4 < x < 8\}$. [2.5.2]
7. The solution set is $\{x|x > 4\} \cup \{x|x < -1\}$. [2.5.2] 8. -98 [3.2.1]
9. The range is $\{0, 1, 5, 8, 16\}$. [3.2.1] 10. 1 [3.2.1] 11. $3h$ [3.2.1] 12. [1.1.2]
13. The equation is $y = -\frac{2}{3}x + \frac{5}{3}$. [3.5.1] 14. The equation is $y = -\frac{3}{2}x + \frac{1}{2}$. [3.6.1] 15. The distance is $2\sqrt{10}$. [3.1.1]
16. The coordinates of the midpoint are $\left(-\frac{1}{2}, 4\right)$. [3.1.1] 17. [3.4.2] 18. [3.7.1]
19. The solution is $(-5, -11)$. [4.1.2] 20. The solution is $(1, 0, -1)$. [4.2.2] 21. 3 [4.3.1] 22. [4.1.1]
The solution is $(2, 0)$.

A FINAL EXAM is included after the last chapter of the text. It is designed to simulate a comprehensive exam covering all the concepts presented in the text. The ANSWERS to the final exam questions are provided in the appendix at the back of the text and include references to the section objectives upon which the questions are based.

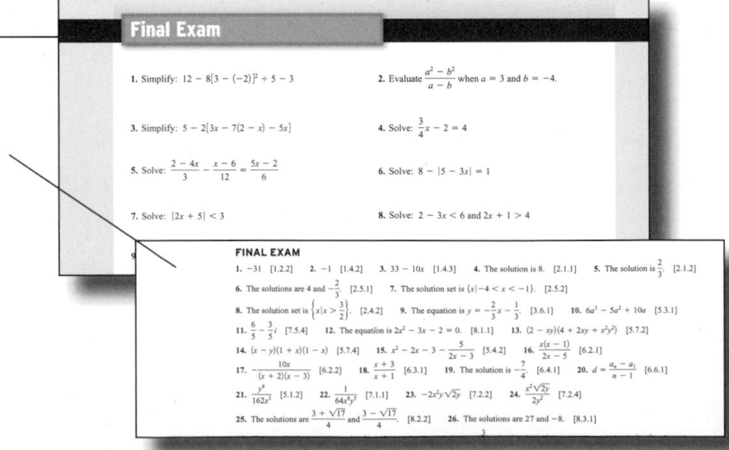

Final Exam

1. Simplify: $12 - 8[3 - (-2)]^2 \div 5 - 3$
2. Evaluate $\frac{a^2 - b^2}{a - b}$ when $a = 3$ and $b = -4$.
3. Simplify: $5 - 2[3x - 7(2 - x) - 5x]$
4. Solve: $\frac{3}{4}x - 2 = 4$
5. Solve: $\frac{2 - 4x}{3} - \frac{x - 6}{12} = \frac{5x - 2}{6}$
6. Solve: $8 - |5 - 3x| = 1$
7. Solve: $|2x + 5| < 3$
8. Solve: $2 - 3x < 6$ and $2x + 1 > 4$

FINAL EXAM

1. -31 [1.2.2] 2. -1 [1.4.2] 3. $33 - 10x$ [1.4.3] 4. The solution is 8. [2.1.1] 5. The solution is $\frac{2}{3}$. [2.1.2]
6. The solutions are 4 and $-\frac{2}{3}$. [2.5.1] 7. The solution set is $\{x| -4 < x < -1\}$. [2.5.2]
8. The solution set is $\left\{x|x > \frac{3}{2}\right\}$. [2.4.2] 9. The equation is $y = -\frac{2}{3}x - \frac{1}{3}$. [3.6.1] 10. $6a^3 - 5a^2 + 10a$ [5.3.1]
11. $\frac{6}{5} - \frac{3}{5}i$ [7.5.4] 12. The equation is $2x^2 - 3x - 2 = 0$. [8.1.1] 13. $(2 - xy)(4 + 2xy + x^2y^2)$ [5.7.2]
14. $(x - y)(1 + x)(1 - x)$ [5.7.4] 15. $x^2 - 2x - 3 - \frac{5}{2x - 3}$ [5.4.2] 16. $\frac{x(x - 1)}{2x - 5}$ [6.2.1]
17. $-\frac{10x}{(x + 2)(x - 3)}$ [6.2.2] 18. $\frac{x + 3}{x + 1}$ [6.3.1] 19. The solution is $-\frac{7}{4}$. [6.4.1] 20. $d = \frac{a_n - a_1}{n - 1}$ [6.6.1]
21. $\frac{y^6}{162x^3}$ [5.1.2] 22. $\frac{1}{64x^3y^3}$ [7.1.1] 23. $-2x^2y\sqrt{2y}$ [7.2.2] 24. $\frac{x^2\sqrt{2y}}{2y^2}$ [7.2.4]
25. The solutions are $\frac{3 + \sqrt{17}}{4}$ and $\frac{3 - \sqrt{17}}{4}$. [8.2.2] 26. The solutions are 27 and -8. [8.3.1]

Other Key Features

MARGINS Within the margins, you can find the following features.

> **Take Note**
> The order of the elements in a set does not matter. For instance, in example (3) at the right, we could have written the domain as $\{4, -5, -1, -7\}$. However, it

TAKE NOTE boxes alert you to concepts that require special attention.

> **Point of Interest**

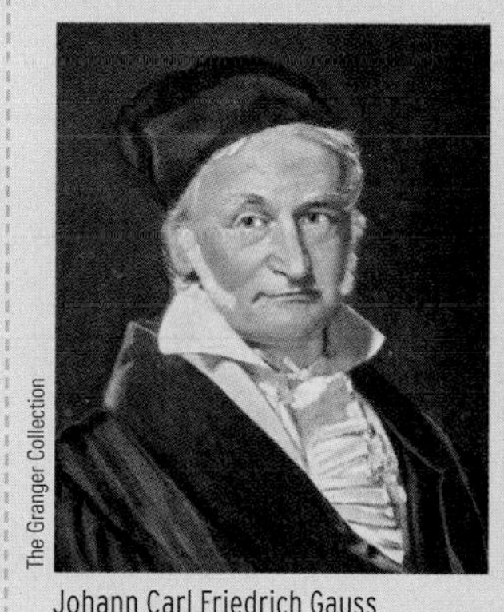

The Granger Collection

> Johann Carl Friedrich Gauss (1777–1855) is considered one of

POINT OF INTEREST boxes, which relate to the topic under discussion, may be historical in nature or may be of general interest.

> **How It's Used**
> Large systems of linear inequalities containing over 100 inequalities have been used to solve application problems in such diverse areas as providing health care and hardening a nuclear missile silo.

NEW! HOW IT'S USED boxes relate to the topic under discussion. These boxes present real-world scenarios that demonstrate the utility of selected concepts from the text.

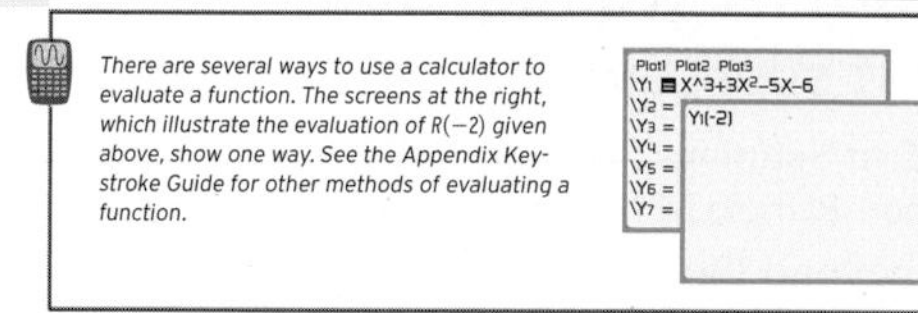

There are several ways to use a calculator to evaluate a function. The screens at the right, which illustrate the evaluation of $R(-2)$ given above, show one way. See the Appendix Keystroke Guide for other methods of evaluating a function.

While the text is not dependent on the use of a calculator, TECHNOLOGY boxes that focus on calculator instruction are included for selected topics. The boxes contain tips for using a graphing calculator.

PROBLEM-SOLVING STRATEGIES The problem-solving approach used throughout the text emphasizes the importance of a well-defined strategy. Model strategies are presented as guides for you to follow as you attempt the parallel Problem that accompanies each numbered Example.

> **STRATEGY** FOR SOLVING A VALUE MIXTURE PROBLEM
> ► For each ingredient in the mixture, write a numerical or variable expression for the amount of the ingredient used, the unit cost of the ingredient, and the value of the amount used. For the mixture, write a numerical or variable expression for the amount, the unit cost of the mixture, and the value of the amount. The results can be recorded in a table.

Instructor Resources

PRINT SUPPLEMENTS

Annotated Instructor's Edition (ISBN: 978-1-111-82696-3)
The Annotated Instructor's Edition features answers to all of the problems in the text, as well as an appendix denoting those problems that can be found in Enhanced WebAssign.

Instructor's Solutions Manual (ISBN: 978-1-133-11236-5)
Author: Patricia M. Parker, *Germanna Community College*
The Instructor's Solutions Manual provides worked-out solutions to all of the problems in the text.

Instructor's Resource Binder with Appendix
(ISBN: 978-1-133-11254-9)
Author: Instructor's Resource Binder by Maria H. Andersen, *Muskegon Community College,* with Appendices by Richard N. Aufmann, *Palomar College,* and Joanne S. Lockwood, *Nashua Community College*
Each section of the main text is discussed in uniquely designed Teaching Guides that contain tips, examples, activities, worksheets, overheads, assessments, and solutions to all worksheets and activities.

ELECTRONIC SUPPLEMENTS

Text-Specific Videos
Author: Dana Mosely
These text-specific instructional videos provide students with visual reinforcement of concepts and explanations. The videos contain easy-to-understand language along with detailed examples and sample problems. A flexible format offers versatility. Topics can be accessed quickly, and lectures can be catered to self-paced, online, or hybrid courses. Closed captioning is provided for the hearing impaired. These videos are available through Enhanced WebAssign and CourseMate.

PowerLecture with Diploma® (ISBN: 978-1-133-11235-8)
This CD-ROM provides you with dynamic media tools for teaching. You can create, deliver, and customize tests (both print and online) in minutes with Diploma's Computerized Testing featuring algorithmic equations. The Solution Builder's online solutions manual easily builds solution sets for homework or exams. Practice Sheets, First-Day-of-Class PowerPoint® lecture slides, art and figures from the book, and a test bank in electronic format are also included on this CD-ROM.

Syllabus Creator (Included on the PowerLecture)
Author: Richard N. Aufmann, *Palomar College,* and Joanne S. Lockwood, *Nashua Community College*
NEW! Easily write, edit, and update your syllabus with the Aufmann/Lockwood Syllabus Creator. This software program enables you to create your new syllabus in several easy steps: first select the required course objectives; then add your contact information, course information, student expectations, grading policy, dates and location of your course, and course outline. You now have your syllabus!

Solution Builder
This online instructor database offers complete worked solutions to all exercises in the text, allowing you to create customized, secure solutions printouts (in PDF format) matched exactly to the problems you assign in class. For more information, visit www.cengage.com/solutionbuilder.

Enhanced WebAssign® (ISBN: 978-0-538-73810-1)
Exclusively from Cengage Learning, Enhanced WebAssign combines the exceptional mathematics content that you know and love with the most powerful online homework solution, WebAssign. Enhanced WebAssign engages students with immediate feedback and rich tutorial content. Interactive eBooks help students develop a deeper conceptual understanding of their subject matter. Online assignments can be built by selecting from thousands of text-specific problems. Assignments can be supplemented with problems from any Cengage Learning textbook.

Enhanced WebAssign: Start Smart Guide for Students
(ISBN: 978-0-495-38479-3)
Author: Brooks/Cole
The Enhanced WebAssign Student Start Smart Guide helps students get up and running quickly with Enhanced WebAssign so that they can study smarter and improve their performance in class.

Printed Access Card for CourseMate with eBook
(ISBN: 978-1-4282-7616-1)

Instant Access Card for CourseMate with eBook
(ISBN: 978-1-4282-7615-4)
Complement your text and course content with study and practice materials. Cengage Learning's Developmental Mathematics CourseMate brings course concepts to life with interactive learning, study, and exam preparation tools that support the printed textbook. Watch student comprehension soar as your class works with the printed textbook and the textbook-specific website. Developmental Mathematics CourseMate goes beyond the book to deliver what you need!

Student Resources

PRINT SUPPLEMENTS

Student Solutions Manual (ISBN: 978-1-133-11237-2)
Author: Patricia M. Parker, *Germanna Community College*
Go beyond the answers—and improve your grade! This manual provides worked-out, step-by-step solutions to the odd-numbered problems in the text. The Student Solutions Manual gives you the information you need to truly understand how the problems are solved.

Student Workbook (ISBN: 978-1-133-11239-6)
Author: Maria H. Andersen, *Muskegon Community College*
Get a head start. The Student Workbook contains assessments, activities, and worksheets for classroom discussions, in-class activities, and group work.

AIM for Success Student Practice Sheets (ISBN: 978-1-133-11238-9)
Author: Christine S. Verity
AIM for Success Student Practice Sheets provide additional practice problems to help you learn the material.

ELECTRONIC SUPPLEMENTS

Text-Specific Videos
Author: Dana Mosely
These text-specific instructional videos provide you with visual reinforcement of concepts and explanations. The videos contain easy-to-understand language along with detailed examples and sample problems. A flexible format offers versatility. Topics can be accessed quickly, and lectures can be catered to self-paced, online, or hybrid courses. Closed captioning is provided for the hearing impaired. These videos are available through Enhanced WebAssign and CourseMate.

Enhanced WebAssign (ISBN: 978-0-538-73810-1)
Enhanced WebAssign (assigned by the instructor) provides instant feedback on homework assignments. This online homework system is easy to use and includes helpful links to textbook sections, video examples, and problem-specific tutorials.

Chapter Test Videos
(Available through Enhanced WebAssign)
Available through Enhanced WebAssign, the chapter test videos provide step-by-step solutions that follow the problem-solving methods used in the text for every end-of-chapter text question. Some solution videos feature interactive questions that provide immediate feedback on your answers.

Enhanced WebAssign: Start Smart Guide for Students
(ISBN: 978-0-495-38479-3)
Author: Brooks/Cole
If your instructor has chosen to package Enhanced WebAssign with your text, this manual will help you get up and running quickly with the Enhanced WebAssign system so that you can study smarter and improve your performance in class.

Printed Access Card for CourseMate with eBook
(ISBN: 978-1-4282-7616-1)

Instant Access Card for CourseMate with eBook
(ISBN: 978-1-4282-7615-4)
The more you study, the greater your success. You can make the most of your study time by accessing everything you need to succeed in one place—online with CourseMate. You can use CourseMate to read the textbook, take notes, review flashcards, watch videos, and take practice quizzes.

Acknowledgments

The authors would like to thank the people who have reviewed the seventh edition and provided many valuable suggestions.

Dimos Arsenidis, *California State University–Long Beach*
Peter Arvanites, *Rockland Community College*
Yugal Behl, *Central New Mexico Community College*
Oiyin Pauline Chow, *Central Pennsylvania Community College*
Mark Harbison, *Sacramento Community College*
Brooke Quinlan, *Hillsborough Community College*
Jean Shutters, *Central Pennsylvania Community College*
Lynn Vazquez, *Ocean County College*
Thomas Edward Wells, *Delta College*
Judith Wood, *Central Florida Community College*
Cathleen M. Zucco-Teveloff, *Rowan University*

Special thanks go to Jean Bermingham for copyediting the manuscript and proofreading the pages, to Patricia M. Parker for preparing the solutions manuals, and to Lauri Semarne for her work in ensuring the accuracy of the text. We would also like to thank the many people at Cengage Learning who worked to guide the manuscript for the eighth edition from development through production.

AIM for Success

CHAPTER

A

Focus on Success

Digital Vision

This important chapter describes study skills that are used by students who have been successful in this course. Chapter A covers a wide range of topics that focus on what you need to do to succeed in this class. It includes a complete guide to the textbook and how to use its features to become a successful student.

OBJECTIVES

A.1
- Get ready
- Motivate yourself
- Develop a "can do" attitude toward math
- Strategies for success
- Time management
- Habits of successful students

A.2
- Get the big picture
- Understand the organization
- Use the interactive method
- Use a strategy to solve word problems
- Ace the test

PREP TEST

Are you ready to succeed in this course?

1. Read this chapter. Answer all of the questions. Write down your answers on paper.
2. Write down your instructor's name.
3. Write down the classroom number.
4. Write down the days and times the class meets.
5. Bring your textbook, a notebook, and a pen or pencil to every class.
6. Be an active participant, not a passive observer.

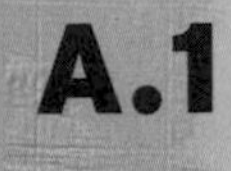

A.1 How to Succeed in This Course

GET READY

We are committed to your success in learning mathematics and have developed many tools and resources to support you along the way.

DO YOU WANT TO EXCEL IN THIS COURSE?

Read on to learn about the skills you'll need and how best to use this book to get the results you want.

We have written this text in an *interactive* style. More about this later but, in short, this means that you are supposed to interact with the text. Do not just read the text! Work along with it. Ready? Let's begin!

WHY ARE YOU TAKING THIS COURSE?

Did you interact with the text, or did you just read the last question? Get some paper and a pencil or pen and answer the question. Really—you will have more success in math and other courses you take if you **actively participate.** Now, **interact.** Write down one reason you are taking this course.

____

____

Of course, we have no idea what you just wrote, but experience has shown us that many of you wrote something along the lines of "I have to take it to graduate" or "It is a prerequisite to another course I have to take" or "It is required for my major." Those reasons are perfectly fine. Every teacher has had to take courses that were not directly related to his or her major.

WHY DO YOU WANT TO SUCCEED IN THIS COURSE?

Think about why you want to succeed in this course. List the reasons here (not in your head . . . on the paper!):

____

____

One reason you may have listed is that math skills are important in order to be successful in your chosen career. That is certainly an important reason. Here are some other reasons.

- Math is a skill that applies across careers, which is certainly a benefit in our world of changing job requirements. A good foundation in math may enable you to more easily make a career change.
- Math can help you learn critical thinking skills, an attribute all employers want.
- Math can help you see relationships between ideas and identify patterns.

MOTIVATE YOURSELF

You'll find many real-life problems in this book, relating to sports, money, cars, music, and more. We hope that these topics will help you understand how mathematics is used in everyday life. To learn all of the necessary skills and to understand how you can apply them to your life outside of this course, **motivate yourself to learn.**

One of the reasons we asked you why you are taking this course was to provide motivation for you to succeed. When there is a reason to do something, that task is easier to accomplish. We understand that you may not want to be taking this course but, to achieve your career goal, this is a necessary step. Let your career goal be your motivation for success.

Take Note
Motivation alone won't lead to success. For example, suppose a person who cannot swim is rowed out to the middle of a lake and thrown overboard. That person has a lot of motivation to swim, but most likely will drown without some help. You'll need motivation *and* learning in order to succeed.

MAKE THE COMMITMENT TO SUCCEED!

With practice, you will improve your math skills. Skeptical? Think about when you first learned to drive a car, ride a skateboard, dance, paint, surf, or any other talent that you now have. You may have felt self-conscious or concerned that you might fail. But with time and practice, you learned the skill.

List a situation in which you accomplished your goal by spending time practicing and perfecting your skills (such as learning to play the piano or to play basketball):

____

____

You do not get "good" at something by doing it once a week. **Practice** is the backbone of any successful endeavor—including math!

DEVELOP A "CAN DO" ATTITUDE TOWARD MATH

You can do math! When you first learned the skills you just listed above, you may not have done them well. With practice, you got better. With practice, you will get better at math. Stay focused, motivated, and committed to success.

We cannot emphasize enough how important it is to overcome the "I Can't Do Math" syndrome. If you listen to interviews of very successful athletes after a particularly bad performance, you will note that they focus on the positive aspects of what they did, not the negative. Sports psychologists encourage athletes always to be positive—to have a "can do" attitude. Develop this attitude toward math and you will succeed.

Change your conversation about mathematics. Do not say "I can't do math," "I hate math," or "Math is too hard." These comments just give you an excuse to fail. You don't want to fail, and we don't want you to fail. Write it down now: **I can do math!**

____

____

STRATEGIES FOR SUCCESS

PREPARE TO SUCCEED

There are a number of things that may be worrisome to you as you begin a new semester. List some of those things now.

Here are some of the concerns expressed by our students.

- **Tuition**
  Will I be able to afford school?
- **Job**
  I must work. Will my employer give me a schedule that will allow me to go to school?
- **Anxiety**
  Will I succeed?
- **Child care**
  What will I do with my kids while I'm in class or when I need to study?
- **Time**
  Will I be able to find the time to attend class and study?
- **Degree goals**
  How long will it take me to finish school and earn my degree?

These are all important and valid concerns. Whatever your concerns, acknowledge them. Choose an education path that allows you to accommodate your concerns. Make sure they don't prevent you from succeeding.

SELECT A COURSE

Many schools offer math assessment tests. These tests evaluate your present math skills. They don't evaluate how smart you are, so don't worry about your score on the test. If you are unsure about where you should start in the math curriculum, these tests can show you where to begin. You are better off starting at a level that is appropriate for you than starting a more advanced class and then dropping it because you can't keep up. Dropping a class is a waste of time and money.

If you have difficulty with math, avoid short courses that compress the class into a few weeks. If you have struggled with math in the past, this environment does not give you the time to process math concepts. Similarly, avoid classes that meet once a week. The time delay between classes makes it difficult to make connections between concepts.

Some career goals require a number of math courses. If that is true of your major, try to take a math course every semester until you complete the requirements. Think about it this way. If you take, say, French I, and then wait two semesters before taking French II, you may forget a lot of material. Math is much the same. You must keep the concepts fresh in your mind.

TIME MANAGEMENT

One of the most important requirements in completing any task is to acknowledge the amount of time it will take to finish the job successfully. Before a construction company starts to build a skyscraper, the company spends months looking at how much time each of the phases of construction will take. This is done so that resources can be allocated when appropriate. For instance, it would not make sense to schedule the electricians to run wiring until the walls are up.

MANAGE YOUR TIME!

We know how busy you are outside of school. Do you have a full-time or a part-time job? Do you have children? Do you visit your family often? Do you play school sports or participate in the school orchestra or theater company? It can be stressful to balance all of the important activities and responsibilities in your life. Creating a time management plan will help you schedule enough time to do everything you need to do. Let's get started.

First, you need a calendar. You can use a daily planner, a calendar for a smartphone, or an online calendar, such as the ones offered by Google, MSN, or Yahoo. It is best to have a calendar on which you can fill in daily activities and be able to see a weekly or monthly view as well.

Start filling in your calendar now, even if it means stopping right here and finding a calendar. Some of the things you might include are:

- The hours each class meets
- Time for driving to and from work or school
- Leisure time, an important aspect of a healthy lifestyle
- Time for study. Plan at least one hour of study for each hour in class. This is a *minimum!*
- Time to eat
- Your work schedule
- Time for extracurricular activities such as sports, music lessons, or volunteer work
- Time for family and friends
- Time for sleep
- Time for exercise

Take Note

Be realistic about how much time you have. One gauge is that working 10 hours per week is approximately equivalent to taking one three-unit course. If your college considers 15 units a full load and you are working 10 hours per week, you should consider taking 12 units. The more you work, the fewer units you should take.

We really hope you did this. If not, please reconsider. One of the best pathways to success is understanding how much time it takes to succeed. When you finish your calendar, if it does not allow you enough time to stay physically and emotionally healthy, rethink some of your school or work activities. We don't want you to lose your job because you have to study math. On the other hand, we don't want you to fail in math because of your job.

If math is particularly difficult for you, consider taking fewer course units during the semesters you take math. This applies equally to any other subject that you may find difficult. There is no rule that you must finish college in four years. It is a myth—discard it now.

Now extend your calendar for the entire semester. Many of the entries will repeat, such as the time a class meets. In your extended calendar, include significant events that may disrupt your normal routine. These might include holidays, family outings, birthdays, anniversaries, or special events such as a concert or a football game. In addition to these events, be sure to include the dates of tests, the date of the final exam, and dates that projects or papers are due. These are all important semester events. Having them on your calendar will remind you that you need to make time for them.

CLASS TIME

To be successful, **attend class.** You should consider your commitment to attend class as serious as your commitment to your job or to keeping an appointment with a dear friend. It is difficult to overstate the importance of attending class. If you miss work, you don't get paid. If you miss class, you are not getting the full benefit of your tuition dollar. You are losing money.

If, by some unavoidable situation, you cannot attend class, find out as soon as possible what was covered in class. You might:

- Ask a friend for notes and the assignment.
- Contact your instructor and get the assignment. Missing class is no excuse for not being prepared for the next class.
- Determine whether there are online resources that you can use to help you with the topics and concepts that were discussed in the class you missed.

Going to class is important. Once you are there, **participate in class.** Stay involved and active. When your instructor asks a question, try to at least mentally answer the question. If you have a question, ask. Your instructor expects questions and wants you to understand the concept being discussed.

HOMEWORK TIME

In addition to attending class, you must **do homework.** Homework is the best way to reinforce the ideas presented in class. You should plan on at least one to two hours of homework and study for each hour you are in class. We've had many students tell us that one to two hours seems like a lot of time. That may be true, but if you want to attain your goals, you must be willing to devote the time to being successful in this math course.

You should schedule study time just as if it were class time. To do this, write down where and when you study best. For instance, do you study best at home, in the library, at the math center, under a tree, or somewhere else? Some psychologists who research successful study strategies suggest that just by varying where you study, you can increase the effectiveness of a study session. While you are considering where you prefer to study, also think about the time of day during which your study period will be most productive. Write down your thoughts.

Look at what you have written, and be sure that you can consistently be in your favorite study environment at the time you have selected. Study and homework are extremely important. Just as you should not miss class, **do not miss study time.**

Before we leave this important topic, we have a few suggestions. If at all possible, create a study hour right after class. The material will be fresh in your mind, and the immediate review, along with your homework, will help reinforce the concepts you are studying.

If you can't study right after class, make sure that you set some time *on the day of the class* to review notes and begin the homework. The longer you wait, the more difficult it is to recall some of the important points covered during class. Studying math in small chunks—one hour a day (perhaps not enough for most of us), every day, is better than seven hours in one sitting. If you are studying for an extended period of time, break up your study session by studying one subject for a while and then moving on to another subject. Try to alternate between similar or related courses. For instance, study math for a while, then science, and then back to math. Or study history for a while, then political science, and then back to history.

Meet some of the people in your class and try to **put together a study group.** The group could meet two or three times a week. During those meetings, you could quiz each other, prepare for a test, try to explain a concept to someone else in the group, or get help on a topic that is difficult for you.

After reading these suggestions, you may want to rethink where and when you study best. If so, do that now. Remember, however, that it is your individual style that is important. Choose what works for *you,* and stick to it.

HABITS OF SUCCESSFUL STUDENTS

There are a number of habits that successful students use. Think about what these might be, and write them down.

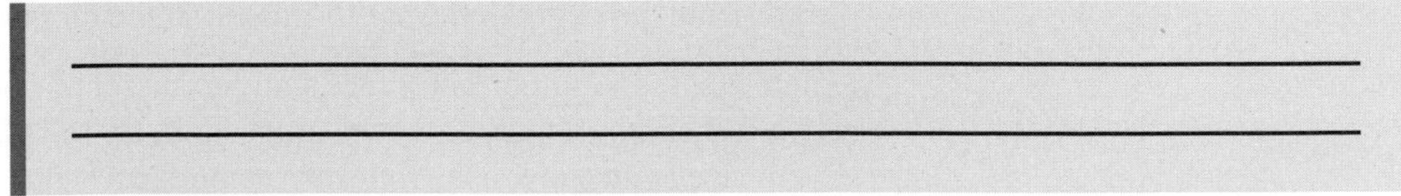

What you have written is very important. The habits you have listed are probably the things you know you must do to succeed. Here is a list of some responses from successful students we have known.

- **Set priorities.** You will encounter many distractions during the semester. Do not allow them to prevent you from reaching your goal.

- **Take responsibility.** Your instructor, this textbook, tutors, math centers, and other resources are there to help you succeed. Ultimately, however, you must choose to learn. You must choose success.
- **Hang out with successful students.** Success breeds success. When you work and study with successful students, you are in an environment that will help you succeed. Seek out people who are committed to their goals.
- **Study regularly.** We have mentioned this before, but it is too important not to be repeated.
- **Self test.** Once every few days, select homework exercises from previous assignments and use them to test your understanding. Try to do these exercises without getting help from examples in the text. These self tests will help you gain confidence that you can do these types of problems on a test given in class.
- **Try different strategies.** If you read the text and are still having difficulty understanding a concept, consider going a step further. Contact the instructor or find a tutor. Many campuses have some free tutorial services. Go to the math or learning center. Consult another textbook. Be active and get the help you need.
- **Make flash cards.** This is one of the strategies that some math students do not think to try. Flash cards are a very important part of learning math. For instance, your instructor may use words or phrases such as *linear, quadratic, exponent, base, rational,* and many others. If you don't know the meanings of these words, you will not know what is being discussed.
- **Plod along.** Your education is not a race. The primary goal is to finish. Taking too many classes and then dropping some does not get you to the end any faster. Take only the classes that you can manage.

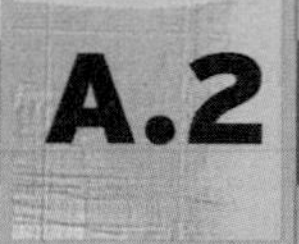

A.2 How to Use This Text to Succeed in This Course

GET THE BIG PICTURE

One of the major resources that you will have access to the entire semester is this textbook. We have written this text with you and your success in mind. The following is a guide to the features of this text that will help you succeed.

Actually, we want you to get the *really* big picture. Take a few minutes to read the table of contents. You may feel some anxiety about all the new concepts you will be learning. Try to think of this as an exciting opportunity to learn math. Now look through the entire book. Move quickly. Don't spend more than a few seconds on each page. Scan titles, look at pictures, and notice diagrams.

Getting this "big picture" view will help you see where this course is going. To reach your goal, it's important to get an idea of the steps you will need to take along the way.

As you look through the book, find topics that interest you. What's your preference? Racing? Sailing? TV? Amusement parks? Find the Index of Applications at the back of the book, and pull out three subjects that interest you. Write those topics here.

____

____

____

UNDERSTAND THE ORGANIZATION

Look again at the Table of Contents. There are 12 chapters in this book. You'll see that every chapter is divided into sections, and each section contains a number of learning objectives. Each learning objective is labeled with a number from 1 to 5. Knowing how this book is organized will help you locate important topics and concepts as you're studying.

Before you start a new objective, take a few minutes to read the Objective Statement for that objective. Then, browse through the objective material. Especially note the words or phrases in bold type—these are important concepts that you'll need to know as you move along in the course. These words are good candidates for flash cards. If possible, include an example of the concept on the flash card, as shown at the left.

Flash Card

Rule for Multiplying Exponential Expressions

If m and n are integers, then $x^m \cdot x^n = x^{m+n}$.

Examples

$x^5 \cdot x^3 = x^{5+3} = x^8$

$a \cdot a^4 = a^{1+4} = a^5$

$z^2 \cdot z^4 \cdot z^5 = z^{2+4+5} = z^{11}$

$(v^4r^3)(v^2r) = v^{4+2}r^{3+1} = v^6r^4$

You will also see important concepts and rules set off in boxes. Here is one about exponents. These rules are also good candidates for flash cards.

RULE FOR MULTIPLYING EXPONENTIAL EXPRESSIONS

If m and n are integers, then $x^m \cdot x^n = x^{m+n}$.

EXAMPLES

1. $x^5 \cdot x^3 = x^{5+3} = x^8$

2. $a \cdot a^4 = a^{1+4}$
$= a^5$ • **Recall that $a = a^1$.**

3. $z^2 \cdot z^4 \cdot z^5 = z^{2+4+5} = z^{11}$

4. $(v^4r^3)(v^2r) = v^{4+2}r^{3+1}$
$= v^6r^4$ • **Add exponents on like bases.**

Leaf through Objective 5.1.1 of Chapter 5. Write down the words in bold and any concepts or rules that are displayed in boxes.

____

____

____

USE THE INTERACTIVE METHOD

As we mentioned earlier, this textbook is based on an interactive approach. We want you to be actively involved in learning mathematics, and have given you many suggestions for getting "hands-on" with this book.

Focus On Look on pages 259–260. See the Focus On? The Focus On introduces a concept (in this case, adding polynomials) and includes a step-by-step solution of the type of exercise you will find in the homework.

Focus on adding polynomials using a horizontal or a vertical format

A. Add $(3x^2 + 2x - 7) + (7x^3 - 3 + 4x^2)$. Use a horizontal format.

Use the Commutative and Associative Properties of Addition to rearrange and group like terms.

$$(3x^2 + 2x - 7) + (7x^3 - 3 + 4x^2)$$
$$= 7x^3 + (3x^2 + 4x^2) + 2x + (-7 - 3)$$

Combine like terms.

$$= 7x^3 + 7x^2 + 2x - 10$$

B. Add $(4x^2 - 3x + 2) + (2x^3 + 4x - 7)$. Use a vertical format.

Write each polynomial in descending order with like terms in columns.

$$\begin{array}{r} 4x^2 - 3x + 2 \\ \underline{2x^3 \qquad\quad + 4x - 7} \end{array}$$

Add the terms in each column.

$$2x^3 + 4x^2 + x - 5$$

Grab paper and a pencil and work along as you're reading through the Focus On. When you're done, get a clean sheet of paper. Write down the problem and try to complete the solution without looking at your notes or at the book. When you're done, check your answer. If you got it right, you're ready to move on.

Look through the text and find three instances of a Focus On. Write the concepts mentioned in each Focus On here.

____

____

____

Example/Problem Pair You'll need hands-on practice to succeed in mathematics. When we show you an example, work it out right beside our solution. Use the Example/Problem pairs to get the practice you need.

Take a look at page 260. Example 5 and Problem 5 are shown here.

EXAMPLE 5 Add $(2x^3 + 5x^2 - 7x + 1) + (-x^3 - 5x^2 + 3x - 6)$ using a vertical format.

Solution

$$\begin{array}{r} 2x^3 + 5x^2 - 7x + 1 \\ \underline{-x^3 - 5x^2 + 3x - 6} \\ x^3 + 0x^2 - 4x - 5 \end{array}$$

- **Write each polynomial in descending order with like terms in columns.**
- **Add the terms in each column.**

$$(2x^3 + 5x^2 - 7x + 1) + (-x^3 - 5x^2 + 3x - 6) = x^3 - 4x - 5$$

Problem 5 Add $(x^3 - x + 2) + (x^2 + x - 6)$ using a horizontal format.

Solution See page S16.

Try Exercise 49, page 264.

You'll see that each Example is fully worked out. Study the Example by carefully working through each step. Then, try to complete the Problem. Use the solution to the Example as a model for solving the Problem. If you get stuck, the solutions to the Problems are provided in the back of the book. There is a page number directly following the Problem that shows you where you can find the completely-worked-out solution. Use the solution to get a hint for the step on which you are stuck. Then, try again!

When you've arrived at your solution, check your work against the solution in the back of the book. Turn to page S16 to see the solution for Problem 5.

Remember that sometimes there is more than one way to solve a problem. But your answer should always match the answer we've given in the back of the book. If you have any questions about whether your method will always work, check with your instructor.

Now note the line that says, "Try Exercise 49, page 264." You should do the Try Exercise from the exercise set to test your understanding of the concepts that have been discussed. When you have finished the exercise, check your answer with the one given in the Answer Section. If you got the answer wrong, try again. If you continue to have difficulty, seek help from a friend, a tutor, or your instructor.

USE A STRATEGY TO SOLVE WORD PROBLEMS

Learning to solve word problems is one of the reasons you are studying math. This is where you combine all of the critical thinking skills you have learned to solve practical problems.

Try not to be intimidated by word problems. Basically, what you need is a strategy that will help you come up with the equation you will need to solve the problem. When you are looking at a word problem, try the following:

- **Read the problem.** This may seem pretty obvious, but we mean really **read** it. Don't just scan it. Read the problem slowly and carefully.
- **Write down what is known and what is unknown.** Now that you have read the problem, go back and write down everything that is known. Next, write down what it is you are trying to find. Write this—don't just think it! Be as specific as you can. For instance, if you are asked to find a distance, don't just write "I need to find the distance." Be specific and write "I need to find the distance between Earth and the moon."
- **Think of a method to find the unknown.** For instance, is there a formula that relates the known and unknown quantities? This is certainly the most difficult step. Eventually, you must write an equation to be solved.
- **Solve the equation.** Be careful as you solve the equation. There is no sense in getting to this point and then making a careless mistake. The unknown in most word problems will include a unit such as feet, dollars, or miles per hour. When you write your answer, include the unit. An answer such as 20 doesn't mean much. Is it 20 feet, 20 dollars, 20 miles per hour, or something else?
- **Check your solution.** Now that you have an answer, go back to the problem and ask yourself whether it makes sense. This is an important step. For instance, if, according to your answer, the cost of a car is \$2.51, you know that something went wrong.

In this text, the solution of every word problem is broken down into two steps, **Strategy** and **Solution.** The Strategy consists of the first three steps discussed above. The Solution is the last two steps. Here is an Example from pages 567–568 of the text. Because you have not yet studied the concepts involved in the problem, you may not be able to solve it. However, note the details given in the Strategy. When you do the Problem following an Example, be sure to include your own Strategy.

EXAMPLE 1 Molybdenum-99 is a radioactive isotope used in medicine. An original 20-microgram sample of molybdenum-99 decays to 18 micrograms in 10 h. Find the half-life of molybdenum-99. Round to the nearest hour.

Strategy A_0, the original amount, is 20 micrograms. A, the final amount, is 18 micrograms. The time is 10 h. To find the half-life, solve the exponential decay equation $A = A_0\left(\frac{1}{2}\right)^{\frac{t}{k}}$ for the half-life, k.

Solution

$$A = A_0\left(\frac{1}{2}\right)^{\frac{t}{k}}$$ • **Use the exponential decay equation.**

$$18 = 20\left(\frac{1}{2}\right)^{\frac{10}{k}}$$ • $A_0 = 20, A = 18, t = 10$

$$\frac{18}{20} = \left(\frac{1}{2}\right)^{\frac{10}{k}}$$ • **Solve for k.**

$$\log\frac{18}{20} = \log\left(\frac{1}{2}\right)^{\frac{10}{k}}$$

$$\log\frac{18}{20} = \frac{10}{k}\log\frac{1}{2}$$

$$k = \frac{10\log\frac{1}{2}}{\log\frac{18}{20}}$$

$$k \approx 65.8$$

The half-life of molybdenum-99 is about 66 h.

Problem 1 The number of words per minute that a student can type will increase with practice and can be approximated by the equation $N = 100[1 - (0.9)^t]$, where N is the number of words typed per minute after t days of practice. In how many days will the student be able to type 60 words per minute? Round to the nearest whole number of days.

Solution See page S33.

Try Exercise 11, page 571.

When you have finished studying a section, **do the exercises your instructor has selected.** Math is not a spectator sport. You must practice every day. Do the homework and do not get behind.

ACE THE TEST

There are a number of features in this text that will help you prepare for a test. These features will help you even more if you do just one simple thing: When you are doing your homework, go back to each previous homework assignment for the current chapter and rework two exercises. That's right—just *two* exercises. You will be surprised at how much better prepared you will be for a test by doing this.

Here are some additional aides to help you **ace the test.**

Chapter Summary Once you've completed a chapter, look at the Chapter Summary. The Chapter Summary is divided into two sections: Key Words and Essential Rules and Procedures. Flip to page 103 to see the Chapter Summary for Chapter 2. The summary shows all of the important topics covered in the chapter. Do you see the reference following each topic? This reference shows you the objective and page in the text where you can find more information on the concept.

Write down one Key Word and one Essential Rule or Procedure. Explain the meaning of the reference 2.1.1, page 56.

____

____

Chapter Review Exercises Turn to page 105 to see the Chapter Review Exercises for Chapter 2. When you do the review exercises, you're giving yourself an important opportunity to test your understanding of the chapter. The answer to each review exercise is given at the back of the book, along with the objective the question relates to. When you're done with the Chapter Review Exercises, check your answers. If you had trouble with any of the questions, you can restudy the objectives and retry some of the exercises in those objectives for extra help.

Go to the Answer Section at the back of the text. Find the answers for the Chapter Review Exercises for Chapter 2. Write down the answer to Exercise 17. What is the meaning of the reference 2.4.2?

____

____

Chapter Test The Chapter Test for each chapter can be found after the Chapter Review Exercises and can be used to help you prepare for your exam. The answer to each test question is given at the back of the book, along with both an objective reference and a reference to a Focus On, Example, or Problem that the question relates to. Think of these tests as "practice runs" for your in-class tests. Take the test in a quiet place, and try to work through it in the same amount of time that will be allowed for your actual exam.

The aids we have mentioned above will help you prepare for a test. You should begin your review *at least* two days before the test—three days is better. These aids will get you ready for the test.

Here are some suggestions to try while you are actually taking the test.

- **Try to relax.** We know that test situations make some students quite nervous or anxious. These feelings are normal. Try to stay calm and focused on what you know. If you have prepared as we have suggested, the answers will begin to come to you.
- **Scan the test.** Get a feeling for the big picture.
- **Read the directions carefully.** Make sure you answer each question fully.
- **Work the problems that are easiest for you first.** This will help you with your confidence and help reduce the nervous feeling you may have.

READY, SET, SUCCEED!

It takes hard work and commitment to succeed, but we know you can do it! Doing well in mathematics is just one step you'll take on your path to success. Good luck. We wish you success.

Review of Real Numbers

CHAPTER 1

Focus on Success

Have you read AIM for Success: Getting Started? It describes study skills used by students who have been successful in their math courses. This feature gives you tips on how to stay motivated, how to manage your time, and how to prepare for exams. It also includes a complete guide to the textbook and how to use its features to be successful in this course. AIM for Success starts on page AIM-1.

Digital Vision

OBJECTIVES

1.1
1. Inequality and absolute value
2. Interval notation and set operations

1.2
1. Operations on integers
2. The Order of Operations Agreement

1.3
1. Operations on rational numbers
2. Order of Operations and complex fractions
3. Decimal notation

1.4
1. The Properties of the Real Numbers
2. Evaluate variable expressions
3. Simplify variable expressions

1.5
1. Translate a verbal expression into a variable expression
2. Application problems

PREP TEST

Are you ready to succeed in this chapter? *Take the Prep Test below to find out if you are ready to learn the new material.*

For Exercises 1 to 8, add, subtract, multiply, or divide.

1. $\frac{5}{12} + \frac{7}{30}$ **2.** $\frac{8}{15} - \frac{7}{20}$

3. $\frac{5}{6} \cdot \frac{4}{15}$ **4.** $\frac{4}{15} \div \frac{2}{5}$

5. $8 + 29.34 + 7.065$ **6.** $92 - 18.37$

7. $2.19(3.4)$ **8.** $32.436 \div 0.6$

9. Which of the following numbers are greater than -8?
(i) -6 (ii) -10 (iii) 0 (iv) 8

10. Match each fraction with its decimal equivalent.

a. $\frac{1}{2}$ **A.** 0.75

b. $\frac{7}{10}$ **B.** 0.89

c. $\frac{3}{4}$ **C.** 0.5

d. $\frac{89}{100}$ **D.** 0.7

1.1 Introduction to Real Numbers

OBJECTIVE 1

Inequality and absolute value

> **Point of Interest**
> The Big Dipper, known to the Greeks as *Ursa Major*, the great bear, is a constellation that can be seen from northern latitudes. The stars of the Big Dipper are Alkaid, Mizar, Alioth, Megrez, Phecda, Merak, and Dubhe. The star at the bend of the handle, Mizar, is actually two stars, Mizar and Alcor. An imaginary line from Merak through Dubhe passes through Polaris, the north star.

It seems to be a human characteristic to put similar items into the same group. For instance, an astronomer places stars in *constellations* and a geologist divides the history of Earth into *eras.*

Mathematicians likewise place objects with similar properties in *sets.* A **set** is a collection of objects called **elements** of the set. Sets are denoted by placing braces around the elements in the set. For instance, the set of the first five letters of the alphabet is {a, b, c, d, e}. The symbol for "is an element of" is $\in$; the symbol for "is not an element of" is $\notin$. For example,

$$a \in \{a, b, c, d, e\} \qquad d \in \{a, b, c, d, e\} \qquad k \notin \{a, b, c, d, e\}$$

The numbers that we use to count things, such as the number of people in a city or the number of different species of flowers, are called *natural numbers.*

Natural numbers = {1, 2, 3, 4, 5, 6, 7, 8, 9, 10, 11, . . .}

Each natural number other than 1 is either a *prime number* or a *composite number.* A **prime number** is a natural number, other than 1, that is evenly divisible only by itself and 1. The first six prime numbers are 2, 3, 5, 7, 11, and 13. A **composite number** is a natural number, other than 1, that is not a prime number. The numbers 4, 6, 8, 9, 10, and 12 are the first six composite numbers.

> **Point of Interest**
> The concept of zero developed very gradually over many centuries. It has been variously denoted by a blank space, by a dot, and finally as 0. Negative numbers, although evident in Chinese manuscripts dating from 200 B.C., were not fully integrated into mathematics until late in the 14th century.

The natural numbers do not have a symbol to denote the concept of "none"—for instance, the number of trees taller than 1000 feet. The *whole numbers* include zero and the natural numbers.

Whole numbers = {0, 1, 2, 3, 4, 5, 6, 7, 8, . . .}

The whole numbers alone do not provide all the numbers that are useful in applications. For instance, a meteorologist needs numbers below zero and above zero.

Integers = {. . . , −5, −4, −3, −2, −1, 0, 1, 2, 3, 4, 5, . . .}

The integers . . . , −5, −4, −3, −2, −1 are **negative integers.** The integers 1, 2, 3, 4, 5, . . . are **positive integers.** Note that the natural numbers and the positive integers are the same set of numbers. The integer zero is neither a positive nor a negative number.

Still other numbers are necessary to solve the variety of application problems that exist. For instance, a landscape architect may need to purchase irrigation pipe that has a diameter of $\frac{5}{8}$ in. Numbers that can be written in the form of a fraction $\frac{p}{q}$, where p and q are integers and $q \neq 0$, are called *rational numbers.*

$$\textbf{Rational numbers} = \left\{\frac{p}{q}, \text{ where } p \text{ and } q \text{ are integers and } q \neq 0\right\}$$

Examples of rational numbers are $\frac{2}{3}$, $-\frac{9}{2}$, and $\frac{5}{1}$. Note that $\frac{5}{1} = 5$, so all integers are rational numbers. The number $\frac{4}{\pi}$ is not a rational number because π is not an integer.

Numbers that can be written as terminating or repeating decimals are rational numbers. For repeating decimals, we place a bar over the digits that repeat.

Terminating decimals	0.5 2.34 −6.20137 7
Repeating decimals	$0.\overline{3} = 0.33\ldots$ $1.2\overline{67} = 1.26767\ldots$ $-4.10\overline{782} = -4.10782782\ldots$

Some numbers cannot be written as terminating or repeating decimals. Such numbers include 0.01001000100001 . . . , $\sqrt{7} \approx 2.6457513$, and $\pi \approx 3.1415927$. These numbers have decimal representations that neither terminate nor repeat. They are called **irrational numbers.** The rational numbers and the irrational numbers taken together are the *real numbers.*

Real numbers = {rational numbers and irrational numbers}

The relationship among the various sets of numbers is shown in the following figure.

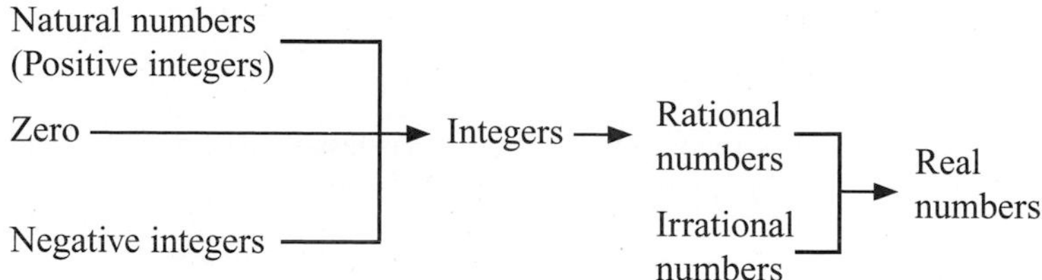

Focus on

identifying the sets to which a number belongs

Determine which of the following numbers are

a. integers **b.** rational numbers **c.** irrational numbers

d. real numbers **e.** prime numbers **f.** composite numbers

$$-1, -3.347, 0, 5, 6.1\overline{01}, \sqrt{48}, 2.2020020002\ldots, 63, \frac{19}{2}, \frac{20}{\sqrt{7}}$$

a. Integers: $-1, 0, 5, 63$

b. Rational numbers: $-1, -3.347, 0, 5, 6.1\overline{01}, 63, \frac{19}{2}$

c. Irrational numbers: $\sqrt{48}, 2.2020020002\ldots, \frac{20}{\sqrt{7}}$

d. Real numbers: $-1, -3.347, 0, 5, 6.1\overline{01}, \sqrt{48}, 2.2020020002\ldots, 63, \frac{19}{2}, \frac{20}{\sqrt{7}}$

e. Prime numbers: 5

f. Composite numbers: 63

The **graph of a real number** is made by placing a heavy dot on a number line directly above the number. The graphs of some real numbers are shown below.

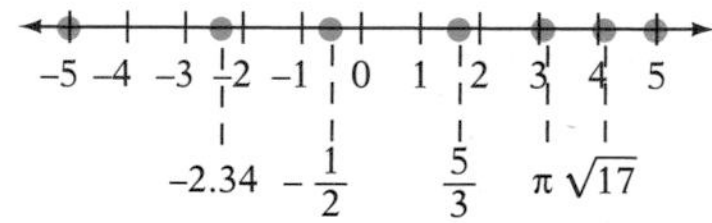

Consider these sentences:

A restaurant's chef prepared a dinner and served *it* to the customers.

A maple tree was planted, and *it* grew 2 ft in one year.

In the first sentence, "it" means dinner; in the second sentence, "it" means tree. In language, the word *it* can stand for many different objects. Similarly, in mathematics, a letter of the alphabet can be used to stand for some number. A letter used in this way is called a **variable.**

It is convenient to use a variable to represent, or stand for, any one of the elements of a set. For instance, the statement "x is an element of the set {0, 2, 4, 6}" means that x can be replaced by 0, 2, 4, or 6. The set {0, 2, 4, 6} is called the **domain** of the variable.

Variables are used in the next defintion.

DEFINITION OF INEQUALITY

If a and b are two real numbers and a is to the left of b on the number line, then a **is less than** b. This is written $a < b$.

If a and b are two real numbers and a is to the right of b on the number line, then a **is greater than** b. This is written $a > b$.

EXAMPLES

1. $-2 < 8$ 2. $-1 > -5$ 3. $0 > -\frac{2}{3}$ 4. $\pi < \sqrt{17}$

The inequality symbols $\leq$ (is less than or equal to) and $\geq$ (is greater than or equal to) are also important. Note the following examples.

$4 \leq 5$ is a true statement because $4 < 5$.

$5 \leq 5$ is a true statement because $5 = 5$.

EXAMPLE 1 Let $y \in \{-5, -3, -1, 1\}$. For which values of y is the inequality $y \geq -1$ a true statement?

Solution Replace y by each element of the set and determine whether the statement is true.

$y \geq -1$	
$-5 \geq -1$	A false statement
$-3 \geq -1$	A false statement
$-1 \geq -1$	A true statement
$1 \geq -1$	A true statement

The inequality is true for -1 and 1.

Problem 1 Let $z \in \{-2, -1, 0, 1, 2\}$. For which values of z is the inequality $z \leq 0$ a true statement?

Solution See page S1.

Try Exercise 25, page 10.

The numbers 5 and -5 are the same distance from zero on the number line but on opposite sides of zero. The numbers 5 and -5 are called **additive inverses,** or **opposites.**

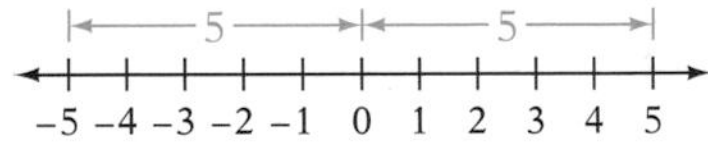

The additive inverse (or opposite) of 5 is -5. The additive inverse of -5 is 5. The symbol for additive inverse is $-$.

$-(4)$ means the additive inverse of positive 4. $\quad -(4) = -4$

$-(-4)$ means the additive inverse of negative 4. $\quad -(-4) = 4$

EXAMPLE 2 Let $a \in \{-12, 0, 4\}$. Determine $-a$, the additive inverse of a, for each element of the set.

Solution

$-a$ • **Write the expression for the additive inverse of *a*.**

$-(-12) = 12$ • **Replace *a* by each element of the set and determine the value of the expression.**

$-(0) = 0$

$-(4) = -4$

Problem 2 Let $v \in \{-8, 0, 9\}$. Determine $-v$, the additive inverse of v, for each element of the set.

Solution See page S1.

Try Exercise 23, page 10.

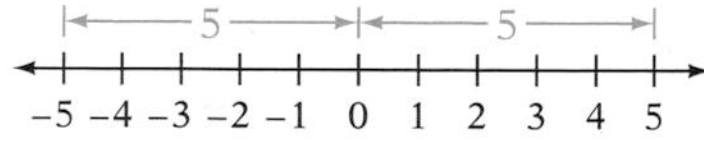

The **absolute value** of a number is a measure of its distance from zero on a number line. The symbol for absolute value is $|\ |$. Note from the figure at the left that the distance from 0 to 5 is 5. Therefore, $|5| = 5$. The figure shows that the distance from 0 to -5 is also 5. Thus $|-5| = 5$.

ABSOLUTE VALUE

The absolute value of a positive number or 0 is the number. The absolute value of a negative number is the additive inverse of the number. This can be written as follows. If a is a real number, then

$$|a| = \begin{cases} a, & a \geq 0 \\ -a, & a < 0 \end{cases}$$

EXAMPLES

1. $|7| = 7$. Because $7 \geq 0$, the absolute value of 7 is the number 7 itself.
2. $|-8| = 8$. Because $-8 < 0$, the absolute value of -8 is the additive inverse of -8. The additive inverse of -8 is 8.
3. $|0| = 0$. The absolute value of 0 is 0. One way to think of this is that the distance from 0 to 0 on the number line is 0.

EXAMPLE 3 Evaluate: $-|-12|$

Solution From the definition of absolute value, $|-12| = 12$. Therefore, $-|-12| = -12$.

Problem 3 Evaluate: $|-23|$

Solution See page S1.

Try Exercise 33, page 10.

OBJECTIVE 2 Interval notation and set operations

The **roster method** of writing a set encloses a list of the elements of the set in braces. We used this method at the beginning of this section to define sets of numbers. Using the roster method, the set of even natural numbers less than 10 is written $\{2, 4, 6, 8\}$. This is an example of a **finite set;** all the elements can be listed. The set of natural numbers, $\{1, 2, 3, 4, \ldots\}$, is an **infinite set;** it is impossible to list all the elements.

The **empty set,** or **null set,** is the set that contains no elements. The symbol $\varnothing$ or $\{\ \}$ is used to represent the empty set.

EXAMPLE 4 Use the roster method to write the set of natural numbers less than 10.

Solution $\{1, 2, 3, 4, 5, 6, 7, 8, 9\}$

Problem 4 Use the roster method to write the set of negative odd integers greater than -8.

Solution See page S1.

Try Exercise 43, page 11.

A second method of representing a set is **set-builder notation.** Set-builder notation can be used to describe almost any set, but it is especially useful when writing infinite sets. In set-builder notation, the set of integers greater than -3 is written

$$\{x \mid x > -3, x \in \text{integers}\}$$

and is read "the set of all x such that x is greater than -3 and x is an element of the integers." This is an infinite set. It is impossible to list all the elements of the set, but the set can be described by using set-builder notation.

The set of real numbers less than 5 is written

$$\{x \mid x < 5, x \in \text{real numbers}\}$$

and is read "the set of all x such that x is less than 5 and x is an element of the real numbers."

Because most of our work is with real numbers, we will generally omit "$x \in$ real numbers" from set-builder notation. Thus we would write $\{x \mid x < 5, x \in \text{real numbers}\}$ as $\{x \mid x < 5\}$, where we assume that x is a real number.

EXAMPLE 5 Use set-builder notation to write the set of real numbers greater than -2.

Solution $\{x \mid x > -2\}$

Problem 5 Use set-builder notation to write the set of integers less than or equal to 7.

Solution See page S1.

Try Exercise 51, page 11.

The graph of a set of real numbers written in set-builder notation can be shown on a number line. The graph of $\{x \mid x > -2\}$ is shown below. The parenthesis on the graph indicates that -2 is not part of the set.

The graph of $\{x \mid x \geq -2\}$ is shown below. The bracket on the graph indicates that -2 is included in the set.

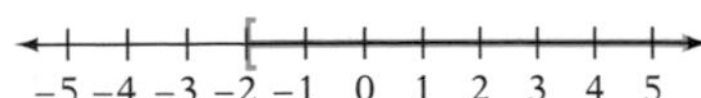

EXAMPLE 6 Graph: $\{x \mid x \leq 3\}$

Solution The set is the real numbers less than or equal to 3.

- **Draw a right bracket at 3, and darken the number line to the left of 3.**

Problem 6 Graph: $\{x|x > -3\}$

Solution See page S1.

Try Exercise 63, page 11.

It is also possible to graph a set of real numbers *between* two given numbers.

Focus on graphing a set of real numbers

Graph: $\{x|0 \leq x < 4\}$

The notation $0 \leq x < 4$ indicates the set of real numbers between 0 and 4, including 0 but not including 4. A bracket is placed at 0 to indicate that 0 is included in the graph; a parenthesis is placed at 4 to indicate that 4 is not part of the graph.

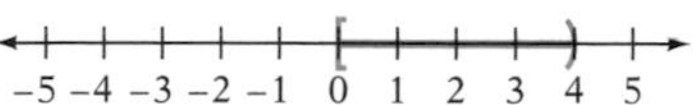

Given two real numbers, an **interval** is the set of all real numbers between the given numbers. The two numbers are the **endpoints** of the interval. For example, the set $\{x|-1 < x < 3\}$ represents the interval of all real numbers between -1 and 3. The endpoints of this interval are -1 and 3.

A **closed interval** includes both endpoints; an **open interval** contains neither endpoint; a **half-open interval** contains one endpoint but not the other. For example, the set $\{x|-1 < x < 3\}$ is an open interval.

Intervals can be represented in set-builder notation or **interval notation.** In interval notation, the brackets or parentheses that are used to graph the set are written with the endpoints of the interval. The set $\{x|0 \leq x < 4\}$ shown above is written $[0, 4)$ in interval notation; 0 and 4 are the endpoints. Here are some additional examples.

Set-builder notation	Interval notation	Graph
$\{x\|-3 \leq x \leq 2\}$	$[-3, 2]$, a closed interval	−5 −4 −3 −2 −1 0 1 2 3 4 5
$\{x\|-3 < x < 2\}$	$(-3, 2)$, an open interval	−5 −4 −3 −2 −1 0 1 2 3 4 5
$\{x\|-3 \leq x < 2\}$	$[-3, 2)$, a half-open interval	−5 −4 −3 −2 −1 0 1 2 3 4 5
$\{x\|-3 < x \leq 2\}$	$(-3, 2]$, a half-open interval	−5 −4 −3 −2 −1 0 1 2 3 4 5

To indicate an interval that extends forever in one or both directions using interval notation, we use the **infinity symbol** ∞ or the **negative infinity symbol** $-\infty$. The infinity symbol is not a number; it is simply a notation used to indicate that the interval is unlimited. In interval notation, a parenthesis is always used to the right of an infinity symbol or to the left of a negative infinity symbol, as shown in the following examples.

Set-builder notation	Interval notation	Graph
$\{x\|x > 1\}$	$(1, \infty)$	−5 −4 −3 −2 −1 0 1 2 3 4 5
$\{x\|x \geq 1\}$	$[1, \infty)$	−5 −4 −3 −2 −1 0 1 2 3 4 5
$\{x\|x < 1\}$	$(-\infty, 1)$	−5 −4 −3 −2 −1 0 1 2 3 4 5
$\{x\|x \leq 1\}$	$(-\infty, 1]$	−5 −4 −3 −2 −1 0 1 2 3 4 5
$\{x\|-\infty < x < \infty\}$	$(-\infty, \infty)$	−5 −4 −3 −2 −1 0 1 2 3 4 5

EXAMPLE 7 Use the given notation or graph to supply the notation or graph that is marked with a question mark.

	Set-builder notation	Interval notation	Graph
A.	$\{x \mid 0 \le x \le 1\}$	?	?
B.	?	$[-3, 4)$	?
C.	?	?	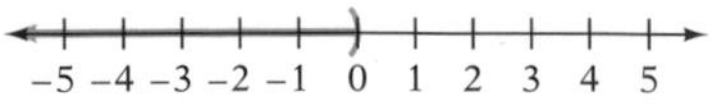

Solution

	Set-builder notation	Interval notation	Graph
A.	$\{x \mid 0 \le x \le 1\}$	$[0, 1]$	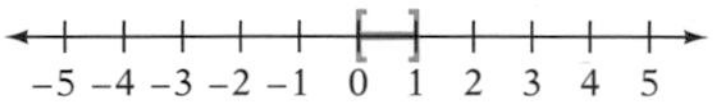
B.	$\{x \mid -3 \le x < 4\}$	$[-3, 4)$	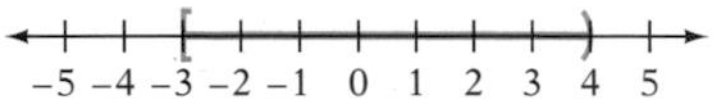
C.	$\{x \mid x < 0\}$	$(-\infty, 0)$	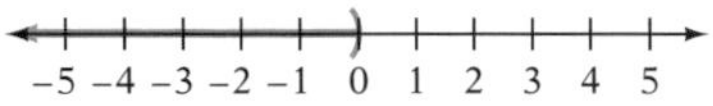

Problem 7 Use the given notation or graph to supply the notation or graph that is marked with a question mark.

	Set-builder notation	Interval notation	Graph
A.	$\{x \mid -2 < x < 0\}$	?	?
B.	?	$(-1, 2]$	?
C.	?	?	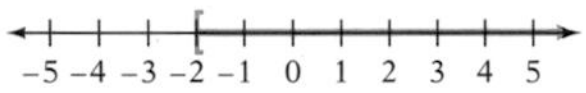

Solution See page S1.

Try Exercises 73, 77, and 93, pages 11, 12.

Just as operations such as addition and multiplication are performed on real numbers, operations are performed on sets. Two operations performed on sets are *union* and *intersection*.

UNION OF TWO SETS

The **union** of two sets, written $A \cup B$, is the set of all elements that belong to either A **or** B. In set-builder notation, this is written

$$A \cup B = \{x \mid x \in A \text{ or } x \in B\}$$

EXAMPLES

1. Given $A = \{2, 3, 4, 5, 6\}$ and $B = \{4, 5, 6, 7, 8\}$, $A \cup B = \{2, 3, 4, 5, 6, 7, 8\}$. Note that the elements 4, 5, and 6, which belong to both sets, are listed only once.
2. Given $C = \{-3, -1, 1, 3\}$ and $D = \{-2, 0, 2\}$, $C \cup D = \{-3, -2, -1, 0, 1, 2, 3\}$.
3. Given $X = \{0, 2, 4, 6, 8\}$ and $Y = \{4, 8\}$, $X \cup Y = \{0, 2, 4, 6, 8\}$.

Point of Interest

The symbols $\in$, $\cup$, and $\cap$ were first used by Giuseppe Peano in *Arithmetices Principia, Nova Exposita (The Principle of Mathematics, a New Method of Exposition)*, published in 1889. The purpose of this book was to deduce the principles of mathematics from pure logic.

INTERSECTION OF TWO SETS

The **intersection** of two sets, written $A \cap B$, is the set of all elements that are common to both A **and** B. In set-builder notation, this is written

$$A \cap B = \{x \mid x \in A \text{ and } x \in B\}$$

EXAMPLES

1. Given $A = \{2, 3, 4, 5, 6\}$ and $B = \{4, 5, 6, 7, 8\}$, $A \cap B = \{4, 5, 6\}$.
2. Given $C = \{-3, -1, 1, 3\}$ and $D = \{-2, 0, 2\}$, $C \cap D = \varnothing$. There are no elements common to both C and D.
3. Given $X = \{0, 2, 4, 6, 8\}$ and $Y = \{4, 8\}$, $X \cap Y = \{4, 8\}$.

Set operations also can be performed on intervals.

EXAMPLE 8 Graph. **A.** $\{x|x \leq -1\} \cup \{x|x > 3\}$ **B.** $(-\infty, 3) \cap [-1, \infty)$

Solution **A.** The set $\{x|x \leq -1\} \cup \{x|x > 3\}$ is the set of real numbers less than or equal to -1 *or* greater than 3. This set can be written $\{x|x \leq -1 \text{ or } x > 3\}$.

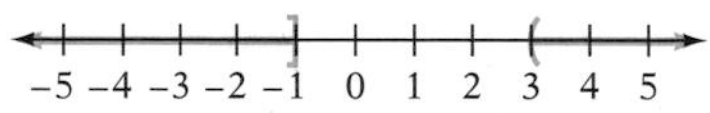

• **The graph of $\{x|x \leq -1$ or $x > 3\}$ contains all the points on the graphs of $x \leq -1$ and $x > 3$.**

B. The set $(-\infty, 3) \cap [-1, \infty)$ is the set of real numbers less than 3 *and* greater than or equal to -1.

The graph of $(-\infty, 3)$ is shown in red, and the graph of $[-1, \infty)$ is shown in blue.

Real numbers that are elements of both $(-\infty, 3)$ *and* $[-1, \infty)$ correspond to the points in the section of overlap; thus, $(-\infty, 3) \cap [-1, \infty) = [-1, 3)$. Note that 3 is not an element of $(-\infty, 3)$. Therefore, 3 is not an element of the intersection of the sets.

Problem 8 Graph.

A. $(-\infty, -1] \cup [2, 4)$

B. $\{x|x \leq 3\} \cap \{x|-3 < x < 5\}$

Solution See page S1.

Try Exercise 103, page 12.

1.1 Exercises

CONCEPT CHECK

Determine which of the numbers are **a.** natural numbers, **b.** whole numbers, **c.** integers, **d.** positive integers, **e.** negative integers. List all that apply.

1. $-14, 9, 0, 53, 7.8, -626$

2. $31, -45, -2, 9.7, 8600, \frac{1}{2}$

Determine which of the numbers are **a.** integers, **b.** rational numbers, **c.** irrational numbers, **d.** real numbers. List all that apply.

3. $-\frac{15}{2}, 0, -3, \pi, 2.\overline{33}, 4.232232223\ldots, \frac{\sqrt{5}}{4}, \sqrt{7}$

4. $-17, 0.3412, \frac{3}{\pi}, -1.010010001\ldots, \frac{27}{91}, 6.1\overline{2}$

5. What is a terminating decimal? Provide an example.

6. What is a repeating decimal? Provide an example.

7. What is the additive inverse of a number?

8. What is the absolute value of a number?

9. Explain the difference between the union of two sets and the intersection of two sets.

10. Explain the difference between $\{x|x < 5\}$ and $\{x|x \leq 5\}$.

1 Inequality and absolute value (See pages 2–5.)

GETTING READY

11. A number such as 0.63633633363333 . . . , whose decimal notation neither ends nor repeats, is an example of an __?__ number.

12. The additive inverse of a negative number is a __?__ number.

13. $y \in \{1, 3, 5, 7, 9\}$ is read "y __?__ the set $\{1, 3, 5, 7, 9\}$."

14. Write the phrase "the opposite of the absolute value of n" in symbols.

Find the additive inverse of each of the following.

15. 27
16. -3
17. $\frac{3}{4}$
18. $\sqrt{17}$
19. 0
20. $-\pi$
21. $-\sqrt{33}$
22. -1.23
23. -91
24. $-\frac{2}{3}$

25. Let $x \in \{-3, 0, 7\}$. For which values of x is $x < 5$ true?

26. Let $z \in \{-4, -1, 4\}$. For which values of z is $z > -2$ true?

27. Let $y \in \{-6, -4, 7\}$. For which values of y is $y > -4$ true?

28. Let $x \in \{-6, -3, 3\}$. For which values of x is $x < -3$ true?

29. Let $w \in \{-2, -1, 0, 1\}$. For which values of w is $w \leq -1$ true?

30. Let $p \in \{-10, -5, 0, 5\}$. For which values of p is $p \geq 0$ true?

31. Let $b \in \{-9, 0, 9\}$. Evaluate $-b$ for each element of the set.

32. Let $a \in \{-3, -2, 0\}$. Evaluate $-a$ for each element of the set.

33. Let $c \in \{-4, 0, 4\}$. Evaluate $|c|$ for each element of the set.

34. Let $q \in \{-3, 0, 7\}$. Evaluate $|q|$ for each element of the set.

35. Let $m \in \{-6, -2, 0, 1, 4\}$. Evaluate $-|m|$ for each element of the set.

36. Let $x \in \{-5, -3, 0, 2, 5\}$. Evaluate $-|x|$ for each element of the set.

37. Are there any real numbers x for which $-x > 0$? If so, describe them.

38. Are there any real numbers y for which $-|y| > 0$? If so, describe them.

2 Interval notation and set operations (See pages 5–9.)

GETTING READY

39. Two ways to write the set of natural numbers less than 5 are $\{1, 2, 3, 4\}$ and $\{n|n < 5, n \in \text{natural numbers}\}$. The first way uses the __?__ method, and the second way uses __?__ notation.

40. The symbol for "union" is __?__. The symbol for "intersection" is __?__.

41. The symbol ∞ is called the __?__ symbol.

42. Replace each question mark with "includes" or "does not include" to make the following statement true. The set $[-4, 7)$ __?__ the number -4 and __?__ the number 7.

Use the roster method to write the set.

43. the integers between -3 and 5

44. the integers between -4 and 0

45. the even natural numbers less than 14

46. the odd natural numbers less than 14

47. the positive-integer multiples of 3 that are less than or equal to 30

48. the negative-integer multiples of 4 that are greater than or equal to -20

49. the negative-integer multiples of 5 that are greater than or equal to -35

50. the positive-integer multiples of 6 that are less than or equal to 36

Use set-builder notation to write the set.

51. the integers greater than 4

52. the integers less than -2

53. the real numbers greater than or equal to -2

54. the real numbers less than or equal to 2

55. the real numbers between 0 and 1

56. the real numbers between -2 and 5

57. the real numbers between 1 and 4, inclusive

58. the real numbers between 0 and 2, inclusive

Graph.

59. $\{x|-1 < x < 5\}$

60. $\{x|1 < x < 3\}$

61. $\{x|0 \leq x \leq 3\}$

62. $\{x|-1 \leq x \leq 1\}$

63. $\{x|x < 2\}$

64. $\{x|x < -1\}$

65. $\{x|x \geq 1\}$

66. $\{x|x \leq -2\}$

Write each interval in set-builder notation.

67. $(0, 8)$

68. $(-2, 4)$

69. $[-5, 7]$

70. $[3, 4]$

71. $[-3, 6)$

72. $(4, 5]$

73. $(-\infty, 4]$

74. $(-\infty, -2)$

75. $(5, \infty)$

76. $[-2, \infty)$

Write each set of real numbers in interval notation.

77. $\{x|-2 < x < 4\}$

78. $\{x|0 < x < 3\}$

79. $\{x|-1 \leq x \leq 5\}$

80. $\{x|0 \leq x \leq 3\}$

81. $\{x|x < 1\}$

82. $\{x|x \leq 6\}$

83. $\{x|-2 \leq x < 6\}$

84. $\{x|x \geq 3\}$

85. $\{x|x \in \text{real numbers}\}$

86. $\{x|x > -1\}$

Graph.

87. $(-2, 5)$

88. $(0, 3)$

89. $[-1, 2]$

90. $[-3, 2]$

91. $(-\infty, 3]$

92. $(-\infty, -1)$

93. $[3, \infty)$

94. $[-2, \infty)$

Find $A \cup B$ and $A \cap B$.

95. $A = \{1, 4, 9\}, B = \{2, 4, 6\}$

96. $A = \{-1, 0, 1\}, B = \{0, 1, 2\}$

97. $A = \{2, 3, 5, 8\}, B = \{9, 10\}$

98. $A = \{1, 3, 5, 7\}, B = \{2, 4, 6, 8\}$

99. $A = \{-4, -2, 0, 2, 4\}, B = \{0, 4, 8\}$

100. $A = \{-3, -2, -1\}, B = \{-2, -1, 0, 1\}$

101. $A = \{1, 2, 3, 4, 5\}, B = \{3, 4, 5\}$

102. $A = \{2, 4\}, B = \{0, 1, 2, 3, 4, 5\}$

Graph.

103. $\{x|x > 1\} \cup \{x|x < -1\}$

104. $\{x|x \leq 2\} \cup \{x|x > 4\}$

105. $\{x|x \leq 2\} \cap \{x|x \geq 0\}$

106. $\{x|x > -1\} \cap \{x|x \leq 4\}$

107. $\{x|x > 1\} \cap \{x|x \geq -2\}$

108. $\{x|x < 4\} \cap \{x|x \leq 0\}$

109. $\{x|x > 2\} \cup \{x|x > 1\}$

110. $\{x|x < -2\} \cup \{x|x < -4\}$

111. $(-\infty, 2] \cup [4, \infty)$

112. $(-3, 4] \cup [-1, 5)$

113. $[-1, 2] \cap [0, 4]$

114. $[-5, 4) \cap (-2, \infty)$

115. $(2, \infty) \cup (-2, 4]$

116. $(-\infty, 2] \cup (4, \infty)$

117. Which set is the empty set?
(i) $\{x|x \in \text{integers}\} \cap \{x|x \in \text{rational numbers}\}$
(ii) $\{-4, -2, 0, 2, 4\} \cup \{-3, -1, 1, 3\}$
(iii) $[5, \infty) \cap (0, 5)$

118. Which set is *not* equivalent to the interval $[-1, 6)$?
(i) $\{x|-1 \leq x < 6\}$
(ii) $\{x|x \geq -1\} \cup \{x|x < 6\}$
(iii) $\{x|x < 6\} \cap \{x|x \geq -1\}$

APPLYING CONCEPTS

Let $R = \{x|x \in \text{real numbers}\}$, $A = \{x|-1 \leq x \leq 1\}$, $B = \{x|0 \leq x \leq 1\}$, $C = \{x|-1 \leq x \leq 0\}$, and $\emptyset$ = empty set. Indicate whether each of the following expressions is equivalent to R, A, B, C, or $\emptyset$.

119. $A \cup B$

120. $A \cup A$

121. $B \cap B$

122. $A \cup C$

123. $A \cap R$

124. $C \cap R$

125. $B \cup R$

126. $A \cup R$

127. $R \cup R$

128. $R \cap \emptyset$

129. The set $B \cap C$ cannot be expressed using R, A, B, C, or $\emptyset$. What real number is represented by $B \cap C$?

130. A student wrote $-3 > x > 5$ as the inequality that represents the real numbers less than -3 or greater than 5. Explain why this notation is incorrect.

Graph the solution set.

131. $|x| < 2$

132. $|x| < 5$

133. $|x| > 3$

134. $|x| > 4$

135. Given that a, b, c, and d are positive real numbers, which of the following will ensure that $\frac{a-b}{c-d} \leq 0$?
(i) $a \geq b$ and $c > d$ (ii) $a \leq b$ and $c > d$ (iii) $a \geq b$ and $c < d$ (iv) $a \leq b$ and $c < d$

PROJECTS OR GROUP ACTIVITIES

A **universal set** U is the set of all elements under consideration. For instance, if our focus were the set of integers, then the universal set would be the set of integers. If our focus were all natural numbers less than 10, then the universal set would be $U = \{1, 2, 3, 4, 5, 6, 7, 8, 9\}$. The complement of a set E, designated by E^c, is the set of elements that belong to the universal set but do not belong to E.

136. Let $U = \{1, 2, 3, 4, 5, 6, 7, 8, 9\}$ and let $E = \{2, 4, 6, 8\}$. Find E^c.

137. Let $U = \{1, 2, 3, 4, 5, 6, 7, 8, 9\}$ and let $E = \{\text{primes less than } 10\}$. Find E^c.

138. Let $U = \{x | x \in \text{natural numbers}\}$ and $E = \{x | x \in \text{odd natural numbers}\}$. Find E^c.

139. Let $U = \{x | x \in \text{real numbers}\}$ and $E = \{x | x \in \text{rational numbers}\}$. Find E^c.

140. If E is a set within the universal set U, find **a.** $E \cup E^c$ and **b.** $E \cap E^c$.

1.2 Operations on Integers

OBJECTIVE 1

Operations on integers

An understanding of the operations on real numbers is necessary to succeed in algebra. Here is a review of the basic operations on real numbers.

Point of Interest

Rules for operating with positive and negative numbers have existed for a long time. Although there are older records of these rules (from the 3rd century), one of the most thorough appears in *The Correct Astronomical System of Brahma*, written by the Indian mathematician Brahmagupta around A.D. 600.

ADDITION OF REAL NUMBERS

Numbers that have the same sign

To add two numbers with the same sign, add the absolute values of the numbers. Then attach the sign of the addends.

Numbers that have different signs

To add two numbers with different signs, find the absolute value of each number. Subtract the smaller of these absolute values from the larger. Then attach the sign of the number with the larger absolute value.

Focus on adding real numbers

Add. **A.** $-65 + (-48)$ **B.** $17 + (-53)$ **C.** $-45 + 81$

A. The signs are the same. Add the absolute values of the numbers.

$$|-65| + |-48| = 65 + 48 = 113$$

Then attach the sign of the addends.

$$-65 + (-48) = -113$$

B. The numbers have different signs. Find the absolute value of each number.

$$|17| = 17 \qquad |-53| = 53$$

Subtract the smaller of these two numbers from the larger one.

$$53 - 17 = 36$$

Attach the sign of the number with larger absolute value. Because $|-53| > |17|$, attach the sign of -53.

$$17 + (-53) = -36$$

C. The numbers have different signs. Find the absolute value of each number.

$$|-45| = 45 \qquad |81| = 81$$

Subtract the smaller of these two numbers from the larger one.

$$81 - 45 = 36$$

Attach the sign of the number with larger absolute value. Because $|81| > |-45|$, attach the sign of 81.

$$-45 + 81 = 36$$

SUBTRACTION OF REAL NUMBERS

To subtract two real numbers, add the opposite of the second number to the first.

Focus on subtracting real numbers

Subtract. **A.** $48 - (-22)$ **B.** $-17 - 37$ **C.** $-25 - (-14)$

Add the opposite of -22.

A. $48 - (-22) = 48 + 22 = 70$

Add the opposite of 37.

B. $-17 - 37 = -17 + (-37) = -54$

Add the opposite of -14.

C. $-25 - (-14) = -25 + 14 = -11$

Many times it is necessary to write a mathematical expression from a verbal one. Here are some words and phrases that are used to indicate addition and subtraction.

Words or phrases for addition		
more than	-3 more than 8	$8 + (-3) = 5$
the sum of	the sum of -5 and -9	$-5 + (-9) = -14$
increased by	-7 increased by 10	$-7 + 10 = 3$
the total of	the total of -23 and 14	$-23 + 14 = -9$
plus	-15 plus -19	$-15 + (-19) = -34$

Words or phrases for subtraction		
minus	12 minus 20	$12 - 20 = 12 + (-20) = -8$
less than	5 less than -9	$-9 - 5 = -9 + (-5) = -14$
less	8 less -9	$8 - (-9) = 8 + 9 = 17$
the difference between	the difference between 3 and -8	$3 - (-8) = 3 + 8 = 11$
decreased by	-7 decreased by 5	$-7 - 5 = -7 + (-5) = -12$

EXAMPLE 1 Solve.

A. What is 27 more than -5?

B. Find the difference between -11 and -8.

Solution **A.** $-5 + 27 = 22$

B. $-11 - (-8) = -11 + 8 = -3$

Problem 1 Solve.

A. What is the sum of -21 and 32?

B. What is -12 less than 7?

Solution See page S1.

Try Exercise 41, page 20.

MULTIPLICATION OR DIVISION OF REAL NUMBERS

Numbers that have the same sign

The product or quotient of two numbers with the same sign is positive.

Numbers that have different signs

The product or quotient of two numbers with different signs is negative.

Take Note

Note in example (4) at the right that we used a fraction bar to denote division. The fraction $\frac{-75}{25}$ is read "−75 divided by 25." The fraction bar is read "divided by."

EXAMPLES

1. $-12(-15) = 180$
2. $-21(14) = -294$
3. $(-65) \div (-5) = 13$
4. $\dfrac{-75}{25} = -3$

Here are some words and phrases that indicate multiplication and division.

Words or phrases for multiplication		
times	5 times −6	$5(-6) = -30$
the product of	the product of −5 and −9	$-5(-9) = 45$
twice	twice −5	$2(-5) = -10$

Words or phrases for division		
the quotient of	the quotient of 15 and −3	$15 \div (-3) = -5$
divided by	−28 divided by −7	$(-28) \div (-7) = 4$

EXAMPLE 2 Solve.

A. Find the quotient of −60 and 12.

B. What is the product of −15 and −5?

Solution **A.** $(-60) \div 12 = -5$

B. $-15(-5) = 75$

Problem 2 Solve.

A. Find −14 times −5.

B. What is −36 divided by 9?

Solution See page S1.

Try Exercise 35, page 20.

Take Note

Recall that for every division problem, there is a related multiplication problem. For example, for the division problem $\frac{12}{3} = 4$, the related multiplication problem is $4 \cdot 3 = 12$.

The relationship between multiplication and division leads to the following properties.

PROPERTIES OF ZERO AND ONE IN DIVISION

1. Any number divided by 1 is the number.

$$\frac{a}{1} = a$$

2. Zero divided by any number other than zero is zero.

$$\frac{0}{a} = 0, a \neq 0$$

3. Division by zero is undefined.
4. Any number other than zero divided by itself is 1.

EXAMPLES

1. $\dfrac{12}{1} = 12 \qquad \dfrac{-5}{1} = -5 \qquad \dfrac{1}{1} = 1$
2. $\dfrac{0}{7} = 0 \qquad \dfrac{0}{-15} = 0$
3. $\frac{4}{0}$ is undefined. $\frac{0}{0}$ is undefined.
4. $\dfrac{12}{12} = 1 \qquad \dfrac{-27}{-27} = 1$

To understand that division by zero is not permitted, suppose $\frac{4}{0} = n$, where n is some number. Because each division problem has a related multiplication problem, $\frac{4}{0} = n$ means $n \cdot 0 = 4$. But $n \cdot 0 = 4$ is impossible because any number times 0 is 0. Therefore, division by 0 is not defined.

Similarly, suppose $\frac{0}{0} = n$. The related multiplication is $0 = n \cdot 0$. The difficulty here is that any number n would make the equation true, so there is no unique answer. Because of this, $\frac{0}{0}$ is not defined.

Repeated multiplication of the same factor can be written using an exponent.

$$2 \cdot 2 \cdot 2 \cdot 2 \cdot 2 \cdot 2 = 2^6 \qquad b \cdot b \cdot b \cdot b \cdot b = b^5$$

(In 2^6 and b^5: 6 and 5 are labeled "Exponent"; 2 and b are labeled "Base".)

The **exponent** indicates how many times the factor, called the **base,** occurs in the multiplication. The multiplication $2 \cdot 2 \cdot 2 \cdot 2 \cdot 2 \cdot 2$ is in **factored form.** The exponential expression 2^6 is in **exponential form.**

2^1 is read "the first power of two" or just "two." ⟶ Usually the exponent 1 is not written.

2^2 is read "the second power of two" or "two squared."

2^3 is read "the third power of two" or "two cubed."

2^4 is read "the fourth power of two."

2^5 is read "the fifth power of two."

b^5 is read "the fifth power of b."

nTH POWER OF *a*

If a is a real number and n is a positive integer, then the ***n*th power of *a*** is the product of n factors of a.

$$a^n = \underbrace{a \cdot a \cdot a \cdot \ldots \cdot a}_{a \text{ as a factor } n \text{ times}}$$

EXAMPLES

1. $5^3 = 5 \cdot 5 \cdot 5 = 125$
2. $(-3)^4 = (-3)(-3)(-3)(-3) = 81$
3. $-3^4 = -(3 \cdot 3 \cdot 3 \cdot 3) = -81$

Note the difference between examples (2) and (3) above. $(-3)^4$ means that we use -3 as a factor 4 times. However, $-3^4 = -(3^4)$. In this case, -3^4 means that we use 3 as a factor 4 times and then find the opposite of the result. Here are a few more examples.

$(-6)^3 = (-6)(-6)(-6) = -216$	Use -6 as a factor 3 times.
$-6^3 = -(6 \cdot 6 \cdot 6) = -216$	Use 6 as a factor 3 times. Then find the opposite.
$(-5)^4 = (-5)(-5)(-5)(-5) = 625$	Use -5 as a factor 4 times.
$-5^4 = -(5 \cdot 5 \cdot 5 \cdot 5) = -625$	Use 5 as a factor 4 times. Then find the opposite.

EXAMPLE 3 Evaluate.

A. $(-7)^3$
B. -4^4
C. $(-2)^4 \cdot (-3)^3$

Solution **A.** $(-7)^3 = (-7)(-7)(-7) =$
B. $-4^4 = -(4 \cdot 4 \cdot 4 \cdot 4) = -256$
C. $(-2)^4 \cdot (-3)^3 = (-2)(-2)(-2)(-2)(-3)(-3)(-3) = -432$

Problem 3 Evaluate.

A. -5^3
B. $(-2)^7$
C. $3^4 \cdot (-2)^2$

Solution See page S1.

Try Exercise 53, page 20.

OBJECTIVE 2 The Order of Operations Agreement

When one expression contains several operations, the Order of Operations Agreement is used to simplify the expression.

ORDER OF OPERATIONS AGREEMENT

Step 1 Perform operations inside grouping symbols. Grouping symbols include parentheses, brackets, the absolute value symbol, and the fraction bar.

Step 2 Simplify exponential expressions.

Step 3 Do multiplications and divisions as they occur from left to right.

Step 4 Do additions and subtractions as they occur from left to right.

EXAMPLES

1. $3(4 - 9) = 3(-5)$	• **Perform operations inside grouping symbols. [Step 1]**
$= -15$	• **Multiply. [Step 3]**
2. $3 \cdot 2^4 = 3 \cdot 16$	• **Simplify exponential expressions. [Step 2]**
$= 48$	• **Multiply. [Step 3]**
3. $12 \div 6 \cdot 3 = 2 \cdot 3$	• **Do multiplications and divisions from left to right. [Step 3]**
$= 6$	
4. $5 - 2 \cdot 4 = 5 - 8$	• **Do multiplications and divisions from left to right. [Step 3]**
$= -3$	• **Do additions and subtractions from left to right. [Step 4]**

EXAMPLE 4 Simplify: $9 \div (6 - 3) - 3(8 - 10)^3$

Solution

$$\begin{aligned} &9 \div (6 - 3) - 3(8 - 10)^3 \\ &= 9 \div 3 - 3(-2)^3 \\ &= 9 \div 3 - 3 \cdot (-8) \\ &= 3 - (-24) \\ &= 27 \end{aligned}$$

- **Perform operations inside grouping symbols.**
- **Simplify exponential expressions.**
- **Do multiplications and divisions from left to right.**
- **Do additions and subtractions from left to right.**

Problem 4 Simplify: $24 - 18 \div 6(3 - 6)^3$

Solution See page S1.

Try Exercise 63, page 21.

EXAMPLE 5 Simplify: $\dfrac{14 - 26}{4} - 18\left(\dfrac{9 - 15}{6}\right) \div 3^2$

Solution

$$\begin{aligned} &\frac{14 - 26}{4} - 18\left(\frac{9 - 15}{6}\right) \div 3^2 \\ &= \frac{-12}{4} - 18\left(\frac{-6}{6}\right) \div 3^2 \\ &= -3 - 18(-1) \div 3^2 \\ &= -3 - 18(-1) \div 9 \\ &= -3 - (-18) \div 9 \\ &= -3 - (-2) \\ &= -1 \end{aligned}$$

- **Perform operations inside grouping symbols. The grouping symbols for this expression are the fraction bars and parentheses. Simplify the numerator in each fraction.**
- **Simplify exponential expressions.**
- **Do multiplications and divisions as they occur from left to right.**
- **Do additions and subtractions as they occur from left to right.**

Problem 5 Simplify: $4 - 2[(25 - 9) \div 2^3]^2$

Solution See page S1.

Try Exercise 69, page 21.

1.2 Exercises

CONCEPT CHECK

1. Explain how to add **a.** two integers with the same sign and **b.** two integers with different signs.
2. Explain how to rewrite $8 - (-12)$ as addition of the opposite.
3. When adding two numbers, is the sum always greater than either of the two numbers being added? If not, give an example.
4. If the product of two numbers is positive, what can be said about the numbers?
5. If the quotient of two numbers is negative, what can be said about the numbers?
6. Is it possible to subtract two negative numbers and get a positive result? If so, give an example.
7. If the product of two numbers is zero, what can be said about the numbers?

8. Write the seventh power of 3 in exponential form.

9. Write the sixth power of -5 in exponential form.

10. Replace the question mark between the two expressions with $=$, $<$, or $>$ to make a true statement.
$8 - 2 \cdot 5$ ___?___ $(8 - 2) \cdot 5$

1 Operations on integers (See pages 13–18.)

GETTING READY

11. Simplify: $-8 - (-3)$
Write subtraction as addition of the opposite. $-8 - (-3) = -8 +$ ___?___
Add. $=$ ___?___

12. Write the division expression $63 \div (-9)$ as the fraction $\frac{?}{?}$. The quotient is ___?___.

13. Without doing the computation, state whether -7^8 simplifies to a positive or a negative number.

14. Which of the following is a positive number?
$-4^5, (-4)^5, -4^6, (-4)^6$

Perform the indicated operation.

15. $2 - (-7)$
16. $192 \div (-32)$
17. $14 + (-26)$
18. $-10(-27)$
19. $29 + (-2)$
20. $24 - (-2)$
21. $-8(29)$
22. $-16(32)$
23. $-20(2)$
24. $210 \div (-30)$
25. $-16 + 33$
26. $-140 \div (-28)$
27. $13 + (-29)$
28. $-28(32)$
29. $-21 - 6$
30. $-16 - 35$
31. $-30(-3)$
32. $34 + (-6)$

33. Find the sum of -4 and -8.
34. What is the product of -5 and 12?
35. What is 48 divided by -12?
36. What number is 5 less than 2?
37. What is the difference between -23 and 41?
38. What is the quotient of -65 and -5?
39. Find 17 decreased by 21.
40. Find -27 increased by 9.
41. Find 16 more than -42.
42. Find the difference between -4 and 14.
43. Find 21 less than -33.
44. What is the total of -21 and -15?

Evaluate.

45. 5^3
46. 3^4
47. -2^3
48. -4^3
49. $(-5)^3$
50. $(-8)^2$
51. $2^2 \cdot 3^4$
52. $4^2 \cdot 3^3$
53. $-2^2 \cdot 3^2$
54. $-3^2 \cdot 5^3$
55. $(-2)^3(-3)^2$
56. $(-4)^3(-2)^3$

Complete Exercises 57 and 58 without using a calculator.

57. State whether the sum, difference, or product is positive or negative.
  a. the sum $567 + (-812)$
  b. the difference $-259 - (-327)$
  c. the product of four positive numbers and three negative numbers
  d. the product of three positive numbers and four negative numbers

58. State whether each quotient is positive, negative, zero, or undefined.

a. $\frac{0}{-91}$ **b.** $-416 \div 52$ **c.** $-\frac{693}{-99}$ **d.** $-87 \div 0$

2 The Order of Operations Agreement (See pages 18–19.)

GETTING READY

59. Simplify: $5^2 - (10 \div 2)^3 \div 5$

$$5^2 - (10 \div 2)^3 \div 5 = 5^2 - (\underline{\ ?\ })^3 \div 5$$
$$= \underline{\ ?\ } - \underline{\ ?\ } \div 5$$
$$= 25 - \underline{\ ?\ }$$
$$= \underline{\ ?\ }$$

- **Perform operations inside grouping symbols.**
- **Simplify exponential expressions.**
- **Do multiplication and division.**
- **Do addition and subtraction.**

60. Why do we need an Order of Operations Agreement?

Simplify.

61. $27 - 12 \div 3 - 5^2$

62. $-3 \cdot 2^3 - 5(1 - 7)$

63. $15 - 3^2(5 - 7) \div 6$

64. $5 - 3(8 \div 4)^2$

65. $4^2 - (5 - 2)^2 \cdot 3$

66. $16 - \frac{5 - 2^4}{3^2 + 2}$

67. $5[(2 - 4) \cdot 3 - 2]$

68. $2[(16 \div 8) - (-2)] + 4$

69. $16 - 4\left(\frac{8 - 2}{3 - 6}\right) \div 2$

70. $25 \div 5\left(\frac{16 + 8}{-2^2 + 8}\right) - 5$

71. $6[3 - (-4 + 2) \div 2]$

72. $12 - 4[2 - (-3 + 5) - 8]$

73. $2(8 - 11) - 12 \div 3 \div 4 + 18 \cdot 3 \div 6$

74. $-3(5 - 9)^3 \div 4 \cdot (-2)$

75. $6 - 4 \cdot 3 - 18 \div (-3)^2$

76. $-9 - 2[4 - (3 - 8)^2] \div 7 - 4$

77. $5 + 3[52 + 4(2 - 5)^3]^2$

78. $-3(5 - 8)^3 + (19 - 7) \div (1 - 3)$

79. $28 \div (7 - 9)^2 \cdot (1 - 3)^4 \div 14$

80. $\frac{9 - 4^2}{(6 - 7)^5} - \frac{3(5 - 7)^2}{2 - 2 \cdot 7}$

81. $\frac{15 - 19}{-2^2} \cdot \frac{3(2 - 7)^2}{3 \cdot 2 - 1} \div \frac{3 \cdot 2 - 2^4}{10 - 4 \cdot 3}$

82. Which expression is equivalent to $32 + 32 \div 4 - 2^3$?
(i) $64 \div 4 - 8$ (ii) $32 + 32 \div (-4)$ (iii) $32 + 8 - 8$ (iv) $32 + 32 \div 8$

83. Which expression is equivalent to $8^2 - 2^2(5 - 3)^3$?
(i) $6^2(2)^3$ (ii) $64 - 4(8)$ (iii) $60(2)^3$ (iv) $64 - 8^3$

APPLYING CONCEPTS

84. What is the tens digit of 11^{22}?

85. What is the ones digit of 7^{18}?

86. What are the last two digits of 5^{33}?

87. What are the last three digits of 5^{234}?

88. Does $(2^3)^4 = 2^{(3^4)}$? If not, which expression is larger?

89. What is the Order of Operations Agreement for a^{b^c}? (*Note:* Even calculators that normally follow the Order of Operations Agreement may not do so for this expression.)

PROJECTS OR GROUP ACTIVITIES

The **proper divisors** of a number are the divisors that are less than the number. For instance, the proper divisors of 12 arc 1, 2, 3, 4, and 6. Using this idea, we can define *deficient numbers, perfect numbers,* and *abundant numbers.*

A **deficient number** is one for which the sum of its proper divisors is less than the number. For instance, 8 is a deficient number. The proper divisors of 8 are 1, 2, and 4. The sum of these numbers is $1 + 2 + 4 = 7$, which is less than 8.

A **perfect number** equals the sum of its proper divisors. For instance, 6 is a perfect number. The proper divisors of 6 are 1, 2, and 3. The sum of these numbers is $1 + 2 + 3 = 6$.

An **abundant number** is one for which the sum of its proper divisors is greater than the number. For instance, 12 is an abundant number. The sum of the divisors is 16, which is greater than 12.

90. Determine whether 20 is deficient, perfect, or abundant.

91. Determine whether 28 is deficient, perfect, or abundant.

92. Is the square of a prime number deficient, perfect, or abundant?

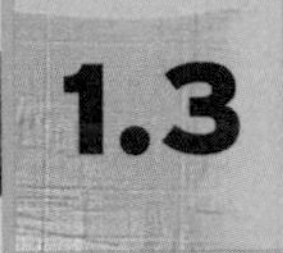

1.3 Operations on Rational Numbers

OBJECTIVE 1 Operations on rational numbers

Our work with fractions frequently requires us to use the least common multiple and greatest common divisor of two or more integers. The **least common multiple (LCM)** of two or more numbers is the smallest number that is a multiple of all the numbers. For instance, 36 is the LCM of 12 and 18 because it is the smallest number that is evenly divisible by both 12 and 18.

Focus on finding the LCM of two numbers

Find the LCM of 12 and 14.

Determine the prime factorization of each number. The common factor is shown in red.

$$12 = 2 \cdot 2 \cdot 3$$
$$14 = 2 \cdot 7$$

Factors of 14

$$\text{LCM} = 2 \cdot 2 \cdot 3 \cdot 7 = 84$$

Factors of 12

- **The LCM is the product of the prime factors of both numbers. Use the common factor only once.**

The LCM of 12 and 14 is 84.

The **greatest common factor (GCF)** of two or more numbers is the greatest number that divides evenly into all the numbers. For instance, the GCF of 12 and 18 is 6, the largest number that divides evenly into 12 and 18.

The GCF can be found by first writing each number as a product of prime factors. The GCF contains the prime factors common to both numbers.

Focus on finding the GCF of two numbers

Find the GCF of 36 and 90.

Determine the prime factorization of each number. The common factors are shown in red.

$$36 = 2 \cdot 2 \cdot 3 \cdot 3$$
$$90 = 2 \cdot 3 \cdot 3 \cdot 5$$

The GCF is the product of the prime factors common to both numbers.

$$\text{GCF} = 2 \cdot 3 \cdot 3 = 18$$

The GCF of 36 and 90 is 18.

How It's Used

Secure websites—those that have URLs that begin with https—use very large prime numbers to encrypt credit card numbers for e-commerce.

The concept of GCF is used when simplifying a rational number. Recall that a rational number is one that can be written in the form $\frac{p}{q}$, where p and q are integers and $q \neq 0$. Examples of rational numbers are $-\frac{5}{9}$ and $\frac{12}{5}$. Because any integer c can be written as $c = \frac{c}{1}$ (for instance, $3 = \frac{3}{1}$), all integers are rational numbers. Also note that if c is a nonzero integer, then $\frac{c}{c} = 1$ (for instance, $\frac{5}{5} = 1$).

A good understanding of the operations on rational numbers is required to succeed in this course.

Take Note

The fractions $\frac{-15}{28}$, $\frac{15}{-28}$, and $-\frac{15}{28}$ all represent the same number. In this text, when an answer is a negative fraction, we will write the negative sign in front of the fraction.

MULTIPLICATION OF FRACTIONS

The product of two fractions is the product of the numerators over the product of the denominators.

$$\frac{a}{b} \cdot \frac{c}{d} = \frac{ac}{bd}$$

EXAMPLES

1. $\frac{2}{3} \cdot \frac{5}{7} = \frac{2 \cdot 5}{3 \cdot 7} = \frac{10}{21}$

2. $\frac{3}{4}\left(-\frac{5}{7}\right) = \frac{3 \cdot (-5)}{4 \cdot 7} = \frac{-15}{28} = -\frac{15}{28}$

A rational number is in **simplest form** when the numerator and denominator contain no common factor greater than 1.

Focus on simplifying a fraction

Simplify: $\frac{30}{45}$

Factor the GCF from the numerator and denominator. The GCF of 30 and 45 is 15. $\quad \frac{30}{45} = \frac{2 \cdot 15}{3 \cdot 15}$

Write as a product of fractions. $\quad = \frac{2}{3} \cdot \frac{15}{15}$

Note that $\frac{15}{15} = 1$. $\quad = \frac{2}{3} \cdot 1$

$$= \frac{2}{3}$$

You can simplify a fraction in steps by dividing the numerator and denominator by a common factor until the fraction is in simplest form.

$$\frac{\overset{10}{\cancel{30}}}{\underset{15}{\cancel{45}}} = \frac{\overset{2}{\cancel{10}}}{\underset{3}{\cancel{15}}} = \frac{2}{3}$$

- **Divide the numerator and denominator by 3. Then divide the resulting numerator and denominator by 5.**

EXAMPLE 1 Multiply: $\left(-\frac{5}{8}\right)\left(-\frac{28}{45}\right)$

Solution $\left(-\frac{5}{8}\right)\left(-\frac{28}{45}\right) = \frac{5 \cdot 28}{8 \cdot 45}$

- **The signs are the same. The product is positive.**

$= \frac{7}{18}$

- **Write the answer in simplest form.**

Problem 1 Multiply: $\frac{10}{21}\left(-\frac{14}{25}\right)$

Solution See page S2.

Try Exercise 25, page 30.

EXAMPLE 2 Evaluate: $\left(-\frac{3}{4}\right)^4$

Solution $\left(-\frac{3}{4}\right)^4 = \left(-\frac{3}{4}\right)\left(-\frac{3}{4}\right)\left(-\frac{3}{4}\right)\left(-\frac{3}{4}\right) = \frac{3 \cdot 3 \cdot 3 \cdot 3}{4 \cdot 4 \cdot 4 \cdot 4} = \frac{81}{256}$

Problem 2 Evaluate: $\left(-\frac{2}{5}\right)^3$

Solution See page S2.

Try Exercise 31, page 31.

The **multiplicative inverse** of a nonzero number a is $\frac{1}{a}$. This number is also called the **reciprocal** of a. For instance, the reciprocal of 2 is $\frac{1}{2}$, and the reciprocal of $-\frac{3}{4}$ is $-\frac{4}{3}$. Division of real numbers is defined in terms of multiplication by the reciprocal.

DIVISION OF FRACTIONS

To divide fractions, multiply by the reciprocal of the divisor.

$$\frac{a}{b} \div \frac{c}{d} = \frac{a}{b} \cdot \frac{d}{c} = \frac{ad}{bc}$$

EXAMPLE 3 Divide: $\frac{3}{8} \div \left(-\frac{9}{16}\right)$

Solution $\frac{3}{8} \div \left(-\frac{9}{16}\right) = -\left(\frac{3}{8} \div \frac{9}{16}\right)$ • **The signs are different. The quotient is negative.**

$= -\left(\frac{3}{8} \cdot \frac{16}{9}\right)$ • **Multiply by the reciprocal of the divisor.**

$= -\frac{2}{3}$ • **Write the answer in simplest form.**

Problem 3 Divide: $\left(-\frac{5}{6}\right) \div \left(-\frac{25}{12}\right)$

Solution See page S2.

Try Exercise 13, page 30.

ADDITION OR SUBTRACTION OF FRACTIONS

The sum or difference of two fractions with the same denominator is the sum or difference of the numerators over the common denominator.

EXAMPLES

1. $\frac{3}{8} + \frac{1}{8} = \frac{3+1}{8} = \frac{4}{8} = \frac{1}{2}$
2. $-\frac{3}{5} + \frac{2}{5} = \frac{-3+2}{5} = \frac{-1}{5} = -\frac{1}{5}$
3. $\frac{4}{7} - \frac{6}{7} = \frac{4-6}{7} = \frac{-2}{7} = -\frac{2}{7}$
4. $\frac{1}{8} - \left(-\frac{5}{8}\right) = \frac{1-(-5)}{8} = \frac{6}{8} = \frac{3}{4}$

To add or subtract rational numbers written as fractions, first rewrite the fractions as equivalent fractions with a common denominator. A common denominator is the product of the denominators of the fractions. The **lowest common denominator (LCD)** is the least common multiple (LCM) of the denominators.

Focus on adding fractions with different denominators

Add: $\frac{5}{6} + \left(-\frac{7}{8}\right)$

The LCM of 6 and 8 is 24. Therefore, the lowest common denominator is 24.

$$\frac{5}{6} + \left(-\frac{7}{8}\right) = \frac{5}{6}\cdot\frac{4}{4} + \left(\frac{-7}{8}\cdot\frac{3}{3}\right)$$

Write each fraction in terms of the lowest common denominator, 24.

$$= \frac{20}{24} + \left(\frac{-21}{24}\right)$$

Add the numerators, and place the sum over the common denominator.

$$= \frac{20 + (-21)}{24}$$

$$= \frac{-1}{24} = -\frac{1}{24}$$

EXAMPLE 4 Subtract: $-\frac{7}{9} - \frac{5}{6}$

Solution

$$
\begin{aligned}
-\frac{7}{9} - \frac{5}{6} &= \frac{-7}{9}\cdot\frac{2}{2} - \frac{5}{6}\cdot\frac{3}{3} \\
&= \frac{-14}{18} - \frac{15}{18} \\
&= \frac{-14 - 15}{18} \\
&= \frac{-29}{18} = -\frac{29}{18}
\end{aligned}
$$

- **The lowest common denominator is 18. Write each fraction in terms of the lowest common denominator.**
- **Subtract the numerators, and place the difference over the common denominator.**

Problem 4 Subtract: $\frac{5}{12} - \left(-\frac{3}{8}\right)$

Solution See page S2.

Try Exercise 23, page 30.

OBJECTIVE 2 Order of Operations and complex fractions

The Order of Operations Agreement can be used to simplify rational expressions.

EXAMPLE 5 Simplify: $-\frac{1}{4} - \frac{7}{5}\left[-\frac{20}{7} + \left(-\frac{5}{4}\right)\right]$

Solution

$$
\begin{aligned}
-\frac{1}{4} - \frac{7}{5}\left[-\frac{20}{7} + \left(-\frac{5}{4}\right)\right] &= -\frac{1}{4} - \frac{7}{5}\left[\frac{-80}{28} + \frac{-35}{28}\right] \\
&= -\frac{1}{4} - \frac{7}{5}\left[\frac{-115}{28}\right] \\
&= -\frac{1}{4} - \left(-\frac{23}{4}\right) \\
&= \frac{22}{4} \\
&= \frac{11}{2}
\end{aligned}
$$

- **Add the fractions in the brackets. The LCD is 28.**
- **Multiply the fractions.**
- **Subtract the fractions.**
- **Write in simplest form.**

Problem 5 Simplify: $\frac{5}{16} - \left(\frac{3}{4}\right)^2 + \frac{5}{8}$

Solution See page S2.

Try Exercise 53, page 31.

EXAMPLE 6 Simplify: $\left(-\frac{3}{4}\right)^2 \div \left(\frac{1}{2} - \frac{2}{3}\right) + \frac{7}{16}$

Solution

$$
\begin{aligned}
&\left(-\frac{3}{4}\right) \div \left(\frac{1}{2} - \frac{2}{3}\right) + \frac{7}{16} \\
&= \left(-\frac{3}{4}\right)^2 \div \left(-\frac{1}{6}\right) + \frac{7}{16} \\
&= \frac{9}{16} \div \left(-\frac{1}{6}\right) + \frac{7}{16}
\end{aligned}
$$

- **Perform operations inside parentheses.** $\frac{1}{2} - \frac{1}{3} = \frac{3}{6} - \frac{4}{6} = -\frac{1}{6}$
- **Simplify exponential expressions.**

$= -\frac{27}{8} + \frac{7}{16}$ • **Perform multiplications and divisions from left to right.**

$= -\frac{47}{16}$ • **Perform additions and subtractions from left to right.**

Problem 6 Simplify: $\left(\frac{4}{3} - \frac{5}{6}\right)^2 + \frac{7}{8} \div \frac{3}{4}$

Solution See page S2.

Try Exercise 55, page 31.

A **complex fraction** is a fraction in which the numerator or denominator contains a fraction. Here are some examples of complex fractions.

$$\text{Main fraction bar} \rightarrow \frac{\frac{1}{6} - \frac{3}{4}}{\frac{2}{5} + \frac{1}{4}} \qquad \frac{\frac{3}{4}}{\frac{1}{2} - \frac{2}{3}} \qquad \frac{\frac{1}{5}}{\frac{3}{4}}$$

The **main fraction bar** separates the numerator and denominator of a complex fraction.

The Order of Operations Agreement can be used to simplify a complex fraction. The main fraction bar is a grouping symbol and can be read as "divided by." For instance, the first complex fraction above can be thought of as

$$\frac{\frac{1}{6} - \frac{3}{4}}{\frac{2}{5} + \frac{1}{4}} = \left(\frac{1}{6} - \frac{3}{4}\right) \div \left(\frac{2}{5} + \frac{1}{4}\right)$$

Looking at the complex fraction in this form gives us a method for simplifying it. Simplify the numerator; simplify the denominator; divide the two results.

Focus on simplifying a complex fraction

Simplify: $\dfrac{\frac{1}{6} - \frac{3}{4}}{\frac{2}{5} + \frac{1}{4}}$

Simplify the numerator and denominator of the complex fraction. $\dfrac{\frac{1}{6} - \frac{3}{4}}{\frac{2}{5} + \frac{1}{4}} = \dfrac{-\frac{7}{12}}{\frac{13}{20}}$

Rewrite the complex fraction as division. $= \left(-\frac{7}{12}\right) \div \frac{13}{20}$

Multiply by the reciprocal of the divisor. $= \left(-\frac{7}{12}\right) \cdot \frac{20}{13}$

Simplify. $= -\frac{35}{39}$

A complex fraction may occur within an expression, as shown in Example 7.

EXAMPLE 7 Simplify: $\frac{1}{2} - \dfrac{\frac{3}{4} + \frac{1}{3}}{\frac{5}{8} - \frac{1}{6}}$

Solution

$$\frac{1}{2} - \dfrac{\frac{3}{4} + \frac{1}{3}}{\frac{5}{8} - \frac{1}{6}} = \frac{1}{2} - \dfrac{\frac{13}{12}}{\frac{11}{24}}$$

• **Simplify the numerator and denominator of the complex fraction.**

$$= \frac{1}{2} - \frac{13}{12} \div \frac{11}{24}$$

• **Rewrite the complex fraction as division.**

$$= \frac{1}{2} - \frac{13}{12} \cdot \frac{24}{11}$$

• **Multiply by the reciprocal.**

$$= \frac{1}{2} - \frac{26}{11}$$

• **Perform multiplications and divisions from left to right.**

$$= \frac{11}{22} - \frac{52}{22} = -\frac{41}{22}$$

• **Perform additions and subtractions from left to right.**

Problem 7 Simplify: $\dfrac{\frac{2}{3} - \frac{3}{4}}{\frac{3}{10} - \frac{1}{5}} + \left(\frac{2}{3}\right)^2$

Solution See page S2.

Try Exercise 67, page 32.

OBJECTIVE 3 Decimal notation

As mentioned in Section 1.1, terminating and repeating decimals can be represented by rational numbers. We can find the decimal representation of a rational number written as a fraction by dividing the numerator by the denominator.

Take Note

The calculator screen below shows the result of 8 ÷ 55. Note that the last digit has been rounded.

Write $\frac{3}{8}$ as a decimal.

Divide 3 by 8.

$$\begin{array}{r} 0.375 \\ 8\overline{)3.000} \\ -2\,4 \\ \hline 60 \\ -56 \\ \hline 40 \\ -40 \\ \hline 0 \end{array}$$

0.375 ← This is a terminating decimal.

0 ← The remainder is zero.

$\frac{3}{8} = 0.375$

Write $\frac{8}{55}$ as a decimal.

Divide 8 by 55.

$$\begin{array}{r} 0.14\overline{545} \\ 55\overline{)8.00000} \\ -5\,5 \\ \hline 2\,50 \\ -2\,20 \\ \hline 300 \\ -275 \\ \hline 250 \\ -220 \\ \hline 300 \\ -275 \\ \hline 25 \end{array}$$

← This is a repeating decimal. Note that the differences shown in red and blue are repeating. When a difference repeats, the repeating digits have been found.

$\frac{8}{55} = 0.1\overline{45}$ ← A bar is placed over the repeating digits.

EXAMPLE 8 Find the decimal representation of $\frac{27}{220}$. Place a bar over the repeating digits.

Solution

$$
\begin{array}{r}
0.122727\overline{27} \\
220\overline{)27.000000} \\
-22\,0 \\
\hline
5\,00 \\
-4\,40 \\
\hline
600 \\
-440 \\
\hline
1600 \\
-1540 \\
\hline
600 \\
-440 \\
\hline
1600 \\
-1540 \\
\hline
60
\end{array}
$$

The difference 60 begins to repeat. We could have stopped here rather than continue.

$\frac{27}{220} = 0.12\overline{27}$

Problem 8 Find the decimal representation of $\frac{5}{7}$. Place a bar over the repeating digits.

Solution See page S2.

Try Exercise 81, page 32.

The rules for addition, subtraction, multiplication, and division of real numbers are used to perform operations on decimals.

Focus on operations on decimals

A. To add two decimals with the same sign, add the absolute values of the decimals. Then attach the sign of the addends.

Add: $-15.23 + (-18.1)$

$-15.23 + (-18.1) = -33.33$

B. To subtract two decimals, write subtraction as addition of the opposite, and then add.

Subtract: $-18.42 - (-9.354)$

$-18.42 - (-9.354) = -18.42 + 9.354$
$= -9.066$

C. The product of two factors with different signs is negative.

Multiply: $(-0.23)(0.04)$

$(-0.23)(0.04) = -0.0092$

D. The quotient of two numbers with the same sign is positive.

Divide: $(-2.835) \div (-1.35)$

$(-2.835) \div (-1.35) = 2.1$

Here is an example using the Order of Operations Agreement with decimals.

EXAMPLE 9 Simplify: $5.4 - 2.7[-6 - (-1.7)]^2$

Solution

$5.4 - 2.7[-6 - (-1.7)]^2$

$= 5.4 - 2.7[-4.3]^2$ • **Perform operations within grouping symbols.**

$= 5.4 - 2.7(18.49)$ • **Simplify exponential expressions.**

$= 5.4 - 49.923$ • **Perform multiplications and divisions from left to right.**

$= -44.523$ • **Perform additions and subtractions from left to right.**

Problem 9 Simplify: $6.4 \div (-0.8) + 1.2(0.3^2 - 0.2)$

Solution See page S2.

Try Exercise 105, page 32.

1.3 Exercises

CONCEPT CHECK

1. What is the LCM of two numbers? How can the LCM be used when adding fractions?
2. What is the GCF of two numbers? How can the GCF be used when simplifying fractions?
3. Are all integers rational numbers?
4. Give two examples of rational numbers that are not integers.
5. Is there a smallest positive integer? Is there a smallest positive rational number?
6. Do all rational numbers have a reciprocal? If not, give an example.

1 Operations on rational numbers (See pages 22–26.)

GETTING READY

7. The least common multiple of the denominators of the fractions $\frac{5}{8}$, $-\frac{1}{28}$, and $\frac{2}{7}$ is ___?___. Write each fraction using the LCM as the denominator:

$$\frac{5}{8} = \frac{?}{56}, -\frac{1}{28} = -\frac{?}{56}, \text{ and } \frac{2}{7} = \frac{?}{56}.$$

8. The reciprocal of $-\frac{8}{27}$ is ___?___. To find the quotient $\frac{2}{3} \div \left(-\frac{8}{27}\right)$, find the product $\frac{2}{3} \cdot$ (___?___). The quotient $\frac{2}{3} \div \left(-\frac{8}{27}\right)$ is ___?___.

Perform the indicated operation.

9. $\frac{4}{5}\left(-\frac{2}{5}\right)$
10. $-\frac{23}{2} + \frac{5}{2}$
11. $-\frac{35}{3} - \left(-\frac{3}{5}\right)$
12. $-\frac{3}{14}\left(-\frac{4}{3}\right)$
13. $-\frac{15}{8} \div \left(-\frac{14}{3}\right)$
14. $-\frac{13}{3} - \frac{19}{6}$
15. $\frac{8}{19} + \left(-\frac{17}{2}\right)$
16. $\frac{39}{5} + \left(-\frac{37}{9}\right)$
17. $\left(-\frac{3}{2}\right)^4$
18. $-\frac{14}{3}\left(\frac{30}{13}\right)$
19. $\left(-\frac{9}{2}\right)^3$
20. $-\frac{14}{3}\left(-\frac{11}{14}\right)$
21. $-\frac{9}{16} - \frac{35}{2}$
22. $-\frac{15}{7} - \frac{15}{2}$
23. $-\frac{11}{6} + \left(-\frac{9}{2}\right)$
24. $-\frac{1}{3} \div \frac{33}{16}$
25. $\frac{8}{3}\left(-\frac{8}{7}\right)$
26. $\left(-\frac{5}{3}\right)^3$
27. $\frac{9}{2} + \left(-\frac{3}{5}\right)$
28. $-\frac{39}{4} + \frac{19}{5}$

29. $\frac{33}{2} \div \left(-\frac{31}{20}\right)$

30. $\frac{19}{4} - \left(-\frac{12}{11}\right)$

31. $\left(-\frac{1}{10}\right)^2$

32. $-\frac{25}{2} - \frac{14}{19}$

33. $\frac{6}{7} \div \left(-\frac{7}{5}\right)$

34. $\frac{29}{14} \div \left(-\frac{38}{3}\right)$

35. $\frac{27}{11}\left(-\frac{5}{2}\right)$

36. $\left(-\frac{2}{5}\right)^3$

37. Find the sum of $-\frac{3}{4}$ and $\frac{5}{6}$.

38. What number is $\frac{2}{3}$ more than $-\frac{11}{15}$?

39. What is the difference between $-\frac{11}{12}$ and $\frac{13}{18}$?

40. What is the product of $-\frac{15}{16}$ and $\frac{8}{3}$?

41. What number is $\frac{2}{3}$ less than $-\frac{4}{5}$?

42. Find $\frac{8}{9}$ decreased by $\frac{5}{6}$.

43. What is the quotient of $\frac{5}{8}$ and $-\frac{7}{5}$?

44. Find $-\frac{8}{3}$ increased by $-\frac{32}{11}$.

45. Find $\frac{13}{12}$ more than $-\frac{5}{8}$.

46. Find the difference between $-\frac{7}{9}$ and $-\frac{5}{6}$.

47. Given any two integers, is it possible to find an integer between the given integers? Explain.

48. Given any two rational numbers, is it possible to find a rational number between the given rational numbers? Explain.

2 Order of Operations and complex fractions (See pages 26–28.)

GETTING READY

49. A complex fraction is a fraction that has one or more ___?___ in its numerator and/or denominator.

50. The complex fraction $\frac{3}{\frac{1}{2}}$ can be rewritten as the division expression ___?___ $\div$ ___?___.

Simplify.

51. $-\frac{3}{4} \div \frac{5}{8} \cdot \left(-\frac{10}{11}\right)$

52. $\frac{1}{2} \div \left(-\frac{4}{5}\right) \div \frac{8}{9}$

53. $\left(\frac{2}{3} - \frac{5}{6}\right)^2 - \frac{5}{12}$

54. $\frac{3}{4} - \frac{5}{3}\left(-\frac{5}{4}\right)$

55. $-\frac{5}{16} \cdot \frac{8}{9} - \frac{5}{6} \div \frac{3}{4}$

56. $-\frac{9}{8}\left(\frac{5}{18} - \frac{3}{4}\right) + \frac{5}{8}$

57. $\frac{1}{2} - \left(\frac{2}{3} \div \frac{5}{9}\right) + \frac{5}{6}$

58. $\frac{3}{4} \div \left[\frac{5}{8} - \frac{5}{12}\right] + 2$

59. $\frac{1}{6} - \frac{5}{4}\left(-\frac{7}{12} + \frac{1}{24}\right)$

60. $\frac{2}{3} - \left[\frac{3}{8} + \frac{5}{6}\right] \div \frac{3}{5}$

61. $\left(-\frac{1}{2}\right)^3 \div \left(-\frac{3}{2} - \frac{1}{4}\right) - \frac{2}{3}$

62. $\left(\frac{7}{8} - \frac{11}{12}\right) \div \left(\frac{3}{4} - \frac{4}{5}\right) - \frac{1}{2}\left(\frac{5}{8} - \frac{2}{3}\right)$

63. $\dfrac{\frac{2}{3}}{\frac{4}{5}}$

64. $\dfrac{-\frac{5}{6}}{\frac{2}{3}}$

65. $\dfrac{\frac{2}{3} - \frac{5}{6}}{\frac{3}{4} - \frac{1}{2}}$

66. $\dfrac{\frac{1}{8}-\frac{1}{12}}{\frac{2}{3}-\frac{3}{4}}$

67. $\dfrac{\frac{2}{3}-\left(\frac{1}{2}\right)^2}{\frac{5}{4}\left(\frac{1}{2}-\frac{3}{4}\right)}$

68. $\dfrac{\left(\frac{2}{3}-\frac{3}{4}\right)^3}{\frac{3}{8}\left(\frac{5}{6}-\frac{5}{12}\right)}$

69. $\dfrac{5}{8}-\dfrac{\frac{2}{3}+\frac{1}{4}}{\frac{5}{6}-\frac{7}{8}}$

70. $\dfrac{1}{2}+\dfrac{\frac{5}{18}-\frac{7}{9}}{\frac{1}{2}+\frac{2}{3}}$

71. $\dfrac{1}{2}-\dfrac{\frac{17}{25}}{4-\frac{3}{5}}+\dfrac{1}{5}$

72. $\dfrac{3}{4}+\dfrac{3-\frac{7}{9}}{\frac{5}{6}}\cdot\dfrac{2}{3}$

73. $\dfrac{\frac{1-2\cdot 3}{4(5-4)}}{\frac{3-5\cdot 2}{3\cdot 5-1}}$

74. $\dfrac{\frac{3(2-5)}{-2(2-6)}}{\frac{1-3(2-5)}{2\cdot 3-1}}$

75. Which of the following complex fractions simplify to integer values?

(i) $\dfrac{3}{\frac{1}{3}}$ (ii) $\dfrac{\frac{1}{3}}{\frac{1}{3}}$ (iii) $\dfrac{\frac{1}{3}}{3}$ (iv) $\dfrac{1}{\frac{1}{3}}$

76. The complex fraction $\dfrac{\frac{2}{5}}{\frac{1}{\ }}$ simplifies to an integer value. What must be true about the number that replaces the question mark?

3 Decimal notation (See pages 28–30.)

Find a terminating or repeating decimal equivalent to each fraction.

77. $\dfrac{5}{8}$ 78. $\dfrac{5}{16}$ 79. $\dfrac{1}{6}$ 80. $\dfrac{7}{12}$

81. $\dfrac{8}{55}$ 82. $\dfrac{19}{99}$ 83. $\dfrac{3}{13}$ 84. $\dfrac{2}{26}$

GETTING READY

85. The sum of 0.773 and −81.5 will have __?__ decimal places.

86. The product of 0.773 and −81.5 will have __?__ decimal places.

Perform the indicated operation.

87. $0.0015 + (-0.0027)$
88. $0.31(-0.1)$
89. $-0.0008 + 3.5$
90. $0.0022(-0.8)$
91. $0.0003 + (-0.39)$
92. $-0.031 \div 3.1$
93. $-0.024(-0.019)$
94. $0.0029 + (-0.003)$
95. $0.000072 \div (-0.004)$
96. $-0.000189 \div (-0.0009)$
97. $-0.0585 \div 4.5$
98. $0.02 + (-0.4)$
99. $3.8(-3.9)$
100. $-0.091 \div (-3.5)$
101. $-0.0026 - (-0.028)$
102. $2.7 - (-0.007)$
103. $-0.000192 \div 0.016$
104. $-0.18 - 0.007$

Simplify.

105. $0.4(1.2 - 2.3)^2 + 5.8$
106. $5.4 - (0.3)^2 \div 0.09$
107. $1.75 \div 0.25 - (1.25)^2$
108. $(3.5 - 4.2)^2 - 3.50 \div 2.5$
109. $25.76 \div (6.54 \div 3.27)^2$
110. $(3.09 - 4.77)^2 - 4.07 \cdot 3.66$

111. Is the decimal representation of $\frac{5}{23}$ a non-terminating, non-repeating decimal? Explain.

112. A student entered the expression at the right into a calculator as shown to find the value of $\frac{2}{3} \div \frac{3}{4}$. Is this correct? If not, why not?

```
2/3/3/4
              .0555555556
Ans▸Frac
                     1/18
```

APPLYING CONCEPTS

Simplify.

113. $2 + \cfrac{2}{2 - \cfrac{2}{2 + 1}}$

114. $1 + \cfrac{2}{3 + \cfrac{4}{5 + \cfrac{6}{7 + 8}}}$

115. $3 - \cfrac{1}{3 - \cfrac{1}{3 - \cfrac{1}{3}}}$

116. $\cfrac{1}{\cfrac{1}{2} - \cfrac{1}{\cfrac{1}{2} - \cfrac{1}{1 - \cfrac{1}{2}}}}$

PROJECTS OR GROUP ACTIVITIES

For Exercises 117 to 122, determine whether the statement is true or false. If the statement is false, give an example to show that it is false.

117. Given any two rational numbers a and b, there is a rational number between a and b.

118. The product of any two rational numbers is a rational number.

119. The sum of any two rational numbers is a rational number.

120. The product of any two irrational numbers is an irrational number.

121. The sum of any two irrational numbers is an irrational number.

122. The sum of a rational number and an irrational number is an irrational number.

123. Find a rational number between $\frac{1}{3}$ and $\frac{1}{2}$.

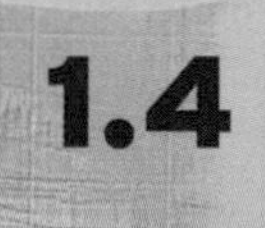

1.4 Variable Expressions

OBJECTIVE 1 The Properties of the Real Numbers

The Properties of the Real Numbers describe the ways in which operations on numbers can be performed. Following is a list of some of the Properties of the Real Numbers and an example of each property.

THE COMMUTATIVE PROPERTY OF ADDITION	EXAMPLE
$a + b = b + a$	$3 + 2 = 2 + 3$ $5 = 5$

THE COMMUTATIVE PROPERTY OF MULTIPLICATION

$$a \cdot b = b \cdot a$$

EXAMPLE

$$(3)(-2) = (-2)(3)$$
$$-6 = -6$$

THE ASSOCIATIVE PROPERTY OF ADDITION

$$(a + b) + c = a + (b + c)$$

EXAMPLE

$$(3 + 4) + 5 = 3 + (4 + 5)$$
$$7 + 5 = 3 + 9$$
$$12 = 12$$

THE ASSOCIATIVE PROPERTY OF MULTIPLICATION

$$(a \cdot b) \cdot c = a \cdot (b \cdot c)$$

EXAMPLE

$$(3 \cdot 4) \cdot 5 = 3 \cdot (4 \cdot 5)$$
$$12 \cdot 5 = 3 \cdot 20$$
$$60 = 60$$

THE ADDITION PROPERTY OF ZERO

$$a + 0 = 0 + a = a$$

EXAMPLE

$$3 + 0 = 0 + 3 = 3$$

THE MULTIPLICATION PROPERTY OF ZERO

$$a \cdot 0 = 0 \cdot a = 0$$

EXAMPLE

$$8 \cdot 0 = 0 \cdot 8 = 0$$

THE MULTIPLICATION PROPERTY OF ONE

$$a \cdot 1 = 1 \cdot a = a$$

EXAMPLE

$$5 \cdot 1 = 1 \cdot 5 = 5$$

THE INVERSE PROPERTY OF ADDITION

$$a + (-a) = (-a) + a = 0$$

EXAMPLE

$$4 + (-4) = (-4) + 4 = 0$$

$-a$ is called the **additive inverse** of a. Also, a is the additive inverse of $-a$. The sum of a number and its additive inverse is 0.

THE INVERSE PROPERTY OF MULTIPLICATION

$$a \cdot \frac{1}{a} = \frac{1}{a} \cdot a = 1, \quad a \neq 0$$

EXAMPLE

$$(4)\left(\frac{1}{4}\right) = \left(\frac{1}{4}\right)(4) = 1$$

$\frac{1}{a}$ is called the **multiplicative inverse** of a. It is also called the **reciprocal** of a. The product of a number and its multiplicative inverse is 1.

THE DISTRIBUTIVE PROPERTY

$$a(b + c) = ab + ac$$

EXAMPLE

$$3(4 + 5) = 3 \cdot 4 + 3 \cdot 5$$
$$3 \cdot 9 = 12 + 15$$
$$27 = 27$$

EXAMPLE 1 Complete the statement by using the Inverse Property of Addition.

$3x + ? = 0$

Solution $3x + (-3x) = 0$

Problem 1 Complete the statement by using the Commutative Property of Multiplication.

$(x)\left(\frac{1}{4}\right) = (?)(x)$

Solution See page S2.

Try Exercise 25, page 39.

EXAMPLE 2 Identify the property that justifies the statement.
$3(x + 4) = 3x + 12$

Solution The Distributive Property

Problem 2 Identify the property that justifies the statement.
$(a + 3b) + c = a + (3b + c)$

Solution See page S2.

Try Exercise 29, page 39.

OBJECTIVE 

Evaluate variable expressions

An expression that contains one or more variables is called a **variable expression.**

A variable expression is shown at the right. The expression has four addends that are called **terms** of the expression. The variable expression has three **variable terms** and one **constant term.**

$$\underbrace{\overbrace{4x^2y \quad - \quad 2z \quad - \quad x}^{}}_{\text{Variable terms}} \quad \underbrace{+ \quad 4}_{\text{Constant term}} \qquad \text{(4 terms)}$$

Each variable term is composed of a **numerical coefficient** and a **variable part.** When the numerical coefficient is 1 or −1, the 1 is usually not written.

Numerical coefficient

$$4x^2y \quad - \quad 2z \quad - \quad 1x \quad + \quad 4$$

Variable part

Replacing the variable in a variable expression by a numerical value and then simplifying the resulting expression is called **evaluating the variable expression.**

Focus on evaluating a variable expression

Evaluate $3 - 2|3x - 2y^2|$ when $x = -1$ and $y = 2$.

$3 - 2|3x - 2y^2|$

Replace each variable with its value.

$3 - 2|3(-1) - 2(2)^2|$

Use the Order of Operations Agreement to simplify the resulting numerical expression.

$= 3 - 2|3(-1) - 2(4)|$
$= 3 - 2|-3 - 8|$
$= 3 - 2|-11|$
$= 3 - 2(11)$
$= 3 - 22$
$= -19$

EXAMPLE 3 Evaluate $a^2 - (ab - c)$ when $a = -2$, $b = 3$, and $c = -4$.

Solution $a^2 - (ab - c)$

$(-2)^2 - [(-2)(3) - (-4)]$ • **Replace each variable in the expression with its value.**

$= (-2)^2 - [-6 - (-4)]$ • **Use the Order of Operations Agreement to simplify the resulting numerical expression.**
$= (-2)^2 - [-2]$
$= 4 - [-2]$
$= 6$

Problem 3 Evaluate $(b - c)^2 \div ab$ when $a = -3$, $b = 2$, and $c = -4$.

Solution See page S3.

Try Exercise 49, page 40.

EXAMPLE 4 The radius of the base of a cylinder is 3 in., and the height is 6 in. Find the volume of the cylinder. Round to the nearest hundredth.

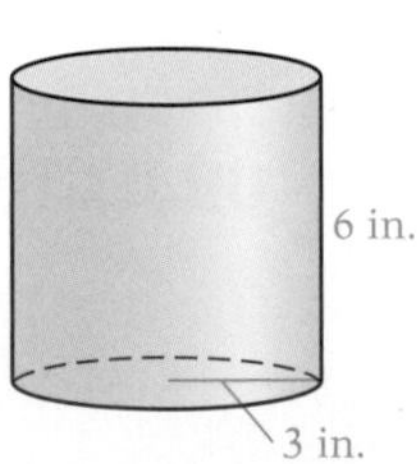

Solution $V = \pi r^2 h$ • Use the formula for the volume of a cylinder.

$V = \pi(3)^2 6$ • Substitute 3 for r and 6 for h.

$V = 54\pi$ • The exact volume of the cylinder is 54π in^3.

$V \approx 169.65$ • An approximate measure can be found by using the π key on a calculator.

The approximate volume of the cylinder is 169.65 in^3.

Take Note
For your reference, geometric formulas are provided at the back of this textbook.

Problem 4 Find the surface area of a right circular cone with a radius of 5 cm and a slant height of 12 cm. Give both the exact area and an approximation to the nearest hundredth.

12 cm

5 cm

Solution See page S3.

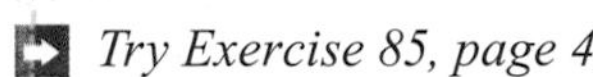
Try Exercise 85, page 41.

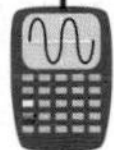

A graphing calculator can be used to evaluate variable expressions. When the value of each variable is stored in the calculator's memory, and a variable expression is then entered into the calculator, the calculator evaluates that variable expression for the values of the variables stored in its memory. See the Appendix for a description of keystroking procedures.

OBJECTIVE 3 Simplify variable expressions

Like terms of a variable expression are terms with the same variable part.

Constant terms are like terms.

Like terms

$4x - 5 + 7x^2 + 3x - 9$

Like terms

To **combine like variable terms,** use the Distributive Property $ba + ca = (b + c)a$ to add the coefficients.

$4x + 3x = (4 + 3)x = 7x$

EXAMPLE 5 Simplify: $2(x + y) + 3(y - 3x)$

Solution $2(x + y) + 3(y - 3x)$

$= 2x + 2y + 3y - 9x$ • Use the Distributive Property to remove parentheses.

$= (2x - 9x) + (2y + 3y)$ • Use the Commutative and Associative Properties of Addition to rearrange and group like terms.

$= -7x + 5y$ • Combine like terms.

Problem 5 Simplify: $(2x + xy - y) - (5x - 7xy + y)$

Solution See page S3.

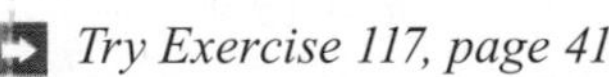
Try Exercise 117, page 41.

EXAMPLE 6 Simplify: $4y - 2[x - 3(x + y) - 5y]$

Solution

$4y - 2[x - 3(x + y) - 5y]$
$= 4y - 2[x - 3x - 3y - 5y]$ • Use the Distributive Property to remove parentheses.
$= 4y - 2[-2x - 8y]$ • Combine like terms.
$= 4y + 4x + 16y$ • Use the Distributive Property to remove brackets.
$= 4x + 20y$ • Combine like terms.

Problem 6 Simplify: $2x - 3[y - 3(x - 2y + 4)]$

Solution See page S3.

Try Exercise 113, page 41.

1.4 Exercises

CONCEPT CHECK

1. Which of the four operations of addition, subtraction, multiplication, and division have a commutative property?
2. Which of the four operations of addition, subtraction, multiplication, and division have an associative property?
3. What is the additive inverse of $-a$?
4. If c is a nonzero number, what is the multiplicative inverse of c?
5. What property of real numbers is illustrated by $2(x + y) = 2x + 2y$?
6. Write any two terms that are like terms with x.
7. Are $2z$ and $2z^2$ like terms? Why or why not?
8. Are $-4a^2b^3c$ and $\frac{2ca^2b^3}{3}$ like terms? Why or why not?

1 The Properties of the Real Numbers (See pages 33–35.)

GETTING READY

9. The fact that two terms can be multiplied in either order is called the __?__ Property of Multiplication.
10. The fact that three or more addends can be added by grouping them in any order is called the __?__ Property of Addition.
11. The Multiplication Property of Zero tells us that the product of a number and zero is __?__.
12. The Addition Property of Zero tells us that the sum of a number and __?__ is the number.

Use the given Property of the Real Numbers to complete the statement.

13. The Commutative Property of Multiplication
$3 \cdot 4 = 4 \cdot ?$

14. The Commutative Property of Addition
$7 + 15 = ? + 7$

15. The Associative Property of Addition
$(3 + 4) + 5 = ? + (4 + 5)$

16. The Associative Property of Multiplication
$(3 \cdot 4) \cdot 5 = 3 \cdot (? \cdot 5)$

17. A Division Property of Zero
$\frac{5}{?}$ is undefined.

18. The Addition Property of Zero
$4 + ? = 4$

19. The Distributive Property
$3(x + 2) = 3x + ?$

20. The Distributive Property
$5(y + 4) = ? \cdot y + 20$

21. A Division Property of Zero
$\frac{?}{-6} = 0$

22. The Inverse Property of Addition
$(x + y) + ? = 0$

23. The Inverse Property of Multiplication
$\frac{1}{mn}(mn) = ?$

24. The Multiplication Property of One
$? \cdot 1 = x$

25. The Associative Property of Multiplication
$2(3x) = ? \cdot x$

26. The Commutative Property of Addition
$ab + bc = bc + ?$

Identify the property that justifies the statement.

27. $\frac{0}{-5} = 0$

28. $-8 + 8 = 0$

29. $(-12)\left(-\frac{1}{12}\right) = 1$

30. $(3 \cdot 4) \cdot 2 = 2 \cdot (3 \cdot 4)$

31. $y + 0 = y$

32. $2x + (5y + 8) = (2x + 5y) + 8$

33. $\frac{-9}{0}$ is undefined.

34. $(x + y)z = xz + yz$

35. $6(x + y) = 6x + 6y$

36. $0 + 2 = 2$

37. $(ab)c = a(bc)$

38. $(x + y) + z = (y + x) + z$

39. The sum of a number n and its additive inverse is multiplied by the reciprocal of the number n. What is the result?

40. The product of a number n and its reciprocal is multiplied by the number n. What is the result?

2 Evaluate variable expressions (See pages 35–37.)

41. Explain the meaning of the phrase "evaluate a variable expression."

42. Explain the difference between the meaning of "the value of the variable" and the meaning of "the value of the variable expression."

GETTING READY

43. Evaluate $n^2|m - n|$ when $n = -5$ and $m = -3$.

$n^2|m - n|$
$= (-5)^2|-3 - (-5)|$ • **Replace n with __?__ and m with __?__.**
$= (-5)^2|$__?__$|$ • **Perform operations inside the grouping symbols.**
$= (-5)^2($__?__$)$ • **Evaluate the absolute value.**
$= ($__?__$)(2)$ • **Simplify the exponential expression.**
$=$ __?__ • **Multiply.**

44. The volume of a cone with height h and base of radius r is $V =$ __?__. If $r = 2$ in. and $h = 3$ in., then the exact volume of the cone is $V = \frac{1}{3}\pi($__?__$)^2($__?__$)$ in^3 = (__?__)π in^3. Use the $\boxed{\pi}$ key on a calculator to approximate the volume to the nearest hundredth: $V \approx$ __?__ in^3.

Evaluate the variable expression when $a = 2$, $b = 3$, $c = -1$, and $d = -4$.

45. $ab + dc$

46. $2ab - 3dc$

47. $b - a(2c + d)$

48. $b^2 - (d - c)^2$

49. $(b - 2a)^2 + c$

50. $(b - d)^2 \div (b - d)$

51. $(bc + a)^2 \div (d - b)$

52. $2d^2 - a^2c^3$

53. $b^2 - 4ac$

54. $\frac{a}{2b} - \frac{2c}{d}$

55. $\frac{3ac}{-4} - c^2$

56. $\frac{a - \frac{1}{a}}{d + \frac{1}{d}}$

57. $\frac{3b - 5c}{3a - c}$

58. $\frac{2d - a}{b - 2c}$

59. $\frac{a - d}{b + c}$

60. $|a^2 + d|$

61. $-a|a + 2d|$

62. $d|b - 2d|$

63. $\frac{2a - 4d}{3b - c}$

64. $\frac{3d - b}{b - 2c}$

65. $-3d \div \left|\frac{ab - 4c}{2b + c}\right|$

66. $-2bc + \left|\frac{bc + d}{ab - c}\right|$

67. $2(d - b) \div (3a - c)$

68. $2d^3 + 4a^2b^3$

69. $-d^2 - c^3a$

70. $a^2c - d^3$

71. $-d^3 + 4ac$

72. b^a

73. 4^{a^2}

74. a^b

Geometry Find the volume of each figure. For calculations involving π, give both the exact value and an approximation to the nearest hundredth.

75.

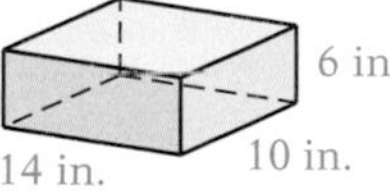

76.

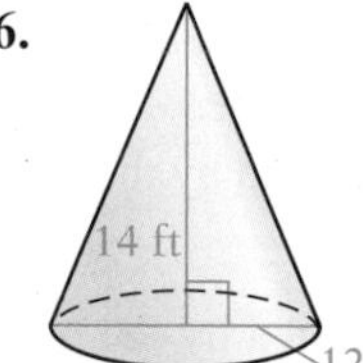

77.

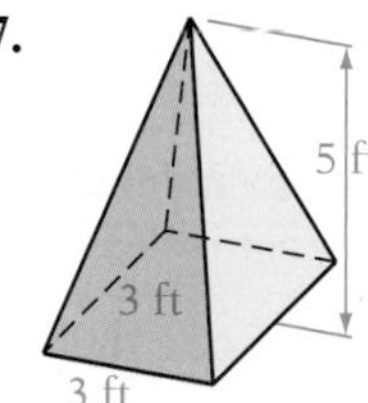

78.

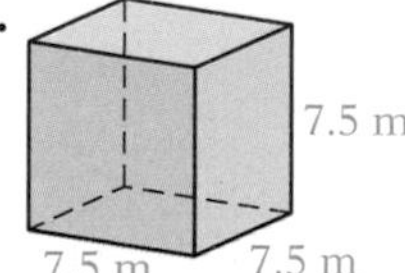

79.

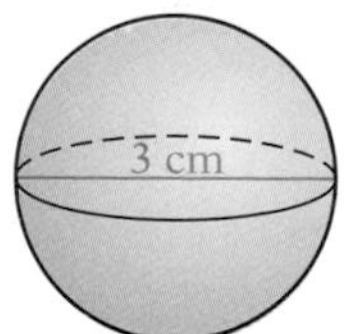

80.

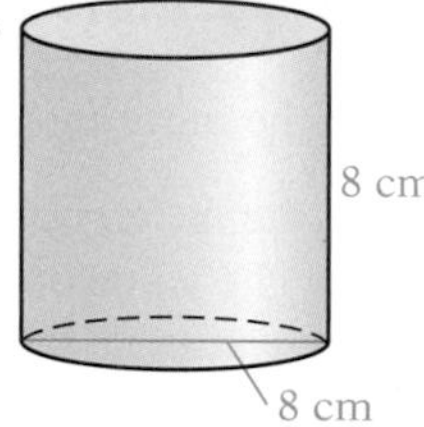

Geometry Find the surface area of each figure. For calculations involving π, give both the exact value and an approximation to the nearest hundredth.

81.

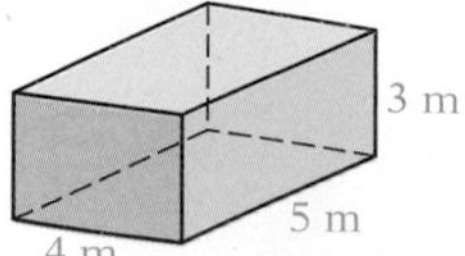

82.

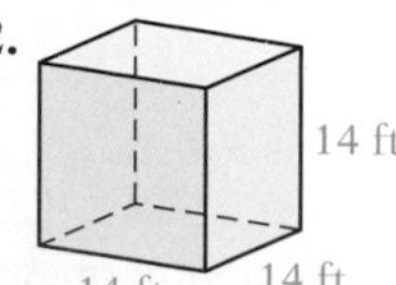

83.

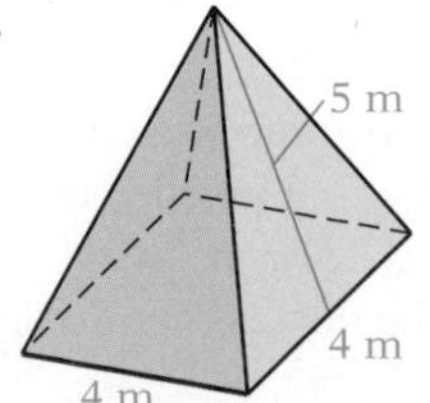

84. 2 cm

85. 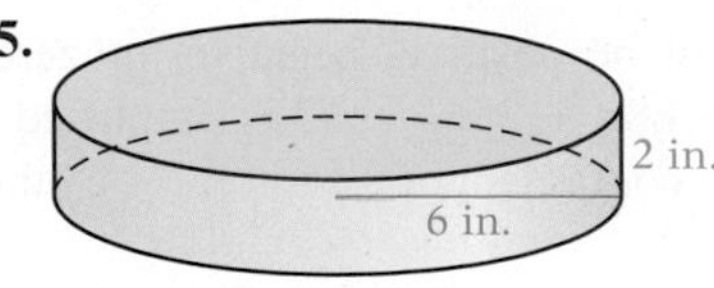

86. 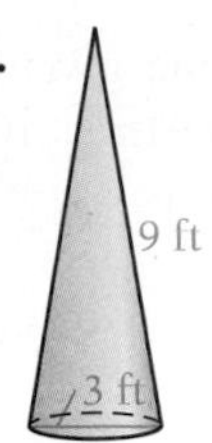

Complete Exercises 87 and 88 without using a calculator.

87. If $\frac{abc}{b - a}$ is evaluated when $a = -38$, $b = -52$, and c is a positive integer, will the result be a positive or a negative number?

88. The formula for the volume of a right circular cylinder with height h and base of radius r is $V = \pi r^2 h$. Suppose a cylinder has a radius of 2 cm and a height of 3 cm.
a. Can the exact volume of the cylinder be V cm^3, where V is a whole number?
b. Can the exact volume of the cylinder be V cm^2, where V is an irrational number?

3 Simplify variable expressions (See pages 37–38.)

89. Explain how the Distributive Property is used to combine like terms.

GETTING READY

90. The four terms of the variable expression $5x - 6y + 4x - 8$ are ___?___, ___?___, ___?___, and ___?___. The terms $5x$ and $4x$ are called ___?___ terms. The coefficient of the y term is ___?___. The constant term is ___?___.

91. Simplify: $-5(y - 5x) + 9y$

$-5(y - 5x) + 9y$
$= (\_\_?\_\_)y + (\_\_?\_\_)x + 9y$ • **Use the Distributive Property to remove parentheses.**
$= (-5y + 9y) + 25x$ • **Use the ___?___ and ___?___ Properties of Addition to rearrange and group like terms.**
$= \_\_?\_\_ + 25x$ • **Combine like terms.**

Simplify.

92. $5x + 7x$
93. $3x + 10x$
94. $-8ab - 5ab$
95. $12\left(\frac{1}{12}x\right)$
96. $\frac{1}{3}(3y)$
97. $-3(x - 2)$
98. $-5(x - 9)$
99. $(x + 2)5$
100. $-(x + y)$
101. $-(-x - y)$
102. $5 + 2(3x - 7)$
103. $7 - 3(4a - 5)$
104. $5v - 3(2 - 4v)$
105. $-3m - 2(4m + 3)$
106. $-3 + 4(2z - 9)$
107. $-5 - 6(2y - 3)$
108. $4x - 3(2y - 5)$
109. $-2a - 3(3a - 7)$
110. $3x - 2(5x - 7)$
111. $2x - 3(x - 2y)$
112. $3[a - 5(5 - 3a)]$
113. $5[-2 - 6(a - 5)]$
114. $3[x - 2(x + 2y)]$
115. $5[y - 3(y - 2x)]$
116. $-2(x - 3y) + 2(3y - 5x)$
117. $4(-a - 2b) - 2(3a - 5b)$
118. $5(3a - 2b) - 3(-6a + 5b)$
119. $-7(2a - b) + 2(-3b + a)$
120. $3x - 2[y - 2(x + 3[2x + 3y])]$
121. $2x - 4[x - 4(y - 2[5y + 3])]$
122. $4 - 2(7x - 2y) - 3(-2x + 3y)$
123. $3x + 8(x - 4) - 3(2x - y)$
124. $\frac{1}{3}[8x - 2(x - 12) + 3]$
125. $\frac{1}{4}[14x - 3(x - 8) - 7x]$

126. State whether the given number will be positive, negative, or zero after the variable expression $31a - 102b + 73 - 88a + 256b - 73$ is simplified.
a. the coefficient of a **b.** the coefficient of b **c.** the constant term

127. State whether the given expression is equivalent to $3[5 - 2(y - 6)]$.
a. $3[3(y - 6)]$ **b.** $15 - 6(y - 6)$

APPLYING CONCEPTS

In each of the following, it is possible that at least one of the Properties of the Real Numbers has been applied incorrectly. If the statement is incorrect, state the incorrect application of the Properties of the Real Numbers and correct the answer. If the statement is correct, state the Property of the Real Numbers that is being used.

128. $-4(5x - y) = -20x + 4y$

129. $4(3y + 1) = 12y + 4$

130. $6 - 6x = 0x = 0$

131. $2 + 3x = (2 + 3)x = 5x$

132. $3a - 4b = 4b - 3a$

133. $2(3y) = (2 \cdot 3)(2y) = 12y$

134. $x^4 \cdot \frac{1}{x^4} = 1, x \neq 0$

135. $-x^2 + y^2 = y^2 - x^2$

PROJECTS OR GROUP ACTIVITIES

Let $A = \{1, 2, 3, 4, 5, 6\}$. If x and y are two numbers in A, then we will define a new product operation on x and y, denoted by $x \otimes y$, as the remainder when xy is divided by 7. For instance, $2 \otimes 5 = 3$ because $2 \cdot 5 = 10$ and, when 10 is divided by 7, the remainder is 3. Although this may seem like a strange way to define an operation, a variation of this operation is used for sending credit card information over the Internet.

136. Find each of the following.
a. $4 \otimes 5$ **b.** $6 \otimes 3$

137. Is this operation commutative?

138. Is the following equation true? $2 \otimes (3 \otimes 5) = (2 \otimes 3) \otimes 5$

139. Try a few more examples of this operation that are similar to Exercise 138. Based on your examples, does it appear that the $\otimes$ operation is associative?

140. Recall that a and b are multiplicative inverses if $ab = 1$. Find the multiplicative inverse of 5 for the $\otimes$ operation.

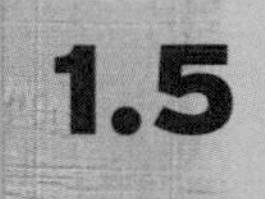

1.5 Verbal Expressions and Variable Expressions

OBJECTIVE 

Translate a verbal expression into a variable expression

One of the major skills required in applied mathematics is the ability to translate a verbal expression into a mathematical expression. As discussed in Sections 1.2 and 1.3, doing so requires recognizing the verbal phrases that translate into mathematical operations. Following is a partial list of the verbal phrases used to indicate the different mathematical operations.

Point of Interest

Mathematical symbolism, as shown on this page, has advanced through various stages: rhetorical, syncoptical, and modern. In the rhetorical stage, all mathematical description was through words. In the syncoptical stage, there was a combination of words and symbols. For instance, "x plano 4 in y" meant $4xy$. The modern stage, which is used today, began in the 17th century. Modern symbolism is also changing. For example, there are advocates of a system of symbolism that would place all operations last. Using this notation, 4 plus 7 would be written 47+ and 6 divided by 4 would be written 64÷.

Addition	more than	8 more than w	$w + 8$
	added to	x added to 9	$9 + x$
	the sum of	the sum of z and 9	$z + 9$
	the total of	the total of r and s	$r + s$
	increased by	x increased by 7	$x + 7$
Subtraction	less than	12 less than b	$b - 12$
	the difference between	the difference between x and 1	$x - 1$
	minus	z minus 7	$z - 7$
	decreased by	17 decreased by a	$17 - a$
Multiplication	times	negative 2 times c	$-2c$
	the product of	the product of x and y	xy
	multiplied by	3 multiplied by n	$3n$
	of	three-fourths of m	$\frac{3}{4}m$
	twice	twice d	$2d$
Division	divided by	v divided by 15	$\frac{v}{15}$
	the quotient of	the quotient of y and 3	$\frac{y}{3}$
	ratio	the ratio of x to 7	$\frac{x}{7}$
Power	the square of or the second power of	the square of x	x^2
	the cube of or the third power of	the cube of r	r^3
	the fifth power of	the fifth power of a	a^5

Be especially careful when translating a phrase that contains the word *sum*, *difference*, *product*, or *quotient*. In the examples at the right, note where the operation symbol is placed.

the *sum* of x and y $x + y$

the *difference* between x and y $x - y$

the *product* of x and y $x \cdot y$

the *quotient* of x and y $\frac{x}{y}$

Focus on translating a verbal expression into a variable expression

A. Translate "five less than twice the difference between a number and seven" into a variable expression. Then simplify.

Assign a variable to the unknown number.	the unknown number: x
Identify the words that indicate the mathematical operations.	5 <u>less than</u> <u>twice</u> the <u>difference</u> <u>between</u> x and 7
Use the identified words to write the variable expression.	$2(x - 7) - 5$
Simplify the expression.	$= 2x - 14 - 5 = 2x - 19$

B. The sum of two numbers is 37. If x represents the smaller number, translate "twice the larger number" into a variable expression.

Write an expression for the larger number by subtracting the smaller number, x, from 37.	larger number: $37 - x$
Identify the words that indicate the mathematical operations.	<u>twice</u> the larger number
Use the identified words to write a variable expression.	$2(37 - x)$

EXAMPLE 1 Translate and simplify "the total of five times a number and twice the difference between the number and three."

Solution the unknown number: n
- **Assign a variable to the unknown number.**

five times the number: $5n$
the difference between the number and three: $n - 3$
twice the difference between the number and three: $2(n - 3)$
- **Use the assigned variable to write an expression for any other unknown quantity.**

$5n + 2(n - 3)$
- **Write the variable expression.**

$= 5n + 2n - 6$
$= 7n - 6$
- **Simplify the variable expression.**

Problem 1 Translate and simplify "a number decreased by the difference between eight and twice the number."

Solution See page S3.

Try Exercise 11, page 46.

EXAMPLE 2 Translate and simplify "fifteen minus one-half the sum of a number and ten."

Solution the unknown number: n
- **Assign a variable to the unknown number.**

the sum of the number and ten: $n + 10$
one-half the sum of the number and ten: $\frac{1}{2}(n + 10)$
- **Use the assigned variable to write an expression for any other unknown quantity.**

$15 - \frac{1}{2}(n + 10)$ • **Write the variable expression.**

$= 15 - \frac{1}{2}n - 5$ • **Simplify the variable expression.**

$= -\frac{1}{2}n + 10$

Problem 2 Translate and simplify "the sum of three-eighths of a number and five-twelfths of the number."

Solution See page S3.

Try Exercise 13, page 46.

OBJECTIVE 2 Application problems

Many of the applications of mathematics require that you identify the unknown quantity, assign a variable to that quantity, and then attempt to express other unknowns in terms of that quantity.

Focus on translating an application problem

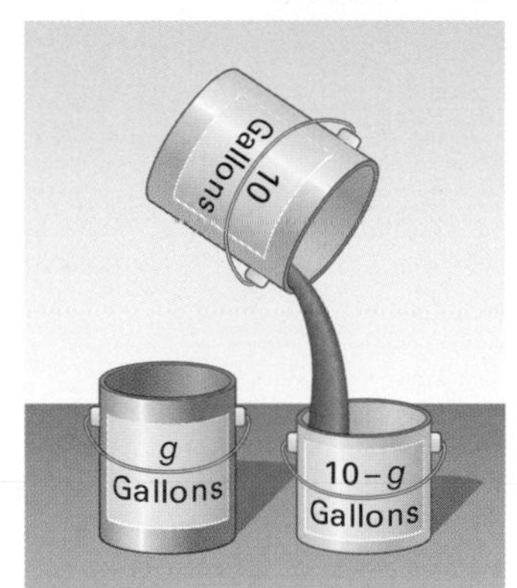

Ten gallons of paint were poured into two containers of different sizes. Express the amount of paint poured into the smaller container in terms of the amount poured into the larger container.

Assign a variable to the amount of paint poured into the larger container. — the number of gallons of paint poured into the larger container: g

Express the amount of paint poured into the smaller container in terms of g (g gallons of paint were poured into the larger container). — the number of gallons of paint poured into the smaller container: $10 - g$

EXAMPLE 3 A cyclist is riding at a rate that is twice the speed of a runner. Express the speed of the cyclist in terms of the speed of the runner.

Solution the speed of the runner: r

the speed of the cyclist is twice r: $2r$

Problem 3 A mixture of candy contains 3 lb more of milk chocolate than of caramel. Express the amount of milk chocolate in the mixture in terms of the amount of caramel in the mixture.

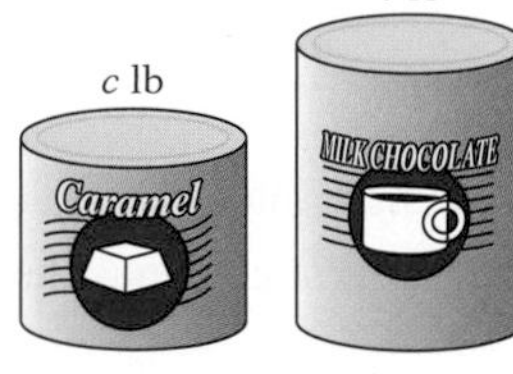

Solution See page S3.

Try Exercise 33, page 47.

EXAMPLE 4 The length of a rectangle is 2 ft more than 3 times the width. Express the length of the rectangle in terms of the width.

Solution the width of the rectangle: W

the length is 2 more than 3 times W: $3W + 2$

Problem 4 The depth of the deep end of a swimming pool is 2 ft more than twice the depth of the shallow end. Express the depth of the deep end in terms of the depth of the shallow end.

Solution See page S3.

Try Exercise 37, page 48.

1.5 Exercises

CONCEPT CHECK

1. Write two verbal expressions that would translate to $y + 6$.
2. Write two verbal expressions that would translate to $5x$.
3. Do the expressions "the difference between x and 2" and "x less than 2" translate to the same variable expression?
4. Do the phrases "ten less than m" and "ten less m" translate to the same variable expression?
5. If the sum of two numbers is 14 and one number is x, express the second number in terms of x.
6. Do the phrases "the ratio of x to y" and "the quotient of x and y" translate to the same variable expression?

1 Translate a verbal expression into a variable expression (See pages 43–45.)

GETTING READY

For each phrase in Exercises 7 to 9, identify the words that indicate mathematical operations.

7. ten more than the product of eight and a number
8. thirteen subtracted from the quotient of negative five and the cube of a number
9. the difference between ten times a number and sixteen times the number
10. The sum of two numbers is 24. To express both numbers in terms of the same variable, let x be one number. Then the other number is ___?___.

Translate into a variable expression. Then simplify.

11. a number minus the sum of the number and two
12. a number decreased by the difference between five and the number
13. the sum of one-third of a number and four-fifths of the number
14. the difference between three-eighths of a number and one-sixth of the number
15. five times the product of eight and a number
16. a number increased by two-thirds of the number
17. the difference between the product of seventeen and a number and twice the number
18. one-half the total of six times a number and twenty-two
19. the difference between the square of a number and the total of twelve and the square of the number
20. eleven more than the square of a number added to the difference between the number and seventeen

21. the sum of five times a number and twelve added to the product of fifteen and the number

22. four less than twice the sum of a number and eleven

23. The sum of two numbers is 15. Using x to represent the smaller of the two numbers, translate "the sum of twice the smaller number and two more than the larger number" into a variable expression. Then simplify.

24. The sum of two numbers is 20. Using x to represent the smaller of the two numbers, translate "the difference between five times the larger number and three less than the smaller number" into a variable expression. Then simplify.

25. The sum of two numbers is 34. Using x to represent the larger of the two numbers, translate "the difference between two more than the smaller number and twice the larger number" into a variable expression. Then simplify.

26. The sum of two numbers is 33. Using x to represent the larger of the two numbers, translate "the difference between six more than twice the smaller number and three more than the larger number" into a variable expression. Then simplify.

27. Which phrase(s) translate(s) into the expression $8n^3 - 5$?
(i) the difference between five and the product of eight and the cube of a number
(ii) five subtracted from the cube of eight and a number
(iii) five less than the product of eight and the cube of a number

28. Which phrase(s) translate(s) into the expression $(5n + 2) + 15$?
(i) fifteen more than the sum of five times a number and two
(ii) the total of two more than the product of five and a number plus fifteen
(iii) fifteen added to two more than the product of five and a number

2 Application problems (See pages 45–46.)

GETTING READY

29. The length of a rectangle is eight more than the width. To express the length and the width in terms of the same variable, let W be the __?__. Then the length is __?__.

30. The width of a rectangle is one-third the length. To express the length and the width in terms of the same variable, let L be the __?__. Then the width is __?__.

31. **Government Spending** See the news clipping at the right. Express the amount of money that the government spent on highways in terms of the amount of money it will spend on high-speed rails.

32. **Demographics** The population of New York City is four times the population of Houston, Texas. Express the population of New York City in terms of the population of Houston, Texas. (*Source: Information Please Almanac*)

33. **Astronomy** The distance from Earth to the sun is approximately 390 times the distance from Earth to the moon. Express the distance from Earth to the sun in terms of the distance from Earth to the moon.

34. **Construction** The longest rail tunnel (from Hanshu to Hokkaido, Japan) is 18.2 mi longer than the longest road tunnel (from Laerdal to Aurland, Norway). Express the length of the longest rail tunnel in terms of the length of the longest road tunnel.

35. **Investments** A financial advisor has invested \$10,000 in two accounts. If one account contains x dollars, express the amount in the second account in terms of x.

36. **Recreation** A fishing line 3 ft long is cut into two pieces, one shorter than the other. Express the length of the shorter piece in terms of the length of the longer piece.

In the News

Help for High-Speed Rails

In an effort to make high-speed rail travel more available in the U.S., the government is funding the beginning of a national high-speed rail network. But car travel still leads the way: last year the federal government spent eight times as much on highways as it will spend on the first round of work on high-speed rails.

Source: Time magazine

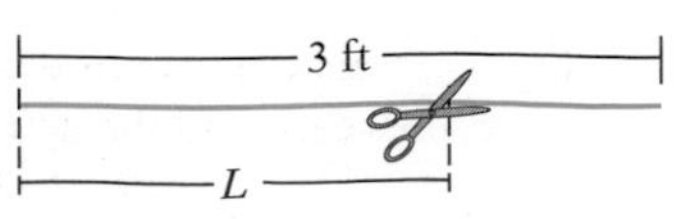

37. **Travel** The total flying time for a round trip between New York and San Diego is 13 h. Because of the jet stream, the time going is not equal to the time returning. Express the flying time between New York and San Diego in terms of the flying time between San Diego and New York.

38. **Carpentry** A 12-foot board is cut into two pieces of different lengths. Express the length of the longer piece in terms of the length of the shorter piece.

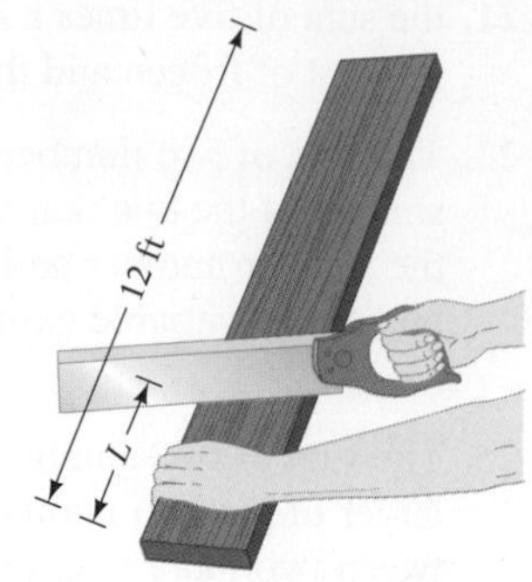

39. **Geometry** The measure of angle A of a triangle is twice the measure of angle B. The measure of angle C is twice the measure of angle A. Write expressions for angle A and angle C in terms of angle B.

40. **Geometry** The length of a rectangle is three more than twice the width. Express the length of the rectangle in terms of the width.

APPLYING CONCEPTS

Some English phrases require more than one variable in order to be translated into a variable expression.

41. **Physics** Translate "the product of one-half the acceleration due to gravity (g) and the time (t) squared" into a variable expression. (This expression gives the distance a dropped object will fall during a certain time interval.)

42. **Physics** Translate "the product of mass (m) and acceleration (a)" into a variable expression. (This expression is used to calculate the force exerted on an accelerating object.)

43. **Physics** Translate "the product of the area (A) and the square of the velocity (v)" into a variable expression. (This expression is used to compute the force a wind exerts on a sail.)

44. **Physics** Translate "the square root of the quotient of the spring constant (k) and the mass (m)" into a variable expression. (This is part of the expression that is used to compute the frequency of oscillation of a mass on the end of a spring.)

PROJECTS OR GROUP ACTIVITIES

For each of the following, write a phrase that would translate into the given expression.

45. $2x + 3$

46. $5y - 4$

47. $2(x + 3)$

48. $5(y - 4)$

CHAPTER 1 Summary

Key Words	Objective and Page Reference	Examples
The **integers** are . . . , −4, −3, −2, −1, 0, 1, 2, 3, ...	[1.1.1, p. 2][1]	
The **negative integers** are the integers . . . , −4, −3, −2, −1. The **positive integers** are the integers 1, 2, 3, 4, ... The positive integers are also called the **natural numbers.**	[1.1.1, p. 2]	
The positive integers and zero are called the **whole numbers.**	[1.1.1, p. 2]	0, 1, 2, 3, 4, . . .
A **rational number** is a number of the form $\frac{p}{q}$, where p and q are integers and q is not equal to zero.	[1.1.1, p. 2]	$\frac{5}{6}$, $\frac{-3}{4}$, and $\frac{6}{1}$ are rational numbers.
An **irrational number** is a number whose decimal representation never terminates or repeats.	[1.1.1, p. 3]	$\sqrt{3}$, π, and 0.21211211121111 . . . are irrational numbers.
The rational numbers and the irrational numbers taken together are called the **real numbers.**	[1.1.1, p. 3]	-3, $\sqrt{5}$, π, $\frac{-6}{7}$, and 0.232332333 . . . are real numbers.
The **additive inverse,** or **opposite,** of a number is the same distance from zero on the number line, but on the opposite side.	[1.1.1, p. 4]	The additive inverse of −3 is 3.
The **absolute value** of a number is the measure of its distance from zero on the number line.	[1.1.1, p. 5]	$\lvert 4 \rvert = 4$ $\lvert -3 \rvert = 3$ $-\lvert -6 \rvert = -6$
A **set** is a collection of objects. The objects in the set are called the **elements** of the set. The **roster method** of writing a set encloses a list of the elements of the set in braces.	[1.1.1, 1.1.2, pp. 2, 5]	The set $A = \{4, 5, 6, 7\}$ contains the elements 4, 5, 6, and 7.
The **empty set,** or **null set,** is the set that contains no elements.	[1.1.2, p. 5]	The empty set is written as $\emptyset$ or $\{\ \}$.
A **finite set** is a set in which the elements can be counted.	[1.1.2, p. 5]	$A = \{1, 2, 3, 4\}$
An **infinite set** is a set in which it is impossible to list all the elements.	[1.1.2, p. 5]	$A = \{1, 2, 3, 4, \ldots\}$
Set-builder notation is used to describe finite and infinite sets.	[1.1.2, p. 6]	In set-builder notation, the set of integers greater than −7 is written $\{x \mid x > -7, x \in \text{integers}\}$

[1] The numbers in brackets are a reference to the objective in which the Key Word or Essential Rule or Procedure is first introduced. For example, the reference [1.1.1] stands for Chapter 1, Section 1, Objective 1. This notation will be used in all Chapter Summaries throughout the text.

Interval notation is an alternative method of representing a set. An interval is **closed** if it includes both **endpoints.** An interval is **open** if it does not include either endpoint. An interval is **half-open** if one endpoint is included and the other endpoint is not included. A bracket is used to indicate that an endpoint is included, and a parenthesis is used to indicate that an endpoint is excluded.	[1.1.2, p. 7]	−5 −4 −3 −2 −1 0 1 2 3 4 5 Closed interval [−4, 5] −5 −4 −3 −2 −1 0 1 2 3 4 5 Open interval (−1, 3) −5 −4 −3 −2 −1 0 1 2 3 4 5 Half-open interval (−3, 4]
The **union** of two sets, written $A \cup B$, is the set that contains all the elements of A and all the elements of B. The elements that are in both set A and set B are listed only once.	[1.1.2, p. 8]	If $A = \{2, 3, 4\}$ and $B = \{4, 5, 6\}$, then $A \cup B = \{2, 3, 4, 5, 6\}$.
The **intersection** of two sets, written $A \cap B$, is the set that contains the elements that are common to both A and B.	[1.1.2, p. 8]	If $A = \{2, 3, 4\}$ and $B = \{4, 5, 6\}$, then $A \cap B = \{4\}$.
The expression a^n is in **exponential form,** where a is the **base** and n is the **exponent.**	[1.2.1, p. 17]	5^3 is in exponential form.
The **multiplicative inverse,** or **reciprocal,** of a nonzero number a is $\frac{1}{a}$.	[1.3.1, p. 24]	The multiplicative inverse of $\frac{3}{5}$ is $\frac{5}{3}$.
A **complex fraction** is a fraction whose numerator or denominator (or both) contains one or more fractions.	[1.3.2, p. 27]	$\dfrac{\frac{3}{5} - 2}{\frac{4}{5}}$ is a complex fraction.
A **variable expression** is an expression that contains one or more variables. The **terms** of a variable expression are the addends of the expression.	[1.4.2, pp. 35–36]	$3xy + x^2 - 4$ is a variable expression. $3xy$, x^2, and -4 are the terms of the expression.
A **variable term** is composed of a **numerical coefficient** and a **variable part.**	[1.4.2, p. 36]	The numerical coefficient of $3xy$ is 3, and the variable part is xy.
Like terms of a variable expression have the same variable part. Constant terms are like terms.	[1.4.3, p. 37]	$3xy$ and $-7xy$ are like terms.

Essential Rules and Procedures	Objective and Page Reference	Examples
To add two numbers with the same sign, add the absolute values of the two numbers, and attach the sign of the addends.	[1.2.1, p. 13]	$-5 + (-6) = -11$ $4 + 2 = 6$
To add two numbers with different signs, find the difference between the absolute values of the two numbers, and attach the sign of the number with the larger absolute value.	[1.2.1, p. 13]	$-4 + 8 = 4$ $6 + (-11) = -5$
To subtract two numbers, add the opposite of the second number to the first.	[1.2.1, p. 14]	$-4 - (-5) = -4 + 5 = 1$ $5 - 12 = 5 + (-12) = -7$

The product of two numbers with the same sign is positive. **The product of two numbers with different signs** is negative.	[1.2.1, p. 15]	$(4)(5) = 20$ $\quad (-3)(-7) = 21$ $(-4)(2) = -8$ $\quad (5)(-4) = -20$
The quotient of two numbers with the same sign is positive. **The quotient of two numbers with different signs** is negative.	[1.2.1, p. 15]	$8 \div 2 = 4$ $\quad (-20) \div (-5) = 4$ $(-10) \div 5 = -2$ $\quad 15 \div (-3) = -5$
Order of Operations Agreement	[1.2.2, p. 18]	$(5 - 2) + 9^2 \div 3$
Step 1 Perform operations inside grouping symbols.		$= 3 + 9^2 \div 3$
Step 2 Simplify exponential expressions.		$= 3 + 81 \div 3$
Step 3 Do multiplication and division as they occur from left to right.		$= 3 + 27$
Step 4 Do addition and subtraction as they occur from left to right.		$= 30$
Properties of the Real Numbers for Addition	[1.4.1, pp. 33–34]	
Commutative: $a + b = b + a$		$3 + 2 = 2 + 3$
Associative: $(a + b) + c = a + (b + c)$		$(4 + 5) + 9 = 4 + (5 + 9)$
Property of Zero: $a + 0 = 0 + a = 0$		$0 + 4 = 4$
Inverse Property: $a + (-a) = (-a) + a = 0$		$8 + (-8) = 0$
Properties of the Real Numbers for Multiplication	[1.4.1, pp. 34–35]	
Commutative: $a \cdot b = b \cdot a$		$3 \cdot 5 = 5 \cdot 3$
Associative: $(a \cdot b) \cdot c = a \cdot (b \cdot c)$		$(2 \cdot 3) \cdot 5 = 2 \cdot (3 \cdot 5)$
Property of Zero: $a \cdot 0 = 0 \cdot a = 0$		$7 \cdot 0 = 0$
Property of One: $a \cdot 1 = 1 \cdot a = a$		$6 \cdot 1 = 6$
Inverse Property: $a \cdot \frac{1}{a} = \frac{1}{a} \cdot a = 1$		$3\left(\frac{1}{3}\right) = 1$
The Distributive Property $a(b + c) = ab + ac$	[1.4.1, p. 35]	$3(x + 5) = 3x + 3(5) = 3x + 15$

CHAPTER 1 Review Exercises

1. Find the additive inverse of $-\frac{3}{4}$.

2. Let $x \in \{-4, -2, 0, 2\}$. For what values of x is $x > -1$ true?

3. Let $p \in \{-4, 0, 7\}$. Evaluate $-|p|$ for each element of the set.

4. Use the roster method to write the set of integers between -3 and 4.

5. Use set-builder notation to write the set of real numbers less than -3.

6. Write $[-2, 3]$ in set-builder notation.

7. Find $A \cup B$ given $A = \{1, 3, 5, 7\}$ and $B = \{2, 4, 6, 8\}$.

8. Find $A \cap B$ given $A = \{0, 1, 2, 3\}$ and $B = \{2, 3, 4, 5\}$.

9. Graph: $[-3, \infty)$

10. Graph: $\{x|x < 1\}$

11. Graph: $\{x|x \le -3\} \cup \{x|x > 0\}$

12. Graph: $(-2, 4]$

13. Graph: $\{x|x > 4\} \cup \{x|-2 \le x < 0\}$

14. Graph: $(-\infty, 2] \cap (0, \infty)$

15. Subtract: $-10 - (-3)$

16. Divide: $-204 \div (-17)$

17. Subtract: $-\frac{3}{8} - \frac{1}{6}$

18. Divide: $-\frac{3}{8} \div \frac{3}{5}$

19. Add: $-4.07 + 2.3$

20. Simplify: $-4^2 - (-3)^2$

21. Simplify: $7 - 3(5 - 9)^2 \div 4 \cdot (-2)$

22. Simplify: $\frac{2}{3} - \frac{1}{3}\left(\frac{1}{2} - \frac{5}{6}\right)^2 \div \frac{8}{9}$

23. Simplify: $\dfrac{\dfrac{6 - 8(5 - 3)}{4 + 2(1 - 6)}}{\dfrac{3^2 - 2^2}{(2 - 5)^2}}$

24. Simplify: $-3.2 + 1.1(4 - 3.8)^2 \div (-2.2)$

25. Use the Distributive Property to complete the statement.
$3(2x - 7y) = 6x - ?$

26. Use the Commutative Property of Multiplication to complete the statement.
$(ab)14 = 14?$

27. Identify the property that justifies the statement.
$(-4) + 4 = 0$

28. Identify the property that justifies the statement.
$2(3x) = (2 \cdot 3)x$

29. Evaluate $b^2 - 4ac$ when $a = 2$, $b = -3$, and $c = -4$.

30. Evaluate $-a^2 - b(2a - 2)^2 + 2b$ when $a = -3$, $b = 4$, and $c = -1$.

31. Simplify: $6 - 2(4a - 2)$

32. Simplify: $-2(x - 3) + 4(2 - x)$

33. Simplify: $4y - 3[x - 2(3 - 2x) - 4y]$

34. Simplify: $5 + 2(4x - 3y) - 3[4 - 2(x - y)]$

35. Translate and simplify "four times the sum of a number and four."

36. Translate and simplify "eight more than twice the difference between a number and two."

37. **Integer Problem** The sum of two numbers is 40. Using x to represent the smaller of the two numbers, translate "the sum of twice the smaller number and five more than the larger number" into a variable expression. Then simplify.

38. **Integer Problem** The sum of two numbers is 9. Using x to represent the larger of the two numbers, translate "the difference between three more than twice the smaller number and one more than the larger number" into a variable expression. Then simplify.

39. **Geometry** The length of a rectangle is 3 ft less than three times the width. Express the length of the rectangle in terms of the width.

40. **Integer Problem** A second integer is five more than four times the first integer. Express the second integer in terms of the first integer.

CHAPTER 1 Test

1. Find the additive inverse of -12.

2. Let $x \in \{-5, 3, 7\}$. For which values of x is $-1 > x$ true?

3. Simplify: $2 - (-12) + 3 - 5$

4. Multiply: $(-2)(-3)(-5)$

5. Divide: $-180 \div 12$

6. Simplify: $|-3 - (-5)|$

7. Simplify: $-5^2 \cdot 4$

8. Simplify: $(-2)^3(-3)^2$

9. Simplify: $\frac{2}{3} - \frac{5}{12} + \frac{4}{9}$

10. Multiply: $\left(-\frac{2}{3}\right)\left(\frac{9}{15}\right)\left(\frac{10}{27}\right)$

11. Simplify: $4.27 - 6.98 + 1.3$

12. Divide: $-15.092 \div 3.08$

13. Simplify: $12 - 4\left(\frac{5^2 - 1}{3}\right) \div 16$

14. Simplify: $8 - 4(2 - 3)^2 \div 2$

15. Evaluate $(a - b)^2 \div (2b + 1)$ when $a = 2$ and $b = -3$.

16. Evaluate $\frac{b^2 - c^2}{a - 2c}$ when $a = 2$, $b = 3$, and $c = -1$.

17. Use the Commutative Property of Addition to complete the statement.
$(3 + 4) + 2 = (? + 3) + 2$

18. Identify the property that justifies the statement.
$-2(x + y) = -2x - 2y$

19. Simplify: $3x - 2(x - y) - 3(y - 4x)$

20. Simplify: $2x - 4[2 - 3(x + 4y) - 2]$

21. Translate and simplify "thirteen decreased by the product of three less than a number and nine."

22. Translate and simplify "one-third of the total of twelve times a number and twenty-seven."

23. Find $A \cup B$ given $A = \{1, 3, 5, 7\}$ and $B = \{2, 3, 4, 5\}$.

24. Find $A \cup B$ given $A = \{-2, -1, 0, 1, 2, 3\}$ and $B = \{-1, 0, 1\}$.

25. Find $A \cap B$ given $A = \{1, 3, 5, 7\}$ and $B = \{5, 7, 9, 11\}$.

26. Find $A \cap B$ given $A = \{-3, -2, -1, 0, 1, 2, 3\}$ and $B = \{-1, 0, 1\}$.

27. Graph: $(-\infty, -1]$

28. Graph: $(3, \infty)$

29. Graph: $\{x|x \leq 3\} \cup \{x|x < -2\}$

30. Graph: $\{x|x < 3\} \cap \{x|x > -2\}$

First-Degree Equations and Inequalities

CHAPTER 2

Focus on Success

Do you have trouble with word problems? Word problems show the variety of ways in which math can be used. The solution of every word problem can be broken down into two steps: Strategy and Solution. The Strategy consists of reading the problem, writing down what is known and unknown, and devising a plan to find the unknown. The Solution often consists of solving an equation and then checking the solution. (See Use a Strategy to Solve Word Problems, page AIM-10.)

OBJECTIVES

2.1
1. Solve equations using the Addition and Multiplication Properties of Equations
2. Solve equations containing parentheses
3. Application problems

2.2
1. Value mixture problems
2. Uniform motion problems

2.3
1. Investment problems
2. Percent mixture problems

2.4
1. Solve inequalities in one variable
2. Solve compound inequalities
3. Application problems

2.5
1. Absolute value equations
2. Absolute value inequalities
3. Application problems

PREP TEST

Are you ready to succeed in this chapter?
Take the Prep Test below to find out if you are ready to learn the new material.

For Exercises 1 to 5, add, subtract, multiply, or divide.

1. $8 - 12$

2. $-9 + 3$

3. $\dfrac{-18}{-6}$

4. $-\dfrac{3}{4}\left(-\dfrac{4}{3}\right)$

5. $-\dfrac{5}{8}\left(\dfrac{4}{5}\right)$

For Exercises 6 to 9, simplify.

6. $3x - 5 + 7x$

7. $6(x - 2) + 3$

8. $n + (n + 2) + (n + 4)$

9. $0.08x + 0.05(400 - x)$

10. Twenty ounces of a snack mixture contains nuts and pretzels. Let n represent the number of ounces of nuts in the mixture. Express the number of ounces of pretzels in the mixture in terms of n.

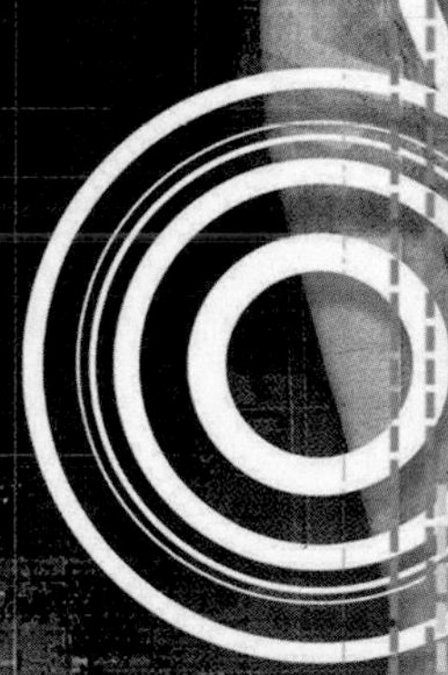

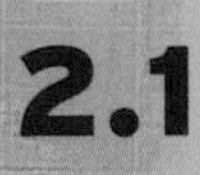

2.1 Equations in One Variable

OBJECTIVE 1 Solve equations using the Addition and Multiplication Properties of Equations

An **equation** expresses the equality of two mathematical expressions. The expressions can be either numerical or variable expressions.

$$\left.\begin{array}{r} 2 + 8 = 10 \\ x + 8 = 11 \\ x^2 + 2y = 7 \end{array}\right\} \text{ Equations}$$

The equation at the right is a **conditional equation.** The equation is *true* if the variable is replaced by 3. The equation is *false* if the variable is replaced by 4.

$x + 2 = 5$	Conditional equation
$3 + 2 = 5$	A true equation
$4 + 2 = 5$	A false equation

The replacement values of the variable that will make an equation true are called the **roots,** or **solutions,** of the equation.

The solution of the equation $x + 2 = 5$ is 3.

The equation at the right is an **identity.** Any replacement for x will result in a true equation.

$x + 2 = x + 2$ Identity

The equation at the right has *no solution* because there is no number that equals itself plus 1. Any replacement value for x will result in a false equation.

$x = x + 1$ No solution

Each of the equations at the right is a **first-degree equation in one variable.** All variables have an exponent of 1.

$$\left.\begin{array}{l} x + 2 = 12 \\ 3y - 2 = 5y \\ 3(a + 2) = 14a \end{array}\right. \text{ First-degree equations}$$

Solving an equation means finding a solution of the equation. The simplest equation to solve is an equation of the form **variable = constant,** because the constant is the solution.

If $x = 3$, then 3 is the solution of the equation because $3 = 3$ is a true equation.

In solving an equation, the goal is to rewrite the given equation in the form *variable = constant*. The Addition Property of Equations can be used to rewrite an equation in this form.

Take Note

The model of an equation as a balance scale applies.

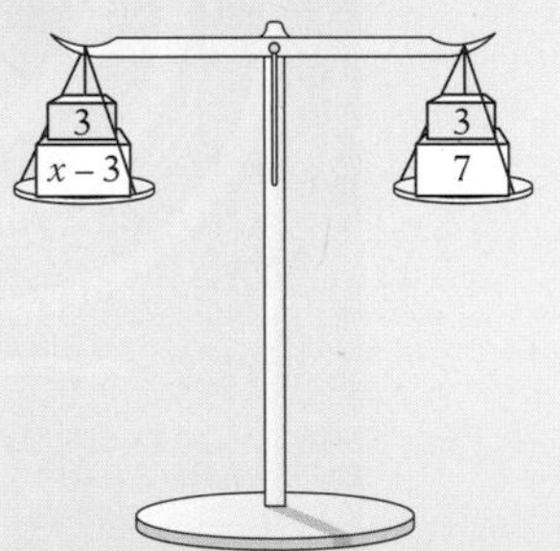

Adding a weight to one side of the equation requires adding the same weight to the other side of the equation so that the equation remains in balance.

THE ADDITION PROPERTY OF EQUATIONS

If a, b, and c are algebraic expressions, then the equation $a = b$ has the same solutions as the equation $a + c = b + c$.

The Addition Property of Equations states that the same quantity can be added to each side of an equation without changing the solution of the equation. This property is used to remove a term from one side of an equation by adding the opposite of that term to each side of the equation.

Focus on solving an equation using the Addition Property of Equations

A. Solve: $x - 3 = 7$

Add the opposite of the constant term -3 to each side of the equation. Simplify.
After simplifying, the equation is in the form *variable* = *constant*.

$$x - 3 = 7$$
$$x - 3 + 3 = 7 + 3$$
$$x + 0 = 10$$
$$x = 10$$

To check the solution, replace the variable with 10. Simplify the left side of the equation. Because $7 = 7$ is a true equation, 10 is a solution.

Check:
$$\begin{array}{c|c} x - 3 & 7 \\ \hline 10 - 3 & 7 \\ 7 & 7 \end{array}$$

The solution is 10.

Take Note

Remember to check the solution.

$$\begin{array}{c|c} x + \frac{7}{12} & \frac{1}{2} \\ \hline -\frac{1}{12} + \frac{7}{12} & \frac{1}{2} \\ \frac{6}{12} & \frac{1}{2} \\ \frac{1}{2} & \frac{1}{2} \end{array}$$

B. Solve: $x + \frac{7}{12} = \frac{1}{2}$

$$x + \frac{7}{12} = \frac{1}{2}$$

Add the opposite of the constant term $\frac{7}{12}$ to each side of the equation. This is equivalent to subtracting $\frac{7}{12}$ from each side.

$$x + \frac{7}{12} - \frac{7}{12} = \frac{1}{2} - \frac{7}{12}$$

Simplify.

$$x + 0 = \frac{6}{12} - \frac{7}{12}$$
$$x = -\frac{1}{12}$$

The solution is $-\frac{1}{12}$.

The Multiplication Property of Equations can also be used to rewrite an equation in the form *variable* = *constant*.

THE MULTIPLICATION PROPERTY OF EQUATIONS

If a, b, and c are algebraic expressions, and $c \neq 0$, then the equation $a = b$ has the same solutions as the equation $ac = bc$.

The Multiplication Property of Equations states that we can multiply each side of an equation by the same nonzero number without changing the solution of the equation. This property is used to remove a coefficient from a variable term in an equation by multiplying each side of the equation by the reciprocal of the coefficient.

Focus on solving an equation using the Multiplication Property of Equations

Take Note

When using the Multiplication Property of Equations, it is usually easier to multiply each side of the equation by the reciprocal of the coefficient when the coefficient is a fraction, as in part (A). Divide each side of the equation by the coefficient when the coefficient is an integer or a decimal, as in part (B).

A. Solve: $-\frac{3}{4}x = 12$

$$-\frac{3}{4}x = 12$$

Multiply each side of the equation by $-\frac{4}{3}$, the reciprocal of $-\frac{3}{4}$.

$$\left(-\frac{4}{3}\right)\left(-\frac{3}{4}\right)x = \left(-\frac{4}{3}\right)12$$

Simplify.

$$1x = -16$$

After simplifying, the equation is in the form *variable* = *constant*.

$$x = -16$$

Check:
$$\begin{array}{c|c} -\frac{3}{4}x & 12 \\ \hline -\frac{3}{4}(-16) & 12 \\ 12 & 12 \end{array}$$

The solution is -16.

B. Solve: $-5x = 9$

$$-5x = 9$$

Multiplying each side of the equation by the reciprocal of -5 is equivalent to dividing each side of the equation by -5.

$$\frac{-5x}{-5} = \frac{9}{-5}$$

Simplify.

$$1x = -\frac{9}{5}$$

$$x = -\frac{9}{5}$$

You should check the solution. The solution is $-\frac{9}{5}$.

In solving an equation, it is often necessary to apply both the Addition and the Multiplication Properties of Equations.

EXAMPLE 1 Solve: $5 - 6x = 9$

Solution

$$5 - 6x = 9$$

$$5 - 5 - 6x = 9 - 5$$ • Subtract 5 from each side of the equation.

$$-6x = 4$$ • Simplify.

$$\frac{-6x}{-6} = \frac{4}{-6}$$ • Divide each side of the equation by -6.

$$x = -\frac{2}{3}$$ • Simplify.

The solution is $-\frac{2}{3}$.

Problem 1 Solve: $\frac{6x}{5} - 3 = -7$

Solution See page S3.

Try Exercise 41, page 61.

EXAMPLE 2 Solve: $3x - 5 = -6x + 2$

Solution

$$3x - 5 = -6x + 2$$

$$3x + 6x - 5 = -6x + 6x + 2$$ • Add $6x$ to each side of the equation.

$$9x - 5 = 2$$

$$9x - 5 + 5 = 2 + 5$$ • Add 5 to each side of the equation.

$$9x = 7$$

$$\frac{9x}{9} = \frac{7}{9}$$ • Divide each side of the equation by 9.

$$x = \frac{7}{9}$$

The solution is $\frac{7}{9}$.

Problem 2 Solve: $3x - 5 = 14 - 5x$

Solution See page S3.

Try Exercise 43, page 61.

OBJECTIVE 2

Solve equations containing parentheses

When an equation contains parentheses, one of the steps in solving the equation requires the use of the Distributive Property.

Focus on solving an equation containing parentheses

Solve: $3(x - 2) + 3 = 2(6 - x)$

$$3(x - 2) + 3 = 2(6 - x)$$

Use the Distributive Property to remove parentheses. $3x - 6 + 3 = 12 - 2x$

Simplify. $3x - 3 = 12 - 2x$

Add $2x$ to each side of the equation. $5x - 3 = 12$

Add 3 to each side of the equation. $5x = 15$

Divide each side of the equation by the coefficient 5. $x = 3$

You should check the solution. The solution is 3.

EXAMPLE 3 Solve: $5(2x - 7) + 2 = 3(4 - x) - 12$

Solution

$$5(2x - 7) + 2 = 3(4 - x) - 12$$
$$10x - 35 + 2 = 12 - 3x - 12$$
$$10x - 33 = -3x$$
$$-33 = -13x$$
$$\frac{33}{13} = x$$

- **Use the Distributive Property.**
- **Simplify.**
- **Subtract 10x from each side of the equation.**
- **Divide each side of the equation by −13.**

The solution is $\frac{33}{13}$.

Problem 3 Solve: $6(5 - x) - 12 = 2x - 3(4 + x)$

Solution See page S3.

Try Exercise 57, page 62.

To solve an equation that contains fractions, first **clear denominators** by multiplying each side of the equation by the least common multiple (LCM) of the denominators.

Focus on solving an equation by clearing denominators

Solve: $\frac{x}{2} - \frac{7}{9} = \frac{x}{6} + \frac{2}{3}$

$$\frac{x}{2} - \frac{7}{9} = \frac{x}{6} + \frac{2}{3}$$

Multiply each side of the equation by 18, the LCM of 2, 9, 6, and 3. $18\left(\frac{x}{2} - \frac{7}{9}\right) = 18\left(\frac{x}{6} + \frac{2}{3}\right)$

Use the Distributive Property to remove parentheses. $\frac{18x}{2} - \frac{18 \cdot 7}{9} = \frac{18x}{6} + \frac{18 \cdot 2}{3}$

Simplify. $9x - 14 = 3x + 12$

Subtract $3x$ from each side of the equation. $6x - 14 = 12$

Add 14 to each side of the equation. $6x = 26$

Divide each side of the equation by the coefficient 6. $x = \frac{13}{3}$

You should check the solution. The solution is $\frac{13}{3}$.

EXAMPLE 4 Solve: $\frac{3x-2}{12} - \frac{x}{9} = \frac{x}{2}$

Solution

$$\frac{3x-2}{12} - \frac{x}{9} = \frac{x}{2}$$
• The LCM of 12, 9, and 2 is 36.

$$36\left(\frac{3x-2}{12} - \frac{x}{9}\right) = 36\left(\frac{x}{2}\right)$$
• Multiply each side of the equation by the LCM of the denominators.

$$\frac{36(3x-2)}{12} - \frac{36x}{9} = \frac{36x}{2}$$
• Use the Distributive Property.

$$3(3x-2) - 4x = 18x$$
• Simplify.

$$9x - 6 - 4x = 18x$$
$$5x - 6 = 18x$$
$$-6 = 13x$$
$$-\frac{6}{13} = x$$

The solution is $-\frac{6}{13}$.

Problem 4 Solve: $\frac{2x-7}{3} - \frac{5x+4}{5} = \frac{-x-4}{30}$

Solution See page S3.

Try Exercise 79, page 62.

OBJECTIVE 3 Application problems

Solving application problems is primarily a skill in translating sentences into equations and then solving the equations. An equation states that two mathematical expressions are equal. Therefore, translating a sentence into an equation requires recognizing the words or phrases that mean *equals.* These phrases include "is," "is equal to," "amounts to," and "represents." Once the sentence is translated into an equation, solve the equation by rewriting the equation in the form *variable* = *constant.*

EXAMPLE 5 A plumber charges \$80 for a service call plus \$1.25 for each additional minute of service over 60 min. If the bill for a plumbing repair job was \$115, how many minutes did the service call take?

Strategy To find the length of the service call in minutes, write and solve an equation using n to represent the total number of minutes of the call. Then $n - 60$ is the number of additional minutes after the first 60 min of the service call. The fixed charge for the 60 min plus the charge for the additional minutes is the total cost of the service call.

Solution

$$80 + 1.25(n - 60) = 115$$
$$80 + 1.25n - 75 = 115$$
$$1.25n + 5 = 115$$
$$1.25n = 110$$
$$n = 88$$

The service call lasted 88 min.

Problem 5 You are making a salary of \$34,500 and receive a 4% raise for next year. Find your next year's salary.

Solution See page S4.

Try Exercise 95, page 63.

2.1 Exercises

CONCEPT CHECK

1. How does an equation differ from an expression?
2. What is the Addition Property of Equations and how is it used?
3. What is the Multiplication Property of Equations and how is it used?

State whether each of the following is a first-degree equation in one variable.

4. $4a - 5 = 0$
5. $2x + 7$
6. $x^2 + 3 = 4$
7. $2 + 3y = 6$
8. $6 - 2(4a - 1)$
9. $5 = 7 - 2$
10. Do all equations have at least one solution?

1 Solve equations using the Addition and Multiplication Properties of Equations (See pages 56-58.)

11. Is 1 a solution of $7 - 3m = 4$?
12. Is 5 a solution of $4y - 5 = 3y$?
13. Is -2 a solution of $6x - 1 = 7x + 1$?
14. Is 3 a solution of $x^2 = 4x - 5$?

GETTING READY

15. To solve the equation $a - 42 = 13$, use the Addition Property of Equations to add __?__ to each side of the equation. The solution is __?__.
16. To solve the equation $12 + x = 5$, __?__ 12 from each side of the equation. The solution is __?__.
17. To solve the equation $-\frac{2}{5}n = 8$, use the Multiplication Property of Equations to multiply each side of the equation by __?__. The solution is __?__.
18. To solve the equation $9 = 18b$, __?__ each side of the equation by 18. The solution is __?__.

Solve and check.

19. $x - 2 = 7$
20. $x - 8 = 4$
21. $a + 3 = -7$
22. $-12 = x - 3$
23. $3x = 12$
24. $8x = 4$
25. $-y = 7$
26. $-x = 0$
27. $\frac{2}{7} + x = \frac{17}{21}$
28. $x + \frac{2}{3} = \frac{5}{6}$
29. $\frac{3a}{7} = -21$
30. $\frac{3t}{8} = -15$
31. $-\frac{5}{12}y = \frac{7}{16}$
32. $-\frac{3}{4}x = -\frac{4}{7}$
33. $b - 14.72 = -18.45$
34. $b + 3.87 = -2.19$
35. $3x + 5x = 12$
36. $2x - 7x = 15$
37. $2x - 4 = 12$
38. $5 - 7a = 19$
39. $16 = 1 - 6x$
40. $7 = 7 - 5x$
41. $-9 = 4x + 3$
42. $2x + 2 = 3x + 5$
43. $2 - 3t = 3t - 4$

44. $3x - 2x + 7 = 12 - 4x$

45. $2x - 9x + 3 = 6 - 5x$

46. $2x - 5 + 7x = 11 - 3x + 4x$

47. $9 + 4x - 12 = -3x + 5x + 8$

48. r is a positive number less than 1. Is the solution of the equation $\frac{10}{9} + x = r$ positive or negative?

49. a is a negative number less than -5. Is the solution of the equation $a = -5b$ less than or greater than 1?

2 Solve equations containing parentheses (See pages 58–60.)

(See pages 58–60.)

GETTING READY

50. Use the Distributive Property to remove the parentheses from the equation $9x - 3(5 - x) = 3(8x + 7)$: $9x -$ __?__ $+$ __?__ $=$ __?__ $+$ __?__

51. To clear denominators from the equation $\frac{x}{7} + \frac{1}{14} = \frac{1}{6}$, multiply each side of the equation by __?__, the least common multiple of the denominators 7, 14, and 6.

Solve and check.

52. $2x + 2(x + 1) = 10$

53. $2x + 3(x - 5) = 15$

54. $2(a - 3) = 2(4 - 2a)$

55. $5(2 - b) = -3(b - 3)$

56. $3 - 2(y - 3) = 4y - 7$

57. $3(y - 5) - 5y = 2y + 9$

58. $4(x - 2) + 2 = 4x - 2(2 - x)$

59. $2x - 3(x - 4) = 2(3 - 2x) + 2$

60. $2(2d + 1) - 3d = 5(3d - 2) + 4d$

61. $-4(7y - 1) + 5y = -2(3y + 4) - 3y$

62. $4[3 + 5(3 - x) + 2x] = 6 - 2x$

63. $2[4 + 2(5 - x) - 2x] = 4x - 7$

64. $2[b - (4b - 5)] = 3b + 4$

65. $-3[x + 4(x + 1)] = x + 4$

66. $4[a - (3a - 5)] = a - 7$

67. $5 - 6[2t - 2(t + 3)] = 8 - t$

68. $-3(x - 2) = 2[x - 4(x - 2) + x]$

69. $3[x - (2 - x) - 2x] = 3(4 - x)$

70. $\frac{2}{9}t - \frac{5}{6} = \frac{1}{12}t$

71. $\frac{3}{4}t - \frac{7}{12} = \frac{1}{6}$

72. $\frac{2}{3}x - \frac{5}{6}x - 3 = \frac{1}{2}x - 5$

73. $\frac{1}{2}x - \frac{3}{4}x + \frac{5}{8} = \frac{3}{2}x - \frac{5}{2}$

74. $\frac{3x - 2}{4} - 3x = 12$

75. $\frac{2a - 9}{5} + 3 = 2a$

76. $\frac{x - 2}{4} - \frac{x + 5}{6} = \frac{5x - 2}{9}$

77. $\frac{2x - 1}{4} + \frac{3x + 4}{8} = \frac{1 - 4x}{12}$

78. $\frac{2}{3}(15 - 6a) = \frac{5}{6}(12a + 18)$

79. $\frac{1}{5}(20x + 30) = \frac{1}{3}(6x + 36)$

80. $\frac{1}{3}(x - 7) + 5 = 6x + 4$

81. $2(y - 4) + 8 = \frac{1}{2}(6y + 20)$

82. $\frac{7}{8}x - \frac{1}{4} = \frac{3}{4}x - \frac{1}{2}$

83. $\frac{1}{2}x - \frac{3}{5} = \frac{2}{5}x + \frac{1}{2}$

84. $-4.2(p + 3.4) = 11.13$

85. $-1.6(b - 2.35) = -11.28$

86. $0.11x + 0.04(700 - x) = 0.06(700)$

87. $x + 0.06(60) = 0.20(x + 20)$

88. Consider the equation $-3[5 - 4(x - 2)] = 5(x - 5)$. How many times would you use the Distributive Property to remove grouping symbols if you solved the equation?

89. Which of the following equations is equivalent to the equation in Exercise 88?
(i) $-15 + 12(x - 2) = 5x - 25$ (ii) $-3[x - 2] = 5(x - 5)$
(iii) $-3[5 - 4x - 8] = 5x - 25$

3 Application problems (See page 60.)

GETTING READY

90. When a sentence is translated into an equation, the word "is" translates into the __?__ sign.

91. Suppose 10 friends go to a restaurant. Some people in the group order the all-you-can-eat buffet, while the rest of the group order the soup-and-sandwich combo. If 2 people order the buffet, then the number of people who order the soup-and-sandwich combo is __?__. If 7 people order the buffet, then the number of people who order the soup-and-sandwich combo is __?__. If n people order the buffet, then an expression that represents the number of people who order the soup-and-sandwich combo is __?__.

92. Temperature The Fahrenheit temperature is 59°. This is 32° more than $\frac{9}{5}$ of the Celsius temperature. Find the Celsius temperature.

93. Temperature The Celsius temperature on a fall morning was 5°. This is $\frac{5}{9}$ the difference between the Fahrenheit temperature and 32°. Find the Fahrenheit temperature.

94. Labor The repair bill on your car is $428.55. The charge for parts was $148.55. A mechanic worked on your car for 4 h. What was the charge per hour for labor?

95. Consumerism A local feed store sells a 100-pound bag of feed for $10.90. If a customer buys more than one bag, each additional bag costs $10.50. A customer bought $84.40 worth of feed. How many 100-pound bags of feed did this customer purchase?

96. Consumerism The Showcase Cinema of Lawrence charges $7.75 for an adult ticket and $4.75 for a child's ticket for all shows before 6:00 P.M. If a family of six pays $34.50 to get into an afternoon show, how many adult tickets and how many children's tickets did the family purchase?

97. Salaries See the news clipping at the right. Find the hourly rate the teachers will be paid for the extra hours they will work.

98. Consumerism The admission charge for a family at a city zoo is $7.50 for the first person and $4.25 for each additional member of the family. How many people are in a family that is charged $28.75 for admission?

99. Federal Income Tax Charlotte's annual federal income tax was $4681.25 plus 25% of her income over $34,000. If she paid $8181.25 in federal income tax, what was her annual income?

100. Federal Income Tax The annual federal income tax for a married couple filing jointly was $9362.50 plus 25% of their income over $68,000. If they paid $10,612.50 in federal income tax, what was their annual income?

In the News

Extra Hours for Teachers

In an effort to improve student performance, teachers at 12 city schools are being asked to work a longer day next year for a relatively small increase in salary. For working an additional 190 h, a teacher currently earning $79,400 per year would see his or her salary increase to $83,500 per year.

Source: The Boston Globe

APPLYING CONCEPTS

Solve.

101. $\frac{1}{\frac{1}{y}} = -9$

102. $8 \div \frac{1}{x} = -3$

103. $\frac{10}{\frac{3}{x}} - 5 = 4x$

104. $\frac{6}{\frac{7}{a}} = -18$

Solve. If the equation has no solution, write "no solution."

105. $2[3(x + 4) - 2(x + 1)] = 5x + 3(1 - x)$

106. $3[4(y + 2) - (y + 5)] = 3(3y + 1)$

107. $\frac{4[(x - 3) + 2(1 - x)]}{5} = x + 1$

108. $\frac{4(x - 5) - (x + 1)}{3} = x - 7$

109. $3(2x + 2) - 4(x - 3) = 2(x + 9)$

110. $2584 \div x = 54\frac{46}{x}$

PROJECTS OR GROUP ACTIVITIES

Recall that an **even integer** is an integer that is divisible by 2. An **odd integer** is an integer that is not divisible by 2.

Consecutive integers are integers that follow one another in order. Examples of consecutive integers are shown at the right.

8, 9, 10
$-3, -2, -1$
$n, n + 1, n + 2$, where n is an integer

Examples of **consecutive even integers** are shown at the right.

16, 18, 20
$-6, -4, -2$
$n, n + 2, n + 4$, where n is an even integer

Examples of **consecutive odd integers** are shown at the right.

11, 13, 15
$-23, -21, -19$
$n, n + 2, n + 4$, where n is an odd integer

111. The sum of three consecutive integers is 33. Find the integers.

112. The sum of three consecutive odd integers is 105. Find the integers.

113. The sum of four consecutive even integers is 92. Find the integers.

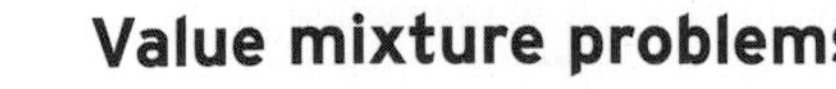

2.2 Value Mixture and Motion Problems

OBJECTIVE 1 Value mixture problems

A **value mixture problem** involves combining two ingredients that have different prices into a single blend. For example, a coffee manufacturer may blend two types of coffee into a single blend.

A solution of a value mixture problem is based on the equation $V = AC$, where V is the value of the ingredient, A is the amount of the ingredient, and C is the cost per unit of the ingredient.

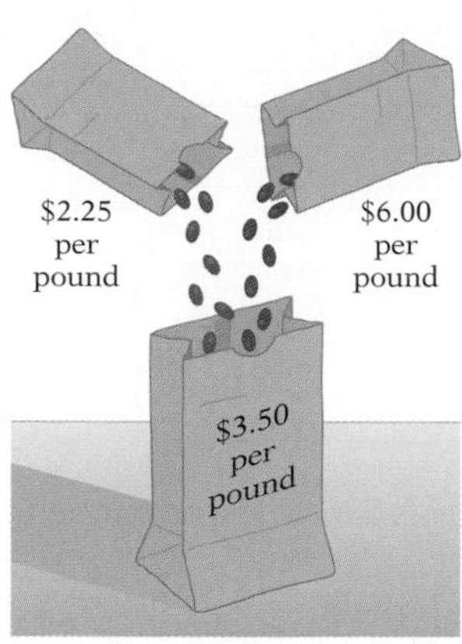

For instance, we can use the value mixture equation to find the value of 12 lb of coffee costing \$5.25 per pound.

$$V = AC$$
$$V = 12(5.25)$$
$$V = 63$$

The value of the coffee is \$63.

Solve: How many pounds of peanuts that cost \$2.25 per pound must be mixed with 40 lb of cashews that cost \$6.00 per pound to make a mixture that costs \$3.50 per pound?

STRATEGY FOR SOLVING A VALUE MIXTURE PROBLEM

▶ For each ingredient in the mixture, write a numerical or variable expression for the amount of the ingredient used, the unit cost of the ingredient, and the value of the amount used. For the mixture, write a numerical or variable expression for the amount, the unit cost of the mixture, and the value of the amount. The results can be recorded in a table.

Amount of peanuts: x

	Amount, A	·	Unit cost, C	=	Value, V
Peanuts	x	·	2.25	=	$2.25x$
Cashews	40	·	6.00	=	$6.00(40)$
Mixture	$x + 40$	·	3.50	=	$3.50(x + 40)$

▶ Determine how the values of the individual ingredients are related. Use the fact that the sum of the values of the ingredients is equal to the value of the mixture.

The sum of the values of the peanuts and the cashews is equal to the value of the mixture.

$$2.25x + 6.00(40) = 3.50(x + 40)$$
$$2.25x + 240 = 3.50x + 140$$
$$-1.25x + 240 = 140$$
$$-1.25x = -100$$
$$x = 80$$

The mixture must contain 80 lb of peanuts.

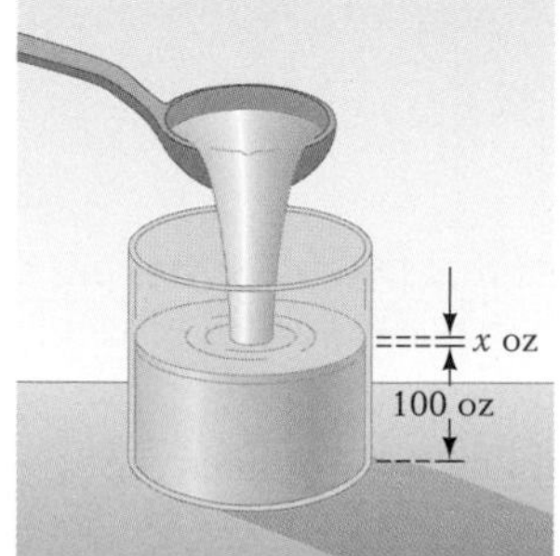

EXAMPLE 1 How many ounces of a gold alloy that costs \$320 an ounce must be mixed with 100 oz of an alloy that costs \$100 an ounce to make a mixture that costs \$160 an ounce?

Strategy ▶ Ounces of the \$320 gold alloy: x

	Amount	Cost	Value
\$320 alloy	x	320	$320x$
\$100 alloy	100	100	$100(100)$
Mixture	$x + 100$	160	$160(x + 100)$

▶ The sum of the values before mixing equals the value after mixing.

Solution

$$320x + 100(100) = 160(x + 100)$$
$$320x + 10{,}000 = 160x + 16{,}000$$
$$160x + 10{,}000 = 16{,}000$$
$$160x = 6000$$
$$x = 37.5$$

The mixture must contain 37.5 oz of the \$320 gold alloy.

Problem 1 A butcher combined hamburger that costs \$4.00 per pound with hamburger that costs \$2.80 per pound. How many pounds of each were used to make a 75-pound mixture that costs \$3.20 per pound?

Solution See page S4.

 *Try Exercise 15, page 72.*

OBJECTIVE 2 Uniform motion problems

Any object that travels at a constant speed in a straight line is said to be in *uniform motion.* **Uniform motion** means that the speed and direction of an object do not change. For instance, a train traveling at a constant speed of 50 mph on a straight track is in uniform motion.

How It's Used

A car's global positioning system (GPS) receiver repeatedly uses the equation $d = rt$ each time it determines the car's location, with r being the speed of light and t representing the time it takes a signal to travel from a GPS satellite to the receiver in the car.

The solution of a uniform motion problem is based on the equation $\boldsymbol{d = rt}$, where d is the distance traveled, r is the rate of travel, and t is the time spent traveling. For instance, suppose a train travels for 2 h at an average speed of 45 mph. Because the time (2 h) and the rate (45 mph) are known, we can find the distance traveled by solving the equation $d = rt$ for d.

$$d = rt$$
$$d = 45(2) \qquad \bullet\ r = 45, t = 2$$
$$d = 90$$

The train travels a distance of 90 mi.

Focus on using the equation $d = rt$

A chef leaves a restaurant and drives to his home, which is 16 mi away. If it takes the chef 20 min to get home, what is the average rate of speed?

Because the answer must be in miles per *hour* and the time is given in *minutes,* convert 20 min to hours: 20 min $= \frac{20}{60}$ h $= \frac{1}{3}$ h.

To find the rate of speed, solve the equation $d = rt$ for r using $d = 16$ and $t = \frac{1}{3}$.

$$d = rt$$
$$16 = r\left(\frac{1}{3}\right)$$
$$16 = \frac{1}{3}r$$
$$(3)16 = (3)\frac{1}{3}r$$
$$48 = r$$

The average rate of speed is 48 mph.

Take Note

The abbreviation ft/s means "feet per second."

If two objects are moving in opposite directions, then the rate at which the distance between them is increasing is the sum of the speeds of the two objects. For instance, in the diagram at the right, two joggers start from the same point and run in opposite directions. The distance between them is changing at a rate of 21 ft/s.

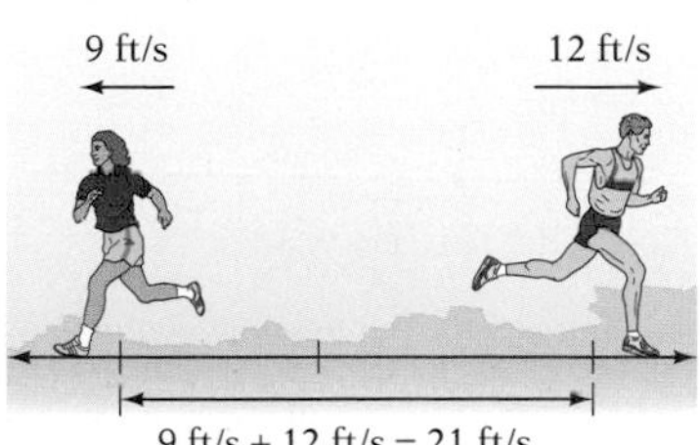

Similarly, if two objects are moving toward each other, the distance between them is decreasing at a rate that is equal to the sum of the speeds. The rate at which the two cyclists at the right are approaching one another is 35 mph.

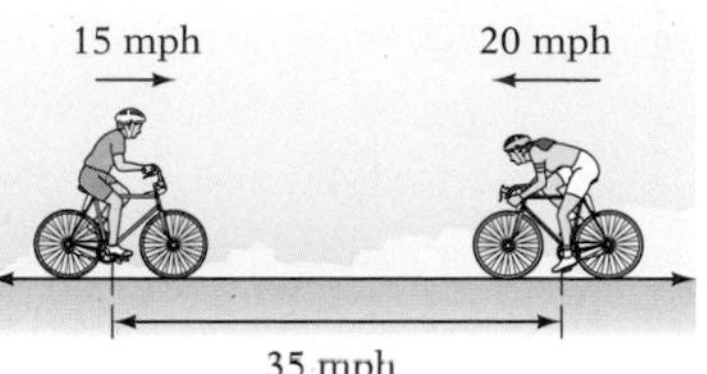

Focus on using the equation $d = rt$

Two cars start from the same point and move in opposite directions. The car moving west is traveling 50 mph, and the car moving east is traveling 65 mph. In how many hours will the cars be 230 mi apart?

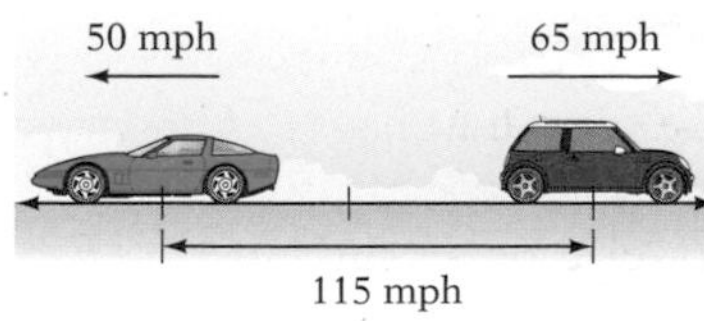

The cars are moving in opposite directions, so the rate at which the distance between them is changing is the sum of the rates of the cars. Thus $r = 115$.

50 mph + 65 mph = 115 mph

The distance is 230 mi. To find the time, solve $d = rt$ for t.

$$d = rt$$
$$230 = 115t$$
$$\frac{230}{115} = \frac{115t}{115}$$
$$2 = t$$

The time is 2 h.

Point of Interest

The jet stream that generally flows from west to east across the United States affects the time it takes a plane to fly from Los Angeles to New York. For instance, on a typical day, the flying time from New York City to Los Angeles is about 40 min longer than the trip from Los Angeles to New York City.

The typical cruising speed of a Boeing 777 jet is 525 mph. However, wind affects the speed of the plane. For example, when a plane is flying from California to New York, the wind is usually in the direction of the plane's flight, thereby increasing the effective speed of the plane. If the speed of the wind is 50 mph, then the effective speed of the plane is the sum of the plane's speed and the wind's speed: 525 mph + 50 mph = 575 mph.

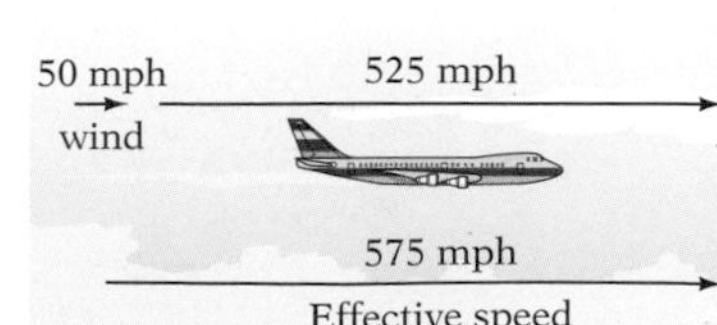

If the plane is traveling from New York to California, the wind is generally in the opposite direction of the plane's flight, so it decreases the speed of the plane. If the speed of the wind is 50 mph, then the effective speed of the plane is the difference between the plane's speed and the wind's speed: 525 mph − 50 mph = 475 mph.

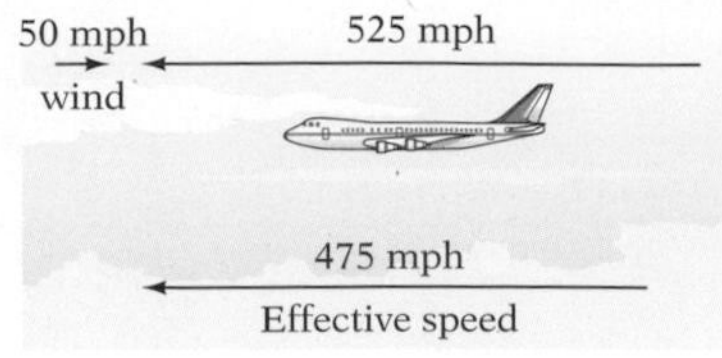

There are other situations in which these concepts may be applied.

EXAMPLE 2 A Mississippi River sightseeing tour takes tourists on a 30-mile trip down the river from the tour dock to a Civil War historical site, where they spend 1 h walking around the grounds. The tour then returns to the tour dock. If the speed of the boat is set to travel at 16 mph in calm water and the rate of the current is 4 mph, find the total time of the trip.

Strategy To determine the total time:

- Find the time spent traveling down the river by solving $d = rt$ for t. The distance is 30 mi. Therefore, $d = 30$.

  Because the boat is traveling with the current, the rate of the boat is the sum of the boat's speed in calm water and the speed of the current: 16 mph + 4 mph = 20 mph. Thus $r = 20$ mph.

- Find the time spent returning to the tour station by solving $d = rt$ for t. The distance is 30 mi. Therefore, $d = 30$. Because the boat is traveling against the current, the rate of the boat is the difference between the boat's speed in calm water and the speed of the current: 16 mph − 4 mph = 12 mph. Thus $r = 12$ mph.

- Add the time spent traveling down the river, the time spent at the historical site (1 h), and the time spent returning.

Solution

$$d = rt$$
$$30 = 20t \quad \bullet\ d = 30, r = 16 + 4 = 20$$
$$\frac{30}{20} = \frac{20t}{20}$$
$$1.5 = t$$

The time spent traveling down the river is 1.5 h.

$$d = rt$$
$$30 = 12t \quad \bullet\ d = 30, r = 16 - 4 = 12$$
$$\frac{30}{12} = \frac{12t}{12}$$
$$2.5 = t$$

The time spent returning to the tour station is 2.5 h.

Total time = 1.5 + 1 + 2.5 = 5

The total time of the trip is 5 h.

Problem 2 A plane that can normally travel at 175 mph in calm air is flying into a headwind of 35 mph. How long will it take the plane to fly 350 mi?

Solution See page S4.

Try Exercise 29, page 73.

For some uniform motion problems, it may be helpful to record the known information in a table. This is illustrated in the following example.

Solve: An executive has an appointment 785 mi from the office. The executive takes a helicopter from the office to the airport and a plane from the airport to the business appointment. The helicopter averages 70 mph, and the plane averages 500 mph. The total time spent traveling is 2 h. Find the distance from the executive's office to the airport.

STRATEGY FOR SOLVING A UNIFORM MOTION PROBLEM

- For each object, write a numerical or variable expression for the distance, rate, and time. The results can be recorded in a table.

The total time of travel is 2 h.

Unknown time in the helicopter: t

Time in the plane: $2 - t$

	Rate, r	$\cdot$	Time, t	$=$	Distance, d
Helicopter	70	$\cdot$	t	$=$	$70t$
Plane	500	$\cdot$	$2 - t$	$=$	$500(2 - t)$

- Determine how the distances traveled by the individual objects are related. For example, the total distance traveled by both objects may be known, or it may be known that the two objects traveled the same distance.

The total distance traveled is 785 mi.

$$70t + 500(2 - t) = 785$$
$$70t + 1000 - 500t = 785$$
$$-430t + 1000 = 785$$
$$-430t = -215$$
$$t = 0.5$$

$70t$ | $500(2 - t)$

Office Airport Appointment

785 mi

The time spent traveling from the office to the airport in the helicopter is 0.5 h. To find the distance between these two locations, substitute 70 for r and 0.5 for t into the equation $rt = d$.

$$rt = d$$
$$70(0.5) = d$$
$$35 = d$$

The distance from the office to the airport is 35 mi.

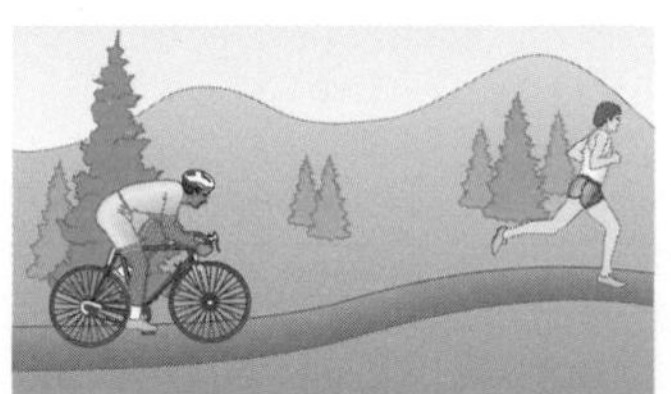

EXAMPLE 3 A long-distance runner started a course running at an average speed of 6 mph. One and one-half hours later, a cyclist traveled the same course at an average speed of 12 mph. How long after the runner started did the cyclist overtake the runner?

Strategy ▶ Unknown time for the cyclist: t
Time for the runner: $t + 1.5$

	Rate	Time	Distance
Runner	6	$t + 1.5$	$6(t + 1.5)$
Cyclist	12	t	$12t$

▶ The runner and the cyclist traveled the same distance.

Solution

$$\begin{aligned} 6(t + 1.5) &= 12t \\ 6t + 9 &= 12t \\ 9 &= 6t \\ \frac{3}{2} &= t \\ t + 1.5 &= 1.5 + 1.5 \\ &= 3 \end{aligned}$$

- **The cyclist traveled for 1.5 h.**
- **Substitute the value of t into the variable expression for the runner's time.**

The cyclist overtook the runner 3 h after the runner started.

Problem 3 Two small planes start from the same point and fly in opposite directions. The first plane is flying 30 mph faster than the second plane. In 4 h, the planes are 1160 mi apart. Find the rate of each plane.

Solution See page S4.

Try Exercise 35, page 74.

2.2 Exercises

CONCEPT CHECK

1. A grocer mixes peanuts that cost \$4 per pound with almonds that cost \$8 per pound. Which of the following choices are *not* possible for the cost of the mixture?
(i) \$4.50 (ii) \$3.75 (iii) \$9.50 (iv) \$6.25 (v) \$6.00

2. A grass seed mixture sells for \$1.75 per pound. The mixture was created by mixing Michigan Premium grass seed that cost \$1.60 per pound with Fescue grass seed that cost \$2.00 per pound. Which of the following is true?
(i) There is more Michigan Premium in the mixture than Fescue.
(ii) There are equal amounts of Michigan Premium and Fescue in the mixture.
(iii) There is more Fescue in the mixture than Michigan Premium.
(iv) There is not enough information.

3. A 10-pound bag of trail mix is made by mixing raisins and nuts. If the bag contains x pounds of raisins, then the amount of nuts is
(i) $(10 - x)$ pounds
(ii) $(x - 10)$ pounds
(iii) $(x + 10)$ pounds

4. Lois and Damien begin at the same time and walk toward each other until they meet. They walk on a straight road that is 2 mi long. Lois walks faster than Damien.
a. Is the distance walked by Lois less than, equal to, or greater than the distance walked by Damien?
b. Is the time walked by Lois less than, equal to, or greater than the time walked by Damien?
c. What is the total distance traveled by Lois and Damien?

5. Margot and Juanita begin at the same time and walk in opposite directions on a straight hiking path. Margot walks at 2 mph and Juanita walks at 3 mph. After 1 h, will they be less than or more than 6 mi apart?

6. Morgan and Emma ride their bikes from Emma's house to the store using the same route. Emma bikes faster than Morgan. Morgan begins biking 5 min before Emma begins, but they arrive at the store at the same time.
a. When they reach the store, is the distance biked by Emma less than, equal to, or greater than the distance biked by Morgan?
b. When they reach the store, is the time spent biking by Emma less than, equal to, or greater than the time spent biking by Morgan?

1 Value mixture problems (See pages 64–66.)

GETTING READY

7. a. The total value of a 20-pound bag of birdseed that costs \$.42 per pound is __?__.
b. The cost per ounce of a 24-ounce box of chocolates that has a total value of \$16.80 is __?__.

8. A dried fruit mixture is made from dried cranberries that cost \$7.20 per pound and dried apricots that cost \$4.50 per pound. Fifteen pounds of the mixture costs \$5.40 per pound. Let x be the amount of dried cranberries in the mixture.
a. Complete the following table.

	Amount, A	·	Unit Cost, C	=	Value, V
Dried cranberries	x	·	__?__	=	__?__
Dried apricots	__?__	·	__?__	=	__?__
Mixture	__?__	·	__?__	=	__?__

b. Use the expressions in the last column of the table in part (a) to write an equation that can be solved to find the number of pounds of dried cranberries in the mixture: __?__ + __?__ = __?__.

9. A coffee merchant mixes a dark roast coffee that cost \$10 per pound with a light roast coffee that cost \$7 per pound. Which of the following could be true about the cost C per pound of the mixture? There may be more than one correct answer.
(i) $C = \$17$ (ii) $C = \$10$ (iii) $C < \$7$ (iv) $C > \$7$
(v) $C > \$10$ (vi) $C < \$10$

10. A snack mix is made from peanuts that cost \$3 per pound and caramel popcorn that costs \$2.20 per pound. The mixture costs \$2.50 per pound. Does the mixture contain more peanuts or more popcorn?

11. A restaurant chef mixes 20 lb of snow peas costing \$1.99 a pound with 14 lb of petite onions costing \$1.19 a pound to make a vegetable medley for the evening meal. Find the cost per pound of the mixture.

12. A coffee merchant combines coffee costing \$5.50 per pound with coffee costing \$3.00 per pound. How many pounds of each should be used to make 40 lb of a blend costing \$4.00 per pound?

13. Adult tickets for a play cost \$5.00, and children's tickets cost \$2.00. For one performance, 460 tickets were sold. Receipts for the performance were \$1880. Find the number of adult tickets sold.

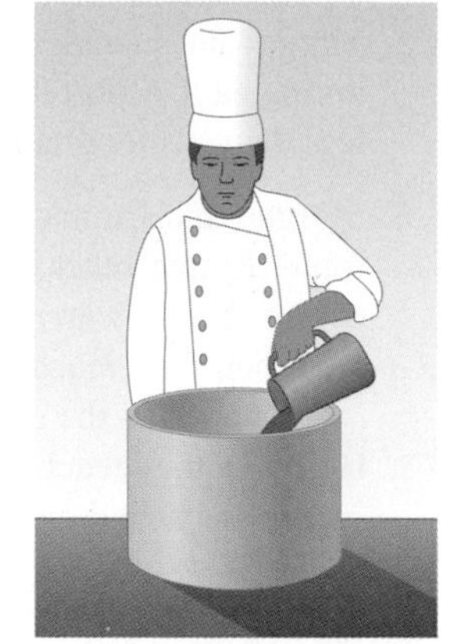

14. The tickets for a local production of Gilbert and Sullivan's *H.M.S. Pinafore* cost \$6.00 for adults and \$3.00 for children and seniors. The total receipts for 505 tickets were \$1977. Find the number of adult tickets sold.

15. Fifty liters of pure maple syrup that costs \$9.50 per liter are mixed with imitation maple syrup that costs \$4.00 per liter. How much imitation maple syrup is needed to make a mixture that costs \$5.00 per liter?

16. To make a flour mix, a miller combined soybeans that cost \$8.50 per bushel with wheat that cost \$4.50 per bushel. How many bushels of each were used to make a mixture of 800 bushels that cost \$5.50 per bushel?

17. A hiking instructor is preparing a trail mix for a group of young hikers. She mixes nuts that cost \$3.99 a pound with pretzels that cost \$1.29 a pound to make a 20-pound mixture that costs \$2.37 a pound. Find the number of pounds of nuts used.

18. A silversmith combined pure silver that costs \$5.20 an ounce with 50 oz of a silver alloy that costs \$2.80 an ounce. How many ounces of the pure silver were used to make an alloy of silver that costs \$4.40 an ounce?

19. A tea mixture was made from 30 lb of tea that costs \$6.00 per pound and 70 lb of tea that costs \$3.20 per pound. Find the cost per pound of the tea mixture.

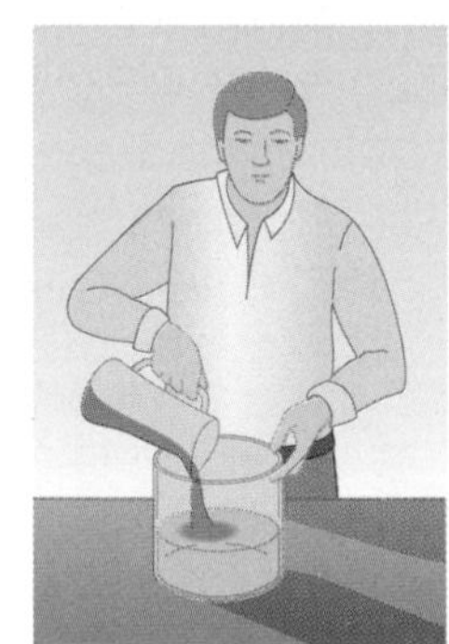

20. Find the cost per ounce of a salad dressing mixture made from 64 oz of olive oil that costs \$8.29 and 20 oz of vinegar that costs \$1.99. Round to the nearest cent.

21. A fruitstand owner combined cranberry juice that costs \$4.20 per gallon with 50 gal of apple juice that costs \$2.10 per gallon. How much cranberry juice was used to make cranapple juice that costs \$3.00 per gallon?

22. Walnuts that cost \$4.05 per kilogram were mixed with cashews that cost \$7.25 per kilogram. How many kilograms of each were used to make a 50-kilogram mixture that costs \$6.25 per kilogram? Round to the nearest tenth.

2 Uniform motion problems (See pages 66–70.)

GETTING READY

23. a. Marvin and Nancy start biking at the same time. Marvin bikes at 12 mph and Nancy bikes at 15 mph. After t hours, Marvin has biked __?__ miles and Nancy has biked __?__ miles.

b. A plane flies at a rate of 380 mph in calm air. The wind is blowing at 20 mph. Flying with the wind, the plane flies __?__ mph. Flying against the wind, the plane flies __?__ mph.

24. Sam and Jose leave a movie theatre and head in opposite directions. Sam bikes and Jose walks. Sam bikes four times faster than Jose walks. After a quarter of an hour, Sam and Jose are 3.75 mi apart. Let r represent Jose's rate.

a. Complete the following table.

	Rate, r	·	Time, t	=	Distance, d
Sam	?	·	?	=	?
Jose	r	·	?	=	?

b. Use the expressions in the last column of the table in part (a) to write an equation that can be solved to find Jose's rate: ___?___ + ___?___ = ___?___.

25. Mike and Mindy live 3 mi apart. They leave their houses at the same time and walk toward each other until they meet. Mindy walks faster than Mike.

a. Is the distance walked by Mike less than, equal to, or greater than the distance walked by Mindy?

b. Is the time spent walking by Mike less than, equal to, or greater than the time spent walking by Mindy?

c. What is the total distance traveled by Mike and Mindy?

26. Eric and James ride their bikes from Eric's house to school. James begins biking 10 min before Eric begins. Eric bikes faster than James and catches up with him just as they reach school.

a. Is the distance biked by James less than, equal to, or greater than the distance biked by Eric?

b. Is the time spent biking by James less than, equal to, or greater than the time spent biking by Eric?

27. It takes a student 30 min to drive from school to work, a distance of 20 mi. At what average rate of speed does the student drive?

28. A cyclist rides at a rate of 16 mph for 45 min. How far does the cyclist travel in that time?

29. The typical cruising speed for a Boeing 747-400 airplane is 30 mph faster than that for a Boeing 737-800. If a Boeing 747-400 can make a trip of 1680 mi in 3 h, how long would it take a Boeing 737-800 to make the same trip? Round to the nearest tenth.

30. Some hotels in Las Vegas use moving sidewalks to transport guests between various destinations. Suppose a hotel guest decides to get on a moving sidewalk that is 195 ft long. If the guest walks at a rate of 5 ft/s and the moving sidewalk moves at a rate of 7 ft/s, how many seconds will it take the guest to walk from one end of the moving sidewalk to the other?

31. Two cyclists start at 1:00 P.M. from opposite ends of a 54-mile race course. The average rate of speed of the first cyclist is 17 mph, and the average rate of speed of the second cyclist is 19 mph. At what time will the two cyclists meet?

32. At 9:00 A.M., Katrina starts down an exercise path and walks at a rate of 4 mph. At 9:30 A.M., Carla leaves from the same place and begins chasing Katrina, catching her at 10:00 A.M. Find Carla's jogging rate.

33. A bicyclist traveling at 18 mph overtakes an in-line skater who is traveling at 10 mph and had a 0.5-hour head start. How far from the starting point does the bicyclist overtake the in-line skater?

34. A helicopter traveling 120 mph overtakes a speeding car traveling 90 mph. The car had a 0.5-hour head start. How far from the starting point did the helicopter overtake the car?

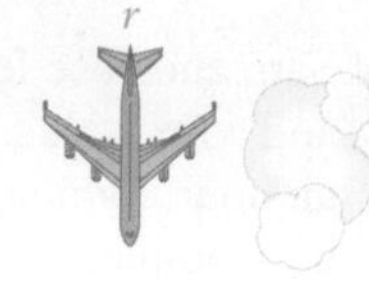

35. Two planes are 1380 mi apart and traveling toward each other. One plane is traveling 80 mph faster than the other plane. The planes pass each other in 1.5 h. Find the speed of each plane.

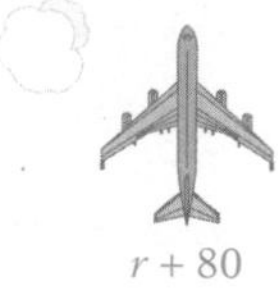

36. Two jet skiers leave the same dock at the same time and travel in opposite directions. One skier is traveling 14 mph slower than the other skier. In half an hour the skiers are 48 mi apart. Find the rate of the slower skier.

37. Two planes start from the same point and fly in opposite directions. The first plane is flying 50 mph slower than the second plane. In 2.5 h, the planes are 1400 mi apart. Find the rate of each plane.

38. A commuter plane provides transportation from an international airport to surrounding cities. One commuter plane averaged 210 mph flying to a city airport and 140 mph returning to the international airport. The total flying time was 4 h. Find the distance between the two airports.

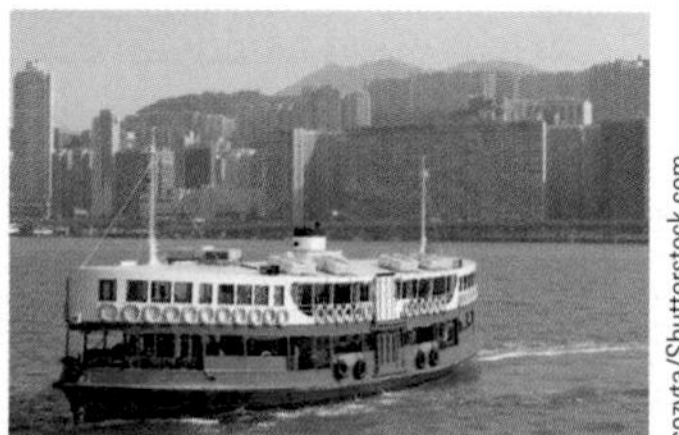
cozyta/Shutterstock.com

39. A ferry leaves a harbor and travels to a resort island at an average speed of 18 mph. On the return trip, because of the fog, the ferry travels at an average speed of 12 mph. The total time for the trip is 6 h. How far is the island from the harbor?

40. Two hikers start from the same point and hike in opposite directions around a lake whose shoreline is 13 mi long. One hiker walks 0.5 mph faster than the other hiker. How fast did each hiker walk if they meet in 2 h?

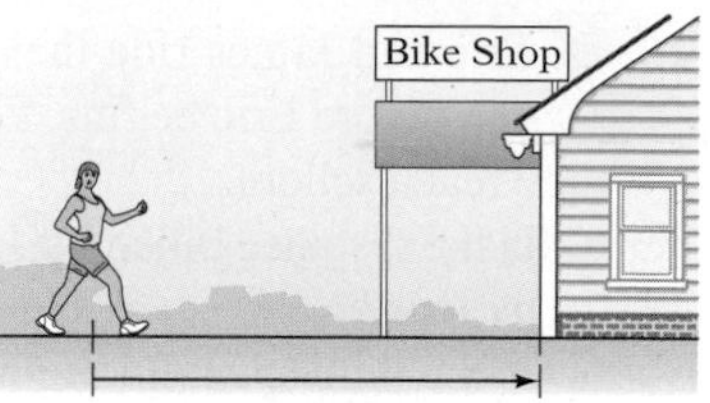

41. Marcella walked from her home to a bicycle repair shop at a rate of 3.5 mph. After picking up her bike, she rode it home along the same route at a rate of 14 mph. Her total travel time was 1 h. How far is Marcella's home from the shop?

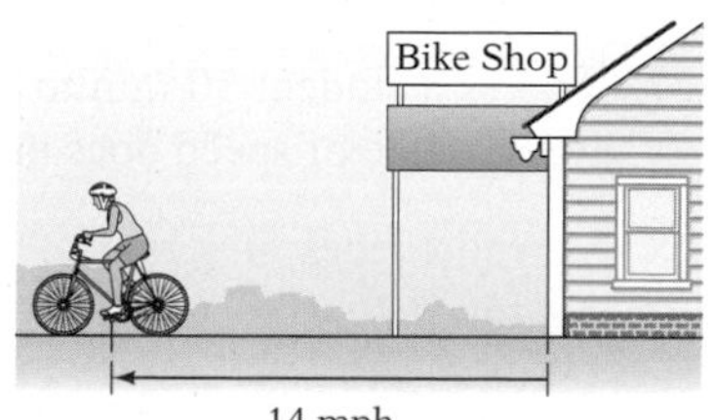

42. An express train leaves Grand Central Station 1 h after a freight train leaves the same station. The express train is traveling 15 mph faster than the freight train. Find the rate at which each train is traveling if the express train overtakes the freight train in 3 h.

43. At noon, a train leaves Washington, D.C., headed for Pittsburgh, Pennsylvania, a distance of 260 mi. The train travels at a speed of 60 mph. At 1 P.M., a second train leaves Pittsburgh headed for Washington, D.C., traveling at 40 mph. How long after the train leaves Pittsburgh will the two trains pass each other?

44. A plane leaves an airport at 3 P.M. At 4 P.M., another plane leaves the same airport traveling in the same direction at a speed that is 150 mph faster than that of the first plane. Four hours after the first plane takes off, the second plane is 250 mi ahead of the first plane. How far did the second plane travel?

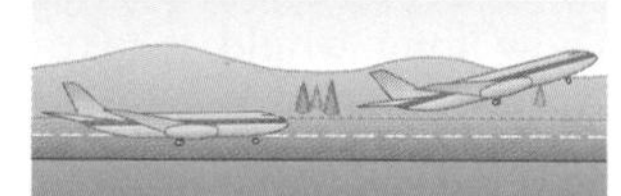

APPLYING CONCEPTS

45. **Space Travel** In 1999, astronomers confirmed the existence of three planets orbiting a star other than the sun. The planets are approximately the size of Jupiter and orbit the star Upsilon Andromeda, which is approximately 260 trillion miles from Earth. How many years after leaving Earth would a spacecraft traveling 18 million miles per hour reach this star? Round to the nearest hundred. (Eighteen million miles per hour is about 1000 times faster than current spacecraft can travel.)

46. **Uniform Motion** If a parade 2 mi long is proceeding at 3 mph, how long will it take a runner, jogging at 6 mph, to travel from the end of the parade to the start of the parade?

47. **Uniform Motion** Two cars are headed directly toward each other at rates of 40 mph and 60 mph. How many miles apart are they 2 min before impact?

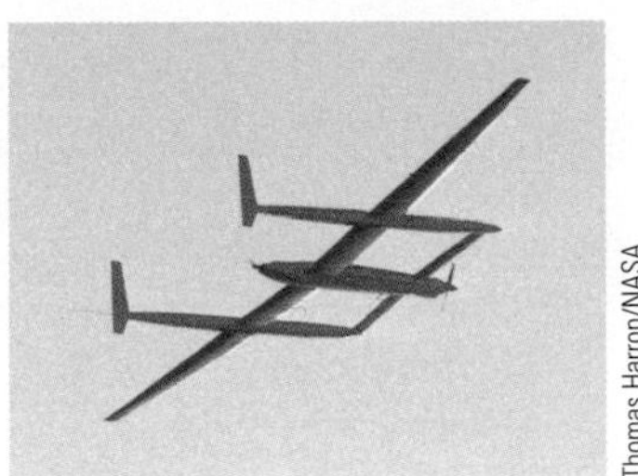
Thomas Harrop/NASA

48. **Aeronautics** In December of 1986, pilots Dick Rutan and Jeana Yeager flew the *Voyager* in the first nonstop, nonrefueled flight around the world. They flew east from Edwards Air Force Base in California on December 14, traveled 24,986.727 mi around the world, and returned to Edwards 216 h 3 min 44 s after their departure.
   a. On what date after their flight did Rutan and Yeager land at Edwards Air Force Base?
   b. What was their average speed in miles per hour? Round to the nearest whole number.
   c. Find the circumference of Earth in a reference almanac, and then calculate the approximate distance above Earth that the flight was flown. Round to the nearest whole number.

49. **Uniform Motion** A car travels at an average speed of 30 mph for 1 mi. Is it possible for the car to increase its speed during the next mile so that its average speed for the 2 mi is 60 mph?

50. A student jogs 1 mi at a rate of 8 mph and jogs back at a rate of 6 mph. Does it seem reasonable that the average rate is 7 mph? Why or why not? Support your answer.

PROJECTS OR GROUP ACTIVITIES

51. A mixture is made from ingredient A, which costs $10 per pound, and ingredient B, which costs $20 per pound.
   a. If equal amounts of ingredients A and B are used, what is the cost of the mixture?
   b. If the mixture is made using less of ingredient A than of B, is the cost of the mixture less than or greater than the cost found in part (a)?

52. Jason and Quan share a bicycle. Jason rides for an agreed-upon distance and then locks up the bike for Quan, who has been walking. Meanwhile, Jason walks on ahead. The friends alternate walking and riding. If both boys walk at a rate of 4 mph and ride at a rate of 12 mph, what part of the time has the bike been locked up when Jason and Quan meet again?

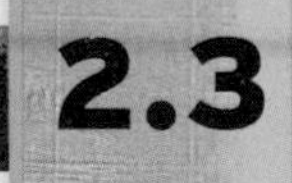

2.3 Applications: Problems Involving Percent

OBJECTIVE 

Investment problems

The annual simple interest that an investment earns is given by the equation $I = Pr$, where I is the simple interest, P is the principal, or the amount invested, and r is the simple interest rate. The solution of an investment problem is based on this equation.

For instance, if the annual interest rate on a $3000 investment is 9%, then we can use the simple interest equation to find the annual simple interest earned on the investment.

$$I = Pr$$
$$I = 3000(0.09)$$
$$I = 270$$

The annual simple interest earned is $270.

Solve: You have a total of \$8000 invested in two simple interest accounts. On one account, a money market fund, the annual simple interest rate is 11.5%. On the second account, a bond fund, the annual simple interest rate is 9.75%. The total annual interest earned by the two accounts is \$823.75. How much do you have invested in each account?

STRATEGY FOR SOLVING A PROBLEM INVOLVING MONEY DEPOSITED IN TWO SIMPLE INTEREST ACCOUNTS

▶ For each amount invested, use the equation $Pr = I$. Write a numerical or variable expression for the principal, the interest rate, and the interest earned. The results can be recorded in a table.

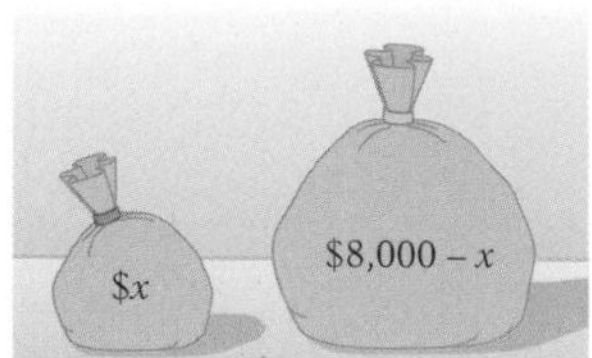

The total amount invested is \$8000.
Amount invested at 11.5%: x
Amount invested at 9.75%: $8000 - x$

	Principal, P	·	Interest rate, r	=	Interest earned, I
Amount at 11.5%	x	·	0.115	=	$0.115x$
Amount at 9.75%	$8000 - x$	·	0.0975	=	$0.0975(8000 - x)$

▶ Determine how the amounts of interest earned on the individual investments are related. For example, the total interest earned by both accounts may be known, or it may be known that the interest earned on one account is equal to the interest earned on the other account.

The total annual interest earned is \$823.75.

$$
\begin{aligned}
0.115x + 0.0975(8000 - x) &= 823.75 \\
0.115x + 780 - 0.0975x &= 823.75 \\
0.0175x + 780 &= 823.75 \\
0.0175x &= 43.75 \\
x &= 2500
\end{aligned}
$$

The amount invested at 9.75% is $8000 - x$.
Replace x with 2500 and evaluate.

$$8000 - x = 8000 - 2500 = 5500$$

The amount invested at 11.5% is \$2500.
The amount invested at 9.75% is \$5500.

EXAMPLE 1 An investment of \$4000 is made at an annual simple interest rate of 4.9%. How much additional money must be invested at an annual simple interest rate of 7.4% so that the total interest earned is 6.4% of the total investment?

Strategy ► Additional amount to be invested at 7.4%: x

	Principal	Rate	Interest
Amount at 4.9%	4000	0.049	0.049(4000)
Amount at 7.4%	x	0.074	$0.074x$
Amount at 6.4%	$4000 + x$	0.064	$0.064(4000 + x)$

► The sum of the amounts of interest earned by the two investments equals the interest earned by the total investment.

Solution

$$0.049(4000) + 0.074x = 0.064(4000 + x)$$
$$196 + 0.074x = 256 + 0.064x$$
$$196 + 0.01x = 256$$
$$0.01x = 60$$
$$x = 6000$$

$6000 must be invested at an annual simple interest rate of 7.4%.

Problem 1 An investment of $3500 is made at an annual simple interest rate of 5.2%. How much additional money must be invested at an annual simple interest rate of 7.5% so that the total interest earned is $575?

Solution See page S4.

 *Try Exercise 15, page 80.*

OBJECTIVE 2 Percent mixture problems

Take Note

As stated at the right, in a 10% hydrogen peroxide solution, 10% of the total solution is hydrogen peroxide. Here we are referring to weight: The solution is 10% hydrogen peroxide *by weight*. For example, if the solution weighs 100 g, then the hydrogen peroxide in the solution weighs 10 g and the water weighs 90% of 100 g, or 90 g.

The amount of a substance in a solution or an alloy can be given as a percent of the total solution or alloy. For example, in a 10% hydrogen peroxide solution, 10% of the total solution is hydrogen peroxide. The remaining 90% is water.

The solution of a percent mixture problem is based on the equation $\boldsymbol{Q = Ar}$, where Q is the quantity of a substance in the solution or alloy, A is the amount of solution or alloy, and r is the percent of concentration.

For instance, we can use the percent mixture equation to find the number of grams of silver in 50 g of a 40% silver alloy.

$$Q = Ar$$
$$Q = 50(0.40)$$
$$Q = 20$$

There are 20 g of silver in the alloy.

Solve: A chemist mixes an 11% acid solution with a 4% acid solution. How many milliliters of each solution should the chemist use to make a 700-milliliter solution that is 6% acid?

STRATEGY FOR SOLVING A PERCENT MIXTURE PROBLEM

► For each solution, use the equation $Ar = Q$. Write a numerical or variable expression for the amount of solution, the percent of concentration, and the quantity of the substance in the solution. The results can be recorded in a table.

The total amount of solution is 700 ml.

Amount of 11% solution: x

Amount of 4% solution: $700 - x$

	Amount of solution, A	$\cdot$	Percent of concentration, r	$=$	Quantity of substance, Q
11% solution	x	$\cdot$	0.11	$=$	$0.11x$
4% solution	$700 - x$	$\cdot$	0.04	$=$	$0.04(700 - x)$
6% solution	700	$\cdot$	0.06	$=$	$0.06(700)$

▶ Determine how the quantities of the substance in the individual solutions are related. Use the fact that the sum of the quantities of the substance being mixed is equal to the quantity of the substance after mixing.

The sum of the amounts of acid in the 11% solution and the 4% solution is equal to the amount of acid in the 6% solution.

$$0.11x + 0.04(700 - x) = 0.06(700)$$
$$0.11x + 28 - 0.04x = 42$$
$$0.07x + 28 = 42$$
$$0.07x = 14$$
$$x = 200$$

The amount of 4% solution is $700 - x$. Replace x by 200 and evaluate.

$$700 - x = 700 - 200 = 500$$

The chemist should use 200 ml of the 11% solution and 500 ml of the 4% solution.

EXAMPLE 2 How many grams of pure acid must be added to 60 g of an 8% acid solution to make a 20% acid solution?

Strategy ▶ Grams of pure acid: x

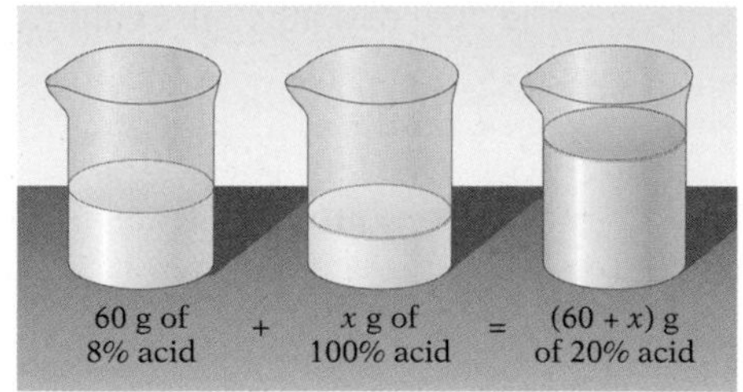

	Amount	Percent	Quantity
Pure acid (100%)	x	1.00	x
8%	60	0.08	$0.08(60)$
20%	$x + 60$	0.20	$0.20(x + 60)$

▶ The sum of the quantities before mixing equals the quantity after mixing.

Solution

$$\begin{aligned} x + 0.08(60) &= 0.20(x + 60) \\ x + 4.8 &= 0.20x + 12 \\ 0.8x + 4.8 &= 12 \\ 0.8x &= 7.2 \\ x &= 9 \end{aligned}$$

To make the 20% acid solution, 9 g of pure acid must be used.

Problem 2 A butcher has some hamburger that is 22% fat and some that is 12% fat. How many pounds of each should be mixed to make 80 lb of hamburger that is 18% fat?

Solution See pages S4–S5.

Try Exercise 39, page 83.

2.3 Exercises

CONCEPT CHECK

1. For the following example, give **a.** the principal, **b.** the interest rate, and **c.** the interest earned.

   The annual simple interest rate on an investment of \$1500 is 4%. Find the annual simple interest earned on the investment.

2. Investment A has an annual interest rate of 5%, and investment B has an annual interest rate of 8%. Which of the following is true?
   (i) The annual rate of return on both investments is less than 5%.
   (ii) The annual rate of return on both investments is between 5% and 8%.
   (iii) The annual rate of return on both investments is more than 8%.

3. A solution that is 10% salt is mixed with a solution that is 20% salt. Which of the following are not possible choices for the concentration of salt in the resulting mixture?
   (i) 10% (ii) 15% (iii) 18% (iv) 20% (v) 25%

4. Pure water is added to a solution of sugar and water that is 20% sugar. Which of the following are not possible choices for the concentration of sugar in the resulting mixture?
   (i) 10% (ii) 15% (iii) 20% (iv) 25% (v) 30%

1 Investment problems (See pages 75–77.)

5. Explain the meaning of each variable in the equation $I = Pr$. Give an example of how this equation is used.

GETTING READY

6. You invest an amount of money at an annual simple interest rate of 6.25%. You invest a second amount, $500 less than the first, at an annual simple interest rate of 6%. Let x represent the amount invested at 6.25%. Complete the following table.

	Principal, P	$\cdot$	Interest rate, r	=	Interest earned, I
Amount at 6.25%	x	$\cdot$	?	=	?
Amount at 6%	?	$\cdot$	?	=	?

7. The total annual interest earned on the investments in Exercise 6 is $115. Use this information and the information in the table in Exercise 6 to write an equation that can be solved to find the amount of money invested at 6.25%: ___?___ + ___?___ = ___?___.

8. Suppose $5000 is invested in two simple interest accounts. On one account the annual simple interest rate is 6%, and on the other the annual simple interest rate is 7%. The total annual interest earned on the two accounts is $330. In the context of this situation, explain each term of the equation

$$0.06x + 0.07(5000 - x) = 330$$

9. Joseph Abruzzio decides to divide a gift of $5000 into two different accounts. He deposits $2000 in an IRA account that earns an annual simple interest rate of 5.5%. The remaining money is placed in an account that earns an annual simple interest rate of 7.25%. How much interest will Joseph earn in one year from the two accounts?

10. Kristi invests $1500 at a 7.25% annual simple interest rate, and her sister Kari invests $2000 at a 6.75% annual simple interest rate. Which of the two will earn the greater amount of interest after one year? How much greater?

11. Deon Brown purchases a municipal bond for $2000. The bond earns an annual simple interest rate of 6.4%. How much must Deon invest in an account that earns 8% annual simple interest so that both investments will earn the same amount of interest in one year?

12. A $5000 investment at an annual simple interest rate of 5.2% earned as much interest after one year as another investment in an account that earned 6.5% annual simple interest. How much was invested at 6.5%?

13. Two investments earn a total annual income of $2825. One investment is in a 6.75% annual simple interest certificate of deposit. The other is in a 7.25% tax-free annual simple interest account. The total amount invested is $40,000. Find the amount invested in the certificate of deposit.

14. Two investments earn an annual income of $765. One investment earns an annual simple interest rate of 8.5%, and the other investment earns an annual simple interest rate of 10.2%. The total amount invested is $8000. How much is invested in each account?

15. An investment club invested $5000 at an annual simple interest rate of 8.4%. How much additional money must be invested at an annual simple interest rate of 10.5% so that the total interest earned will be 9% of the total investment?

16. Two investments earn an annual income of \$465. One investment is a 5.5% tax-free annual simple interest account, and the other is a 4.5% annual simple interest certificate of deposit. The total amount invested is \$9600. How much is invested in each account?

17. Two investments earn an annual income of \$575. One investment earns an annual simple interest rate of 8.5%, and the other investment earns an annual simple interest rate of 6.4%. The total amount invested is \$8000. How much is invested in each account?

18. An investment club invested \$6000 at an annual simple interest rate of 4%. How much additional money must be invested at an annual simple interest rate of 6.5% so that the total annual interest earned will be 5% of the total investment?

19. Dee Pinckney made an investment of \$6000 at an annual simple interest rate of 5.5%. How much additional money must she invest at an annual simple interest rate of 10% in order for the total annual interest earned to be 7% of the investment?

20. An account executive deposited \$42,000 into two simple interest accounts. On the tax-free account, the annual simple interest rate is 3.5%, and on the money market fund, the annual simple interest rate is 4.5%. How much should be invested in each account so that both accounts earn the same annual interest?

21. An investment club invested \$13,600 into two simple interest accounts. On one account, the annual simple interest rate is 4.2%. On the other, the annual simple interest rate is 6%. How much should be invested in each account so that both accounts earn the same annual interest?

22. Orlando Salavarrio, a financial planner, recommended that 25% of a client's investment be placed in a 4% annual simple interest tax-free account, 40% be placed in 6% certificates of deposit, and the remainder be placed in a 9% high-risk investment. The total interest earned from the investments would be \$6550. Find the total amount to be invested.

23. The amount of annual interest earned on the x dollars that Will invested in one simple interest account was $0.055x$, and the amount of annual interest earned on the money that Will invested in another simple interest account was $0.072(6000 - x)$.
a. What were the interest rates on the two accounts?
b. What was the total amount of money Will had invested in the two accounts?

24. Refer to the investments described in Exercise 23. Which of the following could be true about the total amount T of interest earned by Will's two accounts? There may be more than one correct answer.
(i) $T < 330$ (ii) $T < 432$ (iii) $T > 432$ (iv) $T > 330$

2 Percent mixture problems (See pages 77–79.)

GETTING READY

25. The label on a 32-ounce bottle of lemonade says that it contains 10% real lemon juice. The amount of real lemon juice in the bottle is 32(__?__) = __?__ oz.

26. A 25% acid solution is made by adding an amount of 75% acid solution to 50 ml of a 15% acid solution. Let x be the amount of 75% acid solution that is in the mixture.

a. Complete the following table.

	Amount, A	$\cdot$	Percent of concentration, r	$=$	Quantity, Q
15% acid solution	?	$\cdot$	?	$=$	?
75% acid solution	x	$\cdot$	?	$=$	?
Mixture	?	$\cdot$	?	$=$	?

b. Use the expressions in the last column of the table in part (a) to write an equation that can be solved to find the amount of 75% acid solution that is in the mixture: ___?___ + ___?___ = ___?___.

27. A 32-ounce bottle of Orange Ade contains 8 oz of orange juice. At the same percent concentration, how much orange juice is contained in a 40-ounce bottle of Orange Ade?

28. A chemist has 4 L of a solution that contains 0.36 L of acetic acid. The chemist needs to prepare 6 L of a solution with the same percent concentration of acetic acid. How much acetic acid must the 6 L of the solution contain?

29. A 750-milliliter solution is 4% hydrogen peroxide. How much more hydrogen peroxide is contained in an 850-milliliter solution that is 5% hydrogen peroxide?

30. One 8-ounce bar of a silver alloy is 30% silver. The percent concentration of silver in a second bar, which weighs 12 oz, is 35%. Find the total amount of silver in the two bars.

31. A jeweler mixed 15 g of a 60% silver alloy with 45 g of a 20% silver alloy. What is the percent concentration of silver in the resulting alloy?

32. A goldsmith mixed 10 g of a 50% gold alloy with 40 g of a 15% gold alloy. What is the percent concentration of the resulting alloy?

33. A silversmith mixed 25 g of a 70% silver alloy with 50 g of a 15% silver alloy. What is the percent concentration of the resulting alloy?

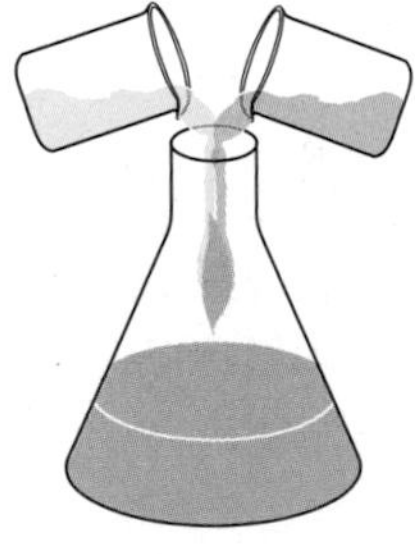

34. A chemist mixed 100 ml of an 8% saline solution with 60 ml of a 5% saline solution. Find the percent concentration of the resulting mixture.

35. How many pounds of a 12% aluminum alloy must be mixed with 400 lb of a 30% aluminum alloy to make a 20% aluminum alloy?

36. How many pounds of a 20% copper alloy must be mixed with 600 lb of a 30% copper alloy to make a 27.5% copper alloy?

37. A hospital staff member mixed a 65% disinfectant solution with a 15% disinfectant solution. How many liters of each were used to make 50 L of a 40% disinfectant solution?

38. A butcher has some hamburger that is 20% fat and some hamburger that is 12% fat. How many pounds of each should be mixed to make 80 lb of hamburger that is 17% fat?

39. How many quarts of water must be added to 5 qt of an 80% antifreeze solution to make a 50% antifreeze solution?

40. Rubbing alcohol is typically diluted with water to 70% strength. If you need 3.5 oz of 45% rubbing alcohol, how many ounces of 70% rubbing alcohol and how much water should you combine?

41. How many ounces of pure water must be added to 60 oz of a 7.5% salt solution to make a 5% salt solution?

42. How much water must be evaporated from 10 gal of a 12% sugar solution in order to obtain a 15% sugar solution?

43. Many fruit drinks are actually only 5% real fruit juice. If you let 2 oz of water evaporate from 12 oz of a drink that is 5% fruit juice, what is the percent concentration of the result?

44. A student mixed 50 ml of a 3% hydrogen peroxide solution with 30 ml of a 12% hydrogen peroxide solution. Find the percent concentration of the resulting mixture. Round to the nearest tenth of a percent.

45. Eighty pounds of a 54% copper alloy are mixed with 200 lb of a 22% copper alloy. Find the percent concentration of the resulting mixture. Round to the nearest tenth of a percent.

46. A druggist mixed 100 cc (cubic centimeters) of a 15% alcohol solution with 50 cc of pure alcohol. Find the percent concentration of the resulting mixture. Round to the nearest tenth of a percent.

47. n ounces of a 30% salt solution are mixed with m ounces of a 50% salt solution. Which relationship between n and m will produce a mixture with a salt concentration of 40%?
(i) $m = 2n$ (ii) $m = n$ (iii) $m = 0.4n$

48. n ounces of a 20% acid solution are mixed with m ounces of a 60% salt solution. Which of the following are possible salt concentrations of the mixture?
(i) 40% (ii) 80% (iii) 15% (iv) 25% (v) 55%

APPLYING CONCEPTS

49. **Investment Problem** A financial manager invested 25% of a client's money in bonds paying 9% annual simple interest, 30% in an 8% annual simple interest account, and the remainder in 9.5% corporate bonds. Find the amount invested in each if the total annual interest earned is \$1785.

50. **Percent Mixture Problem** A silversmith mixed 90 g of a 40% silver alloy with 120 g of a 60% silver alloy. Find the percent concentration of the resulting alloy. Round to the nearest tenth of a percent.

51. **Value Mixture Problem** Find the cost per pound of a tea mixture made from 50 lb of tea that costs \$5.50 per pound and 75 lb of tea that costs \$4.40 per pound.

52. **Percent Mixture Problem** How many kilograms of water must be evaporated from 75 kg of a 15% salt solution to produce a 20% salt solution?

53. **Percent Mixture Problem** A chemist added 20 g of pure acid to a beaker containing an unknown number of grams of pure water. The resulting solution was 25% acid. How many grams of water were in the beaker before the acid was added?

54. **Percent Mixture Problem** A radiator contains 6 L of a 25% antifreeze solution. How much should be drained and replaced with pure antifreeze to produce a 50% antifreeze solution?

55. **Consumer Price Index** The consumer price index (CPI), or the "cost-of-living index," measures the average cost of consumer goods and services. The graph at the right shows the percent increase or decrease in the CPI over the previous year, for the years 2000 through 2010.
   a. During which year shown was the percent increase in the CPI the least?
   b. During which year was the cost of consumer goods and services lowest?

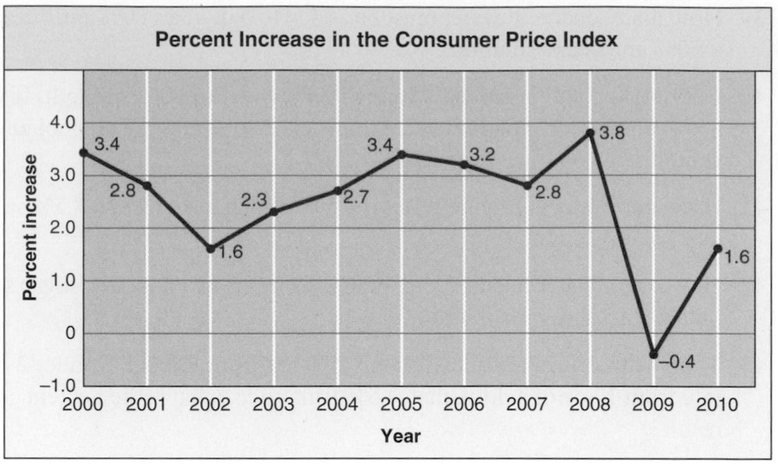

Source: Bureau of Labor Statistics

56. Write a report on series trade discounts. Explain how to convert a series discount to a single-discount equivalent.

PROJECTS OR GROUP ACTIVITIES

57. Suppose there are two 1-gallon bottles on a table. One bottle contains a quart of apple juice, and the other contains a quart of cranberry juice. A tablespoon of juice is removed from the apple juice bottle and added to the cranberry juice bottle. After thoroughly mixing, a tablespoon of the apple-cranberry mixture is removed and mixed into the apple juice bottle. At this time, is the amount of apple juice in the cranberry juice bottle greater than, equal to, or less than the amount of cranberry juice in the apple juice bottle?

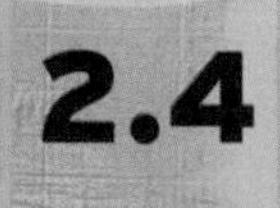

2.4 Inequalities in One Variable

OBJECTIVE 1 Solve inequalities in one variable

The **solution set of an inequality** is a set of numbers, each element of which, when substituted for the variable, results in a true inequality.

The inequality at the right is true if the variable is replaced by 3, for example, or by -1.98, or by $\frac{2}{3}$.

$$x - 1 < 4$$
$$3 - 1 < 4$$
$$-1.98 - 1 < 4$$
$$\frac{2}{3} - 1 < 4$$

There are many values of the variable x that will make the inequality $x - 1 < 4$ true. The solution set of the inequality is any number less than 5. The solution set can be written in set-builder notation as $\{x|x < 5\}$ or in interval notation as $(-\infty, 5)$.

The graph of the solution set of $x - 1 < 4$ is shown at the right.

−5 −4 −3 −2 −1 0 1 2 3 4 5

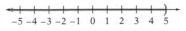

In solving an inequality, use the Addition and Multiplication Properties of Inequalities to rewrite the inequality in the form *variable* < *constant* or *variable* > *constant*.

THE ADDITION PROPERTY OF INEQUALITIES

If $a > b$ and c is a real number, then the inequalities $a > b$ and $a + c > b + c$ have the same solution set.

If $a < b$ and c is a real number, then the inequalities $a < b$ and $a + c < b + c$ have the same solution set.

EXAMPLES

1. Begin with a true inequality. $6 > 2$
   Add -8 to each side. $6 + (-8) > 2 + (-8)$
   Simplify. The inequality is true. $-2 > -6$
2. Begin with a true inequality. $-9 < -3$
   Add 5 to each side. $-9 + 5 < -3 + 5$
   Simplify. The inequality is true. $-4 < 2$

The Addition Property of Inequalities states that the same number can be added to each side of an inequality without changing the solution set of the inequality. This property is also true for an inequality that contains the symbol $\le$ or $\ge$.

The Addition Property of Inequalities is used to remove a term from one side of an inequality by adding the additive inverse of that term to each side of the inequality. Because subtraction is defined in terms of addition, the same number can be subtracted from each side of an inequality without changing the solution set of the inequality.

Focus on using the Addition Property of Inequalities

Solve: $3x - 4 < 2x - 1$

$3x - 4 < 2x - 1$

Subtract $2x$ from each side of the inequality. $x - 4 < -1$

Add 4 to each side of the inequality. $x < 3$

Write the solution set in either set-builder notation or interval notation. The solution set is $\{x|x < 3\}$, or $(-\infty, 3)$.

Take Note

$c > 0$ means that c is a positive number. Note that the inequality symbols are not changed.

THE MULTIPLICATION PROPERTY OF INEQUALITIES

RULE 1

If $a > b$ and $c > 0$, then the inequalities $a > b$ and $ac > bc$ have the same solution set.

If $a < b$ and $c > 0$, then the inequalities $a < b$ and $ac < bc$ have the same solution set.

EXAMPLES

1. Begin with a true inequality. $5 > 2$
   Multiply each side of the inequality by *positive* 3. $3(5) > 3(2)$
   Simplify. The inequality is true. $15 > 6$
2. Begin with a true inequality. $-5 < -3$
   Multiply each side of the inequality by *positive* 4. $4(-5) < 4(-3)$
   Simplify. The inequality is true. $-20 < -12$

Take Note

$c < 0$ means that c is a negative number. Note that the inequality symbols are reversed.

RULE 2

If $a > b$ and $c < 0$, then the inequalities $a > b$ and $ac < bc$ have the same solution set.

If $a < b$ and $c < 0$, then the inequalities $a < b$ and $ac > bc$ have the same solution set.

EXAMPLES

3. Begin with a true inequality. $9 > 3$

Multiply each side of the inequality by *negative* 2 and reverse the inequality symbol. $(-2)9 \quad (-2)3$

Simplify. The inequality is true. $-18 < -6$

4. Begin with a true inequality. $-6 < -4$

Multiply each side of the inequality by *negative* 3 and reverse the inequality symbol. $(-3)(-6) > (-3)(-4)$

Simplify. The inequality is true. $18 > 12$

The Multiplication Property of Inequalities is also true for the symbol $\leq$ or $\geq$. Rule 1 states that when each side of an inequality is multiplied by a positive number, the inequality symbol remains the same. However, Rule 2 states that when each side of an inequality is multiplied by a negative number, the inequality symbol must be reversed.

Because division is defined in terms of multiplication, when each side of an inequality is divided by a positive number, the inequality symbol remains the same. When each side of an inequality is divided by a negative number, the inequality symbol must be reversed.

The Multiplication Property of Inequalities is used to remove a coefficient from one side of an inequality.

Focus on using the Multiplication Property of Inequalities

Solve $-3x > 9$. Write the solution set in interval notation.

$$-3x > 9$$

Divide each side of the inequality by the coefficient -3, and reverse the inequality symbol. Simplify.

$$\frac{-3x}{-3} < \frac{9}{-3}$$

$$x < -3$$

Write the solution set in interval notation. The solution set is $(-\infty, -3)$.

Take Note

Each side of the inequality is divided *by* a negative number; the inequality symbol must be reversed.

EXAMPLE 1 Solve: $x + 3 > 4x + 6$
Write the solution set in set-builder notation.

Solution

$x + 3 > 4x + 6$

$-3x + 3 > 6$ • **Subtract $4x$ from each side of the inequality.**

$-3x > 3$ • **Subtract 3 from each side of the inequality.**

$x < -1$ • **Divide each side of the inequality by -3, and reverse the inequality symbol.**

The solution set is $\{x|x < -1\}$. • **Write the solution set in set-builder notation.**

Problem 1 Solve: $2x - 1 < 6x + 7$
Write the solution set in set-builder notation.

Solution See page S5.

Try Exercise 27, page 91.

When an inequality contains parentheses, the first step in solving the inequality is to use the Distributive Property to remove the parentheses.

EXAMPLE 2 Solve: $5(x - 2) \geq 9x - 3(2x - 4)$
Write the solution set in set-builder notation.

Solution

$5(x - 2) \geq 9x - 3(2x - 4)$

$5x - 10 \geq 9x - 6x + 12$ • **Use the Distributive Property to remove parentheses.**

$5x - 10 \geq 3x + 12$ • **Simplify.**

$2x - 10 \geq 12$ • **Subtract 3x from each side of the inequality.**

$2x \geq 22$ • **Add 10 to each side of the inequality.**

$x \geq 11$ • **Divide each side of the inequality by 2.**

The solution set is $\{x|x \geq 11\}$. • **Write the solution set in set-builder notation.**

Problem 2 Solve: $5x - 2 \leq 4 - 3(x - 2)$
Write the solution set in interval notation.

Solution See page S5.

Try Exercise 33, page 91.

OBJECTIVE 2 Solve compound inequalities

A **compound inequality** is formed by joining two inequalities with a connective word such as "and" or "or." The inequalities shown below are compound inequalities.

$2x < 4$ and $3x - 2 > -8$

$2x + 3 > 5$ or $x + 2 < 5$

The solution set of a compound inequality with the connective word *or* is the *union* of the solution sets of the two inequalities.

Focus on solving a compound inequality with the connective word *or*

Solve: $3x + 1 \geq 10$ or $2x - 3 < 1$

Write the solution using interval notation.

This inequality is read "3*x* plus 1 is greater than or equal to 10 *or* 2*x* minus 3 is less than 1."

Solve each inequality.

$3x + 1 \geq 10$ or $2x - 3 < 1$

$3x \geq 9$ $\quad$ $2x < 4$

$x \geq 3$ $\quad$ $x < 2$

$[3, \infty)$ $\quad$ $(-\infty, 2)$

The inequalities are combined with the word *or*. Find the *union* of the two solution sets.

The solution set is $[3, \infty) \cup (-\infty, 2)$.

EXAMPLE 3 Solve: $3 - 4x > 7$ or $4x + 5 < 9$
Write the solution set in interval notation.

Solution

$$\begin{aligned} 3 - 4x &> 7 \quad \text{or} \quad & 4x + 5 &< 9 \\ -4x &> 4 & 4x &< 4 \\ x &< -1 & x &< 1 \end{aligned}$$

• **Solve each inequality.**

$(-\infty, -1)$ $\quad$ $(-\infty, 1)$

$(-\infty, -1) \cup (-\infty, 1) = (-\infty, 1)$ • **Find the union of the solution sets.**

The solution set is $(-\infty, 1)$.

Problem 3 Solve: $2 - 3x > 11$ or $5 + 2x > 7$
Write the solution set in set-builder notation.

Solution See page S5.

Try Exercise 71, page 92.

The solution set of a compound inequality with the connective word *and* is the set of all elements common to the solution sets of both inequalities. Therefore, it is the *intersection* of the solution sets of the two inequalities.

Focus on solving a compound inequality with the connective word *and*

Solve: $2x < 6$ and $3x + 2 > -4$

Write the solution set in interval notation.

This inequality is read "$2x$ is less than 6 *and* $3x$ plus 2 is greater than -4."

Solve each inequality.

$$\begin{aligned} 2x &< 6 \quad \text{and} \quad & 3x + 2 &> -4 \\ x &< 3 & 3x &> -6 \\ & & x &> -2 \end{aligned}$$

$(-\infty, 3)$ $\quad$ $(-2, \infty)$

Find the intersection of the solution sets. $(-\infty, 3) \cap (-2, \infty) = (-2, 3)$

The solution set is $(-2, 3)$.

EXAMPLE 4 Solve: $11 - 2x > -3$ and $7 - 3x < 4$
Write the solution set in set-builder notation.

Solution

$$\begin{aligned} 11 - 2x &> -3 \quad \text{and} \quad & 7 - 3x &< 4 \\ -2x &> -14 & -3x &< -3 \\ x &< 7 & x &> 1 \end{aligned}$$

• **Solve each inequality.**

$\{x|x < 7\}$ $\quad$ $\{x|x > 1\}$

$\{x|x < 7\} \cap \{x|x > 1\} = \{x|1 < x < 7\}$ • **Find the intersection of the solution sets.**

The solution set is $\{x|1 < x < 7\}$.

Problem 4 Solve: $5 - 4x > 1$ and $6 - 5x < 11$
Write the solution set in interval notation.

Solution See page S5.

Try Exercise 79, page 93.

Some inequalities that use the connective word *and* can be written using a more compact notation. For instance, the inequality $2x + 1 > -3$ *and* $2x + 1 \leq 7$ can be written as $-3 < 2x + 1 \leq 7$. When the compound inequality is written in this form, an alternative method of solving the compound inequality can be used.

Focus on solving a compound inequality of the form $c < ax + b < d$

Solve: $-3 < 2x + 1 < 7$

Write the solution set in set-builder notation.

$$-3 < 2x + 1 < 7$$

Subtract 1 from each of the three parts of the inequality.

$$-3 - 1 < 2x + 1 - 1 < 7 - 1$$
$$-4 < 2x < 6$$

Divide each of the three parts of the inequality by the coefficient 2.

$$\frac{-4}{2} < \frac{2x}{2} < \frac{6}{2}$$
$$-2 < x < 3$$

Write the solution set in set-builder notation.

The solution set is $\{x \mid -2 < x < 3\}$.

EXAMPLE 5 Solve: $1 < 3x - 5 < 4$
Write the solution set in interval notation.

Solution

$$1 < 3x - 5 < 4$$

$$1 + 5 < 3x - 5 + 5 < 4 + 5$$
• Add 5 to each of the three parts of the inequality.

$$6 < 3x < 9$$
• Simplify.

$$\frac{6}{3} < \frac{3x}{3} < \frac{9}{3}$$
• Divide each of the three parts of the inequality by 3.

$$2 < x < 3$$

The solution set is $(2, 3)$.
• Write the solution set in interval notation.

Problem 5 Solve: $-2 \le 5x + 3 \le 13$
Write the solution set in interval notation.

Solution See page S5.

Try Exercise 73, page 92.

OBJECTIVE 3 Application problems

EXAMPLE 6 An average score of 80 to 89 in a history course receives a B grade. A student has grades of 72, 94, 83, and 70 on four exams. Find the range of scores on the fifth exam that will give the student a B for the course.

Strategy To find the range of scores, write and solve an inequality using S to represent the score on the fifth exam.

Solution

$$\frac{72 + 94 + 83 + 70 + S}{5}$$
• The student's average score is the sum of the five scores, divided by 5.

$$80 \le \frac{72 + 94 + 83 + 70 + S}{5} \le 89$$
• Write the inequality that puts the student's average score between 80 and 89, inclusive.

$$80 \le \frac{319 + S}{5} \le 89$$
• Simplify the numerator.

$$5(80) \le 5\left(\frac{319 + S}{5}\right) \le 5(89)$$
• Multiply each part of the inequality by 5.

$$400 \le 319 + S \le 445$$

$$400 - 319 \le 319 - 319 + S \le 445 - 319$$
• Subtract 319 from each part of the inequality.

$$81 \le S \le 126$$

The maximum score on an exam is 100, so eliminate values of S above 100. The range of scores that will give the student a B for the course is $81 \leq S \leq 100$.

Problem 6 Company A rents cars for \$6 a day and 14¢ for every mile driven. Company B rents cars for \$12 a day and 8¢ for every mile driven. You want to rent a car for 5 days. How many miles can you drive a Company A car during the 5 days if it is to cost less than a Company B car?

Solution See page S5.

Try Exercise 99, page 93.

EXAMPLE 7 A coffee merchant mixes some Colombian coffee costing \$8 per pound with 10 lb of French Roast coffee costing \$5 per pound. How many pounds of the Colombian coffee should be used if the merchant wants a blend that costs between \$6 and \$7 per pound?

Strategy Number of pounds of the Colombian coffee: n

The sum of the values of the two ingredients:
$8n + 5(10) = 8n + 50$

Number of pounds of the Colombian and French Roast blend:
$n + 10$

Value of the blend at \$6 per pound: $6(n + 10) = 6n + 60$

Value of the blend at \$7 per pound: $7(n + 10) = 7n + 70$

The sum of the values of the two ingredients is greater than the value of the blend at \$6 per pound and less than the value of the blend at \$7 per pound.

Solution

$$\begin{aligned} 8n + 50 &> 6n + 60 \\ 2n + 50 &> 60 \\ 2n &> 10 \\ n &> 5 \end{aligned} \qquad \text{and} \qquad \begin{aligned} 8n + 50 &< 7n + 70 \\ n + 50 &< 70 \\ n &< 20 \end{aligned}$$

The merchant should use between 5 lb and 20 lb of the Colombian coffee to make a blend costing between \$6 and \$7 per pound.

Problem 7 Eight pounds of walnuts that cost \$4 per pound are mixed with pecans that cost \$7 per pound. How many pounds of pecans should be used to create a mixture that costs between \$5 and \$6 per pound?

Solution See page S5.

Try Exercise 109, page 94.

2.4 Exercises

CONCEPT CHECK

1. State the Addition Property of Inequalities and give numerical examples of its use.

2. State the Multiplication Property of Inequalities and give numerical examples of its use.

3. How does the solution set for $x < 5$ differ from the solution set for $x \le 5$?

4. If $a < b$ and $ac > bc$, what can be said about c?

5. Are the inequalities $5 > x > 1$ and $1 < x < 5$ equivalent?

6. If $-x < 0$, is x a positive number or a negative number?

1 Solve inequalities in one variable (See pages 84-87.)

7. Which numbers are solutions of the inequality $x + 7 \le -3$?
(i) -17 (ii) 8 (iii) -10 (iv) 0

8. Which numbers are solutions of the inequality $2x - 1 > 5$?
(i) 6 (ii) -4 (iii) 3 (iv) 5

GETTING READY

Complete Exercises 9 to 12 by replacing the question mark with "remains the same" or "is reversed."

9. The inequality $x + 5 > 10$ can be solved by subtracting 5 from each side of the inequality. The inequality symbol ___?___.

10. The inequality $\frac{1}{5}x > 10$ can be solved by multiplying each side of the inequality by the reciprocal of $\frac{1}{5}$. The inequality symbol ___?___.

11. The inequality $-5x > 10$ can be solved by dividing each side of the inequality by -5. The inequality symbol ___?___.

12. The inequality $x - 5 > -10$ can be solved by adding 5 to each side of the inequality. The inequality symbol ___?___.

Solve. Write the solution set in set-builder notation.

13. $x - 3 < 2$

14. $x + 4 \ge 2$

15. $4x \le 8$

16. $6x > 12$

17. $-2x > 8$

18. $-3x \le -9$

19. $3x - 1 > 2x + 2$

20. $5x + 2 \ge 4x - 1$

21. $2x - 1 > 7$

22. $4x + 3 \le -1$

23. $6x + 3 > 4x - 1$

24. $7x + 4 < 2x - 6$

25. $8x + 1 \ge 2x + 13$

26. $5x - 4 < 2x + 5$

27. $7 - 2x \ge 1$

28. $3 - 5x \le 18$

29. $4x - 2 < x - 11$

30. $6x + 5 \ge x - 10$

Solve. Write the solution set in interval notation.

31. $x + 7 \ge 4x - 8$

32. $3x + 1 \le 7x - 15$

33. $6 - 2(x - 4) \le 2x + 10$

34. $4(2x - 1) > 3x - 2(3x - 5)$

35. $2(1 - 3x) - 4 > 10 + 3(1 - x)$

36. $2 - 5(x + 1) \ge 3(x - 1) - 8$

37. $\frac{3}{5}x - 2 < \frac{3}{10} - x$

38. $\frac{5}{6}x - \frac{1}{6} \le x - 4$

39. $\frac{1}{3}x - \frac{3}{2} \ge \frac{7}{6} - \frac{2}{3}x$

40. $\frac{7}{12}x - \frac{3}{2} < \frac{2}{3}x + \frac{5}{6}$

41. $\frac{1}{2}x - \frac{3}{4} > \frac{7}{4}x - 2$

42. $\frac{2 - x}{4} - \frac{3}{8} \ge \frac{2}{5}x$

43. $2 - 2(7 - 2x) < 3(3 - x)$

44. $3 + 2(x + 5) \ge x + 5(x + 1) + 1$

State whether the solution set of an inequality in the given form contains only negative numbers, only positive numbers, or both positive and negative numbers.

45. $x + n > a$, where n and a are both positive and $n < a$

46. $nx > a$, where both n and a are negative

47. $nx > a$, where n is negative and a is positive

48. $x - n > -a$, where n and a are both positive and $n < a$

2 Solve compound inequalities (See pages 87–89.)

49. a. Which set operation is used when a compound inequality is combined with the word *or*?

b. Which set operation is used when a compound inequality is combined with the word *and*?

50. Explain why writing $-3 > x > 4$ does not make sense.

GETTING READY

51. To solve the compound inequality $4x \le 4$ or $x + 2 > 8$, divide each side of the first inequality by __?__, and subtract __?__ from each side of the second inequality: $x \le$ __?__ or $x >$ __?__.

52. The solution set in Exercise 51 can be written in interval notation as $(-\infty,$ __?__$] \cup (6,$ __?__$)$.

State whether the inequality describes the empty set, all real numbers, two intervals of real numbers, or one interval of real numbers.

53. $x > -3$ and $x > 2$

54. $x > -3$ or $x < 2$

55. $x < -3$ and $x > 2$

56. $x < -3$ or $x > 2$

Solve. Write the solution set in interval notation.

57. $3x < 6$ and $x + 2 > 1$

58. $x - 3 \le 1$ and $2x \ge -4$

59. $x + 2 \ge 5$ or $3x \le 3$

60. $2x < 6$ or $x - 4 > 1$

61. $-2x > -8$ and $-3x < 6$

62. $\frac{1}{2}x > -2$ and $5x < 10$

63. $\frac{1}{3}x < -1$ or $2x > 0$

64. $\frac{2}{3}x > 4$ or $2x < -8$

65. $x + 4 \ge 5$ and $2x \ge 6$

66. $3x < -9$ and $x - 2 < 2$

67. $-5x > 10$ and $x + 1 > 6$

68. $7x < 14$ and $1 - x < 4$

69. $2x - 3 > 1$ and $3x - 1 < 2$

70. $4x + 1 < 5$ and $4x + 7 > -1$

71. $3x + 7 < 10$ or $2x - 1 > 5$

72. $6x - 2 < -14$ or $5x + 1 > 11$

Solve. Write the solution set in set-builder notation.

73. $-5 < 3x + 4 < 16$

74. $5 < 4x - 3 < 21$

75. $0 \le 2x - 6 \le 4$

76. $-2 \le 3x + 7 \le 1$

77. $4x - 1 > 11$ or $4x - 1 \leq -11$

78. $3x - 5 > 10$ or $3x - 5 < -10$

79. $2x + 3 \geq 5$ and $3x - 1 > 11$

80. $6x - 2 < 5$ or $7x - 5 < 16$

81. $9x - 2 < 7$ and $3x - 5 > 10$

82. $8x + 2 \leq -14$ and $4x - 2 > 10$

83. $3x - 11 < 4$ or $4x + 9 \geq 1$

84. $5x + 12 \geq 2$ or $7x - 1 \leq 13$

85. $3 - 2x > 7$ and $5x + 2 > -18$

86. $1 - 3x < 16$ and $1 - 3x > -16$

87. $5 - 4x > 21$ or $7x - 2 > 19$

88. $6x + 5 < -1$ or $1 - 2x < 7$

89. $3 - 7x \leq 31$ and $5 - 4x > 1$

90. $9 - x \geq 7$ and $9 - 2x < 3$

91. $\frac{2}{3}x - 4 > 5$ or $x + \frac{1}{2} < 3$

92. $\frac{5}{8}x + 2 < -3$ or $2 - \frac{3}{5}x < -7$

93. $-\frac{3}{8} \leq 1 - \frac{1}{4}x \leq \frac{7}{2}$

94. $-2 \leq \frac{2}{3}x - 1 \leq 3$

3 Application problems (See pages 89–90.)

GETTING READY

For Exercises 95 and 96, translate each sentence into an inequality.

95. The minimum value of a number n is 40.

96. A number n is between 150 and 300.

97. **Consumerism** A cellular phone company offers its customers a rate of \$49 for up to 200 min per month of cellular phone time, or a rate of \$25 per month plus \$.40 for each minute of cellular phone time. For how many minutes per month can a customer who chooses the second option use a cellular phone before the charges exceed those of the first option?

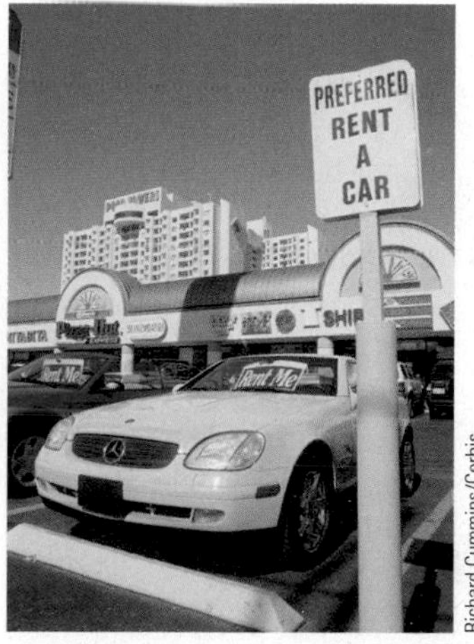

Richard Cummins/Corbis

98. **Consumerism** Suppose PayRite Rental Cars rents compact cars for \$32 per day with unlimited mileage, and Otto Rentals offers compact cars for \$19.99 per day but charges \$.19 for each whole mile beyond 100 mi driven per day. You want to rent a car for one week. How many miles can you drive during the week if Otto Rental is to be less expensive than PayRite?

99. **Consumerism** AirTouch advertises text messaging for \$6.95 per month for up to 400 messages, and \$.10 per message thereafter. A competitor advertises text messaging for \$3.95 per month for up to 400 messages, and \$.15 per message thereafter. For what number of messages per month is the AirTouch plan less expensive?

100. **Consumerism** Heritage National Bank offers two different checking accounts. The first account charges \$3 per month and \$.50 per check after the first 10 checks. The second account charges \$8 per month with unlimited check writing. How many checks can be written per month if the first account is to be less expensive than the second account?

101. **Consumerism** Glendale Federal Bank offers a checking account to small businesses. The charge is \$8 per month plus \$.12 per check after the first 100 checks. A competitor is offering an account for \$5 per month plus \$.15 per check after the first 100 checks. If a business chooses the Glendale Federal Bank account, how many checks does the business write monthly, assuming this account costs less than the competitor's account?

102. **Commissions** George Stoia earns \$1000 per month plus 5% commission on the amount of sales. George's goal is to earn a minimum of \$3200 per month. What amount of sales will enable George to earn \$3200 or more per month?

103. **Hybrid Vehicles** See the news clipping at the right. If a typical city bus has a fuel tank that holds 112 gal of diesel fuel, find the range of miles that the buses in a city's fleet can travel on a full tank of fuel.

In the News

Cities Introducing Hybrid Buses

More and more cities around the country are introducing diesel-electric hybrid buses to their public transportation bus fleets, achieving significant improvement in miles traveled on a single tank of fuel. While a city's conventional diesel buses may average as low as 3.5 mpg, the hybrids can average up to 5 mpg.

Sources: www.boston.com, www.courier-journal.com, www.naplesnews.com, www.sanantonio.bizjournals.com

104. **Test Scores** An average score of 90 or above in an English course receives an A grade. A student has grades of 85, 88, 90, and 98 on four tests. Find the range of scores on the fifth test that will give the student an A grade.

105. **Test Scores** An average score of 80 to 89 in a math course receives a B grade. A student has grades of 82, 78, 75, and 88 on four tests. Find the range of scores on the fifth test that will give the student a B grade.

106. **Value Mixture Problem** A tea merchant mixes some black tea costing \$40 per pound with 4 lb of Earl Grey tea costing \$52 per pound. How many pounds of the black tea should be used if the merchant wants a blend that costs between \$44 and \$48 per pound?

107. **Value Mixture Problem** How many pounds of peanuts costing \$2.50 per pound should be mixed with 50 lb of cashews costing \$6.00 per pound to make a mixture costing between \$4.50 and \$5.00 per pound?

108. **Value Mixture Problem** A chemist mixes a silver alloy AG1 costing \$12 per ounce with another silver alloy AG2 costing \$16 per ounce to make 10 oz of a new alloy. If the chemist wants the new alloy to cost between \$14 and \$15 per ounce, how many ounces of AG1 should the chemist use?

109. **Value Mixture Problem** A miller combined soybeans costing \$8 per bushel with wheat costing \$5 per bushel to make 30 bushels of a new mixture. How many bushels of soybeans should be used to make a 30-bushel mixture costing between \$6 and \$7 per bushel?

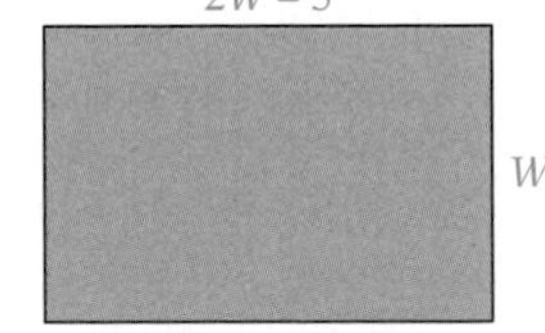

110. **Geometry** The length of a rectangle is 5 cm less than twice the width. Express as an integer the maximum width of the rectangle when the perimeter is less than 60 cm.

111. **Geometry** The length of a rectangle is 2 ft more than four times the width. Express as an integer the maximum width of the rectangle when the perimeter is less than 34 ft.

112. **Geometry** The length of a rectangle is 4 ft more than twice the width. Find, to the nearest whole number, the width of the rectangle if the perimeter is more than 28 ft and less than 40 ft.

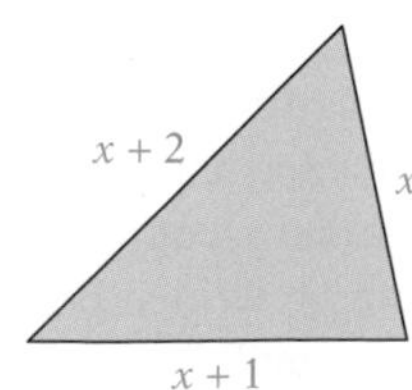

113. **Geometry** One side of a triangle is 1 in. longer than the second side. The third side is 2 in. longer than the second side. Find, to the nearest whole number, the length of the second side of the triangle if the perimeter is more than 15 in. and less than 25 in.

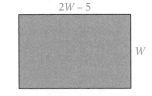

APPLYING CONCEPTS

Use the roster method to list the positive integers that are solutions of the inequality.

114. $2x + 9 \geq 5x - 4$

115. $8x - 7 < 2x + 9$

116. $6 + 4(2 - x) > 7 + 3(x + 5)$

117. $5 + 3(2 + x) > 8 + 4(x - 1)$

118. $3x - 2 > 1$ and $2x - 3 < 5$

119. $-3x < 15$ and $x + 2 < 7$

120. $5 < 7x - 3 \leq 24$

121. $-4 \leq 3x + 8 < 16$

122. Temperature The relationship between Celsius temperature and Fahrenheit temperature is given by the formula $F = \frac{9}{5}C + 32$. If the temperature is between 77°F and 86°F, what is the temperature range in degrees Celsius?

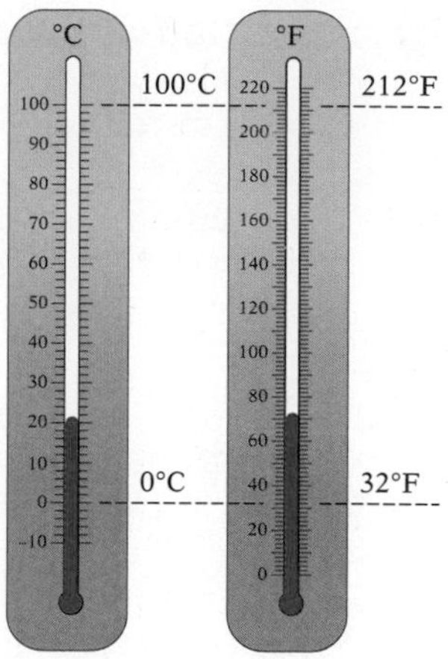

123. Consumerism The charges for a transatlantic telephone call are \$1.56 for the first 3 min and \$.52 for each additional minute or fraction of a minute. What is the largest whole number of minutes a call can last if it is to cost less than \$5.40?

PROJECTS OR GROUP ACTIVITIES

Given the rules for rounding numbers, some possible values for the number 2.7 before it was rounded to the nearest tenth are 2.73, 2.68, 2.65, and 2.749. If V represents the exact value of 2.7 before it was rounded, then the inequality $2.65 \le V < 2.75$ represents all possible values of 2.7 before it was rounded.

124. Suppose a rectangle has a width of 3.4 m and a length of 4.8 m, with each measurement rounded to the nearest tenth of a meter.
  a. Write an inequality for the possible exact measures of the width.
  b. Write an inequality for the possible exact measures of the length.
  c. The area of the rectangle is $A = LW$. Write an inequality for the possible exact areas of the rectangle.

125. Suppose a triangle has a base of 3.26 ft, rounded to the nearest hundredth, and a height of 5.13 ft, rounded to the nearest hundredth.
  a. Write an inequality for the possible exact measures of the base.
  b. Write an inequality for the possible exact measures of the height.
  c. The area of the triangle is $A = \frac{1}{2}bh$. Write an inequality for the possible exact areas of the triangle.

2.5 Absolute Value Equations and Inequalities

OBJECTIVE 1 Absolute value equations

The **absolute value** of a number is its distance from zero on the number line. Distance is always a positive number or zero. Therefore, the absolute value of a number is always a positive number or zero.

The distance from 0 to 3 or from 0 to −3 is 3 units.

$$|3| = 3 \qquad |-3| = 3$$

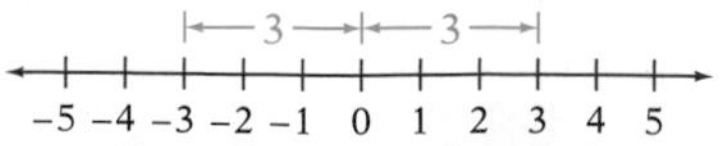

An equation containing a variable within an absolute value symbol is called an **absolute value equation.**

$$\left.\begin{aligned} |x| &= 3 \\ |x + 2| &= 8 \\ |3x - 4| &= 5x - 9 \end{aligned}\right\} \text{Absolute value equations}$$

ABSOLUTE VALUE EQUATIONS

If $a \geq 0$ and $|x| = a$, then $x = a$ or $x = -a$.

EXAMPLES

1. Solve: $|x| = 3$
   $x = 3$ or $x = -3$
   The solutions are 3 or -3.
2. Solve: $|y| = -6$
   The absolute value of a number cannot be negative. The equation has no solution.

EXAMPLE 1 Solve.

A. $|3x| = 15$

B. $|2 - x| = 12$

C. $3 - |2x - 4| = -5$

Solution

A. $|3x| = 15$

$3x = 15$ or $3x = -15$ • Remove the absolute value sign, and write two equations.

$x = 5$ $\quad x = -5$ • Solve for x.

The solutions are 5 and -5.

B. $|2 - x| = 12$

$2 - x = 12$ or $2 - x = -12$ • Remove the absolute value sign, and write two equations.

$-x = 10$ $\quad -x = -14$ • Solve each equation.

$x = -10$ $\quad x = 14$

The solutions are -10 and 14.

C. $3 - |2x - 4| = -5$

$-|2x - 4| = -8$ • Solve for the absolute value.

$|2x - 4| = 8$ • Multiply each side of the equation by -1.

$2x - 4 = 8$ or $2x - 4 = -8$ • Remove the absolute value sign, and write two equations.

$2x = 12$ $\quad 2x = -4$ • Solve each equation.

$x = 6$ $\quad x = -2$

The solutions are 6 and -2.

Take Note

Remember to check the solutions. For Part C:

$3 - \|2x - 4\|$	$= -5$
$3 - \|2(6) - 4\|$	-5
$3 - \|12 - 4\|$	-5
$3 - \|8\|$	-5
$3 - 8$	-5
-5	$= -5$

$3 - \|2x - 4\|$	$= -5$
$3 - \|2(-2) - 4\|$	-5
$3 - \|-4 - 4\|$	-5
$3 - \|-8\|$	-5
$3 - 8$	-5
-5	$= -5$

Problem 1 Solve.

A. $|x| = 25$

B. $|2x - 3| = 5$

C. $5 - |3x + 5| = 3$

Solution See pages S5–S6.

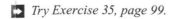

 *Try Exercise 35, page 99.*

OBJECTIVE 2 Absolute value inequalities

Absolute value can be used to represent the distance between two points. For example, the solutions of the absolute value equation $|x - 1| = 3$ are -2 and 4, the numbers whose distance from 1 is 3.

The solutions of the **absolute value inequality** $|x - 1| < 3$ are the numbers whose distance from 1 is *less than* 3. Therefore, the solutions are the numbers greater than -2 *and* less than 4. The solution set is $\{x|-2 < x < 4\}$. *Note:* In this text, solutions to absolute value inequalities will always be written in set-builder notation.

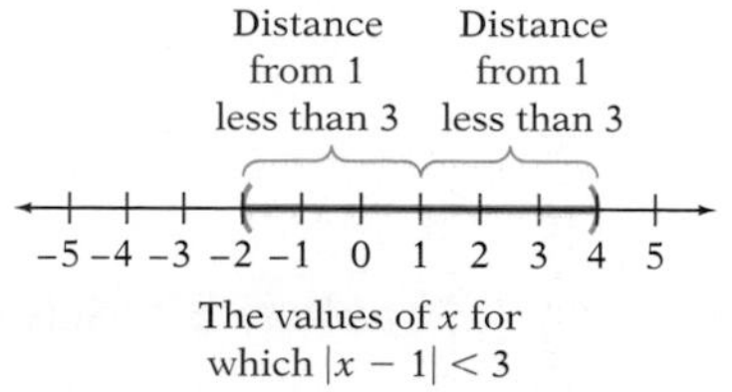

The values of x for which $|x - 1| < 3$

ABSOLUTE VALUE INEQUALITIES OF THE FORM $|ax + b| < c$

To solve an absolute value inequality of the form $|ax + b| < c$, solve the equivalent compound inequality $-c < ax + b < c$.

EXAMPLE 2 Solve: $|4x - 3| < 5$

Solution

$$|4x - 3| < 5$$

$$-5 < 4x - 3 < 5$$ • Solve the equivalent compound inequality.

$$-5 + 3 < 4x - 3 + 3 < 5 + 3$$ • Add 3 to each of the three parts of the inequality.

$$-2 < 4x < 8$$

$$\frac{-2}{4} < \frac{4x}{4} < \frac{8}{4}$$ • Divide each of three parts of the inequality by 4.

$$-\frac{1}{2} < x < 2$$

The solution set is $\left\{x \middle| -\frac{1}{2} < x < 2\right\}$. • Write the solution set.

Problem 2 Solve: $|3x + 2| < 8$

Solution See page S6.

Try Exercise 63, page 100.

The solutions of the absolute value inequality $|x + 1| > 2$ are the numbers whose distance from -1 is *greater than* 2. Therefore, the solutions are the numbers less than -3 *or* greater than 1. The solution set is $\{x|x < -3\} \cup \{x|x > 1\}$.

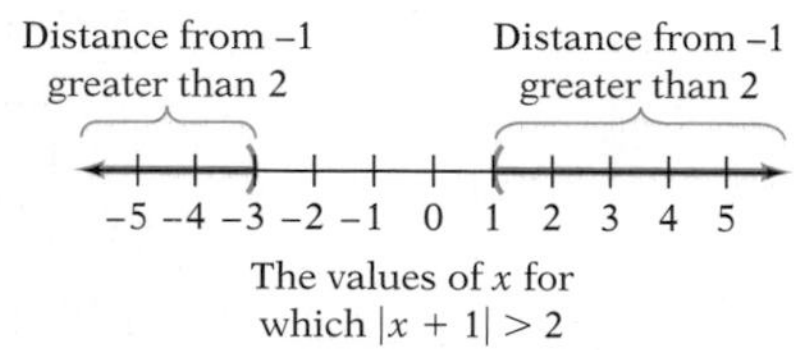

The values of x for which $|x + 1| > 2$

ABSOLUTE VALUE INEQUALITIES OF THE FORM $|ax + b| > c$

To solve an absolute value inequality of the form $|ax + b| > c$, solve the equivalent compound inequality $ax + b < -c$ or $ax + b > c$.

Take Note
Carefully observe the difference between the solution method for $|ax + b| > c$ shown here and that for $|ax + b| < c$ shown in Example 2.

EXAMPLE 3 Solve: $|2x - 1| > 7$

Solution

$|2x - 1| > 7$

$2x - 1 < -7$ or $2x - 1 > 7$ • Solve the equivalent compound inequality.

$2x < -6$ $\quad 2x > 8$

$x < -3$ $\quad x > 4$

$\{x|x < -3\}$ $\quad \{x|x > 4\}$ • Write the two solution sets.

The solution set is $\{x|x < -3\} \cup \{x|x > 4\}$. • Find the union of the solution sets.

Problem 3 Solve: $|5x + 3| > 8$

Solution See page S6.

Try Exercise 61, page 100.

OBJECTIVE 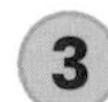

Application problems

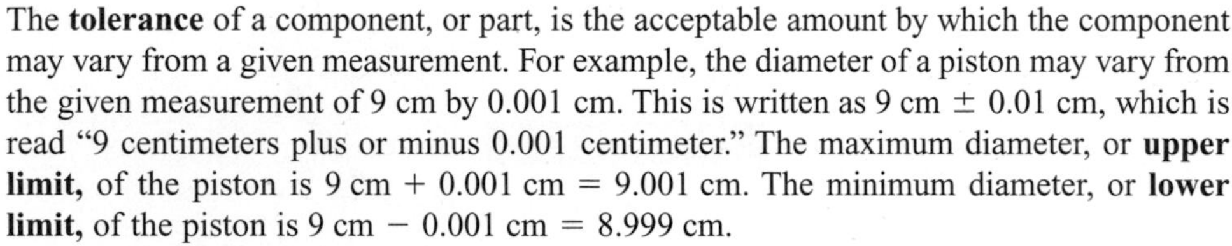

The **tolerance** of a component, or part, is the acceptable amount by which the component may vary from a given measurement. For example, the diameter of a piston may vary from the given measurement of 9 cm by 0.001 cm. This is written as 9 cm ± 0.01 cm, which is read "9 centimeters plus or minus 0.001 centimeter." The maximum diameter, or **upper limit,** of the piston is 9 cm + 0.001 cm = 9.001 cm. The minimum diameter, or **lower limit,** of the piston is 9 cm − 0.001 cm = 8.999 cm.

The lower and upper limits of the diameter could also be found by solving the absolute value inequality $|d - 9| \le 0.001$, where d is the diameter of the piston.

$$|d - 9| \le 0.001$$
$$-0.001 \le d - 9 \le 0.001$$
$$-0.001 + 9 \le d - 9 + 9 \le 0.001 + 9$$
$$8.999 \le d \le 9.001$$

The lower and upper limits of the diameter are 8.999 cm and 9.001 cm.

EXAMPLE 4 A doctor has prescribed 2 cc of medication for a patient. The tolerance is 0.03 cc. Find the lower and upper limits of the amount of medication to be given.

Strategy Let p represent the prescribed amount of medication, T the tolerance, and m the given amount of medication. Solve the absolute value inequality $|m - p| \le T$ for m.

Solution

$|m - p| \le T$

$|m - 2| \le 0.03$ • Substitute the values of p and T into the inequality.

$-0.03 \le m - 2 \le 0.03$ • Solve the equivalent compound inequality.

$1.97 \le m \le 2.03$

The lower and upper limits of the amount of medication to be given to the patient are 1.97 cc and 2.03 cc.

Problem 4 A machinist must make a bushing that has a tolerance of 0.003 in. The diameter of the bushing is 2.55 in. Find the lower and upper limits of the diameter of the bushing.

Solution See page S6.

Try Exercise 85, page 101.

2.5 Exercises

CONCEPT CHECK

1. Explain why the equation $|2x + 1| = -4$ has no solution.

2. For what value of a does the equation $|x - 2| = a$ have only one solution?

3. Does the solution set of $|x| > 3$ represent the union or the intersection of the sets $\{x|x > 3\}$ and $\{x|x < -3\}$?

4. Does the solution set of $|x| < 3$ represent the union or the intersection of the sets $\{x|x < 3\}$ and $\{x|x > -3\}$?

1 Absolute value equations (See pages 95–96.)

GETTING READY

5. Is 2 a solution of $|x - 8| = 6$?

6. Is -2 a solution of $|2x - 5| = 9$?

7. Is -1 a solution of $|3x - 4| = 7$?

8. Is 1 a solution of $|6x - 1| = -5$?

9. If $|x| = 8$, then $x =$ __?__ or $x =$ __?__.

10. If $|x + 5| = 8$, then $x + 5 =$ __?__ or $x + 5 =$ __?__. Subtract 5 from each side of each equation to find that the two solutions of the equation $|x + 5| = 8$ are __?__ and __?__.

Solve.

11. $|x| = 7$
12. $|a| = 2$
13. $|-t| = 3$
14. $|-a| = 7$
15. $|3x| = 12$
16. $|4a| = 28$
17. $|9x| = 12$
18. $|6z| = 3$
19. $|-5y| = 20$
20. $|-8t| = 16$
21. $|x + 2| = 3$
22. $|x + 5| = 2$
23. $|y - 5| = 3$
24. $|y - 8| = 4$
25. $|a - 2| = 0$
26. $|a + 7| = 0$
27. $|x - 2| = -4$
28. $|x + 8| = -2$
29. $|2x - 5| = 4$
30. $|4 - 3x| = 4$
31. $|2 - 5x| = 2$
32. $|2x - 3| = 0$
33. $|5x + 5| = 0$
34. $|x - 2| - 2 = 3$
35. $|x - 9| - 3 = 2$
36. $|3a + 2| - 4 = 4$
37. $|8 - y| - 3 = 1$
38. $|2x - 3| + 3 = 3$
39. $|4x - 7| - 5 = -5$
40. $|6x - 5| - 2 = 4$
41. $|4b + 3| - 2 = 7$
42. $|3t + 2| + 3 = 4$
43. $|5x - 2| + 5 = 7$
44. $2 + |3x - 4| = 5$
45. $5 + |2x + 1| = 8$
46. $5 - |2x + 1| = 5$
47. $9 - |5x + 3| = 1$
48. $8 - |1 - 3x| = -1$

Suppose a and b are positive numbers such that $a < b$. State whether an absolute value equation in the given form has no solution, two negative solutions, two positive solutions, or one positive and one negative solution.

49. $|x - b| = a$

50. $|x - b| = -a$

51. $|x + b| = a$

52. $|x + a| = b$

2 Absolute value inequalities (See pages 96–98.)

GETTING READY

53. If $|x| > 9$, then x __?__ -9 or x __?__ 9.

54. If $|x| < 5$, then __?__ $< x <$ __?__.

Solve.

55. $|x| > 3$

56. $|x| < 5$

57. $|x + 1| > 2$

58. $|x - 2| > 1$

59. $|x - 5| \leq 1$

60. $|x - 4| \leq 3$

61. $|2 - x| \geq 3$

62. $|3 - x| \geq 2$

63. $|2x + 1| < 5$

64. $|3x - 2| < 4$

65. $|5x + 2| > 12$

66. $|7x - 1| > 13$

67. $|4x - 3| \leq -2$

68. $|5x + 1| \leq -4$

69. $|2x + 7| > -5$

70. $|3x - 1| > -4$

71. $|4 - 3x| \geq 5$

72. $|7 - 2x| > 9$

73. $|5 - 4x| \leq 13$

74. $|3 - 7x| < 17$

75. $|6 - 3x| \leq 0$

76. $|10 - 5x| \geq 0$

77. $|2 - 9x| > 20$

78. $|5x - 1| < 16$

Suppose a and b are positive numbers such that $a < b$. State whether an absolute value inequality in the given form has no solution, all negative solutions, all positive solutions, or both positive and negative solutions.

79. $|x + b| < a$

80. $|x + a| < b$

3 Application problems (See page 98.)

GETTING READY

81. The desired dosage of a particular medicine is 50 mg, but it is acceptable for the dosage to vary by 0.5 mg. The number 0.5 is called the __?__ for the desired dosage of 50 mg. The value $50 + 0.5 =$ __?__ mg is called the __?__ limit for the dosage and the value $50 - 0.5 =$ __?__ mg is called the __?__ limit for the dosage.

82. The tolerance for the dosage described in Exercise 81 can be represented by an absolute value inequality by letting D be the actual dosage and stating that the distance between D and 50 must be less than or equal to 0.5: $|D -$ __?__$| \leq$ __?__.

83. Mechanics The diameter of a bushing is 1.75 in. The bushing has a tolerance of 0.008 in. Find the lower and upper limits of the diameter of the bushing.

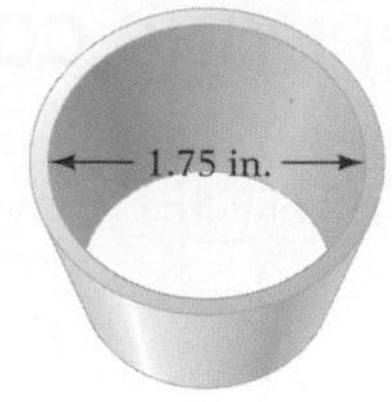

84. Mechanics A machinist must make a bushing that has a tolerance of 0.004 in. The diameter of the bushing is 3.48 in. Find the lower and upper limits of the diameter of the bushing.

85. Medicine A doctor has prescribed 2.5 cc of medication for a patient. The tolerance is 0.2 cc. Find the lower and upper limits of the amount of medication to be given.

86. Automobiles The diameter of a piston for an automobile is $3\frac{5}{16}$ in., with a tolerance of $\frac{1}{64}$ in. Find the lower and upper limits of the diameter of the piston.

87. Political Polling In a poll, the *margin of error* is a measure of the pollsters' confidence in their results. If the pollsters conduct the same poll many times, they expect that 95% of the time they will get results that fall within the margin of error of the reported results. Read the article at the right. For the poll described, the pollsters are 95% sure that the percent of American voters who felt the economy was the most important election issue lies between what lower and upper limits?

In the News

Economy Is Number-One Issue

A *Washington Post*/ABC News poll showed that 41% of American voters felt the economy was the most important election issue. The results of the poll had a margin of error of plus or minus 3 percentage points.

Source: www.washingtonpost.com

88. Aquatic Environments Different species of fish have different requirements for the temperature and pH of the water in which they live. The gold swordtail requires a temperature of 73°F plus or minus 9°F and a pH level of 7.65 plus or minus 0.65. Find the upper and lower limits of **a.** the temperature and **b.** the pH level for the water in which a gold swordtail lives. (*Source:* www.tacomapet.com)

89. Computers A power strip is utilized on a computer to prevent the loss of programming by electrical surges. The power strip is designed to allow 110 volts plus or minus 16.5 volts. Find the lower and upper limits of voltage to the computer.

90. Appliances An electric motor is designed to run on 220 volts plus or minus 25 volts. Find the lower and upper limits of voltage on which the motor will run.

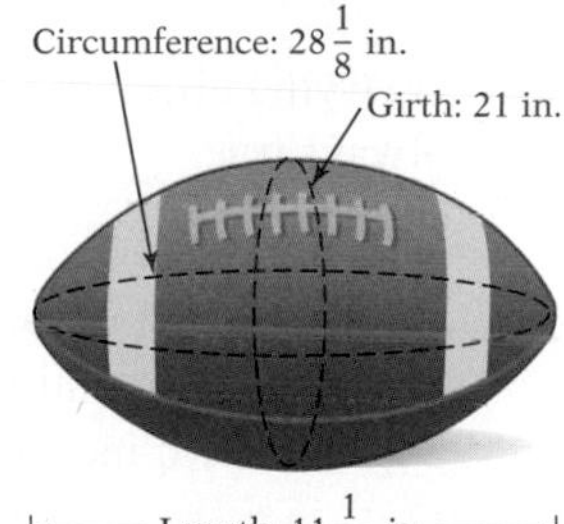

91. Football Manufacturing An NCAA football must conform to the measurements shown in the diagram, with tolerance of $\frac{1}{4}$ in. for the girth, $\frac{3}{8}$ in. for the circumference, and $\frac{5}{32}$ in. for the length. Find the lower and upper limits for **a.** the girth, **b.** the circumference, and **c.** the length of an NCAA football. (*Source:* www.ncaa.org)

92. A dosage of medicine may safely range from 2.8 ml to 3.2 ml. What is the desired dosage of the medicine? What is the tolerance?

93. The tolerance for the diameter, in inches, of a piston is described by the absolute value inequality $|d - 5| \leq 0.01$. What is the desired diameter of the piston? By how much can the actual diameter of the piston vary from the desired diameter?

Electronics The tolerance of the resistors used in electronics is given as a percent of the resistance, which is measured in ohms. For example, if a 20,000-ohm resistor has a tolerance of 4%, then the tolerance is

$$0.04(20{,}000) = 800 \text{ ohms}$$

The lower and upper limits of the resistor can be found by solving the absolute value inequality $|M - 20{,}000| \leq 800$, where M represents the amount of ohms.

94. Find the lower and upper limits of a 29,000-ohm resistor with a 2% tolerance.

95. Find the lower and upper limits of a 15,000-ohm resistor with a 10% tolerance.

APPLYING CONCEPTS

Solve.

96. $\left|\dfrac{2x-5}{3}\right| = 7$ **97.** $\left|\dfrac{3x-2}{4}\right| + 5 = 6$ **98.** $\left|\dfrac{4x-2}{3}\right| > 6$

99. $\left|\dfrac{2x-1}{5}\right| \le 3$ **100.** $|2|x| - 5| = 1$ **101.** $|5|x| - 8| = 2$

For what values of the variable is the equation true? Write the solution set in set-builder notation.

102. $|x+3| = x+3$ **103.** $|y+6| = y+6$ **104.** $|a-4| = 4-a$ **105.** $|b-7| = 7-b$

106. Replace the question mark with $\le$, $\ge$, or $=$.

a. $|x-y| \ ? \ |x| - |y|$ **b.** $\left|\dfrac{x}{y}\right| \ ? \ \dfrac{|x|}{|y|}, y \neq 0$ **c.** $|xy| \ ? \ |x||y|$

107. Write an absolute value inequality to represent all real numbers within 5 units of 2.

108. Write an absolute value inequality to represent all real numbers within k units of j. (Assume $k > 0$.)

PROJECTS OR GROUP ACTIVITIES

109. By trying different values for a and b, determine which of the following appears to be always true.

(i) $|a+b| \le |a| + |b|$ (ii) $|a+b| = |a| + |b|$ (iii) $|a+b| \ge |a| + |b|$

110. By trying different values for a and b, determine which of the following appears to be always true.

(i) $||a| - |b|| \le |a-b|$ (ii) $||a| - |b|| = |a-b|$ (iii) $||a| - |b|| \ge |a-b|$

111. The **Triangle Inequality** states that for all real numbers a and b, $|a+b| \le |a| + |b|$. Prove this inequality. *Hint:* Note that $-|a| \le a \le |a|$ and $-|b| \le b \le |b|$. Now add these two inequalities and use the properties of absolute value.

CHAPTER 2 Summary

Key Words	Objective and Page Reference	Examples
An **equation** expresses the equality of two mathematical expressions.	[2.1.1, p. 56]	$3 + 2 = 5 \quad 2x - 5 = 4$
The simplest equation to solve is an equation of the form **variable = constant.**	[2.1.1, p. 56]	$x = 5 \quad -2 = x$
A **solution, or root, of an equation** is a replacement value for the variable that will make the equation true.	[2.1.1, p. 56]	The solution of $x + 5 = 2$ is -3.
Solving an equation means finding a solution of the equation. The goal is to rewrite the equation in the form **variable = constant,** because the constant is the solution.	[2.1.1, p. 56]	The equation $x = 12$ is in the form *variable* = *constant*. The constant 12 is the solution of the equation.
An equation in which all variables have an exponent of 1 is called a **first-degree equation.**	[2.1.1, p. 56]	$3x - 2 = 5$
Uniform motion means that the speed and direction of an object do not change.	[2.2.2, p. 66]	A car traveling on a straight road at a constant speed of 60 mph is in uniform motion.
The **solution set of an inequality** is a set of numbers, each element of which, when substituted in the inequality, results in a true inequality.	[2.4.1, p. 84]	Any number greater than 4 is a solution of the inequality $x > 4$.
A **compound inequality** is formed by joining two inequalities with a connective word such as "and" or "or."	[2.4.2, p. 87]	$3x > 6$ and $2x + 5 < 7$ $2x + 2 < 3$ or $x + 2 > 4$
An **absolute value equation** is an equation that contains a variable within an absolute value symbol.	[2.5.1, p. 95]	$\lvert x - 2\rvert = 3$
An **absolute value inequality** is an inequality that contains a variable within an absolute value symbol.	[2.5.2, p. 97]	$\lvert x - 4\rvert < 5 \quad \lvert 2x - 3\rvert > 6$

Essential Rules and Procedures	Objective and Page Reference	Examples
The Addition Property of Equations If $a = b$, then $a + c = b + c$.	[2.1.1, p. 56]	$x + 5 = -3$ $x + 5 - 5 = -3 - 5$ $x = -8$
The Multiplication Property of Equations If $a = b$ and $c \neq 0$, then $ac = bc$.	[2.1.1, p. 57]	$\frac{2}{3}x = 4$ $\left(\frac{3}{2}\right)\left(\frac{2}{3}x\right) = \left(\frac{3}{2}\right)4$ $x = 6$

Value Mixture Equation $V = AC$	[2.2.1, p. 64]	A merchant combines coffee that costs \$6 per pound with coffee that costs \$3.20 per pound. How many pounds of each should be used to make 60 lb of a blend that costs \$4.50 per pound? $6x + 3.20(60 - x) = 4.50(60)$
Uniform Motion Equation $d = rt$	[2.2.2, p. 66]	Two planes are 1640 mi apart and traveling toward each other. One plane is traveling 60 mph faster than the other plane. The planes pass each other in 2 h. Find the speed of each plane. $2r + 2(r + 60) = 1640$
Annual Simple Interest Equation $I = Pr$	[2.3.1, p. 75]	An investment of \$4000 is made at an annual simple interest rate of 5%. How much additional money must be invested at an annual simple interest rate of 6.5% so that the total interest earned is \$720? $0.05(4000) + 0.065x = 720$
Percent Mixture Equation $Q = Ar$	[2.3.2, p. 77]	A silversmith mixed 120 oz of an 80% silver alloy with 240 oz of a 30% silver alloy. Find the percent concentration of the resulting silver alloy. $0.80(120) + 0.30(240) = x(360)$
The Addition Property of Inequalities If $a > b$, then $a + c > b + c$. If $a < b$, then $a + c < b + c$.	[2.4.1, p. 85]	$x + 3 > -2$ $x + 3 - 3 > -2 - 3$ $x > -5$
The Multiplication Property of Inequalities **Rule 1** If $a > b$ and $c > 0$, then $ac > bc$. If $a < b$ and $c > 0$, then $ac < bc$.	[2.4.1, pp. 85–86]	$3x > 2$ $\left(\frac{1}{3}\right)(3x) > \left(\frac{1}{3}\right)2$ $x > \frac{2}{3}$
Rule 2 If $a > b$ and $c < 0$, then $ac < bc$. If $a < b$ and $c < 0$, then $ac > bc$.		$-2x < 5$ $\frac{-2x}{-2} > \frac{5}{-2}$ $x > -\frac{5}{2}$
Absolute Value Equations If $a \geq 0$ and $\|x\| = a$, then $x = a$ or $x = -a$.	[2.5.1, p. 96]	If $\|x\| = 5$, then $x = 5$ or $x = -5$.
Absolute Value Inequalities An absolute value inequality of the form $\|ax + b\| < c$ is equivalent to the compound inequality $-c < ax + b < c$.	[2.5.2, p. 97]	If $\|x + 1\| < 3$, then $-3 < x + 1 < 3$.
An absolute value inequality of the form $\|ax + b\| > c$ is equivalent to the compound inequality $ax + b < -c$ or $ax + b > c$.	[2.5.2, p. 97]	If $\|x - 2\| > 5$, then $x - 2 < -5$ or $x - 2 > 5$.

CHAPTER 2 Review Exercises

1. Solve: $x + 4 = -5$

2. Solve: $\frac{2}{3}x = \frac{4}{9}$

3. Solve: $5x + 7 = 2$

4. Solve: $3 - 7x = 5$

5. Solve: $1 - \frac{4x}{3} = 5$

6. Solve: $\frac{2x}{5} + 7 = 1$

7. Solve: $3y - 5 = 3 - 2y$

8. Solve: $3x - 3 + 2x = 7x - 15$

9. Solve: $2(x - 3) = 5(4 - 3x)$

10. Solve: $2x - (3 - 2x) = 4 - 3(4 - 2x)$

11. Solve: $\frac{1}{2}x - \frac{5}{8} = \frac{3}{4}x + \frac{3}{2}$

12. Solve: $\frac{2x - 3}{3} + 2 = \frac{2 - 3x}{5}$

13. Solve: $3x - 7 > -2$
Write the solution set in interval notation.

14. Solve: $2x - 9 < 8x + 15$
Write the solution set in interval notation.

15. Solve: $\frac{2}{3}x - \frac{5}{8} \geq \frac{5}{4}x + 3$
Write the solution set in set-builder notation.

16. Solve: $2 - 3(x - 4) \leq 4x - 2(1 - 3x)$
Write the solution set in set-builder notation.

17. Solve: $-5 < 4x - 1 < 7$
Write the solution set in interval notation.

18. Solve: $5x - 2 > 8$ or $3x + 2 < -4$
Write the solution set in interval notation.

19. Solve: $3x < 4$ and $x + 2 > -1$
Write the solution set in set-builder notation.

20. Solve: $3x - 2 > -4$ or $7x - 5 < 3x + 3$
Write the solution set in set-builder notation.

21. Solve: $2 + |5 - 8x| = 15$

22. Solve: $|5x + 8| = 0$

23. Solve: $6 + |3x - 3| = 2$

24. Solve: $|2x - 5| \leq 3$

25. Solve: $|4x - 5| \geq 3$

26. Solve: $|5x - 4| < -2$

27. Mechanics The diameter of a bushing is 2.75 in. The bushing has a tolerance of 0.003 in. Find the lower and upper limits of the diameter of the bushing.

28. Medicine A doctor has prescribed 2 cc of medication for a patient. The tolerance is 0.25 cc. Find the lower and upper limits of the amount of medication to be given.

29. Metallurgy A silversmith combines 40 oz of pure silver that costs \$16.00 per ounce with 200 oz of a silver alloy that costs \$8.50 per ounce. Find the cost per ounce of the mixture.

30. Food Mixtures A grocer mixed apple juice that costs \$3.20 per gallon with 40 gal of cranberry juice that costs \$5.50 per gallon. How much apple juice was used to make cranapple juice that costs \$4.20 per gallon?

31. Uniform Motion A jogger started on a trail running at a rate of 8 mph. One-half hour later, a cyclist started on the same path cycling toward the jogger at a rate of 12 mph. How long did it take for the cyclist to reach the jogger?

32. Uniform Motion Two planes are 1680 mi apart and traveling toward each other. One plane is traveling 80 mph faster than the other plane. The planes pass each other in 1.75 h. Find the speed of each plane.

33. Investments Two investments earn an annual income of \$635. One investment is earning 10.5% annual simple interest, and the other investment is earning 6.4% annual simple interest. The total investment is \$8000. Find the amount invested in each account.

34. **Metallurgy** An alloy containing 30% tin is mixed with an alloy containing 70% tin. How many pounds of each were used to make 500 lb of an alloy containing 40% tin?

35. **Commissions** A sales executive earns \$800 per month plus a 4% commission on the amount of sales. The executive's goal is to earn at least \$3000 per month. What amount of sales will enable the executive to earn \$3000 or more per month?

36. **Test Scores** An average score of 80 to 90 in a psychology class receives a B grade. A student has grades of 92, 66, 72, and 88 on four tests. Find the range of scores on the fifth test that will give the student a B for the course.

37. **Silver Mixtures** Pure silver costing \$18 per ounce is mixed with 5 oz of a silver alloy costing \$12 per ounce. How many ounces of the pure silver should be mixed with the silver alloy to create a new mixture costing between \$16 and \$17 per ounce?

CHAPTER 2 Test

1. Solve: $x - 2 = -4$

2. Solve: $x + \frac{3}{4} = \frac{5}{8}$

3. Solve: $-\frac{3}{4}y = -\frac{5}{8}$

4. Solve: $3x - 5 = 7$

5. Solve: $\frac{3}{4}y - 2 = 6$

6. Solve: $2x - 3 - 5x = 8 + 2x - 10$

7. Solve: $2[x - (2 - 3x) - 4] = x - 5$

8. Solve: $\frac{2}{3}x - \frac{5}{6}x = 4$

9. Solve: $\frac{2x + 1}{3} - \frac{3x + 4}{6} = \frac{5x - 9}{9}$

10. Solve: $2x - 5 \geq 5x + 4$
Write the solution set in interval notation.

11. Solve: $4 - 3(x + 2) < 2(2x + 3) - 1$
Write the solution set in interval notation.

12. Solve: $3x - 2 > 4$ or $4 - 5x < 14$
Write the solution set in set-builder notation.

13. Solve: $4 - 3x \geq 7$ and $2x + 3 \geq 7$
Write the solution set in set-builder notation.

14. Solve: $|3 - 5x| = 12$

15. Solve: $2 - |2x - 5| = -7$

16. Solve: $|3x - 1| \leq 2$

17. Solve: $|2x - 1| > 3$

18. Solve: $4 + |2x - 3| = 1$

19. **Rental Cars** Agency A rents cars for \$12 per day and 10¢ for every mile driven. Agency B rents cars for \$24 per day with unlimited mileage. How many miles per day can you drive an Agency A car if it is to cost you less than an Agency B car?

20. **Medicine** A doctor prescribed 3 cc of medication for a patient. The tolerance is 0.1 cc. Find the lower and upper limits of the amount of medication to be given.

21. **Food Mixtures** A butcher combines 100 lb of hamburger that costs \$2.60 per pound with 60 lb of hamburger that costs \$4.20 per pound. Find the cost of the hamburger mixture.

22. **Uniform Motion** A jogger runs a distance at a speed of 8 mph and returns the same distance running at a speed of 6 mph. Find the total distance the jogger runs if the total time running is 1 h 45 min.

23. **Investments** An investment of \$12,000 is deposited in two simple interest accounts. On one account, the annual simple interest rate is 7.8%. On the other, the annual simple interest rate is 9%. The total interest earned for one year is \$1020. How much was invested in each account?

24. **Mixture Problem** How many ounces of pure water must be added to 60 oz of an 8% salt solution to make a 3% salt solution?

Cumulative Review Exercises

1. Simplify: $-2^2 \cdot 3^3$

2. Simplify: $4 - (2 - 5)^2 \div 3 + 2$

3. Simplify: $4 \div \dfrac{\frac{3}{8} - 1}{5} \cdot 2$

4. Evaluate $2a^2 - (b - c)^2$ when $a = 2$, $b = 3$, and $c = -1$.

5. Identify the property that justifies the statement. $(2x + 3y) + 2 = (3y + 2x) + 2$

6. Find $A \cap B$ given $A = \{3, 5, 7, 9\}$ and $B = \{3, 6, 9\}$.

7. Simplify: $3x - 2[x - 3(2 - 3x) + 5]$

8. Simplify: $5[y - 2(3 - 2y) + 6]$

9. Solve: $4 - 3x = -2$

10. Solve: $-\frac{5}{6}b = -\frac{5}{12}$

11. Solve: $2x + 5 = 5x + 2$

12. Solve: $\frac{5}{12}x - 3 = 7$

13. Solve: $2[3 - 2(3 - 2x)] = 2(3 + x)$

14. Solve: $3[2x - 3(4 - x)] = 2(1 - 2x)$

15. Solve: $\dfrac{3x - 1}{4} - \dfrac{4x - 1}{12} = \dfrac{3 + 5x}{8}$

16. Solve: $3x - 2 \geq 6x + 7$
Write the solution set in set-builder notation.

17. Solve: $5 - 2x \geq 6$ and $3x + 2 \geq 5$
Write the solution set in set-builder notation.

18. Solve: $4x - 1 > 5$ or $2 - 3x < 8$
Write the solution set in set-builder notation.

19. Solve: $|3 - 2x| = 5$

20. Solve: $3 - |2x - 3| = -8$

21. Solve: $|3x - 5| \leq 4$

22. Solve: $|4x - 3| > 5$

23. Graph: $\{x|x \geq -2\}$

24. Graph: $\{x|x \geq 1\} \cup \{x|x < -2\}$

25. Translate and simplify "the sum of three times a number and six added to the product of three and the number."

26. **Recreation** Tickets for a school play sold for \$2.25 for each adult and \$.75 for each child. The total receipts for 75 tickets were \$128.25. Find the number of adult tickets sold.

27. **Uniform Motion** Two planes are 1400 mi apart and traveling toward each other. One plane is traveling 120 mph faster than the other plane. The planes pass each other in 2.5 h. Find the speed of the faster plane.

28. **Mixture Problem** How many liters of a 12% acid solution must be mixed with 4 L of a 5% acid solution to make an 8% acid solution?

29. **Investments** An investment advisor invested \$10,000 in two accounts. One investment earned 9.8% annual simple interest, and the other investment earned 12.8% annual simple interest. The amount of interest earned in one year was \$1085. How much was invested in the 9.8% account?

30. **Mixture Problem** Ten pounds of trail mix is made by mixing raisins that cost \$3.00 per pound with mixed nuts that cost \$5 per pound. How many pounds of mixed nuts should be used so that the trail mix costs between \$3.70 and \$4.20 per pound?

Linear Functions and Inequalities in Two Variables

CHAPTER 3

Digital Vision

Focus on Success

Are you making attending class a priority? Remember that to be successful, you must attend class. You need to be in class to hear your instructor's explanations and instructions, as well as to ask questions when something is unclear. Most students who miss a class fall behind and then find it very difficult to catch up. (See Class Time, page AIM-5.)

OBJECTIVES

3.1
1. Distance and midpoint formulas
2. Graph an equation in two variables

3.2
1. Evaluate a function
2. Graph of a function
3. Vertical line test

3.3
1. Graph a linear function
2. Graph an equation of the form $Ax + By = C$
3. Application problems

3.4
1. Find the slope of a line given two points
2. Graph a line given a point and the slope
3. Average rate of change

3.5
1. Find the equation of a line given a point and the slope
2. Find the equation of a line given two points
3. Application problems

3.6
1. Find equations of parallel and perpendicular lines

3.7
1. Graph the solution set of an inequality in two variables

PREP TEST

Are you ready to succeed in this chapter?
Take the Prep Test below to find out if you are ready to learn the new material.

1. Simplify: $-4(x - 3)$
2. Simplify: $\sqrt{(-6)^2 + (-8)^2}$
3. Simplify: $\dfrac{3 - (-5)}{2 - 6}$
4. Evaluate $-2x + 5$ for $x = -3$.
5. Evaluate $\dfrac{2r}{r - 1}$ for $r = 5$.
6. Evaluate $2p^3 - 3p + 4$ for $p = -1$.
7. Evaluate $\dfrac{x_1 + x_2}{2}$ for $x_1 = 7$ and $x_2 = -5$.
8. Given $3x - 4y = 12$, find the value of x when $y = 0$.

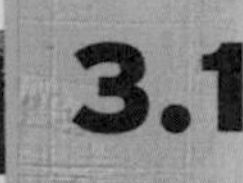

3.1 The Rectangular Coordinate System

OBJECTIVE 1 Distance and midpoint formulas

Point of Interest

A rectangular coordinate system is also called a **Cartesian** coordinate system in honor of Descartes.

Before the 15th century, geometry and algebra were considered separate branches of mathematics. That changed when René Descartes, a French mathematician who lived from 1596 to 1650, founded **analytic geometry.** In this geometry, a *coordinate system* is used to study relationships between variables.

A **rectangular coordinate system** is formed by two number lines, one horizontal and one vertical, that intersect at the zero point of each line. The point of intersection is called the **origin.** The two lines are called **coordinate axes,** or simply **axes.**

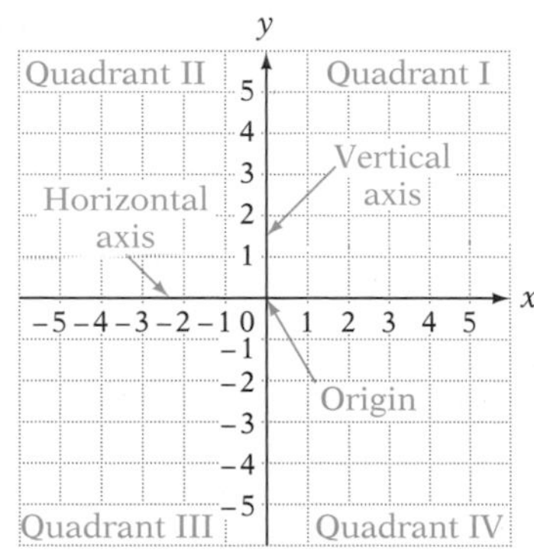

The axes determine a **plane,** which can be thought of as a large, flat sheet of paper. The two axes divide the plane into four regions called **quadrants.** The quadrants are numbered counterclockwise from I to IV.

Each point in the plane can be identified by a pair of numbers called an **ordered pair.** The first number of the pair measures a horizontal distance and is called the **abscissa.** The second number of the pair measures a vertical distance and is called the **ordinate.** The **coordinates** of the point are the numbers in the ordered pair associated with the point. The abscissa is also called the **first coordinate,** or ***x*-coordinate,** of the ordered pair, and the ordinate is also called the **second coordinate,** or ***y*-coordinate,** of the ordered pair.

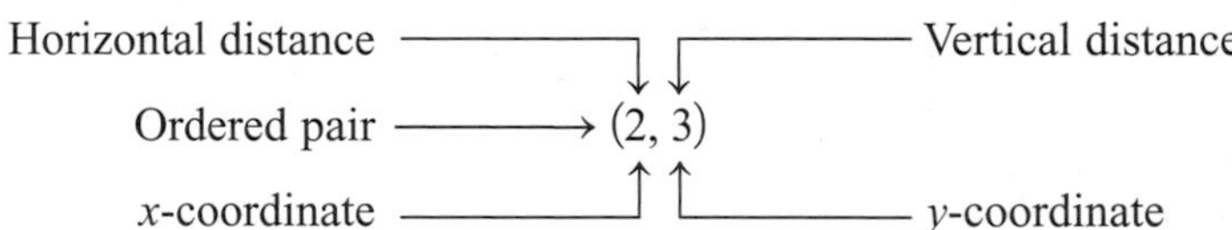

When drawing a rectangular coordinate system, we often label the horizontal axis x and the vertical axis y. In this case, the coordinate system is called an ***xy*-coordinate system.** To graph or plot a point in the xy-coordinate system, place a dot at the location given by the ordered pair. The **graph of an ordered pair** is the dot drawn at the coordinates of the point in the xy-coordinate system. The points whose coordinates are $(3, 4)$ and $(-2.5, -3)$ are graphed in the figure at the right.

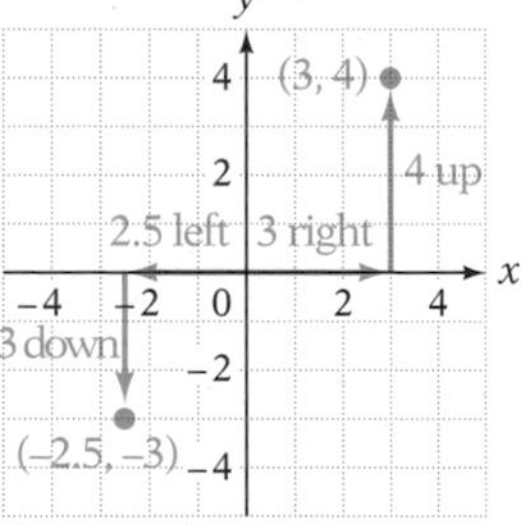

The points whose coordinates are $(3, -1)$ and $(-1, 3)$ are graphed at the right. Note that the graphs are in different locations. The *order* of the coordinates of an ordered pair is important.

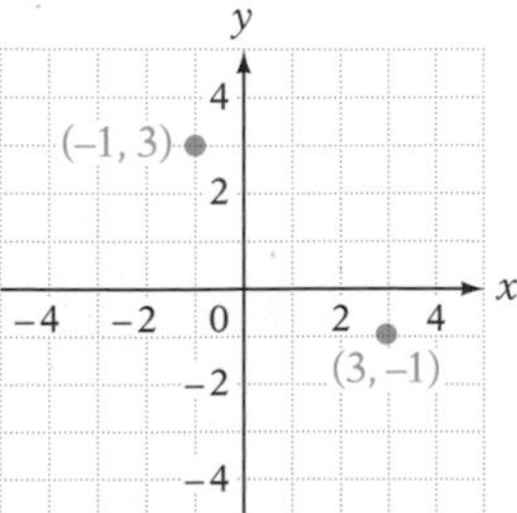

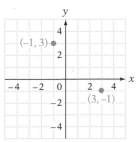

The distance between two points in an xy-coordinate system can be calculated by using the Pythagorean Theorem.

Take Note

A **right triangle** contains one 90° angle. The side opposite the 90° angle is called the **hypotenuse.** The other two sides are called **legs.**

PYTHAGOREAN THEOREM

If a and b are the lengths of the legs of a right triangle and c is the length of the hypotenuse, then $a^2 + b^2 = c^2$.

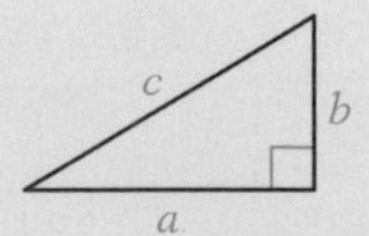

Point of Interest

Although Pythagoras (c. 550 B.C.) is given credit for the Pythagorean Theorem, a clay tablet called Plimpton 322 shows that the Babylonians knew the theorem 1000 years before Pythagoras!

Consider the two points and the right triangle shown at the right. The vertical distance between $P_1(x_1, y_1)$ and $P_2(x_2, y_2)$ is $|y_2 - y_1|$. The horizontal distance between the points $P_1(x_1, y_1)$ and $P_2(x_2, y_2)$ is $|x_2 - x_1|$.

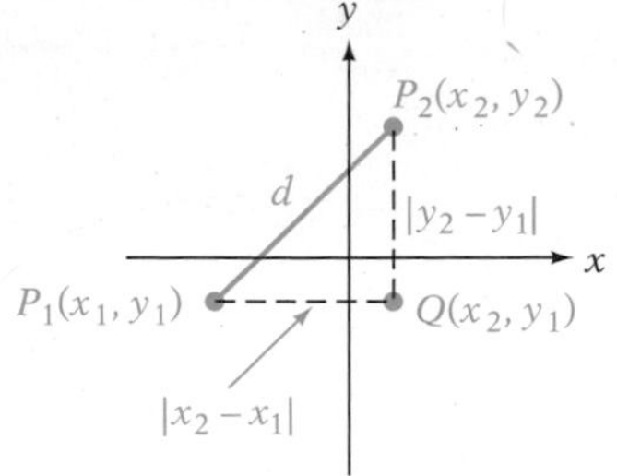

The quantity d^2 is calculated by applying the Pythagorean Theorem to the right triangle. $d^2 = |x_2 - x_1|^2 + |y_2 - y_1|^2$

Because the square of a number is always nonnegative, the absolute value signs are not necessary. $d^2 = (x_2 - x_1)^2 + (y_2 - y_1)^2$

The distance d is the square root of d^2. $d = \sqrt{(x_2 - x_1)^2 + (y_2 - y_1)^2}$

THE DISTANCE FORMULA

If $P_1(x_1, y_1)$ and $P_2(x_2, y_2)$ are two points in the plane, then the distance d between the two points is given by

$$d = \sqrt{(x_2 - x_1)^2 + (y_2 - y_1)^2}$$

EXAMPLE 1 Find the exact distance between the points whose coordinates are $(-3, 2)$ and $(4, -1)$.

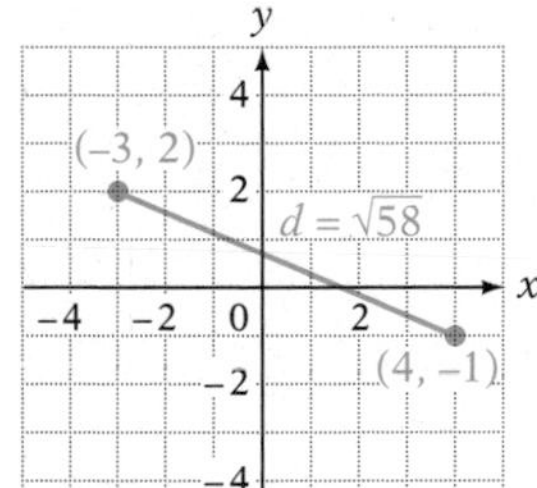

Solution

$d = \sqrt{(x_2 - x_1)^2 + (y_2 - y_1)^2}$ • **Use the distance formula.**

$= \sqrt{[4 - (-3)]^2 + (-1 - 2)^2}$ • $(x_1, y_1) = (-3, 2)$ **and** $(x_2, y_2) = (4, -1)$.

$= \sqrt{7^2 + (-3)^2} = \sqrt{49 + 9}$

$= \sqrt{58}$

The distance between the points is $\sqrt{58}$. See the graph at the left.

Problem 1 Find the exact distance between the points whose coordinates are $(5, -2)$ and $(-4, 3)$.

Solution See page S6.

Try Exercise 15, part (a), page 117.

The **midpoint of a line segment** is equidistant from its endpoints. The coordinates of the midpoint of the line segment P_1P_2 are (x_m, y_m). The intersection of the horizontal line segment through P_1 and the vertical line segment through P_2 is Q, with coordinates (x_2, y_1).

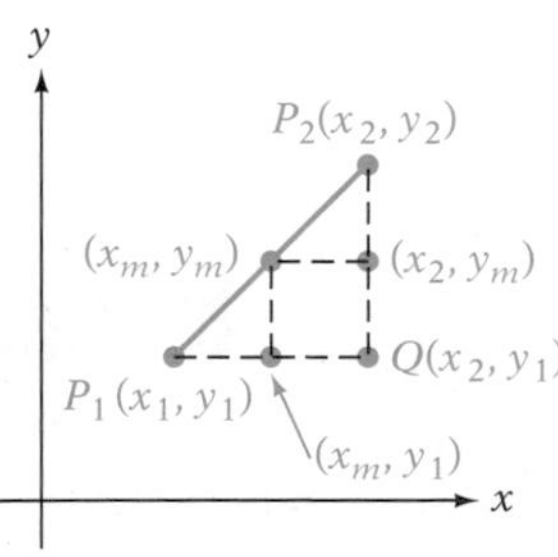

The x-coordinate x_m of the midpoint of the line segment P_1P_2 is the same as the x-coordinate of the midpoint of the line segment P_1Q. It is the average of the x-coordinates of the points P_1 and P_2. $x_m = \dfrac{x_1 + x_2}{2}$

Similarly, the y-coordinate y_m of the midpoint of the line segment P_1P_2 is the same as the y-coordinate of the midpoint of the line segment P_2Q. It is the average of the y-coordinates of the points P_1 and P_2. $y_m = \dfrac{y_1 + y_2}{2}$

THE MIDPOINT FORMULA

If $P_1(x_1, y_1)$ and $P_2(x_2, y_2)$ are the endpoints of a line segment, then the coordinates (x_m, y_m) of the midpoint of the line segment are given by

$$x_m = \frac{x_1 + x_2}{2} \quad \text{and} \quad y_m = \frac{y_1 + y_2}{2}$$

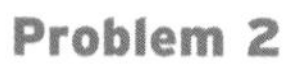

EXAMPLE 2 Find the coordinates of the midpoint of the line segment with endpoints $P_1(-5, 4)$ and $P_2(1, -3)$.

Solution

$$x_m = \frac{x_1 + x_2}{2} \qquad y_m = \frac{y_1 + y_2}{2}$$

• **Use the midpoint formula.**

$$= \frac{-5 + 1}{2} \qquad = \frac{4 + (-3)}{2}$$

• **$(x_1, y_1) = (-5, 4)$ and $(x_2, y_2) = (1, -3)$.**

$$= -2 \qquad = \frac{1}{2}$$

The coordinates of the midpoint are $\left(-2, \frac{1}{2}\right)$. See the graph at the left.

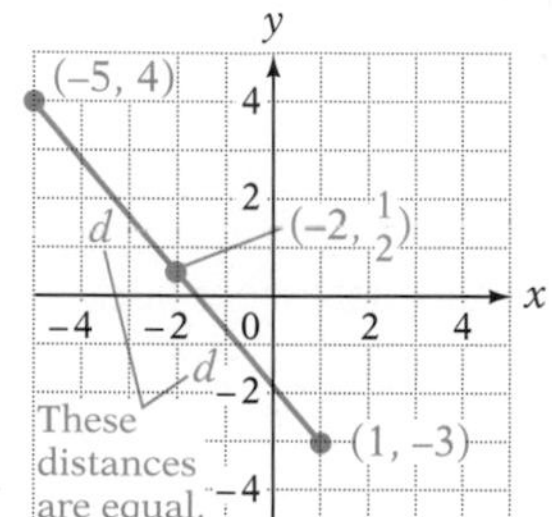

Problem 2 Find the coordinates of the midpoint of the line segment with endpoints $P_1(-3, -5)$ and $P_2(-2, 3)$.

Solution See page S6.

Try Exercise 15, part (b), page 117.

OBJECTIVE 2 Graph an equation in two variables

The *xy*-coordinate system is used to graph equations in *two variables*. Examples of equations in two variables are shown at the right.

$$y = 3x + 7$$
$$y = x^2 - 4x + 3$$
$$x^2 + y^2 = 25$$
$$x = \frac{y}{y^2 + 4}$$

A **solution of an equation in two variables** is an ordered pair (x, y) whose coordinates make the equation a true statement.

Focus on determining whether an ordered pair is a solution of an equation

Is the ordered pair $(-3, 7)$ a solution of the equation $y = -2x + 1$?

Replace x by -3 and replace y by 7. Then simplify.

y	$= -2x + 1$
7	$-2(-3) + 1$
7	$6 + 1$

$7 = 7$

Compare the results. If the resulting equation is a true statement, the ordered pair is a solution of the equation. If it is not a true statement, the ordered pair is not a solution of the equation.

Yes, the ordered pair $(-3, 7)$ is a solution of the equation.

Besides the ordered pair $(-3, 7)$, there are many other ordered-pair solutions of the equation $y = -2x + 1$. For example, $(-5, 11)$, $(0, 1)$, $(-\frac{3}{2}, 4)$ and $(4, -7)$ are also solutions of the equation.

In general, an equation in two variables has an infinite number of solutions. By choosing any value of x and substituting that value into the equation, we can calculate a corresponding value of y.

EXAMPLE 3 Determine the ordered-pair solution of $y = \frac{x}{x-2}$ that corresponds to $x = 4$.

Solution

$$y = \frac{x}{x-2}$$

$$y = \frac{4}{4-2}$$ • **Replace x by 4 and solve for y.**

$$y = 2$$

The ordered-pair solution is $(4, 2)$.

Problem 3 Determine the ordered-pair solution of $y = \frac{3x}{x+1}$ that corresponds to $x = -2$.

Solution See page S6.

Try Exercise 31, page 118.

The ordered-pair solutions of an equation in two variables can be graphed in a rectangular coordinate system.

Focus on graphing some solutions of an equation in two variables

Graph the solutions (x, y) of $y = x^2 - 1$ when x equals $-2, -1, 0, 1$, and 2.

Substitute each value of x into the equation and solve for y. Then graph the resulting ordered pairs by placing a dot at the coordinates of each point. This is sometimes referred to as *plotting the points*. It is convenient to record the ordered-pair solutions in a table similar to the one shown below.

x	$y = x^2 - 1$	y	(x, y)
-2	$y = (-2)^2 - 1 = 3$	3	$(-2, 3)$
-1	$y = (-1)^2 - 1 = 0$	0	$(-1, 0)$
0	$y = 0^2 - 1 = -1$	-1	$(0, -1)$
1	$y = 1^2 - 1 = 0$	0	$(1, 0)$
2	$y = 2^2 - 1 = 3$	3	$(2, 3)$

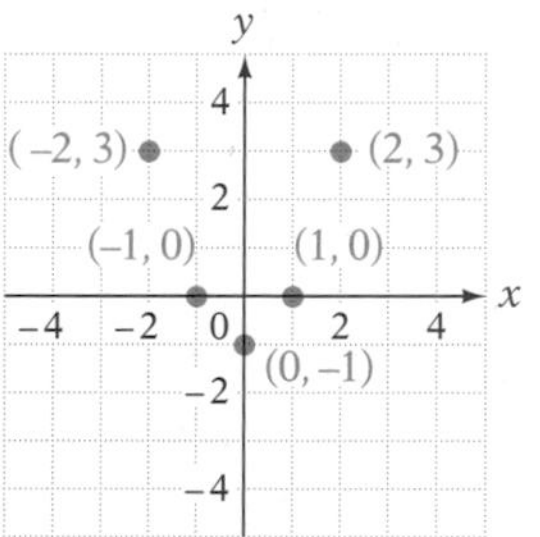

Generally, when we graph an equation in two variables, we include *all* the solutions, not just some selected ones as we did above.

Consider the equation $y = -2x + 1$. We can find ordered pair solutions when $x = -2, -1, 0, 1, 2$, and 3. These are shown in the table on the next page. Figure 1 shows the graph of those solutions.

x	$y = -2x + 1$	y	(x, y)
-2	$y = -2(-2) + 1 = 5$	5	$(-2, 5)$
-1	$y = -2(-1) + 1 = 3$	3	$(-1, 3)$
0	$y = -2(0) + 1 = 1$	1	$(0, 1)$
1	$y = -2(1) + 1 = -1$	-1	$(1, -1)$
2	$y = -2(2) + 1 = -3$	-3	$(2, -3)$
3	$y = -2(3) + 1 = -5$	-5	$(3, -5)$

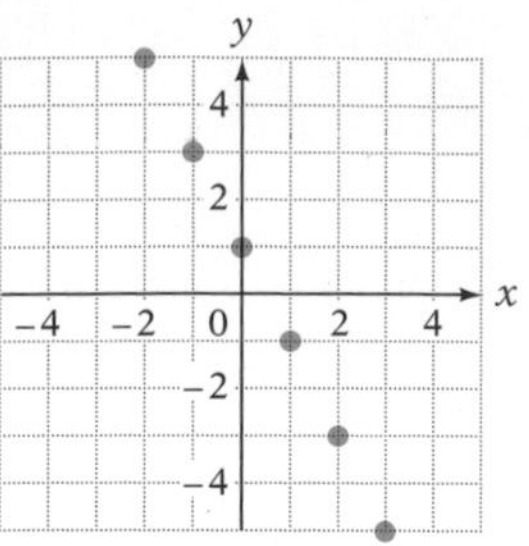

FIGURE 1

If we find additional solutions, such as the ordered pairs that correspond to $x = -1.5$, -0.5, 0.5, 1.5, and 2.5, we get more points, as shown in Figure 2. If we continue to add more and more points, there will be so many dots that the graph will look like the straight line in Figure 3, which is the graph of $y = -2x + 1$. The arrowheads indicate that the graph extends forever in both directions.

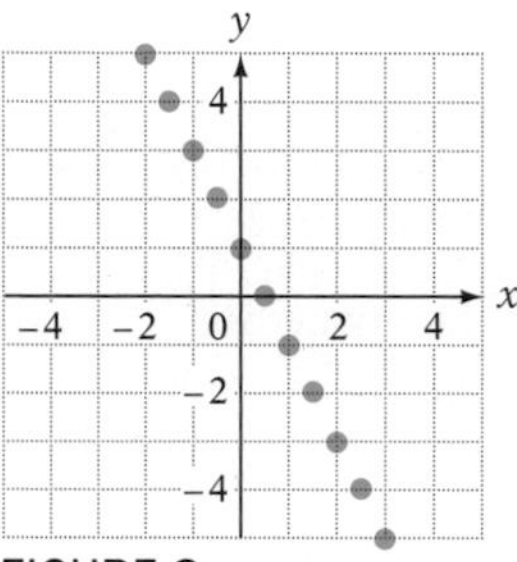

FIGURE 2

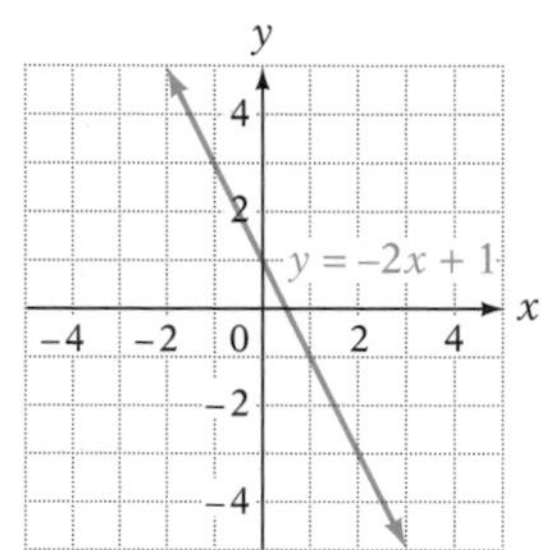

FIGURE 3

The graph of $y = -2x + 1$ is shown again below. As can be seen from the graph, the point with coordinates $(1, 4)$ is not on the graph. And, as shown below, $(1, 4)$ is not a solution of $y = -2x + 1$. The point with coordinates $(2, -3)$ is a point on the graph and is a solution of the equation.

$$\begin{array}{l|l} \multicolumn{2}{c}{y = -2x + 1} \\ \hline 4 & -2(1) + 1 \\ 4 & -2 + 1 \\ \multicolumn{2}{l}{4 \neq -1} \end{array}$$

(1, 4) does *not* represent a point on the graph and is *not* a solution of the equation.

$$\begin{array}{l|l} \multicolumn{2}{c}{y = -2x + 1} \\ \hline -3 & -2(2) + 1 \\ -3 & -4 + 1 \\ \multicolumn{2}{l}{-3 = -3} \end{array}$$

$(2, -3)$ *does* represent a point on the graph and *is* a solution of the equation.

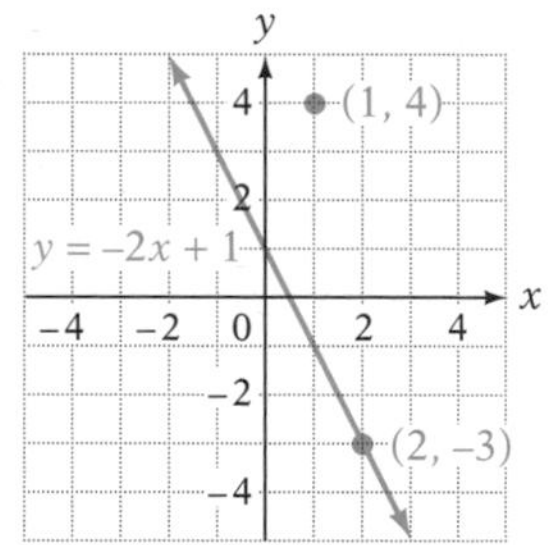

Every ordered pair on the graph of an equation is a solution of the equation, and every ordered-pair solution of an equation gives the coordinates of a point on the graph of the equation.

EXAMPLE 4 Graph $y = \frac{1}{2}x + 1$ by plotting the solutions of the equation when $x = -4, -2, 0, 2$, and 4, and then connecting the points with a smooth graph.

Solution Determine the ordered-pair solutions (x, y) for the given values of x. Plot the points and then connect the points with a smooth graph.

x	$y = \frac{1}{2}x + 1$	y	(x, y)
-4	$y = \frac{1}{2}(-4) + 1 = -1$	-1	$(-4, -1)$
-2	$y = \frac{1}{2}(-2) + 1 = 0$	0	$(-2, 0)$
0	$y = \frac{1}{2}(0) + 1 = 1$	1	$(0, 1)$
2	$y = \frac{1}{2}(2) + 1 = 2$	2	$(2, 2)$
4	$y = \frac{1}{2}(4) + 1 = 3$	3	$(4, 3)$

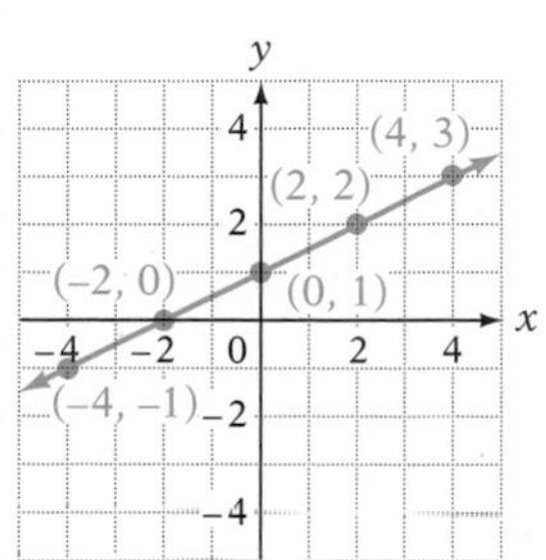

Problem 4 Graph $y = -\frac{1}{2}x + 3$ by plotting the solutions of the equation when $x = -4, -2, 0, 2$, and 4, and then connecting the points with a smooth graph.

Solution See page S6.

Try Exercise 37, page 118.

EXAMPLE 5 Graph $y = x^2 - 2x - 3$ by plotting the solutions of the equation when $x = -2, -1, 0, 1, 2, 3$, and 4, and then connecting the points with a smooth graph.

Solution Determine the ordered pair solutions (x, y) for the given values of x. Plot the points and then connect the points with a smooth graph.

x	$y = x^2 - 2x - 3$	y	(x, y)
-2	$y = (-2)^2 - 2(\quad) - 3 = 5$	5	$(-2, 5)$
-1	$y = (-1)^2 - 2(\quad) - 3 = 0$	0	$(-1, 0)$
0	$y = (0)^2 - 2(\;) - 3 = -3$	-3	$(0, -3)$
1	$y = (1)^2 - 2(\;) - 3 = -4$	-4	$(1, -4)$
2	$y = (2)^2 - 2(\;) - 3 = -3$	-3	$(2, -3)$
3	$y = (3)^2 - 2(\;) - 3 = 0$	0	$(3, 0)$
4	$y = (4)^2 - 2(\;) - 3 = 5$	5	$(4, 5)$

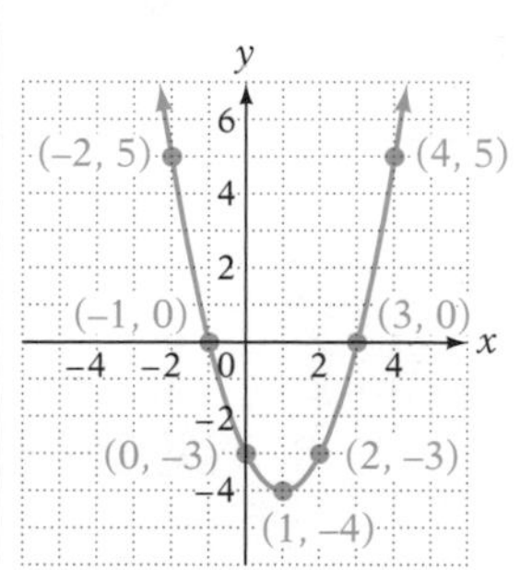

Problem 5 Graph $y = -x^2 + 4$ by plotting the solutions of the equation when $x = -3, -2, -1, 0, 1, 2$, and 3, and then connecting the points with a smooth graph.

Solution See page S7.

Try Exercise 41, page 118.

EXAMPLE 6 Graph $y = |2x - 4| - 1$ by plotting the solutions of the equation when $x = -1, 0, 1, 2, 3, 4$, and 5, and then connecting the points with a smooth graph.

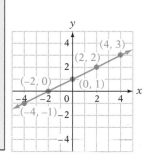

Solution Determine the ordered pair solutions (x, y) for the given values of x. Plot the points and then connect the points with a smooth graph.

x	$y = \|2x - 4\| - 1$	y	(x, y)
-1	$y = \|2(-1) - 4\| - 1 = 5$	5	$(-1, 5)$
0	$y = \|2(0) - 4\| - 1 = 3$	3	$(0, 3)$
1	$y = \|2(1) - 4\| - 1 = 1$	1	$(1, 1)$
2	$y = \|2(2) - 4\| - 1 = -1$	-1	$(2, -1)$
3	$y = \|2(3) - 4\| - 1 = 1$	1	$(3, 1)$
4	$y = \|2(4) - 4\| - 1 = 3$	3	$(4, 3)$
5	$y = \|2(5) - 4\| - 1 = 5$	5	$(5, 5)$

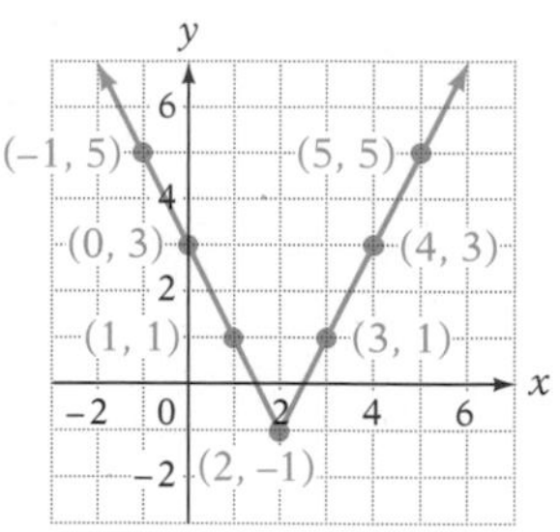

Problem 6 Graph $y = 3 - |x|$ by plotting the solutions of the equation when $x = -3, -2, -1, 0, 1, 2$, and 3, and then connecting the points with a smooth graph.

Solution See page S7.

Try Exercise 45, page 118.

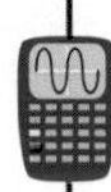

A graphing calculator graphs an equation in two variables much as we have done, by selecting values of x, finding the ordered pair solutions (x, y) for those values of x, and then plotting the points. The graphs of the three preceding examples are shown below. See the Appendix for details on how to create graphs using a graphing calculator.

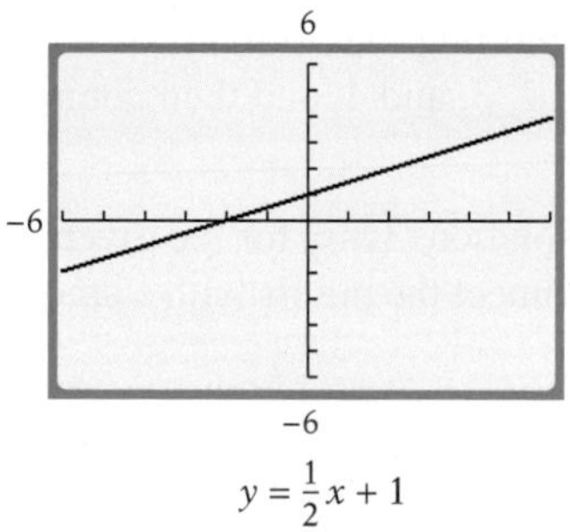

$y = \frac{1}{2}x + 1$

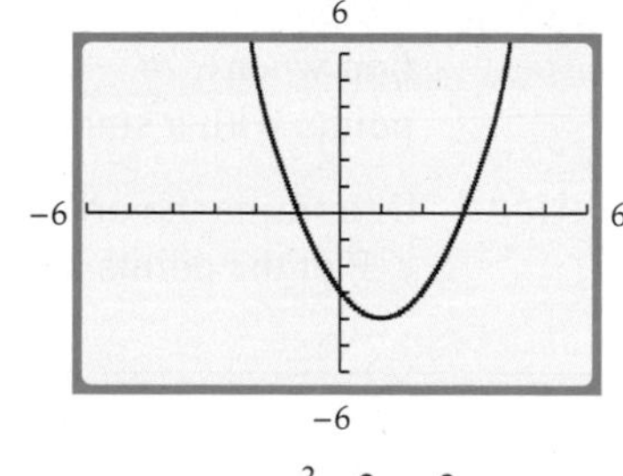

$y = x^2 - 2x - 3$

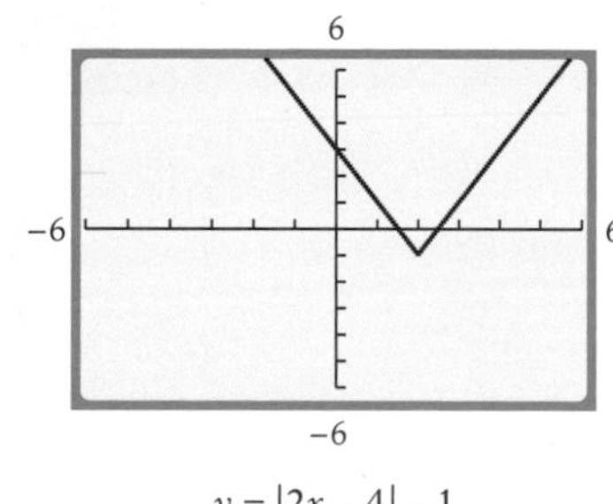

$y = |2x - 4| - 1$

3.1 Exercises

CONCEPT CHECK

1. What is the x-coordinate of a point on the y-axis?
2. What is the y-coordinate of a point on the x-axis?
3. Name the quadrant in which the graph of each of the following ordered pairs is located.

   a. $(-2, 3)$ **b.** $(4, 1)$ **c.** $(-3, -1)$ **d.** $(5, -1)$
4. In which quadrants is the y-coordinate of a point a negative number?
5. In which quadrants is the x-coordinate of a point a positive number?
6. Choose the word or words that make the following sentence true: If M is the midpoint of a line segment between A and B, then the distance between A and M equals/does not equal the distance between B and M.

7. Determine whether each of the following ordered pairs belongs to the graph of $y = -2x + 6$.

a. $(2, 2)$ **b.** $(-3, 0)$ **c.** $(-1, 4)$ **d.** $(3, 0)$

8. Explain why $(2, 4)$ is not a solution of the equation $y = \frac{4}{x - 2}$.

1 Distance and midpoint formulas (See pages 110–112.)

GETTING READY

9. For the points $P_1(3, -3)$ and $P_2(-1, 4)$, identify each of the following values: $x_1 =$ __?__, $x_2 =$ __?__, $y_1 =$ __?__, and $y_2 =$ __?__.

10. Find the length of the line segment between the points P_1 and P_2 given in Exercise 9.

$d = \sqrt{(x_2 - x_1)^2 + (y_2 - y_1)^2}$ • **Use the __?__ Formula.**

$= \sqrt{(\underline{\ ?\ } - \underline{\ ?\ })^2 + (\underline{\ ?\ } - \underline{\ ?\ })^2}$ • **Substitute coordinates.**

$= \sqrt{(\underline{\ ?\ })^2 + (\underline{\ ?\ })^2}$ • **Perform operations inside parentheses.**

$= \sqrt{\underline{\ ?\ } + \underline{\ ?\ }}$ • **Simplify exponential expressions.**

$= \sqrt{\underline{\ ?\ }}$ • **Add.**

The length of the line segment between $P_1(3, -3)$ and $P_2(-1, 4)$ is __?__.

Find **a.** the exact distance between the points P_1 and P_2, and **b.** the coordinates of the midpoint of the line segment with endpoints P_1 and P_2.

11. $P_1(2, 5)$ and $P_2(5, 9)$

12. $P_1(-5, -2)$ and $P_2(7, 3)$

13. $P_1(-2, -9)$ and $P_2(6, 6)$

14. $P_1(-2, 3)$ and $P_2(5, -1)$

15. $P_1(3, 5)$ and $P_2(4, 1)$

16. $P_1(6, -1)$ and $P_2(-3, -2)$

17. $P_1(0, 3)$ and $P_2(-2, 4)$

18. $P_1(-7, -5)$ and $P_2(-2, -1)$

19. $P_1(-3, -5)$ and $P_2(2, -4)$

20. $P_1(3, -5)$ and $P_2(6, 0)$

21. $P_1(5, -2)$ and $P_2(-1, 5)$

22. $P_1(5, -5)$ and $P_2(2, -5)$

23. a, b, c, and d are positive numbers such that $a > c$ and $b < d$. In what quadrant is the midpoint of the line segment joining $P_1(a, b)$ and $P_2(-c, -d)$ located?

24. a, b, c, and d are positive numbers such that $a < c$ and $b < d$. In what quadrant is the midpoint of the line segment joining $P_1(-a, -b)$ and $P_2(c, d)$ located?

25. Draw a line through all points with an abscissa of 2.

26. Draw a line through all points with an abscissa of -3.

27. Draw a line through all points with an ordinate of -3.

28. Draw a line through all points with an ordinate of 4.

2 Graph an equation in two variables (See pages 112–116.)

GETTING READY

29. To find the ordered-pair solution of $y = 4x - 3$ when $x = -1$, replace x by __?__ and then solve for y. The value of y is __?__. The ordered-pair solution is __?__.

30. To find the ordered-pair solution of $y = x^2 + x - 1$ when $x = 2$, replace x by __?__ and then solve for y. The value of y is __?__. The ordered-pair solution is __?__.

31. Determine the ordered-pair solution of $y = |x + 1|$ that corresponds to $x = -5$.

32. Determine the ordered-pair solution of $y = -2|x|$ that corresponds to $x = 3$.

33. Determine the ordered-pair solution of $y = -x^2 + 2$ that corresponds to $x = -1$.

34. Determine the ordered-pair solution of $y = -x^2 - x + 4$ that corresponds to $x = 1$.

35. Determine the ordered-pair solution of $y = x^3 - 2$ that corresponds to $x = 0$.

36. Determine the ordered-pair solution of $y = -x^3 + 1$ that corresponds to $x = -2$.

37. Graph $y = 2x - 3$ by plotting the solutions of the equation when $x = -1, 0, 1, 2, 3,$ and 4, and then connecting the points with a smooth graph.

38. Graph $y = -2x + 1$ by plotting the solutions of the equation when $x = -2, -1, 0, 1,$ and 2, and then connecting the points with a smooth graph.

39. Graph $y = -\frac{2}{3}x + 1$ by plotting the solutions of the equation when $x = -6, -3, 0, 3,$ and 6, and then connecting the points with a smooth graph.

40. Graph $y = \frac{3}{2}x - 2$ by plotting the solutions of the equation when $x = -4, -2, 0, 2,$ and 4, and then connecting the points with a smooth graph.

41. Graph $y = x^2 - 4$ by plotting the solutions of the equation when $x = -3, -2, -1, 0, 1, 2,$ and 3, and then connecting the points with a smooth graph.

42. Graph $y = -x^2 + 5$ by plotting the solutions of the equation when $x = -3, -2, -1, 0, 1, 2,$ and 3, and then connecting the points with a smooth graph.

43. Graph $y = -x^2 + 2x + 3$ by plotting the solutions of the equation when $x = -2, -1, 0, 1, 2, 3,$ and 4, and then connecting the points with a smooth graph.

44. Graph $y = x^2 - 2x - 4$ by plotting the solutions of the equation when $x = -2, -1, 0, 1, 2, 3,$ and 4, and then connecting the points with a smooth graph.

45. Graph $y = |x + 2|$ by plotting the solutions of the equation when $x = -6, -5, -4, -3, -2, -1, 0, 1,$ and 2, and then connecting the points with a smooth graph.

46. Graph $y = -|x| + 2$ by plotting the solutions of the equation when $x = -6, -4, -2, 0, 2, 4,$ and 6, and then connecting the points with a smooth graph.

47. Graph $y = -|x - 1| + 3$ by plotting the solutions of the equation when $x = -3, -1, 1, 3$, and 5, and then connecting the points with a smooth graph.

48. Graph $y = |x + 2| - 4$ by plotting the solutions of the equation when $x = -6, -4, -2, 0, 2$, and 4, and then connecting the points with a smooth graph.

49. There is one value of x for which the equation $y = \frac{3}{x - 1}$ has no ordered-pair solution. What is that value?

50. There is one value of x for which the equation $y = \frac{x}{2x - 4}$ has no ordered-pair solution. What is that value?

APPLYING CONCEPTS

51. Graph $y = x + |x|$ by plotting the solutions of the equation when $x = -7, -5, -3, -1, 0, 1, 2$, and 3, and then connecting the points with a smooth graph.

52. Graph $y = \frac{6}{x^2 + 1}$ by plotting the solutions of the equation when $x = -4, -3, -2, -1, 0, 1, 2, 3$, and 4, and then connecting the points with a smooth graph.

53. Draw a line passing through every point whose abscissa equals its ordinate.

54. Draw a line passing through every point whose ordinate is the additive inverse of its abscissa.

55. Describe the graph of all the ordered pairs (x, y) that are 5 units from the origin.

56. Consider two distinct fixed points in the plane. Describe the graph of all the points (x, y) that are equidistant from these fixed points.

PROJECTS OR GROUP ACTIVITIES

Two points are said to be **symmetric with respect to the *x*-axis** if they have the same x-coordinate and y-coordinates that are opposites. Two points are said to be **symmetric with respect to the *y*-axis** if they have the same y-coordinate and x-coordinates that are opposites.

57. Find the coordinates of the point that is symmetric, with respect to the x-axis, to the point with the given coordinates.

a. (3, 5) **b.** (5, −3) **c.** (−3, 4) **d.** (0, −5)

58. Find the coordinates of the point that is symmetric, with respect to the y-axis, to the point with the given coordinates.

a. (2, 4) **b.** (−3, 2) **c.** (−4, −5) **d.** (4, 0)

3.2 Introduction to Functions

OBJECTIVE 1 Evaluate a function

In mathematics, a *function* is used to describe a relationship between two quantities. Because two quantities are involved, it is natural to use ordered pairs.

DEFINITION OF A FUNCTION

A **function** is a set of ordered pairs in which no two ordered pairs have the same first coordinate. The **domain** of a function is the set of first coordinates of the ordered pairs; the **range** of a function is the set of second coordinates of the ordered pairs.

EXAMPLES

1. $\{(1, 2), (2, 4), (3, 6), (4, 8)\}$
   Domain $= \{1, 2, 3, 4\}$
   Range $= \{2, 4, 6, 8\}$
2. $\{(-1, 0), (0, 0), (1, 0), (2, 0), (3, 0)\}$
   Domain $= \{-1, 0, 1, 2, 3\}$
   Range $= \{0\}$
3. $\{(-5, 5), (-1, 1), (-7, 10), (4, 5)\}$
   Domain $= \{-7, -5, -1, 4\}$
   Range $= \{1, 5, 10\}$

Take Note

The order of the elements in a set does not matter. For instance, in example (3) at the right, we could have written the domain as $\{4, -5, -1, -7\}$. However, it is customary to list the numbers from least to greatest as we did in the example. Also, recall that elements of a set are not repeated. In example (3), the number 5 is included only once in the range.

Now consider the set of ordered pairs $\{(1, 2), (4, 5), (7, 8), (4, 6)\}$. This set of ordered pairs is *not* a function. There are two ordered pairs, $(4, 5)$ and $(4, 6)$, with the same first coordinate. This set of ordered pairs is called a *relation.* A **relation** is any set of ordered pairs. A function is a special type of relation. The concepts of domain and range apply to relations as well as to functions.

There are various ways in which we can describe a function: as a *graph,* as a *table,* and as an *equation.* We will look at each of these.

The bar graph at the right shows the number of people who watched the Super Bowl for the years 2005 to 2010. The jagged line between 0 and 85 on the vertical axis indicates that a portion of the vertical axis has been omitted.

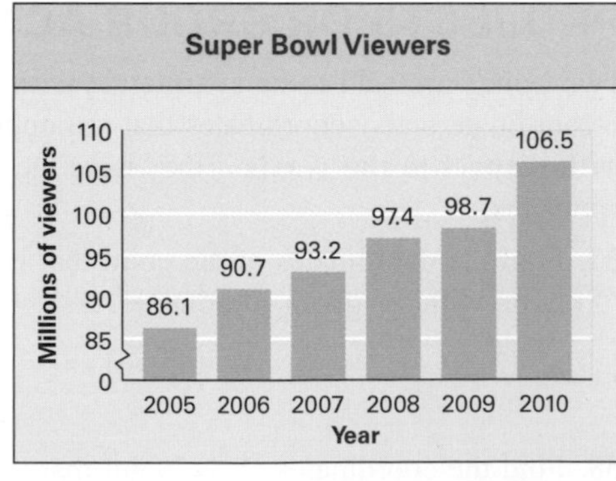

The data in the graph can be written as a set of ordered pairs.

$$\{(2005, 86.1), (2006, 90.7), (2007, 93.2), (2008, 97.4), (2009, 98.7), (2010, 106.5)\}$$

This set is a function. There are no two ordered pairs with the same first coordinate. The domain is $\{2005, 2006, 2007, 2008, 2009, 2010\}$. The range is $\{86.1, 90.7, 93.2, 97.4, 98.7, 106.5\}$. The ordered pair $(2010, 106.5)$ means that in 2010, the number of people who watched the Super Bowl was 106.5 million.

A table is another way of describing a relationship between two quantities. The table at the right shows a grading scale for an exam. This grading scale also can be described with ordered pairs. Some of the possible ordered pairs are (95, A), (72, C), (84, B), (92, A), (61, D), and (88, B).

Exam Score	Grade
90–100	A
80–89	B
70–79	C
60–69	D
0–59	F

The table represents a function. There are no two ordered pairs with the same first coordinate. To understand this, imagine two ordered pairs with the same first coordinate and different second coordinates: say, (92, A) and (92, C). This would mean that one student who scored 92 received an A grade, and a second student who scored 92 received a C grade. Such a result does not happen with a grading scale. The domain of the grading-scale function is $\{0, 1, 2, 3, ..., 97, 98, 99, 100\}$. The range is $\{A, B, C, D, F\}$.

Point of Interest

About 28 s after Kittinger left the gondola, he was at an altitude of 90,000 ft and was traveling about 614 mph. This remains the fastest speed a human has ever attained without the aid of a propulsion system. As he continued to descend, air resistance slowed his speed. At 50,000 ft, his speed was about 250 mph. During the first 30 s, because air resistance was negligible, the equation $s = 16t^2$ was a good approximation of the distance he fell. At lower altitudes, a different equation that considers air resistance would be more appropriate.

A third way to describe a relationship between two quantites is an equation. Consider the situation of one Joseph Kittinger, an Air Force officer who participated in a high-altitude experiment for the Air Force in 1960. He ascended in Excelsior III, a helium-filled balloon, to a height of 102,800 ft above Earth and then jumped from the gondola. The distance s, in feet, he had fallen t seconds after leaving the gondola can be approximated by the equation $s = 16t^2$.

Ordered pairs can be found by substituting values of t in the equation $s = 16t^2$ and solving for s. The ordered pairs for $t = 2, 5, 7$, and 9 are shown in the table at the right.

t	$s = 16t^2$	s	(t, s)
2	$s = 16(2)^2 = 64$	64	(2, 64)
5	$s = 16(5)^2 = 400$	400	(5, 400)
7	$s = 16(7)^2 = 784$	784	(7, 784)
9	$s = 16(9)^2 = 1296$	1296	(9, 1296)

One advantage of the equation is that we can calculate the distance Kittinger fell for any value of t. For instance, when $t = 6.7$, $s = 16(6.7)^2 = 718.24$. The ordered pair (6.7, 718.24) means that 6.7 s after leaving the gondola, Kittinger had fallen 718.24 ft.

Once a value of t is chosen, there is only one possible value of s. Therefore, there are no two ordered pairs with the same first coordinate. The equation represents a function. For instance, when $t = 3$, $s = 16(3)^2 = 144$. 144 is the only possible value of s when $t = 3$.

As mentioned in the Point of Interest at the left, the equation $s = 16t^2$ approximates the distance Kittinger fell during the first 30 s. We can describe the domain of the associated function (all of the times between 0 s and 30 s) using set-builder notation or interval notation.

$$\text{Domain} = \{t \mid 0 \le t \le 30\} \quad \text{or} \quad \text{Domain} = [0, 30]$$

The range of the function is from 0 ft (just starting the descent) to the distance Kittinger fell in 30 s. Substituting 30 into $s = 16t^2$ gives $s = 16(30)^2 = 14{,}400$. Thus

$$\text{Range} = \{s \mid 0 \le s \le 14{,}400\} \quad \text{or} \quad \text{Range} = [0, 14{,}400]$$

EXAMPLE 1 What are the domain and range of the following relation? Is the relation a function?

$$\{(2, 4), (3, 6), (4, 8), (5, 10), (4, 6), (6, 10)\}$$

Solution The domain is $\{2, 3, 4, 5, 6\}$.

- **The domain of the relation is the set of first coordinates of the ordered pairs.**

The range is $\{4, 6, 8, 10\}$.

- **The range of the relation is the set of second coordinates of the ordered pairs.**

The relation is not a function. • **There are two ordered pairs with the same first coordinate and different second coordinates, (4, 8) and (4, 6).**

Problem 1 What are the domain and range of the following relation? Is the relation a function?

$$\{(-2, 6), (-1, 3), (0, 0), (1, -3), (2, -6), (3, -9)\}$$

Solution See page S7.

Try Exercise 9, page 129.

For each element in the domain of a relation, there is a corresponding element in the range of the relation. A possible diagram of the relation in Example 1 is

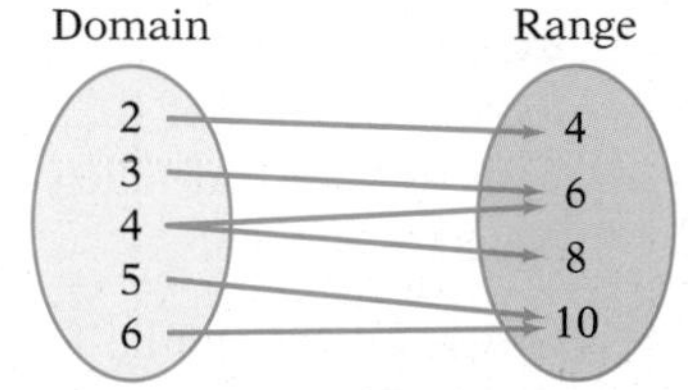

{(2, 4), (3, 6), (4, 8), (5, 10), (4, 6), (6, 10)}

Note from the diagram that there are two arrows from 4 in the domain to different elements, 6 and 8, in the range. This means that there are two ordered pairs, (4, 6) and (4, 8), with the same first coordinate and different second coordinates. The relation is not a function.

Focus on determining whether a diagram represents a function

Write the relation shown in the diagram below as a set of ordered pairs. Is the relation a function?

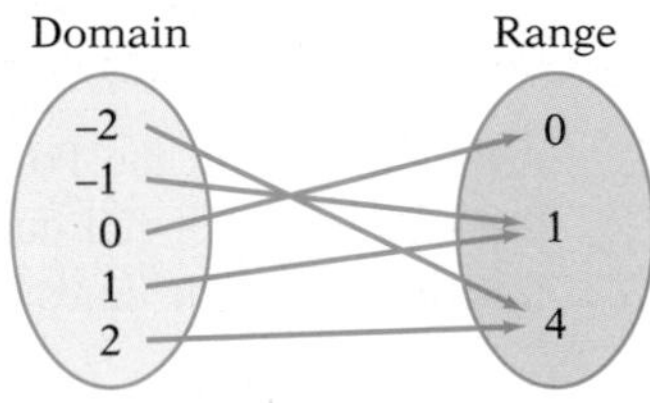

The relation is $\{(-2, 4), (-1, 1), (0, 0), (1, 1), (2, 4)\}$.

There are no ordered pairs with the same first coordinate. The relation is a function.

How It's Used

Functions are used in many applications. For instance, there are functions to calculate π to millions of digits, to predict your chances of winning the lottery, to determine the age of a fossil, and to find a monthly car payment.

Although a function can be described in terms of ordered pairs, in a table, or by a graph, a major focus of this text will be functions represented by equations in two variables.

For instance, the equation $s = 16t^2$ given earlier describes the relationship between the time t Joseph Kittinger fell and the distance s he fell. The ordered pairs can be written as (t, s), where $s = 16t^2$. By substituting $16t^2$ for s, we can also write the ordered pairs as $(t, 16t^2)$. Because the distance he falls *depends* on how long he has been falling, s is called the **dependent variable** and t is called the **independent variable.** For the equation $s = 16t^2$, we say that "distance is a function of time."

A function can be thought of as a rule that pairs one number with another number. For instance, the **square function** pairs a real number with its square. This function can be represented by the equation $y = x^2$. This equation states that, given any element x in the domain, the value of y in the range is the square of x. In this case, the independent variable is x and the dependent variable is y. When $x = 6$, $y = 36$, and one ordered pair of the function is (6, 36). When $x = -7$, $y = 49$. The ordered pair $(-7, 49)$ belongs to the function.

Other ordered pairs of the function are $(-2, 4)$, $\left(\frac{3}{4}, \frac{9}{16}\right)$, and $(-1.1, 1.21)$.

In many cases, you may think of a function as a machine that turns one number into another number. For instance, you can think of the square function machine at the right taking an **input** (an element from the domain) and creating an **output** (an element of the range) that is the square of the input.

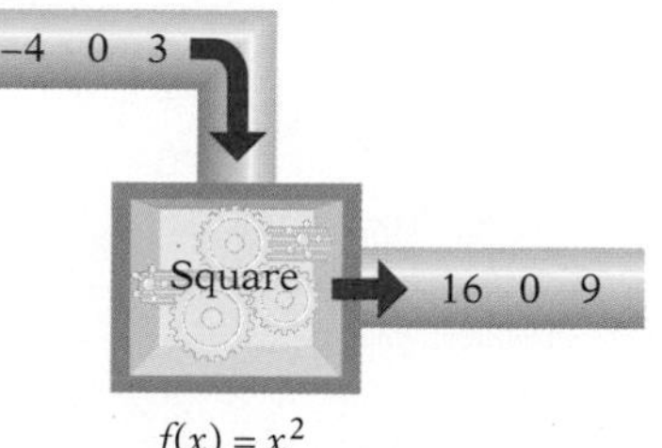

$f(x) = x^2$

Not all equations in two variables define a function. For instance,

$$y^2 = x^2 + 9$$

is not an equation that defines a function. As shown below, the ordered pairs (4, 5) and $(4, -5)$ are solutions of the equation.

$$\begin{array}{c|c} y^2 & x^2 + 9 \\ \hline 5^2 & 4^2 + 9 \\ 25 & 16 + 9 \\ \multicolumn{2}{c}{25 = 25} \end{array}$$

- **Let $(x, y) = (4, 5)$. Replace x by 4 and y by 5.**
- **(4, 5) checks.**

$$\begin{array}{c|c} y^2 & x^2 + 9 \\ \hline (-5)^2 & 4^2 + 9 \\ 25 & 16 + 9 \\ \multicolumn{2}{c}{25 = 25} \end{array}$$

- **Let $(x, y) = (4, -5)$. Replace x by 4 and y by -5.**
- **$(4, -5)$ checks.**

Consequently, there are two ordered pairs, (4, 5) and $(4, -5)$, with the same first coordinate and *different* second coordinates; the equation does not define a function. The phrase "y is a function of x," or a similar phrase with different variables, is used to describe those equations in two variables that define functions.

Take Note

The notation $f(x)$ does not mean f *times* x. The letter f is the *name* of the function, and $f(x)$ is the *value* of the function at x.

Function notation is frequently used for those equations that define functions. Just as x is commonly used as a variable for a number, the letter f is commonly used to name a function. For instance, using the square function, we can write $f(x) = x^2$. The symbol $f(x)$ is read "the value of f at x" or "f of x." The symbol $f(x)$ is the **value of the function** and represents the value of the dependent variable for a given value of the independent variable.

This is the value of the function.
It is the value of the dependent variable.

$$f(x) = x^2$$

The name of the function is f.

This is an algebraic expression that defines the relationship between the dependent variable and the independent variable.

The process of finding the value of $f(x)$ for a given value of x is called **evaluating the function.** For instance, to evaluate $f(x) = x^2$ when x is 4, replace x by 4 and simplify.

$$f(x) = x^2$$

$$f(4) = 4^2$$ • **Replace x by 4. Then simplify.**

$$= 16$$

The *value* of the function is 16 when $x = 4$. An ordered pair of the function is (4, 16).

The letters used to represent a function are somewhat arbitrary. All of the following equations represent the same function. Note that more than one letter can be used.

$f(x) = x^2$ $\quad g(t) = t^2$ $\quad P(v) = v^2$ $\quad SQ(x) = x^2$ • **These equations all represent the square function.**

We write $y = f(x)$ to emphasize the relationship between the independent variable x and the dependent variable y. **Remember:** y and $f(x)$ are different symbols for the same number. Also, the *name* of the function is f; the *value* of the function at x is $f(x)$.

Focus on evaluating a function

Find the value of $R(v) = v^3 + 3v^2 - 5v - 6$ when $v = -2$.

Replace v by -2 and simplify.

$$R(v) = v^3 + 3v^2 - 5v - 6$$
$$R(-2) = (-2)^3 + 3(-2)^2 - 5(-2) - 6$$
$$= -8 + 12 + 10 - 6$$
$$= 8$$

The value of $R(v)$ when $v = -2$ is 8.

Take Note

To evaluate a function, you can substitute open parentheses for the variable in each term. For instance,

$R(v) = v^3 + 3v^2 - 5v - 6$
$R(\ ) = (\ )^3 + 3(\ )^2 - 5(\ ) - 6$

To evaluate the function, fill each pair of parentheses with the same number and then simplify.

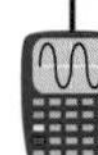

There are several ways to use a calculator to evaluate a function. The screens at the right, which illustrate the evaluation of R(−2) given above, show one way. See the Appendix Keystroke Guide for other methods of evaluating a function.

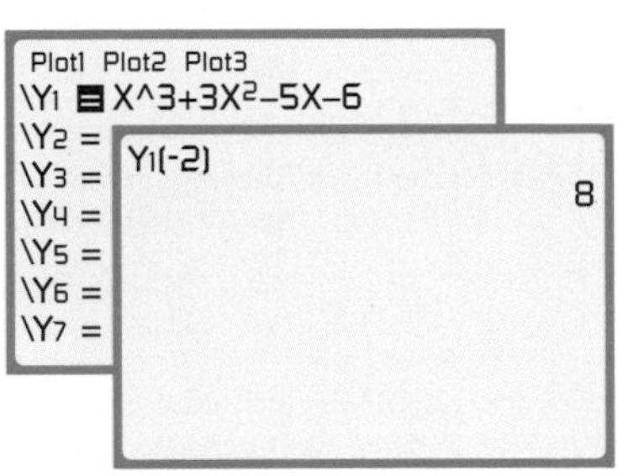

EXAMPLE 2 Let $q(r) = 2r^3 + 5r^2 - 6$.

A. Find $q(-3)$.

B. Find the value of $q(r)$ when $r = 2$.

Solution **A.** $q(r) = 2r^3 + 5r^2 - 6$

$q(-3) = 2(-3)^3 + 5(-3)^2 - 6$ • **Replace *r* by −3.**

$= 2(-27) + 5(9) - 6 = -54 + 45 - 6$ • **Simplify.**

$q(-3) = -15$

B. To find the value of $q(r)$ when $r = 2$ means to evaluate the function when $r = 2$.

$q(r) = 2r^3 + 5r^2 - 6$

$q(2) = 2(2)^3 + 5(2)^2 - 6$ • **Replace *r* by 2.**

$= 2(8) + 5(4) - 6 = 16 + 20 - 6$ • **Simplify.**

$q(2) = 30$

The value of the function when $r = 2$ is 30.

Problem 2 Let $f(z) = z + |z|$.

A. Find $f(-4)$.

B. Evaluate $f(z)$ when $z = 3$.

Solution See page S7.

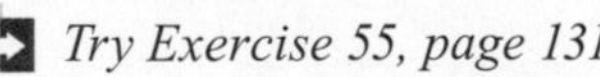
Try Exercise 55, page 131.

It is also possible to evaluate a function at another variable or at a variable expression.

EXAMPLE 3 Let $f(x) = 3 - 2x$. Find $f(3h - 2)$.

Solution

$$f(x) = 3 - 2x$$

$$f(3h - 2) = 3 - 2(3h - 2) \quad \text{• Replace } x \text{ by } 3h - 2.$$

$$= 3 - 6h + 4 \quad \text{• Simplify.}$$

$$f(3h - 2) = -6h + 7$$

Problem 3 Let $r(s) = 3s - 6$. Find $r(2a + 3)$.

Solution See page S7.

Try Exercise 39, page 131.

The range of a function contains all the elements that result from evaluating the function for each element of the domain. If the domain contains an infinite number of elements, as in the case of the falling pilot, it may be difficult to find the range. However, if the domain has only a finite number of elements, then the range can be found by evaluating the function for each element of the domain.

EXAMPLE 4 Find the range of $f(x) = x^2 + 2x - 1$ if the domain is $\{-3, -2, -1, 0, 1\}$.

Solution Evaluate the function for each element of the domain. The range includes the values of $f(-3)$, $f(-2)$, $f(-1)$, $f(0)$, and $f(1)$.

$$f(x) = x^2 + 2x - 1$$

$$f(-3) = (-3)^2 + 2(-3) - 1 = 2$$

$$f(-2) = (-2)^2 + 2(-2) - 1 = -1$$

$$f(-1) = (-1)^2 + 2(-1) - 1 = -2$$

$$f(0) = (0)^2 + 2(0) - 1 = -1$$

$$f(1) = (1)^2 + 2(1) - 1 = 2$$

The range is $\{-2, -1, 2\}$.

Problem 4 Find the range of $f(x) = x^3 + x$ if the domain is $\{-3, -2, -1, 0, 1, 2\}$.

Solution See page S7.

Try Exercise 61, page 131.

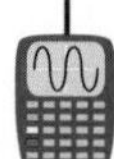

The TABLE feature of a calculator can be used to determine the range values of a function for selected values in the domain.

Plot1 Plot2 Plot3
\Y1 = X²+2X–1
\Y2 =
\Y3 =
\Y4 =
\Y5 =
\Y6 =
\Y7 =

TABLE SETUP
TblStart = -3
ΔTbl = 1
Indpnt: Auto Ask
Depend: Auto Ask

X	Y1
-3	2
-2	-1
-1	-2
0	-1
1	2
2	7
3	14

X = -3

Given an element in the range of a function, we can find a corresponding element of the domain.

EXAMPLE 5 Given $f(x) = 3x + 1$, find a number c in the domain of f such that $f(c) = 7$. Write the corresponding ordered pair of the function.

Solution

$f(x) = 3x + 1$

$f(c) = 3c + 1$ • **Replace x by c.**

$7 = 3c + 1$ • **$f(c) = 7$**

$6 = 3c$ • **Solve for c.**

$2 = c$

The value of c is 2. The corresponding ordered pair is (2, 7).

> **Take Note**
>
> You can check the answer in Example 5 by evaluating $f(x) = 3x + 1$ at $x = 2$.
>
> $f(2) = 3(2) + 1$
>
> $f(2) = 7$

Problem 5 Given $f(x) = 2x - 5$, find a number c in the domain of f such that $f(c) = -3$. Write the corresponding ordered pair of the function.

Solution See page S7.

Try Exercise 69, page 132.

OBJECTIVE 

Graph of a function

The **graph of a function** is the graph of the ordered pairs that belong to the function. Graphing a function is similar to graphing an equation in two variables. First, evaluate the function at selected values of x and plot the corresponding ordered pairs. Then connect the points with a smooth curve to form the graph.

EXAMPLE 6 Graph $h(x) = x^2 - 3$. First evaluate the function when $x = -3, -2, -1, 0, 1, 2$, and 3. Plot the resulting ordered pairs. Then connect the points to form the graph.

Solution Evaluate the function for the given values of x. This will produce some ordered pairs of the function. The results can be recorded in a table. Plot the ordered pairs and then connect the points to form the graph.

x	$y = h(x) = x^2 - 3$	y	(x, y)
-3	$h(-3) = (-3)^2 - 3 = 6$	6	$(-3, 6)$
-2	$h(-2) = (-2)^2 - 3 = 1$	1	$(-2, 1)$
-1	$h(-1) = (-1)^2 - 3 = -2$	-2	$(-1, -2)$
0	$h(0) = (0)^2 - 3 = -3$	-3	$(0, -3)$
1	$h(1) = (1)^2 - 3 = -2$	-2	$(1, -2)$
2	$h(2) = (2)^2 - 3 = 1$	1	$(2, 1)$
3	$h(3) = (3)^2 - 3 = 6$	6	$(3, 6)$

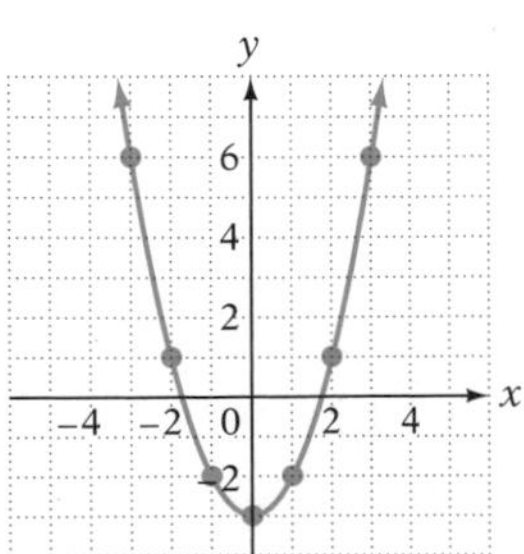

Problem 6 Graph $f(x) = -x^2 - 4x + 2$. First evaluate the function when $x = -5, -4, -3, -2, -1, 0$, and 1. Plot the resulting ordered pairs. Then connect the points to form the graph.

Solution See page S7.

Try Exercise 87, page 133.

The graph of an absolute value function will be used to illustrate various concepts throughout this text. We can graph $f(x) = |x|$ by evaluating the function for various values of x, plotting the resulting ordered pairs, and then connecting the points to produce the graph.

x	$y = f(x) = \|x\|$	y	(x, y)
-3	$f(-3) = \|-3\| = 3$	3	$(-3, 3)$
-2	$f(-2) = \|-2\| = 2$	2	$(-2, 2)$
-1	$f(-1) = \|-1\| = 1$	1	$(-1, 1)$
0	$f(0) = \|0\| = 0$	0	$(0, 0)$
1	$f(1) = \|1\| = 1$	1	$(1, 1)$
2	$f(2) = \|2\| = 2$	2	$(2, 2)$
3	$f(3) = \|3\| = 3$	3	$(3, 3)$

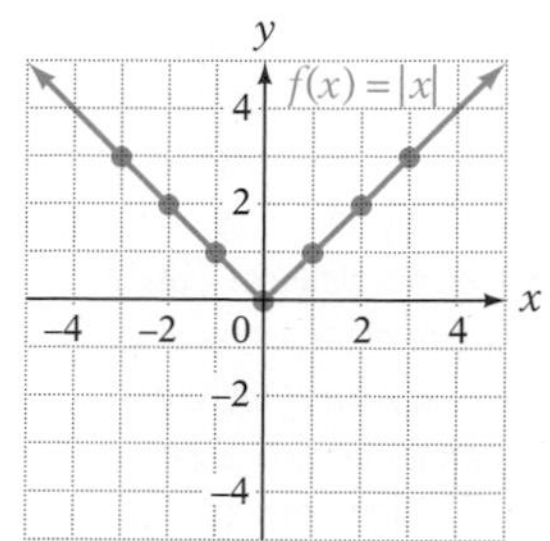

EXAMPLE 7 Graph $f(x) = -|x| + 2$. First evaluate the function when $x = -3, -2, -1, 0, 1, 2,$ and 3. Plot the resulting ordered pairs. Then connect the points to form the graph.

Solution Evaluate the function for the given values of x. Plot the ordered pairs and then connect the points to form a V-shaped graph.

x	$y = f(x) = -\|x\| + 2$	y	(x, y)
-3	$f(-3) = -\|-3\| + 2 = -1$	-1	$(-3, -1)$
-2	$f(-2) = -\|-2\| + 2 = 0$	0	$(-2, 0)$
-1	$f(-1) = -\|-1\| + 2 = 1$	1	$(-1, 1)$
0	$f(0) = -\|0\| + 2 = 2$	2	$(0, 2)$
1	$f(1) = -\|1\| + 2 = 1$	1	$(1, 1)$
2	$f(2) = -\|2\| + 2 = 0$	0	$(2, 0)$
3	$f(3) = -\|3\| + 2 = -1$	-1	$(3, -1)$

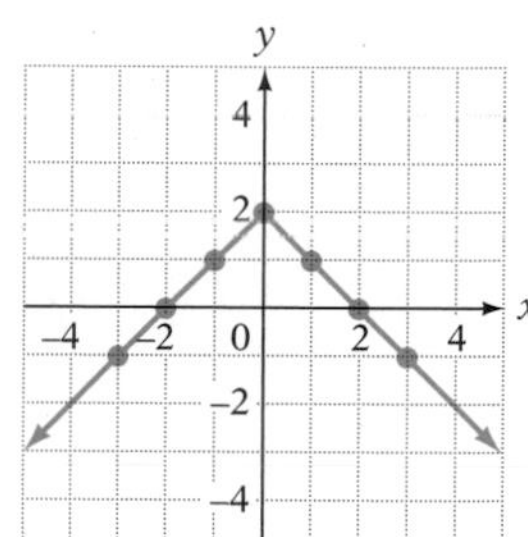

Problem 7 Graph $f(x) = |x - 2|$. First evaluate the function when $x = -1, 0, 1, 2, 3, 4,$ and 5. Plot the resulting ordered pairs. Then connect the points to form the graph.

Solution See page S7.

Try Exercise 97, page 133.

OBJECTIVE  Vertical line test

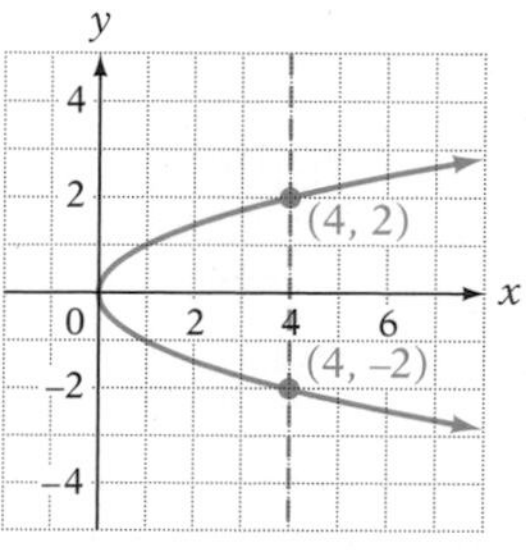

A graph is a visual description of a relation. For the relation graphed at the left, the two ordered pairs $(4, 2)$ and $(4, -2)$ belong to the graph, and those points lie on a vertical line. Because there are two ordered pairs with the same first coordinate, the set of ordered pairs of the graph is not a function. With this observation in mind, we can make use of a quick method to determine whether a graph is the graph of a function.

VERTICAL LINE TEST FOR THE GRAPH OF A FUNCTION

If every vertical line intersects a graph at most once, then the graph is the graph of a function.

This graphical interpretation of a function is often described by saying that each value in the domain of the function is paired with *exactly one* value in the range of the function.

Take Note

For the graph at the far right, note that there are values of x for which there is only one value of y. For instance, when $x = -5$, $y = 4$. For the relation to be a function, however, every value of x in the domain of the function must correspond to exactly one value of y. If there is even one value of x that pairs with two or more values of y, the condition for a function is not met. This is what the vertical line test states: If there is any place on the graph at which a vertical line intersects the graph more than once, the graph is not the graph of a function.

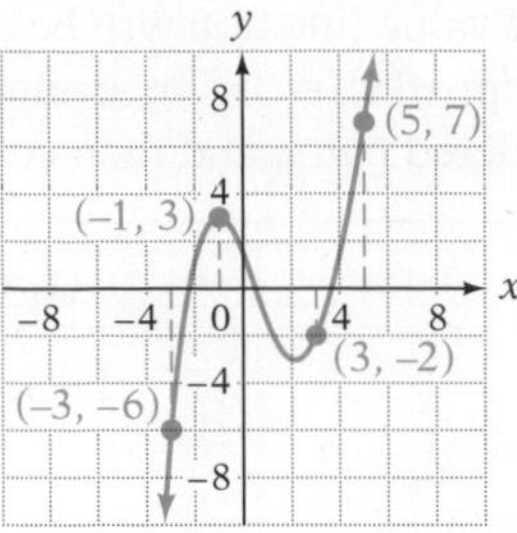

For each x, there is *exactly one* value of y. For instance, when $x = -3$, $y = -6$. This is the graph of a function.

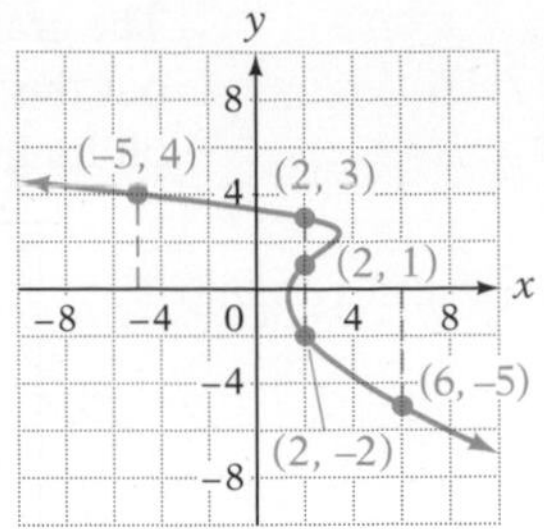

Some values of x can be paired with more than one value of y. For instance, 2 can be paired with -2, 1, and 3. This is not the graph of a function.

EXAMPLE 8 Use the vertical line test to determine whether the graph is the graph of a function.

A.

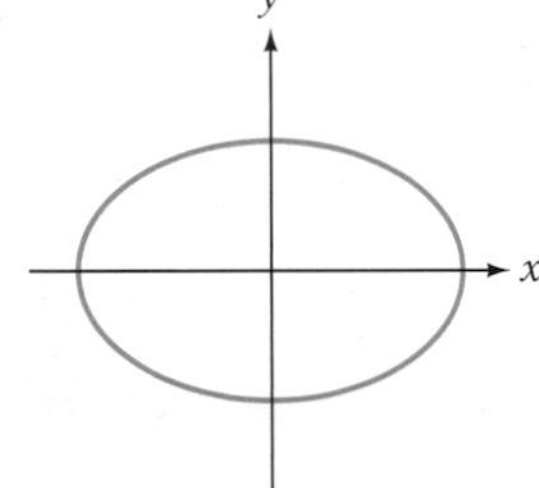

B.

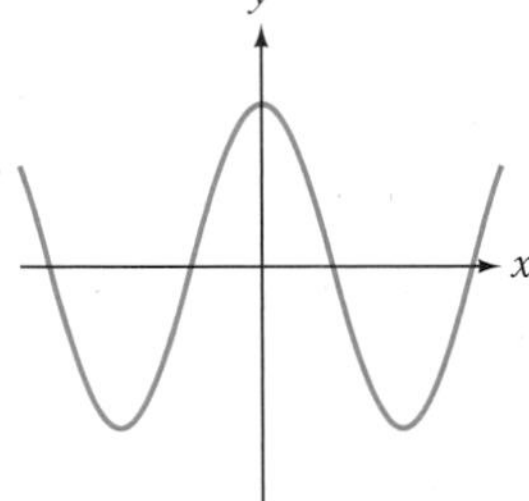

Solution

A. As shown at the right, there are vertical lines that intersect the graph at more than one point. Therefore, the graph is not the graph of a function.

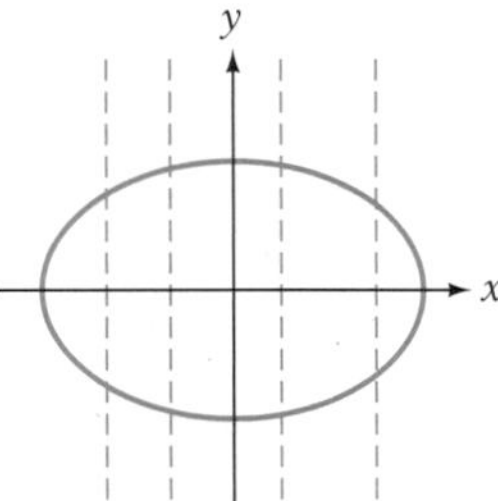

B. For the graph at the right, every vertical line intersects the graph at most once. Therefore, the graph is the graph of a function.

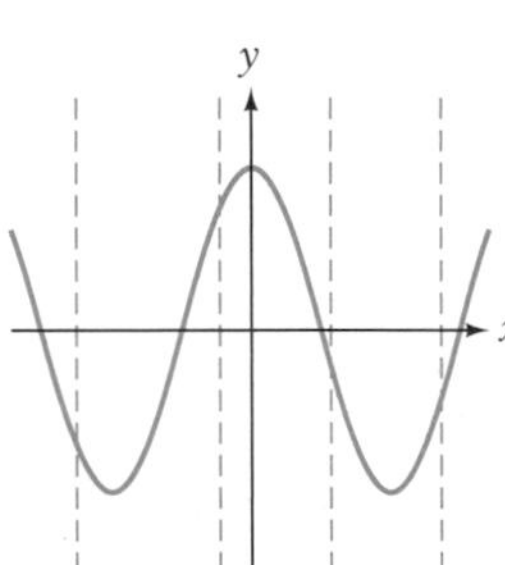

Problem 8 For each of the graphs below, determine whether the graph is the graph of a function.

A.

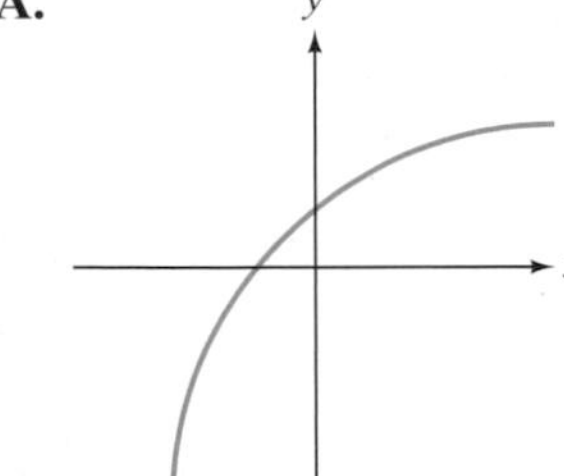

B.

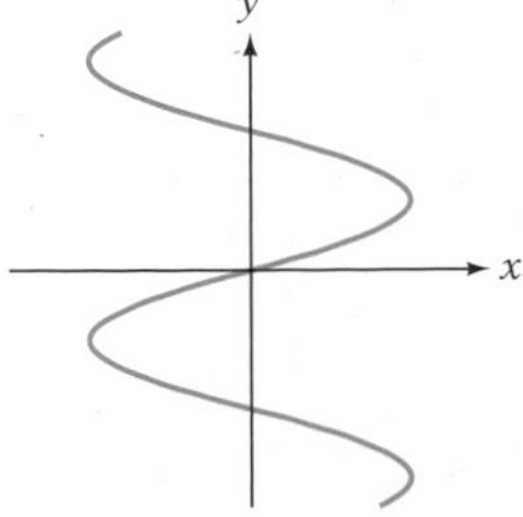

Solution See page S8.

Try Exercise 105, page 134.

3.2 Exercises

CONCEPT CHECK

1. Is every function a relation? Explain your answer.

2. What is the domain of a function? What is the range of a function?

3. Complete the following sentence using *exactly one* or *more than one.* For a function, each element of the domain is paired with __?__ element of the range.

4. Complete the following sentences using *can* or *cannot.* If (5, 2) belongs to a function, then (5, 9) __?__ belong to the function. If (5, 2) belongs to a relation, then (5, 9) __?__ belong to the relation.

5. A function f is defined by $\{(-2, 3), (-1, 0), (0, 2), (3, 4), (5, 7)\}$. What is the value of $f(3)$?

6. What number is not in the domain of the function defined $f(x) = \frac{x}{x-2}$?

7. Does the graph of a vertical line represent a function? Does the graph of a horizontal line represent a function?

8. If f is a function and $f(-2) = 4$, what is an ordered pair that belongs to the function?

1 Evaluate a function (See pages 120–126.)

Determine whether the relation is a function. Give the domain and range of the relation.

9. $\{(-3, 1), (-2, 2), (1, 5), (4, -7)\}$

10. $\{(-5, 4), (-2, 3), (0, 1), (3, 2)\ (7, 11)\}$

11. $\{(1, 5), (2, 5), (3, 5), (4, 5), (5, 5)\}$

12. $\{(1, 0), (10, 1), (100, 2), (1000, 3), (10{,}000, 4)\}$

13. $\left\{\left(-2, -\frac{1}{2}\right), (-1, -1), (1, 1), \left(2, \frac{1}{2}\right), \left(3, \frac{1}{3}\right)\right\}$

14. $\{(0, 0), (1, 1), (4, 2), (9, 3), (1, -1), (4, -2), (9, -3)\}$

15. $\{(2, 3), (4, 5), (6, 7), (8, 9), (6, 8)\}$

16. $\{(-1, 2), (-2, 4), (2, 4), (-2, 5), (4, 9)\}$

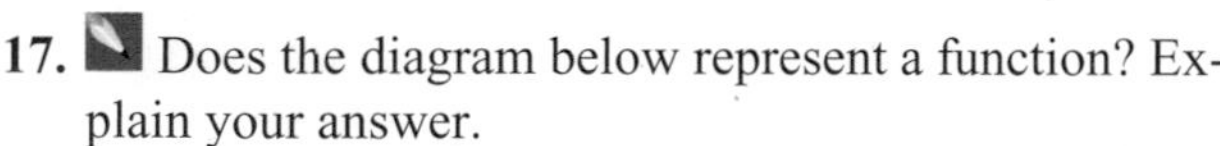
17. Does the diagram below represent a function? Explain your answer.

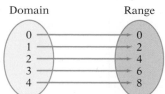

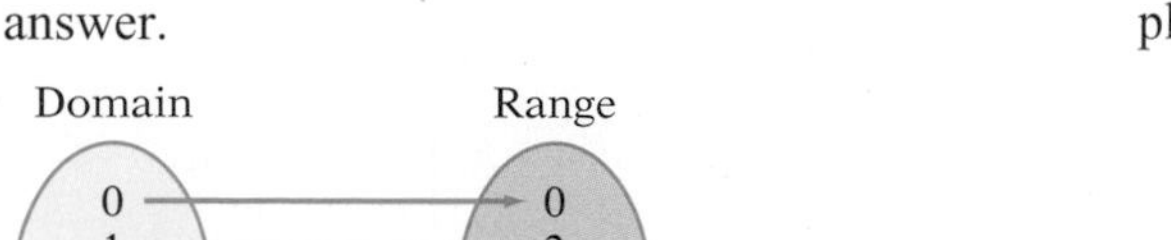

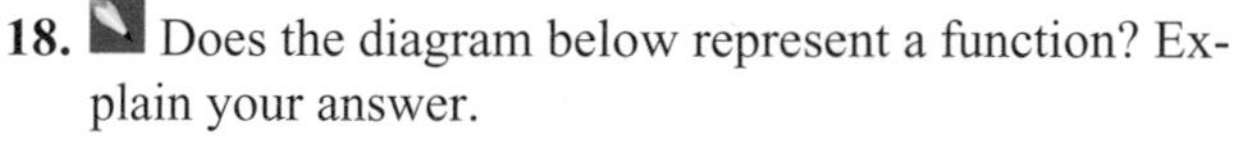
18. Does the diagram below represent a function? Explain your answer.

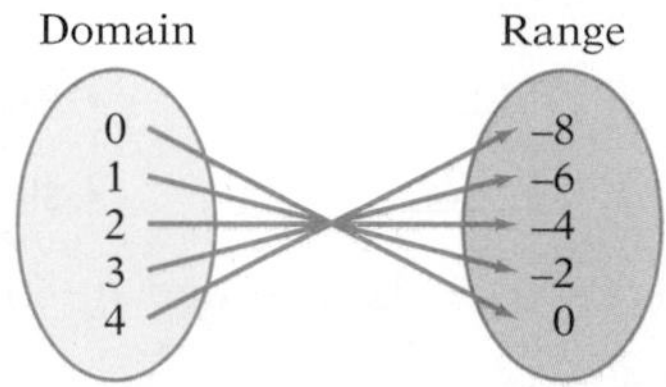

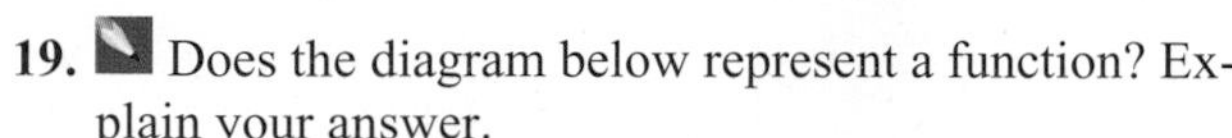
19. Does the diagram below represent a function? Explain your answer.

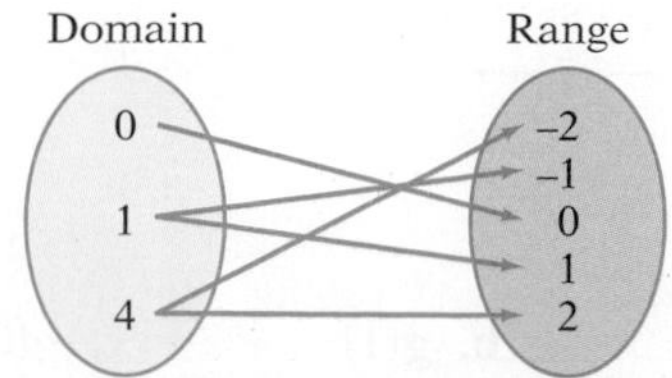

20. Does the diagram below represent a function? Explain your answer.

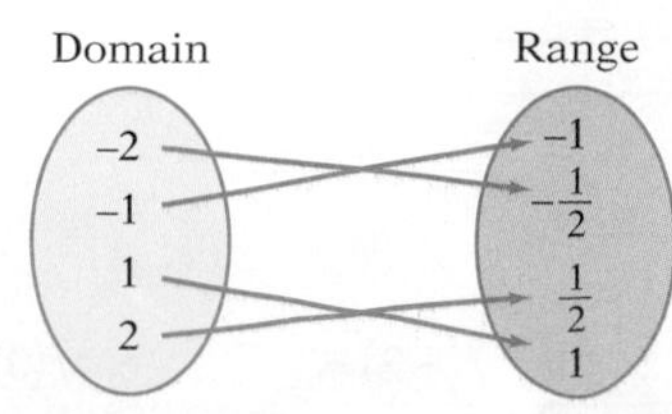

21. Does the diagram below represent a function? Explain your answer.

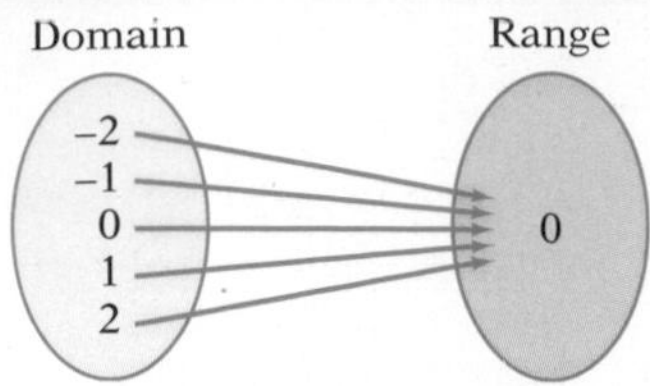

22. Does the diagram below represent a function? Explain your answer.

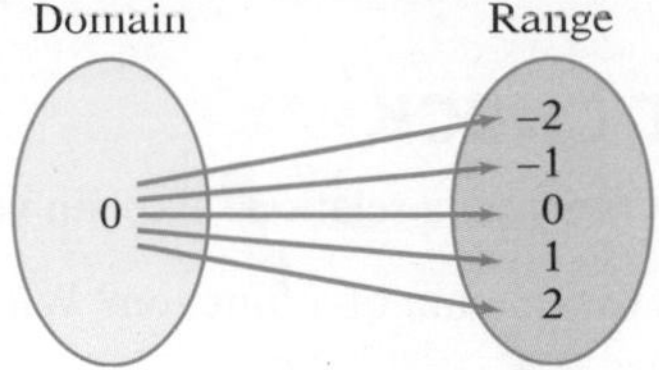

23. **Shipping** Use the table in the news article at the right.
   a. Does the table define a function?
   b. Given $p = 3.15$ lb, find c.
   c. Given $p = 2$ lb, find c.
   d. How much does it cost to send a package that weighs 0.55 lb?

24. **Shipping** The table at the right shows the cost to send an overnight package using United Parcel Service.
   a. Does the table define a function?
   b. Given $x = 3.54$ lb, find y.
   c. Given $x = 2$ lb, find y.
   d. How much does it cost to send a package that weighs 0.35 lb?

Weight, in pounds (x)	Cost (y)
$0 < x \le 1$	\$25.20
$1 < x \le 2$	\$27.40
$2 < x \le 3$	\$29.10
$3 < x \le 4$	\$30.45
$4 < x \le 5$	\$31.75

In the News

Express Mail Rates on the Rise

The U.S. Postal Service is set to raise the shipping costs for Express Mail packages. New rates, like those shown here for Zone 3 packages, go into effect in early January.

Pounds (p)	Cost (c)
$0 < p < 0.5$	\$15.90
$0.5 \le p < 1$	\$20.70
$1 \le p < 2$	\$21.85
$2 \le p < 3$	\$23.20
$3 \le p < 4$	\$24.70

Sources: www.stamps.com, www.usps.com

25. What does it mean to evaluate a function? Explain how to evaluate $f(x) = 3x$ when $x = 2$.

26. What is the value of a function?

27. True or false? If f is a function, then it is possible that $f(0) = -2$ and $f(3) = -2$.

28. True or false? If f is a function, then it is possible that $f(4) = 3$ and $f(4) = 2$.

GETTING READY

29. Given $f(x) = 5x - 7$, find $f(3)$ by completing the following.

$$f(x) = 5x - 7$$
$$f(\underline{\ ?\ }) = 5(\underline{\ ?\ }) - 7 \quad \text{• Replace } x \text{ by 3.}$$
$$f(3) = \underline{\ ?\ } \quad \text{• Simplify.}$$

30. Given $f(x) = x^2 - 3x + 1$, find $f(-2)$ by completing the following.

$$f(x) = x^2 - 3x + 1$$
$$f(\underline{\ ?\ }) = (\underline{\ ?\ })^2 - 3(\underline{\ ?\ }) + 1 \quad \text{• Replace } x \text{ by } -2.$$
$$f(-2) = \underline{\ ?\ } \quad \text{• Simplify.}$$

Evaluate the function.

31. $f(x) = 4x + 5$
   a. $f(2)$ b. $f(-2)$ c. $f(0)$

32. $g(x) = -2x - 7$
   a. $g(-3)$ b. $g(1)$ c. $g(0)$

33. $v(s) = 6 - 3s$

a. $v(3)$ **b.** $v(-2)$ **c.** $v\left(-\frac{2}{3}\right)$

34. $p(z) = 4 - 6z$

a. $p\left(\frac{2}{3}\right)$ **b.** $p(4)$ **c.** $p\left(-\frac{5}{6}\right)$

35. $f(x) = -\frac{3}{2}x - 2$

a. $f(4)$ **b.** $f(-2)$ **c.** $f\left(\frac{4}{3}\right)$

36. $f(x) = \frac{3}{4}x + 1$

a. $f(8)$ **b.** $f(-4)$ **c.** $f\left(-\frac{8}{3}\right)$

37. $p(c) = \frac{1}{2}c - \frac{3}{4}$

a. $p\left(\frac{7}{2}\right)$ **b.** $p\left(-\frac{1}{2}\right)$ **c.** $p\left(\frac{1}{2}\right)$

38. $H(x) = \frac{2}{3} - \frac{5}{6}x$

a. $H\left(-\frac{2}{5}\right)$ **b.** $H(2)$ **c.** $H(0)$

39. $f(x) = 4x - 1$

a. $f(a + 3)$ **b.** $f(2a)$

40. $f(x) = 3x - 2$

a. $f(2 - h)$ **b.** $f(-4h)$

41. $f(x) = 2x^2 - 1$

a. $f(3)$ **b.** $f(-2)$ **c.** $f(0)$

42. $f(x) = 2x^2 + 3$

a. $f(0)$ **b.** $f(-3)$ **c.** $f(2)$

43. $h(t) = 3t^2 - 4t - 5$

a. $h(-2)$ **b.** $h(-1)$ **c.** $h(w)$

44. $p(t) = 3 - 4t - 2t^2$

a. $p(4)$ **b.** $p(-4)$ **c.** $p(0)$

45. $g(x) = x^2 + 2x - 1$

a. $g(1)$ **b.** $g(-3)$ **c.** $g(a)$

46. $g(x) = 2x^2 - 4x + 1$

a. $g(-2)$ **b.** $g(4)$ **c.** $g(z)$

47. $p(t) = 4t^2 - 8t + 3$

a. $p(-2)$ **b.** $p\left(\frac{1}{2}\right)$ **c.** $p(-a)$

48. $r(s) = 3 - 6s - 3s^2$

a. $r(-2)$ **b.** $r\left(-\frac{2}{3}\right)$ **c.** $r(-x)$

49. $f(x) = |x - 3|$

a. $f(-1)$ **b.** $f(5)$ **c.** $f(3)$

50. $h(x) = |2x + 4|$

a. $h(-2)$ **b.** $h(3)$ **c.** $h(-3)$

51. $C(r) = 3|r| - 2$

a. $C(-3)$ **b.** $C(4)$ **c.** $C(0)$

52. $y(x) = 3 - |2x|$

a. $y\left(\frac{1}{2}\right)$ **b.** $y\left(-\frac{3}{2}\right)$ **c.** $y\left(\frac{1}{4}\right)$

53. $K(p) = 5 - 3|p + 2|$

a. $K(-2)$ **b.** $K(-7)$ **c.** $K(3)$

54. $R(s) = 2|1 - s| - 3$

a. $R(-7)$ **b.** $R(4)$ **c.** $R(-1)$

55. Evaluate $s(t) = -16t^2 + 48t$ when $t = 3$.

56. Evaluate $T(s) = s^2 - 4s + 1$ when $s = \frac{3}{4}$.

57. Evaluate $P(x) = 3x^3 - 4x^2 + 6x - 7$ when $x = 2$.

58. Evaluate $R(s) = s^3 - 2s^2 - 5s + 2$ when $s = -3$.

59. Evaluate $R(p) = \frac{3p}{2p - 3}$ when $p = -3$.

60. Evaluate $f(x) = \frac{x + 1}{3x - 1}$ when $x = 3$.

Find the range of the function defined by each equation, for the given domain.

61. $f(x) = 3x - 5$; domain $= \{-3, -2, -1, 0, 1, 2\}$

62. $g(x) = 1 - 2x$; domain $= \{-4, -2, 0, 2, 4, 6\}$

63. $r(t) = \frac{t}{2}$; domain $= \{-3, -2, -1, 0, 1, 2\}$

64. $v(s) = \frac{3}{4}s - 1$; domain $= \{-8, -4, 0, 4, 8\}$

65. $f(x) = x^2 + 3$; domain $= \{-3, -2, -1, 0, 1, 2\}$

66. $r(t) = t^2 - t - 6$; domain $= \{-2, -1, 0, 1, 2, 3\}$

67. $c(n) = n^3 - n - 2$; domain $= \{-3, -2, -1, 0, 1, 2\}$

68. $q(x) = x^3 + 2x^2 - x - 2$; domain $=$ $\{-3, -2, -1, 0, 1, 2\}$

69. Given $f(x) = 2x - 3$, find a number c in the domain of f such that $f(c) = 5$. Write the corresponding ordered pair of the function.

70. Given $f(x) = 3x + 1$, find a number c in the domain of f such that $f(c) = -8$. Write the corresponding ordered pair of the function.

71. Given $f(x) = 1 - 2x$, find a number c in the domain of f such that $f(c) = -7$. Write the corresponding ordered pair of the function.

72. Given $f(x) = -20 - 5x$, find a number c in the domain of f such that $f(c) = -10$. Write the corresponding ordered pair of the function.

73. Given $f(x) = \frac{2}{3}x - 2$, find a number c in the domain of f such that $f(c) = 0$. Write the corresponding ordered pair of the function.

74. Given $f(x) = 3x + 3$, find a number c in the domain of f such that $f(c) = 0$. Write the corresponding ordered pair of the function.

2 Graph of a function (See pages 126–127.)

GETTING READY

75. Let f be the function defined by the equation $f(x) = 3x - 4$. To find the ordered pair of f that has an x-coordinate of -5, evaluate $f(-5)$:

$f(-5) = 3(\underline{\ ?\ }) - 4 = \underline{\ ?\ } - 4 = \underline{\ ?\ }$

The ordered pair of f that has x-coordinate -5 is $\underline{\ ?\ }$.

76. Let g be the function defined by the equation $g(x) = -2|x|$. To find the point on the graph of g that has an x-coordinate of -6, evaluate $g(-6)$:

$g(-6) = -2|\underline{\ ?\ }| = -2(\underline{\ ?\ }) = \underline{\ ?\ }$

The point on the graph of g that has x-coordinate -6 is $\underline{\ ?\ }$.

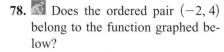

77. Does the ordered pair $(-3, 4)$ belong to the function graphed below?

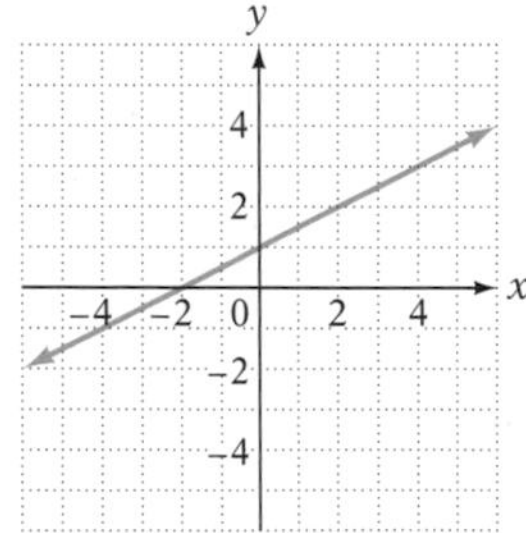

78. Does the ordered pair $(-2, 4)$ belong to the function graphed below?

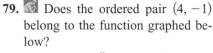

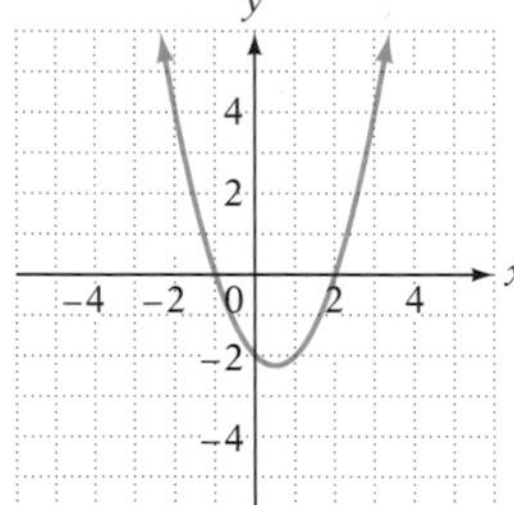

79. Does the ordered pair $(4, -1)$ belong to the function graphed below?

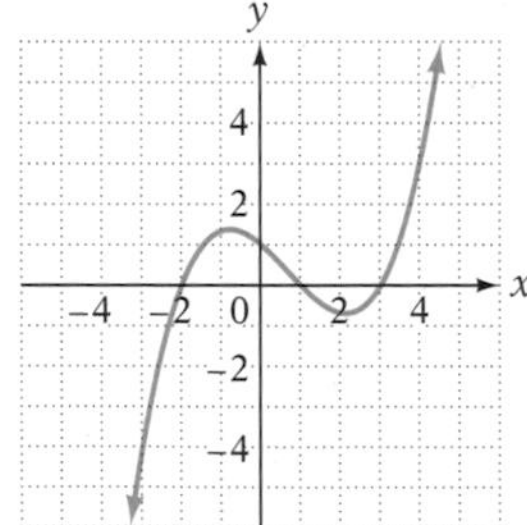

80. If $f(x) = |x| - 5$, does the ordered pair $(-1, -6)$ belong to the function f?

81. If $f(x) = 3x - 4$, does the ordered pair $(-1, -7)$ belong to the function f?

82. If $f(x) = x^2 - 4x$, does the ordered pair $(2, 4)$ belong to the function f?

Graph each function. First evaluate the function at the given values of x. Plot the resulting ordered pairs. Then connect the points to form the graph.

83. $f(x) = x - 3$

$x = -1, 0, 1, 2, 3, 4, 5$

84. $f(x) = -x + 2$

$x = -2, -1, 0, 1, 2, 3, 4$

85. $g(x) = -2x + 2$

$x = -2, -1, 0, 1, 2, 3$

86. $g(x) = 2x - 2$

$x = -2, -1, 0, 1, 2, 3, 4$

87. $h(x) = \frac{1}{2}x - 1$

$x = -6, -4, -2, 0, 2, 4, 6$

88. $h(x) = \frac{2}{3}x + 1$

$x = -6, -3, 0, 3, 6$

89. $f(x) = x^2 - 4$

$x = -3, -2, -1, 0, 1, 2, 3$

90. $g(x) = -x^2 + 5$

$x = -3, -2, -1, 0, 1, 2, 3$

91. $h(x) = x^2 - 4x$

$x = -1, 0, 1, 2, 3, 4, 5$

92. $h(x) = x^2 + 2x$

$x = -4, -3, -2, -1, 0, 1, 2$

93. $g(x) = x^2 + 2x - 3$

$x = -4, -3, -2, -1, 0, 1, 2$

94. $f(x) = -x^2 + 4x + 2$

$x = -1, 0, 1, 2, 3, 4, 5$

95. $f(x) = |x| - 3$

$x = -6, -4, -2, 0, 2, 4, 6$

96. $f(x) = -|x| + 4$

$x = -6, -4, -2, 0, 2, 4, 6$

97. $h(x) = -2|x| + 5$

$x = -3, -2, -1, 0, 1, 2, 3$

98. $h(x) = |2x| - 5$

$x = -5, -3, -1, 0, 1, 3, 5$

99. $f(x) = |x - 2| - 3$

$x = -6, -4, 0, 2, 4, 5, 6$

100. $g(x) = -|x + 2| + 4$

$x = -6, -4, -2, 0, 2, 4, 6$

3 Vertical line test (See pages 127–128.)

GETTING READY

101. Complete the following sentence using *every* or *at least one*. If __?__ vertical line intersects a graph at most once, then the graph is the graph of a function.

102. Complete the following sentence using *x* or *y*. If a vertical line intersects a graph more than once, then at least two points on the graph have the same __?__-coordinate.

Use the vertical line test to determine whether the graph is the graph of a function.

103.

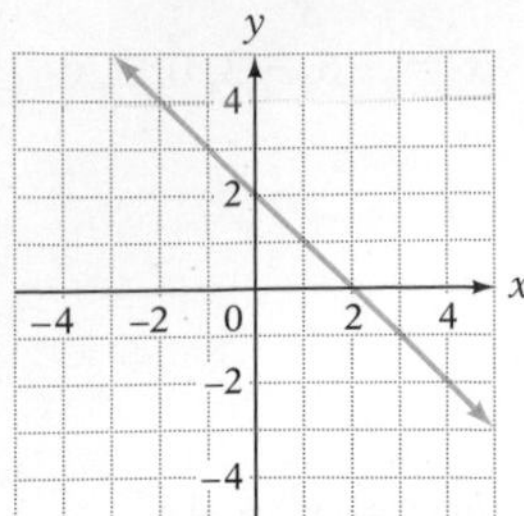

104.

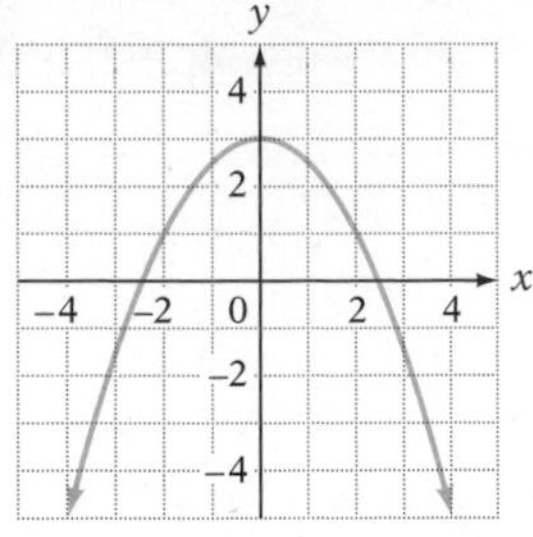

105.

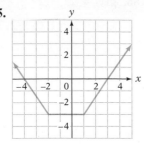

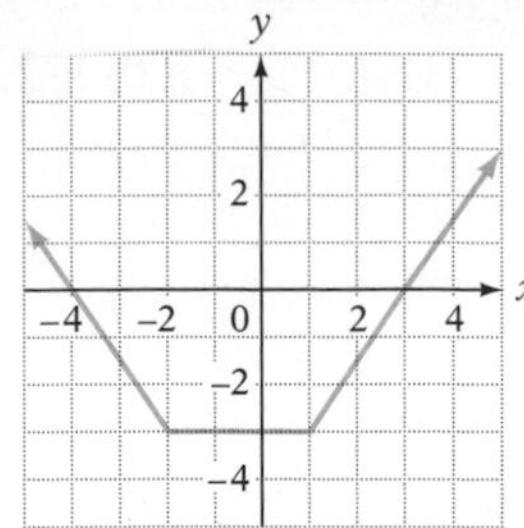

106.

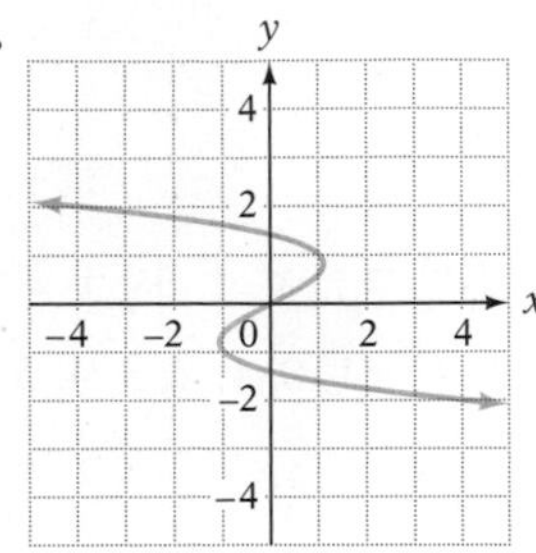

107.

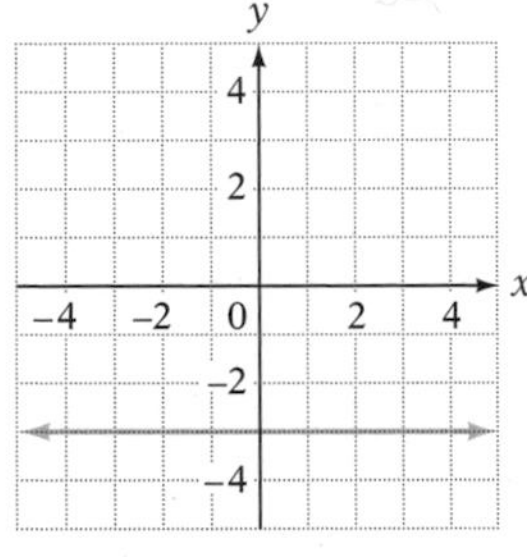

108.

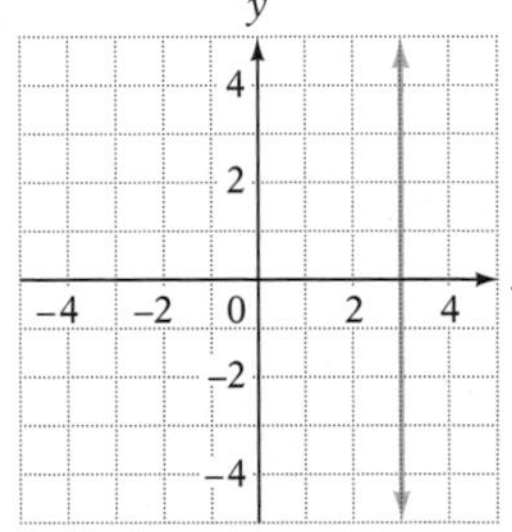

109.

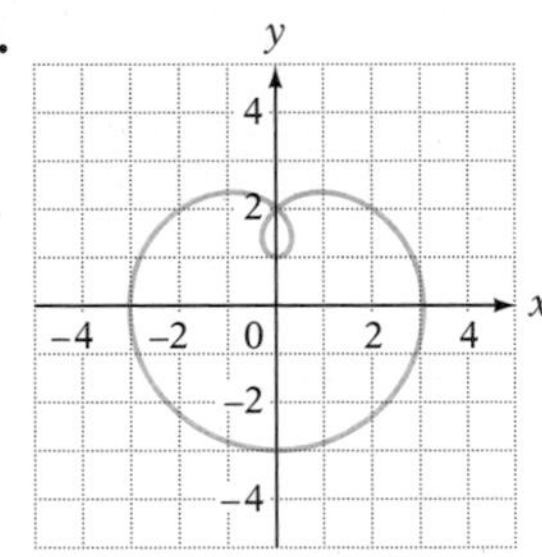

110.

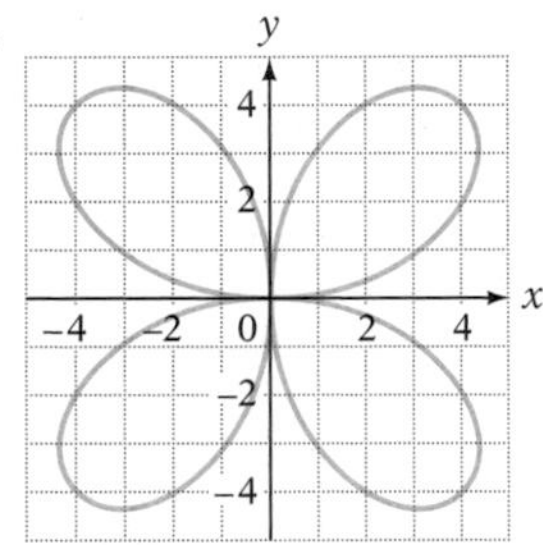

111.

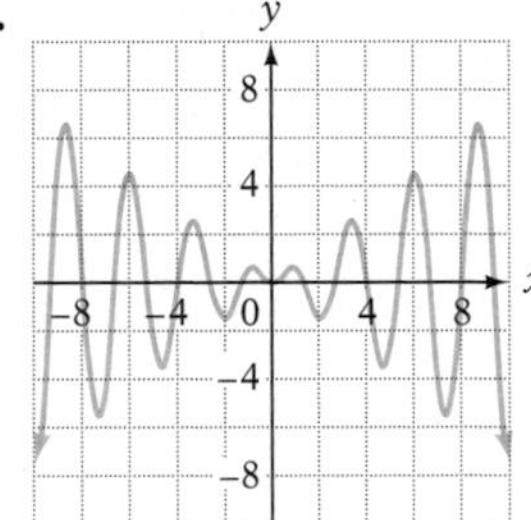

APPLYING CONCEPTS

112. Given $f(x) = |x|$, find all numbers c in the domain of f such that $f(c) = 3$.

113. Given $f(x) = |x - 2|$, find all numbers c in the domain of f such that $f(c) = 4$.

114. Given $f(x) = |x| + 4$, is it possible to find a number c such that $f(c) = 2$? Is 2 in the range of f?

115. Given $f(x) = 4x + 7$, write $f(-2 + h) - f(-2)$ in simplest form.

116. Given $P(x) = 3 - 2x$, write $P(3 + h) - P(3)$ in simplest form.

117. Find the set of ordered pairs (x, y) determined by the equation $|y| = x$, where $x \in \{0, 1, 2, 3\}$. Does the set of ordered pairs define a function? Why or why not?

118. Credit Card Debt The graph below shows the amount of interest a person with credit card debt of $1000 will pay and the time it will take to pay off the debt if that person makes only the minimum monthly payment (and makes no additional charges).

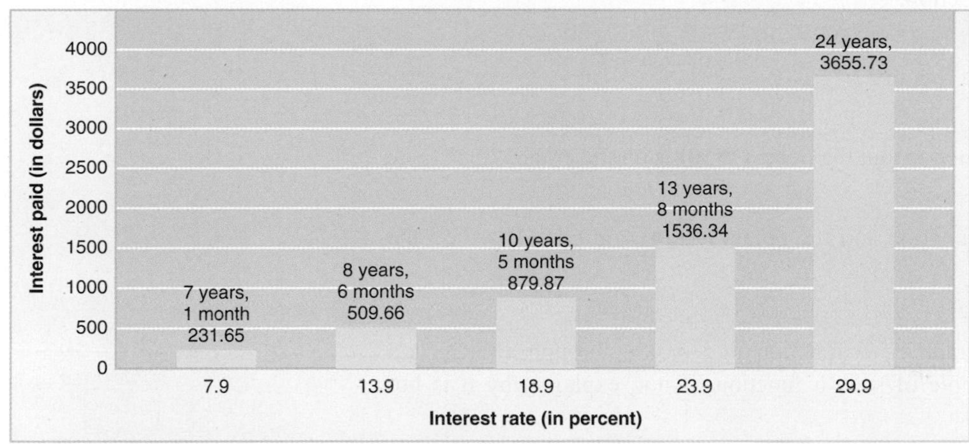

a. How much interest will this person pay if the credit card interest rate is 18.9% (a fairly typical rate)?

b. How many years and months will it take to pay off the debt if the interest rate is 23.9%?

119. Automotive Technology The distance s (in feet) that a car will skid on a certain road surface after the brakes are applied is a function of the car's velocity v (in miles per hour). The function can be approximated by $s = f(v) = 0.017v^2$. How far will a car skid after its brakes are applied if it is traveling 60 mph?

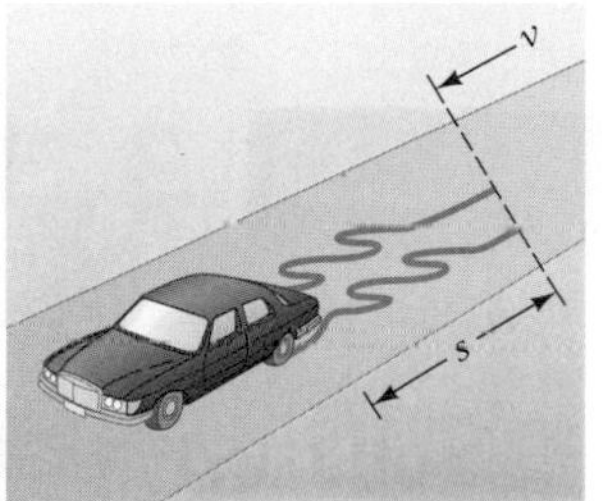

120. Energy The power a windmill can generate is a function of the velocity of the wind. The function can be approximated by $P = f(v) = 0.015v^3$, where P is the power in watts and v is the velocity of the wind in meters per second. How much power will be produced by a windmill when the velocity of the wind is 15 m/s?

Each of the following graphs defines a function. Evaluate the function by estimating the ordinate (which is the value of the function) for the given value of t.

121. Physics The graph at the right shows the temperature T (in degrees Fahrenheit) of a can of cola t hours after it is placed in a refrigerator. Use the graph to estimate the temperature of the cola when:

a. $t = 10$ h

b. $t = 15$ h

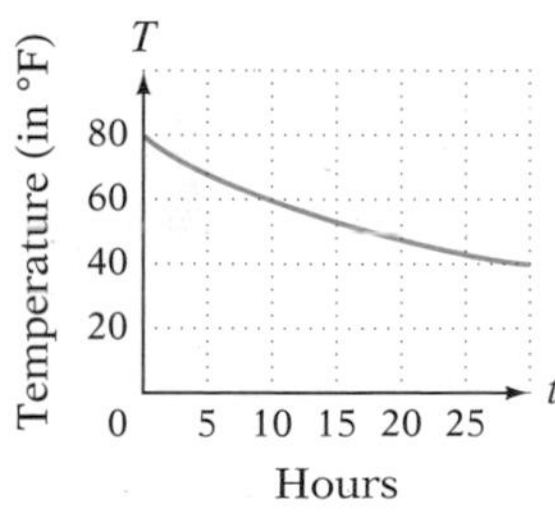

122. Health Science The graph at the right shows the decrease in the heart rate r (in beats per minute) of a runner t minutes after the completion of a race. Use the graph to estimate the heart rate of a runner when:

a. $t = 5$ min

b. $t = 20$ min

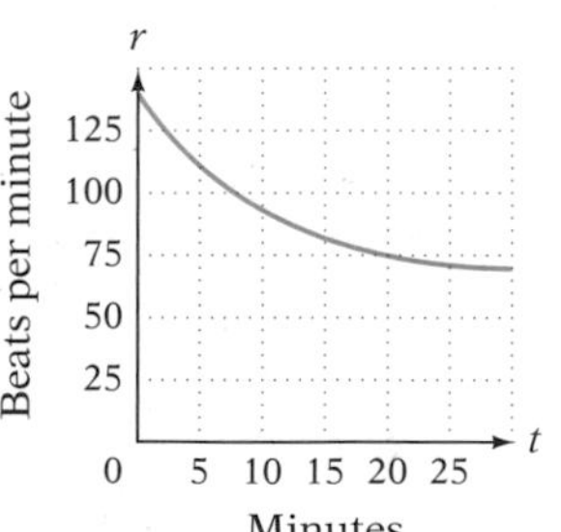

123. Psychology The graph at the right shows what an industrial psychologist has determined to be the average percent score P for an employee taking a performance test t weeks after training begins. Estimate the score an employee would receive on this test when:

a. $t = 4$ weeks

b. $t = 10$ weeks

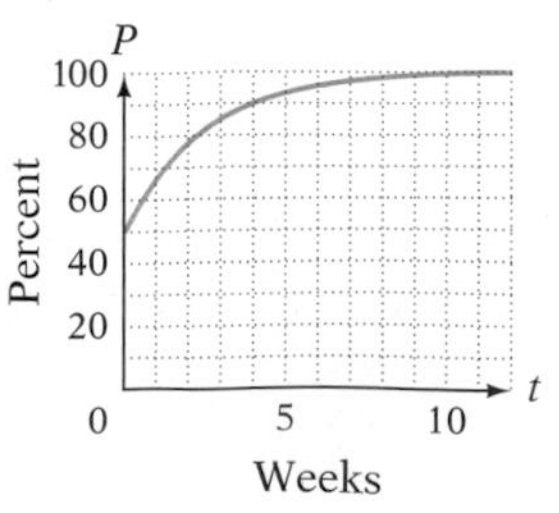

PROJECTS OR GROUP ACTIVITIES

On a coordinate plane, draw a curve that has the indicated property.

124. The curve is the graph of a function.

125. The curve is not the graph of a function.

126. The curve is the graph of a function and the point $P(3, 0)$ is on the graph.

127. The curve is the graph of a function and the points $P(1, 4)$ and $Q(5, 4)$ are on the graph.

128. Is it possible to draw the graph of a function that passes through $P(3, 2)$ and $Q(3, 7)$? If so, draw an example of such a function. If not, explain why it is not possible.

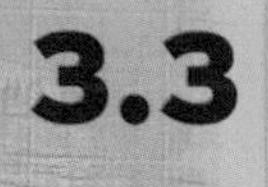

3.3 Linear Functions

OBJECTIVE 1 Graph a linear function

Recall that the ordered pairs of a function can be written as $(x, f(x))$ or (x, y). The **graph of a function** is a graph of the ordered pairs (x, y) that belong to the function. Certain functions have characteristic graphs. A function that can be written in the form $f(x) = mx + b$ (or $y = mx + b$) is called a **linear function** because its graph is a straight line.

Examples of linear functions are shown at the right. Note that the exponent on each variable is 1.

$f(x) = 2x + 5 \quad (m = 2, b = 5)$
$P(t) = 3t - 2 \quad (m = 3, b = -2)$
$y = -2x \quad (m = -2, b = 0)$
$y = -\frac{2}{3}x + 1 \quad \left(m = -\frac{2}{3}, b = 1\right)$
$g(z) = z - 2 \quad (m = 1, b = -2)$

The equation $y = x^2 + 4x + 3$ is not a linear function because it includes a term with a variable squared. The equation $f(x) = \frac{3}{x - 2}$ is not a linear function because a variable occurs in the denominator. Another example of an equation that is not a linear function is $y = \sqrt{x} + 4$; this equation contains a variable within a radical and so is not a linear function.

Consider $f(x) = 2x + 1$. Evaluating the linear function when $x = -3, -2, -1, 0, 1,$ and 2 produces some of the ordered pairs of the function. It is convenient to record the results in a table similar to the one at the right. The graph of the ordered pairs is shown in Figure 1 on the next page.

x	$f(x) = 2x + 1$	y	(x, y)
-3	$f(-3) = 2(-3) + 1 = -5$	-5	$(-3, -5)$
-2	$f(-2) = 2(-2) + 1 = -3$	-3	$(-2, -3)$
-1	$f(-1) = 2(-1) + 1 = -1$	-1	$(-1, -1)$
0	$f(0) = 2(0) + 1 = 1$	1	$(0, 1)$
1	$f(1) = 2(1) + 1 = 3$	3	$(1, 3)$
2	$f(2) = 2(2) + 1 = 5$	5	$(2, 5)$

Evaluating the function when x is not an integer produces more ordered pairs to graph, such as $\left(-\frac{5}{2}, -4\right)$ and $\left(\frac{3}{2}, 4\right)$, as shown in Figure 2 below. Evaluating the function for still other values of x would result in more and more ordered pairs to graph. The result would be so many dots that the graph would look like the straight line shown in Figure 3, which is the graph of $f(x) = 2x + 1$.

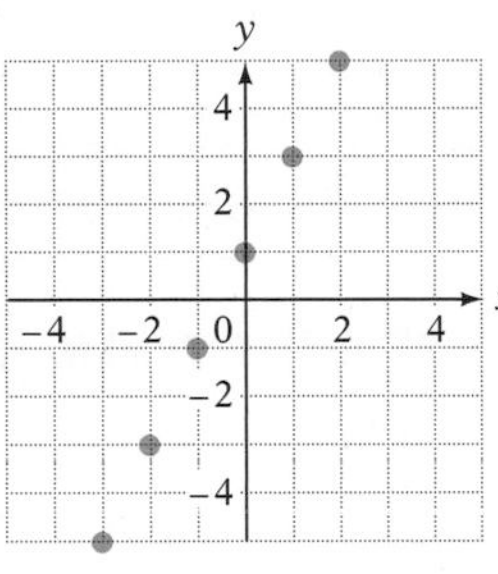

FIGURE 1

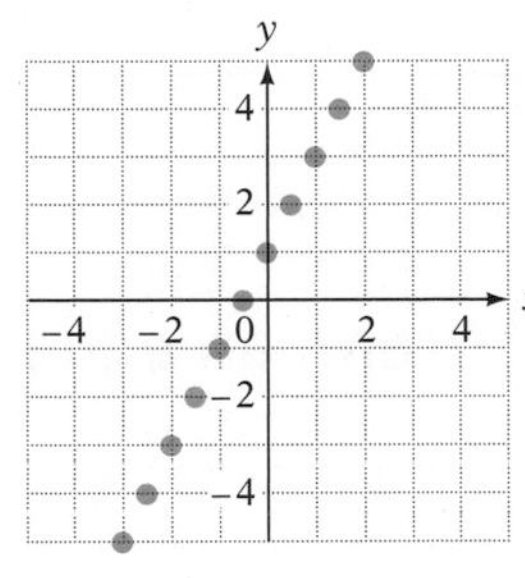

FIGURE 2

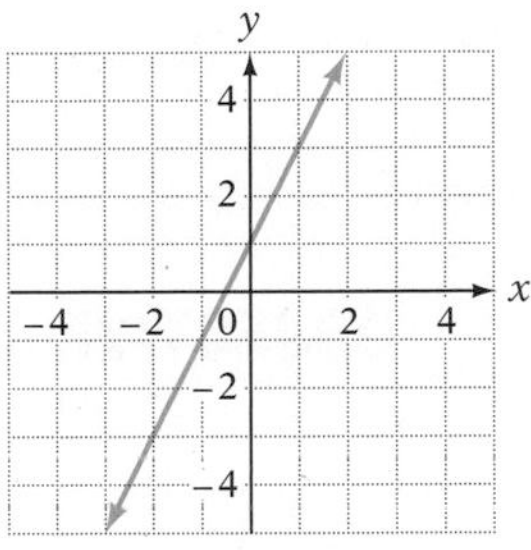

FIGURE 3

No matter what value of x is chosen, $2x + 1$ is a real number. This means that the domain of $f(x) = 2x + 1$ is all real numbers. Therefore, we can use any real number when evaluating the function. Normally, however, values such as π and $\sqrt{5}$ are not used because it is difficult to graph the resulting ordered pairs.

Whether an equation is written as $f(x) = mx + b$ or as $y = mx + b$, the equation represents a linear function, and the graph of the equation is a straight line.

Because the graph of a linear function is a straight line, and a straight line is determined by two points, the graph of a linear function can be drawn by finding only two of the ordered pairs of the function. However, it is recommended that you find at least *three* ordered pairs to ensure accuracy.

Take Note

When the coefficient of x is a fraction, choose values of x that are multiples of the denominator of the fraction. This will result in coordinates that are integers.

EXAMPLE 1 Graph: $f(x) = -\frac{3}{2}x - 3$

Solution

x	$y = f(x) = -\frac{3}{2}x - 3$
0	−3
−2	0
−4	3

- **Find at least three ordered pairs. Because the coefficient of x is a fraction with denominator 2, choosing values of x that are divisible by 2 simplifies the evaluations. The ordered pairs can be displayed in a table.**

- **Graph the ordered pairs and draw a line through the points.**

Problem 1 Graph: $y = \frac{3}{5}x - 4$

Solution See page S8.

Try Exercise 15, page 144.

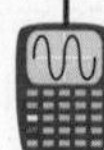

Graphing utilities create graphs by plotting points and then connecting the points to form a curve. Using a graphing utility, enter the equation $y = -\frac{3}{2}x - 3$ and verify the graph drawn in Example 1. (Refer to the Appendix for suggestions on what domain and range to use.) Trace along the graph and verify that $(0, -3)$, $(-2, 0)$, and $(-4, 3)$ are the coordinates of points on the graph. Now enter the equation $y = \frac{3}{5}x - 4$ given in Problem 1. Verify that the ordered pairs you found for this function are coordinates of points on the graph.

OBJECTIVE 2 Graph an equation of the form $Ax + By = C$

A **literal equation** is an equation with more than one variable. Examples of literal equations are $P = 2L + 2W$, $V = LWH$, $d = rt$, and $3x + 2y = 6$.

Linear equations of the form $y = mx + b$ are literal equations. In some cases, a linear equation has the form $Ax + By = C$. In such a case, it may be convenient to solve the equation for y to write the equation in the form $y = mx + b$. To solve for y, we use the same rules and procedures that we use to solve equations with numerical values.

Focus on solving an equation for y

Write $4x - 3y = 6$ in the form $y = mx + b$.

$$4x - 3y = 6$$

Subtract $4x$ from each side of the equation.

$$4x - 4x - 3y = 6 - 4x$$

Simplify.

$$-3y = 6 - 4x$$

Divide each side of the equation by -3.

$$\frac{-3y}{-3} = \frac{6 - 4x}{-3}$$

Simplify.

$$y = \frac{6 - 4x}{-3}$$

Divide each term in the numerator by the denominator.

$$y = \frac{6}{-3} - \frac{4x}{-3}$$

Simplify.

$$y = -2 + \frac{4}{3}x$$

Write the equation in the form $y = mx + b$.

$$y = \frac{4}{3}x - 2$$

We will show two methods of graphing an equation of the form $Ax + By = C$. In the first method, we solve the equation for y and then follow the same procedure used for graphing an equation of the form $y = mx + b$.

EXAMPLE 2 Graph: $3x + 2y = 6$

Solution

$$3x + 2y = 6$$

$$2y = -3x + 6$$ • **Solve the equation for y.**

$$y = -\frac{3}{2}x + 3$$

x	$y = -\frac{3}{2}x + 3$
0	3
2	0
4	−3

- **Find at least three solutions. Choose multiples of 2 for x.**

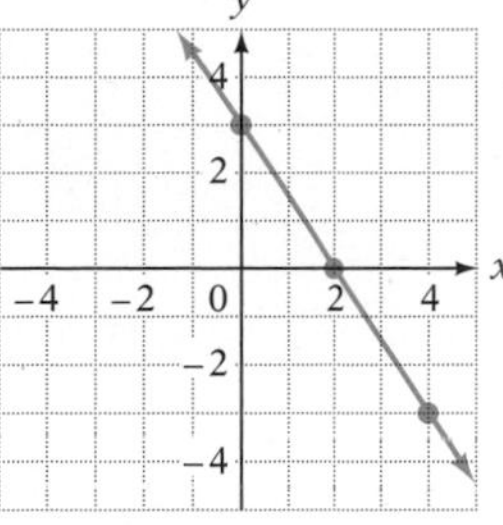

- **Graph the ordered pairs and draw a straight line through the points.**

Problem 2 Graph: $-3x + 2y = 4$

Solution See page S8.

Try Exercise 23, page 145.

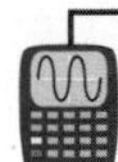

Use a graphing utility to graph the equation $3x + 2y = 6$. First solve the equation for y and then enter the equation $y = -\frac{3}{2}x + 3$. Next trace along the graph and verify that $(0, 3)$, $(2, 0)$, and $(4, -3)$ are coordinates of points on the graph. Follow the same procedure for Problem 2.

An equation in which one of the variables is missing has a graph that is either a horizontal or a vertical line.

The equation $y = -2$ can be written

$$0 \cdot x + y = -2$$

Because $0 \cdot x = 0$ for any value of x, y is -2 for every value of x.

Some of the possible ordered-pair solutions of $y = -2$ are given in the following table. The graph is shown at the right.

x	$y = -2$
−2	−2
0	−2
3	−2

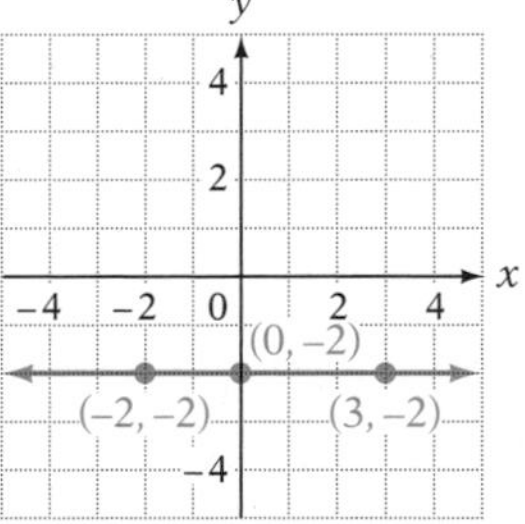

The equation $y = -2$ represents a function. Some of the ordered pairs of this function are $(-2, -2)$, $(0, -2)$, and $(3, -2)$. In function notation, we would write $f(x) = -2$. This function is an example of a *constant function.* No matter what value of x is selected, $f(x) = -2$.

DEFINITION OF A CONSTANT FUNCTION

A function given by $f(x) = b$, where b is a constant, is a **constant function.** The graph of a constant function is a horizontal line passing through $(0, b)$.

Focus on graphing a constant function

Graph: $y + 4 = 0$

Solve for y.

$$y + 4 = 0$$
$$y = -4$$

The graph of $y = -4$ is a horizontal line passing through $(0, -4)$.

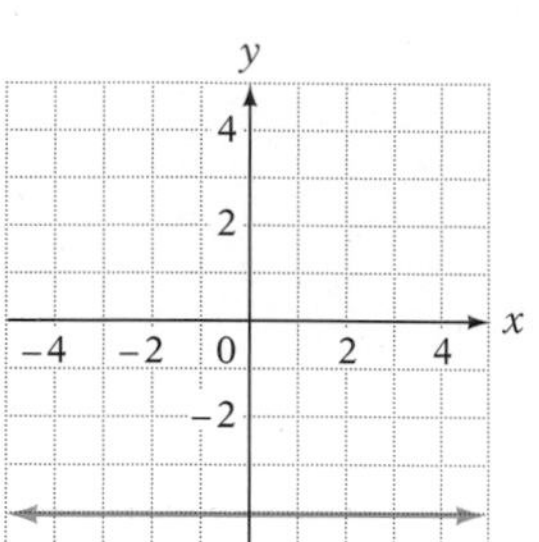

For each value in the domain of a constant function, the value of the function is the same (that is, it is constant). For instance, if $f(x) = 4$, then $f(2) = 4$, $f(3) = 4$, $f(\sqrt{3}) = 4$, $f(\pi) = 4$, and so on. The value of $f(x)$ is 4 for all values of x.

Focus on evaluating a constant function

Evaluate $P(t) = -7$ when $t = 6$.

The value of the constant function is the same for all values of the variable.

$$P(t) = -7$$
$$P(6) = -7$$

For the equation $y = -2$, the coefficient of x is zero. For the equation $x = 2$, the coefficient of y is zero. For instance, the equation $x =$ can be written

$$x + 0 \cdot y = 2$$

No matter what value of y is chosen, $0 \cdot y = 0$, and therefore x is always 2.

Some of the possible ordered-pair solutions are given in the following table. The graph is shown at the right.

x	y
2	6
2	1
2	−4

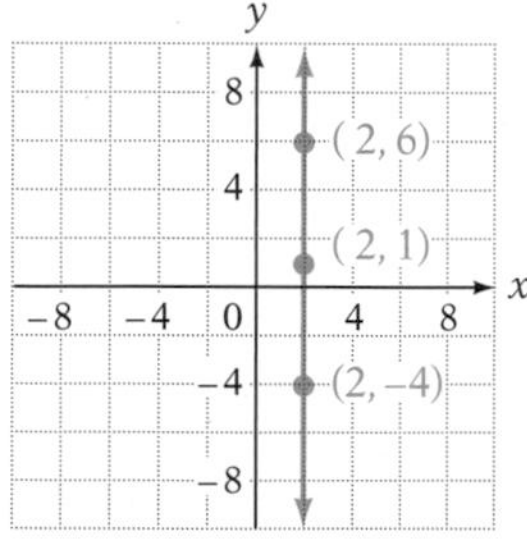

Take Note

The equation $y = b$ represents a function. The equation $x = a$ does *not* represent a function. Remember, not all equations represent functions.

GRAPH OF $x = a$

The graph of $x = a$ is a vertical line passing through the point $(a, 0)$.

Recall that a function is a set of ordered pairs in which no two ordered pairs have the same first coordinate. Because $(2, 6)$, $(2, 1)$, and $(2, -4)$ are ordered-pair solutions of the equation $x = 2$, this equation does not represent a function, and its graph is not the graph of a function.

EXAMPLE 3 Graph: $x = -4$

Solution

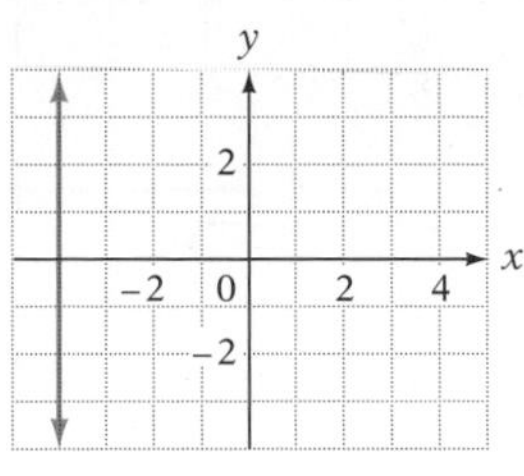

- **The graph of an equation of the form $x = a$ is a vertical line passing through the point whose coordinates are $(a, 0)$. The graph of $x = -4$ passes through $(-4, 0)$.**

Problem 3 Graph: $y - 3 = 0$

Solution See page S8.

Try Exercise 27, page 145.

A second method of graphing straight lines uses the *intercepts* of the graph.

The graph of the equation $x - 2y = 4$ is shown at the right. The graph crosses the x-axis at the point with coordinates $(4, 0)$. This point is called the **x-intercept.** The graph crosses the y-axis at the point with coordinates $(0, -2)$. This point is called the **y-intercept.**

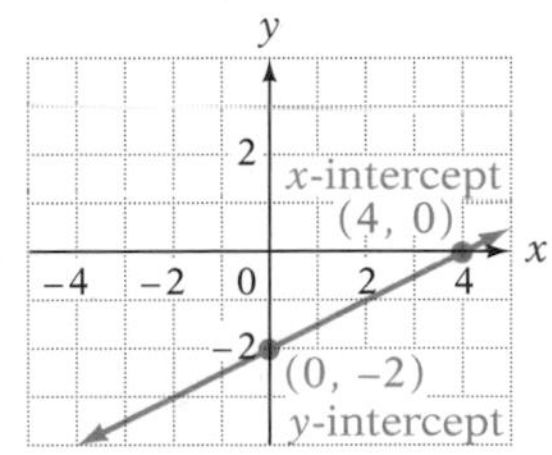

Focus on finding x- and y-intercepts

Take Note

The x-intercept occurs when $y = 0$. The y-intercept occurs when $x = 0$.

Find the coordinates of the x- and y-intercepts of the graph of the equation $3x + 4y = -12$.

To find the x-intercept, let $y = 0$. (Any point on the x-axis has y-coordinate 0.)

$$\begin{aligned} 3x + 4y &= -12 \\ 3x + 4(0) &= -12 \\ 3x &= -12 \\ x &= -4 \end{aligned}$$

The coordinates of the x-intercept are $(-4, 0)$.

To find the y-intercept, let $x = 0$. (Any point on the y-axis has x-coordinate 0.)

$$\begin{aligned} 3x + 4y &= -12 \\ 3(0) + 4y &= -12 \\ 4y &= -12 \\ y &= -3 \end{aligned}$$

The coordinates of the y-intercept are $(0, -3)$.

A linear equation can be graphed by finding the x- and y-intercepts and then drawing a line through the two points.

EXAMPLE 4 Graph $4x - y = 4$ by using the x- and y-intercepts.

Solution

x-intercept:
$$\begin{aligned} 4x - y &= 4 \\ 4x - 0 &= 4 \\ 4x &= 4 \\ x &= 1 \end{aligned}$$

- **To find the x-intercept, let $y = 0$.**

The coordinates of the x-intercept are $(1, 0)$.

y-intercept:
$$\begin{aligned} 4x - y &= 4 \\ 4(0) - y &= 4 \\ -y &= 4 \\ y &= -4 \end{aligned}$$

- **To find the y-intercept, let $x = 0$.**

The coordinates of the y-intercept are $(0, -4)$.

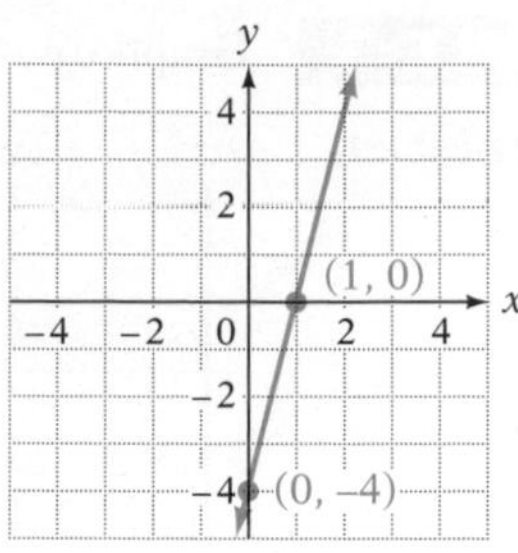

• **Graph the x- and y-intercepts. Draw a line through the two points.**

Problem 4 Graph $3x - y = 2$ by using the x- and y-intercepts.

Solution See page S8.

Try Exercise 37, page 145.

The graph of $f(x) = 2x + 4$ is shown at the right.

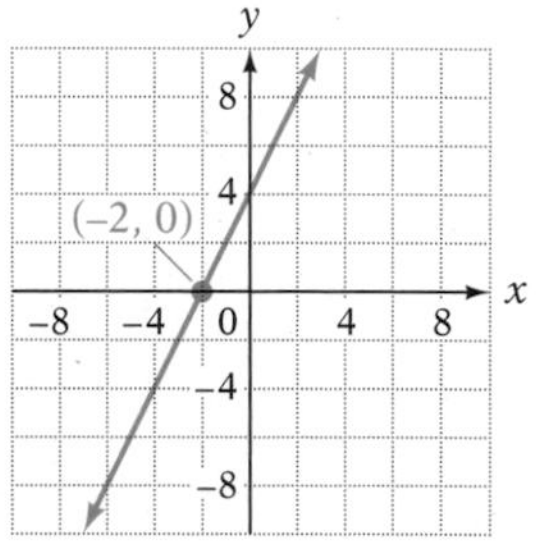

Evaluating the function when $x = -2$, we have

$$f(x) = 2x + 4$$
$$f(-2) = 2(-2) + 4 = -4 + 4$$
$$f(-2) = 0$$

-2 is the x value for which $f(x) = 0$. A value of x for which $f(x) = 0$ is called a *zero* of f.

Note that the x-intercept of the graph has coordinates $(-2, 0)$. The x-coordinate of the x-intercept is -2, the zero of the function.

ZERO OF A FUNCTION

A value of x for which $f(x) = 0$ is called a **zero** of f.

EXAMPLES

1. Let $f(x) = 3x - 6$ and $x = 2$.

$$f(2) = 3(2) - 6 = 0$$

Because $f(2) = 0$, 2 is a zero of f.

2. Let $g(x) = 4x + 8$ and $x = -2$.

$$g(-2) = 4(-2) + 8$$
$$= -8 + 8 = 0$$

Because $g(-2) = 0$, -2 is a zero of g.

3. Let $h(x) = 2x + 1$ and $x = 0$.

$$h(0) = 2(0) + 1$$
$$= 0 + 1$$
$$= 1 \neq 0$$

Because $h(0) \neq 0$, 0 is not a zero of h.

To find a zero of a function f, let $f(x) = 0$ and solve for x.

EXAMPLE 5 Find the zero of $f(x) = -2x - 3$.

Solution

$$f(x) = -2x - 3$$

$$0 = -2x - 3 \quad \text{• To find a zero of a function, let } f(x) = 0.$$

$$2x = -3 \quad \text{• Solve for } x.$$

$$x = -\frac{3}{2}$$

The zero is $-\frac{3}{2}$. The graph of f is shown at the left. Note that the x-coordinate of the x-intercept is the zero of f.

Problem 5 Find the zero of $f(x) = \frac{2}{3}x + 4$.

Solution See page S8.

Try Exercise 59, page 146.

OBJECTIVE 3 Application problems

EXAMPLE 6 On the basis of data from *The Joy of Cooking,* the daily caloric allowance for a woman can be approximated by the equation $C = -7.5A + 2187.5$, where C is the caloric intake and A is the age of the woman. Graph this equation for $25 \le A \le 75$. The point whose coordinates are (45, 1850) is on the graph. Write a sentence that describes the meaning of this ordered pair.

Solution

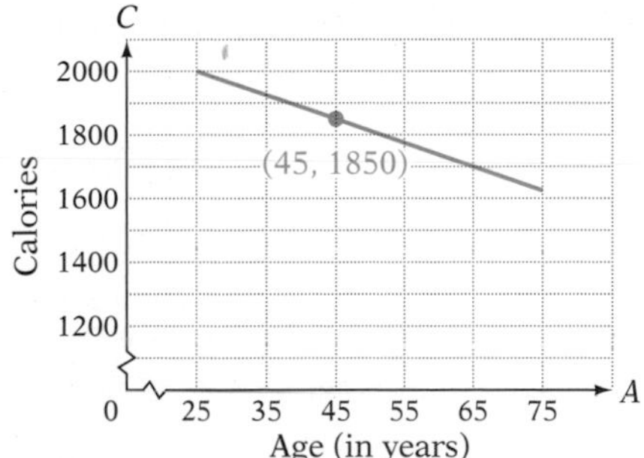

The ordered pair (45, 1850) means that the caloric allowance for a 45-year-old woman is 1850 calories per day.

Problem 6 The height h (in inches) of a person and the length L (in inches) of that person's stride while walking are related. The equation $h = \frac{3}{4}L + 50$ approximates the relationship. Graph this equation for $15 \le L \le 40$. The point whose coordinates are (32, 74) is on the graph. Write a sentence that describes the meaning of this ordered pair.

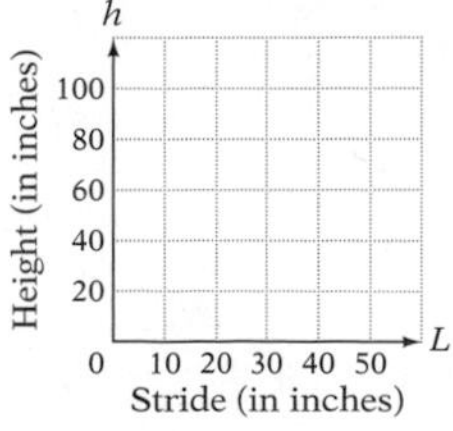

Solution See page S8.

Try Exercise 69, page 146.

3.3 Exercises

CONCEPT CHECK

State whether the graph of the given equation is a straight line.

1. $y = 1 - x$

2. $f(x) = x$

3. $g(x) = 0.2x + 3$

4. $y = 1 - x^2$

5. $f(x) = \frac{2}{x} + 3$

6. $y = 2x + \frac{2}{3}$

7. Explain how to graph a linear function by plotting points.

8. Explain what the graph of an equation represents.

1 Graph a linear function (See pages 136–138.)

GETTING READY

9. Find three ordered pairs on the graph of $y = 2x - 5$ by finding the y values that correspond to x values of -1, 0, and 1.

a. When $x = -1$, $y =$ __?__. An ordered pair on the graph is (__?__, __?__).

b. When $x = 0$, $y =$ __?__. An ordered pair on the graph is (__?__, __?__).

c. When $x = 1$, $y =$ __?__. An ordered pair on the graph is (__?__, __?__).

10. To find ordered pairs on the graph of $y = -\frac{1}{4}x + 1$, it is helpful to choose x values that are divisible by __?__.

Graph.

11. $y = 3x - 4$

12. $y = -2x + 3$

13. $y = -\frac{2}{3}x$

14. $y = \frac{3}{2}x$

15. $y = \frac{2}{3}x - 4$

16. $y = \frac{3}{4}x + 2$

17. $y = -\frac{1}{3}x + 2$

18. $y = -\frac{3}{2}x - 3$

19. $y = \frac{3}{5}x - 1$

20. $y = 3 - \frac{1}{2}x$

21. $f(x) = 4 - \frac{2x}{3}$

22. $g(x) = 2 + \frac{x}{4}$

2 Graph an equation of the form $Ax + By = C$ (See pages 138–143.)

Graph.

23. $2x + y = -3$

24. $2x - y = 3$

25. $x - 4y = 8$

26. $2x + 5y = 10$

27. $y = 1$

28. $y = -2$

29. $x = 4$

30. $x = \quad 3$

31. $2x - 3y = 12$

32. $x - 3y = 0$

33. $3x - 2y = 8$

34. $3x - y = -2$

GETTING READY

35. The x-intercept of a graph of a linear equation is the point where the graph crosses the ___?___. Its ___?___-coordinate is 0.

36. The y-intercept of a graph of a linear equation is the point where the graph crosses the ___?___. Its ___?___-coordinate is 0.

Find the coordinates of the x- and y-intercepts of the graph of each equation. Then graph the equation.

37. $x - 2y = -4$

38. $3x + y = 3$

39. $2x - 3y = 9$

40. $4x - 2y = 5$

41. $2x - y = 4$

42. $2x + y = 3$

43. $3x + 2y = 5$

44. $4x - 3y = 8$

45. $3x + 2y = 4$

46. $2x - 3y = 4$

47. $3x - 5y = 9$

48. $4x - 3y = 6$

49. The y-intercept of the graph of $Ax + By = C$ is below the x-axis. Are the signs of B and C the same or opposite?

50. The x-intercept of the graph of $Ax + By = C$ is to the right of the y-axis. Are the signs of A and C the same or opposite?

51. The x-intercept of the graph of $f(x) = -\frac{4}{3}x - 4$ has coordinates $(-3, 0)$. What is the zero of f?

52. The x-intercept of the graph of $g(x) = -\frac{2}{5}x + 4$ has coordinates $(10, 0)$. What is the zero of g?

From the graph, estimate the zero of the function.

53.
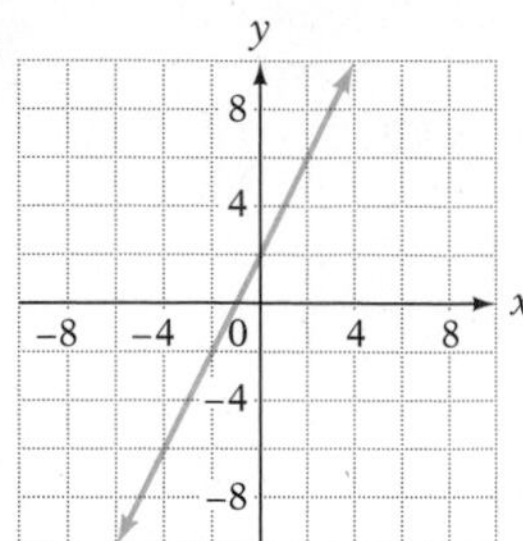

54.
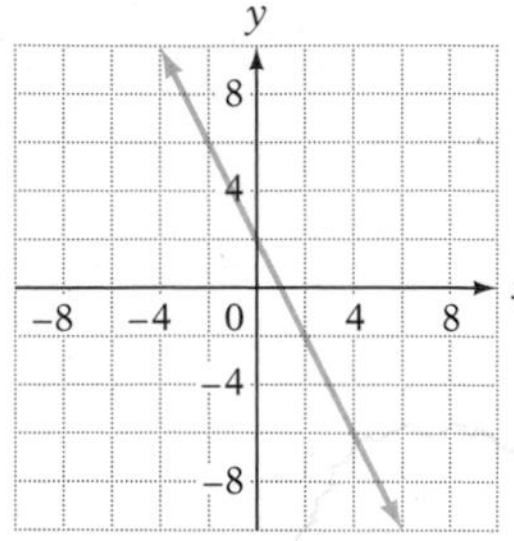

55.
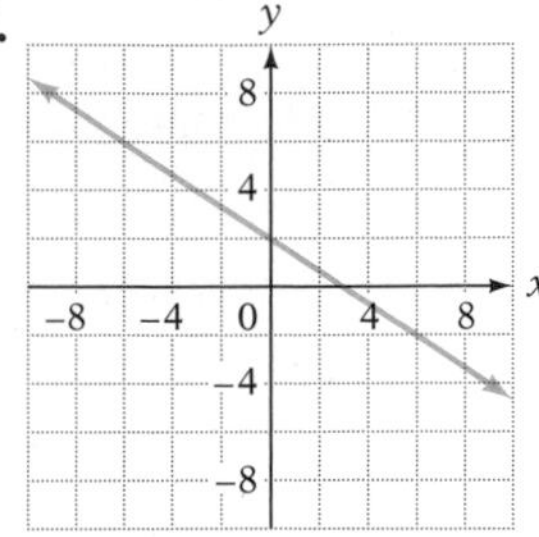

56.
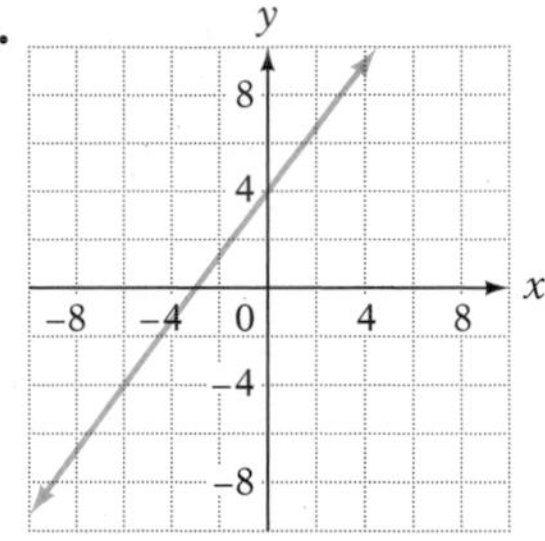

Find the zero of each linear function.

57. $f(x) = 2x - 6$

58. $g(x) = 3x + 6$

59. $g(x) = -2x + 5$

60. $f(x) = -3x - 9$

61. $y(x) = \frac{3}{4}x - 6$

62. $s(t) = -\frac{2}{5}t + 4$

63. $f(t) = \frac{t}{2} - 3$

64. $f(x) = \frac{3x}{2} + 3$

65. $g(x) = 3x + 8$

66. $h(z) = \frac{2}{3}z - 3$

67. $f(x) = -\frac{3}{4}x + 2$

68. $g(x) = \frac{5}{2}x + 4$

3 Application problems (See page 143.)

69. **Amusement Parks** The Kingda Ka roller coaster at Six Flags Great Adventure in Jackson, N.J., has a maximum speed of 188 ft/s. The equation that describes the total number of feet traveled by the roller coaster in t seconds at this speed is given by $D = 188t$. Use the coordinate axes at the right to graph this equation for $0 \le t \le 6$. The point whose coordinates are $(5, 940)$ is on this graph. Write a sentence that describes the meaning of this ordered pair.

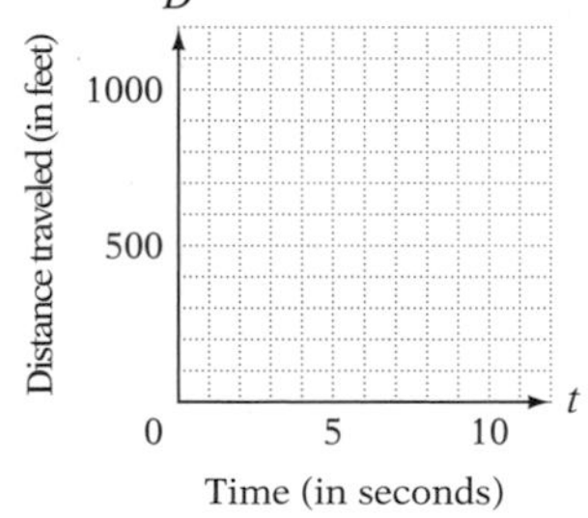

70. **Sports** A tennis pro is paid $40 an hour for a private lesson by members of a condominium group. The equation that describes the total earnings E (in dollars) received by the pro is $E = 40t$, where t is the total number of hours worked. Use the coordinate axes at the right to graph this equation for $0 \le t \le 30$. The point whose coordinates are $(25, 1000)$ is on the graph. Write a sentence that describes the meaning of this ordered pair.

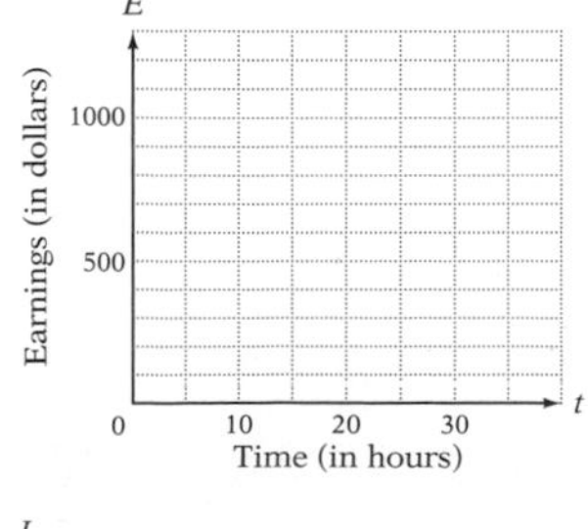

71. **Real Estate** A realtor receives $400 per month plus 6% commission on sales. The equation that describes the total monthly income I (in dollars) of the realtor is $I = 0.06s + 400$, where s is the amount of sales. Use the coordinate axes at the right to graph this equation for $0 \le s \le 150{,}000$. The point whose coordinates are $(60{,}000, 4000)$ is on the graph. Write a sentence that describes the meaning of this ordered pair.

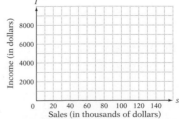

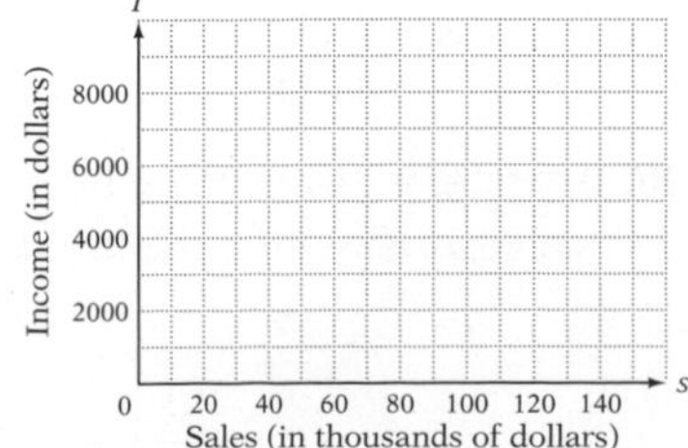

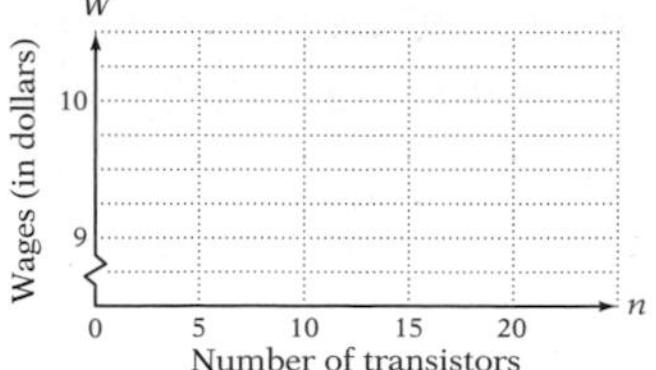

72. **Human Resources** An electronics company pays line workers \$9 per hour plus \$.05 for each transistor produced. The equation that describes the hourly wage W (in dollars) is $W = 0.05n + 9$, where n is the number of transistors produced during any given hour. Use the coordinate axes at the right to graph this equation for $0 \le n \le 20$. The point whose coordinates are $(16, 9.80)$ is on the graph. Write a sentence that describes the meaning of this ordered pair.

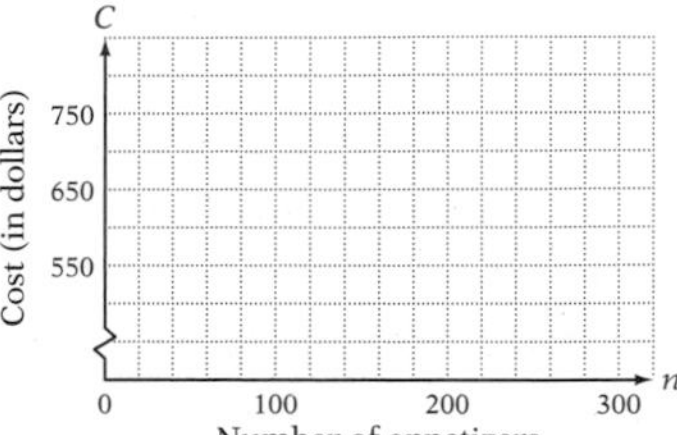

73. **Food Science** A caterer charges a flat rate of \$500 plus \$.95 for each hot appetizer. The equation that describes the cost C (in dollars) of catering a dinner party is $C = 0.95n + 500$, where n is the number of hot appetizers. Use the coordinate axes at the right to graph this equation for $0 \le n \le 300$. The point whose coordinates are $(120, 614)$ is on the graph. Write a sentence that describes the meaning of this ordered pair.

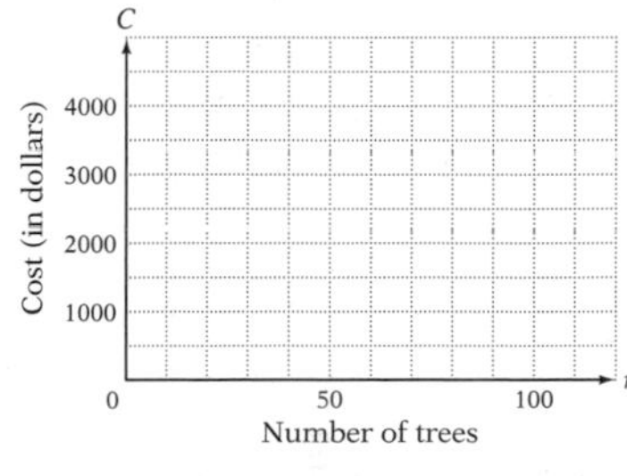

74. **Environmental Science** A tree service charges \$60 plus \$35 for each tree removed. The equation that describes the cost C (in dollars) of tree removal is $C = 35n + 60$, where n is the number of trees removed. Use the coordinate axes at the right to graph this equation for $0 \le n \le 100$. The point whose coordinates are $(50, 1810)$ is on the graph. Write a sentence that describes the meaning of this ordered pair.

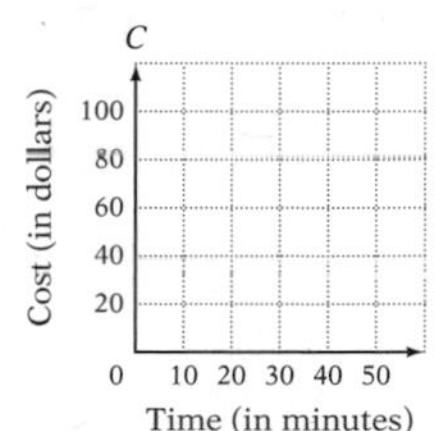

75. **Home Maintenance** An electronics technician charges \$45 plus \$1 per minute to repair defective wiring in a home or apartment. The equation that describes the total cost C (in dollars) to have defective wiring repaired is $C = t + 45$, where t is the number of minutes the technician works. Use the coordinate axes at the right to graph this equation for $0 \le t \le 60$. The point whose coordinates are $(15, 60)$ is on the graph. Write a sentence that describes the meaning of this ordered pair.

76. **Oceanography** Read the news article below about the small submarine *Alvin* used by scientists to explore the ocean.

In the News

***Alvin* Scientists Get New Digs**

In *Alvin,* the original Human Occupied Vehicle (HOV) built in 1964 for deep-sea exploration, scientists can descend at a rate of 30 m/min to a maximum depth of 4500 m. As part of a plan to create an HOV that can dive deeper and stay underwater longer, *Alvin* now, after more than 40 years, has a larger, stronger sphere to house its scientists.

Source: Woods Hole Oceanographic Institute

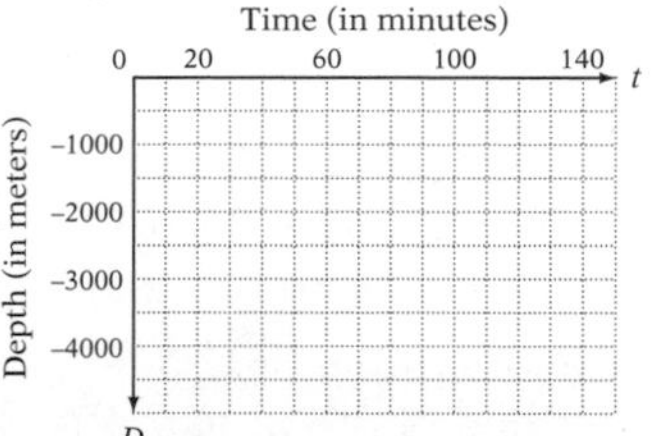

The equation that describes *Alvin's* depth D, in meters, is $D = -30t$, where t is the number of minutes *Alvin* has been descending. Use the coordinate axes at the right to graph this equation for $0 \le t \le 150$. The point whose coordinates are $(65, -1950)$ is on the graph. Write a sentence that describes the meaning of this ordered pair.

APPLYING CONCEPTS

77. **Business** The sale price of an item is a function s of the original price p, where $s(p) = 0.80p$. If an item's original price is \$200, what is the sale price of the item?

78. **Business** The markup of an item is a function m of its cost c, where $m(c) = 0.25c$. If the cost of an item is \$150, what is the markup of the item?

Use a graphing utility to draw the graph of each of the following equations.

79. $f(x) = 1.2x + 2.3$

80. $f(x) = 2.4x + 0.5$

81. $f(x) = \frac{2x}{3} - \frac{5}{3}$

82. $y = -\frac{3x}{4} - \frac{5}{2}$

83. $3x - y = 4$

84. $2x + y = 3$

85. Explain the relationship between the zero of a linear function and the x-intercept of the graph of that function.

86. There is a relationship between the number of times a cricket chirps and the air temperature. The function $f(C) = 7C - 30$, where C is the temperature in degrees Celsius, can be used to approximate the number of times per minute that a cricket chirps. Discuss the domain and range of this function, and explain the significance of its x-intercept.

87. Explain why you cannot graph the equation $4x + 3y = 0$ by using just its intercepts.

PROJECTS OR GROUP ACTIVITIES

The **intercept form** of the equation of a line is given by $\frac{x}{a} + \frac{y}{b} = 1$, where $(a, 0)$ are the coordinates of the x-intercept and $(0, b)$ are the coordinates of the y-intercept.

88. Find the coordinates of the x- and y-intercepts of the graph of $\frac{x}{4} + \frac{y}{5} = 1$.

89. Find the coordinates of the x- and y-intercepts of the graph of $\frac{x}{2} - \frac{y}{3} = 1$.

90. Write the equation $3x + 2y = 6$ in intercept form.

91. Write the equation $\frac{x}{2} + \frac{y}{6} = 1$ in slope-intercept form.

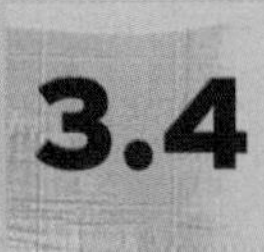

3.4 Slope of a Straight Line

OBJECTIVE  1

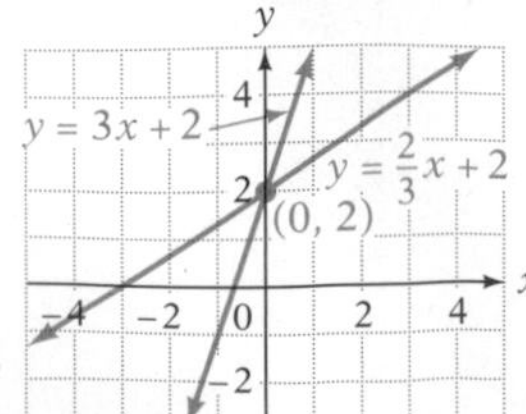

Find the slope of a line given two points

The graphs of $y = 3x + 2$ and $y = \frac{2}{3}x + 2$ are shown at the left. Each graph crosses the y-axis at the point whose coordinates are $(0, 2)$, but the graphs have different slants. The **slope** of a line is a measure of the slant of the line. The symbol for slope is m.

The slope of a line containing two points is the ratio of the change in the y values between the two points to the change in the x values. The line containing the points whose coordinates are $(-1, -3)$ and $(5, 2)$ is shown on the next page.

The change in the y values is the difference between the y-coordinates of the two points.

$$\text{Change in } y = 2 - (-3) = 5$$

The change in the x values is the difference between the x-coordinates of the two points.

$$\text{Change in } x = 5 - (-1) = 6$$

The slope of the line between the two points is the ratio of the change in y to the change in x.

$$\text{Slope} = m = \frac{\text{change in } y}{\text{change in } x} = \frac{5}{6} \qquad m = \frac{2 - (-3)}{5 - (-1)} = \frac{5}{6}$$

In general, if $P_1(x_1, y_1)$ and $P_2(x_2, y_2)$ are two points on a line, then

$$\text{Change in } y = y_2 - y_1 \qquad \text{Change in } x = x_2 - x_1$$

Using these ideas, we can state a formula for slope.

SLOPE FORMULA

The slope of the line containing the two points $P_1(x_1, y_1)$ and $P_2(x_2, y_2)$ is given by

$$m = \frac{y_2 - y_1}{x_2 - x_1}, \quad x_1 \neq x_2$$

Frequently, the Greek letter Δ (delta) is used to designate the change in a variable. Using this notation, we can write the equations for the change in y and the change in x as follows:

$$\text{Change in } y = y_2 - y_1 = \Delta y \qquad \text{Change in } x = x_2 - x_1 = \Delta x$$

With this notation, the slope formula is written $m = \frac{\Delta y}{\Delta x}$.

Focus on finding the slope of a line between two points

Take Note

It does not matter which point is named P_1 and which P_2; the slope will be the same.

A. Find the slope of the line containing the points whose coordinates are $(-2, 0)$ and $(4, 5)$.

Let $P_1 = (-2, 0)$ and $P_2 = (4, 5)$.

$$m = \frac{y_2 - y_1}{x_2 - x_1} = \frac{5 - 0}{4 - (-2)} = \frac{5}{6}$$

The slope is $\frac{5}{6}$.

A line that slants upward to the right always has a **positive slope.**

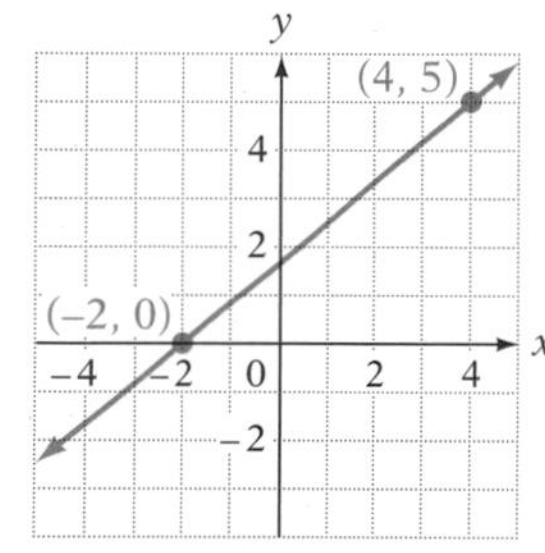

Positive slope

B. Find the slope of the line containing the points whose coordinates are $(-3, 4)$ and $(4, 2)$.

Let $P_1 = (-3, 4)$ and $P_2 = (4, 2)$.

$$m = \frac{y_2 - y_1}{x_2 - x_1} = \frac{2 - 4}{4 - (-3)} = \frac{-2}{7} = -\frac{2}{7}$$

The slope is $-\frac{2}{7}$.

A line that slants downward to the right always has a **negative slope.**

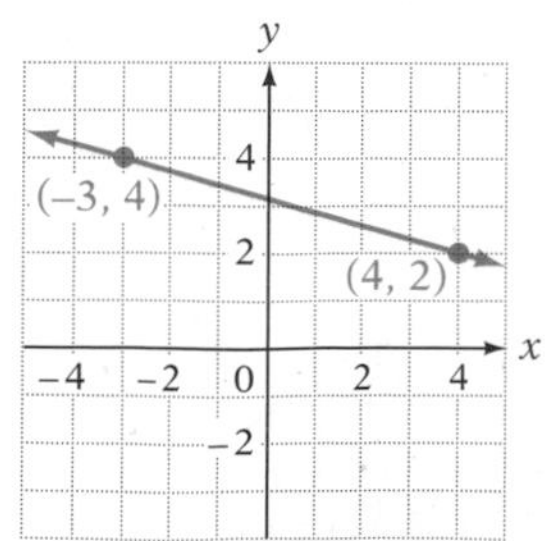

Negative slope

C. Find the slope of the line containing the points whose coordinates are $(-2, 2)$ and $(4, 2)$.

Let $P_1 = (-2, 2)$ and $P_2 = (4, 2)$.

$$m = \frac{y_2 - y_1}{x_2 - x_1} = \frac{2 - 2}{4 - (-2)} = \frac{0}{6} = 0$$

The slope is 0.

A horizontal line has **zero slope.**

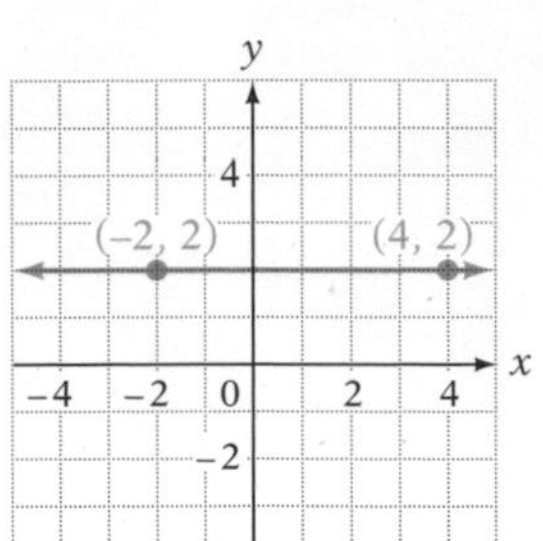

Zero slope

D. Find the slope of the line containing the points whose coordinates are $(1, -2)$ and $(1, 3)$.

Let $P_1 = (1, -2)$ and $P_2 = (1, 3)$.

$$m = \frac{y_2 - y_1}{x_2 - x_1} = \frac{3 - (-2)}{1 - 1} = \frac{5}{0}$$

$\frac{5}{0}$ is not a real number. The slope is undefined.

The slope of a vertical line is **undefined.**

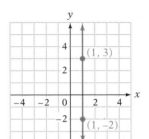

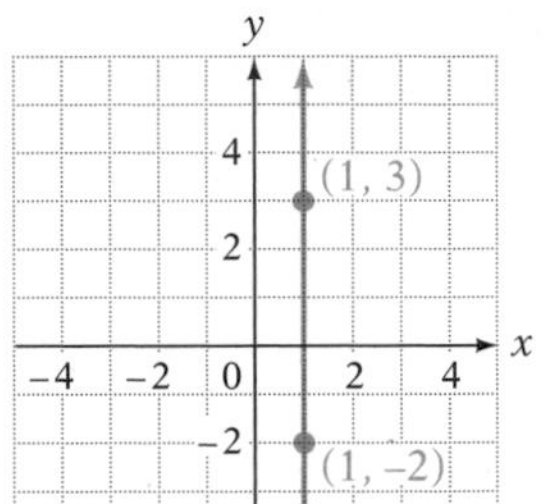

Undefined slope

EXAMPLE 1 Find the slope of the line containing the points whose coordinates are $(2, -5)$ and $(-4, 2)$.

Solution

$$m = \frac{y_2 - y_1}{x_2 - x_1}$$

$$= \frac{2 - (-5)}{-4 - 2} = \frac{7}{-6} = -\frac{7}{6}$$

• **Let $P_1 = (2, -5)$ and $P_2 = (-4, 2)$.**

The slope is $-\frac{7}{6}$.

Problem 1 Find the slope of the line containing the points whose coordinates are $(4, -3)$ and $(2, 7)$.

Solution See page S8.

Try Exercise 11, page 157.

There are many applications of slope. Here are two examples.

The first record for the one-mile run was recorded in 1865 in England. Richard Webster ran the mile in 4 min 36.5 s. His average speed was approximately 19 feet per second.

The graph at the right shows the distance Webster ran during that run. From the graph, note that after 60 s (1 min) he had traveled 1140 ft, and that after 180 s (3 min) he had traveled 3420 ft.

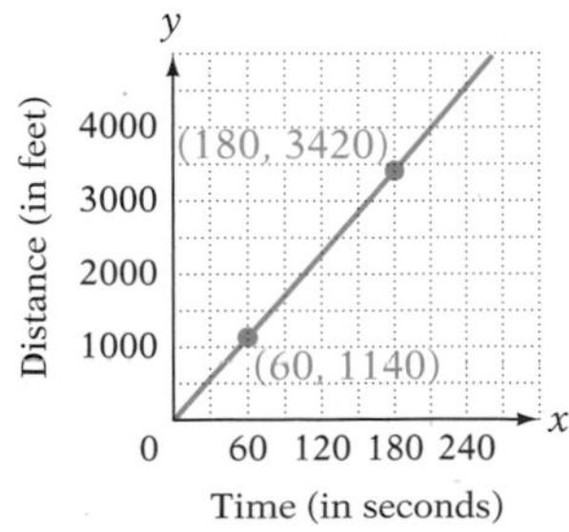

Let the ordered pair $(60, 1140)$ be (x_1, y_1) and the ordered pair $(180, 3420)$ be (x_2, y_2). The slope of the line between these two points is

$$m = \frac{y_2 - y_1}{x_2 - x_1} = \frac{3420 - 1140}{180 - 60} = \frac{2280}{120} = 19$$

Note that the slope of the line is the same as Webster's average speed, 19 feet per second.

The value of a car decreases as the number of miles it has been driven increases. The graph at the right is based on data from the Kelley Blue Book website that show the decline in value of a certain Corvette Z06. The ordered pair (35, 50) means that the value of a car that has been driven 35,000 mi is \$50,000, and the ordered pair (60, 42) means that the value of a car that has been driven 60,000 mi is \$42,000.

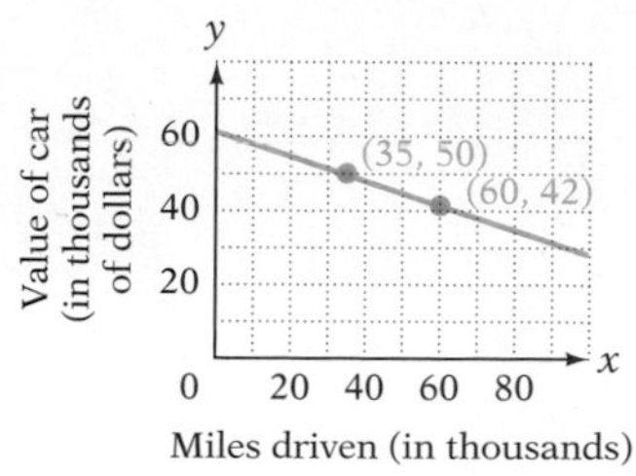

The slope of the line between the two points is

$$m = \frac{y_2 - y_1}{x_2 - x_1}$$

$$= \frac{42 - 50}{60 - 35} = -\frac{8}{25}$$

$$= -0.32$$

Because the units in the numerator of the slope are dollars and the units in the denominator are miles, the units of the slope are dollars per mile. The negative sign indicates that the slope is decreasing. The slope of the line means that the value of the car is decreasing at a rate of \$.32 per mile, or 32 cents per mile.

EXAMPLE 2 The graph shows the relationship between the cost of an item and the sales tax. Find the slope of the line between the two points shown on the graph. Write a sentence that states the meaning of the slope.

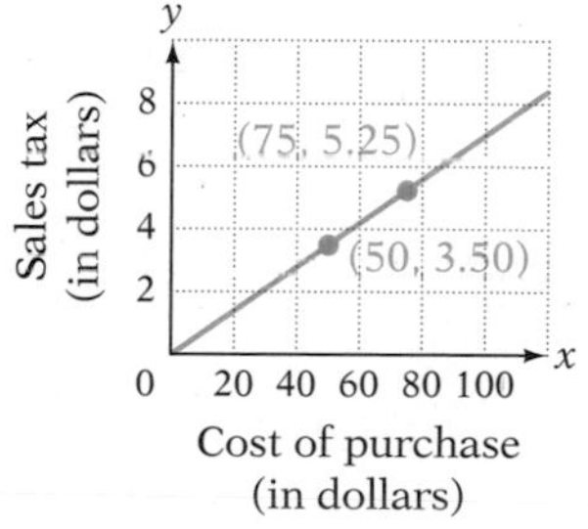

Solution

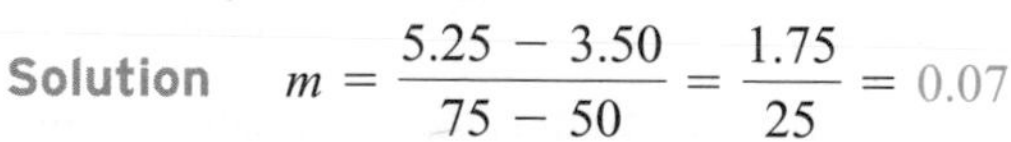

$$m = \frac{5.25 - 3.50}{75 - 50} = \frac{1.75}{25} = 0.07$$

A slope of 0.07 means that the sales tax is \$.07 per dollar.

Problem 2 The graph shows the decrease in the value of a printing press for a period of six years. Find the slope of the line between the two points shown on the graph. Write a sentence that states the meaning of the slope.

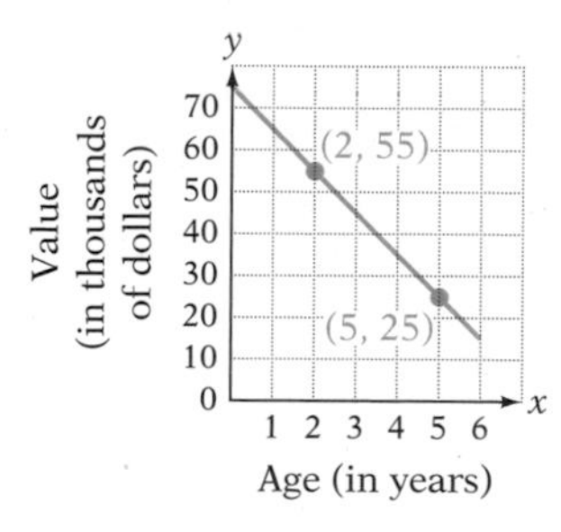

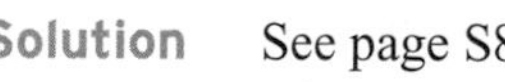

Solution See page S8.

Try Exercise 27, page 157.

OBJECTIVE 2 Graph a line given a point and the slope

The graph of the equation $y = -\frac{3}{4}x + 4$ is shown at the right. The points whose coordinates are $(-4, 7)$ and $(4, 1)$ are on the graph. The slope of the line is

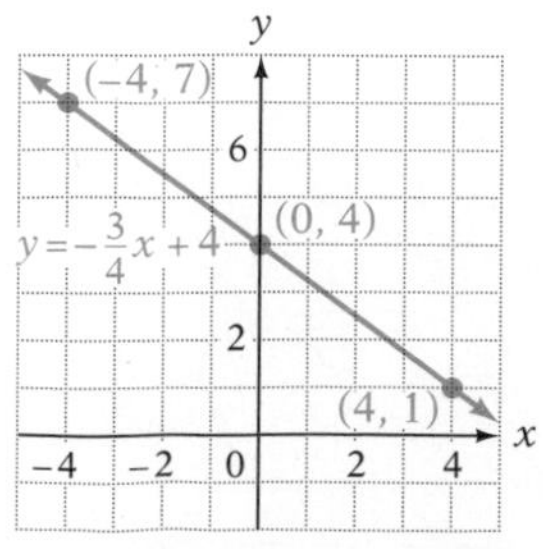

$$m = \frac{7 - 1}{-4 - 4} = \frac{6}{-8} = -\frac{3}{4}$$

Note that the slope of the line has the same value as the coefficient of x.

Recall that the y-intercept is found by replacing x by zero and then solving for y.

$$y = -\frac{3}{4}x + 4$$

$$y = -\frac{3}{4}(0) + 4 = 4$$

The coordinates of the y-intercept are $(0, 4)$. Note that the y-coordinate of the y-intercept is the constant term of the equation

$$y = -\frac{3}{4}x + 4$$

SLOPE-INTERCEPT FORM OF A STRAIGHT LINE

The equation $y = mx + b$ is called the **slope-intercept form** of a straight line. The slope of the line is m, the coefficient of x. The coordinates of the y-intercept are $(0, b)$.

When the equation of a straight line is in the form $y = mx + b$, the graph can be drawn by using the slope and the y-intercept. First locate the y-intercept. Use the slope to find a second point on the line. Then draw a line through the two points.

Focus on graphing a line using the slope and the y-intercept

A. Graph $y = \frac{5}{3}x - 4$ by using the slope and the y-intercept.

The slope is the coefficient of x.

$$m = \frac{5}{3} = \frac{\text{change in } y}{\text{change in } x}$$

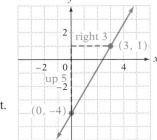

The constant term is the y-coordinate of the y-intercept.

The y-intercept has coordinates $(0, -4)$.

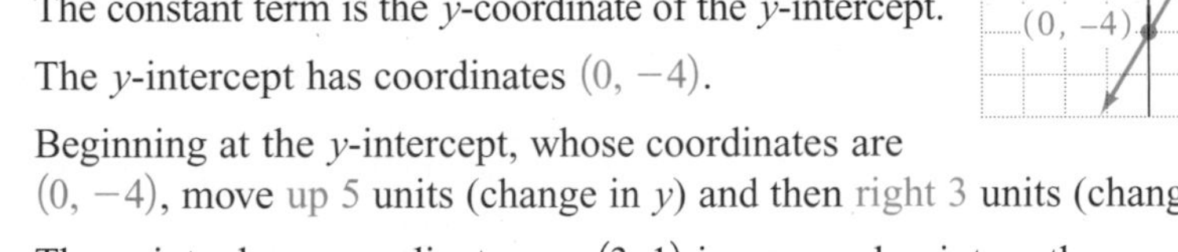

Beginning at the y-intercept, whose coordinates are $(0, -4)$, move up 5 units (change in y) and then right 3 units (change in x).

The point whose coordinates are $(3, 1)$ is a second point on the graph. Draw a line through the points whose coordinates are $(0, -4)$ and $(3, 1)$.

Take Note

When graphing a line by using its slope and y-intercept, *always* start at the y-intercept.

B. Graph $x + 2y = 4$ by using the slope and the y-intercept.

Solve the equation for y.

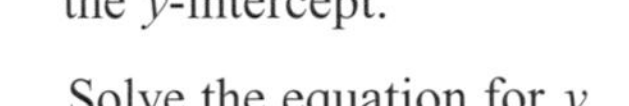

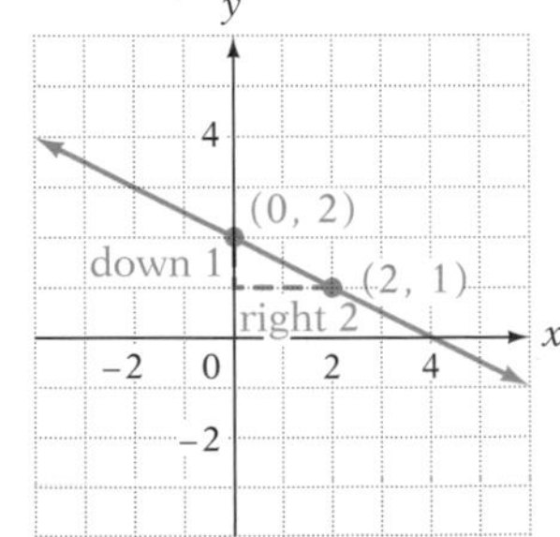

$$x + 2y = 4$$

$$2y = -x + 4$$

$$y = -\frac{1}{2}x + 2$$

Identify the slope and the y-intercept.

The slope is $m = -\frac{1}{2} = \frac{-1}{2}$, and the y-intercept has coordinates $(0, 2)$.

Beginning at the y-intercept, whose coordinates are $(0, 2)$, move down 1 unit (change in y) and then right 2 units (change in x).

The point whose coordinates are $(2, 1)$ is a second point on the graph. Draw a line through the points whose coordinates are $(0, 2)$ and $(2, 1)$.

EXAMPLE 3 Graph $y = -\frac{3}{2}x + 4$ by using the slope and the y-intercept.

Solution y-intercept: (0, 4)

- **Determine the y-intercept from the constant term.**

$$m = -\frac{3}{2} = \frac{-3}{2} = \frac{\text{change in } y}{\text{change in } x}$$

- **Move the negative sign into the numerator of the slope fraction.**

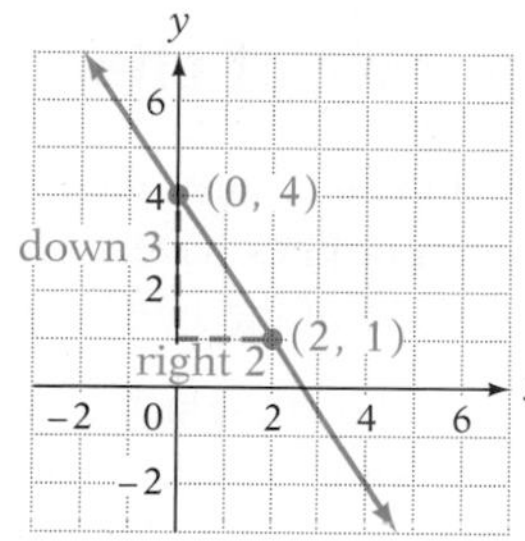

- **Beginning at the y-intercept, whose coordinates are (0, 4), move down 3 units and right 2 units. (2, 1) are the coordinates of a second point on the graph.**
- **Draw a line through the points whose coordinates are (0, 4) and (2, 1).**

Problem 3 Graph $2x + 3y = 6$ by using the slope and the y-intercept.

Solution See page S8.

Try Exercise 37, page 158.

The graph of a line can be drawn when any point on the line and the slope of the line are given.

Focus on graphing a line through a given point and having a given slope

Graph the line that passes through the point $P(-4, -4)$ and has slope 2.

When the slope is an integer, write it as a fraction with a denominator of 1.

$$m = 2 = \frac{2}{1} = \frac{\text{change in } y}{\text{change in } x}$$

Locate $P(-4, -4)$ in the coordinate plane. Beginning at that point, move up 2 units (change in y) and then right 1 unit (change in x).

The point whose coordinates are $(-3, -2)$ is a second point on the graph. Draw a line through the points whose coordinates are $(-4, -4)$ and $(-3, -2)$.

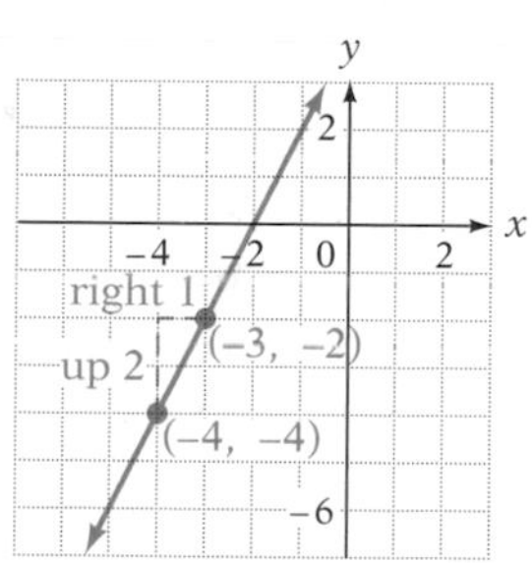

Take Note

This example differs from the others in that a point other than the y-intercept is used. In this case, start at the given point.

EXAMPLE 4 Graph the line that passes through the point $P(-2, 3)$ and has slope $-\frac{4}{3}$.

Solution $$m = -\frac{4}{3} = \frac{-4}{3} = \frac{\text{change in } y}{\text{change in } x}$$

- **Move the negative sign into the numerator of the slope fraction.**

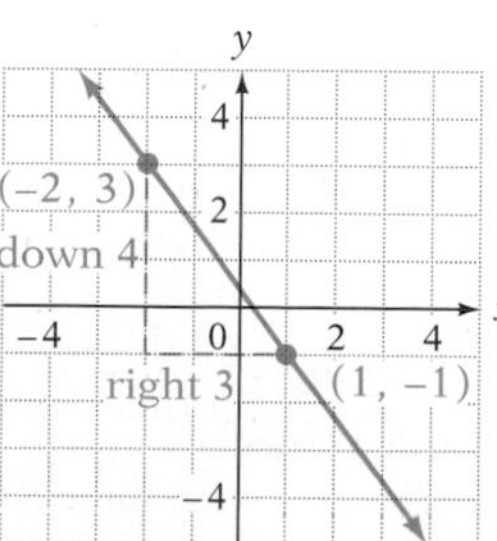

- **Locate (−2, 3). Beginning at that point, move down 4 units and then right 3 units. (1, −1) are the coordinates of a second point on the line.**
- **Draw a line through the points whose coordinates are (−2, 3) and (1, −1).**

Problem 4 Graph the line that passes through the point $P(-3, -2)$ and has slope 3.

Solution See page S9.

Try Exercise 51, page 159.

OBJECTIVE 3 Average rate of change

Recall that slope measures the rate at which one quantity changes with respect to a second quantity. Straight lines have a constant slope. No matter which two points on the line are chosen, the slope of the line between the two points is the same.

If a graph is not a straight line, the slope of the line between two points on the graph may be different from the slope of the line between two other points. Consider the graph of $f(x) = \frac{1}{2}x^2 + 2x - 10$ shown at the right. The slope m_P of the line between P_1 and P_2 is different from the slope m_Q of the line between Q_1 and Q_2.

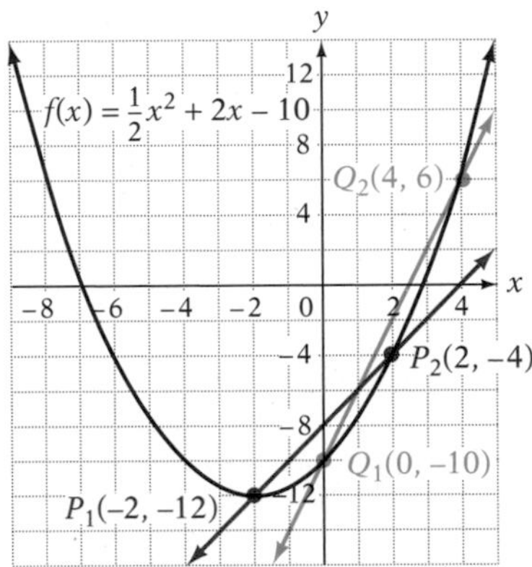

$$m_P = \frac{-4 - (-12)}{2 - (-2)} = \frac{8}{4} = 2$$

$$m_Q = \frac{6 - (-10)}{4 - 0} = \frac{16}{4} = 4$$

In cases such as these, the **average rate of change** between any two points is the slope of the line between the two points.

EXAMPLE 5 Find the average rate of change of $f(x) = 2x^2 - 4x + 5$ between the points whose x-coordinates are $x_1 = 2$ and $x_2 = 4$.

Solution Find the coordinates of each point by finding the y-coordinate for the given x-coordinate.

$$y_1 = f(x_1)$$
$$= 2(2)^2 - 4(2) + 5 = 5 \qquad \bullet\ x_1 = 2$$

The first point is P_1 (2, 5).

$$y_2 = f(x_2)$$
$$= 2(4)^2 - 4(4) + 5 = 21 \qquad \bullet\ x_2 = 4$$

The second point is P_2 (4, 21).

To find the average rate of change between the two points, find the slope of the line between P_1 (2, 5) and P_2 (4, 21).

$$m = \frac{y_2 - y_1}{x_2 - x_1}$$
$$= \frac{21 - 5}{4 - 2} = \frac{16}{2} = 8$$

The average rate of change between the two points is 8. See the graph above.

Problem 5 Find the average rate of change of $y = x^2 + x - 1$ between the points whose x-coordinates are $x_1 = -4$ and $x_2 = -1$.

Solution See page S9.

Try Exercise 55, page 159.

Population of California

Year	Population (in millions)
1850	0.1
1860	0.4
1870	0.6
1880	0.9
1890	1.2
1900	1.5
1910	2.4
1920	3.4
1930	5.7
1940	6.9
1950	10.6
1960	15.7
1970	20.0
1980	23.7
1990	29.8
2000	33.9
2010	37.3

The table at the left shows the population of California for each decade from 1850 to 2010. (*Source:* U.S. Bureau of the Census) The graph below, called a **scatter diagram,** is a graph of the data in the table.

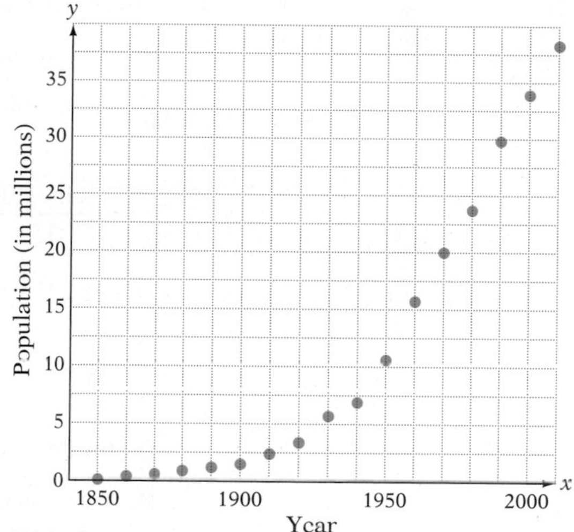

To find the average rate of change of the population between 1980 and 2000, find the slope of the line between the corresponding points on the graph.

$$m = \frac{33.9 - 23.7}{2000 - 1980} = \frac{10.2}{20} = 0.51$$

The average rate of change was 0.51 million, or 510,000, people per year. This means that *on average,* from 1980 to 2000, the population of California increased by 510,000 people per year.

EXAMPLE 6 Find the average annual rate of change in California's population from 1850 to 1950.

Solution In 1850, the population was 0.1 million: (1850, 0.1)

In 1950, the population was 10.6 million: (1950, 10.6)

$$m = \frac{10.6 - 0.1}{1950 - 1850}$$

$$= \frac{10.5}{100} = 0.105$$

The average rate of change in the population of California from 1850 to 1950 was 105,000 people per year.

Problem 6 Find the average annual rate of change in California's population from 1900 to 2000.

Solution See page S9.

Try Exercise 63, part (a), page 159.

The table below shows the median salaries in 1995, 2000, 2005, and 2010 for Boston Red Sox players and New York Yankees players. (*Source:* usatoday.com) Use these data for Example 7 and Problem 7.

Year	1995	2000	2005	2010
Median salary, Boston Red Sox	282,500	2,000,000	2,875,000	3,750,000
Median salary, New York Yankees	531,000	1,350,000	5,833,334	5,500,000

EXAMPLE 7 Find the average annual rate of change in the median salary of Boston Red Sox players between 1995 and 2010. Round to the nearest thousand dollars.

Solution In 1995, the median salary was 282,500: (1995, 282,500)

In 2010, the median salary was 3,750,000: (2010, 3,750,000)

$$m = \frac{3{,}750{,}000 - 282{,}500}{2010 - 1995}$$

$$= \frac{3{,}467{,}500}{15} \approx 231{,}000$$

The average annual rate of change in median salary was approximately \$231,000 per year.

Problem 7 **A.** Find the average annual rate of change in the median salary of New York Yankees players between 1995 and 2010. Round to the nearest thousand dollars.

B. Was this greater than or less than the average annual rate of change in the median salary of Boston Red Sox players between 1995 and 2010?

Solution See page S9.

Try Exercise 63, parts (b) and (c), page 159.

3.4 Exercises

CONCEPT CHECK

Complete each sentence using *increases* or *decreases*.

1. If a line has positive slope, then as x increases, y ___?___.

2. If a line has negative slope, then as x ___?___, y decreases.

Determine the slope and the coordinates of the y-intercept of the graph of the equation.

3. $y = -\frac{3}{4}x + 2$ **4.** $y = x$ **5.** $f(x) = 3 - x$ **6.** $f(x) = \frac{2x}{3} - 1$

1 Find the slope of a line given two points (See pages 148–151.)

GETTING READY

7. To find the slope m of the line containing $P_1(2,\ \ 4)$ and $P_2(3, -5)$, use the slope formula $m = \frac{y_2 - y_1}{x_2 - x_1}$. Identify each x and y value to substitute into the slope formula.

$y_2 =$ ___?___, $y_1 =$ ___?___, $x_2 =$ ___?___, $x_1 =$ ___?___

8. The slope of the line containing $P_1(1, -6)$ and $P_2(5, -10)$ is $\frac{-10 - (\ )}{5 - (\ )} =$ ___?___. This line slants ___?___ (upward or downward) to the right.

Find the slope of the line containing the points P_1 and P_2.

9. $P_1(1, 3), P_2(3, 1)$

10. $P_1(2, 3), P_2(5, 1)$

11. $P_1(-1, 4), P_2(2, 5)$

12. $P_1(3, -2), P_2(1, 4)$

13. $P_1(-1, 3), P_2(-4, 5)$

14. $P_1(-1, -2), P_2(-3, 2)$

15. $P_1(0, 3), P_2(4, 0)$

16. $P_1(-2, 0), P_2(0, 3)$

17. $P_1(2, 4), P_2(2, -2)$

18. $P_1(4, 1), P_2(4, -3)$

19. $P_1(2, 5), P_2(-3, -2)$

20. $P_1(4, 1), P_2(-1, -2)$

21. $P_1(2, 3), P_2(-1, 3)$

22. $P_1(3, 4), P_2(0, 4)$

23. $P_1(0, 4), P_2(-2, 5)$

24. $P_1(-2, 3), P_2(-2, 5)$

25. $P_1(-3, -1), P_2(-3, 4)$

26. $P_1(-2, -5), P_2(-4, -1)$

27. **Uniform Motion** The graph below shows the relationship between the distance traveled by a motorist and the time of travel. Find the slope of the line between the two points shown on the graph. Write a sentence that states the meaning of the slope.

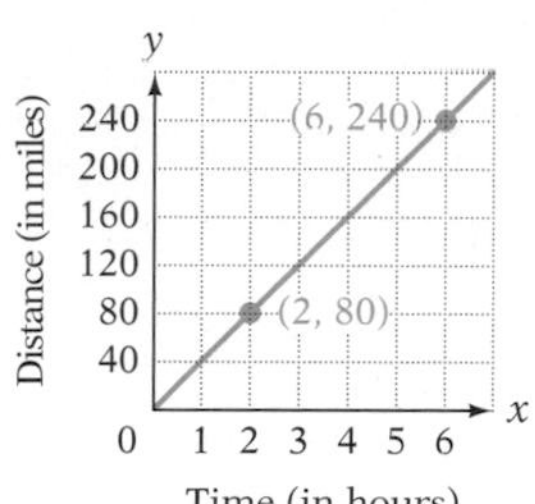

28. **Aeronautics** The graph below shows how the altitude of an airplane above the runway changes after takeoff. Find the slope of the line. Write a sentence that states the meaning of the slope.

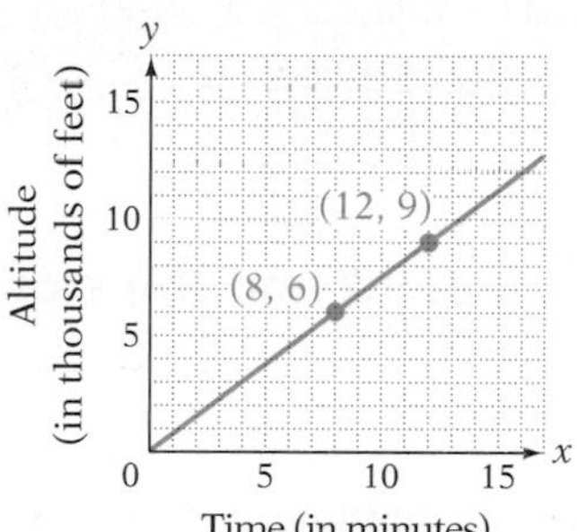

29. **Automotive Technology** The graph below shows how the amount of gas in the tank of a car decreases as the car is driven. Find the slope of the line. Write a sentence that states the meaning of the slope.

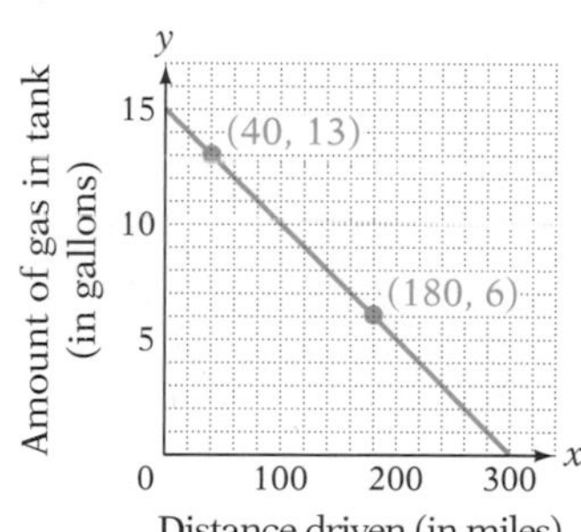

30. **Meteorology** The troposphere extends from the surface of Earth to an elevation of approximately 11 km. The graph below shows the decrease in temperature of the troposphere as altitude increases. Find the slope of the line. Write a sentence that states the meaning of the slope.

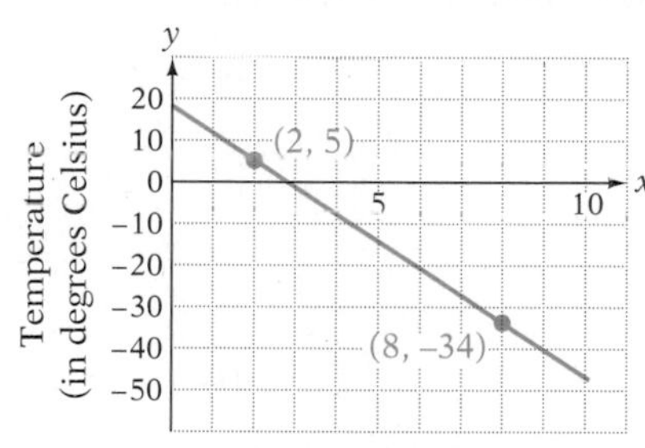

31. **Sports** Lois and Tanya start from the same place on a jogging course. Lois is jogging at 9 km/h, and Tanya is jogging at 6 km/h. The graphs below show the total distance traveled by each jogger and the total distance between Lois and Tanya. Which lines represent which distances?

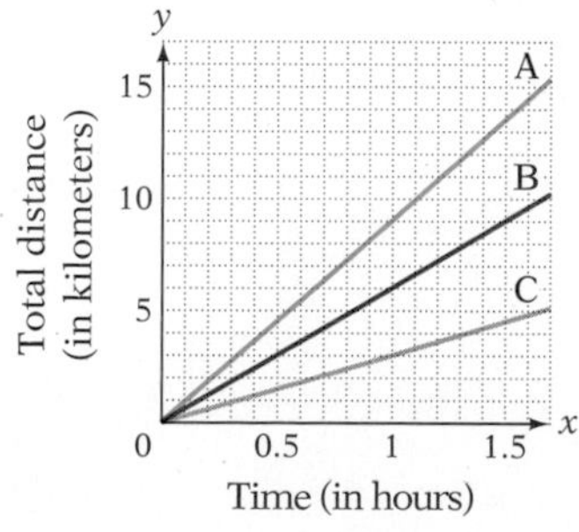

32. **Chemistry** A chemist is filling two cans from a faucet that releases water at a constant rate. Can 1 has a diameter of 20 mm, and can 2 has a diameter of 30 mm. The depth of the water in each can is measured at 5-second intervals. The graph of the results is shown below. On the graph, which line represents the depth of the water for which can?

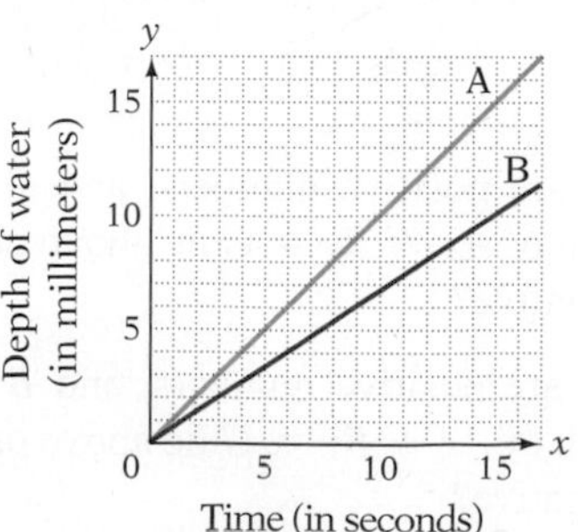

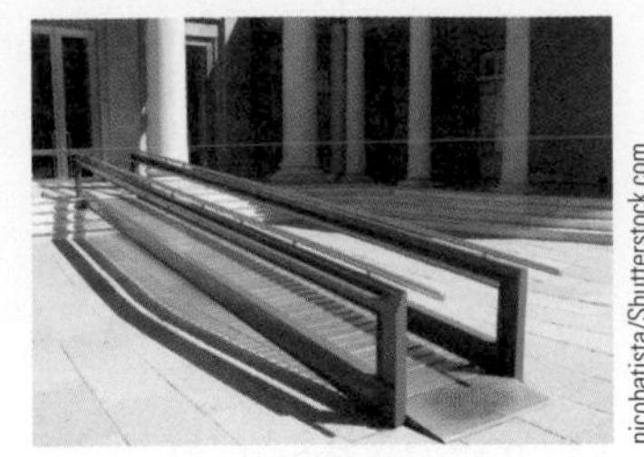
nicobatista/Shutterstock.com

33. **Health Science** The American National Standards Institute (ANSI) states that the slope for a wheelchair ramp must not exceed $\frac{1}{12}$.
   a. Does a ramp that is 6 in. high and 5 ft long meet the requirements of ANSI?
   b. Does a ramp that is 12 in. high and 170 in. long meet the requirements of ANSI?

34. **Solar Energy** Look at the butterfly roof design in the article below. Which side of the roof, the left or the right, has a slope of approximately 1? Is the slope of the other side of the roof greater than 1 or less than 1? (*Note:* Consider both slopes to be positive.)

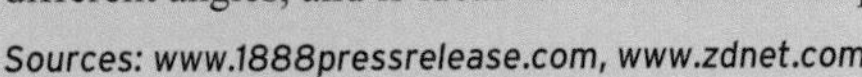

Solar Powered Practice Rooms

A music school in Vermont will construct eight modular buildings with butterfly roofs to serve as its newest practice rooms. The roof design combines two sections that slant toward each other at different angles, and is ideal for the use of solar panels.

Sources: www.1888pressrelease.com, www.zdnet.com

2 Graph a line given a point and the slope (See pages 151–154.)

GETTING READY

35. The slope of the line with equation $y = 5x - 3$ is ___?___, and the coordinates of its y-intercept are ___?___.

36. The slope of the line with equation $2x + y = 7$ is ___?___, and the coordinates of its y-intercept are ___?___.

Graph by using the slope and the y-intercept.

37. $y = \frac{1}{2}x + 2$

38. $y = \frac{2}{3}x - 3$

39. $y = -\frac{3}{2}x$

40. $y = \frac{3}{4}x$

41. $y = -\frac{1}{2}x + 2$

42. $y = \frac{2}{3}x - 1$

43. $x - 3y = 3$

44. $3x + 2y = 8$

45. $4x + y = 2$

46. Suppose A and B are positive numbers, and C is a negative number. Does the y-intercept of the graph of $Ax + By = C$ lie above or below the x-axis? Is the slope of the line positive or negative?

47. Suppose A and C are positive numbers, and B is a negative number. Does the y-intercept of the graph of $Ax + By = C$ lie above or below the x-axis? Is the slope of the line positive or negative?

48. Graph the line that passes through the point $P(-1, -3)$ and has slope $\frac{4}{3}$.

49. Graph the line that passes through the point $P(-2, -3)$ and has slope $\frac{5}{4}$.

50. Graph the line that passes through the point $P(-3, 0)$ and has slope -3.

51. Graph the line that passes through the point $P(2, 0)$ and has slope -1.

3 Average rate of change (See pages 154–156.)

GETTING READY

52. To find the average rate of change between two points on the graph of a nonlinear function, find the ___?___ of the line that contains the two points.

53. To find the average rate of change of a function between two ordered pairs whose x-coordinates are given, first find the ___?___ for each of the given x-coordinates.

Find the average rate of change of the given function between the points with the given x-coordinates.

54. $f(x) = -x^2 + 3x;\ x_1 = 6, x_2 = 9$

55. $f(x) = 2x^2 - x + 3;\ x_1 = 3, x_2 = 5$

56. $f(x) = x^2 - 3x + 1;\ x_1 = -2, x_2 = -1$

57. $f(x) = 2x^2 + 2x + 4;\ x_1 = -6, x_2 = -2$

58. $f(x) = -2x^3 + 6x + 6;\ x_1 = -1, x_2 = 1$

59. $f(x) = x^3 - 7x^2 + 4;\ x_1 = 5, x_2 = 8$

60. $f(x) = 2x^3 - 5x^2 - 3x + 2;\ x_1 = -1, x_2 = 2$

61. $f(x) = 3x^3 + x^2 + 3x - 6;\ x_1 = -3, x_2 = 0$

62. Temperature On November 5 in midstate New Hampshire, the temperature at 6 A.M. was 34°F. At 2 P.M. on the same day, the temperature was 58°F. Find the average rate of change in temperature per hour.

63. Demography The table at the right shows the population of Texas for each decade from 1850 to 2010. (*Source:* U.S. Bureau of the Census)

a. Find the average annual rate of change in the population of Texas from 1900 to 2000.

b. Was the average annual rate of change in the population from 1900 to 1950 greater than or less than the average annual rate of change in the population from 1950 to 2000?

c. Was the average annual rate of change in the population of Texas from 1980 to 2000 greater than or less than the average annual rate of change in the population of California from 1980 to 2000? (See page 155 for the average annual rate of change in the population of California from 1980 to 2000.)

64. Demography Use the table in Exercise 63.

a. During which decade was the average annual rate of change in the population the least?

b. During which decade was the average annual rate of change in the population approximately 300,000 people?

Population of Texas

Year	Population (in millions)
1850	0.2
1860	0.6
1870	0.8
1880	1.6
1890	2.2
1900	3.0
1910	3.9
1920	4.7
1930	5.8
1940	6.4
1950	7.7
1960	9.6
1970	11.2
1980	14.2
1990	17.0
2000	20.9
2010	25.1

65. **Climate Change** Use the information in the article at the right.
   a. Find the goal for the average rate of change in green-house gas emissions from 2020 to 2030.
   b. Find the goal for the average rate of change in green-house gas emissions from 2030 to 2050.

In the News

Reducing Green-House Gasses

Congress has passed a bill that could help the United States reduce green-house gas emissions for decades to come. The goals are for emissions levels to be at 6.0 billion metric tons in 2020, 4.2 billion metric tons in 2030, and 1.2 billion metric tons in 2050.

Source: Time magazine

66. **Cremation** In the United States in 1990, there were 2.15 million deaths and 367,975 cremations. In 2007, there were 2.42 million deaths and 832,340 cremations. (*Source:* Cremation Association of North America)
   a. From 1990 to 2007, what was the average annual rate of change in the number of deaths in the United States? Round to the nearest thousand.
   b. From 1990 to 2007, what was the average annual rate of change in the number of cremations in the United States? Round to the nearest thousand.
   c. From 1990 to 2007, which increased at a greater rate, the number of deaths or the number of cremations?

67. **Traffic Fatalities** The number of pedestrians killed by motor vehicles in 1998 was 5228; in 2008 it was 4378. (*Source:* National Highway Transportation Safety Administration)
   a. Find the average annual rate of change in the number of pedestrian fatalities from 1998 to 2008. (*Note:* A negative average rate of change denotes a decrease.)
   b. Why can the answer to part (a) be considered encouraging?

68. **Energy Prices** The table at the right shows the average price per gallon of gasoline in the United States during January of each year from 2000 to 2010. (*Source:* Energy Information Administration)
   a. Find the average rate of change per year in the price of a gallon of gasoline from 2000 to 2010.
   b. Which was greater, the average rate of change per year from 2000 to 2005 or the average rate of change per year from 2005 to 2010? How much greater?

Year	Price per Gallon
2000	\$1.29
2001	\$1.45
2002	\$1.11
2003	\$1.46
2004	\$1.57
2005	\$1.83
2006	\$2.32
2007	\$2.24
2008	\$3.04
2009	\$1.79
2010	\$2.72

69. **Trademarks** The table below shows the number of applications for trademark registration in the United States in 1990, 1995, 2000, and 2005. (*Source:* International Trademark Association) One reason for the large number of applications is the increase in popularity of the Internet and the desire to register domain names and websites.

Year	1990	1995	2000	2005
Applications (in thousands)	120	175	375	324

   a. Find the average annual rate of change in the number of applications from 1990 to 2000.
   b. How much greater was the average annual rate of change in the number of applications from 1995 to 2000 than from 1990 to 1995?

APPLYING CONCEPTS

70. Let $f(x)$ be the digit in the nth decimal place of the repeating decimal $0.\overline{387}$. For example, $f(3) = 7$ because 7 is the digit in the third decimal place. Find $f(14)$.

Determine the value of k such that the three points whose coordinates are given all lie on the same line.

71. $(3, 2), (4, 6), (5, k)$

72. $(k, 1), (0, -1), (2, -2)$

73. Match each equation with its graph.

i. $y = -2x + 4$

ii. $y = 2x - 4$

iii. $y = 2$

iv. $2x + 4y = 0$

v. $y = \frac{1}{2}x + 4$

vi. $y = -\frac{1}{4}x - 2$

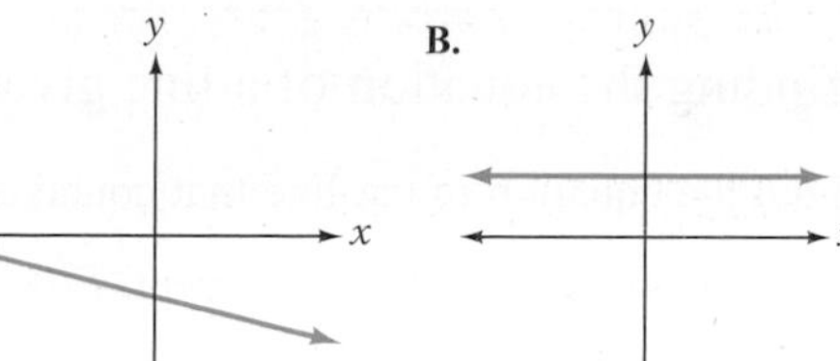
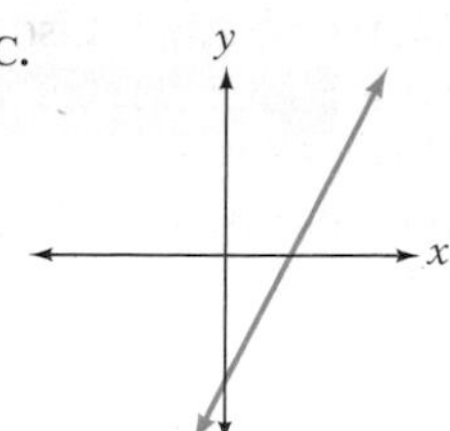
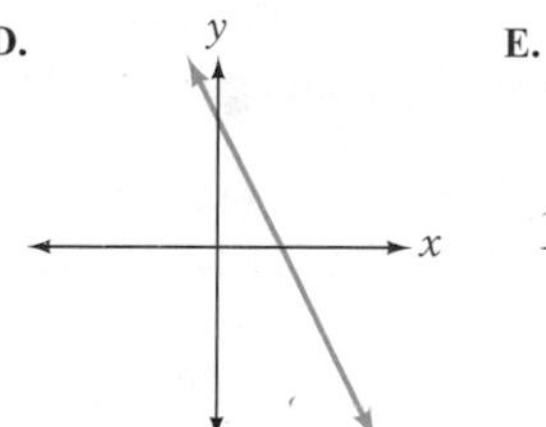
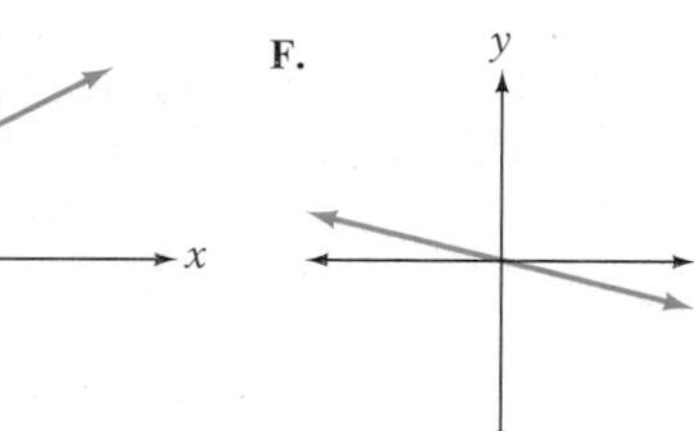

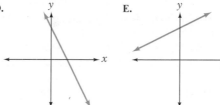

74. 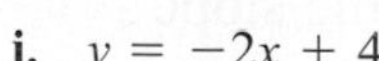  Graph $y = 2x + 3$ and $y = 2x - 1$ on the same coordinate system. Explain how the graphs are different and how they are the same. If b is any real number, how is the graph of $y = 2x + b$ related to the two graphs you have drawn?

PROJECTS OR GROUP ACTIVITIES

Complete each sentence using *increases* or *decreases* in the first blank and a positive number in the second blank.

75. If a line has a slope of 2, then the value of y __?__ by __?__ as the value of x increases by 1.

76. If a line has a slope of -3, then the value of y __?__ by __?__ as the value of x increases by 1.

77. If a line has a slope of $\frac{1}{2}$, then the value of y __?__ by __?__ as the value of x increases by 1.

78. If a line has a slope of $-\frac{2}{3}$, then the value of y __?__ by __?__ as the value of x increases by 1.

79. Explain how you can use the slope of a line to determine whether three given points lie on the same line. Then use your procedure to determine whether the three points whose coordinates are given lie on the same line.

a. $(2, 5), (-1, -1), (3, 7)$

b. $(-1, 5), (0, 3), (-3, 4)$

3.5 Finding Equations of Lines

OBJECTIVE 1 Find the equation of a line given a point and the slope

When the slope of a line and a point on the line are known, the equation of the line can be determined. If the particular point is the y-intercept, use the slope-intercept form, $y = mx + b$, to find the equation.

Focus on finding the equation of a line given the y-intercept and the slope

Find the equation of the line that contains the point $P(0, 3)$ and has slope $\frac{1}{2}$.

The known point is the y-intercept, $P(0, 3)$.

Use the slope-intercept form. $y = mx + b$

Replace m with $\frac{1}{2}$, the given slope, and replace b with 3, the y-coordinate of the y-intercept. $y = \frac{1}{2}x + 3$

The equation of the line is $y = \frac{1}{2}x + 3$.

One method of finding the equation of a line when the slope and any point on the line are known involves using the *point-slope formula.* This formula is derived from the formula for the slope of a line.

Let $P_1(x_1, y_1)$ be the given point on the line, and let $P(x, y)$ be another point on the line.

Use the formula for the slope of a line. $\frac{y - y_1}{x - x_1} = m$

Multiply each side of the equation by $(x - x_1)$. $\frac{y - y_1}{x - x_1}(x - x_1) = m(x - x_1)$

Then simplify. $y - y_1 = m(x - x_1)$

POINT-SLOPE FORMULA

Let m be the slope of a line, and let $P_1(x_1, y_1)$ be a point on the line. The equation of the line can be found by using the **point-slope formula:**

$$y - y_1 = m(x - x_1)$$

Focus on finding the equation of a line given a point and the slope

A. Find the equation of the line that contains the point $P(4, -1)$ and has slope $-\frac{3}{4}$.

Use the point-slope formula. $y - y_1 = m(x - x_1)$

Substitute the slope, $-\frac{3}{4}$, and the coordinates of the given point, $(4, -1)$, into the point-slope formula. Then simplify.

$$y - (-1) = -\frac{3}{4}(x - 4)$$

$$y + 1 = -\frac{3}{4}x + 3$$

Solve for y. $y = -\frac{3}{4}x + 2$

The equation of the line is $y = -\frac{3}{4}x + 2$.

B. Find the equation of the line that passes through the point $P(4, 3)$ and whose slope is undefined.

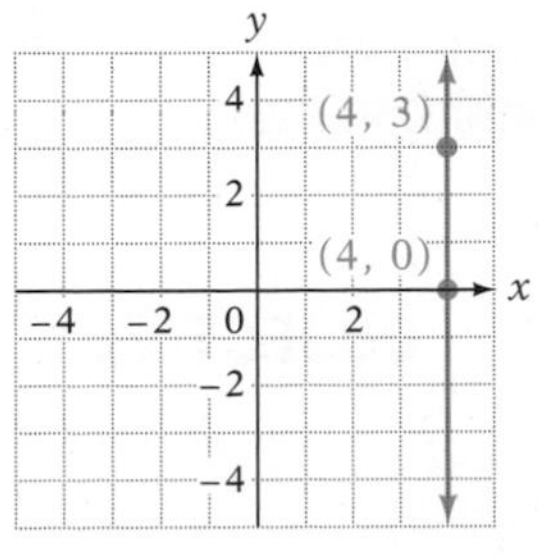

Because the slope is undefined, the point-slope formula cannot be used to find the equation. Instead, recall that when the slope is undefined, the line is vertical, and that the equation of a vertical line is $x = a$, where a is the x-coordinate of the x-intercept. Because the line is vertical and passes through the point $P(4, 3)$, the coordinates of the x-intercept are $(4, 0)$.

The equation of the line is $x = 4$.

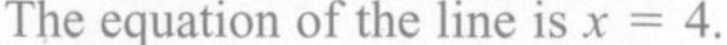

EXAMPLE 1 Find the equation of the line that contains the point $P(-2, 4)$ and has slope 2.

Solution

$y - y_1 = m(x - x_1)$ • Use the point-slope formula.

$y - 4 = 2[x - (-2)]$ • Substitute the slope, 2, and the coordinates of the given point, $(-2, 4)$, into the point-slope formula.

$y - 4 = 2(x + 2)$ • Solve for y.

$y - 4 = 2x + 4$

$y = 2x + 8$

The equation of the line is $y = 2x + 8$.

Problem 1 Find the equation of the line that contains the point $P(4, -3)$ and has slope -3.

Solution See page S9.

Try Exercise 7, page 165.

OBJECTIVE 2 Find the equation of a line given two points

The point-slope formula and the formula for slope are used to find the equation of a line when two points are known.

Focus on finding the equation of a line between two points

Find the equation of the line containing the points $P_1(3, 2)$ and $P_2(-5, 6)$.

To use the point-slope formula, we must know the slope. Use the formula for slope to determine the slope of the line between the two given points.

Let $(x_1, y_1) = (3, 2)$ and $(x_2, y_2) = (-5, 6)$.

$$m = \frac{y_2 - y_1}{x_2 - x_1} = \frac{6 - 2}{-5 - 3} = \frac{4}{-8} = -\frac{1}{2}$$

Now use the point-slope formula with $m = -\frac{1}{2}$ and $(x_1, y_1) = (3, 2)$.

$$y - y_1 = m(x - x_1)$$

$$y - 2 = -\frac{1}{2}(x - 3)$$

Solve for y.

$$y - 2 = -\frac{1}{2}x + \frac{3}{2}$$

$$y = -\frac{1}{2}x + \frac{7}{2}$$

The equation of the line is $y = -\frac{1}{2}x + \frac{7}{2}$.

EXAMPLE 2 Find the equation of the line containing the points $P_1(2, 3)$ and $P_2(4, 1)$.

Solution

$m = \frac{y_2 - y_1}{x_2 - x_1} = \frac{1 - 3}{4 - 2} = \frac{-2}{2} = -1$ • Find the slope. Let $(x_1, y_1) = (2, 3)$ and $(x_2, y_2) = (4, 1)$.

$y - y_1 = m(x - x_1)$ • Substitute the slope and the coordinates of either one of the known points into the point-slope formula.

$y - 3 = -1(x - 2)$

$y - 3 = -x + 2$ • Solve for y.

$y = -x + 5$

The equation of the line is $y = -x + 5$.

Problem 2 Find the equation of the line containing the points $P_1(2, 0)$ and $P_2(5, 3)$.

Solution See page S9.

Try Exercise 37, page 166.

OBJECTIVE 3 Application problems

Linear functions can be used to model a variety of applications in science and business. For each application, data are collected and the independent and dependent variables are selected. Then a linear function is determined that models the data.

EXAMPLE 3 In 2000, there were approximately 50,000 centenarians (people 100 years old or older). Data from the Census Bureau show that this population is expected to increase through the year 2020 at a rate of approximately 4250 centenarians per year. Find a linear function that approximates the population of centenarians in terms of the year. Use your function to approximate the number of centenarians in 2015.

Strategy Select the independent and dependent variables. Because we want to determine the population of centenarians, that quantity is the *dependent* variable, y. The year is the *independent* variable.

From the data, the ordered pair (2000, 50,000) gives the coordinates of a point on the line. The slope of the line is the *rate of increase*, 4250 centenarians per year.

Solution

$$y - y_1 = m(x - x_1)$$ • Use the point-slope formula.

$$y - 50{,}000 = 4250(x - 2000)$$ • $m = 4250$; $(x_1, y_1) = (2000, 50{,}000)$

$$y - 50{,}000 = 4250x - 8{,}500{,}000$$

$$y = 4250x - 8{,}450{,}000$$

The linear function is $f(x) = 4250x - 8{,}450{,}000$.

$$\begin{aligned} f(x) &= 4250x - 8{,}450{,}000 \\ f(2015) &= 4250(2015) - 8{,}450{,}000 \\ &= 8{,}563{,}750 - 8{,}450{,}000 \\ &= 113{,}750 \end{aligned}$$

• Evaluate the function at 2015 to predict the number of centenarians in 2015.

The function gives an estimate of 113,750 centenarians in 2015.

Problem 3 Gabriel Daniel Fahrenheit invented the mercury thermometer in 1717. In terms of readings on this thermometer, water freezes at 32°F and boils at 212°F. In 1742, Anders Celsius invented the Celsius temperature scale. On this scale, water freezes at 0°C and boils at 100°C. Find a linear function that can be used to predict the Celsius temperature when the Fahrenheit temperature is known.

Solution See page S9.

Try Exercise 61, page 167.

3.5 Exercises

CONCEPT CHECK

1. How many lines with a given slope can be drawn through a given point in the plane?

2. Given two points in the plane, how many lines can be drawn through the two points?

3. Can the point-slope formula be used to find the equation of a line with zero slope?

4. Can the point-slope formula be used to find the equation of any line? Explain.

1 Find the equation of a line given a point and the slope (See pages 161–163.)

GETTING READY

5. In the equation of the line that has slope $-\frac{6}{5}$ and y-intercept $P(0, 2)$, m is __?__ and b is __?__. The equation is $y =$ __?__.

6. To find the equation of the line that contains the point $P(-4, 5)$ and has slope 2, use the point-slope formula.

$y - y_1 = m(x - x_1)$

$y - 5 = 2(x - (-4))$ • **Substitute __?__ for y_1, __?__ for m, and __?__ for x_1.**

$y - 5 = 2(x +$ __?__$)$ • **Simplify inside the parentheses.**

$y - 5 =$ __?__ • **Use the Distributive Property on the right side of the equation.**

$y =$ __?__ • **Add 5 to each side of the equation.**

Find the equation of the line that contains the given point and has the given slope.

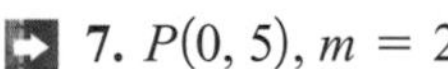 7. $P(0, 5), m = 2$

8. $P(0, 3), m = 1$

9. $P(2, 3), m = \frac{1}{2}$

10. $P(5, 1), m = \frac{2}{3}$

11. $P(3, 0), m = -\frac{5}{3}$

12. $P(-2, 0), m = \frac{3}{2}$

13. $P(-1, 7), m = -3$

14. $P(-2, 4), m = -4$

15. $P(0, 0), m = \frac{1}{2}$

16. $P(0, 0), m = \frac{3}{4}$

17. $P(2, -3), m = 3$

18. $P(4, -5), m = 2$

19. $P(3, 5), m = -\frac{2}{3}$

20. $P(5, 1), m = -\frac{4}{5}$

21. $P(0, -3), m = -1$

22. $P(2, 0), m = \frac{5}{6}$

23. $P(3, -4)$, slope is undefined

24. $P(-2, 5)$, slope is undefined

25. $P(-2, -3), m = 0$

26. $P(-3, -2), m = 0$

27. $P(4, -5), m = -2$

28. $P(-3, 5), m = 3$

29. $P(-5, -1)$, slope is undefined

30. $P(0, 4)$, slope is undefined

31. A student found the equation of the line containing the points $P(-2, 4)$ and $Q(1, 1)$ to be $y = x + 6$. Sketch the line containing $P(-2, 4)$ and $Q(1, 1)$, and use your sketch to explain why the student's values for m and b cannot be correct.

32. A student found the equation of the line containing the points $P(-4, 4)$ and $Q(4, 6)$ to be $y = \frac{1}{4}x + 5$. Sketch the line containing $P(-4, 4)$ and $Q(4,6)$, and use your sketch to explain why the student's values for m and b seem reasonable.

2 Find the equation of a line given two points (See pages 163–164.)

GETTING READY

33. The first step in finding the equation of the line through two given points is to use the coordinates of the points to find the __?__ of the line.

34. Points $P_1(4, 3)$ and $P_2(4, -2)$ have the same __?__-coordinate, so the slope of the line containing P_1 and P_2 is __?__. The equation of the line is __?__.

Find the equation of the line that contains the given points.

35. $P_1(0, 2), P_2(3, 5)$

36. $P_1(0, 4), P_2(1, 5)$

37. $P_1(0, -3), P_2(-4, 5)$

38. $P_1(0, -2), P_2(-3, 4)$

39. $P_1(-1, 3), P_2(2, 4)$

40. $P_1(-1, 1), P_2(4, 4)$

41. $P_1(0, 3), P_2(2, 0)$

42. $P_1(0, 4), P_2(2, 0)$

43. $P_1(-2, -3), P_2(-1, -2)$

44. $P_1(4, 1), P_2(3, -2)$

45. $P_1(2, 3), P_2(5, 5)$

46. $P_1(7, 2), P_2(4, 4)$

47. $P_1(2, 0), P_2(0, -1)$

48. $P_1(0, 4), P_2(-2, 0)$

49. $P_1(3, -4), P_2(-2, -4)$

50. $P_1(-3, 3), P_2(-2, 3)$

51. $P_1(0, 0), P_2(4, 3)$

52. $P_1(2, -5), P_2(0, 0)$

53. $P_1(-2, 5), P_2(-2, -5)$

54. $P_1(3, 2), P_2(3, -4)$

55. $P_1(2, 1), P_2(-2, -3)$

56. $P_1(-3, -2), P_2(1, -4)$

57. $P_1(0, 3), P_2(3, 0)$

58. $P_1(1, -3), P_2(-2, 4)$

3 Application problems (See page 164.)

GETTING READY

59. A plane takes off from an airport that is 500 ft above sea level. The plane climbs at a rate of 1000 ft/min. If $y = mx + b$ is a linear equation that gives the height y (in feet) of the plane above sea level in terms of the time x (in minutes) after take-off, then $m =$ __?__, and an ordered pair that is a solution of the equation is (0, __?__).

60. A plane takes off from an airport that is at sea level and climbs to a height of 12,500 ft in 10 min. If $y = mx + b$ is a linear equation that gives the height y (in feet) of the plane above sea level in terms of the time x (in minutes) after take-off, then two ordered pairs that are solutions of the equation are (0, __?__) and (10, __?__).

61. **Aeronautics** The pilot of a Boeing 757 jet takes off from Boston's Logan Airport, which is at sea level, and climbs to a cruising altitude of 32,000 ft at a constant rate of 1200 ft/min. Find a linear function for the height of the plane in terms of the time after take-off. Use your function to find the height of the plane 11 min after take-off.

62. **Physical Fitness** A jogger running at 9 mph burns approximately 14 calories per minute. Find a linear function for the number of calories burned by the jogger in terms of the number of minutes run. Use your function to find the number of calories burned after jogging for 32 min.

63. **Telecommunications** A cellular phone company offers several different options for using a cellular telephone. One option, for people who plan on using the phone only in emergencies, costs the user $4.95 per month plus $.59 per minute for each minute the phone is used. Find a linear function for the monthly cost of the phone in terms of the number of minutes the phone is used. Use your function to find the monthly cost of using the cellular phone for 13 min in one month.

64. **Aeronautics** An Airbus 320 plane takes off from Denver International Airport, which is 5200 ft above sea level, and climbs to 30,000 ft at a constant rate of 1000 ft/min. Find a linear function for the height of the plane in terms of the time after take-off. Use your function to find the height of the plane 8 min after take-off.

65. **Nutrition** There are approximately 126 calories in a 2-ounce serving of lean hamburger and approximately 189 calories in a 3-ounce serving. Find a linear function for the number of calories in lean hamburger in terms of the size of the serving. Use your function to estimate the number of calories in a 5-ounce serving of lean hamburger.

66. **Chemistry** At sea level, the boiling point of water is 100°C. At an altitude of 2 km, the boiling point of water is 93°C. Find a linear function for the boiling point of water in terms of the altitude above sea level. Use your function to predict the boiling point of water on the top of Mount Everest, which is approximately 8.85 km above sea level. Round to the nearest degree.

67. **Ecology** Use the information in the news clipping at the right. Find a linear function for the percent of trees at 2600 ft that are hardwoods in terms of the year. Use your function to predict the percent of trees at 2600 ft that will be hardwoods in 2020.

68. **Oceanography** Whales, dolphins, and porpoises communicate using high-pitched sounds that travel through the water. The speed at which the sound travels depends on many factors, one of which is the depth of the water. At approximately 1000 m below sea level, the speed of sound is 1480 m/s. Below 1000 m, the speed of sound increases at a constant rate of 0.017 m/s for each additional meter below 1000 m. Find a linear function for the speed of sound in terms of the number of meters below sea level. Use your function to approximate the speed of sound 2500 m below sea level. Round to the nearest meter per second.

69. **Automotive Technology** The gas tank of a certain car contains 16 gal when the driver of the car begins a trip. Each mile driven by the driver decreases the amount of gas in the tank by 0.032 gal. Find a linear function for the number of gallons of gas in the tank in terms of the number of miles driven. Use your function to find the number of gallons in the tank after driving 150 mi.

In the News

Is Global Warming Moving Mountains?

In the mountains of Vermont, maples, beeches, and other hardwood trees that thrive in warm climates are gradually taking over areas that once supported more cold-loving trees, such as balsam and fir. Ecologists report that in 2004, 82% of the trees at an elevation of 2600 ft were hardwoods, as compared to only 57% in 1964.

Source: The Boston Globe

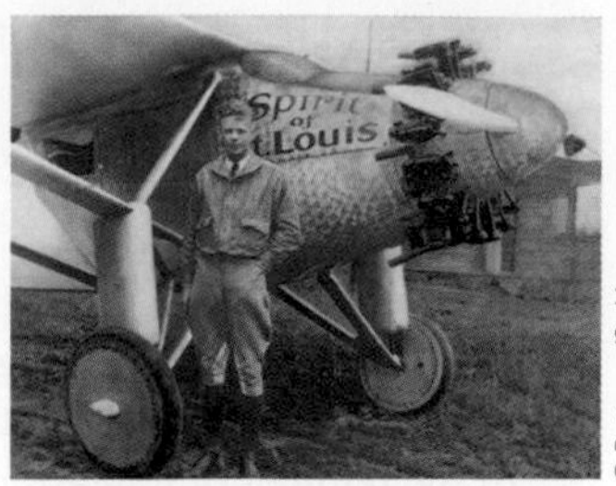

70. **History** In 1927, Charles Lindbergh made history by making the first transatlantic flight from New York to Paris. It took Lindbergh approximately 33.5 h to make the trip. In 1997, the Concorde could make the same trip in approximately 3.3 h. Find a linear function for the time, in hours, to fly across the Atlantic in terms of the year. Use your function to predict how long a flight between the two cities would have taken in 1967. Round your answer to the nearest tenth.

APPLYING CONCEPTS

71. What are the coordinates of the x-intercept of the graph of $y = mx + b$?

72. A line contains the points $P(4, -1)$ and $Q(2, 1)$. Find the coordinates of three other points that are on this line.

73. Given that f is a linear function for which $f(1) = 3$ and $f(-1) = 5$, determine $f(4)$.

74. Find the equation of the line that passes through the midpoint of the line segment between $P_1(2, 5)$ and $P_2(-4, 1)$ and has slope -2.

75. If $y = mx + b$, where m is a given constant, how does the graph of this equation change as the value of b changes?

76. Explain the similarities and differences between the point-slope formula and the slope-intercept form of a straight line.

77. Explain why the point-slope formula cannot be used to find the equation of a line that is parallel to the y-axis.

PROJECTS OR GROUP ACTIVITIES

78. Assume that the maximum speed your car will attain is a linear function of the steepness of the hill it is climbing or descending. If the steepness of the hill is 5° up (the road makes a 5° angle with the horizontal), your car's maximum speed is 77 km/h. If the steepness of the hill is 2° down, your car's maximum speed is 154 km/h. When your car's maximum speed is 99 km/h, how steep is the hill? State your answer in degrees, and note whether the steepness is up or down.

79. Suppose that f is a linear function and that $f(3) = 10$ and $f(-3) = -2$. Find the value of a for which $f(a) = 0$.

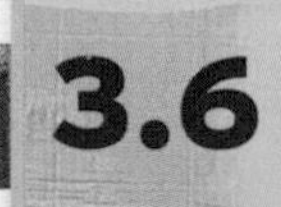

3.6 Parallel and Perpendicular Lines

OBJECTIVE 1 Find equations of parallel and perpendicular lines

Two lines that have the same slope and different y-intercepts do not intersect and are called **parallel lines.**

The slope of each of the lines at the right is $\frac{2}{3}$.

The lines are parallel.

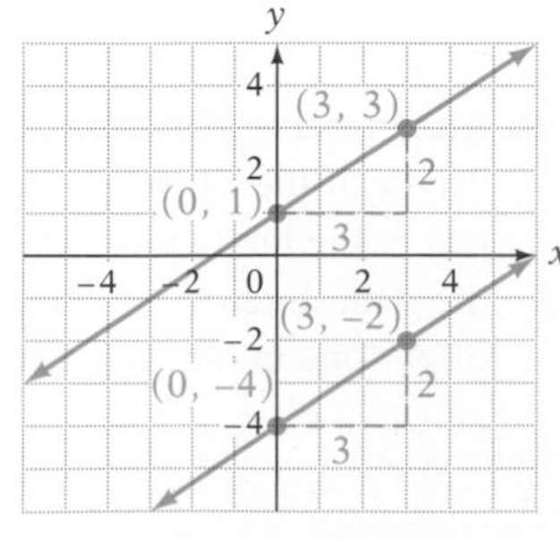

SLOPES OF PARALLEL LINES

Two nonvertical lines with slopes of m_1 and m_2 are parallel if and only if $m_1 = m_2$. Vertical lines are parallel lines.

EXAMPLES

1. The slopes of the graphs of $y = -\frac{2}{3}x + 2$ and $y = -\frac{2}{3}x - 3$ are both $-\frac{2}{3}$. The lines are parallel.
2. The graphs of $x = 2$ and $x = 5$ are vertical lines. The lines are parallel.
3. The slope of the graph of $y = 2x + 3$ is 2. The slope of the graph of $y = -2x + 1$ is -2. The slopes are not equal. The lines are not parallel.

EXAMPLE 1 Are the graphs of $y = -\frac{3}{2}x - 5$ and $2x + 3y = 6$ parallel?

Solution Write, if necessary, each equation in the form $y = mx + b$, and then compare the values of m, the slopes of the lines. If the values of m are equal, the graphs are parallel.

The equation $y = -\frac{3}{2}x - 5$ is in the form $y = mx + b$.

The slope is $-\frac{3}{2}$.

$$2x + 3y = 6$$
$$3y = -2x + 6$$
$$y = -\frac{2}{3}x + 2$$

• **Write the equation $2x + 3y = 6$ in the form $y = mx + b$.**

The slope is $-\frac{2}{3}$.

$$-\frac{2}{3} \neq -\frac{3}{2}$$

• **Compare the slopes.**

Because the slopes are not the same, the graphs are not parallel.

Problem 1 Is the line that contains the points with coordinates $(-2, 1)$ and $(-5, -1)$ parallel to the line that contains the points with coordinates $(1, 0)$ and $(4, 2)$?

Solution See page S9.

Try Exercise 19, page 174.

Focus on finding the equation of a line parallel to a given line

A. Find the equation of the line that contains the point $P(-1, 4)$ and is parallel to the graph of $2x - 3y = 5$.

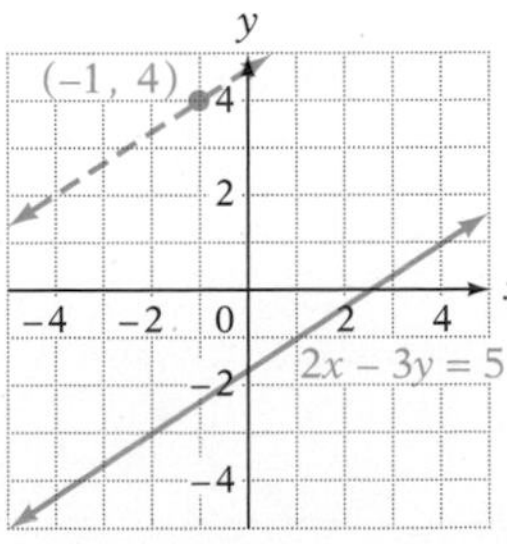

Because the lines are parallel, the slope of the unknown line is the same as the slope of the given line. Solve $2x - 3y = 5$ for y and determine its slope.

$$2x - 3y = 5$$
$$-3y = -2x + 5$$
$$y = \frac{2}{3}x - \frac{5}{3}$$

The slope of the given line is $\frac{2}{3}$. Because the lines are parallel, the slope of the unknown line is also $\frac{2}{3}$.

Use the point-slope formula. $y - y_1 = m(x - x_1)$

Substitute $m = \frac{2}{3}$ and $(x_1, y_1) = (-1, 4)$. $y - 4 = \frac{2}{3}[x - (-1)]$

Solve for y.

$$y - 4 = \frac{2}{3}(x + 1)$$

$$y - 4 = \frac{2}{3}x + \frac{2}{3}$$

$$y = \frac{2}{3}x + \frac{14}{3}$$

The equation of the line is $y = \frac{2}{3}x + \frac{14}{3}$.

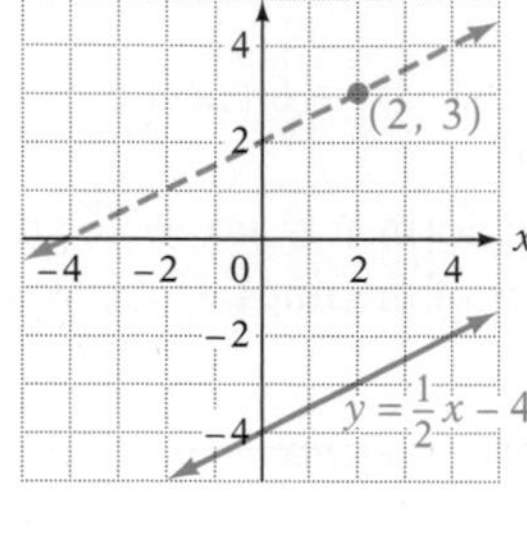

B. Find the equation of the line that contains the point $P(2, 3)$ and is parallel to the graph of $y = \frac{1}{2}x - 4$.

The slope of the given line is $\frac{1}{2}$. Because parallel lines have the same slope, the slope of the unknown line is also $\frac{1}{2}$.

Use the point-slope formula. $y - y_1 = m(x - x_1)$

Substitute $m = \frac{1}{2}$ and $(x_1, y_1) = (2, 3)$. $y - 3 = \frac{1}{2}(x - 2)$

Solve for y.

$$y - 3 = \frac{1}{2}x - 1$$

$$y = \frac{1}{2}x + 2$$

The equation of the line is $y = \frac{1}{2}x + 2$.

EXAMPLE 2 Find the equation of the line that contains the point $P(3, -1)$ and is parallel to the graph of $3x - 2y = 4$.

Solution

$$3x - 2y = 4$$

$$-2y = -3x + 4$$

$$y = \frac{3}{2}x - 2$$

• **Solve the equation for y to find the slope.**

$$m = \frac{3}{2}$$

• **The parallel line has the same slope as the given line.**

$$y - y_1 = m(x - x_1)$$

• **Use the point-slope formula.**

$$y - (-1) = \frac{3}{2}(x - 3)$$

• **$m = \frac{3}{2}$, $(x_1, y_1) = (3, -1)$**

$$y + 1 = \frac{3}{2}x - \frac{9}{2}$$

• **Solve for y.**

$$y = \frac{3}{2}x - \frac{11}{2}$$

The equation of the line is $y = x - \frac{11}{2}$.

Problem 2 Find the equation of the line that contains the point $P(4, 1)$ and is parallel to the graph of $y = \frac{3}{4}x - 1$.

Solution See page S10.

Try Exercise 31, page 175.

Two lines that intersect at right angles, as in Figure 1 below, are **perpendicular lines.** A horizontal line is perpendicular to a vertical line, as shown in Figure 2 below.

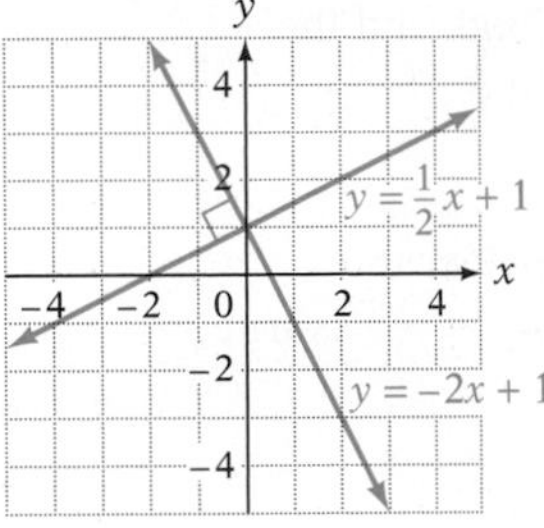

FIGURE 1

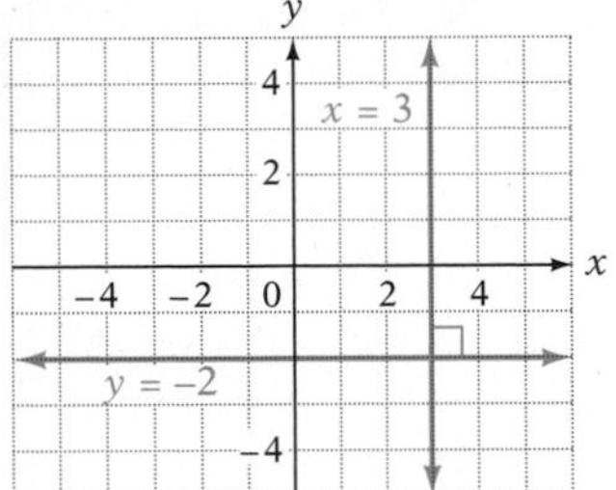

FIGURE 2

SLOPES OF PERPENDICULAR LINES

If m_1 and m_2 are the slopes of two lines, neither of which is vertical, then the lines are perpendicular if and only if $m_1 \cdot m_2 = -1$. A vertical line is perpendicular to a horizontal line.

EXAMPLES

1. The slope of the graph of $y = \frac{3}{4}x + 1$ is $\frac{3}{4}$, and the slope of the graph of $y = -\frac{4}{3}x - 3$ is $-\frac{4}{3}$. The product of the two slopes is $\frac{3}{4} \cdot \left(-\frac{4}{3}\right) = -1$. The lines are perpendicular.
2. The graph of $x = -1$ is a vertical line, and the graph of $y = 5$ is a horizontal line. The lines are perpendicular.
3. The slope of the graph of $y = \frac{1}{2}x + 3$ is $\frac{1}{2}$. The slope of the graph of $y = 2x + 1$ is 2. The product of the slopes is $\frac{1}{2} \cdot 2 = 1 \neq -1$. The lines are not perpendicular.

Focus on determining whether two lines are perpendicular

A. Is the line that contains the points with coordinates (4, 2) and (−2, 5) perpendicular to the line that contains the points with coordinates (−4, 3) and (−3, 5)?

Find the slope of the line that passes through the points with coordinates (4, 2) and (−2, 5).

$$m_1 = \frac{5 - 2}{-2 - 4} = \frac{3}{-6} = -\frac{1}{2}$$

Find the slope of the line that passes through the points with coordinates (−4, 3) and (−3, 5).

$$m_2 = \frac{5 - 3}{-3 - (-4)} = \frac{2}{1} = 2$$

Find the product of the two slopes.

$$m_1 \cdot m_2 = -\frac{1}{2}(2) = -1$$

Because $m_1 \cdot m_2 = -1$, the lines are perpendicular.

B. Are the lines whose equations are $3x + 4y = 8$ and $8x + 6y = 5$ perpendicular?

To determine whether the lines are perpendicular, solve each equation for y and find the slope of each line.

$$3x + 4y = 8 \qquad\qquad 8x + 6y = 5$$
$$4y = -3x + 8 \qquad\qquad 6y = -8x + 5$$
$$y = -\frac{3}{4}x + 2 \qquad\qquad y = -\frac{4}{3}x + \frac{5}{6}$$

Then check whether the equation $m_1 \cdot m_2 = -1$ is true.

$$m_1 = -\frac{3}{4} \qquad\qquad m_2 = -\frac{4}{3}$$
$$m_1 \cdot m_2 = \left(-\frac{3}{4}\right)\left(-\frac{4}{3}\right) = 1$$

Because $m_1 \cdot m_2 = 1 \neq -1$, the lines are not perpendicular.

EXAMPLE 3 Is the line that contains the points with coordinates $(-2, 3)$ and $(-2, -5)$ perpendicular to the line that contains the points with coordinates $(-1, 4)$ and $(2, 4)$?

Solution

$$m_1 = \frac{-5 - 3}{-2 - (-2)} = \frac{-8}{0} \quad \text{undefined}$$

- **Find the slope of the line between the points with coordinates $(-2, 3)$ and $(-2, -5)$. The slope is undefined. The line is vertical.**

$$m_2 = \frac{4 - 4}{2 - (-1)} = \frac{0}{3} = 0$$

- **Find the slope of the line between the points with coordinates $(-1, 4)$ and $(2, 4)$. The slope is zero. The line is horizontal.**

One line is vertical and one is horizontal. The lines are perpendicular.

Problem 3 Are the graphs of $4x - y = -2$ and $x + 4y = -12$ perpendicular?

Solution See page S10.

Try Exercise 21, page 174.

Solving $m_1 \cdot m_2 = -1$ for m_1 gives $m_1 = -\frac{1}{m_2}$. This last equation states that the slopes of perpendicular lines are **negative reciprocals** of each other.

Focus on finding the equations of perpendicular lines

A. Find the equation of the line that contains the point $P(-2, 1)$ and is perpendicular to the graph of $y = -\frac{2}{3}x + 1$.

The slope of the graph of $y = -\frac{2}{3}x + 1$ is $-\frac{2}{3}$. The slope of the line perpendicular to the given line is $\frac{3}{2}$, the negative reciprocal of $-\frac{2}{3}$.

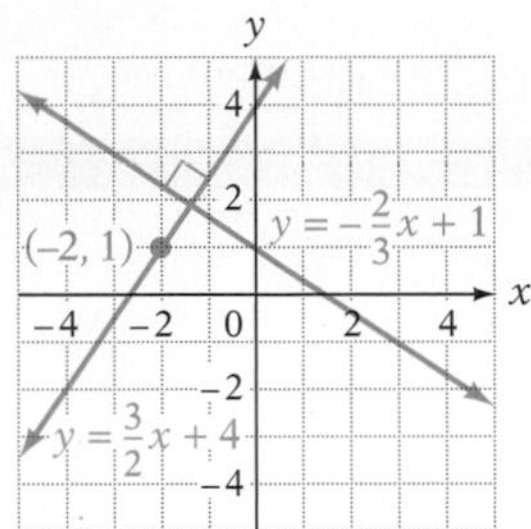

Use the point-slope formula.	$y - y_1 = m(x - x_1)$
Substitute $m = \frac{3}{2}$ and $(x_1, y_1) = (-2, 1)$.	$y - 1 = \frac{3}{2}(x - (-2))$
Solve for y.	$y - 1 = \frac{3}{2}x + 3$
	$y = \frac{3}{2}x + 4$

The equation of the line is $y = \frac{3}{2}x + 4$.

B. Find the equation of the line that contains the point $P(3, -4)$ and is perpendicular to the graph of $2x - y = -3$.

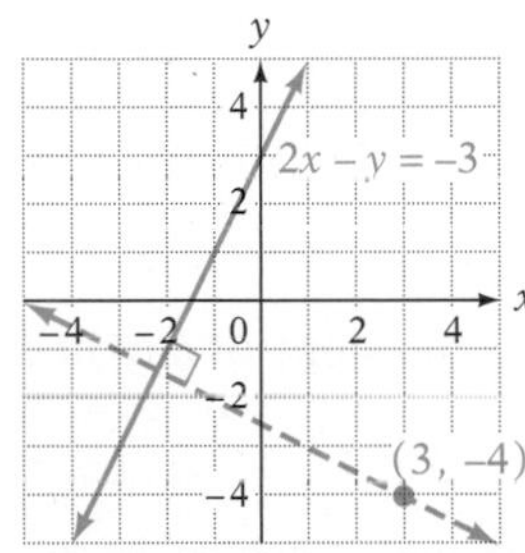

Determine the slope of the given line by solving the equation for y.	$2x - y = -3$
	$-y = -2x - 3$
The slope is 2.	$y = 2x + 3$
The slope of the line perpendicular to the given line is $-\frac{1}{2}$, the negative reciprocal of 2.	
Use the point-slope formula.	$y - y_1 = m(x - x_1)$
Substitute $m = -\frac{1}{2}$ and $(x_1, y_1) = (3, -4)$.	$y - (-4) = -\frac{1}{2}(x - 3)$
Solve for y.	$y + 4 = -\frac{1}{2}x + \frac{3}{2}$
	$y = -\frac{1}{2}x - \frac{5}{2}$

The equation of the line is $y = -\frac{1}{2}x - \frac{5}{2}$.

EXAMPLE 4 Find the equation of the line that contains the point $P(3, -5)$ and is perpendicular to the graph of $y = -3x + 2$.

Solution The slope of $y = -3x + 2$ is -3. The slope of a line perpendicular to the given line is $\frac{1}{3}$, the negative reciprocal of -3.

$y - y_1 = m(x - x_1)$	• **Use the point-slope formula.**
$y - (\ 5) = \frac{1}{3}(x - 3)$	• **Substitute $m = \frac{1}{3}$ and $(x_1, y_1) = (3, -5)$.**
$y + 5 = \frac{1}{3}x - 1$	• **Solve for y.**
$y = \frac{1}{3}x - 6$	

The equation of the line is $y = \frac{1}{3}x - 6$.

Problem 4 Find the equation of the line that contains the point $P(-2, 3)$ and is perpendicular to the graph of $x - 4y = 3$.

Solution See page S10.

Try Exercise 35, page 175.

3.6 Exercises

CONCEPT CHECK

1. Given the slopes of two lines, explain how to determine whether the two lines are parallel.

2. Given the slopes of two lines, how can you determine whether the two lines are perpendicular?

3. Complete the following sentence. Parallel lines have the same __?__.

4. What is the negative reciprocal of $-\frac{3}{4}$?

5. The slope of a line is -5. What is the slope of any line parallel to this line?

6. The slope of a line is $\frac{3}{2}$. What is the slope of any line parallel to this line?

7. The slope of a line is 4. What is the slope of any line perpendicular to this line?

8. The slope of a line is $-\frac{4}{5}$. What is the slope of any line perpendicular to this line?

9. Give the slope of any line that is parallel to the graph of $y = -\frac{1}{3}x + 5$.

10. Give the slope of any line that is perpendicular to the graph of $y = \frac{3}{5}x + 2$.

1 Find equations of parallel and perpendicular lines (See pages 168–173.)

11. Is the graph of $x = -2$ perpendicular to the graph of $y = 3$?

12. Is the graph of $y = \frac{1}{2}$ perpendicular to the graph of $y = -4$?

13. Is the graph of $x = -3$ parallel to the graph of $y = \frac{1}{3}$?

14. Is the graph of $x = 4$ perpendicular to the graph of $y = -4$?

15. Is the graph of $y = \frac{2}{3}x - 4$ parallel to the graph of $y = -\frac{3}{2}x - 4$?

16. Is the graph of $y = -2x + \frac{2}{3}$ parallel to the graph of $y = -2x + 3$?

17. Is the graph of $y = \frac{4}{3}x - 2$ perpendicular to the graph of $y = -\frac{3}{4}x + 2$?

18. Is the graph of $y = \frac{1}{2}x + \frac{3}{2}$ perpendicular to the graph of $y = -\frac{1}{2}x + \frac{3}{2}$?

19. Are the graphs of $2x + 3y = 2$ and $2x + 3y = -4$ parallel?

20. Are the graphs of $2x - 4y = 3$ and $2x + 4y = -3$ parallel?

21. Are the graphs of $x - 4y = 2$ and $4x + y = 8$ perpendicular?

22. Are the graphs of $4x - 3y = 2$ and $4x + 3y = -7$ perpendicular?

23. Is the line that contains the points with coordinates $(3, 2)$ and $(1, 6)$ parallel to the line that contains the points with coordinates $(-1, 3)$ and $(-1, -1)$?

24. Is the line that contains the points with coordinates $(4, -3)$ and $(2, 5)$ parallel to the line that contains the points with coordinates $(-2, -3)$ and $(-4, 1)$?

25. Is the line that contains the points with coordinates $(-3, 2)$ and $(4, -1)$ perpendicular to the line that contains the points with coordinates $(1, 3)$ and $(-2, -4)$?

26. Is the line that contains the points with coordinates $(-1, 2)$ and $(3, 4)$ perpendicular to the line that contains the points with coordinates $(-1, 3)$ and $(-4, 1)$?

27. Is the line that contains the points with coordinates $(-5, 0)$ and $(0, 2)$ parallel to the line that contains the points with coordinates $(5, 1)$ and $(0, -1)$?

28. Is the line that contains the points with coordinates $(3, 5)$ and $(-3, 3)$ perpendicular to the line that contains the points with coordinates $(2, -5)$ and $(-4, 4)$?

29. Find the equation of the line that contains the point $P(-2, -4)$ and is parallel to the graph of $2x - 3y = 2$.

30. Find the equation of the line that contains the point $P(3, 2)$ and is parallel to the graph of $3x + y = -3$.

31. Find the equation of the line that contains the point $P(-4, -1)$ and is parallel to the graph of $y = -\frac{5}{4}x + 3$.

32. Find the equation of the line that contains the point $P(3, -3)$ and is parallel to the graph of $y = \frac{2x}{3} - 1$.

33. Find the equation of the line that contains the point $P(4, 1)$ and is perpendicular to the graph of $y = -3x + 4$.

34. Find the equation of the line that contains the point $P(2, -5)$ and is perpendicular to the graph of $y = \frac{5}{2}x - 4$.

35. Find the equation of the line that contains the point $P(-1, -3)$ and is perpendicular to the graph of $3x - 5y = 2$.

36. Find the equation of the line that contains the point $P(-1, 3)$ and is perpendicular to the graph of $2x + 4y = -1$.

37. l is a line with equation $Ax + By = C$ such that A and B are both positive. Does a line parallel to l slant downward to the right or upward to the right?

38. l is a line with equation $Ax + By = C$ such that A is positive and B is negative. Does a line perpendicular to l slant downward to the right or upward to the right?

APPLYING CONCEPTS

39. If the graphs of $A_1x + B_1y = C_1$ and $A_2x + B_2y = C_2$ are parallel, express $\frac{A_1}{B_1}$ in terms of A_2 and B_2.

40. If the graphs of $A_1x + B_1y = C_1$ and $A_2x + B_2y = C_2$ are perpendicular, express $\frac{A_1}{B_1}$ in terms of A_2 and B_2.

For Exercises 41 and 42, suppose a ball is being twirled at the end of a string and the center of rotation is the origin of a coordinate system. If the string breaks, the initial path of the ball is on a line that is perpendicular to the radius of the circle.

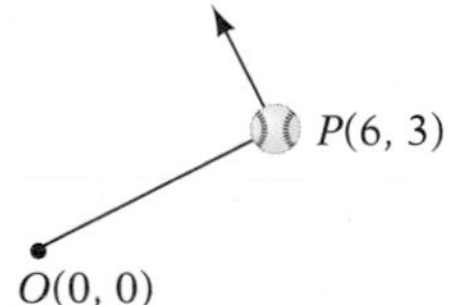

41. Suppose the string breaks when the ball is at the point $P(6, 3)$. Find the equation of the line on which the initial path lies.

42. Suppose the string breaks when the ball is at the point $P(2, 8)$. Find the equation of the line on which the initial path lies.

43. The graphs of $y = -\frac{1}{2}x + 2$ and $y = \frac{2}{3}x - 5$ intersect at the point whose coordinates are $(6, -1)$. Find the equation of a line whose graph intersects the graphs of the given lines to form a right triangle. (*Hint:* There is more than one answer to this question.)

PROJECTS OR GROUP ACTIVITIES

44. Geometry A theorem from geometry states that a line passing through the center of a circle and through a point P on the circle is perpendicular to the tangent line at P. See the figure at the right.

a. If the coordinates of P are $(5, 4)$ and the coordinates of C are $(3, 2)$, what is the equation of the tangent line?

b. What are the coordinates of the x- and y-intercepts of the tangent line?

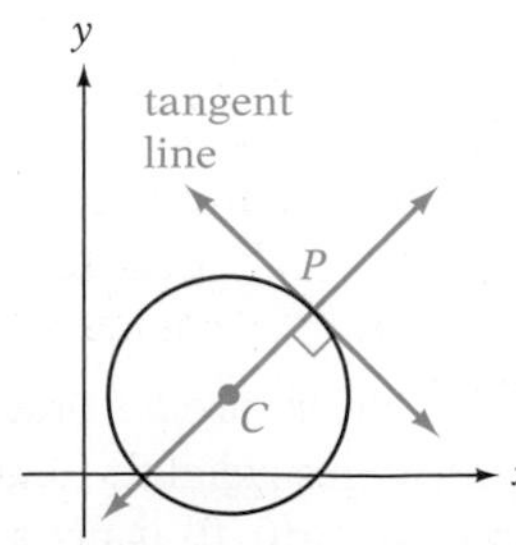

3.7 Inequalities in Two Variables

OBJECTIVE 1 Graph the solution set of an inequality in two variables

The graph of the linear equation $y = x - 1$ separates the plane into three sets: the set of points on the line, the set of points above the line, and the set of points below the line.

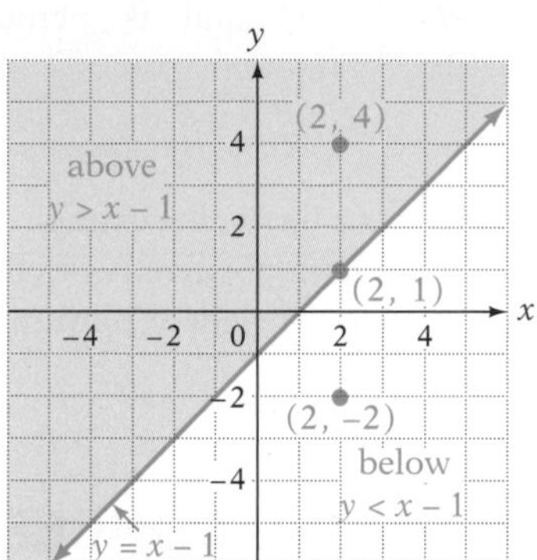

The point whose coordinates are (2, 1) is a solution of $y = x - 1$ and is a point on the line.

The point whose coordinates are (2, 4) is a solution of $y > x - 1$ and is a point above the line.

The point whose coordinates are (2, −2) is a solution of $y < x - 1$ and is a point below the line.

The set of points on the line are the solutions of the equation $y = x - 1$. The set of points above the line are the solutions of the inequality $y > x - 1$. These points form a **half-plane.** The set of points below the line are solutions of the inequality $y < x - 1$. These points also form a half-plane.

An inequality of the form $y > mx + b$ or $Ax + By > C$ is a **linear inequality in two variables.** (The inequality symbol could be replaced by $\geq$, $<$, or $\leq$.) The solution set of a linear inequality in two variables is a half-plane.

The following illustrates the procedure for graphing the solution set of a linear inequality in two variables.

Focus on graphing the solution set of an inequality in two variables

Graph the solution set of $3x - 4y < 12$.

Solve the inequality for y.

$$3x - 4y < 12$$
$$-4y < -3x + 12$$
$$y > \frac{3}{4}x - 3$$

Take Note

When solving the inequality at the right for y, we divide each side of the inequality by −4, so the inequality symbol must be reversed.

Change the inequality $y > \frac{3}{4}x - 3$ to the equality $y = \frac{3}{4}x - 3$, and graph the line.

If the inequality contains $\leq$ or $\geq$, the line belongs to the solution set and is shown by a *solid line.* If the inequality contains $<$ or $>$, the line is not part of the solution set and is shown by a *dashed line.*

If the inequality contains $>$ or $\geq$, shade the upper half-plane. If the inequality contains $<$ or $\leq$, shade the lower half-plane.

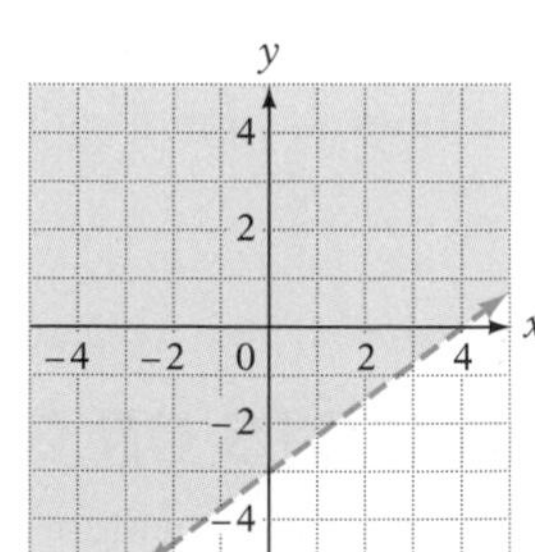

As a check, use the ordered pair (0, 0) to determine whether the correct region of the plane has been shaded. If (0, 0) is a solution of the inequality, then (0, 0) should be in the shaded region. If (0, 0) is not a solution of the inequality, then (0, 0) should not be in the shaded region. *Note:* If the line passes through the point (0, 0), another point, such as (0, 1), must be used as a check.

As shown below, (0, 0) is a solution of the inequality in the preceding Focus On example.

$$y > \frac{3}{4}x - 3$$
$$0 > \frac{3}{4}(0) - 3$$
$$0 > 0 - 3$$
$$0 > -3 \qquad \text{True}$$

Because (0, 0) is a solution of the inequality, (0, 0) should be in the shaded region. The solution set as graphed in the Focus On is correct.

From the graph of $y > \frac{3}{4}x - 3$, note that for a given value of x, more than one value of y can be paired with the value of x. For instance, (4, 1), (4, 3), (5, 1), and $(5, \frac{9}{4})$ are all ordered pairs that belong to the graph.

Because there are ordered pairs with the same first coordinate and different second coordinates, the inequality does not represent a function. The inequality is a relation but not a function.

EXAMPLE 1 Graph the solution set of $x + 2y \leq 4$.

Solution

$$x + 2y \leq 4$$
$$2y \leq -x + 4$$
$$y \leq -\frac{1}{2}x + 2$$

- **Solve the inequality for y.**

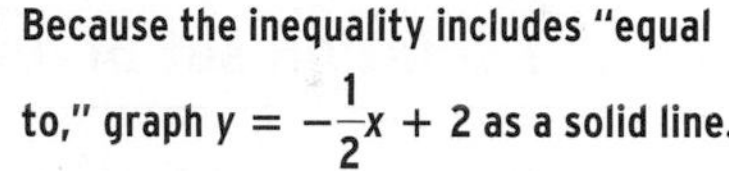
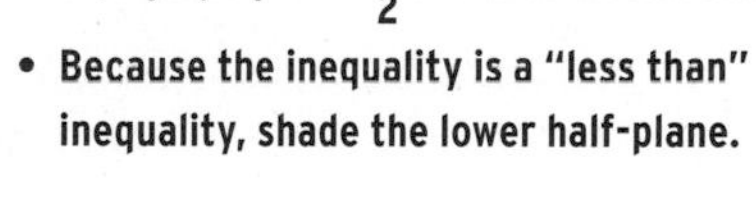

- **Because the inequality includes "equal to," graph $y = -\frac{1}{2}x + 2$ as a solid line.**
- **Because the inequality is a "less than" inequality, shade the lower half-plane.**

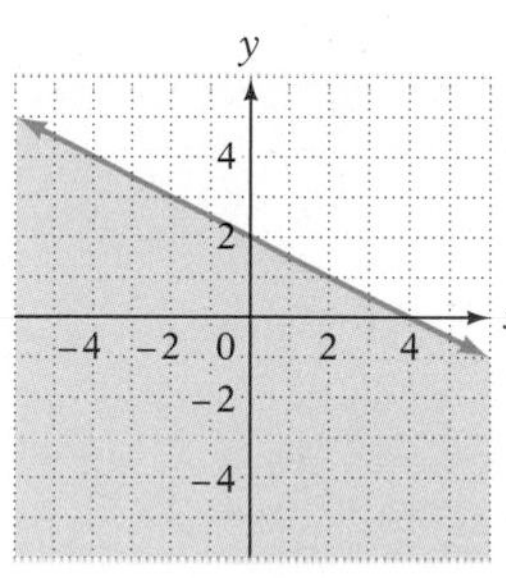

Problem 1 Graph the solution set of $x + 3y > 6$.

Solution See page S10.

Try Exercise 19, page 178.

EXAMPLE 2 Graph the solution set of $x \geq -1$.

Solution

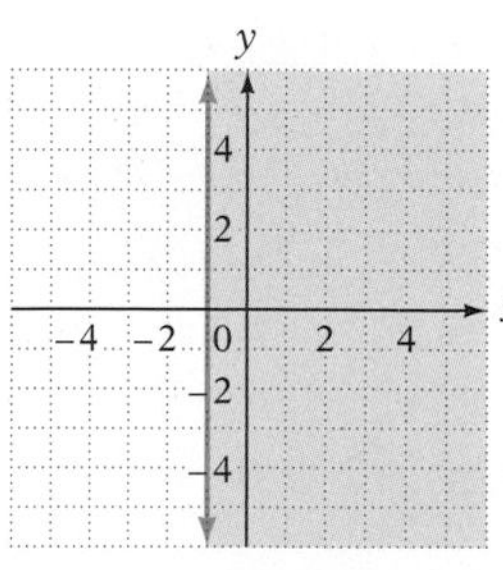

- **Graph $x = -1$ as a solid line.**
- **The point (0, 0) satisfies the inequality. Shade the half-plane to the right of the line.**

Problem 2 Graph the solution set of $y < 2$.

Solution See page S10.

Try Exercise 21, page 178.

3.7 Exercises

CONCEPT CHECK

1. What is a half-plane?

2. Explain a method that you can use to check that the graph of a linear inequality in two variables has been shaded correctly.

3. Is (0, 0) a solution of $y > 2x - 7$?

4. Is (0, 0) a solution of $y < 5x + 3$?

5. Is (0, 0) a solution of $y \leq -\frac{2}{3}x - 8$?

6. Is (0, 0) a solution of $y \geq -\frac{3}{4}x + 9$?

7. Complete the following sentence with *above* or *below*. The solution set of an inequality of the form $y > mx + b$ is all the points in the plane __?__ the graph of $y = mx + b$.

8. Complete the following sentence with *above* or *below*. The solution set of an inequality of the form $y < mx + b$ is all the points in the plane __?__ the graph of $y = mx + b$.

1 Graph the solution set of an inequality in two variables (See pages 176–177.)

Graph the solution set.

9. $y \leq \frac{3}{2}x - 3$

10. $y \geq \frac{4}{3}x - 4$

11. $y < \frac{4}{5}x - 2$

12. $y < \frac{3}{5}x - 3$

13. $y < -\frac{1}{3}x + 2$

14. $y < -\frac{4}{3}x + 3$

15. $x + 3y < 4$

16. $2x - 5y \leq 10$

17. $2x + 3y \geq 6$

18. $3x + 2y < 4$

19. $-x + 2y > -8$

20. $-3x + 2y > 2$

21. $y - 4 < 0$

22. $x + 2 \geq 0$

23. $6x + 5y < 15$

24. $3x - 5y < 10$

25. $-5x + 3y \geq -12$

26. $3x + 4y \geq 12$

27. Given that (0, 0) are the coordinates of a point on the graph of the linear inequality $Ax + By > C$, and that C is not zero, is C positive or negative?

28. Are (0, 0) the coordinates of a point on the graph of the linear inequality $Ax + By > C$, where C is a positive number?

APPLYING CONCEPTS

29. Does the inequality $y < 3x - 1$ represent a function? Explain your answer.

30. Are there any points whose coordinates satisfy both $y \leq x + 3$ and $y \geq -\frac{1}{2}x + 1$? If so, give the coordinates of three such points. If not, explain why not.

31. Are there any points whose coordinates satisfy both $y \leq x - 1$ and $y \geq x + 2$? If so, give the coordinates of three such points. If not, explain why not.

PROJECTS OR GROUP ACTIVITIES

32. Graph the solution set of the inequality $|x| + |y| \leq 5$.

CHAPTER 3 Summary

Key Words	Objective and Page Reference	Examples
A **rectangular coordinate system** is formed by two number lines, one horizontal and one vertical, that intersect at the zero point of each line. The number lines that make up a coordinate system are called the **coordinate axes** or simply the **axes.** The **origin** is the point of intersection of the two coordinate axes. Generally, the horizontal axis is labeled the **x-axis,** and the vertical axis is labeled the **y-axis.** A rectangular coordinate system divides the **plane** determined by the axes into four regions called **quadrants.**	[3.1.1, p. 110]	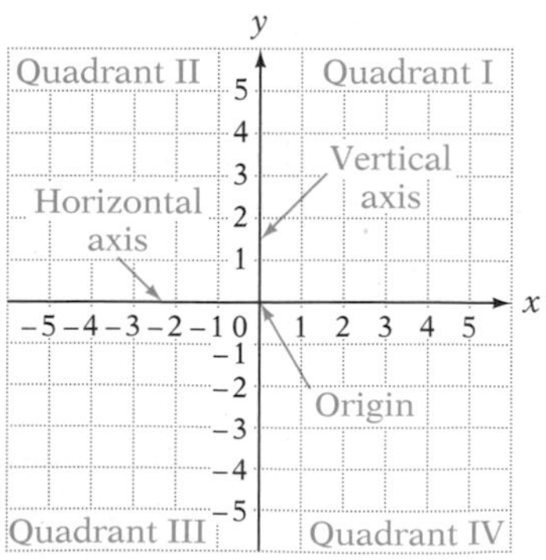

An **ordered pair (x, y)** is used to locate a point in a plane. The first number in an ordered pair is called the **abscissa.** The second number in an ordered pair is called the **ordinate.**

[3.1.1, p. 110]

(3, 4) is an ordered pair.

3 is the abscissa.

4 is the ordinate.

A **function** is a set of ordered pairs in which no two ordered pairs have the same first coordinate.

[3.2.1, p. 120]

$\{(2, 3), (3, 5), (5, 7), (6, 9)\}$

A **relation** is any set of ordered pairs.

[3.2.1, p. 120]

$\{(2, 3), (2, 4), (3, 4), (5, 7)\}$

The **domain** of a relation is the set of first coordinates of all the ordered pairs of the relation. The **range** of a relation is the set of second coordinates of all the ordered pairs of the relation.

[3.2.1, p. 120]

$\{(1, 2), (2, 4), (5, 7), (8, 3)\}$

Domain: $\{1, 2, 5, 8\}$

Range: $\{2, 3, 4, 7\}$

Function notation is used for those equations that represent functions. For the equation at the right, x is the **independent variable** and y is the **dependent variable.** The symbol $f(x)$ is the **value of the function** and represents the value of the dependent variable for a given value of the independent variable. The process of finding the value of $f(x)$ for a given value of x is called **evaluating the function.**

[3.2.1, pp. 122–123]

In function notation, $y = 3x + 7$ is written as $f(x) = 3x + 7$.

$f(4) = 3(4) + 7 = 19$

The **graph of a function** is the graph of the ordered pairs that belong to the function. A **linear function** is one that can be expressed in the form $f(x) = mx + b$. The graph of a linear function is a straight line. The point at which the graph crosses the x-axis is called the **x-intercept,** and the point at which a graph crosses the y-axis is called the **y-intercept.**

[3.2.2, 3.3.1, 3.3.2, pp. 126, 136, and 141]

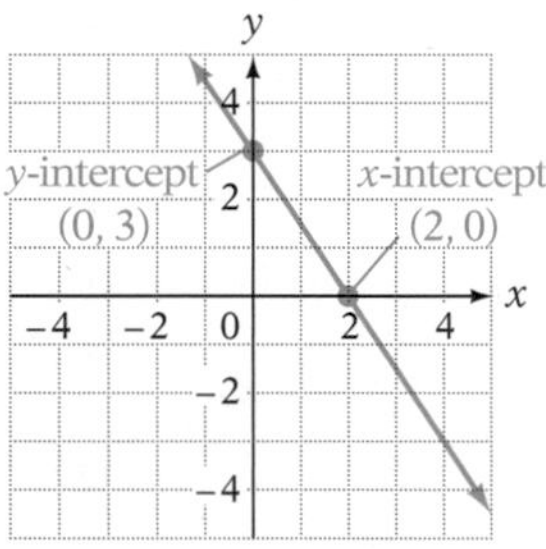

Graph of $f(x) = -\frac{3}{2}x + 3$

The **slope** of a line is a measure of the slant, or tilt, of the line. The symbol for slope is ***m***. A line that slants upward to the right has a **positive slope,** and a line that slants downward to the right has a **negative slope.**

[3.4.1, pp. 148–149]

The graph of $y = 2x - 3$ slants upward to the right. The line has positive slope.

The graph of $y = -5x + 2$ slants downward to the right. The line has negative slope.

A horizontal line has **zero slope.** The slope of a vertical line is **undefined.**

[3.4.1, p. 150]

$y = 3$ is the equation of a line with zero slope.

The slope of the line $x = -3$ is undefined.

An inequality of the form $y > mx + b$ or of the form $Ax + By > C$ is a **linear inequality in two variables.** (The symbol $>$ could be replaced by $\geq$, $<$, or $\leq$.) The solution set of an inequality in two variables is a **half-plane.**

[3.7.1, p. 176]

Essential Rules and Procedures	Objective and Page Reference	Examples
The **distance formula** $d = \sqrt{(x_2 - x_1)^2 + (y_2 - y_1)^2}$ can be used to find the distance between any two points in a plane.	[3.1.1, p. 111]	$(x_1, y_1) = (2, 4), (x_2, y_2) = (-3, 5)$ $d = \sqrt{(-3 - 2)^2 + (5 - 4)^2}$ $= \sqrt{25 + 1}$ $= \sqrt{26}$
The **midpoint formula** $(x_m, y_m) = \left(\frac{x_1 + x_2}{2}, \frac{y_1 + y_2}{2}\right)$ is used to find the midpoint of a line segment.	[3.1.1, p. 112]	$(x_1, y_1) = (2, 3), (x_2, y_2) = (-6, -1)$ $x_m = \frac{2 + (-6)}{2} = -2$ $y_m = \frac{3 + (-1)}{2} = 1$ $(x_m, y_m) = (-2, 1)$
The vertical line test for the graph of a function states that if every vertical line intersects a graph at most once, then the graph is the graph of a function.	[3.2.3, p. 127]	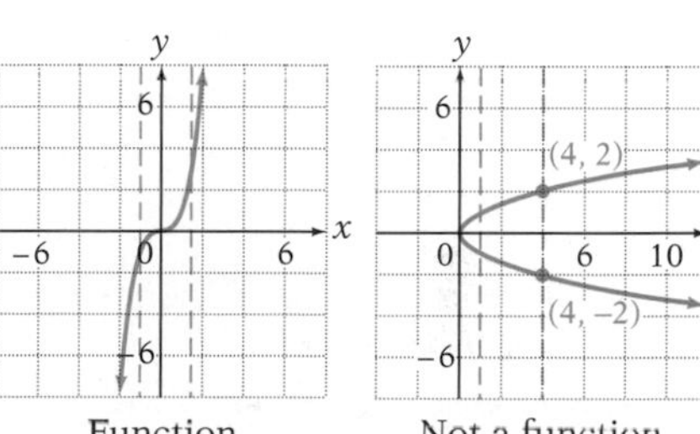 Function Not a function
To find the x-intercept of $Ax + By = C$, let $y = 0$. **To find the y-intercept,** let $x = 0$.	[3.3.2, p. 141]	$3x + 4y = 12$ Let $y = 0$: $3x + 4(0) = 12$, $x = 4$. The coordinates of the x-intercept are (4, 0). Let $x = 0$: $3(0) + 4y = 12$, $y = 3$. The coordinates of the y-intercept are (0, 3).
A value of x for which $f(x) = 0$ is called a **zero of the function.**	[3.3.2, p. 142]	$f(x) = x + 3$ $0 = x + 3$ $-3 = x$ The zero of $f(x) = x + 3$ is -3.
The **slope formula** $m = \frac{y_2 - y_1}{x_2 - x_1}, x_1 \neq x_2$, is used to find the slope of a line when the coordinates of two points on the line are known.	[3.4.1, p. 149]	$(x_1, y_1) = (-3, 2), (x_2, y_2) = (1, 4)$ $m = \frac{4 - 2}{1 - (-3)} = \frac{2}{4} = \frac{1}{2}$
The equation $y = mx + b$ is called the **slope-intercept form of a straight line.**	[3.4.2, p. 152]	$y = -3x + 2$ The slope is -3; the coordinates of the y-intercept are (0, 2).
The **average rate of change** between two points on the graph of a function is the slope of the line between the two points.	[3.4.3, p. 154]	The average rate of change between the points $P(1, 1)$ and $Q(3, 9)$ on the graph of $f(x) = x^2$ is $m = \frac{9 - 1}{3 - 1} = \frac{8}{2} = 4$.

The **point-slope formula,** $y - y_1 = m(x - x_1)$, is used to find the equation of a line when the slope and the coordinates of a point on the line are known.	[3.5.1, p. 162]	The equation of the line that passes through the point $P(4, 2)$ and has slope -3 is $y - y_1 = m(x - x_1)$ $y - 2 = -3(x - 4)$ $y - 2 = -3x + 12$ $y = -3x + 14$
Two lines that have the same slope and different y-intercepts do not intersect and are called **parallel lines.**	[3.6.1, pp. 168–169]	$y = 3x - 4, m_1 = 3$ $y = 3x + 2, m_2 = 3$ Because $m_1 = m_2$, the lines are parallel.
Two lines that intersect at right angles are called **perpendicular lines.** The product of the slopes of two nonvertical perpendicular lines is -1.	[3.6.1, p. 171]	$y = \frac{1}{2}x - 1, m_1 = \frac{1}{2}$ $y = -2x + 2, m_2 = -2$ Because $m_1 \cdot m_2 = -1$, the lines are perpendicular.

CHAPTER 3 Review Exercises

1. Determine the ordered-pair solution of $y = \frac{x}{x - 2}$ that corresponds to $x = 4$.

2. Find the coordinates of the midpoint and the length of the line segment with endpoints $P_1(-2, 4)$ and $P_2(3, 5)$.

3. Graph $f(x) = x^2 - 2$. First evaluate the function when $x = -3, -2, -1, 0, 1, 2$, and 3. Plot the resulting ordered pairs. Then connect the points to form the graph.

4. Graph: $y = -2$

5. Given $P(x) = 3x + 4$, evaluate $P(-2)$ and $P(a)$.

6. What are the domain and range of the relation? Is the relation a function? $\{(-1, 0), (0, 2), (1, 2), (2, 0), (5, 3)\}$

7. Find the range of $f(x) = x^2 - 2$ if the domain is $\{-2, -1, 0, 1, 2\}$.

8. Given $f(x) = 3x - 1$, find a number c in the domain of f such that $f(c) = 5$. Write the corresponding ordered pair of the function.

9. Find the coordinates of the x- and y-intercepts and graph $4x - 6y = 12$.

10. Graph: $y = -2x + 2$

11. Graph: $2x - 3y = -6$

12. Find the slope of the line that contains the points $P_1(3, -2)$ and $P_2(-1, 2)$.

13. Graph $3x + 2y = -4$ by using the x- and y-intercepts.

14. Graph the line that passes through the point $P(-1, 4)$ and has slope $-\frac{1}{3}$.

15. Find the equation of the line that contains the point $P(-3, 4)$ and has slope $\frac{5}{2}$.

16. Find the equation of the line that contains the points $P_1(-2, 4)$ and $P_2(4, -3)$.

17. Find the equation of the line that contains the point $P(3, -2)$ and is parallel to the graph of $y = -3x + 4$.

18. Find the equation of the line that contains the point $P(-2, -4)$ and is parallel to the graph of $2x - 3y = 4$.

19. Find the equation of the line that contains the point $P(2, 5)$ and is perpendicular to the graph of $y = -\frac{2}{3}x + 6$.

20. Find the equation of the line that contains the point $P(-3, -1)$ and is perpendicular to the graph of $4x - 2y = 7$.

21. Find the zero of $f(x) = -3x + 12$.

22. Use the vertical line test to determine whether the graph is the graph of a function.

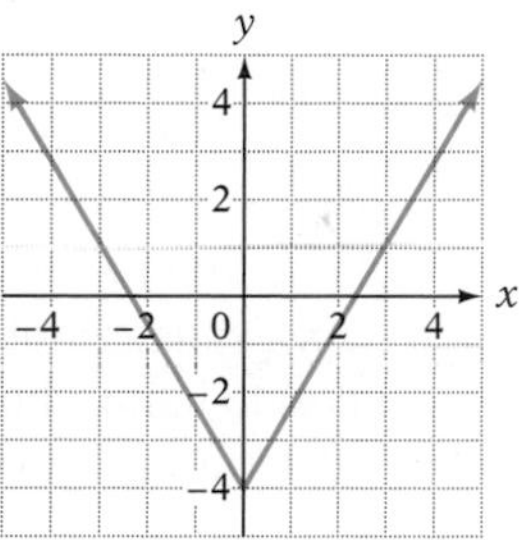

23. Graph the solution set of $y \geq 2x - 3$.

24. Graph the solution set of $3x - 2y < 6$.

25. **Demographics** The table below shows the number of foreign-born people living in the United States for each decade from 1960 to 2000. (*Source:* U.S. Census Bureau)
  a. What was the average annual rate of change in the number of foreign-born people living in the United States from 1980 to 2000? Round to the nearest thousand.
  b. Find the average annual rate of change in the number of foreign-born people living in the United States from 1970 to 1980. Is this more than or less than the average annual rate of change in the number of foreign-born people living in the United States from 1980 to 1990?

Year	1960	1970	1980	1990	2000
Number of foreign born (in millions)	9.7	9.6	14.1	19.8	28.4

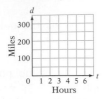

26. **Uniform Motion** A car is traveling at 55 mph. The equation that describes the distance d (in miles) traveled is $d = 55t$, where t is the number of hours driven. Use the coordinate axes below to graph this equation for $0 \leq t \leq 6$. The point with coordinates (4, 220) is on the graph. Write a sentence that describes the meaning of this ordered pair.

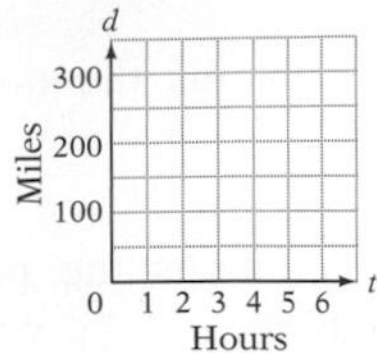

27. **Manufacturing** The graph below shows the relationship between the cost of manufacturing calculators and the number of calculators manufactured. Find the slope of the line between the two points shown on the graph. Write a sentence that states the meaning of the slope.

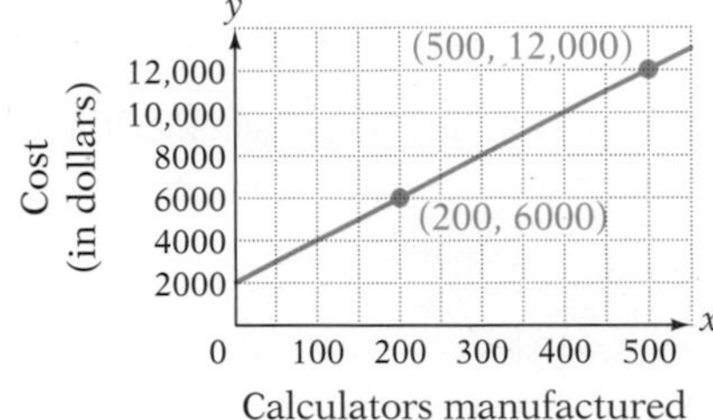

28. **Construction** A building contractor estimates that the cost to build a new home is \$25,000 plus \$80 for each square foot of floor space in the house. Find a linear function that will give the cost to build a house that contains a given number of square feet. Use your function to determine the cost of building a house that contains 2000 ft^2.

CHAPTER 3 Test

1. Graph $f(x) = -x^2 + 2$. First evaluate the function when $x = -3, -2, -1, 0, 1, 2$, and 3. Plot the resulting ordered pairs. Then connect the points to form the graph.

2. Find the ordered-pair solution of $y = 2x + 6$ that corresponds to $x = -3$.

3. Graph: $y = \frac{2}{3}x - 4$

4. Graph: $2x + 3y = -3$

5. Find the equation of the vertical line that contains the point $P(-2, 3)$.

6. Find the coordinates of the midpoint and the length of the line segment with endpoints $P_1(4, 2)$ and $P_2(-5, 8)$.

7. Find the slope of the line that contains the points $P_1(-2, 3)$ and $P_2(4, 2)$.

8. Given $P(x) = 3x^2 - 2x + 1$, evaluate $P(2)$.

9. Graph $2x - 3y = 6$ by using the x- and y-intercepts.

10. Graph the line that passes through the point $P(-2, 3)$ and has slope $-\frac{3}{2}$.

11. Find the equation of the line that contains the point $P(-5, 2)$ and has slope $\frac{2}{5}$.

12. Given $f(x) = 5x - 2$, find a number c in the domain of f such that $f(c) = 3$. Write the corresponding ordered pair of the function.

13. Find the equation of the line that contains the points $P_1(3, -4)$ and $P_2(-2, 3)$.

14. Find the zero of $s(t) = \frac{4}{3}t - 8$.

15. What are the domain and range of the relation? Is the relation a function? $\{(-4, 2), (-2, 2), (0, 0), (3, 5)\}$

16. Find the equation of the line that contains the point $P(1, 2)$ and is parallel to the graph of $y = -\frac{3}{2}x - 6$.

17. Find the equation of the line that contains the point $P(-2, -3)$ and is perpendicular to the graph of $y = -\frac{1}{2}x - 3$.

18. Graph the solution set of $3x - 4y > 8$.

19. Use the vertical line test to determine whether the graph is the graph of a function.

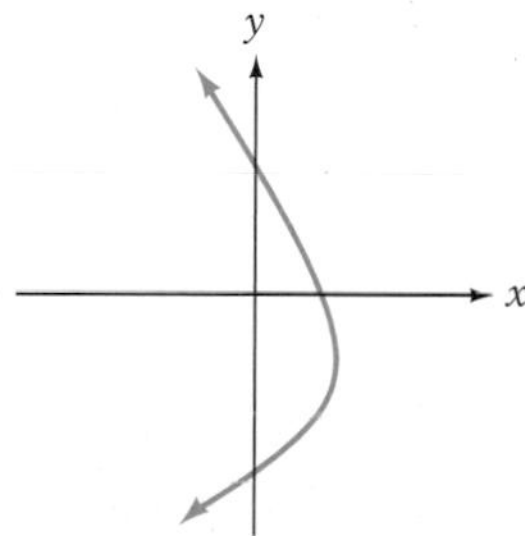

20. **Business** The director of a baseball camp estimates that 100 students will enroll if the tuition is \$250. For each \$20 increase in tuition, 6 fewer students will enroll. Find a linear function that will predict the number of students who will enroll at a given tuition. Use your function to predict enrollment when the tuition is \$300.

21. **Depreciation** The graph at the right shows the relationship between the value of a rental house and the depreciation allowed for income tax purposes. Find the slope of the line between the two points shown on the graph. Write a sentence that states the meaning of the slope.

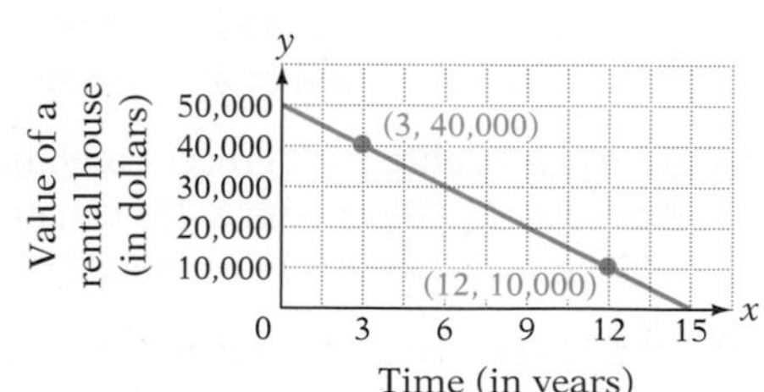

Cumulative Review Exercises

1. Identify the property that justifies the statement. $(x + y) \cdot 2 = 2 \cdot (x + y)$

2. Solve: $3 - \frac{x}{2} = \frac{3}{4}$

3. Solve: $2[y - 2(3 - y) + 4] = 4 - 3y$

4. Solve: $\frac{1 - 3x}{2} + \frac{7x - 2}{6} = \frac{4x + 2}{9}$

5. Solve: $x - 3 < -4$ or $2x + 2 > 3$

6. Solve: $8 - |2x - 1| = 4$

7. Solve: $|3x - 5| < 5$

8. Simplify: $4 - 2(4 - 5)^3 + 2$

9. Evaluate $(a - b)^2 \div ab$ when $a = 4$ and $b = -2$.

10. Graph: $\{x|x < -2\} \cup \{x|x > 0\}$

11. Solve: $3x - 1 < 4$ and $x - 2 > 2$

12. Given $P(x) = x^2 + 5$, evaluate $P(-3)$.

13. Find the ordered-pair solution of $y = -\frac{5}{4}x + 3$ that corresponds to $x = -8$.

14. Find the slope of the line that contains the points $P_1(-1, 3)$ and $P_2(3, -4)$.

15. Find the equation of the line that contains the point $P(-1, 5)$ and has slope $\frac{3}{2}$.

16. Find the equation of the line that contains the points $P_1(4, -2)$ and $P_2(0, 3)$.

17. Find the equation of the line that contains the point $P(2, 4)$ and is parallel to the graph of $y = -\frac{3}{2}x + 2$.

18. Find the equation of the line that contains the point $P(4, 0)$ and is perpendicular to the graph of $3x - 2y = 5$.

19. Graph $3x - 5y = 15$ by using the x- and y-intercepts.

20. Graph the line that passes through the point $P(-3, 1)$ and has slope $-\frac{3}{2}$.

21. Graph the solution set of $3x - 2y \geq 6$.

22. **Uniform Motion** Two planes are 1800 mi apart and are traveling toward each other. One plane is traveling twice as fast as the other plane. The planes will pass each other in 3 h. Find the speed of each plane.

23. **Mixture Problem** A grocer combines coffee that costs $8.00 per pound with coffee that costs $3.00 per pound. How many pounds of each should be used to make 80 lb of a blend that costs $5.00 per pound?

24. **Depreciation** The relationship between the depreciated value of a truck for income tax purposes and its age in years is shown in the graph at the right. Use the x- and y-intercepts to find the slope of the line that represents the depreciated value of the truck. Write a sentence that explains the meaning of the slope.

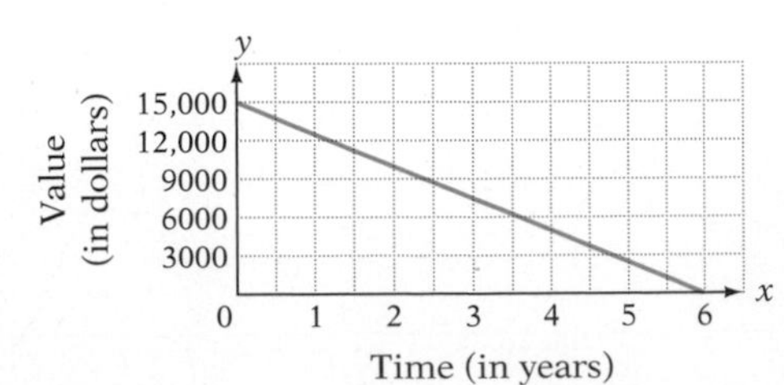

Systems of Equations and Inequalities

CHAPTER 4

Focus on Success

Have you formed or are you part of a study group? Remember that a study group can be a great way to stay focused on succeeding in this course. You can support each other, get help and offer help on homework, and prepare for tests together. (See Homework Time, page AIM-5.)

OBJECTIVES

4.1
1. Solve a system of linear equations by graphing
2. Solve a system of linear equations by the substitution method

4.2
1. Solve a system of two linear equations in two variables by the addition method
2. Solve a system of three linear equations in three variables by the addition method

4.3
1. Evaluate determinants
2. Solve systems of linear equations by using Cramer's Rule
3. Solve systems of linear equations by using matrices

4.4
1. Rate-of-wind and rate-of-current problems
2. Application problems

4.5
1. Graph the solution set of a system of linear inequalities

PREP TEST

Are you ready to succeed in this chapter? *Take the Prep Test below to find out if you are ready to learn the new material.*

1. Simplify: $10\left(\frac{3}{5}x + \frac{1}{2}y\right)$
2. Evaluate $3x + 2y - z$ for $x = -1$, $y = 4$, and $z = -2$.
3. Given $3x - 2z = 4$, find the value of x when $z = -2$.
4. Solve: $3x + 4(-2x - 5) = -5$
5. Solve: $0.45x + 0.06(-x + 4000) = 630$
6. Graph: $y = \frac{1}{2}x - 4$
7. Graph: $3x - 2y = 6$
8. Graph: $y > -\frac{3}{5}x + 1$

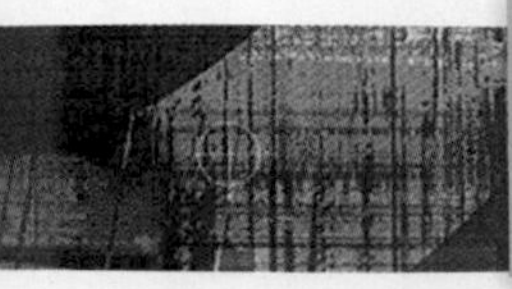

4.1 Solving Systems of Linear Equations by Graphing and by the Substitution Method

OBJECTIVE 1 Solve a system of linear equations by graphing

A **system of equations** is two or more equations considered together. The system at the right is a system of two linear equations in two variables. The graphs of the equations are straight lines.

$$\begin{aligned} 3x + 4y &= 7 \\ 2x - 3y &= 6 \end{aligned}$$

A **solution of a system of equations in two variables** is an ordered pair that is a solution of each equation of the system.

Focus on determining whether an ordered pair is a solution of a system of equations

Is $(3, -2)$ a solution of the system of equations shown at the right?

$$\begin{aligned} 2x - 3y &= 12 \\ 5x + 2y &= 11 \end{aligned}$$

In each equation, replace x by 3 and y by -2.

$2x - 3y = 12$		$5x + 2y = 11$	
$2(3) - 3(-2)$	12	$5(3) + 2(-2)$	11
$6 - (-6)$	12	$15 + (-4)$	11
$12 = 12$		$11 = 11$	

Yes, because $(3, -2)$ is a solution of each equation, it is a solution of the system of equations.

A solution of a system of linear equations can be found by graphing the equations of the system on the same set of coordinate axes. We will now look at three very different systems of linear equations.

Consider the following system of equations:

$$\begin{aligned} x + 2y &= 4 \\ 2x + y &= -1 \end{aligned}$$

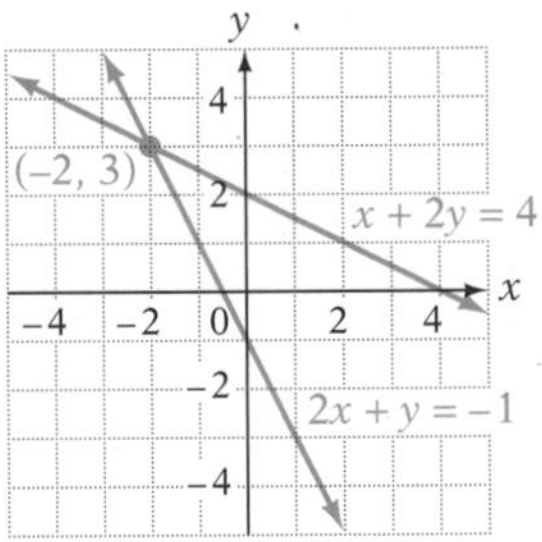

The graphs of the equations in this system are shown at the right. The lines intersect at a single point whose coordinates are $(-2, 3)$. Because this point lies on both lines, its coordinates give the ordered-pair solution of the system of equations. We can check this by substituting -2 for x and 3 for y in each equation of the system.

$x + 2y = 4$		$2x + y = -1$	
$-2 + 2(3)$	4	$2(-2) + 3$	-1
$-2 + 6$	4	$-4 + 3$	-1
$4 = 4$		$-1 = -1$	

• **Replace x by -2 and y by 3.**

The ordered pair $(-2, 3)$ is the solution of the system of equations.

When the graphs of the equations in a system of equations intersect at exactly one point, the system of equations is called an **independent system of equations.** The system of equations

$$\begin{aligned} x + 2y &= 4 \\ 2x + y &= -1 \end{aligned}$$

is an independent system of equations.

Now consider a second system of equations:

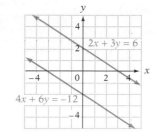

$$2x + 3y = 6$$
$$4x + 6y = -12$$

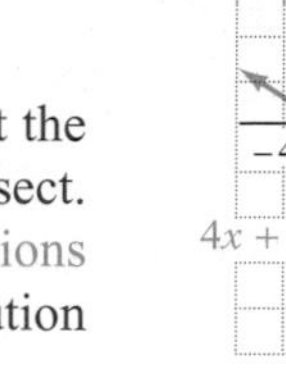

The graphs of the equations in this system are shown at the right. The graphs are parallel and therefore do not intersect. Because the lines do not intersect, the system of equations has no solution. A system of equations that has no solution is called an **inconsistent system of equations.**

Finally, consider the following system of equations:

$$x - 2y = 4$$
$$2x - 4y = 8$$

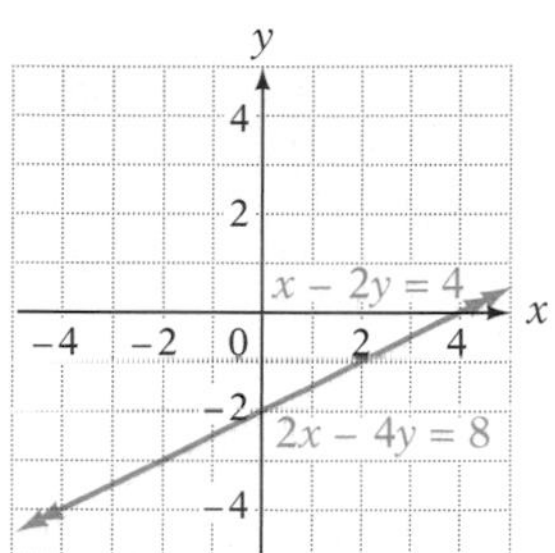

The graphs of the equations in this system are shown at the right. In this case, the graph of one equation lies on top of the graph of the other. The graphs intersect at an infinite number of points, so there are an infinite number of solutions of the system of equations. Because each equation represents the same set of points, the solutions of the system of equations can be written using the ordered-pair solutions of either one of the equations. To do this, choose one of the equations in the system of equations. We have chosen the first equation.

$$x - 2y = 4$$
$$-2y = -x + 4 \qquad \text{• Solve the equation for } y.$$
$$y = \frac{1}{2}x - 2$$

Write the ordered-pair solutions (x, y) as $\left(x, \frac{1}{2}x - 2\right)$. The ordered pairs $\left(x, \frac{1}{2}x - 2\right)$ are the solutions of the system of equations.

When the graphs of the equations in a system of equations intersect at infinitely many points, the system of equations is called a **dependent system of equations.** The system of equations

$$x - 2y = 4$$
$$2x - 4y = 8$$

is a dependent system of equations.

The three systems discussed above illustrate the three possibilities for a system of linear equations in two variables.

Take Note

Keep in mind the distinction among independent, dependent, and inconsistent systems of equations. You should be able to express your understanding of these terms by using graphs.

1. The graphs intersect at one point.
   The solution of the system of equations is the ordered pair (x, y) whose coordinates are the point of intersection.
   The system of equations is *independent.*

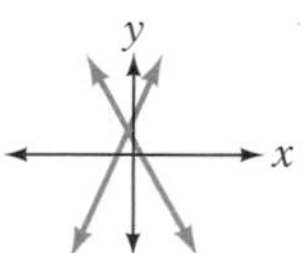

2. The lines are parallel and never intersect.
   The system of equations has no solution.
   The system of equations is *inconsistent.*

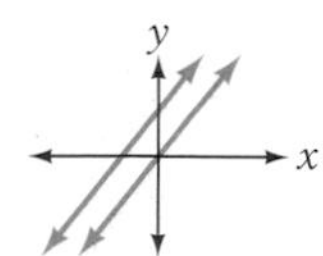

3. The graphs are the same line, and they intersect at infinitely many points.
   There are an infinite number of solutions of the system of equations.
   The system of equations is *dependent.*

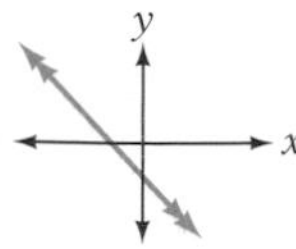

Focus on solving a dependent system of equations

Solve by graphing: $2x - y = 3$
$4x - 2y = 6$

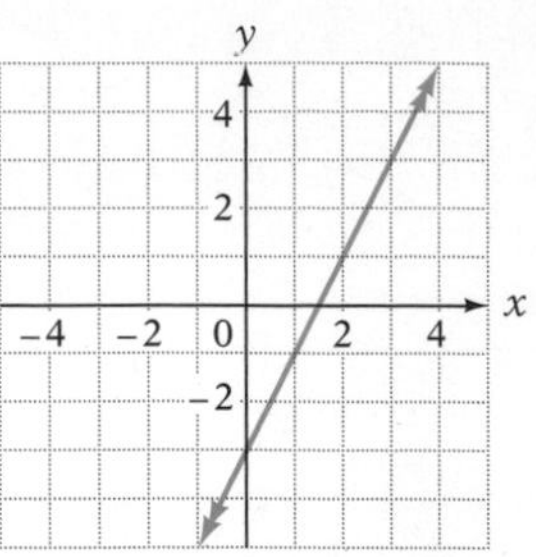

Graph each line.
The system of equations is dependent.
Solve one of the equations for y.

$$\begin{aligned} 2x - y &= 3 \\ -y &= -2x + 3 \\ y &= 2x - 3 \end{aligned}$$

The solutions are the ordered pairs $(x, 2x - 3)$.

EXAMPLE 1 Solve by graphing: $3x - y = 4$
$3x + y = 2$

Solution

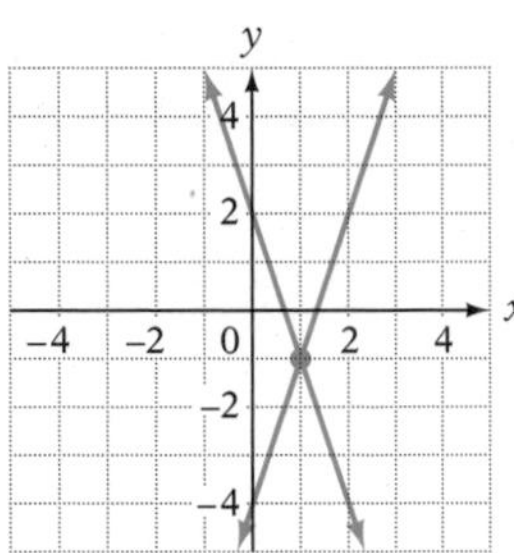

- **Graph each line. The graphs intersect. The coordinates of the point of intersection give the ordered-pair solution of the system.**

The solution is $(1, -1)$.

Problem 1 Solve by graphing: $x + y = 1$
$2x + y = 0$

Solution See page S10.

Try Exercise 19, page 193.

EXAMPLE 2 Solve by graphing.

A. $2x + 3y = 6$
$y = -\frac{2}{3}x + 1$

B. $x - 2y = 6$
$y = \frac{1}{2}x - 3$

Solution **A.**

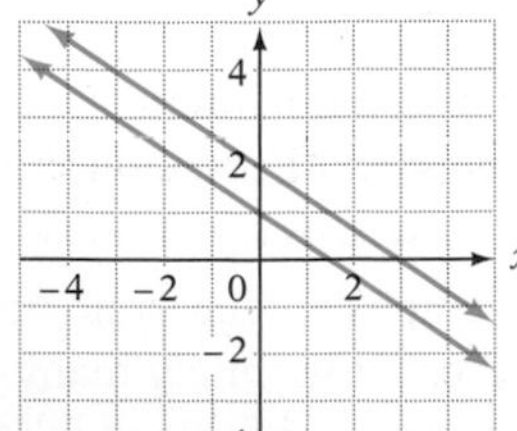

- **Graph each line. The graphs are parallel. The system of equations is inconsistent.**

The system of equations has no solution.

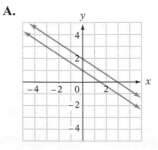

B.

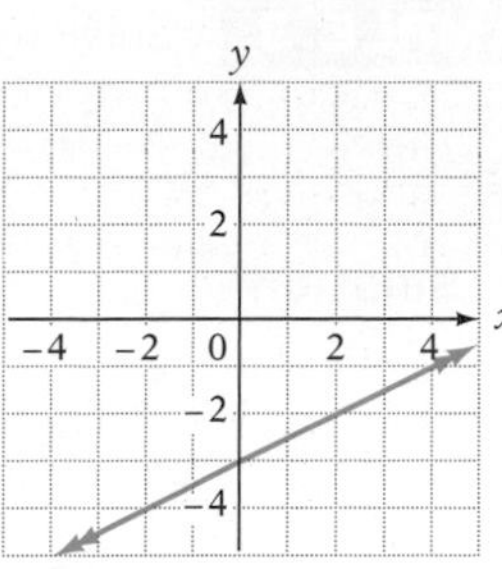

- **Graph each line. The graphs are the same line. The system of equations is dependent.**
- **For this system of equations, one of the equations is already solved for y.**

The solutions are the ordered pairs $(x, \frac{1}{2}x - 3)$.

Problem 2 Solve by graphing.

A. $2x + 5y = 10$

$$y = -\frac{2}{5}x - 2$$

B. $3x - 4y = 12$

$$y = \frac{3}{4}x - 3$$

Solution See page S10.

Try Exercise 23, page 193.

OBJECTIVE 2 Solve a system of linear equations by the substitution method

Solving a system of equations by graphing is based on approximating the coordinates of a point of intersection. An algebraic method called the **substitution method** can be used to find an *exact* solution of a system of equations. To use the substitution method, we must write one of the equations of the system in terms of x or in terms of y.

Focus on solving a system of equations by using the substitution method

Solve by the substitution method: $3x + y = 5$
$4x + 5y = 3$

(1) $3x + y = 5$
(2) $4x + 5y = 3$

Solve equation (1) for y. The result is equation (3).

$3x + y = 5$
(3) $y = -3x + 5$

Use equation (2).
Substitute $-3x + 5$ for y.
Solve for x.

$$4x + 5y = 3$$
$$4x + 5(-3x + 5) = 3$$
$$4x - 15x + 25 = 3$$
$$-11x = -22$$
$$x = 2$$

Now substitute the value of x into equation (3) to find the value of y.

$$y = -3x + 5$$
$$y = -3(2) + 5$$
$$= -6 + 5$$
$$= -1$$

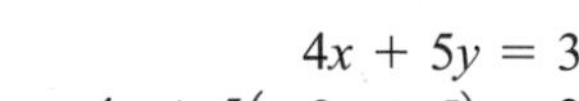

The solution is $(2, -1)$.

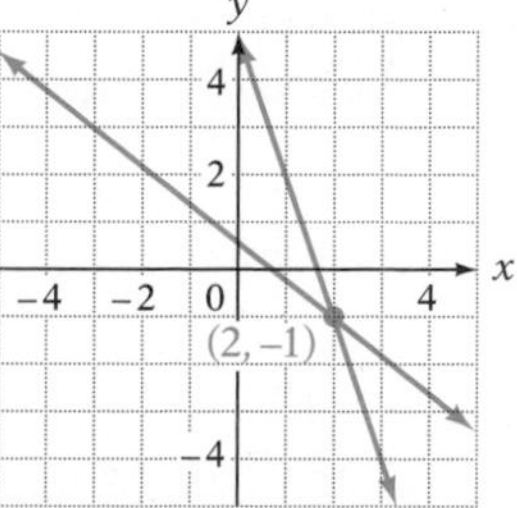

The graph of the system is shown at the left. Note that the graphs intersect at the point whose coordinates are $(2, -1)$, the solution of the system of equations.

Take Note

If, after simplifying, there is no variable in the equation and the resulting equation is *not true*, as in part (B), the system of equations is inconsistent. If the resulting equation is *true*, as in part (C), the system of equations is dependent.

EXAMPLE 3 Solve by substitution.

A. $3x - 5y = 9$
$x = 2y + 4$

B. $3x - 3y = 2$
$y = x + 2$

C. $9x + 3y = 12$
$y = -3x + 4$

Solution

A. (1) $3x - 5y = 9$

(2) $x = 2y + 4$

$3(2y + 4) - 5y = 9$ • Substitute $2y + 4$ for x in equation (1).

$6y + 12 - 5y = 9$ • Solve for y.

$y + 12 = 9$

$y = -3$

$x = 2y + 4$ • Use equation (2).

$x = 2(-3) + 4$ • Substitute -3 for y.

$= -2$ • Simplify.

The solution is $(-2, -3)$.

B. (1) $3x - 3y = 2$

(2) $y = x + 2$ • Equation (2) states that $y = x + 2$.

$3x - 3(x + 2) = 2$ • Substitute $x + 2$ for y in equation (1).

$3x - 3x - 6 = 2$ • Solve for x.

$-6 = 2$

$-6 = 2$ is not a true equation. The system of equations is inconsistent. The system has no solution.

C. (1) $9x + 3y = 12$

(2) $y = -3x + 4$ • Equation (2) states that $y = -3x + 4$.

$9x + 3(-3x + 4) = 12$ • Substitute $-3x + 4$ for y.

$9x - 9x + 12 = 12$ • Solve for x.

$12 = 12$

$12 = 12$ is a true equation. The system of equations is dependent. The solutions are the ordered pairs $(x, -3x + 4)$.

Problem 3 Solve by substitution.

A. $3x - y = 3$
$6x + 3y = -4$

B. $6x - 3y = 6$
$2x - y = 2$

Solution See pages S10–S11.

Try Exercise 57, page 195.

4.1 Exercises

CONCEPT CHECK

Complete each sentence using *exactly one solution, infinitely many solutions,* or *no solution.*

1. An independent system of equations has ___?___. An inconsistent system of equations has ___?___. A dependent system of equations has ___?___.

Determine whether the ordered pair is a solution of the system of equations.

2. $(0, -1)$
$3x - 2y = 2$
$x + 2y = 6$

3. $(2, 1)$
$x + y = 3$
$2x - 3y = 1$

4. $(-3, -5)$
$x + y = -8$
$2x + 5y = -31$

State whether the system of equations represented by the graph is an independent, inconsistent, or dependent system of equations.

5.
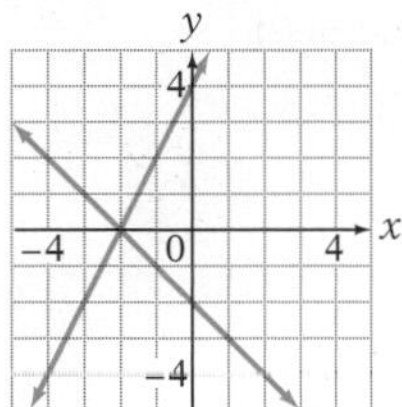

6.
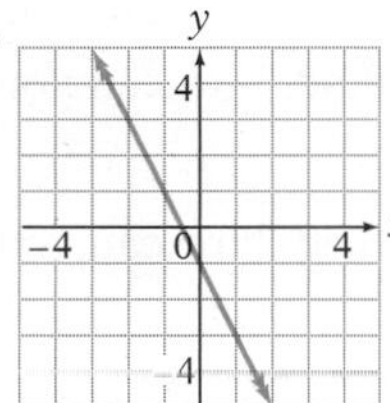

7.
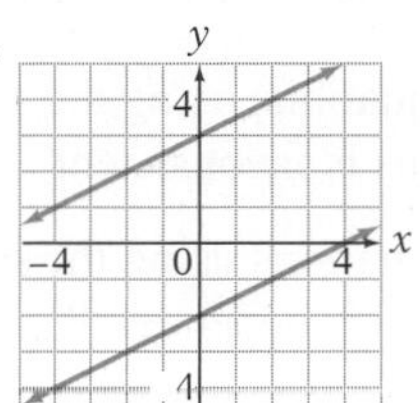

8.
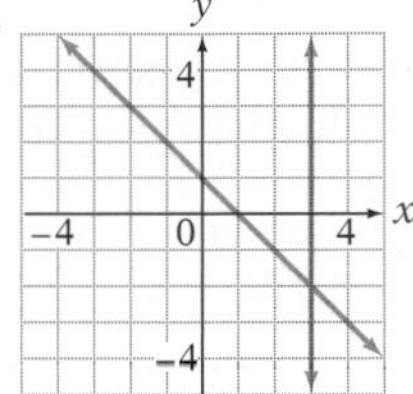

1 Solve a system of linear equations by graphing (See pages 188–191.)

GETTING READY

9. Use the graph shown at the right to solve the system of equations:
$x - 2y = -8$
$3x + y = -3$
The solution is __?__.

10. The equations in the system of equations
$-4x - 2y = 4$
$y = -2x - 2$
have the same graph. The system of equations has an infinite number of ordered-pair solutions in the form $(x, \_\_?\_\_)$.

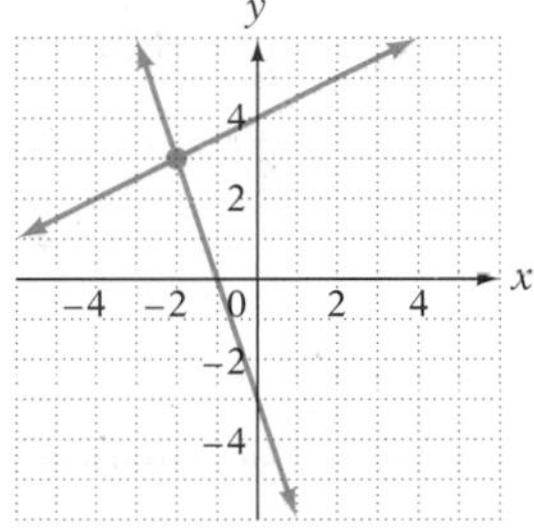

Solve by graphing.

11. $x + y = 2$
$x - y = 4$

12. $x + y = 1$
$3x - y = -5$

13. $x - y = -2$
$x + 2y = 10$

14. $2x - y = 5$
$3x + y = 5$

15. $3x - 2y = 6$
$y = 3$

16. $x = 4$
$3x - 2y = 4$

17. $x = 4$
$y = -1$

18. $x + 2 = 0$
$y - 1 = 0$

19. $y = x - 5$
$2x + y = 4$

20. $2x - 5y = 4$
$y = x + 1$

21. $y = \frac{1}{2}x - 2$
$x - 2y = 8$

22. $2x + 3y = 6$
$y = -\frac{2}{3}x + 1$

23. $2x - 5y = 10$
$y = \frac{2}{5}x - 2$

24. $3x - 2y = 6$
$y = \frac{3}{2}x - 3$

25. $3x - 4y = 12$
$5x + 4y = -12$

26. $2x - 3y = 6$
$2x - 5y = 10$

27. If the following system of equations is an independent system of equations, how are a_1 and a_2 related?

$y = a_1x + b_1$
$y = a_2x + b_2$

28. If a and b are nonzero real numbers, is the following system of equations independent, inconsistent, or dependent?

$x = a$
$y = b$

2 Solve a system of linear equations by the substitution method (See pages 191–192.)

29. When you solve a system of equations using the substitution method, how can you tell whether the system of equations is inconsistent?

30. When you solve a system of equations using the substitution method, how can you tell whether the system of equations is dependent?

GETTING READY

31. Use this system of equations: (1) $y = 2x - 8$
(2) $x = 3$

To solve the system by substitution, substitute __?__ for x in equation (1): $y = 2(3) - 8 =$ __?__.

The solution of the system is (__?__, __?__).

32. Use this system of equations: (1) $y = x - 1$
(2) $3x + y = 3$

To solve the system by substitution, substitute __?__ for y in equation (2): $3x + x - 1 = 3$.

Solving the equation $3x + x - 1 = 3$ for x gives $x =$ __?__.

To find y, substitute this value of x into equation (1): $y =$ __?__ $- 1 =$ __?__.

The solution of the system is (__?__, __?__).

Solve by substitution.

33. $3x - 2y = 4$
$x = 2$

34. $y = -2$
$2x + 3y = 4$

35. $y = 2x - 1$
$x + 2y = 3$

36. $y = -x + 1$
$2x - y = 5$

37. $4x - 3y = 5$
$y = 2x - 3$

38. $3x + 5y = -1$
$y = 2x - 8$

39. $x = 2y + 4$
$4x + 3y = -17$

40. $3x - 2y = -11$
$x = 2y - 9$

41. $5x + 4y = -1$
$y = 2 - 2x$

42. $3x + 2y = 4$
$y = 1 - 2x$

43. $7x - 3y = 3$
$x = 2y + 2$

44. $3x - 4y = 6$
$x = 3y + 2$

45. $2x + 2y = 7$
$y = 4x + 1$

46. $3x + 7y = -5$
$y = 6x - 5$

47. $3x + y = 5$
$2x + 3y = 8$

48. $4x + y = 9$
$3x - 4y = 2$

49. $x + 3y = 5$
$2x + 3y = 4$

50. $x - 4y = 2$
$2x - 5y = 1$

51. $3x + 4y = 14$
$2x + y = 1$

52. $5x + 3y = 8$
$3x + y = 8$

53. $3x + 5y = 0$
$x - 4y = 0$

54. $2x - 7y = 0$
$3x + y = 0$

55. $5x - 3y = -2$
$-x + 2y = -8$

56. $2x + 7y = 1$
$-x + 4y = 7$

57. $x + 3y = 4$
$x = 5 - 3y$

58. $6x + 2y = 7$
$y = -3x + 2$

59. $2x - 4y = 16$
$-x + 2y = -8$

60. $3x - 12y = -24$
$-x + 4y = 8$

61. $3x - y = 10$
$6x - 2y = 5$

62. $6x - 4y = 3$
$3x - 2y = 9$

63. $y = 3x + 2$
$y = 2x + 3$

64. $y = 3x - 7$
$y = 2x - 5$

65. $x = 2y + 1$
$x = 3y - 1$

66. $x = 4y + 1$
$x = -2y - 5$

67. $y = 5x - 1$
$y = 5 - x$

68. $y = 3 - 2x$
$y = 2 - 3x$

APPLYING CONCEPTS

69. For what values of k will the following system of equations be independent?

$2x + 3y = 6$
$2x + ky = 9$

70. If the following system of equations is inconsistent, how are the values of C and D related?

$3x - 4y = C$
$3x - 4y = D$

71. Suppose the following system of equations is an independent system of equations. What is the relationship between $\frac{a_1}{b_1}$ and $\frac{a_2}{b_2}$?

$a_1x + b_1y = c_1$
$a_2x + b_2y = c_2$

72. Suppose the following system of equations is a dependent or inconsistent system of equations. What is the relationship between $\frac{a_1}{b_1}$ and $\frac{a_2}{b_2}$?

$a_1x + b_1y = c_1$
$a_2x + b_2y = c_2$

Use a graphing utility to solve each of the following systems of equations. Round answers to the nearest hundredth.

73. $y = -\frac{1}{2}x + 2$
$y = 2x - 1$

74. $y = 1.2x + 2$
$y = -1.3x - 3$

75. $y = \sqrt{2}x - 1$
$y = -\sqrt{3}x + 1$

76. $y = \pi x - \frac{2}{3}$
$y = -x + \frac{\pi}{2}$

PROJECTS OR GROUP ACTIVITIES

77. Given a system of equations in two variables, how can you, without solving the system of equations, determine whether the system of equations is independent, dependent, or inconsistent?

78. Write a system of equations in two variables that satisfies the given condition.
a. The system of equations has $(-3, 5)$ as its only solution.
b. The system of equations is a dependent system.
c. The system of equations is an inconsistent system.

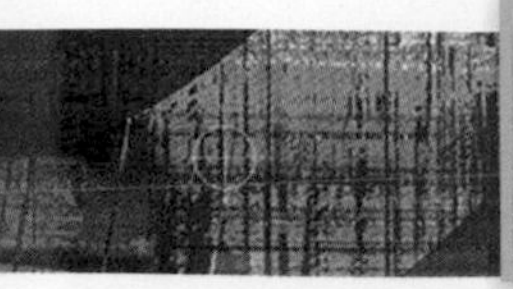

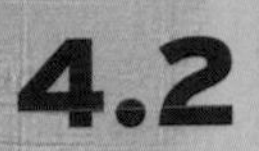

4.2 Solving Systems of Linear Equations by the Addition Method

OBJECTIVE  1 Solve a system of two linear equations in two variables by the addition method

The **addition method** is an alternative method for solving a system of equations. This method is based on the Addition Property of Equations. Use the addition method when it is not convenient to solve one equation for one variable in terms of another variable.

Note, for the system of equations at the right, the effect of adding equation (2) to equation (1). Because $-3y$ and $3y$ are additive inverses, adding the equations results in an equation with only one variable.

$$\begin{array}{lr}(1) & 5x - 3y = 14 \\ (2) & 2x + 3y = -7 \\ \hline & 7x + 0y = 7 \\ & 7x = 7\end{array}$$

The solution of the resulting equation is the first coordinate of the ordered-pair solution of the system.

$$\begin{aligned} 7x &= 7 \\ x &= 1 \end{aligned}$$

The second coordinate is found by substituting the value of x into equation (1) or (2) and then solving for y. Equation (1) is used here.

$$\begin{array}{lr}(1) & 5x - 3y = 14 \\ & 5(1) - 3y = 14 \\ & 5 - 3y = 14 \\ & -3y = 9 \\ & y = -3\end{array}$$

The solution is $(1, -3)$.

Sometimes adding the two equations does not eliminate one of the variables. In this case, use the Multiplication Property of Equations to rewrite one or both of the equations so that when the equations are added, one of the variables is eliminated. To do this, first choose which variable to eliminate. The coefficients of that variable must be additive inverses. Multiply each equation by a constant that will produce coefficients that are additive inverses.

Focus on solving a system of equations by the addition method

A. Solve by the addition method: $\begin{aligned} 3x + 4y &= 2 \\ 2x + 5y &= -1 \end{aligned}$

Number the equations (1) and (2).

$$\begin{array}{lr}(1) & 3x + 4y = 2 \\ (2) & 2x + 5y = -1\end{array}$$

To eliminate x, multiply equation (1) by 2 and equation (2) by -3. Note how the constants are selected. The negative sign is used so that the coefficients will be additive inverses.

$$\begin{aligned} 2(3x + 4y) &= 2(2) \\ -3(2x + 5y) &= -3(-1) \end{aligned}$$

After simplifying, the coefficients of the x-terms are additive inverses.

$$\begin{array}{r} 6x + 8y = 4 \\ -6x - 15y = 3 \\ \hline \end{array}$$

Add the equations. The x-term is eliminated. Solve for y.

$$\begin{aligned} -7y &= 7 \\ y &= -1 \end{aligned}$$

Substitute the value of y into one of the equations, and solve for x. Equation (1) is used here.

$$\begin{array}{lr}(1) & 3x + 4y = 2 \\ & 3x + 4(-1) = 2 \\ & 3x - 4 = 2 \\ & 3x = 6 \\ & x = 2\end{array}$$

The solution is $(2, -1)$.

Point of Interest

There are records of Babylonian mathematicians solving systems of equations 3600 years ago. Here is a system of equations from that time (in our modern notation):

$$\begin{aligned} \frac{2}{3}x &= \frac{1}{2}y + 500 \\ x + y &= 1800 \end{aligned}$$

We say *modern notation* for many reasons. Foremost is the fact that variable notation did not become widespread until the 17th century. There are many other reasons, however. The equals sign had not been invented, 2 and 3 did not look like they do today, and zero had not even been considered as a possible number.

B. Solve by the addition method: $\frac{2}{3}x + \frac{1}{2}y = 4$

$\frac{1}{4}x - \frac{3}{8}y = -\frac{3}{4}$

Number the equations (1) and (2).

(1) $\frac{2}{3}x + \frac{1}{2}y = 4$

(2) $\frac{1}{4}x - \frac{3}{8}y = -\frac{3}{4}$

Clear the fractions. Multiply each equation by the LCM of its denominators.

$6\left(\frac{2}{3}x + \frac{1}{2}y\right) = 6(4)$

$8\left(\frac{1}{4}x - \frac{3}{8}y\right) = 8\left(-\frac{3}{4}\right)$

Simplify. Note that the coefficients of y are additive inverses.

$4x + 3y = 24$

$2x - 3y = -6$

To eliminate y, add the equations. Solve for x.

$6x = 18$

$x = 3$

Substitute the value of x into equation (1), and solve for y.

(1) $\frac{2}{3}x + \frac{1}{2}y = 4$

$\frac{2}{3}(3) + \frac{1}{2}y = 4$

$2 + \frac{1}{2}y = 4$

$\frac{1}{2}y = 2$

$y = 4$

The solution is (3, 4).

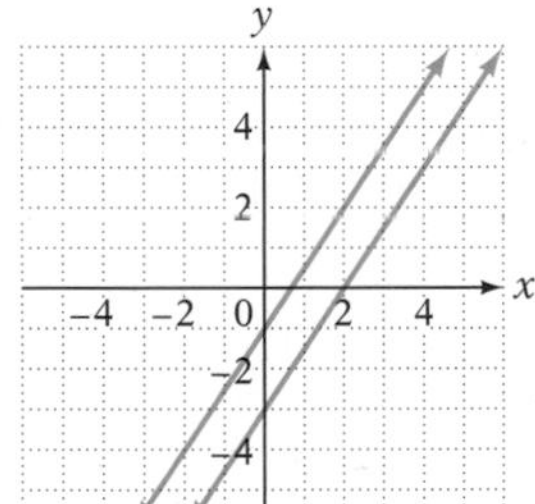

To solve the system of equations at the right, eliminate x. Multiply equation (1) by -2 and add to equation (2).

(1) $3x - 2y = 6$

(2) $6x - 4y = 4$

$-6x + 4y = -12$

$6x - 4y = 4$

$0 = -8$

$0 = -8$ is not a true equation. The system of equations is inconsistent. The system has no solution.

The graph of the equations of the system is shown at the left. Note that the lines are parallel and therefore do not intersect.

EXAMPLE 1 Solve by the addition method.

A. $3x - 2y = 2x + 5$
$2x + 3y = -4$

B. $4x - 8y = 36$
$3x - 6y = 27$

Solution **A.** (1) $3x - 2y = 2x + 5$

(2) $2x + 3y = -4$

$x - 2y = 5$
$2x + 3y = -4$
• **Write equation (1) in the form $Ax + By = C$.**

$-2(x - 2y) = -2(5)$
$2x + 3y = -4$
• **To eliminate x, multiply each side of equation (1) by -2.**

$-2x + 4y = -10$
$2x + 3y = -4$
• **Simplify.**

$7y = -14$
• **Add the equations.**

$y = -2$

$$\begin{aligned} 2x + 3y &= -4 \\ 2x + 3(-2) &= -4 \\ 2x - 6 &= -4 \\ 2x &= 2 \\ x &= 1 \end{aligned}$$

• **Replace y in equation (2) by its value. Solve for x.**

The solution is $(1, -2)$.

B. (1) $4x - 8y = 36$
(2) $3x - 6y = 27$

$$\begin{aligned} 3(4x - 8y) &= 3(36) \\ -4(3x - 6y) &= -4(27) \end{aligned}$$

• **To eliminate x, multiply each side of equation (1) by 3 and each side of equation (2) by -4.**

$$\begin{aligned} 12x - 24y &= 108 \\ -12x + 24y &= -108 \end{aligned}$$

• **Simplify.**

$$0 = 0$$

• **Add the equations.**

Take Note

The result of adding the equations is $0 = 0$. It is not $x = 0$. It is not $y = 0$. There is no variable in the equation $0 = 0$. The equation does not indicate that the solution is $(0, 0)$. Rather, it indicates a dependent system of equations.

$0 = 0$ is a true equation. The system of equations is dependent.

$$\begin{aligned} 4x - 8y &= 36 \\ -8y &= -4x + 36 \\ y &= \tfrac{1}{2}x - \tfrac{9}{2} \end{aligned}$$

• **Solve equation (1) for y.**

The solutions are the ordered pairs $\left(x, \frac{1}{2}x - \frac{9}{2}\right)$.

Problem 1 Solve by the addition method.

A. $2x + 5y = 6$
$3x - 2y = 6x + 2$

B. $2x + y = 5$
$4x + 2y = 6$

Solution See page S11.

Try Exercise 11, page 202.

OBJECTIVE 2 Solve a system of three linear equations in three variables by the addition method

An equation of the form $\boldsymbol{Ax + By + Cz = D}$, where A, B, and C are coefficients and D is a constant, is a **linear equation in three variables.** Two examples of this type of equation are shown below.

$$2x + 4y - 3z = 7 \qquad x - 6y + z = -3$$

Graphing an equation in three variables requires a third coordinate axis perpendicular to the xy-plane. The third axis is commonly called the z-axis. The result is a three-dimensional coordinate system called the ***xyz*-coordinate system.** To help visualize a three-dimensional coordinate system, think of a corner of a room: the floor is the xy-plane, one wall is the yz-plane, and the other wall is the xz-plane. A three-dimensional coordinate system is shown at the right.

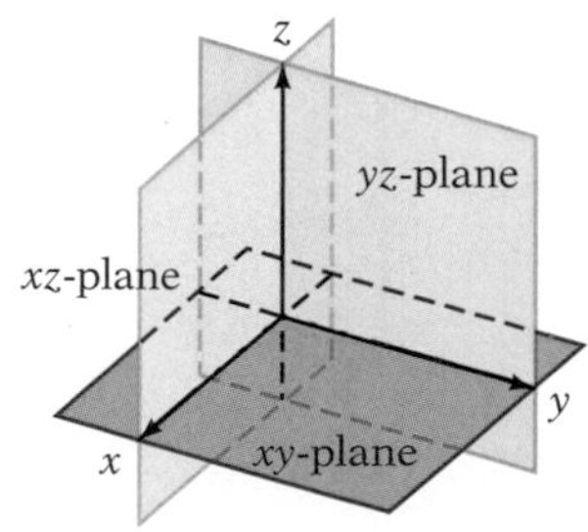

Each point in an xyz-coordinate system is the graph of an **ordered triple** (x, y, z). Graphing an ordered triple requires three moves, the first along the x-axis, the second parallel to the y-axis, and the third parallel to the z-axis. The graphs of the points with coordinates $(-4, 2, 3)$ and $(3, 4, -2)$ are shown at the right.

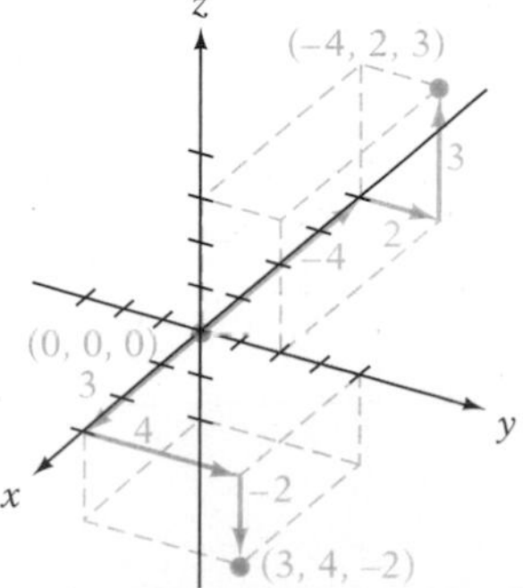

The graph of a linear equation in three variables is a plane. That is, if all the solutions of a linear equation in three variables were plotted in an *xyz*-coordinate system, the graph would look like a large piece of paper extending infinitely. The graph of $x + y + z = 3$ is shown at the right.

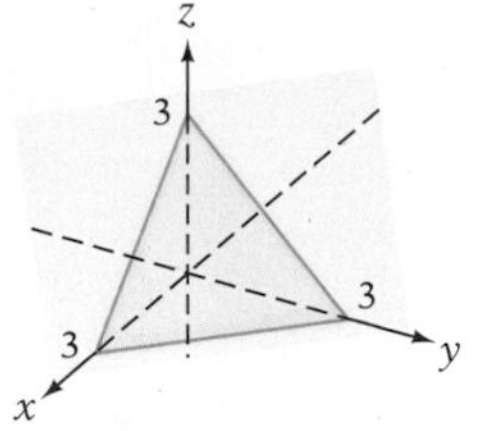

Just as a solution of an equation in two variables is an ordered pair (x, y), a **solution of an equation in three variables** is an ordered triple (x, y, z). For example, $(2, 1, -3)$ is a solution of the equation $2x - y - 2z = 9$. The ordered triple $(1, 3, 2)$ is not a solution.

A **system of linear equations in three variables** is shown at the right. A **solution of a system of equations in three variables** is an ordered triple that is a solution of each equation of the system.

$$\begin{aligned} x - 2y + z &= 7 \\ 3x + y - 2z &= 4 \\ 2x - 3y + 5z &= 19 \end{aligned}$$

The ordered triple $(3, -1, 2)$ is a solution of the system of equations. As shown below, it is a solution of each equation in the system.

$$\begin{array}{r|l} \multicolumn{2}{c}{x - 2y + z = 7} \\ \hline 3 - 2(-1) + 2 & 7 \\ 7 = & 7 \end{array} \qquad \begin{array}{r|l} \multicolumn{2}{c}{3x + y - 2z = 4} \\ \hline 3(3) + (-1) - 2(2) & 4 \\ 4 = & 4 \end{array} \qquad \begin{array}{r|l} \multicolumn{2}{c}{2x - 3y + 5z = 19} \\ \hline 2(3) - 3(-1) + 5(2) & 19 \\ 19 = & 19 \end{array}$$

For a system of three equations in three variables to have a solution, the graphs of the equations must be three planes that intersect at a single point, be three planes that intersect along a common line, or all be the same plane. These situations are shown in the figures that follow.

The three planes shown in Figure A below intersect at a point. A system of equations represented by planes that intersect at a point is independent.

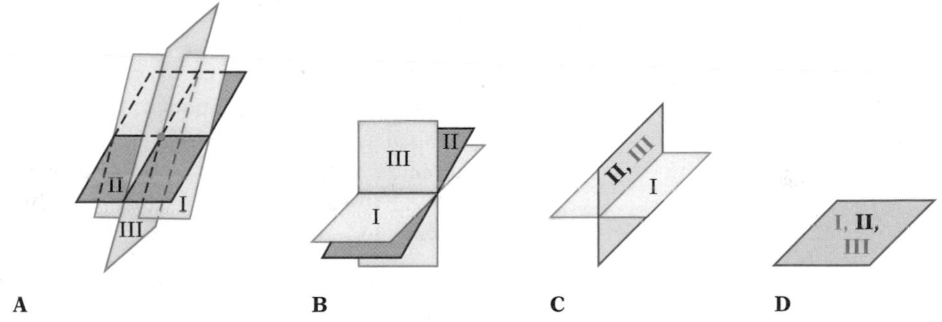

A B C D

Graph of an Independent System of Equations

Graphs of Dependent Systems of Equations

The three planes shown in Figures B and C above intersect along a common line. In Figure D, the three planes are all the same plane. The systems of equations represented by the planes in Figures B, C, and D are dependent systems.

The systems of equations represented by the planes in the four figures below are inconsistent systems.

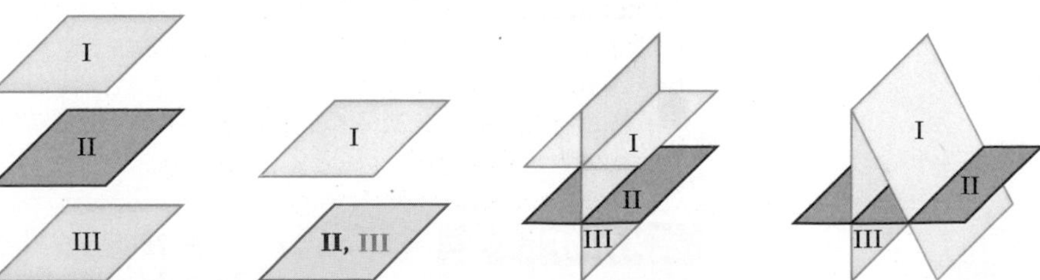

Graphs of Inconsistent Systems of Equations

A system of linear equations in three variables can be solved by using the addition method. First, eliminate one variable from any two of the given equations. Then eliminate the same variable from any other two equations. The result will be a system of two equations in two variables. Solve this system by the addition method.

Point of Interest

In the early 1980s, Stephen Hoppe became interested in winning Monopoly strategies. Finding these strategies required solving a system that contained 123 equations in 123 variables!

Focus on solving a system of equations in three variables

A. Solve: $x + 4y - z = 10$
$3x + 2y + z = 4$
$2x - 3y + 2z = -7$

$$\begin{aligned} (1)\quad x + 4y - z &= 10 \\ (2)\quad 3x + 2y + z &= 4 \\ (3)\quad 2x - 3y + 2z &= -7 \end{aligned}$$

Eliminate z from equations (1) and (2) by adding the two equations. The result is equation (4).

$$\begin{aligned} x + 4y - z &= 10 \\ 3x + 2y + z &= 4 \\ \hline (4)\quad 4x + 6y &= 14 \end{aligned}$$

Eliminate z from equations (1) and (3). Multiply equation (1) by 2 and add it to equation (3). The result is equation (5).

$$\begin{aligned} 2x + 8y - 2z &= 20 \\ 2x - 3y + 2z &= -7 \\ \hline (5)\quad 4x + 5y &= 13 \end{aligned}$$

Solve the system of two equations in two variables, equations (4) and (5).

$$\begin{aligned} (4)\quad 4x + 6y &= 14 \\ (5)\quad 4x + 5y &= 13 \end{aligned}$$

Eliminate x. Multiply equation (5) by -1 and add it to equation (4).

$$\begin{aligned} 4x + 6y &= 14 \\ -4x - 5y &= -13 \\ \hline y &= 1 \end{aligned}$$

Substitute the value of y into equation (4) or (5), and solve for x. Equation (4) is used here.

$$\begin{aligned} (4)\quad 4x + 6y &= 14 \\ 4x + 6(1) &= 14 \\ 4x + 6 &= 14 \\ 4x &= 8 \\ x &= 2 \end{aligned}$$

Substitute the value of y and the value of x into one of the equations in the original system, and solve for z. Equation (2) is used here.

$$\begin{aligned} (2)\quad 3x + 2y + z &= 4 \\ 3(2) + 2(1) + z &= 4 \\ 6 + 2 + z &= 4 \\ 8 + z &= 4 \\ z &= -4 \end{aligned}$$

The solution is $(2, 1, -4)$.

B. Solve: $2x - 3y - z = 1$
$x + 4y + 3z = 2$
$4x - 6y - 2z = 5$

$$\begin{aligned} (1)\quad 2x - 3y - z &= 1 \\ (2)\quad x + 4y + 3z &= 2 \\ (3)\quad 4x - 6y - 2z &= 5 \end{aligned}$$

Eliminate x from equations (1) and (2). Multiply equation (2) by -2 and add to equation (1).

$$\begin{aligned} 2x - 3y - z &= 1 \\ -2x - 8y - 6z &= -4 \\ \hline (4)\quad -11y - 7z &= -3 \end{aligned}$$

Eliminate x from equations (1) and (3). Multiply equation (1) by -2 and add to equation (3). Equation (5) is not a true equation.

$$\begin{aligned} -4x + 6y + 2z &= -2 \\ 4x - 6y - 2z &= 5 \\ \hline (5)\quad 0 &= 3 \end{aligned}$$

The system of equations has no solution.

EXAMPLE 2 Solve: $3x - y + 2z = 1$
$2x + 3y + 3z = 4$
$x + y - 4z = -9$

Solution

$$\begin{aligned} (1)\quad 3x - y + 2z &= 1 \\ (2)\quad 2x + 3y + 3z &= 4 \\ (3)\quad x + y - 4z &= -9 \end{aligned}$$

• Number the equations (1), (2), and (3).

$$\begin{aligned} 3x - y + 2z &= 1 \\ x + y - 4z &= -9 \\ \hline 4x - 2z &= -8 \end{aligned}$$

• Eliminate y. Add equations (1) and (3).

$$(4)\quad 2x - z = -4$$

• Simplify the resulting equation by multiplying each side of the equation by $\frac{1}{2}$.

$$\begin{aligned} 9x - 3y + 6z &= 3 \\ 2x + 3y + 3z &= 4 \\ \hline (5)\quad 11x + 9z &= 7 \end{aligned}$$

• Multiply equation (1) by 3 and add to equation (2).

$$\begin{aligned} (4)\quad 2x - z &= -4 \\ (5)\quad 11x + 9z &= 7 \end{aligned}$$

• Solve the system of two equations, equations (4) and (5).

$$\begin{aligned} 18x - 9z &= -36 \\ 11x + 9z &= 7 \\ \hline 29x &= -29 \\ x &= -1 \end{aligned}$$

• Multiply equation (4) by 9 and add to equation (5). Solve for x.

$$\begin{aligned} (4)\quad 2x - z &= -4 \\ 2(-1) - z &= -4 \\ -2 - z &= -4 \\ -z &= -2 \\ z &= 2 \end{aligned}$$

• Replace x by -1 in equation (4). Solve for z.

$$\begin{aligned} (3)\quad x + y - 4z &= -9 \\ -1 + y - 4(2) &= -9 \\ -1 + y - 8 &= -9 \\ -9 + y &= -9 \\ y &= 0 \end{aligned}$$

• Replace x by -1 and z by 2 in equation (1), (2), or (3). Equation (3) is used here. Solve for y.

The solution is $(-1, 0, 2)$.

Problem 2 Solve:

$$\begin{aligned} x - y + z &= 6 \\ 2x + 3y - z &= 1 \\ x + 2y + 2z &= 5 \end{aligned}$$

Solution See page S11.

Try Exercise 49, page 203.

4.2 Exercises

CONCEPT CHECK

1. Can a system of linear equations in two variables have exactly two solutions?

2. Suppose you add two equations in a system of equations, and the result is $x = 0$. Does this mean that the system of equations is inconsistent? Why or why not?

3. If you begin to solve the system of equations

$$\begin{aligned} x + y &= 6 \\ x - y &= 4 \end{aligned}$$

by adding the two equations, the result is $2x = 10$, which means that $x = 5$. Does this mean that 5 is a solution of the system of equations? Why or why not?

4. Suppose you add two equations in a system of equations, and the result is $4 = 6$. What can you conclude about the system of equations?

5. What are the values of A, B, and C if $(3, -2, 1)$ is a solution of the following system of equations?

$$\begin{aligned} x + y - 3z &= A \\ 2x - y + 4z &= B \\ x + 3y - z &= C \end{aligned}$$

1 Solve a system of two linear equations in two variables by the addition method (See pages 196–198.)

GETTING READY

6. Use this system of equations: (1) $-4x - y = 5$
(2) $x - 2y = 4$

a. To eliminate x from the system of equations by using the addition method, multiply each side of equation (__?__) by __?__.

b. To eliminate y from the system of equations by using the addition method, multiply each side of equation (__?__) by __?__.

Solve by the addition method.

7. $x - y = 5$
$x + y = 7$

8. $x + y = 1$
$2x - y = 5$

9. $3x + y = 4$
$x + y = 2$

10. $x - 3y = 4$
$x + 5y = -4$

11. $3x + y = 7$
$x + 2y = 4$

12. $x - 2y = 7$
$3x - 2y = 9$

13. $3x - y = 4$
$6x - 2y = 8$

14. $x - 2y = -3$
$-2x + 4y = 6$

15. $2x + 5y = 9$
$4x - 7y = -16$

16. $8x - 3y = 21$
$4x + 5y = -9$

17. $4x - 6y = 5$
$2x - 3y = 7$

18. $3x + 6y = 7$
$2x + 4y = 5$

19. $3x - 5y = 7$
$x - 2y = 3$

20. $3x + 4y = 25$
$2x + y = 10$

21. $3x + 2y = 16$
$2x - 3y = -11$

22. $2x - 5y = 13$
$5x + 3y = 17$

23. $4x + 4y = 5$
$2x - 8y = -5$

24. $3x + 7y = 16$
$4x - 3y = 9$

25. $5x + 4y = 0$
$3x + 7y = 0$

26. $3x - 4y = 0$
$4x - 7y = 0$

27. $3x - 6y = 6$
$9x - 3y = 8$

28. $4x - 8y = 5$
$8x + 2y = 1$

29. $5x + 2y = 2x + 1$
$2x - 3y = 3x + 2$

30. $3x + 3y = y + 1$
$x + 3y = 9 - x$

31. $\frac{2}{3}x - \frac{1}{2}y = 3$
$\frac{1}{3}x - \frac{1}{4}y = \frac{3}{2}$

32. $\frac{3}{4}x + \frac{1}{3}y = -\frac{1}{2}$
$\frac{1}{2}x - \frac{5}{6}y = -\frac{7}{2}$

33. $\frac{2}{5}x - \frac{1}{3}y = 1$
$\frac{3}{5}x + \frac{2}{3}y = 5$

34. $\frac{5}{6}x + \frac{1}{3}y = \frac{4}{3}$
$\frac{2}{3}x - \frac{1}{2}y = \frac{11}{6}$

35. $\frac{3}{4}x + \frac{2}{5}y = -\frac{3}{20}$
$\frac{3}{2}x - \frac{1}{4}y = \frac{3}{4}$

36. $\frac{2}{5}x - \frac{1}{2}y = \frac{13}{2}$
$\frac{3}{4}x - \frac{1}{5}y = \frac{17}{2}$

37. $4x - 5y = 3y + 4$
$2x + 3y = 2x + 1$

38. $5x - 2y = 8x - 1$
$2x + 7y = 4y + 9$

39. $2x + 5y = 5x + 1$
$3x - 2y = 3y + 3$

40. When you solve a system of equations using the addition method, how can you tell whether the system of equations is inconsistent?

41. When you solve a system of equations using the addition method, how can you tell whether the system of equations is dependent?

Suppose A, B, C, and D are nonzero real numbers. State whether the given system of equations is independent, inconsistent, or dependent.

42. $Ax + By = 0$
$Cx + Dy = 0, D \neq B$

43. $Ax + By = C$
$DAx + DBy = DC$

2 Solve a system of three linear equations in three variables by the addition method (See pages 198–201.)

GETTING READY

44. Look at the system of equations in Exercise 46. State two different pairs of equations that can be added together to eliminate the variable z.

45. Look at the system of equations in Exercise 48. To eliminate z from the first and second equations, multiply the second equation by ___?___ and then add the equations.

Solve by the addition method.

46. $x + 2y - z = 1$
$2x - y + z = 6$
$x + 3y - z = 2$

47. $x + 3y + z = 6$
$3x + y - z = -2$
$2x + 2y - z = 1$

48. $2x - y + 2z = 7$
$x + y + z = 2$
$3x - y + z = 6$

49. $x - 2y + z = 6$
$x + 3y + z = 16$
$3x - y - z = 12$

50. $3x + y = 5$
$3y - z = 2$
$x + z = 5$

51. $2y + z = 7$
$2x - z = 3$
$x - y = 3$

52. $x - y + z = 1$
$2x + 3y - z = 3$
$-x + 2y - 4z = 4$

53. $2x + y - 3z = 7$
$x - 2y + 3z = 1$
$3x + 4y - 3z = 13$

54. $2x + 3z = 5$
$3y + 2z = 3$
$3x + 4y = -10$

55. $3x + 4z = 5$
$2y + 3z = 2$
$2x - 5y = 8$

56. $2x + 4y - 2z = 3$
$x + 3y + 4z = 1$
$x + 2y - z = 4$

57. $x - 3y + 2z = 1$
$x - 2y + 3z = 5$
$2x - 6y + 4z = 3$

58. $2x + y - z = 5$
$x + 3y + z = 14$
$3x - y + 2z = 1$

59. $3x - y - 2z = 11$
$2x + y - 2z = 11$
$x + 3y - z = 8$

60. $3x + y - 2z = 2$
$x + 2y + 3z = 13$
$2x - 2y + 5z = 6$

61. $4x + 5y + z = 6$
$2x - y + 2z = 11$
$x + 2y + 2z = 6$

62. $2x - y + z = 6$
$3x + 2y + z = 4$
$x - 2y + 3z = 12$

63. $3x + 2y - 3z = 8$
$2x + 3y + 2z = 10$
$x + y - z = 2$

64. $3x - 2y + 3z = -4$
$2x + y - 3z = 2$
$3x + 4y + 5z = 8$

65. $3x - 3y + 4z = 6$
$4x - 5y + 2z = 10$
$x - 2y + 3z = 4$

66. $3x - y + 2z = 2$
$4x + 2y - 7z = 0$
$2x + 3y - 5z = 7$

67. $2x + 2y + 3z = 13$
$-3x + 4y - z = 5$
$5x - 3y + z = 2$

68. $2x - 3y + 7z = 0$
$x + 4y - 4z = -2$
$3x + 2y + 5z = 1$

69. $5x + 3y - z = 5$
$3x - 2y + 4z = 13$
$4x + 3y + 5z = 22$

APPLYING CONCEPTS

Solve. (*Hint:* First multiply both sides of each equation in the system by a multiple of 10 so that the coefficients and constants are integers.)

70. $0.2x - 0.3y = 0.5$
$0.3x - 0.2y = 0.5$

71. $0.4x - 0.9y = -0.1$
$0.3x + 0.2y = 0.8$

72. $1.25x - 0.25y = -1.5$
$1.5x + 2.5y = 1$

73. $2.25x + 1.5y = 3$
$1.75x + 2.25y = 1.25$

74. $1.5x + 2.5y + 1.5z = 8$
$0.5x - 2y - 1.5z = -1$
$2.5x - 1.5y + 2z = 2.5$

75. $1.6x - 0.9y + 0.3z = 2.9$
$1.6x + 0.5y - 0.1z = 3.3$
$0.8x - 0.7y + 0.1z = 1.5$

Solve.

76. The point of intersection of the graphs of the equations $Ax + 3y = 6$ and $2x + By = -4$ is $(3, -2)$. Find A and B.

77. The point of intersection of the graphs of the equations $Ax + 3y + 2z = 8$, $2x + By - 3z = -12$, and $3x - 2y + Cz = 1$ is $(3, -2, 4)$. Find A, B, and C.

78. The distance between a point and a line is the perpendicular distance from the point to the line. Find the distance between the point with coordinates $(3, 1)$ and the line with equation $y = x$.

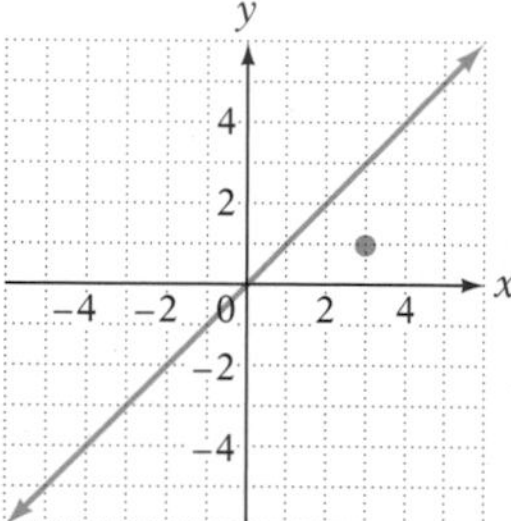

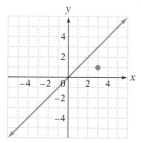

PROJECTS OR GROUP ACTIVITIES

79. The intersection of two distinct planes is a line. Let L be the line of intersection of the planes with equations $2x + y - z = 13$ and $x - 2y + z = 5$. If the point with coordinates $(x, 3, z)$ is on line L, find the value of $x - z$.

80. Describe the graph of each of the following equations in an xyz-coordinate system.

a. $x = 3$ **b.** $y = 4$ **c.** $z = 2$ **d.** $y = x$

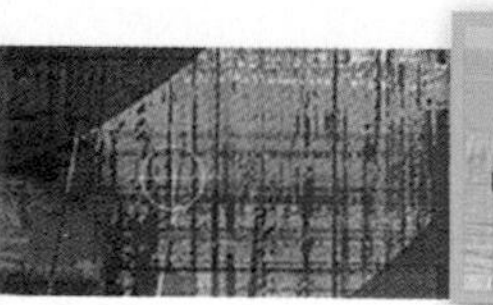

4.3 Solving Systems of Equations by Using Determinants and by Using Matrices

OBJECTIVE  1 Evaluate determinants

A **matrix** is a rectangular array of numbers. Each number in the matrix is called an **element** of the matrix. The matrix at the right, with three rows and four columns, is called a 3×4 ("3 by 4") matrix.

$$A = \begin{bmatrix} 1 & -3 & 2 & 4 \\ 0 & 4 & -3 & 2 \\ 6 & -5 & 4 & -1 \end{bmatrix}$$

A matrix of m rows and n columns is said to be of **order** $m \times n$. Matrix A above has order 3×4. The notation a_{ij} refers to the element of a matrix in the ith row and the jth column. For matrix A, $a_{23} = -3$, $a_{31} = 6$, and $a_{13} = 2$.

A **square matrix** is one that has the same number of rows as columns. A 2×2 matrix and a 3×3 matrix are shown at the right.

$$\begin{bmatrix} -1 & 3 \\ 5 & 2 \end{bmatrix} \quad \begin{bmatrix} 4 & 0 & 1 \\ 5 & -3 & 7 \\ 2 & 1 & 4 \end{bmatrix}$$

Associated with every square matrix is a number called its **determinant.**

Take Note

Matrices are denoted by brackets. Determinants are denoted by vertical bars.

DETERMINANT OF A 2 × 2 MATRIX

The determinant of a 2 × 2 matrix $\begin{bmatrix} a_{11} & a_{12} \\ a_{21} & a_{22} \end{bmatrix}$ is written $\begin{vmatrix} a_{11} & a_{12} \\ a_{21} & a_{22} \end{vmatrix}$. The value of this determinant is given by the formula

$$\begin{vmatrix} a_{11} & a_{12} \\ a_{21} & a_{22} \end{vmatrix} = a_{11}a_{22} - a_{21}a_{12}$$

EXAMPLES

1. Evaluate the determinant $\begin{vmatrix} 3 & 4 \\ -1 & 2 \end{vmatrix}$.

$$\begin{vmatrix} 3 & 4 \\ -1 & 2 \end{vmatrix} = 3(2) - 4(-1) = 6 + 4 = 10$$

2. Evaluate the determinant $\begin{vmatrix} -3 & -5 \\ 6 & 10 \end{vmatrix}$.

$$\begin{vmatrix} -3 & -5 \\ 6 & 10 \end{vmatrix} = -3(10) - (-5)(6) = -30 + 30 = 0$$

The value of the determinant of a matrix of order larger than 2 × 2 can be found by using smaller determinants within the large one. The smaller determinants are called *minors.*

MINOR OF AN ELEMENT OF A DETERMINANT

The **minor of an element** a_{ij} of a determinant is the determinant that remains after row i and column j have been removed.

EXAMPLES

1. Find the minor of −3 for the determinant $\begin{vmatrix} 2 & -3 & 4 \\ 0 & 4 & 8 \\ -1 & 3 & 6 \end{vmatrix}$.

$$\begin{vmatrix} 2 & -3 & 4 \\ 0 & 4 & 8 \\ -1 & 3 & 6 \end{vmatrix}$$

• **Remove the row and column containing −3.**

The minor of −3 is $\begin{vmatrix} 0 & 8 \\ -1 & 6 \end{vmatrix}$.

2. Find the minor of 0 for the determinant $\begin{vmatrix} 2 & -3 & 3 & 1 \\ 5 & -2 & 1 & -7 \\ 4 & 7 & 0 & 9 \\ 3 & -1 & 6 & 10 \end{vmatrix}$.

$$\begin{vmatrix} 2 & -3 & 3 & 1 \\ 5 & -2 & 1 & -7 \\ 4 & 7 & 0 & 9 \\ 3 & -1 & 6 & 10 \end{vmatrix}$$

• **Remove the row and column containing 0.**

The minor of 0 is $\begin{vmatrix} 2 & -3 & 1 \\ 5 & -2 & -7 \\ 3 & -1 & 10 \end{vmatrix}$.

Related to the minor of an element of a determinant is the *cofactor* of the element.

DEFINITION OF A COFACTOR

The **cofactor of an element** a_{ij} of a determinant is $(-1)^{i+j}$ times the minor of a_{ij}.

EXAMPLES

Use the determinant $\begin{vmatrix} 3 & -2 & 1 \\ 2 & -5 & -4 \\ 0 & 3 & 1 \end{vmatrix}$.

1. Find the cofactor of -2.
   Because -2 is in the first row and the second column, $i = 1$ and $j = 2$. Therefore, $i + j = 1 + 2 = 3$, and $(-1)^{i+j} = (-1)^3 = -1$.

   The cofactor of -2 is $(-1)\begin{vmatrix} 2 & -4 \\ 0 & 1 \end{vmatrix}$.

2. Find the cofactor of -5.
   Because -5 is in the second row and the second column, $i = 2$ and $j = 2$. Therefore, $i + j = 2 + 2 = 4$, and $(-1)^{i+j} = (-1)^4 = 1$.

   The cofactor of -5 is $(1)\begin{vmatrix} 3 & 1 \\ 0 & 1 \end{vmatrix}$.

Note from these examples that the cofactor of an element is -1 times the minor of that element or 1 times the minor of that element, depending on whether the sum $i + j$ is an odd or an even integer.

The value of a 3×3 or larger determinant can be found by **expanding by cofactors** of *any* row or *any* column. This process involves multiplying each element of the selected row or column by its cofactor and then finding the sum of the results.

Focus on finding the value of a 3×3 determinant

Evaluate the determinant $\begin{vmatrix} 2 & -3 & 2 \\ 1 & 3 & -1 \\ 0 & -2 & 2 \end{vmatrix}$.

Expand by cofactors of the first row. Multiply each element of the first row by its cofactor and then add the products.

$$\begin{vmatrix} 2 & -3 & 2 \\ 1 & 3 & -1 \\ 0 & -2 & 2 \end{vmatrix} = 2(-1)^{1+1}\begin{vmatrix} 3 & -1 \\ -2 & 2 \end{vmatrix} + (-3)(-1)^{1+2}\begin{vmatrix} 1 & -1 \\ 0 & 2 \end{vmatrix} + 2(-1)^{1+3}\begin{vmatrix} 1 & 3 \\ 0 & -2 \end{vmatrix}$$

$$= 2(1)\begin{vmatrix} 3 & -1 \\ -2 & 2 \end{vmatrix} + (-3)(-1)\begin{vmatrix} 1 & -1 \\ 0 & 2 \end{vmatrix} + 2(1)\begin{vmatrix} 1 & 3 \\ 0 & -2 \end{vmatrix}$$

$$= 2(6 - 2) + 3(2 - 0) + 2(-2 - 0)$$

$$= 2(4) + 3(2) + 2(-2) = 8 + 6 + (-4)$$

$$= 10$$

Now consider finding the value of the determinant shown above by expanding by cofactors of the second column.

$$\begin{vmatrix} 2 & -3 & 2 \\ 1 & 3 & -1 \\ 0 & -2 & 2 \end{vmatrix} = -3(-1)^{1+2}\begin{vmatrix} 1 & -1 \\ 0 & 2 \end{vmatrix} + 3(-1)^{2+2}\begin{vmatrix} 2 & 2 \\ 0 & 2 \end{vmatrix} + (-2)(-1)^{3+2}\begin{vmatrix} 2 & 2 \\ 1 & -1 \end{vmatrix}$$

$$= -3(-1)\begin{vmatrix} 1 & -1 \\ 0 & 2 \end{vmatrix} + 3(1)\begin{vmatrix} 2 & 2 \\ 0 & 2 \end{vmatrix} + (-2)(-1)\begin{vmatrix} 2 & 2 \\ 1 & -1 \end{vmatrix}$$

$$= 3(2 - 0) + 3(4 - 0) + 2(-2 - 2)$$

$$= 3(2) + 3(4) + 2(-4) = 6 + 12 + (-8)$$
$$= 10$$

Note that the value of the determinant is the same whether the first row or the second column is used to expand by cofactors. *Any row or column* can be used to evaluate a determinant by expanding by cofactors. In order to simplify the calculations, choose the row or column with the most zeros.

EXAMPLE 1 Evaluate the determinant. **A.** $\begin{vmatrix} 3 & -2 \\ 6 & -4 \end{vmatrix}$ **B.** $\begin{vmatrix} -2 & 3 & 1 \\ 4 & -2 & 0 \\ 1 & -2 & 3 \end{vmatrix}$

Solution **A.** $\begin{vmatrix} 3 & -2 \\ 6 & -4 \end{vmatrix} = 3(-4) - (6)(-2) = -12 + 12 = 0$

B. There is a zero in row 2, column 3. Expand by cofactors of either row 2 or column 3. We will use row 2.

$$\begin{vmatrix} -2 & 3 & 1 \\ 4 & -2 & 0 \\ 1 & -2 & 3 \end{vmatrix} = 4(-1)^{2+1}\begin{vmatrix} 3 & 1 \\ -2 & 3 \end{vmatrix} + (-2)(-1)^{2+2}\begin{vmatrix} -2 & 1 \\ 1 & 3 \end{vmatrix} + 0(-1)^{2+3}\begin{vmatrix} -2 & 3 \\ 1 & -2 \end{vmatrix}$$
$$= 4(-1)\begin{vmatrix} 3 & 1 \\ -2 & 3 \end{vmatrix} + (-2)(1)\begin{vmatrix} -2 & 1 \\ 1 & 3 \end{vmatrix} + 0(-1)\begin{vmatrix} -2 & 3 \\ 1 & -2 \end{vmatrix}$$
$$= -4(9 - (-2)) - 2(-6 - 1) + 0$$
$$= -4(11) - 2(-7) = -44 + 14$$
$$= -30$$

Take Note

The reason for choosing a row or column with a zero element can be seen at the right. Because 0 times any number equals 0, it is not necessary to evaluate the minor of the zero element.

Problem 1 Evaluate the determinant.

A. $\begin{vmatrix} -1 & -4 \\ 3 & -5 \end{vmatrix}$ **B.** $\begin{vmatrix} 2 & -1 & 3 \\ 4 & -1 & 1 \\ 2 & 2 & 1 \end{vmatrix}$

Solution See page S11.

Try Exercise 21, page 217.

OBJECTIVE 2 Solve systems of linear equations by using Cramer's Rule

The connection between determinants and systems of equations can be understood by solving a general system of linear equations.

Solve: $a_{11}x + a_{12}y = b_1$
$a_{21}x + a_{22}y = b_2$

(1) $a_{11}x + a_{12}y = b_1$
(2) $a_{21}x + a_{22}y = b_2$

Eliminate y. Multiply equation (1) by a_{22} and equation (2) by $-a_{12}$.

$$a_{11}a_{22}x + a_{12}a_{22}y = b_1a_{22}$$
$$-a_{21}a_{12}x - a_{12}a_{22}y = -b_2a_{12}$$

Add the equations.

$$(a_{11}a_{22} - a_{21}a_{12})x = b_1a_{22} - b_2a_{12}$$

Solve for x. Assume $a_{11}a_{22} - a_{21}a_{12} \neq 0$.

$$x = \frac{b_1a_{22} - b_2a_{12}}{a_{11}a_{22} - a_{21}a_{12}}$$

Note that the denominator for x, $a_{11}a_{22} - a_{21}a_{12}$, is the determinant of the coefficients of x and y. This is called the **coefficient determinant.**

$$a_{11}a_{22} - a_{21}a_{12} = \begin{vmatrix} a_{11} & a_{12} \\ a_{21} & a_{22} \end{vmatrix}$$

Coefficients of x (first column)

Coefficients of y (second column)

The numerator for x, $b_1a_{22} - b_2a_{12}$, is the determinant obtained by replacing the first column in the coefficient determinant by the constants b_1 and b_2. This is called the **numerator determinant.**

$$b_1a_{22} - b_2a_{12} = \begin{vmatrix} b_1 & a_{12} \\ b_2 & a_{22} \end{vmatrix}$$

Constants of the equations →

Following a similar procedure and eliminating x, we can also express the y-coordinate of the solution in determinant form. These results are summarized in Cramer's Rule.

Point of Interest

Cramer's Rule is named after Gabriel Cramer, who used it in a book he published in 1750. However, this rule had already been published in 1683 by the Japanese mathematician Seki Kown. That publication occurred seven years before Cramer's birth.

CRAMER'S RULE FOR TWO EQUATIONS IN TWO VARIABLES

The solution of the system of equations

$$\begin{aligned} a_{11}x + a_{12}y &= b_1 \\ a_{21}x + a_{22}y &= b_2 \end{aligned}$$

is given by $x = \dfrac{D_x}{D}$ and $y = \dfrac{D_y}{D}$, where

$$D = \begin{vmatrix} a_{11} & a_{12} \\ a_{21} & a_{22} \end{vmatrix},\ D_x = \begin{vmatrix} b_1 & a_{12} \\ b_2 & a_{22} \end{vmatrix},\ D_y = \begin{vmatrix} a_{11} & b_1 \\ a_{21} & b_2 \end{vmatrix},\ \text{and } D \neq 0.$$

EXAMPLE 2 Solve by using Cramer's Rule: $\begin{aligned} 2x - 3y &= 8 \\ 5x + 6y &= 11 \end{aligned}$

Solution

$D = \begin{vmatrix} 2 & -3 \\ 5 & 6 \end{vmatrix} = 27$ • **Evaluate the coefficient determinant.**

$D_x = \begin{vmatrix} 8 & -3 \\ 11 & 6 \end{vmatrix} = 81$ • **Evaluate each numerator determinant.**

$D_y = \begin{vmatrix} 2 & 8 \\ 5 & 11 \end{vmatrix} = -18$

$x = \dfrac{D_x}{D} = \dfrac{81}{27} = 3$ • **Use Cramer's Rule to find the x- and y-coordinates of the solution.**

$y = \dfrac{D_y}{D} = \dfrac{-18}{27} = -\dfrac{2}{3}$

The solution is $\left(3, -\dfrac{2}{3}\right)$.

Problem 2 Solve by using Cramer's Rule: $\begin{aligned} 6x - 6y &= 5 \\ 2x - 10y &= -1 \end{aligned}$

Solution See page S12.

Try Exercise 37, page 218.

Consider the system of equations $\begin{aligned} 6x - 9y &= 5 \\ 4x - 6y &= 4 \end{aligned}$. As shown below, the value of the determinant of the coefficients, D, is zero.

$$D = \begin{vmatrix} 6 & -9 \\ 4 & -6 \end{vmatrix} = 6(-6) - (-9)4 = -36 + 36 = 0$$

Therefore, the fractions $\dfrac{D_x}{D}$ and $\dfrac{D_y}{D}$ are undefined. When $D = 0$, the system of equations is dependent if both D_x and D_y are zero. The system of equations is inconsistent if $D = 0$ and either D_x or D_y is not zero.

A procedure similar to that followed for two equations in two variables can be used to extend Cramer's Rule to three equations in three variables.

CRAMER'S RULE FOR THREE EQUATIONS IN THREE VARIABLES

The solution of the system of equations

$$\begin{aligned} a_{11}x + a_{12}y + a_{13}z &= b_1 \\ a_{21}x + a_{22}y + a_{23}z &= b_2 \\ a_{31}x + a_{32}y + a_{33}z &= b_3 \end{aligned}$$

is given by $x = \dfrac{D_x}{D}$, $y = \dfrac{D_y}{D}$, and $z = \dfrac{D_z}{D}$, where

$$D = \begin{vmatrix} a_{11} & a_{12} & a_{13} \\ a_{21} & a_{22} & a_{23} \\ a_{31} & a_{32} & a_{33} \end{vmatrix}, D_x = \begin{vmatrix} b_1 & a_{12} & a_{13} \\ b_2 & a_{22} & a_{23} \\ b_3 & a_{32} & a_{33} \end{vmatrix}, D_y = \begin{vmatrix} a_{11} & b_1 & a_{13} \\ a_{21} & b_2 & a_{23} \\ a_{31} & b_3 & a_{33} \end{vmatrix},$$

$$D_z = \begin{vmatrix} a_{11} & a_{12} & b_1 \\ a_{21} & a_{22} & b_2 \\ a_{31} & a_{32} & b_3 \end{vmatrix}, \text{ and } D \neq 0.$$

EXAMPLE 3 Solve by using Cramer's Rule:

$$\begin{aligned} 3x - y + z &= 5 \\ x + 2y - 2z &= -3 \\ 2x + 3y + z &= 4 \end{aligned}$$

Solution

$$D = \begin{vmatrix} 3 & -1 & 1 \\ 1 & 2 & -2 \\ 2 & 3 & 1 \end{vmatrix} = 28$$

• **Evaluate the coefficient determinant.**

$$D_x = \begin{vmatrix} 5 & -1 & 1 \\ -3 & 2 & -2 \\ 4 & 3 & 1 \end{vmatrix} = 28$$

• **Evaluate each numerator determinant.**

$$D_y = \begin{vmatrix} 3 & 5 & 1 \\ 1 & -3 & -2 \\ 2 & 4 & 1 \end{vmatrix} = 0$$

$$D_z = \begin{vmatrix} 3 & -1 & 5 \\ 1 & 2 & -3 \\ 2 & 3 & 4 \end{vmatrix} = 56$$

$$x = \frac{D_x}{D} = \frac{28}{28} = 1$$

$$y = \frac{D_y}{D} = \frac{0}{28} = 0$$

$$z = \frac{D_z}{D} = \frac{56}{28} = 2$$

• **Use Cramer's Rule to find the *x*-, *y*-, and *z*-coordinates of the solution.**

The solution is (1, 0, 2).

Problem 3 Solve by using Cramer's Rule:

$$\begin{aligned} 2x - y + z &= -1 \\ 3x + 2y - z &= 3 \\ x + 3y + z &= -2 \end{aligned}$$

Solution See page S12.

Try Exercise 45, page 218.

OBJECTIVE 3 Solve systems of linear equations by using matrices

Consider the 3×4 matrix below.

$$\begin{bmatrix} 1 & 4 & -3 & 6 \\ -2 & 5 & 2 & 0 \\ -1 & 3 & 7 & -4 \end{bmatrix}$$

The elements $a_{11}, a_{22}, a_{33}, \ldots, a_{nn}$ form the **main diagonal** of a matrix. The elements 1, 5, and 7 form the main diagonal of the matrix above.

By considering only the coefficients and constants for the following system of equations, we can form the corresponding 3×4 **augmented matrix.**

> **Take Note**
> For augmented matrices, it is common practice to show a vertical line between the coefficients of the variables and the constant terms.

System of Equations **Augmented Matrix**

$$\begin{aligned} 3x - 2y + z &= 2 \\ x \quad\quad - 3z &= -2 \\ 2x - y + 4z &= 5 \end{aligned} \qquad \left[\begin{array}{ccc|c} 3 & -2 & 1 & 2 \\ 1 & 0 & -3 & -2 \\ 2 & -1 & 4 & 5 \end{array}\right]$$

Note that when a term is missing from one of the equations of the system, the coefficient of that term is 0, and a 0 is entered in the matrix.

A system of equations can be written from an augmented matrix.

$$\left[\begin{array}{ccc|c} 2 & -1 & 4 & 1 \\ 1 & 1 & 0 & 3 \\ 3 & -2 & -1 & 5 \end{array}\right] \qquad \begin{aligned} 2x - y + 4z &= 1 \\ x + y \quad\quad &= 3 \\ 3x - 2y - z &= 5 \end{aligned}$$

EXAMPLE 4 Write the augmented matrix for the system of equations.

$$\begin{aligned} 2x - 3y &= 4 \\ x + 5y &= 0 \end{aligned}$$

Solution The coefficients of x, 2 and 1, are the first column. The coefficients of y, -3 and 5, are the second column. The constant terms are the third column.

The augmented matrix is $\left[\begin{array}{cc|c} 2 & -3 & 4 \\ 1 & 5 & 0 \end{array}\right]$.

Problem 4 Write the system of equations that corresponds to the augmented matrix $\left[\begin{array}{ccc|c} 2 & -3 & 1 & 4 \\ 1 & 0 & -2 & 3 \\ 0 & 1 & 2 & -3 \end{array}\right]$.

Solution See page S12.

Try Exercises 55 and 61, page 219.

A system of equations can be solved by writing the system in matrix form and then performing operations on the matrix similar to those performed on the equations of the system. These operations are called **elementary row operations.**

ELEMENTARY ROW OPERATIONS

1. Interchange two rows.
2. Multiply all the elements in a row by the same nonzero number.
3. Replace a row by the sum of that row and a multiple of any other row.

Take Note

For each example of a row operation at the right, we are using the same system of equations. The solution of this system of equations is (2, 1). You should verify that (2, 1) is also a solution of the new system of equations. This means that these row operations have not changed the solution.

Each of these elementary row operations has as its basis the operations that can be performed on a system of equations. These operations do not change the solution of the system of equations. Here are some examples of each row operation.

1. Interchange two rows.

Original System		This notation means to interchange rows 1 and 2.		**New System**
$2x - 3y = 1$ $4x + 5y = 13$	$\left[\begin{array}{rr\|r} 2 & -3 & 1 \\ 4 & 5 & 13 \end{array}\right]$	$R_1 \leftrightarrow R_2$	$\left[\begin{array}{rr\|r} 4 & 5 & 13 \\ 2 & -3 & 1 \end{array}\right]$	$4x + 5y = 13$ $2x - 3y = 1$
The solution is (2, 1).				The solution is (2, 1).

2. Multiply all the elements in a row by the same nonzero number.

Original System		This notation means to multiply row 2 by 3.		**New System**
$2x - 3y = 1$ $4x + 5y = 13$	$\left[\begin{array}{rr\|r} 2 & -3 & 1 \\ 4 & 5 & 13 \end{array}\right]$	$3R_2 \rightarrow$	$\left[\begin{array}{rr\|r} 2 & -3 & 1 \\ 12 & 15 & 39 \end{array}\right]$	$2x - 3y = 1$ $12x + 15y = 39$
The solution is (2, 1).				The solution is (2, 1).

Take Note

Note that for the third elementary row operation, the row being multiplied is *not* changed. For the example at the right, row 1 is not changed. The row being added to is replaced by the result of the operation. It is not absolutely necessary to follow this convention, but for consistency, we will always use this method.

3. Replace a row by the sum of that row and a multiple of any other row.

Original System		This notation means to replace row 2 by the sum of that row and −2 times row 1.		**New System**
$2x - 3y = 1$ $4x + 5y = 13$	$\left[\begin{array}{rr\|r} 2 & -3 & 1 \\ 4 & 5 & 13 \end{array}\right]$	$-2R_1 + R_2 \rightarrow$	$\left[\begin{array}{rr\|r} 2 & -3 & 1 \\ 0 & 11 & 11 \end{array}\right]$	$2x - 3y = 1$ $11y = 11$
The solution is (2, 1).				The solution is (2, 1).

Note that we replace the row that follows the addition symbol. See the Take Note at the left.

EXAMPLE 5 Let $A = \left[\begin{array}{rrr|r} 1 & 3 & -4 & 6 \\ 3 & 2 & 0 & -1 \\ -2 & -5 & 3 & 4 \end{array}\right]$. Perform the following elementary row operations on A.

A. $R_1 \leftrightarrow R_3$ **B.** $-2R_3$ **C.** $2R_3 + R_1$

Solution **A.** $R_1 \leftrightarrow R_3$ means to interchange row 1 and row 3.

$$\left[\begin{array}{rrr|r} 1 & 3 & -4 & 6 \\ 3 & 2 & 0 & -1 \\ -2 & -5 & 3 & 4 \end{array}\right] \quad R_1 \leftrightarrow R_3 \quad \left[\begin{array}{rrr|r} -2 & -5 & 3 & 4 \\ 3 & 2 & 0 & -1 \\ 1 & 3 & -4 & 6 \end{array}\right]$$

B. $-2R_3$ means to multiply row 3 by -2.

$$\left[\begin{array}{rrr|r} 1 & 3 & -4 & 6 \\ 3 & 2 & 0 & -1 \\ -2 & -5 & 3 & 4 \end{array}\right] \quad -2R_3 \rightarrow \quad \left[\begin{array}{rrr|r} 1 & 3 & -4 & 6 \\ 3 & 2 & 0 & -1 \\ 4 & 10 & -6 & -8 \end{array}\right]$$

C. $2R_3 + R_1$ means to multiply row 3 by 2 and then add the result to row 1. The result replaces row 1.

$$\left[\begin{array}{rrr|r} 1 & 3 & -4 & 6 \\ 3 & 2 & 0 & -1 \\ -2 & -5 & 3 & 4 \end{array}\right] \quad 2R_3 + R_1 \rightarrow \quad \left[\begin{array}{rrr|r} -3 & -7 & 2 & 14 \\ 3 & 2 & 0 & -1 \\ -2 & -5 & 3 & 4 \end{array}\right]$$

Problem 5 Let $B = \left[\begin{array}{rrr|r} 1 & 8 & -2 & 3 \\ 2 & -3 & 4 & 1 \\ 3 & 5 & -7 & 3 \end{array}\right]$. Perform the following elementary row operations on B.

A. $R_2 \leftrightarrow R_3$ **B.** $3R_2$ **C.** $-3R_1 + R_3$

Solution See page S12.

Try Exercise 67, page 219.

Elementary row operations are used to solve a system of equations. The goal is to use the elementary row operations to rewrite the augmented matrix with 1's down the main diagonal and 0's to the left of the 1's in all rows except the first. This is called the **row echelon form** of the matrix. Here are some examples of the row echelon form.

$$\left[\begin{array}{rr|r} 1 & 3 & -2 \\ 0 & 1 & 3 \end{array}\right] \quad \left[\begin{array}{rrr|r} 1 & -2 & 3 & 1 \\ 0 & 1 & 2.5 & -4 \\ 0 & 0 & 1 & 2 \end{array}\right] \quad \left[\begin{array}{rrr|r} 1 & 4 & \frac{1}{2} & -3 \\ 0 & 1 & 3 & 0 \\ 0 & 0 & 1 & -\frac{2}{3} \end{array}\right]$$

We will follow a very definite procedure to rewrite an augmented matrix in row echelon form.

STEPS FOR REWRITING A 2 × 3 AUGMENTED MATRIX IN ROW ECHELON FORM

The order in which the elements of the augmented matrix below are changed is as follows:

Step 1: Change a_{11} to a 1.

Step 2: Change a_{21} to a 0.

Step 3: Change a_{22} to a 1.

$$\left[\begin{array}{cc|c} a_{11} & a_{12} & a_{13} \\ a_{21} & a_{22} & a_{23} \end{array}\right]$$

Focus on writing a 2 × 3 augmented matrix in row echelon form

Write the matrix $\left[\begin{array}{rr|r} 3 & -6 & 12 \\ 2 & 1 & -3 \end{array}\right]$ in row echelon form.

Step 1: Change a_{11} to 1. One way to do this is to multiply row 1 by the reciprocal of a_{11}.

$$\left[\begin{array}{rr|r} 3 & -6 & 12 \\ 2 & 1 & -3 \end{array}\right] \quad \tfrac{1}{3}R_1 \rightarrow \quad \left[\begin{array}{rr|r} 1 & -2 & 4 \\ 2 & 1 & -3 \end{array}\right]$$

Step 2: Change a_{21} to 0 by multiplying row 1 by the opposite of a_{21} and then adding to row 2.

$$\left[\begin{array}{rr|r} 1 & -2 & 4 \\ 2 & 1 & -3 \end{array}\right] \quad -2R_1 + R_2 \rightarrow \quad \left[\begin{array}{rr|r} 1 & -2 & 4 \\ 0 & 5 & -11 \end{array}\right]$$

Step 3: Change a_{22} to 1 by multiplying row 2 by the reciprocal of a_{22}.

$$\left[\begin{array}{cc|c} 1 & -2 & 4 \\ 0 & 5 & -11 \end{array}\right] \xrightarrow{\frac{1}{5}R_2} \left[\begin{array}{cc|c} 1 & -2 & 4 \\ 0 & 1 & -\frac{11}{5} \end{array}\right]$$

A row echelon form of the matrix is $\left[\begin{array}{cc|c} 1 & -2 & 4 \\ 0 & 1 & -\frac{11}{5} \end{array}\right]$.

The row echelon form of a matrix is not unique and depends on the elementary row operations that are used. For instance, suppose we again start with $\left[\begin{array}{cc|c} 3 & -6 & 12 \\ 2 & 1 & -3 \end{array}\right]$ and follow the elementary row operations below.

$$\left[\begin{array}{cc|c} 3 & -6 & 12 \\ 2 & 1 & -3 \end{array}\right] \xrightarrow{-1R_2 + R_1} \left[\begin{array}{cc|c} 1 & -7 & 15 \\ 2 & 1 & -3 \end{array}\right] \xrightarrow{-2R_1 + R_2} \left[\begin{array}{cc|c} 1 & -7 & 15 \\ 0 & 15 & -33 \end{array}\right] \xrightarrow{\frac{1}{15}R_2} \left[\begin{array}{cc|c} 1 & -7 & 15 \\ 0 & 1 & -\frac{11}{5} \end{array}\right]$$

The row echelon forms $\left[\begin{array}{cc|c} 1 & -7 & 15 \\ 0 & 1 & -\frac{11}{5} \end{array}\right]$ and $\left[\begin{array}{cc|c} 1 & -2 & 4 \\ 0 & 1 & -\frac{11}{5} \end{array}\right]$ are different. Row echelon form is not unique.

STEPS FOR REWRITING A 3 × 4 AUGMENTED MATRIX IN ROW ECHELON FORM

The order in which the elements of the augmented matrix below are changed is as follows:

Step 1: Change a_{11} to a 1.

Step 2: Change a_{21} and a_{31} to 0's.

Step 3: Change a_{22} to a 1.

Step 4: Change a_{32} to a 0.

Step 5: Change a_{33} to a 1.

$$\left[\begin{array}{ccc|c} a_{11} & a_{12} & a_{13} & a_{14} \\ a_{21} & a_{22} & a_{23} & a_{24} \\ a_{31} & a_{32} & a_{33} & a_{34} \end{array}\right]$$

EXAMPLE 6 Write $\left[\begin{array}{ccc|c} 2 & 1 & 3 & -1 \\ 1 & 3 & 5 & -1 \\ -3 & -1 & 1 & 2 \end{array}\right]$ in row echelon form.

Solution **Step 1:** Change a_{11} to 1 by interchanging row 1 and row 2. *Note:* We could have chosen to multiply row 1 by $\frac{1}{2}$. The sequence of steps to get to row echelon form is not unique.

$$\left[\begin{array}{ccc|c} 2 & 1 & 3 & -1 \\ 1 & 3 & 5 & -1 \\ -3 & -1 & 1 & 2 \end{array}\right] \quad R_1 \leftrightarrow R_2 \quad \left[\begin{array}{ccc|c} 1 & 3 & 5 & -1 \\ 2 & 1 & 3 & -1 \\ -3 & -1 & 1 & 2 \end{array}\right]$$

Step 2: Change a_{21} to 0 by multiplying row 1 by the opposite of a_{21} and then adding to row 2.

$$\left[\begin{array}{ccc|c} 1 & 3 & 5 & -1 \\ 2 & 1 & 3 & -1 \\ -3 & -1 & 1 & 2 \end{array}\right] \xrightarrow{-2R_1 + R_2} \left[\begin{array}{ccc|c} 1 & 3 & 5 & -1 \\ 0 & -5 & -7 & 1 \\ -3 & -1 & 1 & 2 \end{array}\right]$$

Change a_{31} to 0 by multiplying row 1 by the opposite of a_{31} and then adding to row 3.

$$\left[\begin{array}{ccc|c} 1 & 3 & 5 & -1 \\ 0 & -5 & -7 & 1 \\ -3 & -1 & 1 & 2 \end{array}\right] \quad 3R_1 + R_3 \rightarrow \quad \left[\begin{array}{ccc|c} 1 & 3 & 5 & -1 \\ 0 & -5 & -7 & 1 \\ 0 & 8 & 16 & -1 \end{array}\right]$$

Step 3: Change a_{22} to 1 by multiplying row 2 by the reciprocal of a_{22}.

$$\left[\begin{array}{ccc|c} 1 & 3 & 5 & -1 \\ 0 & -5 & -7 & 1 \\ 0 & 8 & 16 & -1 \end{array}\right] \quad -\tfrac{1}{5}R_2 \rightarrow \quad \left[\begin{array}{ccc|c} 1 & 3 & 5 & -1 \\ 0 & 1 & \frac{7}{5} & -\frac{1}{5} \\ 0 & 8 & 16 & -1 \end{array}\right]$$

Step 4: Change a_{32} to 0 by multiplying row 2 by the opposite of a_{32} and then adding to row 3.

$$\left[\begin{array}{ccc|c} 1 & 3 & 5 & -1 \\ 0 & 1 & \frac{7}{5} & -\frac{1}{5} \\ 0 & 8 & 16 & -1 \end{array}\right] \quad -8R_2 + R_3 \rightarrow \quad \left[\begin{array}{ccc|c} 1 & 3 & 5 & -1 \\ 0 & 1 & \frac{7}{5} & -\frac{1}{5} \\ 0 & 0 & \frac{24}{5} & \frac{3}{5} \end{array}\right]$$

Step 5: Change a_{33} to 1 by multiplying row 3 by the reciprocal of a_{33}.

$$\left[\begin{array}{ccc|c} 1 & 3 & 5 & -1 \\ 0 & 1 & \frac{7}{5} & -\frac{1}{5} \\ 0 & 0 & \frac{24}{5} & \frac{3}{5} \end{array}\right] \quad \tfrac{5}{24}R_3 \rightarrow \quad \left[\begin{array}{ccc|c} 1 & 3 & 5 & -1 \\ 0 & 1 & \frac{7}{5} & -\frac{1}{5} \\ 0 & 0 & 1 & \frac{1}{8} \end{array}\right]$$

A row echelon form of the matrix is $\left[\begin{array}{ccc|c} 1 & 3 & 5 & -1 \\ 0 & 1 & \frac{7}{5} & -\frac{1}{5} \\ 0 & 0 & 1 & \frac{1}{8} \end{array}\right]$.

Problem 6 Write $\left[\begin{array}{ccc|c} 1 & -3 & 2 & 1 \\ -4 & 14 & 0 & -2 \\ 2 & -5 & -3 & 16 \end{array}\right]$ in row echelon form.

Solution See pages S12–S13.

Try Exercise 73, page 219.

If an augmented matrix is in row echelon form, the corresponding system of equations can be solved by substitution. For instance, consider the following matrix in row echelon form and the corresponding system of equations.

$$\left[\begin{array}{ccc|c} 1 & -3 & 4 & 7 \\ 0 & 1 & 3 & -6 \\ 0 & 0 & 1 & -1 \end{array}\right] \qquad \begin{aligned} x - 3y + 4z &= 7 \\ y + 3z &= -6 \\ z &= -1 \end{aligned}$$

From the last equation of the system above, we have $z = -1$. Substitute this value into the second equation and solve for y. Thus $y = -3$.

$$\begin{aligned} y + 3z &= -6 \\ y + 3(-1) &= -6 \\ y - 3 &= -6 \\ y &= -3 \end{aligned}$$

Substitute $y = -3$ and $z = -1$ in the first equation of the system and solve for x. Thus $x = 2$.

$$\begin{aligned} x - 3y + 4z &= 7 \\ x - 3(-3) + 4(-1) &= 7 \\ x + 9 - 4 &= 7 \\ x &= 2 \end{aligned}$$

The solution of the system of equations is $(2, -3, -1)$.

The process of solving a system of equations by using elementary row operations is called the **Gaussian elimination method.**

Point of Interest

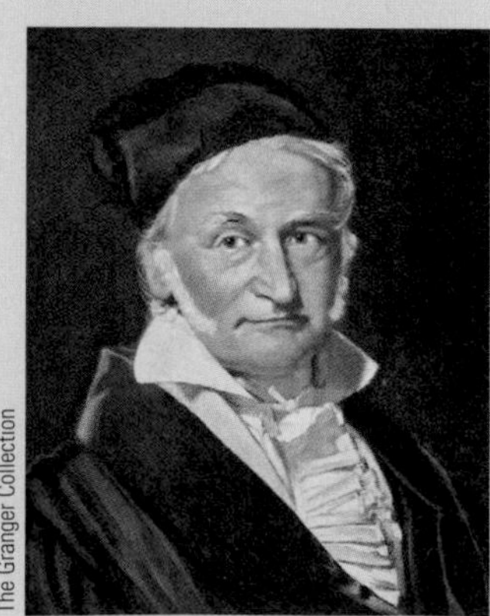

The Granger Collection

Johann Carl Friedrich Gauss (1777–1855) is considered one of the greatest mathematicians of all time. He contributed not only to mathematics but to astronomy and physics as well. A unit of magnetism called a gauss was named in his honor.

EXAMPLE 7 Solve by using the Gaussian elimination method:

$2x - 5y = 19$
$3x + 4y = -6$

Solution Write the augmented matrix and then use elementary row operations to rewrite the matrix in row echelon form.

$$\left[\begin{array}{cc|c} 2 & -5 & 19 \\ 3 & 4 & -6 \end{array}\right] \xrightarrow[\frac{1}{2}R_1]{\text{Change } a_{11} \text{ to } 1.} \left[\begin{array}{cc|c} 1 & -\frac{5}{2} & \frac{19}{2} \\ 3 & 4 & -6 \end{array}\right]$$

• Multiply R_1 by the reciprocal of a_{11}.

$$\left[\begin{array}{cc|c} 1 & -\frac{5}{2} & \frac{19}{2} \\ 3 & 4 & -6 \end{array}\right] \xrightarrow[-3R_1 + R_2]{\text{Change } a_{21} \text{ to } 0.} \left[\begin{array}{cc|c} 1 & -\frac{5}{2} & \frac{19}{2} \\ 0 & \frac{23}{2} & -\frac{69}{2} \end{array}\right]$$

• Multiply R_1 by the opposite of a_{21} and then add R_2.

$$\left[\begin{array}{cc|c} 1 & -\frac{5}{2} & \frac{19}{2} \\ 0 & \frac{23}{2} & -\frac{69}{2} \end{array}\right] \xrightarrow[\frac{2}{23}R_2]{\text{Change } a_{22} \text{ to } 1.} \left[\begin{array}{cc|c} 1 & -\frac{5}{2} & \frac{19}{2} \\ 0 & 1 & -3 \end{array}\right]$$

• Multiply R_2 by the reciprocal of a_{22}.

(1) $x - \frac{5}{2}y = \frac{19}{2}$
(2) $y = -3$

• Write the system of equations corresponding to the row echelon form of the matrix.

$$x - \frac{5}{2}(-3) = \frac{19}{2}$$
$$x + \frac{15}{2} = \frac{19}{2}$$
$$x = 2$$

• Substitute -3 for y in equation (1) and solve for x.

The solution is $(2, -3)$.

Problem 7 Solve by using the Gaussian elimination method:
$4x - 5y = 17$
$3x + 2y = 7$

Solution See page S13.

Try Exercise 79, page 220.

The Gaussian elimination method can be extended to systems of equations with more than two variables.

EXAMPLE 8 Solve by using the Gaussian elimination method:

$x + 2y - z = 9$
$2x - y + 2z = -1$
$-2x + 3y - 2z = 7$

Solution Write the augmented matrix and then use elementary row operations to rewrite the matrix in row echelon form.

$$\left[\begin{array}{ccc|c} 1 & 2 & -1 & 9 \\ 2 & -1 & 2 & -1 \\ -2 & 3 & -2 & 7 \end{array}\right] \xrightarrow[-2R_1 + R_2]{a_{11} \text{ is 1. Change } a_{21} \text{ to } 0.} \left[\begin{array}{ccc|c} 1 & 2 & -1 & 9 \\ 0 & -5 & 4 & -19 \\ -2 & 3 & -2 & 7 \end{array}\right]$$

$$\left[\begin{array}{ccc|c} 1 & 2 & -1 & 9 \\ 0 & -5 & 4 & -19 \\ -2 & 3 & -2 & 7 \end{array}\right] \xrightarrow[2R_1 + R_3]{\text{Change } a_{31} \text{ to } 0.} \left[\begin{array}{ccc|c} 1 & 2 & -1 & 9 \\ 0 & -5 & 4 & -19 \\ 0 & 7 & -4 & 25 \end{array}\right]$$

$$\begin{bmatrix} 1 & 2 & -1 & | & 9 \\ 0 & -5 & 4 & | & -19 \\ 0 & 7 & -4 & | & 25 \end{bmatrix} \xrightarrow[-\frac{1}{5}R_2]{\text{Change } a_{22} \text{ to } 1.} \begin{bmatrix} 1 & 2 & -1 & | & 9 \\ 0 & 1 & -\frac{4}{5} & | & \frac{19}{5} \\ 0 & 7 & -4 & | & 25 \end{bmatrix}$$

$$\begin{bmatrix} 1 & 2 & -1 & | & 9 \\ 0 & 1 & -\frac{4}{5} & | & \frac{19}{5} \\ 0 & 7 & -4 & | & 25 \end{bmatrix} \xrightarrow[-7R_2 + R_3]{\text{Change } a_{32} \text{ to } 0.} \begin{bmatrix} 1 & 2 & -1 & | & 9 \\ 0 & 1 & -\frac{4}{5} & | & \frac{19}{5} \\ 0 & 0 & \frac{8}{5} & | & -\frac{8}{5} \end{bmatrix}$$

$$\begin{bmatrix} 1 & 2 & -1 & | & 9 \\ 0 & 1 & -\frac{4}{5} & | & \frac{19}{5} \\ 0 & 0 & \frac{8}{5} & | & -\frac{8}{5} \end{bmatrix} \xrightarrow[\frac{5}{8}R_2]{\text{Change } a_{33} \text{ to } 1.} \begin{bmatrix} 1 & 2 & -1 & | & 9 \\ 0 & 1 & -\frac{4}{5} & | & \frac{19}{5} \\ 0 & 0 & 1 & | & -1 \end{bmatrix}$$

• **This is row echelon form.**

(1) $x + 2y - z = 9$

(2) $y - \frac{4}{5}z = \frac{19}{5}$

(3) $z = -1$

• **Write the system of equations corresponding to the row echelon form of the matrix.**

$$y - \frac{4}{5}(-1) = \frac{19}{5}$$
$$y + \frac{4}{5} = \frac{19}{5}$$
$$y = 3$$

• **Substitute −1 for *z* in equation (2) and solve for *y*.**

$$x + 2y - z = 9$$
$$x + 2(3) - (-1) = 9$$
$$x + 7 = 9$$
$$x = 2$$

• **Substitute −1 for *z* and 3 for *y* in equation (1) and solve for *x*.**

The solution is (2, 3, −1).

Problem 8 Solve by using the Gaussian elimination method:

$$2x + 3y + 3z = -2$$
$$x + 2y - 3z = 9$$
$$3x - 2y - 4z = 1$$

Solution See page S13.

Try Exercise 91, page 220.

4.3 Exercises

CONCEPT CHECK

1. What is a matrix? What is a square matrix?
2. What is the difference in notation between a matrix and a determinant?

For Exercises 3 to 5, use the matrix $A = \begin{bmatrix} 2 & -1 & 3 & 2 \\ -3 & 0 & 2 & 1 \\ 4 & 5 & 3 & 1 \end{bmatrix}$.

3. What is the order of matrix A?
4. What is the value of a_{32} in matrix A?
5. What are the elements of the main diagonal of matrix A?

6. What is the minor of element a_{ij} of a determinant?

7. When the value of the determinant of the coefficients of a system of equations is zero, what can be said about the system of equations?

8. What is an augmented matrix?

9. What are the three elementary row operations?

10. Which of the following matrices are in row echelon form?

(i) $\begin{bmatrix} 1 & -3 & 4 \\ 0 & 1 & 1 \end{bmatrix}$ (ii) $\begin{bmatrix} 0 & 1 & -1 & 0 \\ 1 & 0 & 3 & -5 \\ 0 & 0 & 1 & 6 \end{bmatrix}$ (iii) $\begin{bmatrix} 1 & 0 & -3 & 1 \\ 0 & 1 & 4 & 9 \\ 0 & 0 & 1 & 7 \end{bmatrix}$

1 Evaluate determinants (See pages 204–207.)

GETTING READY

11. The determinant of $\begin{bmatrix} -1 & -1 \\ 5 & 8 \end{bmatrix}$ is written $\begin{vmatrix} -1 & -1 \\ 5 & 8 \end{vmatrix}$.

The value of this determinant is $(-1)(\underline{\ ?\ }) - 5(\underline{\ ?\ }) = \underline{\ ?\ }$.

12. For the determinant $\begin{vmatrix} 4 & 1 & -7 \\ 0 & -2 & 6 \\ 5 & 0 & -3 \end{vmatrix}$, the minor of 6 is $\begin{vmatrix} \underline{\ ?\ } & \underline{\ ?\ } \\ \underline{\ ?\ } & \underline{\ ?\ } \end{vmatrix}$.

13. For the determinant $\begin{vmatrix} 2 & 5 & 7 \\ -3 & 4 & -1 \\ 0 & 8 & 4 \end{vmatrix}$, find the minor of 7 and of 8.

14. For the determinant $\begin{vmatrix} 5 & -3 & 4 \\ 2 & 0 & -1 \\ 3 & 7 & 6 \end{vmatrix}$, find the cofactor of 3 and of -1.

Evaluate the determinant.

15. $\begin{vmatrix} 2 & -1 \\ 3 & 4 \end{vmatrix}$

16. $\begin{vmatrix} 5 & 1 \\ -1 & 2 \end{vmatrix}$

17. $\begin{vmatrix} 6 & -2 \\ -3 & 4 \end{vmatrix}$

18. $\begin{vmatrix} -3 & 5 \\ 1 & 7 \end{vmatrix}$

19. $\begin{vmatrix} 3 & 6 \\ 2 & 4 \end{vmatrix}$

20. $\begin{vmatrix} 5 & -10 \\ 1 & -2 \end{vmatrix}$

21. $\begin{vmatrix} 1 & -1 & 2 \\ 3 & 2 & 1 \\ 1 & 0 & 4 \end{vmatrix}$

22. $\begin{vmatrix} 4 & 1 & 3 \\ 2 & -2 & 1 \\ 3 & 1 & 2 \end{vmatrix}$

23. $\begin{vmatrix} 3 & -1 & 2 \\ 0 & 1 & 2 \\ 3 & 2 & -2 \end{vmatrix}$

24. $\begin{vmatrix} 4 & 5 & -2 \\ 3 & -1 & 5 \\ 2 & 1 & 4 \end{vmatrix}$

25. $\begin{vmatrix} 4 & 2 & 6 \\ -2 & 1 & 1 \\ 2 & 1 & 3 \end{vmatrix}$

26. $\begin{vmatrix} 3 & 6 & -3 \\ 4 & -1 & 6 \\ -1 & -2 & 3 \end{vmatrix}$

2 Solve systems of linear equations by using Cramer's Rule (See pages 207–209.)

GETTING READY

For Exercises 27 to 30, use this system of equations:

$$\begin{aligned} 2x - 3y &= 5 \\ -5x + 4y &= -2 \end{aligned}$$

27. a. The coefficient determinant for the system is $D = \begin{vmatrix} \underline{\ ?\ } & \underline{\ ?\ } \\ \underline{\ ?\ } & \underline{\ ?\ } \end{vmatrix}$.

b. The value of the coefficient determinant is $D = \underline{\ ?\ }$.

28. **a.** The numerator determinant for x is $D_x = \begin{vmatrix} ? & ? \\ ? & ? \end{vmatrix}$.

b. The value of the numerator determinant for x is $D_x = $ __?__.

29. **a.** The numerator determinant for y is $D_y = \begin{vmatrix} ? & ? \\ ? & ? \end{vmatrix}$.

b. The value of the numerator determinant for y is $D_y = $ __?__.

30. Use your answers to Exercises 27(b), 28(b), and 29(b) to find the solution of the system of equations.

$x = \dfrac{D_x}{D} = \dfrac{?}{?} = $ __?__ and $y = \dfrac{D_y}{D} = \dfrac{?}{?} = $ __?__,

so the solution is (__?__, __?__).

Solve by using Cramer's Rule, if possible.

31. $2x - 5y = 26$
$5x + 3y = 3$

32. $3x + 7y = 15$
$2x + 5y = 11$

33. $x - 4y = 8$
$3x + 7y = 5$

34. $5x + 2y = -5$
$3x + 4y = 11$

35. $2x + 3y = 4$
$6x - 12y = -5$

36. $5x + 4y = 3$
$15x - 8y = -21$

37. $2x + 5y = 6$
$6x - 2y = 1$

38. $7x + 3y = 4$
$5x - 4y = 9$

39. $-2x + 3y = 7$
$4x - 6y = 9$

40. $9x + 6y = 7$
$3x + 2y = 4$

41. $2x - 5y = -2$
$3x - 7y = -3$

42. $8x + 7y = -3$
$2x + 2y = 5$

43. $2x - y + 3z = 9$
$x + 4y + 4z = 5$
$3x + 2y + 2z = 5$

44. $3x - 2y + z = 2$
$2x + 3y + 2z = -6$
$3x - y + z = 0$

45. $3x - y + z = 11$
$x + 4y - 2z = -12$
$2x + 2y - z = -3$

46. $x + 2y + 3z = 8$
$2x - 3y + z = 5$
$3x - 4y + 2z = 9$

47. $4x - 2y + 6z = 1$
$3x + 4y + 2z = 1$
$2x - y + 3z = 2$

48. $x - 3y + 2z = 1$
$2x + y - 2z = 3$
$3x - 9y + 6z = -3$

49. $5x - 4y + 2z = 4$
$3x - 5y + 3z = -4$
$3x + y - 5z = 12$

50. $2x + 4y + z = 7$
$x + 3y - z = 1$
$3x + 2y - 2z = 5$

3 Solve systems of linear equations by using matrices (See pages 210–216.)

GETTING READY

For Exercises 51 to 53, use the augmented matrix at the right, for a system of three equations in three variables x, y, and z.

$$\left[\begin{array}{rrr|r} 1 & -2 & 3 & 2 \\ 2 & -3 & 1 & 1 \\ 3 & -1 & 2 & 9 \end{array}\right]$$

51. The coefficient of y in the second equation of the system of equations associated with the augmented matrix is __?__.

52. The constant for the third equation of the system of equations associated with the augmented matrix is __?__.

53. When the augmented matrix is in row echelon form, all the elements of the main diagonal will be __?__. All the elements to the left of the main diagonal will be __?__.

54. Is the matrix $\left[\begin{array}{ccc|c} 1 & -2 & 3 & 1 \\ 0 & 1 & 2 & 8 \\ 0 & 1 & 0 & 1 \end{array}\right]$ in row echelon form? Why or why not?

Write the augmented matrix for the given system of equations.

55. $2x + y = 4$
$x - 3y = -10$

56. $-x + 6y = -3$
$7x - y = -20$

57. $3x + y - 3z = -5$
$x - y + 2z = 6$
$-x + 4y - z = 0$

58. $x - 5y + z = 23$
$9x - y - 4z = 0$
$x + y - z = 5$

Write the system of equations for the given augmented matrix.

59. $\left[\begin{array}{cc|c} 8 & -1 & -15 \\ 2 & 1 & -3 \end{array}\right]$

60. $\left[\begin{array}{cc|c} 1 & 2 & -17 \\ 0 & 1 & -9 \end{array}\right]$

61. $\left[\begin{array}{ccc|c} 1 & -4 & -1 & 0 \\ 0 & 1 & 6 & 15 \\ 0 & 0 & 1 & 2 \end{array}\right]$

62. $\left[\begin{array}{ccc|c} -8 & -1 & 3 & 10 \\ 7 & -2 & 9 & 0 \\ -4 & 1 & -10 & -5 \end{array}\right]$

Use matrices A, B, and C shown below. Perform the given row operation on the indicated matrix.

$$A = \left[\begin{array}{cc|c} 3 & 4 & -2 \\ 2 & -3 & 5 \end{array}\right] \quad B = \left[\begin{array}{ccc|c} 2 & -3 & 4 & 1 \\ 5 & -1 & 3 & -2 \\ 1 & 2 & -3 & 7 \end{array}\right] \quad C = \left[\begin{array}{ccc|c} 1 & 3 & -2 & 1 \\ 0 & 4 & -3 & 2 \\ -2 & 1 & -2 & 2 \end{array}\right]$$

63. $R_1 \leftrightarrow R_2$; matrix B

64. $R_1 \leftrightarrow R_3$; matrix C

65. $\frac{1}{3}R_1$; matrix A

66. $-2R_2$; matrix A

67. $2R_1 + R_3$; matrix C

68. $-R_2 + R_1$; matrix B

Write each matrix in row echelon form.

69. $\left[\begin{array}{cc|c} 2 & -4 & 1 \\ 3 & -7 & -1 \end{array}\right]$

70. $\left[\begin{array}{cc|c} 4 & 2 & -2 \\ 7 & 4 & -1 \end{array}\right]$

71. $\left[\begin{array}{cc|c} 5 & -2 & 3 \\ -7 & 3 & 1 \end{array}\right]$

72. $\left[\begin{array}{cc|c} 2 & 5 & -4 \\ 3 & 1 & 2 \end{array}\right]$

73. $\left[\begin{array}{ccc|c} 1 & 4 & 1 & -2 \\ 3 & 11 & -1 & 2 \\ 2 & 3 & 1 & 4 \end{array}\right]$

74. $\left[\begin{array}{ccc|c} 1 & 2 & 2 & -1 \\ -4 & -10 & -1 & 3 \\ 3 & 4 & 2 & -2 \end{array}\right]$

75. $\left[\begin{array}{ccc|c} -2 & 6 & -1 & 3 \\ 1 & -2 & 2 & 1 \\ 3 & -6 & 7 & 6 \end{array}\right]$

76. $\left[\begin{array}{ccc|c} 2 & 6 & 10 & 3 \\ 3 & 8 & 15 & 0 \\ 1 & 2 & 3 & -1 \end{array}\right]$

77. What is the solution of the system of equations that has $\left[\begin{array}{ccc|c} 1 & -1 & 3 & -2 \\ 0 & 1 & -1 & 1 \\ 0 & 0 & 1 & 3 \end{array}\right]$ as the row echelon form of the augmented matrix for the system of equations?

78. What is the solution of the system of equations that has $\left[\begin{array}{ccc|c} 1 & -3 & 2 & 4 \\ 0 & 1 & -2 & 3 \\ 0 & 0 & 1 & -1 \end{array}\right]$ as the row echelon form of the augmented matrix for the system of equations?

Solve by using the Gaussian elimination method.

79. $3x + y = 6$, $2x - y = -1$

80. $2x + y = 3$, $x - 4y = 6$

81. $x - 3y = 8$, $3x - y = 0$

82. $2x + 3y = 16$, $x - 4y = -14$

83. $y = 4x - 10$, $2y = 5x - 11$

84. $2y = 4 - 3x$, $y = 1 - 2x$

85. $2x - y = -4$, $y = 2x - 8$

86. $3x - 2y = -8$, $y = \frac{3}{2}x - 2$

87. $4x - 3y = -14$, $3x + 4y = 2$

88. $5x + 2y = 3$, $3x + 4y = 13$

89. $5x + 4y + 3z = -9$, $x - 2y + 2z = -6$, $x - y - z = 3$

90. $x - y - z = 0$, $3x - y + 5z = -10$, $x + y - 4z = 12$

91. $5x - 5y + 2z = 8$, $2x + 3y - z = 0$, $x + 2y - z = 0$

92. $2x + y - 5z = 3$, $3x + 2y + z = 15$, $5x - y - z = 5$

93. $2x + 3y + z = 5$, $3x + 3y + 3z = 10$, $4x + 6y + 2z = 5$

94. $x - 2y + 3z = 2$, $2x + y + 2z = 5$, $2x - 4y + 6z = -4$

95. $3x + 2y + 3z = 2$, $6x - 2y + z = 1$, $3x + 4y + 2z = 3$

96. $2x + 3y - 3z = -1$, $2x + 3y + 3z = 3$, $4x - 4y + 3z = 4$

97. $5x - 5y - 5z = 2$, $5x + 5y - 5z = 6$, $10x + 10y + 5z = 3$

98. $3x - 2y + 2z = 5$, $6x + 3y - 4z = -1$, $3x - y + 2z = 4$

99. $4x + 4y - 3z = 3$, $8x + 2y + 3z = 0$, $4x - 4y + 6z = -3$

APPLYING CONCEPTS

Solve for x.

100. $\begin{vmatrix} 3 & 2 \\ 4 & x \end{vmatrix} = -11$

101. $\begin{vmatrix} 1 & 0 & 2 \\ 4 & 3 & -1 \\ 0 & 2 & x \end{vmatrix} = -24$

102. $\begin{vmatrix} -2 & 1 & 3 \\ 0 & x & 4 \\ -1 & 2 & -3 \end{vmatrix} = -24$

Complete the statement.

103. If all the elements in one row or one column of a 2×2 matrix are zeros, the value of the determinant of the matrix is __?__.

104. If all the elements in one row or one column of a 3×3 matrix are zeros, the value of the determinant of the matrix is __?__.

105. a. The value of the determinant $\begin{vmatrix} x & x & a \\ y & y & b \\ z & z & c \end{vmatrix}$ is __?__.

b. If two columns of a 3×3 matrix contain identical elements, the value of the determinant is __?__.

106. Suppose that applying elementary row operations to an augmented matrix produces the matrix $\left[\begin{array}{ccc|c} 1 & 2 & -3 & 5 \\ 0 & 1 & -2 & -2 \\ 0 & 0 & 0 & -3 \end{array}\right]$. Explain why the original system of equations has no solution.

PROJECTS OR GROUP ACTIVITIES

107. **Surveying** Surveyors use a formula to find the area of a plot of land. The *surveyor's area formula* states that if the vertices $P_1(x_1, y_1), P_2(x_2, y_2), P_3(x_3, y_3), \ldots, P_n(x_n, y_n)$ of a simple polygon are listed counterclockwise around the perimeter, then the area of the polygon is

$$A = \frac{1}{2}\left\{\begin{vmatrix} x_1 & x_2 \\ y_1 & y_2 \end{vmatrix} + \begin{vmatrix} x_2 & x_3 \\ y_2 & y_3 \end{vmatrix} + \begin{vmatrix} x_3 & x_4 \\ y_3 & y_4 \end{vmatrix} + \cdots + \begin{vmatrix} x_n & x_1 \\ y_n & y_1 \end{vmatrix}\right\}$$

Use the surveyor's area formula to find the area of the polygon with vertices $P_1(9, -3)$, $P_2(26, 6)$, $P_3(18, 21)$, $P_4(16, 10)$, and $P_5(1, 11)$. Measurements are given in feet.

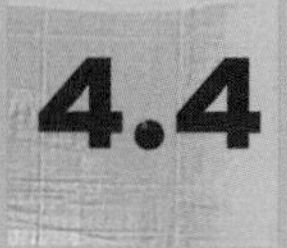

4.4 Application Problems

OBJECTIVE 1 Rate-of-wind and rate-of-current problems

Solving motion problems that involve an object moving with or against a wind or current normally requires two variables.

Solve: A motorboat traveling with the current can go 24 mi in 2 h. Against the current, it takes 3 h to go the same distance. Find the rate of the motorboat in calm water and the rate of the current.

STRATEGY FOR SOLVING RATE-OF-WIND AND RATE-OF-CURRENT PROBLEMS

- Choose one variable to represent the rate of the object in calm conditions and a second variable to represent the rate of the wind or current. Using these variables, express the rate of the object with and against the wind or current. Use the equation $rt = d$ to write expressions for the distance traveled by the object. The results can be recorded in a table.

Rate of the boat in calm water: x
Rate of the current: y

	Rate	·	Time	=	Distance
With current	$x + y$	·	2	=	$2(x + y)$
Against current	$x - y$	·	3	=	$3(x - y)$

- Determine how the expressions for the distance are related, and write a system of equations.

The distance traveled with the current is 24 mi. $\quad 2(x + y) = 24$
The distance traveled against the current is 24 mi. $\quad 3(x - y) = 24$

The figure at the right depicts the system of equations

$$2(x + y) = 24$$
$$3(x - y) = 24$$

where x is the rate of the boat in calm water, and y is the rate of the current.

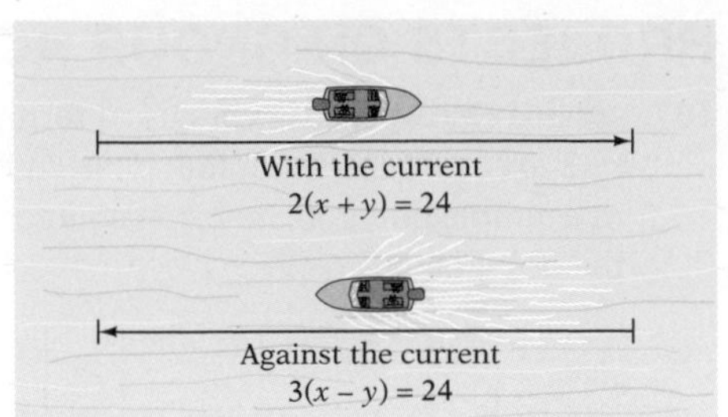

Solve the system of equations.

$$2(x + y) = 24 \quad \rightarrow \quad \frac{1}{2} \cdot 2(x + y) = \frac{1}{2} \cdot 24 \quad \rightarrow \quad x + y = 12$$
$$3(x - y) = 24 \quad \rightarrow \quad \frac{1}{3} \cdot 3(x - y) = \frac{1}{3} \cdot 24 \quad \rightarrow \quad x - y = 8$$
$$2x = 20$$
$$x = 10$$

Replace x by 10 in the equation $x + y = 12$. Solve for y.

$$x + y = 12$$
$$10 + y = 12$$
$$y = 2$$

The rate of the boat in calm water is 10 mph.
The rate of the current is 2 mph.

EXAMPLE 1 Flying with the wind, a plane flew 1000 mi in 5 h. Flying against the wind, the plane could fly only 500 mi in the same amount of time. Find the rate of the plane in calm air and the rate of the wind.

Strategy ▶ Rate of the plane in still air: p
Rate of the wind: w

	Rate	Time	Distance
With wind	$p + w$	5	$5(p + w)$
Against wind	$p - w$	5	$5(p - w)$

▶ The distance traveled with the wind is 1000 mi.
The distance traveled against the wind is 500 mi.

Solution

$$5(p + w) = 1000$$
$$5(p - w) = 500$$

• **Write a system of equations.**

$$p + w = 200$$
$$p - w = 100$$

• **Multiply each side of both equations by $\frac{1}{5}$.**

$$2p = 300$$

• **Add the equations.**

$$p = 150$$

• **Solve for p.**

$$p + w = 200$$
$$150 + w = 200$$
$$w = 50$$

• **Substitute the value of p into one of the equations and solve for w.**

The rate of the plane in calm air is 150 mph.
The rate of the wind is 50 mph.

Problem 1 A rowing team rowing with the current traveled 18 mi in 2 h. Against the current, the team rowed 10 mi in 2 h. Find the rate of the rowing team and the rate of the current.

Solution See page S13.

Try Exercise 9, page 226.

OBJECTIVE 2 Application problems

The application problems in this objective are varieties of those problems solved earlier in the text. Each of the strategies for the problems in this objective will result in a system of equations.

Solve: A store owner purchased twenty 60-watt light bulbs and thirty fluorescent lights for a total cost of \$40. A second purchase, at the same prices, included thirty 60-watt bulbs and ten fluorescent lights for a total cost of \$25. Find the cost of a 60-watt bulb and of a fluorescent light.

STRATEGY FOR SOLVING AN APPLICATION PROBLEM IN TWO VARIABLES

- Choose one variable to represent one of the unknown quantities and a second variable to represent the other unknown quantity. Write numerical or variable expressions for all the remaining quantities. These results can be recorded in two tables, one for each of the conditions.

Cost of 60-watt bulb: b
Cost of a fluorescent light: f

First purchase:

	Amount	·	Unit cost	=	Value
60-watt	20	·	b	=	$20b$
Fluorescent	30	·	f	=	$30f$

Second purchase:

	Amount	·	Unit cost	=	Value
60-watt	30	·	b	=	$30b$
Fluorescent	10	·	f	=	$10f$

- Determine a system of equations. The strategies presented in Chapter 2 can be used to determine the relationships between the expressions in the tables. Each table will give one equation of the system.

The total of the first purchase was \$40. $20b + 30f = 40$
The total of the second purchase was \$25. $30b + 10f = 25$

Solve the system of equations.

$$\begin{aligned} 20b + 30f &= 40 \\ 30b + 10f &= 25 \end{aligned} \quad \longrightarrow \quad \begin{aligned} 3(20b + 30f) &= 3 \cdot 40 \\ -2(30b + 10f) &= -2 \cdot 25 \end{aligned} \quad \longrightarrow \quad \begin{aligned} 60b + 90f &= 120 \\ -60b - 20f &= -50 \\ \hline 70f &= 70 \\ f &= 1 \end{aligned}$$

Replace f by 1 in the equation $20b + 30f = 40$.
Solve for b.

$$\begin{aligned} 20b + 30f &= 40 \\ 20b + 30(1) &= 40 \\ 20b + 30 &= 40 \\ 20b &= 10 \\ b &= 0.5 \end{aligned}$$

The cost of a 60-watt bulb was \$.50.
The cost of a fluorescent light was \$1.00.

EXAMPLE 2 A metallurgist has two alloys of stainless steel. Alloy I is 14% chromium and 6% nickel, and Alloy II is 18% chromium and 8% nickel. How many kilograms of each alloy should the metallurgist use to make a new stainless steel compound that contains 23 kg of chromium and 10 kg of nickel?

Strategy ▶ Kilograms of Alloy I: x
Kilograms of Alloy II: y

Kilograms of chromium:

	Amount of alloy	Percent	Quantity of chromium
Chromium in Alloy I	x	0.14	$0.14x$
Chromium in Alloy II	y	0.18	$0.18y$
Chromium in new alloy			23

Kilograms of nickel:

	Amount of alloy	Percent	Quantity of nickel
Nickel in Alloy I	x	0.06	$0.06x$
Nickel in Alloy II	y	0.08	$0.08y$
Nickel in new alloy			10

▶ The sum of the quantities of chromium in the two alloys equals the amount of chromium in the new alloy: $0.14x + 0.18y = 23$.

The sum of the quantities of nickel in the two alloys equals the amount of nickel in the new alloy: $0.06x + 0.08y = 10$.

Solution

$$
\begin{aligned}
(1)\quad 0.14x + 0.18y &= 23 && \text{• Write a system of equations.}\\
(2)\quad 0.06x + 0.08y &= 10\\
100(0.14x + 0.18y) &= 100(23) && \text{• Multiply each equation by 100 to remove decimals.}\\
100(0.06x + 0.08y) &= 100(10)\\
(3)\quad 14x + 18y &= 2300 && \text{• Simplify.}\\
(4)\quad 6x + 8y &= 1000\\
6(14x + 18y) &= 6(2300) && \text{• Multiply equation (3) by 6.}\\
-14(6x + 8y) &= -14(1000) && \text{• Multiply equation (4) by } -14.\\
84x + 108y &= 13{,}800 && \text{• Simplify.}\\
-84x - 112y &= -14{,}000\\
-4y &= -200 && \text{• Add the equations.}\\
y &= 50 && \text{• Solve for } y.\\
0.14x + 0.18y &= 23 && \text{• Substitute the value of } y \text{ into equation (1).}\\
0.14x + 0.18(50) &= 23\\
0.14x + 9 &= 23 && \text{• Solve for } x.\\
0.14x &= 14\\
x &= 100
\end{aligned}
$$

The metallurgist should use 100 kg of Alloy I and 50 kg of Alloy II.

Problem 2 A citrus fruit grower purchased 25 orange trees and 20 grapefruit trees for \$290. The next week, at the same prices, the grower bought 20 orange trees and 30 grapefruit trees for \$330. Find the cost of an orange tree and the cost of a grapefruit tree.

Solution See page S14.

Try Exercise 23, page 227.

Example 3 is an application problem that requires more than two variables. The solution of this problem illustrates how the substitution method can be used to solve a system of linear equations in three variables.

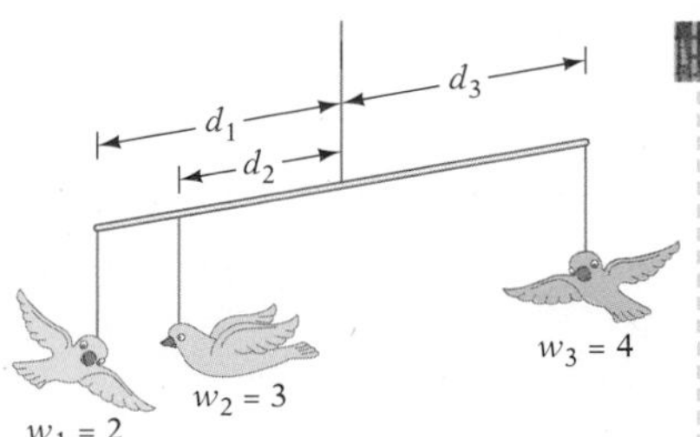

EXAMPLE 3 An artist is creating a mobile in which three objects will be suspended from a light rod that is 18 in. long, as shown at the left. The weight, in ounces, of each object is shown in the diagram. For the mobile to balance, the objects must be positioned so that $w_1d_1 + w_2d_2 = w_3d_3$. The artist wants d_1 to be 1.5 times d_2. Find the distances d_1, d_2, and d_3 such that the mobile will balance.

Strategy There are three unknowns for this problem. Use the information in the problem to write three equations with d_1, d_2, and d_3 as the variables.

The length of the rod is 18 in. Therefore, $d_1 + d_3 = 18$. Using $w_1d_1 + w_2d_2 = w_3d_3$, we have $2d_1 + 3d_2 = 4d_3$. The artist wants d_1 to be 1.5 times d_2. Thus $d_1 = 1.5d_2$.

Solution

(1) $d_1 + d_3 = 18$ • **Write a system of equations.**

(2) $2d_1 + 3d_2 = 4d_3$

(3) $d_1 = 1.5d_2$

(4) $1.5d_2 + d_3 = 18$ • **To solve by substitution, first use equation (3) to replace d_1 in equation (1) and in equation (2).**

(5) $2(1.5d_2) + 3d_2 = 4d_3$

$6d_2 = 4d_3$ • **Simplify equation (5).**

(6) $1.5d_2 = d_3$ • **Divide each side of equation (5) by 4.**

$1.5d_2 + 1.5d_2 = 18$ • **Replace d_3 by $1.5d_2$ in equation (4) and solve for d_2.**

$3d_2 = 18$

$d_2 = 6$

From equation (6), $d_3 = 1.5d_2 = 1.5(6) = 9$.

Substituting the value of d_3 into equation (1), we have $d_1 = 9$.

The distances are $d_1 = 9$ in., $d_2 = 6$ in., and $d_3 = 9$ in.

Problem 3 A science museum charges \$10 for an admission ticket, but members receive a discount of \$3, and students are admitted for half price. Last Saturday, 750 tickets were sold for a total of \$5400. If 20 more student tickets than full-price tickets were sold, how many of each type of ticket were sold?

Solution See page S14.

Try Exercise 35, page 228.

4.4 Exercises

CONCEPT CHECK

1. The speed of a plane in calm air is 500 mph. If the plane is flying into a 50-mile-per-hour headwind, what is the speed of the plane relative to an observer on the ground?

2. The speed of a boat in calm water is x miles per hour, and the speed of the current is y miles per hour. What is the rate of the boat going with the current?

3. A contractor bought 50 yd of nylon carpet for x dollars per yard and 100 yd of wool carpet for y dollars per yard. Express the total cost of the two purchases in terms of x and y.

4. A chemist has x grams of an alloy that is 20% silver and 10% gold, and y grams of an alloy that is 25% silver and 30% gold. Express the number of grams of gold in the two alloys in terms of x and y.

1 Rate-of-wind and rate-of-current problems (See pages 221–222.)

GETTING READY

5. A boat travels down a river for 3 h (traveling with the current), then turns around and takes 5 h to return (traveling against the current). Let b be the rate of the boat, in miles per hour, in calm water, and let c be the rate of the current, in miles per hour. Complete the following table.

	Rate	·	Time	=	Distance
With current	?	·	?	=	?
Against current	?	·	?	=	?

6. Suppose the distance traveled down the river by the boat in Exercise 5 is 45 mi. Use the expressions in the last column of the table in Exercise 5 to write a system of equations that can be solved to find the rate of the boat in calm water and the rate of the current.

7. Traveling with the wind, a plane flies m miles in h hours. Traveling against the wind, the plane flies n miles in h hours. Is n less than, equal to, or greater than m?

8. Traveling against the current, it takes a boat h hours to go m miles. Traveling with the current, the boat takes k hours to go m miles. Is h less than, equal to, or greater than k?

9. Flying with the wind, a small plane flew 320 mi in 2 h. Against the wind, the plane could fly only 280 mi in the same amount of time. Find the rate of the plane in calm air and the rate of the wind.

10. A jet plane flying with the wind flew 2100 mi in 4 h. Against the wind, the plane could fly only 1760 mi in the same amount of time. Find the rate of the plane in calm air and the rate of the wind.

11. A cabin cruiser traveling with the current went 48 mi in 3 h. Against the current, it took 4 h to travel the same distance. Find the rate of the cabin cruiser in calm water and the rate of the current.

12. A motorboat traveling with the current went 48 mi in 2 h. Against the current, it took 3 h to travel the same distance. Find the rate of the boat in calm water and the rate of the current.

13. Flying with the wind, a pilot flew 450 mi between two cities in 2.5 h. The return trip against the wind took 3 h. Find the rate of the plane in calm air and the rate of the wind.

14. A turbo-prop plane flying with the wind flew 600 mi in 2 h. Flying against the wind, the plane required 3 h to travel the same distance. Find the rate of the wind and the rate of the plane in calm air.

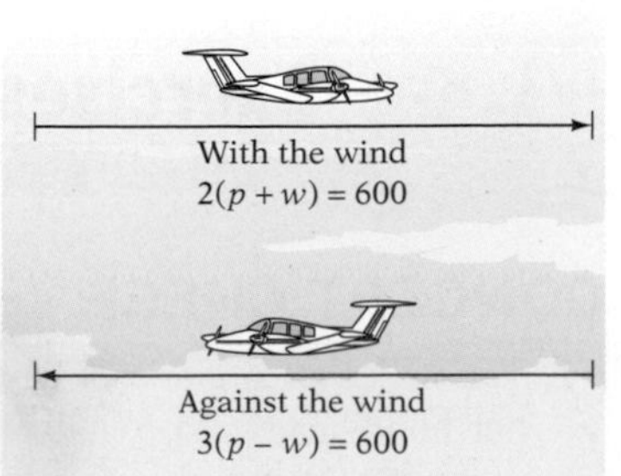

15. A motorboat traveling with the current went 88 km in 4 h. Against the current, the boat could go only 64 km in the same amount of time. Find the rate of the boat in calm water and the rate of the current.

16. A rowing team rowing with the current traveled 18 km in 2 h. Rowing against the current, the team rowed 12 km in the same amount of time. Find the rate of the rowing team in calm water and the rate of the current.

17. A plane flying with a tailwind flew 360 mi in 3 h. Against the wind, the plane required 4 h to fly the same distance. Find the rate of the plane in calm air and the rate of the wind.

18. Flying with the wind, a plane flew 1000 mi in 4 h. Against the wind, the plane required 5 h to fly the same distance. Find the rate of the plane in calm air and the rate of the wind.

19. A motorboat traveling with the current went 54 mi in 3 h. Against the current, it took 3.6 h to travel the same distance. Find the rate of the boat in calm water and the rate of the current.

20. A plane traveling with the wind flew 3625 mi in 6.25 h. Against the wind, the plane required 7.25 h to fly the same distance. Find the rate of the plane in calm air and the rate of the wind.

2 Application problems (See pages 223–225.)

GETTING READY

21. For one project, a builder buys 200 ft² of hardwood flooring and 300 ft² of wall-to-wall carpet. For another project, the builder buys 350 ft² of hardwood flooring and 100 ft² of wall-to-wall carpet. Let h be the cost per square foot of hardwood flooring, and let w be the cost per square foot of wall-to-wall carpet. Complete the following tables by replacing the question marks.

First Project	Amount	·	Unit cost	=	Value
Hardwood flooring	?	·	?	=	?
Wall-to-wall carpet	?	·	?	=	?

Second Project	Amount	·	Unit cost	=	Value
Hardwood flooring	?	·	?	=	?
Wall-to-wall carpet	?	·	?	=	?

22. Suppose the builder in Exercise 21 spent \$2300 on the flooring and carpet for the first project and \$2750 on the flooring and carpet for the second project. Use the expressions in the last columns of the tables in Exercise 21 to write a system of equations that can be solved to find the cost per square foot of hardwood flooring and the cost per square foot of wall-to-wall carpet.

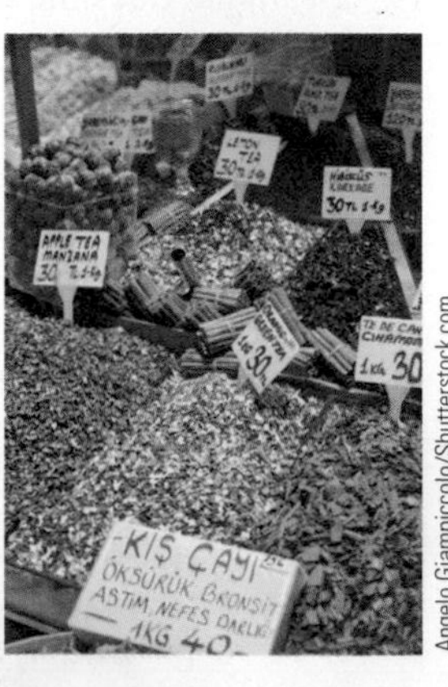

Angelo Giampiccolo/Shutterstock.com

23. **Construction** A carpenter purchased 50 ft of redwood and 90 ft of pine for a total cost of \$31.20. A second purchase, at the same prices, included 200 ft of redwood and 100 ft of pine for a total cost of \$78. Find the cost per foot of redwood and of pine.

24. **Business** A merchant mixed 10 lb of cinnamon tea with 5 lb of spice tea. The 15-pound mixture cost \$40. A second mixture included 12 lb of the cinnamon tea and 8 lb of the spice tea. The 20-pound mixture cost \$54. Find the cost per pound of the cinnamon tea and of the spice tea.

25. **Home Economics** During one month, a homeowner used 400 units of electricity and 120 units of gas for a total cost of $147.20. The next month, 350 units of electricity and 200 units of gas were used for a total cost of $144. Find the cost per unit of gas.

26. **Construction** A contractor buys 20 yd^2 of nylon carpet and 28 yd^2 of wool carpet for $1360. A second purchase, at the same prices, includes 15 yd^2 of nylon carpet and 20 yd^2 of wool carpet for $990. Find the cost per square yard of the wool carpet.

27. **Manufacturing** A company manufactures both mountain bikes and trail bikes. The cost of materials for a mountain bike is $70, and the cost of materials for a trail bike is $50. The cost of labor to manufacture a mountain bike is $80, and the cost of labor to manufacture a trail bike is $40. During a week in which the company has budgeted $2500 for materials and $2600 for labor, how many mountain bikes does the company plan to manufacture?

28. **Manufacturing** A company manufactures both liquid crystal display (LCD) and plasma televisions. The cost of materials for an LCD television is $125, and the cost of materials for a plasma TV is $150. The cost of labor to manufacture one LCD television is $80, and the cost of labor to manufacture one plasma television is $85. How many of each type of television can the manufacturer produce during a week in which $18,000 has been budgeted for materials and $10,750 has been budgeted for labor?

Fuel Economy Use the information in the article at the right for Exercises 29 and 30.

29. One week, the owner of a hybrid car drove 394 mi and spent $34.74 on gasoline. How many miles did the owner drive in the city? On the highway?

30. Gasoline for one week of driving cost the owner of a hybrid car $26.50. The owner would have spent $51.50 for gasoline to drive the same number of miles in a traditional car. How many miles did the owner drive in the city? On the highway?

In the News

Hybrids Easier on the Pocketbook?

A hybrid car can make up for its high sticker price with savings at the pump. At current gas prices, here's a look at the cost per mile for one company's hybrid and traditional cars.

Gasoline Cost per Mile

Car Type	City ($/mi)	Highway ($/mi)
Hybrid	0.09	0.08
Traditional	0.18	0.13

Source: www.fueleconomy.gov

31. **Health Science** A pharmacist has two vitamin-supplement powders. The first powder is 25% vitamin B_1 and 15% vitamin B_2. The second is 15% vitamin B_1 and 20% vitamin B_2. How many milligrams of each of the two powders should the pharmacist use to make a mixture that contains 117.5 mg of vitamin B_1 and 120 mg of vitamin B_2?

32. **Chemistry** A chemist has two alloys, one of which is 10% gold and 15% lead and the other of which is 30% gold and 40% lead. How many grams of each of the two alloys should be used to make an alloy that contains 60 g of gold and 88 g of lead?

33. **Business** On Monday, a computer manufacturing company sent out three shipments. The first order, which contained a bill for $114,000, was for 4 Model II, 6 Model VI, and 10 Model IX computers. The second shipment, which contained a bill for $72,000, was for 8 Model II, 3 Model VI, and 5 Model IX computers. The third shipment, which contained a bill for $81,000, was for 2 Model II, 9 Model VI, and 5 Model IX computers. What does the manufacturer charge for a Model VI computer?

34. **Health Care** A relief organization supplies blankets, cots, and lanterns to victims of fires, floods, and other natural disasters. One week the organization purchased 15 blankets, 5 cots, and 10 lanterns for a total cost of $1250. The next week, at the same prices, the organization purchased 20 blankets, 10 cots, and 15 lanterns for a total cost of $2000. The next week, at the same prices, the organization purchased 10 blankets, 15 cots, and 5 lanterns for a total cost of $1625. Find the cost of one blanket, the cost of one cot, and the cost of one lantern.

35. **Investments** An investor has a total of $18,000 deposited in three different accounts, which earn annual interest of 9%, 7%, and 5%. The amount deposited in the 9% account is twice the amount in the 5% account. If the three accounts earn total annual interest of $1340, how much money is deposited in each account?

36. Investments A financial planner invested \$33,000 of a client's money, part at 9%, part at 12%, and the remainder at 8%. The total annual interest income from these three investments was \$3290. The amount invested at 12% was \$5000 less than the combined amounts invested at 9% and 8%. Find the amount invested at each rate.

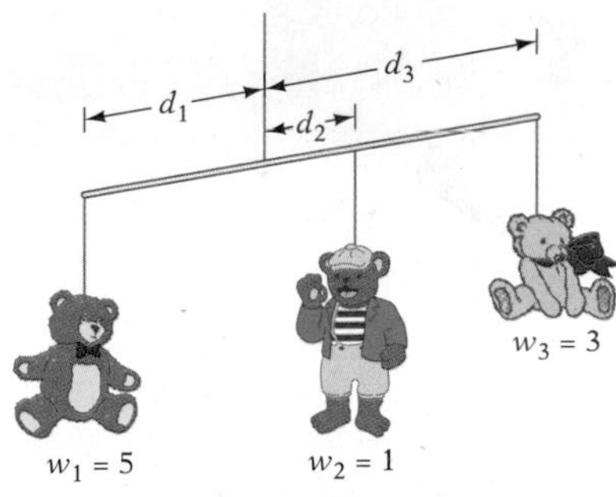

37. Art A mobile is made by suspending three objects from a light rod that is 15 in. long, as shown at the right. The weight, in ounces, of each object is shown in the diagram. For the mobile to balance, the objects must be positioned so that $w_1d_1 = w_2d_2 + w_3d_3$. The artist wants d_3 to be three times d_2. Find the distances d_1, d_2, and d_3 such that the mobile will balance.

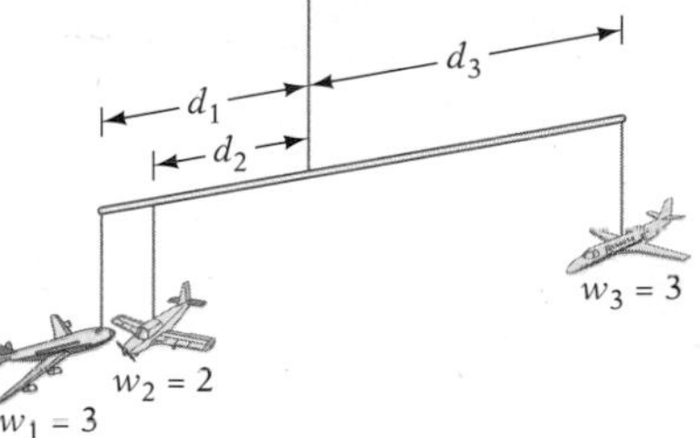

38. Art A mobile is made by suspending three objects from a light rod that is 20 in. long, as shown at the right. The weight, in ounces, of each object is shown in the diagram. For the mobile to balance, the objects must be positioned so that $w_1d_1 + w_2d_2 = w_3d_3$. The artist wants d_3 to be twice d_2. Find the distances d_1, d_2, and d_3 such that the mobile will balance.

APPLYING CONCEPTS

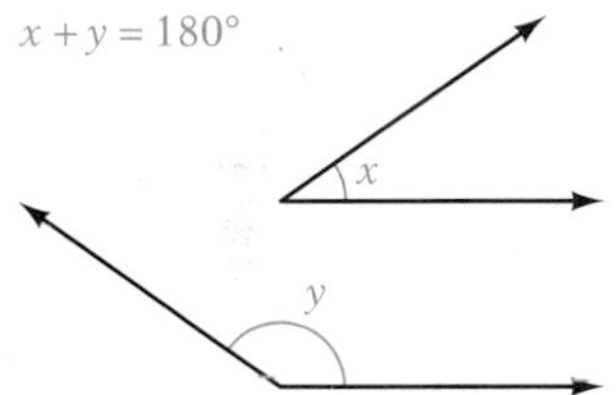

39. Geometry Two angles are supplementary. The measure of the larger angle is 40° more than three times the measure of the smaller angle. Find the measures of the two angles. (Supplementary angles are two angles whose sum is 180°.)

40. Geometry Two angles are complementary. The measure of the larger angle is 9° more than eight times the measure of the smaller angle. Find the measures of the two angles. (Complementary angles are two angles whose sum is 90°.)

PROJECTS OR GROUP ACTIVITIES

41. A plane is flying the 3500 mi from New York City to London. The speed of the plane in calm air is 375 mph, and there is a 50-mile-per-hour tailwind. The *point of no return* is the point at which the flight time required to return to New York City is the same as the flight time to travel on to London. For this flight, how far from New York is the point of no return? Round to the nearest whole number.

4.5 Solving Systems of Linear Inequalities

OBJECTIVE 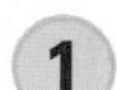

Graph the solution set of a system of linear inequalities

Two or more inequalities considered together are called a **system of inequalities.** The **solution set of a system of inequalities** is the intersection of the solution sets of the individual inequalities. To graph the solution set of a system of inequalities, first graph the solution set of each inequality. Recall that the solution set of an inequality is a half-plane. The solution set of the system of inequalities is the region of the plane represented by the intersection of the two shaded half-planes.

Focus on solving a system of linear inequalities

A. Graph the solution set: $2x - y \leq 3$
$3x + 2y > 8$

Solve each inequality for y.

$$\begin{aligned} 2x - y &\leq 3 \\ -y &\leq -2x + 3 \\ y &\geq 2x - 3 \end{aligned} \qquad \begin{aligned} 3x + 2y &> 8 \\ 2y &> -3x + 8 \\ y &> -\frac{3}{2}x + 4 \end{aligned}$$

Graph $y = 2x - 3$ as a solid line. Because the inequality is $\geq$, shade above the line.

Graph $y = -\frac{3}{2}x + 4$ as a dashed line.

Because the inequality is $>$, shade above the line.

The solution set of the system is the region of the plane that represents the intersection of the solution sets of the individual inequalities.

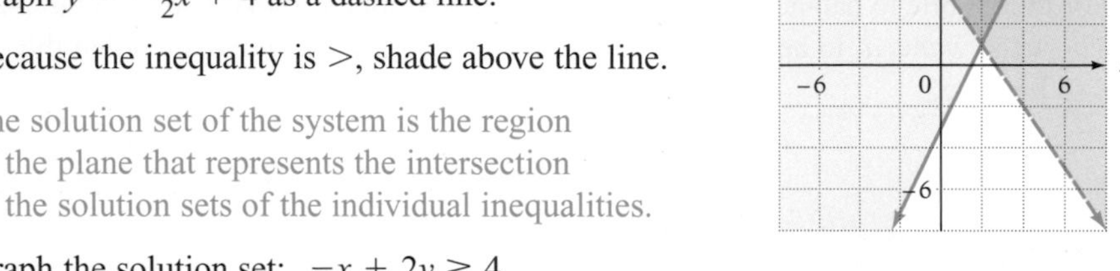

Take Note

You can use a test point to check that the correct region has been designated as the solution set. We can see from the graph that the point with coordinates (2, 4) is in the solution set, and as shown below, it is a solution of each inequality in the system. This indicates that the solution set as graphed is correct.

$$\begin{aligned} 2x - y &\leq 3 \\ 2(2) - (4) &\leq 3 \\ 0 &\leq 3 \quad \text{True} \end{aligned}$$

$$\begin{aligned} 3x + 2y &> 8 \\ 3(2) + 2(4) &> 8 \\ 14 &> 8 \quad \text{True} \end{aligned}$$

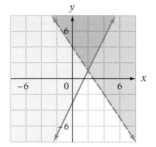

B. Graph the solution set: $-x + 2y \geq 4$
$x - 2y \geq 6$

Solve each inequality for y.

$$\begin{aligned} -x + 2y &\geq 4 \\ 2y &\geq x + 4 \\ y &\geq \frac{1}{2}x + 2 \end{aligned} \qquad \begin{aligned} x - 2y &\geq 6 \\ -2y &\geq -x + 6 \\ y &\leq \frac{1}{2}x - 3 \end{aligned}$$

Shade above the solid line graph of $y = \frac{1}{2}x + 2$.

Shade below the solid line graph of $y = \frac{1}{2}x - 3$.

Because the solution sets of the two inequalities do not intersect, the solution set of the system is the empty set.

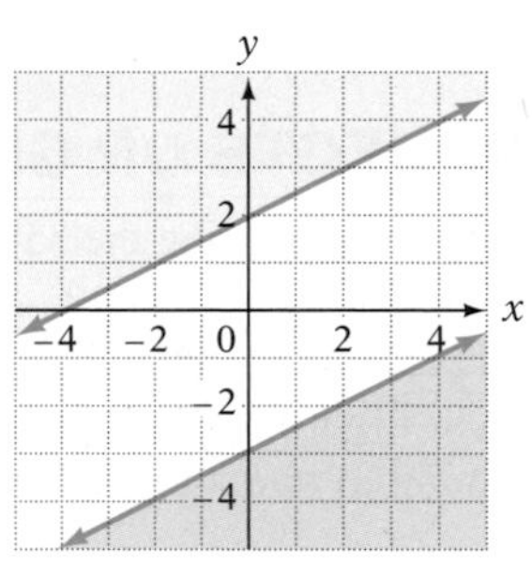

How It's Used

Large systems of linear inequalities containing over 100 inequalities have been used to solve application problems in such diverse areas as providing health care and hardening a nuclear missile silo.

EXAMPLE 1 Graph the solution set.

A. $y \geq x - 1$
$y < -2x$

B. $2x + 3y > 9$
$y < -\frac{2}{3}x + 1$

Solution **A.**

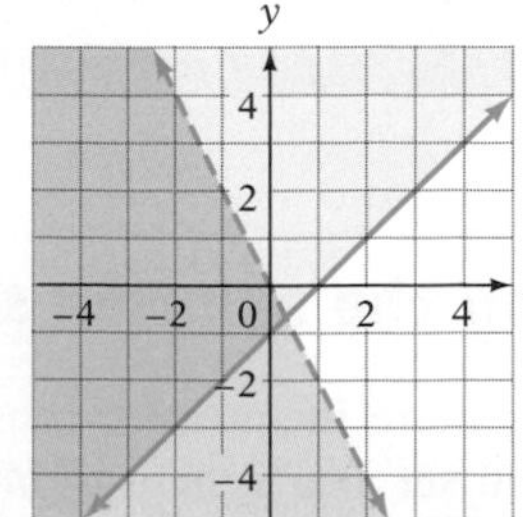

- **Shade above the solid line graph of $y = x - 1$.**
- **Shade below the dashed line graph of $y = -2x$.**

The solution set of the system is the intersection of the solution sets of the individual inequalities.

B. $2x + 3y > 9$ • **Solve $2x + 3y > 9$ for y.**

$$3y > -2x + 9$$

$$y > -\frac{2}{3}x + 3$$

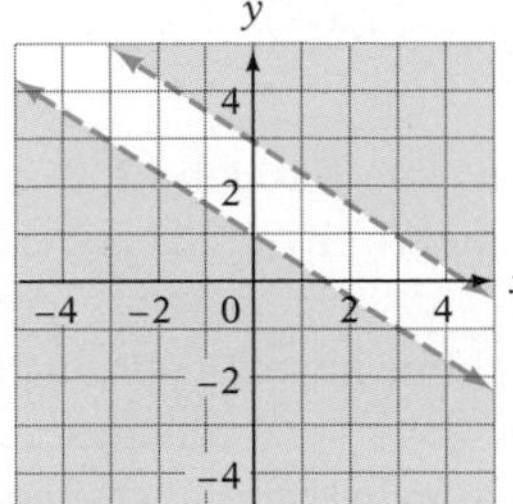

• **Shade above the dashed line graph of $y = -\frac{2}{3}x + 3$.**

• **Shade below the dashed line graph of $y = -\frac{2}{3}x + 1$.**

The solution set of the system is the empty set because the solution sets of the two inequalities do not intersect.

Problem 1 Graph the solution set.

A. $y \geq 2x - 3$ $y > -3x$

B. $3x + 4y > 12$ $y < \frac{3}{4}x - 1$

Solution See page S14.

Try Exercise 7, page 232.

4.5 Exercises

CONCEPT CHECK

1. Which ordered pair is a solution of the system of inequalities shown at the right? $2x - y < 4$, $x - 3y \geq 6$
   (i) $(5, 1)$ (ii) $(-3, -5)$
2. Which ordered pair is a solution of the system of inequalities shown at the right? $3x - 2y \geq 6$, $x + y < 5$
   (i) $(-2, 3)$ (ii) $(3, -2)$
3. Is the solution set of a system of inequalities the union or the intersection of the solution sets of the individual inequalities?
4. For which system of inequalities is the solution set the empty set?
   (i) $y > 3x - 1$, $y < 3x - 3$ (ii) $y > 3x - 1$, $y < 3x + 3$

1 Graph the solution set of a system of linear inequalities (See pages 229–231.)

GETTING READY

5. Use the graph showing the four regions of the plane that are determined by the intersecting lines that are the graphs of $y = 2x + 2$ and $y = -x + 5$. Complete parts (a), (b), and (c) by replacing each question mark with a region letter from the graph.
   a. Points in the solution set of the inequality $y \leq 2x + 2$ are located in regions __?__ and __?__.

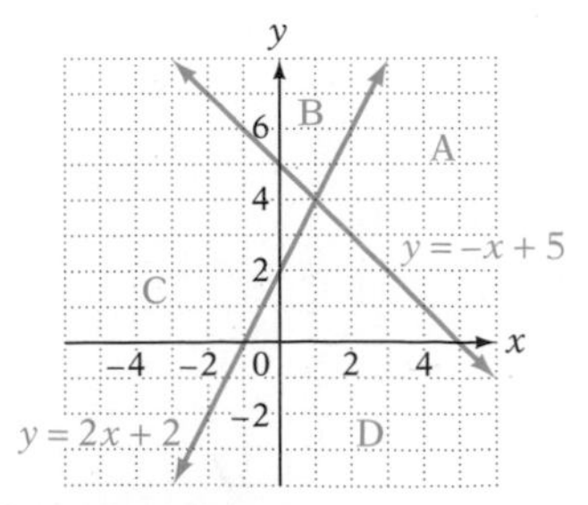

b. Points in the solution set of the inequality $y \geq -x + 5$ are located in regions __?__ and __?__.

c. Points in the solution set of the system of inequalities shown below are located in region __?__.

$y \leq 2x + 2$
$y \geq -x + 5$

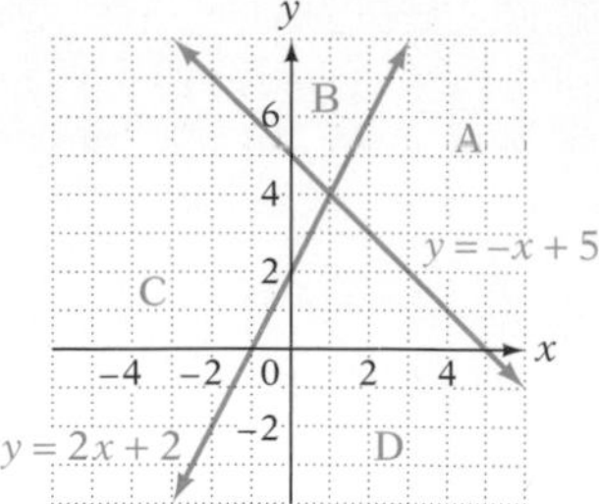

6. Use the graph showing the vertical line through (2, 0) and the horizontal line through (0, 4). Complete parts (a) and (b) by replacing each question mark with *greater than* or *less than.*

a. Points to the right of the vertical line have x-coordinates __?__ 2.

b. Points below the horizontal line have y-coordinates __?__ 4.

c. Write the system of inequalities for which the shaded region is the solution set.

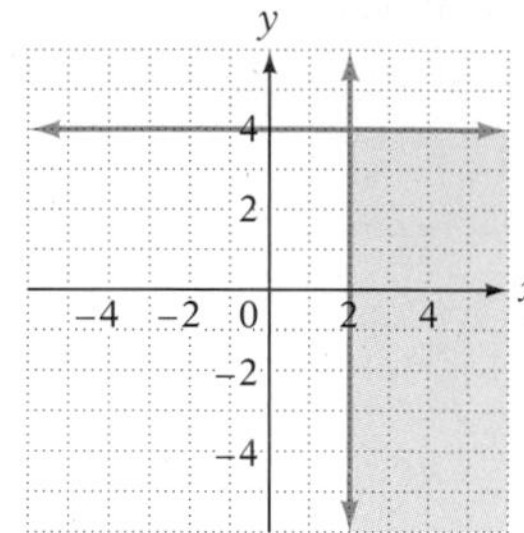

Graph the solution set.

7. $y \leq x - 3$
$y \leq -x + 5$

8. $y > 2x - 4$
$y < -x + 5$

9. $y > 3x - 3$
$y \geq -2x + 2$

10. $y \leq -\frac{1}{2}x + 3$
$y \geq x - 3$

11. $2x + y \geq -2$
$6x + 3y \leq 6$

12. $x + y \geq 5$
$3x + 3y \leq 6$

13. $3x - 2y < 6$
$y \leq 3$

14. $x \leq 2$
$3x + 2y > 4$

15. $y > 2x - 6$
$x + y < 0$

16. $x < 3$
$y < -2$

17. $x + 1 \geq 0$
$y - 3 \leq 0$

18. $5x - 2y \geq 10$
$3x + 2y \geq 6$

19. $2x + y \geq 4$
$3x - 2y < 6$

20. $3x - 4y < 12$
$x + 2y < 6$

21. $x - 2y \leq 6$
$2x + 3y \leq 6$

22. $x - 3y > 6$
$2x + y > 5$

23. $x - 2y \leq 4$
$3x + 2y \leq 8$

24. $3x - 2y < 0$
$5x + 3y > 9$

For Exercises 25 to 28, a and b are positive numbers such that $a > b$. Describe the solution set of each system of inequalities.

25. $x + y < a$
$x + y < b$

26. $x + y > a$
$x + y < b$

27. $x + y < a$
$x + y > b$

28. $x + y > a$
$x + y > b$

APPLYING CONCEPTS

Graph the solution set.

29. $2x + 3y \le 15$
$3x - y \le 6$
$y \ge 0$

30. $x + y \le 6$
$x - y \le 2$
$x \ge 0$

31. $x - y \le 5$
$2x - y \ge 6$
$y \ge 0$

32. $x - 3y \le 6$
$5x - 2y \ge 4$
$y \ge 0$

33. $2x - y \le 4$
$3x + y < 1$
$y \le 0$

34. $x - y \le 4$
$2x + 3y > 6$
$x \ge 0$

PROJECTS OR GROUP ACTIVITIES

A set of points in a plane is a **convex set** if each line segment connecting a pair of points in the set is contained completely within the set.

35. Which of the following are convex sets?

(i)

(ii)

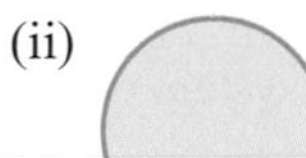

(iii)

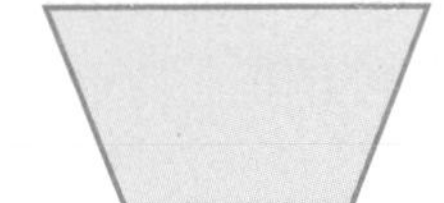

(iv)

36. Graph the system of inequalities given below. Is the solution set a convex set?

$x + y \le 10$
$2x + y \le 15$
$x \ge 0, y \ge 0$

CHAPTER 4 Summary

Key Words	Objective and Page Reference	Examples
A **system of equations** is two or more equations considered together. A **solution of a system of equations in two variables** is an ordered pair that is a solution of each equation of the system.	[4.1.1, p. 188]	The solution of the system $x + y = 2$ $x - y = 4$ is the ordered pair $(3, -1)$. $(3, -1)$ is the only ordered pair that is a solution of both equations.

When the graphs of a system of equations intersect at exactly one point, the system is called an **independent system of equations.** [4.1.1, p. 188]

The system of equations

$$6x - 5y = -3$$
$$5x + 4y = 22$$

is independent. The solution is (2, 3).

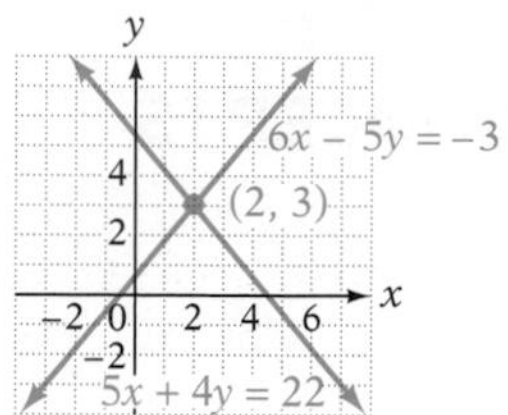

When the graphs of a system of equations do not intersect, the system has no solution and is called an **inconsistent system of equations.** [4.1.1, p. 189]

The system of equations

$$x + 2y = 6$$
$$x + 2y = -2$$

is inconsistent. The system has no solution.

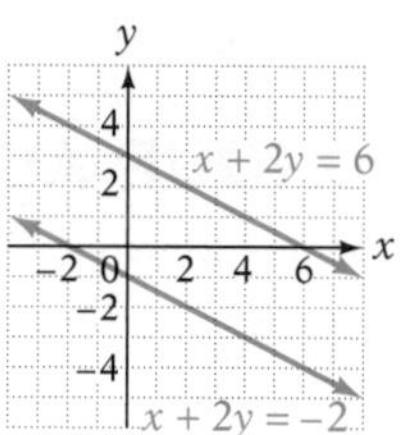

When the graphs of a system of equations coincide, the system is called a **dependent system of equations.** [4.1.1, p. 189]

The system of equations

$$2x - y = 4$$
$$4x - 2y = 8$$

is dependent. The solutions are the ordered pairs $(x, 2x - 4)$.

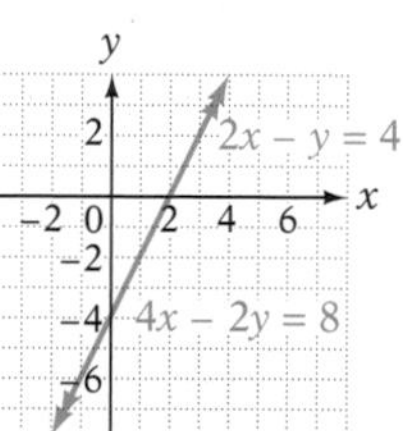

An equation of the form $Ax + By + Cz = D$ is called a **linear equation in three variables.** [4.2.2, p. 198]

$3x + 2y - 5z = 12$ is a linear equation in three variables.

A **solution of a system of equations in three variables** is an ordered triple that is a solution of each equation of the system. [4.2.2, p. 199]

The ordered triple (1, 2, 1) is a solution of the system of equations.

$$3x + y - 3z = 2$$
$$-x + 2y + 3z = 6$$
$$2x + 2y - 2z = 4$$

A **matrix** is a rectangular array of numbers. [4.3.1, p. 204]

$\begin{bmatrix} 2 & 3 & 6 \\ -1 & 2 & 4 \end{bmatrix}$ is a 2×3 matrix.

A **square matrix** has the same number of rows as columns.

[4.3.1, p. 204]

$\begin{bmatrix} 2 & 3 \\ 4 & -1 \end{bmatrix}$ is a square matrix.

A **determinant** is a number associated with a square matrix. The determinant of a 2 × 2 matrix

$$\begin{bmatrix} a_{11} & a_{12} \\ a_{21} & a_{22} \end{bmatrix} \text{ is } \begin{vmatrix} a_{11} & a_{12} \\ a_{21} & a_{22} \end{vmatrix} = a_{11}a_{22} - a_{21}a_{12}.$$

[4.3.1, pp. 204–205]

$$\begin{vmatrix} 2 & -3 \\ -4 & 5 \end{vmatrix} = 2(5) - (-4)(-3) = 10 - 12 = -2$$

The **minor of an element** a_{ij} of a determinant is the determinant that remains after row i and column j have been removed.

[4.3.1, p. 205]

$\begin{vmatrix} 2 & 1 & 3 \\ 4 & 6 & 2 \\ 1 & 8 & 3 \end{vmatrix}$ The minor of 4 is $\begin{vmatrix} 1 & 3 \\ 8 & 3 \end{vmatrix}$.

The **cofactor of an element** a_{ij} of a determinant is $(-1)^{i+j}$ times the minor of a_{ij}.

[4.3.1, p. 206]

In the determinant above, 4 is a_{21}. The cofactor of 4 is

$$(-1)^{2+1}\begin{vmatrix} 1 & 3 \\ 8 & 3 \end{vmatrix} = -\begin{vmatrix} 1 & 3 \\ 8 & 3 \end{vmatrix}$$

The evaluation of the determinant of a 3 × 3 or larger matrix is accomplished by **expanding by cofactors.**

[4.3.1, p. 206]

Expand by cofactors of the first column.

$$\begin{vmatrix} 2 & 1 & 3 \\ 4 & 6 & 2 \\ 1 & 8 & 3 \end{vmatrix} = 2\begin{vmatrix} 6 & 2 \\ 8 & 3 \end{vmatrix} - 4\begin{vmatrix} 1 & 3 \\ 8 & 3 \end{vmatrix} + 1\begin{vmatrix} 1 & 3 \\ 6 & 2 \end{vmatrix}$$

An **augmented matrix** is a matrix consisting of the coefficients and constant terms of a system of equations.

[4.3.3, p. 210]

The augmented matrix associated with the system

$$\begin{aligned} 4x - 3y + z &= 7 \\ x + y - 5z &= 14 \\ 2x + 3z &= -4 \end{aligned}$$

is

$$\left[\begin{array}{rrr|r} 4 & -3 & 1 & 7 \\ 1 & 1 & -5 & 14 \\ 2 & 0 & 3 & -4 \end{array}\right]$$

The process of solving a system of equations by using elementary row operations is called the **Gaussian elimination method.**

[4.3.3, p. 214]

Inequalities considered together are called a **system of inequalities.** The **solution set of a system of inequalities** is the intersection of the solution sets of the individual inequalities.

[4.5.1, p. 229]

$x + y > 3$
$x - y > -2$

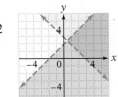

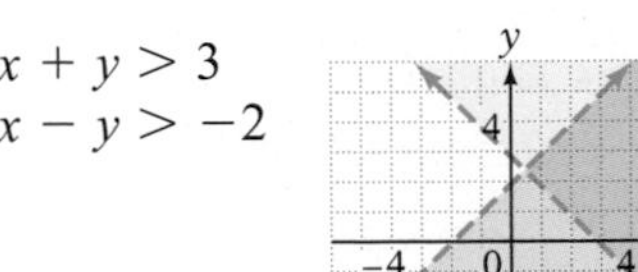

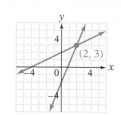

Essential Rules and Procedures	Objective and Page Reference	Examples
A system of equations can be solved by:		
1. Graphing	[4.1.1, pp. 188–189]	$y = \frac{1}{2}x + 2$ $y = \frac{5}{2}x - 2$ 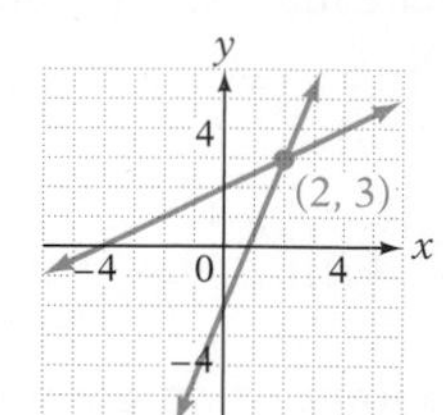
2. The substitution method	[4.1.2, pp. 191–192]	(1) $2x - 3y = 4$ (2) $y = -x + 2$ Substitute the expression for y into equation (1). $2x - 3(-x + 2) = 4$ $5x - 6 = 4$ $5x = 10$ $x = 2$ Substitute for x in equation (2). $y = -x + 2$ $y = -(2) + 2 = 0$ The solution is (2, 0).
3. The addition method	[4.2.1 and 4.2.2, pp. 196–197, 200]	(1) $-2x + 3y = 6$ (2) $2x - 5y = -2$ $-2y = 4$ $y = -2$ • **Add the equations.** (1) $-2x + 3y = 6$ • **Replace y in equation (1).** $-2x + 3(-2) = 6$ $-2x - 6 = 6$ $-2x = 12$ $x = -6$ The solution is (−6, −2).
Cramer's Rule	[4.3.2, pp. 207, 208]	
For two variables: $a_{11}x + a_{12}y = b_1$ $a_{21}x + a_{22}y = b_2$		$2x - y = 6$ $x + 3y = 4$
$x = \frac{D_x}{D}, y = \frac{D_y}{D}$, where $D = \begin{vmatrix} a_{11} & a_{12} \\ a_{21} & a_{22} \end{vmatrix}$,		$D = \begin{vmatrix} 2 & -1 \\ 1 & 3 \end{vmatrix} = 7$,
$D_x = \begin{vmatrix} b_1 & a_{12} \\ b_2 & a_{22} \end{vmatrix}, D_y = \begin{vmatrix} a_{11} & b_1 \\ a_{21} & b_2 \end{vmatrix}$,		$D_x = \begin{vmatrix} 6 & -1 \\ 4 & 3 \end{vmatrix} = 22$,
and $D \neq 0$.		$D_y = \begin{vmatrix} 2 & 6 \\ 1 & 4 \end{vmatrix} = 2$
		$x = \frac{D_x}{D} = \frac{22}{7}, y = \frac{D_y}{D} = \frac{2}{7}$

Elementary row operations are used to rewrite a matrix in **row echelon form:** 1. Interchange two rows. 2. Multiply all the elements in a row by the same nonzero number. 3. Replace a row by the sum of that row and a multiple of any other row.	[4.3.3, pp. 210, 212]	$\left[\begin{array}{ccc|c} 1 & 4 & -2 & 3 \\ 0 & 1 & 3 & -1 \\ 0 & 0 & 1 & 6 \end{array}\right]$ Row echelon form

CHAPTER 4 Review Exercises

1. Solve by substitution: $\begin{aligned} 2x - 6y &= 15 \\ x &= 3y + 8 \end{aligned}$

2. Solve by substitution: $\begin{aligned} 3x + 12y &= 18 \\ x + 4y &= 6 \end{aligned}$

3. Solve by the addition method: $\begin{aligned} 3x + 2y &= 2 \\ x + y &= 3 \end{aligned}$

4. Solve by the addition method: $\begin{aligned} 5x - 15y &= 30 \\ x - 3y &= 6 \end{aligned}$

5. Solve by the addition method: $\begin{aligned} 3x + y &= 13 \\ 2y + 3z &= 5 \\ x + 2z &= 11 \end{aligned}$

6. Solve by the addition method:
$\begin{aligned} 3x - 4y - 2z &= 17 \\ 4x - 3y + 5z &= 5 \\ 5x - 5y + 3z &= 14 \end{aligned}$

7. Evaluate the determinant: $\begin{vmatrix} 6 & 1 \\ 2 & 5 \end{vmatrix}$

8. Evaluate the determinant: $\begin{vmatrix} 1 & 5 & -2 \\ -2 & 1 & 4 \\ 4 & 3 & -8 \end{vmatrix}$

9. Solve by using Cramer's Rule: $\begin{aligned} 2x - y &= 7 \\ 3x + 2y &= 7 \end{aligned}$

10. Solve by using Cramer's Rule: $\begin{aligned} 3x - 4y &= 10 \\ 2x + 5y &= 15 \end{aligned}$

11. Solve by using Cramer's Rule: $\begin{aligned} x + y + z &= 0 \\ x + 2y + 3z &= 5 \\ 2x + y + 2z &= 3 \end{aligned}$

12. Solve by using Cramer's Rule: $\begin{aligned} x + 3y + z &= 6 \\ 2x + y - z &= 12 \\ x + 2y - z &= 13 \end{aligned}$

13. Solve by the addition method: $\begin{aligned} x - 2y + z &= 7 \\ 3x - z &= -1 \\ 3y + z &= 1 \end{aligned}$

14. Solve by using Cramer's Rule: $\begin{aligned} 3x - 2y &= 2 \\ -2x + 3y &= 1 \end{aligned}$

15. Solve by using the Gaussian elimination method:
$\begin{aligned} 2x - 2y - 6z &= 1 \\ 4x + 2y + 3z &= 1 \\ 2x - 3y - 3z &= 3 \end{aligned}$

16. Evaluate the determinant: $\begin{vmatrix} 3 & -2 & 5 \\ 4 & 6 & 3 \\ 1 & 2 & 1 \end{vmatrix}$

17. Solve by using Cramer's Rule: $\begin{aligned} 4x - 3y &= 17 \\ 3x - 2y &= 12 \end{aligned}$

18. Solve by using the Gaussian elimination method:
$\begin{aligned} 3x + 2y - z &= -1 \\ x + 2y + 3z &= -1 \\ 3x + 4y + 6z &= 0 \end{aligned}$

19. Solve by graphing: $x + y = 3$
$3x - 2y = -6$

20. Solve by graphing: $2x - y = 4$
$y = 2x - 4$

21. Graph the solution set: $x + 3y < 6$
$2x - y > 4$

22. Graph the solution set: $2x + 4y \geq 8$
$x + y \leq 3$

23. Uniform Motion A cabin cruiser traveling with the current went 60 mi in 3 h. Against the current, it took 5 h to travel the same distance. Find the rate of the cabin cruiser in calm water and the rate of the current.

24. Uniform Motion A plane flying with the wind flew 600 mi in 3 h. Flying against the wind, the plane required 4 h to travel the same distance. Find the rate of the plane in calm air and the rate of the wind.

25. Recreation At a movie theater, admission tickets are $5 for children and $8 for adults. The receipts for one Friday evening were $2500. The next day, there were three times as many children as the preceding evening and only half as many adults as the night before, yet the receipts were still $2500. Find the number of children who attended the movie Friday evening.

26. Investments An investor has a total of $25,000 deposited in three different accounts, which earn annual interest rates of 8%, 6%, and 4%. The amount deposited in the 8% account is twice the amount in the 6% account. If the three accounts earn total annual interest of $1520, how much money is deposited in each account?

CHAPTER 4 Test

1. Solve by substitution: $3x + 2y = 4$
$x = 2y - 1$

2. Solve by substitution: $5x + 2y = -23$
$2x + y = -10$

3. Solve by substitution: $y = 3x - 7$
$y = -2x + 3$

4. Solve by using the Gaussian elimination method:
$3x + 4y = -2$
$2x + 5y = 1$

5. Solve by the addition method: $4x - 6y = 5$
$6x - 9y = 4$

6. Solve by the addition method:
$3x - y = 2x + y - 1$
$5x + 2y = y + 6$

7. Solve by the addition method:
$2x + 4y - z = 3$
$x + 2y + z = 5$
$4x + 8y - 2z = 7$

8. Solve by using the Gaussian elimination method:
$x - y - z = 5$
$2x + z = 2$
$3y - 2z = 1$

9. Evaluate the determinant: $\begin{vmatrix} 3 & -1 \\ -2 & 4 \end{vmatrix}$

10. Evaluate the determinant: $\begin{vmatrix} 1 & -2 & 3 \\ 3 & 1 & 1 \\ 2 & -1 & -2 \end{vmatrix}$

11. Solve by using Cramer's Rule: $x - y = 3$
$2x + y = -4$

12. Solve by using Cramer's Rule:
$x - y + z = 2$
$2x - y - z = 1$
$x + 2y - 3z = -4$

13. Solve by using Cramer's Rule:
$3x + 2y + 2z = 2$
$x - 2y - z = 1$
$2x - 3y - 3z = -3$

14. Solve by graphing: $2x - 3y = -6$
$2x - y = 2$

15. Solve by graphing: $x - 2y = -5$
$3x + 4y = -15$

16. Graph the solution set: $2x - y < 3$
$4x + 3y < 11$

17. Graph the solution set: $x + y > 2$
$2x - y < -1$

18. **Uniform Motion** A plane flying with the wind went 350 mi in 2 h. The return trip, flying against the wind, took 2.8 h. Find the rate of the plane in calm air and the rate of the wind.

19. **Purchasing** A clothing manufacturer purchased 60 yd of cotton and 90 yd of wool for a total cost of $1800. Another purchase, at the same prices, included 80 yd of cotton and 20 yd of wool for a total cost of $1000. Find the cost per yard of the cotton and of the wool.

Cumulative Review Exercises

1. Solve: $\frac{3}{2}x - \frac{3}{8} + \frac{1}{4}x = \frac{7}{12}x - \frac{5}{6}$

2. Find the equation of the line that contains the points $P_1(2, -1)$ and $P_2(3, 4)$.

3. Simplify: $3[x - 2(5 - 2x) - 4x] + 6$

4. Evaluate $a + bc \div 2$ when $a = 4$, $b = 8$, and $c = -2$.

5. Solve: $2x - 3 < 9$ or $5x - 1 < 4$

6. Solve: $|x - 2| - 4 < 2$

7. Solve: $|2x - 3| > 5$

8. Given $f(x) = 3x^3 - 2x^2 + 1$, evaluate $f(-3)$.

9. Find the range of $f(x) = 3x^2 - 2x$ if the domain is $\{-2, -1, 0, 1, 2\}$.

10. Given $F(x) = x^2 - 3$, find $F(2)$.

11. Given $f(x) = 3x - 4$, write $f(2 + h) - f(2)$ in simplest form.

12. Graph: $\{x|x \leq 2\} \cap \{x|x > -3\}$

13. Find the equation of the line that contains the point $P(-2, 3)$ and has slope $-\frac{2}{3}$.

14. Find the equation of the line that contains the point $P(-1, 2)$ and is perpendicular to the graph of $2x - 3y = 7$.

15. Find the distance between the points with coordinates $(-4, 2)$ and $(2, 0)$.

16. Find the coordinates of the midpoint of the line segment connecting the points $P_1(-4, 3)$ and $P_2(3, 5)$.

17. Graph $2x - 5y = 10$ by using the slope and y-intercept.

18. Graph the solution set of the inequality $3x - 4y \geq 8$.

19. Solve by substitution: $\begin{aligned} 3x - 2y &= 7 \\ y &= 2x - 1 \end{aligned}$

20. Solve by the addition method: $\begin{aligned} 3x + 2z &= 1 \\ 2y - z &= 1 \\ x + 2y &= 1 \end{aligned}$

21. Evaluate the determinant: $\begin{vmatrix} 2 & -5 & 1 \\ 3 & 1 & 2 \\ 6 & -1 & 4 \end{vmatrix}$

22. Solve by graphing: $\begin{aligned} 5x - 2y &= 10 \\ 3x + 2y &= 6 \end{aligned}$

23. Solve by using Cramer's Rule: $\begin{aligned} 4x - 3y &= 17 \\ 3x - 2y &= 12 \end{aligned}$

24. Graph the solution set: $\begin{aligned} 3x - 2y &\geq 4 \\ x + y &< 3 \end{aligned}$

25. **Mixture Problem** How many milliliters of pure water must be added to 100 ml of a 4% salt solution to make a 2.5% salt solution?

26. **Uniform Motion** Flying with the wind, a small plane required 2 h to fly 150 mi. Against the wind, it took 3 h to fly the same distance. Find the rate of the wind.

27. **Purchasing** A restaurant manager buys 100 lb of hamburger and 50 lb of steak for a total cost of \$490. A second purchase, at the same prices, includes 150 lb of hamburger and 100 lb of steak. The total cost is \$860. Find the price of one pound of steak.

28. **Electronics** Find the lower and upper limits of a 12,000-ohm resistor with a 15% tolerance.

29. **Compensation** The graph at the right shows the relationship between the monthly income, in dollars, and sales, in thousands of dollars, of an account executive. Find the slope of the line between the two points shown on the graph. Write a sentence that states the meaning of the slope.

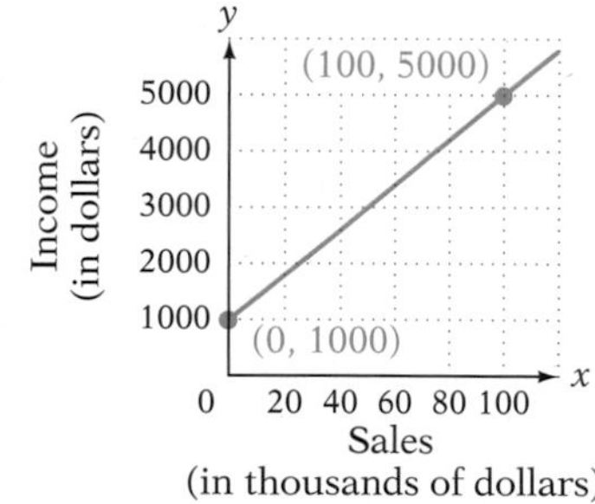

Polynomials and Exponents

CHAPTER 5

Digital Vision

Focus on Success

Did you read Ask the Authors at the front of this text? If you did, then you know that the authors' advice is that you practice, practice, practice—and then practice some more. The more time you spend doing math outside of class, the more successful you will be in this course. (See Make the Commitment to Succeed, page AIM-3.)

OBJECTIVES

5.1
1. Multiply monomials
2. Divide monomials and simplify expressions with negative exponents
3. Scientific notation
4. Application problems

5.2
1. Evaluate polynomial functions
2. Add or subtract polynomials

5.3
1. Multiply a polynomial by a monomial
2. Multiply two polynomials
3. Multiply polynomials that have special products
4. Application problems

5.4
1. Divide a polynomial by a monomial
2. Divide polynomials
3. Synthetic division
4. Evaluate a polynomial using synthetic division

5.5
1. Factor a monomial from a polynomial
2. Factor by grouping

5.6
1. Factor trinomials of the form $x^2 + bx + c$
2. Factor trinomials of the form $ax^2 + bx + c$

5.7
1. Factor the difference of two perfect squares and factor perfect-square trinomials
2. Factor the sum or the difference of two cubes
3. Factor trinomials that are quadratic in form
4. Factor completely

5.8
1. Solve equations by factoring
2. Application problems

PREP TEST

Are you ready to succeed in this chapter? *Take the Prep Test below to find out if you are ready to learn the new material.*

1. Simplify: $-4(3y)$
2. Simplify: $(-2)^3$
3. Simplify: $-4a - 8b + 7a$
4. Simplify: $3x - 2[y - 4(x + 1) + 5]$
5. Simplify: $-(x - y)$
6. Write 40 as a product of prime numbers.
7. Find the GCF of 16, 20, and 24.
8. Evaluate $x^3 - 2x^2 + x + 5$ for $x = -2$.
9. Solve: $3x + 1 = 0$

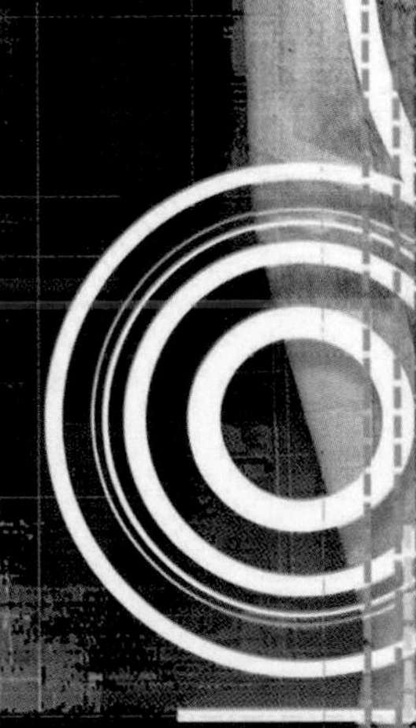

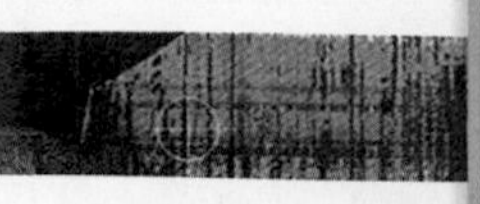

5.1 Exponential Expressions

OBJECTIVE 1 Multiply monomials

Point of Interest

Around A.D. 250, the monomial $3x^2$ shown at the right would have been written Δ^Y3, or at least approximately like that. In A.D. 250, the symbol for 3 was not the one we use today.

A **monomial** is a number, a variable, or a product of a number and variables.

The examples at the right are monomials. The **degree of a monomial** is the sum of the exponents of the variables.

x	degree 1 $(x = x^1)$
$3x^2$	degree 2
$4x^2y$	degree 3
$6x^3y^4z^2$	degree 9
x^n	degree n
6	degree 0

The degree of a nonzero constant term is zero.

The expression $5\sqrt{x}$ is not a monomial because $\sqrt{x}$ cannot be written as a product of variables. The expression $\frac{x}{y}$ is not a monomial because it is a quotient of variables.

The expression x^4 is an exponential expression. The exponent, 4, indicates the number of times the base, x, occurs as a factor.

The product of exponential expressions with the *same* base can be simplified by writing each expression in factored form and writing the result with an exponent.

$$x^3 \cdot x^4 = \underbrace{\overbrace{(x \cdot x \cdot x)}^{3\text{ factors}} \cdot \overbrace{(x \cdot x \cdot x \cdot x)}^{4\text{ factors}}}_{7\text{ factors}}$$
$$= x^7$$

Note that adding the exponents results in the same product.

$$x^3 \cdot x^4 = x^{3+4} = x^7$$

RULE FOR MULTIPLYING EXPONENTIAL EXPRESSIONS

If m and n are integers, then $x^m \cdot x^n = x^{m+n}$.

EXAMPLES

1. $x^5 \cdot x^3 = x^{5+3} = x^8$
2. $a \cdot a^4 = a^{1+4} = a^5$ • **Recall that $a = a^1$.**
3. $z^2 \cdot z^4 \cdot z^5 = z^{2+4+5} = z^{11}$
4. $(v^4r^3)(v^2r) = v^{4+2}r^{3+1} = v^6r^4$ • **Add exponents on like bases.**

Focus on simplifying the product of exponential expressions

Simplify: $(-4x^5y)(-3x^2y^3)$

Use the Commutative and Associative Properties to rearrange and group factors.

$$(-4x^5y)(-3x^2y^3) = [(-4)(-3)](x^5 \cdot x^2)(y \cdot y^3)$$

Multiply the coefficients. Multiply variables with like bases by adding the exponents.

$$= 12x^{5+2}y^{1+3}$$
$$= 12x^7y^4$$

EXAMPLE 1 Simplify: $(-2a^3b^2c)(4a^4c^7)$

Solution $(-2a^3b^2c)(4a^4c^7) = (-2 \cdot 4)a^{3+4}b^2c^{1+7}$
$= -8a^7b^2c^8$

• Multiply the coefficients. Multiply variables with like bases by adding the exponents.

Problem 1 Simplify: $(7xy^3)(-5x^2y^2)(-xy^2)$

Solution See page S14.

Try Exercise 11, page 250.

As shown below, the power of a monomial can be simplified by writing the power in factored form and then using the Rule for Multiplying Exponential Expressions.

Write in factored form.	$(a^2)^3 = a^2 \cdot a^2 \cdot a^2$	$(x^3y^4)^2 = (x^3y^4)(x^3y^4)$
Use the Rule for Multiplying Exponential Expressions.	$= a^{2+2+2}$ $= a^6$	$= x^{3+3}y^{4+4}$ $= x^6y^8$
The expression can also be simplified by multiplying each exponent inside the parentheses by the exponent outside the parentheses.	$(a^2)^3 = a^{2 \cdot 3} = a^6$	$(x^3y^4)^2 = x^{3 \cdot 2}y^{4 \cdot 2} = x^6y^8$

RULE FOR SIMPLIFYING A POWER OF AN EXPONENTIAL EXPRESSION

If m and n are integers, then $(x^m)^n = x^{mn}$.

EXAMPLES

1. $(x^5)^3 = x^{5 \cdot 3} = x^{15}$
2. $(y^8)^2 = y^{8 \cdot 2} = y^{16}$

A rule related to the one above applies to the power of a monomial.

RULE FOR SIMPLIFYING POWERS OF PRODUCTS

If m, n, and p are integers, then $(x^my^n)^p = x^{mp}y^{np}$.

EXAMPLES

1. $(x^4y^3)^5 = x^{4 \cdot 5}y^{3 \cdot 5} = x^{20}y^{15}$
2. $(2x^4)^3 = 2^{1 \cdot 3}x^{4 \cdot 3}$ • $2 = 2^1$
   $= 2^3x^{12} = 8x^{12}$

EXAMPLE 2 Simplify. **A.** $(-2a^4b^2)^3$ **B.** $(-2x^2y^3)(-3xy^2)^4$

Solution **A.** $(-2a^4b^2)^3 = (-2)^{1 \cdot 3}a^{4 \cdot 3}b^{2 \cdot 3}$ • Multiply the exponents.
$= (-2)^3a^{12}b^6 = -8a^{12}b^6$

B. $(-2x^2y^3)(-3xy^2)^4$
$= (-2x^2y^3)[(-3)^{1 \cdot 4}x^{1 \cdot 4}y^{2 \cdot 4}]$ • Use the Rule for Simplifying Powers of Products.
$= (-2x^2y^3)[(-3)^4x^4y^8]$
$= (-2x^2y^3)[81x^4y^8]$
$= -162x^6y^{11}$ • Use the Rule for Multiplying Exponential Expressions.

Problem 2 Simplify. **A.** $(-x^2y^4)^5$ **B.** $(2a^3b^2)(-3a^4b)^3$

Solution See pages S14–S15.

Try Exercise 25, page 250.

OBJECTIVE 2

Divide monomials and simplify expressions with negative exponents

The quotient of two exponential expressions with the same base can be simplified by writing each expression in factored form, dividing by common factors, and writing the result with an exponent.

$$\frac{x^5}{x^2} = \frac{\overset{1}{\cancel{x}} \cdot \overset{1}{\cancel{x}} \cdot x \cdot x \cdot x}{\underset{1}{\cancel{x}} \cdot \underset{1}{\cancel{x}}} = x^3$$

Subtracting the exponents gives the same result.

$$\frac{x^5}{x^2} = x^{5-2} = x^3$$

To divide two monomials with the same base, subtract the exponents of the like bases.

RULE FOR DIVIDING EXPONENTIAL EXPRESSIONS

If m and n are integers and $x \neq 0$, then $\dfrac{x^m}{x^n} = x^{m-n}$.

EXAMPLES

1. $\dfrac{x^7}{x^5} = x^{7-5} = x^2$
2. $\dfrac{a^5b^7}{a^4b} = a^{5-4}b^{7-1}$ • **Subtract exponents on like bases.**

   $= ab^6$

EXAMPLE 3 Simplify. **A.** $\dfrac{6x^5y^3}{8x^2y}$ **B.** $\dfrac{-6a^7b^5}{2a^5b^4}$

Solution **A.** $\dfrac{6x^5y^3}{8x^2y} = \dfrac{3x^{5-2}y^{3-1}}{4}$ • **Write $\frac{6}{8}$ in simplest form. Subtract the exponents on the like bases.**

$= \dfrac{3x^3y^2}{4}$

B. $\dfrac{-6a^7b^5}{2a^5b^4} = -3a^{7-5}b^{5-4}$ • **Write $\frac{-6}{2}$ in simplest form. Subtract the exponents on the like bases.**

$= -3a^2b$

Problem 3 Simplify. **A.** $\dfrac{-8x^7y^5}{16xy^4}$ **B.** $\dfrac{5a^5b^3}{8a^2b}$

Solution See page S15.

Try Exercise 47, page 251.

Consider the expression $\dfrac{x^4}{x^4}$, $x \neq 0$. This expression can be simplified by subtracting exponents or by dividing by common factors.

$$\frac{x^4}{x^4} = x^{4-4} = x^0 \qquad \frac{x^4}{x^4} = \frac{\overset{1}{\cancel{x}} \cdot \overset{1}{\cancel{x}} \cdot \overset{1}{\cancel{x}} \cdot \overset{1}{\cancel{x}}}{\underset{1}{\cancel{x}} \cdot \underset{1}{\cancel{x}} \cdot \underset{1}{\cancel{x}} \cdot \underset{1}{\cancel{x}}} = 1$$

The equations $\frac{x^4}{x^4} = x^0$ and $\frac{x^4}{x^4} = 1$ suggest the following definition of x^0.

DEFINITION OF ZERO AS AN EXPONENT

If $x \neq 0$, then $x^0 = 1$. The expression 0^0 is not defined.

EXAMPLES

Assume the value of a variable is not equal to zero.

1. $(-12)^0 = 1$
2. $a^0 = 1$
3. $(3y)^0 = 1$
4. $3y^0 = 3 \cdot 1 = 3$

Consider the expression $\frac{x^4}{x^6}$, $x \neq 0$. This expression can be simplified by subtracting exponents or by dividing by common factors.

$$\frac{x^4}{x^6} = x^{4-6} = x^{-2} \qquad \frac{x^4}{x^6} = \frac{\overset{1}{\cancel{x}} \cdot \overset{1}{\cancel{x}} \cdot \overset{1}{\cancel{x}} \cdot \overset{1}{\cancel{x}}}{\underset{1}{\cancel{x}} \cdot \underset{1}{\cancel{x}} \cdot \underset{1}{\cancel{x}} \cdot \underset{1}{\cancel{x}} \cdot x \cdot x} = \frac{1}{x^2}$$

The equations $\frac{x^4}{x^6} = x^{-2}$ and $\frac{x^4}{x^6} = \frac{1}{x^2}$ suggest that $x^{-2} = \frac{1}{x^2}$.

DEFINITION OF A NEGATIVE EXPONENT

If $x \neq 0$ and n is a positive integer, then $x^{-n} = \frac{1}{x^n}$ and $\frac{1}{x^{-n}} = x^n$.

EXAMPLES

Assume the value of a variable is not equal to zero.

1. $3^{-2} = \frac{1}{3^2} = \frac{1}{9}$
2. $2x^{-5} = \frac{2}{x^5}$
3. $(2x)^{-5} = \frac{1}{(2x)^5} = \frac{1}{32x^5}$
4. $\frac{5}{a^{-4}} = 5a^4$

An exponential expression is in simplest form when it contains only positive exponents.

Focus on writing an exponential expression with negative exponents in simplest form

Use the Definition of a Negative Exponent to write the following in simplest form. Assume that $x \neq 0$ and $y \neq 0$.

A. $2^{-3}x = \frac{1}{2^3} \cdot x = \frac{1}{8} \cdot x = \frac{x}{8}$

B. $\frac{x^{-2}}{y^{-3}} = x^{-2} \cdot \frac{1}{y^{-3}} = \frac{1}{x^2} \cdot y^3 = \frac{y^3}{x^2}$

C. $\frac{2x^{-4}}{y} = \frac{2}{y} \cdot x^{-4} = \frac{2}{y} \cdot \frac{1}{x^4} = \frac{2}{x^4 y}$

For the above examples, we assumed that $x \neq 0$ and $y \neq 0$. This was done because division by zero is not defined. In this textbook, we will assume that values of the variables are chosen so that division by zero does not occur. Using this convention, we generally will not state the restrictions on variables.

Zero and negative exponents may occur when using the Rule for Dividing Exponential Expressions.

EXAMPLE 4 Simplify. **A.** $\dfrac{9x^5y^2}{12x^5y^7}$ **B.** $\dfrac{5x^{-2}y^4}{10x^5y^{-1}}$

Solution **A.** $\dfrac{9x^5y^2}{12x^5y^7} = \dfrac{3x^{5-5}y^{2-7}}{4} = \dfrac{3x^0y^{-5}}{4}$

- **Use the Rule for Dividing Exponential Expressions.**

$= \dfrac{3}{4y^5}$

- **Use the Definitions of Zero and Negative Exponents to write in simplest form.**

B. $\dfrac{5x^{-2}y^4}{10x^5y^{-1}} = \dfrac{x^{-2-5}y^{4-(-1)}}{2} = \dfrac{x^{-7}y^5}{2}$

- **Use the Rule for Dividing Exponential Expressions.**

$= \dfrac{y^5}{2x^7}$

- **Use the Definition of a Negative Exponent to write in simplest form.**

Problem 4 Simplify. **A.** $\dfrac{18x^{-6}y}{9x^{-6}y^7}$ **B.** $\dfrac{6x^8y^{-3}}{4x^{-2}y^4}$

Solution See page S15.

Try Exercise 75, page 251.

Consider the expression $\left(\dfrac{x^3}{y^4}\right)^2$, $y \neq 0$. This expression can be simplified by squaring $\dfrac{x^3}{y^4}$ or by multiplying each exponent in the quotient by the exponent outside the parentheses.

$$\left(\frac{x^3}{y^4}\right)^2 = \left(\frac{x^3}{y^4}\right)\left(\frac{x^3}{y^4}\right) = \frac{x^3 \cdot x^3}{y^4 \cdot y^4} = \frac{x^{3+3}}{y^{4+4}} = \frac{x^6}{y^8} \qquad \left(\frac{x^3}{y^4}\right)^2 = \frac{x^{3\cdot 2}}{y^{4\cdot 2}} = \frac{x^6}{y^8}$$

RULE FOR SIMPLIFYING POWERS OF QUOTIENTS

If m, n, and p are integers and $y \neq 0$, then $\left(\dfrac{x^m}{y^n}\right)^p = \dfrac{x^{mp}}{y^{np}}$.

EXAMPLES

1. $\left(\dfrac{a^4}{b^5}\right)^3 = \dfrac{a^{4\cdot 3}}{b^{5\cdot 3}} = \dfrac{a^{12}}{b^{15}}$ **2.** $\left(\dfrac{r^3}{t}\right)^5 = \dfrac{r^{3\cdot 5}}{t^{1\cdot 5}} = \dfrac{r^{15}}{t^5}$

EXAMPLE 5 Simplify. **A.** $\left(\dfrac{r^{-2}}{t^4}\right)^3$ **B.** $\left(\dfrac{a^4}{b^3}\right)^{-2}$

Solution **A.** $\left(\dfrac{r^{-2}}{t^4}\right)^3 = \dfrac{r^{-2\cdot 3}}{t^{4\cdot 3}} = \dfrac{r^{-6}}{t^{12}}$

- **Use the Rule for Simplifying Powers of Quotients.**

$= \dfrac{1}{r^6t^{12}}$

- **Use the Definition of a Negative Exponent to simplify.**

B. $\left(\dfrac{a^4}{b^3}\right)^{-2} = \dfrac{a^{4(-2)}}{b^{3(-2)}} = \dfrac{a^{-8}}{b^{-6}}$

- **Use the Rule for Simplifying Powers of Quotients.**

$= \dfrac{b^6}{a^8}$

- **Use the Definition of a Negative Exponent to simplify.**

Problem 5 Simplify. **A.** $\left(\frac{x^2}{y^{-3}}\right)^4$ **B.** $\left(\frac{x^5}{y^4}\right)^{-3}$

Solution See page S15.

Try Exercise 73, page 251.

The rules for simplifying exponential expressions and powers of exponential expressions are restated below for convenience.

RULES OF EXPONENTS

If m, n, and p are integers, and $x \neq 0$ and $y \neq 0$, then

$x^m \cdot x^n = x^{m+n}$ $(x^m)^n = x^{mn}$ $(x^m y^n)^p = x^{mp} y^{np}$

$\frac{x^m}{x^n} = x^{m-n}$ $\left(\frac{x^m}{y^n}\right)^p = \frac{x^{mp}}{y^{np}}$ $x^{-n} = \frac{1}{x^n}$

$x^0 = 1$

EXAMPLE 6 Simplify. **A.** $(3x^2y^{-3})(6x^{-4}y^5)$ **B.** $\left(\frac{3a^2b^{-2}c^{-1}}{27a^{-1}b^2c^{-4}}\right)^{-2}$

Solution **A.** $(3x^2y^{-3})(6x^{-4}y^5)$

$= 18x^{2+(-4)}y^{-3+5}$
$= 18x^{-2}y^2$
- Use the Rule for Multiplying Exponential Expressions.

$= \frac{18y^2}{x^2}$
- Use the Definition of a Negative Exponent to rewrite the expression without negative exponents.

B. $\left(\frac{3a^2b^{-2}c^{-1}}{27a^{-1}b^2c^{-4}}\right)^{-2} = \left(\frac{a^3b^{-4}c^3}{9}\right)^{-2}$
- Simplify inside the parentheses by using the Rule for Dividing Exponential Expressions.

$= \frac{a^{-6}b^8c^{-6}}{9^{-2}}$
- Multiply each exponent inside the parentheses by the exponent outside the parentheses.

$= \frac{9^2b^8}{a^6c^6}$
- Use the Definition of a Negative Exponent to rewrite the expression without negative exponents.

$= \frac{81b^8}{a^6c^6}$
- Simplify.

Problem 6 Simplify. **A.** $(2x^{-5}y)(5x^4y^{-3})$ **B.** $\left(\frac{2^{-1}x^2y^{-3}}{4x^{-2}y^{-5}}\right)^{-2}$

Solution See page S15.

Try Exercise 77, page 252.

OBJECTIVE 3 Scientific notation

> **Point of Interest**
>
> Astronomers measure the distance to some stars by using the *parsec*. One parsec is approximately 1.92×10^{13} mi.

Very large and very small numbers are encountered in the fields of science and engineering. For example, the mass of the electron is 0.0000000000000000000000000009 g. Numbers such as this are difficult to read and write, so a more convenient system for writing them has been developed. It is called **scientific notation.**

To express a number in scientific notation, write the number as the product of a number between 1 and 10 and a power of 10. The form for scientific notation is $a \times 10^n$, where $1 \leq a < 10$ and n is an integer.

> **Take Note**
>
> There are two steps in writing a number in scientific notation: (1) determine the number between 1 and 10, and (2) determine the exponent on 10.

For numbers greater than 10, move the decimal point to the right of the first digit. The exponent n is positive and equal to the number of places the decimal point has been moved.

$965{,}000 = 9.65 \times 10^5$

$3{,}600{,}000 = 3.6 \times 10^6$

$92{,}000{,}000{,}000 = 9.2 \times 10^{10}$

For numbers less than 1, move the decimal point to the right of the first nonzero digit. The exponent n is negative. The absolute value of the exponent is equal to the number of places the decimal point has been moved.

$0.0002 = 2 \times 10^{-4}$

$0.0000000974 = 9.74 \times 10^{-8}$

$0.000000000086 = 8.6 \times 10^{-11}$

EXAMPLE 7 Write 0.000041 in scientific notation.

Solution $0.000041 = 4.1 \times 10^{-5}$ • **The decimal point must be moved 5 places to the right. The exponent is negative.**

Problem 7 Write 942,000,000 in scientific notation.

Solution See page S15.

Try Exercise 93, page 252.

Converting a number written in scientific notation to decimal notation requires moving the decimal point.

When the exponent is positive, move the decimal point to the right the same number of places as the exponent.

$1.32 \times 10^4 = 13{,}200$

$1.4 \times 10^8 = 140{,}000{,}000$

When the exponent is negative, move the decimal point to the left the same number of places as the absolute value of the exponent.

$1.32 \times 10^{-2} = 0.0132$

$1.4 \times 10^{-4} = 0.00014$

EXAMPLE 8 Write 3.3×10^7 in decimal notation.

Solution $3.3 \times 10^7 = 33{,}000{,}000$ • **Move the decimal point 7 places to the right.**

Problem 8 Write 2.7×10^{-5} in decimal notation.

Solution See page S15.

Try Exercise 101, page 252.

Numerical calculations involving numbers that have more digits than a handheld calculator is able to handle can be performed using scientific notation.

EXAMPLE 9 Simplify: $\dfrac{2{,}400{,}000{,}000 \times 0.0000063}{0.00009 \times 480}$

Solution

$$\frac{2{,}400{,}000{,}000 \times 0.0000063}{0.00009 \times 480}$$

$$= \frac{2.4 \times 10^9 \times 6.3 \times 10^{-6}}{9 \times 10^{-5} \times 4.8 \times 10^2}$$

• Write the numbers in scientific notation.

$$= \frac{(2.4)(6.3) \times 10^{9+(-6)-(-5)-2}}{(9)(4.8)}$$

• Use the Rules for Multiplying and Dividing Exponential Expressions.

$$= 0.35 \times 10^6$$

• Simplify.

$$= 3.5 \times 10^5$$

• Write in scientific notation.

Problem 9 Simplify: $\dfrac{5{,}600{,}000 \times 0.000000081}{900 \times 0.000000028}$

Solution See page S15.

Try Exercise 105, page 252.

OBJECTIVE 4 Application problems

EXAMPLE 10  How many miles does light travel in 1 day? The speed of light is 186,000 mi/s. Write the answer in scientific notation.

Strategy To find the distance traveled:

- ▶ Write the speed of light in scientific notation.
- ▶ Write the number of seconds in 1 day in scientific notation.
- ▶ Use the equation $d = rt$, where r is the speed of light and t is the number of seconds in 1 day.

Solution $186{,}000 = 1.86 \times 10^5$

$24 \cdot 60 \cdot 60 = 86{,}400 = 8.64 \times 10^4$

$d = rt$
$d = (1.86 \times 10^5)(8.64 \times 10^4)$
$d = 1.86 \times 8.64 \times 10^9$
$d = 16.0704 \times 10^9$
$d = 1.60704 \times 10^{10}$

Light travels 1.60704×10^{10} mi in 1 day.

Problem 10 A computer can do an arithmetic operation in 1×10^{-7} s. How many arithmetic operations can the computer perform in 1 min? Write the answer in scientific notation.

Solution See page S15.

Try Exercise 125, page 253.

5.1 Exercises

CONCEPT CHECK

1. Which of the following are monomials?
   (i) 7 (ii) $-3x^2y^3$ (iii) $2x + 3$ (iv) $\frac{xy^2}{3}$ (v) $\frac{3y}{x}$

2. What is the Rule for Multiplying Exponential Expressions?

3. Does the Rule for Simplifying Powers of Products apply to $(a^2 + b^3)^4$? Why or why not?

4. If x is a positive number, is x^{-3} a positive or a negative number?

5. What is the value of $\frac{1}{2^{-5}}$?

6. If $a \neq 0$, what are the values of $-a^0$ and $(-a)^0$?

1 Multiply monomials (See pages 242–244.)

GETTING READY

7. Of the two expressions $x^4 + y^5$ and x^4y^5, the one that is a monomial is __?__. The degree of this monomial is __?__.

8. a. Use the Rule for Multiplying Exponential Expressions to multiply: $(a^8)(a^5) =$ __?__.

   b. Use the Rule for Simplifying a Power of an Exponential Expression to simplify: $(a^8)^5 =$ __?__.

   c. Use the Rule for Simplifying Powers of Products to simplify: $(a^8b)^5 =$ __?__.

9. State whether the expression can be simplified using one or more of the rules mentioned in Exercise 8.
   a. $(ab)^2$ b. $(a + b)^2$ c. $(x^2 + y^3)^2$ d. $(x^2y^3)^2$

10. a. True or false? $x^3y^4 = xy^{12}$
    b. True or false? $x^3y^4 = xy^7$
    c. True or false? $(x^3y^4)^2 = x^9y^{16}$
    d. True or false? $(x^3 + y^4)^2 = x^6 + y^8$

Simplify.

11. $(-6r^2t^5)(-4rt)$
12. $(5x^2y^4)(3y^2z^3)$
13. $(-b^3c^5)(-a^2c^4)$
14. $(z^3)^4$
15. $(b^5)^2$
16. $(a^7b^3)^2$
17. $(xy^6)^4$
18. $(r^3s)^5$
19. $(x^2y^4)^4$
20. $(2x^3)^5$
21. $(3a)^4$
22. $(-5x^3)^3$
23. $(-4c^5)^4$
24. $(6x^3y^4)^2$
25. $(-2x^5y)^6$
26. $(-3x^2y^3)^4$
27. $(-2ab^2)^3$
28. $(xy^2z^3)^4$
29. $(-2x^2yz^3)^5$
30. $(-3r^3s^2t)^6$
31. $(2xy)(-3x^2yz)(x^2y^3z^3)$

32. $(x^2z^4)(2xyz^4)(-3x^3y^2)$
33. $(3b^5)(2ab^2)(-2ab^2c^2)$
34. $(-c^3)(-2a^2bc)(3a^2b)$
35. $(-2x^2y^3z)(3x^2yz^4)$
36. $(2a^2b)^3(-3ab^4)^2$
37. $(-3ab^3)^3(-2^2a^2b)^2$
38. $(4ab)^2(-2ab^2c^3)^3$
39. $(-2ab^2)(-3a^4b^5)^3$

2 Divide monomials and simplify expressions with negative exponents (See pages 244–247.)

40. Explain how to divide two exponential expressions with the same base.

41. If a variable has a negative exponent, how can you rewrite it with a positive exponent?

GETTING READY

42. As long as x is not zero, x^0 is defined to be equal to ___?___. Using this definition, $5^0 =$ ___?___, $(2x^2)^0 =$ ___?___, and $-4x^0 =$ ___?___.

43. Simplify: $\left(\dfrac{a^{-3}}{a^5}\right)^2$

$\left(\dfrac{a^{-3}}{a^5}\right)^2 = (\underline{\quad ? \quad})^2$ • **The bases inside the parentheses are the same. Subtract the exponents.**

$= \underline{\quad ? \quad}$ • **Multiply the exponent inside the parentheses by the exponent outside the parentheses.**

$= \underline{\quad ? \quad}$ • **Use the Definition of a Negative Exponent to rewrite the expression with a positive exponent.**

Simplify.

44. $\dfrac{a^8}{a^5}$
45. $\dfrac{15a^7}{5a^4}$
46. $\dfrac{x^5y^{10}}{x^2y}$
47. $\dfrac{a^7b}{a^2b^4}$

48. $\dfrac{3x^9y^4}{-6x^3y^3}$
49. $\dfrac{12ab^3c^6}{18bc^5}$
50. $\dfrac{x^3}{x^{12}}$
51. $\dfrac{x^3y^6}{x^3y^3}$

52. 2^{-3}
53. $\dfrac{1}{3^{-5}}$
54. $\dfrac{1}{x^{-4}}$
55. $\dfrac{1}{y^{-3}}$

56. $\dfrac{2x^{-2}}{y^4}$
57. $\dfrac{a^3}{4b^{-2}}$
58. $x^{-4}x^4$
59. $x^{-3}x^{-5}$

60. $(3x^{-2})^2$
61. $(5x^2)^{-3}$
62. $\dfrac{x^{-3}}{x^2}$
63. $\dfrac{x^4}{x^{-5}}$

64. $a^{-2} \cdot a^4$
65. $a^{-5} \cdot a^7$
66. $(x^2y^{-4})^2$
67. $(x^3y^5)^{-2}$

68. $(2a^{-1})^{-2}(2a^{-1})^4$
69. $(3a)^{-3}(9a^{-1})^{-2}$
70. $(x^{-2}y)^2(xy)^{-2}$
71. $(x^{-1}y^2)^{-3}(x^2y^{-4})^{-3}$

72. $\dfrac{6^2a^{-2}b^3}{3ab^4}$
73. $\left(\dfrac{x^2}{y}\right)^{-4}$
74. $\dfrac{-48ab^{10}}{32a^4b^3}$
75. $\dfrac{a^2b^3c^7}{a^6bc^5}$

76. $\dfrac{(-4x^2y^3)^2}{(2xy^2)^3}$

77. $\dfrac{(-3a^2b^3)^2}{(-2ab^4)^3}$

78. $\left(\dfrac{x^{-3}y^{-4}}{x^{-2}y}\right)^{-2}$

79. $\left(\dfrac{a^{-2}b}{a^3b^{-4}}\right)^2$

80. $\dfrac{(2a^{-3}b^{-2})^3}{(a^{-4}b^{-1})^{-2}}$

81. $\dfrac{(3x^{-2}y)^{-2}}{(4xy^{-2})^{-1}}$

82. $\left(\dfrac{4^{-2}xy^{-3}}{x^{-3}y}\right)^3\left(\dfrac{8^{-1}x^{-2}y}{x^4y^{-1}}\right)^{-2}$

83. $\left(\dfrac{9ab^{-2}}{8a^{-2}b}\right)^{-2}\left(\dfrac{3a^{-2}b}{2a^2b^{-2}}\right)^3$

84. $[(xy^{-2})^3]^{-2}$

85. $[(x^{-2}y^{-1})^2]^{-3}$

86. $\left[\left(\dfrac{x}{y^2}\right)^{-2}\right]^3$

87. $\left[\left(\dfrac{a^2}{b}\right)^{-1}\right]^2$

88. True or false?

a. $\left(\dfrac{a}{b}\right)^{-1} = \dfrac{b}{a}$

b. $(a-b)^{-1} = \dfrac{1}{a} - \dfrac{1}{b}$

89. True or false?

a. $\dfrac{a^n}{b^m} = \left(\dfrac{a}{b}\right)^{n-m}$

b. $\dfrac{a^n}{a^m} = a^{n-m}$

3 Scientific notation (See pages 248–249.)

GETTING READY

90. A number is in scientific notation if it is written as the product of a number between ___?___ and ___?___ and a power of ___?___.

91. To write the number 0.00000078 in scientific notation, move the decimal point ___?___ places to the ___?___. This means the exponent on 10 is ___?___.

Write in scientific notation.

92. 0.00000467

93. 0.00000005

94. 0.00000000017

95. 4,300,000

96. 200,000,000,000

97. 9,800,000,000

Write in decimal notation.

98. 1.23×10^{-7}

99. 6.2×10^{-12}

100. 8.2×10^{15}

101. 6.34×10^5

102. 3.9×10^{-2}

103. 4.35×10^9

Simplify. Write the answer in scientific notation.

104. $(3 \times 10^{-12})(5 \times 10^{16})$

105. $(8.9 \times 10^{-5})(3.2 \times 10^{-6})$

106. $(0.0000065)(3{,}200{,}000{,}000{,}000)$

107. $(480{,}000)(0.0000000096)$

108. $\dfrac{9 \times 10^{-3}}{6 \times 10^5}$

109. $\dfrac{2.7 \times 10^4}{3 \times 10^{-6}}$

110. $\dfrac{0.0089}{500{,}000{,}000}$

111. $\dfrac{4800}{0.00000024}$

112. $\dfrac{0.00056}{0.000000000004}$

113. $\dfrac{0.000000346}{0.0000005}$

114. $\dfrac{(3.2 \times 10^{-11})(2.9 \times 10^{15})}{8.1 \times 10^{-3}}$

115. $\dfrac{(6.9 \times 10^{27})(8.2 \times 10^{-13})}{4.1 \times 10^{15}}$

116. $\dfrac{(0.00000004)(84{,}000)}{(0.0003)(1{,}400{,}000)}$

117. $\dfrac{(720)(0.0000000039)}{(26{,}000{,}000{,}000)(0.018)}$

For Exercises 118 and 119, $a \times 10^n$ and $b \times 10^m$ are numbers written in scientific notation. State whether the expression is greater than or less than 1.

118. $\dfrac{(a \times 10^n)}{(b \times 10^m)}$, where $n > m$

119. $(a \times 10^n)(b \times 10^m)$, where $n < 0$ and $m < 0$

4 Application problems (See page 249.)

GETTING READY

120. Write the number of seconds in 5 h in scientific notation:

$$5 \text{ h} = (5 \text{ h})(\underline{\quad ? \quad} \text{ min/h})(\underline{\quad ? \quad} \text{ s/min})$$
$$= \underline{\quad ? \quad} \text{ s}$$
$$= 1.8 \times 10^{\underline{?}} \text{ s}$$

121. The speed of light is 3×10^5 km/s. Use your answer to Exercise 120 to find how many kilometers light travels in 5 h:

$$d = (3 \times 10^5)(1.8 \times 10^4) \text{ km} = \underline{\quad ? \quad} \times 10^{\underline{?}} \text{ km}$$

Solve. Write the answer in scientific notation.

122. **Physics** How many meters does light travel in 8 h? The speed of light is 3×10^8 m/s.

123. **Physics** How many meters does light travel in 1 day? The speed of light is 3×10^8 m/s.

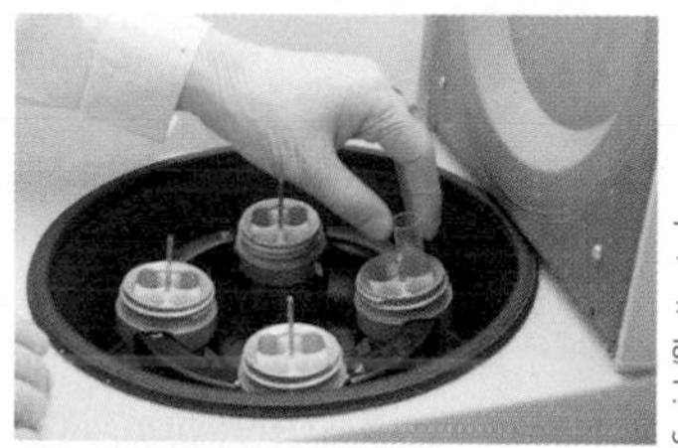

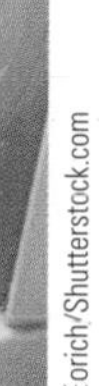

124. **Physics** A high-speed centrifuge makes 4×10^8 revolutions each minute. Find the time in seconds for the centrifuge to make one revolution.

125. **Physics** The mass of an electron is 9.109×10^{-31} kg. The mass of a proton is 1.673×10^{-27} kg. How many times heavier is a proton than an electron?

126. **Geology** The mass of Earth is 5.9×10^{24} kg. The mass of the sun is 2×10^{30} kg. How many times heavier is the sun than Earth?

127. **Astronomy** Use the information in the article below to determine the average number of miles traveled per day by *Phoenix* on its trip to Mars.

In the News

A Mars Landing for *Phoenix*

At 7:53 P.M., a safe landing on the surface of Mars brought an end to the *Phoenix* spacecraft's 296-day, 422-million-mile journey to the Red Planet.

Source: The Los Angeles Times

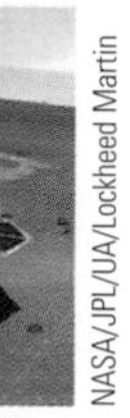

128. **Astronomy** It took 11 min for the commands from a computer on Earth to travel to the *Phoenix* Mars Lander, a distance of 119 million miles. How fast did the signals from Earth to Mars travel?

129. **Forestry** Use the information in the article at the right. If every burned acre of Yellowstone Park had 12,000 lodgepole pine seedlings growing on it 1 year after the fire, how many new seedlings would be growing?

130. **Forestry** Use the information in the article at the right. Find the number of seeds released by the lodgepole pine trees for each surviving seedling.

131. **Astronomy** How long does it take light to travel to Earth from the sun? The sun is 9.3×10^7 mi from Earth, and light travels 1.86×10^5 mi/s.

132. **Astronomy** The diameter of Neptune is 3×10^4 mi. Use the formula $SA = 4\pi r^2$ to find the surface area of Neptune in square miles.

133. **Biology** The radius of a cell is 1.5×10^{-4} mm. Use the formula $V = \frac{4}{3}\pi r^3$ to find the volume of the cell.

134. **Chemistry** One gram of hydrogen contains 6.023×10^{23} atoms. Find the weight of one atom of hydrogen.

135. **Astronomy** Our galaxy is estimated to be 5.6×10^{19} mi across. How long (in hours) would it take a spaceship to cross the galaxy traveling at 25,000 mph?

In the News

Forest Fires Spread Seeds

Forest fires may be feared by humans, but not by the lodgepole pine, a tree that uses the intense heat of a fire to release its seeds from their cones. After a blaze that burned 12,000,000 acres of Yellowstone National Park, scientists counted 2 million lodgepole pine seeds on one acre of the park. One year later, the scientists returned to that acre to find 12,000 lodgepole pine seedlings growing.

Source: National Public Radio

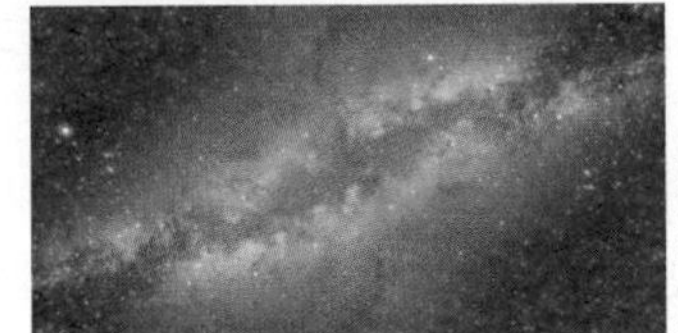
Viktar Malyshchyts/Shutterstock.com

APPLYING CONCEPTS

136. Evaluate 3^{x^2} for the given values of x.

a. $x = 2$ **b.** $x = 3$ **c.** $x = -1$ **d.** $x = -2$

137. Evaluate 2^{-x^2} for the given values of x.

a. $x = 2$ **b.** $x = 3$ **c.** $x = -1$ **d.** $x = -2$

Simplify each expression. Assume that m and n are positive integers and that $x \neq 0$.

138. $x^{3n}x^{4n}$ 139. $(-2x^ny^m)(3x^ny^{2m})$ 140. $(3x^m)^4$

141. $\dfrac{x^{5n}}{x^{2n}}$ 142. $\dfrac{x^ny^{5m}}{x^{3n}y^m}$ 143. $\left(\dfrac{3}{x^{3n}}\right)^{-2}$

Simplify.

144. $\dfrac{4m^4}{n^{-2}} + \left(\dfrac{n^{-1}}{m^2}\right)^{-2}$ 145. $\dfrac{5x^3}{y^{-6}} + \left(\dfrac{x^{-1}}{y^2}\right)^{-3}$

146. $\left(\dfrac{3a^{-2}b}{a^{-4}b^{-1}}\right)^2 \div \left(\dfrac{a^{-1}b}{9a^2b^3}\right)^{-1}$ 147. $\left(\dfrac{2m^3n^{-2}}{4m^4n}\right)^{-2} \div \left(\dfrac{mn^5}{m^{-1}n^3}\right)^3$

148. True or false? $(2 + 3)^{-2} = 2^{-2} + 3^{-2}$

149. If a and b are positive real numbers and $a < b$, what is the relationship between a^{-1} and b^{-1}?

PROJECTS OR GROUP ACTIVITIES

150. Gather three coins: a penny, a nickel, and a dime. Draw three dots on a piece of paper, and place the coins in a pile from largest (in size) to smallest on one of the dots. Now transfer the coins to one of the other dots using the following rules. Move one coin at a time. Never place a larger coin on a smaller one. How many moves are required? Write this number as an exponential expression with a base of 2.

151. Repeat Exercise 150 with four coins: a penny, a nickel, a dime, and a quarter. How many moves are required to move the pile of coins to one of the other dots? Write this number as an exponential expression with a base of 2.

152. Repeat Exercise 150 with five coins: a penny, a nickel, a dime, a quarter, and a half-dollar (use a cardboard disk if you do not have a half-dollar coin). How many moves are required to move the pile of coins to one of the other dots? Write this number as an exponential expression with a base of 2.

5.2 Introduction to Polynomials

OBJECTIVE 1 Evaluate polynomial functions

A **polynomial** is a variable expression in which the terms are monomials.

A polynomial of one term is a **monomial.** $5x$

A polynomial of two terms is a **binomial.** $5x^2y + 6x$

A polynomial of three terms is a **trinomial.** $3x^2 + 9xy - 5y$

Polynomials with more than three terms do not have special names.

Here are some additional examples of polynomials.

A polynomial with four terms $3x^4 - 6x^2 + 5x - 9$

The coefficients may be any real numbers.
The exponents on variables must be positive integers. $\sqrt{3}x^3 - \pi x^2 + \frac{5}{2}x - 7$

Focus on identifying a polynomial

Determine whether the given expression is a polynomial. Explain.

A. $3x^4 + 2x^{-2} - 3$ **B.** $x^4 - 3x - \sqrt{7}$ **C.** $\frac{1}{x} + 2$

A. No. Polynomials must have positive integer exponents on variables.

B. Yes. The exponents on the variables are positive integers.

C. No. Because $\frac{1}{x} + 2 = x^{-1} + 2$, the exponent on the variable is a negative integer.

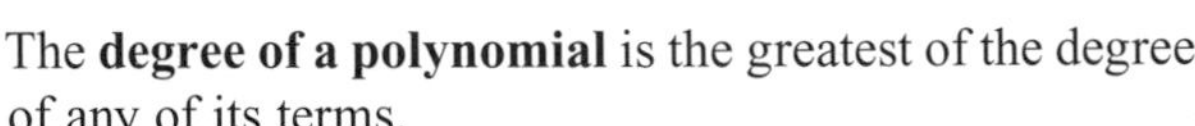

The **degree of a polynomial** is the greatest of the degrees of any of its terms.

$3x + 2$ degree 1
$3x^2 + 2x - 4$ degree 2
$4x^3y^2 + 6x^4$ degree 5

The terms of a polynomial in one variable are usually arranged so that the exponents on the variable decrease from left to right. This is called **descending order.**

$2x^2 - x + 8$

$3y^3 - 3y^2 + y - 12$

For a polynomial in more than one variable, "descending order" may refer to any one of the variables.

The polynomial at the right is shown first in descending order of the x variable and then in descending order of the y variable.

$2x^2 + 3xy + 5y^2$

$5y^2 + 3xy + 2x^2$

The **linear function** given by $f(x) = mx + b$ is an example of a polynomial function. It is a polynomial function of degree 1. A second-degree polynomial function, called a **quadratic function,** is given by the equation $f(x) = ax^2 + bx + c, a \neq 0$. A third-degree polynomial function is called a **cubic function.** In general, a **polynomial function** is an expression whose terms are monomials.

The **leading coefficient** of a polynomial function is the coefficient of the term with the greatest exponent on a variable. The constant term is the term without a variable. For the polynomial function $P(x) = 7x^4 - 3x^2 + 2x - 4$, the leading coefficient is 7 and the constant term is -4. The degree of the polynomial is 4.

EXAMPLE 1 Find the leading coefficient, the constant term, and the degree of the polynomial $P(x) = 5x^6 - 4x^5 - 3x^2 + 7$.

Solution The term with the greatest exponent is $5x^6$. The term without a variable is 7.

The leading coefficient is 5, the constant term is 7, and the degree is 6.

Problem 1 Find the leading coefficient, the constant term, and the degree of the polynomial $R(x) = -3x^4 + 3x^3 + 3x^2 - 2x - 12$.

Solution See page S15.

Try Exercise 9, page 262.

To **evaluate a polynomial function,** replace the variable by its value and simplify.

Focus on evaluating a polynomial function

Given $P(x) = x^3 - 3x^2 + 4$, evaluate $P(-3)$.

Substitute -3 for x.
Simplify.

$$P(x) = x^3 - 3x^2 + 4$$
$$P(-3) = (-3)^3 - 3(-3)^2 + 4$$
$$P(-3) = -27 - 27 + 4$$
$$P(-3) = -50$$

How It's Used

Economists use polynomial functions to model cost, revenue, and profit, and to make predictions about economic trends.

EXAMPLE 2 To overcome the resistance of the wind and the tires on the road, the horsepower (hp), P, required by a cyclist to keep a certain bicycle moving at v miles per hour is given by $P(v) = 0.00003v^3 + 0.00211v$. How much horsepower must the cyclist supply to keep this bicycle moving at 20 mph?

Strategy To find the horsepower, evaluate the function when $v = 20$.

Solution

$$P(v) = 0.00003v^3 + 0.00211v$$
$$P(20) = 0.00003(20)^3 + 0.00211(20) \quad \text{• Replace } v \text{ by 20.}$$
$$= 0.2822 \quad \text{• Simplify.}$$

The cyclist must supply 0.2822 hp.

Problem 2 The velocity of air that is expelled during a cough can be modeled by $v(r) = 600r^2 - 1000r^3$, where v is the velocity of the air in centimeters per second and r is the radius of the trachea in centimeters. What is the velocity of expelled air during a cough when the radius of the trachea is 0.4 cm?

Solution See page S15.

Try Exercise 33, page 263.

The graph of a linear function is a straight line and can be found by plotting just two points. The graph of a polynomial function of degree greater than 1 is a curve. Consequently, many points may have to be found before an accurate graph can be drawn.

Evaluating the quadratic function given by the equation $f(x) = x^2 - x - 6$ when $x = -3$, $-2, -1, 0, 1, 2, 3$, and 4 gives the points shown in Figure 1. For instance, $f(-3) = 6$, so $(-3, 6)$ is graphed; $f(2) = -4$, so $(2, -4)$ is graphed; and $f(4) = 6$, so $(4, 6)$ is graphed. Evaluating the function at noninteger values of x, such as $x = -\frac{3}{2}$ and $x = \frac{5}{2}$, produces more points to graph, as shown in Figure 2. Connecting the points with a smooth curve results in Figure 3, which is the graph of f.

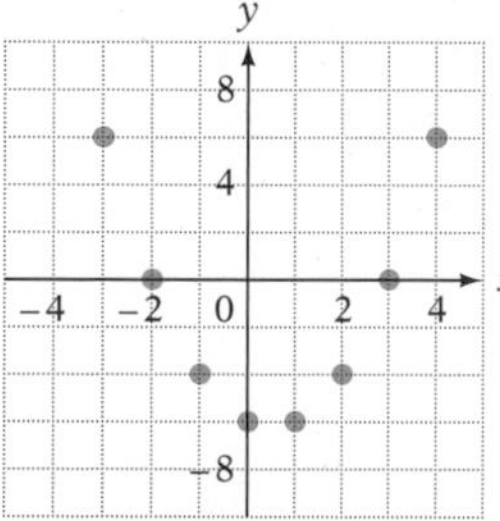

FIGURE 1

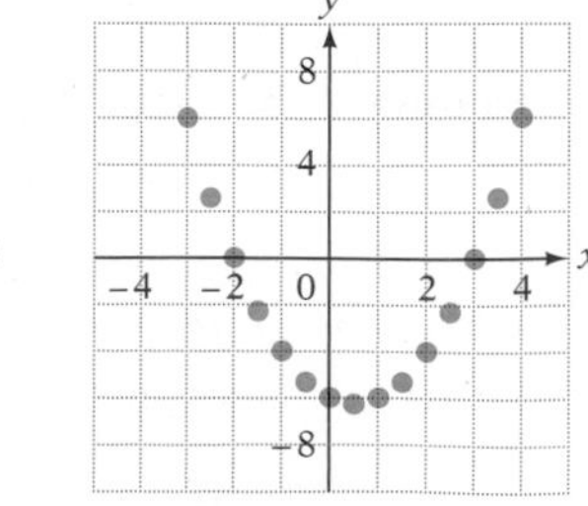

FIGURE 2

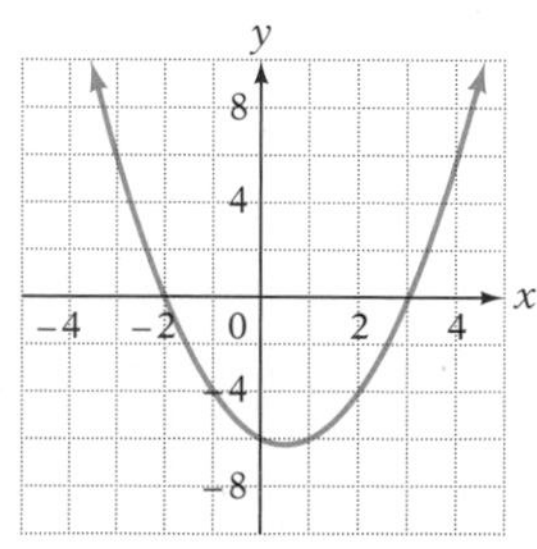

FIGURE 3

Here is an example of graphing a cubic function, $P(x) = x^3 - 2x^2 - 5x + 6$. Evaluating the function when $x = -2, -1, 0, 1, 2, 3$, and 4 gives the graph in Figure 4. Evaluating at some noninteger values gives the graph in Figure 5. Finally, connecting the dots with a smooth curve gives the graph in Figure 6.

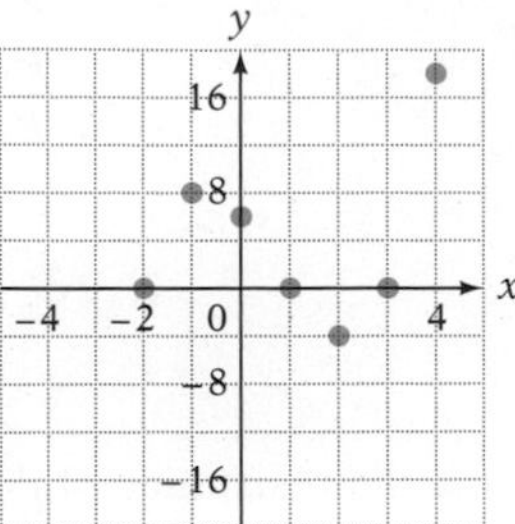
FIGURE 4

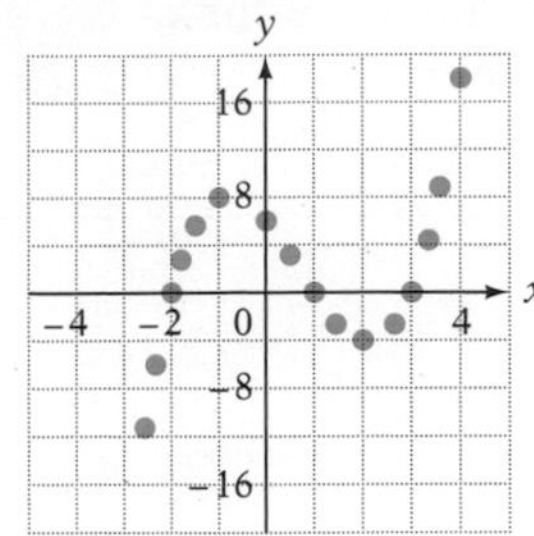
FIGURE 5

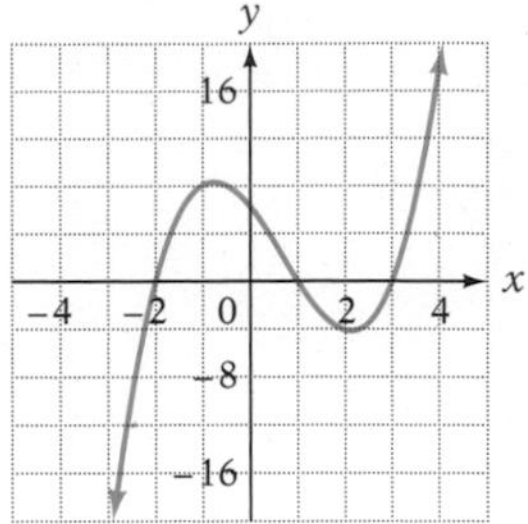
FIGURE 6

The graphs of $f(x) = x^2 - x - 6$ (Figure 3) and $P(x) = x^3 - 2x^2 - 5x + 6$ (Figure 6) are repeated at the right. Note that by the vertical line test, these graphs are graphs of functions.

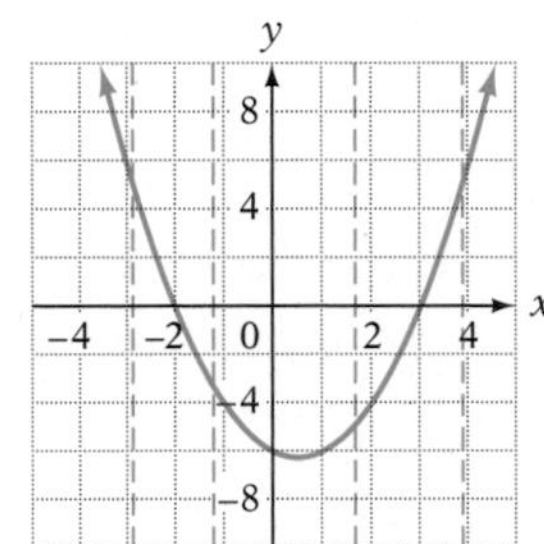

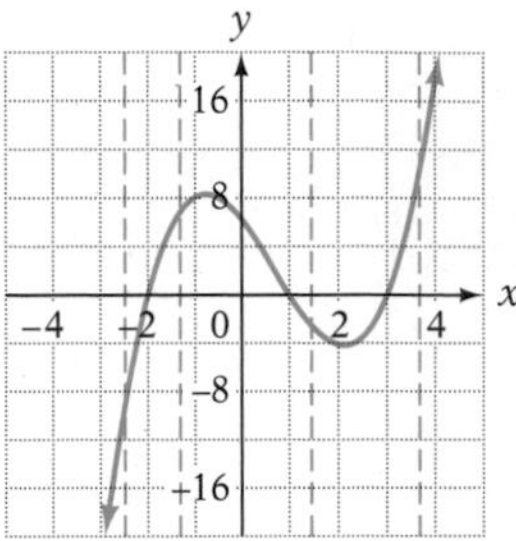

Recall that the domain of a function is the set of first coordinates of the ordered pairs of the function. Because a polynomial function can be evaluated for any real number, the domain of a polynomial function is the set of real numbers.

EXAMPLE 3 Graph: $f(x) = x^2 - 2$

Solution

x	$y = x^2 - 2$
−3	7
−2	2
−1	−1
0	−2
1	−1
2	2
3	7

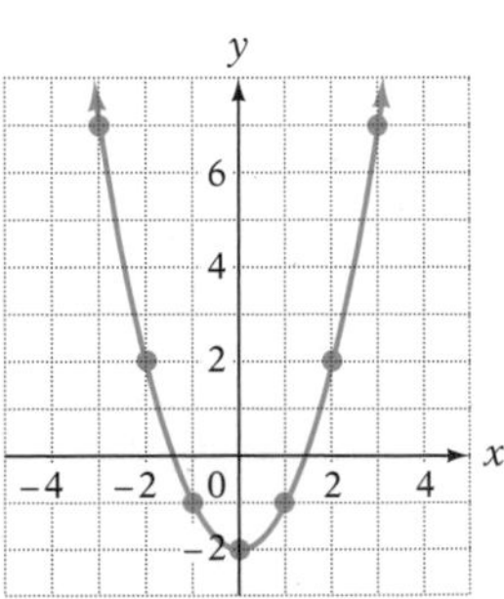

Problem 3 Graph: $f(x) = x^2 - 2x$

Solution See page S15.

Try Exercise 27, page 262.

EXAMPLE 4 Graph: $f(x) = x^3 - 1$

Solution

x	$y = x^3 - 1$
-2	-9
-1	-2
0	-1
1	0
2	7

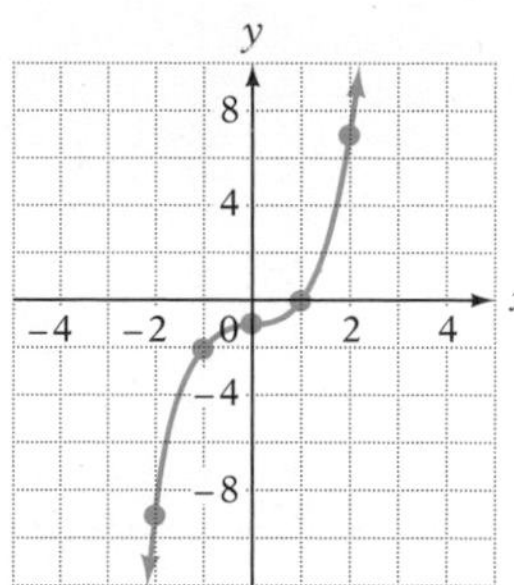

Problem 4 Graph: $F(x) = -x^3 + 1$

Solution See page S16.

Try Exercise 29, page 262.

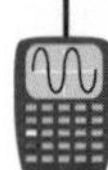

It may be necessary to plot a large number of points before drawing the graph in Example 4. Graphing calculators create graphs by plotting a large number of points and then connecting the points to form a curve. Using a graphing calculator, enter the equation $y = x^3 - 1$. Graph the equation using XMIN = −4.7, XMAX = 4.7, YMIN = −10, and YMAX = 10. Now use the TRACE feature to move around the graph and verify that (−2, −9), (−1, −2), (0, −1), (1, 0), and (2, 7) are the coordinates of points on the graph.

Follow the same procedure to graph the equation in Problem 4. The graph is shown at the right, with the cursor at the point (2, −7). For assistance, refer to the Appendix Keystroke Guide.

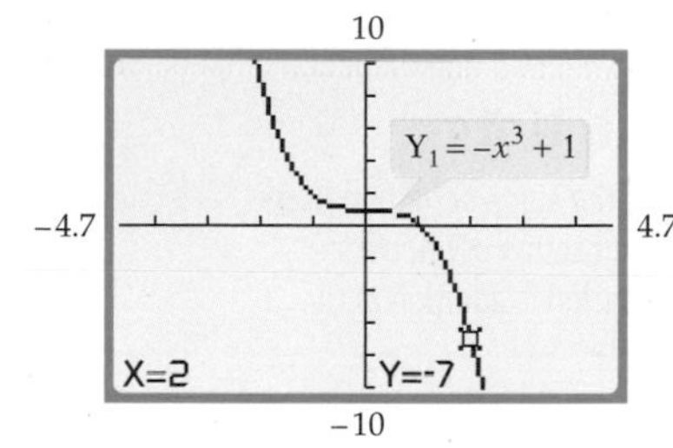

OBJECTIVE 2 Add or subtract polynomials

Polynomials can be added by combining like terms. Either a vertical or a horizontal format can be used.

Focus on adding polynomials using a horizontal or a vertical format

A. Add $(3x^2 + 2x - 7) + (7x^3 - 3 + 4x^2)$. Use a horizontal format.

Use the Commutative and Associative Properties of Addition to rearrange and group like terms.

$$(3x^2 + 2x - 7) + (7x^3 - 3 + 4x^2)$$
$$= 7x^3 + (3x^2 + 4x^2) + 2x + (-7 - 3)$$

Combine like terms.

$$= 7x^3 + 7x^2 + 2x - 10$$

B. Add $(4x^2 - 3x + 2) + (2x^3 + 4x - 7)$. Use a vertical format.

Write each polynomial in descending order with like terms in columns.

$$\begin{array}{r} 4x^2 - 3x + 2 \\ 2x^3 \qquad\quad + 4x - 7 \\ \hline \end{array}$$

Add the terms in each column.

$$2x^3 + 4x^2 + x - 5$$

EXAMPLE 5 Add $(2x^3 + 5x^2 - 7x + 1) + (-x^3 - 5x^2 + 3x - 6)$ using a vertical format.

Solution

$$\begin{array}{r} 2x^3 + 5x^2 - 7x + 1 \\ -x^3 - 5x^2 + 3x - 6 \\ \hline x^3 + 0x^2 - 4x - 5 \end{array}$$

- **Write each polynomial in descending order with like terms in columns.**
- **Add the terms in each column.**

$(2x^3 + 5x^2 - 7x + 1) + (-x^3 - 5x^2 + 3x - 6) = x^3 - 4x - 5$

Problem 5 Add $(x^3 - x + 2) + (x^2 + x - 6)$ using a horizontal format.

Solution See page S16.

Try Exercise 49, page 264.

Take Note

This is the same definition used for subtraction of integers: Subtraction is addition of the opposite.

The additive inverse of the polynomial $x^2 + 5x - 4$ is $-(x^2 + 5x - 4)$.

To simplify the additive inverse of a polynomial, change the sign of every term inside the parentheses.

$$-(x^2 + 5x - 4) = -x^2 - 5x + 4$$

To subtract two polynomials, add the additive inverse of the second polynomial to the first.

Focus on subtracting polynomials using a horizontal or a vertical format

A. Subtract $(3x^2 - 7xy + y^2) - (-4x^2 + 7xy - 3y^2)$. Use a horizontal format.

Rewrite the subtraction as addition of the additive inverse.

$$(3x^2 - 7xy + y^2) - (-4x^2 + 7xy - 3y^2)$$
$$= (3x^2 - 7xy + y^2) + (4x^2 - 7xy + 3y^2)$$

Combine like terms.

$$= 7x^2 - 14xy + 4y^2$$

B. Subtract $(6x^3 - 3x + 7) - (3x^2 - 5x + 12)$. Use a vertical format.

Rewrite the subtraction as addition of the additive inverse.

$$(6x^3 - 3x + 7) - (3x^2 - 5x + 12)$$
$$= (6x^3 - 3x + 7) + (-3x^2 + 5x - 12)$$

Write each polynomial in descending order with like terms in columns. Add the terms in each column.

$$\begin{array}{r} 6x^3 \qquad\quad - 3x + 7 \\ - 3x^2 + 5x - 12 \\ \hline 6x^3 - 3x^2 + 2x - 5 \end{array}$$

EXAMPLE 6 Subtract $(3x^2 - 2x + 4) - (7x^2 + 3x - 12)$ using a vertical format.

Solution $(3x^2 - 2x + 4) - (7x^2 + 3x - 12)$
$(3x^2 - 2x + 4) + (-7x^2 - 3x + 12)$

• Rewrite the subtraction as addition of the additive inverse.

$$\begin{array}{r} 3x^2 - 2x + 4 \\ -7x^2 - 3x + 12 \\ \hline -4x^2 - 5x + 16 \end{array}$$

• Write each polynomial in descending order in a vertical format.
• Combine like terms in each column.

Problem 6 Subtract $(-5x^2 + 2x - 3) - (6x^2 + 3x - 7)$ using a vertical format.

Solution See page S16.

Try Exercise 55, page 264.

Function notation may be used when adding or subtracting polynomials. For instance, if $P(x) = 3x^2 - 2x + 4$ and $R(x) = -5x^3 + 4x + 7$, then

$$\begin{aligned} P(x) + R(x) &= (3x^2 - 2x + 4) + (-5x^3 + 4x + 7) \\ &= -5x^3 + 3x^2 + 2x + 11 \end{aligned}$$

$$\begin{aligned} P(x) - R(x) &= (3x^2 - 2x + 4) - (-5x^3 + 4x + 7) \\ &= (3x^2 - 2x + 4) + (5x^3 - 4x - 7) \\ &= 5x^3 + 3x^2 - 6x - 3 \end{aligned}$$

EXAMPLE 7 Given $P(x) = -3x^2 + 2x - 6$ and $R(x) = 4x^3 - 3x + 4$, find $S(x) = P(x) + R(x)$.

Solution
$$\begin{aligned} S(x) &= P(x) + R(x) \\ &= (-3x^2 + 2x - 6) + (4x^3 - 3x + 4) \\ &= 4x^3 - 3x^2 - x - 2 \end{aligned}$$

Problem 7 Given $P(x) = 4x^3 - 3x^2 + 2$ and $R(x) = -2x^2 + 2x - 3$, find $D(x) = P(x) - R(x)$.

Solution See page S16.

Try Exercise 63, page 264.

5.2 Exercises

CONCEPT CHECK

1. Identify each of the following as a monomial, a binomial, a trinomial, or none of these.

a. $-3x^4 + 1$ **b.** $2x - 7$
c. $3x^2y^5z$ **d.** $1 - 4x - x^2$
e. $5z^4 - 2z^{-2} + 4$ **f.** 7

2. Write each polynomial in descending order.
a. $3x - 7x^2 + 5$ **b.** $3x^4 - 7 - 2x + 4x^2$

3. What is the domain of a polynomial function?

4. Write the additive inverse of $4x^3 - 7x + 8$.

1 Evaluate polynomial functions (See pages 255–259.)

5. Explain how to determine the degree of a polynomial in one variable. Give two examples of a third-degree polynomial.

6. Explain how to evaluate a polynomial function.

GETTING READY

For Exercises 7 and 8, use the polynomial function $P(x) = -3x^2 - 2x + 10$.

7. The degree of P is __?__. The coefficient -3 is called the __?__ coefficient. The term 10 is called the __?__ term.

8. a. To evaluate $P(-2)$, substitute __?__ for __?__, and simplify:
$P(\_\_?\_\_) = -3(\_\_?\_\_)^2 - 2(\_\_?\_\_) + 10$
$= \_\_?\_\_ + \_\_?\_\_ + 10 = \_\_?\_\_$

b. Use the results of part (a) to name a point that is on the graph of $y = P(x)$.

For Exercises 9 to 20, determine whether the function is a polynomial function. For those functions that are polynomial functions, identify **a.** the leading coefficient, **b.** the constant term, and **c.** the degree.

9. $P(x) = -x^2 + 3x + 8$

10. $P(x) = 3x^4 - 3x - 7$

11. $R(x) = \frac{x}{x + 1}$

12. $R(x) = \frac{3x^2 - 2x + 1}{x}$

13. $f(x) = \sqrt{x} - x^2 + 2$

14. $f(x) = x^2 - \sqrt{x + 2} - 8$

15. $g(x) = 3x^5 - 2x^2 + \pi$

16. $g(x) = -4x^5 - 3x^2 + x - \sqrt{7}$

17. $P(x) = 3x^2 - 5x^3 + 2$

18. $P(x) = x^2 - 5x^4 - x^6$

19. $R(x) = 14$

20. $R(x) = \frac{1}{x} + 2$

21. Given $P(x) = 3x^2 - 2x - 8$, evaluate $P(3)$.

22. Given $P(x) = -3x^2 - 5x + 8$, evaluate $P(-5)$.

23. Given $R(x) = 2x^3 - 3x^2 + 4x - 2$, evaluate $R(2)$.

24. Given $R(x) = -x^3 + 2x^2 - 3x + 4$, evaluate $R(-1)$.

25. Given $f(x) = x^4 - 2x^2 - 10$, evaluate $f(-1)$.

26. Given $f(x) = x^5 - 2x^3 + 4x$, evaluate $f(2)$.

Graph.

27. $P(x) = x^2 - 1$

28. $P(x) = 2x^2 + 3$

29. $R(x) = x^3 + 2$

30. $R(x) = x^4 + 1$

31. $f(x) = x^3 - 2x$

32. $f(x) = x^2 - x - 2$

33. **Oceanography** The length of a water wave close to land depends on several factors, one of which is the depth of the water. However, the length L (in meters) of deep-water waves (those far away from land) can be approximated by $L(s) = 0.641s^2$, where s is the speed of the wave in meters per second. Find the length of a deep-water wave that has a speed of 6 m/s. Round to the nearest tenth of a meter.

34. **Physics** The distance s (in feet) that an object on the moon will fall in t seconds is given by $s(t) = 2.735t^2$. How far will an object on the moon fall in 3 s?

35. **Sports** The total number of softball games T that must be scheduled in a league that has n teams such that each team plays every other team twice is given by $T(n) = n^2 - n$. What is the total number of games that must be scheduled for a league that has 8 teams?

36. **Geometry** A diagonal of a polygon is a line from a vertex to a nonadjacent vertex. Seven of the possible diagonals for a decagon (10-sided figure) are shown at the right. The total number T of diagonals for an n-sided polygon is given by $T(n) = \frac{1}{2}n^2 - \frac{3}{2}n$. Find the total number of diagonals for a decagon.

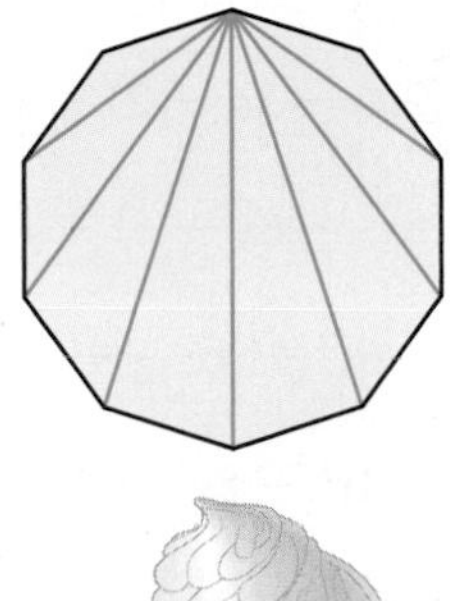

37. **Food Science** Baked Alaska is a dessert that is made by putting a 1-inch meringue coating around a hemisphere of ice cream. The amount of meringue that is needed depends on the radius of the hemisphere of ice cream and is given by $M(r) = 6.14r^2 + 6.14r + 2.094$, where $M(r)$ is the volume of meringue needed (in cubic inches) and r is the radius of the ice cream in inches. Find the amount of meringue that is needed for a Baked Alaska that has a mound of ice cream with a 6-inch radius. Round to the nearest whole number.

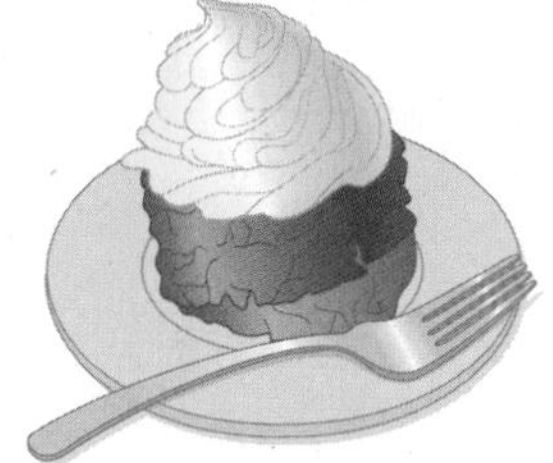

For Exercises 38 and 39, use graphs A, B, and C shown below.

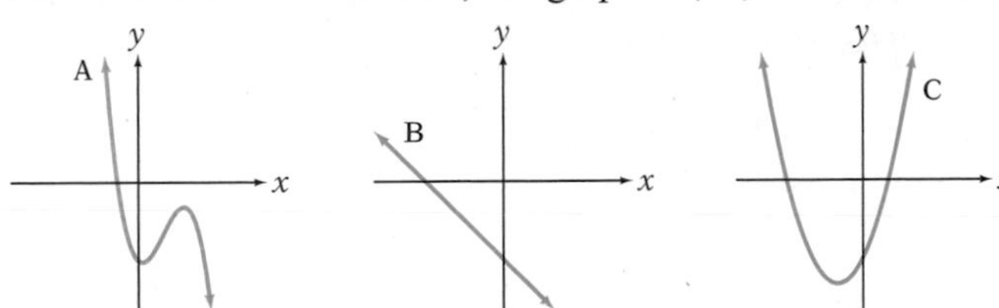

38. Which graph could be the graph of $P(x) = x^2 + 2x - 3$?

39. Which graph could be the graph of $R(x) = x^3 + 3x^2 - x - 3$?

40. Suppose $f(x) = x^2 + 1$ and $g(x) = x^3 + x$. Is $f(x) > g(x)$ or is $f(x) < g(x)$ when $0 < x < 1$? Explain your answer.

2 Add or subtract polynomials (See pages 259–261.)

GETTING READY

41. The additive inverse of $8x^2 + 3x - 5$ is $-(8x^2 + 3x - 5) =$ ___?___.

42. $(4x^2 + 6x - 1) - (3x^2 - 5x + 2) = (4x^2 + 6x - 1) + ($___?___$)$

Simplify. Use a vertical format.

43. $(5x^2 + 2x - 7) + (x^2 - 8x + 12)$

44. $(3x^2 - 2x + 7) + (-3x^2 + 2x - 12)$

45. $(x^2 + 8 - 3x) - (2x^2 - 3x + 7)$

46. $(2x^2 + 3x - 7) - (5x^2 - 1 - 8x)$

47. $(8x^3 - 4x^2 + 2x - 10) - (-9x^3 - 5x^2 + 4x - 4)$

48. $(7x^3 + x^2 - 5x + 4) - (-5x^2 + 2x + 9)$

49. $(-7x^3 + 2x^2 - 3x + 4) + (7x^3 - 10x^2 + 9x + 5)$

50. $(-10x^3 + 9x^2 + x + 3) + (-10x^3 + 3x^2 + 4x + 1)$

51. $(6x^2 + 2x + 3) - (-7x^2 + 10x - 4)$

52. $(9x^2 - 5) + (x^2 + 3x + 10)$

Simplify. Use a horizontal format.

53. $(3y^2 - 7y) + (2y^2 - 8y + 2)$

54. $(-2y^2 - 4y - 12) + (5y^2 - 5y)$

55. $(2a^2 - 3a - 7) - (-5a^2 - 2a - 9)$

56. $(3a^2 - 9a) - (-5a^2 + 7a - 6)$

57. $(8x^2 + 6x + 6) - (2x^2 - 7x + 6)$

58. $(-10x^2 + 7x + 8) - (-3x^2 - 6x - 9)$

59. $(-6x^2 + 4x - 9) + (9x^3 + 2x^2 + 7x + 1)$

60. $(-6x^2 - 8x + 8) + (-10x^3 + 2x^2 - 2x - 10)$

61. $(-3x^3 - 8x^2 + 2x - 2) - (4x - 8)$

62. $(7x^2 - 5x - 3) - (-4x^3 - 3x^2 + 6x - 4)$

63. Given $P(x) = 3x^3 - 4x^2 - x + 1$ and $R(x) = 2x^3 + 5x - 8$, find $P(x) + R(x)$.

64. Given $P(x) = 5x^3 - 3x - 7$ and $R(x) = 2x^3 - 3x^2 + 8$, find $P(x) - R(x)$.

65. Given $P(x) = 3x^4 - 3x^3 - x^2$ and $R(x) = 3x^3 - 7x^2 + 2x$, find $S(x) = P(x) + R(x)$.

66. Given $P(x) = 3x^4 - 2x + 1$ and $R(x) = 3x^5 - 5x - 8$, find $S(x) = P(x) + R(x)$.

67. Given $P(x) = x^2 + 2x + 1$ and $R(x) = 2x^3 - 3x^2 + 2x - 7$, find $D(x) = P(x) - R(x)$.

68. Given $P(x) = 2x^4 - 2x^2 + 1$ and $R(x) = 3x^3 - 2x^2 + 3x + 8$, find $D(x) = P(x) - R(x)$.

69. If $P(x)$ is a third-degree polynomial and $Q(x)$ is a fourth-degree polynomial, what can be said about the degree of $P(x) + Q(x)$? Give some examples of polynomials that support your answer.

70. If $P(x)$ is a fifth-degree polynomial and $Q(x)$ is a fourth-degree polynomial, what can be said about the degree of $P(x) - Q(x)$? Give some examples of polynomials that support your answer.

APPLYING CONCEPTS

Two polynomials are equal if the coefficients of like powers are equal. In Exercises 71 and 72, use this definition of equality of polynomials to find the value of k that makes the equation an identity.

71. $(2x^3 + 3x^2 + kx + 5) - (x^3 + x^2 - 5x - 2) = x^3 + 2x^2 + 3x + 7$

72. $(6x^3 + kx^2 - 2x - 1) - (4x^3 - 3x^2 + 1) = 2x^3 - x^2 - 2x - 2$

73. If $P(-1) = -3$ and $P(x) = 4x^4 - 3x^2 + 6x + c$, find the value of c.

74. If $P(2) = 3$ and $P(x) = 2x^3 - 4x^2 - 2x + c$, find the value of c.

75. Graph $f(x) = x^2$, $g(x) = x^2 - 3$, and $h(x) = x^2 + 4$ on the same coordinate grid. From the graphs, make a conjecture about the shape and location of $k(x) = x^2 - 2$. Test your conjecture by graphing k.

76. Graph $f(x) = x^2$, $g(x) = (x - 3)^2$, and $h(x) = (x + 4)^2$ on the same coordinate grid. From the graphs, make a conjecture about the shape and location of $k(x) = (x - 2)^2$. Test your conjecture by graphing k.

77. **Sports** The deflection D (in inches) of a beam that is uniformly loaded is given by the polynomial function $D(x) = 0.005x^4 - 0.1x^3 + 0.5x^2$, where x is the distance in feet from one end of the beam. See the figure at the right. The maximum deflection occurs when x is the midpoint of the beam. Determine the maximum deflection for the beam in the diagram.

78. **Engineering** Construction of the Golden Gate Bridge, which is a suspension bridge, was completed in 1937. The length of the main span of the bridge is 4200 ft. The height, in feet, of the suspension cables above the roadway varies from 0 ft at the center of the bridge to 525 ft at the towers that support the cables.

Distance from the center of the bridge	Height of the cables above the roadway
0 ft	0 ft
1050 ft	150 ft
2100 ft	525 ft

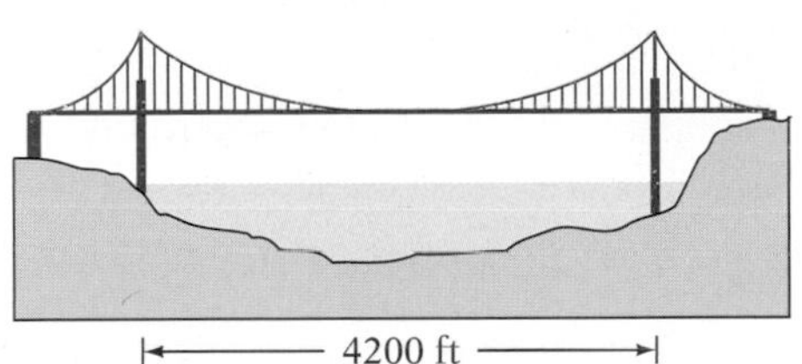

The function that approximately models the data is $f(x) = \frac{1}{8820}x^2 + 25$, where x is the distance from the center of the bridge and $f(x)$ is the height of the cables above the roadway. Use this model to approximate the height of the cables at a distance of **a.** 1000 ft from the center of the bridge and **b.** 1500 ft from the center of the bridge. Round to the nearest tenth.

79. Explain the similarities and differences among the graphs of $f(x) = x^2$, $g(x) = (x - 3)^2$, and $h(x) = x^2 - 3$.

PROJECTS OR GROUP ACTIVITIES

These exercises are a continuation of the Projects or Group Activities exercises from Section 5.1. If you have not completed those exercises, do so now.

80. Suppose n disks are arranged from largest to smallest on one of your three dots. Looking at the pattern of your answers to Exercises 150 to 152 from Section 5.1, write an exponential expression for the number of moves required to transfer the n disks from one dot to another dot.

81. An ancient myth suggests that three shamans were given 64 golden disks arranged from largest to smallest on one of three diamond spires. The shamans were asked to move the golden disks to one of the other spires using the rules stated in Exercise 150 of Section 5.1. If the shamans could make one move each second, how many years would it take them to move all the golden disks to another spire? Write your answer in scientific notation.

82. An estimate for the age of the universe is 1.37×10^{10} years. If the shamans described in Exercise 81 began their task 1.37×10^{10} years ago, approximately what percent of their task is complete?

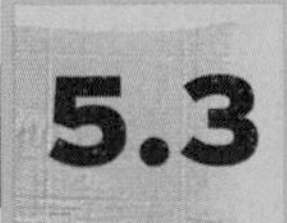

5.3 Multiplication of Polynomials

OBJECTIVE 1 Multiply a polynomial by a monomial

To multiply a polynomial by a monomial, use the Distributive Property and the Rule for Multiplying Exponential Expressions.

EXAMPLE 1 Multiply.
A. $-5x(x^2 - 2x + 3)$ **B.** $x^2 - x[3 - x(x - 2) + 3]$

Solution **A.** $-5x(x^2 - 2x + 3)$

$= -5x(x^2) - (-5x)(2x) + (-5x)(3)$ • Use the Distributive Property.

$= -5x^3 + 10x^2 - 15x$ • Use the Rule for Multiplying Exponential Expressions.

B. $x^2 - x[3 - x(x - 2) + 3]$

$= x^2 - x[3 - x^2 + 2x + 3]$ • Use the Distributive Property to remove the inner grouping symbols.

$= x^2 - x[6 - x^2 + 2x]$ • Combine like terms.

$= x^2 - 6x + x^3 - 2x^2$ • Use the Distributive Property to remove the brackets.

$= x^3 - x^2 - 6x$ • Combine like terms, and write the polynomial in descending order.

Problem 1 Multiply.
A. $-4y(y^2 - 3y + 2)$ **B.** $x^2 - 2x[x - x(4x - 5) + x^2]$

Solution See page S16.

Try Exercise 15, page 271.

OBJECTIVE  2 Multiply two polynomials

The product of two polynomials is the polynomial obtained by multiplying each term of one polynomial by each term of the other polynomial and then combining like terms. The degree of the product is the sum of the degrees of the individual polynomials.

Focus on multiplying polynomials

Multiply: $(3x^2 - 4x - 2)(2x + 3)$

Use the Distributive Property to multiply the trinomial by each term of the binomial.

$$(3x^2 - 4x - 2)(2x + 3) = (3x^2 - 4x - 2)2x + (3x^2 - 4x - 2)3$$
$$= (6x^3 - 8x^2 - 4x) + (9x^2 - 12x - 6)$$
$$= 6x^3 + x^2 - 16x - 6$$

The degree of $3x^2 - 4x - 2$ is 2; the degree of $2x + 3$ is 1; the degree of the product is 3, the sum of 2 and 1.

A convenient method of multiplying two polynomials is to use a vertical format similar to that used for multiplication of whole numbers. Here is this method for the product of the polynomials shown above.

$$\begin{array}{rrrrl} & 3x^2 & - \; 4x & - \; 2 & \\ & & 2x & + \; 3 & \\ \hline & 9x^2 & - \; 12x & - \; 6 & = (3x^2 - 4x - 2)3 \\ 6x^3 & - \; 8x^2 & - \; 4x & & = (3x^2 - 4x - 2)2x \\ \hline 6x^3 & + \; x^2 & - \; 16x & - \; 6 & \end{array}$$

EXAMPLE 2 Multiply: $(4a^3 - 3a + 7)(a - 5)$

Solution

$$\begin{array}{r} 4a^3 - 3a + 7 \\ a - 5 \\ \hline -20a^3 \quad + 15a - 35 \\ 4a^4 \quad - 3a^2 + 7a \quad \\ \hline 4a^4 - 20a^3 - 3a^2 + 22a - 35 \end{array}$$

Problem 2 Multiply: $(-2b^2 + 5b - 4)(-3b + 2)$

Solution See page S16.

Try Exercise 37, page 272.

It is frequently necessary to find the product of two binomials. The product can be found by using a method called **FOIL**, which is based on the Distributive Property. The letters FOIL stand for **F**irst, **O**uter, **I**nner, and **L**ast.

Focus on multiplying binomials using the FOIL method

Multiply: $(3x - 2)(2x + 5)$

Multiply the First terms.	$(3x - 2)(2x + 5)$	$3x \cdot 2x = 6x^2$
Multiply the Outer terms.	$(3x - 2)(2x + 5)$	$3x \cdot 5 = 15x$
Multiply the Inner terms.	$(3x - 2)(2x + 5)$	$-2 \cdot 2x = -4x$
Multiply the Last terms.	$(3x - 2)(2x + 5)$	$-2 \cdot 5 = -10$

F O I L

Add the products. $(3x - 2)(2x + 5) = 6x^2 + 15x - 4x - 10$

Combine like terms. $= 6x^2 + 11x - 10$

FOIL is not really a different way of multiplying. It is based on the Distributive Property. The product $(3x - 2)(2x + 5)$ is shown again below.

$$\begin{aligned} &(3x - 2)(2x + 5) \\ &= 3x(2x + 5) - 2(2x + 5) \\ &= 6x^2 + 15x - 4x - 10 \\ &= 6x^2 + 11x - 10 \end{aligned}$$

FOIL is just an efficient way to multiply binomials.

EXAMPLE 3 Multiply.

A. $(6x - 5)(3x - 4)$ **B.** $(2x + 3y)(4x - 5y)$

C. $(2x - 1)(x + 3)(3x - 4)$

Solution **A.** $(6x - 5)(3x - 4)$

$= 6x(3x) + 6x(-4) + (-5)(3x) + (-5)(-4)$ • **Use FOIL.**

$= 18x^2 - 24x - 15x + 20$ • **Simplify.**

$= 18x^2 - 39x + 20$

B. $(2x + 3y)(4x - 5y)$

$= 2x(4x) + 2x(-5y) + 3y(4x) + 3y(-5y)$ • **Multiply using FOIL.**

$= 8x^2 - 10xy + 12xy - 15y^2$ • **Simplify.**

$= 8x^2 + 2xy - 15y^2$

Take Note

The product $(2x^2 + 5x - 3)(3x - 4)$ could have been found using a vertical format.

$$\begin{array}{r} 2x^2 + 5x - 3 \\ 3x - 4 \\ \hline -8x^2 - 20x + 12 \\ 6x^3 + 15x^2 - 9x \phantom{{}+12} \\ \hline 6x^3 + 7x^2 - 29x + 12 \end{array}$$

C. $(2x - 1)(x + 3)(3x - 4)$

$= [2x(x) + 2x(3) + (-1)x + (-1)3](3x - 4)$ • Use FOIL to multiply $(2x - 1)(x + 3)$.

$= [2x^2 + 6x - x - 3](3x - 4)$ • Simplify.

$= [2x^2 + 5x - 3](3x - 4)$

$= (2x^2 + 5x - 3)3x - (2x^2 + 5x - 3)4$ • Use the Distributive Property to multiply $(2x^2 + 5x - 3)$ by $3x - 4$.

$= (6x^3 + 15x^2 - 9x) - (8x^2 + 20x - 12)$ • Use the Distributive Property again.

$= 6x^3 + 7x^2 - 29x + 12$ • Simplify.

Problem 3 Multiply.

A. $(3x + 4)(5x - 2)$ **B.** $(x - 3y)(2x - 7y)$

C. $(x + 3)(x - 4)(2x - 3)$

Solution See page S16.

Try Exercise 77, page 272.

OBJECTIVE 3 Multiply polynomials that have special products

A formula for the product of the sum and difference of the same terms can be found by using FOIL.

PRODUCT OF THE SUM AND DIFFERENCE OF *a* AND *b*

$$(a + b)(a - b) = a^2 - ab + ab - b^2 = a^2 - b^2$$

Square of a → a^2

Square of b → b^2

EXAMPLES

1. $(x + 4)(x - 4) = x^2 - 4^2 = x^2 - 16$
2. $(3x + 5)(3x - 5) = (3x)^2 - 5^2 = 9x^2 - 25$

We can also find a formula for the square of a binomial.

SQUARE OF A BINOMIAL

$$(a + b)^2 = (a + b)(a + b) = a^2 + ab + ab + b^2$$
$$= a^2 + 2ab + b^2$$

Square of a → a^2

Twice ab → $2ab$

Square of b → b^2

$$(a - b)^2 = (a - b)(a - b) = a^2 - ab - ab + b^2$$
$$= a^2 - 2ab + b^2$$

Square of a → a^2

Twice ab → $2ab$

Square of b → b^2

EXAMPLES

1. $(x + 5)^2 = x^2 + 2(x)(5) + 5^2 = x^2 + 10x + 25$
2. $(2x - 3)^2 = (2x)^2 - 2(2x)(3) + 3^2 = 4x^2 - 12x + 9$

EXAMPLE 4 Simplify.

A. $(3x - 2y)(3x + 2y)$ **B.** $(5x^2 + 6)(5x^2 - 6)$
C. $(5x + 4y)^2$ **D.** $(x^3 - 5)^2$

Solution

A. $(3x - 2y)(3x + 2y) = (3x)^2 - (2y)^2 = 9x^2 - 4y^2$
• **This is the product of the sum and difference of 3*x* and 2*y*.**

B. $(5x^2 + 6)(5x^2 - 6) = (5x^2)^2 - (6)^2 = 25x^4 - 36$
• **This is the product of the sum and difference of $5x^2$ and 6.**

C. $(5x + 4y)^2 = (5x)^2 + 2(5x)(4y) + (4y)^2 = 25x^2 + 40xy + 16y^2$
• **This is the square of a binomial.**

D. $(x^3 - 5)^2 = (x^3)^2 - 2(x^3)(5) + 5^2 = x^6 - 10x^3 + 25$
• **This is the square of a binomial.**

Problem 4 Simplify. **A.** $(4x^2 - y)(4x^2 + y)$ **B.** $(3x^2 + 5y^2)^2$

Solution See page S16.

Try Exercises 113 and 117, page 273.

OBJECTIVE 4 Application problems

EXAMPLE 5 The length of a rectangle is $(2x + 3)$ ft. The width is $(x - 5)$ ft. Find the area of the rectangle in terms of the variable x.

Strategy To find the area, replace the variables L and W in the equation $A = LW$ by the given values, and solve for A.

$x - 5$

$2x + 3$

Solution

$A = LW$
$A = (2x + 3)(x - 5)$
$A = 2x^2 - 10x + 3x - 15$
$A = 2x^2 - 7x - 15$

The area is $(2x^2 - 7x - 15)$ ft^2.

Problem 5 The base of a triangle is $(2x + 6)$ ft. The height is $(x - 4)$ ft. Find the area of the triangle in terms of the variable x.

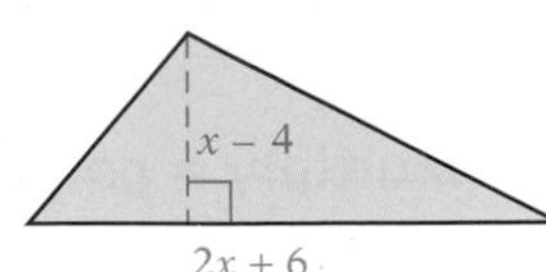

Solution See page S16.

Try Exercise 129, page 274.

EXAMPLE 6 The corners are cut from a rectangular piece of cardboard measuring 8 in. by 12 in. The sides are folded up to make a box. Find the volume of the box in terms of the variable x, where x is the length of the side of the square cut from each corner of the rectangle.

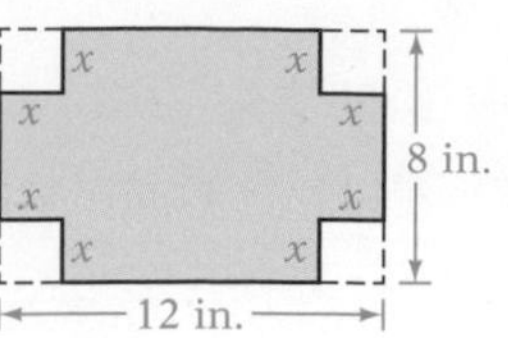

Strategy Length of the box: $12 - 2x$
Width of the box: $8 - 2x$
Height of the box: x
To find the volume, replace the variables L, W, and H in the equation $V = LWH$, and solve for V.

Solution
$$V = LWH$$
$$V = (12 - 2x)(8 - 2x)x$$
$$V = (96 - 24x - 16x + 4x^2)x$$
$$V = (96 - 40x + 4x^2)x$$
$$V = 96x - 40x^2 + 4x^3$$
$$V = 4x^3 - 40x^2 + 96x$$

The volume is $(4x^3 - 40x^2 + 96x)$ in^3.

Problem 6 Find an expression in terms of the variable x for the volume of the rectangular solid shown in the diagram. All dimensions are given in feet.

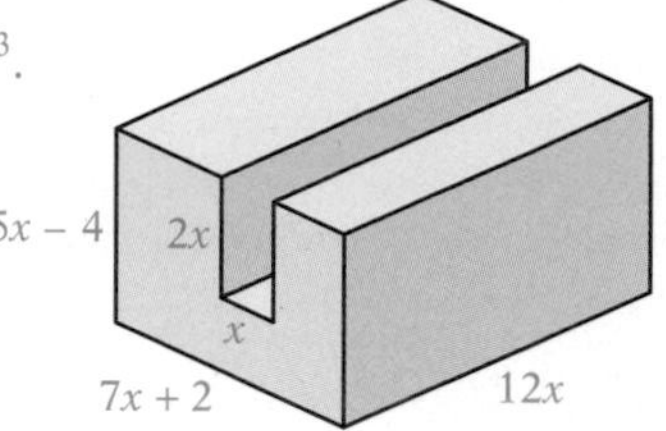

Solution See page S16.

Try Exercise 133, page 274.

5.3 Exercises

CONCEPT CHECK

1. If p is a polynomial of degree 2 and q is a polynomial of degree 3, what is the degree of the product of the two polynomials?

2. What is FOIL?

3. True or false? $(x - y)(x + y) = x^2 - y^2$

4. True or false? $(x + y)^2 = x^2 + y^2$

1 Multiply a polynomial by a monomial (See pages 265–266.)

5. When is the Distributive Property used?

6. Explain how to multiply a monomial and a polynomial by using the Distributive Property.

GETTING READY

7. Multiply: $-3y^3(y^2 - 10)$

$-3y^3(y^2 - 10) = \underline{\quad ? \quad}(y^2) - (\underline{\quad ? \quad})(10)$
- Use the $\underline{\quad ? \quad}$ Property to multiply each term of $(y^2 - 10)$ by $-3y^3$.

$= \underline{\quad ? \quad} + \underline{\quad ? \quad}$
- In the first term, the bases are the same. Add the exponents.

8. Multiply: $a^2b(2a + b)$

$a^2b(2a + b) = (\underline{\quad ? \quad})(2a) + (\underline{\quad ? \quad})(b)$
- Use the Distributive Property to multiply each term of $(2a + b)$ by a^2b.

$= \underline{\quad ? \quad} + \underline{\quad ? \quad}$
- In each term, add the exponents on like bases.

Simplify.

9. $2x(x - 3)$

10. $2a(2a + 4)$

11. $3x^2(2x^2 - x)$

12. $-4y^2(4y - 6y^2)$

13. $3xy(2x - 3y)$

14. $-4ab(5a - 3b)$

15. $2b + 4b(2 - b)$

16. $x - 2x(x - 2)$

17. $3a^2 - 2a(3 - a)$

18. $-2y(3 - y) + 2y^2$

19. $-2a^2(3a^2 - 2a + 3)$

20. $4b(3b^3 - 12b^2 - 6)$

21. $3b(3b^4 - 3b^2 + 8)$

22. $(2x^2 - 3x - 7)(-2x^2)$

23. $-5x^2(4 - 3x + 3x^2 + 4x^3)$

24. $-2y^2(3 - 2y - 3y^2 + 2y^3)$

25. $-2x^2y(x^2 - 3xy + 2y^2)$

26. $3ab^2(3a^2 - 2ab + 4b^2)$

27. $2y^2 - y[3 - 2(y - 4) - y]$

28. $3x^2 - x[x - 2(3x - 4)]$

29. $2y - 3[y - 2y(y - 3) + 4y]$

30. $4a^2 - 2a[3 - a(2 - a + a^2)]$

For Exercises 31–34, state whether the given expression is equivalent to $2x - x(5x - 3)$.

31. $2x - 5x^2 - 3$

32. $x(5x - 3)$

33. $5x^2 - 3x$

34. $5x^2 + 5x$

2 Multiply two polynomials (See pages 266–268.)

GETTING READY

35. For the product $(3x - 2)(x + 5)$:

The First terms are $\underline{\quad ? \quad}$ and $\underline{\quad ? \quad}$.
The Outer terms are $\underline{\quad ? \quad}$ and $\underline{\quad ? \quad}$.
The Inner terms are $\underline{\quad ? \quad}$ and $\underline{\quad ? \quad}$.
The Last terms are $\underline{\quad ? \quad}$ and $\underline{\quad ? \quad}$.

36. Use FOIL to multiply $(y + 4)(2y - 7)$:
The product of the First terms is $y \cdot 2y =$ ___?___.
The product of the Outer terms is $y(-7) =$ ___?___.
The product of the Inner terms is $4 \cdot 2y =$ ___?___.
The product of the Last terms is $4(-7) =$ ___?___.
The sum of these four products is ___?___.

Multiply.

37. $(8x^2 + x + 1)(3x - 8)$

38. $(-x^2 - 2x + 2)(6x - 6)$

39. $(-6x^2 + x + 4)(x - 4)$

40. $(-2x^2 - 4x - 8)(8x - 6)$

41. $(-4x^2 + 4x + 2)(3x - 3)$

42. $(4x^2 + 3x + 1)(6x - 6)$

43. $(5x^3 + x^2 - 8x + 6)(-7x - 3)$

44. $(-3x^3 - 6x^2 + 3x - 5)(x + 3)$

45. $(4x^3 - 4x^2 - 5x + 5)(x + 8)$

46. $(-x^3 - 5x^2 + 7x + 8)(7x - 8)$

47. $(4x^3 + 5x - 4)(x - 1)$

48. $(-3x^3 + 3x + 5)(3x - 8)$

49. $(-x^3 + 3x^2 + 4)(5x + 8)$

50. $(-3x^3 - 4x^2 - 5)(-3x - 2)$

51. $(3x^2 + 5x - 3)(8x^2 + 5x + 6)$

52. $(4x^2 - 5x - 4)(-7x^2 + 7x + 5)$

53. $(-2x^2 - 4x + 3)(x^2 + 4x - 5)$

54. $(-3x^2 + 5x - 4)(-4x^2 - 4x + 7)$

55. $(2x + 1)(x - 3)(x + 4)$

56. $(x + 4)(x - 5)(2x + 3)$

57. $(x - 3)(3x + 1)(x + 3)$

58. $(2x + 5)(x - 1)(3x + 2)$

59. $(2x - 7)(2x - 1)(x + 4)$

60. $(3x + 4)(x - 3)(4x + 3)$

Multiply using FOIL.

61. $(x + 13)(x + 15)$

62. $(x + 2)(x - 10)$

63. $(x - 13)(x - 8)$

64. $(x + 10)(x + 4)$

65. $(x - 4)(x + 12)$

66. $(x - 14)(x - 11)$

67. $(x + 9)(x + 3)$

68. $(x - 12)(x + 4)$

69. $(x + 8)(x - 4)$

70. $(x + 3)(x + 2)$

71. $(4x + 4)(3x + 3)$

72. $(-2x - 2)(-6x - 7)$

73. $(3x + 3)(-3x - 4)$

74. $(6x + 8)(x + 1)$

75. $(x + 6)(-5x - 1)$

76. $(5x - 2)(8x + 8)$

77. $(3x - 8)(3x + 3)$

78. $(-7x - 1)(7x + 3)$

79. $(-x - 2)(-2x - 1)$

80. $(x - 6)(7x + 7)$

81. $(-8x - 2)(3x + 5)$

82. $(7x + 8)(3x + 3)$

83. $(x + 2y)(x - 4y)$

84. $(2a - b)(a + b)$

85. $(3a + b)(a - 3b)$

86. $(5x - 2y)(2x - 3y)$

87. $(7x - y)(x + 5y)$

88. $(2a + 9b)(3a - 7b)$

89. $(xy + 4)(xy - 5)$

90. $(3xy - 2)(2xy + 3)$

91. $(ab + 4)(3ab - 8)$

92. $(x^2 + 3)(x^2 + 6)$

93. $(5a^2 - 6)(2a^2 + 3)$

94. $(2x^3 - 5)(3x^3 + 2)$

95. If the constant terms of two binomials are both negative, will the constant term of the product of the two binomials also be negative?

96. Suppose a and b are positive numbers such that $a < b$. Will the coefficient of the x term of the product $(ax - b)(x + 1)$ be positive or negative?

3 Multiply polynomials that have special products (See pages 268–269.)

GETTING READY

97. The product $(a + b)^2 = a^2 + 2ab + b^2$, so
$(4x + 1)^2 = (\underline{\ ?\ })^2 + 2(\underline{\ ?\ })(\underline{\ ?\ }) + (\underline{\ ?\ })^2$
$= \underline{\ ?\ } + \underline{\ ?\ } + \underline{\ ?\ }$.

98. The product $(a + b)(a - b) = a^2 - b^2$, so
$(2x + 3)(2x - 3) = (\underline{\ ?\ })^2 - (\underline{\ ?\ })^2 = \underline{\ ?\ } - \underline{\ ?\ }$.

Simplify.

99. $(b - 7)(b + 7)$
100. $(a - 4)(a + 4)$
101. $(b - 11)(b + 11)$
102. $(3x - 2)(3x + 2)$
103. $(5x - 4y)^2$
104. $(3a + 5b)^2$
105. $(x^2 + y^2)^2$
106. $(x^2 - 3)^2$
107. $(2a - 3b)(2a + 3b)$
108. $(10 + b)(10 - b)$
109. $(x^2 + 1)(x^2 - 1)$
110. $(5x - 7y)(5x + 7y)$
111. $(5a - 9b)(5a + 9b)$
112. $(x^2 + y^2)(x^2 - y^2)$
113. $(3x + 7y)(3x - 7y)$
114. $(4y + 1)(4y - 1)$
115. $(6 - x)(6 + x)$
116. $(2x^2 - 3y^2)^2$
117. $(3a - 4b)^2$
118. $(2x^2 + 5)^2$
119. $(3x^3 + 2)^2$

For Exercises 120 to 123, state whether the coefficient of the x term of the product is positive, negative, or zero.

120. $(ax + b)^2$, where $a > 0$ and $b > 0$
121. $(ax + b)(ax - b)$, where $a > 0$ and $b > 0$
122. $(ax + b)(ax + b)$, where $a > 0$ and $b < 0$
123. $(ax + b)^2$, where $a < 0$ and $b < 0$

4 Application problems (See pages 269–270.)

GETTING READY

124. Use the figure at the right, which shows a rectangular piece of cardboard with the corners cut out. The sides can be folded up to make a box. State each dimension of the box in terms of x.

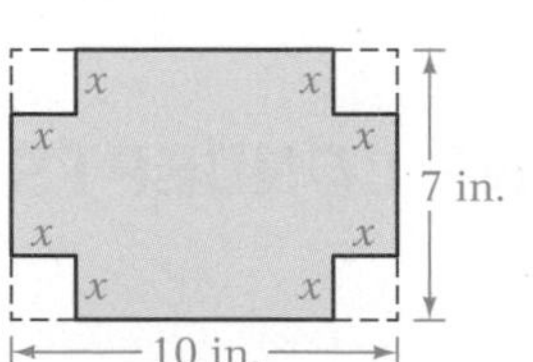

a. Length of the box = $\underline{\ ?\ }$
b. Width of the box = $\underline{\ ?\ }$
c. Height of the box = $\underline{\ ?\ }$

125. a. Use the figure at the right. We can use the properties of rectangles to show that both unlabeled sides of the figure have length x. The unlabeled vertical side has length $4x - \underline{\ ?\ } = x$. The unlabeled horizontal side has length $x + 3 - \underline{\ ?\ } = x$.

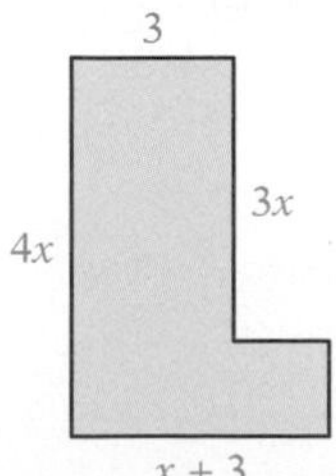

b. Divide the figure into two rectangles by extending the unlabeled horizontal side. The area of the upper rectangle is $3x(\underline{\ ?\ }) = \underline{\ ?\ }$. The area of the lower rectangle is $(x + 3)(\underline{\ ?\ }) = \underline{\ ?\ }$. The area of the whole figure is $\underline{\ ?\ } + \underline{\ ?\ } = \underline{\ ?\ }$.

126. **Geometry** The length of a rectangle is $(3x + 3)$ ft. The width is $(x - 4)$ ft. Find the area of the rectangle in terms of the variable x.

127. **Geometry** The base of a triangle is $(x + 2)$ ft. The height is $(2x - 3)$ ft. Find the area of the triangle in terms of the variable x.

128. **Geometry** Find an expression in terms of x for the area of the figure shown below. All dimensions given are in meters.

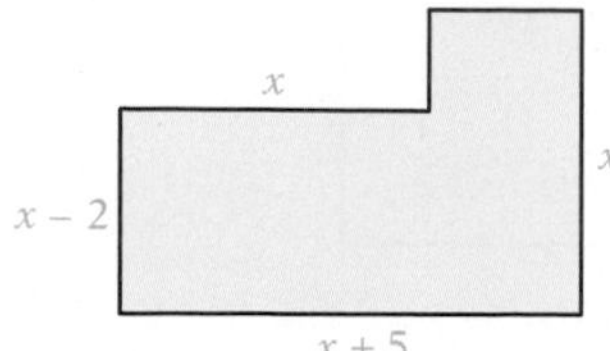

129. **Geometry** Find an expression in terms of x for the area of the figure shown below. All dimensions given are in feet.

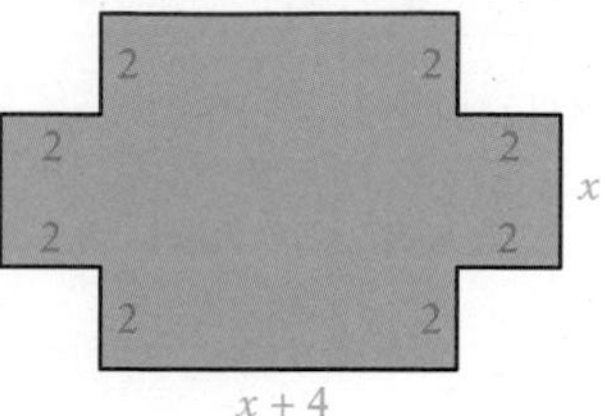

130. **Construction** A trough is made from a rectangular piece of metal by folding up the sides as shown in the figure below. What is the volume of the trough in terms of x, the length of a side folded up?

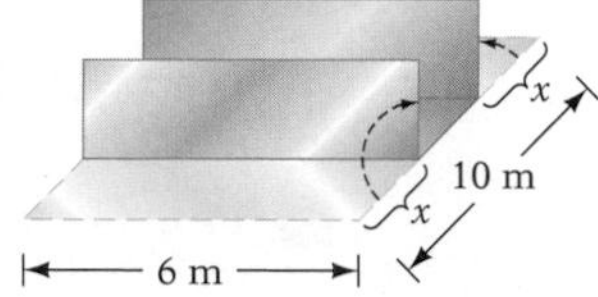

131. **Construction** A square sheet of cardboard measuring 18 in. by 18 in. is used to make an open box by cutting squares of equal size from the corners and folding up the sides. Find the volume of the box in terms of the variable x, where x is the length of the side of a square cut from the cardboard.

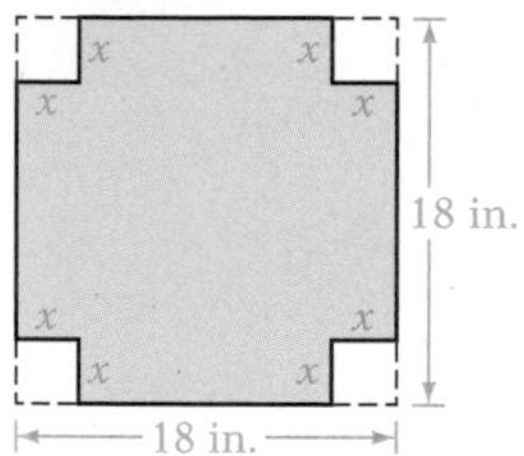

132. **Geometry** The length of the side of a cube is $(x - 2)$ cm. Find the volume of the cube in terms of the variable x.

133. **Geometry** The length of a box is $(2x + 3)$ cm, the width is $(x - 5)$ cm, and the height is x cm. Find the volume of the box in terms of the variable x.

134. **Geometry** Find an expression in terms of x for the volume of the figure shown below. All dimensions given are in inches.

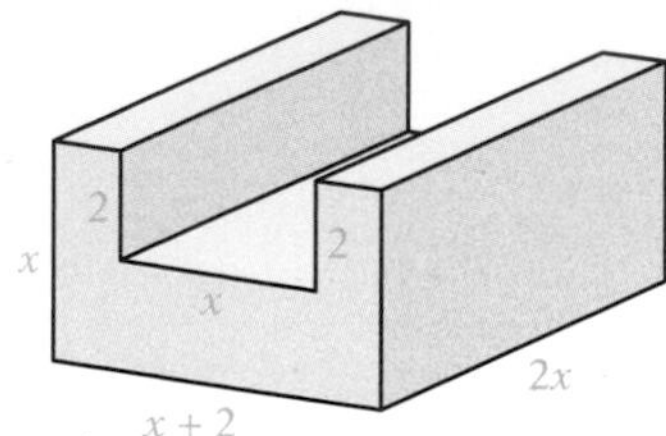

135. **Geometry** Find an expression in terms of x for the volume of the figure shown below. All dimensions given are in centimeters.

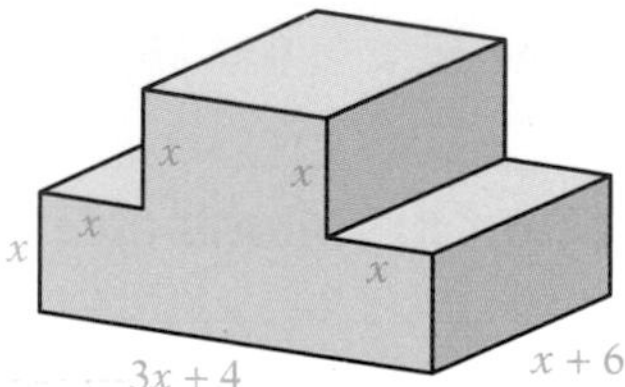

APPLYING CONCEPTS

Simplify.

136. $(a - b)^2 - (a + b)^2$

137. $(x + 2y)^2 + (x + 2y)(x - 2y)$

138. $(4y + 3)^2 - (3y + 4)^2$

139. $2x^2(3x^3 + 4x - 1) - 5x^2(x^2 - 3)$

140. $(2b + 3)(b - 4) + (3 + b)(3 - 2b)$

141. $(3x - 2y)^2 - (2x - 3y)^2$

Multiply.

142. $(x^n + 3)(x^n + 6)$

143. $(2x^n - 3)(x^n + 4)$

144. $(3x^n + 5)(2x^n - 6)$

145. $(x^n + 7)(x^n - 7)$

146. $(2x^n + 5)(2x^n - 5)$

147. $(x^n - y^n)(x^n + y^n)$

148. $(x^n - 4)^2$ **149.** $(3x^n + 5)^2$ **150.** $(x^n + y^n)^2$

For what value of k is the given equation an identity?

151. $(5x - k)(3x + k) = 15x^2 + 4x - k^2$

152. $(kx - 7)(kx + 2) = k^2x^2 + 5x - 14$

153. What polynomial, when divided by $2x - 3$, has a quotient of $x + 7$?

154. What polynomial, when divided by $x - 4$, has a quotient of $2x + 3$?

PROJECTS OR GROUP ACTIVITIES

155. Write two polynomials whose product is a polynomial of degree 3.

156. Write two polynomials whose product is a polynomial of degree 4.

157. What are the possibilities for the degrees of two polynomials whose product is a polynomial of degree 5?

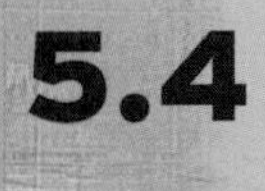

5.4 Division of Polynomials

OBJECTIVE 1 Divide a polynomial by a monomial

Note that $\frac{9x + 6x}{3}$ can be simplified by first adding the terms in the numerator and then dividing by 3.

$$\frac{9x + 6x}{3} = \frac{15x}{3} = 5x$$

The expression $\frac{9x + 6x}{3}$ can also be simplified by first dividing each term in the numerator by the denominator and then adding the results.

$$\frac{9x + 6x}{3} = \frac{9x}{3} + \frac{6x}{3} = 3x + 2x = 5x$$

To divide a polynomial by a monomial, divide each term of the polynomial by the monomial.

Focus on dividing a polynomial by a monomial

Divide: $\frac{8x^3 - 4x^2 + 6x}{2x}$

Divide each term of the polynomial $8x^3 - 4x^2 + 6x$ by the monomial $2x$.

$$\frac{8x^3 - 4x^2 + 6x}{2x} = \frac{8x^3}{2x} - \frac{4x^2}{2x} + \frac{6x}{2x}$$

Simplify.

$$= 4x^2 - 2x + 3$$

EXAMPLE 1 Divide: $\dfrac{6p^4 - 9p^3 - 12p^2}{3p^2}$

Solution $\dfrac{6p^4 - 9p^3 - 12p^2}{3p^2} = \dfrac{6p^4}{3p^2} - \dfrac{9p^3}{3p^2} - \dfrac{12p^2}{3p^2}$ • **Divide each term of the polynomial by $3p^2$.**

$= 2p^2 - 3p - 4$ • **Simplify.**

Problem 1 Divide: $\dfrac{15y^3 + 10y^2 - 25y}{-5y}$

Solution See page S17.

Try Exercise 11, page 281.

EXAMPLE 2 Divide: $\dfrac{18a^2b + 27a^2b^2 - 9ab^2}{9a^2b}$

Solution $\dfrac{18a^2b + 27a^2b^2 - 9ab^2}{9a^2b}$

$= \dfrac{18a^2b}{9a^2b} + \dfrac{27a^2b^2}{9a^2b} - \dfrac{9ab^2}{9a^2b}$ • **Divide each term of the polynomial by $9a^2b$.**

$= 2 + 3b - \dfrac{b}{a}$ • **Simplify.**

Problem 2 Divide: $\dfrac{12x^2y^2 - 16xy^2 - 8x}{4x^2y}$

Solution See page S17.

Try Exercise 21, page 281.

OBJECTIVE 2 Divide polynomials

The division method illustrated in Objective 1 is appropriate only when the divisor is a monomial. To divide two polynomials when the divisor is not a monomial, use a method similar to that used for division of whole numbers.

To check division of polynomials, use

(Quotient × divisor) + remainder = dividend

Focus on dividing two polynomials

A. Divide: $(x^2 + 5x - 7) \div (x + 3)$

Step 1

$$\begin{array}{r} x \phantom{{}+5x-7} \\ x+3\overline{)x^2 + 5x - 7} \\ \underline{x^2 + 3x} \phantom{{}-7} \\ 2x - 7 \end{array}$$

Think: $x\overline{)x^2} = \dfrac{x^2}{x} = x$

Multiply: $x(x + 3) = x^2 + 3x$

Subtract: $(x^2 + 5x) - (x^2 + 3x) = 2x$

Bring down -7.

Step 2

$$\begin{array}{r} x + 2 \\ x + 3\overline{)x^2 + 5x - 7} \\ \underline{x^2 + 3x} \\ 2x - 7 \\ \underline{2x + 6} \\ -13 \end{array}$$

Think: $x\overline{)2x} = \frac{2x}{x} = 2$

Multiply: $2(x + 3) = 2x + 6$

Subtract: $(2x - 7) - (2x + 6) = -13$

The remainder is -13.

Check: (Quotient × divisor) + remainder = dividend

$$(x + 2)(x + 3) + (-13) = x^2 + 3x + 2x + 6 - 13 = x^2 + 5x - 7$$

$$(x^2 + 5x - 7) \div (x + 3) = x + 2 - \frac{13}{x + 3}$$

B. Divide: $\frac{6 - 6x^2 + 4x^3}{2x + 3}$

Arrange the terms of each polynomial in descending order. Note that there is no x term in $4x^3 - 6x^2 + 6$. Insert a zero as $0x$ for the missing term so that like terms will be in the same columns.

$$\begin{array}{r} 2x^2 - 6x + 9 \\ 2x + 3\overline{)4x^3 - 6x^2 + 0x + 6} \\ \underline{4x^3 + 6x^2} \\ -12x^2 + 0x \\ \underline{-12x^2 - 18x} \\ 18x + 6 \\ \underline{18x + 27} \\ -21 \end{array}$$

$$\frac{4x^3 - 6x^2 + 6}{2x + 3} = 2x^2 - 6x + 9 - \frac{21}{2x + 3}$$

EXAMPLE 3 Divide. **A.** $\frac{12x^2 - 11x + 10}{4x - 5}$ **B.** $\frac{x^3 + 1}{x + 1}$

Solution **A.**

$$\begin{array}{r} 3x + 1 \\ 4x - 5\overline{)12x^2 - 11x + 10} \\ \underline{12x^2 - 15x} \\ 4x + 10 \\ \underline{4x - 5} \\ 15 \end{array}$$

$$\frac{12x^2 - 11x + 10}{4x - 5} = 3x + 1 + \frac{15}{4x - 5}$$

B.

$$\begin{array}{r} x^2 - x + 1 \\ x + 1\overline{)x^3 + 0x^2 + 0x + 1} \\ \underline{x^3 + x^2} \\ -x^2 + 0x \\ \underline{-x^2 - x} \\ x + 1 \\ \underline{x + 1} \\ 0 \end{array}$$

$$\frac{x^3 + 1}{x + 1} = x^2 - x + 1$$

Problem 3 Divide. **A.** $\frac{15x^2 + 17x - 20}{3x + 4}$ **B.** $\frac{3x^3 + 8x^2 - 6x + 2}{3x - 1}$

Solution See page S17.

Try Exercise 25, page 282.

OBJECTIVE 3 Synthetic division

Synthetic division is a shorter method of dividing a polynomial by a binomial of the form $x - a$. This method of dividing uses only the coefficients of the variable terms and the constant term.

Both long division and synthetic division are used below to simplify the expression $(3x^2 - 10x + 7) \div (x - 2)$.

LONG DIVISION

Compare the coefficients in this problem worked by long division with the coefficients in the same problem worked by synthetic division.

$$\begin{array}{r} 3x - 4 \\ x - 2\overline{)3x^2 - 10x + 7} \\ \underline{3x^2 - 6x} \\ -4x + 7 \\ \underline{-4x + 8} \\ -1 \end{array}$$

$$(3x^2 - 10x + 7) \div (x - 2) = 3x - 4 - \frac{1}{x - 2}$$

SYNTHETIC DIVISION

Identify the value of a.

$x - a = x - 2$, so $a = 2$.
Bring down the 3.

Value of a — Coefficients of the dividend

$$\begin{array}{r|rrr} 2 & 3 & -10 & 7 \\ & \downarrow & & \\ \hline & 3 & & \end{array}$$

Multiply $2 \cdot 3$ and add the product (6) to -10. The result is -4.

$$\begin{array}{r|rrr} 2 & 3 & -10 & 7 \\ & & 6 & \\ \hline & 3 & -4 & \end{array}$$

Multiply $2(-4)$ and add the product (-8) to 7. The result is -1.

$$\begin{array}{r|rrr} 2 & 3 & -10 & 7 \\ & & 6 & -8 \\ \hline & 3 & -4 & -1 \end{array}$$

Coefficients of the quotient — Remainder

Use the coefficients of the quotient and the remainder to write the result of the division. The degree of the first term of the quotient is one degree less than the degree of the first term of the dividend.

$$(3x^2 - 10x + 7) \div (x - 2)$$
$$= 3x - 4 - \frac{1}{x - 2}$$

Check: $(3x - 4)(x - 2) - 1 = 3x^2 - 6x - 4x + 8 - 1 = 3x^2 - 10x + 7$

Focus on dividing polynomials using synthetic division

Divide: $(2x^3 + 3x^2 - 4x + 8) \div (x + 3)$

Identify the value of a. $x - a = x + 3 = x - (-3)$, so $a = -3$.

Write down the value of a and the coefficients of the dividend.

Bring down the 2. Multiply $-3 \cdot 2$ and add the product (-6) to 3. Continue until all the coefficients have been used.

$$\begin{array}{r|rrrr} -3 & 2 & 3 & -4 & 8 \\ & & -6 & 9 & -15 \\ \hline & 2 & -3 & 5 & -7 \end{array}$$

Coefficients of the quotient: 2, −3, 5 Remainder: −7

Write the quotient. The degree of the quotient is one degree less than the degree of the dividend.

$$(2x^3 + 3x^2 - 4x + 8) \div (x + 3) = 2x^2 - 3x + 5 - \frac{7}{x + 3}$$

EXAMPLE 4 Divide.

A. $(5x^2 - 3x + 7) \div (x - 1)$

B. $(3x^4 - 8x^2 + 2x + 1) \div (x + 2)$

Solution

A.

$$\begin{array}{r|rrr} 1 & 5 & -3 & 7 \\ & & 5 & 2 \\ \hline & 5 & 2 & 9 \end{array}$$

• $x - a = x - 1$; $a = 1$

$$(5x^2 - 3x + 7) \div (x - 1) = 5x + 2 + \frac{9}{x - 1}$$

B.

$$\begin{array}{r|rrrrr} -2 & 3 & 0 & -8 & 2 & 1 \\ & & -6 & 12 & -8 & 12 \\ \hline & 3 & -6 & 4 & -6 & 13 \end{array}$$

• Insert a zero for the missing x^3 term.

• $x - a = x + 2$; $a = -2$

$$(3x^4 - 8x^2 + 2x + 1) \div (x + 2) = 3x^3 - 6x^2 + 4x - 6 + \frac{13}{x + 2}$$

Problem 4 Divide.

A. $(6x^2 + 8x - 5) \div (x + 2)$

B. $(2x^4 - 3x^3 - 8x^2 - 2) \div (x - 3)$

Solution See page S17.

Try Exercise 67, page 283.

OBJECTIVE 4 Evaluate a polynomial using synthetic division

A polynomial can be evaluated by using synthetic division. Consider the polynomial $P(x) = 2x^4 - 3x^3 + 4x^2 - 5x + 1$. One way to evaluate the polynomial when $x = 2$ is to replace x by 2 and then simplify the numerical expression.

$$P(x) = 2x^4 - 3x^3 + 4x^2 - 5x + 1$$
$$P(2) = 2(2)^4 - 3(2)^3 + 4(2)^2 - 5(2) + 1$$
$$P(2) = 2(16) - 3(8) + 4(4) - 5(2) + 1$$
$$P(2) = 32 - 24 + 16 - 10 + 1 = 15$$

Now use synthetic division to divide $(2x^4 - 3x^3 + 4x^2 - 5x + 1) \div (x - 2)$.

$$\begin{array}{r|rrrrr} 2 & 2 & -3 & 4 & -5 & 1 \\ & & 4 & 2 & 12 & 14 \\ \hline & 2 & 1 & 6 & 7 & 15 \end{array}$$

Coefficients of the quotient (2, 1, 6, 7); Remainder (15)

Note that the remainder is 15, which is the same value as $P(2)$. This is not a coincidence. The following theorem states that this situation is always true.

REMAINDER THEOREM

If the polynomial $P(x)$ is divided by $x - a$, the remainder is $P(a)$.

Focus on evaluating a polynomial using the Remainder Theorem

Evaluate $P(x) = x^4 - 3x^2 + 4x - 5$ when $x = -2$ by using the Remainder Theorem.

The value at which the polynomial is evaluated: -2

A 0 is inserted for the x^3 term.

$$\begin{array}{r|rrrrr} -2 & 1 & 0 & -3 & 4 & -5 \\ & & -2 & 4 & -2 & -4 \\ \hline & 1 & -2 & 1 & 2 & -9 \end{array}$$

$-9 \leftarrow$ The remainder is $P(-2)$.

$P(-2) = -9$

EXAMPLE 5 Use the Remainder Theorem to evaluate $P(-2)$ when $P(x) = x^3 - 3x^2 + x + 3$.

Solution Use synthetic division with $a = -2$.

$$\begin{array}{r|rrrr} -2 & 1 & -3 & 1 & 3 \\ & & -2 & 10 & -22 \\ \hline & 1 & -5 & 11 & -19 \end{array}$$

By the Remainder Theorem, $P(-2) = -19$.

Problem 5 Use the Remainder Theorem to evaluate $P(3)$ when $P(x) = 2x^3 - 4x - 5$.

Solution See page S17.

Try Exercise 87, page 284.

5.4 Exercises

CONCEPT CHECK

1. Two polynomials are divided. The degree of the dividend is 5, and the degree of the divisor is 1. What is the degree of the quotient?

2. Suppose r is the remainder when two polynomials are divided. Is the degree of r less than, equal to, or greater than the degree of the divisor?

3. Suppose you are going to divide a polynomial $P(x)$ by $x + 5$. What is the value of a that you will write to the left of the vertical line?

4. What is the Remainder Theorem?

1 Divide a polynomial by a monomial (See pages 275–276.)

GETTING READY

5. Replace each question mark to make a true statement.

$$\frac{9y^2 + 6y}{3y} = \frac{9y^2}{?} + \frac{6y}{?}$$
$$= \underline{\quad ? \quad} + \underline{\quad ? \quad}$$

6. Replace each question mark to make a true statement.

$$\frac{8x^3 - 12x^2}{4x^2} = \frac{8x^3}{?} - \frac{12x^2}{?}$$
$$= \underline{\quad ? \quad} - \underline{\quad ? \quad}$$

7. Every division equation has a related multiplication equation. For instance, $\frac{12}{3} = 4$ means that $12 = 3 \cdot 4$. What is the related multiplication equation for $\frac{12x^2 + 6x}{3x} = 4x + 2$?

8. How can multiplication be used to check that $\frac{6x^3 - 12x^2 - 2x}{2x} = 3x^2 - 6x - 1$?

Divide.

9. $\frac{4a - 8}{4}$

10. $\frac{8x - 12y}{2}$

11. $\frac{6w^2 + 4w}{2w}$

12. $\frac{8z^3 - 6z^2}{-2z}$

13. $\frac{3t^3 - 9t^2 + 12t}{3t}$

14. $\frac{10a^3 - 20a^2 + 15a}{5a}$

15. $\frac{-2xy^2 + 4x^2y}{2xy}$

16. $\frac{8a^3b^2 - 12a^2b^3}{4a^2b}$

17. $\frac{8v^3 - 6v^2 + 12v}{4v}$

18. $\frac{12x^4 - 9x^3 + 8x^2}{6x^2}$

19. $\frac{16x^3 - 24x^2 + 48x}{-8x^2}$

20. $\frac{12y^2 - 9y + 6}{-3y}$

21. $\frac{12x^2y^2 - 16x^2y + 20xy^2}{4x^2y}$

22. $\frac{6uv^3 + 9u^2v^2 - 3u^3v}{3uv^2}$

2 Divide polynomials (See pages 276–277.)

23. Explain how you can check the result of dividing two polynomials.

24. Explain how the degree of the quotient of two polynomials is related to the degrees of the dividend and the divisor.

Divide by using long division.

25. $(x^2 + 3x - 40) \div (x - 5)$

26. $(x^2 - 14x + 24) \div (x - 2)$

27. $(6x^2 + 5x - 6) \div (3x - 2)$

28. $(12x^2 + 13x - 14) \div (3x - 2)$

29. $(10x^2 + 9x - 5) \div (2x - 1)$

30. $(18x^2 - 3x + 2) \div (3x + 2)$

31. $(2x^2 + 3x - 5) \div (x + 2)$

32. $(6x^2 - 7x - 8) \div (2x - 3)$

33. $(x^3 - x^2 - 10x - 8) \div (x + 2)$

34. $(x^3 - 2x^2 - 3x - 20) \div (x - 4)$

35. $(2x^3 - 3x^2 - 24x - 10) \div (2x + 5)$

36. $(12 - 5x^2 + 6x^3 - 13x) \div (3x - 4)$

37. $(3x^2 + 3x + x^3 - 5) \div (x + 4)$

38. $(4x^3 + 2x^2 - 16x - 12) \div (3 + 2x)$

39. $(x^3 - 3x^2 + 2) \div (x - 3)$

40. $(x^3 + 4x^2 - 8) \div (x + 4)$

41. $(x^3 - 7x - 5) \div (2 + x)$

42. $(2x^3 + x - 2) \div (x - 3)$

43. $(2x^3 + 5x^2 - 6) \div (x + 3)$

44. $(3x^3 - 15x^2 + 8) \div (x - 3)$

45. $(8x^3 - 9) \div (2x - 3)$

46. $(64x^3 + 4) \div (4x + 2)$

47. $\dfrac{3x^3 - 8x^2 - 33x - 10}{3x + 1}$

48. $\dfrac{8x^3 - 38x^2 + 49x - 10}{4x - 1}$

49. $\dfrac{4 - 7x + 5x^2 - x^3}{x - 3}$

50. $\dfrac{4 + 6x - 3x^2 + 2x^3}{2x + 1}$

51. $\dfrac{16x^2 - 13x^3 + 2x^4 - 9x + 20}{x - 5}$

52. $\dfrac{x + 3x^4 - x^2 + 5x^3 - 2}{x + 2}$

53. $\dfrac{6x^3 + 2x^2 + x + 4}{2x^2 - 3}$

54. $\dfrac{9x^3 + 6x^2 + 2x + 1}{3x^2 + 2}$

55. True or false? When a tenth-degree polynomial is divided by a second-degree polynomial, the quotient is a fifth-degree polynomial.

56. True or false? When a polynomial of degree $3n$ is divided by a polynomial of degree n, the degree of the quotient polynomial is $2n$.

3 Synthetic division (See pages 278–279.)

GETTING READY

57. Synthetic division is a shorter method of dividing a polynomial by a __?__ of the form __?__.

58. Use this display of the synthetic division of two polynomials in x:

$$\begin{array}{r|rrr} -2 & 4 & -1 & 3 \\ & & -8 & 18 \\ \hline & 4 & -9 & 21 \end{array}$$

a. The dividend is the polynomial ___?___.
b. The divisor is $x -$ ___?___ = ___?___.
c. The "−8" comes from multiplying ___?___ and ___?___.
d. The "−9" comes from adding ___?___ and ___?___.
e. The quotient is the polynomial ___?___ and the remainder ___?___.

Divide by using synthetic division.

59. $(2x^2 - 6x - 8) \div (x + 1)$

60. $(3x^2 + 19x + 20) \div (x + 5)$

61. $(x^2 + 5x - 9) \div (x + 4)$

62. $(x^2 - 7x - 4) \div (x - 3)$

63. $(3x^2 - 4) \div (x - 1)$

64. $(4x^2 - 8) \div (x - 2)$

65. $(2x^2 + 24) \div (2x + 4)$

66. $(3x^2 - 15) \div (x + 3)$

67. $(2x^3 - x^2 + 6x + 9) \div (x + 1)$

68. $(3x^3 + 10x^2 + 6x - 4) \div (x + 2)$

69. $(6x - 3x^2 + x^3 - 9) \div (x + 2)$

70. $(5 - 5x + 4x^2 + x^3) \div (x - 3)$

71. $(x^3 + x - 2) \div (x + 1)$

72. $(x^3 + 2x + 5) \div (x - 2)$

73. $\dfrac{3x^4 + 3x^3 - x^2 + 3x + 2}{x + 1}$

74. $\dfrac{4x^4 + 12x^3 - x^2 - x + 2}{x + 3}$

75. $\dfrac{16x^2 - 13x^3 + 2x^4 - 9x + 20}{x - 5}$

76. $\dfrac{2x^3 - x^2 - 10x + 15 + x^4}{x - 2}$

77. $\dfrac{2x^4 - x^2 + 2}{x - 3}$

78. $\dfrac{x^4 - 3x^3 - 30}{x + 2}$

79. $\dfrac{x^3 + 125}{x + 5}$

80. $\dfrac{x^3 + 343}{x + 7}$

81. Suppose you know the number to the left of the vertical line in a synthetic division display of $P(x) \div Q(x)$, and you know the numbers below the horizontal line. Do you have enough information to determine $P(x)$?

82. Suppose you know the number to the left of the vertical line in a synthetic division display of $P(x) \div Q(x)$, and you know the numbers above the horizontal line. Do you have enough information to determine $P(x)$?

4 Evaluate a polynomial using synthetic division (See page 280.)

GETTING READY

83. Finish the synthetic division of $P(x) = 3x^3 - 2x^2 + x - 9$ by $x - 2$.

$$\begin{array}{r|rrrr} 2 & 3 & -2 & 1 & -9 \\ & & 6 & ? & ? \\ \hline & 3 & 4 & ? & ? \end{array}$$

The value of $P(2)$ is ___?___.

84. Use the synthetic division display to state one point that you know is on the graph of $P(x) = x^3 + x^2 - 9x - 9$.

$$\begin{array}{r|rrrr} 1 & 1 & 1 & -9 & -9 \\ & & 1 & 2 & -7 \\ \hline & 1 & 2 & -7 & -16 \end{array}$$

Use the Remainder Theorem to evaluate the polynomial.

85. $P(x) = 2x^2 - 3x - 1$; $P(3)$

86. $Q(x) = 3x^2 - 5x - 1$; $Q(2)$

87. $R(x) = x^3 - 2x^2 + 3x - 1$; $R(4)$

88. $F(x) = x^3 + 4x^2 - 3x + 2$; $F(3)$

89. $P(z) = 2z^3 - 4z^2 + 3z - 1$; $P(-2)$

90. $R(t) = 3t^3 + t^2 - 4t + 2$; $R(-3)$

91. $Q(x) = x^4 + 3x^3 - 2x^2 + 4x - 9$; $Q(2)$

92. $Y(z) = z^4 - 2z^3 - 3z^2 - z + 7$; $Y(3)$

93. $F(x) = 2x^4 - x^3 - 2x - 5$; $F(-3)$

94. $Q(x) = x^4 - 2x^3 + 4x - 2$; $Q(-2)$

95. $P(x) = x^3 - 3$; $P(5)$

96. $S(t) = 4t^3 + 5$; $S(-4)$

97. $R(t) = 4t^4 - 3t^2 + 5$; $R(-3)$

98. $P(z) = 2z^4 + z^2 - 3$; $P(-4)$

99. $Q(x) = x^5 - 4x^3 - 2x^2 + 5x - 2$; $Q(2)$

100. $T(x) = 2x^5 + 4x^4 - x^2 + 4$; $T(3)$

APPLYING CONCEPTS

Divide.

101. $(x^4 - 2x^3 + 7x^2 - 6x + 12) \div (x^2 + 3)$

102. $(x^4 + 3x^3 - 6x^2 - 12x + 8) \div (x^2 - 4)$

103. $(x^4 + 2x^3 + 2x^2 + x - 20) \div (x^2 + x - 4)$

104. $(x^4 - x^3 - 9x^2 + 19x - 10) \div (x^2 - 3x + 2)$

For what value of k will the remainder be zero?

105. $(x^3 - 3x^2 - x + k) \div (x - 3)$

106. $(x^3 - 2x^2 + x + k) \div (x - 2)$

107. $(x^2 + kx - 6) \div (x - 3)$

108. $(x^3 + kx + k - 1) \div (x - 1)$

109. When $x^2 + x + 2$ is divided by a polynomial, the quotient is $x + 4$, and remainder is 14. Find the polynomial.

PROJECTS OR GROUP ACTIVITIES

The Factor Theorem is a result of the Remainder Theorem. The Factor Theorem states that a polynomial $P(x)$ has a factor $(x - c)$ if and only if $P(c) = 0$. In other words, a remainder of zero means that the divisor is a factor of the dividend.

110. Determine whether $x + 5$ is a factor of $P(x) = x^4 + x^3 - 21x^2 - x + 20$.

111. Judging from your answer to Exercise 110, is -5 a zero of $P(x)$? Explain your answer.

112. Explain why $P(x) = 4x^4 + 7x^2 + 12$ has no factor of the form $x - c$, where c is a real number.

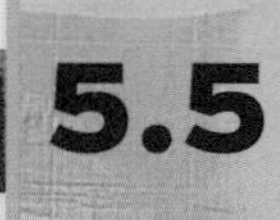

5.5 Introduction to Factoring

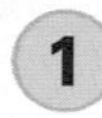

OBJECTIVE 1 Factor a monomial from a polynomial

The greatest common factor (GCF) of two or more exponential expressions with the same base is the exponential expression with the smallest exponent.

$2^2, 2^5, 2^3$	x^5, x^7, x	z^4, z^3, z^7
GCF $= 2^2$	GCF $= x$	GCF $= z^3$

The GCF of two or more monomials is the product of each common factor with its smallest exponent. If there are no common factors, the GCF is 1.

EXAMPLE 1 Find the GCF of the monomials.
A. $6x^6y^3, 4x^2y^6$ **B.** $18x^4y^2z^2, 24xz^3, 36x^2y^2z$

Solution **A.** The common numerical factor is 2. The smallest exponent on 2 is 1.

$6x^6y^3 = 2 \cdot 3 \cdot x^6 \cdot y^3$

$4x^2y^6 = 2^2 \cdot x^2 \cdot y^6$

The common variables are x and y. The smallest exponent on x is 2; the smallest exponent on y is 3. The common variable expression is x^2y^3.

The GCF of $6x^6y^3$ and $4x^2y^6$ is $2x^2y^3$.

B. The common numerical factors are 2 and 3, with 1 being the smallest exponent on each. The common numerical factor is $2 \cdot 3 = 6$.

$18x^4y^2z^2 = 2 \cdot 3^2 \cdot x^4 \cdot y^2 \cdot z^2$

$24xz^3 = 2^3 \cdot 3 \cdot x \cdot z^3$

$36x^2y^2z = 2^2 \cdot 3^2 \cdot x^2 \cdot y^2 \cdot z$

The common variable factors are x and z, with 1 being the smallest exponent on each. The common variable factor is xz.

The GCF of $18x^4y^2z^2$, $24xz^3$, and $36x^2y^2z$ is $6xz$.

Problem 1 Find the GCF of the monomials.
A. $16x^6yz^4, 9y^2z^5$ **B.** $36x^5y^3, 12x^2y^4z^5, 15x^2y^4$

Solution See page S17.

Try Exercise 15, page 288.

To **factor a polynomial** means to write the polynomial as a product of other polynomials.

Take Note
You can determine the monomial factors in the parentheses by dividing each term of the polynomial by the GCF.
$\frac{6x^3}{3x} = 2x^2, \frac{12x^2}{3x} = 4x, \frac{3x}{3x} = 1$

EXAMPLE 2 Factor. **A.** $6x^3 - 12x^2 - 3x$ **B.** $36x^3y^2 + 24x^3y^3 - 60x^4y$

Solution **A.** The GCF of $6x^3$, $12x^2$, and $3x$ is $3x$.

$$6x^3 - 12x^2 - 3x = 3x(2x^2) - 3x(4x) - 3x(1)$$
$$= 3x(2x^2 - 4x - 1)$$

- **Rewrite each term of the polynomial as a product that has the GCF as one factor.**

B. The GCF of $36x^3y^2$, $24x^3y^3$, and $60x^4y$ is $12x^3y$.

$$36x^3y^2 + 24x^3y^3 - 60x^4y$$
$$= 12x^3y(3y) + 12x^3y(2y^2) - 12x^3y(5x)$$
$$= 12x^3y(3y + 2y^2 - 5x)$$

• Rewrite each term of the polynomial as a product that has the GCF as one factor.

Problem 2 Factor. **A.** $6x^6z^3 + 8x^5z^2$ **B.** $16x^5y^4z^3 + 30xy^6z^3 - 30xy^6z$

Solution See page S17.

Try Exercise 25, page 288.

OBJECTIVE 2 Factor by grouping

For the examples below, the binomials in parentheses are **binomial factors.**

$$3rs(2r - 5s)$$
$$-2xy(3x + 7)$$

If the terms of an expression contain a common binomial factor, the Distributive Property can be used to factor the common binomial factor from the expression.

Focus on factoring a common binomial factor from an expression

Factor: $4a(2b + 3) - 5(2b + 3)$

The common binomial factor is $(2b + 3)$. Use the Distributive Property to write the expression as a product of factors.

$$4a(2b + 3) - 5(2b + 3)$$
$$= (2b + 3)(4a - 5)$$

Consider the binomial $y - x$. Factoring -1 from this binomial gives

$$y - x = -(x - y)$$

This equation is sometimes used to factor a common binomial from an expression.

Focus on factoring a common binomial factor from an expression

Factor: $6r(r - s) - 7(s - r)$

$$6r(r - s) - 7(s - r)$$

Rewrite the expression as a sum of terms that have a common binomial factor. Use $s - r = -(r - s)$.

$$= 6r(r - s) + 7(r - s)$$

Write the expression as a product of factors.

$$= (r - s)(6r + 7)$$

Some polynomials can be **factored by grouping** terms so that a common binomial factor is found.

Focus on factoring by grouping

Factor: $8y^2 + 4y - 6ay - 3a$

$8y^2 + 4y - 6ay - 3a$

Group the first two terms and the last two terms. Note that

$$-6ay - 3a = -(6ay + 3a)$$

$= (8y^2 + 4y) - (6ay + 3a)$

Factor the GCF from each group.

$= 4y(2y + 1) - 3a(2y + 1)$

Write the expression as a product of factors.

$= (2y + 1)(4y - 3a)$

EXAMPLE 3 Factor. **A.** $3x(y - 4) - 2(4 - y)$ **B.** $xy - 4x - 2y + 8$

Solution **A.** $3x(y - 4) - 2(4 - y)$

$= 3x(y - 4) + 2(y - 4)$

- **Write the expression as a sum of terms that have a common factor. Note that $4 - y = -(y - 4)$.**

$= (y - 4)(3x + 2)$

- **Factor out the common binomial factor.**

B. $xy - 4x - 2y + 8$

$= (xy - 4x) - (2y - 8)$

- **Group the first two terms and the last two terms. Note that $-2y + 8 = -(2y - 8)$.**

$= x(y - 4) - 2(y - 4)$

- **Factor the GCF from each group.**

$= (y - 4)(x - 2)$

- **Factor out the common binomial factor.**

Problem 3 Factor. **A.** $6a(2b + 5) - 7(5 + 2b)$ **B.** $3rs - 2r - 3s + 2$

Solution See page S17.

Try Exercises 49 and 55, page 289.

EXAMPLE 4 Factor: $x^3 + 4x^2 - 3x - 12$

Solution $x^3 + 4x^2 - 3x - 12$

$= (x^3 + 4x^2) - (3x + 12)$

- **Group the first two terms and the last two terms. Note: $-3x - 12 = -(3x + 12)$.**

$= x^2(x + 4) - 3(x + 4)$

- **Factor the GCF from each group of terms in parentheses.**

$= (x + 4)(x^2 - 3)$

- **Factor out the common binomial factor.**

Problem 4 Factor: $2x^3 - 4x^2 + 3x - 6$

Solution See page S17.

Try Exercise 51, page 289.

5.5 Exercises

CONCEPT CHECK

1. What is the GCF of two or more monomials?

2. What does it mean to factor a polynomial?

3. Of the two expressions $15x^2 - 10x$ and $5x(3x - 2)$, which one is written in *factored form*?

4. The expression $5x(3x - 2)$ has a monomial factor and a binomial factor. The monomial factor is ___?___ and the binomial factor is ___?___.

1 Factor a monomial from a polynomial (See pages 285–286.)

Find the GCF of the monomials.

5. $18x^4y^3$, $12x^4y^3$
6. $4xy^3$, $12y^2$
7. $10y^4$, $8x^5$
8. $12x^5y^3$, $8x^5y^6$
9. $18x^3y^2$, $12x^2$
10. $20x^4y^5$, $15x^6y^2$
11. $12xy^6$, $6x^6$
12. $24y^3$, $12y$
13. $20x^3y^6z^2$, $36x^3yz^4$
14. $8x^4y^3z^4$, $24xz^5$, $12xy^4z^4$
15. $10x^5y^5z^3$, $18xz$, $9x^2y^3z$
16. $24x^4yz$, $10xy^5z^3$, $18xy^2z^6$
17. $12x^3yz$, $12x^4y^5z^2$, $8x^4y^6$
18. $12x^6yz^4$, $24y^4z^5$, $10x^2y^3z^3$
19. $15y^4$, $8y^5z^6$, $4x^3z$
20. $16y^2z$, $8y^4z^3$, $15y^6z^6$
21. $12x^3y^4z^4$, $25x^6y^4$, $12x^2y$
22. $8x^2z^3$, $12x^5yz^2$, $4x^6y^6z^2$

GETTING READY

Complete each factorization.

23. $15x^2y - 9xy^2 = 3xy($___?___$)$

24. $4x^5y^2 + 6x^2y^2 =$ ___?___$(2x^3 + 3)$

Factor the polynomial.

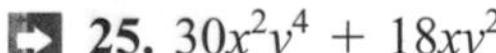

25. $30x^2y^4 + 18xy^2$
26. $18x^5y^5 + 9xy$
27. $12x^2y^6 + 20$
28. $30x^6y - 8x^3y^3$
29. $18x^6y^3 - 10xy^3$
30. $8x^3y^4 - 6y^5$
31. $8x^2y + 6y$
32. $24x^3y^4 - 20xy$
33. $10x^6y^5z^3 - 10x^4y^4z^3 + 6x^3y^4z^6$
34. $12x^5y^2z^2 - 30x^4z^3 + 24y^5z^3$
35. $20x^5z^4 + 15x^4y^3z^6 + 8x^4z^6$
36. $4x^5y^4z^2 + 25x^3y^2z^6 - 15z^3$
37. $18x^4y^3z^2 - 10x^2y^4 + 8x^2y^3z^3$
38. $30x^5y^6z + 36x^5y^2z + 24x^2yz^5$
39. $24x^5yz + 36xy^3 - 30y^5$
40. $24x^5y^5z^2 + 25x^4y^3z - 18x^2z^3$
41. $30x^5 + 20y^4z^3 - 25y^3$
42. $-6x^6y^6 + 18x^5y^4z^6 - 10x^2z^2$

43. If n is a positive integer, what is the GCF of x^{3n} and x^{2n}?

44. If n is a positive integer, what is the GCF of x^{n+3} and x^{n+2}?

2 Factor by grouping (See pages 286–287.)

GETTING READY

45. The common factor of the two terms of the expression $3x(x - 5) - 4(x - 5)$ is __?__.

46. Factor -1 from the binomial: $3 - y = -1($__?__$)$.

Factor.

47. $x(a + 2) - 2(a + 2)$

48. $3(x + y) + a(x + y)$

49. $a(x - 2) - b(2 - x)$

50. $3(a - 7) - b(7 - a)$

51. $x^2 + 3x + 2x + 6$

52. $x^2 - 5x + 4x - 20$

53. $xy + 4y - 2x - 8$

54. $ab + 7b - 3a - 21$

55. $ax + bx - ay - by$

56. $2ax - 3ay - 2bx + 3by$

57. $x^2y - 3x^2 - 2y + 6$

58. $a^2b + 3a^2 + 2b + 6$

59. $6 + 2y + 3x^2 + x^2y$

60. $15 + 3b - 5a^2 - a^2b$

61. $2ax^2 + bx^2 - 4ay - 2by$

62. $4a^2x + 2a^2y - 6bx - 3by$

63. $x^3 + x^2 + 2x + 2$

64. $y^3 - y^2 + 3y - 3$

65. $2x^3 - x^2 + 4x - 2$

66. $2y^3 - y^2 + 6y - 3$

67. $x^3 + 6x^2 - 6x - 36$

68. $2x^3 + 14x^2 - 3x - 21$

69. $3x^3 + 7x^2 - 21x - 49$

70. $6x^3 + 18x^2 - 8x - 24$

APPLYING CONCEPTS

The GCF of two or more monomials is the product of each common factor with its *smallest* exponent. This definition applies to expressions with negative exponents as well. The GCF of $4x^{-2}$ and $6x^{-4}$ is $2x^{-4}$ because, if you compare the exponents, $-4 < -2$.

Factor.

71. $4x^{-2} + 6x^{-4}$

72. $9x^{-3} - 6x^{-1}$

73. $12x^{-1}y^{-2} - 18x^{-2}y^{-1}$

74. $4x^{-2} + 8x^2$

Factoring by grouping can be extended to expressions with more than four terms. For the following, group the first three terms and the last three terms, and then factor by grouping.

75. $ac + ad + 2a + bc + bd + 2b$

76. $ac + ad - a - bc - bd + b$

PROJECTS OR GROUP ACTIVITIES

77. Make up a polynomial with two terms for which $3x^2y$ is the greatest common factor of the terms.

78. Make up a polynomial with three terms for which $2a^2b^3$ is the greatest common factor of the terms.

79. Write an expression for which $2a + b$ is a common binomial factor.

5.6 Factoring Trinomials

OBJECTIVE 1 Factor trinomials of the form $x^2 + bx + c$

Quadratic trinomials of the form $x^2 + bx + c$ are shown at the right. Note that the coefficient of x^2 is 1; b and c are nonzero integers.

$x^2 + 9x + 20$ $\quad b = 9, c = 20$
$x^2 + 2x - 48$ $\quad b = 2, c = -48$
$x^2 - 5x + 6$ $\quad b = -5, c = 6$

To **factor a quadratic trinomial,** we try to write the trinomial as the product of two binomials. For instance,

Trinomial		Factored form
$x^2 + 9x + 20$	=	$(x + 4)(x + 5)$
$x^2 + 2x - 48$	=	$(x - 6)(x + 8)$
$x^2 - 5x + 6$	=	$(x - 3)(x - 2)$

The method by which factors of a trinomial of the form $x^2 + bx + c$ are found is based on FOIL. Consider the following binomial products, noting the relationship between the constant terms of the binomials and the terms of the trinomial.

Sum of binomial constants — Product of binomial constants

$$\begin{aligned}
& \qquad\qquad\quad \text{F} \qquad \text{O} \qquad \text{I} \qquad \text{L} \\
(x + 4)(x + 5) &= x \cdot x + 5x + 4x + 4 \cdot 5 &&= x^2 + 9x + 20 \\
(x - 6)(x + 8) &= x \cdot x + 8x - 6x + (-6) \cdot 8 &&= x^2 + 2x - 48 \\
(x - 3)(x - 2) &= x \cdot x - 2x - 3x + (-3)(-2) &&= x^2 - 5x + 6
\end{aligned}$$

Observe two important points from these examples:

1. The constant term of the trinomial is the product of the constant terms of the binomials. The coefficient of x in the trinomial is the sum of the constant terms of the binomials.
2. When the constant term of the trinomial is positive, the constant terms of the binomials have the same sign. When the constant term of the trinomial is negative, the constant terms of the binomials have opposite signs.

Focus on factoring a trinomial of the form $x^2 + bx + c$

A. Factor: $x^2 - 4x - 12$

The constant term is negative. Find two factors of -12, one positive and one negative, whose sum is -4. You can make a list of factors as we have done at the right. Although we have listed all the possibilities, once the correct sum is found, you may stop.

Factors of -12	Sum of the factors
$1, -12$	-11
$-1, 12$	11
$2, -6$	-4
$-2, 6$	4
$3, -4$	-1
$-3, 4$	1

The product of 2 and -6 is -12; the sum of 2 and -6 is -4.

$x^2 - 4x - 12 = (x + 2)(x - 6)$

Check: $(x + 2)(x - 6) = x^2 - 6x + 2x - 12 = x^2 - 4x - 12$

B. Factor: $z^2 - 11z + 18$

The constant term is positive. Find two factors of 18, both negative (the coefficient of z is negative), whose sum is -11.

Factors of 18	Sum of the factors
$-1, -18$	-19
$-2, -9$	-11
$-3, -6$	-9

The product of -2 and -9 is 18; the sum of -2 and -9 is -11.

$z^2 - 11z + 18 = (z - 2)(z - 9)$

Check: $(z - 2)(z - 9) = z^2 - 9z - 2z + 18 = z^2 - 11z + 18$

Take Note

Be sure to check proposed factorizations. For instance, for the example at the right, we might have tried $(x - 3)(x + 5y^2)$. However,

$(x - 3)(x + 5y^2)$
$= x^2 + 5xy^2 - 3x - 15y^2$

The first and last terms are correct, but the sum of the two middle terms does not equal the middle term of the original polynomial.

C. Factor: $x^2 + 2xy - 15y^2$

The term $-15y^2$ is negative. Find two factors of -15, one positive and one negative, whose sum is 2. From the table, the product of -3 and 5 is -15; the sum of -3 and 5 is 2. Because the last term of $x^2 + 2xy - 15y^2$ contains y^2, we use $-3y$ and $5y$. Note that $(-3y)5y = -15y^2$.

Factors of -15	Sum of the factors
$1, -15$	-14
$-1, 15$	14
$3, -5$	-2
$-3, 5$	2

$x^2 + 2xy - 15y^2 = (x - 3y)(x + 5y)$

Check: $(x - 3y)(x + 5y) = x^2 + 5xy - 3xy - 15y^2 = x^2 + 2xy - 15y^2$

EXAMPLE 1 Factor. **A.** $x^2 + 8x + 12$ **B.** $x^2 + 5x - 84$

Solution **A.** $x^2 + 8x + 12$

$2(6) = 12$ • **Find two factors of 12 whose sum is 8. The factors are 2 and 6.**
$2 + 6 = 8$

$x^2 + 8x + 12 = (x + 2)(x + 6)$ • **Factor the trinomial.**

Check: $(x + 2)(x + 6) = x^2 + 6x + 2x + 12$
$= x^2 + 8x + 12$

B. $x^2 + 5x - 84$

$(-7)(12) = -84$ • **Find two factors of −84 whose sum is 5. The factors are −7 and 12.**
$-7 + 12 = 5$

$x^2 + 5x - 84 = (x + 12)(x - 7)$ • **Factor the trinomial.**

Check: $(x + 12)(x - 7) = x^2 - 7x + 12x - 84$
$= x^2 + 5x - 84$

Problem 1 Factor. **A.** $x^2 + 13x + 42$ **B.** $x^2 - x - 20$

Solution See page S17.

Try Exercise 11, page 297.

When factoring any polynomial, the first step *always* should be to check whether the terms have a common factor.

EXAMPLE 2 Factor: $2x^2y - 4xy - 70y$

Solution $2y$ is a common monomial factor.

$$2x^2y - 4xy - 70y = 2y(x^2 - 2x - 35)$$
$$= 2y(x - 7)(x + 5)$$

- **Factor out the common factor $2y$.**
- **Now factor $x^2 - 2x - 35$ by finding two factors of -35 whose sum is -2. The factors are -7 and 5.**

Check:

$$2y(x - 7)(x + 5) = 2y(x^2 + 5x - 7x - 35) = 2y(x^2 - 2x - 35)$$
$$= 2x^2y - 4xy - 70y$$

Problem 2 Factor: $6x^3y + 6x^2y - 36xy$

Solution See page S17.

Try Exercise 27, page 297.

Not all trinomials factor using only integers.

Consider $x^2 + 4x + 21$. We must find two positive factors of 21 whose sum is 4. From the table at the right, there are no such integers.

Factors of 21	Sum of the factors
1, 21	22
3, 7	10

The polynomial $x^2 + 4x + 21$ is said to be **nonfactorable over the integers.**

EXAMPLE 3 Factor: $x^2 + 2x - 4$

Solution Find two factors of -4 whose sum is 2. From the table below, there are no such integers. $x^2 + 2x - 4$ is nonfactorable over the integers.

Factors of -4	Sum of the factors
$-1, 4$	3
$1, -4$	-3
$-2, 2$	0

Problem 3 Factor: $x^2 + 5x - 1$

Solution See page S17.

Try Exercise 25, page 297.

OBJECTIVE 2 Factor trinomials of the form $ax^2 + bx + c$

There are various methods of factoring trinomials of the form $ax^2 + bx + c$, where $a \neq 1$. Factoring by using trial factors and factoring by grouping will be discussed in this objective. Factoring by using trial factors is illustrated first.

To use the trial factor method, use the factors of a and the factors of c to write all the possible binomial factors of the trinomial. Then use FOIL to determine the correct factorization. To reduce the number of trial factors that must be considered, remember the following:

1. Use the signs of the constant term and the coefficient of x in the trinomial to determine the signs of the binomial factors. If the constant term is positive, the signs of the binomial factors will be the same as the sign of the coefficient of x in the trinomial. If the constant term is negative, the constant terms in the binomials will have opposite signs.
2. If the terms of the trinomial do not have a common factor, then the terms in either one of the binomial factors will not have a common factor.

Focus on factoring a trinomial by trial factors

Factor: $4x^2 + 31x - 8$

The terms of the trinomial do not have a common factor; therefore, the terms of the binomial factors will not have a common factor.

Because the constant term, c, of the trinomial is negative (-8), the constant terms of the binomial factors will have opposite signs.

Find the factors of a (4) and the factors of c (-8).

Factors of 4	Factors of −8
1, 4	1, −8
2, 2	−1, 8
	2, −4
	−2, 4

Take Note

The terms of the binomial factor $4x - 8$ have a common factor of 4: $4x - 8 = 4(x - 2)$. Because the terms of $4x^2 + 31x - 8$ do not have a common factor, none of the terms of the binomial factors can have a common factor.

Using these factors, write trial factors, and use FOIL to check the middle term of each trinomial.

Remember that if the terms of the trinomial do not have a common factor, then the terms of a binomial factor cannot have a common factor. Such trial factors need not be checked.

The correct factors have been found.

Trial Factors	Middle Term
$(x + 1)(4x - 8)$	Common factor
$(x - 1)(4x + 8)$	Common factor
$(x + 2)(4x - 4)$	Common factor
$(x - 2)(4x + 4)$	Common factor
$(2x + 1)(2x - 8)$	Common factor
$(2x - 1)(2x + 8)$	Common factor
$(2x + 2)(2x - 4)$	Common factor
$(2x - 2)(2x + 4)$	Common factor
$(4x + 1)(x - 8)$	$-32x + x = -31x$
$(4x - 1)(x + 8)$	$32x - x = 31x$

$4x^2 + 31x - 8 = (4x - 1)(x + 8)$

Other trial factors need not be checked.

The example above illustrates that many of the trial factors may have common factors and thus need not be tried. For the remainder of this chapter, the trial factors with a common factor will not be listed.

EXAMPLE 4 Factor. **A.** $2x^2 - 21x + 10$ **B.** $6x^2 + 17x - 10$

Solution **A.** $2x^2 - 21x + 10$

Factors of 2	Factors of 10
1, 2	−1, −10
	−2, −5

• **Use negative factors of 10.**

Trial Factors	Middle Term
$(x - 2)(2x - 5)$	$-5x - 4x = -9x$
$(2x - 1)(x - 10)$	$-20x - x = -21x$

• **Write trial factors. Use FOIL to check the middle term.**

$2x^2 - 21x + 10 = (2x - 1)(x - 10)$

B. $6x^2 + 17x - 10$

Factors of 6	Factors of -10
1, 6 2, 3	1, -10 -1, 10 2, -5 -2, 5

• Find the factors of a (6) and the factors of c (-10).

Trial Factors	Middle Term
$(x + 2)(6x - 5)$	$-5x + 12x = 7x$
$(x - 2)(6x + 5)$	$5x - 12x = -7x$
$(2x + 1)(3x - 10)$	$-20x + 3x = -17x$
$(2x - 1)(3x + 10)$	$20x - 3x = 17x$

• Write trial factors. Use FOIL to check the middle term.

$$6x^2 + 17x - 10 = (2x - 1)(3x + 10)$$

Problem 4 Factor. **A.** $4x^2 + 15x - 4$ **B.** $10x^2 + 39x + 14$

Solution See page S18.

Try Exercise 45, page 298.

Trinomials of the form $ax^2 + bx + c$ can also be factored by grouping. This method is an extension of the method discussed in Section 5.5, Objective 2.

To factor $ax^2 + bx + c$, first find the factors of $a \cdot c$ whose sum is b. Use the two factors to rewrite the middle term of the trinomial as the sum of two terms. Then factor by grouping to write the factorization of the trinomial.

Focus on factoring a trinomial by grouping

A. Factor: $3x^2 + 11x + 8$

Find the product of $a = 3$ and $c = 8$.

$$3(8) = 24$$

Now find two factors of 24 whose sum is 11.

Factors of 24	Sum of the factors
1, 24	25
2, 12	14
3, 8	11
4, 6	10

$3(8) = 24 \qquad 3 + 8 = 11$

The factors are 3 and 8.

Use these factors to rewrite $11x$ as $3x + 8x$.

$$3x^2 + 11x + 8 = 3x^2 + 3x + 8x + 8$$

Factor by grouping.

$$= (3x^2 + 3x) + (8x + 8)$$
$$= 3x(x + 1) + 8(x + 1)$$
$$= (x + 1)(3x + 8)$$

Check: $(x + 1)(3x + 8) = 3x^2 + 8x + 3x + 8 = 3x^2 + 11x + 8$

B. Factor: $4x^2 - 12x - 7$

Find the product of $a = 4$ and $c = -7$.

$$4(-7) = -28$$

Factors of -28	Sum of the factors
$1, -28$	-27
$-1, 28$	27
$2, -14$	-12
$-2, 14$	12
$4, -7$	-3
$-4, 7$	3

Now find two factors of -28 whose sum is -12.

$$2(-14) = -24$$
$$2 + (-14) = -12$$

The factors are 2 and -14.

Use these factors to rewrite $-12x$ as $-14x + 2x$.

$$4x^2 - 12x - 7 = 4x^2 - 14x + 2x - 7$$

Factor by grouping.

$$= (4x^2 - 14x) + (2x - 7)$$
$$= 2x(2x - 7) + (2x - 7)$$
$$= (2x - 7)(2x + 1)$$

Check: $(2x - 7)(2x + 1) = 4x^2 + 2x - 14x - 7 = 4x^2 - 12x - 7$

C. Factor: $3x^2 - 11x + 4$

Find the product of $a = 3$ and $c = 4$.

$$3(4) = 12$$

Factors of 12	Sum of the factors
$-1, -12$	-13
$-2, -6$	-8
$-3, -4$	-7

Now find two factors of 12 whose sum is -11. As the table shows, there are no such factors.

$3x^2 - 11x + 4$ is nonfactorable over the integers.

EXAMPLE 5 Factor. **A.** $2x^2 - 21x + 10$ **B.** $10 - 17x - 6x^2$

Solution **A.** $2x^2 - 21x + 10$

$a = 2, c = 10; ac = 20$. Find two factors of 20 whose sum is -21.

$-1(-20) = 20 \qquad -1 + (-20) = -21$

The factors are -1 and -20.

$$2x^2 - 21x + 10 = 2x^2 - x - 20x + 10$$

• **Use these factors to rewrite $-21x$ as $-x - 20x$.**

$$= (2x^2 - x) - (20x - 10)$$
$$= x(2x - 1) - 10(2x - 1)$$

• **Factor by grouping. Note that $-20x + 10 = -(20x - 10)$.**

$$= (2x - 1)(x - 10)$$

• **The factorization checks.**

B. $10 - 17x - 6x^2$

$a = -6, c = 10; ac = -60$. Find two factors of -60 whose sum is -17.

$3(-20) = -60 \qquad 3 + (-20) = -17$

The factors are 3 and -20.

$10 - 17x - 6x^2 = 10 + 3x - 20x - 6x^2$ • **Use these factors to rewrite $-17x$ as $3x - 20x$.**

$= (10 + 3x) - (20x + 6x^2)$ • **Factor by grouping. Note that $-20x - 6x^2 = -(20x + 6x^2)$.**

$= (10 + 3x) - 2x(10 + 3x)$

$= (10 + 3x)(1 - 2x)$ • **The factorization checks.**

Problem 5 Factor. **A.** $6x^2 + 7x - 20$ **B.** $2 - x - 6x^2$

Solution See page S18.

Try Exercise 47, page 298.

Either method of factoring discussed in this objective will always lead to a correct factorization of trinomials of the form $ax^2 + bx + c$ that are factorable.

A polynomial is factored completely when it is written as a product of factors that are nonfactorable over the integers.

EXAMPLE 6 Factor. **A.** $30y + 2xy - 4x^2y$ **B.** $12x^3y + 14x^2y - 6xy$

Solution **A.** $30y + 2xy - 4x^2y$ • **The GCF of $30y$, $2xy$, and $4x^2y$ is $2y$.**

$= 2y(15 + x - 2x^2)$ • **Factor out the GCF.**

$= 2y(5 + 2x)(3 - x)$ • **Factor the trinomial.**

B. $12x^3y + 14x^2y - 6xy$ • **The GCF of $12x^3y$, $14x^2y$, and $6xy$ is $2xy$.**

$= 2xy(6x^2 + 7x - 3)$

$= 2xy(3x - 1)(2x + 3)$

Problem 6 Factor. **A.** $3a^3b^3 + 3a^2b^2 - 60ab$ **B.** $40a - 10a^2 - 15a^3$

Solution See page S18.

Try Exercise 71, page 298.

5.6 Exercises

CONCEPT CHECK

1. Which of the following is the correct factorization of $4x^2 + 9x - 28$?
   (i) $(2x + 7)(2x - 4)$ (ii) $(4x - 7)(x + 4)$ (iii) $(4x + 28)(x - 1)$

2. For each of the following, find two factors of c whose sum is b.
   a. $b = -7, c = 12$ **b.** $b = -10, c = -24$
   c. $b = 7, c = -18$ **d.** $b = 25, c = 84$

3. For each of the following, find two factors of $a \cdot c$ whose sum is b.

a. $a = 2, b = -11, c = 9$ **b.** $a = 3, b = 4, c = -4$

c. $a = -1, b = -1, c = 20$ **d.** $a = 4, b = 5, c = -6$

4. What does it mean for a polynomial to be *nonfactorable over the integers*?

1 Factor trinomials of the form $x^2 + bx + c$ (See pages 290–292.)

GETTING READY

5. a. Find two numbers whose product is -21 and whose sum is 4.

b. Find two numbers whose product is 20 and whose sum is -9.

6. To factor a quadratic trinomial of the form $x^2 + bx + c$ into two binomials $(x + n)$ and $(x + m)$, look for constants n and m whose product is ___?___ and whose sum is ___?___.

Factor.

7. $x^2 - 8x + 15$ **8.** $x^2 + 12x + 20$ **9.** $a^2 + 12a + 11$

10. $a^2 + a - 72$ **11.** $b^2 + 2b - 35$ **12.** $a^2 + 7a + 6$

13. $y^2 - 16y + 39$ **14.** $x^2 + x - 132$ **15.** $a^2 - 15a + 56$

16. $x^2 + 15x + 50$ **17.** $x^2 + 4x - 5$ **18.** $a^2 - 3ab + 2b^2$

19. $a^2 + 11ab + 30b^2$ **20.** $a^2 + 8ab - 33b^2$ **21.** $x^2 - 14xy + 24y^2$

22. $x^2 + 5xy + 6y^2$ **23.** $y^2 + 2xy - 63x^2$ **24.** $x^2 - 5x + 7$

25. $x^2 - 7x - 12$ **26.** $5x^6 - 50x^5 + 105x^4$ **27.** $x^4y^4 - 4x^3y^4 - 21x^2y^4$

28. $x^2y^3 - xy^3 - 2y^3$ **29.** $6x^6 - 60x^5 + 54x^4$ **30.** $4x^4y^4 + 12x^3y^4 + 8x^2y^4$

31. $3x^5y^4 - 24x^4y^4 + 45x^3y^4$ **32.** $5x^2 + 15x - 140$ **33.** $3x^6y - 45x^5y + 168x^4y$

34. $4x^3y^3 + 56x^2y^3 + 180xy^3$ **35.** $2x^5y^4 + 4x^4y^4 - 30x^3y^4$ **36.** $30 + 17a - 20a^2$

37. $4x^4 - 45x^2 + 80$

38. Suppose that b and c are nonzero and that n and m are positive constants such that $x^2 + bx + c = (x - n)(x + m)$.

a. Is c positive or negative?

b. If b is negative, is n less than, equal to, or greater than m?

39. Suppose that b and c are nonzero and that n and m are positive constants such that $x^2 + bx + c = (x - n)(x - m)$.

a. Is c positive or negative?

b. Is b positive or negative?

c. Is b less than, equal to, or greater than c?

2 Factor trinomials of the form $ax^2 + bx + c$ (See pages 292–296.)

If you want to factor $ax^2 + bx + c$ by grouping, you look for factors of ac whose sum is b. In Exercises 40 to 43, information is given about the signs of b and c in the trinomial $ax^2 + bx + c$, $a > 0$. In each case, state whether the factors of ac should be two positive numbers, two negative numbers, or one positive and one negative number.

40. $b < 0$ and $c < 0$

41. $b < 0$ and $c > 0$

42. $b > 0$ and $c < 0$

43. $b > 0$ and $c > 0$

Factor.

44. $2x^2 + 7x + 3$

45. $2x^2 - 11x - 40$

46. $6y^2 + 5y - 6$

47. $4y^2 - 15y + 9$

48. $6b^2 - b - 35$

49. $2a^2 + 13a + 6$

50. $3y^2 - 22y + 39$

51. $12y^2 - 13y - 72$

52. $6a^2 - 26a + 15$

53. $5x^2 + 26x + 5$

54. $4a^2 - a - 5$

55. $11x^2 - 122x + 11$

56. $11y^2 - 47y + 12$

57. $12x^2 - 17x + 5$

58. $12x^2 - 40x + 25$

59. $8y^2 - 18y + 9$

60. $4x^2 + 9x + 10$

61. $6a^2 - 5a - 2$

62. $10x^2 - 29x + 10$

63. $2x^2 + 5x + 12$

64. $4x^2 - 6x + 1$

65. $40 - 3x - x^2$

66. $42 + x - x^2$

67. $24 + 13x - 2x^2$

68. $6 - 7x - 5x^2$

69. $120x^4 - 306x^3 + 162x^2$

70. $25x^2y^3 - 30xy^3 - 27y^3$

71. $32x^4y^3 + 36x^3y^3 - 56x^2y^3$

72. $14x^6y + 26x^5y - 48x^4y$

73. $24x^3 - 96x^2 + 42x$

74. $20x^4y - 24x^3y + 4x^2y$

75. $144x^3y^4 + 28x^2y^5 - 60xy^6$

76. $108x^2y^3 - 63xy^4 + 9y^5$

APPLYING CONCEPTS

Find all integers k such that the trinomial can be factored over the integers.

77. $x^2 + kx - 6$

78. $x^2 + kx + 8$

79. $2x^2 - kx - 5$

80. $2x^2 - kx + 3$

81. $2x^2 + kx - 3$

82. $3x^2 + kx + 5$

83. What are the positive integer values of k for which $x^2 + 6x + k$ factors over the integers?

84. In Exercise 83, suppose we remove the restriction that k be a positive integer. For how many integer values of k will $x^2 + 6x + k$ factor over the integers?

85. If $x^2 + bx + c$ factors as $(x + m)(x + n)$, then $b =$ __?__ and $c =$ __?__.

PROJECTS OR GROUP ACTIVITIES

86. The area of a rectangle is $(3x^2 + x - 2)$ ft². The length and width of the rectangle are the factors of the trinomial.

a. Express the dimensions of the rectangle in terms of x.

b. Can x be less than or equal to zero?

c. What is the smallest value of x for which the area of the rectangle is greater than zero?

d. Assuming the length is greater than the width, which factor is the length and which factor is the width? *Hint:* Your answer will depend on the value of x.

5.7 Special Factoring

OBJECTIVE 1 Factor the difference of two perfect squares and factor perfect-square trinomials

The product of a term and itself is called a **perfect square.** The exponents on variable parts of perfect squares are always even numbers.

Term		*Perfect Square*
5	$5 \cdot 5 =$	25
x	$x \cdot x =$	x^2
$3y^4$	$3y^4 \cdot 3y^4 =$	$9y^8$

The **square root of a perfect square** is one of the two equal factors of the perfect square. $\sqrt{\ }$ is the symbol for square root. To find the exponent of the square root of a variable term, divide the exponent by 2. For the examples at the right, assume that x and y represent positive numbers.

$$\sqrt{25} = 5$$
$$\sqrt{x^2} = x$$
$$\sqrt{9y^8} = 3y^4$$

The factors of the difference of two perfect squares are the sum and difference of the square roots of the perfect squares.

FACTOR THE DIFFERENCE OF TWO SQUARES

$$a^2 - b^2 = (a + b)(a - b)$$

EXAMPLES

1. $x^2 - 9 = x^2 - 3^2 = (x + 3)(x - 3)$
2. $z^2 - 49 = z^2 - 7^2 = (z + 7)(z - 7)$

An expression such as $a^2 + b^2$ is the *sum of two squares.* The sum of two squares does not factor over the integers. For instance, $x^2 + 9$ is the sum of two squares. It does not factor over the integers.

EXAMPLE 1 Factor. **A.** $25x^2 - 81$ **B.** $12x^3 - 147x$

Solution **A.** $25x^2 - 81 = (5x)^2 - 9^2$

- **Write $25x^2 - 81$ as the difference of two squares.**

$= (5x + 9)(5x - 9)$

- **Use $a^2 - b^2 = (a + b)(a - b)$ to factor.**

B. $12x^3 - 147x = 3x(4x^2 - 49)$

- **Factor out the GCF.**

$= 3x[(2x)^2 - 7^2]$

- **Write $4x^2 - 49$ as the difference of two squares.**

$= 3x(2x + 7)(2x - 7)$

- **Use $a^2 - b^2 = (a + b)(a - b)$ to factor.**

Problem 1 Factor. **A.** $81x^2 - 4y^2$ **B.** $20x^3y^2 - 45xy^2$

Solution See page S18.

Try Exercise 19, page 304.

The square of a binomial is a **perfect-square trinomial.** Here are two examples.

$$(a + b)^2 = a^2 + 2ab + b^2$$
$$(a - b)^2 = a^2 - 2ab + b^2$$

To factor a perfect-square trinomial, write the trinomial as the square of a binomial.

FACTOR A PERFECT-SQUARE TRINOMIAL

$$a^2 + 2ab + b^2 = (a + b)^2$$
$$a^2 - 2ab + b^2 = (a - b)^2$$

EXAMPLES

1. $x^2 + 6x + 9 = (x + 3)^2$
2. $x^2 - 12x + 36 = (x - 6)^2$

Focus on factoring a perfect-square trinomial

Factor: $4x^2 - 12x + 9$

Because $4x^2$ is a perfect square and 9 is a perfect square, try factoring $4x^2 - 12x + 9$ as the square of a binomial.

$$4x^2 - 12x + 9 \stackrel{?}{=} (2x - 3)^2$$

Check: $(2x - 3)^2 = (2x - 3)(2x - 3) = 4x^2 - 6x - 6x + 9 = 4x^2 - 12x + 9$

The factorization checks. Thus $4x^2 - 12x + 9 = (2x - 3)^2$.

It is important to check a proposed factorization, as we did above. For instance, consider factoring $x^2 + 13x + 36$.

Because x^2 is a perfect square and 36 is a perfect square, try factoring $x^2 + 13x + 36$ as the square of a binomial.

$$x^2 + 13x + 36 \stackrel{?}{=} (x + 6)^2$$

Check: $(x + 6)^2 = (x + 6)(x + 6) = x^2 + 6x + 6x + 36 = x^2 + 12x + 36$

In this case, the proposed factorization of $x^2 + 13x + 36$ does *not* check. Try another factorization. The numbers 4 and 9 are factors of 36 whose sum is 13.

$$x^2 + 13x + 36 = (x + 4)(x + 9)$$

EXAMPLE 2 Factor: $4x^2 - 20x + 25$

Solution $4x^2 - 20x + 25$ • $4x^2 = (2x)^2$ and $25 = 5^2$.

$= (2x - 5)^2$ • Check that $(2x - 5)^2 = 4x^2 - 20x + 25$.

Problem 2 Factor: $9x^2 + 12x + 4$

Solution See page S18.

Try Exercise 23, page 304.

OBJECTIVE 2 Factor the sum or the difference of two cubes

The product of the same three factors is called a **perfect cube.** The exponents on the variable parts of perfect cubes are always divisible by 3.

Term		*Perfect Cube*
2	$2 \cdot 2 \cdot 2 =$	8
$3y$	$3y \cdot 3y \cdot 3y =$	$27y^3$
x^2	$x^2 \cdot x^2 \cdot x^2 =$	x^6

The **cube root of a perfect cube** is one of the three equal factors of the perfect cube. $\sqrt[3]{}$ is the symbol for cube root. To find the exponent of the cube root of a perfect-cube variable term, divide the exponent by 3.

$\sqrt[3]{8} = 2$

$\sqrt[3]{27y^3} = 3y$

$\sqrt[3]{x^6} = x$

The following rules are used to factor the sum or difference of two perfect cubes.

FACTOR THE SUM OR DIFFERENCE OF TWO CUBES

$$a^3 + b^3 = (a + b)(a^2 - ab + b^2)$$
$$a^3 - b^3 = (a - b)(a^2 + ab + b^2)$$

EXAMPLES

1. $x^3 + 27 = x^3 + 3^3 = (x + 3)(x^2 - 3x + 9)$
2. $z^3 - 64 = z^3 - 4^3 = (z - 4)(z^2 + 4z + 16)$

EXAMPLE 3 Factor: **A.** $8x^3 + 125$ **B.** $3x^4y - 81xy^4$

Solution **A.** $8x^3 + 125 = (2x)^3 + 5^3$ • $8x^3 + 125$ is the sum of two cubes.

$= (2x + 5)[(2x)^2 - 2x(5) + 5^2]$ • Factor using the Sum of Two Cubes formula.

$= (2x + 5)(4x^2 - 10x + 25)$

B. $3x^4y - 81xy^4$

$= 3xy(x^3 - 27y^3)$ • $3xy$ is a common factor.

$= 3xy[x^3 - (3y)^3]$ • $x^3 - 27y^3$ is the difference of two cubes.

$= 3xy(x - 3y)[(x)^2 + x(3y) + (3y)^2]$ • Factor using the Difference of Two Cubes formula.

$= 3xy(x - 3y)(x^2 + 3xy + 9y^2)$

Problem 3 Factor: **A.** $a^3 + 64b^3$ **B.** $32x^4y^3 - 108xy^3$

Solution See page S18.

Try Exercise 65, page 305.

OBJECTIVE 3 Factor trinomials that are quadratic in form

Certain trinomials can be expressed as quadratic trinomials by making suitable variable substitutions.

Take Note

An expression is quadratic in form if it can be written as $a(\ )^2 + b(\ ) + c$, where the same expression is placed in both sets of parentheses.

The expression $2x^6 - 7x^3 + 4$ is quadratic in form because

$$2x^6 - 7x^3 + 4 = 2(x^3)^2 - 7(x^3) + 4$$

The expression $5x^2y^2 + 3xy - 6$ is quadratic in form because

$$5x^2y^2 + 3xy - 6 = 5(xy)^2 + 3(xy) - 6$$

TRINOMIALS THAT ARE QUADRATIC IN FORM

A trinomial is **quadratic in form** if it can be written as $au^2 + bu + c$.

EXAMPLES

1. $2x^6 - 7x^3 + 4$
   Let $u = x^3$. Then $u^2 = (x^3)^2 = x^6$.
   $2x^6 - 7x^3 + 4 \Rightarrow 2u^2 - 7u + 4$
   $2x^6 - 7x^3 + 4$ is quadratic in form.
2. $5x^2y^2 + 3xy - 6$
   Let $u = xy$. Then $u^2 = (xy)^2 = x^2y^2$.
   $5x^2y^2 + 3xy - 6 \Rightarrow 5u^2 + 3u - 6$
   $5x^2y^2 + 3xy - 6$ is quadratic in form.

Focus on factoring a polynomial that is quadratic in form

Factor: $x^4 - 2x^2 - 15$

Let $u = x^2$.	$x^4 - 2x^2 - 15 = (x^2)^2 - 2(x^2) - 15$
Replace x^2 by u.	$= u^2 - 2u - 15$
Factor.	$= (u - 5)(u + 3)$
Replace u by x^2.	$= (x^2 - 5)(x^2 + 3)$

EXAMPLE 4 Factor. **A.** $6x^2y^2 - xy - 12$ **B.** $2x^4 + 5x^2 - 12$

Solution **A.** $6x^2y^2 - xy - 12$
$= (3xy + 4)(2xy - 3)$
- **Let $u = xy$.**
- **Factor $6u^2 - u - 12$.**

B. $2x^4 + 5x^2 - 12$
$= (x^2 + 4)(2x^2 - 3)$
- **Let $u = x^2$.**
- **Factor $2u^2 + 5u - 12$.**

Problem 4 Factor. **A.** $6x^2y^2 - 19xy + 10$ **B.** $3x^4 + 4x^2 - 4$

Solution See page S18.

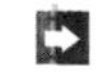 *Try Exercise 89, page 305.*

OBJECTIVE  4 Factor completely

Take Note

Remember that you may have to factor more than once in order to factor the polynomial completely.

When factoring a polynomial completely, ask the following questions about the polynomial.

1. Is there a common factor? If so, factor out the GCF.
2. If the polynomial is a binomial, is it the difference of two perfect squares, the sum of two perfect cubes, or the difference of two perfect cubes? If so, factor.
3. If the polynomial is a trinomial, is it a perfect-square trinomial or the product of two binomials? If so, factor.

4. If the polynomial has four terms, can it be factored by grouping? If so, factor.
5. Is each factor nonfactorable over the integers? If not, factor.

EXAMPLE 5 Factor. **A.** $x^2y + 2x^2 - y - 2$ **B.** $x^6 - y^6$

Solution **A.**

$$\begin{aligned} &x^2y + 2x^2 - y - 2 \\ &= (x^2y + 2x^2) - (y + 2) && \text{• Factor by grouping.} \\ &= x^2(y + 2) - (y + 2) \\ &= (y + 2)(x^2 - 1) \\ &= (y + 2)(x + 1)(x - 1) && \text{• Factor the difference of two perfect squares.} \end{aligned}$$

B.

$$\begin{aligned} &x^6 - y^6 \\ &= (x^3)^2 - (y^3)^2 && \text{• Write } x^6 - y^6 \text{ as the difference of two squares.} \\ &= (x^3 - y^3)(x^3 + y^3) && \text{• Factor the difference of two squares.} \\ &= (x - y)(x^2 + xy + y^2)(x + y)(x^2 - xy + y^2) && \text{• Factor the difference of two cubes and the sum of two cubes.} \end{aligned}$$

Problem 5 Factor. **A.** $4x - 4y - x^3 + x^2y$ **B.** $x^4 - 8x^2 - 9$

Solution See page S18.

Try Exercise 107, page 306.

5.7 Exercises

CONCEPT CHECK

1. State whether each expression is a perfect square.
   a. $2x^4$ **b.** $25x^4y^8$ **c.** $64x^6$ **d.** $9x^5$

2. State whether each expression is a perfect cube.
   a. $8x^6$ **b.** $16x^3$ **c.** $64x^6$ **d.** $27x^9y^6$

3. State whether each expression is the difference of perfect squares.
   a. $4x^2 - 9$ **b.** $x^2 + 9$ **c.** $x^2y^2 - 81$ **d.** $16x^2 - 8$

4. State whether each expression is the sum or difference of perfect cubes.
   a. $8x^3 + 27$ **b.** $(x + 8)^3$ **c.** $(x - 27)^3$ **d.** $x^3 - 64$

5. State whether each expression is quadratic in form.
   a. $2x^4 + x^2 - 6$ **b.** $4x^4 + 9x + 25$ **c.** $x^4 - 9$ **d.** $x^2y^2 - 2xy + 3$

6. Complete the following sentence: When factoring a polynomial, the first step is to find the __?__ of the terms of the polynomial.

1 Factor the difference of two perfect squares and factor perfect-square trinomials (See pages 299–301.)

GETTING READY

Which of the expressions are perfect squares?

7. 4; 8; $25x^6$; $12y^{10}$; $100x^4y^4$

8. 9; 18; $15a^8$; $49b^{12}$; $64a^{16}b^2$

Name the square root of the expression.

9. $16z^8$

10. $36d^{10}$

11. $81a^4b^6$

12. $25m^2n^{12}$

13. a. The binomial $16x^2 - 1$ is in the form $a^2 - b^2$, where $a =$ __?__ and $b =$ __?__.

b. Use the formula $a^2 - b^2 = (a + b)(a - b)$ to factor $16x^2 - 1$: $16x^2 - 1 =$ (__?__)(__?__).

14. a. The trinomial $25y^2 - 20y + 4$ is in the form $a^2 - 2ab + b^2$, where $a =$ __?__ and $b =$ __?__.

b. Use the formula $a^2 - 2ab + b^2 = (a - b)^2$ to factor $25y^2 - 20y + 4$: $25y^2 - 20y + 4 =$ (__?__)^2.

15. Explain how to factor the difference of two perfect squares.

16. What is a perfect-square trinomial?

Factor.

17. $x^2 - 16$

18. $y^2 - 49$

19. $4x^2 - 1$

20. $81x^2 - 4$

21. $b^2 - 2b + 1$

22. $a^2 + 14a + 49$

23. $16x^2 - 40x + 25$

24. $49x^2 + 28x + 4$

25. $x^2 + 4$

26. $a^2 + 16$

27. $x^2 + 6xy + 9y^2$

28. $4x^2y^2 + 12xy + 9$

29. $4x^2 - y^2$

30. $49a^2 - 16b^4$

31. $16x^2 - 121$

32. $49y^2 - 36$

33. $1 - 9a^2$

34. $16 - 81y^2$

35. $4a^2 + 4a - 1$

36. $9x^2 + 12x - 4$

37. $b^2 + 7b + 14$

38. $y^2 - 5y + 25$

39. $25a^2 - 40ab + 16b^2$

40. $4a^2 - 36ab + 81b^2$

41. $-45x^3y + 120x^2y - 80xy$

42. $16x^2y^3 - 9y^3$

43. $54x^5y^4 - 144x^4y^4 + 96x^3y^4$

44. $12x^3y + 36x^2y + 27xy$

45. $45x^4y^3 + 30x^3y^3 + 5x^2y^3$

46. $-54x^4y^4 - 144x^3y^4 - 96x^2y^4$

47. $9x^4 - 4x^6$

48. $3x^3y^4 - 75x^5y^4$

49. $5x^3y - 45x^5y$

50. $16x^3y^4 - 36xy^4$

51. Which expressions are equivalent to $x^{16} - 81$?
(i) $(x^4 + 9)(x^4 - 9)$ (ii) $(x^8 + 9)(x^8 - 9)$ (iii) $(x^4 + 9)(x^2 + 3)(x^2 - 3)$ (iv) $(x^8 + 9)(x^4 + 3)(x^4 - 3)$

52. Given that a and b are positive numbers, which expressions can be factored?
(i) $a^2x^2 - 2abx + b^2$ (ii) $-a^2x^2 - 2abx - b^2$ (iii) $-a^2x^2 + 2abx - b^2$ (iv) $a^2x^2 - 2abx - b^2$

2 Factor the sum or the difference of two cubes (See page 301.)

GETTING READY

Which of the expressions are perfect cubes?

53. 4; 8; x^9; a^8b^8; $27c^{15}d^{18}$

54. 9; 27; y^{12}; m^3n^6; $64mn^9$

Name the cube root of the expression.

55. $8x^9$

56. $27y^{15}$

57. $64a^6b^{18}$

58. $125c^{12}d^3$

59. a. The binomial $125x^3 + 8$ is in the form $a^3 + b^3$, where $a =$ __?__ and $b =$ __?__.

b. Use the formula $a^3 + b^3 = (a + b)(a^2 - ab + b^2)$ to factor $125x^3 + 8$: $125x^3 + 8 =$ (__?__)(__?__).

60. a and b are positive numbers. Which expressions can be factored?
(i) $a^3x^3 - b^3y^3$ (ii) $a^9x^3 + b^9y^3$ (iii) $a^8x^3 + b^8y^3$ (iv) $a^6x^3 - b^3y^6$

Factor.

61. $x^3 - 27$

62. $y^3 + 125$

63. $8x^3 - 1$

64. $64a^3 + 27$

65. $x^3 - 8y^3$

66. $27a^3 + b^3$

67. $64x^3 + 1$

68. $1 - 125b^3$

69. $27x^3 - 8y^3$

70. $64x^3 + 27y^3$

71. $x^3y^3 + 64$

72. $8x^3y^3 + 27$

73. $24x^6y - 3x^3y$

74. $32x^3y - 4y$

75. $54x^6y + 2x^3y$

76. $24x^7 - 3x^4$

77. $81x^3y + 3y$

78. $81x^5y^3 + 24x^2y^3$

3 Factor trinomials that are quadratic in form (See page 302.)

GETTING READY

79. The trinomial $8x^4 + 2x^2 - 3$ can be written in the quadratic form $8u^2 + 2u - 3$, where $u =$ __?__.

80. The trinomial $5x^4y^2 - 17x^2y - 12$ can be written in the quadratic form $5u^2 - 17u - 12$, where $u =$ __?__.

81. What does it mean for a polynomial to be quadratic in form?

82. Do all polynomials that are quadratic in form factor? If not, give an example of one that does not factor.

Factor.

83. $x^2y^2 - 8xy + 15$

84. $x^2y^2 - 8xy - 33$

85. $x^4 - 9x^2 + 18$

86. $y^4 - 6y^2 - 16$

87. $b^4 - 13b^2 - 90$

88. $a^4 + 14a^2 + 45$

89. $x^4y^4 - 8x^2y^2 + 12$

90. $a^4b^4 + 11a^2b^2 - 26$

91. $3x^2y^2 - 14xy + 15$

92. $5x^2y^2 - 59xy + 44$

93. $6a^2b^2 - 23ab + 21$

94. $10a^2b^2 + 3ab - 7$

95. $2x^4 - 13x^2 - 15$

96. $3x^4 + 20x^2 + 32$

97. $6x^5y + x^3y - 12xy$

98. $36x^6y^3 + 45x^4y^3 - 54x^2y^3$

99. $16x^6y^2 - 80x^4y^2 + 100x^2y^2$

100. $60x^7y^3 + 33x^5y^3 - 9x^3y^3$

101. $12x^6y^3 + 2x^4y^3 - 4x^2y^3$

102. $72x^7 - 12x^5 - 4x^3$

4 Factor completely (See pages 302–303.)

GETTING READY

103. When factoring a polynomial, you should look first for a ___?___ factor.

104. To factor a polynomial with four terms, try factoring by ___?___.

Factor.

105. $12x^2 - 36x + 27$

106. $5x^2 + 10x + 5$

107. $27a^4 - a$

108. $3x^4 - 81x$

109. $20x^2 - 5$

110. $7x^2 - 28$

111. $y^5 + 6y^4 - 55y^3$

112. $y^4 - 10y^3 + 21y^2$

113. $16x^4 - 81$

114. $x^4 - 16$

115. $16a - 2a^4$

116. $8x^5 - 98x^3$

117. $x^3 + 2x^2 - x - 2$

118. $x^3y^3 - x^3$

119. $2x^3 + 4x^2 - 3x - 6$

120. $2x^3 - 3x^2 - 8x + 12$

121. $x^3 + x^2 - 16x - 16$

122. $3x^3 - 3x^2 + 4x - 4$

123. $x^4 - 2x^3 - 35x^2$

124. $4x^3 + 8x^2 - 9x - 18$

125. $4x^2 + 4x - 1$

126. $8x^4 - 40x^3 + 50x^2$

127. $6x^5 + 74x^4 + 24x^3$

128. $x^4 - y^4$

129. $16a^4 - b^4$

130. $a^4 - 25a^2 - 144$

131. $x^4 - 5x^2 - 4$

132. $16a^4 - 2a$

133. $3b^5 - 24b^2$

134. $a^4b^2 - 8a^3b^3 - 48a^2b^4$

135. $x^4y^2 - 5x^3y^3 + 6x^2y^4$

136. $24a^2b^2 - 14ab^3 - 90b^4$

137. $16x^3y + 4x^2y^2 - 42xy^3$

138. $x^3 - 2x^2 - 4x + 8$

139. $x^3 - 2x^2 - x + 2$

140. $8xb - 8x - 4b + 4$

141. The factored form of a polynomial $P(x)$ is $x^n(x - a)(x + b)$. What is the degree of the polynomial P?

142. The factored form of a polynomial $P(x)$ is $x^n(x^n + a)(x^n + b)$. What is the degree of the polynomial P?

APPLYING CONCEPTS

Find all integers k such that the trinomial is a perfect-square trinomial.

143. $4x^2 - kx + 25$

144. $9x^2 - kx + 1$

145. $16x^2 + kxy + y^2$

146. $49x^2 + kxy + 64y^2$

Factor.

147. $(a - b)^3 - b^3$

148. $a^3 + (a + b)^3$

Factor the polynomial. Assume that n is a positive integer.

149. $x^{2n} - 49$

150. $x^{2n} + x^n - 6$

151. $x^{3n} + 8$

152. $x^{4n} + 2x^{2n} - 8$

153. $x^{4n} - 16$

154. $x^{4n} - 5x^{2n} + 4$

155. Factor $x^4 + 64$. [*Suggestion:* Add and subtract $16x^2$ so that the expression becomes $(x^4 + 16x^2 + 64) - 16x^2$. Now factor the difference of two squares.]

156. Using the strategy of Exercise 155, factor $x^4 + 4$. (*Suggestion:* Add and subtract $4x^2$.)

157. Can a third-degree polynomial have factors $x - 1$, $x + 1$, $x - 3$, and $x + 4$? Why or why not?

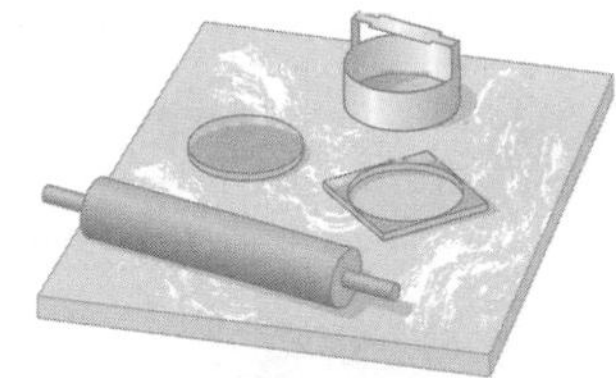

158. Food Science A large circular cookie is cut from a square piece of dough. The diameter of the cookie is x centimeters. The piece of dough is x centimeters on a side and 1 cm deep. In terms of x, how many cubic centimeters of dough are left over? Use the approximation 3.14 for π.

PROJECTS OR GROUP ACTIVITIES

If you have not completed Exercises 110 to 112 from Section 5.4, you should complete them now.

159. Given that 3 is a zero of $P(x) = x^3 - x^2 - 3x - 9$, determine the factorization over the integers of $x^3 - x^2 - 3x - 9$.

160. Given that -3 and 2 are zeros of $P(x) = x^4 + 2x^3 - 4x^2 - 5x - 6$, determine the factorization over the integers of $x^4 + 2x^3 - 4x^2 - 5x - 6$.

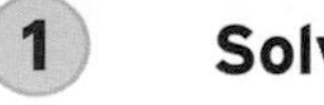

5.8 Solving Equations by Factoring

OBJECTIVE 1

Solve equations by factoring

If the product of two factors is zero, then one or both of the factors must be zero.

> **PRINCIPLE OF ZERO PRODUCTS**
>
> If the product of two factors is zero, then at least one of the factors is equal to zero. If $ab = 0$, then $a = 0$ or $b = 0$.

Take Note

The Principle of Zero Products contains the phrase *at least one*. This phrase means that *one or both* of the factors could be zero. Note how this concept is used in example (3) at the right.

EXAMPLES

1. Suppose $5x = 0$. The factors are 5 and x. The product equals zero, so at least one of the factors must be zero. Because $5 \neq 0$, we know that $x = 0$.
2. Suppose $-3(x - 5) = 0$. The factors are -3 and $x - 5$. The product equals zero, so at least one of the factors must be zero. Because $-3 \neq 0$, we know that $x - 5 = 0$, which means that $x = 5$.
3. Suppose $(x - 5)(x + 4) = 0$. The factors are $x - 5$ and $x + 4$. The product equals 0, so $x - 5 = 0$ or $x + 4 = 0$. If $x - 5 = 0$, then $x = 5$. If $x + 4 = 0$, then $x = -4$.

An equation of the form $ax^2 + bx + c = 0$, $a \neq 0$, is a **quadratic equation.** A quadratic equation is in **standard form** when the polynomial is written in descending order and is equal to zero.

Some quadratic equations can be solved by factoring and then using the Principle of Zero Products.

Focus on solving a quadratic equation by using the Principle of Zero Products

Solve: $2x^2 - x = 1$

	$2x^2 - x = 1$
Write the equation in standard form.	$2x^2 - x - 1 = 0$
Factor the trinomial.	$(2x + 1)(x - 1) = 0$
Use the Principle of Zero Products.	$2x + 1 = 0 \qquad x - 1 = 0$
Solve each equation for x.	$2x = -1 \qquad x = 1$
	$x = -\frac{1}{2}$
The solutions check.	The solutions are $-\frac{1}{2}$ and 1.

EXAMPLE 1 Solve. **A.** $x(x + 2) = 15$ **B.** $(x + 4)(x - 3) = 3x - 4$

Solution **A.**

$$x(x + 2) = 15$$
$$x^2 + 2x = 15$$ • Multiply $x(x + 2)$.
$$x^2 + 2x - 15 = 0$$ • Write the equation in standard form.

$$(x + 5)(x - 3) = 0$$ • Factor the trinomial.

$$x + 5 = 0 \qquad x - 3 = 0$$
$$x = -5 \qquad x = 3$$ • Use the Principle of Zero Products. The product of $x + 5$ and $x - 3$ equals 0. Therefore, $x + 5 = 0$ or $x - 3 = 0$.

The solutions are -5 and 3.

B.

$$(x + 4)(x - 3) = 3x - 4$$
$$x^2 + x - 12 = 3x - 4$$ • Multiply $(x + 4)(x - 3)$.
$$x^2 - 2x - 8 = 0$$ • Write the equation in standard form.

$$(x + 2)(x - 4) = 0$$ • Factor the trinomial.

$$x + 2 = 0 \qquad x - 4 = 0$$
$$x = -2 \qquad x = 4$$ • Use the Principle of Zero Products. The product of $x + 2$ and $x - 4$ equals 0. Therefore, $x + 2 = 0$ or $x - 4 = 0$.

The solutions are -2 and 4.

Problem 1 Solve. **A.** $4x^2 + 11x = 3$ **B.** $(x - 2)(x + 5) = 8$

Solution See pages S18–S19.

Try Exercise 23, page 312.

The Principle of Zero Products can be extended to more than two factors. For example, if $abc = 0$, then $a = 0$, $b = 0$, or $c = 0$.

Focus on solving a third-degree equation by using the Principle of Zero Products

Solve: $x^3 - x^2 - 25x + 25 = 0$

Factor by grouping.

$$x^3 - x^2 - 25x + 25 = 0$$
$$(x^3 - x^2) - (25x - 25) = 0$$
$$x^2(x - 1) - 25(x - 1) = 0$$
$$(x - 1)(x^2 - 25) = 0$$
$$(x - 1)(x + 5)(x - 5) = 0$$

Use the Principle of Zero Products. $x - 1 = 0$ $x + 5 = 0$ $x - 5 = 0$

Solve each equation for x. $x = 1$ $x = -5$ $x = 5$

The solutions are -5, 5, and 1.

The Principle of Zero Products is used to find elements in the domain of a quadratic function that correspond to a given element in the range.

EXAMPLE 2 Given that -1 is in the range of the function defined by $f(x) = x^2 - 3x - 5$, find two values of c for which $f(c) = -1$.

Solution

$f(c) = -1$

$c^2 - 3c - 5 = -1$ • $f(c) = c^2 - 3c - 5$

$c^2 - 3c - 4 = 0$ • Write in standard form.

$(c - 4)(c + 1) = 0$ • Solve for c by factoring.

$c - 4 = 0$ $c + 1 = 0$ • Use the Principle of Zero Products.

$c = 4$ $c = -1$

The values of c are -1 and 4.

Take Note

You can check the values of c by evaluating the function.

$f(x) = x^2 - 3x - 5$
$f(4) = 4^2 - 3(4) - 5$
$= -1$ ✓
$f(-1) = (-1)^2 - 3(-1) - 5$
$= -1$ ✓

Problem 2 Given that 4 is in the range of the function defined by $s(t) = t^2 - t - 2$, find two values of c for which $s(c) = 4$.

Solution See page S19.

Try Exercise 51, page 312.

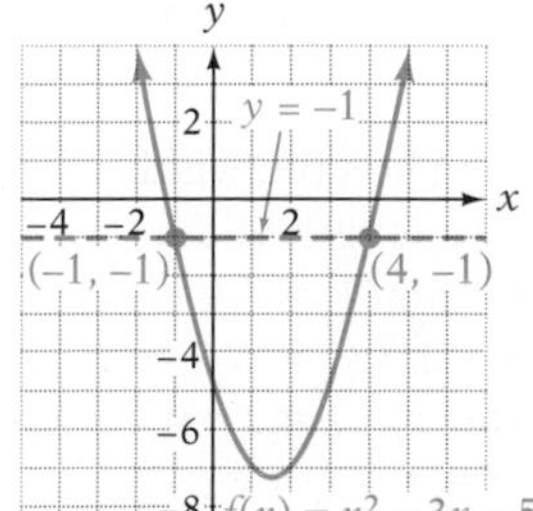

In Example 2, there are two values in the domain that can be paired with the range element -1. The two values are -1 and 4. Two ordered pairs that belong to the function are $(-1, -1)$ and $(4, -1)$. *Remember:* A function can have different first elements paired with the same second element. A function cannot have the same first element paired with different second elements.

The graph of $f(x) = x^2 - 3x - 5$ is shown at the left.

Recall that a *zero* of a function f is a number c for which $f(c) = 0$. The zeros of some quadratic functions can be found by using the Principle of Zero Products. Finding the zeros is similar to solving Example 2, where the range element is 0.

EXAMPLE 3 Find the zeros of $f(x) = x^2 - 2x - 8$.

Solution We must find the values of c for which $f(c) = 0$.

$$f(c) = 0$$

$$c^2 - 2c - 8 = 0 \qquad \bullet\ f(x) = x^2 - 2x - 8.\ \text{Thus } f(c) = c^2 - 2c - 8.$$

$$(c + 2)(c - 4) = 0 \qquad \bullet\ \text{Factor and use the Principle of Zero Products.}$$

$$c + 2 = 0 \qquad c - 4 = 0$$

$$c = -2 \qquad c = 4$$

The zeros of f are -2 and 4.

Problem 3 Find the zeros of $s(t) = t^2 + 5t - 14$.

Solution See page S19.

Try Exercise 59, page 312.

A graphing calculator can be used to find the zeros in Example 3. First graph the function. Some sample screens are shown below. See the Appendix for keystroke help.

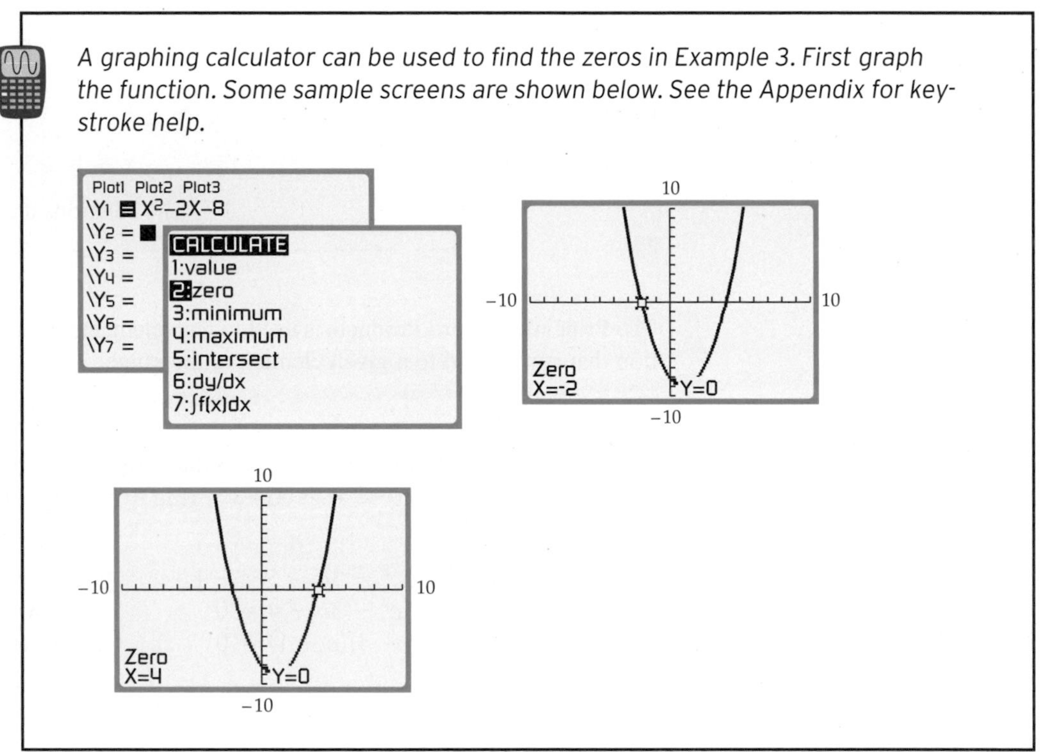

OBJECTIVE 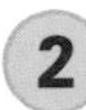 Application problems

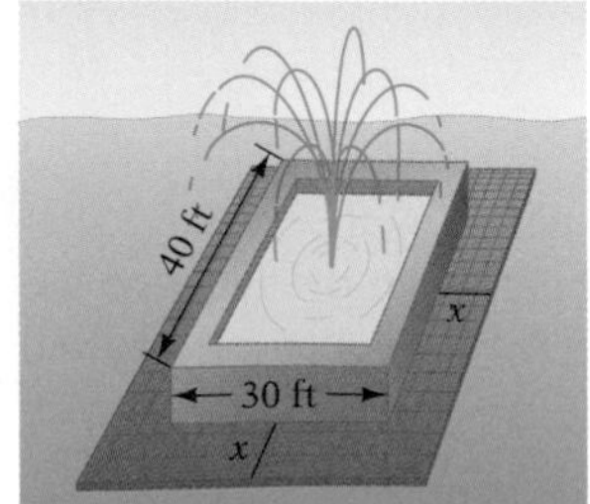

EXAMPLE 4 An architect wants to design a fountain to be placed on the front lawn of a new art museum. The base of the fountain is to be 30 ft by 40 ft, with a uniform brick walkway around the fountain. The total area of the fountain and the walkway is 2576 ft^2. How wide is the walkway?

Strategy Let x represent the width of the brick walkway. Then the width of the fountain and walkway is $30 + 2x$, and the length of the fountain and walkway is $40 + 2x$. Use the formula $A = LW$, where $A = 2576$, to write an equation.

Solution

$$\begin{aligned} A &= LW \\ 2576 &= (40 + 2x)(30 + 2x) && \bullet \text{ Replace } A, L, \text{ and } W \text{ by their values.} \\ 2576 &= 1200 + 140x + 4x^2 && \bullet \text{ Multiply.} \\ 0 &= 4x^2 + 140x - 1376 && \bullet \text{ Write in standard form.} \\ 0 &= x^2 + 35x - 344 && \bullet \text{ Divide each side by 4.} \\ 0 &= (x - 8)(x + 43) && \bullet \text{ Factor and use the Principle of Zero Products.} \end{aligned}$$

$$x - 8 = 0 \qquad x + 43 = 0$$
$$x = 8 \qquad x = -43$$

A walkway -43 ft wide would not make sense.

The walkway must be 8 ft wide.

Take Note

Sometimes a solution of a quadratic equation does not make sense in terms of the problem situation. Eliminate such a solution.

Problem 4 A diagonal of a polygon is a line from one vertex of the polygon to a nonadjacent vertex. The number of diagonals D of a polygon with n sides is given by $D = \frac{n(n-3)}{2}$. Find the number of sides for a polygon that has 20 diagonals.

Solution See page S19.

Try Exercise 67, page 313.

5.8 Exercises

CONCEPT CHECK

1. What is the Principle of Zero Products?

2. True or false? By the Principle of Zero Products, if $(x + 2)(x - 3) = 4$, then $x + 2 = 4$ or $x - 3 = 4$.

3. State whether the given equation is a quadratic equation.
   a. $3x^2 + 6x = 0$ **b.** $x^2 - 3x + 2$ **c.** $3x + 6 = x^2$
   d. $2x - 4 = 0$ **e.** $x^2 = 25$ **f.** $x^4 - 2x^2 - 3 = 0$

4. Write each equation in standard form, with the coefficient of x^2 a positive number.
   a. $2x^2 + 3 = -6x$ **b.** $3x = 6 - x^2$ **c.** $x(x + 1) - 3 = 0$ **d.** $(x + 2)(x - 4) = 9$

1 Solve equations by factoring (See pages 307–310.)

5. What is a quadratic equation? How does it differ from a linear equation? Give an example of each type of equation.

6. How is the Principle of Zero Products used to solve some quadratic equations?

GETTING READY

7. Let $f(x) = x^2 + x - 15$. To find two values of c for which $f(c) = 5$, solve the equation $c^2 + c - 15 =$ ___?___.

8. Solve: $3x^2 - 5x - 2 = 0$

$3x^2 - 5x - 2 = 0$ • **The equation is in __?__ form.**

$(x - \_\_?\_\_)(3x + \_\_?\_\_) = 0$ • **Factor the trinomial.**

$x - 2 = \_\_?\_\_$ $3x + 1 = \_\_?\_\_$ • **Use the Principle of Zero Products.**

$x = \_\_?\_\_$ $x = \_\_?\_\_$ • **Solve each equation for *x*.**

Solve.

9. $(y + 4)(y + 6) = 0$
10. $(a - 5)(a - 2) = 0$
11. $x(x - 7) = 0$
12. $3z(2z + 5) = 0$
13. $(2x + 3)(x - 7) = 0$
14. $(4a - 1)(a + 9) = 0$
15. $b^2 - 49 = 0$
16. $9t^2 - 16 = 0$
17. $y^2 + 4y - 5 = 0$
18. $x^2 + x - 6 = 0$
19. $2b^2 - 5b - 12 = 0$
20. $a^2 - 8a + 16 = 0$
21. $x^2 - 9x = 0$
22. $3a^2 - 12a = 0$
23. $z^2 - 3z = 28$
24. $b^2 - 4b = 32$
25. $3t^2 + 13t = 10$
26. $2x^2 - 5x = 12$
27. $5b^2 - 17b = -6$
28. $4y^2 - 19y = 5$
29. $8x^2 - 10x = 3$
30. $6a^2 + a = 2$
31. $y(y - 2) = 35$
32. $z(z - 1) = 20$
33. $y(3y - 2) = 8$
34. $x(2x - 5) = 12$
35. $3a^2 - 4a = 20 - 15a$
36. $2b^2 - 6b = b - 3$
37. $(y + 5)(y - 7) = -20$
38. $(x + 2)(x - 6) = 20$
39. $(b + 5)(b + 10) = 6$
40. $(a - 9)(a - 1) = -7$
41. $(a - 1)^2 = 3a - 5$
42. $(b - 2)^2 + b^2 = 10$
43. $x^3 + 4x^2 - x - 4 = 0$
44. $x^3 + x^2 - 9x - 9 = 0$
45. $x^3 - 4x^2 - 25x + 100 = 0$
46. $2x^3 + 3x^2 - 18x - 27 = 0$
47. $3x^3 + 2x^2 - 48x - 32 = 0$
48. $2x^3 + x^2 - 8x - 4 = 0$
49. $9x^3 - 9x^2 - 25x + 25 = 0$
50. $12x^3 - 8x^2 - 3x + 2 = 0$

Find the values c in the domain of f for which $f(c)$ is the indicated value.

51. $f(x) = x^2 - 3x + 3;\ f(c) = 1$
52. $f(x) = x^2 + 4x - 2;\ f(c) = 3$
53. $f(x) = 2x^2 - x - 5;\ f(c) = -4$
54. $f(x) = 6x^2 - 5x - 9;\ f(c) = -3$

Find the zeros of the function.

55. $f(x) = x^2 + 3x - 4$
56. $g(x) = x^2 + 2x - 15$
57. $s(t) = t^2 - 4t - 12$
58. $f(t) = t^2 + 3t - 18$
59. $f(x) = 2x^2 - 5x + 3$
60. $g(x) = 3x^2 - 7x - 6$

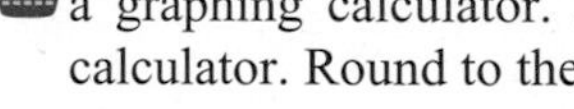

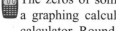

The zeros of some quadratic equations that do not factor easily can be found by using a graphing calculator. Find the zeros of the following equations by using a graphing calculator. Round to the nearest hundredth.

61. $f(x) = x^2 + 3x - 1$

62. $f(x) = -x^2 - 2x + 4$

63. $f(x) = 2x^2 - 3x - 1$

64. $f(x) = -2x^2 + 4x + 1$

2 Application problems (See pages 310–311.)

65. Geometry The number of possible diagonals D in a polygon with n sides is given by $D = \frac{n(n-3)}{2}$. Find the number of sides of a polygon with 54 diagonals.

66. Geometry The length of a rectangle is 5 in. more than twice the width. The area is 168 in^2. Find the width and length of the rectangle.

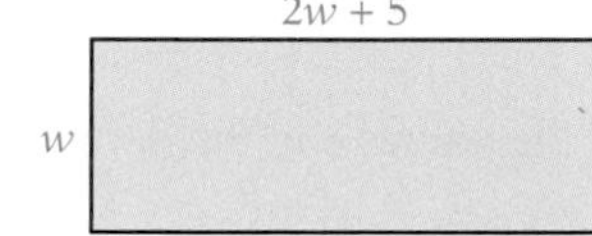

67. Publishing Read the article at the right. If the length of the rectangular e-reader is 5 cm less than twice the width, find the length and width of the e-reader.

In the News

Pocket-Sized E-Reader

Sony has introduced a small e-reader that might truly fit in your pocket, its size being a rectangle of only 150 cm^2.

Source: dailyator.com

68. Geometry The width of a rectangle is 5 ft less than the length. The area of the rectangle is 300 ft^2. Find the length and width of the rectangle.

69. Geometry The length of the base of a triangle is three times the height. The area of the triangle is 24 cm^2. Find the base and height of the triangle.

70. Geometry The height of a triangle is 4 cm more than twice the length of the base. The area of the triangle is 35 cm^2. Find the height of the triangle.

71. Geometry The length of a rectangle is 6 cm, and the width is 3 cm. If both the length and the width are increased by equal amounts, the area of the rectangle is increased by 70 cm^2. Find the length and width of the larger rectangle.

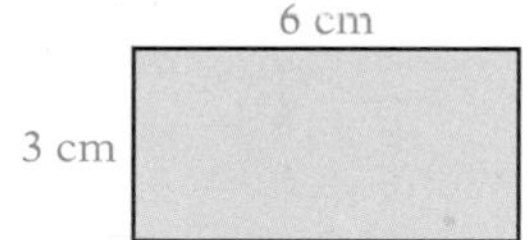

72. Geometry The width of a rectangle is 4 cm, and the length is 8 cm. If both the width and the length are increased by equal amounts, the area of the rectangle is increased by 64 cm^2. Find the length and width of the larger rectangle.

73. Construction A trough is made from a rectangular piece of metal by folding up the sides as shown in the figure at the right. What should the value of x be so that the volume is 72 m^3?

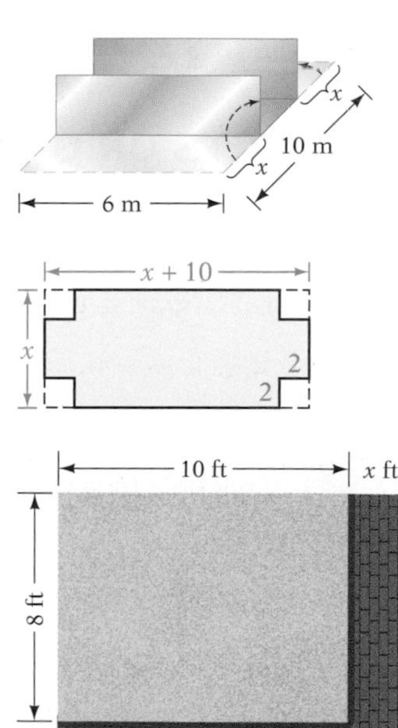

74. Construction A rectangular piece of cardboard is 10 in. longer than it is wide. Squares 2 in. on a side are to be cut from each corner and then the sides folded up to make an open box with a volume of 112 in^3. Find the length and width of the piece of cardboard.

75. Construction A homeowner is constructing a brick walkway around an existing 8-foot by 10-foot concrete patio, as shown in the diagram at the right. How wide should the brick walkway be if the total area of the patio and walkway must be 143 ft^2?

76. Physics The height h, in feet, of a ball above the ground t seconds after being thrown upward with a velocity of 48 ft/s is given by $h = -16t^2 + 48t + 3$. After how many seconds will the ball be 35 ft above the ground?

77. Physics The height h, in feet, of a ball above the ground t seconds after being thrown upward at a velocity of 64 ft/s is given by $h = -16t^2 + 64t + 3$. After how many seconds will the ball be 63 ft above the ground?

78. Space Science A small rocket is launched with an acceleration of 10 m/s^2. The velocity v of the rocket after it has traveled s meters is given by $v^2 = 20s$. Find the velocity of this rocket after it has traveled 500 m.

APPLYING CONCEPTS

79. Find $3n^2 + 2n - 1$ if $n(n + 6) = 16$.

80. Find $4y^2 - y + 3$ if $(y - 1)(y + 2) = 4$.

81. Find $-a^2 + 5a - 2$ if $6a(a - 1) = 36$.

Find the values c in the domain of f for which $f(c)$ is the indicated value.

82. $f(x) = x^3 + 9x^2 - x - 14$; $f(c) = -5$

83. $f(x) = x^3 + 3x^2 - 4x - 11$; $f(c) = 1$

84. Geometry The perimeter of a rectangular garden is 44 m. The area of the garden is 120 m^2. Find the length and width of the garden.

85. Geometry The sides of a rectangular box have areas of 16 cm^2, 20 cm^2, and 80 cm^2. Find the volume of the box.

86. Write an equation whose solutions are 3, 2, and -1.

87. The following seems to show that $1 = 2$. Explain the error.

$a = b$

$a^2 = ab$ • **Multiply each side of the equation by a.**

$a^2 - b^2 = ab - b^2$ • **Subtract b^2 from each side of the equation.**

$(a - b)(a + b) = b(a - b)$ • **Factor.**

$a + b = b$ • **Divide each side by $a - b$.**

$b + b = b$ • **Because $a = b$, substitute b for a.**

$2b = b$

$2 = 1$ • **Divide both sides by b.**

PROJECTS OR GROUP ACTIVITIES

Solve for x.

88. $x^2 - 9ax + 14a^2 = 0$

89. $x^2 + 9xy - 36y^2 = 0$

90. $3x^2 - 4cx + c^2 = 0$

91. $2x^2 + 3bx + b^2 = 0$

92. Is it possible for a third-degree equation to have 1, 2, 3, and 4 as solutions? Why or why not?

CHAPTER 5 Summary

Key Words	Objective and Page Reference	Examples
A **monomial** is a number, a variable, or a product of numbers and variables.	[5.1.1, p. 242]	$2, x, 3x, -4x^2$, and $23x^2y^3$ are monomials.
A number written in **scientific notation** is a number written in the form $a \times 10^n$, where $1 \leq a < 10$ and n is an integer.	[5.1.3, p. 248]	$0.000000023 = 2.3 \times 10^{-8}$
A **polynomial** is a variable expression in which the terms are monomials.	[5.2.1, p. 255]	$x^4 - 2xy - 32x + 8$ is a polynomial. The terms are x^4, $-2xy$, $-32x$, and 8.
A polynomial of two terms is a **binomial.**	[5.2.1, p. 255]	$3x - y$ is a binomial.
A polynomial of three terms is a **trinomial.**	[5.2.1, p. 255]	$x^2 - 2x + 4$ is a trinomial.
The **degree of a polynomial** is the greatest of the degrees of any of its terms.	[5.2.1, p. 256]	The degree of the polynomial $x^3 + 3x^2y^2 - 4xy - 3$ is 4.
Synthetic division is a shorter method of dividing a polynomial by a binomial of the form $x - a$. This method uses only the coefficients of the variable terms.	[5.4.3, p. 278]	$\begin{array}{r\|rrrr} 2 & 3 & 0 & -9 & -5 \\ & & 6 & 12 & 6 \\ \hline & 3 & 6 & 3 & 1 \end{array}$ $(3x^3 - 9x - 5) \div (x - 2)$ $= 3x^2 + 6x + 3 + \dfrac{1}{x-2}$
To **factor a polynomial** means to write the polynomial as the product of other polynomials.	[5.5.1, p. 285]	$x^2 + 5x + 6 = (x + 2)(x + 3)$
A **quadratic trinomial** is a polynomial of the form $ax^2 + bx + c$, where a and b are nonzero coefficients and c is a nonzero constant.	[5.6.1, 5.6.2, pp. 290, 292]	$3x^2 - 2x + 8$: $a = 3$, $b = -2$, $c = 8$
To **factor a quadratic trinomial** means to express the trinomial as the product of two binomials.	[5.6.1, p. 290]	$x^2 + 2x - 8 = (x - 2)(x + 4)$
A polynomial is **nonfactorable over the integers** if it does not factor using only integers.	[5.6.1, p. 292]	$x^2 + x + 1$ is nonfactorable.
The product of a term and itself is a **perfect square.**	[5.7.1, p. 299]	$(5x)(5x) = 25x^2$; $25x^2$ is a perfect square.
The **square root of a perfect square** is one of the two equal factors of the perfect square.	[5.7.1, p. 299]	$\sqrt{25x^2} = 5x$
The product of the same three factors is called a **perfect cube.**	[5.7.2, p. 301]	$(2x)(2x)(2x) = 8x^3$; $8x^3$ is a perfect cube.
The **cube root of a perfect cube** is one of the three equal factors of the perfect cube.	[5.7.2, p. 301]	$\sqrt[3]{8x^3} = 2x$

A trinomial is **quadratic in form** if it can be written as $au^2 + bu + c$.	[5.7.3, p. 302]	$x^4 - 3x^2 + 2$ is quadratic in form because $x^4 - 3x^2 + 2 = u^2 - 3u + 2$, where $u = x^2$.
A **quadratic equation** is an equation of the form $ax^2 + bx + c = 0, a \neq 0$. A quadratic equation is in **standard form** when the polynomial is in descending order and equal to zero.	[5.8.1, p. 308]	$3x^2 + 3x + 8 = 0$ is a quadratic equation in standard form.

Essential Rules and Procedures

Essential Rules and Procedures	Objective and Page Reference	Examples
Rule for Multiplying Exponential Expressions If m and n are integers, then $x^m \cdot x^n = x^{m+n}$.	[5.1.1, p. 242]	$x^2 \cdot x^5 = x^{2+5} = x^7$
Rule for Simplifying a Power of an Exponential Expression If m and n are integers, then $(x^m)^n = x^{mn}$.	[5.1.1, p. 243]	$(x^4)^3 = x^{4 \cdot 3} = x^{12}$
Rule for Simplifying Powers of Products If m, n, and p are integers, then $(x^m y^n)^p = x^{mp}y^{np}$.	[5.1.1, p. 243]	$(x^2y^5)^3 = x^{2\cdot 3}y^{5\cdot 3} = x^6y^{15}$
Rule for Dividing Exponential Expressions If m and n are integers and $x \neq 0$, then $\frac{x^m}{x^n} = x^{m-n}$.	[5.1.2, p. 244]	$\frac{x^{12}}{x^7} = x^{12-7} = x^5$
Definition of Zero as an Exponent If $x \neq 0$, then $x^0 = 1$.	[5.1.2, p. 245]	$5^0 = 1 \quad (2xy^4)^0 = 1$
Definition of a Negative Exponent If $n > 0$ and $x \neq 0$, then $x^{-n} = \frac{1}{x^n}$ and $\frac{1}{x^{-n}} = x^n$.	[5.1.2, p. 245]	$x^{-3} = \frac{1}{x^3} \quad \frac{1}{x^{-4}} = x^4$
Rule for Simplifying Powers of Quotients If m, n, and p are integers and $y \neq 0$, then $\left(\frac{x^m}{y^n}\right)^p = \frac{x^{mp}}{y^{np}}$.	[5.1.2, p. 246]	$\left(\frac{x^2}{y^4}\right)^5 = \frac{x^{2\cdot 5}}{y^{4\cdot 5}} = \frac{x^{10}}{y^{20}}$
To divide a polynomial by a monomial, divide each term of the polynomial by the monomial.	[5.4.1, p. 275]	$\frac{12x^5 + 8x^3 - 6x}{4x^2} = 3x^3 + 2x - \frac{3}{2x}$
Remainder Theorem If the polynomial $P(x)$ is divided by $x - a$, the remainder is $P(a)$.	[5.4.4, p. 280]	$P(x) = x^3 - x^2 + x - 1$ $\begin{array}{r\|rrrr} -2 & 1 & -1 & 1 & -1 \\ & & -2 & 6 & -14 \\ \hline & 1 & -3 & 7 & -15 \end{array}$ $P(-2) = -15$

Factor the Difference of Two Squares $a^2 - b^2 = (a - b)(a + b)$	[5.7.1, p. 299]	$x^2 - 9 = (x - 3)(x + 3)$
Factor a Perfect-Square Trinomial $a^2 + 2ab + b^2 = (a + b)^2$ $(a^2 - 2ab + b^2) = (a - b)^2$	[5.7.1, p. 300]	$4x^2 + 12x + 9 = (2x + 3)^2$ $x^2 - 10x + 25 = (x - 5)^2$
Factor the Sum or Difference of Two Cubes $a^3 + b^3 = (a + b)(a^2 - ab + b^2)$ $a^3 - b^3 = (a - b)(a^2 + ab + b^2)$	[5.7.2, p. 301]	$x^3 + 64 = (x + 4)(x^2 - 4x + 16)$ $8b^3 - 1 = (2b - 1)(4b^2 + 2b + 1)$
The Principle of Zero Products If $ab = 0$, then $a = 0$ or $b = 0$.	[5.8.1, p. 307]	$(x - 4)(x + 2) = 0$ $x - 4 = 0 \quad x + 2 = 0$ $x = 4 \quad x = -2$

CHAPTER 5 Review Exercises

1. Add: $(3x^2 - 2x - 6) + (-x^2 - 3x + 4)$

2. Subtract: $(5x^2 - 8xy + 2y^2) - (x^2 - 3y^2)$

3. Multiply: $(5x^2yz^4)(2xy^3z^{-1})(7x^{-2}y^{-2}z^3)$

4. Multiply: $(2x^{-1}y^2z^5)^4(-3x^3yz^{-3})$

5. Simplify: $\dfrac{3x^4yz^{-1}}{-12xy^3z^2}$

6. Simplify: $\dfrac{(2a^4b^{-3}c^2)^3}{(2a^3b^2c^{-1})^4}$

7. Write 93,000,000 in scientific notation.

8. Write 2.54×10^{-3} in decimal notation.

9. Simplify: $\dfrac{3 \times 10^{-3}}{1.5 \times 10^3}$

10. Given $P(x) = 2x^3 - x + 7$, evaluate $P(-2)$.

11. Graph: $y = x^2 + 1$

12. Identify **a.** the leading coefficient, **b.** the constant term, and **c.** the degree of the polynomial $P(x) = 3x^5 - 3x^2 + 7x + 8$.

13. Use the Remainder Theorem to evaluate $P(x) = -2x^3 + 2x^2 - 4$ when $x = -3$.

14. Divide: $\dfrac{6x^2y^3 - 18x^3y^2 + 12x^4y}{3x^2y}$

15. Divide: $\dfrac{15x^2 + 2x - 2}{3x - 2}$

16. Divide: $\dfrac{12x^2 - 16x - 7}{6x + 1}$

17. Divide: $\dfrac{4x^3 + 27x^2 + 10x + 2}{x + 6}$

18. Divide: $\dfrac{x^3 - 2x + 2}{x - 2}$

19. Multiply: $4x^2y(3x^3y^2 + 2xy - 7y^3)$

20. Simplify: $5x^2 - 4x[x - (3x + 2) + x]$

21. Multiply: $(x + 6)(x^3 - 3x^2 - 5x + 1)$

22. Multiply: $(x - 4)(3x + 2)(2x - 3)$

23. Multiply: $(5a + 2b)(5a - 2b)$

24. Simplify: $(4x - 3y)^2$

25. Factor: $18a^5b^2 - 12a^3b^3 + 30a^2b$

26. Factor: $x(y - 3) + 4(3 - y)$

27. Factor: $2ax + 4bx - 3ay - 6by$

28. Factor: $x^2 + 12x + 35$

29. Factor: $12 + x - x^2$

30. Factor: $x^2 - 16x + 63$

31. Factor: $6x^2 - 31x + 18$

32. Factor: $24x^2 + 61x - 8$

33. Factor: $x^2y^2 - 9$

34. Factor: $4x^2 + 12xy + 9y^2$

35. Factor: $27x^3 + 8$

36. Factor: $64a^3 - 27b^3$

37. Factor: $x^4 - 13x^2 + 36$

38. Factor: $15x^4 + x^2 - 6$

39. Factor: $36x^8 - 36x^4 + 5$

40. Factor: $21x^4y^4 + 23x^2y^2 + 6$

41. Factor: $3a^6 - 15a^4 - 18a^2$

42. Factor: $3a^4b - 3ab^4$

43. Solve: $x^3 - x^2 - 6x = 0$

44. Solve: $6x^2 + 60 = 39x$

45. Solve: $x^3 + 5x^2 - 4x - 20 = 0$

46. Solve: $y^3 + y^2 - 36y - 36 = 0$

47. Find the zeros of $f(x) = x^2 - 5x - 6$.

48. Find two values of c in the domain of $f(x) = x^2 + 4x - 9$ such that $f(c) = 3$.

49. **Astronomy** The most distant object visible from Earth without the aid of a telescope is the Great Galaxy of Andromeda. It takes light from the Great Galaxy of Andromeda 2.2×10^6 years to travel to Earth. Light travels about 5.9×10^{12} mph. How far from Earth is the Great Galaxy of Andromeda?

50. **Solar Power** Light from the sun supplies Earth with 2.4×10^{14} horsepower. Earth receives only 2.2×10^{-7} of the power generated by the sun. How much power is generated by the sun?

51. **Geometry** The length of a rectangle is $(5x + 3)$ cm. The width is $(2x - 7)$ cm. Find the area of the rectangle in terms of the variable x.

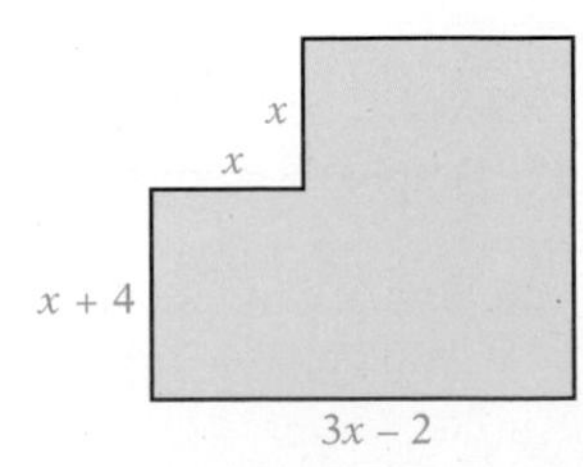

52. **Geometry** Find the area of the figure shown at the right. All dimensions given are in inches.

53. **Geometry** The length of a rectangle is 2 m more than twice the width. The area of the rectangle is 60 m^2. Find the length of the rectangle.

54. **Physics** The height h, in feet, of a ball above the ground t seconds after being batted upward at a velocity of 96 ft/s is given by $h = -16t^2 + 96t + 3$. After how many seconds will the ball be 143 ft above the ground?

CHAPTER 5 Test

1. Subtract: $(6x^3 - 7x^2 + 6x - 7) - (4x^3 - 3x^2 + 7)$

2. Simplify: $(-4a^2b)^3(-ab^4)$

3. Simplify: $\dfrac{(2a^{-4}b^2)^3}{4a^{-2}b^{-1}}$

4. Write the number 0.000000501 in scientific notation.

5. Write the number of seconds in 1 week in scientific notation.

6. Simplify: $(2x^{-3}y)^{-4}$

7. Simplify: $-5x[3 - 2(2x - 4) - 3x]$

8. Multiply: $(3a + 4b)(2a - 7b)$

9. Multiply: $(x^2 + 3x - 5)(2x + 3)$

10. Simplify: $(3z - 5)^2$

11. Divide: $\dfrac{25x^2y^2 - 15x^2y + 20xy^2}{5x^2y}$

12. Divide: $(4x^3 + x - 15) \div (2x - 3)$

13. Divide: $(x^3 - 5x^2 + 5x + 5) \div (x - 3)$

14. Given $P(x) = 3x^2 - 8x + 1$, evaluate $P(2)$.

15. Use the Remainder Theorem to evaluate $P(x) = -x^3 + 4x - 8$ when $x = -2$.

16. Factor: $6a^4 - 13a^2 - 5$

17. Factor: $12x^3 + 12x^2 - 45x$

18. Factor: $16x^2 - 25$

19. Factor: $16t^2 + 24t + 9$

20. Factor: $27x^3 - 8$

21. Factor: $6x^2 - 4x - 3xa + 2a$

22. Find the zeros of $g(x) = 2x^2 - 5x - 12$.

23. Solve: $6x^2 = x + 1$

24. Solve: $6x^3 + x^2 - 6x - 1 = 0$

25. **Geometry** The length of a rectangle is $(5x + 1)$ ft. The width is $(2x - 1)$ ft. Find the area of the rectangle in terms of the variable x.

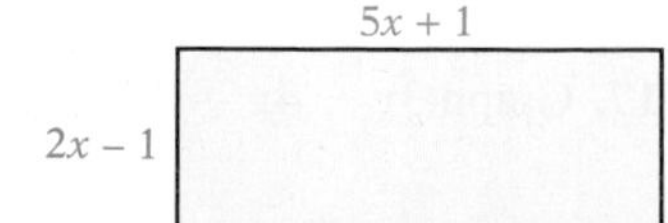

26. **Space Travel** A space vehicle travels 2.4×10^5 mi from Earth to the moon at an average velocity of 2×10^4 mph. How long does it take the space vehicle to reach the moon?

Cumulative Review Exercises

1. Simplify: $8 - 2[-3 - (-1)]^2 + 4$

2. Evaluate $\frac{2a - b}{b - c}$ when $a = 4$, $b = -2$, and $c = 6$.

3. Identify the property that justifies the statement $2x + (-2x) = 0$.

4. Simplify: $2x - 4[x - 2(3 - 2x) + 4]$

5. Solve: $\frac{2}{3} - y = \frac{5}{6}$

6. Solve: $8x - 3 - x = -6 + 3x - 8$

7. Divide: $\frac{x^3 - 3}{x - 3}$

8. Solve: $3 - |2 - 3x| = -2$

9. Given $P(x) = 3x^2 - 2x + 2$, evaluate $P(-2)$.

10. Evaluate $f(x) = \frac{x + 1}{x + 2}$ when $x = -3$.

11. Find the range of the function given by $F(x) = 3x^2 - 4$ if the domain is $\{-2, -1, 0, 1, 2\}$.

12. Find the slope of the line that contains the points $P_1(-2, 3)$ and $P_2(4, 2)$.

13. Find the equation of the line that contains the point $P(-1, 2)$ and has slope $-\frac{3}{2}$.

14. Find the equation of the line that contains the point $P(-2, 4)$ and is perpendicular to the graph of $3x + 2y = 4$.

15. Solve by using Cramer's Rule:
$$\begin{aligned} 2x - 3y &= 2 \\ x + y &= -3 \end{aligned}$$

16. Solve by the addition method:
$$\begin{aligned} x - y + z &= 0 \\ 2x + y - 3z &= -7 \\ -x + 2y + 2z &= 5 \end{aligned}$$

17. Graph $3x - 4y = 12$ by using the x- and y-intercepts.

18. Graph the solution set of $-3x + 2y < 6$.

19. Solve by graphing: $\begin{aligned} x - 2y &= 3 \\ -2x + y &= -3 \end{aligned}$

20. Graph the solution set: $\begin{aligned} 2x + y &< 3 \\ -6x + 3y &\geq 4 \end{aligned}$

21. Simplify: $(4a^{-2}b^3)(2a^3b^{-1})^{-2}$

22. Simplify: $\dfrac{(5x^3y^{-3}z)^{-2}}{y^4z^{-2}}$

23. Divide: $(x^3 + 2x^2 - 4x + 1) \div (x + 3)$

24. Multiply: $(2x + 3)(2x^2 - 3x + 1)$

25. Factor: $-4x^3 + 14x^2 - 12x$

26. Factor: $a(x - y) - b(y - x)$

27. Factor: $x^4 - 16$

28. Factor: $2x^3 - 16$

29. Biology A biologist mixes a 4% acid solution with an 8% acid solution to make 10 L of a 5% acid solution. How many liters of each solution does the biologist use?

30. Metallurgy How many ounces of pure gold that costs \$360 per ounce must be mixed with 80 oz of an alloy that costs \$120 per ounce to make a mixture that costs \$200 per ounce?

31. Uniform Motion Two cyclists are 25 mi apart and are traveling toward each other. One cyclist is traveling at $\frac{2}{3}$ the rate of the other cyclist. They pass in 2 h. Find the rate of each cyclist.

32. Investments If \$3000 is invested at an annual simple interest rate of 7.5%, how much additional money must be invested at an annual simple interest rate of 10% so that the total interest earned is 9% of the total investment?

33. Travel The graph below shows the relationship between the distance traveled in miles and the time of travel in hours. Find the slope of the line between the two points on the graph. Write a sentence that states the meaning of the slope.

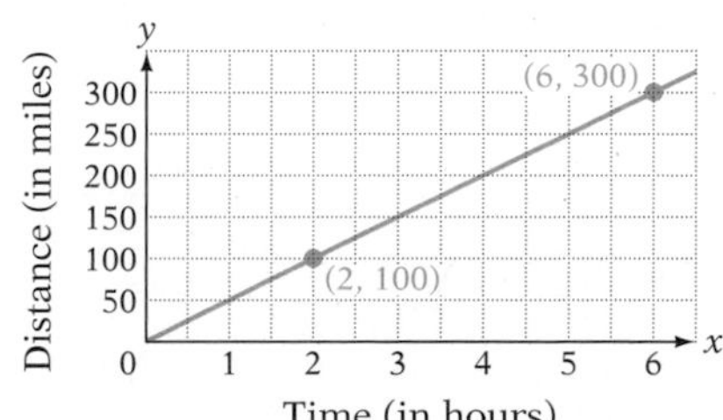

CHAPTER 6

Rational Expressions

Focus on Success

Are you using the features of this text to learn the concepts being presented? The Focus On feature includes a step-by-step solution to a type of exercise you will be working in your homework assignment. A numbered Example provides you with a fully worked out solution. After studying the Example, try completing the Problem that follows it. A complete solution to the Problem is given in the back of the text. Next, try the exercise listed after the Problem. You will be reinforcing the concepts you are learning at each step. (See Use the Interactive Method, page AIM-8.)

OBJECTIVES

PREP TEST

Are you ready to succeed in this chapter? *Take the Prep Test below to find out if you are ready to learn the new material.*

1. Find the LCM of 10 and 25.

Add, subtract, multiply, or divide.

2. $-\frac{3}{8} \cdot \frac{4}{9}$

3. $-\frac{4}{5} \div \frac{8}{15}$

4. $-\frac{5}{6} + \frac{7}{8}$

5. $-\frac{3}{8} - \left(-\frac{7}{12}\right)$

6. Simplify: $\dfrac{\frac{2}{3} - \frac{1}{4}}{\frac{1}{8} - 2}$

7. Evaluate $\dfrac{2x - 3}{x^2 - x + 1}$ for $x = 2$.

8. Solve: $4(2x + 1) = 3(x - 2)$

9. Solve: $10\left(\frac{t}{2} + \frac{t}{5}\right) = 10(1)$

10. Two planes start from the same point and fly in opposite directions. The first plane is flying 20 mph slower than the second plane. In 2 h, the planes are 480 mi apart. Find the rate of each plane.

Digital Vision

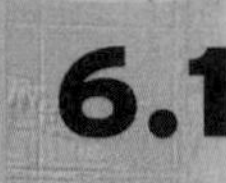

6.1 Introduction to Rational Functions

OBJECTIVE 1 Find the domain of a rational function

An expression in which the numerator and denominator are polynomials is called a **rational expression.** Examples of rational expressions are shown at the right.

$$\frac{9}{z} \qquad \frac{3x+4}{2x^2+1} \qquad \frac{x^3-x+1}{x^2-3x-5}$$

The expression $\frac{\sqrt{x}+3}{x}$ is not a rational expression because $\sqrt{x}+3$ is not a polynomial.

A function that is written in terms of a rational expression is a **rational function.** Each of the following equations represents a rational function.

$$f(x) = \frac{x^2+3}{2x-1} \qquad g(t) = \frac{3}{t^2-4} \qquad R(z) = \frac{z^2+3z-1}{z^2+z-12}$$

To evaluate a rational function, replace the variable by its value. Then simplify.

EXAMPLE 1 Given $f(x) = \frac{3x-4}{x^2-2x+1}$, find $f(-3)$.

Solution

$$f(x) = \frac{3x-4}{x^2-2x+1}$$

$$f(-3) = \frac{3(-3)-4}{(-3)^2-2(-3)+1} \qquad \text{• Substitute } -3 \text{ for } x.$$

$$f(-3) = \frac{-9-4}{9+6+1}$$

$$f(-3) = \frac{-13}{16}$$

$$f(-3) = -\frac{13}{16}$$

Problem 1 Given $f(x) = \frac{3-5x}{x^2+5x+6}$, find $f(2)$.

Solution See page S19.

Try Exercise 11, page 327.

DOMAIN OF A RATIONAL FUNCTION

Because division by zero is not defined, the **domain of a rational function** must exclude those numbers for which the value of the denominator is zero.

EXAMPLES

1. $f(x) = \frac{x+3}{x-4}$

   The denominator equals 0 when $x - 4 = 0$, or $x = 4$. The domain of f is all real numbers except 4. This is written $\{x \mid x \neq 4\}$.

2. $f(x) = \dfrac{x^2 - 6}{4x + 8}$

The denominator equals 0 when $4x + 8 = 0$. Solve the equation for x.

$$4x + 8 = 0$$
$$4x = -8$$
$$x = -2$$

The domain of f is all real numbers except -2. This is written $\{x \mid x \neq -2\}$.

3. $f(x) = \dfrac{2x}{x^2 + 4}$

The dominator equals 0 when $x^2 + 4 = 0$. However, $x^2 \geq 0$ for all real numbers. Therefore, $x^2 + 4 > 0$ for all real numbers, and the denominator is never 0. The domain is all real numbers, or $\{x \mid x \in \text{real numbers}\}$.

EXAMPLE 2 Find the domain of $f(x) = \dfrac{2x - 6}{x^2 - 3x - 4}$.

Solution The domain must exclude values of x for which $x^2 - 3x - 4 = 0$. Solve this equation for x.

$$x^2 - 3x - 4 = 0$$
$$(x + 1)(x - 4) = 0 \qquad \text{• Factor the trinomial.}$$
$$x + 1 = 0 \qquad x - 4 = 0 \qquad \text{• Use the Principle of Zero Products to set each factor equal to zero.}$$
$$x = -1 \qquad x = 4$$

The solutions are -1 and 4. The domain of f must exclude these values.

The domain of f is $\{x \mid x \neq -1, x \neq 4\}$.

Take Note

In Chapter 8, we will discuss how to solve this problem when the polynomial does not factor.

Problem 2 Find the domain of $g(x) = \dfrac{5 - x}{x^2 - 4}$.

Solution See page S19.

Try Exercise 21, page 327.

OBJECTIVE 2 Simplify rational expressions

A rational expression is in **simplest form** when the numerator and denominator have no common factors other than 1. The Multiplication Property of One is used to write a rational expression in simplest form, as shown below.

$$\frac{x^2 - 25}{x^2 + 13x + 40} = \frac{(x - 5)(x + 5)}{(x + 8)(x + 5)}$$
$$= \frac{(x - 5)}{(x + 8)} \cdot \frac{(x + 5)}{(x + 5)}$$
$$= \frac{x - 5}{x + 8} \cdot 1 = \frac{x - 5}{x + 8}, x \neq -8, x \neq -5$$

The requirement $x \neq -8, x \neq -5$ must be included because division by 0 is not allowed.

The preceding simplification is usually shown with slashes to indicate that a common factor has been removed:

$$\frac{x^2 - 25}{x^2 + 13x + 40} = \frac{(x-5)\overset{1}{\cancel{(x+5)}}}{(x+8)\underset{1}{\cancel{(x+5)}}} = \frac{x-5}{x+8}, x \neq -8, x \neq -5$$

We will show a simplification with slashes. We will also omit the restrictions that prevent division by zero. Nonetheless, those restrictions *always* are implied.

EXAMPLE 3 Simplify.

A. $\dfrac{x^2 - 16}{x^2 + 11x + 28}$ B. $\dfrac{12 + 5x - 2x^2}{2x^2 - 3x - 20}$

Solution A. $\dfrac{x^2 - 16}{x^2 + 11x + 28} = \dfrac{(x+4)(x-4)}{(x+4)(x+7)}$

- **Factor the numerator and the denominator.**

$= \dfrac{\overset{1}{\cancel{(x+4)}}(x-4)}{\underset{1}{\cancel{(x+4)}}(x+7)}$

- **Divide by the common factors.**

$= \dfrac{x-4}{x+7}$

- **Write the answer in simplest form.**

B. $\dfrac{12 + 5x - 2x^2}{2x^2 - 3x - 20} = \dfrac{(4-x)(3+2x)}{(x-4)(2x+5)}$

- **Factor the numerator and the denominator.**

$= \dfrac{\overset{-1}{\cancel{(4-x)}}(3+2x)}{\underset{1}{\cancel{(x-4)}}(2x+5)}$

- **Divide by the common factors. Remember that $4 - x = -(x - 4)$. Therefore, $\dfrac{4-x}{x-4} = \dfrac{-(x-4)}{x-4} = \dfrac{-1}{1} = -1$.**

$= -\dfrac{2x+3}{2x+5}$

- **Write the answer in simplest form.**

Take Note

Recall that

$(b - a) = -(a - b)$

Therefore,

$(4 - x) = -(x - 4)$

In general,

$$\frac{b-a}{a-b} = \frac{-\overset{1}{\cancel{(a-b)}}}{\underset{1}{\cancel{a-b}}} = -1$$

Problem 3 Simplify.

A. $\dfrac{6x^4 - 24x^3}{12x^3 - 48x^2}$ B. $\dfrac{20x - 15x^2}{15x^3 - 5x^2 - 20x}$

Solution See page S19.

Try Exercise 53, page 329.

6.1 Exercises

CONCEPT CHECK

1. State whether the function is a rational function.

a. $f(x) = \dfrac{1}{x}$ b. $g(x) = \dfrac{1}{\sqrt{x}}$ c. $h(x) = \dfrac{-2}{x^2 + 1}$

2. If $\frac{p(x)}{q(x)}$ is a rational expression in simplest form, and 2 is a zero of $p(x)$ and 3 is a zero of $q(x)$, then which statement is true?

(i) 2 is not in the domain of the expression.
(ii) 3 is not in the domain of the expression.
(iii) Both 2 and 3 are not in the domain of the expression.

3. What is the common factor of $x^2 + 5x - 6$ and $x^2 + 7x + 6$?

4. Explain why the following simplification is incorrect.

$$\frac{3\overset{1}{\cancel{x}} + 7}{2\underset{1}{\cancel{x}}} = \frac{10}{2} = 5.$$

1 Find the domain of a rational function (See pages 324–325.)

5. What is a rational function? Give an example of a rational function.

6. What values are excluded from the domain of a rational function?

GETTING READY

Complete Exercises 7 and 8 by replacing the question marks. Use the rational function $f(x) = \frac{x + 2}{1 - 2x}$.

7. To find $f(-3)$, substitute -3 for x: $f(-3) = \frac{? + 2}{1 - 2(?)} = \frac{?}{1 + ?} = $ ___?___.

8. Find the domain of the function. The domain must exclude values of x for which the denominator is ___?___.

$1 - 2x =$ ___?___ • **Set the denominator equal to zero.**

$-2x =$ ___?___ • **Subtract 1 from each side of the equation.**

$x =$ ___?___ • **Divide each side of the equation by −2.**

The domain must exclude the value ___?___, so the domain is $\{x | x \neq$ ___?___$\}$.

9. Given $f(x) = \frac{2}{x - 3}$, find $f(4)$.

10. Given $f(x) = \frac{-7}{5 - x}$, find $f(-2)$.

11. Given $f(x) = \frac{x - 2}{x + 4}$, find $f(-2)$.

12. Given $f(x) = \frac{x - 3}{2x - 1}$, find $f(3)$.

13. Given $f(x) = \frac{x - 2}{2x^2 + 3x + 8}$, find $f(3)$.

14. Given $f(x) = \frac{x^2}{3x^2 - 3x + 5}$, find $f(4)$.

15. Given $f(x) = \frac{x^2 - 2x}{x^3 - x + 4}$, find $f(-1)$.

16. Given $f(x) = \frac{8 - x^2}{x^3 - x^2 + 4}$, find $f(-3)$.

Find the domain of the function.

17. $H(x) = \frac{4}{x - 3}$

18. $G(x) = \frac{-2}{x + 2}$

19. $f(x) = \frac{x}{x + 4}$

20. $g(x) = \frac{3x}{x - 5}$

21. $R(x) = \frac{5x}{3x + 9}$

22. $p(x) = \frac{-2x}{6 - 2x}$

23. $q(x) = \dfrac{x-2}{(x-4)(x+2)}$

24. $h(x) = \dfrac{2x+1}{(x+1)(x+5)}$

25. $V(x) = \dfrac{x^2}{(2x+5)(3x-6)}$

26. $F(x) = \dfrac{x^2-1}{(4x+8)(3x-1)}$

27. $f(x) = \dfrac{x^2+1}{x}$

28. $g(x) = \dfrac{2x^3-x-1}{x^2}$

29. $k(x) = \dfrac{x+1}{x^2+1}$

30. $P(x) = \dfrac{2x+3}{2x^2+3}$

31. $f(x) = \dfrac{2x-1}{x^2+x-6}$

32. $G(x) = \dfrac{3-4x}{x^2+4x-5}$

33. $A(x) = \dfrac{5x+2}{x^2+2x-24}$

34. $h(x) = \dfrac{3x}{x^2-4}$

35. $f(x) = \dfrac{4x-7}{3x^2+12}$

36. $g(x) = \dfrac{x^2+x+1}{5x^2+1}$

37. $G(x) = \dfrac{x^2+1}{6x^2-13x+6}$

38. $A(x) = \dfrac{5x-7}{x(x-2)(x-3)}$

39. $f(x) = \dfrac{x^2+8x+4}{2x^3+9x^2-5x}$

40. $H(x) = \dfrac{x^4-1}{2x^3+2x^2-24x}$

41. The function $f(x) = -\dfrac{1}{x+1}$ is not defined at $x = -1$. Is the value of the function when $x > -1$ positive or negative?

42. The function $f(x) = \dfrac{1}{(x-1)(x+2)}$ is undefined for $x = 1$ and $x = -2$. Is the value of the function when $-2 < x < 1$ positive or negative?

2 Simplify rational expressions (See pages 325–326.)

43. When is a rational expression in simplest form?

44. Are the rational expressions $\dfrac{x(x-2)}{2(x-2)}$ and $\dfrac{x}{2}$ equal for all values of x? Why or why not?

GETTING READY

45. To simplify a rational expression, first write the numerator and denominator in __?__ form. Then simplify the rational expression by dividing its numerator and denominator by any __?__ factors.

46. Simplify: $\dfrac{x^2-1}{3x^2+3x}$

$\dfrac{x^2-1}{3x^2+3x} = \dfrac{(x-?)(x+?)}{(?)(x+1)}$ • **Factor the numerator and denominator.**

$= \dfrac{?}{?}$ • **Divide the numerator and denominator by the common factor __?__.**

Simplify.

47. $\dfrac{4-8x}{4}$

48. $\dfrac{8y+2}{2}$

49. $\dfrac{6x^2-2x}{2x}$

50. $\dfrac{3y-12y^2}{3y}$

51. $\dfrac{8x^2(x-3)}{4x(x-3)}$

52. $\dfrac{16y^4(y+8)}{12y^3(y+8)}$

53. $\dfrac{2x-6}{3x-x^2}$

54. $\dfrac{3a^2-6a}{12-6a}$

55. $\dfrac{6x^3-15x^2}{12x^2-30x}$

56. $\dfrac{-36a^2-48a}{18a^3+24a^2}$

57. $\dfrac{a^2+4a}{4a-16}$

58. $\dfrac{3x-6}{x^2+2x}$

59. $\dfrac{16x^3-8x^2+12x}{4x}$

60. $\dfrac{3x^3y^3-12x^2y^2+15xy}{3xy}$

61. $\dfrac{-10a^4-20a^3+30a^2}{-10a^2}$

62. $\dfrac{-7a^5-14a^4+21a^3}{-7a^3}$

63. $\dfrac{x^2-7x+12}{x^2-9x+20}$

64. $\dfrac{x^2-x-20}{x^2-2x-15}$

65. $\dfrac{x^2-xy-2y^2}{x^2-3xy+2y^2}$

66. $\dfrac{2x^2+7xy-4y^2}{4x^2-4xy+y^2}$

67. $\dfrac{6-x-x^2}{3x^2-10x+8}$

68. $\dfrac{3x^2+10x-8}{8-14x+3x^2}$

69. $\dfrac{14-19x-3x^2}{3x^2-23x+14}$

70. $\dfrac{x^2+x-12}{x^2-x-12}$

71. $\dfrac{a^2-7a+10}{a^2+9a+14}$

72. $\dfrac{x^4-y^4}{x^2+y^2}$

73. $\dfrac{x^3+y^3}{3x^3-3x^2y+3xy^2}$

74. $\dfrac{3x^3+3x^2+3x}{9x^3-9}$

75. $\dfrac{x^3-4xy^2}{3x^3-2x^2y-8xy^2}$

76. $\dfrac{4a^2-8ab+4b^2}{4a^2-4b^2}$

77. $\dfrac{4x^3-14x^2+12x}{24x+4x^2-8x^3}$

78. $\dfrac{6x^3-15x^2-75x}{150x+30x^2-12x^3}$

79. $\dfrac{x^4+3x^2+2}{x^4-1}$

80. $\dfrac{x^4-2x^2-3}{x^4+2x^2+1}$

81. $\dfrac{x^2y^2+4xy-21}{x^2y^2-10xy+21}$

82. $\dfrac{6x^2y^2+11xy+4}{9x^2y^2+9xy-4}$

83. True or false? $\dfrac{4x^2-8x}{16-8x}=\dfrac{x^2}{4}$

84. True or false? $\dfrac{4x+1}{8x+2}=\dfrac{1}{2}$

APPLYING CONCEPTS

Simplify each expression. Assume that n is a positive integer.

85. $\dfrac{a^{2n}-a^n-2}{a^{2n}+3a^n+2}$

86. $\dfrac{a^{2n}+a^n-12}{a^{2n}-2a^n-3}$

87. Evaluate $h(x)=\frac{x+2}{x-3}$ when $x = 2.9, 2.99, 2.999,$ and 2.9999. On the basis of your evaluations, complete the following sentence. As x becomes closer to 3, do the values of $h(x)$ increase or decrease?

88. Evaluate $h(x) = \frac{x + 2}{x - 3}$ when $x = 3.1, 3.01, 3.001$, and 3.0001. On the basis of your evaluations, complete the following sentence. As x becomes closer to 3, do the values of $h(x)$ increase or decrease?

89. Suppose that $F(x) = \frac{g(x)}{h(x)}$ and that, for some real number a, $g(a) = 0$ and $h(a) = 0$. Is $F(x)$ in simplest form? Explain your answer.

90. Why can the numerator and denominator of a rational expression be divided by their common factors? What conditions must be placed on the values of the variables when a rational expression is simplified?

PROJECTS OR GROUP ACTIVITIES

91. Photography The focal length of a camera lens is the distance from the lens to the point at which parallel rays of light come to a focus.

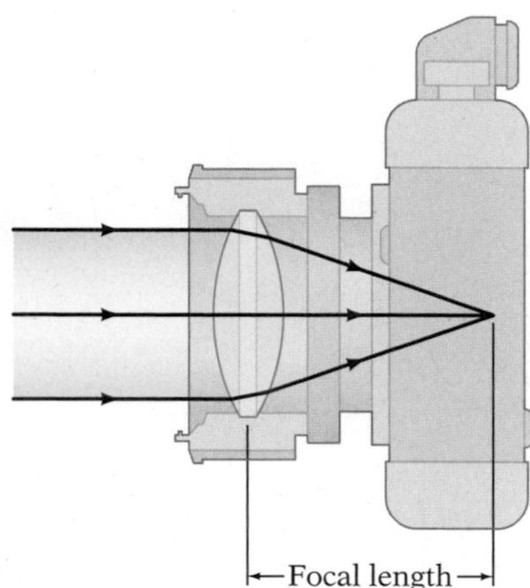

The relationship among the focal length (F), the distance between the object and the lens (x, in meters), and the distance between the lens and the film (y, in millimeters) is given by $\frac{1}{F} = \frac{1}{x} + \frac{1}{y}$. A camera used by a professional photographer has a dial that allows the focal length to be set at a constant value. Suppose a photographer chooses a focal length of 50 mm. Substituting this value into the equation, solving for y, and using $y = f(x)$ notation yield $f(x) = \frac{50x}{x - 50}$.

a. Graph this equation for $50 < x \leq 6000$.
b. The point whose coordinates are (2000, 51), to the nearest integer, is on the graph of the function. Give an interpretation of this ordered pair.
c. Give a reason for choosing the domain such that $x > 50$.
d. Photographers refer to *depth of field* as the range of distances at which an object remains in focus. Use the graph to explain why the depth of field is larger for objects that are far from the lens than for objects that are close to the lens.

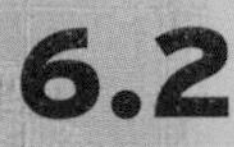

6.2 Operations on Rational Expressions

OBJECTIVE 1 Multiply and divide rational expressions

MULTIPLY RATIONAL EXPRESSIONS

The product of two rational expressions is the product of the numerators of the expressions over the product of the denominators of the expressions.

EXAMPLES

1. $\dfrac{2x^2}{3y} \cdot \dfrac{6y^2}{x^3} = \dfrac{12x^2y^2}{3x^3y}$ • **Multiply the numerators. Multiply the denominators.**

$= \dfrac{4y}{x}$ • **Write the answer in simplest form.**

2. $\dfrac{x+2}{2x-6} \cdot \dfrac{5x-15}{3x+6} = \dfrac{x+2}{2(x-3)} \cdot \dfrac{5(x-3)}{3(x+2)}$ • **Factor the numerators and denominators.**

$= \dfrac{5\overset{1}{\cancel{(x+2)}}\overset{1}{\cancel{(x-3)}}}{6\underset{1}{\cancel{(x-3)}}\underset{1}{\cancel{(x+2)}}}$ • **Multiply the numerators. Multiply the denominators.**

$= \dfrac{5}{6}$ • **Write the answer in simplest form.**

Focus on simplifying the product of two rational expressions

Simplify: $\dfrac{x^2-2x}{2x^2+x-15} \cdot \dfrac{2x^2-x-10}{x^2-4}$

$$\frac{x^2-2x}{2x^2+x-15} \cdot \frac{2x^2-x-10}{x^2-4}$$

Factor the numerators and denominators.

$$= \frac{x(x-2)}{(x+3)(2x-5)} \cdot \frac{(x+2)(2x-5)}{(x+2)(x-2)}$$

Multiply the numerators. Multiply the denominators.

$$= \frac{x(x-2)(x+2)(2x-5)}{(x+3)(2x-5)(x+2)(x-2)}$$

Divide by the common factors.

$$= \frac{x\overset{1}{\cancel{(x-2)}}\overset{1}{\cancel{(x+2)}}\overset{1}{\cancel{(2x-5)}}}{(x+3)\underset{1}{\cancel{(2x-5)}}\underset{1}{\cancel{(x+2)}}\underset{1}{\cancel{(x-2)}}}$$

Write the answer in simplest form.

$$= \frac{x}{x+3}$$

EXAMPLE 1 Multiply: $\frac{2x^2 - 5x - 3}{x^2 - 7x + 12} \cdot \frac{20 - x - x^2}{2x^2 + 13x + 6}$

Solution $\frac{2x^2 - 5x - 3}{x^2 - 7x + 12} \cdot \frac{20 - x - x^2}{2x^2 + 13x + 6}$

$= \frac{(2x + 1)(x - 3)}{(x - 3)(x - 4)} \cdot \frac{(5 + x)(4 - x)}{(2x + 1)(x + 6)}$ • Factor the numerators and denominators.

$= \frac{\overset{1}{\cancel{(2x + 1)}}\overset{1}{\cancel{(x - 3)}}(5 + x)\overset{-1}{\cancel{(4 - x)}}}{\underset{1}{\cancel{(x - 3)}}\underset{1}{\cancel{(x - 4)}}\underset{1}{\cancel{(2x + 1)}}(x + 6)}$ • Multiply the rational expressions. Recall that $\frac{4 - x}{x - 4} = \frac{-\overset{1}{(x - 4)}}{\underset{1}{x - 4}} = -1$.

$= -\frac{x + 5}{x + 6}$ • Write the answer in simplest form.

Problem 1 Multiply: $\frac{12 + 5x - 3x^2}{x^2 + 2x - 15} \cdot \frac{2x^2 + x - 45}{3x^2 + 4x}$

Solution See page S20.

Try Exercise 15, page 338.

The **reciprocal of a rational expression** is the rational expression with the numerator and denominator interchanged.

Rational expression			Reciprocal
	$\frac{a}{b}$	$\frac{b}{a}$	
	$\frac{a^2 - 2y}{4}$	$\frac{4}{a^2 - 2y}$	

DIVISION OF RATIONAL EXPRESSIONS

To divide rational expressions, multiply the dividend by the reciprocal of the divisor.

EXAMPLES

1. $\frac{5a^2b}{7x^2y} \div \frac{10a^3b^2}{9xy^2} = \frac{5a^2b}{7x^2y} \cdot \frac{9xy^2}{10a^3b^2}$ • Multiply by the reciprocal of the divisor.

$= \frac{45a^2bxy^2}{70a^3b^2x^2y}$ • Multiply.

$= \frac{9y}{14abx}$ • Write the answer in simplest form.

2. $\frac{3x + 15}{5x^2} \div \frac{6x + 30}{4x} = \frac{3x + 15}{5x^2} \cdot \frac{4x}{6x + 30}$ • Multiply by the reciprocal of the divisor.

$= \frac{3(x + 5)}{5x^2} \cdot \frac{4x}{6(x + 5)}$ • Factor the numerators and denominators.

$= \frac{12x\overset{1}{\cancel{(x + 5)}}}{30x^2\underset{1}{\cancel{(x + 5)}}} = \frac{2}{5x}$ • Multiply. Then write the answer in simplest form.

EXAMPLE 2 Divide.

A. $\frac{12x^2y^2 - 24xy^2}{5z^2} \div \frac{4x^3y - 8x^2y}{3z^4}$

B. $\dfrac{3y^2 - 10y + 8}{3y^2 + 8y - 16} \div \dfrac{2y^2 - 7y + 6}{2y^2 + 5y - 12}$

Solution **A.** $\dfrac{12x^2y^2 - 24xy^2}{5z^2} \div \dfrac{4x^3y - 8x^2y}{3z^4}$

$= \dfrac{12x^2y^2 - 24xy^2}{5z^2} \cdot \dfrac{3z^4}{4x^3y - 8x^2y}$ • **Multiply by the reciprocal of the divisor.**

$= \dfrac{12xy^2(x - 2)}{5z^2} \cdot \dfrac{3z^4}{4x^2y(x - 2)}$ • **Factor the numerators and denominators.**

$= \dfrac{36xy^2z^4\overset{1}{\cancel{(x - 2)}}}{20x^2yz^2\underset{1}{\cancel{(x - 2)}}} = \dfrac{9yz^2}{5x}$ • **Multiply. Then write the answer in simplest form.**

B. $\dfrac{3y^2 - 10y + 8}{3y^2 + 8y - 16} \div \dfrac{2y^2 - 7y + 6}{2y^2 + 5y - 12}$

$= \dfrac{3y^2 - 10y + 8}{3y^2 + 8y - 16} \cdot \dfrac{2y^2 + 5y - 12}{2y^2 - 7y + 6}$ • **Multiply by the reciprocal of the divisor.**

$= \dfrac{(y - 2)(3y - 4)}{(3y - 4)(y + 4)} \cdot \dfrac{(y + 4)(2y - 3)}{(y - 2)(2y - 3)}$ • **Factor the numerators and denominators.**

$= \dfrac{\overset{1}{\cancel{(y - 2)}}\overset{1}{\cancel{(3y - 4)}}\overset{1}{\cancel{(y + 4)}}\overset{1}{\cancel{(2y - 3)}}}{\underset{1}{\cancel{(3y - 4)}}\underset{1}{\cancel{(y + 4)}}\underset{1}{\cancel{(y - 2)}}\underset{1}{\cancel{(2y - 3)}}} = 1$ • **Multiply. Then write the answer in simplest form.**

Problem 2 Divide.

A. $\dfrac{6x^2 - 3xy}{10ab^4} \div \dfrac{16x^2y^2 - 8xy^3}{15a^2b^2}$

B. $\dfrac{6x^2 - 7x + 2}{3x^2 + x - 2} \div \dfrac{4x^2 - 8x + 3}{5x^2 + x - 4}$

Solution See page S20.

Try Exercise 23, page 338.

OBJECTIVE 2 Add and subtract rational expressions

ADD OR SUBTRACT RATIONAL EXPRESSIONS

To add two rational expressions *with the same denominator*, add the numerators and place the sum over the common denominator. To subtract two rational expressions *with the same denominator*, subtract the numerators and place the difference over the common denominator.

EXAMPLES

1. $\dfrac{2a + b}{a^2 - b^2} + \dfrac{a - 4b}{a^2 - b^2} = \dfrac{(2a + b) + (a - 4b)}{a^2 - b^2}$ • **The denominators are the same. Add the numerators.**

$= \dfrac{3a - 3b}{a^2 - b^2} = \dfrac{3\overset{1}{\cancel{(a - b)}}}{(a + b)\underset{1}{\cancel{(a - b)}}}$ • **Simplify. Write the fraction in simplest form.**

$= \dfrac{3}{a + b}$

2. $\dfrac{7x - 12}{2x^2 + 5x - 12} - \dfrac{3x - 6}{2x^2 + 5x - 12}$

$= \dfrac{(7x - 12) - (3x - 6)}{2x^2 + 5x - 12}$

• The denominators are the same. Subtract the numerators.

$= \dfrac{7x - 12 - 3x + 6}{2x^2 + 5x - 12}$

• Simplify. Write the fraction in simplest form.

$= \dfrac{4x - 6}{2x^2 + 5x - 12}$

$= \dfrac{2(2x - 3)}{(2x - 3)(x + 4)}$

$= \dfrac{2\overset{1}{\cancel{(2x - 3)}}}{\underset{1}{\cancel{(2x - 3)}}(x + 4)} = \dfrac{2}{x + 4}$

For each example above, the denominators of the two expressions were the same. If the denominators of two expressions are not the same, both rational expressions must be expressed in terms of a *common denominator*. A good common denominator to use is the least common multiple (LCM) of the denominators, also called the least common denominator (LCD).

The **LCM of two or more polynomials** is the polynomial of least degree that contains the factors of each polynomial. To find the LCM, first factor each polynomial completely. The LCM is the product of each factor the greatest number of times it occurs in any one factorization.

To find the LCM of $3x^2 + 15x$ and $6x^4 + 24x^3 - 30x^2$, factor each polynomial.

$$3x^2 + 15x = 3x(x + 5)$$
$$6x^4 + 24x^3 - 30x^2 = 6x^2(x^2 + 4x - 5) = 6x^2(x - 1)(x + 5)$$

The LCM is the product of the LCM of the numerical coefficients and each variable factor the greatest number of times it occurs in any one factorization.

$$\text{LCM} = 6x^2(x - 1)(x + 5)$$

Focus on writing rational expressions in terms of the LCD

Write the fractions $\dfrac{x + 2}{x^2 - 2x}$ and $\dfrac{5x}{3x - 6}$ in terms of the LCD (the LCM of the denominators).

Factor each denominator.

$$x^2 - 2x = x(x - 2)$$
$$3x - 6 = 3(x - 2)$$

Find the LCD.

The LCD is $3x(x - 2)$.

For each fraction, multiply the numerator and denominator by the factor whose product with the denominator is the LCD.

$$\frac{x + 2}{x^2 - 2x} = \frac{x + 2}{x(x - 2)} \cdot \frac{3}{3} = \frac{3x + 6}{3x(x - 2)}$$
$$\frac{5x}{3x - 6} = \frac{5x}{3(x - 2)} \cdot \frac{x}{x} = \frac{5x^2}{3x(x - 2)}$$

Focus on adding rational expressions

Add: $\dfrac{3x}{2x-3} + \dfrac{3x+6}{2x^2+x-6}$

Factor the denominators.

$2x - 3$ does not factor.
$2x^2 + x - 6 = (2x-3)(x+2)$

The LCD is $(2x-3)(x+2)$.

$$\frac{3x}{2x-3} + \frac{3x+6}{2x^2+x-6}$$

Rewrite each fraction in terms of the LCD.

$$= \frac{3x}{2x-3}\cdot\frac{x+2}{x+2} + \frac{3x+6}{(2x-3)(x+2)}$$

$$= \frac{3x^2+6x}{(2x-3)(x+2)} + \frac{3x+6}{(2x-3)(x+2)}$$

Add the fractions.

$$= \frac{(3x^2+6x)+(3x+6)}{(2x-3)(x+2)}$$

$$= \frac{3x^2+9x+6}{(2x-3)(x+2)}$$

$$= \frac{3(x^2+3x+2)}{(2x-3)(x+2)}$$

Factor the numerator to determine whether there are common factors in the numerator and denominator.

$$= \frac{3(x+2)(x+1)}{(2x-3)(x+2)}$$

$$= \frac{3\overset{1}{\cancel{(x+2)}}(x+1)}{(2x-3)\underset{1}{\cancel{(x+2)}}} = \frac{3(x+1)}{2x-3}$$

Take Note

Note the steps involved in adding or subtracting rational expressions:

1. Find the LCD.
2. Rewrite each fraction in terms of the LCD.
3. Add or subtract the rational expressions.
4. Simplify the resulting sum or difference.

EXAMPLE 3 Add: $\dfrac{4b}{a^2} + \dfrac{2a}{b^2} + \dfrac{3}{ab}$

Solution The LCD is a^2b^2. • **Find the LCD.**

$$\frac{4b}{a^2} + \frac{2a}{b^2} + \frac{3}{ab} = \frac{4b}{a^2}\cdot\frac{b^2}{b^2} + \frac{2a}{b^2}\cdot\frac{a^2}{a^2} + \frac{3}{ab}\cdot\frac{ab}{ab}$$

• **Write each fraction in terms of the LCD.**

$$= \frac{4b^3}{a^2b^2} + \frac{2a^3}{a^2b^2} + \frac{3ab}{a^2b^2}$$

• **Simplify.**

$$= \frac{4b^3 + 2a^3 + 3ab}{a^2b^2}$$

• **Add the numerators. Place the sum over the common denominator.**

Problem 3 Subtract: $\dfrac{5y}{x} - \dfrac{9}{xy} - \dfrac{4}{y^2}$

Solution See page S20.

Try Exercise 53, page 340.

EXAMPLE 4 Subtract: $\frac{x}{2x-4} - \frac{4-x}{x^2-2x}$

Solution $2x - 4 = 2(x - 2)$
$x^2 - 2x = x(x - 2)$

The LCD is $2x(x - 2)$. • Find the LCD.

$$\frac{x}{2x-4} - \frac{4-x}{x^2-2x}$$

$$= \frac{x}{2(x-2)} \cdot \frac{x}{x} - \frac{4-x}{x(x-2)} \cdot \frac{2}{2}$$ • Write each fraction in terms of the LCD.

$$= \frac{x^2}{2x(x-2)} - \frac{8-2x}{2x(x-2)}$$

$$= \frac{x^2-(8-2x)}{2x(x-2)}$$ • Subtract the fractions.

$$= \frac{x^2+2x-8}{2x(x-2)}$$ • Simplify.

$$= \frac{(x+4)(x-2)}{2x(x-2)}$$

$$= \frac{(x+4)\overset{1}{\cancel{(x-2)}}}{2x\underset{1}{\cancel{(x-2)}}} = \frac{x+4}{2x}$$ • Divide by the common factors.

Problem 4 Add: $\frac{a-3}{a^2-5a} + \frac{a-9}{a^2-25}$

Solution See page S20.

Try Exercise 71, page 340.

EXAMPLE 5 Simplify. **A.** $\frac{4}{x-3} + 3 - \frac{2x}{x-1}$

B. $\frac{6x-23}{2x^2+x-6} + \frac{3x}{2x-3} - \frac{5}{x+2}$

Solution

A. The LCD is $(x - 3)(x - 1)$. • Find the LCD.

$$\frac{4}{x-3} + 3 - \frac{2x}{x-1}$$

$$= \frac{4}{x-3} \cdot \frac{x-1}{x-1} + \frac{3}{1} \cdot \frac{(x-3)(x-1)}{(x-3)(x-1)} - \frac{2x}{x-1} \cdot \frac{x-3}{x-3}$$ • Write each fraction in terms of the LCD.

$$= \frac{4x-4}{(x-3)(x-1)} + \frac{3x^2-12x+9}{(x-3)(x-1)} - \frac{2x^2-6x}{(x-3)(x-1)}$$ • Simplify.

$$= \frac{(4x-4)+(3x^2-12x+9)-(2x^2-6x)}{(x-3)(x-1)}$$ • Write the sum and difference over the common denominator.

$$= \frac{x^2-2x+5}{(x-3)(x-1)}$$ • Write the answer in simplest form.

B. $2x^2 + x - 6 = (2x - 3)(x + 2)$

The LCD is $(2x - 3)(x + 2)$. • Find the LCD.

$$\frac{6x-23}{2x^2+x-6} + \frac{3x}{2x-3} - \frac{5}{x+2}$$

$$= \frac{6x - 23}{(2x - 3)(x + 2)} + \frac{3x}{2x - 3} \cdot \frac{x + 2}{x + 2} - \frac{5}{x + 2} \cdot \frac{2x - 3}{2x - 3}$$

• Write each fraction in terms of the LCD.

$$= \frac{6x - 23}{(2x - 3)(x + 2)} + \frac{3x^2 + 6x}{(2x - 3)(x + 2)} - \frac{10x - 15}{(2x - 3)(x + 2)}$$

• Simplify.

$$= \frac{(6x - 23) + (3x^2 + 6x) - (10x - 15)}{(2x - 3)(x + 2)}$$

• Write the sum and difference over the common denominator.

$$= \frac{6x - 23 + 3x^2 + 6x - 10x + 15}{(2x - 3)(x + 2)} = \frac{3x^2 + 2x - 8}{(2x - 3)(x + 2)}$$

• Simplify the numerator.

$$= \frac{(3x - 4)\overset{1}{\cancel{(x + 2)}}}{(2x - 3)\underset{1}{\cancel{(x + 2)}}} = \frac{3x - 4}{2x - 3}$$

• Write the answer in simplest form.

Problem 5 Simplify. **A.** $\frac{1}{x + 2} - \frac{x}{x - 3} + 2$

B. $\frac{x - 1}{x - 2} - \frac{7 - 6x}{2x^2 - 7x + 6} + \frac{4}{2x - 3}$

Solution See page S20.

Try Exercise 91, page 341.

6.2 Exercises

CONCEPT CHECK

1. When multiplying rational expressions with different denominators, is it necessary to find a common denominator for the expressions?

2. When adding rational expressions with different denominators, is it necessary to find a common denominator for the expressions?

3. What is the reciprocal of $\frac{x^2}{x - 1}$?

4. What is the error in the following division?

$$\frac{x + 1}{x^2 - 4} \div \frac{x + 3}{x - 2} = \frac{x^2 - 4}{x + 1} \cdot \frac{x + 3}{x - 2} = \frac{\overset{1}{\cancel{(x - 2)}}(x + 2)(x + 3)}{(x + 1)\underset{1}{\cancel{(x - 2)}}} = \frac{(x + 2)(x + 3)}{x + 1}$$

1 Multiply and divide rational expressions (See pages 331–333.)

5. Explain how to multiply two rational expressions.

6. Explain how to divide two rational expressions.

GETTING READY

7. The first step in the process of multiplying two rational expressions is to write each numerator and denominator in __?__ form.

8. Complete: $\frac{x - 2}{2 - x} = $ __?__

Multiply or divide.

9. $\dfrac{27a^2b^5}{16xy^2} \cdot \dfrac{20x^2y^3}{9a^2b}$

10. $\dfrac{15x^2y^4}{24ab^3} \cdot \dfrac{28a^2b^4}{35xy^4}$

11. $\dfrac{3x - 15}{4x^2 - 2x} \cdot \dfrac{20x^2 - 10x}{15x - 75}$

12. $\dfrac{2x^2 + 4x}{8x^2 - 40x} \cdot \dfrac{6x^3 - 30x^2}{3x^2 + 6x}$

13. $\dfrac{x^2y^3}{x^2 - 4x - 5} \cdot \dfrac{2x^2 - 13x + 15}{x^4y^3}$

14. $\dfrac{2x^2 - 5x + 3}{x^6y^3} \cdot \dfrac{x^4y^4}{2x^2 - x - 3}$

15. $\dfrac{x^2 - 3x + 2}{x^2 - 8x + 15} \cdot \dfrac{x^2 + x - 12}{8 - 2x - x^2}$

16. $\dfrac{x^2 + x - 6}{12 + x - x^2} \cdot \dfrac{x^2 + x - 20}{x^2 - 4x + 4}$

17. $\dfrac{12 + x - 6x^2}{6x^2 + 29x + 28} \cdot \dfrac{2x^2 + x - 21}{4x^2 - 9}$

18. $\dfrac{x^2 + 5x + 4}{4 + x - 3x^2} \cdot \dfrac{3x^2 + 2x - 8}{x^2 + 4x}$

19. $\dfrac{x^3 - y^3}{2x^2 + xy - 3y^2} \cdot \dfrac{2x^2 + 5xy + 3y^2}{x^2 + xy + y^2}$

20. $\dfrac{x^4 - 5x^2 + 4}{3x^2 - 4x - 4} \cdot \dfrac{3x^2 - 10x - 8}{x^2 - 4}$

21. $\dfrac{6x^2y^4}{35a^2b^5} \div \dfrac{12x^3y^3}{7a^4b^5}$

22. $\dfrac{12a^4b^7}{13x^2y^2} \div \dfrac{18a^5b^6}{26xy^3}$

23. $\dfrac{2x - 6}{6x^2 - 15x} \div \dfrac{4x^2 - 12x}{18x^3 - 45x^2}$

24. $\dfrac{4x^2 - 4y^2}{6x^2y^2} \div \dfrac{3x^2 + 3xy}{2x^2y - 2xy^2}$

25. $\dfrac{2x^2 - 2y^2}{14x^2y^4} \div \dfrac{x^2 + 2xy + y^2}{35xy^3}$

26. $\dfrac{8x^3 + 12x^2y}{4x^2 - 9y^2} \div \dfrac{16x^2y^2}{4x^2 - 12xy + 9y^2}$

27. $\dfrac{2x^2 - 5x - 3}{2x^2 + 7x + 3} \div \dfrac{2x^2 - 3x - 20}{2x^2 - x - 15}$

28. $\dfrac{3x^2 - 10x - 8}{6x^2 + 13x + 6} \div \dfrac{2x^2 - 9x + 10}{4x^2 - 4x - 15}$

29. $\dfrac{x^2 - 8x + 15}{x^2 + 2x - 35} \div \dfrac{15 - 2x - x^2}{x^2 + 9x + 14}$

30. $\dfrac{2x^2 + 13x + 20}{8 - 10x - 3x^2} \div \dfrac{6x^2 - 13x - 5}{9x^2 - 3x - 2}$

31. $\dfrac{2x^2 - 13x + 21}{2x^2 + 11x + 15} \div \dfrac{2x^2 + x - 28}{3x^2 + 4x - 15}$

32. $\dfrac{2x^2 - 13x + 15}{2x^2 - 3x - 35} \div \dfrac{6x^2 + x - 12}{6x^2 + 13x - 28}$

33. $\dfrac{14 + 17x - 6x^2}{3x^2 + 14x + 8} \div \dfrac{4x^2 - 49}{2x^2 + 15x + 28}$

34. $\dfrac{16x^2 - 9}{6 - 5x - 4x^2} \div \dfrac{16x^2 + 24x + 9}{4x^2 + 11x + 6}$

35. $\dfrac{6x^2 + 6x}{3x + 6x^2 + 3x^3} \div \dfrac{x^2 - 1}{1 - x^3}$

36. $\dfrac{x^3 + y^3}{2x^3 + 2x^2y} \div \dfrac{3x^3 - 3x^2y + 3xy^2}{6x^2 - 6y^2}$

37. Which expression is *not* equivalent to $\dfrac{x^2 + 4x - 5}{x^2 - x - 6}$?

(i) $\dfrac{x - 1}{x + 2} \div \dfrac{x - 3}{x + 5}$ (ii) $\dfrac{x - 1}{x - 3} \cdot \dfrac{x + 5}{x + 2}$

(iii) $\dfrac{x - 3}{x - 1} \div \dfrac{x + 5}{x + 2}$ (iv) $\dfrac{x + 5}{2x - 6} \cdot \dfrac{2x - 2}{x + 2}$

38. The simplified form of the product $\left(\dfrac{x - 1}{x}\right)^8\left(\dfrac{x^5}{x - 5}\right)^2$ has a factor of the form x^n in the numerator. What is n?

2 Add and subtract rational expressions (See pages 333–337.)

GETTING READY

39. To find the LCM of $x^2 - 6x + 9$ and $x^2 - 9$, first factor each polynomial:

$x^2 - 6x + 9 =$ __?__ • **$x^2 - 6x + 9$ factors as a perfect-square trinomial.**

$x^2 - 9 =$ __?__ • **$x^2 - 9$ factors as the difference of two squares.**

The LCM will use the factor $x - 3$ __?__ time(s) and the factor $x + 3$ __?__ time(s). The LCM is __?__.

40. True or false? The LCM of $(xy)^3$ and xy^3 is $(xy)^3$.

41. How many factors of $x - 4$ are in the LCM of each pair of expressions?

a. $x^2 - x - 12$ and $x^2 - 8x + 16$ **b.** $x^2 + x - 12$ and $x^2 + 8x + 16$
c. $x^2 - 16$ and $x^2 - x - 12$

42. How many factors of a are in the LCM of $(a^2b)^3$ and a^4b^4? How many factors of b?

Write each fraction in terms of the LCD.

43. $\dfrac{3}{4x^2y}, \dfrac{17}{12xy^4}$

44. $\dfrac{5}{16a^3b^3}, \dfrac{7}{30a^5b}$

45. $\dfrac{3x}{2x-3}, \dfrac{5x}{2x+3}$

46. $\dfrac{2}{7y-3}, \dfrac{-3}{7y+3}$

47. $\dfrac{2x}{x^2-9}, \dfrac{x+1}{x-3}$

48. $\dfrac{3x}{x^2-16}, \dfrac{2x}{4x-16}$

49. $\dfrac{3x}{x^2-1}, \dfrac{5x}{x^2-2x+1}$

50. $\dfrac{2x}{x^2+x-6}, \dfrac{-4x}{x^2+5x+6}$

GETTING READY

51. Subtract: $\dfrac{5}{x-4} - \dfrac{2}{x+3}$

$\dfrac{5}{x-4} - \dfrac{2}{x+3}$ • **The LCD is __?__.**

$= \dfrac{5}{x-4} \cdot \dfrac{x+3}{x+3} - \dfrac{2}{x+3} \cdot \dfrac{x-4}{x-4}$ • **Multiply the numerator and denominator of each fraction by the factor whose product with the denominator is the LCD.**

$= \dfrac{?}{(x-4)(x+3)} - \dfrac{?}{(x-4)(x+3)}$

$= \dfrac{(?) - (?)}{(x-4)(x+3)} = \dfrac{5x + 15 - ? + ?}{(x-4)(x+3)}$

$= \dfrac{? + ?}{(x-4)(x+3)}$

52. True or false? $\dfrac{1}{x-3} + \dfrac{1}{3-x} = 0$

Add or subtract.

53. $\frac{3}{2xy} - \frac{7}{2xy} - \frac{9}{2xy}$

54. $-\frac{3}{4x^2} + \frac{8}{4x^2} - \frac{3}{4x^2}$

55. $\frac{x}{x^2 - 3x + 2} - \frac{2}{x^2 - 3x + 2}$

56. $\frac{3x}{3x^2 + x - 10} - \frac{5}{3x^2 + x - 10}$

57. $\frac{3}{2x^2y} - \frac{8}{5x} - \frac{9}{10xy}$

58. $\frac{2}{5ab} - \frac{3}{10a^2b} + \frac{4}{15ab^2}$

59. $\frac{2x - 1}{12x} - \frac{3x + 4}{9x}$

60. $\frac{3x - 4}{6x} - \frac{2x - 5}{4x}$

61. $\frac{3x + 2}{4x^2y} - \frac{y - 5}{6xy^2}$

62. $\frac{2y - 4}{5xy^2} + \frac{3 - 2x}{10x^2y}$

63. $\frac{3}{x + 3} - \frac{4}{x - 2}$

64. $\frac{5}{x - 4} + \frac{2}{x + 6}$

65. $\frac{2x}{x - 3} - \frac{3x}{x - 5}$

66. $\frac{3a}{a - 2} - \frac{5a}{a + 1}$

67. $\frac{x - 2}{x - 3} + \frac{x + 7}{x - 5}$

68. $\frac{x - 1}{x - 5} - \frac{x + 4}{x + 2}$

69. $\frac{2x + 1}{2x - 3} - \frac{x - 5}{3x + 2}$

70. $\frac{x - 5}{2x + 5} - \frac{x + 6}{3x - 1}$

71. $\frac{3}{x + 5} + \frac{2x + 7}{x^2 - 25}$

72. $\frac{x}{4 - x} - \frac{4}{x^2 - 16}$

73. $\frac{2}{x} - 3 - \frac{10}{x - 4}$

74. $\frac{6a}{a - 3} - 5 + \frac{3}{a}$

75. $\frac{1}{2x - 3} - \frac{5}{2x} + 1$

76. $\frac{5}{x} - \frac{5x}{5 - 6x} + 2$

77. $\frac{3}{x^2 - 1} + \frac{2x}{x^2 + 2x + 1}$

78. $\frac{1}{x^2 - 6x + 9} - \frac{1}{x^2 - 9}$

79. $\frac{x}{x + 3} - \frac{3 - x}{x^2 - 9}$

80. $\frac{1}{x + 2} - \frac{3x}{x^2 + 4x + 4}$

81. $\frac{2x - 3}{x + 5} - \frac{x^2 - 4x - 19}{x^2 + 8x + 15}$

82. $\frac{-3x^2 + 8x + 2}{x^2 + 2x - 8} - \frac{2x - 5}{x + 4}$

83. $\frac{2x - 2}{4x^2 - 9} - \frac{5}{3 - 2x}$

84. $\frac{x^2 + 4}{4x^2 - 36} - \frac{13}{x + 3}$

85. $\frac{x - 2}{x + 1} - \frac{3 - 12x}{2x^2 - x - 3}$

86. $\frac{3x - 4}{4x + 1} + \frac{3x + 6}{4x^2 + 9x + 2}$

87. $\frac{x + 1}{x^2 + x - 6} - \frac{x + 2}{x^2 + 4x + 3}$

88. $\frac{x + 1}{x^2 + x - 12} - \frac{x - 3}{x^2 + 7x + 12}$

89. $\dfrac{x-1}{2x^2+11x+12}+\dfrac{2x}{2x^2-3x-9}$

90. $\dfrac{x-2}{4x^2+4x-3}+\dfrac{3-2x}{6x^2+x-2}$

91. $\dfrac{x}{x-3}-\dfrac{2}{x+4}-\dfrac{14}{x^2+x-12}$

92. $\dfrac{x^2}{x^2+x-2}+\dfrac{3}{x-1}-\dfrac{4}{x+2}$

93. $\dfrac{x^2+6x}{x^2+3x-18}-\dfrac{2x-1}{x+6}+\dfrac{x-2}{3-x}$

94. $\dfrac{2x^2-2x}{x^2-2x-15}-\dfrac{2}{x+3}+\dfrac{x}{5-x}$

APPLYING CONCEPTS

Simplify.

95. $\dfrac{25x-x^3}{x^4-1}\cdot\dfrac{3-x-4x^2}{2x^2+7x-15}\div\dfrac{4x^3-23x^2+15x}{3-5x+2x^2}$

96. $\left(\dfrac{y-2}{3y+1}-\dfrac{y+2}{3y-1}\right)\cdot\left(\dfrac{3y+1}{y}-\dfrac{3y-1}{y^2}\right)$

97. $\left(\dfrac{x+1}{2x-1}-\dfrac{x-1}{2x+1}\right)\cdot\left(\dfrac{2x-1}{x}-\dfrac{2x-1}{x^2}\right)$

98. Replace the question mark with one rational expression that will make the equation true.
$\dfrac{3}{x-2}+?=\dfrac{2}{x+3}$

99. Replace the question mark with one rational expression that will make the equation true.
$\dfrac{2x-1}{x^2-9}\div ?=\dfrac{x}{x+3}$

100. If $\dfrac{Ax+B}{2x^2-5x+2}=\dfrac{10}{x-2}-\dfrac{11}{2x-1}$, find A and B.

101. For adding or subtracting fractions, any common denominator will do. Explain the advantages and disadvantages of using the LCD.

PROJECTS OR GROUP ACTIVITIES

102. **Manufacturing** Manufacturers who package their products in cans would like to design the can such that the minimum amount of aluminum is needed. If a soft drink can contains 12 oz (355 cm^3), the function that relates the surface area of the can (the amount of aluminum needed) to the radius of the bottom of the can is given by the equation $f(r) = 2\pi r^2 + \frac{710}{r}$, where r is measured in centimeters.

a. Express the right side of this equation with a common denominator.
b. Graph the equation for $0 < r \le 7.5$.
c. The point whose coordinates are (7, 409), to the nearest integer, is on the graph of f. Write a sentence that gives an interpretation of this ordered pair.
d. Use a graphing utility to determine the radius of the can that has a minimum surface area. Round to the nearest tenth.
e. The height of the can is determined from the equation $h = \dfrac{355}{\pi r^2}$. Use the answer to part (d) to determine the height of the can that has a minimum surface area. Round to the nearest tenth.
f. Determine the minimum surface area. Round to the nearest tenth.

103. **Manufacturing** A manufacturer wants to make square tissues and package them in a box. The manufacturer has determined that to be competitive, the box needs to hold 175 tissues, which means that the volume of the box will be 132 in^3. The amount of cardboard (surface area) that will be necessary to build this box is given by $f(x) = 2x^2 + \frac{528}{x}$, where x is the height of the box in inches.

a. Express the right side of this equation with a common denominator.
b. Graph the equation for $1.5 < x \leq 10$.
c. The point whose coordinates are (4, 164) is on the graph. Write a sentence that explains the meaning of this point.
d. With a graphing utility, determine, to the nearest tenth, the height of the box that uses the minimum amount of cardboard.
e. Determine the minimum amount of cardboard. Round to the nearest tenth.

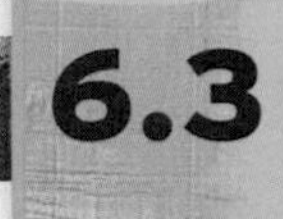

6.3 Complex Fractions

OBJECTIVE 1 Simplify complex fractions

A **complex fraction** is a fraction in which the numerator or denominator contains one or more fractions. Examples of complex fractions are shown below.

$$\frac{5}{2+\frac{1}{2}} \qquad \frac{5+\frac{1}{y}}{5-\frac{1}{y}} \qquad \frac{x+4+\frac{1}{x+2}}{x-2+\frac{1}{x+2}}$$

To simplify a complex fraction, rewrite the complex fraction so that no fraction remains in the numerator or denominator. Write the resulting fraction in simplest form.

Focus on simplifying a complex fraction

Simplify: $\dfrac{\frac{1}{x}+\frac{1}{y}}{\frac{1}{x}-\frac{1}{y}}$

Multiply the numerator and denominator of the complex fraction by the LCD of the fractions. The LCD of $\frac{1}{x}$ and $\frac{1}{y}$ is xy.

$$\frac{\frac{1}{x}+\frac{1}{y}}{\frac{1}{x}-\frac{1}{y}} = \frac{\frac{1}{x}+\frac{1}{y}}{\frac{1}{x}-\frac{1}{y}} \cdot \frac{xy}{xy}$$

Use the Distributive Property, and then simplify each product.

$$= \frac{\frac{1}{x}\cdot xy + \frac{1}{y}\cdot xy}{\frac{1}{x}\cdot xy - \frac{1}{y}\cdot xy} = \frac{y+x}{y-x}$$

How It's Used

Complex fractions arise in some physics equations, such as the lens equation

$$f = \frac{1}{\frac{1}{p}+\frac{1}{q}}$$

that relates the focal length of a lens to the distances of an object and its image from the lens.

Note that after the numerator and denominator of the complex fraction have been multiplied by the LCD, no fraction remains in the numerator or denominator.

The method just shown of simplifying a complex fraction by multiplying the numerator and denominator by the LCD is used in Example 1. However, a different approach is to rewrite the numerator and denominator of the complex fraction as single fractions and then divide the numerator by the denominator. Here, the example shown on the preceding page is simplified by using this alternative method.

Take Note

Recall that the fraction bar can be read "divided by."

Rewrite the numerator and denominator of the complex fraction as single fractions.

$$\frac{\frac{1}{x}+\frac{1}{y}}{\frac{1}{x}-\frac{1}{y}} = \frac{\frac{1}{x}\cdot\frac{y}{y}+\frac{1}{y}\cdot\frac{x}{x}}{\frac{1}{x}\cdot\frac{y}{y}-\frac{1}{y}\cdot\frac{x}{x}} = \frac{\frac{y}{xy}+\frac{x}{xy}}{\frac{y}{xy}-\frac{x}{xy}} = \frac{\frac{y+x}{xy}}{\frac{y-x}{xy}}$$

Divide the numerator of the complex fraction by the denominator.

$$= \frac{y+x}{xy} \div \frac{y-x}{xy} = \frac{y+x}{xy}\cdot\frac{xy}{y-x}$$

Multiply the fractions. Simplify.

$$= \frac{(y+x)xy}{xy(y-x)} = \frac{y+x}{y-x}$$

Note that this is the same result we got before.

EXAMPLE 1 Simplify. **A.** $\dfrac{2-\frac{11}{x}+\frac{15}{x^2}}{3-\frac{5}{x}-\frac{12}{x^2}}$ **B.** $\dfrac{2x-1+\frac{7}{x+4}}{3x-8+\frac{17}{x+4}}$

Solution **A.**

$$\frac{2-\frac{11}{x}+\frac{15}{x^2}}{3-\frac{5}{x}-\frac{12}{x^2}} = \frac{2-\frac{11}{x}+\frac{15}{x^2}}{3-\frac{5}{x}-\frac{12}{x^2}}\cdot\frac{x^2}{x^2}$$

• **Multiply the numerator and denominator by the LCD, x^2.**

$$= \frac{2\cdot x^2-\frac{11}{x}\cdot x^2+\frac{15}{x^2}\cdot x^2}{3\cdot x^2-\frac{5}{x}\cdot x^2-\frac{12}{x^2}\cdot x^2}$$

• **Distributive Property**

$$= \frac{2x^2-11x+15}{3x^2-5x-12}$$

$$= \frac{(2x-5)(x-3)}{(3x+4)(x-3)} = \frac{2x-5}{3x+4}$$

B.

$$\frac{2x-1+\frac{7}{x+4}}{3x-8+\frac{17}{x+4}}$$

$$= \frac{2x-1+\frac{7}{x+4}}{3x-8+\frac{17}{x+4}}\cdot\frac{x+4}{x+4}$$

• **Multiply the numerator and denominator by the LCD, $x + 4$.**

$$= \frac{(2x-1)(x+4)+\frac{7}{x+4}(x+4)}{(3x-8)(x+4)+\frac{17}{x+4}(x+4)}$$

• **Distributive Property**

$$= \frac{2x^2+7x-4+7}{3x^2+4x-32+17} = \frac{2x^2+7x+3}{3x^2+4x-15}$$

$$= \frac{(2x+1)(x+3)}{(3x-5)(x+3)} = \frac{2x+1}{3x-5}$$

Problem 1 A. $\dfrac{3 + \dfrac{16}{x} + \dfrac{16}{x^2}}{6 + \dfrac{5}{x} - \dfrac{4}{x^2}}$ B. $\dfrac{2x + 5 + \dfrac{14}{x - 3}}{4x + 16 + \dfrac{49}{x - 3}}$

Solution See page S21.

Try Exercise 23, page 345.

6.3 Exercises

CONCEPT CHECK

1. What is a complex fraction?

2. What is the general goal of simplifying a complex fraction?

3. What is the LCD of the fractions in the complex fraction $\dfrac{1 - \dfrac{3}{x - 3}}{\dfrac{2}{x} + 1}$?

4. True or false? $\dfrac{c^{-1}}{a^{-1} + b^{-1}} = \dfrac{\dfrac{1}{c}}{\dfrac{1}{a} + \dfrac{1}{b}}$

1 Simplify complex fractions (See pages 342–344.)

GETTING READY

For Exercises 5 to 8, use the complex fraction $\dfrac{1 - \dfrac{2}{a}}{1 + \dfrac{1}{a^2}}$.

5. To simplify the complex fraction, multiply the numerator and denominator of the complex fraction by the LCD of the fractions __?__ and __?__. The LCD is __?__.

6. When you multiply the numerator of the complex fraction by a^2, the numerator of the complex fraction simplifies to __?__.

7. When you multiply the denominator of the complex fraction by a^2, the denominator of the complex fraction simplifies to __?__.

8. The simplified form of the complex fraction is __?__.

Simplify.

9. $\dfrac{2 - \dfrac{1}{3}}{4 + \dfrac{11}{3}}$

10. $\dfrac{3 + \dfrac{5}{2}}{8 - \dfrac{7}{2}}$

11. $\dfrac{3 - \dfrac{2}{3}}{5 + \dfrac{5}{6}}$

12. $\dfrac{1 + \dfrac{1}{x}}{1 - \dfrac{1}{x^2}}$

13. $\dfrac{\dfrac{1}{y^2} - 1}{1 + \dfrac{1}{y}}$

14. $\dfrac{a - 2}{\dfrac{4}{a} - a}$

15. $\dfrac{\dfrac{25}{a} - a}{5 + a}$

16. $\dfrac{\dfrac{1}{a^2} - \dfrac{1}{a}}{\dfrac{1}{a^2} + \dfrac{1}{a}}$

17. $\dfrac{\dfrac{1}{b} + \dfrac{1}{2}}{\dfrac{4}{b^2} - 1}$

18. $\dfrac{2 - \dfrac{4}{x + 2}}{5 - \dfrac{10}{x + 2}}$

19. $\dfrac{4 + \dfrac{12}{2x - 3}}{5 + \dfrac{15}{2x - 3}}$

20. $\dfrac{\dfrac{x}{x + 1} - \dfrac{1}{x}}{\dfrac{x}{x + 1} + \dfrac{1}{x}}$

21. $\dfrac{\dfrac{2a}{a - 1} - \dfrac{3}{a}}{\dfrac{1}{a - 1} + \dfrac{2}{a}}$

22. $\dfrac{\dfrac{-5}{b - 5} - 3}{\dfrac{10}{b - 5} + 6}$

23. $\dfrac{1 - \dfrac{1}{x} - \dfrac{6}{x^2}}{1 - \dfrac{4}{x} + \dfrac{3}{x^2}}$

24. $\dfrac{1 - \dfrac{3}{x} - \dfrac{10}{x^2}}{1 + \dfrac{11}{x} + \dfrac{18}{x^2}}$

25. $\dfrac{\dfrac{15}{x^2} - \dfrac{2}{x} - 1}{\dfrac{4}{x^2} - \dfrac{5}{x} + 4}$

26. $\dfrac{1 - \dfrac{2x}{3x - 4}}{x - \dfrac{32}{3x - 4}}$

27. $\dfrac{1 - \dfrac{12}{3x + 10}}{x - \dfrac{8}{3x + 10}}$

28. $\dfrac{x - 1 + \dfrac{2}{x - 4}}{x + 3 + \dfrac{6}{x - 4}}$

29. $\dfrac{x - 5 - \dfrac{18}{x + 2}}{x + 7 + \dfrac{6}{x + 2}}$

30. $\dfrac{x - 4 + \dfrac{9}{2x + 3}}{x + 3 - \dfrac{5}{2x + 3}}$

31. $\dfrac{\dfrac{1}{a} - \dfrac{3}{a - 2}}{\dfrac{2}{a} + \dfrac{5}{a - 2}}$

32. $\dfrac{\dfrac{2}{b} - \dfrac{5}{b + 3}}{\dfrac{3}{b} + \dfrac{3}{b + 3}}$

33. $\dfrac{\dfrac{1}{y^2} - \dfrac{1}{xy} - \dfrac{2}{x^2}}{\dfrac{1}{y^2} - \dfrac{3}{xy} + \dfrac{2}{x^2}}$

34. $\dfrac{\dfrac{2}{b^2} - \dfrac{5}{ab} - \dfrac{3}{a^2}}{\dfrac{2}{b^2} + \dfrac{7}{ab} + \dfrac{3}{a^2}}$

35. $\dfrac{\dfrac{x - 1}{x + 1} - \dfrac{x + 1}{x - 1}}{\dfrac{x - 1}{x + 1} + \dfrac{x + 1}{x - 1}}$

36. $\dfrac{\dfrac{y}{y + 2} - \dfrac{y}{y - 2}}{\dfrac{y}{y + 2} + \dfrac{y}{y - 2}}$

37. The denominator of a certain complex fraction is the reciprocal of its numerator. Which of the following is the simplified form of this complex fraction?
   (i) 1
   (ii) the square of the complex fraction
   (iii) the square of the numerator of the complex fraction
   (iv) the reciprocal of the complex fraction

38. Determine whether the given expression is equivalent to the reciprocal of the complex fraction $\dfrac{1}{1 - \dfrac{1}{a}}$.

a. $\dfrac{a}{a - 1}$ b. $\dfrac{a - 1}{a}$ c. $1 - \dfrac{1}{a}$

APPLYING CONCEPTS

Simplify.

39. $\frac{x^{-1}}{y^{-1}} + \frac{y}{x}$

40. $\frac{x^{-1} + y^{-1}}{x^{-1} - y^{-1}}$

41. $\frac{x - \frac{1}{x}}{1 + \frac{1}{x}}$

42. $\frac{x^{-1} + y}{x^{-1} - y}$

43. $2 - \frac{2}{2 - \frac{2}{c - 1}}$

44. $1 - \frac{1}{1 - \frac{1}{b - 2}}$

45. $3 - \frac{2}{1 - \frac{2}{3 - \frac{2}{x}}}$

46. $\frac{\frac{1}{x + h} - \frac{1}{x}}{h}$

47. $\frac{\frac{1}{(x + h)^2} - \frac{1}{x^2}}{h}$

48. **Electronics** The total resistance R of three resistors in parallel is given by the formula $R = \frac{1}{\frac{1}{R_1} + \frac{1}{R_2} + \frac{1}{R_3}}$. Find the total parallel resistance when $R_1 = 2$ ohms, $R_2 = 4$ ohms, and $R_3 = 8$ ohms.

PROJECTS OR GROUP ACTIVITIES

49. **Car Loans** The interest rate on a car loan affects the monthly payment. The function that relates the monthly payment for a 5-year loan (60-month loan) to the monthly interest rate is given by

$$P(x) = \frac{Cx}{\left[1 - \frac{1}{(x + 1)^{60}}\right]}$$

where x is the monthly interest rate (as a decimal), C is the loan amount, and $P(x)$ is the monthly payment.

a. Simplify the complex fraction.

b. The point whose coordinates are approximately (0.005, 386.66) is on the graph of this equation. Write a sentence that gives an interpretation of this ordered pair.

c.  Use a graphing utility to determine the monthly payment for a car loan of \$20,000 at an annual interest rate of 8%. Round to the nearest dollar.

6.4 Rational Equations

OBJECTIVE 1 Solve fractional equations

To solve an equation containing fractions, **clear denominators** by multiplying each side of the equation by the LCD of the fractions. Then solve for the variable.

Focus on solving a rational equation

Solve: $\dfrac{3x}{x-5} = 5 - \dfrac{5}{x-5}$

$$\frac{3x}{x-5} = 5 - \frac{5}{x-5}$$

Multiply each side of the equation by the LCD.

$$(x-5)\left(\frac{3x}{x-5}\right) = (x-5)\left(5 - \frac{5}{x-5}\right)$$

Use the Distributive Property on the right side of the equation.

$$3x = (x-5)5 - (x-5)\left(\frac{5}{x-5}\right)$$

Simplify.

$$3x = 5x - 25 - 5$$
$$3x = 5x - 30$$

Solve for x.

$$-2x = -30$$
$$x = 15$$

15 checks as a solution.

The solution is 15.

Occasionally, a value of the variable that appears to be a solution will make one of the denominators zero. Such a solution is called an **extraneous solution.** In such a case, the equation has no solution for that value of the variable.

Focus on a rational equation with no solution

Solve: $\dfrac{3x}{x-3} = 2 + \dfrac{9}{x-3}$

Take Note

If each side of an equation is multiplied by a variable expression, it is essential that the solutions be checked. As this example shows, a proposed solution to an equation may not check when it is substituted into the original equation.

$$\frac{3x}{x-3} = 2 + \frac{9}{x-3}$$

Multiply each side of the equation by the LCD.

$$(x-3)\left(\frac{3x}{x-3}\right) = (x-3)\left(2 + \frac{9}{x-3}\right)$$

Use the Distributive Property on the right side of the equation.

$$3x = (x-3)2 + (x-3)\left(\frac{9}{x-3}\right)$$
$$3x = 2x - 6 + 9$$
$$3x = 2x + 3$$

Solve for x.

$$x = 3$$

Substituting 3 into the original equation results in division by zero.

$$\frac{3x}{x-3} = 2 + \frac{9}{x-3}$$
$$\frac{3(3)}{3-3} = 2 + \frac{9}{3-3}$$

Because division by zero is not defined, the equation has no solution.

$$\frac{9}{0} = 2 + \frac{9}{0}$$

Multiplying each side of an equation by a variable expression may produce an equation with different solutions from the original equation. Thus, any time you multiply each side of an equation by a variable expression, you must check the resulting solution.

EXAMPLE 1 Solve. **A.** $\frac{1}{4} = \frac{5}{x+5}$ **B.** $\frac{2x}{x-2} = \frac{1}{3x-4} + 2$

Solution **A.**

$$\frac{1}{4} = \frac{5}{x+5}$$

$$4(x+5)\frac{1}{4} = 4(x+5)\left(\frac{5}{x+5}\right)$$

• **Multiply each side of the equation by the LCD.**

$$x + 5 = 4(5)$$

$$x + 5 = 20$$

$$x = 15$$

15 checks as a solution.

The solution is 15.

B.

$$\frac{2x}{x-2} = \frac{1}{3x-4} + 2$$

$$(x-2)(3x-4)\left(\frac{2x}{x-2}\right) = (x-2)(3x-4)\left(\frac{1}{3x-4} + 2\right)$$

$$(3x-4)2x = (x-2)(3x-4)\left(\frac{1}{3x-4}\right) + (x-2)(3x-4)2$$

$$6x^2 - 8x = x - 2 + 6x^2 - 20x + 16$$

$$6x^2 - 8x = 6x^2 - 19x + 14$$

$$11x = 14$$

$$x = \frac{14}{11}$$

$\frac{14}{11}$ checks as a solution.

The solution is $\frac{14}{11}$.

Problem 1 Solve. **A.** $\frac{5}{2x-3} = \frac{-2}{x+1}$ **B.** $\frac{4x+1}{2x-1} = 2 + \frac{3}{x-3}$

Solution See page S21.

Try Exercise 23, page 353.

OBJECTIVE 2 Work problems

Rate of work is that part of a task that is completed in one unit of time. If a mason can build a retaining wall in 12 h, then in 1 h the mason can build $\frac{1}{12}$ of the wall. The mason's rate of work is $\frac{1}{12}$ of the wall each hour. If an apprentice can build the wall in x hours, the rate of work for the apprentice is $\frac{1}{x}$ of the wall each hour. In solving a work problem, the goal is to determine the time it takes to complete a task. The basic equation that is used to solve work problems is

Rate of work × Time worked = Part of task completed

For example, if a pipe can fill a tank in 5 h, then in 2 h the pipe will fill $\frac{1}{5} \times 2 = \frac{2}{5}$ of the tank. In t hours, the pipe will fill $\frac{1}{5} \times t = \frac{t}{5}$ of the tank.

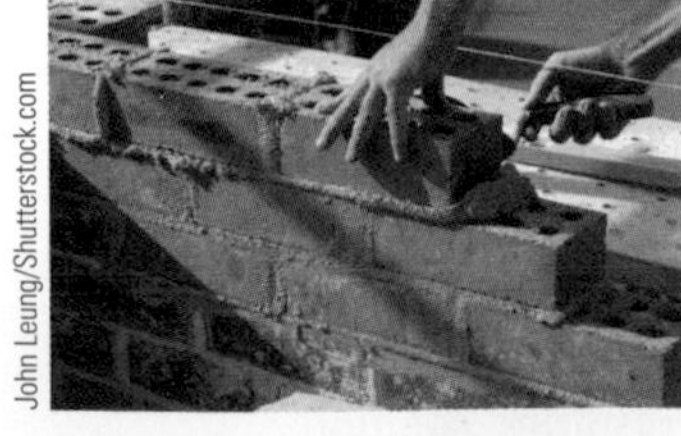

Point of Interest

The following problem was recorded in the *Jiuzhang*, a Chinese text that dates to the Han dynasty (about 200 B.C. to A.D. 200). "A reservoir has 5 channels bringing water to it. The first channel can fill the reservoir in $\frac{1}{3}$ day, the second in 1 day, the third in $2\frac{1}{2}$ days, the fourth in 3 days, and the fifth in 5 days. If all channels are open, how long does it take to fill the reservoir?" This problem is the earliest known work problem.

Solve: A mason can build a wall in 10 h. An apprentice can build a wall in 15 h. How long would it take them to build the wall if they worked together?

STRATEGY FOR SOLVING A WORK PROBLEM

- For each person or machine, write a numerical or variable expression for the rate of work, the time worked, and the part of the task completed. The results can be recorded in a table.

Unknown time to build the wall working together: t

	Rate of work	·	Time worked	=	Part of task completed
Mason	$\frac{1}{10}$	·	t	=	$\frac{t}{10}$
Apprentice	$\frac{1}{15}$	·	t	=	$\frac{t}{15}$

- Determine how the parts of the task completed are related. Use the fact that the sum of the parts of the task completed must equal 1, the complete task.

The sum of the part of the task completed by the mason and the part of the task completed by the apprentice is 1.

$$\frac{t}{10} + \frac{t}{15} = 1$$

$$30\left(\frac{t}{10} + \frac{t}{15}\right) = 30(1)$$

$$3t + 2t = 30$$

$$5t = 30$$

$$t = 6$$

Working together, they would build the wall in 6 h.

EXAMPLE 2 An electrician requires 12 h to wire a house. The electrician's apprentice can wire a house in 16 h. After working alone on a job for 4 h, the electrician quits, and the apprentice completes the task. How long does it take the apprentice to finish wiring the house?

Strategy

- Time required for the apprentice to finish wiring the house: t

	Rate	·	Time	=	Part
Electrician	$\frac{1}{12}$	·	4	=	$\frac{4}{12}$
Apprentice	$\frac{1}{16}$	·	t	=	$\frac{t}{16}$

- The sum of the part of the task completed by the electrician and the part of the task completed by the apprentice is 1.

Solution

$$\frac{4}{12} + \frac{t}{16} = 1$$
$$\frac{1}{3} + \frac{t}{16} = 1$$
$$48\left(\frac{1}{3} + \frac{t}{16}\right) = 48(1)$$
$$16 + 3t = 48$$
$$3t = 32$$
$$t = \frac{32}{3} = 10\frac{2}{3}$$

It takes the apprentice $10\frac{2}{3}$ h to finish wiring the house.

Fills tank in x hours | Fills tank in 9 hours

Fills $\frac{1}{x}$ of the tank each hour | Fills $\frac{1}{9}$ of the tank each hour

Problem 2 Two water pipes can fill a tank with water in 6 h. The larger pipe working alone can fill the tank in 9 h. How long would it take the smaller pipe, working alone, to fill the tank?

Solution See page S21.

Try Exercise 55, page 355.

OBJECTIVE 

Uniform motion problems

A car that travels constantly in a straight line at 55 mph is in uniform motion. **Uniform motion** means that the speed of an object does not change.

The basic equation used to solve uniform motion problems is

Distance = Rate × Time

An alternative form of this equation can be written by solving the equation for time. This form of the equation is used to solve the following problem.

$$\frac{\textbf{Distance}}{\textbf{Rate}} = \textbf{Time}$$

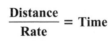

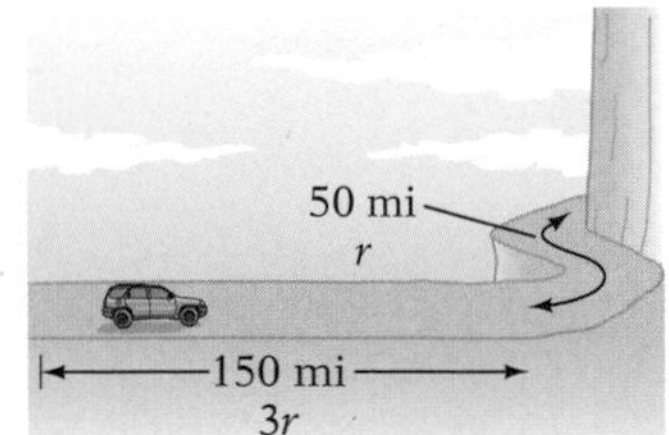

Solve: A motorist drove 150 mi on country roads before driving 50 mi on mountain roads. The rate of speed on the country roads was three times the rate on the mountain roads. The time spent traveling the 200 mi was 5 h. Find the rate of the motorist on the country roads.

STRATEGY FOR SOLVING A UNIFORM MOTION PROBLEM

▶ For each object, write a numerical or variable expression for the distance, rate, and time. The results can be recorded in a table.

The unknown rate of speed on the mountain roads: r
The rate of speed on the country roads: $3r$

	Distance	÷	Rate	=	Time
Country roads	150	÷	$3r$	=	$\frac{150}{3r}$
Mountain roads	50	÷	r	=	$\frac{50}{r}$

▶ Determine how the times traveled by each object are related. For example, it may be known that the times are equal, or the total time may be known.

The total time for the trip is 5 h.

$$\frac{150}{3r} + \frac{50}{r} = 5$$
$$\frac{50}{r} + \frac{50}{r} = 5$$
$$r\left(\frac{50}{r} + \frac{50}{r}\right) = r(5)$$
$$50 + 50 = 5r$$
$$100 = 5r$$
$$20 = r$$

The rate of speed on the country roads was $3r$. Replace r with 20 and evaluate.

$$3r = 3(20) = 60$$

The rate of speed on the country roads was 60 mph.

EXAMPLE 3 A marketing executive traveled 810 mi on a corporate jet in the same amount of time that it took to travel an additional 162 mi by helicopter. The rate of the jet was 360 mph greater than the rate of the helicopter. Find the rate of the jet.

Strategy ▶ Rate of the helicopter: r
Rate of the jet: $r + 360$

r
162 mi

$r + 360$
810 mi

	Distance	÷	Rate	=	Time
Jet	810	÷	$r + 360$	=	$\frac{810}{r + 360}$
Helicopter	162	÷	r	=	$\frac{162}{r}$

▶ The time traveled by jet is equal to the time traveled by helicopter.

Solution

$$\frac{810}{r + 360} = \frac{162}{r}$$
$$r(r + 360)\left(\frac{810}{r + 360}\right) = r(r + 360)\left(\frac{162}{r}\right)$$
$$810r = (r + 360)162$$
$$810r = 162r + 58{,}320$$
$$648r = 58{,}320$$
$$r = 90$$

• **The rate of the helicopter was 90 mph.**

$r + 360 = 90 + 360 = 450$

• **Substitute the value of r into the variable expression for the rate of the jet.**

The rate of the jet was 450 mph.

Problem 3 A plane can fly at a rate of 150 mph in calm air. Traveling with the wind, the plane flew 700 mi in the same amount of time it took to fly 500 mi against the wind. Find the rate of the wind.

Solution See pages S21–S22.

Try Exercise 69, page 356.

6.4 Exercises

CONCEPT CHECK

1. What numbers are not possible solutions of the rational equation $\frac{x-1}{x+1} + \frac{x-2}{x} = 4$?

2. Each side of an equation is multiplied by $x - 3$. The new equation (i) has the same solutions as the original equation, (ii) may have the same solutions as the original equation, or (iii) does not have the same solutions as the original equation.

3. If Hana can paint a wall in 30 min and Miya can paint a wall in 45 min, who has the greater rate of work?

4. If you drove 500 mi in r hours, how many hours did it take to make the trip?

1 Solve fractional equations (See pages 346–348.)

GETTING READY

For Exercises 5 to 8, use the equation $\frac{8}{x} = \frac{5}{x+3}$.

5. The first step in solving the equation is to clear denominators by multiplying each side of the equation by the LCM of the denominators __?__ and __?__. The LCD is __?__.

6. a. When you multiply the left side of the equation by the LCD $x(x + 3)$, the left side simplifies to __?__.

   b. When you multiply the right side of the equation by the LCD $x(x + 3)$, the right side simplifies to __?__.

7. Use your answers to Exercise 6 to write the equation that results from clearing denominators: __?__ = __?__. The solution of this equation is __?__.

8. The solution of a rational equation must be checked in the original equation. To check the solution you found in Exercise 7 in the original equation, substitute __?__ for x and simplify:

   $\frac{8}{-8} = \frac{5}{-8+3}$ simplifies to __?__ = __?__, which shows that the solution checks.

Solve.

9. $\frac{x}{2} + \frac{5}{6} = \frac{x}{3}$

10. $\frac{x}{5} - \frac{2}{9} = \frac{x}{15}$

11. $1 - \frac{3}{y} = 4$

12. $7 + \frac{6}{y} = 5$

13. $\frac{8}{2x - 1} = 2$

14. $3 = \frac{18}{3x - 4}$

15. $\frac{4}{x - 4} = \frac{2}{x - 2}$

16. $\frac{x}{3} = \frac{x + 1}{7}$

17. $\frac{x - 2}{5} = \frac{1}{x + 2}$

18. $\frac{x + 4}{10} = \frac{6}{x - 3}$

19. $\frac{3}{x - 2} = \frac{4}{x}$

20. $\frac{5}{x} = \frac{2}{x + 3}$

21. $\frac{3}{x - 4} + 2 = \frac{5}{x - 4}$

22. $\frac{5}{y + 3} - 2 = \frac{7}{y + 3}$

23. $\frac{8}{x - 5} = \frac{3}{x}$

24. $\frac{16}{2 - x} = \frac{4}{x}$

25. $5 + \frac{8}{a - 2} = \frac{4a}{a - 2}$

26. $\frac{-4}{a - 4} = 3 - \frac{a}{a - 4}$

27. $\frac{x}{2} + \frac{20}{x} = 7$

28. $3x = \frac{4}{x} - \frac{13}{2}$

29. $\frac{6}{x - 5} = \frac{1}{x}$

30. $\frac{8}{x - 2} = \frac{4}{x + 1}$

31. $\frac{x}{x + 2} = \frac{6}{x + 5}$

32. $\frac{x}{x - 2} = \frac{3}{x - 4}$

33. $-\frac{5}{x + 7} + 1 = \frac{4}{x + 7}$

34. $5 - \frac{2}{2x - 5} = \frac{3}{2x - 5}$

35. $\frac{2}{4y^2 - 9} + \frac{1}{2y - 3} = \frac{3}{2y + 3}$

36. $\frac{5}{x - 2} - \frac{2}{x + 2} = \frac{3}{x^2 - 4}$

37. $\frac{5}{x^2 - 7x + 12} = \frac{2}{x - 3} + \frac{5}{x - 4}$

38. $\frac{9}{x^2 + 7x + 10} = \frac{5}{x + 2} - \frac{3}{x + 5}$

When a proposed solution of a rational equation does not check in the original equation, it is because the proposed solution results in an expression involving division by zero. For each equation, state the values of x that, when substituted into the original equation, would result in division by zero.

39. $\frac{x^2 + 24}{x^2 + x - 12} + \frac{x}{4} = \frac{2x}{x - 3}$

40. $\frac{1}{x - 2} - 1 = \frac{2}{x^2 - 2x}$

2 Work problems (See pages 348–350.)

41. If one person can complete a task in 2 h and another person can complete the same task in 3 h, will it take more or less than 2 h to complete the task when both people are working? Explain your answer.

42. If a gardener can mow a lawn in 20 min, what portion of the lawn can the gardener mow in 1 min?

GETTING READY

43. It takes the hot water hose 6 min to fill a washing machine. It takes the cold water hose 4 min to fill the machine. Let t represent the amount of time it would take for the hot and cold water hoses to fill the machine with warm water. Complete the following table.

	Rate of work	·	Time worked	=	Part of task completed
Hot water hose	?	·	?	=	?
Cold water hose	?	·	?	=	?

44. Refer to Exercise 43. When the washing machine is full, the "part of task completed" is the whole task, so the sum of the parts completed by the hot water hose and the cold water hose is ___?___. Use this fact and the expressions in the table in Exercise 43 to write an equation that can be solved to find the amount of time it takes to fill the washer with warm water: ___?___ + ___?___ = ___?___.

45. A large biotech firm uses two computers to process the daily results of its research studies. One computer can process the data in 2 h; the other computer takes 3 h to do the same job. How long would it take to process the data if both computers were used?

46. Two college students have started their own business building computers from kits. Working alone, one student can build a computer in 20 h. When the second student helps, the students can build a computer in 7.5 h. How long would it take the second student, working alone, to build the computer?

47. One solar heating panel can raise the temperature of water 1 degree in 30 min. A second solar heating panel can raise the temperature 1 degree in 45 min. How long would it take to raise the temperature of the water 1 degree with both solar panels operating?

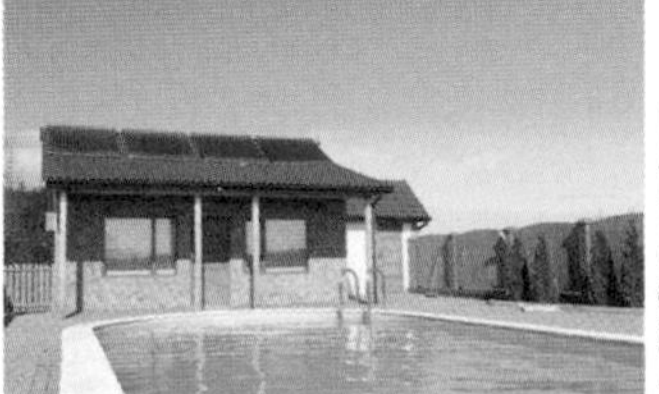
luiggi33/Shutterstock.com

48. One member of a gardening team can landscape a new lawn in 36 h. The other member of the team can do the job in 45 h. How long would it take to landscape a lawn if both gardeners worked together?

49. One member of a telephone crew can wire new telephone lines in 5 h. It takes 7.5 h for the other member of the crew to do the job. How long would it take to wire new telephone lines if both members of the crew worked together?

50. As the spring flood waters began to recede in Texas, a young family was faced with pumping the water from their basement. One pump they were using could dispose of 9000 gal in 3 h. A second pump could dispose of the same number of gallons in 4.5 h. How many hours would it take to dispose of 9000 gal if both pumps were working?

51. A new machine can package transistors four times as fast as an older machine. Working together, the machines can package the transistors in 8 h. How long would it take the new machine, working alone, to package the transistors?

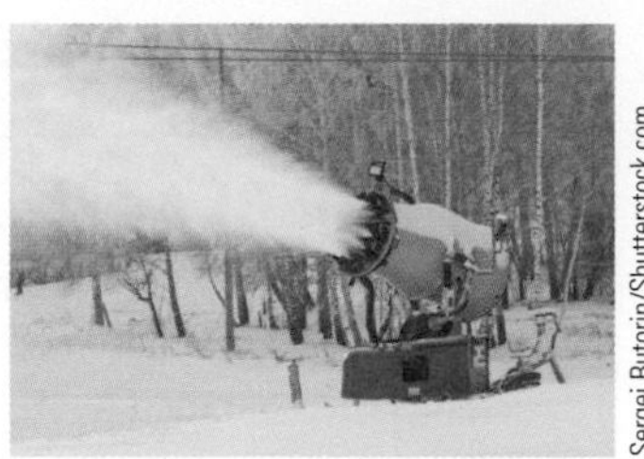
Sergei Butorin/Shutterstock.com

52. A ski resort can manufacture enough snow to open its steepest run in 12 h, whereas naturally falling snow would have to last for 36 h to provide enough snow. If the resort makes snow at the same time that it is snowing naturally, how long will it be until the run can be opened?

53. The larger of two printers being used to print the payroll for a major corporation requires 40 min to print the payroll. After both printers have been operating for 10 min, the larger printer malfunctions. The smaller printer requires 50 more minutes to complete the payroll. How long would it take the smaller printer, working alone, to print the payroll?

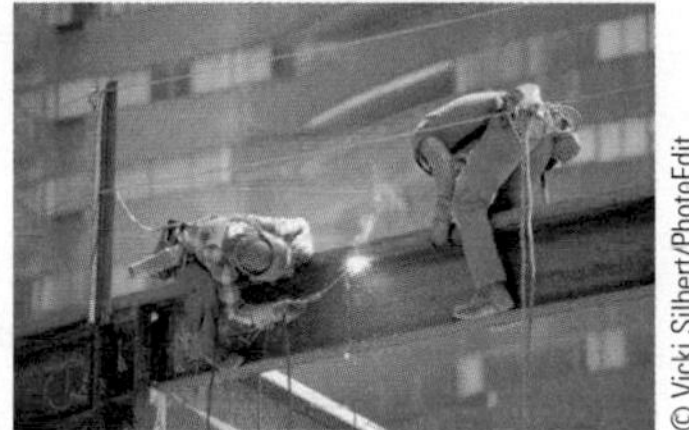
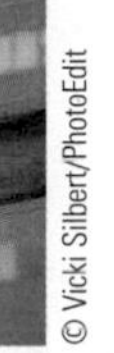
© Vicki Silbert/PhotoEdit

54. A welder requires 25 h to do a job. After the welder and an apprentice work on a job for 10 h, the welder quits. The apprentice finishes the job in 17 h. How long would it take the apprentice, working alone, to do the job?

55. A roofer requires 12 h to shingle a roof. After the roofer and an apprentice work on a roof for 3 h, the roofer moves on to another job. The apprentice requires 12 more hours to finish the job. How long would it take the apprentice, working alone, to do the job?

56. An experienced bricklayer can work twice as fast as an apprentice bricklayer. After the bricklayers worked together on a job for 8 h, the experienced bricklayer quit. The apprentice required 12 more hours to finish the job. How long would it take the experienced bricklayer, working alone, to do the job?

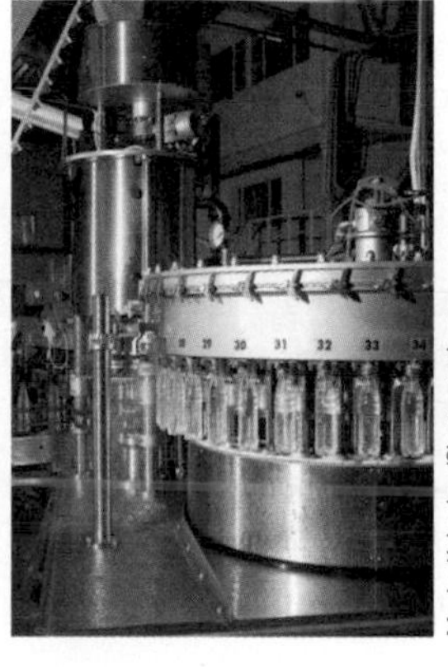
Maxim Kalmykov/Shutterstock.com

57. Two clerks are addressing envelopes for a mayoral candidate's political campaign. One clerk can address one envelope every 30 s, whereas it takes 40 s for the second clerk to address one envelope. How many minutes will it take the clerks, working together, to address 140 envelopes?

58. Three machines are filling water bottles. The machines can fill the daily quota of water bottles in 12 h, 15 h, and 20 h, respectively. How long would it take to fill the daily quota of water bottles with all three machines working?

59. With both hot and cold water running, a bathtub can be filled in 10 min. The drain will empty the tub in 15 min. A child turns both faucets on and leaves the drain open. How long will it be until the bathtub starts to overflow?

60. The inlet pipe of a water tank can fill the tank in 30 min. The outlet pipe can empty the tank in 20 min. How long would it take to empty a full tank with both pipes open?

61. An oil tank has two inlet pipes and one outlet pipe. One inlet pipe can fill the tank in 12 h, and the other inlet pipe can fill the tank in 20 h. The outlet pipe can empty the tank in 10 h. How long would it take to fill the tank with all three pipes open?

62. Water from a tank is being used for irrigation while the tank is being filled. The two inlet pipes can fill the tank in 6 h and 12 h, respectively. The outlet pipe can empty the tank in 24 h. How long would it take to fill the tank with all three pipes open?

63. It takes Jane n minutes to weed a row of a garden, and it takes Paul m minutes to weed a row of the garden, where $m > n$. Let t be the amount of time it takes Jane and Paul to weed a row of the garden together. Is t less than n, between n and m, or greater than m?

64. Ernie and Mike painted a fence together in a hours. It would have taken Ernie b hours to paint the fence by himself. What fraction of the fence did Ernie paint? What fraction of the fence did Mike paint?

3 Uniform motion problems (See pages 350–352.)

GETTING READY

65. a. A plane can fly 470 mph in calm air. In the time it takes the plane to fly 1320 mi against a headwind, it could fly 1500 mi if it were flying with the wind. Let w represent the rate of the wind. Complete the following table.

	Distance	÷	Rate	=	Time
Against the wind	?	÷	?	=	?
With the wind	?	÷	?	=	?

b. Use the relationship between the expressions in the last column of the table to write an equation that can be solved to find the rate of the wind: ? = ?.

66. Use the equation from part (b) of Exercise 65. The first step in solving this equation is to multiply each side of the equation by ?.

67. Two skaters take off for an afternoon of rollerblading in Central Park. The first skater can cover 15 mi in the same amount of time it takes the second skater, traveling 3 mph slower than the first skater, to cover 12 mi. Find the rate of each rollerblader.

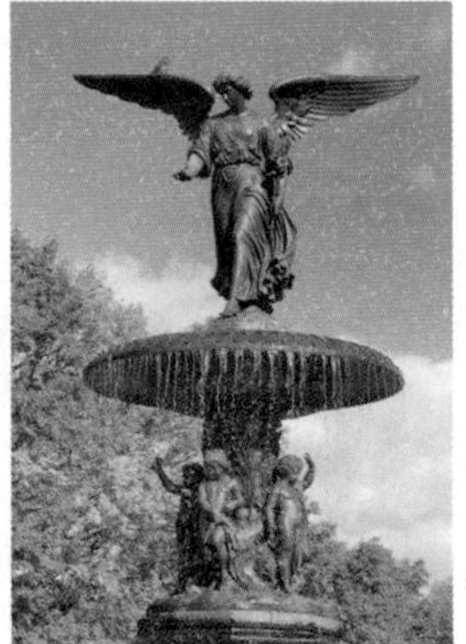

Bethesda Fountain in Central Park

68. A commercial jet travels 1620 mi in the same amount of time it takes a corporate jet to travel 1260 mi. The rate of the commercial jet is 120 mph greater than the rate of the corporate jet. Find the rate of each jet.

69. A passenger train travels 295 mi in the same amount of time it takes a freight train to travel 225 mi. The rate of the passenger train is 14 mph greater than the rate of the freight train. Find the rate of each train.

70. The rate of a bicyclist is 7 mph more than the rate of a long-distance runner. The bicyclist travels 30 mi in the same amount of time it takes the runner to travel 16 mi. Find the rate of the runner.

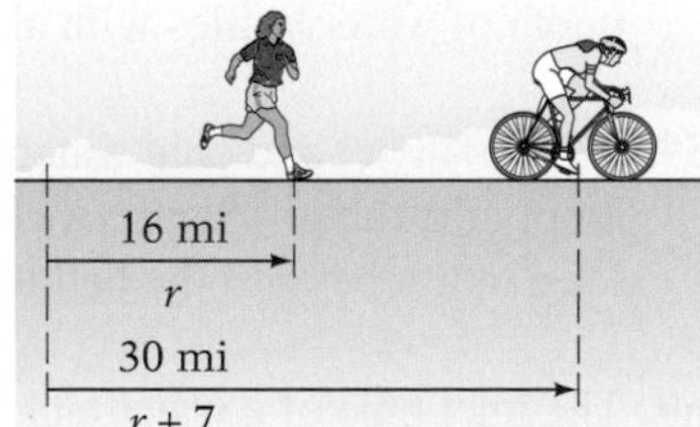

71. A cyclist rode 40 mi before having a flat tire and then walking 5 mi to a service station. The cycling rate was four times the walking rate. The time spent cycling and walking was 5 h. Find the rate at which the cyclist was riding.

72. Use the information in the article below. Find the average rate of the bullet train. Round to the nearest mile per hour.

In the News

First Bullet Train Will Travel in 2015

America's first "bullet train" is expected to be riding the rails in Florida in 2015. Averaging about 55 mph faster than a conventional train, the bullet train will cover the 88 mi from Orlando to Tampa in the same amount of time it would take a conventional train to travel 38 mi.

Source: Time magazine

73. A motorist drove 72 mi before running out of gas and then walking 4 mi to a gas station. The driving rate of the motorist was 12 times the walking rate. The time spent driving and walking was 2.5 h. Find the rate at which the motorist walked.

74. A cyclist and a jogger start from a town at the same time and head for a destination 18 mi away. The rate of the cyclist is twice the rate of the jogger. The cyclist arrives 1.5 h before the jogger. Find the rate of the cyclist.

75. A business executive can travel the 480 ft between two terminals of an airport by walking on a moving sidewalk in the same amount of time required to walk 360 ft without using the moving sidewalk. If the rate of the moving sidewalk is 2 ft/s, find the rate at which the executive can walk.

76. An insurance representative traveled 735 mi by commercial jet and then an additional 105 mi by helicopter. The rate of the jet was four times the rate of the helicopter. The entire trip took 2.2 h. Find the rate of the jet.

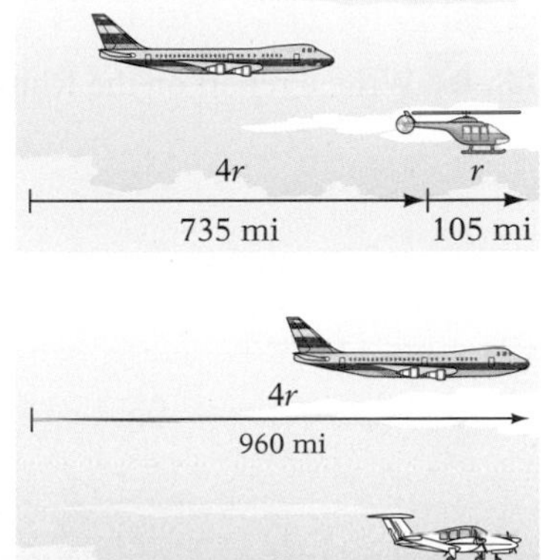

77. A single-engine plane and a commercial jet leave an airport at 10 A.M. and head for an airport 960 mi away. The rate of the jet is four times the rate of the single-engine plane. The single-engine plane arrives 4 h after the jet. Find the rate of each plane.

78. Marlys can row a boat 3 mph faster than she can swim. She is able to row 10 mi in the same amount of time it takes her to swim 4 mi. Find Marlys's swimming rate.

79. A cruise ship can sail 28 mph in calm water. Sailing with the Gulf Stream, the ship can sail 170 mi in the same amount of time it takes to sail 110 mi against the Gulf Stream. Find the rate of the Gulf Stream.

80. A commercial jet can fly 500 mph in calm air. Traveling with the jet stream, the plane flew 2420 mi in the same amount of time it would take to fly 1580 mi against the jet stream. Find the rate of the jet stream.

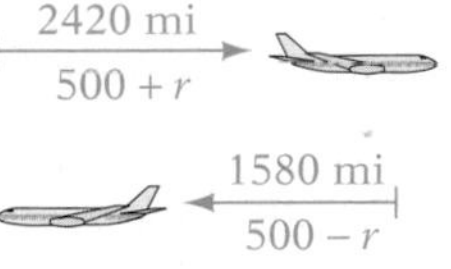

81. A tour boat used for river excursions can travel 7 mph in calm water. The amount of time it takes to travel 20 mi with the current is the same as the amount of time it takes to travel 8 mi against the current. Find the rate of the current.

82. A canoe can travel 8 mph in still water. Traveling with the current of a river, the canoe can travel 15 mi in the same amount of time it takes to travel 9 mi against the current. Find the rate of the current.

APPLYING CONCEPTS

83. Uniform Motion Problem Because of weather conditions, a bus driver reduced the usual speed along a 165-mile bus route by 5 mph. The bus arrived only 15 min later than its usual arrival time. How fast does the bus usually travel?

84. Uniform Motion Problem By increasing your speed by 10 mph, you can drive the 200-mile trip to your hometown in 40 min less time than the trip usually takes. How fast do you usually drive?

85. Uniform Motion Problem If a parade is 1 mi long and is proceeding at 3 mph, how long will it take a runner, jogging at 5 mph, to run from the beginning of the parade to the end and then back to the beginning?

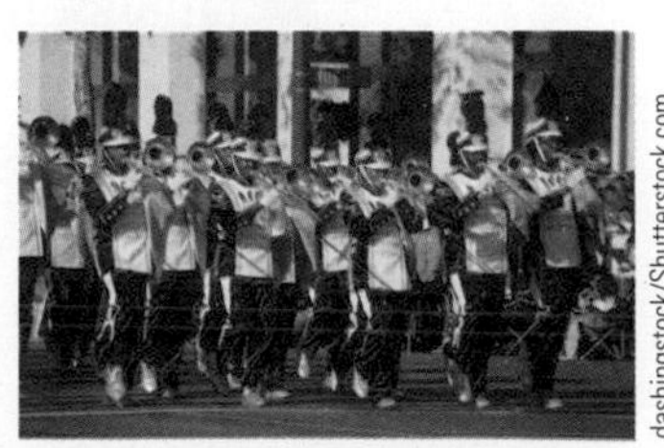
dashingstock/Shutterstock.com

86. Work Problem One pipe can fill a tank in 3 h, a second pipe can fill the tank in 4 h, and a third pipe can fill the tank in 6 h. How long would it take to fill the tank with all three pipes operating?

PROJECTS OR GROUP ACTIVITIES

87. Egyptian mathematicians worked with *unit fractions.*

a. What are unit fractions?

b. Write $\frac{3}{8}$ as the sum of two unit fractions.

c. Write $\frac{3}{5}$ as the sum of two unit fractions.

d. Write $\frac{7}{12}$ as the sum of two unit fractions in two different ways.

88.  Write a report on the Rhind Papyrus.

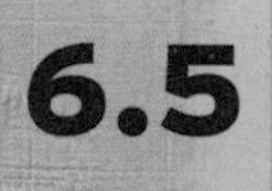

6.5 Proportions and Variation

OBJECTIVE 1 Proportions

Quantities such as 3 feet, 5 liters, and 2 miles are number quantities written with units. In these examples, the units are feet, liters, and miles.

A **ratio** is the quotient of two quantities that have the same unit.

The weekly wages of a painter are \$800. The painter spends \$150 a week for food. The ratio of wages spent for food to total weekly wages is written as shown below.

$\frac{\$150}{\$800} = \frac{150}{800} = \frac{3}{16}$ A ratio is in simplest form when the two numbers do not have a common factor. The units are not written.

A **rate** is the quotient of two quantities that have different units.

A car travels 120 mi on 3 gal of gas. The miles-to-gallon rate is written as shown below.

$\frac{120 \text{ mi}}{3 \text{ gal}} = \frac{40 \text{ mi}}{1 \text{ gal}}$ A rate is in simplest form when the two numbers do not have a common factor. The units are written as part of the rate.

A **proportion** is an equation that states that two ratios or rates are equal. For example, $\frac{90 \text{ km}}{4 \text{ L}} = \frac{45 \text{ km}}{2 \text{ L}}$ and $\frac{3}{4} = \frac{x+2}{16}$ are proportions.

Note that a proportion is a special kind of fractional equation. Many application problems can be solved by using proportions.

Focus on solving a proportion

The sales tax on a car that costs \$24,000 is \$1320. Find the sales tax on a car that costs \$29,000.

Write a proportion using x to represent the unknown sales tax.

$$\frac{1320}{24{,}000} = \frac{x}{29{,}000}$$

Simplify the left side.

$$\frac{11}{200} = \frac{x}{29{,}000}$$

Multiply each side of the equation by the denominators.

$$(200)(29{,}000)\left(\frac{11}{200}\right) = (200)(29{,}000)\left(\frac{x}{29{,}000}\right)$$
$$(29{,}000)(11) = 200x$$
$$319{,}000 = 200x$$
$$1595 = x$$

The sales tax on the $29,000 car is $1595.

EXAMPLE 1 A stock investment of 50 shares pays a dividend of $106. At this rate, how many additional shares are needed to earn a dividend of $424?

Strategy To find the additional number of shares that are required, write and solve a proportion using x to represent the additional number of shares. Then $50 + x$ is the total number of shares of stock.

Solution

$$\frac{106}{50} = \frac{424}{50 + x}$$

$$\frac{53}{25} = \frac{424}{50 + x}$$ • Simplify the left side.

$$25(50 + x)\frac{53}{25} = 25(50 + x)\frac{424}{50 + x}$$ • Multiply each side by the denominators.

$$(50 + x)53 = (25)424$$
$$2650 + 53x = 10{,}600$$
$$53x = 7950$$
$$x = 150$$

An additional 150 shares of stock are required.

Problem 1 Two pounds of cashews cost $5.80. At this rate, how much would 15 lb of cashews cost?

Solution See page S22.

Try Exercise 27, page 364.

OBJECTIVE 2

Variation problems

A **direct variation** is a special function that can be expressed as the equation $y = kx$, where k is a constant. The equation $y = kx$ is read "y varies directly as x" or "y is directly proportional to x." The constant k is called the **constant of variation** or the **constant of proportionality.**

The circumference (C) of a circle varies directly as the diameter (d). The direct variation equation is written $C = \pi d$. The constant of variation is π.

A nurse earns $28 per hour. The nurse's total wage (w) is directly proportional to the number of hours (h) worked. The equation of variation is $w = 28h$. The constant of proportionality is 28.

In general, a direct variation equation can be written in the form $y = kx^n$, where n is a positive number. For example, the equation $y = kx^2$ is read "y varies directly as the square of x."

The area (A) of a circle varies directly as the square of the radius (r) of the circle. The direct variation equation is $A = \pi r^2$. The constant of variation is π.

Given that V varies directly as r, and $V = 20$ when $r = 4$, the constant of variation can be found by writing the basic direct variation equation, replacing V and r by the given values, and solving for the constant of variation.

$$V = kr$$
$$20 = k \cdot 4$$
$$5 = k$$

The direct variation equation can then be written by substituting the value of k into the basic direct variation equation.

$$V = 5r$$

Focus on solving a direct variation problem

The tension (T) in a spring varies directly as the distance (x) it is stretched. If $T =$ 8 lb when $x = 2$ in., find T when $x = 4$ in.

Write the basic direct variation equation. $T = kx$

Replace T and x by the given values. $8 = k \cdot 2$

Solve for the constant of variation. $4 = k$

Write the direct variation equation. $T = 4x$

To find T when $x = 4$ in., substitute 4 for x in the direct variation equation and solve for T.

$$T = 4x$$
$$T = 4 \cdot 4$$
$$T = 16$$

When $x = 4$ in., the tension is 16 lb.

EXAMPLE 2 The amount (A) of medication prescribed for a person varies directly with the person's weight (W). For a person who weighs 50 kg, 2 ml of medication are prescribed. How many milliliters of medication are required for a person who weighs 75 kg?

Strategy To find the required amount of medication:

- Write the basic direct variation equation, replace the variables by the given values, and solve for k.
- Write the direct variation equation, replacing k by its value. Substitute 75 for W, and solve for A.

Solution

$$A = kW$$
$$2 = k \cdot 50$$
$$\frac{1}{25} = k$$

$$A = \frac{1}{25}W$$ • **This is the direct variation equation.**

$$A = \frac{1}{25} \cdot 75 = 3$$ • **Replace W by 75.**

The required amount of medication is 3 ml.

Problem 2 The distance (s) a body falls from rest varies directly as the square of the time (t) of the fall. An object falls 64 ft in 2 s. How far will it fall in 5 s?

Solution See page S22.

Try Exercise 37, page 365.

An **inverse variation** is a function that can be expressed as the equation $y = \frac{k}{x}$, where k is a constant. The equation $y = \frac{k}{x}$ is read "y varies inversely as x" or "y is inversely proportional to x."

In general, an inverse variation equation can be written $y = \frac{k}{x^n}$, where n is a positive number. For example, the equation $y = \frac{k}{x^2}$ is read "y varies inversely as the square of x."

Given that P varies inversely as the square of x, and $P = 5$ when $x = 2$, the variation constant can be found by writing the basic inverse variation equation, replacing P and x by the given values, and solving for the constant of variation.

$$P = \frac{k}{x^2}$$
$$5 = \frac{k}{2^2}$$
$$5 = \frac{k}{4}$$
$$20 = k$$

The inverse variation equation can then be found by substituting the value of k into the basic inverse variation equation.

$$P = \frac{20}{x^2}$$

Focus on solving an inverse variation problem

The length (L) of a rectangle with fixed area is inversely proportional to the width (w). If $L = 6$ ft when $w = 2$ ft, find L when $w = 3$ ft.

Write the basic inverse variation equation.

$$L = \frac{k}{w}$$

Replace L and w by the given values.

$$6 = \frac{k}{2}$$

Solve for the constant of variation.

$$12 = k$$

Write the inverse variation equation.

$$L = \frac{12}{w}$$

To find L when $w = 3$ ft, substitute 3 for w in the inverse variation equation and solve for L.

$$L = \frac{12}{w}$$
$$L = \frac{12}{3}$$
$$L = 4$$

When $w = 3$ ft, the length is 4 ft.

EXAMPLE 3 A company that produces personal computers has determined that the number of computers it can sell (s) is inversely proportional to the price (P) of the computer. Two thousand computers can be sold when the price is \$900. How many computers can be sold when the price of a computer is \$800?

Strategy To find the number of computers:

- ▶ Write the basic inverse variation equation, replace the variables by the given values, and solve for k.
- ▶ Write the inverse variation equation, replacing k by its value. Substitute 800 for P, and solve for s.

Solution

$$s = \frac{k}{P}$$

$$2000 = \frac{k}{900}$$

$$1{,}800{,}000 = k$$

$$s = \frac{1{,}800{,}000}{P}$$

• This is the inverse variation equation.

$$s = \frac{1{,}800{,}000}{800} = 2250$$

• Replace P by 800.

At a price of \$800, 2250 computers can be sold.

Problem 3 The resistance (R) to the flow of electric current in a wire of fixed length is inversely proportional to the square of the diameter (d) of the wire. If a wire of diameter 0.01 cm has a resistance of 0.5 ohm, what is the resistance in a wire that is 0.02 cm in diameter?

Solution See page S22.

Try Exercise 51, page 366.

A **combined variation** is a variation in which two or more types of variation occur at the same time. For example, in chemistry, the volume (V) of a gas varies directly as the temperature (T) and inversely as the pressure (P). This combined variation is written $V = \frac{kT}{P}$. A combined variation is the subject of Example 4.

A **joint variation** is a variation in which a variable varies directly as the product of two or more other variables. A joint variation can be expressed as the equation $z = kxy$, where k is a constant. The equation $z = kxy$ is read "z varies jointly as x and y." For example, the area (A) of a triangle varies jointly as the base (b) and the height (h). The joint variation equation is written $A = \frac{1}{2}bh$. The constant of variation is $\frac{1}{2}$. Problem 4 involves both combined and joint variation.

EXAMPLE 4 The pressure (P) of a gas varies directly as the temperature (T) and inversely as the volume (V). When $T = 50°$ and $V = 275$ in^3, $P = 20$ lb/in^2. Find the pressure of a gas when $T = 60°$ and $V = 250$ in^3.

Strategy To find the pressure:

- ▶ Write the basic combined variation equation, replace the variables by the given values, and solve for k.
- ▶ Write the combined variation equation, replacing k by its value. Substitute 60 for T and 250 for V, and solve for P.

Solution

$$P = \frac{kT}{V}$$

$$20 = \frac{k(50)}{275}$$

$$110 = k$$

$$P = \frac{110T}{V}$$

• This is the combined variation equation.

$$P = \frac{110(60)}{250} = 26.4$$

• Replace T by 60 and V by 250.

The pressure is 26.4 lb/in^2.

Problem 4 The strength (s) of a rectangular beam varies jointly as its width (W) and the square of its depth (d) and inversely as its length (L). If the strength of a beam 2 in. wide, 12 in. deep, and 12 ft long is 1200 lb, find the strength of a beam 4 in. wide, 8 in. deep, and 16 ft long.

Solution See page S22.

Try Exercise 57, page 367.

6.5 Exercises

CONCEPT CHECK

1. If y varies directly as x and k is the constant of variation, which equation is true?

(i) $x + y = k$ (ii) $x - y = k$ (iii) $xy = k$ (iv) $\frac{y}{x} = k$

2. If y varies inversely as x and k is the constant of variation, which equation is true?

(i) $x + y = k$ (ii) $x - y = k$ (iii) $xy = k$ (iv) $\frac{y}{x} = k$

3. Write "y varies jointly as x and z" as an equation, where k is the constant of proportionality.

4. Write "y varies directly as the square of x and inversely as z" as an equation, where k is the constant of proportionality.

1 Proportions (See pages 358–359.)

5. How does a ratio differ from a rate?

6. What is a proportion?

GETTING READY

7. The scale on a map shows that a distance of 2 cm on the map represents an actual distance of 5 mi. This rate can be expressed as the quotient $\frac{?}{?}$ or as the quotient $\frac{?}{?}$.

8. Let x be the distance, in miles, represented by 13 cm on the map described in Exercise 7. Use a proportion to find x.

$\frac{5\text{ mi}}{2\text{ cm}} = \frac{?\text{ mi}}{?\text{ cm}}$ • **Write the proportion so that each side is the ratio of miles to centimeters.**

$(2)(13)\left(\frac{5}{2}\right) = (2)(13)\left(\frac{x}{13}\right)$ • **Multiply each side of the proportion by __?__ and __?__.**

$\underline{\quad ? \quad} = (\underline{\quad ? \quad})x$ • **Simplify.**

$\underline{\quad ? \quad} = x$ • **Divide each side of the equation by __?__.**

13 cm on the map represents __?__ mi.

Solve the proportion.

9. $\frac{x+1}{10} = \frac{2}{5}$

10. $\frac{4}{x+2} = \frac{3}{4}$

11. $\frac{x}{4} = \frac{x-2}{8}$

12. $\frac{8}{x-5} = \frac{3}{x}$

13. $\frac{8}{x-2} = \frac{4}{x+1}$

14. $\frac{4}{x-4} = \frac{2}{x-2}$

15. $\frac{8}{3x-2} = \frac{2}{2x+1}$

16. $\frac{x-2}{x-5} = \frac{2x}{2x+5}$

17. True or false? If $\frac{a}{b} = \frac{c}{d}$, then $\frac{b}{a} = \frac{d}{c}$.

18. True or false? If $\frac{a}{b} = \frac{c}{d}$, then $\frac{a}{c} = \frac{b}{d}$.

19. **Environmental Science** In a wildlife preserve, 60 ducks are captured, tagged, and then released. Later, 200 ducks are examined, and three of the 200 ducks are found to have tags. Estimate the number of ducks in the preserve.

20. **Political Science** A pre-election survey showed that 7 out of every 12 voters would vote in an election. At this rate, how many people would be expected to vote in a city of 210,000?

21. **Architecture** On an architectural drawing, $\frac{1}{4}$ in. represents 1 ft. Find the dimensions of a room that measures $4\frac{1}{4}$ in. by $5\frac{1}{2}$ in. on the drawing.

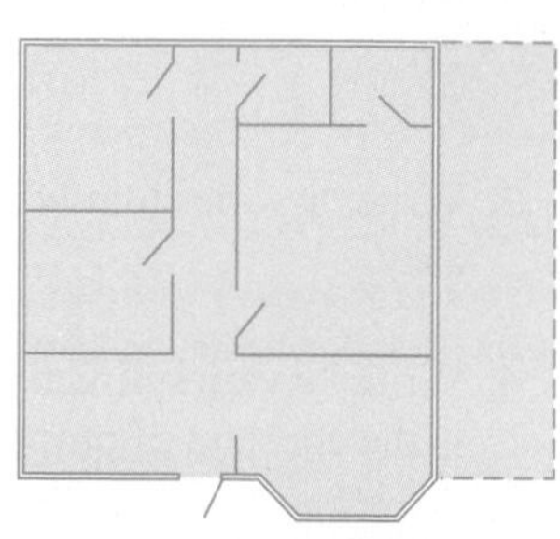

22. **Construction** A contractor estimates that 15 ft^2 of window space will be allowed for every 160 ft^2 of floor space. Using this estimate, how much window space will be allowed for 3200 ft^2 of floor space?

23. **Energy** Walking 4 mi in 2 h will burn 650 calories. Walking at the same rate, how many miles would a person need to walk to lose 1 lb? (The burning of 3500 calories is equivalent to the loss of 1 lb.) Round to the nearest hundredth.

24. **Construction** To tile 24 ft^2 of area, 120 ceramic tiles are needed. At this rate, how many tiles are needed to tile 300 ft^2?

25. **Health Science** Three-fourths of an ounce of a medication are required for a 120-pound adult. At the same rate, how many additional ounces of medication are required for a 200-pound adult?

26. **Biology** Use the information in the article at the right. If the distance the nerve message traveled from the elephant's hind leg to its brain was 9 ft, how far did the nerve message travel from the shrew's hind leg to its brain? Round to the nearest inch.

27. **Environmental Science** Six ounces of an insecticide are mixed with 15 gal of water to make a spray for spraying an orange grove. At the same rate, how much additional insecticide is required to be mixed with 100 gal of water?

28. **Ecology** In an attempt to estimate the number of Siberian tigers in a preserve, a wildlife management team captured, tagged, and released 50 Siberian tigers. A few months later, the team captured 150 Siberian tigers, of which 30 had tags. Estimate the number of Siberian tigers in the region.

In the News

Reaction Times, Large and Small

A study of reaction times in elephants and shrews shows that across the animal kingdom, from large creatures to small, nerve messages travel to and from the brain at the same rate. Scientists found that an elephant reacted to a touch on its hind leg in 100 milliseconds, whereas a shrew needed only 1 millisecond to react to a similar touch.

Source: smithsonianmag.com

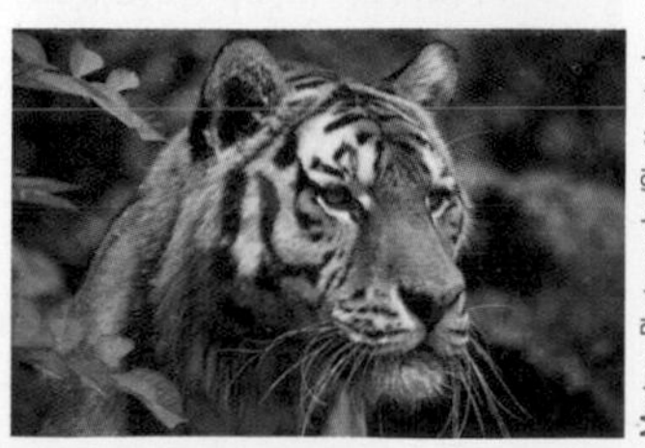

Mertens Photography/Shutterstock.com

29. **Internet Speeds** A certain wireless Internet connection will download a 4-megabyte file in 2 s. At this rate, how many seconds would it take to download a 5-megabyte file?

30. **Construction** A contractor estimated that 30 ft^3 of cement are required to make a concrete floor that measures 90 ft^2. Using this estimate, find how many additional cubic feet of cement would be required to make a concrete floor that measures 120 ft^2.

31. **Graphic Arts** The accompanying picture of a whale uses a scale of 1 in. represents 48 ft. Estimate the actual length of the whale.

32. **Physics** An object that weighs 100 lb on Earth would weigh 90.5 lb on Venus. At this rate, what is the weight on Venus of an object that weighs 150 lb on Earth?

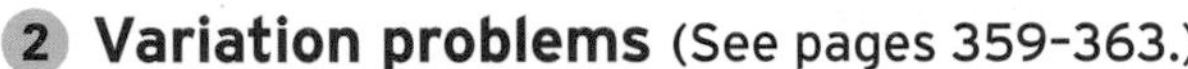

2 Variation problems (See pages 359–363.)

33. What is a direct variation?

34. What is an inverse variation?

35. A ball whose circumference is 30 cm weighs 250 g. An 800-gram ball made of the same material has a circumference of 96 cm. Is the relationship between the circumference and weight of a ball made of this material a direct variation or an inverse variation?

36. A person who weighs 150 lb on Earth weighs only 149.25 lb at 10 mi above Earth and 148.5 lb at 20 mi above Earth. Is the relationship between weight and distance above Earth a direct variation or an inverse variation?

37. **Business** The profit (P) realized by a company varies directly as the number of products it sells (s). If a company makes a profit of \$2500 on the sale of 20 products, what is the profit when the company sells 300 products?

38. **Farming** The number of bushels of wheat (b) produced by a farm is directly proportional to the number of acres (A) planted in wheat. If a 25-acre farm yields 1125 bushels of wheat, what is the yield of a farm that has 220 acres of wheat?

39. **Oil Spill** Read the article at the right about the Deepwater Horizon oil spill of 2010. If the well leaked oil at the same rate throughout the duration of the spill, then the total amount of oil leaked would be directly proportional to the number of days the oil had been leaking. Using the data in the article, estimate how many barrels of oil leaked during the 86 days of the spill. Round to the nearest tenth of a million.

In the News

Oil Leak Finally Capped

On June 15, 56 days after the oil spill, government estimates of the total amount of oil leaked into the Gulf of Mexico stood at 3.1 million barrels. One month later, the well has been successfully capped, and finally, after 86 days, oil no longer leaks into the ocean waters.

Source: nytimes.com

40. **Art** Leonardo da Vinci observed that a person's height varies directly as the width of the person's shoulders. If a person 70 in. tall has a shoulder width of 17.5 in., what is the height of a person whose shoulder width is 16 in.?

41. **Art** Leonardo da Vinci observed that the length of a person's face varies directly as the length of the person's chin. If a person whose face length is 9 in. has a chin length of 1.5 in., what is the length of a person's face whose chin length is 1.7 in.?

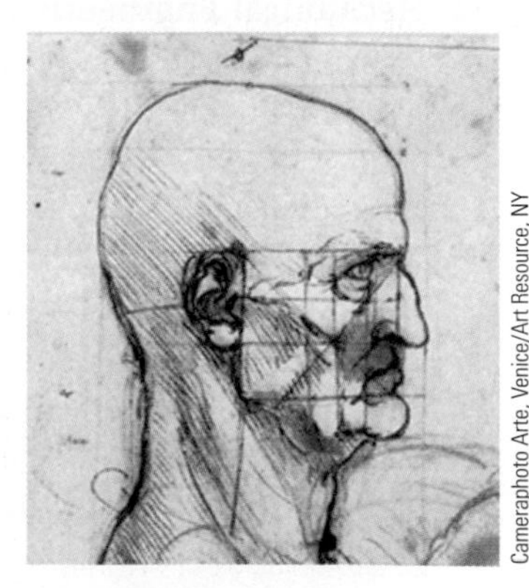

Cameraphoto Arte, Venice/Art Resource, NY

42. **Scuba Diving** The pressure (p) on a diver in the water varies directly as the depth (d). If the pressure is 3.6 lb/in^2 when the depth is 8 ft, what is the pressure when the depth is 30 ft?

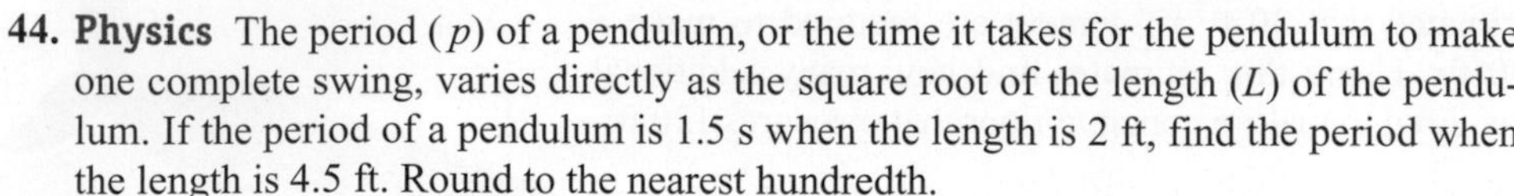

43. **Physics** The distance (d) a spring will stretch varies directly as the force (f) applied to the spring. If a force of 5 lb is required to stretch a spring 2 in., what force is required to stretch the spring 5 in.?

44. **Physics** The period (p) of a pendulum, or the time it takes for the pendulum to make one complete swing, varies directly as the square root of the length (L) of the pendulum. If the period of a pendulum is 1.5 s when the length is 2 ft, find the period when the length is 4.5 ft. Round to the nearest hundredth.

45. **Optics** The distance (d) a person can see to the horizon from a point above the surface of Earth varies directly as the square root of the height (H). If, from a height of 500 ft, the horizon is 19 mi away, how far is the horizon from a point that is 800 ft high? Round to the nearest hundredth.

46. **Automotive Technology** The stopping distance (s) of a car varies directly as the square of its speed (v). If a car traveling 30 mph requires 60 ft to stop, find the stopping distance for a car traveling 45 mph. Round to the nearest tenth.

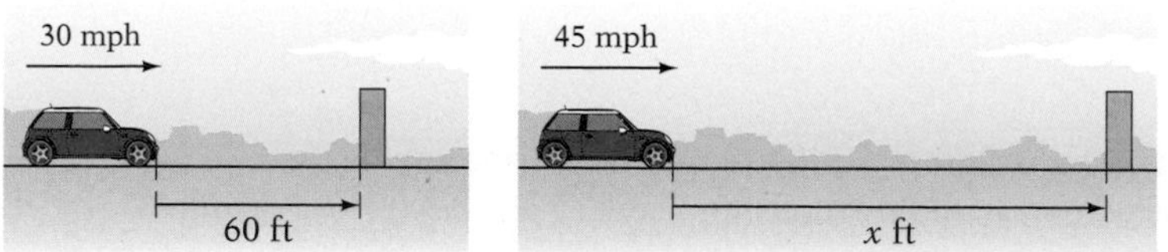

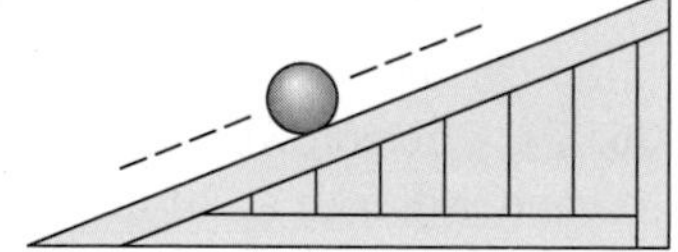

47. **Physics** The distance (s) a ball will roll down an inclined plane is directly proportional to the square of the time (t). If the ball rolls 5 ft in 1 s, how far will it roll in 4 s?

48. **Physics** The pressure (p) in a liquid varies jointly as the depth (d) and the density (D) of the liquid. If the pressure is 37.5 lb/in^2 when the depth is 100 in. and the density is 1.2 lb/in^2, find the pressure when the density remains the same and the depth is 60 in.

49. **Chemistry** At a constant temperature, the pressure (P) of a gas varies inversely as the volume (V). If the pressure is 25 lb/in^2 when the volume is 400 ft^3, find the pressure when the volume is 150 ft^3.

50. **Electricity** The current (I) in a wire varies directly as the voltage (v) and inversely as the resistance (r). If the current is 27.5 amps when the voltage is 110 volts and the resistance is 4 ohms, find the current when the voltage is 195 volts and the resistance is 12 ohms.

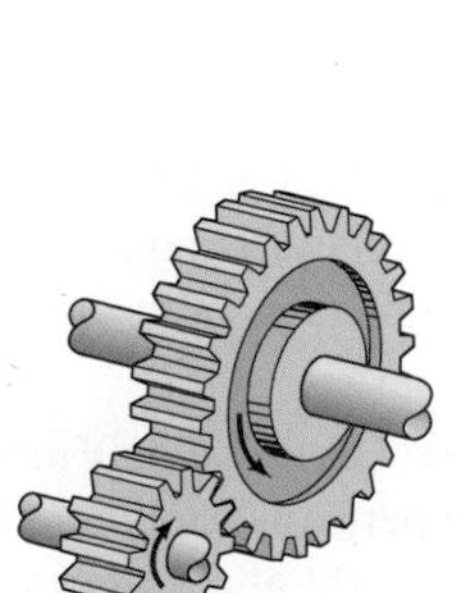

51. **Mechanical Engineering** The speed (v) of a gear varies inversely as the number of teeth (t). If a gear that has 48 teeth makes 20 revolutions per minute, how many revolutions per minute will a gear that has 30 teeth make?

52. **Physics** The intensity (I) of a light source is inversely proportional to the square of the distance (d) from the source. If the intensity is 8 lumens at a distance of 6 m, what is the intensity when the distance is 4 m?

53. **Magnetism** The repulsive force (f) between the north poles of two magnets is inversely proportional to the square of the distance (d) between them. If the repulsive force is 18 lb when the distance is 3 in., find the repulsive force when the distance is 1.2 in.

54. **Music** The frequency of vibration (f) of a guitar string varies directly as the square root of the tension (T) and inversely as the length (L) of the string. If the frequency is 40 vibrations per second when the tension is 25 lb and the length of the string is 3 ft, find the frequency when the tension is 36 lb and the string is 4 ft long.

55. **Electricity** The resistance (R) of a wire varies directly as the length (L) of the wire and inversely as the square of the diameter (d). If the resistance is 9 ohms in 50 ft of wire that has a diameter of 0.05 in., find the resistance in 50 ft of a similar wire that has a diameter of 0.02 in.

56. **Electricity** The power (P) in an electric circuit is directly proportional to the product of the current (I) and the square of the resistance (R). If the power is 100 watts when the current is 4 amps and the resistance is 5 ohms, find the power when the current is 2 amps and the resistance is 10 ohms.

57. **Sailing** The wind force (w) on a vertical surface varies jointly as the area (A) of the surface and the square of the wind velocity (v). When the wind is blowing at 30 mph, the force on an area of 10 ft^2 is 45 lb. Find the force on this area when the wind is blowing at 60 mph.

APPLYING CONCEPTS

58. In the direct variation equation $y = kx$, what is the effect on y when x doubles?

59. In the inverse variation equation $y = \frac{k}{x}$, what is the effect on x when y doubles?

Complete each sentence using the words *directly* or *inversely*.

60. If a varies ________ as b and c, then abc is constant.

61. If a varies directly as b and inversely as c, then c varies ________ as b and ________ as a.

62. If the area of a rectangle is held constant, the length of the rectangle varies ________ as the width.

63. If the width of a rectangle is held constant, the area of the rectangle varies ________ as the length.

PROJECTS OR GROUP ACTIVITIES

64. **Galileo's Experiments** Galileo Galilei (1564–1642) is purported to have dropped balls from the leaning tower of Pisa to demonstrate that balls of different weights reach the ground at the same time. There is, however, no account by Galileo that he actually did this. Nonetheless, he did demonstrate that balls of different weights, rolled down a slightly inclined board with a track notched from its center, will travel the length of the board in the same amount of time. From this experiment, Galileo concluded that balls of different weights, dropped from the same height, would reach the ground at the same time.

The table on the next page gives data, found using more modern techniques, from an experiment similar to the one performed by Galileo.

Distance (cm)	Time (s)	Distance (cm)	Time (s)
50	1.12	300	2.74
100	1.58	350	2.96
150	1.94	400	3.16
200	2.24	450	3.35
250	2.50	500	3.54

a. Find the ratio of any two distances and the ratios of the squares of the times for those distances. Do this for at least five different pairs of distances and times. How are the ratios related?

b. From his experiments, Galileo concluded that the ratio of the distances traveled by a ball equaled the ratio of the squares of the times the ball traveled down the ramp. Let (d_1, t_1) and (d_2, t_2) be any two pairs of data from the table. Write Galileo's conclusion as a proportion.

c. From the result in part (b), Galileo concluded that the distance an object falls varies directly as the square of the time it falls. Write this result as an equation.

6.6 Literal Equations

OBJECTIVE 1 Solve literal equations

A **literal equation** is an equation that contains more than one variable. Examples of literal equations are shown at the right.

$$3x - 2y = 4$$
$$v^2 = v_0^2 + 2as$$

Formulas are used to express relationships among physical quantities. A **formula** is a literal equation that states a rule about measurements. Examples of formulas are shown below.

$$s = vt - 16t^2 \quad \text{(Physics)}$$
$$c^2 = a^2 + b^2 \quad \text{(Geometry)}$$
$$A = P(1 + r)^t \quad \text{(Business)}$$

The Addition and Multiplication Properties of Equations can be used to solve a literal equation for one of the variables. The goal is to rewrite the equation so that the variable being solved for is alone on one side of the equation, and all the other numbers and variables are on the other side.

Focus on solving a literal equation for one variable

Solve $C = \frac{5}{9}(F - 32)$ for F.

$$C = \frac{5}{9}(F - 32)$$

Use the Distributive Property to remove parentheses.

$$C = \frac{5}{9}F - \frac{160}{9}$$

Multiply each side of the equation by 9. $9C = 5F - 160$

Add 160 to each side of the equation. $9C + 160 = 5F$

Divide each side of the equation by the coefficient 5. $\dfrac{9C + 160}{5} = F$

EXAMPLE 1 **A.** Solve $A = P + Prt$ for P. **B.** Solve $\dfrac{S}{S - C} = R$ for C.

Solution **A.**

$$A = P + Prt$$

$$A = (1 + rt)P$$

- **Factor P from $P + Prt$.**

$$\frac{A}{1 + rt} = \frac{(1 + rt)P}{1 + rt}$$

- **Divide each side of the equation by $1 + rt$.**

$$\frac{A}{1 + rt} = P$$

B.

$$\frac{S}{S - C} = R$$

$$(S - C)\frac{S}{S - C} = (S - C)R$$

- **Multiply each side of the equation by $S - C$.**

$$S = SR - CR$$

$$CR + S = SR$$

- **Add CR to each side of the equation.**

$$CR = SR - S$$

- **Subtract S from each side of the equation.**

$$C = \frac{SR - S}{R}$$

- **Divide each side of the equation by R.**

Problem 1 **A.** Solve $\dfrac{1}{R_1} + \dfrac{1}{R_2} = \dfrac{1}{R}$ for R. **B.** Solve $t = \dfrac{r}{r + 1}$ for r.

Solution See page S22.

Try Exercise 7, page 370.

6.6 Exercises

CONCEPT CHECK

1. What is a literal equation?

2. Are the literal equations $\dfrac{1}{A} + \dfrac{1}{B} = \dfrac{1}{C}$ and $A + B = C$ equivalent?

3. How do you solve $a(b + 1) = b$ for a?

4. What is the first step in solving $M = NP + N$ for N?

1 Solve literal equations (See pages 368-369.)

GETTING READY

5. Solve $P = \dfrac{R - C}{n}$ for C.

$$P = \frac{R - C}{n}$$

$$(\underline{\quad ? \quad})P = (\underline{\quad ? \quad})\left(\frac{R - C}{n}\right)$$ • **Multiply each side of the equation by ___?___.**

$$nP = \underline{\quad ? \quad}$$ • **Simplify.**

$$nP - R = -C$$ • **Subtract ___?___ from each side.**

$$\underline{\quad ? \quad} = C$$ • **Multiply each side by −1.**

6. Only two steps are needed to solve $P = \dfrac{R - C}{n}$ for n. The first step is to ___?___ each side of the equation by ___?___. The second step is to ___?___ each side of the equation by ___?___. The result is $n =$ ___?___.

Solve the formula for the variable given.

7. $P = 2L + 2W; W$ (Geometry)

8. $F = \dfrac{9}{5}C + 32; C$ (Temperature conversion)

9. $S = C - rC; C$ (Business)

10. $A = P + Prt; t$ (Business)

11. $PV = nRT; R$ (Chemistry)

12. $A = \dfrac{1}{2}bh; h$ (Geometry)

13. $F = \dfrac{Gm_1m_2}{r^2}; m_2$ (Physics)

14. $\dfrac{P_1V_1}{T_1} = \dfrac{P_2V_2}{T_2}; P_2$ (Chemistry)

15. $I = \dfrac{E}{R + r}; R$ (Physics)

16. $S = V_0t - 16t^2; V_0$ (Physics)

17. $A = \dfrac{1}{2}h(b_1 + b_2); b_2$ (Geometry)

18. $V = \dfrac{1}{3}\pi r^2h; h$ (Geometry)

19. $\dfrac{1}{R} = \dfrac{1}{R_1} + \dfrac{1}{R_2}; R_2$ (Physics)

20. $\dfrac{1}{f} = \dfrac{1}{a} + \dfrac{1}{b}; b$ (Physics)

21. $a_n = a_1 + (n - 1)d; d$ (Mathematics)

22. $P = \dfrac{R - C}{n}; n$ (Business)

23. $S = 2WH + 2WL + 2LH; H$ (Geometry)

24. $S = 2\pi r^2 + 2\pi rH; H$ (Geometry)

Solve for x.

25. $ax + by + c = 0$

26. $x = ax + b$

27. $ax + b = cx + d$

28. $y - y_1 = m(x - x_1)$

29. $\frac{a}{x} = \frac{b}{c}$

30. $\frac{1}{x} + \frac{1}{a} = b$

31. $\frac{1}{a} + \frac{1}{b} = \frac{1}{x}$

32. $a(a - x) = b(b - x)$

33. When asked to solve $ax + b = cx$ for x, one student's result was $x = \frac{b}{c - a}$. Another student's result was $x = -\frac{b}{a - c}$. Are their answers equivalent?

34. When asked to solve $I = \frac{E}{R + r}$ for r, one student's result was $r = \frac{E}{I} - R$. Another student's result was $r = \frac{E - IR}{I}$. Are their answers equivalent?

APPLYING CONCEPTS

Solve for x.

35. $\frac{x}{x + y} = \frac{2x}{4y}$

36. $\frac{2x}{x - 2y} = \frac{x}{2y}$

37. $\frac{x - y}{2x} = \frac{x - 3y}{5y}$

38. $\frac{x - y}{x} = \frac{2x}{9y}$

PROJECTS OR GROUP ACTIVITIES

39. Car Loans The spreadsheet application Microsoft® Excel offers many different formulas. One formula you may find useful is the PMT formula. This formula will give you the monthly payment for, say, a car loan. Start Excel and follow these steps.

Step 1 In cell A1, enter the *monthly* interest rate. For instance, if the annual interest rate is 8%, type = 0.08/12. The equals sign tells Excel to calculate the result. The interest rate is entered as a decimal.

Step 2 In A2, type the number of *months* of the loan. For instance, if the loan is 5 years, enter 60 months. You could also type = 12*5.

Step 3 In A3, type the amount of the loan. For instance, if the loan amount is $10,000, enter 10000. Do not enter the dollar sign or the comma.

Step 4 In A4, type =PMT(A1, A2, A3) and hit ENTER. The payment amount will be shown.

Suppose you can get a 5-year car loan at an annual interest rate of 8%. If you don't want your car payment to exceed $250 per month, how much, to the nearest $100, can you afford to borrow?

CHAPTER 6 Summary

Key Words	Objective and Page Reference	Examples
A fraction in which the numerator and denominator are polynomials is called a **rational expression.**	[6.1.1, p. 324]	$\frac{x^2+2x+1}{x-3}$ is a rational expression.
A rational expression is in **simplest form** when the numerator and denominator have no common factors other than 1.	[6.1.2, p. 325]	$\frac{x^2+3x+2}{(x+1)^2} = \frac{(x+1)(x+2)}{(x+1)(x+1)} = \frac{x+2}{x+1}$
The **reciprocal of a rational expression** is the rational expression with the numerator and denominator interchanged.	[6.2.1, p. 332]	The reciprocal of $\frac{a^2-b^2}{x-y}$ is $\frac{x-y}{a^2-b^2}$.
The **least common multiple (LCM) of two or more polynomials** is the simplest polynomial that contains the factors of each polynomial.	[6.2.2, p. 334]	$4x^2 - 12x = 4x(x-3)$ $3x^3 - 21x^2 + 36x = 3x(x-3)(x-4)$ $\text{LCM} = 12x(x-3)(x-4)$
A **complex fraction** is a fraction in which the numerator or denominator contains one or more fractions.	[6.3.1, p. 342]	$\dfrac{\frac{1}{x}+\frac{1}{y}}{\frac{1}{x}}$ is a complex fraction.
A **ratio** is the quotient of two quantities that have the same unit. When a ratio is in simplest form, the units are not written.	[6.5.1, p. 358]	$\frac{\$35}{\$100}$ written as a ratio in simplest form is $\frac{7}{20}$.
A **rate** is the quotient of two quantities that have different units.	[6.5.1, p. 358]	$\frac{65 \text{ mi}}{2 \text{ gal}}$ is a rate.
A **proportion** is an equation that states the equality of two ratios or rates.	[6.5.1, p. 358]	$\frac{75}{3} = \frac{x}{42}$ is a proportion.
A **direct variation** is a special function that can be expressed as the equation $y = kx^n$, where k is a constant called the **constant of variation** or the **constant of proportionality.**	[6.5.2, p. 359]	$C = \pi d$ is a formula relating the circumference and diameter of a circle. π is the constant of proportionality.
An **inverse variation** is a function that can be expressed as the equation $y = \frac{k}{x^n}$, where k is a constant.	[6.5.2, p. 361]	$I = \frac{k}{d^2}$ gives the intensity of a light source at a distance d from the source.
A **combined variation** is a variation in which two or more types of variation occur at the same time.	[6.5.2, p. 362]	$V = \frac{kT}{P}$ is a formula that states that the volume of a gas is directly proportional to the temperature and inversely proportional to the pressure.
A **joint variation** is a variation in which a variable varies directly as the product of two or more variables. A joint variation can be expressed as the equation $z = kxy$, where k is a constant.	[6.5.2, p. 362]	$C = kAT$ is a formula for the cost of insulation, where A is the area to be insulated and T is the thickness of the insulation.
A **formula** is a literal equation that states a rule about measurements.	[6.6.1, p. 368]	$P = 2L + 2W$ is the formula for the perimeter of a rectangle.

Essential Rules and Procedures	Objective and Page Reference	Examples
Rule for Multiplying Fractions $\frac{a}{b} \cdot \frac{c}{d} = \frac{ac}{bd}$	[6.2.1, p. 331]	$\frac{x^2 + 2x - 8}{x^2 - 2x} \cdot \frac{3x - 12}{x^2 - 16}$ $= \frac{(x + 4)(x - 2)}{x(x - 2)} \cdot \frac{3(x - 4)}{(x + 4)(x - 4)}$ $= \frac{(x + 4)(x - 2) \cdot 3(x - 4)}{x(x - 2)(x + 4)(x - 4)}$ $= \frac{3}{x}$
Rule for Dividing Fractions $\frac{a}{b} \div \frac{c}{d} = \frac{a}{b} \cdot \frac{d}{c}$	[6.2.1, p. 332]	$\frac{3x}{x - 5} \div \frac{3}{x + 4} = \frac{3x}{x - 5} \cdot \frac{x + 4}{3}$ $= \frac{3x(x + 4)}{3(x - 5)} = \frac{x(x + 4)}{x - 5}$ $= \frac{x^2 + 4x}{x - 5}$
Rule for Adding Fractions with the Same Denominator $\frac{a}{c} + \frac{b}{c} = \frac{a + b}{c}$	[6.2.2, pp. 333–334]	$\frac{2x - 7}{x^2 + 4} + \frac{x + 2}{x^2 + 4} = \frac{(2x - 7) + (x + 2)}{x^2 + 4}$ $= \frac{3x - 5}{x^2 + 4}$
Rule for Subtracting Fractions with the Same Denominator $\frac{a}{c} - \frac{b}{c} = \frac{a - b}{c}$	[6.2.2, pp. 333–334]	$\frac{5x - 6}{x + 2} - \frac{2x + 4}{x + 2} = \frac{(5x - 6) - (2x + 4)}{x + 2}$ $= \frac{3x - 10}{x + 2}$
Equation for Work Problems Rate of work $\times$ Time worked $=$ Part of task completed	[6.4.2, p. 348]	A roofer requires 24 h to shingle a roof. An apprentice can shingle the roof in 36 h. How long would it take to shingle the roof if both roofers worked together? The equation used to solve this problem is $\frac{t}{24} + \frac{t}{36} = 1$ where t is the time it takes for the roofers to shingle the roof together.
Equation for Uniform Motion Problems Distance $=$ Rate $\times$ Time or $\frac{\text{Distance}}{\text{Rate}} = \text{Time}$	[6.4.3, p. 350]	A motorcycle travels 195 mi in the same amount of time it takes a car to travel 159 mi. The rate of the motorcycle is 12 mph greater than the rate of the car. Find the rate of the car. The equation used to solve this problem is $\frac{195}{r + 12} = \frac{159}{r}$ where r is the rate of the car.

CHAPTER 6 Review Exercises

1. Given $P(x) = \frac{x}{x - 3}$, find $P(4)$.

2. Given $P(x) = \frac{x^2 - 2}{3x^2 - 2x + 5}$, find $P(-2)$.

3. Find the domain of $g(x) = \frac{2x}{x - 3}$.

4. Find the domain of $f(x) = \frac{2x - 7}{3x^2 + 3x - 18}$.

5. Find the domain of $F(x) = \frac{x^2 - x}{3x^2 + 4}$.

6. Simplify: $\frac{6a^5 + 4a^4 - 2a^3}{2a^3}$

7. Simplify: $\frac{16 - x^2}{x^3 - 2x^2 - 8x}$

8. Simplify: $\frac{x^3 - 27}{x^2 - 9}$

9. Multiply: $\frac{a^6b^4 + a^4b^6}{a^5b^4 - a^4b^4} \cdot \frac{a^2 - b^2}{a^4 - b^4}$

10. Multiply: $\frac{x^3 - 8}{x^3 + 2x^2 + 4x} \cdot \frac{x^3 + 2x^2}{x^2 - 4}$

11. Multiply: $\frac{16 - x^2}{6x - 6} \cdot \frac{x^2 + 5x + 6}{x^2 - 8x + 16}$

12. Divide: $\frac{x^2 - 5x + 4}{x^2 - 2x - 8} \div \frac{x^2 - 4x + 3}{x^2 + 8x + 12}$

13. Divide: $\frac{27x^3 - 8}{9x^3 + 6x^2 + 4x} \div \frac{9x^2 - 12x + 4}{9x^2 - 4}$

14. Divide: $\frac{3 - x}{x^2 + 3x + 9} \div \frac{x^2 - 9}{x^3 - 27}$

15. Add: $\frac{5}{3a^2b^3} + \frac{7}{8ab^4}$

16. Subtract: $\frac{3x^2 + 2}{x^2 - 4} - \frac{9x - x^2}{x^2 - 4}$

17. Simplify: $\frac{8}{9x^2 - 4} + \frac{5}{3x - 2} - \frac{4}{3x + 2}$

18. Simplify: $\frac{6x}{3x^2 - 7x + 2} - \frac{2}{3x - 1} + \frac{3x}{x - 2}$

19. Simplify: $\frac{x}{x - 3} - 4 - \frac{2x - 5}{x + 2}$

20. Simplify: $\dfrac{x - 6 + \frac{6}{x - 1}}{x + 3 - \frac{12}{x - 1}}$

21. Simplify: $\dfrac{x + \frac{3}{x - 4}}{3 + \frac{x}{x - 4}}$

22. Solve: $\frac{5x}{2x - 3} + 4 = \frac{3}{2x - 3}$

23. Solve: $\frac{x}{x - 3} = \frac{2x + 5}{x + 1}$

24. Solve: $\frac{6}{x - 3} - \frac{1}{x + 3} = \frac{51}{x^2 - 9}$

25. Solve: $\frac{30}{x^2 + 5x + 4} + \frac{10}{x + 4} = \frac{4}{x + 1}$

26. Solve $I = \frac{1}{R}V$ for R.

27. Solve $Q = \dfrac{N - S}{N}$ for N.

28. Solve $S = \dfrac{a}{1 - r}$ for r.

29. **Fuel Efficiency** A car uses 4 tanks of fuel to travel 1800 mi. At this rate, how many tanks of fuel would be required for a trip of 3000 mi?

30. **Cartography** On a certain map, 2.5 in. represents 10 mi. How many miles would be represented by 12 in.?

31. **Work Problem** An electrician requires 65 min to install a ceiling fan. The electrician and an apprentice, working together, take 40 min to install the fan. How long would it take the apprentice, working alone, to install the ceiling fan?

32. **Work Problem** The inlet pipe can fill a tub in 24 min. The drain pipe can empty the tub in 15 min. How long would it take to empty a full tub with both pipes open?

33. **Work Problem** Three students can paint a dormitory room in 8 h, 16 h, and 16 h, respectively. How long would it take to do the job if all three students worked together?

34. **Uniform Motion** A canoeist can travel 10 mph in calm water. The amount of time it takes to travel 60 mi with the current is the same as the amount of time it takes to travel 40 mi against the current. Find the rate of the current.

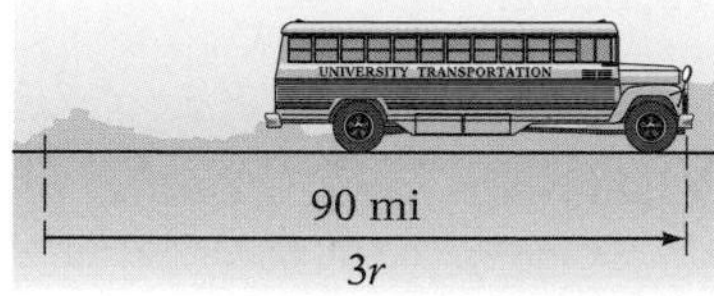

35. **Uniform Motion** A bus and a cyclist leave a school at 8 A.M. and head for a stadium 90 mi away. The rate of the bus is three times the rate of the cyclist. The cyclist arrives 4 h after the bus. Find the rate of the bus.

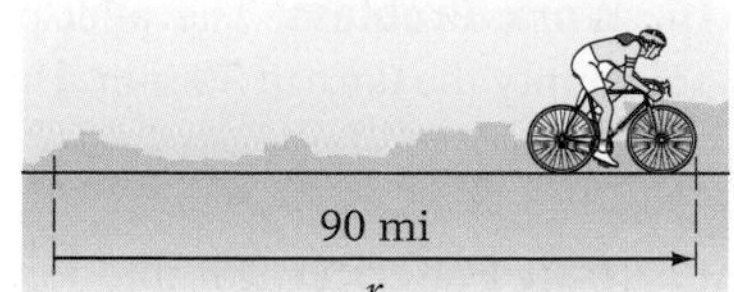

36. **Uniform Motion** A tractor travels 10 mi in the same amount of time it takes a car to travel 15 mi. The rate of the tractor is 15 mph less than the rate of the car. Find the rate of the tractor.

37. **Physics** The pressure (p) of wind on a flat surface varies jointly as the area (A) of the surface and the square of the wind's velocity (v). If the pressure on 22 ft^2 of surface is 10 lb when the wind's velocity is 10 mph, find the pressure on the same surface when the wind's velocity is 20 mph.

38. **Light** The illumination (I) produced by a light varies inversely as the square of the distance (d) from the light. If the illumination produced 10 ft from a light is 12 lumens, find the illumination 2 ft from the light.

39. **Electricity** The electrical resistance (r) of a cable varies directly as its length (l) and inversely as the square of its diameter (d). If a cable 16,000 ft long and $\frac{1}{4}$ in. in diameter has a resistance of 3.2 ohms, what is the resistance of a cable that is 8000 ft long and $\frac{1}{2}$ in. in diameter?

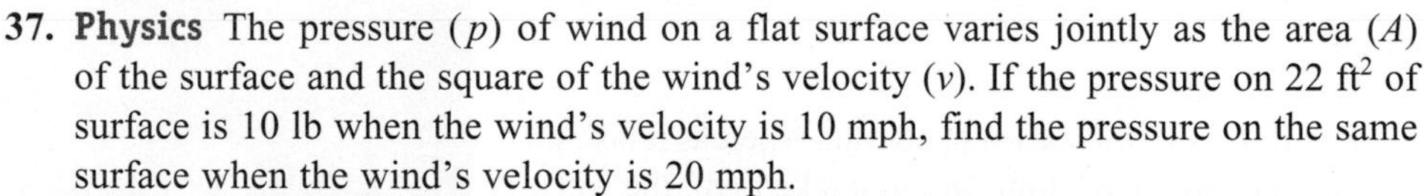

CHAPTER 6 Test

1. Simplify: $\dfrac{v^3 - 4v}{2v^2 - 5v + 2}$

2. Simplify: $\dfrac{2a^2 - 8a + 8}{4 + 4a - 3a^2}$

3. Multiply: $\dfrac{3x^2 - 12}{5x - 15} \cdot \dfrac{2x^2 - 18}{x^2 + 5x + 6}$

4. Given $P(x) = \dfrac{3 - x^2}{x^3 - 2x^2 + 4}$, find $P(-1)$.

5. Divide: $\dfrac{2x^2 - x - 3}{2x^2 - 5x + 3} \div \dfrac{3x^2 - x - 4}{x^2 - 1}$

6. Multiply: $\dfrac{x^2 - x - 2}{x^2 + x} \cdot \dfrac{x^2 - x}{x^2 - 4}$

7. Simplify: $\dfrac{2}{x^2} + \dfrac{3}{y^2} - \dfrac{5}{2xy}$

8. Simplify: $\dfrac{3x}{x - 2} - 3 + \dfrac{4}{x + 2}$

9. Find the domain of $f(x) = \dfrac{3x^2 - x + 1}{x^2 - 9}$.

10. Subtract: $\dfrac{x + 2}{x^2 + 3x - 4} - \dfrac{2x}{x^2 - 1}$

11. Simplify: $\dfrac{1 - \dfrac{1}{x} - \dfrac{12}{x^2}}{1 + \dfrac{6}{x} + \dfrac{9}{x^2}}$

12. Simplify: $\dfrac{1 - \dfrac{1}{x + 2}}{1 - \dfrac{3}{x + 4}}$

13. Solve: $\dfrac{3}{x + 1} = \dfrac{2}{x}$

14. Solve: $\dfrac{4x}{2x - 1} = 2 - \dfrac{1}{2x - 1}$

15. Solve $ax = bx + c$ for x.

16. **Work Problem** The inlet pipe can fill a water tank in 48 min. The outlet pipe can empty the tank in 30 min. How long would it take to empty a full tank with both pipes open?

17. **Wall Papering** An interior designer uses two rolls of wallpaper for every 45 ft² of wall space in an office. At this rate, how many rolls of wallpaper are needed for an office that has 315 ft² of wall space?

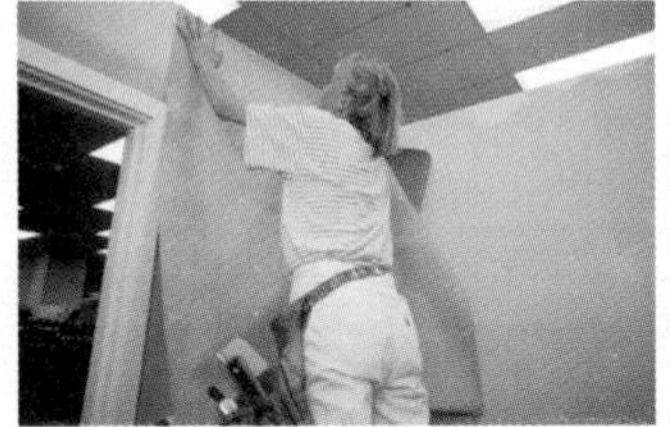

Bonnie Kamin/PhotoEdit, Inc.

18. **Work Problem** One landscaper can till the soil for a lawn in 30 min, whereas it takes a second landscaper 15 min to do the same job. How long would it take to till the soil for the lawn if both landscapers worked together?

19. **Uniform Motion** A cyclist travels 20 mi in the same amount of time it takes a hiker to walk 6 mi. The rate of the cyclist is 7 mph greater than the rate of the hiker. Find the rate of the cyclist.

20. **Physics** The stopping distance (s) of a car varies directly as the square of the speed (v) of the car. For a car traveling at 50 mph, the stopping distance is 170 ft. Find the stopping distance of a car that is traveling at 30 mph.

Cumulative Review Exercises

1. Simplify: $8 - 4[-3 - (-2)]^2 \div 5$

2. Solve: $\dfrac{2x - 3}{6} - \dfrac{x}{9} = \dfrac{x - 4}{3}$

3. Solve: $5 - |x - 4| = 2$

4. Find the domain of $f(x) = \dfrac{x}{x - 3}$.

5. Given $P(x) = \frac{x - 1}{2x - 3}$, find $P(-2)$.

6. Write 0.000000035 in scientific notation.

7. Solve: $\frac{x}{x + 1} = 1$

8. Solve: $(9x - 1)(x - 4) = 0$

9. Simplify: $\frac{(2a^{-2}b^3)}{(4a)^{-1}}$

10. Solve: $x - 3(1 - 2x) \geq 1 - 4(2 - 2x)$

11. Multiply: $(2a^2 - 3a + 1)(-2a^2)$

12. Factor: $2x^2 + 3x - 2$

13. Factor: $x^3y^3 - 27$

14. Simplify: $\frac{x^4 + x^3y - 6x^2y^2}{x^3 - 2x^2y}$

15. Find the equation of the line that contains the point $P(-2, -1)$ and is parallel to the graph of $3x - 2y = 6$.

16. Simplify: $(x - x^{-1})^{-1}$

17. Graph $-3x + 5y = -15$ by using the x- and y-intercepts.

18. Graph the solution set: $\begin{aligned} x + y &\leq 3 \\ -2x + y &> 4 \end{aligned}$

19. Divide: $\frac{4x^3 + 2x^2 - 10x + 1}{x - 2}$

20. Divide: $\frac{16x^2 - 9y^2}{16x^2y - 12xy^2} \div \frac{4x^2 - xy - 3y^2}{12x^2y^2}$

21. Find the domain of $f(x) = \frac{x^2 + 6x + 7}{3x^2 + 5}$.

22. Subtract: $\frac{5x}{3x^2 - x - 2} - \frac{2x}{x^2 - 1}$

23. Evaluate the determinant: $\begin{vmatrix} 6 & 5 \\ 2 & -3 \end{vmatrix}$

24. Simplify: $\frac{x - 4 + \frac{5}{x + 2}}{x + 2 - \frac{1}{x + 2}}$

25. Solve: $\begin{aligned} x + y + z &= 3 \\ -2x + y + 3z &= 2 \\ 2x - 4y + z &= -1 \end{aligned}$

26. Given $f(x) = x^2 - 3x + 3$ and $f(c) = 1$, find c.

27. Solve the proportion: $\frac{2}{x - 3} = \frac{5}{2x - 3}$

28. Solve: $\frac{3}{x^2 - 36} = \frac{2}{x - 6} - \frac{5}{x + 6}$

29. Multiply: $(a + 5)(a^3 - 3a + 4)$

30. Solve $I = \dfrac{E}{R + r}$ for r.

31. Are the graphs of $4x + 3y = 12$ and $3x + 4y = 16$ perpendicular?

32. Graph: $f(x) = 1 - x + x^3$

33. **Food Mixture** How many pounds of almonds that cost \$5.40 per pound must be mixed with 50 lb of peanuts that cost \$2.60 per pound to make a mixture that costs \$4.00 per pound?

34. **Elections** A pre-election survey showed that three out of five voters would vote in an election. At this rate, how many people would be expected to vote in a city of 125,000?

35. **Work Problem** A new computer can work six times as fast as an older computer. Working together, the computers can complete a job in 12 min. How long would it take the new computer, working alone, to do the job?

36. **Uniform Motion** A plane can fly at a rate of 300 mph in calm air. Traveling with the wind, the plane flew 900 mi in the same amount of time it took to fly 600 mi against the wind. Find the rate of the wind.

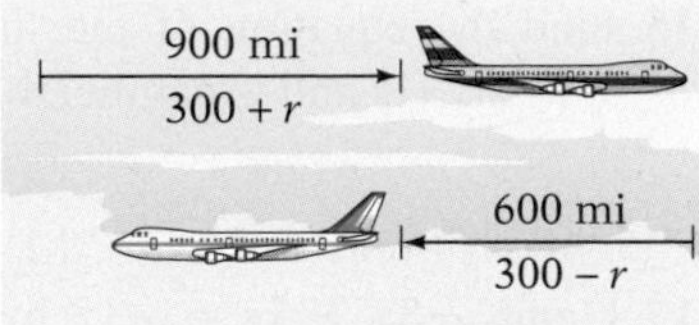

37. **Music** The frequency of vibration (f) in an open pipe organ varies inversely as the length (L) of the pipe. If the air in a pipe 2 m long vibrates 60 times per minute, find the frequency in a pipe that is 1.5 m long.

Rational Exponents and Radicals

CHAPTER 7

Digital Vision

Focus on Success

Do you get nervous before taking a math test? The more prepared you are, the less nervous you will be. There are a number of features in this text that will help you to be prepared. We suggest that you start with the Chapter Summary. The Chapter Summary describes the important topics covered in the chapter. The reference following each topic shows you the objective and page in the text where you can find more information on the concept. Do the Chapter Review Exercises to test your understanding of the material in the chapter. If you have trouble with any of the questions, restudy the objectives the questions are taken from, and retry some of the exercises in those objectives. Take the Chapter Test in a quiet place, working on it as if it were an actual exam. (See Ace the Test, page AIM-11).

OBJECTIVES

7.1
1. Simplify expressions with rational exponents
2. Write exponential expressions as radical expressions and radical expressions as exponential expressions
3. Simplify radical expressions that are roots of perfect powers

7.2
1. Simplify radical expressions
2. Add and subtract radical expressions
3. Multiply radical expressions
4. Divide radical expressions

7.3
1. Find the domain of a radical function
2. Graph a radical function

7.4
1. Solve equations containing one or more radical expressions
2. Application problems

7.5
1. Simplify complex numbers
2. Add and subtract complex numbers
3. Multiply complex numbers
4. Divide complex numbers

PREP TEST

Are you ready to succeed in this chapter? *Take the Prep Test below to find out if you are ready to learn the new material.*

1. Complete: $48 = ? \cdot 3$

Simplify.

2. 2^5

3. $6\left(\frac{3}{2}\right)$

4. $\frac{1}{2} - \frac{2}{3} + \frac{1}{4}$

5. $(3 - 7x) - (4 - 2x)$

6. $\frac{3x^5y^6}{12x^4y}$

7. $(3x - 2)^2$

Multiply.

8. $(2 + 4x)(5 - 3x)$

9. $(6x - 1)(6x + 1)$

10. Solve: $x^2 - 14x - 5 = 10$

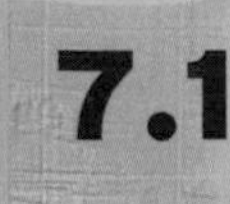

7.1 Rational Exponents and Radical Expressions

OBJECTIVE 1 Simplify expressions with rational exponents

Point of Interest

Nicolas Chuquet (ca. 1475), a French physician, wrote an algebra text in which he used a notation for expressions with fractional exponents. He wrote $R^2 6$ to mean $6^{\frac{1}{2}}$ and $R^3 15$ to mean $15^{\frac{1}{3}}$. This was an improvement over earlier notations that used words for these expressions.

In this section, the definition of an exponent is extended beyond integers so that any rational number can be used as an exponent. The definition is expressed in such a way that the Rules of Exponents hold true for rational exponents.

Consider the expression $(a^{\frac{1}{n}})^n$ for $a > 0$ and n a positive integer. Now simplify, assuming that the Rule for Simplifying a Power of an Exponential Expression is true.

$$(a^{\frac{1}{n}})^n = a^{\frac{1}{n} \cdot n} = a^1 = a$$

Because $(a^{\frac{1}{n}})^n = a$, the number $a^{\frac{1}{n}}$ is the number whose nth power is a.

DEFINITION OF $a^{\frac{1}{n}}$

If n is a positive integer, then $a^{\frac{1}{n}}$ is the number whose nth power is a.

EXAMPLES

1. $9^{\frac{1}{2}} = 3$ because $3^2 = 9$.
2. $64^{\frac{1}{3}} = 4$ because $4^3 = 64$.
3. $(-32)^{\frac{1}{5}} = -2$ because $(-2)^5 = -32$.

Take Note

Suppose that $(-9)^{\frac{1}{2}} = x$. Then by the definition of $a^{\frac{1}{n}}$, $x^2 = -9$. However, the square of any real number is not a negative number. Therefore, $(-9)^{\frac{1}{2}}$ is not a real number.

If a is a negative number and n is an even integer, then $a^{\frac{1}{n}}$ is not a real number. For instance, $(-9)^{\frac{1}{2}}$ is not a real number. We shall discuss such numbers later in this chapter.

As shown at the left, expressions that contain rational exponents do not always represent real numbers when the base of the exponential expression is a negative number. For this reason, all variables in this chapter represent positive numbers unless otherwise stated.

Using the definition of $a^{\frac{1}{n}}$ and the Rules of Exponents, it is possible to define any exponential expression that contains a rational exponent.

DEFINITION OF $a^{\frac{m}{n}}$

If m and n are positive integers and $a^{\frac{1}{n}}$ is a real number, then

$$a^{\frac{m}{n}} = (a^{\frac{1}{n}})^m$$

EXAMPLE 1 Simplify. **A.** $27^{\frac{2}{3}}$ **B.** $32^{-\frac{2}{5}}$

Solution **A.** $27^{\frac{2}{3}} = (3^3)^{\frac{2}{3}}$ • **Rewrite 27 as 3^3.**

$= 3^{3(\frac{2}{3})}$ • **Use the Rule for Simplifying a Power of an Exponential Expression.**

$= 3^2$

$= 9$ • **Simplify.**

Take Note

Note that $32^{-\frac{2}{5}} = \frac{1}{4}$, a positive number. A negative exponent does not affect the sign of a number.

B. $32^{-\frac{2}{5}} = (2^5)^{-\frac{2}{5}}$ • Rewrite 32 as 2^5.

$= 2^{-2}$ • Use the Rule for Simplifying a Power of an Exponential Expression.

$= \frac{1}{2^2}$ • Use the Definition of a Negative Exponent.

$= \frac{1}{4}$ • Simplify.

Problem 1 Simplify. **A.** $64^{\frac{2}{3}}$ **B.** $16^{-\frac{3}{4}}$

Solution See page S23.

Try Exercise 21, page 385.

EXAMPLE 2 Simplify. **A.** $b^{\frac{1}{2}} \cdot b^{\frac{2}{3}} \cdot b^{-\frac{1}{4}}$ **B.** $(-64x^6y^{-\frac{3}{2}})^{\frac{4}{3}}$ **C.** $\left(\frac{8a^3b^{-4}}{64a^{-9}b^2}\right)^{\frac{2}{3}}$

Solution **A.** $b^{\frac{1}{2}} \cdot b^{\frac{2}{3}} \cdot b^{-\frac{1}{4}} = b^{\frac{1}{2}+\frac{2}{3}-\frac{1}{4}}$ • Use the Rule for Multiplying Exponential Expressions.

$= b^{\frac{6}{12}+\frac{8}{12}-\frac{3}{12}}$

$= b^{\frac{11}{12}}$

B. $(-64x^6y^{-\frac{3}{2}})^{\frac{4}{3}} = [(-4)^3x^6y^{-\frac{3}{2}}]^{\frac{4}{3}}$ • Rewrite −64 as $(-4)^3$.

$= (-4)^{3(\frac{4}{3})}x^{6(\frac{4}{3})}y^{-\frac{3}{2}(\frac{4}{3})}$ • Use the Rule for Simplifying Powers of Products.

$= (-4)^4x^8y^{-2}$ • Simplify.

$= \frac{256x^8}{y^2}$

C. $\left(\frac{8a^3b^{-4}}{64a^{-9}b^2}\right)^{\frac{2}{3}} = \left(\frac{a^{12}}{8b^6}\right)^{\frac{2}{3}}$ • Simplify $\frac{8}{64} = \frac{1}{8}$ and use the Rule for Dividing Exponential Expressions.

$= \left(\frac{a^{12}}{2^3b^6}\right)^{\frac{2}{3}}$ • Rewrite 8 as 2^3.

$= \frac{a^8}{2^2b^4}$ • Use the Rule for Simplifying Powers of Quotients.

$= \frac{a^8}{4b^4}$

Problem 2 Simplify. **A.** $\frac{x^{\frac{1}{2}}y^{-\frac{5}{4}}}{x^{-\frac{4}{3}}y^{\frac{1}{3}}}$ **B.** $(x^{\frac{3}{4}}y^{\frac{1}{2}}z^{-\frac{2}{3}})^{-\frac{4}{3}}$ **C.** $\frac{(3x^{\frac{3}{4}}y^{-\frac{1}{4}})^4}{(9x^2y^4)^{\frac{1}{2}}}$

Solution See page S23.

Try Exercise 65, page 386.

OBJECTIVE 2 Write exponential expressions as radical expressions and radical expressions as exponential expressions

Recall that $a^{\frac{1}{n}}$ is the number whose nth power is a. We can also say that $a^{\frac{1}{n}}$ is the ***n*th root of *a*.**

nTH ROOT OF a

If a is a real number and n is a positive integer, then $\sqrt[n]{a} = a^{\frac{1}{n}}$.

In the expression $\sqrt[n]{a}$, the symbol $\sqrt{}$ is called the **radical,** n is the **index** of the radical, and a is the **radicand.**

EXAMPLES

1. $\sqrt[3]{7} = 7^{\frac{1}{3}}$
2. $\sqrt[5]{x} = x^{\frac{1}{5}}$

When $n = 2$, the radical expression represents a square root and the index 2 is usually not written.

An exponential expression with a rational exponent can be written as a radical expression.

Take Note

$a^{\frac{m}{n}} = \sqrt[n]{a^m}$

WRITE $a^{\frac{m}{n}}$ AS A RADICAL EXPRESSION

If $a^{\frac{1}{n}}$ is a real number, then $a^{\frac{m}{n}} = \sqrt[n]{a^m}$.

EXAMPLES

1. $15^{\frac{2}{3}} = \sqrt[3]{15^2}$
2. $x^{\frac{4}{5}} = \sqrt[5]{x^4}$
3. $\sqrt[5]{2^3} = 2^{\frac{3}{5}}$
4. $\sqrt[6]{z^5} = z^{\frac{5}{6}}$

EXAMPLE 3 Rewrite the exponential expression as a radical expression.

A. $(3x)^{\frac{2}{3}}$ **B.** $-2x^{\frac{2}{3}}$

Solution **A.** $(3x)^{\frac{2}{3}} = \sqrt[3]{(3x)^2} = \sqrt[3]{9x^2}$

• **The denominator of the rational exponent is the index of the radical. The numerator is the power of the radicand.**

B. $-2x^{\frac{2}{3}} = -2(x^2)^{\frac{1}{3}} = -2\sqrt[3]{x^2}$

• **The −2 is not raised to the power.**

Problem 3 Rewrite the exponential expression as a radical expression.

A. $(2x^3)^{\frac{3}{4}}$ **B.** $-5a^{\frac{5}{6}}$

Solution See page S23.

Try Exercise 95, page 387.

EXAMPLE 4 Rewrite the radical expression as an exponential expression.

A. $\sqrt[7]{x^5}$ **B.** $\sqrt[3]{a^3 + b^3}$

Solution **A.** $\sqrt[7]{x^5} = (x^5)^{\frac{1}{7}} = x^{\frac{5}{7}}$

• **The index of the radical is the denominator of the rational exponent. The power of the radicand is the numerator of the rational exponent.**

B. $\sqrt[3]{a^3 + b^3} = (a^3 + b^3)^{\frac{1}{3}}$

• **Note that $(a^3 + b^3)^{\frac{1}{3}} \neq a + b$.**

Problem 4 Rewrite the radical expression as an exponential expression.

A. $\sqrt[3]{3ab}$ **B.** $\sqrt[4]{x^4 + y^4}$

Solution See page S23.

Try Exercise 109, page 387.

OBJECTIVE 3 Simplify radical expressions that are roots of perfect powers

Every positive number has two square roots, one a positive and one a negative number. For example, because $(5)^2 = 25$ and $(-5)^2 = 25$, there are two square roots of 25: 5 and -5.

The symbol $\sqrt{\ }$ is used to indicate the positive or **principal square root.** To indicate the negative square root of a number, a negative sign is placed in front of the radical.

$\sqrt{25} = 5$

$-\sqrt{25} = -5$

The square root of zero is zero. $\sqrt{0} = 0$

The square root of a negative number is not a real number, because the square of a real number must be positive. $\sqrt{-25}$ is not a real number.

The square root of a squared negative number is a positive number. $\sqrt{(-5)^2} = \sqrt{25} = 5$

For any real number a, $\sqrt{a^2} = |a|$ and $-\sqrt{a^2} = -|a|$. If a is a positive real number, then $\sqrt{a^2} = a$ and $(\sqrt{a})^2 = a$.

The cube root of a positive number is positive. $\sqrt[3]{8} = 2$, because $2^3 = 8$.

The cube root of a negative number is negative. $\sqrt[3]{-8} = -2$, because $(-2)^3 = -8$.

For any real number a, $\sqrt[3]{a^3} = a$.

The following holds true for finding the nth root of a real number.

If n is an even integer, then $\sqrt[n]{a^n} = |a|$ and $-\sqrt[n]{a^n} = -|a|$. If n is an odd integer, then $\sqrt[n]{a^n} = a$.

For example,

$$\sqrt[6]{y^6} = |y| \qquad -\sqrt[12]{x^{12}} = -|x| \qquad \sqrt[5]{b^5} = b$$

For the remainder of this chapter, we will assume that variable expressions inside a radical represent positive numbers. Therefore, it is not necessary to use the absolute value signs.

Focus on simplifying a root of a perfect power

A. Simplify: $\sqrt[4]{x^4y^8}$

The radicand is a perfect fourth power because the exponents on the variables are divisible by 4.

Write the radical expression as an exponential expression. $\sqrt[4]{x^4y^8} = (x^4y^8)^{\frac{1}{4}}$

Use the Rule for Simplifying Powers of Products. $= xy^2$

B. Simplify: $\sqrt[3]{125c^9d^6}$

The radicand is a perfect cube because 125 is a perfect cube $(125 = 5^3)$ and all the exponents on the variables are divisible by 3.

Write the radical as an exponential expression. $\sqrt[3]{125c^9d^6} = (5^3c^9d^6)^{\frac{1}{3}}$

Use the Rule for Simplifying Powers of Products. $= 5c^3d^2$

Take Note

Note that when the index is an even natural number, the nth root requires absolute value symbols.

$\sqrt[6]{y^6} = |y|$ but $\sqrt[5]{y^5} = y$

Because we stated that variables within radicals represent *positive* numbers, we will omit the absolute value symbols when writing an answer.

Note that a variable expression is a perfect power if the exponents on the factors are evenly divisible by the index of the radical.

The chart below shows roots of perfect powers. Knowledge of these roots is very helpful in simplifying radical expressions.

Square Roots		*Cube Roots*	*Fourth Roots*	*Fifth Roots*
$\sqrt{1} = 1$	$\sqrt{36} = 6$	$\sqrt[3]{1} = 1$	$\sqrt[4]{1} = 1$	$\sqrt[5]{1} = 1$
$\sqrt{4} = 2$	$\sqrt{49} = 7$	$\sqrt[3]{8} = 2$	$\sqrt[4]{16} = 2$	$\sqrt[5]{32} = 2$
$\sqrt{9} = 3$	$\sqrt{64} = 8$	$\sqrt[3]{27} = 3$	$\sqrt[4]{81} = 3$	$\sqrt[5]{243} = 3$
$\sqrt{16} = 4$	$\sqrt{81} = 9$	$\sqrt[3]{64} = 4$	$\sqrt[4]{256} = 4$	
$\sqrt{25} = 5$	$\sqrt{100} = 10$	$\sqrt[3]{125} = 5$	$\sqrt[4]{625} = 5$	

If a number is not a perfect power, its root can only be approximated; examples include $\sqrt{5}$ and $\sqrt[3]{3}$. These numbers are **irrational numbers.** Their decimal representations never terminate or repeat.

$$\sqrt{5} = 2.2360679\ldots \qquad \sqrt[3]{3} = 1.4422495\ldots$$

Focus on simplifying a root of a perfect power

Simplify: $\sqrt[5]{-243x^5y^{15}}$

From the chart, 243 is a perfect fifth power, and each exponent is divisible by 5. Therefore, the radicand is a perfect fifth power.

Divide each exponent by 5. $\sqrt[5]{-243x^5y^{15}} = -3xy^3$

Take Note

From the chart, $\sqrt[5]{243} = 3$, which means that $3^5 = 243$. From this we know that $(-3)^5 = -243$, which means $\sqrt[5]{-243} = -3$.

EXAMPLE 5 Simplify.

A. $\sqrt{49x^2}$ **B.** $\sqrt[3]{-125a^6b^9}$ **C.** $-\sqrt[4]{16a^4b^8}$

Solution

A. $\sqrt{49x^2} = 7x$
- The radicand is a perfect square. Divide the exponent by 2.

B. $\sqrt[3]{-125a^6b^9} = -5a^2b^3$
- The radicand is a perfect cube. Divide each exponent by 3.

C. $-\sqrt[4]{16a^4b^8} = -2ab^2$
- The radicand is a perfect fourth power. Divide each exponent by 4.

Problem 5 Simplify.

A. $\sqrt{121x^{10}}$ **B.** $\sqrt[3]{-8x^{12}y^3}$ **C.** $-\sqrt[4]{81x^{12}y^8}$

Solution See page S23.

Try Exercise 145, page 387.

7.1 Exercises

CONCEPT CHECK

Find the value of a.

1. $a^{\frac{1}{3}} = 4$ **2.** $a^{\frac{1}{4}} = 3$ **3.** $a^{\frac{1}{5}} = -2$ **4.** $a^{\frac{1}{3}} = -5$

Determine whether the given number is a rational or an irrational number.

5. $\sqrt{25}$ **6.** $\sqrt[4]{16}$ **7.** $\sqrt[3]{25}$ **8.** $\sqrt{32}$

Simplify.

9. $\sqrt{100}$ **10.** $\sqrt[3]{64}$ **11.** $\sqrt[4]{81}$ **12.** $\sqrt[5]{32}$

1 Simplify expressions with rational exponents (See pages 380-381.)

GETTING READY

13. Simplify: $64^{-\frac{3}{2}}$

$64^{-\frac{3}{2}} = (\underline{\ \ ?\ \ }^2)^{-\frac{3}{2}}$ • Write 64 as $\underline{(?)^2}$.

$= 8^{\underline{\ ?\ }}$ • Use the Rule for Simplifying the Power of an Exponential Expression: multiply the exponents.

$= \frac{1}{?}$ • Use the Definition of a Negative Exponent: $a^{-n} = \frac{1}{a^n}$.

$= \underline{\ \ \ ?\ \ \ }$ • Evaluate.

14. Simplify: $(a^{\frac{1}{3}} \cdot a^{\frac{1}{2}})^2$

$(a^{\frac{1}{3}} \cdot a^{\frac{1}{2}})^2 = (a^{\underline{\ ?\ }})^2$ • Use the Rule for Multiplying Exponential Expressions: add the exponents.

$= a^{\underline{\ ?\ }}$ • Use the Rule for Simplifying the Power of an Exponential Expression: multiply the exponents.

15. Which of the following is not a real number?
(i) $-25^{\frac{1}{2}}$ (ii) $25^{-\frac{1}{2}}$ (iii) $-25^{-\frac{1}{2}}$ (iv) $(-25)^{\frac{1}{2}}$

16. Which of the following are equivalent to $16^{-\frac{1}{2}}$?
(i) -8 (ii) $\frac{1}{4}$ (iii) -4 (iv) 4^{-1}

Simplify.

17. $8^{\frac{1}{3}}$ **18.** $16^{\frac{1}{2}}$ **19.** $9^{\frac{3}{2}}$

20. $25^{\frac{3}{2}}$ **21.** $27^{-\frac{2}{3}}$ **22.** $64^{-\frac{1}{3}}$

23. $32^{\frac{2}{5}}$ **24.** $16^{\frac{3}{4}}$ **25.** $\left(\frac{25}{49}\right)^{-\frac{3}{2}}$

26. $\left(\frac{8}{27}\right)^{-\frac{2}{3}}$ **27.** $x^{\frac{1}{2}}x^{\frac{1}{2}}$ **28.** $a^{\frac{1}{3}}a^{\frac{5}{3}}$

29. $y^{-\frac{1}{4}}y^{\frac{3}{4}}$ **30.** $x^{\frac{2}{5}} \cdot x^{-\frac{4}{5}}$ **31.** $x^{-\frac{2}{3}} \cdot x^{\frac{3}{4}}$

32. $x \cdot x^{-\frac{1}{2}}$ **33.** $a^{\frac{1}{3}} \cdot a^{\frac{3}{4}} \cdot a^{-\frac{1}{2}}$ **34.** $y^{-\frac{1}{6}} \cdot y^{\frac{2}{3}} \cdot y^{\frac{1}{2}}$

35. $\dfrac{a^{\frac{1}{2}}}{a^{\frac{3}{2}}}$ **36.** $\dfrac{b^{\frac{1}{3}}}{b^{\frac{4}{3}}}$ **37.** $\dfrac{y^{-\frac{3}{4}}}{y^{\frac{1}{4}}}$

38. $\dfrac{x^{-\frac{3}{5}}}{x^{\frac{1}{5}}}$ **39.** $\dfrac{y^{\frac{2}{3}}}{y^{-\frac{5}{6}}}$ **40.** $\dfrac{b^{\frac{3}{4}}}{b^{-\frac{3}{2}}}$

41. $(x^2)^{-\frac{1}{2}}$ **42.** $(a^8)^{-\frac{3}{4}}$ **43.** $(x^{-\frac{2}{3}})^6$

44. $(y^{-\frac{5}{6}})^{12}$ **45.** $(a^{-\frac{1}{2}})^{-2}$ **46.** $(b^{-\frac{2}{3}})^{-6}$

47. $(x^{-\frac{3}{8}})^{-\frac{4}{5}}$ **48.** $(y^{-\frac{3}{2}})^{-\frac{2}{9}}$ **49.** $(x^8y^2)^{\frac{1}{2}}$

50. $(a^3b^9)^{\frac{2}{3}}$ **51.** $(x^4y^2z^6)^{\frac{3}{2}}$ **52.** $(a^8b^4c^4)^{\frac{3}{4}}$

53. $(x^{-3}y^6)^{-\frac{1}{3}}$ **54.** $(a^2b^{-6})^{-\frac{1}{2}}$ **55.** $(x^{-2}y^{\frac{1}{3}})^{-\frac{3}{4}}$

56. $(a^{-\frac{2}{3}}b^{\frac{2}{3}})^{\frac{3}{2}}$ **57.** $(6x^{\frac{3}{2}}y^{\frac{2}{5}})(3x^{\frac{3}{2}}y)$ **58.** $(-2xy)(-5x^{\frac{5}{4}}y)$

59. $(-4xy^{\frac{2}{5}})(2x^{\frac{3}{4}}y^2)$ **60.** $(-2x^{\frac{3}{4}}y^{\frac{2}{5}})(-4x^{\frac{4}{5}}y^{\frac{3}{4}})$ **61.** $(6x^{\frac{4}{5}}y^{\frac{3}{5}})(5xy^{\frac{3}{4}})$

62. $(x^2y^{\frac{2}{5}})(-4x^{\frac{4}{3}}y^{\frac{2}{3}})$ **63.** $\dfrac{36x^2y^{\frac{2}{3}}}{12x^{\frac{4}{5}}y^2}$ **64.** $\dfrac{24x^{\frac{1}{2}}y^{\frac{3}{5}}}{6x^{\frac{4}{3}}y^{\frac{3}{4}}}$

65. $\dfrac{12x^{\frac{5}{3}}y^{\frac{3}{4}}}{-18xy^{\frac{3}{4}}}$ **66.** $\dfrac{24x^{\frac{5}{2}}y^{\frac{3}{5}}}{18xy^2}$ **67.** $\dfrac{-20x^2y^{\frac{1}{2}}}{-12x^{-\frac{1}{2}}y^{\frac{4}{5}}}$

68. $\dfrac{18x^{\frac{1}{2}}y^{\frac{1}{2}}}{-8x^{-\frac{5}{2}}y^{\frac{5}{3}}}$ **69.** $(a^{\frac{2}{3}}b^2)^6(a^3b^3)^{\frac{1}{3}}$ **70.** $(x^3y^{-\frac{1}{2}})^{-2}(x^{-3}y^2)^{\frac{1}{6}}$

71. $(16m^{-2}n^4)^{-\frac{1}{2}}(mn^{\frac{1}{2}})$ **72.** $(27m^3n^{-6})^{\frac{1}{3}}(m^{-\frac{1}{3}}n^{\frac{5}{6}})^6$ **73.** $\left(\dfrac{x^{\frac{1}{2}}y^{-\frac{3}{4}}}{y^{\frac{2}{3}}}\right)^{-6}$

74. $\left(\dfrac{x^{\frac{1}{2}}y^{-\frac{5}{4}}}{y^{-\frac{3}{4}}}\right)^{-4}$ **75.** $\left(\dfrac{2^{-6}b^{-3}}{a^{-\frac{1}{2}}}\right)^{-\frac{2}{3}}$ **76.** $\left(\dfrac{49c^{\frac{5}{3}}}{a^{-\frac{1}{4}}b^{\frac{5}{6}}}\right)^{-\frac{3}{2}}$

77. $\dfrac{(x^{-2}y^4)^{\frac{1}{2}}}{(x^{\frac{1}{2}})^4}$ **78.** $\dfrac{(x^{-3})^{\frac{1}{3}}}{(x^9y^6)^{\frac{1}{6}}}$ **79.** $a^{-\frac{1}{4}}(a^{\frac{5}{4}} - a^{\frac{9}{4}})$

80. $x^{\frac{4}{3}}(x^{\frac{2}{3}} + x^{-\frac{1}{3}})$ **81.** $y^{\frac{2}{3}}(y^{\frac{1}{3}} + y^{-\frac{2}{3}})$ **82.** $b^{-\frac{2}{5}}(b^{-\frac{3}{5}} - b^{\frac{7}{5}})$

83. $a^{\frac{1}{6}}(a^{\frac{5}{6}} - a^{-\frac{7}{6}})$

2 Write exponential expressions as radical expressions and radical expressions as exponential expressions (See pages 381–382.)

GETTING READY

84. In the radical expression $\sqrt[n]{a}$, n is called the ___?___ and a is called the ___?___.

85. To rewrite $\sqrt[5]{a^4}$ as an exponential expression, find the exponent for the base a. The rational exponent is $\dfrac{\text{power of the radicand}}{\text{index}} = \dfrac{?}{?}$, so $\sqrt[5]{a^4} = a^{\underline{?}}$.

86. True or false? $8x^{\frac{1}{3}} = 2\sqrt[3]{x}$

87. True or false? $\sqrt[4]{x^3} = (\sqrt[4]{x})^3$

Rewrite the exponential expression as a radical expression.

88. $3^{\frac{1}{4}}$
89. $5^{\frac{1}{2}}$
90. $a^{\frac{3}{2}}$
91. $b^{\frac{4}{3}}$
92. $(2t)^{\frac{5}{2}}$
93. $(3x)^{\frac{2}{3}}$
94. $-2x^{\frac{2}{3}}$
95. $-3a^{\frac{2}{5}}$
96. $(a^2b^4)^{\frac{3}{5}}$
97. $(a^3b^7)^{\frac{3}{2}}$
98. $(4x + 3)^{\frac{3}{4}}$
99. $(3x - 2)^{\frac{1}{3}}$
100. $x^{-\frac{2}{3}}$

Rewrite the radical expression as an exponential expression.

101. $\sqrt{14}$
102. $\sqrt{7}$
103. $\sqrt[3]{x}$
104. $\sqrt[4]{y}$
105. $\sqrt[3]{x^4}$
106. $\sqrt[4]{a^3}$
107. $\sqrt[5]{b^3}$
108. $\sqrt[4]{b^5}$
109. $\sqrt[3]{2x^2}$
110. $\sqrt[5]{4y^7}$
111. $-\sqrt{3x^5}$
112. $-\sqrt[4]{4x^5}$
113. $3x\sqrt[3]{y^2}$

3 Simplify radical expressions that are roots of perfect powers (See pages 383–384.)

GETTING READY

114. $16x^{12}y^{20}$ is a perfect fourth power because $16 = (\underline{\ ?\ })^4$, $x^{12} = (\underline{\ ?\ })^4$, and $y^{20} = (\underline{\ ?\ })^4$.

115. $32a^{10}b^{15}$ is a perfect fifth power because $32 = (\underline{\ ?\ })^5$, $a^{10} = (\underline{\ ?\ })^5$, and $b^{15} = (\underline{\ ?\ })^5$.

State whether the expression simplifies to a positive number, a negative number, or a number that is not a real number.

116. $-\sqrt[3]{8x^{15}}$
117. $-\sqrt{9x^8}$
118. $\sqrt{-4x^{12}}$
119. $-\sqrt[3]{-27x^9}$

Simplify.

120. $\sqrt{x^{16}}$
121. $\sqrt{y^{14}}$
122. $-\sqrt{x^8}$
123. $-\sqrt{a^6}$
124. $\sqrt{x^2y^{10}}$
125. $\sqrt{a^{14}b^6}$
126. $\sqrt{25x^6}$
127. $\sqrt{121y^{12}}$
128. $\sqrt[3]{x^3y^9}$
129. $\sqrt[3]{a^6b^{12}}$
130. $-\sqrt[3]{x^{15}y^3}$
131. $-\sqrt[3]{a^9b^9}$
132. $\sqrt[3]{27a^9}$
133. $\sqrt[3]{125b^{15}}$
134. $\sqrt[3]{-8x^3}$
135. $\sqrt[3]{-a^6b^9}$
136. $\sqrt{16a^4b^{12}}$
137. $\sqrt{25x^8y^2}$
138. $\sqrt[3]{27x^9}$
139. $\sqrt[3]{8a^{21}b^6}$
140. $\sqrt[3]{-64x^9y^{12}}$
141. $\sqrt[3]{-27a^3b^{15}}$
142. $\sqrt[4]{x^{16}}$
143. $\sqrt[4]{y^{12}}$
144. $\sqrt[4]{16x^{12}}$
145. $\sqrt[4]{81a^{20}}$
146. $-\sqrt[4]{x^8y^{12}}$

147. $-\sqrt[4]{a^{16}b^4}$

148. $\sqrt[5]{x^{20}y^{10}}$

149. $\sqrt[5]{a^5b^{25}}$

150. $\sqrt[4]{81x^4y^{20}}$

151. $\sqrt[4]{16a^8b^{20}}$

152. $\sqrt[5]{32a^5b^{10}}$

153. $\sqrt[5]{-32x^{15}y^{20}}$

154. $\sqrt[5]{243x^{10}y^{40}}$

155. $\sqrt[3]{\dfrac{27b^3}{a^9}}$

APPLYING CONCEPTS

Simplify.

156. $\sqrt{(4x + 1)^2}$

157. $\sqrt{(2x + 3)^2}$

158. $\sqrt{x^2 + 4x + 4}$

159. $\sqrt{x^2 + 2x + 1}$

160. $\sqrt{\sqrt[3]{y^6}}$

161. $\sqrt[3]{\sqrt{x^6}}$

162. $\sqrt{\sqrt{16x^{12}}}$

163. $\sqrt[5]{\sqrt[3]{b^{15}}}$

164. $\sqrt[3]{\sqrt{64x^{36}y^{30}}}$

165. $\sqrt[5]{\sqrt{a^{10}b^{20}}}$

PROJECTS OR GROUP ACTIVITIES

A **continued fraction** is a complex fraction of the form

$$a_0 + \cfrac{b_0}{a_1 + \cfrac{b_1}{a_2 + \cfrac{b_2}{a_3 + \dots}}}$$

where the complex fraction continues indefinitely. By stopping the indefinite process at some point, we can approximate the square root of a natural number.

166. Show that $\sqrt{2} \approx 1 + \cfrac{1}{2 + \cfrac{1}{2 + \cfrac{1}{2 + \cfrac{1}{2}}}}$

167. Show that $\sqrt{3} \approx 1 + \cfrac{1}{1 + \cfrac{1}{2 + \cfrac{1}{1 + \cfrac{1}{2 + 1}}}}$

7.2 Operations on Radical Expressions

OBJECTIVE 1 Simplify radical expressions

A radical expression is not in simplest form if the radicand contains a factor that is a perfect power of the index. Here are some examples.

$\sqrt{32}$ is not in simplest form. $32 = 16 \cdot 2$, so $16 = 4^2$ is a perfect square factor of 32.

$\sqrt[3]{24}$ is not in simplest form. $24 = 8 \cdot 3$, so $8 = 2^3$ is a perfect cube factor of 24.

$\sqrt[5]{160}$ is not in simplest form. $160 = 32 \cdot 5$, so $32 = 2^5$ is a perfect fifth-power factor of 160.

$\sqrt[4]{72}$ is in simplest form. There is no perfect fourth-power factor of 72.

The Product Property of Radicals is used to write a radical expression in simplest form.

PRODUCT PROPERTY OF RADICALS

If $\sqrt[n]{a}$ and $\sqrt[n]{b}$ are real numbers, then $\sqrt[n]{a} \cdot \sqrt[n]{b} = \sqrt[n]{ab}$.

Focus on simplifying a radical expression

A. Simplify: $\sqrt{72}$

Write the radicand as the product of a perfect square and a factor that does not contain a perfect square. $\quad \sqrt{72} = \sqrt{36 \cdot 2}$

Use the Product Property of Radicals. $\quad = \sqrt{36} \cdot \sqrt{2}$

Simplify $\sqrt{36}$. $\quad = 6\sqrt{2}$

B. Simplify: $\sqrt[3]{72x^4}$

Write the radicand as the product of a perfect cube and a factor that does not contain a perfect cube. $\quad \sqrt[3]{72x^4} = \sqrt[3]{8x^3 \cdot 9x}$

Use the Product Property of Radicals. $\quad = \sqrt[3]{8x^3} \cdot \sqrt[3]{9x}$

Simplify $\sqrt[3]{8x^3}$. $\quad = 2x\sqrt[3]{9x}$

Take Note

You might think to begin with $\sqrt{72} = \sqrt{9 \cdot 8}$. Although 9 is a perfect square, 8 contains 4 as a perfect square factor. You would eventually reduce $\sqrt{72}$ to simplest form, but it would take more steps. Begin by factoring out the *greatest* perfect *n*th-power factor from the radicand.

EXAMPLE 1 Simplify: $\sqrt[4]{32x^6y^9z^2}$

Solution

$\sqrt[4]{32x^6y^9z^2} = \sqrt[4]{16x^4y^8 \cdot 2x^2yz^2}$ • **Write the radicand as the product of a perfect fourth power and a factor that does not contain a perfect fourth power.**

$= \sqrt[4]{16x^4y^8} \cdot \sqrt[4]{2x^2yz^2}$ • **Use the Product Property of Radicals.**

$= 2xy^2\sqrt[4]{2x^2yz^2}$ • **Simplify.**

Problem 1 Simplify: $\sqrt[5]{128x^7}$

Solution See page S23.

Try Exercise 37, page 397.

OBJECTIVE 2 Add and subtract radical expressions

The Distributive Property is used to simplify the sum or difference of radical expressions that have the same radicand and the same index. For example,

$$3\sqrt{5} + 8\sqrt{5} = (3 + 8)\sqrt{5} = 11\sqrt{5}$$

$$2\sqrt[3]{3x} - 9\sqrt[3]{3x} = (2 - 9)\sqrt[3]{3x} = -7\sqrt[3]{3x}$$

Radical expressions that are in simplest form and have unlike radicands or different indices cannot be simplified by the Distributive Property. The following expressions cannot be simplified by the Distributive Property.

$3\sqrt[4]{2} - 6\sqrt[4]{3}$ • **The radicands are different.**

$2\sqrt[4]{4x} + 3\sqrt[3]{4x}$ • **The indices are different.**

Focus on adding or subtracting radical expressions

A. Subtract: $\sqrt[3]{81} - \sqrt[3]{192}$

Use the Product Property of Radicals to simplify each radical expression.

$$\sqrt[3]{81} - \sqrt[3]{192} = \sqrt[3]{27 \cdot 3} - \sqrt[3]{64 \cdot 3}$$
$$= \sqrt[3]{27} \cdot \sqrt[3]{3} - \sqrt[3]{64} \cdot \sqrt[3]{3}$$

Use the Distributive Property to combine like terms.

$$= 3\sqrt[3]{3} - 4\sqrt[3]{3} = (3 - 4)\sqrt[3]{3}$$
$$= -\sqrt[3]{3}$$

B. Add: $\sqrt{48a^3} + a\sqrt{75a}$

Use the Product Property of Radicals to simplify each radical expression.

$$\sqrt{48a^3} + a\sqrt{75a} = \sqrt{16a^2 \cdot 3a} + a\sqrt{25 \cdot 3a}$$
$$= \sqrt{16a^2} \cdot \sqrt{3a} + a\sqrt{25} \cdot \sqrt{3a}$$
$$= 4a\sqrt{3a} + 5a\sqrt{3a}$$

Use the Distributive Property to combine like terms.

$$= 9a\sqrt{3a}$$

Some radical expressions cannot be simplified to a single term. For instance,

$$\sqrt[4]{16a^5} + \sqrt[4]{81a} = \sqrt[4]{16a^4 \cdot a} + \sqrt[4]{81 \cdot a}$$
$$= \sqrt[4]{16a^4} \cdot \sqrt[4]{a} + \sqrt[4]{81} \cdot \sqrt[4]{a}$$
$$= 4a\sqrt[4]{a} + 3\sqrt[4]{a}$$

$4a\sqrt[4]{a}$ and $3\sqrt[4]{a}$ are not like terms, so the expression is in simplest form.

EXAMPLE 2

A. Add: $5\sqrt{20a^6b^3} + 4b\sqrt{125a^6b}$

B. Subtract: $2x\sqrt[3]{16y^7} - 4y\sqrt[3]{16x^3y^4}$

Solution

A. $5\sqrt{20a^6b^3} + 4b\sqrt{125a^6b}$

$= 5\sqrt{4a^6b^2 \cdot 5b} + 4b\sqrt{25a^6 \cdot 5b}$ • **Use the Product Property of Radicals to simplify each radical expression.**

$= 5\sqrt{4a^6b^2} \cdot \sqrt{5b} + 4b\sqrt{25a^6} \cdot \sqrt{5b}$

$= 5(2a^3b)\sqrt{5b} + 4b(5a^3)\sqrt{5b}$ • $\sqrt{4a^6b^2} = 2a^3b$; $\sqrt{25a^6} = 5a^3$

$= 10a^3b\sqrt{5b} + 20a^3b\sqrt{5b}$ • **Simplify.**

$= 30a^3b\sqrt{5b}$ • **Combine like terms.**

B. $2x\sqrt[3]{16y^7} - 4y\sqrt[3]{16x^3y^4}$

$= 2x\sqrt[3]{8y^6 \cdot 2y} - 4y\sqrt[3]{8x^3y^3 \cdot 2y}$ • **Use the Product Property of Radicals to simplify each radical expression.**

$= 2x\sqrt[3]{8y^6} \cdot \sqrt[3]{2y} - 4y\sqrt[3]{8x^3y^3} \cdot \sqrt[3]{2y}$

$= 2x(2y^2)\sqrt[3]{2y} - 4y(2xy)\sqrt[3]{2y}$ • $\sqrt[3]{8y^6} = 2y^2$; $\sqrt[3]{8x^3y^3} = 2xy$

$= 4xy^2\sqrt[3]{2y} - 8xy^2\sqrt[3]{2y}$ • **Simplify.**

$= -4xy^2\sqrt[3]{2y}$ • **Combine like terms.**

Problem 2 **A.** Add: $4\sqrt{45x^4y^3} + 2xy\sqrt{20x^2y}$
B. Subtract: $5a\sqrt[4]{32b^5} - 7b\sqrt[4]{162a^4b}$

Solution See page S23.

Try Exercise 77, page 398.

OBJECTIVE 3

Multiply radical expressions

The Product Property of Radicals is used to multiply radical expressions with the same index.

EXAMPLE 3 Multiply. **A.** $(3\sqrt{5})^2$ **B.** $\sqrt{2xy}\sqrt{6x}$ **C.** $\sqrt[3]{9a^2b}\sqrt[3]{18a^5b^2}$

Solution **A.** $(3\sqrt{5})^2 = (3\sqrt{5})(3\sqrt{5})$
$= 9\sqrt{25}$
- Use the Product Property of Radicals to multiply the radicands.

$= 9(5) = 45$
- Simplify.

B. $\sqrt{2xy}\sqrt{6x} = \sqrt{12x^2y}$
$= \sqrt{4x^2}\sqrt{3y}$
- Use the Product Property of Radicals.

$= 2x\sqrt{3y}$
- Simplify.

C. $\sqrt[3]{9a^2b}\sqrt[3]{18a^5b^2} = \sqrt[3]{162a^7b^3}$
$= \sqrt[3]{27a^6b^3}\sqrt[3]{6a}$
- Use the Product Property of Radicals.

$= 3a^2b\sqrt[3]{6a}$
- Simplify.

Problem 3 Multiply. **A.** $(-3\sqrt{7})^2$ **B.** $\sqrt{15xy^3}\sqrt{5xy}$
C. $\sqrt[3]{20ab^4}\sqrt[3]{2a^4b^2}$

Solution See page S23.

Try Exercise 99, page 398.

EXAMPLE 4 Multiply: $\sqrt{2x}(\sqrt{8x} - \sqrt{3})$

Solution $\sqrt{2x}(\sqrt{8x} - \sqrt{3}) = \sqrt{16x^2} - \sqrt{6x}$
- Use the Distributive Property to remove parentheses.

$= 4x - \sqrt{6x}$
- Simplify.

Problem 4 Multiply: $\sqrt{5b}(\sqrt{3b} - \sqrt{10})$

Solution See page S23.

Try Exercise 111, page 398.

When each of the radical expressions being multiplied contains two terms, use FOIL.

EXAMPLE 5 Multiply. **A.** $(2\sqrt{3} - 5)(7\sqrt{3} + 2)$ **B.** $(3\sqrt{x} - 2\sqrt{y})^2$

Solution **A.** $(2\sqrt{3} - 5)(7\sqrt{3} + 2)$
$= 14\sqrt{9} + 4\sqrt{3} - 35\sqrt{3} - 10$
- Use FOIL.

$= 14(3) - 31\sqrt{3} - 10$
$= 42 - 31\sqrt{3} - 10$
- Simplify $\sqrt{9}$ and combine like terms.

$= 32 - 31\sqrt{3}$

B. $(3\sqrt{x} - 2\sqrt{y})^2$
$= (3\sqrt{x} - 2\sqrt{y})(3\sqrt{x} - 2\sqrt{y})$
$= 9\sqrt{x^2} - 6\sqrt{xy} - 6\sqrt{xy} + 4\sqrt{y^2}$ • **Use FOIL.**
$= 9x - 12\sqrt{xy} + 4y$ • **Simplify $\sqrt{x^2}$ and $\sqrt{y^2}$. Combine like terms.**

Problem 5 Multiply. **A.** $(2\sqrt{5} + 3)(4\sqrt{5} - 7)$
B. $(2\sqrt{a} + 3\sqrt{b})(\sqrt{a} - 2\sqrt{b})$

Solution See page S23.

Try Exercise 131, page 399.

OBJECTIVE 4 Divide radical expressions

The Quotient Property of Radicals is used to divide radical expressions with the same index.

QUOTIENT PROPERTY OF RADICALS

If $\sqrt[n]{a}$ and $\sqrt[n]{b}$ are real numbers, and $b \neq 0$, then $\dfrac{\sqrt[n]{a}}{\sqrt[n]{b}} = \sqrt[n]{\dfrac{a}{b}}$.

Focus on simplifying a radical expression using the Quotient Property of Radicals

A. Simplify: $\sqrt[3]{\dfrac{81x^5}{y^6}}$

Use the Quotient Property of Radicals. $\sqrt[3]{\dfrac{81x^5}{y^6}} = \dfrac{\sqrt[3]{81x^5}}{\sqrt[3]{y^6}}$

Simplify each radical expression. $= \dfrac{\sqrt[3]{27x^3}\sqrt[3]{3x^2}}{\sqrt[3]{y^6}}$

$= \dfrac{3x\sqrt[3]{3x^2}}{y^2}$

B. Simplify: $\dfrac{\sqrt{5a^4b^7c^2}}{\sqrt{ab^3c}}$

Use the Quotient Property of Radicals. $\dfrac{\sqrt{5a^4b^7c^2}}{\sqrt{ab^3c}} = \sqrt{\dfrac{5a^4b^7c^2}{ab^3c}}$

Simplify the radicand. $= \sqrt{5a^3b^4c}$

Use the Product Property of Radicals. $= \sqrt{a^2b^4}\sqrt{5ac}$

Simplify. $= ab^2\sqrt{5ac}$

SIMPLEST FORM OF A RADICAL EXPRESSION

A radical expression is in simplest form when *all* of the following conditions are met.

1. The radicand contains no factor greater than 1 that is a perfect power of the index.
2. There is no fraction under the radical sign.
3. No radical remains in the denominator of the radical expression.

EXAMPLES

1. $\sqrt[3]{40}$ is not in simplest form. $8 = 2^3$ is a perfect cube factor of 40.
2. $\sqrt{\frac{2}{3}}$ is not in simplest form. There is a fraction under the radical sign.
3. $\frac{5}{\sqrt{6}}$ is not in simplest form. There is a radical in the denominator.

Condition 3 for the simplest form of a radical expression requires that no radical remain in the denominator of the radical expression. The procedure used to remove a radical from the denominator is called **rationalizing the denominator.**

Focus on rationalizing the denominator of a radical expression

Take Note

To find the factor used to multiply the numerator and denominator, look for an expression that will produce a perfect power of the index.

For part (B), think $\sqrt[3]{2} \cdot \sqrt[3]{?} = \sqrt[3]{2^3}$. Replacing the question mark with $2^2 = 4$ gives $\sqrt[3]{2} \cdot \sqrt[3]{2^2} = \sqrt[3]{2^3}$. Therefore, we multiply the numerator and denominator by $\sqrt[3]{4}$.

For part (C), write $4 = 2^2$ and think $\sqrt[5]{2^2x^3} \cdot \sqrt[5]{?} = \sqrt[5]{2^5x^5}$. Replacing the question mark with $2^3x^2 = 8x^2$ gives $\sqrt[5]{4x^3} \cdot \sqrt[5]{8x^2} = \sqrt[5]{32x^5} = 2x$.

A. Simplify: $\frac{5}{\sqrt{6}}$

The radical in the denominator is a square root. Multiply the numerator and denominator by an expression that produces a perfect square. Because $\sqrt{6} \cdot \sqrt{6} = \sqrt{36} = 6$, we use $\sqrt{6}$.

Multiply the numerator and denominator by $\sqrt{6}$. Because $\frac{\sqrt{6}}{\sqrt{6}} = 1$, the value of the expression is not changed.

$$\frac{5}{\sqrt{6}} = \frac{5}{\sqrt{6}} \cdot \frac{\sqrt{6}}{\sqrt{6}}$$

Simplify.

$$= \frac{5\sqrt{6}}{\sqrt{36}} = \frac{5\sqrt{6}}{6}$$

B. Simplify: $\frac{6}{\sqrt[3]{2}}$

The radical in the denominator is a cube root. Multiply the numerator and denominator by an expression that produces a perfect cube. Because $\sqrt[3]{2} \cdot \sqrt[3]{4} = \sqrt[3]{8} = 2$, we use $\sqrt[3]{4}$.

Multiply the numerator and denominator by $\sqrt[3]{4}$. Because $\frac{\sqrt[3]{4}}{\sqrt[3]{4}} = 1$, the value of the expression is not changed.

$$\frac{6}{\sqrt[3]{2}} = \frac{6}{\sqrt[3]{2}} \cdot \frac{\sqrt[3]{4}}{\sqrt[3]{4}}$$

Simplify.

$$= \frac{6\sqrt[3]{4}}{\sqrt[3]{8}} = \frac{6\sqrt[3]{4}}{2}$$
$$= 3\sqrt[3]{4}$$

C. Simplify: $\frac{3x}{\sqrt[5]{4x^3}}$

The radical in the denominator is a fifth root. Multiply the numerator and denominator by an expression that produces a perfect fifth power. Because $\sqrt[5]{4x^3} \cdot \sqrt[5]{8x^2} = \sqrt[5]{32x^5} = 2x$, we use $\sqrt[5]{8x^2}$.

Multiply the numerator and denominator by $\sqrt[5]{8x^2}$. Because $\frac{\sqrt[5]{8x^2}}{\sqrt[5]{8x^2}} = 1$, the value of the expression is not changed.

$$\frac{3x}{\sqrt[5]{4x^3}} = \frac{3x}{\sqrt[5]{4x^3}} \cdot \frac{\sqrt[5]{8x^2}}{\sqrt[5]{8x^2}}$$
$$= \frac{3x\sqrt[5]{8x^2}}{\sqrt[5]{32x^5}} = \frac{3x\sqrt[5]{8x^2}}{2x}$$

Simplify.

$$= \frac{3\sqrt[5]{8x^2}}{2}$$

EXAMPLE 6 Simplify. **A.** $\sqrt{\frac{2}{3}}$ **B.** $\frac{5}{\sqrt{5x}}$ **C.** $\frac{3x}{\sqrt[3]{4x}}$

Solution **A.** $\sqrt{\frac{2}{3}} = \frac{\sqrt{2}}{\sqrt{3}}$ • **Use the Quotient Property of Radicals.**

$= \frac{\sqrt{2}}{\sqrt{3}} \cdot \frac{\sqrt{3}}{\sqrt{3}} = \frac{\sqrt{6}}{\sqrt{9}}$ • **Rationalize the denominator by multiplying the numerator and denominator by $\sqrt{3}$.**

$= \frac{\sqrt{6}}{3}$ • **Simplify.**

B. $\frac{5}{\sqrt{5x}} = \frac{5}{\sqrt{5x}} \cdot \frac{\sqrt{5x}}{\sqrt{5x}}$ • **Multiply the numerator and denominator by $\sqrt{5x}$.**

$= \frac{5\sqrt{5x}}{\sqrt{25x^2}}$

$= \frac{5\sqrt{5x}}{5x}$

$= \frac{\sqrt{5x}}{x}$

C. $\frac{3x}{\sqrt[3]{4x}} = \frac{3x}{\sqrt[3]{4x}} \cdot \frac{\sqrt[3]{2x^2}}{\sqrt[3]{2x^2}}$ • **Multiply the numerator and denominator by $\sqrt[3]{2x^2}$. Then $\sqrt[3]{4x} \cdot \sqrt[3]{2x^2} = \sqrt[3]{8x^3}$. $8x^3$ is a perfect cube.**

$= \frac{3x\sqrt[3]{2x^2}}{\sqrt[3]{8x^3}}$

$= \frac{3x\sqrt[3]{2x^2}}{2x}$

$= \frac{3\sqrt[3]{2x^2}}{2}$

Take Note

In part (C), note that multiplying by $\frac{\sqrt[3]{4x}}{\sqrt[3]{4x}}$ will not rationalize the denominator:

$$\frac{3x}{\sqrt[3]{4x}} \cdot \frac{\sqrt[3]{4x}}{\sqrt[3]{4x}} = \frac{3x\sqrt[3]{4x}}{\sqrt[3]{16x^2}}$$

Because $16x^2$ is not a perfect cube, the denominator still contains a radical expression.

Problem 6 Simplify. **A.** $\sqrt{\frac{3}{7}}$ **B.** $\frac{y}{\sqrt{3y}}$ **C.** $\frac{3}{\sqrt[3]{3x^2}}$

Solution See pages S23–S24.

Try Exercise 159, page 400.

To simplify a fraction that has a square-root radical expression with two terms in the denominator, multiply the numerator and denominator by the *conjugate* of the denominator.

DEFINITION OF CONJUGATE

The **conjugate** of $a + b$ is $a - b$, and the conjugate of $a - b$ is $a + b$. The product of conjugates is $(a + b)(a - b) = a^2 - b^2$.

EXAMPLES

1. The conjugate of $3 + \sqrt{7}$ is $3 - \sqrt{7}$. The product of the conjugates is
$$(3 + \sqrt{7})(3 - \sqrt{7}) = 3^2 - (\sqrt{7})^2 = 9 - 7 = 2$$
2. The conjugate of $\sqrt{5} - 6$ is $\sqrt{5} + 6$. The product of the conjugates is
$$(\sqrt{5} - 6)(\sqrt{5} + 6) = (\sqrt{5})^2 - 6^2 = 5 - 36 = -31$$
3. The conjugate of $-2 + 3\sqrt{2}$ is $-2 - 3\sqrt{2}$. The product of the conjugates is
$$(-2 + 3\sqrt{2})(-2 - 3\sqrt{2}) = (-2)^2 - (3\sqrt{2})^2 = 4 - (9 \cdot 2) = 4 - 18 = -14$$
4. The conjugate of $\sqrt{x} - \sqrt{y}$ is $\sqrt{x} + \sqrt{y}$. The product of the conjugates is
$$(\sqrt{x} - \sqrt{y})(\sqrt{x} + \sqrt{y}) = (\sqrt{x})^2 - (\sqrt{y})^2 = x - y$$

Focus on simplifying a radical expression by multiplying the numerator and denominator by the conjugate of the denominator

A. Simplify: $\dfrac{12}{4 - 3\sqrt{2}}$

Multiply the numerator and denominator by the conjugate of the denominator.

$$\frac{12}{4 - 3\sqrt{2}} = \frac{12}{4 - 3\sqrt{2}} \cdot \frac{4 + 3\sqrt{2}}{4 + 3\sqrt{2}}$$

Use $(a - b)(a + b) = a^2 - b^2$.

$$= \frac{12(4 + 3\sqrt{2})}{4^2 - (3\sqrt{2})^2}$$

Simplify. Note that $(3\sqrt{2})^2 = 9 \cdot 2 = 18$.

$$= \frac{48 + 36\sqrt{2}}{16 - 18} = \frac{48 + 36\sqrt{2}}{-2}$$

$$= -24 - 18\sqrt{2}$$

B. Simplify: $\dfrac{3}{\sqrt{x} + 4}$

Multiply the numerator and denominator by the conjugate of the denominator.

$$\frac{3}{\sqrt{x} + 4} = \frac{3}{\sqrt{x} + 4} \cdot \frac{\sqrt{x} - 4}{\sqrt{x} - 4}$$

Use $(a + b)(a - b) = a^2 - b^2$.

$$= \frac{3(\sqrt{x} - 4)}{(\sqrt{x})^2 - 4^2}$$

Simplify. Note that $(\sqrt{x})^2 = x$.

$$= \frac{3\sqrt{x} - 12}{x - 16}$$

EXAMPLE 7 Simplify. **A.** $\dfrac{4 - 3\sqrt{5}}{3 + 2\sqrt{5}}$ **B.** $\dfrac{\sqrt{3} + \sqrt{y}}{\sqrt{3} - \sqrt{y}}$

Solution **A.**

$$\frac{4 - 3\sqrt{5}}{3 + 2\sqrt{5}} = \frac{4 - 3\sqrt{5}}{3 + 2\sqrt{5}} \cdot \frac{3 - 2\sqrt{5}}{3 - 2\sqrt{5}}$$

- **Multiply the numerator and denominator by the conjugate of the denominator.**

$$= \frac{12 - 8\sqrt{5} - 9\sqrt{5} + 6\sqrt{25}}{3^2 - (2\sqrt{5})^2}$$

- **Use FOIL.**
- **Use $(a + b)(a - b) = a^2 - b^2$.**

$$= \frac{12 - 17\sqrt{5} + 30}{9 - 20}$$

- **Simplify. $6\sqrt{25} = 6(5) = 30$; $(2\sqrt{5})^2 = 4(5) = 20$**

$$= \frac{42 - 17\sqrt{5}}{-11} = -\frac{42 - 17\sqrt{5}}{11}$$

B.

$$\frac{\sqrt{3} + \sqrt{y}}{\sqrt{3} - \sqrt{y}} = \frac{\sqrt{3} + \sqrt{y}}{\sqrt{3} - \sqrt{y}} \cdot \frac{\sqrt{3} + \sqrt{y}}{\sqrt{3} + \sqrt{y}}$$

- **Multiply the numerator and denominator by the conjugate of the denominator.**

$$= \frac{\sqrt{9} + \sqrt{3y} + \sqrt{3y} + \sqrt{y^2}}{(\sqrt{3})^2 - (\sqrt{y})^2}$$

- **Use FOIL.**
- **Use $(a - b)(a + b) = a^2 - b^2$.**

$$= \frac{3 + 2\sqrt{3y} + y}{3 - y}$$

- **Simplify.**

Problem 7 Simplify. **A.** $\frac{4 + \sqrt{2}}{3 - \sqrt{3}}$ **B.** $\frac{\sqrt{2} + \sqrt{x}}{\sqrt{2} - \sqrt{x}}$

Solution See page S24.

Try Exercise 179, page 400.

7.2 Exercises

CONCEPT CHECK

Determine whether the radical expression is in simplest form. If it is not in simplest form, explain why not.

1. $\sqrt{75}$ **2.** $3\sqrt{6}$ **3.** $\frac{\sqrt{5}}{5}$

4. $\sqrt[3]{16}$ **5.** $\frac{1}{\sqrt{2}}$ **6.** $\sqrt[4]{\frac{3}{4}}$

Write the conjugate of the given expression.

7. $3 - \sqrt{5}$ **8.** $2\sqrt{7} + 6$ **9.** $-\sqrt{3} + \sqrt{5}$ **10.** $5 - \sqrt{x}$

Write the expression whose product with the given radical expression produces a perfect power of the index. This operation is used in rationalizing the denominator of some rational expressions.

11. $\sqrt{7}$ **12.** $\sqrt[3]{25}$ **13.** $\sqrt[4]{x^3}$ **14.** $\sqrt[5]{8x^2}$

1 Simplify radical expressions (See pages 388–389.)

GETTING READY

15. Simplify: $\sqrt{50}$

$\sqrt{50} = \sqrt{\underline{\quad ? \quad} \cdot 2}$ • Write 50 as the product of a perfect square factor and a factor that does not contain a perfect square.

$= \sqrt{\underline{\quad ? \quad}}\sqrt{\underline{\quad ? \quad}}$ • Use the Product Property of Radicals to write the expression as the product of two square roots.

$= \underline{\quad ? \quad}\sqrt{2}$ • Simplify $\sqrt{25}$.

16. Simplify: $\sqrt[3]{16x^{14}}$

$\sqrt[3]{16x^{14}} = \sqrt[3]{(\underline{\quad ? \quad})(2x^2)}$ • Write $16x^4$ as the product of a perfect cube and a factor that does not contain a perfect cube.

$= \sqrt[3]{\underline{\quad ? \quad}}\sqrt[3]{2x^2}$ • Use the Product Property of Radicals.

$= \underline{\quad ? \quad}\sqrt[3]{2x^2}$ • Simplify.

State whether the expression is in simplest form.

17. a. $2\sqrt[5]{x^7}$ **b.** $\sqrt[3]{36y^2}$

18. a. $9a^2\sqrt{54ab}$ **b.** $2b\sqrt[4]{8b^3}$

Explain why the simplification is incorrect.

19. $\sqrt{32} = \sqrt{4 \cdot 8} = \sqrt{4}\sqrt{8} = 2\sqrt{8}$

20. $\sqrt[3]{128} = \sqrt[3]{8 \cdot 16} = \sqrt[3]{8}\sqrt[3]{16} = 2\sqrt[3]{16}$

Simplify.

21. $\sqrt{18}$ **22.** $\sqrt{40}$ **23.** $\sqrt{98}$ **24.** $\sqrt{128}$

25. $\sqrt[3]{72}$ **26.** $\sqrt[3]{54}$ **27.** $\sqrt[3]{16}$ **28.** $\sqrt[3]{128}$

29. $\sqrt[4]{48}$ **30.** $\sqrt[4]{162}$ **31.** $\sqrt[5]{96}$ **32.** $\sqrt[5]{729}$

33. $\sqrt{x^4y^3z^5}$ **34.** $\sqrt{x^3y^6z^9}$ **35.** $\sqrt{8a^3b^8}$

36. $\sqrt{24a^9b^6}$ **37.** $\sqrt{45x^2y^3z^5}$ **38.** $\sqrt{60xy^7z^{12}}$

39. $\sqrt[3]{125x^2y^4}$ **40.** $\sqrt[4]{16x^9y^5}$ **41.** $\sqrt[3]{-216x^5y^9}$

42. $\sqrt[3]{a^8b^{11}c^{15}}$ **43.** $\sqrt[3]{a^5b^8}$ **44.** $\sqrt[4]{64x^8y^{10}}$

2 Add and subtract radical expressions (See pages 389-391.)

GETTING READY

45. The Distributive Property can be used to add or subtract radicals that have the same __?__ and the same __?__.

46. Simplify: $5\sqrt[5]{x^{12}} + 2x\sqrt[5]{x^7}$

$5\sqrt[5]{x^{12}} + 2x\sqrt[5]{x^7}$

$= 5\sqrt[5]{\underline{\quad ? \quad}}\sqrt[5]{x^2} + 2x\sqrt[5]{\underline{\quad ? \quad}}\sqrt[5]{x^2}$ • Use the Product Property of Radicals.

$= 5(\underline{\quad ? \quad})\sqrt[5]{x^2} + 2x(\underline{\quad ? \quad})\sqrt[5]{x^2}$ • Simplify the perfect fifth roots.

$= 5x^2\sqrt[5]{x^2} + \underline{\quad ? \quad}\sqrt[5]{x^2}$ • Multiply.

$= (\underline{\quad ? \quad} + \underline{\quad ? \quad})\sqrt[5]{x^2}$ • Use the Distributive Property.

$= \underline{\quad ? \quad}\sqrt[5]{x^2}$ • Add.

47. State whether the expression can be simplified.

a. $5x\sqrt{x} + x\sqrt[3]{x}$ **b.** $\sqrt[5]{10y} - \sqrt[5]{y}$

c. $\sqrt[3]{2 + a} + \sqrt[3]{2 + a}$ **d.** $\sqrt{8x^2} + \sqrt{2}$

48. State whether the expression is equivalent to $\sqrt{ab} + \sqrt{ab}$.

a. $2\sqrt{ab}$ **b.** $\sqrt{2ab}$ **c.** $2ab$ **d.** $\sqrt{4ab}$

Simplify.

49. $\sqrt{2} + \sqrt{2}$ **50.** $\sqrt{5} - \sqrt{5}$

51. $4\sqrt[3]{7} - \sqrt[3]{7}$ **52.** $3\sqrt[3]{11} - 8\sqrt[3]{11}$

53. $\sqrt{8} - \sqrt{32}$ **54.** $\sqrt{27} - \sqrt{75}$

55. $3\sqrt{108} - 2\sqrt{18} - 3\sqrt{48}$ **56.** $2\sqrt{50} - 3\sqrt{125} + \sqrt{98}$

57. $\sqrt[3]{128} + \sqrt[3]{250}$ **58.** $\sqrt[3]{16} - \sqrt[3]{54}$

59. $4\sqrt[3]{-54} + \sqrt[3]{250}$

60. $3\sqrt[3]{24} - 6\sqrt[3]{-192}$

61. $4\sqrt[4]{16} - 9\sqrt[4]{81}$

62. $-7\sqrt[4]{162} + 2\sqrt[4]{512}$

63. $\sqrt{128x} - \sqrt{98x}$

64. $\sqrt{48x} + \sqrt{147x}$

65. $2\sqrt{2x^3} + 4x\sqrt{8x}$

66. $5y\sqrt{8y} + 2\sqrt{50y^3}$

67. $x\sqrt{75xy} - \sqrt{27x^3y}$

68. $3\sqrt{8x^2y^3} - 2x\sqrt{32y^3}$

69. $\sqrt{27a} - \sqrt{8a}$

70. $\sqrt{18b} + \sqrt{75b}$

71. $2\sqrt{32x^2y^3} - xy\sqrt{98y}$

72. $6y\sqrt{x^3y} - 2\sqrt{x^3y^3}$

73. $7b\sqrt{a^5b^3} - 2ab\sqrt{a^3b^3}$

74. $2a\sqrt{27ab^5} + 3b\sqrt{3a^3b}$

75. $2\sqrt[3]{3a^4} - 3a\sqrt[3]{81a}$

76. $2b\sqrt[3]{16b^2} + \sqrt[3]{128b^5}$

77. $3\sqrt[3]{x^5y^7} - 8xy\sqrt[3]{x^2y^4}$

78. $2x\sqrt[3]{-16x^3y^7} + 5y\sqrt[3]{250x^6y^4}$

79. $3\sqrt[4]{32a^5} - a\sqrt[4]{162a}$

80. $-a\sqrt[4]{16ab^5} - 2b\sqrt[4]{81a^5b}$

81. $2a\sqrt[4]{16ab^5} + 3b\sqrt[4]{256a^5b}$

82. $-4a\sqrt[4]{32a^5b^8} + 9b\sqrt[4]{162a^9b^4}$

83. $\sqrt{4x^7y^5} + 9x^2\sqrt{x^3y^5} - 5xy\sqrt{x^5y^3}$

84. $2x\sqrt{8xy^2} - 3y\sqrt{32x^3} + \sqrt{8x^3y^2}$

85. $3x\sqrt[3]{8xy^4} - 7y\sqrt[3]{64x^4y} + \sqrt[3]{125x^4y^4}$

86. $\sqrt[3]{54xy^3} - 5\sqrt[3]{2xy^3} + \sqrt[3]{128xy^3}$

3 Multiply radical expressions (See pages 391–392.)

GETTING READY

87. Multiply: $\sqrt{3}(\sqrt{6} - 6\sqrt{x})$

$\sqrt{3}(\sqrt{6} - 6\sqrt{x}) = \sqrt{3}\cdot\sqrt{6} - \sqrt{3}\cdot 6\sqrt{x}$ • Use the Distributive Property.

$= \sqrt{\underline{\quad ? \quad}} - 6\sqrt{\underline{\quad ? \quad}}$ • The indices are the same. Multiply the radicands.

$= \underline{\quad ? \quad}\sqrt{2} - 6\sqrt{3x}$ • Simplify $\sqrt{18}$.

88. Multiply: $\sqrt[3]{a^2}\sqrt[3]{a^5}$

$\sqrt[3]{a^2}\sqrt[3]{a^5} = \sqrt[3]{\underline{\quad ? \quad}} = \sqrt[3]{\underline{\quad ? \quad}\cdot a} = \sqrt[3]{\underline{\quad ? \quad}}\cdot\sqrt[3]{a}$

$= \underline{\quad ? \quad}\cdot\sqrt[3]{a}$

Multiply.

89. $\sqrt{8}\sqrt{32}$

90. $\sqrt{14}\sqrt{35}$

91. $3\sqrt{6}\cdot 5\sqrt{3}$

92. $-2\sqrt{14}\cdot 5\sqrt{21}$

93. $\sqrt[3]{4}\sqrt[3]{8}$

94. $\sqrt[3]{6}\sqrt[3]{36}$

95. $-2\sqrt[3]{20}\cdot 5\sqrt[3]{6}$

96. $(3\sqrt[4]{18})(-2\sqrt[4]{27})$

97. $\sqrt{x^2y^5}\sqrt{xy}$

98. $\sqrt{a^3b}\sqrt{ab^4}$

99. $\sqrt{2x^2y}\sqrt{32xy}$

100. $\sqrt{5x^3y}\sqrt{10x^3y^4}$

101. $\sqrt[3]{x^2y}\sqrt[3]{16x^4y^2}$

102. $\sqrt[3]{4a^2b^3}\sqrt[3]{8ab^5}$

103. $\sqrt[4]{12ab^3}\sqrt[4]{4a^5b^2}$

104. $\sqrt[4]{36a^2b^4}\sqrt[4]{12a^5b^3}$

105. $2\sqrt{14xy}\cdot 4\sqrt{7x^2y}\cdot 3\sqrt{8xy^2}$

106. $2\sqrt{3x^2}\cdot 3\sqrt{12xy^3}\cdot\sqrt{6x^3y}$

107. $\sqrt[3]{2a^2b}\sqrt[3]{4a^3b^2}\sqrt[3]{8a^5b^6}$

108. $\sqrt[3]{8ab}\sqrt[3]{4a^2b^3}\sqrt[3]{9ab^4}$

109. $\sqrt{3}(\sqrt{27} - \sqrt{3})$

110. $\sqrt{10}(\sqrt{10} - \sqrt{5})$

111. $\sqrt{x}(\sqrt{x} - \sqrt{2})$

112. $\sqrt{y}(\sqrt{y} - \sqrt{5})$

113. $\sqrt{2x}(\sqrt{8x} - \sqrt{32})$

114. $\sqrt{3a}(\sqrt{27a^2} - \sqrt{a})$

115. $(\sqrt{5} - 5)(2\sqrt{5} + 2)$

116. $(\sqrt{2} - 3)(\sqrt{2} + 4)$

117. $(3 - 4\sqrt{7})(2 + 3\sqrt{7})$

118. $(2\sqrt{5} + 4)(5 - 3\sqrt{5})$

119. $(2\sqrt{3} + \sqrt{6})(3\sqrt{3} - 5\sqrt{6})$

120. $(\sqrt{10} - 2\sqrt{5})(2\sqrt{10} + 3\sqrt{5})$

121. $(4\sqrt{5} + 2)^2$

122. $(2 - 3\sqrt{7})^2$

123. $(\sqrt{x} + 4)(\sqrt{x} - 7)$

124. $(\sqrt{a} - 2)(\sqrt{a} - 3)$

125. $(2\sqrt{z} + 3)(3\sqrt{z} - 4)$

126. $(5 - 2\sqrt{x})(2 + 3\sqrt{x})$

127. $(\sqrt{x} - y)(\sqrt{x} + y)$

128. $(\sqrt{y} - 2)(\sqrt{y} + 2)$

129. $(2\sqrt{3x} - \sqrt{y})(2\sqrt{3x} + \sqrt{y})$

130. $(\sqrt{2x} - 3\sqrt{y})(\sqrt{2x} + 3\sqrt{y})$

131. $(\sqrt{x} - 3)^2$

132. $(\sqrt{2x} + 4)^2$

133. $(1 - 5\sqrt{x})^2$

134. $(3\sqrt{x} + 5)^2$

135. $(\sqrt[3]{x} - 4)(\sqrt[3]{x} + 5)$

136. $(\sqrt[3]{a} + 2)(\sqrt[3]{a} + 3)$

137. True or false? $(\sqrt{a} - 1)(\sqrt{a} + 1) > a$

138. True or false? $\sqrt{a}(\sqrt{a} - a) = a(1 - \sqrt{a})$

4 Divide radical expressions (See pages 392–396.)

GETTING READY

139. Divide: $\dfrac{\sqrt[4]{80x^{10}}}{\sqrt[4]{5x}}$

$\dfrac{\sqrt[4]{80x^{10}}}{\sqrt[4]{5x}} = \sqrt[4]{\dfrac{80x^{10}}{5x}}$ • **Use the __?__ Property of Radicals.**

$= \sqrt[4]{16x^{?}}$ • **Simplify the radicand.**

$= \sqrt[4]{16x^{?}}\sqrt[4]{x}$

$=$ __?__ $\sqrt[4]{x}$ • **Simplify the perfect fourth power.**

140. a. Two expressions in the form $a + b$ and $a - b$ are called __?__.

b. The conjugate of $\sqrt{y} - 5$ is __?__.

c. The product of conjugates of the form $a + b$ and $a - b$ is __?__.

d. $(\sqrt{y} - 5)(\sqrt{y} + 5) =$ (__?__)^2 $-$ (__?__)^2 $=$ __?__ $-$ __?__.

141. State whether the expression is in simplest form.

a. $\dfrac{\sqrt[3]{a^2}}{b}$ **b.** $\dfrac{\sqrt{a^3}}{b}$ **c.** $\sqrt[4]{\dfrac{a}{b}}$ **d.** $\dfrac{\sqrt{a}}{\sqrt{a} + b}$

142. By what expression should the numerator and denominator be multiplied in order to rationalize the denominator?

a. $\dfrac{1}{\sqrt{6}}$ **b.** $\dfrac{7}{\sqrt[3]{2y^5}}$ **c.** $\dfrac{8x}{\sqrt[4]{27x}}$ **d.** $\dfrac{4}{\sqrt{3} - x}$

Simplify.

143. $\dfrac{\sqrt{32x^2}}{\sqrt{2x}}$

144. $\dfrac{\sqrt{60y^4}}{\sqrt{12y}}$

145. $\dfrac{\sqrt{42a^3b^5}}{\sqrt{14a^2b}}$

146. $\dfrac{\sqrt{65ab^4}}{\sqrt{5ab}}$

147. $\dfrac{1}{\sqrt{5}}$

148. $\dfrac{1}{\sqrt{2}}$

149. $\frac{14}{3\sqrt{7}}$

150. $\frac{10}{9\sqrt{5}}$

151. $\sqrt{\frac{5}{8}}$

152. $\sqrt{\frac{7}{18}}$

153. $\frac{1}{\sqrt{2x}}$

154. $\frac{2}{\sqrt{3y}}$

155. $\frac{5}{\sqrt{5x}}$

156. $\frac{9}{\sqrt{3a}}$

157. $\sqrt{\frac{x}{5}}$

158. $\sqrt{\frac{y}{2}}$

159. $\frac{3}{\sqrt[3]{2}}$

160. $\frac{5}{\sqrt[3]{9}}$

161. $\frac{3}{\sqrt[3]{4x^2}}$

162. $\frac{5}{\sqrt[3]{3y}}$

163. $\frac{\sqrt{40x^3y^2}}{\sqrt{80x^2y^3}}$

164. $\frac{\sqrt{15a^2b^5}}{\sqrt{30a^5b^3}}$

165. $\frac{\sqrt{24a^2b}}{\sqrt{18ab^4}}$

166. $\frac{\sqrt{12x^3y}}{\sqrt{20x^4y}}$

167. $\frac{3}{\sqrt[4]{8x^3}}$

168. $\frac{-3}{\sqrt[4]{27y^2}}$

169. $\frac{4}{\sqrt[5]{16a^2}}$

170. $\frac{a}{\sqrt[5]{81a^4}}$

171. $\frac{2x}{\sqrt[5]{64x^3}}$

172. $\frac{3y}{\sqrt[4]{32y^2}}$

173. $\frac{2}{\sqrt{5}+2}$

174. $\frac{5}{2-\sqrt{7}}$

175. $\frac{8}{4+\sqrt{6}}$

176. $\frac{11}{7-3\sqrt{3}}$

177. $\frac{\sqrt{3}}{5-\sqrt{15}}$

178. $\frac{2\sqrt{5}}{3+2\sqrt{5}}$

179. $\frac{3}{\sqrt{y}-2}$

180. $\frac{-7}{\sqrt{x}-3}$

181. $\frac{\sqrt{2}-\sqrt{3}}{\sqrt{2}+\sqrt{3}}$

182. $\frac{\sqrt{3}+\sqrt{4}}{\sqrt{2}+\sqrt{3}}$

183. $\frac{4-\sqrt{2}}{2-\sqrt{3}}$

184. $\frac{3-\sqrt{x}}{3+\sqrt{x}}$

185. $\frac{\sqrt{3}-\sqrt{5}}{\sqrt{2}+\sqrt{5}}$

186. $\frac{\sqrt{2}+\sqrt{3}}{\sqrt{3}-\sqrt{2}}$

187. $\frac{\sqrt{a}+a\sqrt{b}}{\sqrt{a}-a\sqrt{b}}$

188. $\frac{\sqrt{3}-3\sqrt{y}}{\sqrt{3}+3\sqrt{y}}$

189. $\frac{3\sqrt{xy}+2\sqrt{xy}}{\sqrt{x}-\sqrt{y}}$

190. $\frac{2\sqrt{x}+3\sqrt{y}}{\sqrt{x}-4\sqrt{y}}$

APPLYING CONCEPTS

Simplify.

191. $(\sqrt{8}-\sqrt{2})^3$

192. $(\sqrt{27}-\sqrt{3})^3$

193. $(\sqrt{2}-3)^3$

194. $(\sqrt{5}+2)^3$

195. $\frac{3}{\sqrt{y+1}+1}$

196. $\frac{2}{\sqrt{x+4}+2}$

In some cases, it is necessary to rationalize the numerator of a radical expression rather than the denominator. In Exercises 197 and 198, rationalize the numerator.

197. $\dfrac{\sqrt{4+h}-2}{h}$

198. $\dfrac{\sqrt{9+h}-3}{h}$

199. By what factor must you multiply a number in order to double its square root? to triple its square root? to double its cube root? to triple its cube root? Explain.

PROJECTS OR GROUP ACTIVITIES

The factorization formulas for the sum and difference of two perfect cubes can be used to simplify some radical expressions containing cube roots. Recall:

$$a^3 + b^3 = (a + b)(a^2 - ab + b^2)$$
$$a^3 - b^3 = (a - b)(a^2 + ab + b^2)$$

200. Using the sum of perfect cubes formula, show that

$$(3 + \sqrt[3]{2})(9 - 3\sqrt[3]{2} + \sqrt[3]{4}) = 29$$

201. Using the difference of perfect cubes formula, show that

$$(\sqrt[3]{x} - \sqrt[3]{y})(\sqrt[3]{x^2} + \sqrt[3]{x}\sqrt[3]{y} + \sqrt[3]{y^2}) = x - y$$

202. Simplify $\dfrac{1}{3+\sqrt[3]{2}}$ by multiplying the numerator and denominator by $9 - 3\sqrt[3]{2} + \sqrt[3]{4}$. (*Hint:* See Exercise 200.)

203. Simplify: $\dfrac{4}{\sqrt[3]{5}-1}$

7.3 Radical Functions

OBJECTIVE 

Find the domain of a radical function

A **radical function** is one that contains a fractional exponent or a variable underneath a radical. Examples of radical functions are shown at the right.

$$f(x) = 3\sqrt[4]{x^5} - 7$$
$$g(x) = 3x - 2x^{\frac{1}{2}} + 5$$

Note that these are *not* polynomial functions because polynomial functions do not contain variables raised to a fractional power or variable radical expressions.

The domain of a radical function is a set of real numbers for which the radical expression is a real number. For example, -9 is one number that would be excluded from the domain of $f(x) = \sqrt{x+5}$ because

$$f(-9) = \sqrt{-9+5} = \sqrt{-4}, \text{ which is not a real number.}$$

Focus on finding the domain of a radical function

A. State the domain of $f(x) = \sqrt{x+5}$ in set-builder notation.

The value of $\sqrt{x+5}$ is a real number when $x + 5$ is greater than or equal to zero: $x + 5 \geq 0$. Solving this inequality for x results in $x \geq -5$.

The domain of f is $\{x|x \geq -5\}$.

B. State the domain of $F(x) = \sqrt[3]{2x - 6}$ in interval notation.

Because the cube root of a real number is a real number, $\sqrt[3]{2x - 6}$ is a real number for all values of x (for instance, $F(-1) = \sqrt[3]{2(-1) - 6} = \sqrt[3]{-8} = -2$). Therefore, in interval notation, the domain of F is $(-\infty, \infty)$.

These last two examples suggest the following:

If the index of a radical expression is an even number, the radicand must be greater than or equal to zero to ensure that the value of the radical expression will be a real number. If the index of a radical expression is an odd number, the radicand may be a positive or a negative number.

EXAMPLE 1 State the domain of each function in set-builder notation.

A. $V(x) = \sqrt[4]{6 - 4x}$ **B.** $R(x) = \sqrt[5]{x + 4}$

Solution **A.** $6 - 4x \geq 0$

$-4x \geq -6$

$x \leq \dfrac{3}{2}$

• **V contains an even root. Therefore, the radicand must be greater than or equal to zero.**

The domain is $\{x | x \leq \frac{3}{2}\}$.

B. $R(x) = \sqrt[5]{x + 4}$

• **Because R contains an odd root, the radicand may be positive or negative.**

The domain is $\{x | x \in \text{real numbers}\}$.

Problem 1 State the domain of each function in interval notation.

A. $Q(x) = \sqrt[3]{6x + 12}$ **B.** $T(x) = (3x + 9)^{\frac{1}{2}}$

Solution See page S24.

Try Exercise 15, page 405.

OBJECTIVE 2 Graph a radical function

The graph of a radical function is produced in the same manner as the graph of any other function. The function is evaluated at several values in the domain of the function, and the resulting ordered pairs are graphed. Ordered pairs must be graphed until an accurate graph can be drawn.

Focus on graphing a radical function

Graph: $f(x) = \sqrt{x + 2}$

Because f contains an even root, the radicand must be greater than or equal to zero. To determine the domain of f, solve the inequality $x + 2 \geq 0$. The solution is $x \geq -2$, so the domain is $\{x | x \geq -2\}$. Now determine ordered pairs of the function by choosing values of x from the domain. Some possible choices are shown in the following table.

Take Note

By first determining the domain of the function, you discover the possible values of x that may be used when evaluating the function. Any value of x in the domain can be used. We used "nice" values—values of x that result in a perfect-square radicand. However, any value of x can be used. For instance, if $x = 3$, then

$$f(3) = \sqrt{5} \approx 2.24$$

and the ordered pair (3, 2.24) belongs to the graph. This ordered pair is just a little more difficult to graph.

x	$f(x) = \sqrt{x+2}$	y
-2	$f(-2) = \sqrt{-2+2} = \sqrt{0} = 0$	0
-1	$f(-1) = \sqrt{-1+2} = \sqrt{1} = 1$	1
2	$f(2) = \sqrt{2+2} = \sqrt{4} = 2$	2
7	$f(7) = \sqrt{7+2} = \sqrt{9} = 3$	3

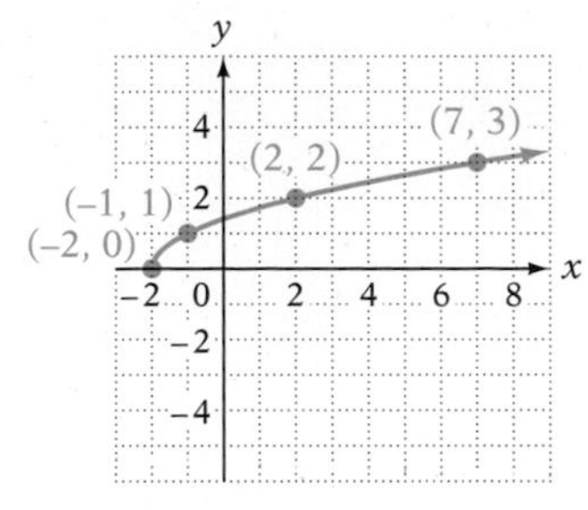

EXAMPLE 2 Graph: $H(x) = \sqrt[3]{x}$

Solution Because H contains only an odd root, the domain of H is all real numbers. Choose some values of x in the domain of H, and evaluate the function for those values. Some possible choices are given in the following table.

x	$y = \sqrt[3]{x}$
-8	-2
-1	-1
0	0
1	1
8	2

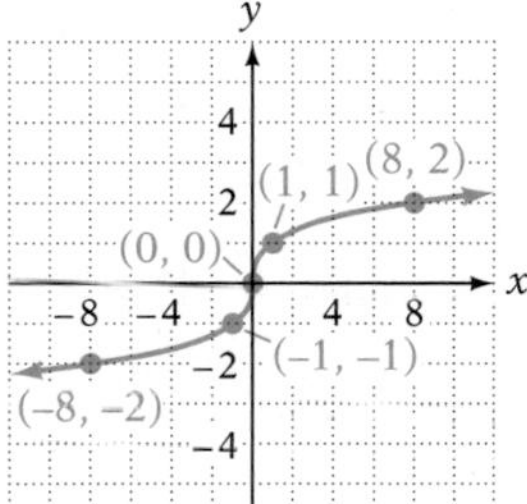

Problem 2 Graph: $F(x) = \sqrt{x-2}$

Solution See page S24.

Try Exercise 39, page 405.

The graphs of $f(x) = \sqrt{x+2}$ and $H(x) = \sqrt[3]{x}$ are shown below. Note that each graph satisfies the vertical line test for the graph of a function.

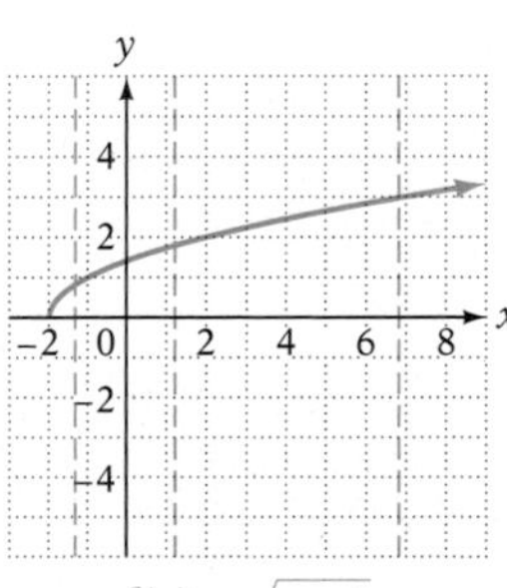
$f(x) = \sqrt{x+2}$

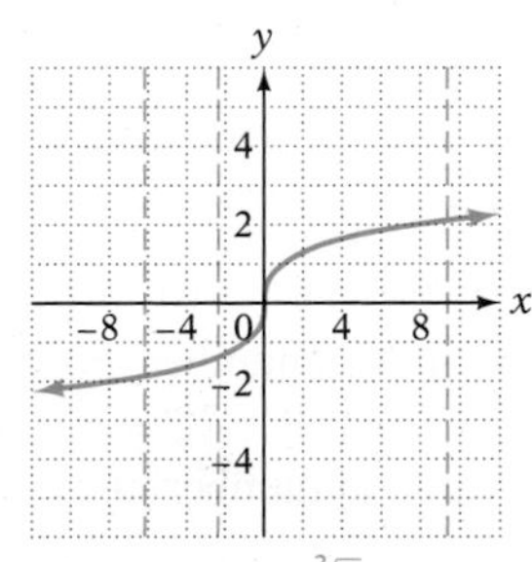
$H(x) = \sqrt[3]{x}$

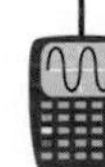
A graphing utility can be used to graph radical functions. See the Appendix for instructions on how to enter a radical function on a graphing utility.

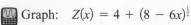

EXAMPLE 3 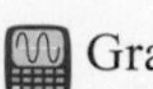 Graph: $Z(x) = 4 + (8 - 6x)^{\frac{1}{2}}$

Solution Because Z involves an even root, the radicand must be greater than or equal to zero. The domain is $\{x|x \leq \frac{4}{3}\}$.

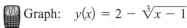

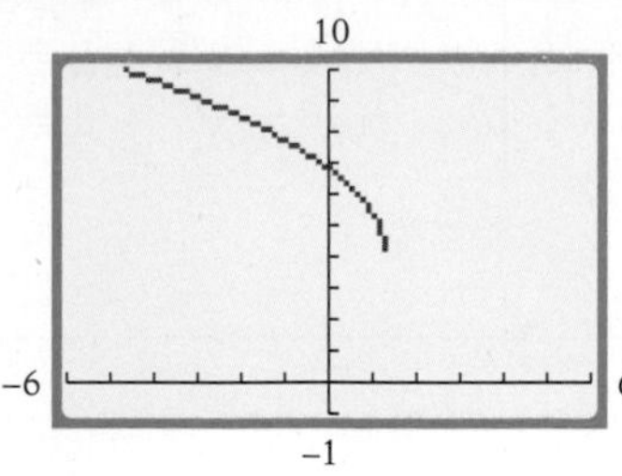

Problem 3 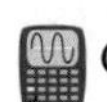 Graph: $y(x) = 2 - \sqrt[3]{x - 1}$

Solution See page S24.

Try Exercise 49, page 406.

7.3 Exercises

CONCEPT CHECK

State whether -5 is in the domain of each of the following.

1. $f(x) = \sqrt[3]{x + 5}$
2. $g(x) = \sqrt{x + 5}$
3. $k(x) = \sqrt[3]{x - 4}$
4. $h(x) = \sqrt{x - 4}$
5. $r(x) = \sqrt[3]{x} + 5$
6. $v(x) = \sqrt{x} + 5$

1 Find the domain of a radical function (See pages 401–402.)

7. What is a radical function?
8. What is the difference between a radical function and a polynomial function?

GETTING READY

9. a. When n is an even number, the domain of $f(x) = x^{\frac{1}{n}}$ is $\{x \mid x$ __?__$\}$.

   b. When n is an odd number, the domain of $f(x) = x^{\frac{1}{n}}$ is $\{x \mid x$ __?__$\}$.

10. To find the domain of $f(x) = 8 + \sqrt[4]{2x - 1}$, note that the index of the radical is even, so the function is defined only when __?__ ≥ 0. Solving this inequality for x gives $x \geq$ __?__. In interval notation, the domain of f is [__?__, __?__).

11. State whether the domain of f is $(-\infty, 0]$, $[0, \infty)$, $(-\infty, \infty)$, or the empty set.

    a. $f(x) = \sqrt{-x}$ b. $f(x) = \sqrt{-x^2}$

    c. $f(x) = -\sqrt{x^2}$ d. $f(x) = \sqrt{(-x)^2}$

12. **a.** True or false? The functions $f(x) = \sqrt[4]{3x - 4}$ and $g(x) = \sqrt[3]{3x - 4}$ have the same domain.
b. True or false? $f(x) = \sqrt{3x^2}$ is a radical function.

State the domain of each function in set-builder notation.

13. $f(x) = 2x^{\frac{1}{3}}$

14. $r(x) = -3\sqrt[5]{2x}$

15. $g(x) = -2\sqrt{x + 1}$

16. $h(x) = 3x^{\frac{1}{4}} - 2$

17. $f(x) = 2x\sqrt{x} - 3$

18. $y(x) = -3\sqrt[3]{1 + x}$

19. $C(x) = -3x^{\frac{3}{4}} + 1$

20. $G(x) = 6x^{\frac{2}{5}} + 5$

21. $F(x) = 4(3x - 6)^{\frac{1}{2}}$

State the domain of each function in interval notation.

22. $f(x) = -2(4x - 12)^{\frac{1}{2}}$

23. $g(x) = 2(2x - 10)^{\frac{2}{3}}$

24. $J(x) = 4 - (3x - 3)^{\frac{2}{5}}$

25. $V(x) = x - \sqrt{12 - 4x}$

26. $Y(x) = -6 + \sqrt{6 - x}$

27. $h(x) = 3\sqrt[4]{(x - 2)^3}$

28. $g(x) = \frac{2}{3}\sqrt[4]{(4 - x)^3}$

29. $f(x) = x - (4 - 6x)^{\frac{1}{2}}$

30. $F(x) = (9 + 12x)^{\frac{1}{2}} - 4$

2 Graph a radical function (See pages 402–404.)

GETTING READY

For Exercises 31 to 34, use the function $f(x) = -\sqrt{x - 5}$.

31. The domain of f is $\{x \mid x \underline{\quad ? \quad}\}$.

32. For what value of x will the radicand be equal to the perfect square 9?

33. $f(14) = -\sqrt{\underline{\quad ? \quad} - 5} = -\sqrt{\underline{\quad ? \quad}} = \underline{\quad ? \quad}$

34. Use the results of Exercise 33 to state a point that is on the graph of $f(x) = -\sqrt{x - 5}$.

Graph.

35. $F(x) = \sqrt{x}$

36. $G(x) = -\sqrt{x}$

37. $h(x) = -\sqrt[3]{x}$

38. $K(x) = \sqrt[3]{x + 1}$

39. $f(x) = -\sqrt{x + 2}$

40. $g(x) = \sqrt{x - 1}$

41. $S(x) = -\sqrt[4]{x}$

42. $C(x) = (x + 2)^{\frac{1}{4}}$

43. $F(x) = (x - 2)^{\frac{1}{2}}$

44. $f(x) = -(x - 1)^{\frac{1}{2}}$

45. $Q(x) = (x - 3)^{\frac{1}{3}}$

46. $H(x) = (-x)^{\frac{1}{3}}$

Graph.

47. $f(x) = 2x^{\frac{2}{5}} - 1$

48. $h(x) = 3x^{\frac{2}{5}} + 2$

49. $g(x) = 3 - (5 - 2x)^{\frac{1}{2}}$

50. $y(x) = 1 + (4 - 8x)^{\frac{1}{2}}$

51. $V(x) = \sqrt[5]{(x - 2)^2}$

52. $A(x) = x\sqrt{3x - 9}$

53. $F(x) = 2x - 3\sqrt[3]{x} - 1$

54. $f(x) = 2x - \sqrt{x} - 1$

55. $f(x) = 3x\sqrt{4x + 8}$

56. **a.** Which graph below could be the graph of $f(x) = -x^{\frac{1}{2}}$?

b. Which graph below could be the graph of $f(x) = (-x)^{\frac{1}{2}}$?

c. Which graph below could be the graph of $f(x) = -(-x)^{\frac{1}{2}}$?

d. Which graph below could be the graph of $f(x) = x^{\frac{1}{2}}$?

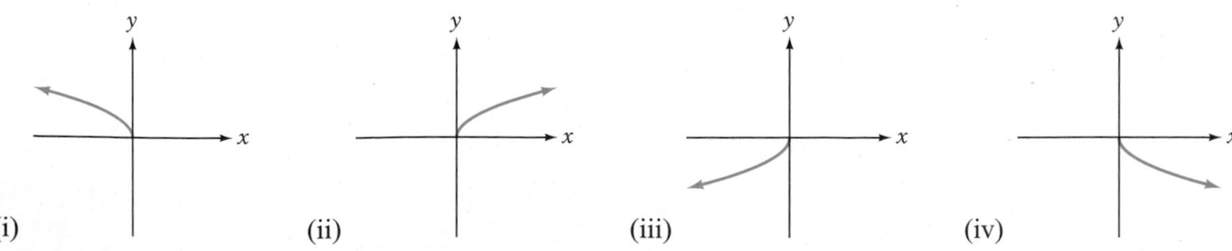

57. State whether the graph of the given function will be above the x-axis, below the x-axis, or both above and below the x-axis.

a. $f(x) = -x^{\frac{1}{3}}$ **b.** $f(x) = -\sqrt{x + 2}$ **c.** $f(x) = \sqrt[4]{3x}$ **d.** $f(x) = (-2x)^{\frac{1}{2}}$

APPLYING CONCEPTS

58. Sports Many major league baseball parks have a symmetric design, as shown in the figure at the right. One decision that the designer must make is the shape of the outfield. One possible design uses the function

$$f(x) = k + (400 - k)\sqrt{1 - \frac{x^2}{a^2}}$$

to determine the shape of the outfield.

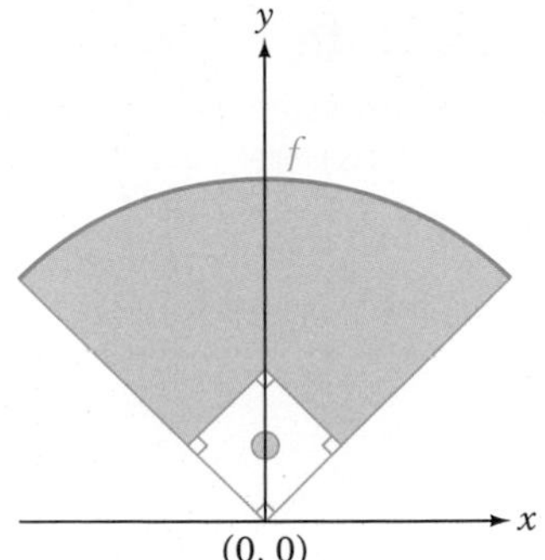

a. Graph this equation for $k = 0$, $a = 287$, and $-240 \le x \le 240$.

b. What is the maximum value of this function for the given interval?

c. The equation of the right-field foul line is $y = x$. Where does the foul line intersect the graph of f? That is, find the point on the graph of f for which $y = x$.

d. If the units on the axes are feet, what is the distance from home plate along the foul line to the base of the right-field wall?

59. Currency According to the Bureau of Engraving and Printing, the average life spans of different denominations of currency are as shown in the table at the right. The function that approximately models these data is $f(x) = 1.38x^{\frac{2}{5}}$, where x is the denomination of the bill and $f(x)$ is the average life span in years.

Currency	Average Life Span
\$1 bill	1.8 years
\$5 bill	2 years
\$10 bill	3 years
\$20 bill	4 years
\$50 bill	9 years
\$100 bill	9 years

a. In the context of the problem, what is the domain of the function $f(x) = 1.38x^{\frac{2}{5}}$? Why is zero not in the domain? Why are negative numbers not in the domain?

b. If 2 were in the domain of the function, what life span would the model predict for the \$2 bill? Round to the nearest tenth. Do you think this estimate is reasonable?

PROJECTS OR GROUP ACTIVITIES

60. Writing the fraction $\frac{2}{4}$ in lowest terms as $\frac{1}{2}$, it appears that $(x^2)^{\frac{1}{4}} = x^{\frac{1}{2}}$. Using $-10 \le x \le 10$, graph $f(x) = (x^2)^{\frac{1}{4}}$ and then graph $g(x) = x^{\frac{1}{2}}$. Are the graphs the same? If they are, try graphing $g(x) = x^{\frac{1}{2}}$ first and then $f(x) = (x^2)^{\frac{1}{4}}$. Explain why the graphs are not the same. In your explanation, tell why $x^{\frac{2}{4}} = x^{\frac{1}{2}}$ is not always a true statement.

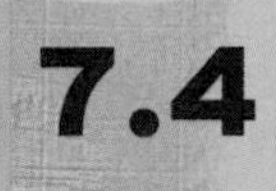

7.4 Solving Equations Containing Radical Expressions

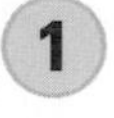

OBJECTIVE 1 Solve equations containing one or more radical expressions

An equation that contains a variable expression in a radicand is a **radical equation**.

$$\left.\begin{array}{l}\sqrt[3]{2x - 5} + x = 7 \\ \sqrt{x + 1} - \sqrt{x} = 4\end{array}\right\}\text{Radical equations}$$

Take Note
You might assume that the converse of this property is also true. That is, if $a^2 = b^2$, then $a = b$. But this is not the case. For example, if $a = -4$ and $b = 4$, then $(-4)^2 = 4^2$ but $-4 \neq 4$.

The following property is used to solve a radical equation.

PROPERTY OF RAISING EACH SIDE OF AN EQUATION TO A POWER

If two numbers are equal, then the same powers of the numbers are equal.

If $a = b$, then $a^n = b^n$.

EXAMPLE 1 Solve. **A.** $\sqrt{3x-2} - 8 = -3$ **B.** $\sqrt[3]{3x-1} = -4$

Solution **A.**

$$\sqrt{3x-2} - 8 = -3$$
$$\sqrt{3x-2} = 5$$

• Rewrite the equation so that the radical is alone on one side of the equation.

$$(\sqrt{3x-2})^2 = 5^2$$

• Square each side of the equation.

$$3x - 2 = 25$$
$$3x = 27$$
$$x = 9$$

• Solve the resulting equation.

Check: • Check the solution.

$$\sqrt{3x-2} - 8 = -3$$

$\sqrt{3 \cdot 9 - 2} - 8$	-3
$\sqrt{27 - 2} - 8$	-3
$\sqrt{25} - 8$	-3
$5 - 8$	-3

$$-3 = -3$$

The solution is 9.

B.

$$\sqrt[3]{3x-1} = -4$$
$$(\sqrt[3]{3x-1})^3 = (-4)^3$$

• Cube each side of the equation.

$$3x - 1 = -64$$
$$3x = -63$$
$$x = -21$$

• Solve the resulting equation.

Check: • Check the solution.

$$\sqrt[3]{3x-1} = -4$$

$\sqrt[3]{3(-21) - 1}$	-4
$\sqrt[3]{-63 - 1}$	-4
$\sqrt[3]{-64}$	-4

$$-4 = -4$$

The solution is -21.

Problem 1 Solve. **A.** $\sqrt{4x+5} - 12 = -5$ **B.** $\sqrt[4]{x-8} = 3$

Solution See page S24.

Try Exercise 23, page 411.

When you raise both sides of an equation to an even power, the resulting equation may have a solution that is an extraneous solution of the original equation. Therefore, it is necessary to check all proposed solutions of a radical equation.

EXAMPLE 2 Solve. **A.** $x + 2\sqrt{x - 1} = 9$ **B.** $\sqrt{x + 7} = \sqrt{x} + 1$

Solution **A.**

$$x + 2\sqrt{x - 1} = 9$$
$$2\sqrt{x - 1} = 9 - x$$

- **Rewrite the equation with the radical alone on one side of the equation.**

$$(2\sqrt{x - 1})^2 = (9 - x)^2$$
$$4(x - 1) = 81 - 18x + x^2$$
$$4x - 4 = 81 - 18x + x^2$$

- **Square each side of the equation.**

$$0 = x^2 - 22x + 85$$

- **Write the quadratic equation in standard form.**

$$0 = (x - 5)(x - 17)$$

- **Factor.**

$$x - 5 = 0 \qquad x - 17 = 0$$
$$x = 5 \qquad x = 17$$

- **Use the Principle of Zero Products.**

Check:

$x + 2\sqrt{x - 1}$	$= 9$
$5 + 2\sqrt{5 - 1}$	9
$5 + 2\sqrt{4}$	9
$5 + 2 \cdot 2$	9
$5 + 4$	9
$9 = 9$	

$x + 2\sqrt{x - 1}$	$= 9$
$17 + 2\sqrt{17 - 1}$	9
$17 + 2\sqrt{16}$	9
$17 + 2 \cdot 4$	9
$17 + 8$	9
$25 \neq 9$	

17 does not check as a solution. It is an extraneous solution of the equation.

The solution is 5.

Take Note

You must check the proposed solutions to radical equations. The proposed solutions of the equation at the right were 5 and 17. However, 17 did not check as a solution. 17 is an extraneous solution.

B.

$$\sqrt{x + 7} = \sqrt{x} + 1$$

- **A radical appears on each side of the equation.**

$$(\sqrt{x + 7})^2 = (\sqrt{x} + 1)^2$$

- **Square each side of the equation.**

$$x + 7 = x + 2\sqrt{x} + 1$$
$$6 = 2\sqrt{x}$$

- **Simplify the resulting equation.**

$$3 = \sqrt{x}$$

- **The equation contains a radical.**

$$3^2 = (\sqrt{x})^2$$
$$9 = x$$

- **Square each side of the equation.**

Check:

- **Check the solution.**

$\sqrt{x + 7}$	$= \sqrt{x} + 1$
$\sqrt{9 + 7}$	$\sqrt{9} + 1$
$\sqrt{16}$	$3 + 1$
$4 = 4$	

The solution is 9.

Take Note

Note that

$(\sqrt{x} + 1)^2$

$= (\sqrt{x} + 1)(\sqrt{x} + 1)$

$= x + 2\sqrt{x} + 1$

Problem 2 Solve. **A.** $x + 3\sqrt{x + 2} = 8$ **B.** $\sqrt{x + 5} = 5 - \sqrt{x}$

Solution See pages S24–S25.

Try Exercise 55, page 411.

OBJECTIVE 2 Application problems

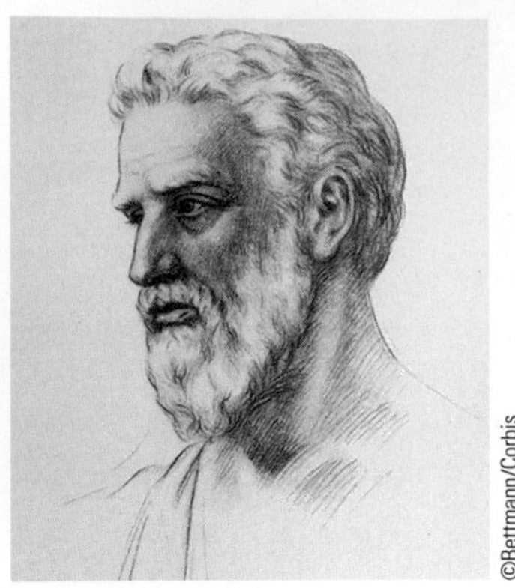

Pythagoras
(c. 580 B.C.–529 B.C.)

A right triangle contains one 90° angle. The side opposite the 90° angle is called the **hypotenuse.** The other two sides are called **legs.**

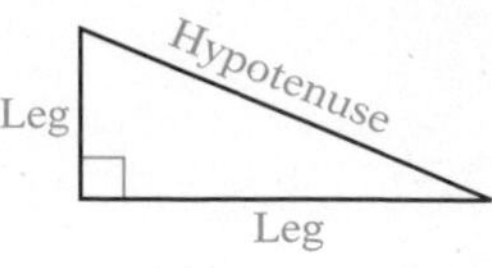

Pythagoras, a Greek mathematician, is credited with the discovery that the square of the hypotenuse of a right triangle is equal to the sum of the squares of the two legs. This is called the Pythagorean Theorem.

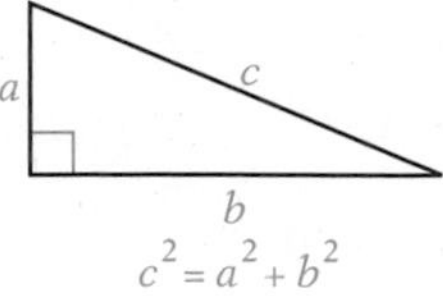

$c^2 = a^2 + b^2$

PYTHAGOREAN THEOREM

The square of the hypotenuse c of a right triangle is equal to the sum of the squares of the two legs, a and b.

$$c^2 = a^2 + b^2$$

EXAMPLE 3 A ladder 20 ft long is leaning against a building. How high on the building will the ladder reach when the bottom of the ladder is 8 ft from the building? Round to the nearest tenth.

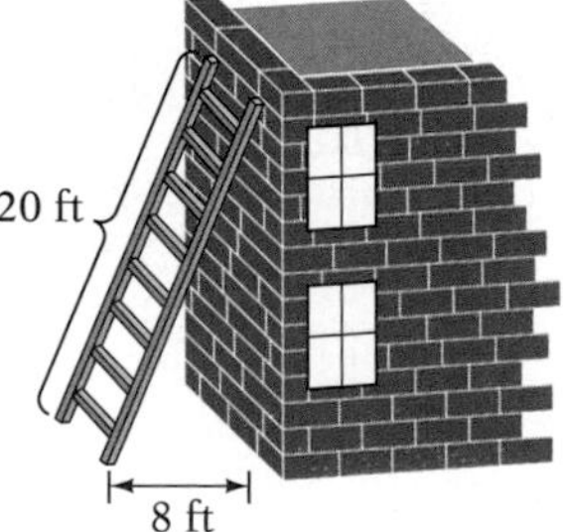

Strategy To find the distance, use the Pythagorean Theorem. The hypotenuse is the length of the ladder. One leg is the distance from the bottom of the ladder to the base of the building. The distance along the building from the ground to the top of the ladder is the unknown leg.

Solution

$$\begin{aligned} c^2 &= a^2 + b^2 \\ 20^2 &= 8^2 + b^2 \\ 400 &= 64 + b^2 \\ 336 &= b^2 \\ \sqrt{336} &= \sqrt{b^2} \\ \sqrt{336} &= b \\ 18.3 &\approx b \end{aligned}$$

The distance is 18.3 ft.

Problem 3 Find the diagonal of a rectangle that is 6 cm in length and 3 cm in width. Round to the nearest tenth.

Solution See page S25.

Try Exercise 67, page 413.

7.4 Exercises

CONCEPT CHECK

Determine whether the statement is sometimes true or always true.

1. If $a^2 = b^2$, then $a = b$.

2. If $a^3 = b^3$, then $a = b$.

3. If $a^4 = b^4$, then $a = b$.

4. If $a^5 = b^5$, then $a = b$.

State whether 2 is a solution of the given equation.

5. $\sqrt{x+2} - \sqrt{4x+1} = 1$

6. $\sqrt{3x+3} - \sqrt{x-1} = 2$

1 Solve equations containing one or more radical expressions (See pages 407–409.)

GETTING READY

7. Which of the equations $\sqrt{2}x + 1 = 7$ and $\sqrt{2x+1} = 7$ is a radical equation?

8. Solve: $\sqrt[3]{x-3} = -4$

$$\sqrt[3]{x-3} = -4$$

$$(\sqrt[3]{x-3})^{?} = (-4)^{?}$$ • Cube each side of the equation.

$$\underline{\ ?\ } = \underline{\ ?\ }$$ • Simplify.

$$x = \underline{\ ?\ }$$ • Add ___?___ to each side of the equation.

9. What is the first step when solving $\sqrt{x} + 3 = 9$?

10. Why is it necessary to check the proposed solutions of a radical equation?

Solve.

11. $\sqrt{3x} = 12$

12. $\sqrt{5x} = 10$

13. $\sqrt[3]{4x} = -2$

14. $\sqrt[3]{6x} = -3$

15. $\sqrt{2x} = -4$

16. $\sqrt{5x} = -5$

17. $\sqrt{3x-2} = 5$

18. $\sqrt{5x-4} = 9$

19. $\sqrt{3-2x} = 7$

20. $\sqrt{9-4x} = 4$

21. $7 = \sqrt{1-3x}$

22. $6 = \sqrt{8-7x}$

23. $\sqrt[3]{4x-1} = 2$

24. $\sqrt[3]{5x+2} = 3$

25. $\sqrt[3]{1-2x} = -3$

26. $\sqrt[3]{3-2x} = -2$

27. $\sqrt{4x-3} - 5 = 0$

28. $\sqrt{x-2} = 4$

29. $\sqrt[3]{x-3} + 5 = 0$

30. $\sqrt[3]{x-2} = 3$

31. $\sqrt[4]{2x-9} = 3$

32. $\sqrt[4]{4x+1} = 2$

33. $\sqrt{3x-5} - 5 = 3$

34. $\sqrt{2x-3} - 2 = 1$

35. $\sqrt[3]{x-4} + 7 = 5$

36. $\sqrt[3]{2x-3} + 5 = 2$

37. $\sqrt{2x+4} = 3 - \sqrt{2x}$

38. $\sqrt{x+1} = 2 - \sqrt{x}$

39. $\sqrt{x+2} - \sqrt{x-1} = 1$

40. $\sqrt{3x+1} + \sqrt{3x-6} = 7$

41. $\sqrt{2x-1} - \sqrt{2x+6} = 1$

42. $\sqrt{x-1} - \sqrt{x+2} = 3$

43. $\sqrt{4x+5} - \sqrt{4x-11} = 2$

44. $\sqrt{3-2x} + \sqrt{10-2x} = 7$

45. $\sqrt{x^2-4x-1} + 3 = x$

46. $\sqrt{x^2+3x-2} - x = 1$

47. $\sqrt{x^2-2x+1} = 3$

48. $\sqrt{x^2-3x-1} = 3$

49. $\sqrt{4x+1} - \sqrt{2x+4} = 1$

50. $\sqrt{2x+5} - \sqrt{3x-2} = 1$

51. $\sqrt{5x+4} - \sqrt{3x+1} = 1$

52. $\sqrt{5x-1} - \sqrt{3x-2} = 1$

53. $3\sqrt{x-2} + 2 = x$

54. $4\sqrt{x+1} - x = 1$

55. $x + 2\sqrt{x+1} = 7$

56. Without actually solving the equations, identify which equation has no solution.
(i) $\sqrt{4x - 8} = -5$ (ii) $\sqrt[3]{4x - 8} = -5$
(iii) $\sqrt{4x} - 8 = -5$ (iv) $-\sqrt{4x - 8} = -5$

57. How many times will the Property of Raising Each Side of an Equation to a Power be used to solve the equation? Do not solve.
a. $\sqrt{x - 8} = \sqrt{2x - 1}$ **b.** $\sqrt{x - 8} = \sqrt{2x} - 1$

2 Application problems (See page 410.)

GETTING READY

58. The equation $v = 8\sqrt{d}$ gives the speed v, in feet per second, of a dropped object after the object has fallen d feet. A stone is dropped from a bridge. To find the distance the stone has fallen when it reaches a speed of 56 ft/s, replace __?__ with 56 and solve for __?__.

59. A 15-foot ladder leans against the side of a building with its bottom d feet from the building. The ladder reaches a height of h feet. Which of the following distances is *not* a possible value for h?
(i) 2 ft (ii) 8 ft (iii) 14 ft (iv) 20 ft

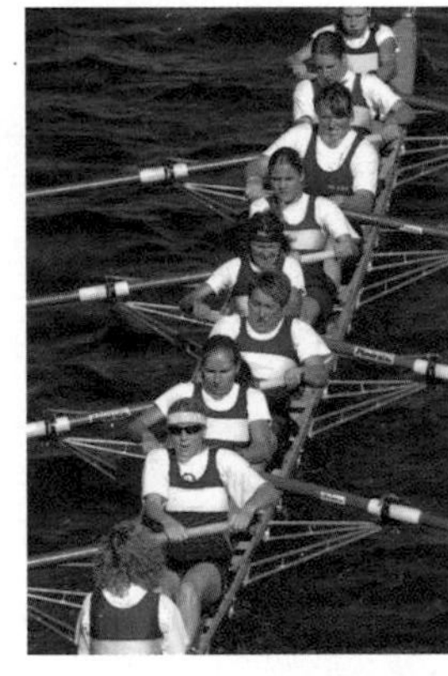
Vladimir Wrangel/Shutterstock.com

60. Sports The equation $s = 16.97\sqrt[9]{n}$ can be used to predict the maximum speed s (in feet per second) of n rowers on a scull. Find, to the nearest whole number, how many rowers are needed to travel at 20 ft/s. Does doubling the number of rowers double the maximum speed of the scull?

61. Health Science The number of calories an animal uses per day (called the metabolic rate of the animal) can be approximated by $M = 126.4\sqrt[4]{W^3}$, where M is the metabolic rate and W is the weight of the animal in pounds. Find, to the nearest hundred pounds, the weight of an elephant whose metabolic rate is 60,000 calories per day.

62. Astronomy The time T (in days) that it takes for a planet to revolve around the sun can be approximated by the equation $T = 0.407\sqrt{d^3}$, where d is the mean distance of the planet from the sun in millions of miles. It takes Venus approximately 226 days to complete one revolution of the sun. To the nearest million miles, what is the mean distance of Venus from the sun?

63. Astronomy The time T (in days) that it takes for a moon of Saturn to revolve around Saturn can be approximated by the equation $T = 0.373\sqrt{d^3}$, where d is the mean distance of the moon from Saturn in units of 100,000 km. It takes the moon Tethys approximately 1.89 days to complete one revolution of Saturn. To the nearest 1000 km, what is the mean distance of Tethys from Saturn?

64. Construction The maximum velocity of a roller coaster depends on the vertical drop from the top of the highest hill to the bottom of that hill. The formula $v = 8\sqrt{h}$ gives the relationship between maximum velocity v (in feet per second) and height h (in feet). The maximum velocity of the Magnum XL-200 roller coaster in Sandusky, Ohio, is approximately 114 ft/s. How tall, to the nearest foot, is the highest hill for this roller coaster?

65. Meteorology The sustained wind velocity v (in meters per second) in a hurricane is given by $v = 6.3\sqrt{1013 - p}$, where p is the air pressure in millibars (mb). Read the article at the right. What was the air pressure when Julia's winds were blowing at the velocity given in the article? What happens to wind speed in a hurricane as air pressure decreases?

In the News

A Special Category of Hurricane

It hasn't happened since 1926—two Category 4 hurricanes existing at the same time over the Atlantic. Now, on the same day, Hurricanes Igor and Julia both have been rated Category 4, with Julia blowing the fiercer winds, at speeds of up to 69.3 m/s (155 mph).

Source: www.examiner.com/weather-in-baltimore

66. Construction A 12-foot ladder is leaning against a building. How high on the building will the ladder reach when the bottom of the ladder is 4 ft from the building? Round to the nearest tenth.

67. **Construction** A 26-foot ladder is leaning against a building. How far is the bottom of the ladder from the wall when the ladder reaches a height of 24 ft on the building?

68. **Uniform Motion** Two joggers start from the same point and travel in the directions shown in the figure at the right. One jogger is traveling 8 ft/s, and the second jogger is traveling 9 ft/s. What is the distance between the joggers after 1 min? Round to the nearest tenth.

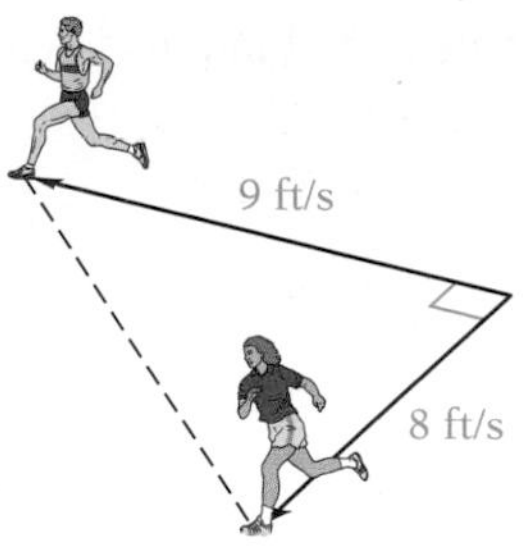

69. **Uniform Motion** Two joggers leave the same point at the same time, one jogging east at 3 m/s and the other jogging south at 3.5 m/s. What is the distance between the joggers after 3 min? Round to the nearest hundredth.

70. **Uniform Motion** Solve the following problem, which appears in a math text written around A.D. 1200. "Two birds start flying from the tops of two towers 50 ft apart at the same time and at the same rate. One tower is 30 ft high, and the other tower is 40 ft high. The birds reach a grass seed on the ground at exactly the same time. How far is the grass seed from the 40-foot tower?"

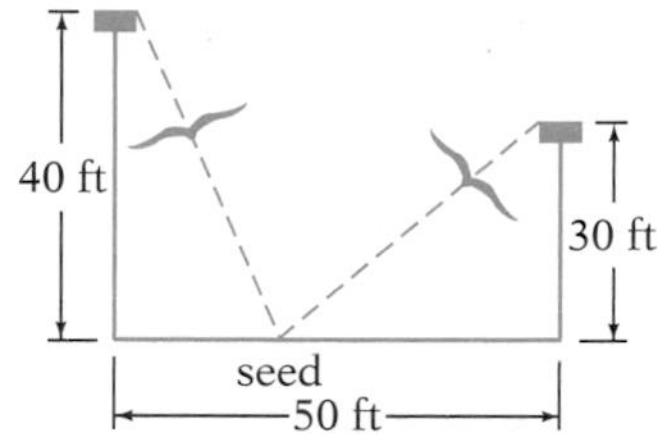

71. **Oceanography** How far above the water would a submarine periscope have to be to locate a ship 3.5 mi away? The equation for the distance in miles that the lookout can see is $d = \sqrt{1.5h}$, where h is the height in feet above the surface of the water. Round to the nearest hundredth.

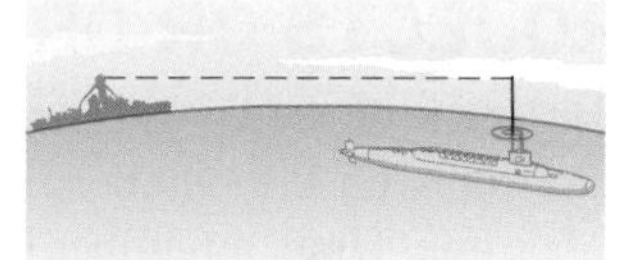

72. **Oceanography** How far above the water would a submarine periscope have to be to locate a ship 3.2 mi away? The equation for the distance in miles that the lookout can see is $d = \sqrt{1.5h}$, where h is the height in feet above the surface of the water. Round to the nearest hundredth.

73. **Physics** An object is dropped from a high building. Find the distance the object has fallen when its speed reaches 120 ft/s. Use the equation $v = 8\sqrt{d}$, where v is the speed of the object and d is the distance.

74. **Physics** An object is dropped from a bridge. Find the distance the object has fallen when its speed reaches 80 ft/s. Use the equation $v = 8\sqrt{d}$, where v is the speed of the object and d is the distance.

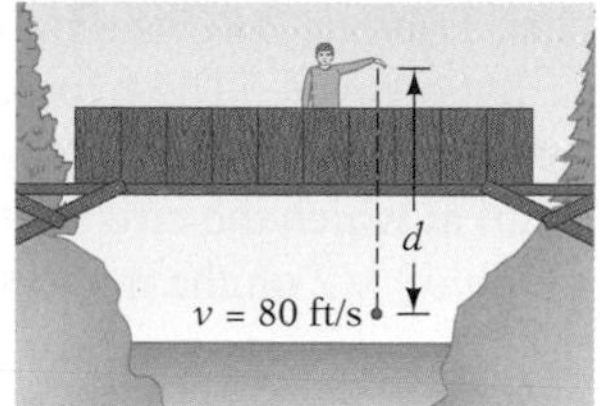

75. **Automotive Technology** Find the distance required for a car to reach a velocity of 48 ft/s when the acceleration is 12 ft/s^2. Use the equation $v = \sqrt{2as}$, where v is the velocity, a is the acceleration, and s is the distance.

76. **Automotive Technology** Find the distance required for a car to reach a velocity of 60 m/s when the acceleration is 10 m/s^2. Use the equation $v = \sqrt{2as}$, where v is the velocity, a is the acceleration, and s is the distance.

77. **Satellites** Read the article at the right. At what height above Earth's surface is the A-Train in orbit? Use the equation $v = \sqrt{\dfrac{4 \times 10^{14}}{h + 6.4 \times 10^6}}$, where v is the speed of the satellites in meters per second and h is the height above Earth's surface in meters. Round to the nearest hundred thousand.

In the News

OCO to Get a Second Shot

A congressional committee has specified that funds in NASA's budget go toward a second attempt to add an Orbiting Carbon Observatory (OCO) to the "A-Train" of satellites that orbit Earth at about 7500 m/s, providing scientists with a wealth of data that can be used to study climate change.

Source: Jet Propulsion Laboratory

78. **Clocks** Find the length of a pendulum that makes one swing in 3 s. The equation for the time of one swing of a pendulum is $T = 2\pi\sqrt{\dfrac{L}{32}}$, where T is the time in seconds and L is the length in feet. Round to the nearest hundredth.

APPLYING CONCEPTS

79. **Geometry** A box has a base that measures 4 in. by 6 in. The height of the box is 3 in. Find the greatest distance between two corners. Round to the nearest hundredth.

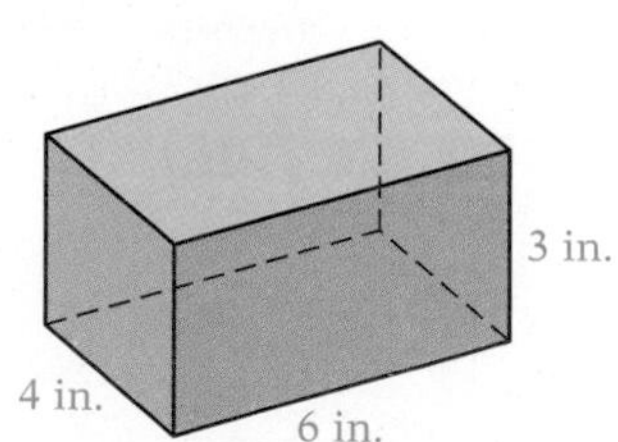

80. Geometry Find the length of the side labeled x.

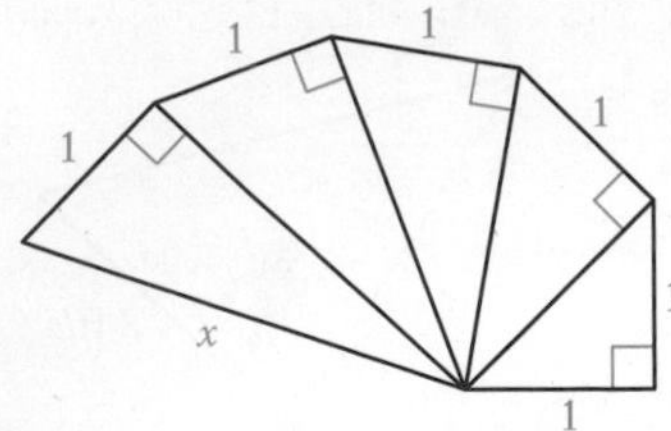

81. If a and b are nonnegative real numbers, is the equation always true, sometimes true, or never true?

$$\sqrt{a^2 + b^2} = a + b$$

Write a paragraph that supports your answer.

82. If a and b are both positive real numbers, and $a > b$, is a^b less than, equal to, or greater than b^a? Does your answer depend on a and b?

PROJECTS OR GROUP ACTIVITIES

Note: You will need a ruler and a compass for this activity. In this activity, you will graph square roots on the number line. We will begin by explaining how to graph $\sqrt{2}$ on the number line. Draw a number line from -1 to 2. Leave 1 inch between adjacent numbers. Starting at 0, construct triangle ABC. Leg $\overline{AC}$, from 0 to 1 on the number line, is 1 unit long. Leg $\overline{BC}$ is perpendicular to $\overline{AC}$ and equal in length to $\overline{AC}$. Draw $\overline{AB}$ from point A to point B. Triangle ABC is a right triangle. Use the Pythagorean Theorem to find the length of the hypotenuse. (The hypotenuse is $\sqrt{2}$ units.) Place the point of your compass at A (0 on the number line) and the compass pencil at point B. Draw a circle with radius AB. Label the point at which the circle intersects the number line as point D. Draw a dot at D. This is the graph of $\sqrt{2}$ on the number line.

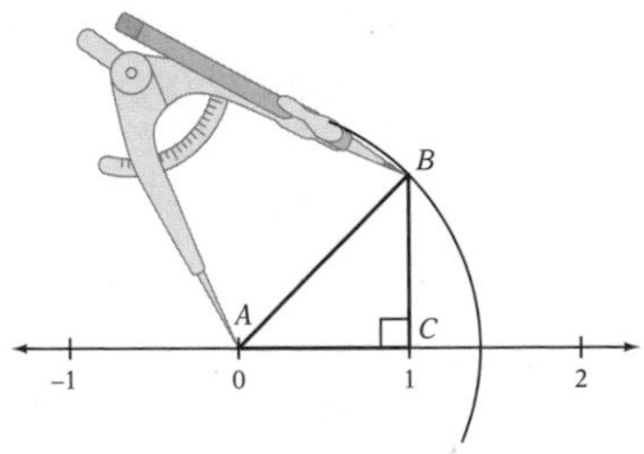

83. Use the procedure described above to graph $\sqrt{5}$ and $\sqrt{8}$ on the number line.

7.5 Complex Numbers

OBJECTIVE 1

Simplify complex numbers

The radical expression $\sqrt{-4}$ is not a real number because there is no real number whose square is -4. However, the solution of an algebraic equation is sometimes the square root of a negative number.

During the late 17th century, a new number, called an *imaginary number,* was defined so that a negative number would have a square root. The letter i was chosen to represent the number whose square is -1.

$$i^2 = -1$$

An imaginary number is defined in terms of i.

DEFINITION OF AN IMAGINARY NUMBER

If a is a positive real number, then the principal square root of $-a$ is the **imaginary number** $i\sqrt{a}$. This can be written

$$\sqrt{-a} = i\sqrt{a}$$

When $a = 1$, we have $\sqrt{-1} = i$.

EXAMPLES

1. $\sqrt{-16} = i\sqrt{16} = 4i$ **2.** $\sqrt{-21} = i\sqrt{21}$

It is customary to write i in front of a radical to avoid confusing $\sqrt{a}\,i$ with $\sqrt{ai}$.

EXAMPLE 1 Simplify: $3\sqrt{-20}$

Solution $3\sqrt{-20} = 3i\sqrt{20} = 3i(2\sqrt{5}) = 6i\sqrt{5}$

Problem 1 Simplify: $-5\sqrt{-80}$

Solution See page S25.

Try Exercise 23, page 421.

The set containing the real numbers and the imaginary numbers is called the set of complex numbers.

DEFINITION OF A COMPLEX NUMBER

A **complex number** is a number of the form $\boldsymbol{a + bi}$, where a and b are real numbers and $i = \sqrt{-1}$. The number a is the **real part** of the complex number, and b is the **imaginary part** of the complex number. A complex number written as $a + bi$ is in **standard form.**

EXAMPLES

1. $3 + 4i$ Real part is 3; imaginary part is 4.

2. $5 - 2i\sqrt{7}$ Real part is 5; imaginary part is $-2\sqrt{7}$.

3. 5 Real part is 5; imaginary part is 0, because $5 = 5 + 0i$.

4. $-4i$ Real part is 0; imaginary part is -4, because $-4i = 0 - 4i$.

5. $\dfrac{2 + 3i}{5} = \dfrac{2}{5} + \dfrac{3}{5}i$ Real part is $\dfrac{2}{5}$; imaginary part is $\dfrac{3}{5}$.

Complex numbers $a + bi$
- Real numbers $a + 0i$ — A **real number** is a complex number in which $b = 0$.
- Imaginary numbers $0 + bi$ — An **imaginary number** is a complex number in which $a = 0$.

How It's Used

Complex numbers have applications in electrical engineering. The *impedance* of an alternating current circuit is a complex number that measures the circuit's resistance to the flow of electricity.

EXAMPLE 2 Write $\frac{4+\sqrt{-20}}{6}$ in standard form.

Solution

$$\frac{4+\sqrt{-20}}{6} = \frac{4+i\sqrt{20}}{6}$$ • Write $\sqrt{-20}$ as $i\sqrt{20}$.

$$= \frac{4+2i\sqrt{5}}{6}$$ • Simplify the radical.

$$= \frac{\overset{1}{\cancel{2}}(2+i\sqrt{5})}{\underset{1}{\cancel{2}}\cdot 3} = \frac{2+i\sqrt{5}}{3}$$ • Factor and simplify.

$$= \frac{2}{3} + \frac{\sqrt{5}}{3}i$$ • Write in standard form.

Problem 2 Write $\frac{12-\sqrt{-72}}{3}$ in standard form.

Solution See page S25.

Try Exercise 33, page 421.

OBJECTIVE 2 Add and subtract complex numbers

ADDITION AND SUBTRACTION OF COMPLEX NUMBERS

To add two complex numbers, add the real parts and add the imaginary parts. To subtract two complex numbers, subtract the real parts and subtract the imaginary parts.

$$(a+bi)+(c+di) = (a+c)+(b+d)i$$
$$(a+bi)-(c+di) = (a-c)+(b-d)i$$

EXAMPLE 3 Add or subtract.
A. $(3+2i)+(6-5i)$ **B.** $(-2+6i)-(4-3i)$

Solution

A. $(3+2i)+(6-5i)$
$= (3+6)+(2-5)i$ • Add the real parts and add the imaginary parts.
$= 9-3i$

B. $(-2+6i)-(4-3i)$
$= (-2-4)+[6-(-3)]i$ • Subtract the real parts and subtract the imaginary parts.
$= -6+9i$

Problem 3 Add or subtract.
A. $(5-7i)+(2-i)$ **B.** $(-4+2i)-(6-8i)$

Solution See page S25.

Try Exercise 45, page 421.

OBJECTIVE 3 Multiply complex numbers

When we multiply complex numbers, the term i^2 is frequently a part of the product. Recall that $i^2 = -1$.

Focus on multiplying imaginary numbers

A. Multiply: $(-3i)(5i)$

Multiply the imaginary numbers. $(-3i)(5i) = -15i^2$

Replace i^2 by -1. Then simplify. $= -15(-1) = 15$

B. Multiply: $\sqrt{-6} \cdot \sqrt{-24}$

Write the imaginary numbers in terms of i. $\sqrt{-6} \cdot \sqrt{-24} = i\sqrt{6} \cdot i\sqrt{24}$

Multiply the imaginary numbers. $= i^2\sqrt{144}$

Replace i^2 by -1. Then simplify. $= (-1)(12) = -12$

Take Note

Part (B) illustrates an important point. When working with the square root of a negative number, always rewrite the expression in terms of *i* before continuing.

EXAMPLE 4 Multiply.

A. $(3 - 4i)(2 + 5i)$ **B.** $\left(\frac{9}{10} + \frac{3}{10}i\right)\left(1 - \frac{1}{3}i\right)$

C. $(4 + 5i)(4 - 5i)$ **D.** $(6 + i)^2$

Solution **A.** $(3 - 4i)(2 + 5i)$

$= 6 + 15i - 8i - 20i^2$ • **Use the FOIL method.**

$= 6 + 7i - 20i^2$ • **Combine like terms.**

$= 6 + 7i - 20(-1)$ • **Replace i^2 by -1.**

$= 26 + 7i$ • **Write the answer in the form $a + bi$.**

B. $\left(\frac{9}{10} + \frac{3}{10}i\right)\left(1 - \frac{1}{3}i\right)$

$= \frac{9}{10} - \frac{3}{10}i + \frac{3}{10}i - \frac{1}{10}i^2$ • **Use the FOIL method.**

$= \frac{9}{10} - \frac{1}{10}i^2$ • **Combine like terms.**

$= \frac{9}{10} - \frac{1}{10}(-1)$ • **Replace i^2 by -1.**

$= \frac{9}{10} + \frac{1}{10} = 1$ • **Simplify.**

C. $(4 + 5i)(4 - 5i)$

$= 16 - 20i + 20i - 25i^2$ • **Use the FOIL method.**

$= 16 - 25i^2$

$= 16 - 25(-1)$

$= 16 + 25 = 41$

D. $(6 + i)^2 = 36 + 12i + i^2$ • **$(6 + i)^2 = (6 + i)(6 + i)$**

$= 36 + 12i + (-1)$

$= 35 + 12i$

Problem 4 Multiply.

A. $(4 - 3i)(2 - i)$ **B.** $(3 - i)\left(\frac{3}{10} + \frac{1}{10}i\right)$

C. $(3 + 6i)(3 - 6i)$ **D.** $(1 + 5i)^2$

Solution See page S25.

Try Exercise 79, page 422.

OBJECTIVE  Divide complex numbers

A fraction containing one or more complex numbers is in simplest form when no imaginary number remains in the denominator.

Focus on dividing by an imaginary number

A. Divide: $\frac{7}{i}$

Multiply the expression by 1 in the form $\frac{i}{i}$. $\quad \frac{7}{i} = \frac{7}{i} \cdot \frac{i}{i} = \frac{7i}{i^2}$

Replace i^2 by -1. Then simplify. $\quad = \frac{7i}{-1} = -7i$

B. Divide: $\frac{4 - 5i}{2i}$

Multiply the expression by 1 in the form $\frac{i}{i}$. $\quad \frac{4 - 5i}{2i} = \frac{4 - 5i}{2i} \cdot \frac{i}{i}$

$= \frac{4i - 5i^2}{2i^2}$

Replace i^2 by -1. $\quad = \frac{4i - 5(-1)}{2(-1)}$

Simplify. $\quad = \frac{5 + 4i}{-2}$

Write the answer in the form $a + bi$. $\quad = -\frac{5}{2} - 2i$

EXAMPLE 5 Divide. **A.** $\frac{5}{4i}$ **B.** $\frac{2 + 7i}{-14i}$

Solution **A.** $\frac{5}{4i} = \frac{5}{4i} \cdot \frac{i}{i} = \frac{5i}{4i^2}$

- Multiply the expression by 1 in the form $\frac{i}{i}$.

$= \frac{5i}{4(-1)} = -\frac{5}{4}i$

- Replace i^2 by -1. Then simplify.

B. $\frac{2 + 7i}{-14i} = \frac{2 + 7i}{-14i} \cdot \frac{i}{i}$

- Multiply the expression by 1 in the form $\frac{i}{i}$.

$= \frac{2i + 7i^2}{-14i^2}$

$= \frac{2i + 7(-1)}{-14(-1)} = \frac{-7 + 2i}{14}$

- Replace i^2 by -1. Then simplify.

$= -\frac{1}{2} + \frac{1}{7}i$

- Write the answer in the form $a + bi$.

Problem 5 Divide. **A.** $\frac{-3}{6i}$ **B.** $\frac{2 - 3i}{4i}$

Solution See page S25.

Try Exercise 97, page 423.

CONJUGATE OF A COMPLEX NUMBER

The **conjugate** of $a + bi$ is $a - bi$, and the conjugate of $a - bi$ is $a + bi$. The product of the conjugates is $(a + bi)(a - bi) = a^2 + b^2$.

EXAMPLES

1. The conjugate of $2 + 5i$ is $2 - 5i$. The product of the conjugates is $(2 + 5i)(2 - 5i) = 2^2 + 5^2 = 29$.
2. The conjugate of $3 - 4i$ is $3 + 4i$. The product of the conjugates is $(3 - 4i)(3 + 4i) = 3^2 + 4^2 = 25$.
3. The conjugate of $-5 + i$ is $-5 - i$. The product of the conjugates is $(-5 + i)(-5 - i) = (-5)^2 + 1^2 = 26$.

The conjugate of a complex number is used to divide complex numbers when the denominator is of the form $a + bi$.

Focus on dividing two complex numbers

A. Divide: $\dfrac{3}{2 - 3i}$

Multiply the numerator and denominator by $2 + 3i$, the conjugate of the denominator.

$$\frac{3}{2 - 3i} = \frac{3}{2 - 3i} \cdot \frac{2 + 3i}{2 + 3i}$$

$$= \frac{3(2 + 3i)}{2^2 + 3^2}$$

Write the answer in the form $a + bi$.

$$= \frac{6 + 9i}{13} = \frac{6}{13} + \frac{9}{13}i$$

B. Divide: $\dfrac{5 - 2i}{1 + 2i}$

Multiply the numerator and denominator by $1 - 2i$, the conjugate of the denominator.

$$\frac{5 - 2i}{1 + 2i} = \frac{5 - 2i}{1 + 2i} \cdot \frac{1 - 2i}{1 - 2i}$$

$$= \frac{5 - 10i - 2i + 4i^2}{1^2 + 2^2}$$

Replace i^2 by -1. Then simplify.

$$= \frac{5 - 12i + 4(-1)}{1 + 4}$$

$$= \frac{5 - 12i - 4}{5}$$

Write the answer in the form of $a + bi$.

$$= \frac{1 - 12i}{5} = \frac{1}{5} - \frac{12}{5}i$$

EXAMPLE 6 Divide. **A.** $\dfrac{13i}{3 + 2i}$ **B.** $\dfrac{5 + 3i}{4 + 2i}$

Solution **A.**

$$\frac{13i}{3 + 2i} = \frac{13i}{3 + 2i} \cdot \frac{3 - 2i}{3 - 2i}$$

- **Multiply the numerator and denominator by $3 - 2i$, the conjugate of the denominator.**

$$= \frac{13i(3 - 2i)}{3^2 + 2^2} = \frac{39i - 26i^2}{9 + 4}$$

$$= \frac{39i - 26(-1)}{13} = \frac{26 + 39i}{13}$$

- **Replace i^2 by -1. Then simplify.**

$$= 2 + 3i$$

- **Write the answer in the form $a + bi$.**

$$\textbf{B.}\quad \frac{5+3i}{4+2i} = \frac{5+3i}{4+2i}\cdot\frac{4-2i}{4-2i}$$

- **Multiply the numerator and denominator by $4-2i$, the conjugate of the denominator.**

$$= \frac{20-10i+12i-6i^2}{4^2+2^2}$$

$$= \frac{20+2i-6(-1)}{16+4}$$

- **Replace i^2 by -1. Then simplify.**

$$= \frac{20+2i+6}{20}$$

$$= \frac{26+2i}{20} = \frac{13}{10} + \frac{1}{10}i$$

- **Write the answer in the form $a + bi$.**

Problem 6 Divide. **A.** $\dfrac{5}{1+3i}$ **B.** $\dfrac{2+5i}{3-2i}$

Solution See page S25.

Try Exercise 107, page 423.

7.5 Exercises

CONCEPT CHECK

1. What is an imaginary number? What is a complex number?
2. Do all real numbers belong to the set of complex numbers? Do all imaginary numbers belong to the set of real numbers?

For each complex number, identify the real part and the imaginary part.

3. $5 - 7i$ **4.** $6i$ **5.** 9 **6.** $\dfrac{1+3i}{2}$

1 Simplify complex numbers (See pages 414–416.)

GETTING READY

7. $\sqrt{-1} = i$, so $i^2 =$ __?__.

8. a. The real part of the complex number $7 - 9i$ is __?__.
   b. The imaginary part of the complex number $7 - 9i$ is __?__.

9. Simplify: $\sqrt{-54}$
   $\sqrt{-54} =$ __?__ $\sqrt{54} = i\sqrt{(\_?\_)(\_?\_)} =$ __?__ $\sqrt{6}$

10. Simplify: $\sqrt{-100x^{16}}$
   $\sqrt{-100x^{16}} =$ __?__ $\sqrt{100x^{16}} =$ __?__ i

11. Is $10 - 3i$ equivalent to $10 + \sqrt{-9}$ or to $10 - \sqrt{-9}$?

12. Which of the following are equivalent to $\sqrt{25} + \sqrt{-25}$?
   (i) $5 - 5i$ (ii) $5 + 5i$ (iii) $\sqrt{25} + i\sqrt{25}$ (iv) $\sqrt{25} - i\sqrt{25}$

Simplify.

13. $\sqrt{-4}$
14. $\sqrt{-64}$
15. $5\sqrt{-100}$
16. $6\sqrt{-64}$
17. $-4\sqrt{-9}$
18. $-5\sqrt{-81}$
19. $\sqrt{-98}$
20. $\sqrt{-72}$
21. $\sqrt{-27}$
22. $\sqrt{-75}$
23. $-4\sqrt{-50}$
24. $-3\sqrt{-20}$
25. $-4\sqrt{-98}$
26. $9\sqrt{-90}$
27. $5 + \sqrt{-49}$
28. $-6 - \sqrt{-25}$
29. $-7 - \sqrt{-80}$
30. $10 + \sqrt{-63}$
31. $11 - 3\sqrt{-96}$
32. $-15 + 4\sqrt{-40}$
33. $\dfrac{14 - 3\sqrt{-49}}{7}$
34. $\dfrac{-15 + 3\sqrt{-25}}{5}$
35. $\dfrac{-6 - 5\sqrt{-4}}{8}$
36. $\dfrac{15 + 2\sqrt{-81}}{9}$
37. $\dfrac{6 + 4\sqrt{-45}}{3}$
38. $\dfrac{-20 + 7\sqrt{-125}}{10}$
39. $\dfrac{12 - 5\sqrt{-27}}{18}$

2 Add and subtract complex numbers (See page 416.)

GETTING READY

40. Add: $(7 - 9i) + (-3 + 4i)$

$(7 - 9i) + (-3 + 4i)$
$= (7 + \underline{\quad ? \quad}) + (-9 + \underline{\quad ? \quad})i$ • Add the real parts. Add the imaginary parts.
$= \underline{\quad ? \quad}$ • Simplify.

41. Subtract: $12 - (6 - 8i)$

$12 - (6 - 8i) = (12 - \underline{\quad ? \quad}) - (\underline{\quad ? \quad})$ • Subtract the real parts. Subtract the imaginary part.
$= \underline{\quad ? \quad} + \underline{\quad ? \quad}$ • Simplify.

Add or subtract.

42. $(2 + 4i) + (6 - 5i)$
43. $(6 - 9i) + (4 + 2i)$
44. $(-2 - 4i) - (6 - 8i)$
45. $(3 - 5i) - (8 - 2i)$
46. $(-5 + 4i) + (-9 + 3i)$
47. $(-7 - 10i) + (4 - 6i)$
48. $(6 - 2i) - (8 - 3i)$
49. $(-7 + 5i) - (7 - 2i)$
50. $(3 - 6i) - (3 + 5i)$
51. $(-2 + 7i) + (5 - 7i)$
52. $(5 - 3i) + 2i$
53. $(6 - 8i) + 4i$
54. $(7 + 2i) + (-7 - 2i)$
55. $(8 - 3i) + (-8 + 3i)$
56. $(9 + 4i) + 6$
57. $(4 + 6i) - 7$

For Exercises 58 to 61, use the complex numbers $m = a + bi$ and $n = c - di$, where a, b, c, and d are all positive real numbers.

58. True or false? If $a = c$ and $b = d$, then $m - n = 2bi$.

59. True or false? If $m - n$ is a real number, then $b + d = 0$.

60. Suppose $a > c$ and $b > d$. Is the imaginary part of $m + n$ positive or negative?

61. Suppose $a < c$ and $b < d$. Is the real part of $m - n$ positive or negative?

3 Multiply complex numbers (See pages 416–417.)

GETTING READY

62. Multiply: $-7i(5 - 8i)$

$-7i(5 - 8i) = (\underline{\ ?\ })(5) - (\underline{\ ?\ })(8i)$ • Use the Distributive Property.

$= \underline{\ ?\ } + \underline{\ ?\ }\, i^2$ • Simplify.

$= -35i + 56(\underline{\ ?\ })$ • $i^2 = \underline{\ ?\ }$

$= \underline{\ ?\ }$ • Write the answer in $a + bi$ form.

63. Multiply: $(1 + 4i)(2 - 3i)$

$(1 + 4i)(2 - 3i) = 2 - 3i + \underline{\ ?\ } - \underline{\ ?\ }$ • Use FOIL.

$= 2 - 3i + 8i - 12(\underline{\ ?\ })$ • $i^2 = \underline{\ ?\ }$

$= \underline{\ ?\ }$ • Combine like terms.

Multiply.

64. $(7i)(-9i)$

65. $(-6i)(-4i)$

66. $\sqrt{-2}\,\sqrt{-8}$

67. $\sqrt{-5}\,\sqrt{-45}$

68. $\sqrt{-3}\,\sqrt{-6}$

69. $\sqrt{-5}\,\sqrt{-10}$

70. $2i(6 + 2i)$

71. $-3i(4 - 5i)$

72. $(3 - 6i)(-7 - 9i)$

73. $(-4 + 4i)(-7 - 4i)$

74. $(3 + 3i)(-2 + 3i)$

75. $(1 - 3i)(9 + 6i)$

76. $(5 - 2i)(3 + i)$

77. $(2 - 4i)(2 - i)$

78. $(6 + 5i)(3 + 2i)$

79. $(4 - 7i)(2 + 3i)$

80. $(1 - i)\left(\frac{1}{2} + \frac{1}{2}i\right)$

81. $\left(\frac{4}{5} - \frac{2}{5}i\right)\left(1 + \frac{1}{2}i\right)$

82. $\left(\frac{6}{5} + \frac{3}{5}i\right)\left(\frac{2}{3} - \frac{1}{3}i\right)$

83. $(2 - i)\left(\frac{2}{5} + \frac{1}{5}i\right)$

84. $(-1 - 3i)(-3 + 9i)$

85. $(2 + 3i)(-4 + 6i)$

86. $(5 + i)(1 + 5i)$

87. $(4 - i)(-1 + 4i)$

88. $(6 + i)^2$

89. $(4 - 3i)^2$

90. $(5 - 2i)^2$

91. $(-1 + i)^2$

92. True or false? For all real numbers a and b, the product $(a + bi)(a - bi)$ is a positive real number.

93. True or false? The product of two imaginary numbers is always a negative real number.

4 Divide complex numbers (See pages 418–420.)

GETTING READY

94. Simplify: $\dfrac{1-2i}{-5i}$

$\dfrac{1-2i}{-5i} = \dfrac{1-2i}{-5i} \cdot \dfrac{?}{?}$ • To remove i from the denominator, multiply the expression by 1 in the form $\frac{i}{i}$.

$= \dfrac{i - (\ ?\ )}{-5i^2}$ • Multiply the numerators. Multiply the denominators.

$= \dfrac{i - 2(\ ?\ )}{-5(\ ?\ )}$ • Replace i^2 by -1.

$= \dfrac{i + (\ ?\ )}{?}$ • Simplify.

$= \underline{\ \ ?\ \ } + \underline{\ \ ?\ \ }\,i$ • Write in the form $a + bi$.

95. To simplify $\dfrac{2i}{2-3i}$, multiply the numerator and denominator by ___?___.

Divide.

96. $\dfrac{3}{i}$

97. $\dfrac{4}{5i}$

98. $\dfrac{2-3i}{-4i}$

99. $\dfrac{16+5i}{-3i}$

100. $\dfrac{4}{5+i}$

101. $\dfrac{6}{5+2i}$

102. $\dfrac{2}{2-i}$

103. $\dfrac{5}{4-i}$

104. $\dfrac{7i}{4-3i}$

105. $\dfrac{6i}{5-9i}$

106. $\dfrac{8i}{8+2i}$

107. $\dfrac{-2i}{3+5i}$

108. $\dfrac{1-3i}{3+i}$

109. $\dfrac{2+12i}{5+i}$

110. $\dfrac{-18+4i}{-1-2i}$

111. $\dfrac{-27-i}{-1-3i}$

112. $\dfrac{-34+31i}{-2-5i}$

113. $\dfrac{41-3i}{1+5i}$

114. $\dfrac{2-3i}{3+i}$

115. $\dfrac{3+5i}{1-i}$

116. True or false? The quotient of two imaginary numbers is an imaginary number.

117. True or false? The reciprocal of an imaginary number is an imaginary number.

APPLYING CONCEPTS

Note the pattern when successive powers of i are simplified. Use the pattern for Exercises 118 to 126.

$i^1 = i$ $\qquad$ $i^5 = i \cdot i^4 = i(1) = i$

$i^2 = -1$ $\qquad$ $i^6 = i^2 \cdot i^4 = -1$

$i^3 = i^2 \cdot i = -i$ $\qquad$ $i^7 = i^3 \cdot i^4 = -i$

$i^4 = i^2 \cdot i^2 = (-1)(-1) = 1$ $\qquad$ $i^8 = i^4 \cdot i^4 = 1$

118. When the exponent on i is a multiple of 4, the power equals ________.

119. i^6 **120.** i^9 **121.** i^{57} **122.** i^{65}

123. i^{-6} **124.** i^{-34} **125.** i^{-58} **126.** i^{-180}

127. a. Is $3i$ a solution of $2x^2 + 18 = 0$?
b. Is $-3i$ a solution of $2x^2 + 18 = 0$?

128. a. Is $1 + 3i$ a solution of $x^2 - 2x - 10 = 0$?
b. Is $1 - 3i$ a solution of $x^2 - 2x - 10 = 0$?

PROJECTS OR GROUP ACTIVITIES

Complex numbers can be graphed in a plane. The graph is called an **Argand diagram**, as shown below. The horizontal axis is the real axis, and the vertical axis is the imaginary axis.

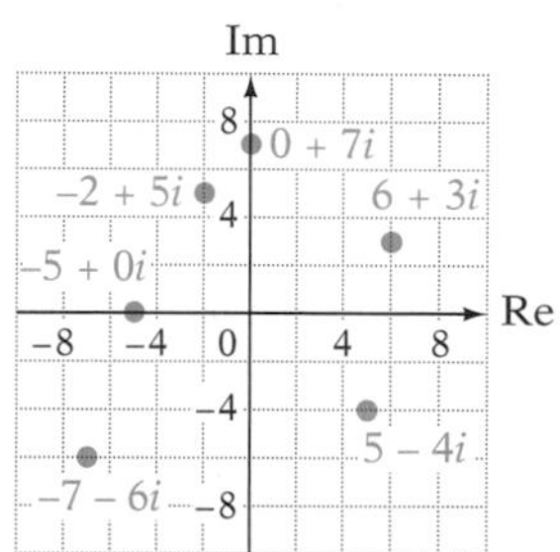

129. Use an Argand diagram like the one above to graph the complex numbers $2 - 5i$, $3 + 5i$, $-4 - 2i$, $-6 + 4i$, $3i$, and 4.

130. The absolute value of the complex number $z = a + bi$ is $|z| = \sqrt{a^2 + b^2}$. Find the absolute value of each of the complex numbers in Exercise 129.

131. Give a geometric interpretation of the absolute value of a complex number.

CHAPTER 7 Summary

Key Words	Objective and Page Reference	Examples
If n is a positive integer, then $a^{\frac{1}{n}}$ is the **nth root of a**, or the number whose nth power is a. The expression $\sqrt[n]{a}$ is another notation for the nth root of a. The symbol $\sqrt{\ }$ is called the **radical**, n is the **index**, and a is the **radicand.**	[7.1.1, p. 380, and 7.1.2, pp. 381–382]	$16^{\frac{1}{4}} = 2$ because $2^4 = 16$. $\sqrt[3]{125} = 5$ because $5^3 = 125$.
If m and n are positive integers, and $a^{\frac{1}{n}}$ is a real number, then the **rational exponent** $\frac{m}{n}$ is defined as follows. $(a^{\frac{m}{n}}) = (a^{\frac{1}{n}})^m$	[7.1.1, p. 380]	$125^{\frac{2}{3}} = (125^{\frac{1}{3}})^2 = 5^2 = 25$
The symbol $\sqrt{\ }$ is used to indicate the positive or **principal square root** of a number.	[7.1.3, p. 383]	$\sqrt{16} = 4$ $-\sqrt{16} = -4$

The **conjugate** of $a + b$ is $a - b$, and the conjugate of $a - b$ is $a + b$. The product of conjugates is $(a + b)(a - b) = a^2 - b^2$.	[7.2.4, p. 394]	The conjugate of $3 + \sqrt{5}$ is $3 - \sqrt{5}$. The product of the conjugates is $(3+\sqrt{5})(3-\sqrt{5}) = 3^2 - (\sqrt{5})^2 = 9 - 5 = 4$
A **radical function** is a function that contains a fractional exponent on a variable or a variable underneath a radical.	[7.3.1, p. 401]	$f(x) = (x - 1)^{\frac{1}{2}}$ $g(x) = \sqrt[3]{x - 2}$
A **radical equation** is an equation that contains a variable expression in the radicand.	[7.4.1, p. 407]	$\sqrt[3]{x + 4} - 1 = 7$ $\sqrt{x} + \sqrt{2x + 1} = 5$
The symbol *i* is used to represent the number whose square is -1.	[7.5.1, p. 415]	$i^2 = -1$
If a is a positive real number, then the principal square root of $-a$ is the **imaginary number** $i\sqrt{a}$. This is written $\sqrt{-a} = i\sqrt{a}$.	[7.5.1, p. 415]	$\sqrt{-37} = i\sqrt{37}$ $\sqrt{-81} = i\sqrt{81} = 9i$ $\sqrt{-1} = i$
A **complex number** is a number of the form $a + bi$, where a and b are real numbers and $i = \sqrt{-1}$. The number a is the **real part** of the complex number; b is the **imaginary part** of the complex number. A complex number written as $a + bi$ is in **standard form.**	[7.5.1, p. 415]	$5 + 2i$ Real part is 5; imaginary part is 2. $6 - 2i$ Real part is 6; imaginary part is -2. 17 Real part is 17; imaginary part is 0. $5i$ Real part is 0; imaginary part is 5.

Essential Rules and Procedures	Objective and Page Reference	Examples
Write an exponential expression as a radical expression, and write a radical expression as an exponential expression $a^{\frac{m}{n}} = \sqrt[n]{a^m}$	[7.1.2, p. 382]	$x^{\frac{2}{3}} = (\sqrt[3]{x})^2$ $\sqrt[4]{3x} = (3x)^{\frac{1}{4}}$
Simplify a radical expression that is a perfect power Divide each exponent by the index of the radical.	[7.1.3, p. 383]	$\sqrt[3]{27x^6} = \sqrt[3]{3^3x^6} = 3x^2$
Product Property of Radicals If n is a positive integer, and $\sqrt[n]{a}$ and $\sqrt[n]{b}$ are real numbers, then $\sqrt[n]{a} \cdot \sqrt[n]{b} = \sqrt[n]{ab}$.	[7.2.1, p. 389]	$\sqrt{7} \cdot \sqrt{5} = \sqrt{7 \cdot 5} = \sqrt{35}$ $\sqrt[4]{9} \cdot \sqrt[4]{7} = \sqrt[4]{9 \cdot 7} = \sqrt[4]{63}$
Simplify a radical expression that is not a perfect power Write the radicand as the product of a perfect nth-power factor and a factor that does not contain a perfect nth power.	[7.2.1, p. 389]	$\sqrt{12x^3y^2} = \sqrt{4x^2y^2(3x)} = 2xy\sqrt{3x}$
Add or subtract radical expressions Use the Distributive Property to add or subtract like radicals as you would like terms.	[7.2.2, p. 389]	$5\sqrt{7} - 8\sqrt{7} = -3\sqrt{7}$

Multiply radical expressions [7.2.3, p. 391]

Use the Product Property of Radicals to multiply radicals with the same index.

Use FOIL to multiply radical expressions of two terms.

$\sqrt[3]{6x} \cdot \sqrt[3]{4x^2} = \sqrt[3]{24x^3} = 2x\sqrt[3]{3}$

$$\begin{aligned}(2 + \sqrt{x})(3 - \sqrt{x}) \\ &= 6 - 2\sqrt{x} + 3\sqrt{x} - x \\ &= 6 + \sqrt{x} - x\end{aligned}$$

Quotient Property of Radicals [7.2.4, p. 392]

If n is a positive integer, and $\sqrt[n]{a}$ and $\sqrt[n]{b}$ are real numbers, then $\dfrac{\sqrt[n]{a}}{\sqrt[n]{b}} = \sqrt[n]{\dfrac{a}{b}}$.

$$\frac{\sqrt{42}}{\sqrt{6}} = \sqrt{\frac{42}{6}} = \sqrt{7}$$

$$\frac{\sqrt[3]{63}}{\sqrt[3]{7}} = \sqrt[3]{\frac{63}{7}} = \sqrt[3]{9}$$

Divide radical expressions [7.2.4, pp. 392–393]

Use the Quotient Property of Radicals to divide radicals with the same index.

Rationalize the denominator by multiplying the numerator and denominator by the conjugate of the denominator. Use $(a + b)(a - b) = a^2 - b^2$.

$$\frac{\sqrt[3]{8x^5}}{\sqrt[3]{2x^3}} = \sqrt[3]{\frac{8x^5}{2x^3}} = \sqrt[3]{4x^2}$$

$$\begin{aligned}\frac{2}{3 + \sqrt{5}} &= \frac{2}{3 + \sqrt{5}} \cdot \frac{3 - \sqrt{5}}{3 - \sqrt{5}} \\ &= \frac{2(3 - \sqrt{5})}{9 - 5} = \frac{6 - 2\sqrt{5}}{4} \\ &= \frac{3 - \sqrt{5}}{2}\end{aligned}$$

Graph of a radical function [7.3.2, pp. 402–403]

The domain of a radical function with an even index is determined by solving the inequality that states that the radicand is greater than or equal to zero.

The domain of $f(x) = \sqrt{4x + 20}$ is $\{x | x \geq -5\}$.

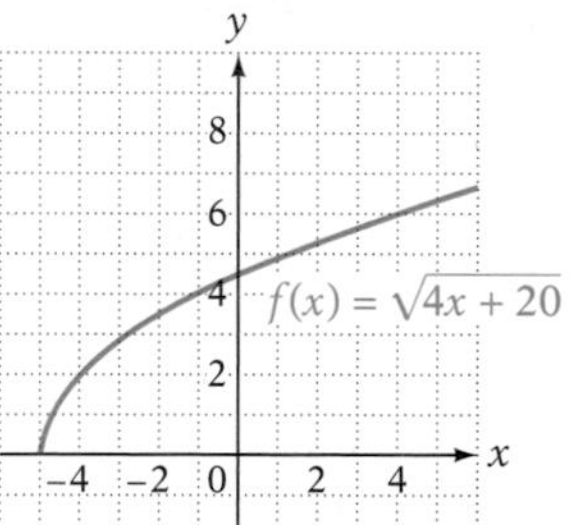

Solve a radical equation [7.4.1, p. 408]

Rewrite the equation so that the radical is alone on one side of the equation. Then raise each side of the equation to the power given by the index of the radical.

$$\begin{aligned}\sqrt[3]{2x} - 2 &= 4 \\ \sqrt[3]{2x} &= 6 \\ (\sqrt[3]{2x})^3 &= 6^3 \\ 2x &= 216 \\ x &= 108\end{aligned}$$

Pythagorean Theorem [7.4.2, p. 410]

If a and b are the lengths of the legs of a right triangle, and c is the length of the hypotenuse, then $a^2 + b^2 = c^2$.

(Right triangle with legs 3 and 4 and hypotenuse 5.) $3^2 + 4^2 = 5^2$

Write a complex number in standard form [7.5.1, p. 415]

Write the number in the form $a + bi$.

$$\frac{4 + \sqrt{-12}}{2} = \frac{4 + 2i\sqrt{3}}{2} = 2 + i\sqrt{3}$$

Add or subtract complex numbers [7.5.2, p. 416]

Add (or subtract) the real parts. Add (or subtract) the imaginary parts.

$(3 + 4i) + (2 - 7i) = 5 - 3i$

Multiply complex numbers [7.5.3, pp. 416–417]

Multiply as you would real numbers, and replace i^2 by -1.

$$\begin{aligned}(3 + i)(2 - 3i) &= 6 - 9i + 2i - 3i^2 \\ &= 6 - 7i - 3(-1) \\ &= 9 - 7i\end{aligned}$$

Divide complex numbers [7.5.4, p. 418]

Multiply the numerator and denominator by the conjugate of the denominator. If the denominator is an imaginary number, multiply the numerator and denominator by i.

$$\frac{4 + i}{2 - i} = \frac{4 + i}{2 - i} \cdot \frac{2 + i}{2 + i} = \frac{7 + 6i}{5} = \frac{7}{5} + \frac{6}{5}i$$

CHAPTER 7 Review Exercises

1. Simplify: $81^{-\frac{1}{4}}$

2. Simplify: $\dfrac{x^{-\frac{3}{2}}}{x^{\frac{7}{2}}}$

3. Simplify: $(a^{16})^{-\frac{5}{8}}$

4. Simplify: $(16x^{-4}y^{12})(100x^6y^{-2})^{\frac{1}{2}}$

5. Simplify: $\dfrac{(3x^{\frac{3}{4}}y^{\frac{1}{2}})^4}{(x^{-2}y^4)^{\frac{1}{2}}}$

6. Rewrite $3x^{\frac{3}{4}}$ as a radical expression.

7. Rewrite $7y\sqrt[3]{x^2}$ as an exponential expression.

8. Simplify: $\sqrt[4]{81a^8b^{12}}$

9. Simplify: $-\sqrt{49x^6y^{16}}$

10. Simplify: $\sqrt[3]{-8a^6b^{12}}$

11. Simplify: $\sqrt{18a^3b^6}$

12. Simplify: $\sqrt[5]{-64a^8b^{12}}$

13. Simplify: $\sqrt[4]{x^6y^8z^{10}}$

14. Add: $\sqrt{54} + \sqrt{24}$

15. Subtract: $4\sqrt[3]{8x^6y^5} - 3xy\sqrt[3]{27x^3y^2}$

16. Subtract: $\sqrt{50a^4b^3} - ab\sqrt{18a^2b}$

17. Simplify: $4x\sqrt{12x^2y} + \sqrt{3x^4y} - x^2\sqrt{27y}$

18. Multiply: $\sqrt{32}\,\sqrt{50}$

19. Multiply: $\sqrt[3]{16x^4y}\,\sqrt[3]{4xy^5}$

20. Multiply: $\sqrt{3x}(3 + \sqrt{3x})$

21. Multiply: $(5 - \sqrt{6})^2$

22. Multiply: $(\sqrt{3} + 8)(\sqrt{3} - 2)$

23. Multiply: $(3\sqrt{a} + 5\sqrt{b})(2\sqrt{a} - 6\sqrt{b})$

24. Simplify: $\dfrac{\sqrt{125x^6}}{\sqrt{5x^3}}$

25. Simplify: $\dfrac{8}{\sqrt{3y}}$

26. Simplify: $\dfrac{6}{\sqrt[3]{9x}}$

27. Simplify: $\dfrac{6}{7 - 3\sqrt{5}}$

28. Simplify: $\dfrac{1 + 2\sqrt{3}}{2 - 5\sqrt{3}}$

29. Simplify: $\dfrac{\sqrt{x} + \sqrt{y}}{\sqrt{x} - \sqrt{y}}$

30. State the domain of $f(x) = \sqrt[4]{3x - 2}$ in set-builder notation.

31. State the domain of $f(x) = \sqrt[3]{x - 5}$ in set-builder notation.

32. Graph: $g(x) = 3x^{\frac{1}{3}}$

33. Graph: $f(x) = \sqrt{x - 3}$

34. Solve: $\sqrt[3]{9x} = -6$

35. Solve: $\sqrt[3]{3x - 5} = 2$

36. Solve: $\sqrt{4x + 9} + 10 = 11$

37. Solve: $\sqrt{3x + 1} + \sqrt{3x + 10} = 9$

38. Simplify: $\sqrt{-36}$

39. Simplify: $\sqrt{-50}$

40. Simplify: $\dfrac{14 - 3\sqrt{-72}}{21}$

41. Add: $(5 + 2i) + (4 - 3i)$

42. Subtract: $(-8 + 3i) - (4 - 7i)$

43. Add: $(3 - 9i) + 7$

44. Multiply: $(8i)(2i)$

45. Multiply: $i(3 - 7i)$

46. Multiply: $(6 - 5i)(4 + 3i)$

47. Divide: $\dfrac{-6}{i}$

48. Divide: $\dfrac{5 + 2i}{3i}$

49. Divide: $\dfrac{7}{2 - i}$

50. Divide: $\dfrac{5 + 9i}{1 - i}$

51. **Geometry** Find the width of a rectangle that has a diagonal of 13 in. and a length of 12 in.

52. **Wind Power** The velocity of the wind determines the amount of power generated by a windmill. A typical equation for this relationship is $v = 4.05\sqrt[3]{P}$, where v is the velocity in miles per hour and P is the power in watts. Find the amount of power generated by a wind of 20 mph. Round to the nearest whole number.

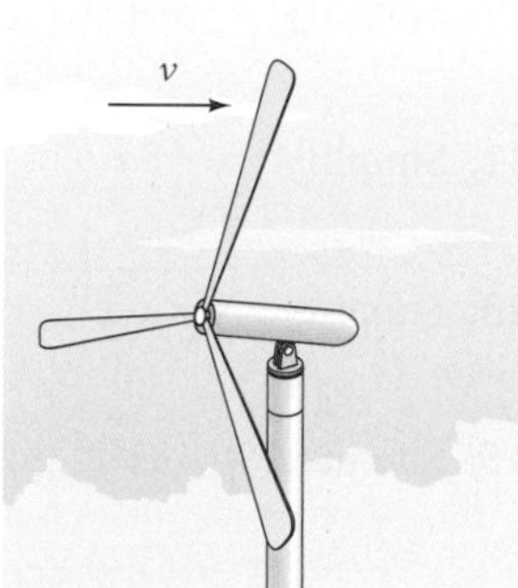

53. **Physics** Find the distance required for a car to reach a velocity of 88 ft/s when the acceleration is 16 ft/s^2. Use the equation $v = \sqrt{2as}$, where v is the velocity, a is the acceleration, and s is the distance.

54. **Building Maintenance** A 12-foot ladder is leaning against a building. How far from the building is the bottom of the ladder when the top of the ladder touches the building 10 ft above the ground? Round to the nearest hundredth.

CHAPTER 7 Test

1. Simplify: $\dfrac{r^{\frac{2}{3}}r^{-1}}{r^{-\frac{1}{2}}}$

2. Simplify: $\dfrac{(2x^{\frac{1}{3}}y^{-\frac{2}{3}})^6}{(x^{-4}y^8)^{\frac{1}{4}}}$

3. Simplify: $\left(\dfrac{4a^4}{b^2}\right)^{-\frac{3}{2}}$

4. Rewrite $3y^{\frac{2}{5}}$ as a radical expression.

5. Rewrite $\frac{1}{2}\sqrt[4]{x^3}$ as an exponential expression.

6. State the domain of $f(x) = \sqrt{4 - x}$ in set-builder notation.

7. State the domain of $f(x) = (2x - 3)^{\frac{1}{3}}$ in interval notation.

8. Simplify: $\sqrt[3]{27a^4b^3c^7}$

9. Add: $\sqrt{18a^3} + a\sqrt{50a}$

10. Subtract: $\sqrt[3]{54x^7y^3} - x\sqrt[3]{128x^4y^3} - x^2\sqrt[3]{2xy^3}$

11. Multiply: $\sqrt{3x}(\sqrt{x} - \sqrt{25x})$

12. Multiply: $(2\sqrt{3} + 4)(3\sqrt{3} - 1)$

13. Multiply: $(\sqrt{a} - 3\sqrt{b})(2\sqrt{a} + 5\sqrt{b})$

14. Multiply: $(2\sqrt{x} + \sqrt{y})^2$

15. Simplify: $\dfrac{\sqrt{32x^5y}}{\sqrt{2xy^3}}$

16. Simplify: $\dfrac{4 - 2\sqrt{5}}{2 - \sqrt{5}}$

17. Simplify: $\dfrac{\sqrt{x}}{\sqrt{x} - \sqrt{y}}$

18. Multiply: $(\sqrt{-8})(\sqrt{-2})$

19. Subtract: $(5 - 2i) - (8 - 4i)$

20. Multiply: $(2 + 5i)(4 - 2i)$

21. Divide: $\dfrac{2 + 3i}{1 - 2i}$

22. Add: $(2 + i) + (2 - i)$

23. Solve: $\sqrt{x + 12} - \sqrt{x} = 2$

24. Solve: $\sqrt[3]{2x - 2} + 4 = 2$

25. **Guy Wires** A guy wire is attached to a point 30 ft above the ground on a lamppost. The wire is anchored to the ground at a point 6 ft from the base of the post. Find the length of the guy wire. Round to the nearest tenth.

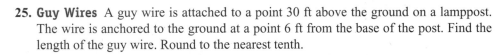

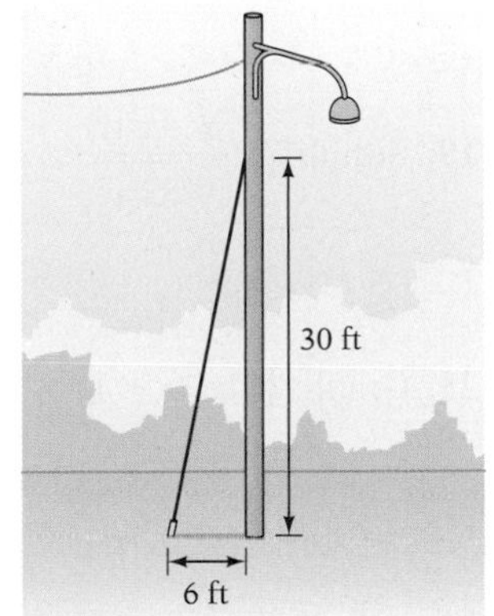

Cumulative Review Exercises

1. Identify the property that justifies the statement. $(a + 2)b = ab + 2b$

2. Simplify: $2x - 3[x - 2(x - 4) + 2x]$

3. Find $A \cap B$ given $A = \{2, 4, 6\}$ and $B = \{1, 3, 5\}$.

4. Solve: $\sqrt[3]{2x - 5} + 3 = 6$

5. Solve: $5 - \dfrac{2}{3}x = 4$

6. Solve: $2[4 - 2(3 - 2x)] = 4(1 - x)$

7. Solve: $3x - 4 \le 8x + 1$
Write the solution in set-builder notation.

8. Solve: $5 < 2x - 3 < 7$
Write the solution in set-builder notation.

9. Solve: $|7 - 3x| > 1$

10. Factor: $64a^2 - b^2$

11. Factor: $x^5 + 2x^3 - 3x$

12. Solve: $3x^2 + 13x - 10 = 0$

13. Graph: $g(x) = \sqrt{1 - x}$

14. Is the line with equation $x - 2y = 4$ perpendicular to the line with equation $2x + y = 4$?

15. Simplify: $(3^{-1}x^3y^{-5})(3^{-1}y^{-2})^{-2}$

16. Simplify: $\left(\dfrac{x^{-\frac{1}{2}}y^{\frac{3}{4}}}{y^{-\frac{5}{4}}}\right)^4$

17. Subtract: $\sqrt{20x^3} - x\sqrt{45x}$

18. Multiply: $(\sqrt{5} - 3)(\sqrt{5} - 2)$

19. Simplify: $\dfrac{\sqrt[3]{4x^5y^4}}{\sqrt[3]{8x^2y^5}}$

20. Divide: $\dfrac{3i}{2 - i}$

21. Graph: $\{x|x > -1\} \cap \{x|x \le 3\}$

22. State the domain of $g(x) = \sqrt[5]{2x + 5}$ in set-builder notation.

23. Given $f(x) = 3x^2 - 2x + 1$, evaluate $f(-3)$.

24. Find an equation of the line that passes through the points $P_1(2, 3)$ and $P_2(-1, 2)$.

25. Evaluate the determinant: $\begin{vmatrix} 1 & 2 & -3 \\ 0 & -1 & 2 \\ 3 & 1 & -2 \end{vmatrix}$

26. Solve by using Cramer's Rule: $\begin{aligned} 2x - y &= 4 \\ -2x + 3y &= 5 \end{aligned}$

27. Find the slope and y-intercept of the line with equation $3x - 2y = -6$, and then graph the line.

28. Graph the solution set of $3x + 2y \le 4$.

29. Investments An investment of \$2500 is made at an annual simple interest rate of 7.2%. How much additional money must be invested at an annual simple interest rate of 8.4% so that the total interest earned is \$516?

30. Geometry The width of a rectangle is 6 ft less than the length. The area of the rectangle is 72 ft². Find the length and width of the rectangle.

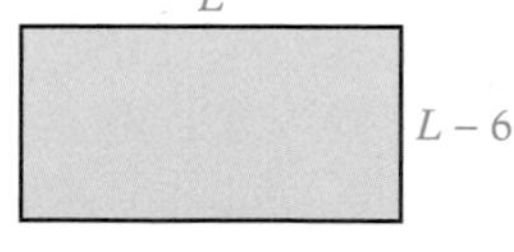

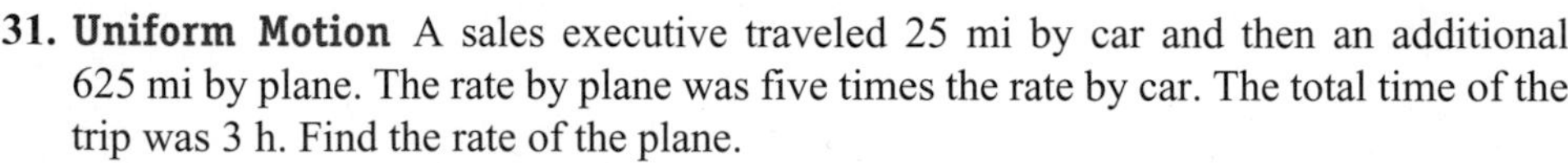

31. Uniform Motion A sales executive traveled 25 mi by car and then an additional 625 mi by plane. The rate by plane was five times the rate by car. The total time of the trip was 3 h. Find the rate of the plane.

32. Astronomy How long does it take light to travel to Earth from the moon when the moon is 232,500 mi from Earth? Light travels 1.86×10^5 mi/s.

33. Periscopes How far above the water would a submarine periscope have to be to locate a ship 7 mi away? The equation for the distance in miles that the lookout can see is $d = \sqrt{1.5h}$, where h is the height in feet above the surface of the water. Round to the nearest tenth.

34. Investments The graph shows the amount invested and the annual interest income from an investment. Find the slope of the line between the two points shown on the graph. Then write a sentence that states the meaning of the slope.

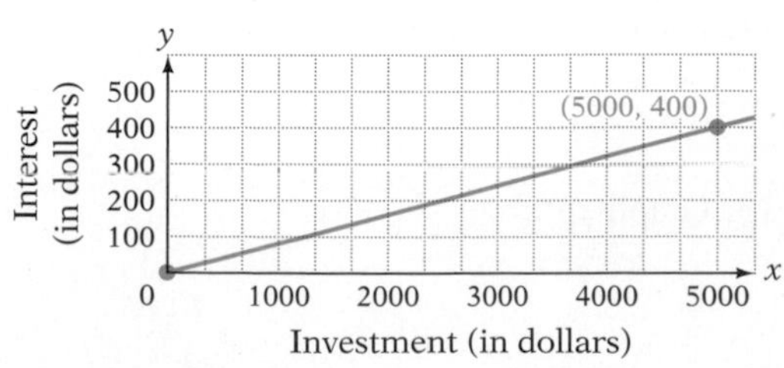

CHAPTER 8

Quadratic Equations and Inequalities

Focus on Success

Did you make a time management plan when you started this course? If not, you can still benefit from doing so. Create a schedule that gives you enough time to do everything you need to do. We want you to schedule enough time to study math each week so that you will successfully complete this course. Once you have determined the hours during which you will study, consider your study time a commitment that you cannot break. (See Time Management, page AIM-4.)

OBJECTIVES

8.1
1. Solve quadratic equations by factoring
2. Solve quadratic equations by taking square roots

8.2
1. Solve quadratic equations by completing the square
2. Solve quadratic equations by using the quadratic formula

8.3
1. Equations that are quadratic in form
2. Radical equations
3. Fractional equations

8.4
1. Application problems

8.5
1. Graph a quadratic function
2. Find the x-intercepts of a parabola

8.6
1. Minimum and maximum problems
2. Applications of minimum and maximum

8.7
1. Solve nonlinear inequalities

PREP TEST

Are you ready to succeed in this chapter?
Take the Prep Test below to find out if you are ready to learn the new material.

1. Simplify: $\sqrt{18}$
2. Simplify: $\sqrt{-9}$
3. Simplify: $\dfrac{3x-2}{x-1} - 1$
4. Evaluate $b^2 - 4ac$ when $a = 2$, $b = -4$, and $c = 1$.
5. Is $4x^2 + 28x + 49$ a perfect-square trinomial?
6. Factor: $4x^2 - 4x + 1$
7. Factor: $9x^2 - 4$
8. Graph $\{x|x < -1\} \cap \{x|x < 4\}$
9. Solve: $x(x-1) = x + 15$
10. Solve: $\dfrac{4}{x-3} = \dfrac{16}{x}$

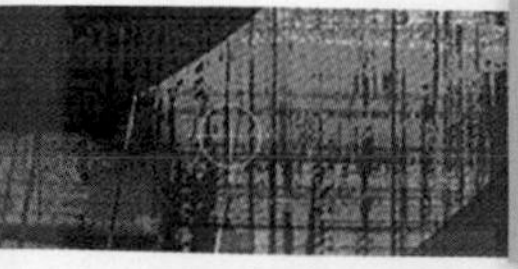

8.1 Solving Quadratic Equations by Factoring or by Taking Square Roots

OBJECTIVE 1 Solve quadratic equations by factoring

A **quadratic equation** is an equation of the form $ax^2 + bx + c = 0$, where a and b are coefficients, c is a constant, and $a \neq 0$.

$$3x^2 - x + 2 = 0, \quad a = 3, \quad b = -1, \quad c = 2$$
$$-x^2 + 4 = 0, \quad a = -1, \quad b = 0, \quad c = 4$$
$$6x^2 - 5x = 0, \quad a = 6, \quad b = -5, \quad c = 0$$

A quadratic equation is in **standard form** when the polynomial is in descending order and equal to zero.

Because the degree of the polynomial $ax^2 + bx + c$ is 2, a quadratic equation is also called a **second-degree equation.**

The Principle of Zero Products states that if the product of two factors is zero, then at least one of the factors equals zero.

Take Note

The Principle of Zero Products contains the phrase *at least one*. This phrase means *one or both* of the factors could be zero. Note how this concept is used in example (3) at the right.

PRINCIPLE OF ZERO PRODUCTS

If the product of two factors is zero, then at least one of the factors equals zero. If $ab = 0$, then $a = 0$ or $b = 0$.

EXAMPLES

1. Suppose $3x = 0$. The factors are 3 and x. The product equals zero, so at least one of the factors must be zero. Because $3 \neq 0$, we know that $x = 0$.
2. Suppose $-4(x - 4) = 0$. The factors are -4 and $x - 4$. The product equals zero, so at least one of the factors must be zero. Beause $-4 \neq 0$, we know that $x - 4 = 0$, which means $x = 4$.
3. Suppose $(x - 2)(x + 3) = 0$. The factors are $x - 2$ and $x + 3$. The product equals zero, so $x - 2 = 0$ or $x + 3 = 0$. If $x - 2 = 0$, then $x = 2$. If $x + 3 = 0$, then $x = -3$.

The Principle of Zero Products can be used to solve some quadratic equations.

EXAMPLE 1 Solve by factoring: $2x^2 - 3x = 2$

Solution

$$2x^2 - 3x = 2$$

$$2x^2 - 3x - 2 = 0$$
• **Write the equation in standard form.**

$$(2x + 1)(x - 2) = 0$$
• **Factor the trinomial.**

$$2x + 1 = 0 \qquad x - 2 = 0$$
$$2x = -1 \qquad x = 2$$
$$x = -\frac{1}{2}$$
• **Use the Principle of Zero Products. The product of 2x + 1 and x − 2 is 0. Therefore, at least one of the factors is zero.**

The solutions are $-\frac{1}{2}$ and 2.

Problem 1 Solve by factoring: $3x^2 + 14x + 8 = 0$

Solution See page S25.

Try Exercise 29, page 436.

EXAMPLE 2 Solve by factoring: $(x + 1)(2x - 1) = 2x + 2$

Solution

$(x + 1)(2x - 1) = 2x + 2$

$2x^2 + x - 1 = 2x + 2$ • **Multiply the factors on the left side.**

$2x^2 - x - 3 = 0$ • **Write the equation in standard form.**

$(x + 1)(2x - 3) = 0$ • **Factor.**

$x + 1 = 0 \qquad 2x - 3 = 0$ • **Use the Principle of Zero Products.**

$x = -1 \qquad x = \frac{3}{2}$

The solutions are -1 and $\frac{3}{2}$.

Problem 2 Solve by factoring: $(x + 5)(x - 1) = 7$

Solution See page S26.

Try Exercise 43, page 437.

When a quadratic equation has two solutions that are the same number, the solution is called a **double root** of the equation.

Focus on solving an equation with a double root

Solve by factoring: $x^2 - 6x + 9 = 0$

$x^2 - 6x + 9 = 0$

Factor. $(x - 3)(x - 3) = 0$

Use the Principle of Zero Products. $x - 3 = 0 \qquad x - 3 = 0$

Solve each equation for x. $x = 3 \qquad x = 3$

3 is a double root of $x^2 - 6x + 9 = 0$.

The Principle of Zero Products also can be used to write an equation that has specific roots. For instance, suppose r and s are given as solutions of an equation. Then one possible equation is $(x - r)(x - s) = 0$, as shown below.

$(x - r)(x - s) = 0$

Use the Principle of Zero Products. $x - r = 0 \qquad x - s = 0$

Solve for x. $x = r \qquad x = s$

The solutions are r and s.

Given two solutions r and s and the equation $(x - r)(x - s) = 0$, we can find a quadratic equation that has the given solutions.

Focus on writing a quadratic equation given its solutions

Write a quadratic equation that has solutions -5 and 4.

$$(x - r)(x - s) = 0$$

Replace r by -5 and s by 4. $\quad [x - (-5)](x - 4) = 0$

Simplify. $\quad (x + 5)(x - 4) = 0$

Multiply the binomials. $\quad x^2 + x - 20 = 0$

A quadratic equation with solutions -5 and 4 is $x^2 + x - 20 = 0$.

EXAMPLE 3 Write a quadratic equation that has integer coefficients and has solutions $\frac{2}{3}$ and $\frac{1}{2}$.

Solution

$$(x - r)(x - s) = 0$$

$$\left(x - \frac{2}{3}\right)\left(x - \frac{1}{2}\right) = 0$$ • **Replace r by $\frac{2}{3}$ and s by $\frac{1}{2}$.**

$$x^2 - \frac{7}{6}x + \frac{1}{3} = 0$$ • **Multiply the binomials.**

$$6\left(x^2 - \frac{7}{6}x + \frac{1}{3}\right) = 6 \cdot 0$$ • **Multiply each side of the equation by 6, the LCD.**

$$6x^2 - 7x + 2 = 0$$

A quadratic equation with solutions $\frac{2}{3}$ and $\frac{1}{2}$ is $6x^2 - 7x + 2 = 0$.

Problem 3 Write a quadratic equation that has integer coefficients and has solutions $-\frac{2}{3}$ and $\frac{1}{6}$.

Solution See page S26.

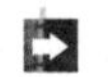 *Try Exercise 59, page 437.*

OBJECTIVE 2 Solve quadratic equations by taking square roots

Recall that if x is a variable that can be positive or negative, then $\sqrt{x^2} = |x|$. This fact is used to solve a quadratic equation by taking square roots.

Focus on solving a quadratic equation by taking square roots

A. Solve: $x^2 = 16$

$$x^2 = 16$$

Take the square root of each side of the equation. $\quad \sqrt{x^2} = \sqrt{16}$

$\sqrt{x^2} = |x|$, $\sqrt{16} = 4$ $\quad |x| = 4$

Solve the absolute value equation. The notation $x = \pm 4$ means $x = 4$ or $x = -4$. $\quad x = \pm 4$

The solutions are -4 and 4.

B. Solve by taking square roots: $3x^2 - 54 = 0$

$$3x^2 - 54 = 0$$

Solve for x^2.

$$3x^2 = 54$$

$$x^2 = 18$$

Take the square root of each side of the equation.

$$\sqrt{x^2} = \sqrt{18}$$

Simplify.

$$|x| = 3\sqrt{2}$$

Solve the absolute value equation.

$$x = \pm 3\sqrt{2}$$

The solutions are $-3\sqrt{2}$ and $3\sqrt{2}$.

C. Solve by taking square roots: $x^2 + 18 = 6$

$$x^2 + 18 = 6$$

Solve for x^2.

$$x^2 = -12$$

Take the square root of each side of the equation.

$$\sqrt{x^2} = \sqrt{-12}$$

Simplify.

$$|x| = i\sqrt{12} = 2i\sqrt{3}$$

$$x = \pm 2i\sqrt{3}$$

The solutions are $-2i\sqrt{3}$ and $2i\sqrt{3}$.

An equation containing the square of a binomial can be solved by taking square roots.

EXAMPLE 4 Solve by taking square roots: $3(x - 2)^2 + 12 = 0$

Solution $3(x - 2)^2 + 12 = 0$

$$3(x - 2)^2 = -12$$ • **Solve for $(x - 2)^2$.**

$$(x - 2)^2 = -4$$

$$\sqrt{(x - 2)^2} = \sqrt{-4}$$ • **Take the square root of each side of the equation. Then simplify.**

$$|x - 2| = 2i$$

$$x - 2 = \pm 2i$$

$x - 2 = 2i$ $\quad$ $x - 2 = -2i$ • **Solve for x.**

$x = 2 + 2i$ $\quad$ $x = 2 - 2i$

The solutions are $2 + 2i$ and $2 - 2i$.

Problem 4 Solve by taking square roots: $5(x + 4)^2 + 7 = 17$

Solution See page S26.

Try Exercise 103, page 438.

8.1 Exercises

CONCEPT CHECK

1. Which of the following are *not* quadratic equations?

(i) $3x - 5 = 0$ (ii) $3x^2 = 4x - 8$ (iii) $x^2 + 4x - 5$ (iv) $0 = x^2 - 6x + 12$

Write the equation in standard form with the coefficient of x^2 positive.

2. $4x = 2x^2 - 3$
3. $x^2 + 3x = 7$
4. $3x = -2x^2$
5. $4x^2 = 15$

Determine whether the statement is always true, sometimes true, or never true.

6. A quadratic equation has two real roots.
7. If $(x - 3)(x + 5) = 8$, then $x - 3 = 8$ or $x + 5 = 8$.
8. A solution of $x^2 + x = 5$ is 2.

Solve the equation.

9. $5(x + 3) = 0$
10. $(x - 4)(x + 2) = 0$
11. $x(x - 5) = 0$
12. $(2x + 3)(3x - 4) = 0$

1 Solve quadratic equations by factoring (See pages 432–434.)

13. Explain why the restriction $a \neq 0$ is given in the definition of a quadratic equation.

14. What does the Principle of Zero Products state? How is it used to solve a quadratic equation?

GETTING READY

15. Solve: $x^2 = 6x + 27$

$$x^2 = 6x + 27$$

$$x^2 - 6x - 27 = 0$$

• **Write the equation in __?__ form by subtracting 6x and 27 from each side of the equation.**

$(x + \_\_?\_\_)(x - \_\_?\_\_) = 0$

• **Factor the trinomial.**

$x + 3 = \_\_?\_\_$ or $x - 9 = \_\_?\_\_$

• **Use the Principle of Zero Products.**

$x = \_\_?\_\_$ $\quad$ $x = \_\_?\_\_$

16. a. The factored form of a quadratic equation that has solutions -4 and 7 is (__?__)(__?__) = 0.

b. The standard form of the equation in part (a) is __?__ = 0.

Solve by factoring.

17. $x^2 - 4x = 0$
18. $y^2 + 6y = 0$
19. $9z^2 - 18z = 0$
20. $4y^2 + 20y = 0$
21. $s^2 - s - 6 = 0$
22. $v^2 + 4v - 5 = 0$
23. $t^2 - 25 = 0$
24. $p^2 - 81 = 0$
25. $y^2 - 6y + 9 = 0$
26. $x^2 + 10x + 25 = 0$
27. $r^2 - 3r = 10$
28. $p^2 + 5p = 6$
29. $v^2 + 10 = 7v$
30. $t^2 - 16 = 15t$
31. $2x^2 - 9x - 18 = 0$
32. $3y^2 - 4y - 4 = 0$
33. $4z^2 - 9z + 2 = 0$
34. $2s^2 - 9s + 9 = 0$
35. $3w^2 + 11w = 4$
36. $2r^2 + r = 6$
37. $6x^2 = 23x + 18$
38. $6x^2 = 7x - 2$
39. $4 - 15u - 4u^2 = 0$
40. $3 - 2y - 8y^2 = 0$

41. $4s(s + 3) = s - 6$

42. $3v(v - 2) = 11v + 6$

43. $(2x - 3)(x + 4) = 4x - 2$

44. $(3x - 2)(x + 5) = 20x - 4$

45. $(3x - 4)(x + 4) = x^2 - 3x - 28$

46. $(3v - 2)(2v + 1) = 3v^2 - 11v - 10$

47. Given that $ax^2 + bx + c = 0$ is a quadratic equation that can be solved by factoring, state whether the equation has two positive solutions, two negative solutions, or one positive and one negative solution.
a. $a > 0, b < 0, c < 0$ **b.** $a > 0, b > 0, c > 0$ **c.** $a > 0, b < 0, c > 0$

48. Which equation has a double root?

(i) $x^2 = a^2$ (ii) $x^2 + 2ax = a^2$ (iii) $x^2 + 2ax = -a^2$ (iv) $x^2 - 2ax = a^2$

Write a quadratic equation that has integer coefficients and has as solutions the given pair of numbers.

49. 2 and 5

50. -2 and -4

51. 6 and -1

52. -2 and 5

53. 3 and -3

54. 5 and -5

55. 4 and 4

56. 2 and 2

57. 0 and 5

58. 0 and -2

59. 3 and $\frac{1}{2}$

60. $-\frac{1}{2}$ and 5

61. $\frac{5}{3}$ and -2

62. $-\frac{3}{2}$ and -1

63. $-\frac{2}{3}$ and $\frac{2}{3}$

64. $-\frac{1}{2}$ and $\frac{1}{2}$

65. $\frac{6}{5}$ and $-\frac{1}{2}$

66. $\frac{3}{4}$ and $-\frac{3}{2}$

67. $-\frac{1}{4}$ and $-\frac{1}{2}$

68. $-\frac{5}{6}$ and $-\frac{2}{3}$

69. $\frac{3}{5}$ and $-\frac{1}{10}$

For Exercises 70 to 73, r and s are solutions of the equation $ax^2 + bx + c = 0$, where a is positive. In each case, state whether b is positive or negative and whether c is positive or negative.

70. $r > 0, s > 0$

71. $r < 0, s < 0$

72. $r > 0, s < 0, r > |s|$

73. $r > 0, s < 0, r < |s|$

2 Solve quadratic equations by taking square roots (See pages 434–435.)

GETTING READY

74. The notation $x = \pm 6$ means $x =$ __?__ or $x =$ __?__.

75. Solve by taking square roots: $4x^2 + 100 = 0$

$4x^2 + 100 = 0$

$4x^2 =$ __?__ • Subtract __?__ from each side of the equation.

$x^2 =$ __?__ • Divide each side of the equation by __?__.

$\sqrt{x^2} = \sqrt{-25}$ • Take the __?__ of each side of the equation.

$|x| =$ __?__ • Simplify $\sqrt{-25}$.

$x = \pm$ __?__ • Solve for x.

Solve by taking square roots.

76. $x^2 = 64$

77. $y^2 = 49$

78. $r^2 - 36 = 0$

79. $s^2 - 4 = 0$

80. $9x^2 - 16 = 0$

81. $4x^2 - 81 = 0$

82. $s^2 - 32 = 0$

83. $v^2 - 48 = 0$

84. $4r^2 - 80 = 0$

85. $\frac{x^2}{2} - 16 = 0$

86. $(x + 2)^2 = 25$

87. $(x - 1)^2 = 36$

88. $(x - 5)^2 - 18 = 0$

89. $(x + 4)^2 - 50 = 0$

90. $3(x + 2)^2 - 36 = 0$

91. $7(y - 3)^2 - 56 = 0$

92. $\left(y + \frac{1}{3}\right)^2 = \frac{4}{9}$

93. $\left(x - \frac{2}{5}\right)^2 = \frac{9}{25}$

94. $2\left(x + \frac{3}{5}\right)^2 = \frac{8}{25}$

95. $3\left(x - \frac{5}{3}\right)^2 = \frac{4}{3}$

96. $z^2 + 16 = 0$

97. $y^2 + 49 = 0$

98. $t^2 + 27 = 0$

99. $z^2 + 18 = 0$

100. $(x - 2)^2 = -4$

101. $(x + 5)^2 = -25$

102. $(r - 2)^2 + 28 = 0$

103. $(z + 1)^2 + 12 = 0$

104. $2\left(x + \frac{1}{2}\right)^2 + 50 = 0$

105. $3\left(y - \frac{2}{3}\right)^2 + 4 = 0$

106. $\left(z - \frac{3}{4}\right)^2 + \frac{9}{2} = 0$

107. $\left(v + \frac{4}{5}\right)^2 + \frac{9}{5} = 0$

For Exercises 108 to 111, a and b are positive real numbers. In each case, state how many real or complex number solutions the equation has.

108. $(x - a)^2 + b = 0$

109. $x^2 + a = 0$

110. $(x - a)^2 = 0$

111. $(x - a)^2 - b = 0$

APPLYING CONCEPTS

Write an equation with integer coefficients that has the given numbers as solutions.

112. $1 - \sqrt{5}$ and $1 + \sqrt{5}$

113. $4 - 2\sqrt{3}$ and $4 + 2\sqrt{3}$

114. $-3i$ and $3i$

115. $-i\sqrt{3}$ and $i\sqrt{3}$

116. $2 - i$ and $2 + i$

117. $1 - 5i$ and $1 + 5i$

PROJECTS OR GROUP ACTIVITIES

118. One solution of $2x^2 - 5x + c = 0$ is 4. What is the other solution?

119. For how many integer values of b does the equation $x^2 - bx - 16 = 0$ have integer solutions?

120. Marc and Elena tried to solve the same equation of the form $ax^2 + bx + c = 0$. Unfortunately, Marc copied the value of a incorrectly and came up with the solutions -2 and $-\frac{3}{2}$. Elena copied the value of b incorrectly and came up with the solutions 2 and 3. If neither student had made an error in copying the equation, what would be the correct equation?

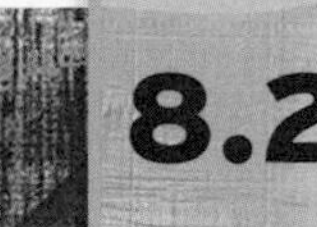

8.2 Solving Quadratic Equations by Completing the Square and by Using the Quadratic Formula

OBJECTIVE 1 Solve quadratic equations by completing the square

Recall that a perfect-square trinomial is the square of a binomial. Some examples of perfect-square trinomials follow.

Perfect-square trinomial		Square of a binomial
$x^2 + 8x + 16$	$=$	$(x + 4)^2$
$x^2 - 10x + 25$	$=$	$(x - 5)^2$
$x^2 - 5x + \frac{25}{4}$	$=$	$\left(x - \frac{5}{2}\right)^2$

For each perfect-square trinomial, the square of $\frac{1}{2}$ the coefficient of x equals the constant term.

$$\left(\frac{1}{2}\text{ coefficient of } x\right)^2 = \text{Constant term}$$

$x^2 + 8x + 16,$	$\left(\frac{1}{2} \cdot 8\right)^2 = 16$
$x^2 - 10x + 25,$	$\left[\frac{1}{2}(-10)\right]^2 = 25$
$x^2 - 5x + \frac{25}{4},$	$\left(\frac{1}{2}(-5)\right)^2 = \frac{25}{4}$

To **complete the square** on $x^2 + bx$, add $\left(\frac{1}{2}b\right)^2$ to $x^2 + bx$.

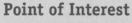

Focus on completing the square

A. Complete the square on $x^2 - 12x$. Write the resulting perfect-square trinomial as the square of a binomial.

Find the constant term. $\left[\frac{1}{2}(-12)\right]^2 = (-6)^2 = 36$

Complete the square on $x^2 - 12x$ by adding the constant term. $x^2 - 12x + 36$

Write the resulting perfect-square trinomial as the square of a binomial. $x^2 - 12x + 36 = (x - 6)^2$

B. Complete the square on $z^2 + 3z$. Write the resulting perfect-square trinomial as the square of a binomial.

Find the constant term. $\left(\frac{1}{2} \cdot 3\right)^2 = \left(\frac{3}{2}\right)^2 = \frac{9}{4}$

Complete the square on $z^2 + 3z$ by adding the constant term. $z^2 + 3z + \frac{9}{4}$

Write the resulting perfect-square trinomial as the square of a binomial. $z^2 + 3z + \frac{9}{4} = \left(z + \frac{3}{2}\right)^2$

Point of Interest

Early attempts to solve quadratic equations were primarily geometric. The Persian mathematician al-Khwarizmi (c. A.D. 800) essentially completed a square of $x^2 + 12x$ as shown below.

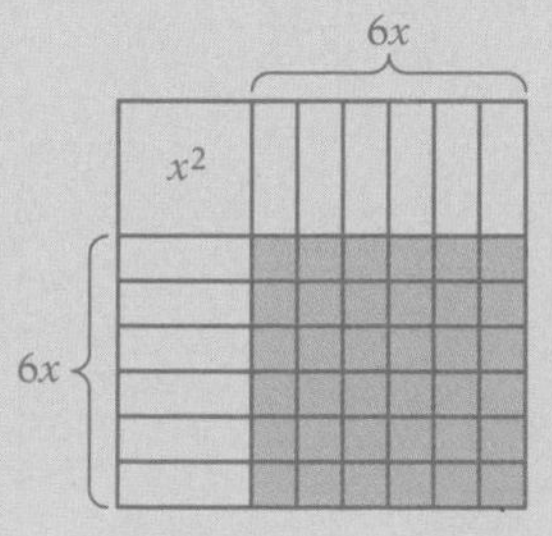

Not all quadratic equations can be solved by factoring, but any quadratic equation can be solved by completing the square. Add to each side of the equation the term that completes the square. Rewrite the equation in the form $(x + a)^2 = b$. Then take the square root of each side of the equation.

Focus on solving a quadratic equation by completing the square

Solve by completing the square: $x^2 - 4x - 14 = 0$

$$x^2 - 4x - 14 = 0$$

Add 14 to each side of the equation.

$$x^2 - 4x = 14$$

Add the constant term that completes the square on $x^2 - 4x$ to each side of the equation.

$$x^2 - 4x + 4 = 14 + 4$$

$\left[\frac{1}{2}(-4)\right]^2 = 4$

Factor the perfect-square trinomial.

$$(x - 2)^2 = 18$$

Take the square root of each side of the equation.

$$\sqrt{(x - 2)^2} = \sqrt{18}$$

Simplify.

$$|x - 2| = 3\sqrt{2}$$

Solve for x.

$$x - 2 = \pm 3\sqrt{2}$$

$$x - 2 = 3\sqrt{2} \qquad x - 2 = -3\sqrt{2}$$

$$x = 2 + 3\sqrt{2} \qquad x = 2 - 3\sqrt{2}$$

The solutions are $2 + 3\sqrt{2}$ and $2 - 3\sqrt{2}$.

Take Note

You should check your solutions.

Check:

$$\begin{array}{r|l} x^2 - 4x - 14 = 0 & \\ \hline (2 + 3\sqrt{2})^2 - 4(2 + 3\sqrt{2}) - 14 & 0 \\ 4 + 12\sqrt{2} + 18 - 8 - 12\sqrt{2} - 14 & 0 \\ 0 = 0 & \end{array}$$

$$\begin{array}{r|l} x^2 - 4x - 14 = 0 & \\ \hline (2 - 3\sqrt{2})^2 - 4(2 - 3\sqrt{2}) - 14 & 0 \\ 4 - 12\sqrt{2} + 18 - 8 + 12\sqrt{2} - 14 & 0 \\ 0 = 0 & \end{array}$$

The solutions of the equation $x^2 - 4x - 14 = 0$ are $2 + 3\sqrt{2}$ and $2 - 3\sqrt{2}$. These are the exact solutions. However, in some situations it may be preferable to have decimal approximations of the solutions of a quadratic equation. Approximate solutions can be found by using a calculator and then rounding to the desired degree of accuracy.

$$2 + 3\sqrt{2} \approx 6.243 \text{ and } 2 - 3\sqrt{2} \approx -2.243$$

To the nearest thousandth, the approximate solutions of the equation $x^2 - 4x - 14 = 0$ are 6.243 and -2.243.

When a, the coefficient of the x^2 term, is not 1, divide each side of the equation by a before completing the square.

Focus on solving a quadratic equation by completing the square

Solve: $2x^2 - x - 2 = 0$

$$2x^2 - x - 2 = 0$$

$$2x^2 - x = 2$$

Divide each side of the equation by the coefficient of x^2.

$$\frac{2x^2 - x}{2} = \frac{2}{2}$$

The coefficient of the x^2 term is now 1.

$$x^2 - \frac{1}{2}x = 1$$

Take Note

This example illustrates all the steps required in solving a quadratic equation by completing the square.

1. Write the equation in the form $ax^2 + bx = -c$.
2. Divide each side of the equation by a.
3. Complete the square on $x^2 + \frac{b}{a}x$. Add to both sides of the equation the number that completes the square.
4. Factor the perfect-square trinomial.
5. Take the square root of each side of the equation.
6. Solve the resulting equation for x.
7. Check the solutions.

Add the term that completes the square on $x^2 - \frac{1}{2}x$ to each side of the equation.

$$x^2 - \frac{1}{2}x + \frac{1}{16} = 1 + \frac{1}{16}$$

Factor the perfect-square trinomial.

$$\left(x - \frac{1}{4}\right)^2 = \frac{17}{16}$$

Take the square root of each side of the equation.

$$\sqrt{\left(x - \frac{1}{4}\right)^2} = \sqrt{\frac{17}{16}}$$

Simplify.

$$\left|x - \frac{1}{4}\right| = \frac{\sqrt{17}}{4}$$

$$x - \frac{1}{4} = \pm\frac{\sqrt{17}}{4}$$

Solve for x.

$$x - \frac{1}{4} = \frac{\sqrt{17}}{4} \qquad x - \frac{1}{4} = -\frac{\sqrt{17}}{4}$$

$$x = \frac{1}{4} + \frac{\sqrt{17}}{4} \qquad x = \frac{1}{4} - \frac{\sqrt{17}}{4}$$

The solutions are $\frac{1 + \sqrt{17}}{4}$ and $\frac{1 - \sqrt{17}}{4}$.

EXAMPLE 1 Solve by completing the square.

A. $4x^2 - 8x + 1 = 0$ **B.** $x^2 + 4x + 5 = 0$

Solution **A.** $4x^2 - 8x + 1 = 0$

$$4x^2 - 8x = -1$$

• **Subtract 1 from each side of the equation.**

$$\frac{4x^2 - 8x}{4} = \frac{-1}{4}$$

• **The coefficient of the x^2 term must be 1. Divide each side of the equation by 4.**

$$x^2 - 2x = -\frac{1}{4}$$

$$x^2 - 2x + 1 = -\frac{1}{4} + 1$$

• **Complete the square.** $\left[\frac{1}{2}(-2)\right]^2 = 1$

$$(x - 1)^2 = \frac{3}{4}$$

• **Factor the perfect-square trinomial.**

$$\sqrt{(x - 1)^2} = \sqrt{\frac{3}{4}}$$

• **Take the square root of each side of the equation.**

$$|x - 1| = \frac{\sqrt{3}}{2}$$

• **Simplify.**

$$x - 1 = \pm\frac{\sqrt{3}}{2}$$

$$x - 1 = \frac{\sqrt{3}}{2} \qquad x - 1 = -\frac{\sqrt{3}}{2}$$

• **Solve for x.**

$$x = 1 + \frac{\sqrt{3}}{2} \qquad x = 1 - \frac{\sqrt{3}}{2}$$

$$x = \frac{2 + \sqrt{3}}{2} \qquad x = \frac{2 - \sqrt{3}}{2}$$

The solutions are $\frac{2 + \sqrt{3}}{2}$ and $\frac{2 - \sqrt{3}}{2}$.

B. $x^2 + 4x + 5 = 0$

$x^2 + 4x = -5$ • Subtract 5 from each side of the equation.

$x^2 + 4x + 4 = -5 + 4$ • Complete the square.

$(x + 2)^2 = -1$ • Factor the perfect-square trinomial.

$\sqrt{(x + 2)^2} = \sqrt{-1}$ • Take the square root of each side of the equation. Simplify.

$|x + 2| = i$

$x + 2 = \pm i$

$x + 2 = i$ $\quad$ $x + 2 = -i$ • Solve for x.

$x = -2 + i$ $\quad$ $x = -2 - i$

The solutions are $-2 + i$ and $-2 - i$.

Problem 1 Solve by completing the square.

A. $4x^2 - 4x - 1 = 0$ **B.** $2x^2 + x - 5 = 0$

Solution See page S26.

Try Exercise 15, page 447.

OBJECTIVE 2 Solve quadratic equations by using the quadratic formula

A general formula known as the **quadratic formula** can be derived by applying the method of completing the square to the standard form of a quadratic equation. This formula can be used to solve any quadratic equation.

The solution of the equation $ax^2 + bx + c = 0$ by completing the square is shown below.

$$ax^2 + bx + c = 0$$

Subtract the constant term from each side of the equation.

$$ax^2 + bx + c - c = 0 - c$$

$$ax^2 + bx = -c$$

Divide each side of the equation by a, the coefficient of x^2.

$$\frac{ax^2 + bx}{a} = \frac{-c}{a}$$

$$x^2 + \frac{b}{a}x = -\frac{c}{a}$$

Complete the square by adding $\left(\frac{1}{2} \cdot \frac{b}{a}\right)^2$ to each side of the equation.

$$x^2 + \frac{b}{a}x + \left(\frac{1}{2} \cdot \frac{b}{a}\right)^2 = \left(\frac{1}{2} \cdot \frac{b}{a}\right)^2 - \frac{c}{a}$$

$$x^2 + \frac{b}{a}x + \frac{b^2}{4a^2} = \frac{b^2}{4a^2} - \frac{c}{a}$$

Simplify the right side of the equation.

$$x^2 + \frac{b}{a}x + \frac{b^2}{4a^2} = \frac{b^2}{4a^2} - \left(\frac{c}{a} \cdot \frac{4a}{4a}\right)$$

$$x^2 + \frac{b}{a}x + \frac{b^2}{4a^2} = \frac{b^2}{4a^2} - \frac{4ac}{4a^2}$$

$$x^2 + \frac{b}{a}x + \frac{b^2}{4a^2} = \frac{b^2 - 4ac}{4a^2}$$

Factor the perfect-square trinomial on the left side of the equation.

$$\left(x + \frac{b}{2a}\right)^2 = \frac{b^2 - 4ac}{4a^2}$$

Take the square root of each side of the equation.

$$\sqrt{\left(x + \frac{b}{2a}\right)^2} = \sqrt{\frac{b^2 - 4ac}{4a^2}}$$

$$\left|x + \frac{b}{2a}\right| = \frac{\sqrt{b^2 - 4ac}}{2a}$$

$$x + \frac{b}{2a} = \pm\frac{\sqrt{b^2 - 4ac}}{2a}$$

Solve for x. $x + \frac{b}{2a} = \frac{\sqrt{b^2 - 4ac}}{2a}$ $\qquad$ $x + \frac{b}{2a} = -\frac{\sqrt{b^2 - 4ac}}{2a}$

$$x = -\frac{b}{2a} + \frac{\sqrt{b^2 - 4ac}}{2a} \qquad x = -\frac{b}{2a} - \frac{\sqrt{b^2 - 4ac}}{2a}$$

$$= \frac{-b + \sqrt{b^2 - 4ac}}{2a} \qquad = \frac{-b - \sqrt{b^2 - 4ac}}{2a}$$

Point of Interest

Although mathematicians have studied quadratic equations since around 500 B.C., it was not until the 18th century that the formula was written as it is today. Of further note, the word *quadratic* has the same Latin root as does the word *square*.

QUADRATIC FORMULA

The solutions of $ax^2 + bx + c = 0$, $a \neq 0$, are

$$\frac{-b + \sqrt{b^2 - 4ac}}{2a} \quad \text{and} \quad \frac{-b - \sqrt{b^2 - 4ac}}{2a}$$

The quadratic formula is frequently written in the form

$$x = \frac{-b \pm \sqrt{b^2 - 4ac}}{2a}$$

Focus on solving a quadratic equation with rational solutions by using the quadratic formula

Solve by using the quadratic formula: $2x^2 + 5x + 3 = 0$

The equation $2x^2 + 5x + 3 = 0$ is in standard form. $a = 2$, $b = 5$, $c = 3$

$$x = \frac{-b \pm \sqrt{b^2 - 4ac}}{2a}$$

Replace a, b, and c in the quadratic formula with these values.

$$= \frac{-(5) \pm \sqrt{(5)^2 - 4(2)(3)}}{2(2)}$$

$$= \frac{-5 \pm \sqrt{25 - 24}}{4}$$

$$= \frac{-5 \pm \sqrt{1}}{4} = \frac{-5 \pm 1}{4}$$

$$x = \frac{-5 + 1}{4} = \frac{-4}{4} = -1 \qquad x = \frac{-5 - 1}{4} = \frac{-6}{4} = -\frac{3}{2}$$

The solutions are -1 and $-\frac{3}{2}$.

When a quadratic equation has rational number solutions (as does the one above), it may be easier to solve the equation by factoring, as shown below.

$$2x^2 + 5x + 3 = 0$$

$$(2x + 3)(x + 1) = 0 \qquad \text{• Factor.}$$

$$2x + 3 = 0 \qquad x + 1 = 0 \qquad \text{• Use the Principle of Zero Products.}$$

$$2x = -3 \qquad x = -1$$

$$x = -\frac{3}{2}$$

The solutions are $-\frac{3}{2}$ and -1.

Focus on solving a quadratic equation with irrational solutions by using the quadratic formula

Solve $3x^2 = 4x + 6$ by using the quadratic formula.

Write the equation in standard form by subtracting $4x$ and 6 from each side of the equation.

$$3x^2 = 4x + 6$$
$$3x^2 - 4x - 6 = 0$$
$$a = 3, b = -4, c = -6$$
$$x = \frac{-b \pm \sqrt{b^2 - 4ac}}{2a}$$

Replace a, b, and c in the quadratic formula with their values.

$$= \frac{-(-4) \pm \sqrt{(-4)^2 - 4(3)(-6)}}{2(3)}$$
$$= \frac{4 \pm \sqrt{16 - (-72)}}{6}$$
$$= \frac{4 \pm \sqrt{88}}{6} = \frac{4 \pm 2\sqrt{22}}{6}$$
$$= \frac{\cancel{2}(2 \pm \sqrt{22})}{\cancel{2} \cdot 3} = \frac{2 \pm \sqrt{22}}{3}$$

The solutions are $\frac{2 + \sqrt{22}}{3}$ and $\frac{2 - \sqrt{22}}{3}$.

Focus on solving a quadratic equation with complex number solutions by using the quadratic formula

Solve by using the quadratic formula: $4x^2 = 8x - 13$

Write the equation in standard form.

$$4x^2 = 8x - 13$$
$$4x^2 - 8x + 13 = 0$$
$$a = 4, b = -8, c = 13$$
$$x = \frac{-b \pm \sqrt{b^2 - 4ac}}{2a}$$

Replace a, b, and c in the quadratic formula by their values.

$$= \frac{-(-8) \pm \sqrt{(-8)^2 - 4 \cdot 4 \cdot 13}}{2 \cdot 4}$$

Simplify.

$$= \frac{8 \pm \sqrt{64 - 208}}{8}$$
$$= \frac{8 \pm \sqrt{-144}}{8}$$

Write the answer in the form $a + bi$.

$$= \frac{8 \pm 12i}{8} = \frac{2 \pm 3i}{2} = 1 \pm \frac{3}{2}i$$

The solutions are $1 + \frac{3}{2}i$ and $1 - \frac{3}{2}i$.

Take Note

Remember to check your solutions.

Check:

$4x^2 = 8x - 13$	
$4\left(1 + \frac{3}{2}i\right)^2$	$8\left(1 + \frac{3}{2}i\right) - 13$
$4\left(1 + 3i - \frac{9}{4}\right)$	$8 + 12i - 13$
$4\left(-\frac{5}{4} + 3i\right)$	$-5 + 12i$

$-5 + 12i = -5 + 12i$

$4x^2 = 8x - 13$	
$4\left(1 - \frac{3}{2}i\right)^2$	$8\left(1 - \frac{3}{2}i\right) - 13$
$4\left(1 - 3i - \frac{9}{4}\right)$	$8 - 12i - 13$
$4\left(-\frac{5}{4} - 3i\right)$	$-5 - 12i$

$-5 - 12i = -5 - 12i$

EXAMPLE 2 Solve by using the quadratic formula.

A. $4x^2 + 12x + 9 = 0$ **B.** $2x^2 - x + 5 = 0$

Solution **A.** $4x^2 + 12x + 9 = 0$

• $a = 4, b = 12, c = 9$

$$x = \frac{-b \pm \sqrt{b^2 - 4ac}}{2a}$$

$$= \frac{-12 \pm \sqrt{12^2 - 4 \cdot 4 \cdot 9}}{2 \cdot 4}$$

• Replace a, b, and c in the quadratic formula by their values. Then simplify.

$$= \frac{-12 \pm \sqrt{144 - 144}}{8}$$

$$= \frac{-12 \pm \sqrt{0}}{8} = \frac{-12}{8} = -\frac{3}{2}$$

• The equation has a double root.

The solution is $-\frac{3}{2}$.

B. $2x^2 - x + 5 = 0$

$$x = \frac{-b \pm \sqrt{b^2 - 4ac}}{2a}$$

• $a = 2, b = -1, c = 5$

$$= \frac{-(-1) \pm \sqrt{(-1)^2 - 4 \cdot 2 \cdot 5}}{2 \cdot 2}$$

• Replace a, b, and c in the quadratic formula by their values. Then simplify.

$$= \frac{1 \pm \sqrt{1 - 40}}{4}$$

$$= \frac{1 \pm \sqrt{-39}}{4} = \frac{1 \pm i\sqrt{39}}{4}$$

The solutions are $\frac{1}{4} + \frac{\sqrt{39}}{4}i$ and $\frac{1}{4} - \frac{\sqrt{39}}{4}i$.

Problem 2 Solve by using the quadratic formula.

A. $x^2 + 6x - 9 = 0$ **B.** $4x^2 = 4x - 1$

Solution See pages S26–S27.

Try Exercise 91, page 448.

In the quadratic formula, the quantity $b^2 - 4ac$ is called the **discriminant.** When a, b, and c are real numbers, the discriminant determines whether a quadratic equation will have a double root, two real number solutions that are not equal, or two complex number solutions.

EFFECT OF THE DISCRIMINANT ON THE SOLUTIONS OF A QUADRATIC EQUATION

1. If $b^2 - 4ac = 0$, the equation has one real number solution, a double root.
2. If $b^2 - 4ac > 0$, the equation has two real number solutions that are not equal.
3. If $b^2 - 4ac < 0$, the equation has two complex number solutions.

EXAMPLE 3 Use the discriminant to determine whether $4x^2 - 2x + 5 = 0$ has one real number solution, two real number solutions, or two complex number solutions.

Solution $b^2 - 4ac = (-2)^2 - 4(4)(5)$ • $a = 4, b = -2, c = 5$

$= 4 - 80$

$= -76$

$-76 < 0$ • **The discriminant is less than 0.**

The equation has two complex number solutions.

Problem 3 Use the discriminant to determine whether $3x^2 - x - 1 = 0$ has one real number solution, two real number solutions, or two complex number solutions.

Solution See page S27.

Try Exercise 113, page 449.

8.2 Exercises

CONCEPT CHECK

1. State whether each of the following is a perfect-square trinomial.

a. $x^2 + 6x - 9$ **b.** $x^2 - 12x + 36$

c. $x^2 + 9x + \frac{81}{4}$ **d.** $x^2 - 5x + \frac{25}{2}$

2. Completing the square is a method for solving what type of equation?

3. Can every quadratic equation be solved by using the quadratic formula?

4. Complete. When using the quadratic formula, first write the equation in __?__ form. The value of a is the coefficient of __?__, the value of b is the coefficient of __?__, and c is the __?__ term.

5. After being written in standard form, which of the following equations can be solved by using the quadratic formula?

(i) $3x + 2 = x^2$ (ii) $2(3x + 1) = 4x - 2$

(iii) $2x(x - 1) = 4$ (iv) $(2x + 1)^2 = x - 1$

6. Complete. If the discriminant of a quadratic equation is positive, then the equation will have two __?__ solutions. If the discriminant is negative, the equation will have two __?__ solutions.

1 Solve quadratic equations by completing the square

(See pages 439–442.)

GETTING READY

7. a. To complete the square on $x^2 + 14x$, find the constant c that makes $x^2 + 14x + c$ a __?__ trinomial. The constant term c will be the square of half the coefficient of __?__, so $c = \left[\frac{1}{2}(\underline{\ ?\ })\right]^2 = \underline{\ ?\ }$.

b. Complete the square on $x^2 + 14x$, and write the result as the square of a binomial:

$x^2 + 14x + \underline{\ ?\ } = (\underline{\ ?\ })^2$

8. Rewrite the equation $3x^2 - 6x + 2 = 0$ in the form $(x - m)^2 = n$.

$3x^2 - 6x + 2 = 0$

$3x^2 - 6x = \underline{\ ?\ }$ • **Subtract 2 from each side of the equation.**

$x^2 - (\underline{\ ?\ })x = \underline{\ ?\ }$ • **To make the coefficient of x^2 equal to 1, divide each side of the equation by 3.**

$x^2 - 2x + \underline{\ ?\ } = -\frac{2}{3} + \underline{\ ?\ }$ • **Add $\left[\frac{1}{2}(\underline{\ ?\ })\right]^2$ to each side of the equation.**

$(x - \underline{\ ?\ })^2 = \underline{\ ?\ }$ • **Factor the left side of the equation. Simplify the right side of the equation.**

9. What is the next step when using the method of completing the square to solve $x^2 + 6x = 4$?

10. Are there some quadratic equations that cannot be solved by completing the square? Why or why not?

Solve by completing the square.

11. $x^2 - 4x - 5 = 0$

12. $w^2 - 2w - 24 = 0$

13. $z^2 - 6z + 9 = 0$

14. $u^2 + 10u + 25 = 0$

15. $r^2 + 4r - 7 = 0$

16. $s^2 + 6s - 1 = 0$

17. $x^2 - 6x + 7 = 0$

18. $y^2 + 8y + 13 = 0$

19. $z^2 - 2z + 2 = 0$

20. $t^2 - 4t + 8 = 0$

21. $t^2 - t - 1 = 0$

22. $u^2 - u - 7 = 0$

23. $y^2 - 6y = 4$

24. $w^2 + 4w = 2$

25. $x^2 = 8x - 15$

26. $z^2 = 4z - 3$

27. $v^2 = 4v - 13$

28. $x^2 = 2x - 17$

29. $p^2 + 6p = -13$

30. $x^2 + 4x = -20$

31. $y^2 - 2y = 17$

32. $x^2 + 10x = 7$

33. $z^2 = z + 4$

34. $r^2 = 3r - 1$

35. $x^2 + 13 = 2x$

36. $x^2 + 27 = 6x$

37. $2y^2 + 3y + 1 = 0$

38. $2t^2 + 5t - 3 = 0$

39. $4r^2 - 8r = -3$

40. $4u^2 - 20u = -9$

41. $4x^2 - 4x + 5 = 0$

42. $4t^2 - 4t + 17 = 0$

43. $9x^2 - 6x + 2 = 0$

44. $9y^2 - 12y + 13 = 0$

45. $2s^2 = 4s + 5$

46. $3u^2 = 6u + 1$

47. $y - 2 = (y - 3)(y + 2)$

48. $8s - 11 = (s - 4)(s - 2)$

49. $6t - 2 = (2t - 3)(t - 1)$

50. $2z + 9 = (2z + 3)(z + 2)$

51. $(x - 4)(x + 1) = x - 3$

52. $(y - 3)^2 = 2y + 10$

Solve by completing the square. Approximate the solutions to the nearest thousandth.

53. $z^2 + 2z = 4$

54. $t^2 - 4t = 7$

55. $2x^2 = 4x - 1$

56. $3y^2 = 5y - 1$

57. $4z^2 + 2z - 1 = 0$

58. $4w^2 - 8w = 3$

59. True or false? If $c > 4$, the solutions of $x^2 + 4x + c = 0$ are real numbers.

60. True or false? $x^2 + 2bx = -b^2$ has one real solution.

2 Solve quadratic equations by using the quadratic formula

(See pages 442–446.)

61. Write the quadratic formula. What does each variable in the formula represent?

62. Write the expression that appears under the radical symbol in the quadratic formula. What is this quantity called? What can it be used to determine?

63. Suppose you must solve the quadratic equation $x^2 = 3x + 5$. Does it matter whether you rewrite the equation as $x^2 - 3x - 5 = 0$ or as $0 = -x^2 + 3x + 5$ before you begin?

64. Which method of solving a quadratic equation do you prefer: completing the square or using the quadratic formula? Why?

GETTING READY

65. To write the equation $x^2 = 6x - 10$ in standard form, subtract ___?___ from and add ___?___ to each side of the equation. The resulting equation is ___?___ $= 0$. Then $a =$ ___?___, $b =$ ___?___, and $c =$ ___?___.

66. Use the values of a, b, and c that you found in Exercise 65 to evaluate the discriminant for the equation $x^2 = 6x - 10$:

$$b^2 - 4ac = (\underline{\ ?\ })^2 - 4(\underline{\ ?\ })(\underline{\ ?\ })$$
$$= \underline{\ ?\ } - \underline{\ ?\ } = \underline{\ ?\ }$$

67. Based on the value of the discriminant you found in Exercise 66, the equation $x^2 = 6x - 10$ must have two ___?___ number solutions.

68. Using the quadratic formula to solve $x^2 = 6x - 10$ leads to the result $\frac{6 \pm \sqrt{-4}}{2}$. Simplify this result.

$$\frac{6 \pm \sqrt{-4}}{2} = \frac{6 \pm ?}{2}$$
- **Rewrite $\sqrt{-4}$ as the imaginary number ___?___.**

$$= \underline{\ ?\ } \pm \underline{\ ?\ }$$
- **Divide each term of the numerator by the denominator.**

Solve by using the quadratic formula.

69. $x^2 - 3x - 10 = 0$
70. $y^2 + 5y - 36 = 0$
71. $v^2 = 24 - 5v$
72. $t^2 = 2t + 35$
73. $8s^2 = 10s + 3$
74. $12t^2 = 5t + 2$
75. $v^2 - 2v - 7 = 0$
76. $t^2 - 2t - 11 = 0$
77. $y^2 - 8y - 20 = 0$
78. $x^2 = 14x - 24$
79. $v^2 = 12v - 24$
80. $2z^2 - 2z - 1 = 0$
81. $4x^2 - 4x - 7 = 0$
82. $2p^2 - 8p + 5 = 0$
83. $2s^2 - 3s + 1 = 0$
84. $4w^2 - 4w - 1 = 0$
85. $3x^2 + 10x + 6 = 0$
86. $3v^2 = 6v - 2$
87. $6w^2 = 19w - 10$
88. $z^2 + 2z + 2 = 0$
89. $p^2 - 4p + 5 = 0$
90. $y^2 - 2y + 5 = 0$
91. $x^2 + 6x + 13 = 0$
92. $2w^2 - 2w + 5 = 0$

93. $4v^2 + 8v + 3 = 0$

94. $2x^2 + 6x + 5 = 0$

95. $2y^2 + 2y + 13 = 0$

96. $4t^2 - 6t + 9 = 0$

97. $3v^2 + 6v + 1 = 0$

98. $2r^2 = 4r - 11$

99. $3y^2 = 6y - 5$

100. $2x(x - 2) = x + 12$

101. $10y(y + 4) = 15y - 15$

102. $(3s - 2)(s + 1) = 2$

103. $(2t + 1)(t - 3) = 9$

Solve by using the quadratic formula. Approximate the solutions to the nearest thousandth.

104. $x^2 - 6x - 6 = 0$

105. $p^2 - 8p + 3 = 0$

106. $r^2 - 2r = 4$

107. $w^2 + 4w = 1$

108. $3t^2 = 7t + 1$

109. $2y^2 = y + 5$

Use the discriminant to determine whether the quadratic equation has one real number solution, two real number solutions, or two complex number solutions.

110. $2z^2 - z + 5 = 0$

111. $3y^2 + y + 1 = 0$

112. $9x^2 - 12x + 4 = 0$

113. $4x^2 + 20x + 25 = 0$

114. $2v^2 - 3v - 1 = 0$

115. $3w^2 + 3w - 2 = 0$

116. $2p^2 + 5p + 1 = 0$

117. $2t^2 + 9t + 3 = 0$

118. $5z^2 + 2 = 0$

119. True or false? If $a > 0$ and $c < 0$, the solutions of $ax^2 + bx + c = 0$ are real numbers.

120. True or false? If $a > 0$, $b = 0$, and $c > 0$, the solutions of $ax^2 + bx + c = 0$ are real numbers.

APPLYING CONCEPTS

Solve.

121. $\sqrt{2}x^2 + 5x - 3\sqrt{2} = 0$

122. $\sqrt{3}w^2 + w - 2\sqrt{3} = 0$

123. $t^2 - t\sqrt{3} + 1 = 0$

124. $y^2 + y\sqrt{7} + 2 = 0$

Solve for x.

125. $x^2 - ax - 2a^2 = 0$

126. $x^2 - ax - 6a^2 = 0$

127. $x^2 - 2x - y = 0$

128. $x^2 - 4xy - 4 = 0$

For what values of p does the quadratic equation have two real number solutions that are not equal? Write the answer in set-builder notation.

129. $x^2 - 6x + p = 0$

130. $x^2 + 10x + p = 0$

For what values of p does the quadratic equation have two complex number solutions? Write the answer in set-builder notation.

131. $x^2 - 2x + p = 0$

132. $x^2 + 4x + p = 0$

133. **Sports** The height h (in feet) of a baseball above the ground t seconds after it is hit can be approximated by the equation $h = -16t^2 + 70t + 4$. Using this equation, determine how long it will take for the ball to hit the ground. Round to the nearest hundredth. (*Hint:* The ball hits the ground when $h = 0$.)

134. **Sports** After a baseball is hit, there are two equations that can be considered. One gives the height h (in feet) of the ball above the ground t seconds after it is hit. The second gives the distance s (in feet) of the ball from home plate t seconds after it is hit. A model of this situation is given by $h = -16t^2 + 70t + 4$ and $s = 44.5t$. Using this model, determine whether the ball will clear a fence 325 ft from home plate. Round to the nearest tenth.

	U.S. Forest Service	Lande
s_j	0.34	0.11
s_s	0.97	0.71
s_a	0.97	0.94
f	0.24	0.24

135. **Conservation** The National Forest Management Act of 1976 specifies that harvesting timber in national forests must be accomplished in conjunction with environmental considerations. One such consideration is providing a habitat for the spotted owl. One model of the survival of the spotted owl requires solving the equation $x^2 - s_a x - s_j s_s f = 0$ for x. Different values of s_a, s_j, s_s, and f are given in the table at the right. (*Source:* Charles Biles and Barry Noon, "The Spotted Owl!" from *The Journal of Undergraduate Mathematics and Its Application,* Vol. 11, No. 2, 1990. Reprinted with permission.) The values are particularly important because they relate to the survival of the owl. If $x > 1$, then the model predicts a growth in the population; if $x = 1$, the population remains steady; if $x < 1$, the population decreases. The important solution of the equation is the larger of the two roots of the equation.

a. Determine the larger root of this equation for values provided by the U.S. Forest Service. Round to the nearest hundredth. Does the model predict that the population will increase, remain steady, or decrease?

b. Determine the larger root of this equation for the values provided by R. Lande in *Oecologia* (Vol. 75, 1988). Round to the nearest hundredth. Does the model predict that the population will increase, remain steady, or decrease?

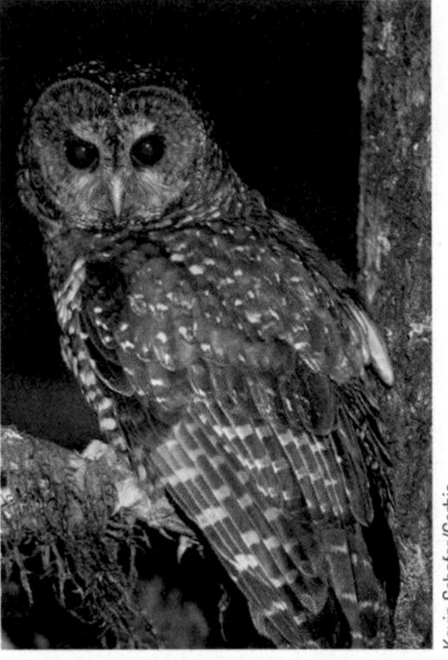
Kevin Schafer/Corbis

136. **Agriculture** An agricultural research group determined that the amount of grass harvested from a pasture could be modeled by the equation $A = -0.001532x^2 + 0.3358x - 7.5323$, where x is the amount, in 100,000 kg per hectare, of fertilizer that is applied to the pasture. According to this model, how many kilograms per hectare of fertilizer must be applied to a pasture to produce 800,000 kg per hectare of grass? Round to the nearest thousand.

PROJECTS OR GROUP ACTIVITIES

The discriminant of a quadratic polynomial can also be defined as the *square of the difference* between the solutions of the equation. For instance, the solutions of the equation $x^2 - x - 6 = 0$ are -2 and 3. The square of the difference between the solutions is $(-2 - 3)^2 = 25$, which is the same value of the discriminant that we would get by using $b^2 - 4ac$. Use this alternative definition to find the discriminant for each of the following equations.

137. $x^2 + 3x - 18 = 0$

138. $x^2 - 6x + 9 = 0$

139. $x^2 - 5x + 2 = 0$

140. $x^2 + 2x + 3 = 0$

8.3 Equations That Are Reducible to Quadratic Equations

OBJECTIVE 1 Equations that are quadratic in form

Certain equations that are not quadratic equations can be expressed in quadratic form by making suitable substitutions. An equation is **quadratic in form** if it can be written as $au^2 + bu + c = 0$.

To see that the equation at the right is quadratic in form, let $x^2 = u$. Replace x^2 by u. The equation is quadratic in form.

$$x^4 - 4x^2 - 5 = 0$$
$$(x^2)^2 - 4(x^2) - 5 = 0$$
$$u^2 - 4u - 5 = 0$$

To see that the equation at the right is quadratic in form, let $y^{\frac{1}{2}} = u$. Replace $y^{\frac{1}{2}}$ by u. The equation is quadratic in form.

$$y - y^{\frac{1}{2}} - 6 = 0$$
$$(y^{\frac{1}{2}})^2 - (y^{\frac{1}{2}}) - 6 = 0$$
$$u^2 - u - 6 = 0$$

The key to recognizing equations that are quadratic in form is that when the equation is written in standard form, the exponent on one variable term is $\frac{1}{2}$ the exponent on the other variable term.

Focus on solving an equation that is quadratic in form

Solve: $z + 7z^{\frac{1}{2}} - 18 = 0$

The equation $z + 7z^{\frac{1}{2}} - 18 = 0$ is quadratic in form.

To solve this equation, let $z^{\frac{1}{2}} = u$.

Solve for u by factoring.

$$z + 7z^{\frac{1}{2}} - 18 = 0$$
$$(z^{\frac{1}{2}})^2 + 7(z^{\frac{1}{2}}) - 18 = 0$$
$$u^2 + 7u - 18 = 0$$
$$(u - 2)(u + 9) = 0$$
$$u - 2 = 0 \qquad u + 9 = 0$$
$$u = 2 \qquad u = -9$$

Replace u by $z^{\frac{1}{2}}$.

$$z^{\frac{1}{2}} = 2 \qquad z^{\frac{1}{2}} = -9$$

Solve for z by squaring each side of the equation.

$$(z^{\frac{1}{2}})^2 = 2^2 \qquad (z^{\frac{1}{2}})^2 = (-9)^2$$
$$z = 4 \qquad z = 81$$

Check the solutions. When each side of the equation has been squared, the resulting equation may have an extraneous solution.

Check:

$$\begin{array}{c|c} z + 7z^{\frac{1}{2}} - 18 & 0 \\ \hline 4 + 7(4)^{\frac{1}{2}} - 18 & 0 \\ 4 + 7 \cdot 2 - 18 & \\ 4 + 14 - 18 & \end{array}$$
$$0 = 0$$

4 checks as a solution.

$$\begin{array}{c|c} z + 7z^{\frac{1}{2}} - 18 & 0 \\ \hline 81 + 7(81)^{\frac{1}{2}} - 18 & 0 \\ 81 + 7 \cdot 9 - 18 & \\ 81 + 63 - 18 & \end{array}$$
$$126 \neq 0$$

81 does not check as a solution.

Write the solution.

The solution is 4.

EXAMPLE 1 Solve: **A.** $x^4 + x^2 - 12 = 0$ **B.** $x^{\frac{2}{3}} - 2x^{\frac{1}{3}} - 3 = 0$

Solution **A.**

$$x^4 + x^2 - 12 = 0$$
$$(x^2)^2 + (x^2) - 12 = 0$$
$$u^2 + u - 12 = 0$$
$$(u - 3)(u + 4) = 0$$

- **The equation is quadratic in form.**
- **Let $x^2 = u$.**
- **Solve for u by factoring.**

$$u - 3 = 0 \qquad u + 4 = 0$$
$$u = 3 \qquad u = -4$$
$$x^2 = 3 \qquad x^2 = -4$$
$$\sqrt{x^2} = \sqrt{3} \qquad \sqrt{x^2} = \sqrt{-4}$$
$$x = \pm\sqrt{3} \qquad x = \pm 2i$$

- **Replace u by x^2.**
- **Solve for x by taking square roots.**

The solutions are $\sqrt{3}$, $-\sqrt{3}$, $2i$, and $-2i$.

B.

$$x^{\frac{2}{3}} - 2x^{\frac{1}{3}} - 3 = 0$$
$$(x^{\frac{1}{3}})^2 - 2(x^{\frac{1}{3}}) - 3 = 0$$
$$u^2 - 2u - 3 = 0$$
$$(u - 3)(u + 1) = 0$$

- **The equation is quadratic in form.**
- **Let $x^{\frac{1}{3}} = u$.**
- **Solve for u by factoring.**

$$u - 3 = 0 \qquad u + 1 = 0$$
$$u = 3 \qquad u = -1$$
$$x^{\frac{1}{3}} = 3 \qquad x^{\frac{1}{3}} = -1 \qquad \text{• Replace } u \text{ by } x^{\frac{1}{3}}.$$
$$(x^{\frac{1}{3}})^3 = 3^3 \qquad (x^{\frac{1}{3}})^3 = (-1)^3 \qquad \text{• Solve for } x \text{ by cubing both sides of the equation.}$$
$$x = 27 \qquad x = -1$$

The solutions are 27 and -1.

Problem 1 Solve. **A.** $x - 5x^{\frac{1}{2}} + 6 = 0$ **B.** $4x^4 + 35x^2 - 9 = 0$

Solution See page S27.

Try Exercise 11, page 455.

OBJECTIVE 

Radical equations

Certain equations containing a radical can be solved by first solving the equation for the radical expression and then squaring each side of the equation.

Remember that when each side of an equation has been squared, the resulting equation may have an extraneous solution. Therefore, the solutions of a radical equation must be checked.

Focus on solving a radical equation

Solve: $\sqrt{x + 2} + 4 = x$

Solve for the radical expression.
$$\sqrt{x + 2} + 4 = x$$
$$\sqrt{x + 2} = x - 4$$

Square each side of the equation.
$$(\sqrt{x + 2})^2 = (x - 4)^2$$

Simplify.
$$x + 2 = x^2 - 8x + 16$$

Write the equation in standard form.
$$0 = x^2 - 9x + 14$$

Solve for x by factoring.
$$0 = (x - 7)(x - 2)$$
$$x - 7 = 0 \qquad x - 2 = 0$$
$$x = 7 \qquad x = 2$$

Check the solution.

Check:
$$\sqrt{x + 2} + 4 = x$$
$$\sqrt{7 + 2} + 4 \mid 7$$
$$\sqrt{9} + 4$$
$$3 + 4$$
$$7 = 7$$

$$\sqrt{x + 2} + 4 = x$$
$$\sqrt{2 + 2} + 4 \mid 2$$
$$\sqrt{4} + 4$$
$$2 + 4$$
$$6 \neq 2$$

2 does not check as a solution.

Write the solution. The solution is 7.

EXAMPLE 2 Solve: $\sqrt{3x + 7} - x = 3$

Solution
$$\sqrt{3x + 7} - x = 3$$
$$\sqrt{3x + 7} = x + 3 \qquad \text{• Solve for the radical expression.}$$
$$(\sqrt{3x + 7})^2 = (x + 3)^2 \qquad \text{• Square each side of the equation.}$$
$$3x + 7 = x^2 + 6x + 9 \qquad \text{• Simplify.}$$
$$0 = x^2 + 3x + 2 \qquad \text{• Write the equation in standard form.}$$
$$0 = (x + 2)(x + 1) \qquad \text{• Factor.}$$

$x + 2 = 0 \qquad x + 1 = 0$ • **Use the Principle of Zero Products.**

$x = -2 \qquad x = -1$

Check:

$$\begin{array}{c|c} \sqrt{3x+7} - x = 3 & \\ \hline \sqrt{3(-2)+7} - (-2) & 3 \\ \sqrt{1} + 2 & \\ 1 + 2 & \\ 3 = 3 & \end{array} \qquad \begin{array}{c|c} \sqrt{3x+7} - x = 3 & \\ \hline \sqrt{3(-1)+7} - (-1) & 3 \\ \sqrt{4} + 1 & \\ 2 + 1 & \\ 3 = 3 & \end{array}$$

-2 and -1 check as solutions. The solutions are -2 and -1.

Problem 2 Solve: $\sqrt{2x+1} + x = 7$

Solution See page S27.

Try Exercise 43, page 456.

If an equation contains more than one radical, the procedure of solving for the radical expression and squaring each side of the equation may have to be repeated.

EXAMPLE 3 Solve: $\sqrt{2x+5} - \sqrt{x+2} = 1$

Solution

$\sqrt{2x+5} - \sqrt{x+2} = 1$

$\sqrt{2x+5} = \sqrt{x+2} + 1$ • **Solve for one of the radical expressions.**

$(\sqrt{2x+5})^2 = (\sqrt{x+2} + 1)^2$ • **Square each side of the equation.**

$2x + 5 = x + 2 + 2\sqrt{x+2} + 1$ • **Simplify.**

$2x + 5 = x + 2\sqrt{x+2} + 3$

$x + 2 = 2\sqrt{x+2}$ • **Solve for the radical expression.**

$(x+2)^2 = (2\sqrt{x+2})^2$ • **Square each side of the equation.**

$x^2 + 4x + 4 = 4(x+2)$ • **Simplify.**

$x^2 + 4x + 4 = 4x + 8$

$x^2 - 4 = 0$ • **Write the equation in standard form.**

$(x+2)(x-2) = 0$ • **Factor.**

$x + 2 = 0 \qquad x - 2 = 0$ • **Use the Principle of Zero Products.**

$x = -2 \qquad x = 2$

The solutions are -2 and 2. • **As shown at the left, -2 and 2 check as solutions.**

Take Note

You must check the solutions in Example 3.

Check:

$$\begin{array}{c|c} \sqrt{2x+5} - \sqrt{x+2} = 1 & \\ \hline \sqrt{2(-2)+5} - \sqrt{(-2)+2} & 1 \\ \sqrt{1} - \sqrt{0} & \\ 1 - 0 & \\ 1 = 1 & \end{array}$$

$$\begin{array}{c|c} \sqrt{2x+5} - \sqrt{x+2} = 1 & \\ \hline \sqrt{2(2)+5} - \sqrt{(2)+2} & 1 \\ \sqrt{9} - \sqrt{4} & \\ 3 - 2 & \\ 1 = 1 & \end{array}$$

Problem 3 Solve: $\sqrt{2x-1} + \sqrt{x} = 2$

Solution See page S27.

Try Exercise 53, page 456.

OBJECTIVE 3 Fractional equations

After each side of a fractional equation has been multiplied by the LCD, the resulting equation is sometimes a quadratic equation. The solutions to the resulting equation must be checked, because multiplying each side of an equation by a variable expression may produce an equation that has a solution that is not a solution of the original equation.

Focus on solving a fractional equation

Solve: $\frac{1}{r} + \frac{1}{r+1} = \frac{3}{2}$

$$\frac{1}{r} + \frac{1}{r+1} = \frac{3}{2}$$

Multiply each side of the equation by the LCD.

$$2r(r+1)\left(\frac{1}{r} + \frac{1}{r+1}\right) = 2r(r+1) \cdot \frac{3}{2}$$

$$2(r+1) + 2r = r(r+1) \cdot 3$$

$$2r + 2 + 2r = (r^2 + r) \cdot 3$$

$$4r + 2 = 3r^2 + 3r$$

Write the equation in standard form.

$$0 = 3r^2 - r - 2$$

$$0 = (3r+2)(r-1)$$

Solve for r by factoring.

$$3r + 2 = 0 \qquad r - 1 = 0$$

$$3r = -2 \qquad r = 1$$

$$r = -\frac{2}{3}$$

$-\frac{2}{3}$ and 1 check as solutions.

Write the solutions.

The solutions are $-\frac{2}{3}$ and 1.

EXAMPLE 4 Solve: $\frac{18}{2a-1} + 3a = 17$

Solution

$$\frac{18}{2a-1} + 3a = 17$$

• The LCD is $2a - 1$.

$$(2a-1)\left(\frac{18}{2a-1} + 3a\right) = (2a-1)17$$

$$(2a-1)\frac{18}{2a-1} + (2a-1)(3a) = (2a-1)17$$

$$18 + 6a^2 - 3a = 34a - 17$$

$$6a^2 - 37a + 35 = 0$$

• Write the equation in standard form.

$$(6a-7)(a-5) = 0$$

• Solve for a by factoring.

$$6a - 7 = 0 \qquad a - 5 = 0$$

$$6a = 7 \qquad a = 5$$

$$a = \frac{7}{6}$$

$\frac{7}{6}$ and 5 check as solutions.

The solutions are $\frac{7}{6}$ and 5.

Problem 4 Solve: $3y + \frac{25}{3y-2} = -8$

Solution See page S27.

Try Exercise 61, page 457.

8.3 Exercises

CONCEPT CHECK

1. State whether each equation is quadratic in form.

a. $x^5 + 7x^3 + 12 = 0$ **b.** $x^4 - x^2 - 1 = 0$

c. $2x^{\frac{1}{2}} + 3x^{\frac{1}{4}} - 5 = 0$ **d.** $4 - 3x^3 + 2x^6 = 0$

2. Simplify each of the following.

a. $(\sqrt{x-5})^2$ **b.** $(\sqrt{x} - 5)^2$

c. $(\sqrt{2x} + 3)^2$ **d.** $(\sqrt{2x+3})^2$

3. What is the first step in solving the equation $\sqrt{x+1} + x = 5$?

4. Complete. Squaring each side of an equation may introduce an __?__ solution.

1 Equations that are quadratic in form (See pages 450–452.)

GETTING READY

5. The equation $x^8 + 4x^4 - 5 = 0$ is equivalent to the quadratic equation $u^2 + 4u - 5 = 0$, where $u =$ __?__.

6. The equation $p - 2p^{\frac{1}{2}} - 8 = 0$ is equivalent to the quadratic equation $u^2 - 2u - 8 = 0$, where $u =$ __?__.

7. What does it mean for an equation to be quadratic in form?

8. Explain how to show that $x^4 - 2x^2 - 3 = 0$ is quadratic in form.

Solve.

9. $x^4 - 13x^2 + 36 = 0$

10. $y^4 - 5y^2 + 4 = 0$

11. $z^4 - 6z^2 + 8 = 0$

12. $t^4 - 12t^2 + 27 = 0$

13. $p - 3p^{\frac{1}{2}} + 2 = 0$

14. $v - 7v^{\frac{1}{2}} + 12 = 0$

15. $x - x^{\frac{1}{2}} - 12 = 0$

16. $w - 2w^{\frac{1}{2}} - 15 = 0$

17. $z^4 + 3z^2 - 4 = 0$

18. $y^4 + 5y^2 - 36 = 0$

19. $x^4 + 12x^2 - 64 = 0$

20. $x^4 - 81 = 0$

21. $p + 2p^{\frac{1}{2}} - 24 = 0$

22. $v + 3v^{\frac{1}{2}} - 4 = 0$

23. $y^{\frac{2}{3}} - 9y^{\frac{1}{3}} + 8 = 0$

24. $z^{\frac{2}{3}} - z^{\frac{1}{3}} - 6 = 0$

25. $x^6 - 9x^3 + 8 = 0$

26. $y^6 + 9y^3 + 8 = 0$

27. $z^8 - 17z^4 + 16 = 0$

28. $v^4 - 15v^2 - 16 = 0$

29. $p^{\frac{2}{3}} + 2p^{\frac{1}{3}} - 8 = 0$

30. $w^{\frac{2}{3}} + 3w^{\frac{1}{3}} - 10 = 0$

31. $2x - 3x^{\frac{1}{2}} + 1 = 0$

32. $3y - 5y^{\frac{1}{2}} - 2 = 0$

33. Which equations can be solved by solving the quadratic equation $u^2 - 8u - 20 = 0$?

(i) $x^{10} - 8x^5 - 20 = 0$ (ii) $x^{16} - 8x^4 - 20 = 0$

(iii) $x^{\frac{1}{10}} - 8x^{\frac{1}{5}} - 20 = 0$ (iv) $x^{\frac{2}{5}} - 8x^{\frac{1}{5}} - 20 = 0$

34. Each of the following equations can be solved by solving the quadratic equation $u^2 + 5u + 6 = 0$. Which equations could possibly end up with extraneous solutions?

(i) $x^{\frac{1}{4}} + 5x^{\frac{1}{8}} + 6 = 0$ (ii) $x^8 + 5x^4 + 6 = 0$

(iii) $x + 5x^{\frac{1}{2}} + 6 = 0$ (iv) $x^{\frac{2}{3}} + 5x^{\frac{1}{3}} + 6 = 0$

2 Radical equations (See pages 452–453.)

GETTING READY

35. The first step in the solution of the equation $\sqrt{a+1} + 7 = a + 2$ is to solve for __?__ by subtracting __?__ from each side of the equation. The second step is to __?__ each side of the equation.

36. Solving the equation given in Exercise 35 yields $a = 8$ and $a = 3$. Checking these solutions in the original equation shows that the solution __?__ is extraneous.

37. Look at the equations to be solved in Exercises 39 to 56. List the exercise numbers of the equations for which the first step of the solution will be to square both sides of the equation.

38. Look at the equations to be solved in Exercises 39 to 56. List the exercise numbers of the equations whose solutions will require squaring both sides two separate times.

Solve.

39. $\sqrt{3w + 3} = w + 1$

40. $\sqrt{2s + 1} = s - 1$

41. $\sqrt{p + 11} = 1 - p$

42. $x - 7 = \sqrt{x - 5}$

43. $\sqrt{2y - 1} = y - 2$

44. $\sqrt{t + 8} = 2t + 1$

45. $\sqrt{x + 1} + x = 5$

46. $\sqrt{x - 4} + x = 6$

47. $x = \sqrt{x} + 6$

48. $\sqrt{4y + 1} - y = 1$

49. $\sqrt{10x + 5} - 2x = 1$

50. $\sqrt{3s + 4} + 2s = 12$

51. $\sqrt{2x - 1} = 1 - \sqrt{x - 1}$

52. $\sqrt{y} + 1 = \sqrt{y + 5}$

53. $\sqrt{x - 1} - \sqrt{x} = -1$

54. $\sqrt{5 - 2x} = \sqrt{2 - x} + 1$

55. $\sqrt{t + 3} + \sqrt{2t + 7} = 1$

56. $\sqrt{x + 6} + \sqrt{x + 2} = 2$

3 Fractional equations (See pages 453–454.)

GETTING READY

57. Use the equation $\frac{4}{y} = \frac{y - 6}{y - 4}$.

a. The first step in solving the equation is to clear denominators by multiplying each side of the equation by the LCM of the denominators __?__ and __?__. The LCD is __?__.

b. When you multiply the left side of the equation by the LCD $y(y - 4)$, the left side simplifies to __?__.

c. When you multiply the right side of the equation by the LCD $y(y - 4)$, the right side simplifies to __?__.

d. Use your answers to parts (b) and (c) to write the equation that results from clearing denominators: __?__ = __?__.

58. a. Using the results of Exercise 57, we know we can solve the rational equation $\frac{4}{y} = \frac{y-6}{y-4}$ by solving the quadratic equation $4y - 16 = y^2 - 6y$. Written in standard form, this equation is $0 = \underline{\quad ? \quad}$.

b. Solve $0 = y^2 - 10y + 16$ by factoring: $0 = (\underline{\ ?\ })(\underline{\ ?\ })$, so $y = \underline{\ ?\ }$ or $y = \underline{\ ?\ }$.

c. Both solutions from part (b) check in the original equation, because neither solution makes one of the denominators equal to $\underline{\ ?\ }$.

Solve.

59. $x = \frac{10}{x-9}$

60. $z = \frac{5}{z-4}$

61. $\frac{y-1}{y+2} + y = 1$

62. $\frac{2p-1}{p-2} + p = 8$

63. $\frac{3r+2}{r+2} - 2r = 1$

64. $\frac{2v+3}{v+4} + 3v = 4$

65. $\frac{2}{2x+1} + \frac{1}{x} = 3$

66. $\frac{3}{s} - \frac{2}{2s-1} = 1$

67. $\frac{16}{z-2} + \frac{16}{z+2} = 6$

68. $\frac{2}{y+1} + \frac{1}{y-1} = 1$

69. $\frac{t}{t-2} + \frac{2}{t-1} = 4$

70. $\frac{4t+1}{t+4} + \frac{3t-1}{t+1} = 2$

71. $\frac{5}{2p-1} + \frac{4}{p+1} = 2$

72. $\frac{3w}{2w+3} + \frac{2}{w+2} = 1$

73. $\frac{x+2}{x-4} + \frac{2x-4}{x+1} = 8$

APPLYING CONCEPTS

Solve.

74. $\frac{x^4}{4} + 1 = \frac{5x^2}{4}$

75. $\frac{x+2}{3} + \frac{2}{x-2} = 3$

76. $\frac{x^2}{4} + \frac{x}{2} + \frac{1}{8} = 0$

77. $\frac{x^4}{3} - \frac{8x^2}{3} = 3$

78. $\frac{x^4}{8} + \frac{x^2}{4} = 3$

79. $\sqrt{x^4 + 4} = 2x$

80. $(\sqrt{x} - 2)^2 - 5\sqrt{x} + 14 = 0$ (*Hint:* Let $u = \sqrt{x} - 2$.)

81. $(\sqrt{x} + 3)^2 - 4\sqrt{x} - 17 = 0$ (*Hint:* $u = \sqrt{x} + 3$.)

PROJECTS OR GROUP ACTIVITIES

82. **Sports** According to the Compton's Interactive Encyclopedia, the minimum dimensions of a football used in the National Football Association games are 10.875 in. long and 20.75 in. in circumference at the center. A possible model for the cross section of a football is given by $y = \pm 3.3041\sqrt{1 - \frac{x^2}{29.7366}}$, where x is the distance from the center of the football and y is the radius of the football at x.

a. What is the domain of the equation?

b. Graph $y = 3.3041\sqrt{1 - \frac{x^2}{29.7366}}$ and $y = -3.3041\sqrt{1 - \frac{x^2}{29.7366}}$ on the same coordinate axes. Explain why the $\pm$ symbol occurs in the equation.

c. Determine the radius of the football when x is 3 in. Round to the nearest ten-thousandth.

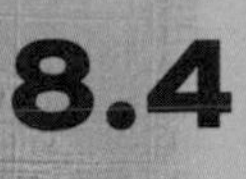

8.4 Applications of Quadratic Equations

OBJECTIVE 1

Application problems

The application problems in this section are similar to problems that were solved earlier in the text. Each of the strategies for the problems in this section will result in a quadratic equation.

Solve: A small pipe takes 16 min longer to empty a tank than does a larger pipe. Working together, the pipes can empty the tank in 6 min. How long would it take each pipe, working alone, to empty the tank?

STRATEGY FOR SOLVING AN APPLICATION PROBLEM

▶ Determine the type of problem. Is it a uniform motion problem, a geometry problem, an integer problem, or a work problem?

The problem is a work problem.

▶ Choose a variable to represent the unknown quantity. Write numerical or variable expressions for all the remaining quantities. These results can be recorded in a table.

The unknown time of the larger pipe: t
The unknown time of the smaller pipe: $t + 16$

	Rate of work	$\cdot$	Time worked	$=$	Part of task completed
Larger pipe	$\frac{1}{t}$	$\cdot$	6	$=$	$\frac{6}{t}$
Smaller pipe	$\frac{1}{t+16}$	$\cdot$	6	$=$	$\frac{6}{t+16}$

▶ Determine how the quantities are related. If necessary, review the strategies presented in earlier chapters.

The sum of the parts of the task completed must equal 1.

$$\frac{6}{t} + \frac{6}{t+16} = 1$$

$$t(t+16)\left(\frac{6}{t} + \frac{6}{t+16}\right) = t(t+16)\cdot 1$$

$$(t+16)6 + 6t = t^2 + 16t$$

$$6t + 96 + 6t = t^2 + 16t$$

$$0 = t^2 + 4t - 96$$

$$0 = (t+12)(t-8)$$

$$t + 12 = 0 \qquad t - 8 = 0$$

$$t = -12 \qquad t = 8$$

The solution $t = -12$ is not possible because time cannot be a negative number.

The time for the smaller pipe is $t + 16$. Replace t by 8 and evaluate.

$$t + 16 = 8 + 16 = 24$$

The larger pipe requires 8 min to empty the tank.
The smaller pipe requires 24 min to empty the tank.

EXAMPLE 1 A kicker punts a football at an angle of 60° with the ground. Assuming no air resistance, the height h, in feet, of the punted football x feet from where it was kicked can be given by $h = -0.0065x^2 + 1.73x + 4$. How far is the football from the kicker when the height of the football is 70 ft? Round to the nearest tenth.

Strategy To find the football's distance from the kicker when it is 70 ft above the ground, solve the equation $h = -0.0065x^2 + 1.73x + 4$ for x when $h = 70$.

Solution

$$h = -0.0065x^2 + 1.73x + 4$$

$$70 = -0.0065x^2 + 1.73x + 4$$ • **Replace h by 70.**

$$0 = -0.0065x^2 + 1.73x - 66$$ • **Write in standard form.**

$$x = \frac{-1.73 \pm \sqrt{1.73^2 - 4(-0.0065)(-66)}}{2(-0.0065)}$$ • **Solve by using the quadratic formula.**

$$x = \frac{-1.73 \pm \sqrt{1.2769}}{-0.013} \approx \frac{-1.73 \pm 1.13}{-0.013}$$

$$x \approx \frac{-1.73 + 1.13}{-0.013} \qquad x \approx \frac{-1.73 - 1.13}{-0.013}$$

$$x \approx 46.2 \qquad x \approx 220$$

When the football is 70 ft high, it is either 46.2 ft or 220 ft from the kicker.

The flight of the football is shown below. Note that it is 70 ft above the ground twice, when $x = 46.2$ ft and when $x = 220$ ft from the kicker.

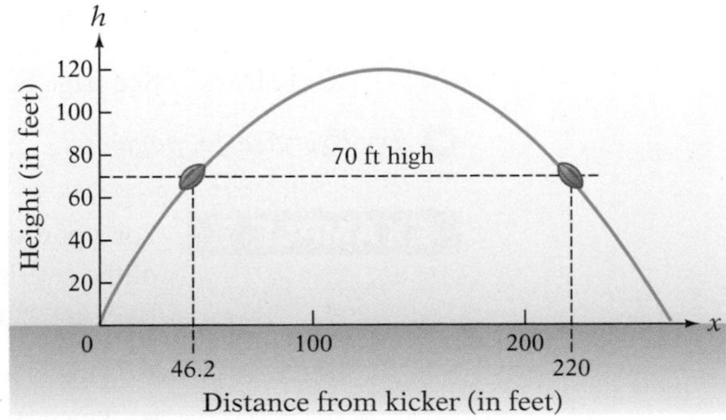

Problem 1 For the punt in Example 1, the height h, in feet, of the football t seconds after it has been kicked is given by $h = -16t^2 + 60.6t + 4$. Assuming the ball is not caught and lands on the ground, what is the "hang time"—the amount of time the ball is in the air? Round to the nearest tenth.

Solution See pages S27–S28.

Try Exercise 7, page 461.

EXAMPLE 2 A swimming pool is being emptied using two hoses. The smaller hose takes 2 h longer to empty the pool than does the larger hose. After the valves on both hoses have been opened for 1 h, the larger hose is turned off. It takes the smaller hose 1 more hour to empty the pool. How long would it take the larger hose, working alone, to empty the pool?

Strategy
- ▶ This is a work problem.
- ▶ The unknown time for the larger hose working alone: t
- ▶ The unknown time for the smaller hose working alone: $t + 2$
- ▶ The larger hose operates for 1 h; the smaller hose operates for 2 h.

	Rate	·	Time	=	Part
Larger hose	$\frac{1}{t}$	·	1	=	$\frac{1}{t}$
Smaller hose	$\frac{1}{t+2}$	·	2	=	$\frac{2}{t+2}$

- ▶ The sum of the part of the task completed by the larger hose and the part completed by the smaller hose is 1.

Solution

$$\frac{1}{t} + \frac{2}{t+2} = 1$$

$$t(t+2)\left(\frac{1}{t} + \frac{2}{t+2}\right) = t(t+2) \cdot 1$$
• **Multiply each side of the equation by the LCD.**

$$(t+2) + 2t = t^2 + 2t$$
• **Simplify.**

$$0 = t^2 - t - 2$$
• **Write the quadratic equation in standard form.**

$$0 = (t+1)(t-2)$$
• **Factor.**

$$t + 1 = 0 \qquad t - 2 = 0$$
$$t = -1 \qquad t = 2$$
• **Use the Principle of Zero Products.**

Because time cannot be negative, $t = -1$ is not possible.

It would take the larger hose, working alone, 2 h to empty the pool.

Problem 2 It takes Seth 3 h longer to wash the windows in a house than it does Tessa. Working together, they can wash the windows in 2 h. How long would it take Tessa, working alone, to wash the windows in the house?

Solution See page S28.

Try Exercise 13, page 462.

EXAMPLE 3 For a portion of the Snake River as it winds through Idaho, the rate of the river's current is 4 mph. A tour guide can row 5 mi down this river and back in 3 h. Find the rowing rate of the guide in calm water.

Strategy
- ▶ This is a uniform motion problem.
- ▶ The unknown rowing rate of the guide: r

	Distance	÷	Rate	=	Time
Down river	5	÷	$r + 4$	=	$\frac{5}{r+4}$
Up river	5	÷	$r - 4$	=	$\frac{5}{r-4}$

- ▶ The total time of the trip was 3 h.

Solution
$$\frac{5}{r+4} + \frac{5}{r-4} = 3$$

$$(r+4)(r-4)\left(\frac{5}{r+4} + \frac{5}{r-4}\right) = (r+4)(r-4) \cdot 3$$
• Multiply each side of the equation by the LCD.

$$5(r-4) + 5(r+4) = 3r^2 - 48$$
• Simplify.

$$10r = 3r^2 - 48$$

$$0 = 3r^2 - 10r - 48$$
• Write the quadratic equation in standard form.

$$0 = (3r+8)(r-6)$$
• Factor.

$$3r + 8 = 0 \qquad r - 6 = 0$$
$$r = -\frac{8}{3} \qquad r = 6$$
• Use the Principle of Zero Products.

Because the rate cannot be negative, $r = -\frac{8}{3}$ is not possible.
The tour guide's rowing rate in calm water is 6 mph.

Problem 3 The rate of a jet in calm air is 250 mph. Flying with the wind, the jet can fly 1200 mi in 2 h less time than is required to make the return trip against the wind. Find the rate of the wind.

Solution See page S28.

Try Exercise 21, page 463.

8.4 Exercises

CONCEPT CHECK

1. If the work on a project takes t hours to complete, what portion of the job is completed in 1 h?

2. Suppose one person can do a job in 2 h and a second person can do the same job in 3 h. If they work together, will they complete the job in less than 2 h, in between 2 h and 3 h, or in more than 3 h?

3. Suppose one person can do a task in 3 h and a second person can do the same task in 5 h. If they work together, how much of the task will be completed in 1 h?

4. Let r be the rowing rate of a person in calm water. If the rate of a river's current is 2 mph, express the rate rowing down the river (with the current) and the rate rowing up the river (against the current) as variable expressions.

5. Let r be the rate at which a plane can fly in calm air. Write a variable expression for the time it would take the plane to fly 500 mi into a headwind of 50 mph.

6. Let r be the rate at which a plane can fly in calm air. Write a variable expression for the time it would take the plane to fly 500 mi with a tailwind of 50 mph.

1 Application problems (See pages 458–461.)

7. **Physics** The height of a projectile fired upward is given by the formula $s = v_0t - 16t^2$, where s is the height, v_0 is the initial velocity, and t is the time. Find the time for a projectile to return to Earth if it has an initial velocity of 200 ft/s.

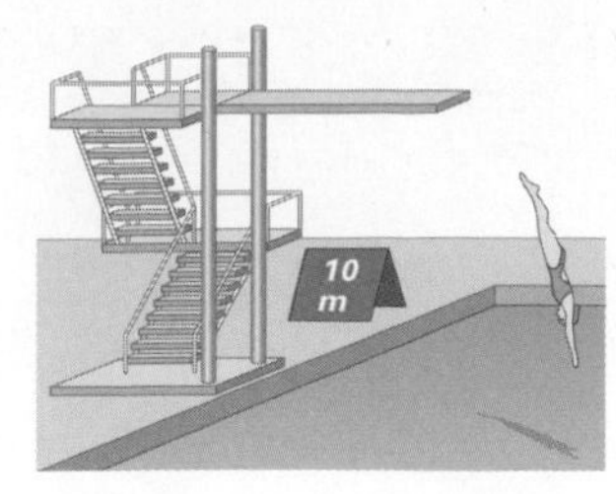

8. **Sports** A diver jumps from a platform that is 10 m high. The height of the diver t seconds after jumping is given by $h = -4.9t^2 + 3.2t + 10.5$. To the nearest hundredth of a second, how long after jumping from the platform will the diver enter the water?

9. **Automotive Technology** The distance d (in meters) required to stop a car traveling at v kilometers per hour is $d = 0.019v^2 + 0.69v$. Approximate, to the nearest tenth, the maximum speed at which a driver can be driving and still be able to stop within 150 m.

10. **Physics** The height of a projectile fired upward is given by the formula $s = v_0t - 16t^2$, where s is the height, v_0 is the initial velocity, and t is the time. Find the times at which a projectile with an initial velocity of 128 ft/s will be 64 ft above the ground. Round to the nearest hundredth of a second.

11. **Aeronautics** A model rocket is launched with an initial velocity of 200 ft/s. The height h of the rocket t seconds after the launch is given by $h = -16t^2 + 200t$. How many seconds after it is launched will the rocket be 300 ft above the ground? Round to the nearest hundredth of a second.

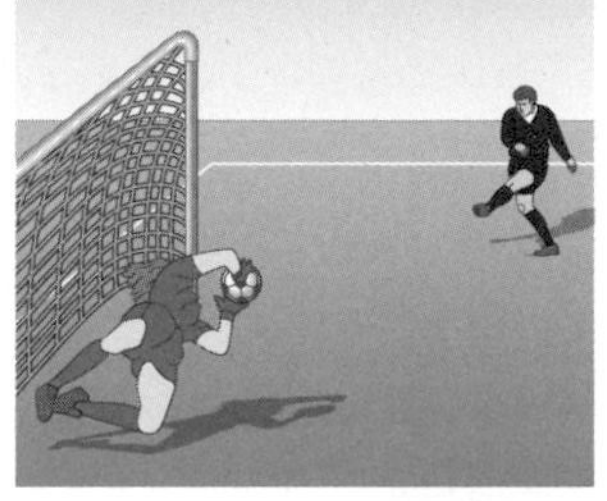

12. **Sports** A penalty kick in soccer is made from a penalty mark that is 36 ft from a goal that is 8 ft high. A possible equation for the flight of a penalty kick is $h = -0.002x^2 + 0.35x$, where h is the height (in feet) of the ball x feet from the penalty mark. Assuming that the flight of the kick is toward the goal and that the ball is not touched by the goalie, will the ball land in the net?

13. **Petroleum Engineering** A small pipe can fill an oil tank in 6 min more time than it takes a larger pipe to fill the same tank. Working together, both pipes can fill the tank in 4 min. How long would it take each pipe, working alone, to fill the tank?

14. **Metallurgy** A small heating unit takes 8 h longer to melt a piece of iron than does a larger unit. Working together, the heating units can melt the iron in 3 h. How long would it take each heating unit, working alone, to melt the iron?

15. **Parallel Processing** Parallel processing is the simultaneous use of more than one computer to run a program. Suppose one computer, working alone, takes 4 h longer than a second computer to run a program that determines whether a number is prime. After the computers work together for 1 h, the faster computer crashes. The slower computer continues for another 2 h before completing the program. How long would it take the faster computer, working alone, to complete the program?

16. **Construction** It takes an apprentice carpenter 2 h longer to install a section of wood floor than it does a more experienced carpenter. After the carpenters work together for 2 h, the experienced carpenter leaves for another job. It then takes the apprentice carpenter 2 h longer to complete the installation. How long would it take the apprentice carpenter, working alone, to install the floor? Round to the nearest tenth.

17. **Landscaping** It takes a small sprinkler 16 min longer to soak a lawn than it takes a large sprinkler. Working together, the sprinklers can soak the lawn in 6 min. How long would it take each sprinkler, working alone, to soak the lawn?

18. **Payroll** It takes one printer, working alone, 6 h longer to print a weekly payroll than it takes a second printer. Working together, the two printers can print the weekly payroll in 4 h. How long would it take each printer, working alone, to print the payroll?

19. **Recreation** A cruise ship made a trip of 100 mi in 8 h. The ship traveled the first 40 mi at a constant rate before increasing its speed by 5 mph. Another 60 mi was traveled at the increased speed. Find the rate of the cruise ship for the first 40 mi.

20. **Sports** A cyclist traveled 60 mi at a constant rate before reducing the speed by 2 mph. Another 40 mi was traveled at the reduced speed. The total time for the 100-mile trip was 9 h. Find the rate of the cyclist during the first 60 mi.

21. **Aeronautics** The rate of a single-engine plane in calm air is 100 mph. Flying with the wind, the plane can fly 240 mi in 1 h less time than is required to make the return trip of 240 mi. Find the rate of the wind.

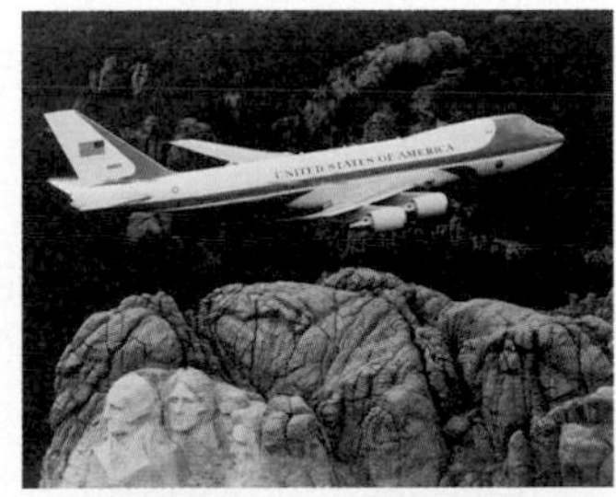
USAF

22. **Transportation** A car travels 120 mi. A second car, traveling 10 mph faster than the first car, makes the same trip in 1 h less time. Find the speed of each car.

23. **Air Force One** The United States Air Force uses the designation VC-25 for the plane on which the president of the United States flies. When the president is on the plane, its call sign is Air Force One. The plane's speed in calm air is 630 mph. Flying with the jet stream, the plane can fly from Washington, D.C., to London, a distance of approximately 3660 mi, in 1.75 h less time than is required to make the return trip. Find the rate of the jet stream. Round to the nearest tenth of a mile per hour.

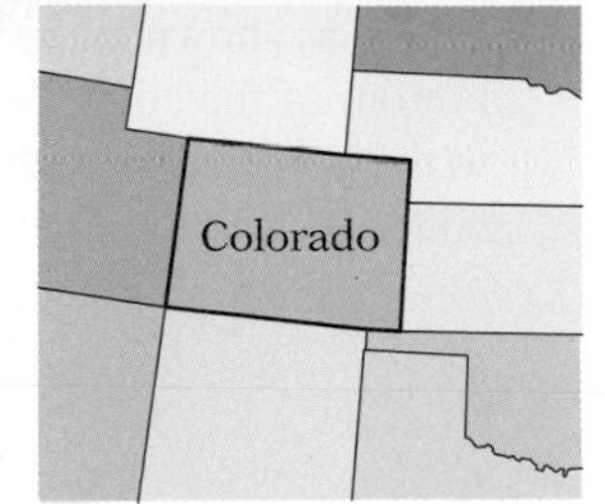

24. **Geography** The state of Colorado is almost perfectly rectangular, with its north border 111 mi longer than its west border. The state encompasses 104,000 mi^2. Estimate the dimensions of Colorado. Round to the nearest mile.

25. **Geometry** An open box is formed from a rectangular piece of cardboard whose length is 8 cm more than its width by cutting squares whose sides are 2 cm in length from each corner and then folding up the sides. Find the dimensions of the box if its volume is 256 cm^3.

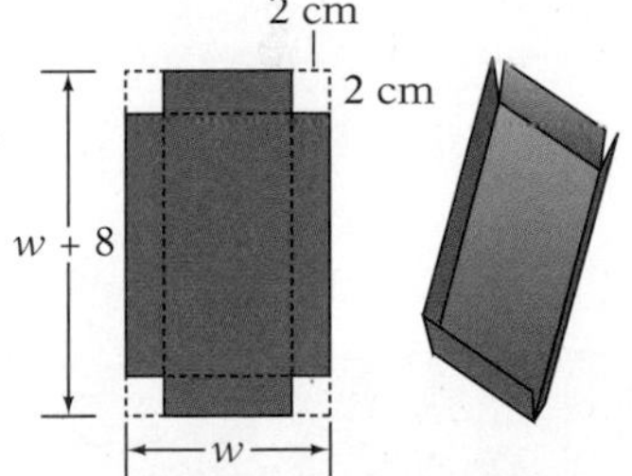

26. **Geometry** A square piece of cardboard is formed into a box by cutting 10-centimeter squares from each of the four corners and then folding up the sides, as shown in the figure. If the volume of the box is to be 49,000 cm^3, what size square piece of cardboard is needed?

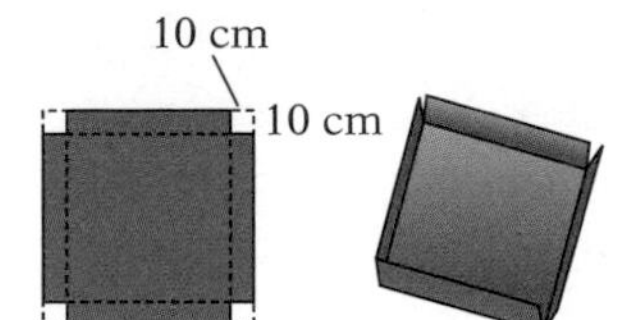

27. **Sports** Read the article at the right.
   a. The screen behind the east end zone is a rectangle with length 13 ft less than three times its width. Find the length and width of the east-end-zone screen.
   b. The screen behind the west end zone is a rectangle with length 1 ft less than twice its width. Find the length and width of the west-end-zone screen.

28. **Fencing** A dog trainer has 80 ft of fencing that will be used to create a rectangular work area for dogs. If the trainer wants to enclose an area of 300 ft^2, what will be the dimensions of the work area?

In the News

Dolphins on the Big Screen

The Miami Dolphins' stadium, Landshark Stadium, is home to one of the world's largest digital displays—a 6850-square-foot screen installed behind the east end zone. A smaller screen, of area 4950 ft^2, sits behind the west end zone.

Source: Business Wire

29. **Construction** A homeowner hires a mason to lay a brick border around a cement patio that measures 8 ft by 10 ft. If the total area of the patio and border is 168 ft^2, what is the width of the border?

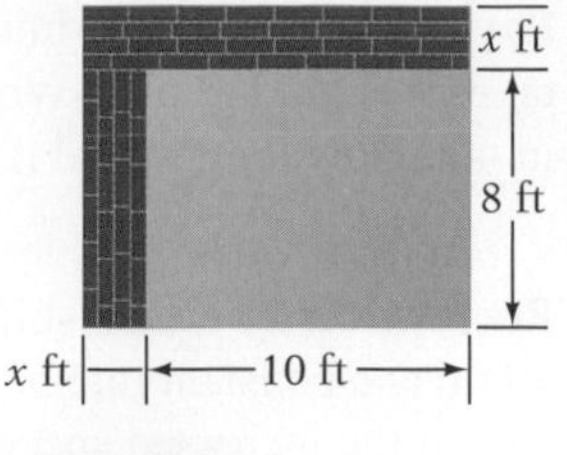

APPLYING CONCEPTS

30. **Manufacturing** The surface area of the ice cream cone shown at the right is given by $A = \pi r^2 + \pi rs$, where r is the radius of the circular top of the ice cream cone and s is the slant height of the cone. If the area of the cone is 11.25π in^2 and the slant height of the cone is 6 in., find the radius of the cone.

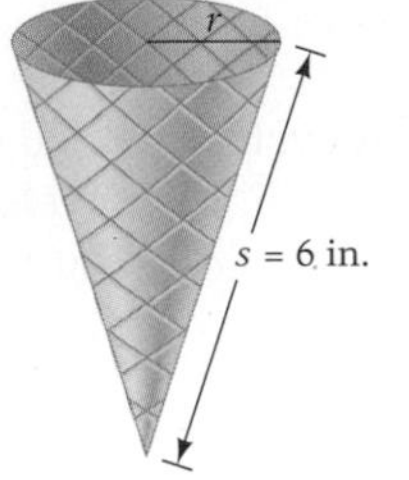

PROJECTS OR GROUP ACTIVITIES

31. **Automotive** A car with good tread can stop in less distance than a car with poor tread. A model for the stopping distance d, in feet, of a car with good tread on dry cement is $d = 0.04v^2 + 0.5v$, where v is the speed of the car in miles per hour. If the driver must be able to stop within 60 ft, what is the maximum safe speed of the car? Round to the nearest whole number.

32. **Physics** When the valve in a certain water tank is opened, the depth d, in centimeters, of the water t seconds after the valve is opened can be approximated by $d = 0.0000177t^2 - 0.0532t + 400$. How long after the valve is opened will the depth of the water be 100 cm? Round to the nearest tenth.

33. **Geometry** A perfectly spherical scoop of mint chocolate chip ice cream is placed in a cone as shown at the right. How far is the bottom of the scoop of ice cream from the bottom of the cone? (*Hint:* A line segment from the center of the scoop of ice cream to the point at which the ice cream touches the cone is perpendicular to the edge of the cone.) Round to the nearest tenth.

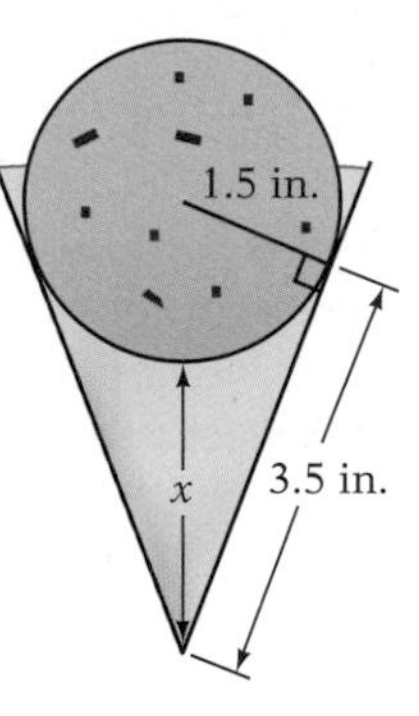

8.5 Properties of Quadratic Functions

OBJECTIVE  1 Graph a quadratic function

QUADRATIC FUNCTION

A **quadratic function** is a function that can be expressed by the equation $f(x) = ax^2 + bx + c, a \neq 0$.

EXAMPLES

1. $f(x) = 2x^2 - 3x + 4$ $\quad a = 2, b = -3, c = 4$
2. $g(x) = x^2 + 4x$ $\quad a = 1, b = 4, c = 0$
3. $h(x) = 6 - x^2$ $\quad a = -1, b = 0, c = 6$

The graph of a quadratic function can be drawn by finding ordered pairs that belong to the function.

Focus on graphing a quadratic function

Graph: $f(x) = x^2 - 2x - 3$

Evaluate the function for various values of x. Find enough ordered pairs to determine the shape of the graph.

x	$f(x) = x^2 - 2x - 3$	$f(x)$	(x, y)
-2	$f(-2) = (-2)^2 - 2(-2) - 3$	5	$(-2, 5)$
-1	$f(-1) = (-1)^2 - 2(-1) - 3$	0	$(-1, 0)$
0	$f(0) = (0)^2 - 2(0) - 3$	-3	$(0, -3)$
1	$f(1) = (1)^2 - 2(1) - 3$	-4	$(1, -4)$
2	$f(2) = (2)^2 - 2(2) - 3$	-3	$(2, -3)$
3	$f(3) = (3)^2 - 2(3) - 3$	0	$(3, 0)$
4	$f(4) = (4)^2 - 2(4) - 3$	5	$(4, 5)$

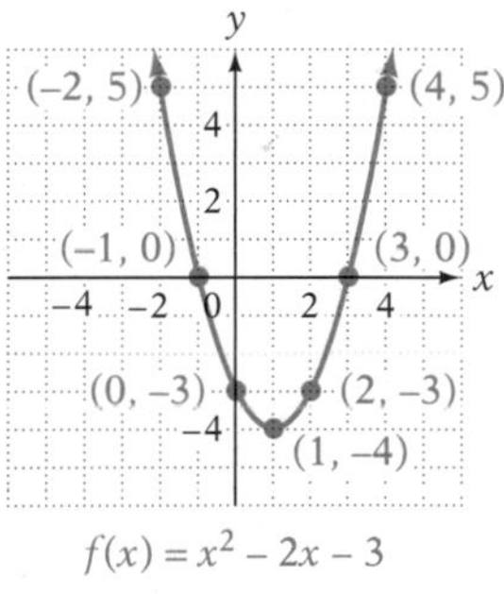

$f(x) = x^2 - 2x - 3$

The graph of a quadratic function, called a **parabola,** is "cup" shaped, as shown above. The cup can open up or down. The parabola opens up when $a > 0$ and opens down when $a < 0$.

Point of Interest

The suspension cables for some bridges, such as the Golden Gate bridge, have the shape of a parabola. Parabolic shapes are also used for the mirrors in telescopes and in the design of certain antennas.

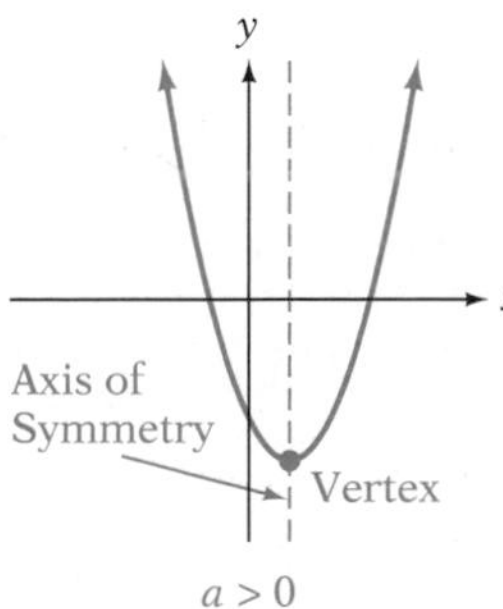

$a > 0$

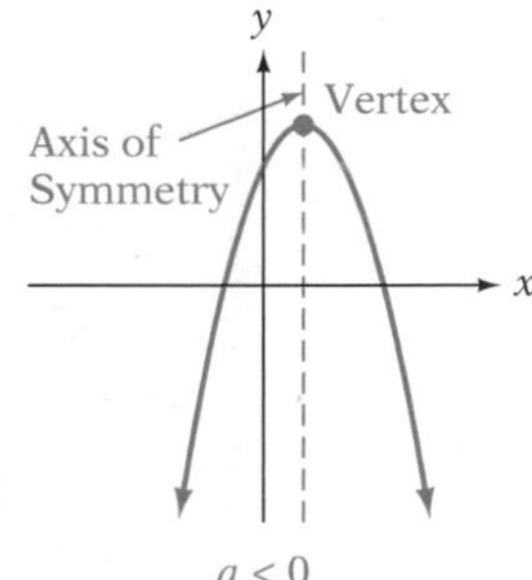

$a < 0$

The **vertex** of a parabola is the point with the least y-coordinate when $a > 0$ and the point with the greatest y-coordinate when $a < 0$. The line that passes through the vertex and is parallel to the y-axis is called the **axis of symmetry.** To understand the axis of symmetry, think of folding the graph along that line. The two halves of the graph would match up.

Focus on finding the coordinates of the vertex and the equation of the axis of symmetry of a parabola

Find the coordinates of the vertex and the equation of the axis of symmetry of the graph of $F(x) = x^2 + 4x + 3$.

To find the coordinates of the vertex, complete the square. Group the variable terms.

$$F(x) = x^2 + 4x + 3$$
$$F(x) = (x^2 + 4x) + 3$$

Complete the square of $x^2 + 4x$. Add and subtract $\left[\frac{1}{2}(4)\right]^2 = 4$ to and from $x^2 + 4x$.	$F(x) = (x^2 + 4x + 4) - 4 + 3$
Factor and combine like terms.	$F(x) = (x + 2)^2 - 1$
Because a, the coefficient of x^2, is positive ($a = 1$), the parabola opens up and the vertex is the point with the least y-coordinate. Because $(x + 2)^2 \geq 0$ for all values of x, the least y-coordinate occurs when $(x + 2)^2 = 0$. When $x = -2$, this expression equals zero.	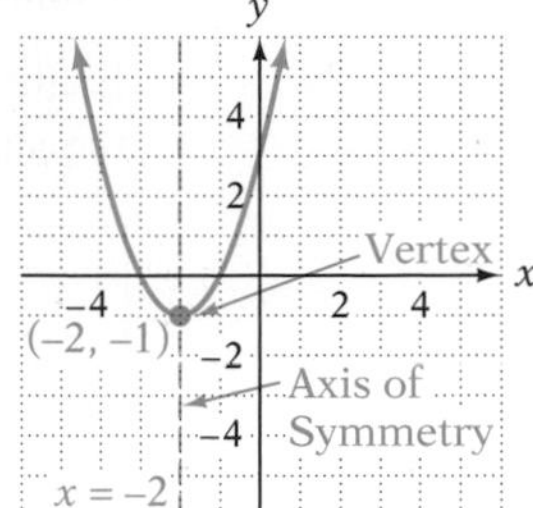 The x-coordinate of the vertex is -2.
To find the y-coordinate of the vertex, evaluate the function at $x = -2$.	$F(x) = (x + 2)^2 - 1$ $F(-2) = (-2 + 2)^2 - 1 = -1$ The y-coordinate of the vertex is -1.

From these results, the coordinates of the vertex are $(-2, -1)$. The equation of the axis of symmetry is $x = -2$.

By following the process illustrated in the above example and completing the square of $f(x) = ax^2 + bx + c$, we can find a formula for the coordinates of the vertex of a parabola.

VERTEX AND AXIS OF SYMMETRY OF A PARABOLA

Let $f(x) = ax^2 + bx + c$, $a \neq 0$, be the equation of a parabola. The coordinates of the vertex are $\left(-\frac{b}{2a}, f\left(-\frac{b}{2a}\right)\right)$. The equation of the axis of symmetry is $x = -\frac{b}{2a}$.

EXAMPLE

Let $f(x) = -2x^2 + 8x + 5$. For this equation, $a = -2$, $b = 8$, and $c = 5$.

x-coordinate of the vertex: $-\frac{b}{2a} = -\frac{8}{2(-2)} = 2$

y-coordinate of the vertex: $f\left(-\frac{b}{2a}\right) = f(2) = -2(2)^2 + 8(2) + 5 = 13$

The coordinates of the vertex are $(2, 13)$.

The equation of axis of symmetry is $x = 2$.

EXAMPLE 1 Find the coordinates of the vertex and the equation of the axis of symmetry for the parabola with equation $y = x^2 + 2x - 3$. Then sketch its graph.

Solution Find the x-coordinate of the vertex. $a = 1$ and $b = 2$.

$$x = -\frac{b}{2a} = -\frac{2}{2(1)} = -1$$

Find the y-coordinate of the vertex by replacing x with -1 and solving for y.

$$y = x^2 + 2x - 3$$
$$y = (-1)^2 + 2(-1) - 3$$
$$y = 1 - 2 - 3 = -4$$

The coordinates of the vertex are $(-1, -4)$. The equation of the axis of symmetry is $x = -1$.

Find some ordered-pair solutions of the equation and record these in a table. Because the graph is symmetric with respect to the line with equation $x = -1$, choose values of x greater than -1.

x	$y = x^2 + 2x - 3$
0	-3
1	0
2	5

Graph the ordered-pair solutions on a rectangular coordinate system. Use symmetry to locate points of the graph on the other side of the axis of symmetry. Remember that corresponding points on the graph are the same distance from the axis of symmetry.

Draw a parabola through the points.

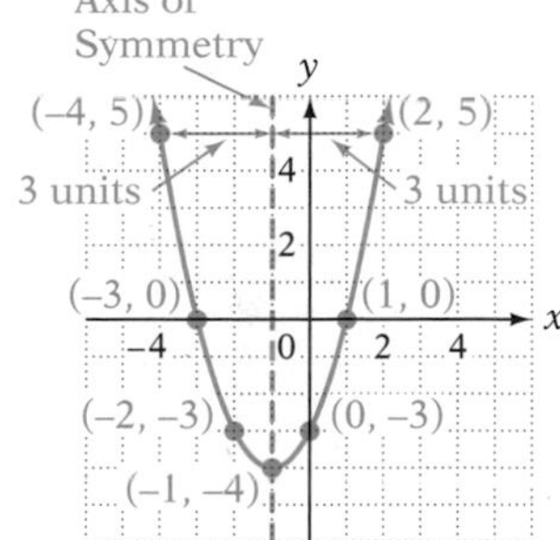

Problem 1 Find the coordinates of the vertex and the equation of the axis of symmetry for the parabola with equation $y = x^2 - 2$. Then sketch its graph.

Solution See page S28.

Try Exercise 23, page 473.

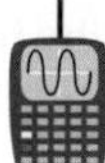

Using a graphing utility, graph the equation $y = x^2 + 2x - 3$ with a domain of $[-4.7, 4.7]$. Trace along the graph and verify that $(-1, -4)$ are the coordinates of the vertex.

Because $f(x) = ax^2 + bx + c$ is a real number for all real numbers x, the domain of a quadratic function is all real numbers. The range of a quadratic function can be determined from the y-coordinate of the vertex.

EXAMPLE 2 Graph $f(x) = -2x^2 - 4x + 3$. State the domain and range of f.

Solution Because a is negative ($a = -2$), the graph of f will open down. The x-coordinate of the vertex is

$$x = -\frac{b}{2a} = -\frac{-4}{2(-2)} = -1.$$

The y-coordinate of the vertex is
$f(-1) = -2(-1)^2 - 4(-1) + 3 = 5$.
The coordinates of the vertex are $(-1, 5)$.

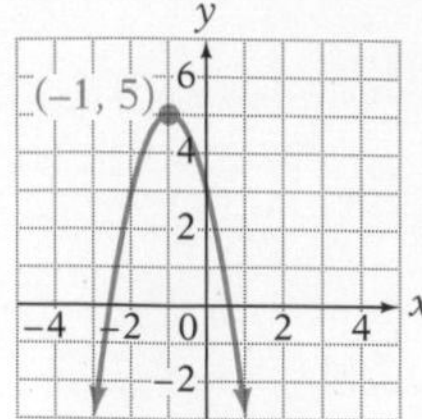

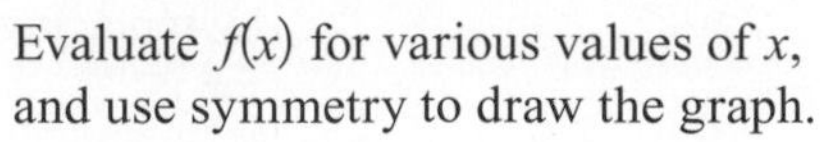

Evaluate $f(x)$ for various values of x, and use symmetry to draw the graph.

Because $f(x) = -2x^2 - 4x + 3$ is a real number for all values of x, the domain of f is $\{x|x \in \text{real numbers}\}$. The vertex of the parabola is the highest point on the graph. Because the y-coordinate at that point is 5, the range of f is $\{y|y \leq 5\}$.

Problem 2 Graph $g(x) = x^2 + 4x - 2$. State the domain and range of the function.

Solution See page S28.

Try Exercise 47, page 474.

OBJECTIVE 2 Find the x-intercepts of a parabola

Recall that a point at which a graph crosses the x- or y-axis is called an *intercept* of the graph. The x-intercepts of the graph of an equation occur when $y = 0$; the y-intercepts occur when $x = 0$.

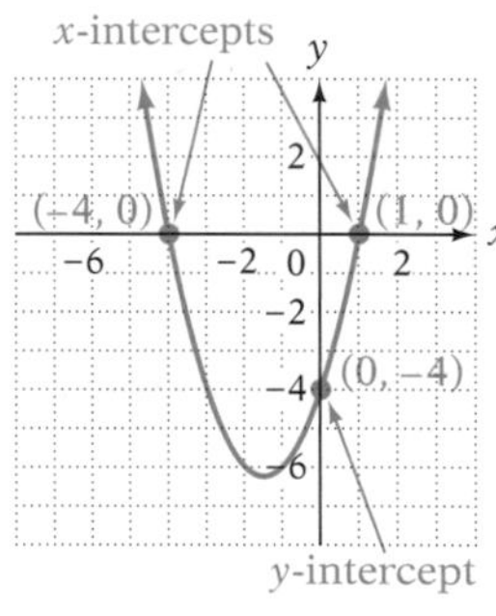

The graph of $y = x^2 + 3x - 4$ is shown at the right. The points whose coordinates are $(-4, 0)$ and $(1, 0)$ are x-intercepts of the graph.

EXAMPLE 3 Find the coordinates of the x-intercepts of the parabola whose equation is $y = 2x^2 - x - 6$.

Solution

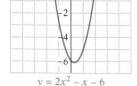

$y = 2x^2 - x - 6$

$0 = 2x^2 - x - 6$ • **To find the x-intercepts, let $y = 0$.**

$0 = (2x + 3)(x - 2)$ • **Factor.**

$2x + 3 = 0 \qquad x - 2 = 0$ • **Use the Principle of Zero Products.**

$x = -\dfrac{3}{2} \qquad x = 2$

The coordinates of the x-intercepts are $\left(-\frac{3}{2}, 0\right)$ and $(2, 0)$. See the graph at the left.

$y = 2x^2 - x - 6$

Problem 3 Find the coordinates of the x-intercepts of the parabola whose equation is $f(x) = x^2 + 2x - 8$.

Solution See page S28.

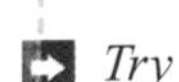

Try Exercise 61, page 475.

If $ax^2 + bx + c = 0$ has a double root, then the graph of $y = ax^2 + bx + c$ intersects the x-axis at one point. In that case, the graph is said to be **tangent** to the x-axis.

Focus on finding the x-intercept of a parabola tangent to the x-axis

Find the coordinates of the x-intercept of the parabola whose equation is $y = 4x^2 - 4x + 1$.

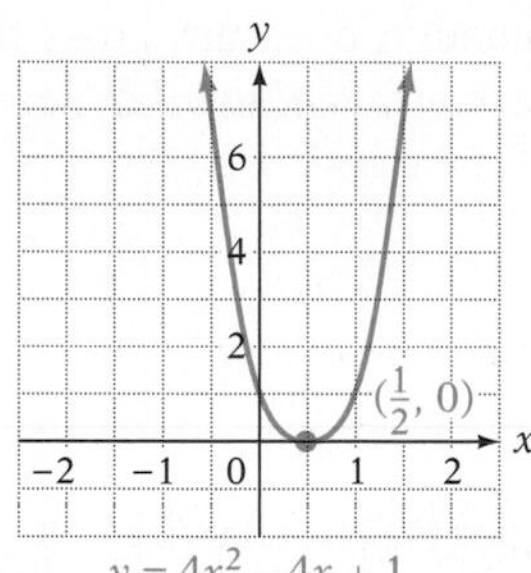

To find the x-intercept, let $y = 0$.	$y = 4x^2 - 4x + 1$ $0 = 4x^2 - 4x + 1$
Solve for x by factoring and using the Principle of Zero Products.	$0 = (2x - 1)(2x - 1)$ $2x - 1 = 0 \qquad 2x - 1 = 0$ $2x = 1 \qquad 2x = 1$ $x = \frac{1}{2} \qquad x = \frac{1}{2}$

The coordinates of the x-intercept are $\left(\frac{1}{2}, 0\right)$. See the graph at the left.

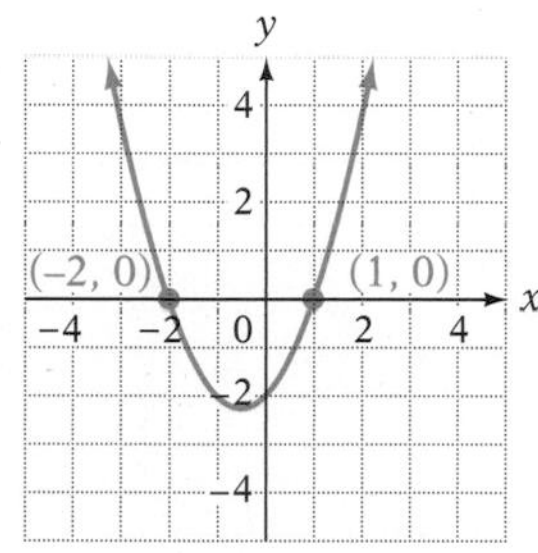

Recall that a *zero* of a function f is a number c for which $f(c) = 0$. For the graph of $f(x) = x^2 + x - 2$ shown at the left, the x-coordinates of the x-intercepts are zeros of f.

$f(x) = x^2 + x - 2$	$f(x) = x^2 + x - 2$
$f(-2) = (-2)^2 + (-2) - 2$	$f(1) = 1^2 + 1 - 2$
$f(-2) = 4 + (-2) - 2$	$f(1) = 1 + 1 - 2$
$f(-2) = 0$	$f(1) = 0$
-2 is a zero of f.	1 is a zero of f.

There is an important connection between the x-intercepts and the real zeros of a function. Because the numbers on the x-axis are *real* numbers, the x-coordinate of an x-intercept of the graph of a function is a *real zero* of the function.

EXAMPLE 4 Find the zeros of $f(x) = x^2 - 2x - 1$.

Solution

$$f(x) = x^2 - 2x - 1$$
$$0 = x^2 - 2x - 1$$

- **To find the zeros, let $f(x) = 0$ and solve for x.**

$$x = \frac{-b \pm \sqrt{b^2 - 4ac}}{2a}$$

- **Use the quadratic formula.**

$$= \frac{-(-2) \pm \sqrt{(-2)^2 - 4(1)(-1)}}{2(1)}$$

- **$a = 1$, $b = -2$, and $c = -1$.**

$$= \frac{2 \pm \sqrt{4 + 4}}{2}$$
$$= \frac{2 \pm \sqrt{8}}{2}$$
$$= \frac{2 \pm 2\sqrt{2}}{2}$$
$$= 1 \pm \sqrt{2}$$

The zeros of the function are $1 - \sqrt{2}$ and $1 + \sqrt{2}$.

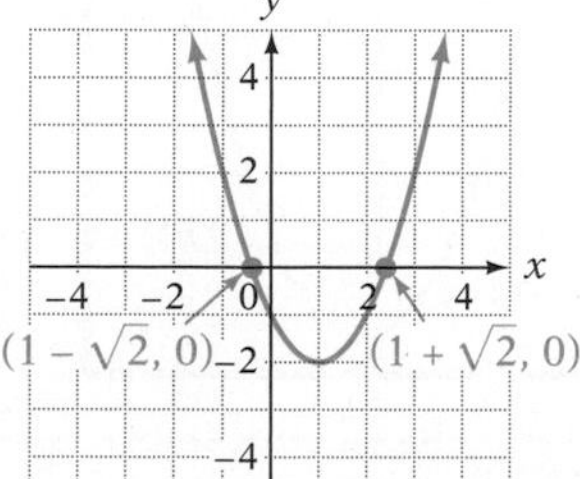

The graph of $f(x) = x^2 - 2x - 1$ is shown at the left, along with the x-intercepts of the graph. Note that the x-coordinates of the x-intercepts are the real zeros of the function.

Problem 4 Find the zeros of $f(x) = 2x^2 + 3x - 1$.

Solution See page S29.

Try Exercise 81, page 475.

The x-axis consists only of real numbers. If the graph of a function does not cross the x-axis, it will not have any x-intercepts and therefore will not have any *real number* zeros. In this case, the function has *complex number* zeros.

EXAMPLE 5 Find the zeros of $f(x) = -x^2 + 2x - 2$.

Solution

$$f(x) = -x^2 + 2x - 2$$

$$0 = -x^2 + 2x - 2$$

- **To find the zeros, let $f(x) = 0$ and solve for x.**

$$x = \frac{-b \pm \sqrt{b^2 - 4ac}}{2a}$$

- **Use the quadratic formula.**

$$= \frac{-2 \pm \sqrt{2^2 - 4(-1)(-2)}}{2(-1)}$$

- **$a = -1, b = 2, c = -2$**

$$= \frac{-2 \pm \sqrt{4 - 8}}{-2}$$

$$= \frac{-2 \pm \sqrt{-4}}{-2} = \frac{-2 \pm 2i}{-2}$$

$$= 1 \pm i$$

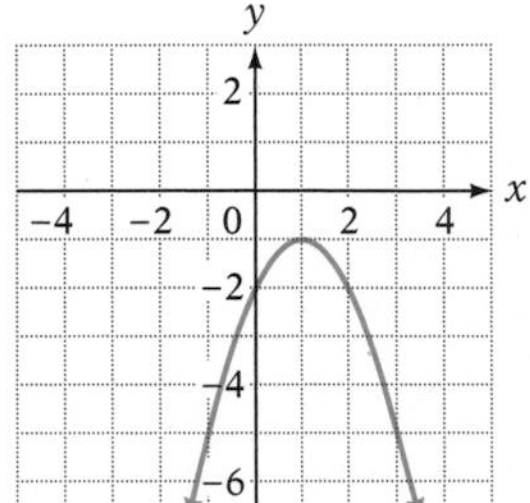

The complex zeros of the function are $1 - i$ and $1 + i$.

The graph of $f(x) = -x^2 + 2x - 2$ is shown at the left. Note that the graph does not cross the x-axis. The function has no real zeros.

Problem 5 Find the zeros of $f(x) = 4x^2 - 8x + 5$.

Solution See page S29.

 *Try Exercise 89, page 476.*

Example 5 shows that a function can have a zero without crossing the x-axis. The zero of the function, in this case, is a complex number. If the graph of a function crosses the x-axis, the function has a real zero at the x-coordinate of the x-intercept. If the graph of a function never crosses the x-axis, the function has only complex number zeros.

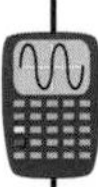

The zeros of a quadratic function can always be determined exactly by solving a quadratic equation. However, it is possible to approximate the real zeros by graphing the function and determining the x-intercepts.

Focus on estimating the real zeros of a function with a graphing calculator

Use a graphing calculator to estimate, to the nearest tenth, the real zeros of $f(x) = 2x^2 - x - 5$.

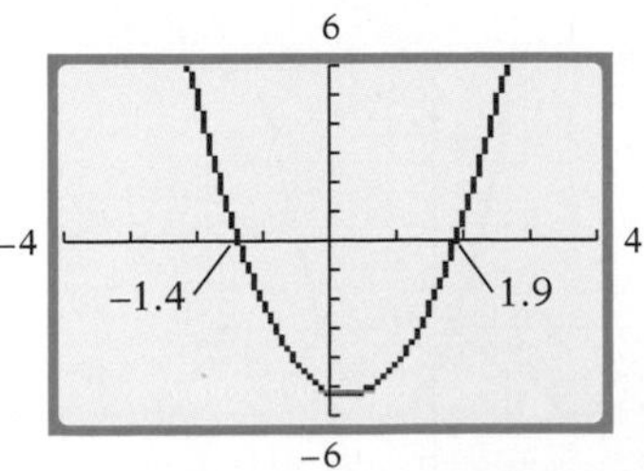

Graph the function. Use the features of the graphing calculator to estimate the zeros.

The approximate values of the zeros are -1.4 and 1.9.

Recall that the discriminant of $ax^2 + bx + c$ is the expression $b^2 - 4ac$ and that this expression can be used to determine whether $ax^2 + bx + c = 0$ has zero, one, or two real number solutions. Because there is a connection between the solutions of $ax^2 + bx + c = 0$ and the x-intercepts of the graph of $y = ax^2 + bx + c$, the discriminant can be used to determine the number of x-intercepts of a parabola.

EFFECT OF THE DISCRIMINANT ON THE NUMBER OF x-INTERCEPTS OF A PARABOLA WITH EQUATION $y = ax^2 + bx + c$

1. If $b^2 - 4ac = 0$, the parabola has one x-intercept.
2. If $b^2 - 4ac > 0$, the parabola has two x-intercepts.
3. If $b^2 - 4ac < 0$, the parabola has no x-intercepts.

EXAMPLE 6 Use the discriminant to determine the number of x-intercepts of the parabola with the given equation.

A. $y = 2x^2 - x + 2$ **B.** $f(x) = -x^2 + 4x - 4$

Solution **A.** $y = 2x^2 - x + 2$

$b^2 - 4ac$ • **Evaluate the discriminant.**

$(-1)^2 - 4(2)(2)$ • $a = 2, b = -1, c = 2$

$= 1 - 16$

$= -15$

The discriminant is negative.
The parabola has no x-intercepts.

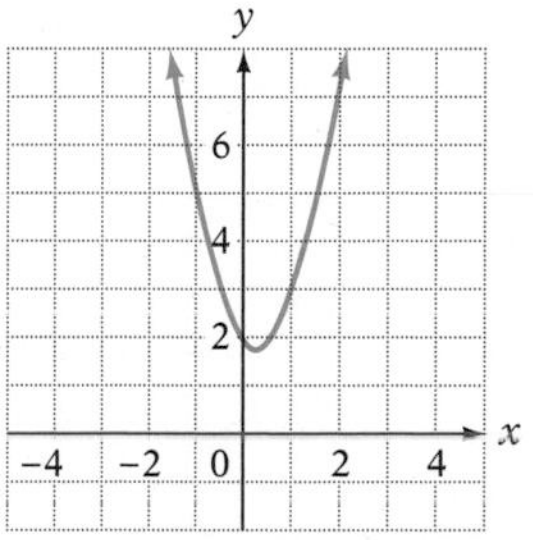

$y = 2x^2 - x + 2$

B. $f(x) = -x^2 + 4x - 4$

$b^2 - 4ac$ • **Evaluate the discriminant.**

$(4)^2 - 4(-1)(-4)$ • $a = -1, b = 4, c = -4$

$= 16 - 16$

$= 0$

The discriminant is zero.
The parabola has one x-intercept.

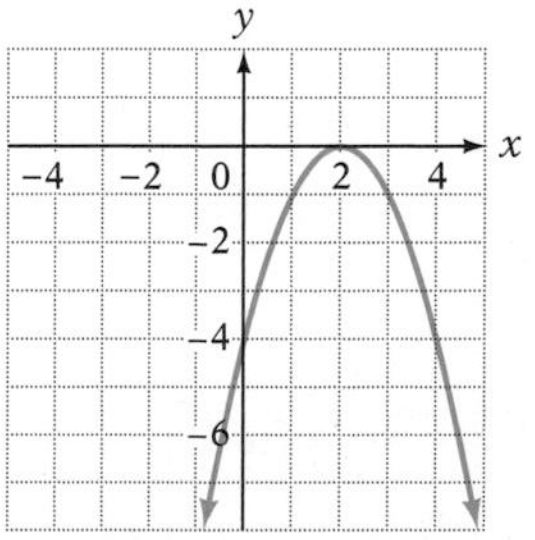

$f(x) = -x^2 + 4x - 4$

Problem 6 Use the discriminant to determine the number of x-intercepts of the parabola with equation $y = x^2 - x - 6$.

Solution See page S29.

Try Exercise 101, page 476.

8.5 Exercises

CONCEPT CHECK

1. State whether each function is a quadratic function.

   a. $f(x) = 1 - 2x - x^2$ **b.** $g(x) = 3x + 2$

   c. $h(x) = \frac{2}{x^2} + 3x + 1$ **d.** $v(x) = x^2$

2. The coordinates of the vertex of the graph of $f(x) = x^2 - 2x + 3$ are $(1, 2)$. What is the equation of the axis of symmetry?

3. The x-coordinate of the vertex of the graph of $f(x) = -x^2 + 6x - 3$ is 3. What is the y-coordinate of the vertex?

4. State whether the given number is a zero of $f(x) = x^2 + 4x - 5$.

   a. 0 **b.** 1 **c.** -5 **d.** -2

5. If $2 - i$ and $2 + i$ are the zeros of a quadratic function, how many x-intercepts does the graph of the function have?

For Exercises 6 to 9, the equation of the axis of symmetry of a parabola is given, along with the coordinates of a point on the parabola. Use this information to find the coordinates of another point on the parabola.

6. $x = 0$; $(3, 4)$
7. $x = 2$; $(1, 1)$
8. $x = -3$; $(0, -2)$
9. $x = -1$; $(-5, 0)$

10. The zeros of the function $f(x) = x^2 - 2x - 3$ are -1 and 3. What are the coordinates of the x-intercepts of the graph of the equation $y = x^2 - 2x - 3$?

11. If the value of the discriminant of a quadratic function is 4, how many x-intercepts does the graph of the function have?

12. Match each equation with its graph.

    i. $f(x) = x^2$
    ii. $f(x) = x^2 - 1$
    iii. $f(x) = -x^2$
    iv. $f(x) = -x^2 + 3$
    v. $f(x) = x^2 + 3x - 2$
    vi. $f(x) = -x^2 - 3x + 2$

A.

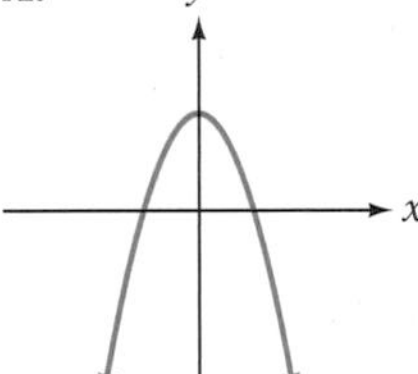

B.

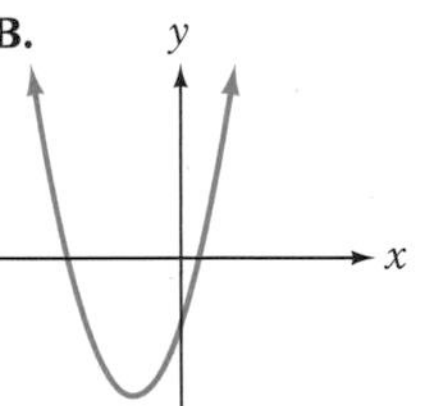

C.

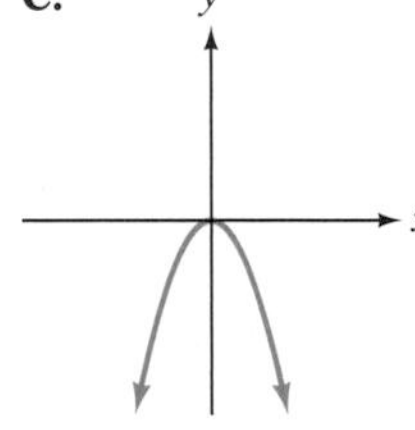

D.

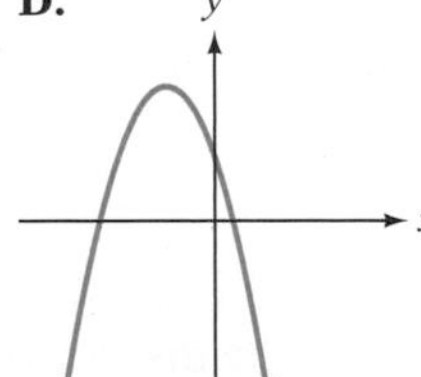

E. **F.**

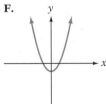

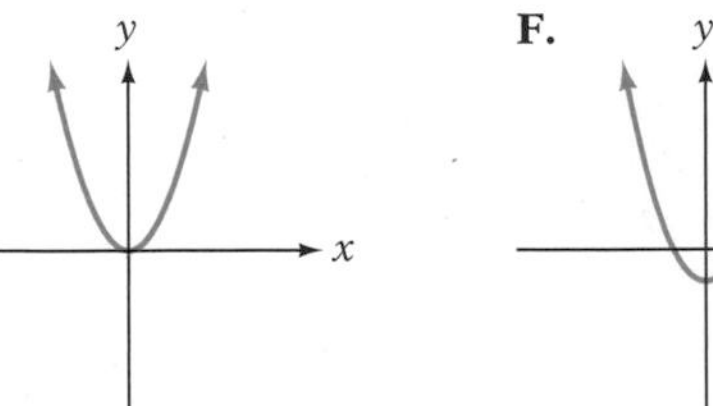

1 Graph a quadratic function (See pages 464–468.)

13. What is a quadratic function?

14. What are the vertex and axis of symmetry of the graph of a parabola?

15. The axis of symmetry of a parabola is the line with equation $x = -5$. What is the x-coordinate of the vertex of the parabola?

16. The axis of symmetry of a parabola is the line with equation $x = 8$. What is the x-coordinate of the vertex of the parabola?

17. The coordinates of the vertex of a parabola are $(7, -9)$. What is the equation of the axis of symmetry of the parabola?

18. The coordinates of the vertex of a parabola are $(-4, 10)$. What is the equation of the axis of symmetry of the parabola?

GETTING READY

19. Use the quadratic function $f(x) = 3x^2 - 1$.

a. The function is in the standard form $f(x) = ax^2 + bx + c$, where $a = \underline{\ ?\ }$, $b = \underline{\ ?\ }$, and $c = \underline{\ ?\ }$.

b. The graph of the function is a $\underline{\ ?\ }$. Because $a > 0$, the graph opens $\underline{\ ?\ }$.

c. The x-coordinate of the vertex is $x = -\dfrac{b}{2a} = -\dfrac{?}{2(?)} = \underline{\ ?\ }$.

d. The y-coordinate of the vertex is $f(0) = 3(\underline{\ ?\ })^2 - 1 = \underline{\ ?\ }$. The coordinates of the vertex of the parabola are $(\underline{\ ?\ }, \underline{\ ?\ })$.

20. Use the results of Exercise 19.

a. The axis of symmetry of the graph of $f(x) = 3x^2 - 1$ is the vertical line with equation $x = \underline{\ ?\ }$.

b. The coordinates of one point on the graph of $f(x) = 3x^2 - 1$ are $(2, 11)$. This point is $\underline{\ ?\ }$ units to the right of the axis of symmetry. The corresponding point 2 units to the left of the axis of the symmetry has coordinates $(\underline{\ ?\ }, \underline{\ ?\ })$.

Find the coordinates of the vertex and the equation of the axis of symmetry for the parabola given by each equation. Then sketch its graph.

21. $y = x^2 - 2$

22. $y = x^2 + 2$

23. $y = -x^2 + 3$

24. $y = -x^2 - 1$

25. $y = \frac{1}{2}x^2$

26. $y = 2x^2$

27. $y = 2x^2 - 1$

28. $y = -\frac{1}{2}x^2 + 2$

29. $y = x^2 - 2x$

30. $y = x^2 + 2x$

31. $y = -2x^2 + 4x$

32. $y = \frac{1}{2}x^2 - x$

33. $y = x^2 - x - 2$

34. $y = x^2 - 3x + 2$

35. $y = 2x^2 - x - 5$

36. $y = 2x^2 - x - 3$

37. $y = -2x^2 + 4x + 1$

38. $y = -x^2 - 3x + 3$

Graph the function. State the domain and range of the function.

39. $f(x) = 2x^2 - 4x - 5$

40. $f(x) = 2x^2 + 8x + 3$

41. $f(x) = -2x^2 - 3x + 2$

42. $f(x) = 2x^2 - 7x + 3$

43. $f(x) = x^2 - 4x + 4$

44. $f(x) = -x^2 + 6x - 9$

45. $f(x) = x^2 + 4x - 3$

46. $f(x) = x^2 - 2x - 2$

47. $f(x) = -x^2 - 4x - 5$

48. $f(x) = -x^2 + 4x + 1$

For Exercises 49 to 52, use the quadratic function $f(x) = ax^2 + bx + c$.

49. True or false? If $a < 0$, then the range of f is $\left\{y \middle| y \geq f\left(-\frac{b}{2a}\right)\right\}$.

50. True or false? If $b = 0$, then the axis of symmetry for the graph of f is the y-axis.

51. If $a > 0$ and $b < 0$, is the axis of symmetry for the graph of f to the left or to the right of the y-axis?

52. True or false? If $b = 2a$, then the point with x-coordinate -2 has the same y-coordinate as the point with x-coordinate 0.

2 Find the x-intercepts of a parabola (See pages 468–471.)

GETTING READY

53. If 3 is a zero of the function $y = f(x)$, then $f(\underline{\ ?\ }) = \underline{\ ?\ }$, and $(\underline{\ ?\ }, \underline{\ ?\ })$ are the coordinates of an $\underline{\ ?\ }$-intercept of the graph of $y = f(x)$.

54. Use the equation $y = 2x^2 - 3x + 5$. The discriminant is

$$b^2 - 4ac = (\underline{\ ?\ })^2 - 4(\underline{\ ?\ })(\underline{\ ?\ })$$
$$= \underline{\ ?\ } - \underline{\ ?\ } = \underline{\ ?\ }$$

Because the discriminant is negative, the graph of $y = 2x^2 - 3x + 5$ has $\underline{\ ?\ }$ x-intercepts.

55. How can you find the x-intercepts of the graph of a quadratic function?

56. How can the discriminant be used to determine the number of x-intercepts of the graph of a quadratic function?

Find the coordinates of the x-intercepts of the parabola given by each equation.

57. $y = x^2 - 4$

58. $y = x^2 - 9$

59. $y = 2x^2 - 4x$

60. $y = 3x^2 + 6x$

61. $y = x^2 - x - 2$

62. $y = x^2 - 2x - 8$

63. $y = 2x^2 - 5x - 3$

64. $y = 4x^2 + 11x + 6$

65. $y = 3x^2 - 19x - 14$

66. $y = 6x^2 + 7x + 2$

67. $y = 9x^2 - 12x + 4$

68. $y = x^2 - 2$

69. $y = 9x^2 - 2$

70. $y = 2x^2 - x - 1$

71. $y = 4x^2 - 4x - 15$

72. $y = x^2 + 2x - 1$

73. $y = x^2 + 4x - 3$

74. $y = x^2 + 6x + 10$

75. $y = -x^2 - 4x - 5$

76. $y = x^2 - 2x - 2$

77. $y = -x^2 - 2x + 1$

78. $y = -x^2 + 4x + 1$

Find the zeros of the function.

79. $f(x) = 2x^2 - 3x$

80. $f(x) = -3x^2 + 4x$

81. $f(x) = x^2 + 3x + 2$

82. $f(x) = -3x^2 + 4x - 1$

83. $f(x) = 9x^2 + 12x + 4$

84. $f(x) = x^2 - 6x + 9$

85. $f(x) = x^2 - 16$

86. $f(x) = 2x^2 - 4$

87. $f(x) = -x^2 + 3x + 8$

88. $f(x) = -2x^2 + x + 5$

89. $f(x) = 2x^2 + 3x + 2$

90. $f(x) = 3x^2 - x + 4$

Graph the function. Estimate the real zeros of the function to the nearest tenth.

91. $f(x) = x^2 + 3x - 1$

92. $f(x) = x^2 - 2x - 4$

93. $f(x) = 2x^2 - 3x - 7$

94. $f(x) = -2x^2 - x + 2$

95. $f(x) = x^2 + 6x + 12$

96. $f(x) = x^2 - 3x + 9$

Use the discriminant to determine the number of x-intercepts of the parabola with the given equation.

97. $y = 2x^2 + x + 1$

98. $y = 2x^2 + 2x - 1$

99. $y = -x^2 - x + 3$

100. $y = -2x^2 + x + 1$

101. $y = x^2 - 8x + 16$

102. $y = x^2 - 10x + 25$

103. $y = -3x^2 - x - 2$

104. $y = -2x^2 + x - 1$

105. $y = 4x^2 - x - 2$

106. $y = 2x^2 + x + 4$

107. $y = -2x^2 - x - 5$

108. $y = -3x^2 + 4x - 5$

109. $y = x^2 + 8x + 16$

110. $y = x^2 - 12x + 36$

111. $y = x^2 + x - 3$

For Exercises 112 to 115, $f(x) = ax^2 + bx + c$ is a quadratic function whose graph is a parabola that has only one x-intercept, $P(n, 0)$. Determine whether each statement is true or false.

112. $P(n, 0)$ is also the vertex of the parabola.

113. $ax^2 + bx + c$ is a perfect-square trinomial.

114. If $a > 0$, then for all real numbers x, $f(x) \geq 0$.

115. The point with x-coordinate $n - m$ has the same y-coordinate as the point with x-coordinate $n + m$.

APPLYING CONCEPTS

Find the value of k such that the graph of the equation contains the given point.

116. $y = x^2 - 3x + k$; $(2, 5)$

117. $y = x^2 + 2x + k$; $(-3, 1)$

118. $y = 2x^2 + kx - 3$; $(4, -3)$

119. $y = 3x^2 + kx - 6$; $(-2, 4)$

120. One zero of the function $f(x) = 2x^2 - 5x + k$ is 4. What is the other zero?

121. The zeros of the function $f(x) = mx^2 + nx + 6$ are -2 and 3. What are the zeros of the function $g(x) = nx^2 + mx - 6$?

Recall that we can find a zero of a quadratic function $f(x) = ax^2 + bx + c$ by finding a value of x in the domain of f such that $f(x) = 0$. We can extend this idea to find a value x in the domain of f such that $f(x) = c$, where c is any number in the range of f. Here is an example.

Find two values of x in the domain of $f(x) = x^2 + 3x - 1$ such that $f(x) = 9$.

$$f(x) = 9$$

$$x^2 + 3x - 1 = 9 \quad \text{• Replace } f(x) \text{ by } x^2 + 3x - 1.$$

$$x^2 + 3x - 10 = 0 \quad \text{• Solve the quadratic equation.}$$

$$(x + 5)(x - 2) = 0$$

$$x + 5 = 0 \qquad x - 2 = 0$$

$$x = -5 \qquad x = 2$$

The values of x for which $f(x) = 9$ are -5 and 2.

For Exercises 122 to 125, find two values of x in the domain of f such that $f(x)$ has the given value.

122. $f(x) = x^2 + 3x - 2$; $f(x) = 2$

123. $f(x) = -x^2 + 6x - 3$; $f(x) = 5$

124. $f(x) = 2x^2 + x - 4$; $f(x) = -1$

125. $f(x) = 2x^2 + 5x - 1$; $f(x) = -3$

PROJECTS OR GROUP ACTIVITIES

An equation of the form $y = ax^2 + bx + c$ can be written in the form $y = a(x - h)^2 + k$, where (h, k) are the coordinates of the vertex of the parabola. Use the process of completing the square to rewrite each equation in the form $y = a(x - h)^2 + k$. Find the coordinates of the vertex. (*Hint:* Review the Focus On example on pages 465–466.)

126. $y = x^2 - 2x - 2$

127. $y = x^2 - 4x + 7$

128. $y = x^2 - x - 3$

129. $y = x^2 + x + 2$

Using $y = a(x - h)^2 + k$ as the equation of a parabola with vertex at $V(h, k)$, find the equation of the parabola that has the given characteristics. Write the final equation in the form $y = ax^2 + bx + c$.

130. Vertex $V(0, -3)$; the graph passes through $P(3, -2)$

131. Vertex $V(1, 2)$; the graph passes through $P(2, 5)$

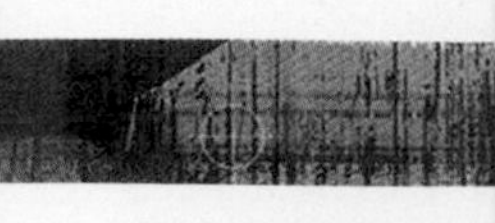

8.6 Applications of Quadratic Functions

OBJECTIVE 1 Minimum and maximum problems

How It's Used

Calculus is a branch of mathematics that demonstrates, among other things, how to find the maximum or minimum of functions other than quadratic functions. These are very important problems in applied mathematics. For instance, an automotive engineer wants to design a car whose shape will *minimize* the effect of air flow. The same engineer tries to *maximize* the efficiency of a car's engine. Similarly, an economist may try to determine what business practices will *minimize* cost and *maximize* profit.

The graph of $f(x) = x^2 - 2x + 3$ is shown at the right. Because a is positive, the parabola opens up. The vertex of the parabola is the lowest point on the parabola. It is the point that has the minimum y-coordinate. Therefore, the value of the function at this point is a **minimum.**

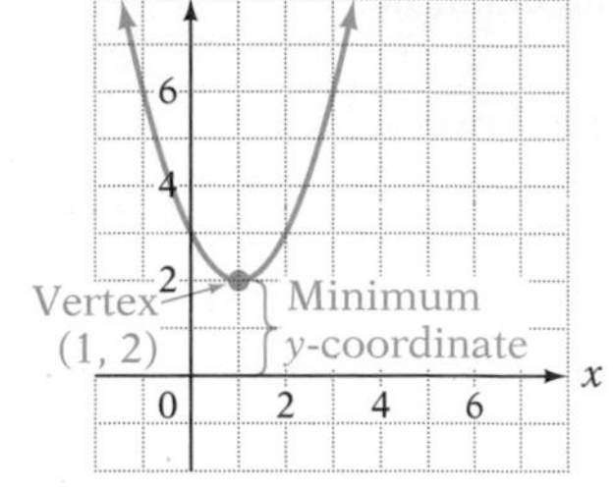

The graph of $f(x) = -x^2 + 2x + 1$ is shown at the right. Because a is negative, the parabola opens down. The vertex of the parabola is the highest point on the parabola. It is the point that has the maximum y-coordinate. Therefore, the value of the function at this point is a **maximum.**

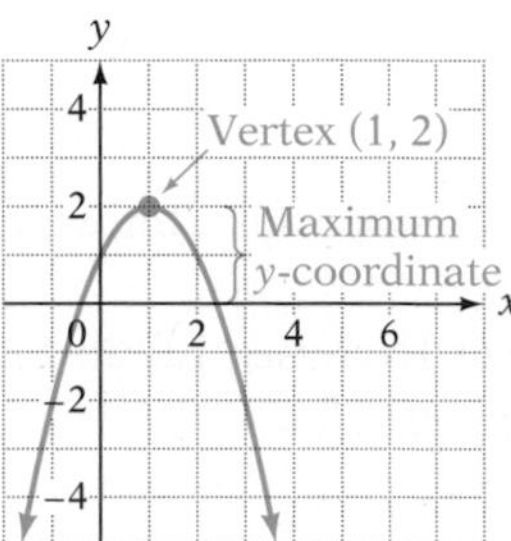

To find the minimum or maximum value of a quadratic function, first find the x-coordinate of the vertex. Then evaluate the function at that value.

EXAMPLE 1 Find the maximum or minimum value of the function $f(x) = -2x^2 + 4x + 3$.

Solution $x = -\dfrac{b}{2a} = -\dfrac{4}{2(-2)} = 1$ • **Find the x-coordinate of the vertex. $a = -2, b = 4$**

$f(x) = -2x^2 + 4x + 3$

$f(1) = -2(1)^2 + 4(1) + 3$ • **Evaluate the function at $x = 1$.**

$f(1) = 5$

Because $a < 0$, the graph of f opens down. Therefore, the function has a maximum value.

The maximum value of the function is 5. See the graph at the left.

$f(x) = -2x^2 + 4x + 3$

Problem 1 Find the maximum or minimum value of the function $f(x) = 2x^2 - 3x + 1$.

Solution See page S29.

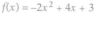

Try Exercise 11, page 481.

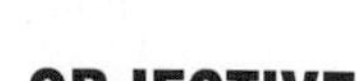

OBJECTIVE 2 Applications of minimum and maximum

Focus on solving a maximization problem

A mason is forming a rectangular floor for a storage shed. The perimeter of the rectangle is 44 ft. What dimensions of the rectangle will give the floor a maximum area? What is the maximum area?

We are given the perimeter of the rectangle, and we want to find the dimensions of the rectangle that will yield the maximum area for the floor. Use the equation for the perimeter of a rectangle.

$$P = 2L + 2W$$

$P = 44$ $$44 = 2L + 2W$$

Divide both sides of the equation by 2. $$22 = L + W$$

Solve the equation for W. $$22 - L = W$$

Now use the equation for the area of a rectangle. Use substitution to express the area in terms of L.

$$A = LW$$

From the equation above, $W = 22 - L$.
Substitute $22 - L$ for W.

$$A = L(22 - L)$$
$$A = 22L - L^2$$

The area of the rectangle is $22L - L^2$. To find the length of the rectangle, find the L-coordinate of the vertex of the function $f(L) = -L^2 + 22L$.

For the function $f(L) = -L^2 + 22L$, $a = -1$ and $b = 22$.

$$L = -\frac{b}{2a} = -\frac{22}{2(-1)} = 11$$

The length of the rectangle is 11 ft.

To find the width, replace L in $22 - L$ by 11, the L-coordinate of the vertex, and evaluate.

$$W = 22 - L$$
$$W = 22 - 11 = 11$$

The width of the rectangle is 11 ft.

The dimensions of the rectangle that would give the floor a maximum area are 11 ft by 11 ft.

To find the maximum area of the floor, evaluate the function $f(L) = -L^2 + 22L$ at 11, the L-coordinate of the vertex.

$$f(L) = -L^2 + 22L$$
$$f(11) = -(11)^2 + 22(11)$$
$$= -121 + 242 = 121$$

The maximum area of the floor is 121 ft^2.

The graph of the function $f(L) = -L^2 + 22L$ is shown at the right. Note that the vertex of the parabola is (11, 121). For any value of L less than 11, the area of the floor will be less than 121 ft^2. For any value of L greater than 11, the area of the floor will be less than 121 ft^2. 121 is the maximum value of the function, and the maximum value occurs when $L = 11$.

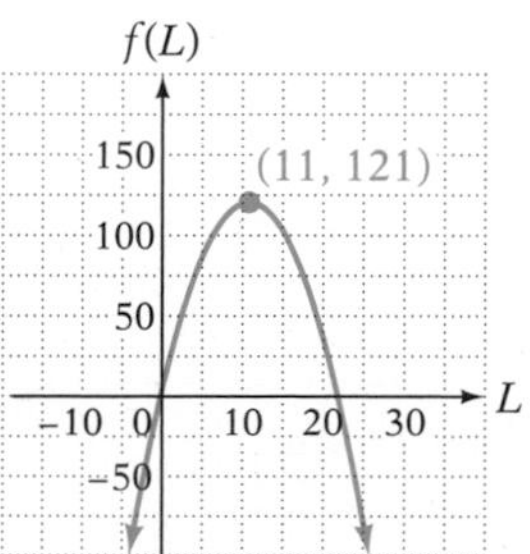

EXAMPLE 2 A mining company has determined that the cost c, in dollars per ton, of mining a mineral is given by

$$c(x) = 0.2x^2 - 2x + 12$$

where x is the number of tons of the mineral that are mined. Find the number of tons of the mineral that should be mined to minimize the cost. What is the minimum cost?

Strategy ▶ To find the number of tons that will minimize the cost, find the x-coordinate of the vertex.

▶ To find the minimum cost, evaluate the function at the x-coordinate of the vertex.

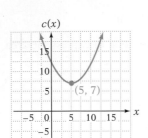

Solution $x = -\dfrac{b}{2a} = -\dfrac{-2}{2(0.2)} = 5$

To minimize the cost, 5 tons should be mined.

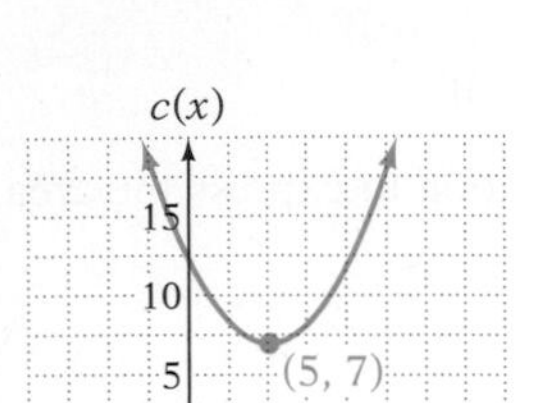

$c(x) = 0.2x^2 - 2x + 12$
$c(5) = 0.2(5)^2 - 2(5) + 12 = 5 - 10 + 12 = 7$

The minimum cost per ton is \$7.

Note: The graph of the function $c(x) = 0.2x^2 - 2x + 12$ is shown at the left. The vertex of the parabola is $(5, 7)$. For any value of x less than 5, the cost per ton is greater than \$7. For any value of x greater than 5, the cost per ton is greater than \$7. 7 is the minimum value of the function, and the minimum value occurs when $x = 5$.

Problem 2 The height s, in feet, of a ball thrown straight up is given by $s(t) = -16t^2 + 64t$, where t is the time in seconds. Find the time it takes the ball to reach its maximum height. What is the maximum height?

Solution See page S29.

Try Exercise 35, page 483.

EXAMPLE 3 Find two numbers whose difference is 10 and whose product is a minimum.

Strategy ▶ Let x and y represent the two numbers.

▶ Express y in terms of x.

$y - x = 10$ • **The difference of the numbers is 10.**
$y = x + 10$ • **Solve for y.**

▶ Express the product of the numbers in terms of x.

$xy = x(x + 10)$ • **$y = x + 10$**
$f(x) = x^2 + 10x$ • **The function f represents the product of the two numbers.**

▶ To find one of the two numbers, find the x-coordinate of the vertex of $f(x) = x^2 + 10x$.

▶ To find the other number, replace x in $x + 10$ by the x-coordinate of the vertex and evaluate.

Solution $x = -\dfrac{b}{2a} = -\dfrac{10}{2(1)} = -5$

$x + 10 = -5 + 10 = 5$

The numbers are -5 and 5.

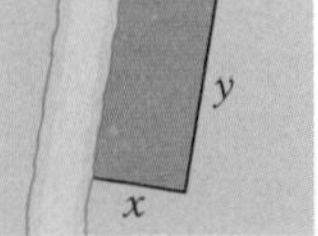

Problem 3 A rectangular area is being fenced along a stream to enclose a picnic area. If 100 ft of fencing is available, what dimensions of the rectangle will produce the maximum area for picnicking? See the figure at the left.

Solution See page S29.

Try Exercise 41, page 483.

8.6 Exercises

CONCEPT CHECK

1. For each of the following, determine whether the function has a minimum value, a maximum value, or neither.

   a. $f(x) = -2x^2 + 3x - 7$ **b.** $f(x) = 3x + 5$ **c.** $f(x) = x^2$

2. The coordinates of the vertex of a parabola that opens downward are $(2, -3)$. What is the maximum value of the function?

3. The x-coordinate of the vertex of the graph of $f(x) = 3x^2 - 6x + 3$ is 2. What is the minimum value of f?

4. Can a quadratic function have both a minimum and a maximum value?

1 Minimum and maximum problems (See page 478.)

5. What is the minimum value or the maximum value of a quadratic function?

6. Describe how to find the minimum or maximum value of a quadratic function.

GETTING READY

For Exercises 7 and 8, use the quadratic function $f(x) = -8x^2 + 8x$.

7. Because $a =$ __?__, the graph of f opens __?__. The vertex is the highest point on the graph, so its __?__-coordinate is the maximum value of $f(x)$.

8. The vertex of the graph of f occurs at $x = -\frac{b}{2a} = -\frac{?}{2(?)} =$ __?__. The maximum value of $f(x)$ is

   $f(\_?\_) = -8(\_?\_)^2 + 8(\_?\_) = \_?\_ + \_?\_ = \_?\_$

9. Does the function have a minimum value or a maximum value?

   a. $f(x) = -x^2 + 6x - 1$ **b.** $f(x) = 2x^2 - 4$ **c.** $f(x) = -5x^2 + x$

10. Does the function have a minimum value or a maximum value?

    a. $f(x) = 3x^2 - 2x + 4$ **b.** $f(x) = -x^2 + 9$ **c.** $f(x) = 6x^2 - 3x$

Find the minimum or maximum value of the quadratic function.

11. $f(x) = x^2 - 2x + 3$
12. $f(x) = 2x^2 + 4x$
13. $f(x) = -2x^2 + 4x - 3$
14. $f(x) = -2x^2 + 4x - 5$
15. $f(x) = -2x^2 - 3x + 4$
16. $f(x) = -2x^2 - 3x$
17. $f(x) = 2x^2 + 3x - 8$
18. $f(x) = 3x^2 + 3x - 2$
19. $f(x) = -3x^2 + x - 6$

20. $f(x) = -x^2 - x + 2$

21. $f(x) = x^2 - 5x + 3$

22. $f(x) = 3x^2 + 5x + 2$

23. Which of the following parabolas has the greatest minimum value?

(i) $y = x^2 - 2x - 3$ (ii) $y = x^2 - 10x + 20$ (iii) $y = 3x^2 - 6$

24. Which of the following parabolas has the greatest maximum value?

(i) $y = -2x^2 + 2x - 1$ (ii) $y = -x^2 + 8x - 2$ (iii) $y = -4x^2 + 3$

25. The vertex of a parabola that opens up is $P(-4, 7)$. Does the function have a maximum or a minimum value? What is the maximum or minimum value of the function?

26. The vertex of a parabola that opens down is $P(3, -5)$. Does the function have a maximum or a minimum value? What is the maximum or minimum value of the function?

For Exercises 27 and 28, use the following information.

f is a quadratic function whose graph opens up and has vertex $P(n, p)$.
g is a quadratic function whose graph opens down and has vertex $Q(n, q)$.

27. True or false? If $p > q$, then for all values of x, $f(x) > g(x)$.

28. True or false? If $p < q$, then for all values of x, $f(x) < g(x)$.

2 Applications of minimum and maximum (See pages 478–480.)

GETTING READY

Complete Exercises 29 and 30 given that the approximate height h, in meters, of a ball t seconds after it is thrown vertically upward from a height of 2 m, at an initial speed of 30 m/s, is given by the function $h(t) = -5t^2 + 30t + 2$.

29. To find out how many seconds it takes for the ball to reach its maximum height, find the __?__-coordinate of the vertex of the graph of h:

$$t = -\frac{b}{2a} = -\frac{?}{-2(?)} = \underline{\quad ? \quad}$$

The ball reaches its maximum height after __?__ s.

30. Use the result of Exercise 29. Find the maximum height of the ball:

$$h(\underline{\ ?\ }) = -5(\underline{\ ?\ })^2 + 30(\underline{\ ?\ }) + 2$$
$$= \underline{\ ?\ } + \underline{\ ?\ } + 2 = \underline{\ ?\ }$$

The maximum height of the ball is __?__ m.

Solve.

31. **Physics** The height s, in feet, of a rock thrown upward at an initial speed of 64 ft/s from a cliff 50 ft above an ocean beach is given by the function $s(t) = -16t^2 + 64t + 50$, where t is the time in seconds. Find the maximum height above the beach that the rock will attain.

32. **Sports** An event in the Summer Olympics is 10-meter springboard diving. In this event, the height s, in meters, of a diver above the water t seconds after jumping is given by $s(t) = -4.9t^2 + 7.8t + 10$. What is the maximum height that the diver will be above the water? Round to the nearest tenth.

33. **Space Travel** Read the article at the right. NASA uses an airplane similar to the one described in the article to prepare American astronauts for their work in the weightless environment of space. Suppose the height h, in feet, of NASA's airplane is modeled by the equation $h(t) = -6.63t^2 + 431t + 25{,}000$, where t is the number of seconds elapsed after the plane enters its parabolic path. Find the maximum height of the NASA airplane. Round to the nearest thousand feet.

In the News

Zero-G Tourism

This summer, why not head to the moon? Zero Gravity Corporation now offers rides in an airplane that produces a weightless environment by flying in a series of parabolic paths. The first parabolic path mimics the gravity of Mars for about 35 s. The next two paths give you the feeling of being on the moon, and after that you will ride 12 parabolas of true weightlessness.

Source: www.space.com/news

34. **Manufacturing** A manufacturer of microwave ovens believes that the revenue R, in dollars, the company receives is related to the price P, in dollars, of an oven by the function $R(P) = 125P - 0.25P^2$. What price will give the maximum revenue?

35. **Business** A tour operator believes that the profit P, in dollars, from selling x tickets is given by $P(x) = 40x - 0.25x^2$. Using this model, find the maximum profit that the tour operator can expect.

36. **Art** The Buckingham Fountain in Chicago shoots water from a nozzle at the base of the fountain. The height h, in feet, of the water above the ground t seconds after it leaves the nozzle is given by the function $h(t) = -16t^2 + 90t + 15$. What is the maximum height of the water? Round to the nearest tenth.

37. **Structural Engineering** The suspension cable that supports a small footbridge hangs in the shape of a parabola. The height h, in feet, of the cable above the bridge is given by the function $h(x) = 0.25x^2 - 0.8x + 25$, where x is the distance in feet from one end of the bridge. What is the minimum height of the cable above the bridge?

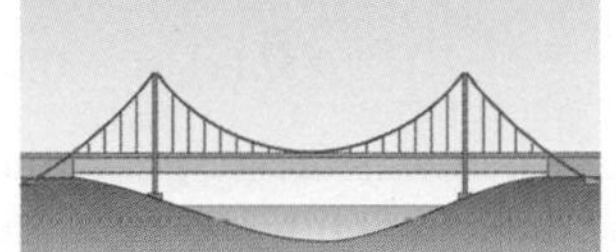

38. **Telescopes** An equation of the thickness h, in inches, of the mirror in the telescope at the Palomar Mountain Observatory in California is given by the function $h(x) = 0.000379x^2 - 0.0758x + 24$, where x is measured from the edge of the mirror. Find the minimum thickness of the mirror.

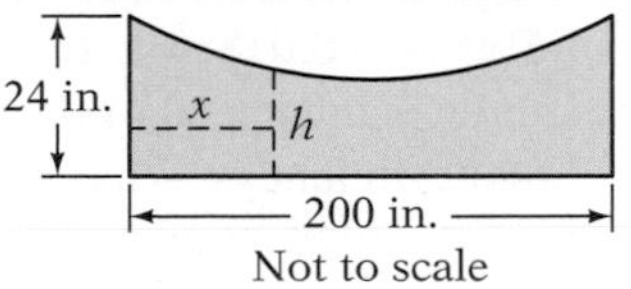

39. **Fire Science** The height s, in feet, of water squirting from a fire hose nozzle is given by the equation $s(x) = -\frac{1}{30}x^2 + 2x + 5$, where x is the horizontal distance from the nozzle. How high on a building 40 ft from the fire hose will the water land?

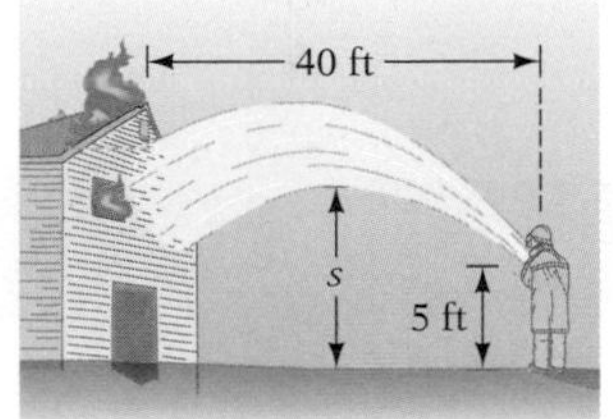

40. **Automotive Technology** On wet concrete, the stopping distance s, in feet, of a car traveling v miles per hour is given by $s(v) = 0.055v^2 + 1.1v$. At what speed could a car be traveling and still stop at a stop sign 44 ft away?

41. **Number Problem** Find two numbers whose sum is 20 and whose product is a maximum.

42. **Number Problem** Find two numbers whose difference is 14 and whose product is a minimum.

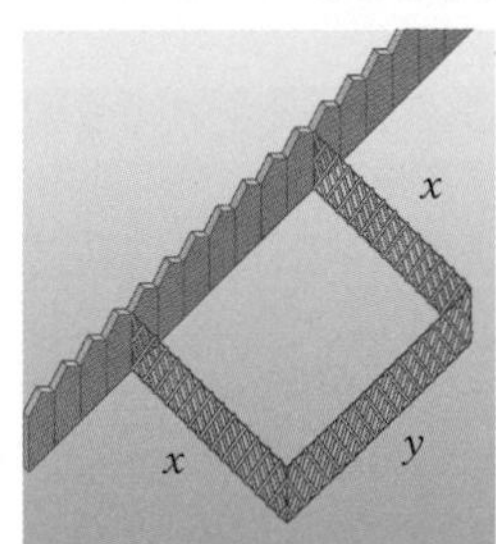

43. **Ranching** A rancher has 200 ft of fencing with which to build a rectangular corral alongside an existing fence. Determine the dimensions of the corral that will maximize the enclosed area.

44. **Outdoor Dining** The parks and recreation department for a city wants to fence a rectangular eating area for an existing snack shop at a beach, as shown in the diagram at the right. If the city has 400 ft of fencing, what dimensions of the rectangle will maximize the eating area?

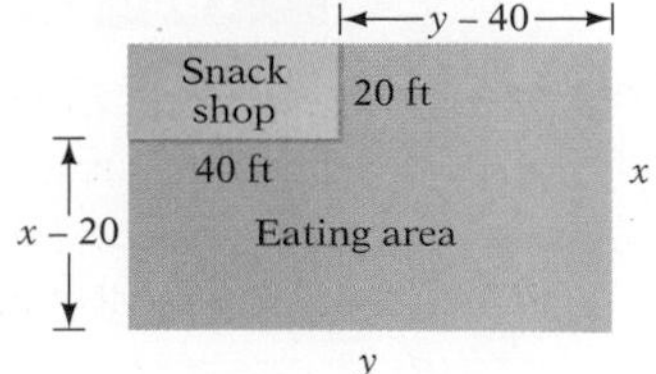

APPLYING CONCEPTS

45. Show that the minimum value of

$$S(x) = (2 - x)^2 + (5 - x)^2 + (4 - x)^2 + (7 - x)^2$$

occurs when x is the average of the numbers 2, 5, 4, and 7.

Use a graphing utility to find the minimum or maximum value of the function. Round to the nearest tenth. See the Graphing Calculator Appendix for assistance.

46. $f(x) = x^4 + 2x^3 + 1$

47. $f(x) = x^4 - 2x^2 + 4$

48. $f(x) = -x^8 + x^6 - x^4 + 5x^2 + 7$

49. $f(x) = -x^6 + x^4 - x^3 + x$

50. On the basis of Exercises 46–49, make a conjecture about the relationship between the sign of the leading coefficient of a polynomial function of even degree and whether that function will have a minimum or a maximum value.

PROJECTS OR GROUP ACTIVITIES

51. Football Some football fields are built in a parabolic mound shape so that water will drain off the field. A model for the shape of such a field is given by

$$h(x) = -0.00023475x^2 + 0.0375x$$

where h is the height of the field in feet at a distance of x feet from a sideline. What is the maximum height of the field? Round to the nearest tenth.

52. Fuel Efficiency The fuel efficiency of an average car is given by the equation $E(v) = -0.018v^2 + 1.476v + 3.4$, where E is the fuel efficiency in miles per gallon and v is the speed of the car in miles per hour. What speed will yield the maximum fuel efficiency? What is the maximum fuel efficiency?

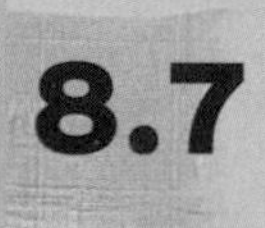

8.7 Nonlinear Inequalities

OBJECTIVE 1 Solve nonlinear inequalities

A **quadratic inequality in one variable** is an inequality that can be written in the form $ax^2 + bx + c < 0$ or $ax^2 + bx + c > 0$, where $a \neq 0$. The symbols $\leq$ and $\geq$ can also be used.

Quadratic inequalities can be solved by algebraic means. However, it is often easier to use a graphical method to solve these inequalities. The graphical method is used in the example that follows.

Focus on solving a quadratic inequality

Solve and graph the solution set of $x^2 - x - 6 < 0$.

Factor the trinomial.

$$x^2 - x - 6 < 0$$
$$(x - 3)(x + 2) < 0$$

On a number line, draw vertical lines at the numbers that make each factor equal to zero.

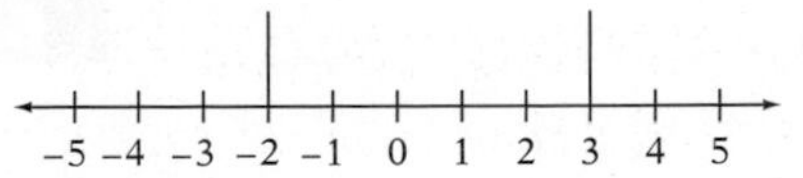

$x - 3 = 0 \qquad x + 2 = 0$
$x = 3 \qquad\quad x = -2$

For each factor, place plus signs above the number line for those regions where the factor is positive, and place negative signs where the factor is negative. $x - 3$ is positive for $x > 3$, and $x + 2$ is positive for $x > -2$.

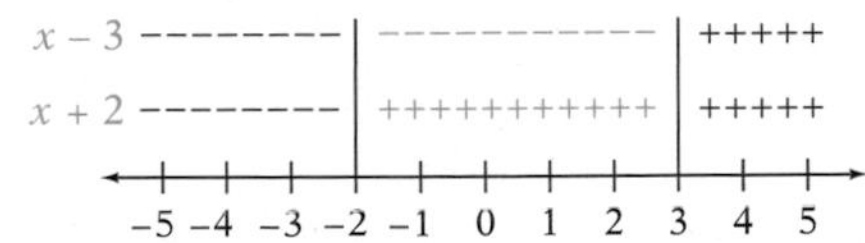

Because $x^2 - x - 6 < 0$, the solution set will be the regions where one factor is positive and the other factor is negative.

Write the solution set using set-builder notation or interval notation.

The solution set is $\{x | -2 < x < 3\}$, or $(-2, 3)$.

The graph of the solution set of the inequality $x^2 - x - 6 < 0$ is shown at the right.

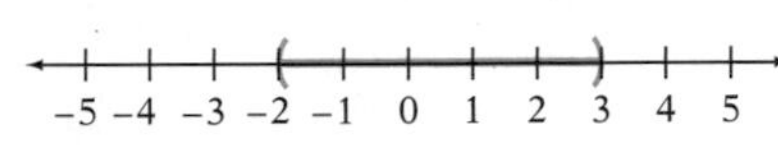

Take Note

For each factor, choose a number in each region. For example, when $x = -4$, $x - 3$ is negative; when $x = 1$, $x - 3$ is negative; and when $x = 4$, $x - 3$ is positive. When $x = -4$, $x + 2$ is negative; when $x = 1$, $x + 2$ is positive; and when $x = 4$, $x + 2$ is positive.

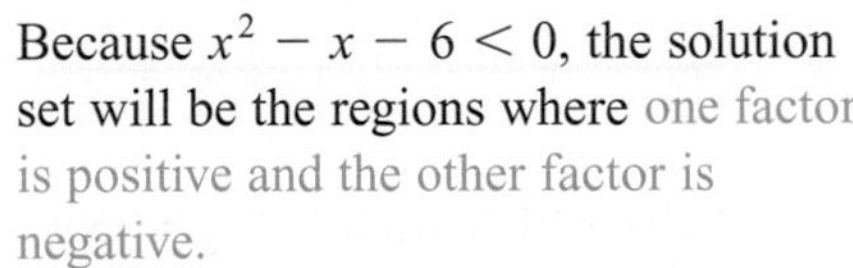

This method of solving quadratic inequalities can be used on any polynomial that can be factored into linear factors.

Focus on solving a cubic inequality

Solve and graph the solution set of $x^3 - 4x^2 - 4x + 16 > 0$. Write the solution set using set-builder notation.

Factor the polynomial by grouping.

$$x^3 - 4x^2 - 4x + 16 > 0$$
$$x^2(x - 4) - 4(x - 4) > 0$$
$$(x^2 - 4)(x - 4) > 0$$
$$(x - 2)(x + 2)(x - 4) > 0$$

On a number line, identify for each factor the regions where the factor is positive and those where the factor is negative.

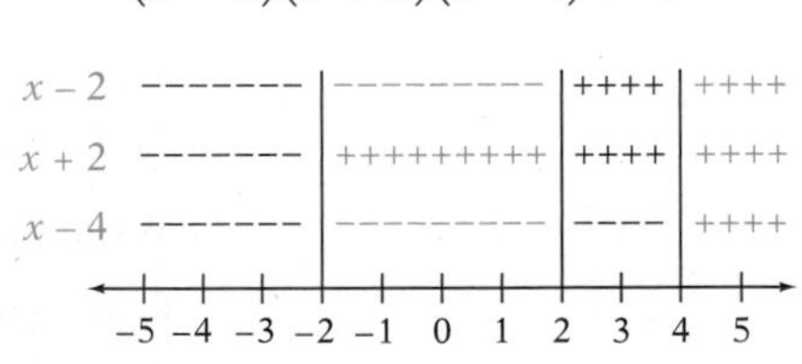

There are two regions where the product of the three factors is positive.

Write the solution set.

The solution set is $\{x | -2 < x < 2\} \cup \{x | x > 4\}$.

The graph of the solution set of the inequality $x^3 - 4x^2 - 4x + 16 > 0$ is shown at the right.

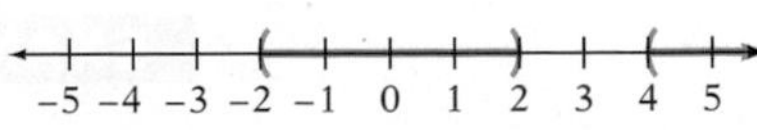

EXAMPLE 1 Solve and graph the solution set of $2x^2 - x - 3 \geq 0$. Write the solution set using interval notation.

Solution $2x^2 - x - 3 \geq 0$

$(2x - 3)(x + 1) \geq 0$

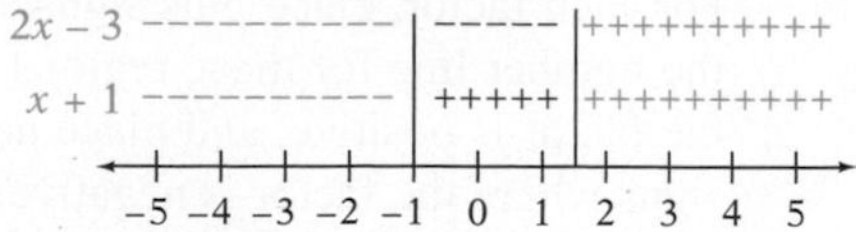

The solution set is $(-\infty, -1] \cup [\frac{3}{2}, \infty)$.

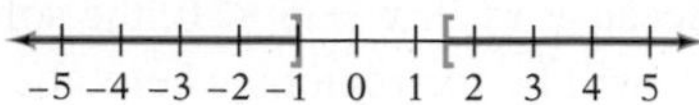

Problem 1 Solve and graph the solution set of $2x^2 - x - 10 \leq 0$. Write the solution set using set-builder notation.

Solution See page S29.

Try Exercise 23, page 488.

The graphical method can be used to solve rational inequalities.

Focus on solving a rational inequality

Solve: $\frac{2x - 5}{x - 4} \leq 1$. Write the solution set using interval notation.

$$\frac{2x - 5}{x - 4} \leq 1$$

Rewrite the inequality so that zero appears on the right side of the inequality.

$$\frac{2x - 5}{x - 4} - 1 \leq 0$$

Then simplify.

$$\frac{2x - 5}{x - 4} - \frac{x - 4}{x - 4} \leq 0$$

$$\frac{x - 1}{x - 4} \leq 0$$

On a number line, identify for each factor of the numerator and each factor of the denominator the regions where the factor is positive and where the factor is negative.

x − 1 --------------- ++++++ +++
x − 4 --------------- ------ +++
−5 −4 −3 −2 −1 0 1 2 3 4 5

The region where the quotient of the two factors is negative is between 1 and 4.

Write the solution set. The solution set is $[1, 4)$.

Note that 1 is part of the solution set, but 4 is not part of the solution set because the denominator of the rational expression is zero when $x = 4$.

EXAMPLE 2 Solve and graph the solution set of $\frac{x + 4}{x - 3} \geq 0$. Write the solution set using set-builder notation.

Solution $\frac{x + 4}{x - 3} \geq 0$

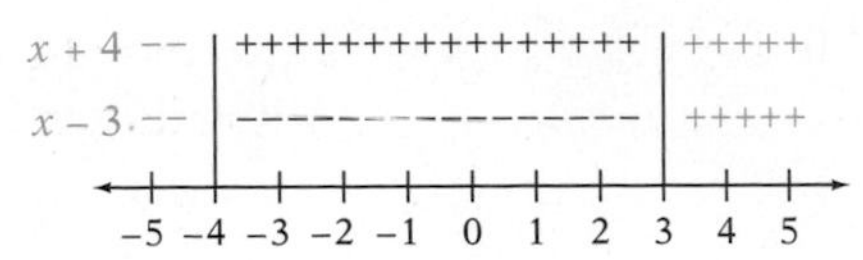

The solution set is $\{x|x > 3\} \cup \{x|x \leq -4\}$.

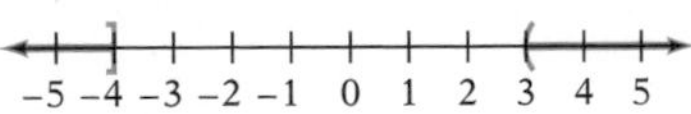

Problem 2 Solve and graph the solution set of $\frac{x}{x-2} \leq 0$. Write the solution set using interval notation.

Solution See page S29.

Try Exercise 15, page 488.

8.7 Exercises

CONCEPT CHECK

1. Suppose $\frac{x-2}{x-3} \geq 0$. Is 3 an element of the solution set? Is 2 an element of the solution set?
2. If $xy > 0$, what must be true about the values of x and y?
3. If $xy \leq 0$, what must be true about the values of x and y?
4. Complete: If $-2x > 0$, then x is a __?__ number.

1 Solve nonlinear inequalities (See pages 484–487.)

GETTING READY

5. To use a number line diagram to solve the inequality $(x + 3)(x - 8) > 0$, begin by drawing vertical lines through __?__ and __?__.
6. When x is between -4 and 7, the numerator of $\frac{x-7}{x+4}$ is negative and the denominator is __?__. The quotient of a negative number and a positive number is __?__, so the expression $\frac{x-7}{x+4}$ is __?__ for all values of x between -4 and 7.

Solve and graph the solution set. Write the solution set using set-builder notation.

7. $(x - 4)(x + 2) > 0$
8. $(x + 1)(x - 3) > 0$
9. $x^2 - 3x + 2 \geq 0$
10. $x^2 + 5x + 6 > 0$
11. $x^2 - x - 12 < 0$
12. $x^2 + x - 20 < 0$

13. $(x - 1)(x + 2)(x - 3) < 0$

14. $(x + 4)(x - 2)(x - 1) \geq 0$

15. $\dfrac{x - 4}{x + 2} > 0$

16. $\dfrac{x + 2}{x - 3} > 0$

17. $\dfrac{x - 3}{x + 1} \leq 0$

18. $\dfrac{x - 1}{x} > 0$

19. $\dfrac{(x - 1)(x + 2)}{x - 3} \leq 0$

20. $\dfrac{(x + 3)(x - 1)}{x - 2} \geq 0$

21. Which inequalities can be solved using this diagram?

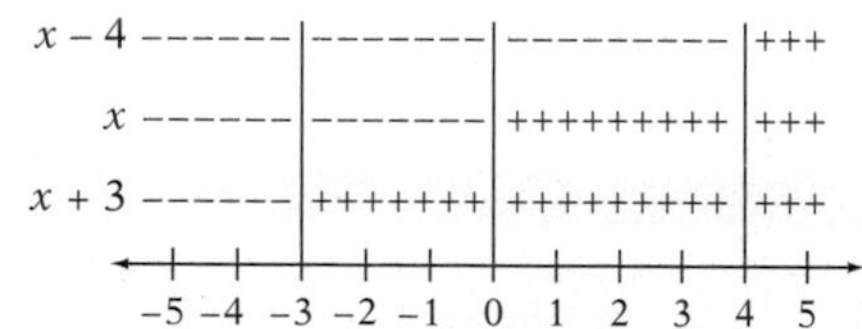

(i) $x(x + 3)(x - 4) > 0$ (ii) $x + 3 < x(x - 4)$

(iii) $0 > \dfrac{x}{3 - x}$ (iv) $\dfrac{x + 3}{x(x - 4)} < 0$

22. a, b, and c are positive numbers such that $a < b < c$. In which of the following intervals is the product $(x - a)(x - b)(x - c)$ always negative?

(i) $\{x \mid a < x < c\}$ (ii) $\{x \mid x > c\}$

(iii) $\{x \mid x < a\}$ (iv) $\{x \mid b < x < c\}$

Solve. Write the solution set using interval notation.

23. $x^2 - 16 > 0$

24. $x^2 - 4 \geq 0$

25. $x^2 - 4x + 4 > 0$

26. $x^2 + 6x + 9 > 0$

27. $x^2 - 9x \leq 36$

28. $x^2 + 4x > 21$

29. $2x^2 - 5x + 2 \geq 0$

30. $4x^2 - 9x + 2 < 0$

31. $4x^2 - 8x + 3 < 0$

32. $2x^2 + 11x + 12 \geq 0$

33. $(x - 6)(x + 3)(x - 2) \leq 0$

34. $(x + 5)(x - 2)(x - 3) > 0$

35. $(2x - 1)(x - 4)(2x + 3) > 0$

36. $(x - 2)(3x - 1)(x + 2) \leq 0$

37. $x^3 + 3x^2 - x - 3 \leq 0$

38. $x^3 + x^2 - 9x - 9 < 0$

39. $x^3 - x^2 - 4x + 4 \geq 0$

40. $2x^3 + 3x^2 - 8x - 12 \geq 0$

41. $\dfrac{3x}{x - 2} > 1$

42. $\dfrac{2x}{x + 1} < 1$

43. $\frac{2}{x+1} \ge 2$

44. $\frac{3}{x-1} < 2$

45. $\frac{x}{(x-1)(x+2)} \ge 0$

46. $\frac{x-2}{(x+1)(x-1)} \le 0$

47. $\frac{1}{x} < 2$

48. $\frac{x}{2x-1} \ge 1$

APPLYING CONCEPTS

Graph the solution set.

49. $(x+2)(x-3)(x+1)(x+4) > 0$

50. $(x-1)(x+3)(x-2)(x-4) \ge 0$

51. $(x^2+2x-8)(x^2-2x-3) < 0$

52. $(x^2+2x-3)(x^2+3x+2) \ge 0$

53. $(x^2+1)(x^2-3x+2) > 0$

54. $(x^2-9)(x^2+5x+6) \le 0$

55. $\frac{x^2(3-x)(2x+1)}{(x+4)(x+2)} \ge 0$

56. $\frac{1}{x} + x > 2$

57. $3x - \frac{1}{x} \le 2$

58. $x^2 - x < \frac{1-x}{x}$

PROJECTS OR GROUP ACTIVITIES

59. You shoot an arrow into the air with an initial velocity of 70 m/s. The distance up, in meters, is given by $d = rt - 5t^2$, where t is the number of seconds since the arrow was shot and r is the initial velocity. Find the interval of time during which the arrow will be more than 200 m high.

CHAPTER 8 Summary

Key Words	Objective and Page Reference	Examples
A **quadratic equation** is an equation of the form $ax^2 + bx + c = 0$, where $a \neq 0$. A quadratic equation is also called a **second-degree equation.**	[8.1.1, p. 432]	$3x^2 + 4x - 7 = 0$ and $x^2 - 1 = 0$ are quadratic equations.

A quadratic equation is in **standard form** when the polynomial is in descending order and equal to zero. [8.1.1, p. 432]

$x^2 - 5x + 6 = 0$ is a quadratic equation in standard form.

When a quadratic equation has two solutions that are the same number, the solution is called a **double root** of the equation. [8.1.1, p. 433]

$$x^2 - 4x + 4 = 0$$
$$(x - 2)(x - 2) = 0$$
$$x - 2 = 0 \qquad x - 2 = 0$$
$$x = 2 \qquad x = 2$$

2 is a double root.

The graph of a **quadratic function** $f(x) = ax^2 + bx + c$ is a **parabola** that opens up when $a > 0$ and down when $a < 0$. The **axis of symmetry** is a line that passes through the **vertex** of the parabola. [8.5.1, pp. 464–465]

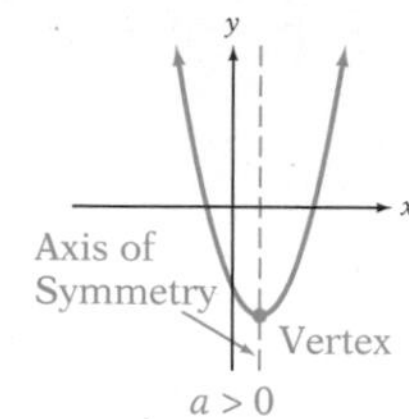

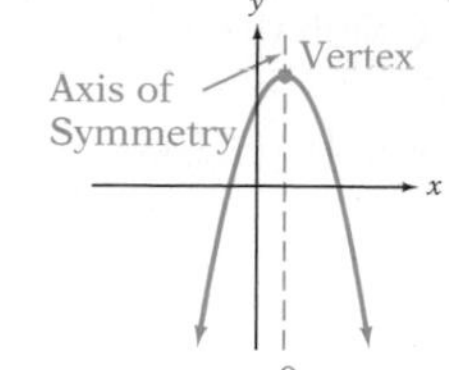

The graph of $f(x) = ax^2 + bx + c$ has a **minimum** value if $a > 0$ and a **maximum** value if $a < 0$. [8.6.1, p. 478]

A **quadratic inequality in one variable** is one that can be written in the form $ax^2 + bx + c > 0$ or $ax^2 + bx + c < 0$, where $a \neq 0$. The symbols $\leq$ and $\geq$ can also be used. [8.7.1, p. 484]

$3x^2 + 5x - 8 \leq 0$ is a quadratic inequality.

Essential Rules and Procedures	Objective and Page Reference	Examples

Some quadratic equations can be solved by factoring and using the **Principle of Zero Products.** [8.1.1, p. 432]

If $ab = 0$, then $a = 0$ or $b = 0$.

$$(x - 3)(x + 4) = 0$$
$$x - 3 = 0 \qquad x + 4 = 0$$
$$x = 3 \qquad x = -4$$

A quadratic equation can be solved by **taking the square root of each side of the equation.** [8.1.2, p. 434]

$$(x + 2)^2 - 9 = 0$$
$$(x + 2)^2 = 9$$
$$\sqrt{(x + 2)^2} = \sqrt{9}$$
$$x + 2 = \pm\sqrt{9}$$
$$x + 2 = \pm 3$$
$$x + 2 = 3 \qquad x + 2 = -3$$
$$x = 1 \qquad x = -5$$

A quadratic equation can be solved by **completing the square.** [8.2.1, pp. 439–440]

$$x^2 + 4x - 1 = 0$$
$$x^2 + 4x = 1$$
$$x^2 + 4x + 4 = 1 + 4$$
$$(x + 2)^2 = 5$$
$$\sqrt{(x + 2)^2} = \sqrt{5}$$
$$x + 2 = \pm\sqrt{5}$$
$$x + 2 = \sqrt{5} \qquad x + 2 = -\sqrt{5}$$
$$x = -2 + \sqrt{5} \qquad x = -2 - \sqrt{5}$$

A quadratic equation $ax^2 + bx + c = 0$ can be solved by using the **quadratic formula.** $x = \dfrac{-b \pm \sqrt{b^2 - 4ac}}{2a}$	[8.2.2, p. 443]	$2x^2 - 3x + 4 = 0$ $a = 2, b = -3, c = 4$ $x = \dfrac{-(-3) \pm \sqrt{(-3)^2 - 4(2)(4)}}{2(2)}$ $= \dfrac{3 \pm \sqrt{9 - 32}}{4} = \dfrac{3 \pm \sqrt{-23}}{4}$ $= \dfrac{3 \pm i\sqrt{23}}{4} = \dfrac{3}{4} \pm \dfrac{\sqrt{23}}{4}i$
For an equation of the form $ax^2 + bx + c = 0$, the quantity $b^2 - 4ac$ is called the **discriminant.**	[8.2.2, p. 445]	
If $b^2 - 4ac = 0$, the equation has a double root.		$x^2 + 8x + 16 = 0$ has a double root because $b^2 - 4ac = 8^2 - 4(1)(16) = 0$
If $b^2 - 4ac > 0$, the equation has two real number solutions that are not equal.		$2x^2 + 3x - 5 = 0$ has two unequal real number solutions because $b^2 - 4ac = 3^2 - 4(2)(-5) = 49$
If $b^2 - 4ac < 0$, the equation has two complex number solutions.		$3x^2 + 2x + 4 = 0$ has two complex number solutions because $b^2 - 4ac = 2^2 - 4(3)(4) = -44$
Vertex and axis of symmetry of a parabola The coordinates of the vertex of a parabola are $\left(-\frac{b}{2a}, f\left(-\frac{b}{2a}\right)\right)$. The equation of the axis of symmetry is $x = -\frac{b}{2a}$.	[8.5.1, p. 466]	$f(x) = x^2 - 2x - 4 \quad a = 1, b = -2$ $-\dfrac{b}{2a} = -\dfrac{-2}{2(1)} = 1$ $f(1) = 1^2 - 2(1) - 4 = -5$ The coordinates of the vertex are $(1, -5)$. The equation of the axis of symmetry is $x = 1$.
The **x-intercepts** of the graph of a quadratic equation occur when $y = 0$.	[8.5.2, p. 468]	$x^2 - 5x + 6 = 0$ $(x - 3)(x - 2) = 0$ $x - 3 = 0 \quad x - 2 = 0$ $x = 3 \quad x = 2$ The coordinates of the x-intercepts are $(3, 0)$ and $(2, 0)$.

CHAPTER 8 Review Exercises

1. Solve: $2x^2 - 3x = 0$

2. Solve for x: $6x^2 + 9x = 6$

3. Solve: $x^2 = 48$

4. Solve: $\left(x + \dfrac{1}{2}\right)^2 + 4 = 0$

5. Find the minimum value of the function $f(x) = x^2 - 7x + 8$.

6. Find the maximum value of the function $f(x) = -2x^2 + 4x + 1$.

7. Write a quadratic equation that has integer coefficients and has solutions $\frac{1}{3}$ and -3.

8. Solve: $2x^2 + 9x = 5$

9. Solve: $2(x + 1)^2 - 36 = 0$

10. Solve: $x^2 + 6x + 10 = 0$

11. Solve: $\frac{2}{x - 4} + 3 = \frac{x}{2x - 3}$

12. Solve: $x^4 - 6x^2 + 8 = 0$

13. Solve: $\sqrt{2x - 1} + \sqrt{2x} = 3$

14. Solve: $2x^{\frac{2}{3}} + 3x^{\frac{1}{3}} - 2 = 0$

15. Solve: $\sqrt{3x - 2} + 4 = 3x$

16. Solve: $x^2 - 6x - 2 = 0$

17. Solve: $\frac{2x}{x - 4} + \frac{6}{x + 1} = 11$

18. Solve: $2x^2 - 2x = 1$

19. Solve: $2x = 4 - 3\sqrt{x - 1}$

20. Solve: $3x = \frac{9}{x - 2}$

21. Solve: $\frac{3x + 7}{x + 2} + x = 3$

22. Solve: $\frac{x - 2}{2x + 3} - \frac{x - 4}{x} = 2$

23. Solve: $1 - \frac{x + 4}{2 - x} = \frac{x - 3}{x + 2}$

24. Find the equation of the axis of symmetry of the parabola with equation $y = -x^2 + 6x - 5$.

25. Find the coordinates of the vertex of the parabola with equation $y = -x^2 + 3x - 2$.

26. Find the zeros of $f(x) = x^2 - 8x - 4$.

27. Use the discriminant to determine the number of x-intercepts of the parabola with equation $y = 3x^2 - 2x - 4$.

28. Find the coordinates of the x-intercepts of the parabola with equation $y = 4x^2 + 12x + 4$.

29. Find the coordinates of the x-intercepts of the parabola with equation $y = -2x^2 - 3x + 2$.

30. Find the zeros of $f(x) = 3x^2 + 2x + 2$.

31. Solve: $(x + 3)(2x - 5) < 0$. Write the solution set in interval notation.

32. Solve: $(x - 2)(x + 4)(2x + 3) \le 0$. Write the solution set in set-builder notation.

33. Solve and graph the solution set of $\frac{x - 2}{2x - 3} \ge 0$. Write the solution set in interval notation.

34. Solve and graph the solution set of $\frac{(2x - 1)(x + 3)}{x - 4} \le 0$. Write the solution set in set-builder notation.

35. Graph $f(x) = x^2 + 2x - 4$. State the domain and range.

36. Find the coordinates of the vertex and the equation of the axis of symmetry of the parabola with equation $y = x^2 - 2x + 3$. Then graph the equation.

37. Geometry The length of a rectangle is 2 cm more than twice the width. The area of the rectangle is 60 cm^2. Find the length and width of the rectangle.

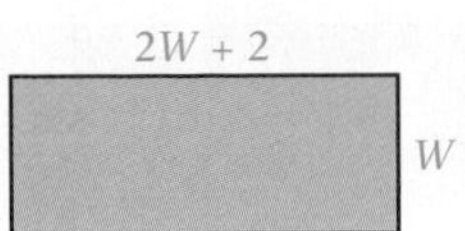

38. Work Problem An older computer requires 12 min longer to print the payroll than does a newer computer. Together the computers can print the payroll in 8 min. Find the time for the newer computer, working alone, to complete the payroll.

39. Uniform Motion A car travels 200 mi. A second car, traveling 10 mph faster than the first car, makes the same trip in 1 h less time. Find the speed of each car.

CHAPTER 8 Test

1. Solve: $2x^2 + x = 6$

2. Solve: $12x^2 + 7x - 12 = 0$

3. Find the maximum value of the function $f(x) = -x^2 + 8x - 7$.

4. Write the quadratic equation that has integer coefficients and has solutions $-\frac{1}{3}$ and 3.

5. Solve: $2(x + 3)^2 - 36 = 0$

6. Solve: $x^2 + 4x - 1 = 0$

7. Find the zeros of $g(x) = x^2 + 3x - 8$.

8. Solve: $3x^2 - x + 8 = 0$

9. Solve: $\frac{2x}{x - 1} + \frac{3}{x + 2} = 1$

10. Solve: $2x + 7x^{\frac{1}{2}} - 4 = 0$

11. Solve: $x^4 - 11x^2 + 18 = 0$

12. Solve: $\sqrt{2x + 1} + 5 = 2x$

13. Find the zeros of $f(x) = x^2 - 4x + 5$.

14. Use the discriminant to determine the number of x-intercepts of the parabola with equation $y = 3x^2 + 2x - 4$.

15. Find the coordinates of the x-intercepts of the parabola with equation $y = 2x^2 + 5x - 12$.

16. Find the equation of the axis of symmetry of the parabola with equation $y = 2x^2 + 6x + 3$.

17. Graph $y = \frac{1}{2}x^2 + x - 4$. State the domain and range.

18. Solve and graph the solution set of $\frac{2x - 3}{x + 4} \leq 0$. Write the answer in set-builder notation.

19. Work Problem A small pipe takes 6 min longer to fill a tank than does a larger pipe. Working together, the pipes can fill the tank in 4 min. How long would it take each pipe, working alone, to fill the tank?

20. Uniform Motion The rate of a river's current is 2 mph. A canoeist rowed 6 mi down the river and back in 4 h. Find the rowing rate in calm water.

Cumulative Review Exercises

1. Evaluate $2a^2 - b^2 \div c^2$ when $a = 3$, $b = -4$, and $c = -2$.

2. Solve: $\frac{2x - 3}{4} - \frac{x + 4}{6} = \frac{3x - 2}{8}$

3. Find the slope of the line that contains the points with coordinates $(3, -4)$ and $(-1, 2)$.

4. Find the equation of the line that contains the point $P(1, 2)$ and is parallel to the graph of $x - y = 1$.

5. Factor: $-3x^3y + 6x^2y^2 - 9xy^3$

6. Factor: $6x^2 - 7x - 20$

7. Factor: $ax + ay - 2x - 2y$

8. Divide: $(3x^3 - 13x^2 + 10) \div (3x - 4)$

9. Multiply: $\frac{x^2 + 2x + 1}{8x^2 + 8x} \cdot \frac{4x^3 - 4x^2}{x^2 - 1}$

10. Find the distance between the points $P_1(-2, 3)$ and $P_2(2, 5)$.

11. Solve $S = \frac{n}{2}(a + b)$ for b.

12. Multiply: $-2i(7 - 4i)$

13. Simplify: $a^{-\frac{1}{2}}(a^{\frac{1}{2}} - a^{\frac{3}{2}})$

14. Simplify: $\frac{\sqrt[3]{8x^4y^5}}{\sqrt[3]{16xy^6}}$

15. Solve: $\frac{x}{x + 2} - \frac{4x}{x + 3} = 1$

16. Solve: $\frac{x}{2x + 3} - \frac{3}{4x^2 - 9} = \frac{x}{2x - 3}$

17. Solve: $x^4 - 6x^2 + 8 = 0$

18. Solve: $\sqrt{3x + 1} - 1 = x$

19. Solve: $|3x - 2| < 8$

20. Find the coordinates of the x- and y-intercepts of the graph of $6x - 5y = 15$.

21. Graph the solution set: $\begin{aligned} x + y &\le 3 \\ 2x - y &< 4 \end{aligned}$

22. Solve the system of equations:

$$\begin{aligned} x + y + z &= 2 \\ -x + 2y - 3z &= -9 \\ x - 2y - 2z &= -1 \end{aligned}$$

23. Given $f(x) = \frac{2x - 3}{x^2 - 1}$, find $f(-2)$.

24. Find the domain of the function $f(x) = \frac{x - 2}{x^2 - 2x - 15}$.

25. Solve and graph the solution set of $x^3 + x^2 - 6x < 0$. Write the answer in set-builder notation.

26. Solve and graph the solution set of $\frac{(x - 1)(x - 5)}{x + 3} \ge 0$. Write the answer in interval notation.

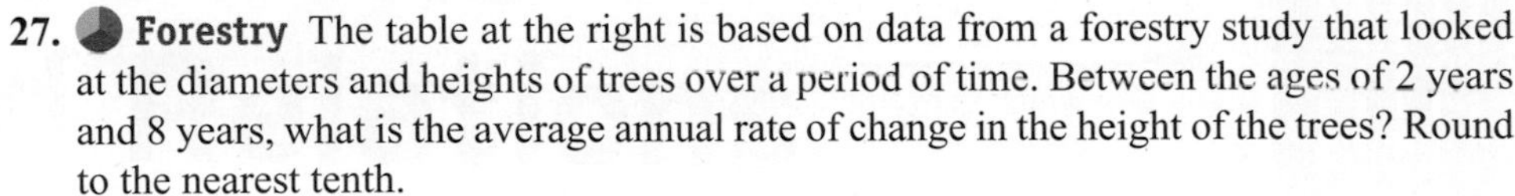

27. **Forestry** The table at the right is based on data from a forestry study that looked at the diameters and heights of trees over a period of time. Between the ages of 2 years and 8 years, what is the average annual rate of change in the height of the trees? Round to the nearest tenth.

Age (in years)	Height (in feet)	Diameter (in inches)
2	7	0.8
3	12	2.0
4	20	3.8
5	22	4.2
6	30	6.9
7	37	7.1
8	42	8.1

28. **Mechanics** A piston rod for an automobile is $9\frac{3}{8}$ in. with a tolerance of $\frac{1}{64}$ in. Find the lower and upper limits of the length of the piston rod.

29. **Geometry** The base of a triangle is $(x + 8)$ ft. The height is $(2x - 4)$ ft. Find the area of the triangle in terms of the variable x.

30. **Depreciation** The graph shows the relationship between the cost of a building and the depreciation allowed for income tax purposes. Find the slope of the line between the two points shown on the graph. Write a sentence that states the meaning of the slope.

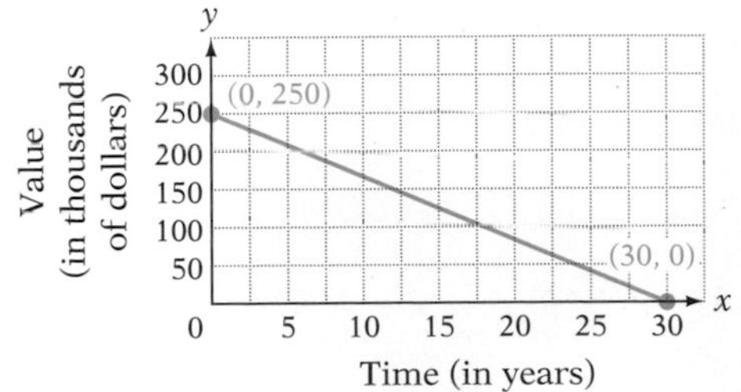

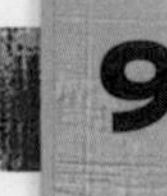

9.1 Translations of Graphs

OBJECTIVE 1 Graph by using translations

The graphs of $f(x) = |x|$ and $g(x) = |x| + 2$ are shown in Figure 1. Note that for a given x-coordinate, the y-coordinate on the graph of g is 2 units higher than that on the graph of f. The graph of g is said to be a **vertical translation,** or **vertical shift,** of the graph of f.

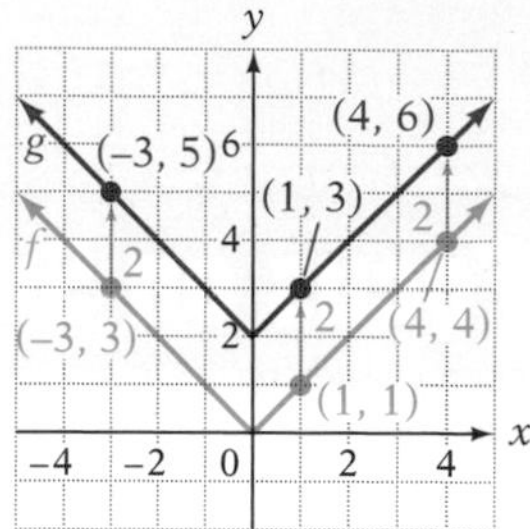

FIGURE 1

Note that because $f(x) = |x|$,

$$g(x) = |x| + 2$$
$$= f(x) + 2$$

Thus $g(x)$ is 2 units more than $f(x)$.

Take Note

Remember that $g(x)$ and $f(x)$ are y-coordinates of an ordered pair. The fact that $g(x)$ is 2 units more than $f(x)$ means that, for a given value of x, the y values for g are 2 units higher than the y values for f.

Now consider the graphs of the functions $f(x) = |x|$ and $h(x) = |x| - 3$, shown in Figure 2. Note that for a given x-coordinate, the y-coordinate on the graph of h is 3 units lower than that on the graph of f. The graph of h is a vertical translation of the graph of f.

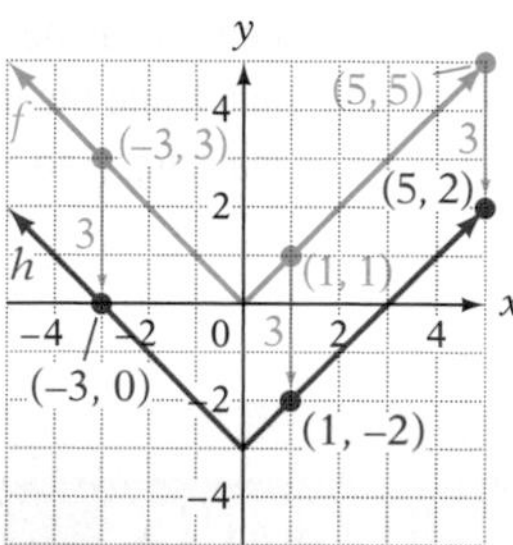

FIGURE 2

In Figure 1, the graph of $g(x) = |x| + 2$ is the graph of f shifted *up* 2 units, whereas in Figure 2, the graph of $h(x) = |x| - 3$ is the graph of f shifted *down* 3 units.

VERTICAL TRANSLATION

If f is a function and c is a positive constant, then

the graph of $y = f(x) + c$ is the graph of $y = f(x)$ shifted up c units.

the graph of $y = f(x) - c$ is the graph of $y = f(x)$ shifted down c units.

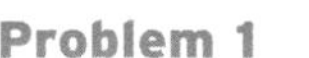

EXAMPLE 1 Given the graph of the function $y = f(x)$ shown at the right in blue, graph $g(x) = f(x) + 3$ by using a vertical translation.

Solution The graph of $g(x) = f(x) + 3$ is the graph of $y = f(x)$ shifted 3 units up. This graph is shown in red at the right.

Problem 1 Given the graph of the function $y = f(x)$ shown at the right, graph $g(x) = f(x) - 3$ by using a vertical translation.

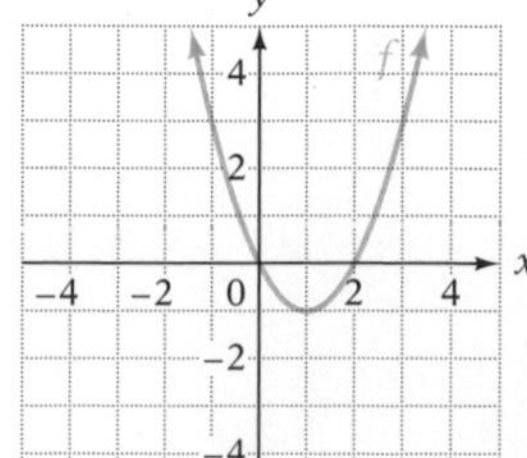

Solution See page S30.

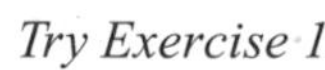

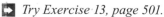

Try Exercise 13, page 501.

The graphs of $f(x) = |x|$ and $g(x) = |x - 2|$ are shown in Figure 3. Note that the graph of g is the graph of f shifted 2 units to the right. The graph of g is a **horizontal translation,** or **horizontal shift,** of the graph of f. In this situation, each x-coordinate is moved to the right 2 units, but the y-coordinates are unchanged.

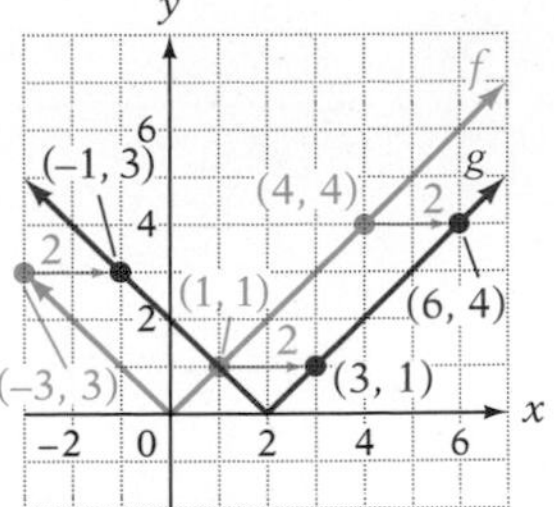

FIGURE 3

The graphs of $f(x) = |x|$ and $h(x) = |x + 3|$ are shown in Figure 4. In this case, the graph of f is translated 3 units to the left.

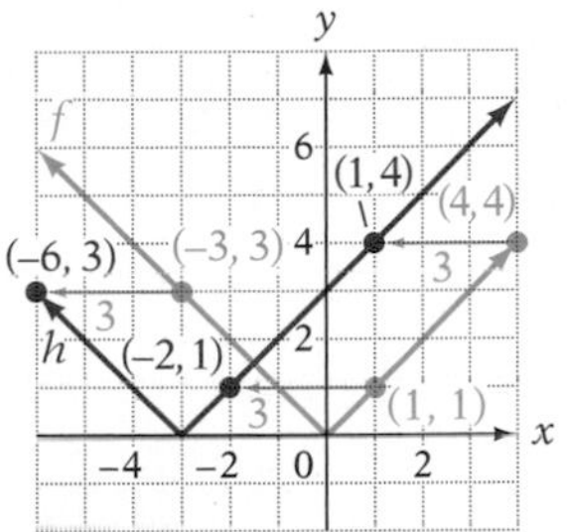

FIGURE 4

Point of Interest

Translations of a figure or pattern may be applied in the design of wallpaper, in the background of a Web page, and in the use of shadows to create the illusion of a letter raised off a page.

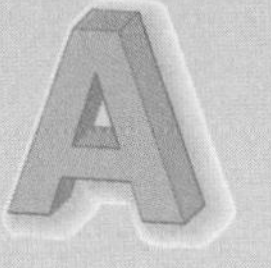

HORIZONTAL TRANSLATION

If f is a function and c is a positive constant, then

the graph of $y = f(x - c)$ is the graph of $y = f(x)$ shifted to the right c units.

the graph of $y = f(x + c)$ is the graph of $y = f(x)$ shifted to the left c units.

EXAMPLE 2 Given the graph of the function $y = F(x)$ shown at the right in blue, graph $G(x) = F(x + 2)$ by using a horizontal translation.

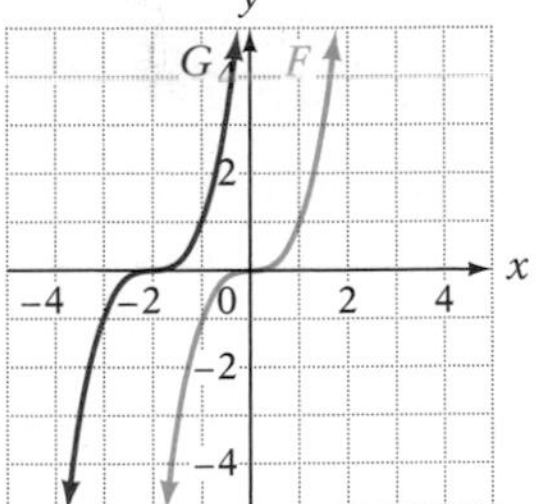

Solution The graph of G (shown in red) is the graph of F shifted horizontally 2 units to the left.

Problem 2 Given the graph of the function $y = f(x)$ shown at the right, graph $g(x) = f(x - 3)$ by using a horizontal translation.

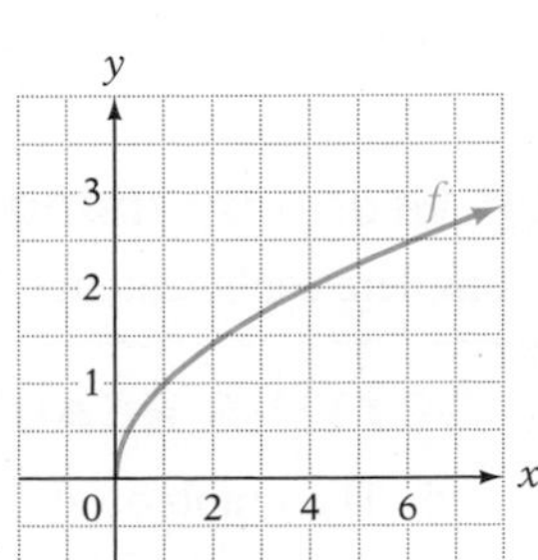

Solution See page S30.

Try Exercise 17, page 501.

It is possible for a graph to involve both a horizontal and a vertical translation.

EXAMPLE 3 Given the graph of the function $y = f(x)$ shown at the right in blue, graph $A(x) = f(x + 1) - 3$ by using both a horizontal and a vertical translation.

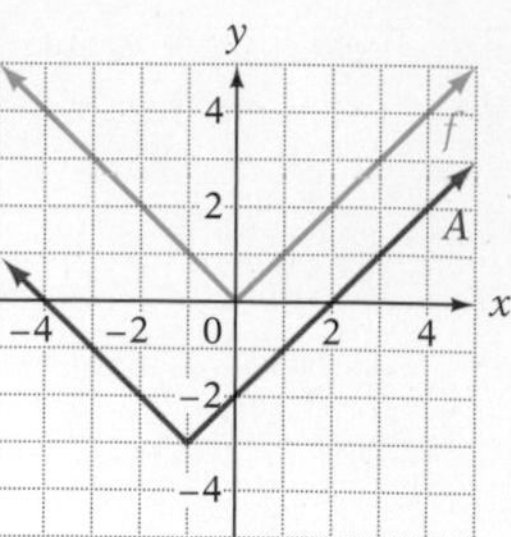

Solution The graph of A includes a horizontal shift of 1 unit to the left and a vertical shift of 3 units down. The graph of A is shown in red at the right.

Problem 3 Given the graph of the function $y = g(x)$ shown at the right, graph $V(x) = g(x - 2) + 1$ by using both a horizontal and a vertical translation.

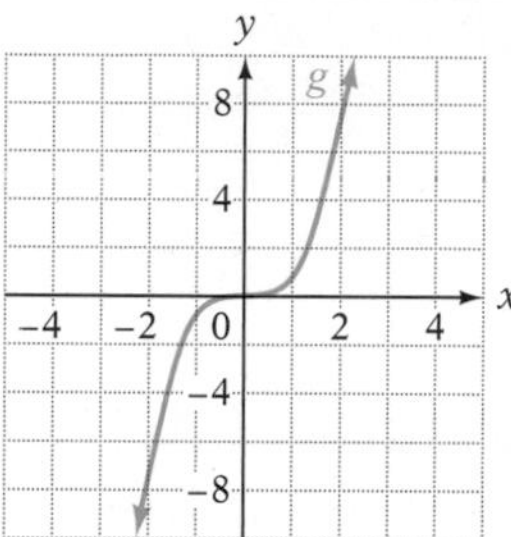

Solution See page S30.

Try Exercise 25, page 502.

9.1 Exercises

CONCEPT CHECK

Complete each sentence.

1. If c is a positive constant, then the graph of $y = f(x + c)$ is the graph of $y = f(x)$ shifted c units __?__.

2. If c is a positive constant, then the graph of $y = f(x - c)$ is the graph of $y = f(x)$ shifted c units __?__.

3. If c is a positive constant, then the graph of $y = f(x) + c$ is the graph of $y = f(x)$ shifted c units __?__.

4. If c is a positive constant, then the graph of $y = f(x) - c$ is the graph of $y = f(x)$ shifted c units __?__.

1 Graph by using translations (See pages 498–500.)

Complete each sentence by giving the number of units and the direction of the shift.

GETTING READY

5. The graph of $g(x) = f(x) - 5$ is the graph of $y = f(x)$ shifted __?__.

6. The graph of $g(x) = f(x - 3)$ is the graph of $y = f(x)$ shifted __?__.

7. The graph of $g(x) = f(x + 7)$ is the graph of $y = f(x)$ shifted __?__.

8. The graph of $g(x) = f(x - 2) + 4$ is the graph of $y = f(x)$ shifted __?__ and __?__.

For Exercises 9 to 34, use translations to draw the graphs. Many of these exercises refer to Figures 1 to 3 below.

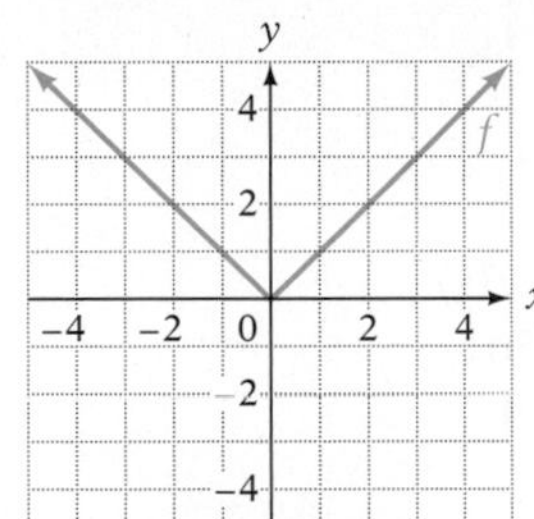
FIGURE 1

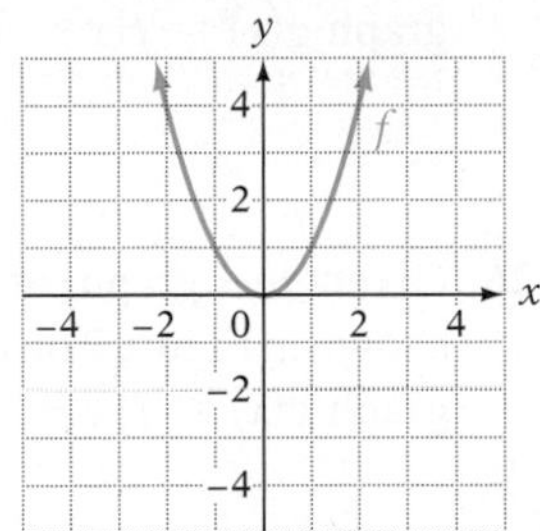
FIGURE 2

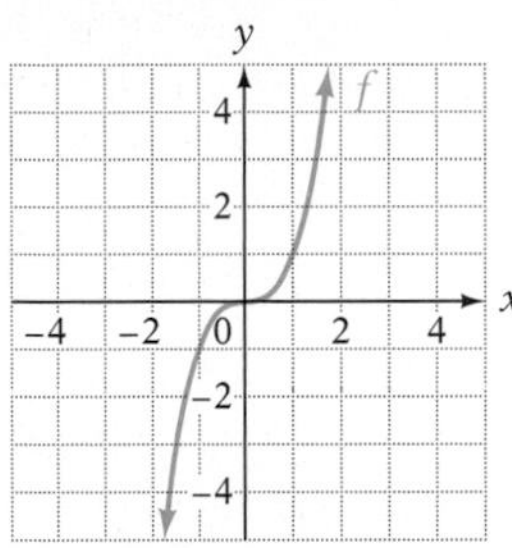
FIGURE 3

9. Given the graph of $y = f(x)$ shown in Figure 1 above, graph $g(x) = f(x) + 4$.

10. Given the graph of $y = f(x)$ shown in Figure 1 above, graph $g(x) = f(x) - 3$.

11. Given the graph of $y = f(x)$ shown in Figure 2 above, graph $g(x) = f(x) - 2$.

12. Given the graph of $y = f(x)$ shown in Figure 2 above, graph $g(x) = f(x) + 1$.

13. Given the graph of $y = f(x)$ shown in Figure 3 above, graph $g(x) = f(x) + 1$.

14. Given the graph of $y = f(x)$ shown in Figure 3 above, graph $g(x) = f(x) - 2$.

15. Given the graph of $y = f(x)$ shown in Figure 2 above, graph $g(x) = f(x - 3)$.

16. Given the graph of $y = f(x)$ shown in Figure 2 above, graph $g(x) = f(x + 2)$.

17. Given the graph of $y = f(x)$ shown in Figure 1 above, graph $g(x) = f(x - 1)$.

18. Given the graph of $y = f(x)$ shown in Figure 1 above, graph $g(x) = f(x + 4)$.

19. Given the graph of $y = f(x)$ shown in Figure 3 above, graph $g(x) = f(x - 4)$.

20. Given the graph of $y = f(x)$ shown in Figure 3 above, graph $g(x) = f(x + 3)$.

21. Given the graph of $y = f(x)$ shown below, graph $g(x) = f(x) + 2$.

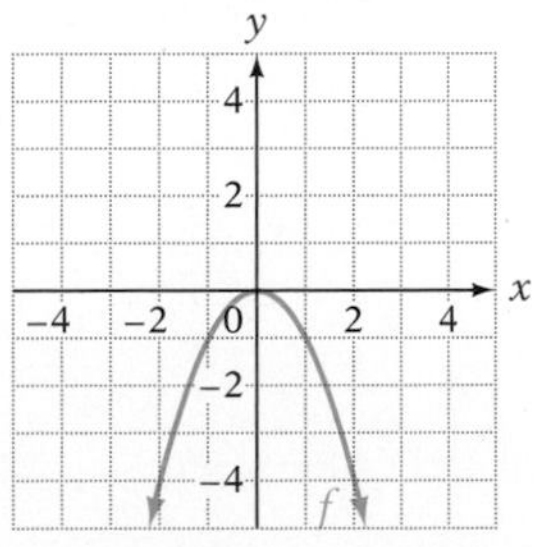

22. Given the graph of $y = f(x)$ shown below, graph $g(x) = f(x) - 2$.

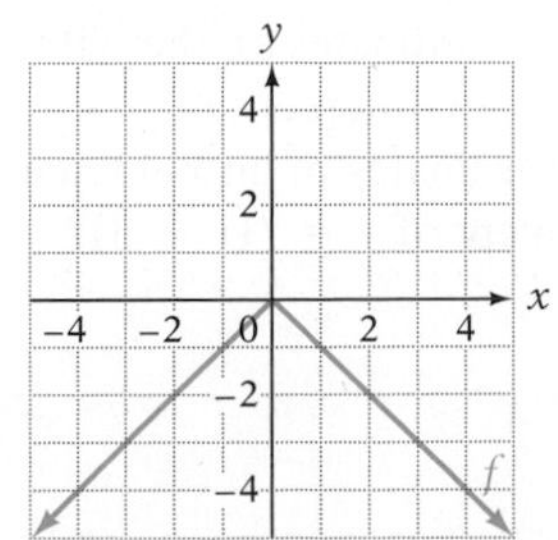

23. Given the graph of $y = f(x)$ shown in Figure 3 on page 501, graph $g(x) = f(x + 1) - 2$.

24. Given the graph of $y = f(x)$ shown in Figure 3 on page 501, graph $g(x) = f(x - 2) + 1$.

25. Given the graph of $y = f(x)$ shown in Figure 2 on page 501, graph $g(x) = f(x - 3) - 4$.

26. Given the graph of $y = f(x)$ shown in Figure 2 on page 501, graph $g(x) = f(x + 2) + 4$.

27. Given the graph of $y = f(x)$ shown below, graph $g(x) = f(x - 1) + 2$.

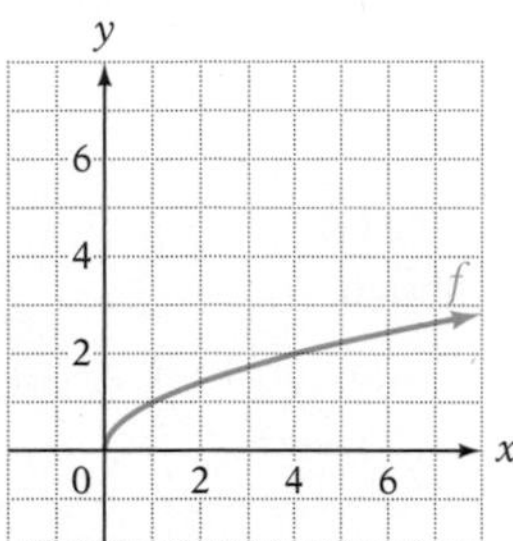

28. Given the graph of $y = f(x)$ shown below, graph $g(x) = f(x + 1) - 3$.

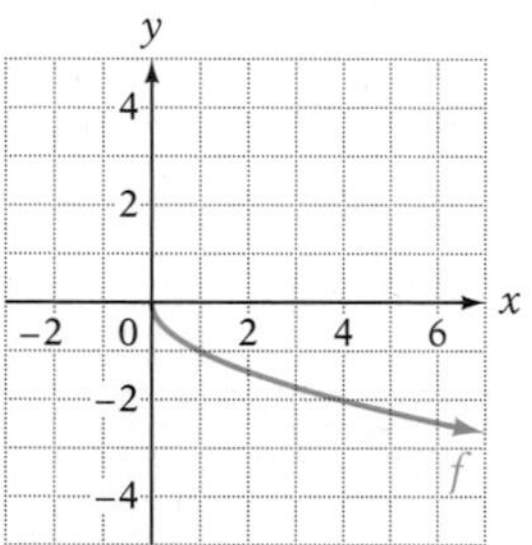

29. Given the graph of $y = f(x)$ shown below, graph $g(x) = f(x) - 1$.

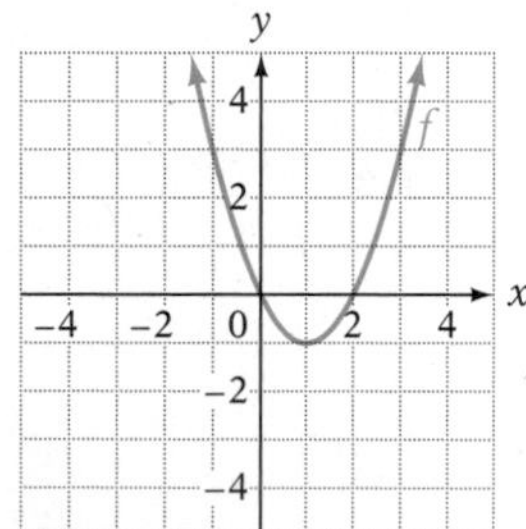

30. Given the graph of $y = f(x)$ shown below, graph $g(x) = f(x) + 2$.

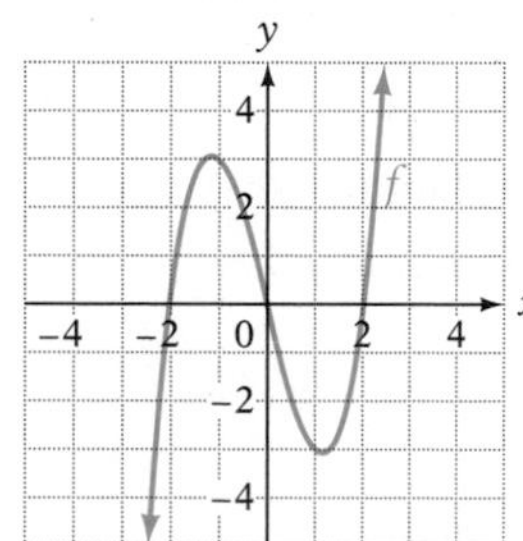

31. Given the graph of $y = f(x)$ shown in Exercise 21, graph $g(x) = f(x + 1) - 2$.

32. Given the graph of $y = f(x)$ shown in Exercise 21, graph $g(x) = f(x - 1) + 1$.

33. Given the graph of $y = f(x)$ shown in Exercise 28, graph $g(x) = f(x - 2) + 1$.

34. Given the graph of $y = f(x)$ shown in Exercise 27, graph $g(x) = f(x + 2) - 3$.

35. **a.** If (0, 7) are the coordinates of the y-intercept of $y = f(x) - 2$, then what are the coordinates of the y-intercept of $y = f(x)$?

b. If (5, 0) are the coordinates of an x-intercept of $y = f(x)$, then what are the coordinates of an x-intercept of $y = f(x - 3)$?

36. The graph of $y = f(x)$ is shown at the right. Which graph is the graph of $y = f(x + 2) - 3$?

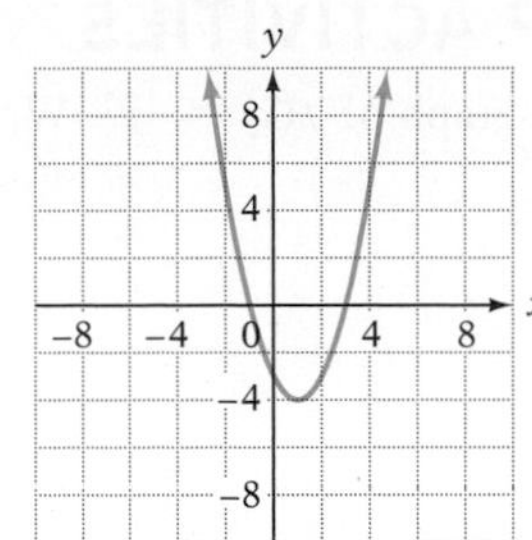

(i)

(ii)

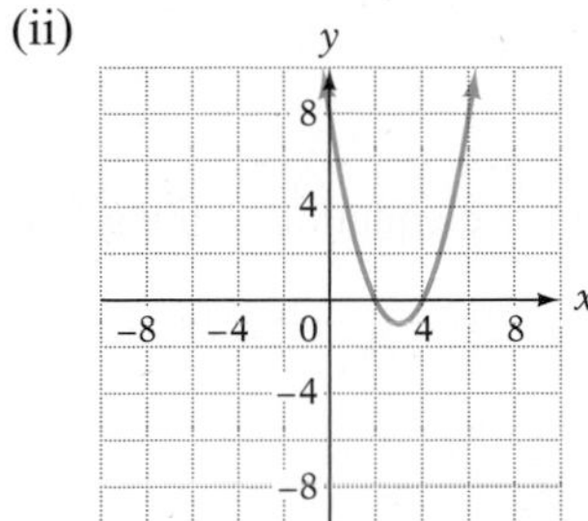

(iii)

APPLYING CONCEPTS

In Exercises 37 to 42, use translations to draw the graph. These exercises refer to Figures 4 to 6 below.

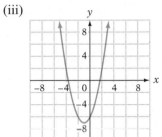

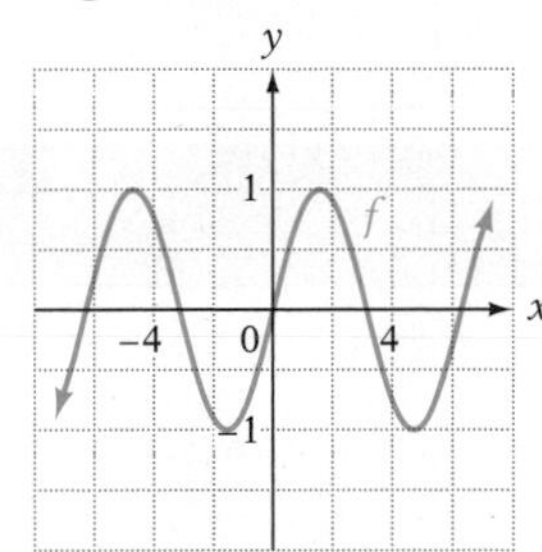

FIGURE 4

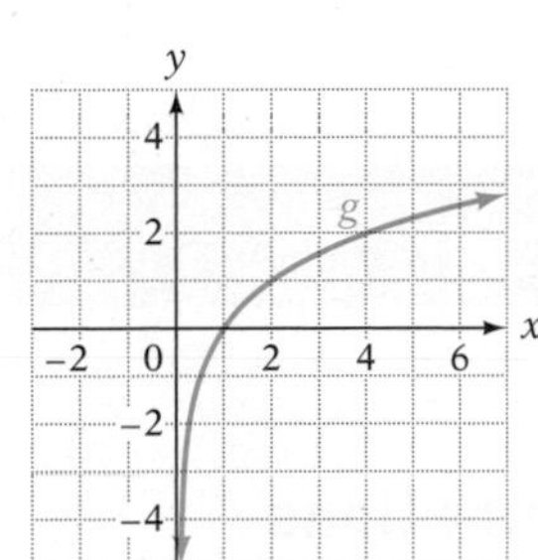

FIGURE 5

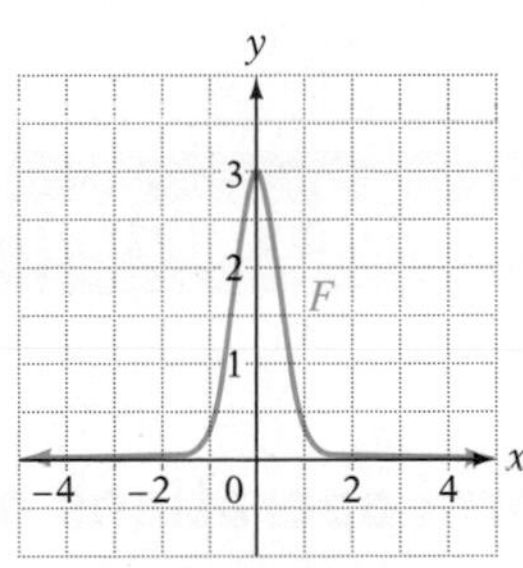

FIGURE 6

37. Given the graph of $y = f(x)$ shown in Figure 4 above, graph $y = f(x) + 1$.

38. Given the graph of $y = f(x)$ shown in Figure 4 above, graph $y = f(x + 1)$.

39. Given the graph of $y = g(x)$ shown in Figure 5 above, graph $y = g(x) - 3$.

40. Given the graph of $y = g(x)$ shown in Figure 5 above, graph $y = g(x) + 1$.

41. Given the graph of $y = F(x)$ shown in Figure 6 above, graph $y = F(x + 3)$.

42. Given the graph of $y = F(x)$ shown in Figure 6 above, graph $y = F(x - 1)$.

PROJECTS OR GROUP ACTIVITIES

43. The graph is a translation of the graph of $f(x) = |x|$. Write the equation of the graph.

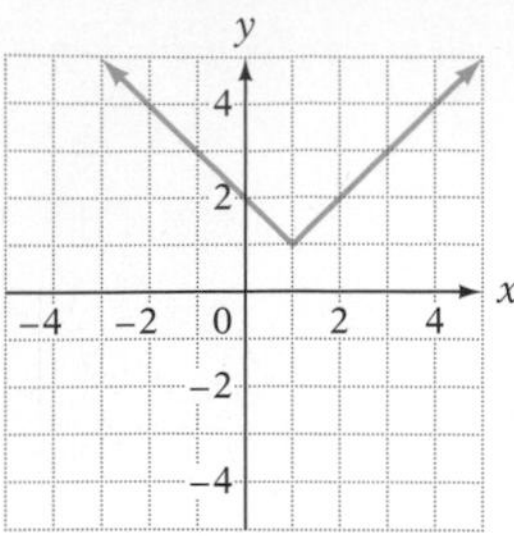

44. The graph is a translation of the graph of $f(x) = x^2$. Write the equation of the graph.

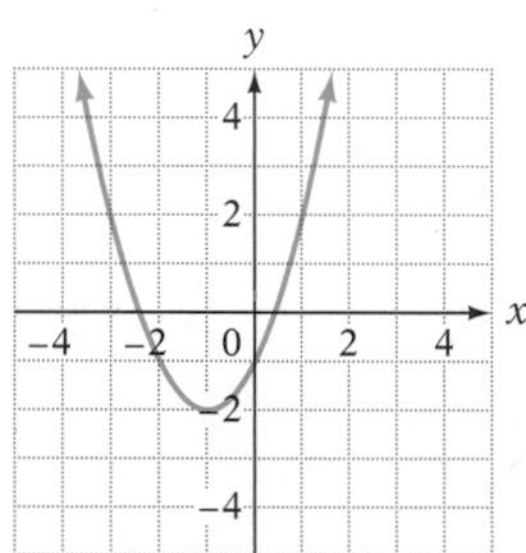

9.2 Algebra of Functions

OBJECTIVE 1 Perform operations on functions

The operations of addition, subtraction, multiplication, and division of functions are defined as follows:

OPERATIONS ON FUNCTIONS

If f and g are functions and x is an element of the domain of each function, then

$$(f + g)(x) = f(x) + g(x)$$
$$(f - g)(x) = f(x) - g(x)$$
$$(f \cdot g)(x) = f(x) \cdot g(x)$$
$$\left(\frac{f}{g}\right)(x) = \frac{f(x)}{g(x)},\ g(x) \neq 0$$

Focus on using operations on functions

A. Given $f(x) = x^2 + 1$ and $g(x) = 3x - 2$, find $(f + g)(3)$.

$$\begin{aligned}(f + g)(x) &= f(x) + g(x)\\ (f + g)(3) &= f(3) + g(3)\\ &= [(3)^2 + 1] + [3(3) - 2]\\ &= 10 + 7 = 17\end{aligned}$$

B. Given $f(x) = 2x + 1$ and $g(x) = x^2 + 1$, find $\left(\frac{f}{g}\right)(-2)$.

$$\left(\frac{f}{g}\right)(x) = \frac{f(x)}{g(x)}$$

$$\left(\frac{f}{g}\right)(-2) = \frac{f(-2)}{g(-2)}$$

$$= \frac{2(-2) + 1}{(-2)^2 + 1}$$

$$= -\frac{3}{5}$$

C. Given $f(x) = 3x - 2$ and $g(x) = 2x + 1$, find $(f \cdot g)(x)$.

$$(f \cdot g)(x) = f(x) \cdot g(x)$$

$$= (3x - 2) \cdot (2x + 1)$$

$$= 6x^2 - x - 2$$

You cannot evaluate the quotient of two functions at any value that results in a denominator of zero. For instance, consider trying to evaluate $\left(\frac{f}{g}\right)(1)$, given $f(x) = 2x^2 - 5x + 6$ and $g(x) = x^2 - 1$.

$$\left(\frac{f}{g}\right)(1) = \frac{f(1)}{g(1)}$$

$$= \frac{2(1)^2 - 5(1) + 6}{(1)^2 - 1}$$

$$= \frac{3}{0} \leftarrow \text{This is not a real number.}$$

Because $\frac{3}{0}$ is not defined, the expression $\left(\frac{f}{g}\right)(1)$ cannot be evaluated.

EXAMPLE 1 Given $f(x) = x^2 - x + 1$ and $g(x) = x^3 - 4$, find $(f - g)(3)$.

Solution

$$(f - g)(3) = f(3) - g(3)$$

$$= (3^2 - 3 + 1) - (3^3 - 4)$$

$$= 7 - 23$$

$$(f - g)(3) = -16$$

Problem 1 Given $f(x) = x^2 + 2x$ and $g(x) = 5x - 2$, find $(f + g)(-2)$.

Solution See page S30.

Try Exercise 7, page 510.

EXAMPLE 2 Given $f(x) = 3x^2 + 4x - 7$ and $g(x) = x^2 - 4x + 2$, vfind $(f - g)(x)$.

Solution

$$(f - g)(x) = f(x) - g(x)$$

$$= (3x^2 + 4x - 7) - (x^2 - 4x + 2)$$

$$(f - g)(x) = 2x^2 + 8x - 9$$

Problem 2 Given $f(x) = x^2 + 3$ and $g(x) = 3x - 5$, find $(f \cdot g)(x)$.

Solution See page S30.

Try Exercise 9, page 510.

EXAMPLE 3 Given $f(x) = x^2 + 4x + 4$ and $g(x) = x^3 - 2$, find $\left(\frac{f}{g}\right)(3)$.

Solution

$$\left(\frac{f}{g}\right)(3) = \frac{f(3)}{g(3)}$$
$$= \frac{3^2 + 4(3) + 4}{3^3 - 2}$$
$$= \frac{25}{25}$$
$$\left(\frac{f}{g}\right)(3) = 1$$

Problem 3 Given $f(x) = x^2 - 4$ and $g(x) = x^2 + 2x + 1$, find $\left(\frac{f}{g}\right)(4)$.

Solution See page S30.

Try Exercise 27, page 510.

OBJECTIVE 2

Find the composition of two functions

Composition of functions is another way in which functions can be combined. This method of combining functions uses the output of one function as the input for a second function.

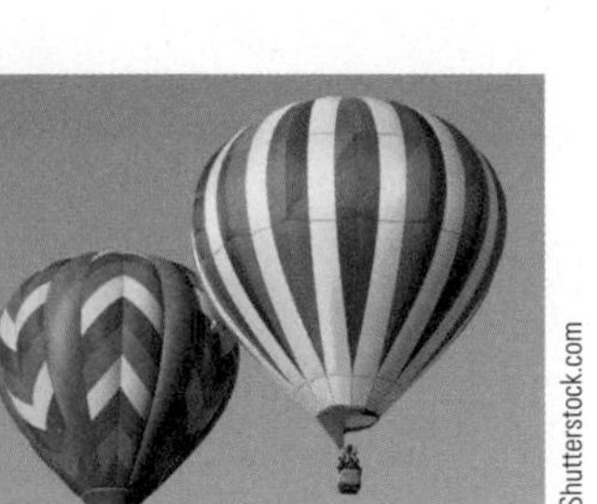
Leene/Shutterstock.com

Suppose that a propane heater is used to heat the air in a hot-air balloon and that the shape of the balloon can be approximated by a sphere. The propane expands the air in the balloon in such a way that the radius of the balloon, in feet, is given by $r(t) = 0.3t$, where t is the time, in minutes, that the propane heater has been running.

The volume of the balloon depends on its radius and is given by $V(r) = \frac{4}{3}\pi r^3$. To find the volume of the balloon 20 min after the heater is turned on, we first find the radius of the balloon and then use that number to find the volume of the balloon.

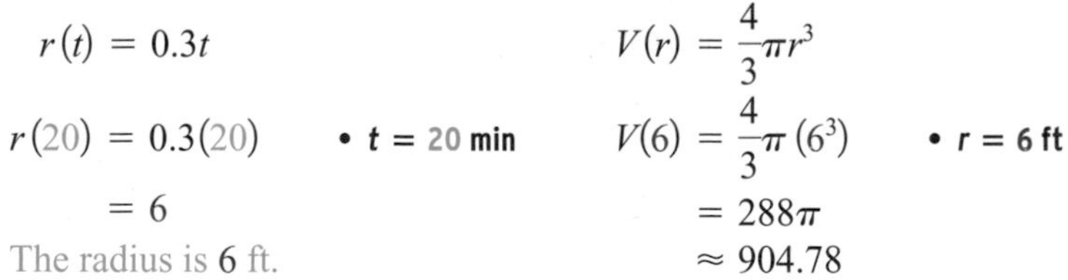

$$r(t) = 0.3t$$
$$r(20) = 0.3(20) \quad \bullet\ t = 20 \text{ min}$$
$$= 6$$

The radius is 6 ft.

$$V(r) = \frac{4}{3}\pi r^3$$
$$V(6) = \frac{4}{3}\pi(6^3) \quad \bullet\ r = 6 \text{ ft}$$
$$= 288\pi$$
$$\approx 904.78$$

The volume is approximately 904.78 ft^3.

There is an alternative way to solve this problem. Because the volume of the balloon depends on the radius and the radius depends on the time, there is a relationship between the volume and the time. We can determine that relationship by evaluating the formula for the volume using $r(t) = 0.3t$ as our input. This will give the volume of the balloon as a function of time.

$$V(r) = \frac{4}{3}\pi r^3$$
$$V[r(t)] = \frac{4}{3}\pi[r(t)]^3 \quad \bullet\ \text{Replace } r \text{ by } r(t).$$
$$= \frac{4}{3}\pi[0.3t]^3 \quad \bullet\ r(t) = 0.3t$$
$$= 0.036\pi t^3$$

The volume of the balloon as a function of time is $V(t) = 0.036\pi t^3$. To find the volume of the balloon 20 min after the heater is turned on, evaluate this function at $t = 20$.

$$\begin{aligned} V(t) &= 0.036\pi t^3 \\ V(20) &= 0.036\pi(20)^3 \qquad \bullet\ t = 20 \text{ min} \\ &= 288\pi \\ &\approx 904.78 \end{aligned}$$

This is exactly the same result we calculated earlier.

The function $V(t) = 0.036\pi t^3$ is referred to as the *composition* of V with r. The notation $V \circ r$ is used to denote this composition of functions. That is,

$$(V \circ r)(t) = 0.036\pi t^3$$

How It's Used

The process of repeatedly composing a function with itself is called *iteration*, and is the basis of *chaos theory*. Chaos theory can be used to study topics as diverse as population growth, the weather, and electrical circuits.

DEFINITION OF THE COMPOSITION OF TWO FUNCTIONS

Let f and g be two functions such that $g(x)$ is in the domain of f for all x in the domain of g. Then the **composition** of the two functions, denoted by $f \circ g$, is the function whose value at x is given by $(f \circ g)(x) = f[g(x)]$.

The function defined by $(f \circ g)(x)$ is called the **composite** of f and g. We read $(f \circ g)(x)$ as "f circle g of x," and we read $f[g(x)]$ as "f of g of x."

Consider the functions $f(x) = 2x$ and $g(x) = x^2$. Then $(f \circ g)(3) = f[g(3)]$ means to evaluate the function f at $g(3)$.

$$g(3) = 3^2 = 9$$

$$f[g(3)] = f(9) = 2(9) = 18$$

The function machine that is illustrated at the right shows the composition of $g(x) = x^2$ and $f(x) = 2x$. Note that the composite function, $(f \circ g)(x) = f[g(x)]$, uses the output of the square function $g(x) = x^2$ as the input of the double function $f(x) = 2x$.

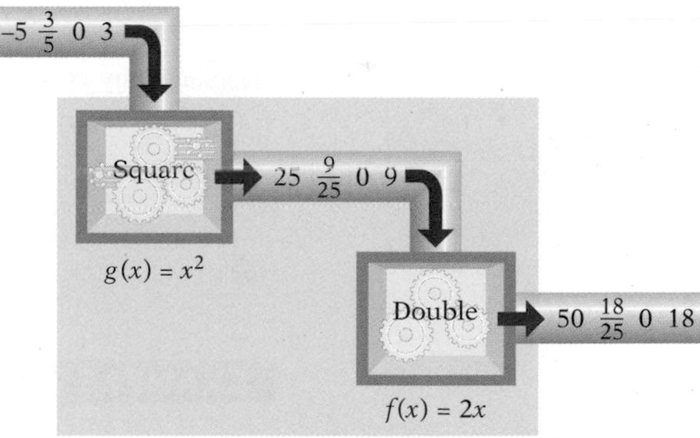

$(f \circ g)(x) = f[g(x)]$

The requirement in the definition of the composition of two functions that $g(x)$ be in the domain of f for all x in the domain of g is important. For instance, let

$$f(x) = \frac{1}{x-1} \quad \text{and} \quad g(x) = 3x - 5$$

When $x = 2$,

$$g(2) = 3(2) - 5 = 1$$

$$f[g(2)] = f(1) = \frac{1}{1-1} = \frac{1}{0} \leftarrow \text{This is not a real number.}$$

In this case, $g(2)$ is not in the domain of f. Thus the composition is not defined at 2.

Focus on finding a composition of functions at a given value

A. Given $f(x) = x^3 + 3x - 4$ and $g(x) = 3x - 1$, find $f[g(1)]$.

Find $g(1)$.

$g(1) = 2$

$$g(x) = 3x - 1$$
$$g(1) = 3(1) - 1 = 2$$
$$f(x) = x^3 + 3x - 4$$

Find $f[g(1)]$.

Replace $g(1)$ by 2.

$$\begin{aligned} f[g(1)] &= [g(1)]^3 + 3[g(1)] - 4 \\ &= 2^3 + 3(2) - 4 \\ &= 8 + 6 - 4 \\ f[g(1)] &= 10 \end{aligned}$$

B. Given $f(x) = x^3 + 3x - 4$ and $g(x) = 3x - 1$, find $g[f(1)]$.

Find $f(1)$.

$f(1) = 0$

$$f(x) = x^3 + 3x - 4$$
$$f(1) = 1^3 + 3(1) - 4 = 0$$
$$g(x) = 3x - 1$$

Find $g[f(1)]$.

Replace $f(1)$ by 0.

$$\begin{aligned} g[f(1)] &= 3[f(1)] - 1 \\ &= 3(0) - 1 \\ g[f(1)] &= -1 \end{aligned}$$

Note from the two examples above that $f[g(1)] \neq g[f(1)]$. In general, composition of functions is *not* a commutative operation: $f[g(x)] \neq g[f(x)]$.

Focus on finding a general composition of functions

Given $f(x) = x^2 + 4x - 1$ and $g(x) = 2x + 3$, find $f[g(x)]$.

$$f(x) = x^2 + 4x - 1$$

Replace x by $g(x)$.

$g(x) = 2x + 3$

Simplify.

$$\begin{aligned} f[g(x)] &= [g(x)]^2 + 4[g(x)] - 1 \\ &= [2x + 3]^2 + 4[2x + 3] - 1 \\ &= [4x^2 + 12x + 9] + [8x + 12] - 1 \\ f[g(x)] &= 4x^2 + 20x + 20 \end{aligned}$$

EXAMPLE 4 Given $f(x) = x^2 - 1$ and $g(x) = 3x + 4$, find each composite function.

A. $f[g(2)]$ **B.** $g[f(x)]$

Solution **A.**

$$g(x) = 3x + 4$$
$$g(2) = 3(2) + 4 = 10 \quad \text{• Find } g(2).$$
$$f(x) = x^2 - 1$$
$$f[g(2)] = [g(2)]^2 - 1 \quad \text{• Find } f[g(2)].$$
$$= 10^2 - 1 \quad \text{• Replace } g(2) \text{ by } 10.$$
$$= 100 - 1$$
$$f[g(2)] = 99$$

B.

$$g(x) = 3x + 4$$
$$g[f(x)] = 3[f(x)] + 4 \quad \text{• Replace } x \text{ by } f(x).$$
$$= 3[x^2 - 1] + 4 \quad \text{• } f(x) = x^2 - 1$$
$$= 3x^2 - 3 + 4 \quad \text{• Simplify.}$$
$$g[f(x)] = 3x^2 + 1$$

Problem 4 Given $g(x) = 3x - 2$ and $h(x) = x^2 + 1$, evaluate each composite function.

A. $g[h(0)]$ **B.** $h[g(x)]$

Solution See page S30.

Try Exercise 41, page 510.

9.2 Exercises

CONCEPT CHECK

1. Is it possible to add $f(x) = \sqrt{x-2}$ and $g(x) = x^2$ when **a.** $x = 2$ or **b.** $x = 1$? Explain your answer.

2. If f and g are functions, which of the following are true for all x in the domain of both f and g?
   (i) $(f+g)(x) = (g+f)(x)$ (ii) $(f-g)(x) = (g-f)(x)$
   (iii) $(f \cdot g)(x) = (g \cdot f)(x)$ (iv) $\left(\frac{f}{g}\right)(x) = \left(\frac{g}{f}\right)(x)$

3. To evaluate $(f \circ g)(x)$, x must be in the domain of __?__ and $g(x)$ must be in the domain of __?__.

4. True of false? For any two functions f and g, $(f \circ g)(x) = (g \circ f)(x)$.

1 Perform operations on functions (See pages 504–506.)

GETTING READY

5. Given $f(x) = x^2 + 2x$ and $g(x) = 3x - 1$, find $(f-g)(-2)$.

$(f-g)(-2)$
$= f(-2) - g(-2)$ • $(f-g)(x) = f(x) - g(x)$
$= [(\_\_?\_\_)^2 + 2(\_\_?\_\_)] - [3(\_\_?\_\_) - 1]$ • In each function, replace x with -2.
$= [\_\_?\_\_ - \_\_?\_\_] - [\_\_?\_\_ - 1]$ • Evaluate the expressions inside the grouping symbols.
$= \_\_?\_\_ - (\_\_?\_\_)$
$= \_\_?\_\_$ • Subtract.

6. Given $f(x) = 2x^3 - x$ and $g(x) = x^2 + 1$, find $(f \cdot g)(-1)$.

$(f \cdot g)(-1)$
$= f(-1) \cdot g(-1)$ • $(f \cdot g)(x) = f(x) \cdot g(x)$
$= [2(\_\_?\_\_)^3 - (\_\_?\_\_)] \cdot [(\_\_?\_\_)^2 + 1]$ • In each function, replace x with -1.
$= [\_\_?\_\_ + \_\_?\_\_] \cdot [\_\_?\_\_ + 1]$ • Evaluate the expressions inside the grouping symbols.
$= \_\_?\_\_ \cdot \_\_?\_\_$
$= \_\_?\_\_$ • Multiply.

Given $f(x) = 2x^2 - 3$ and $g(x) = -2x + 4$, find:

7. $(f - g)(2)$
8. $(f + g)(0)$
9. $(f + g)(x)$
10. $(f - g)(x)$
11. $(f \cdot g)(2)$
12. $\left(\frac{f}{g}\right)(-1)$
13. $(f \cdot g)(x)$
14. $\left(\frac{g}{f}\right)(x)$
15. $(g \cdot g)(x)$

Given $f(x) = 2x^2 + 3x - 1$ and $g(x) = 2x - 4$, find:

16. $(f + g)(-3)$
17. $(f - g)(4)$
18. $(f \cdot g)(-2)$
19. $\left(\frac{f}{g}\right)(2)$
20. $\left(\frac{g}{f}\right)(2)$
21. $(f + g)(x)$
22. $(g - f)(x)$
23. $(g \cdot f)(x)$
24. $\left(\frac{g}{f}\right)(x)$

Given $f(x) = x^2 + 3x - 5$ and $g(x) = x^3 - 2x + 3$, find:

25. $(f - g)(2)$
26. $(f \cdot g)(-3)$
27. $\left(\frac{f}{g}\right)(-2)$

Use the functions $f(x) = \sqrt{x}$, $g(x) = \sqrt{x^2}$, and $h(x) = \sqrt{a - x}$, where $a > 0$. State whether the given expression is *defined* or *undefined*.

28. $(f \cdot g)(a)$
29. $(f \cdot g)(-a)$
30. $\left(\frac{f}{h}\right)(a)$
31. $\left(\frac{h}{g}\right)(a)$

2 Find the composition of two functions (See pages 506–509.)

GETTING READY

For Exercises 32 and 33, use the functions $f(x) = x^2 - 2x$ and $g(x) = 4x$.

32. Evaluate $g[f(-1)]$.

$f(-1) = (\underline{\ ?\ })^2 - 2(\underline{\ ?\ }) = \underline{\ ?\ }$ • **First evaluate $f(-1)$.**

$g[f(-1)] = g[\underline{\ ?\ }]$ • **Replace $f(-1)$ with its value, 3.**

$= 4[\underline{\ ?\ }]$ • **Replace x with 3 in $g(x) = 4x$.**

$= \underline{\ ?\ }$ • **Evaluate.**

33. Find $f[g(x)]$.

$f[g(x)] = f(4x)$ • **$g(x) = \underline{\ ?\ }$**

$= (\underline{\ ?\ })^2 - 2(\underline{\ ?\ })$ • **Replace x in $f(x) = x^2 - 2x$ with $4x$.**

$= \underline{\ ?\ } - \underline{\ ?\ }$ • **Simplify.**

34. Explain the meaning of the notation $f[g(2)]$.
35. What is the meaning of the notation $(f \circ g)(x)$?

Given $f(x) = 2x - 3$ and $g(x) = 4x - 1$, find each composite function.

36. $f[g(0)]$
37. $g[f(0)]$
38. $f[g(2)]$
39. $g[f(-2)]$
40. $f[g(x)]$
41. $g[f(x)]$

Given $g(x) = x^2 + 3$ and $h(x) = x - 2$, find each composite function.

42. $g[h(0)]$ **43.** $h[g(0)]$ **44.** $g[h(4)]$ **45.** $h[g(-2)]$ **46.** $g[h(x)]$ **47.** $h[g(x)]$

Given $f(x) = x^2 + x + 1$ and $h(x) = 3x + 2$, find each composite function.

48. $f[h(0)]$ **49.** $h[f(0)]$ **50.** $f[h(-1)]$ **51.** $h[f(-2)]$ **52.** $f[h(x)]$ **53.** $h[f(x)]$

Given $f(x) = x - 2$ and $g(x) = x^3$, find each composite function.

54. $f[g(2)]$ **55.** $f[g(-1)]$ **56.** $g[f(2)]$ **57.** $g[f(-1)]$ **58.** $f[g(x)]$ **59.** $g[f(x)]$

State whether the given pair of functions will produce the composite function $(f \circ g)(x) = 3x^2 - 3$.

60. $f(x) = x^2 - 1$ and $g(x) = 3x$ **61.** $f(x) = x^2$ and $g(x) = 3x - 3$

62. $f(x) = 3x - 3$ and $g(x) = x^2$ **63.** $f(x) = 3x$ and $g(x) = x^2 - 1$

APPLYING CONCEPTS

Given $g(x) = x^2 - 1$, find:

64. $g(2 + h)$ **65.** $g(3 + h) - g(3)$ **66.** $g(-1 + h) - g(-1)$

67. $\dfrac{g(1 + h) - g(1)}{h}$ **68.** $\dfrac{g(-2 + h) - g(-2)}{h}$ **69.** $\dfrac{g(a + h) - g(a)}{h}$

Given $f(x) = 2x$, $g(x) = 3x - 1$, and $h(x) = x - 2$, find:

70. $f(g[h(2)])$ **71.** $g(h[f(1)])$ **72.** $h(g[f(-1)])$

73. $f(h[g(0)])$ **74.** $f(g[h(x)])$ **75.** $g(f[h(x)])$

PROJECTS OR GROUP ACTIVITIES

The graphs of the functions f and g are shown below. Use these graphs to determine the values of the following composite functions.

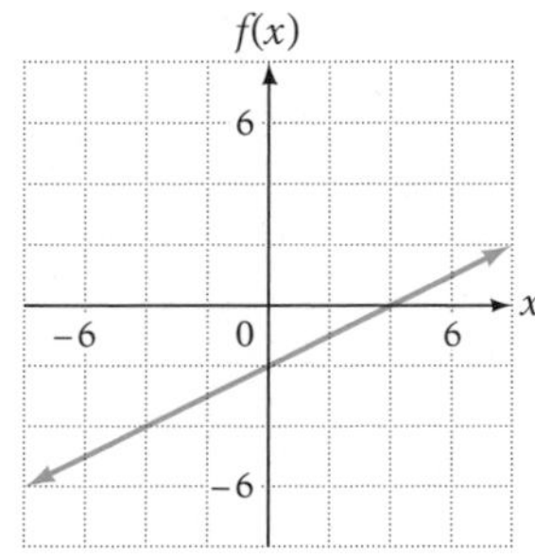

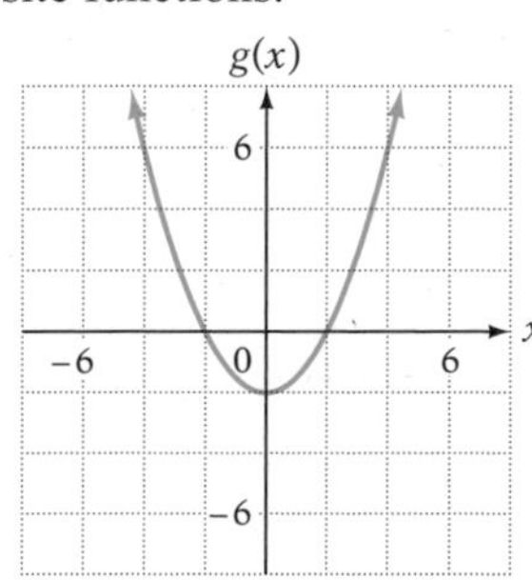

76. $f[g(0)]$ **77.** $f[g(2)]$ **78.** $f[g(4)]$ **79.** $f[g(-2)]$

80. $f[g(-4)]$ **81.** $g[f(0)]$ **82.** $g[f(4)]$ **83.** $g[f(-4)]$

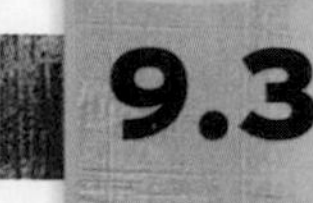

9.3 One-to-One and Inverse Functions

OBJECTIVE  Determine whether a function is one-to-one

Recall that a function is a set of ordered pairs in which no two ordered pairs that have the same first component have different second components. This means that given any x, there is only one y that can be paired with that x. A **one-to-one function** satisfies the additional condition that given any y, there is only one x that can be paired with the given y. One-to-one functions are commonly expressed by writing 1–1.

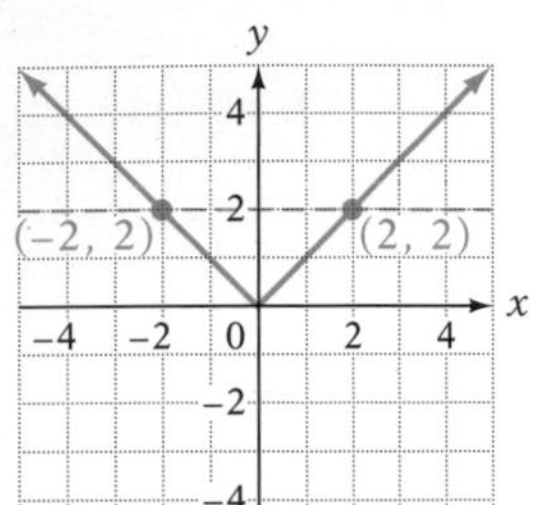

The function given by the equation $y = |x|$ is not a 1–1 function since, given $y = 2$, there are two possible values of x, 2 and -2, that can be paired with the given y-value. The graph at the left illustrates that a horizontal line intersects the graph more than once.

Just as the vertical line test can be used to determine whether a graph represents a function, a **horizontal line test** can be used to determine whether the graph of a function represents a 1–1 function.

HORIZONTAL LINE TEST

A graph of a function is the graph of a 1–1 function if any horizontal line intersects the graph at no more than one point.

The graph of a function is shown at the right. Since any horizontal line will intersect the graph at no more than one point, the graph is the graph of a 1–1 function.

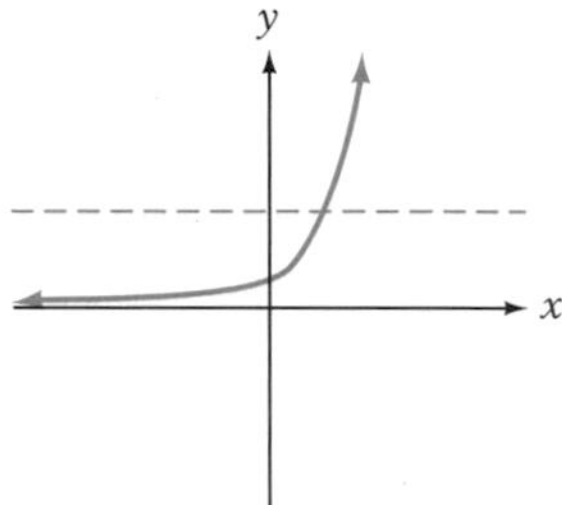

The graph of a quadratic function is shown at the right. Note that a horizontal line can intersect the graph at more than one point. Therefore, this graph is not the graph of a 1–1 function. In general, $f(x) = ax^2 + bx + c, a \neq 0$, is not a 1–1 function.

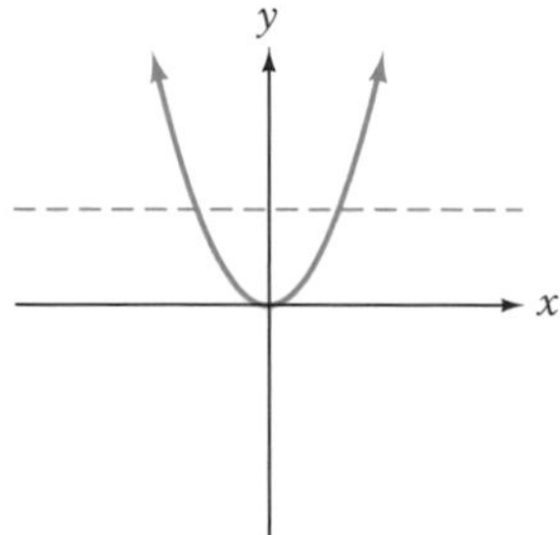

EXAMPLE 1 Determine whether the graph represents the graph of a 1–1 function.

A.

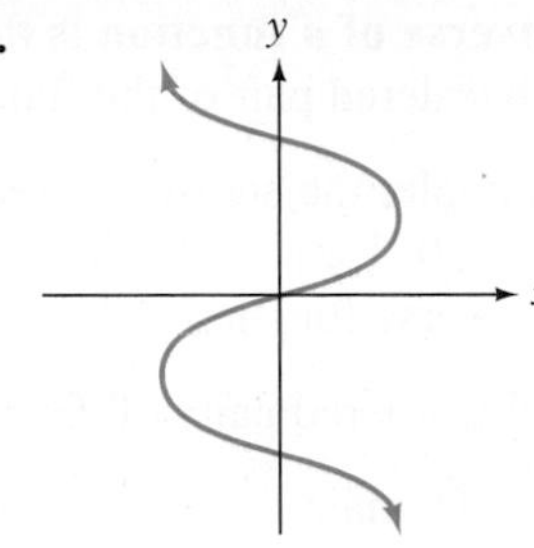

B.

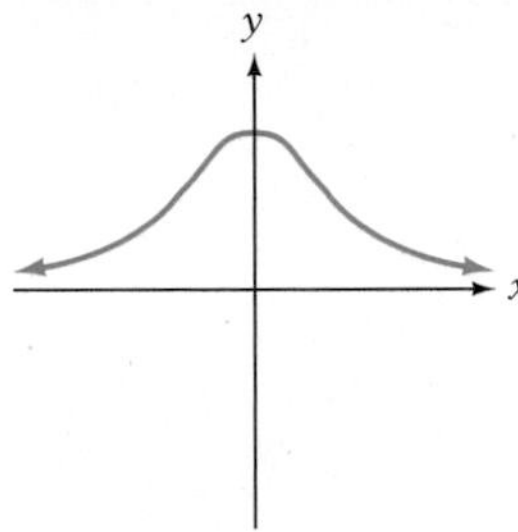

Solution **A.**

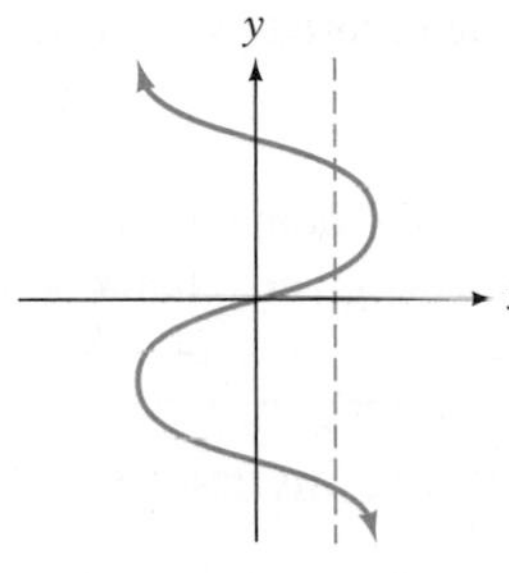

- **A vertical line can intersect the graph at more than one point. The graph does not represent a function.**

This is not the graph of a 1–1 function.

B.

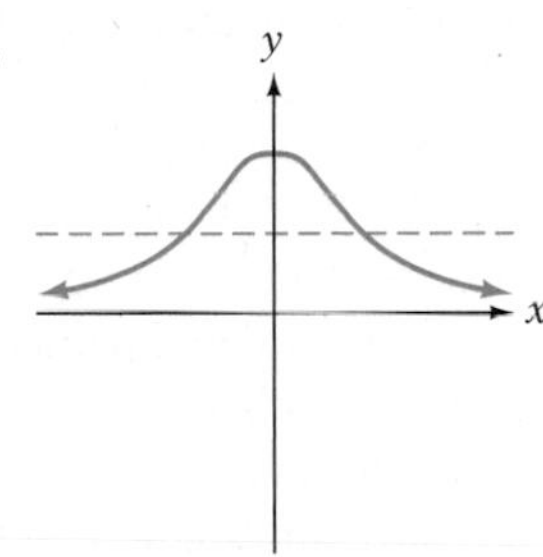

- **A horizontal line can intersect the curve at more than one point.**

This is not the graph of a 1–1 function.

Problem 1 Determine whether the graph represents the graph of a 1–1 function.

A.

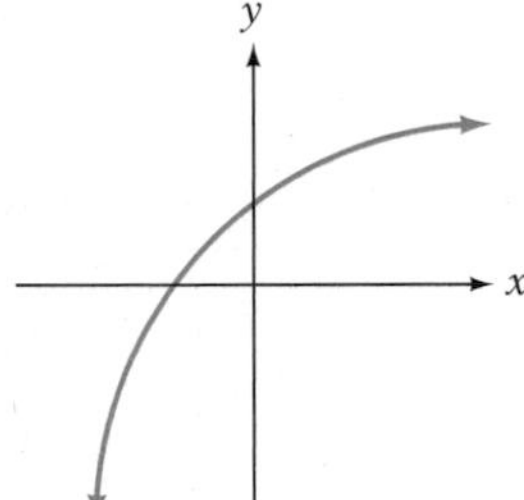

B.

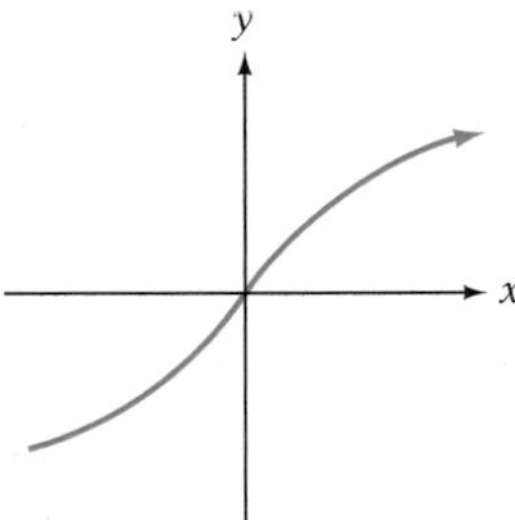

Solution See page S30.

Try Exercise 13, page 517.

OBJECTIVE 2 Find the inverse of a function

The **inverse of a function** is the set of ordered pairs formed by reversing the coordinates of each ordered pair of the function.

For example, the set of ordered pairs of the function defined by $f(x) = 2x$ with domain $\{-2, -1, 0, 1, 2\}$ is $\{(-2, -4), (-1, -2), (0, 0), (1, 2), (2, 4)\}$. The set of ordered pairs of the inverse function is $\{(-4, -2), (-2, -1), (0, 0), (2, 1), (4, 2)\}$.

From the ordered pairs of f, we have

$$\text{Domain} = \{-2, -1, 0, 1, 2\} \quad \text{and} \quad \text{Range} = \{-4, -2, 0, 2, 4\}$$

From the ordered pairs of the inverse function, we have

$$\text{Domain} = \{-4, -2, 0, 2, 4\} \quad \text{and} \quad \text{Range} = \{-2, -1, 0, 1, 2\}$$

Note that the domain of the inverse function is the range of the original function, and the range of the inverse function is the domain of the original function.

Focus on finding the inverse of a function given as a set of ordered pairs

Find the inverse of the function $\{(-1, 3), (1, -2), (3, -3), (5, 4), (6, 5)\}$.

Reverse the coordinates of each ordered pair. $\{(3, -1), (-2, 1), (-3, 3), (4, 5), (5, 6)\}$

Now consider the function defined by $g(x) = x^2$ with domain $\{-2, -1, 0, 1, 2\}$. The set of ordered pairs of this function is $\{(-2, 4), (-1, 1), (0, 0), (1, 1), (2, 4)\}$. Reversing the coordinates of the ordered pairs gives $\{(4, -2), (1, -1), (0, 0), (1, 1), (4, 2)\}$. These ordered pairs do not satisfy the condition of a function because there are ordered pairs with the same first coordinate and different second coordinates. This example illustrates that not all functions have an inverse function.

The graphs of $f(x) = 2x$ and $g(x) = x^2$ with the set of real numbers as the domain are shown below.

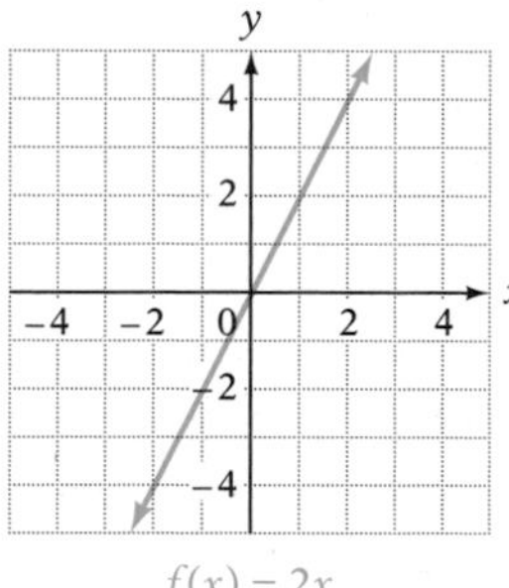

$f(x) = 2x$

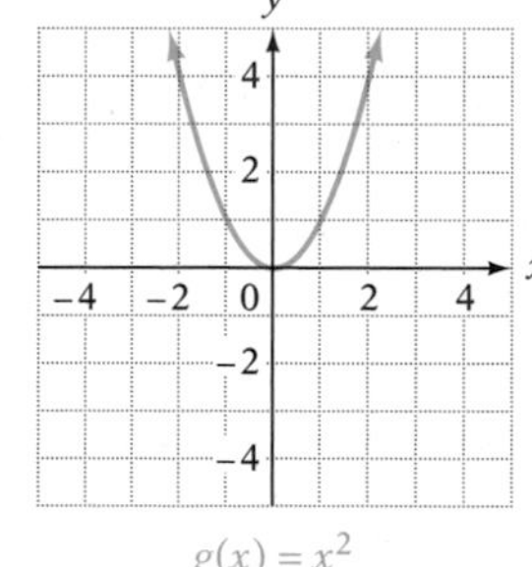

$g(x) = x^2$

By the horizontal line test, f is a 1–1 function but g is not.

Take Note

It is important to note that f^{-1} is the symbol for the *inverse* function and does not mean "reciprocal."

$$f^{-1}(x) \neq \frac{1}{f(x)}$$

CONDITION FOR AN INVERSE FUNCTION

A function f has an inverse function if and only if f is a 1–1 function.

The symbol f^{-1} is used to denote the inverse of a 1–1 function f. The symbol $f^{-1}(x)$ is read "f inverse of x."

$f^{-1}(x)$ does *not* denote the reciprocal of $f(x)$ but is the notation for the inverse of a 1–1 function.

To find the inverse of a function, interchange x and y. Then solve for y.

Focus on finding the inverse of a function given by an equation

Find the inverse of the function defined by $f(x) = 3x + 6$.

Take Note

If the ordered pairs of f are given by (x, y), then the ordered pairs of f^{-1} are given by (y, x). That is, x and y are interchanged. This is the reason for Step 2 at the right.

$f(x) = 3x + 6$

Replace $f(x)$ by y. $\quad y = 3x + 6$

Interchange x and y. $\quad x = 3y + 6$

Solve for y. $\quad x - 6 = 3y$

$\frac{1}{3}x - 2 = y$

Replace y by $f^{-1}(x)$. $\quad f^{-1}(x) = \frac{1}{3}x - 2$

The inverse of the function is given by $f^{-1}(x) = \frac{1}{3}x - 2$.

EXAMPLE 2 Find the inverse of the function defined by the equation $f(x) = 2x - 4$.

Solution

$f(x) = 2x - 4$

$y = 2x - 4$ • **Replace f(x) by y.**

$x = 2y - 4$ • **Interchange x and y.**

$2y = x + 4$ • **Solve for y.**

$y = \frac{1}{2}x + 2$

$f^{-1}(x) = \frac{1}{2}x + 2$

Problem 2 Find the inverse of the function defined by the equation $f(x) = 4x + 2$.

Solution See page S30.

Try Exercise 39, page 519.

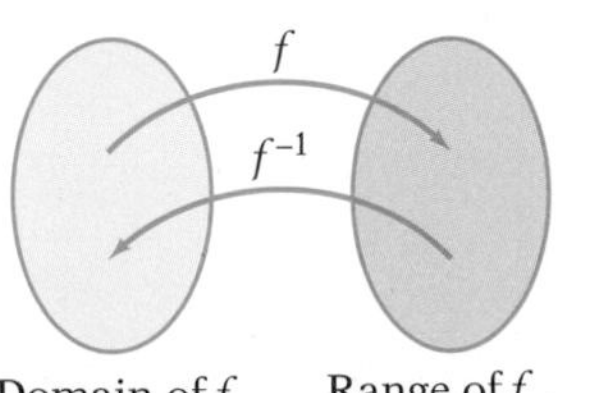

Domain of f Range of f
Range of f^{-1} Domain of f^{-1}

The fact that the ordered pairs of the inverse of a function are the reverse of those of the original function has a graphical interpretation. The function shown at the left below includes the points with coordinates $(-2, 0)$, $(-1, 2)$, $(1, 4)$, and $(5, 6)$. In the graph in the middle, the points with the reverse coordinates are plotted. The inverse function is graphed by drawing a smooth curve through these points, as shown in the rightmost figure.

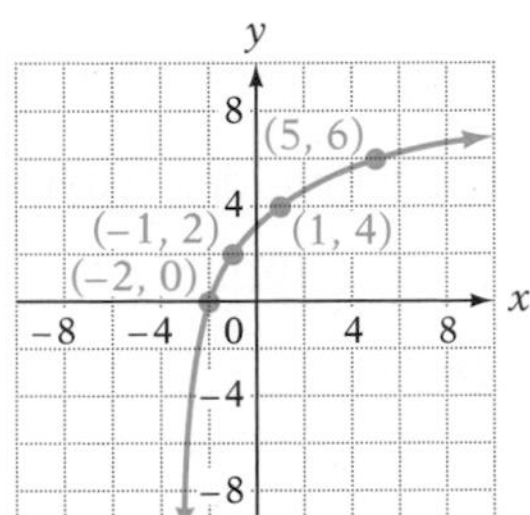

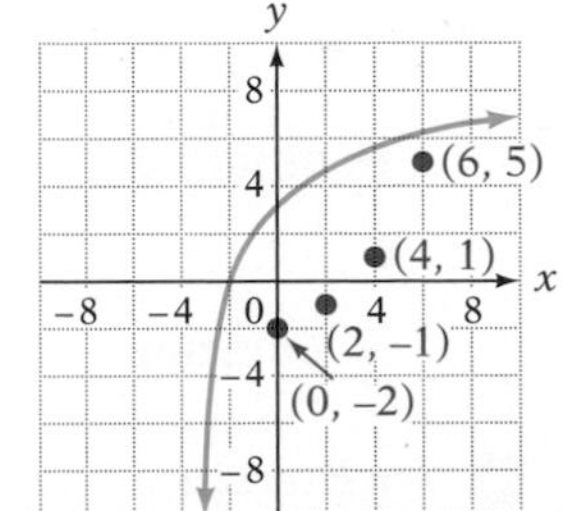

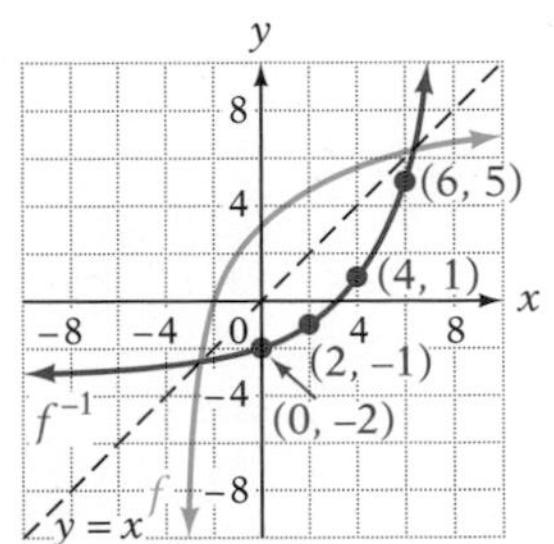

Note that the dashed graph of $y = x$ is shown in the rightmost figure. If two functions are inverses of each other, then their graphs are mirror images with respect to the graph of the equation $y = x$.

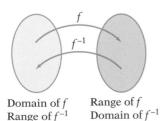

A special property relates the composition of a function and its inverse.

COMPOSITION OF INVERSE FUNCTIONS PROPERTY

$$f^{-1}[f(x)] = x \quad \text{and} \quad f[f^{-1}(x)] = x$$

This property can be used to determine whether two functions are inverses of each other.

Focus on determining whether two functions are inverses

Are $f(x) = 2x - 4$ and $g(x) = \frac{1}{2}x + 2$ inverses of each other?

To determine whether the functions are inverses, check that they satisfy the Composition of Inverse Functions Property.

$$f[g(x)] = 2\left(\frac{1}{2}x + 2\right) - 4 \qquad g[f(x)] = \frac{1}{2}(2x - 4) + 2$$

$$= x + 4 - 4 \qquad\qquad = x - 2 + 2$$

$$= x \qquad\qquad = x$$

Because $f[g(x)] = x$ and $g[f(x)] = x$, the functions are inverses of each other.

EXAMPLE 3 Are the functions defined by the equations $f(x) = -2x + 3$ and $g(x) = -\frac{1}{2}x + \frac{3}{2}$ inverses of each other?

Solution

$$f[g(x)] = f\left(-\frac{1}{2}x + \frac{3}{2}\right)$$

$$= -2\left(-\frac{1}{2}x + \frac{3}{2}\right) + 3$$

$$= x - 3 + 3$$

$$= x$$

$$g[f(x)] = g(-2x + 3)$$

$$= -\frac{1}{2}(-2x + 3) + \frac{3}{2}$$

$$= x - \frac{3}{2} + \frac{3}{2}$$

$$= x$$

- **Check that the functions f and g satisfy the Composition of Inverse Functions Property.**
- **$f[g(x)] = x$**
- **$g[f(x)] = x$**

The functions are inverses of each other.

Problem 3 Are the functions defined by the equations $h(x) = 4x + 2$ and $g(x) = \frac{1}{4}x - \frac{1}{2}$ inverses of each other?

Solution See page S30.

Try Exercise 57, page 519.

The function given by the equation $f(x) = \frac{1}{2}x^2$ does not have an inverse that is a function. Two of the ordered-pair solutions of this function are $(4, 8)$ and $(-4, 8)$.

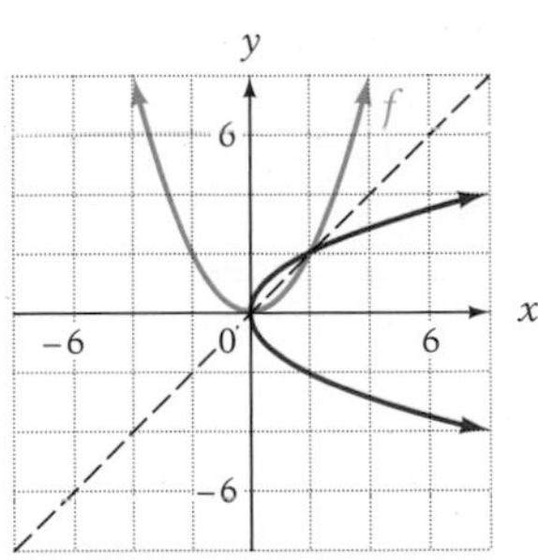

The graph of $f(x) = \frac{1}{2}x^2$ is shown at the left. This graph does not pass the horizontal line test for the graph of a 1–1 function. The mirror image of the graph of f with respect to the graph of $y = x$ is also shown. This graph does not pass the vertical line test for the graph of a function.

A quadratic function with the real numbers as its domain does not have an inverse function.

9.3 Exercises

CONCEPT CHECK

Determine whether the set of ordered pairs has an inverse function.

1. $\{(0, 4), (2, 5), (4, 8), (3, 1), (1, 6)\}$
2. $\{(-3, 2), (-2, 3), (-1, 4), (1, 1), (0, -3)\}$
3. $\{(-1, 10), (0, 9), (1, 11), (2, 8), (3, 10)\}$
4. Do all functions have an inverse function?
5. Does $f(x) = mx + b, m \neq 0$, have an inverse function?
6. Does $f(x) = ax^2, a \neq 0$, have an inverse function?

1 Determine whether a function is one-to-one (See pages 512–513.)

7. What is a 1–1 function?
8. What is the horizontal line test?

GETTING READY

9. The function $y = x^2$ is not a 1–1 function because, for example, the x values 2 and __?__ are both paired with the same y value, __?__.
10. Suppose the horizontal line $y = 3$ intersects the graph of the function $y = f(x)$ at the points with coordinates $(-2, 3)$ and $(2, 3)$. The function $y = f(x)$ is not a 1–1 function because two different x values, __?__ and __?__, are paired with the same y value, __?__.

Determine whether the graph represents the graph of a 1–1 function.

11.

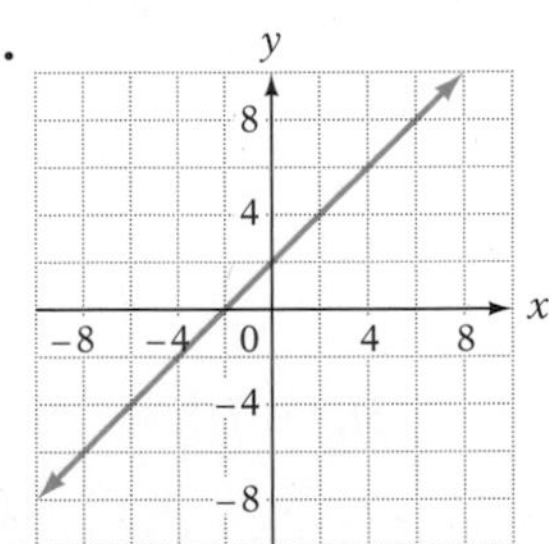

12.

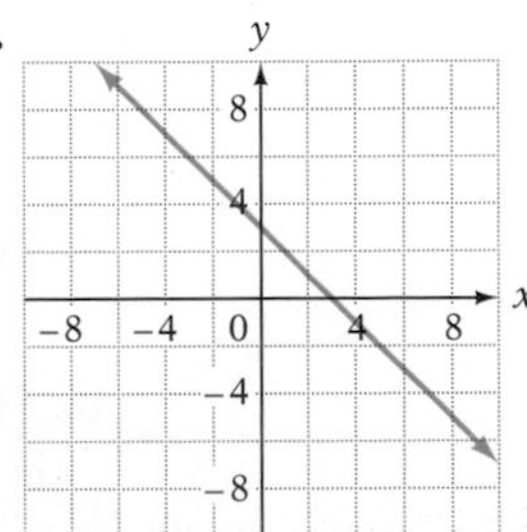

13.

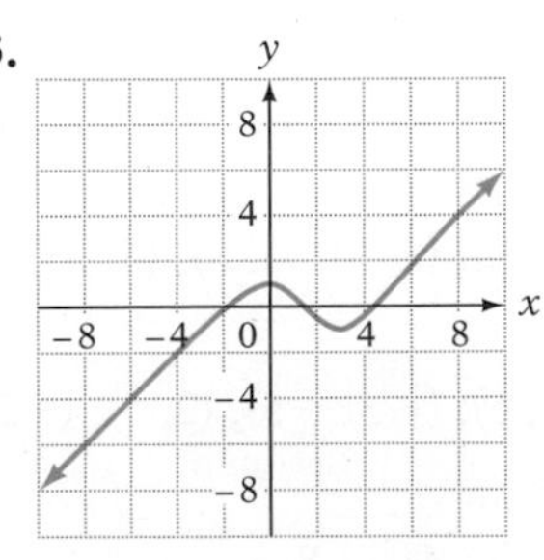

14.

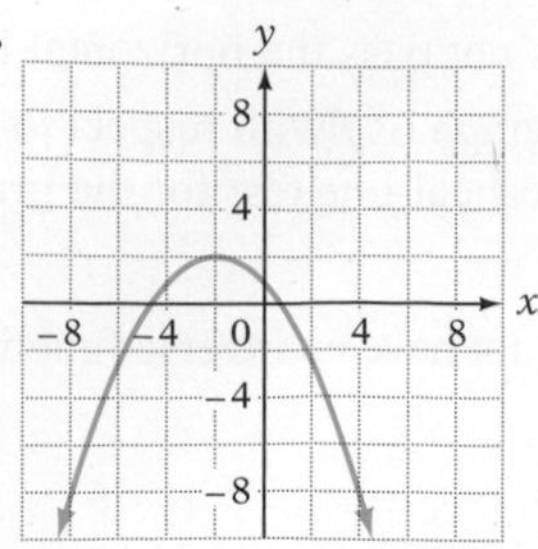

15.

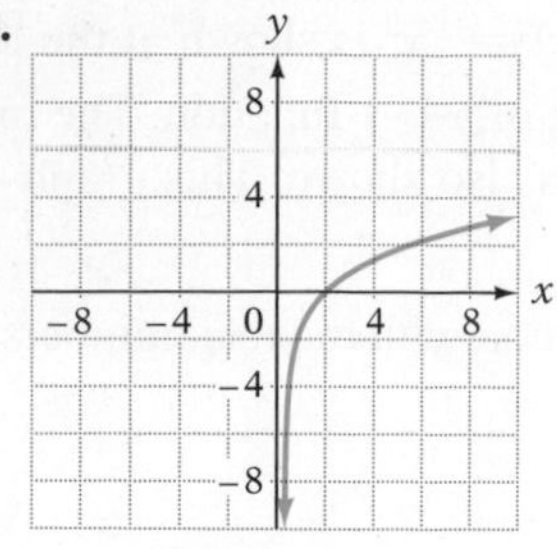

16.

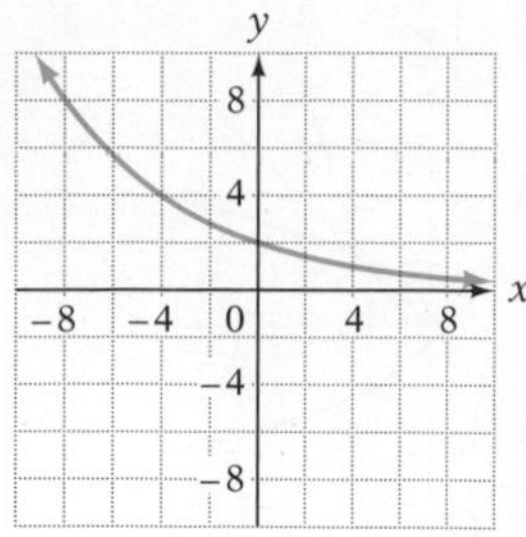

17.

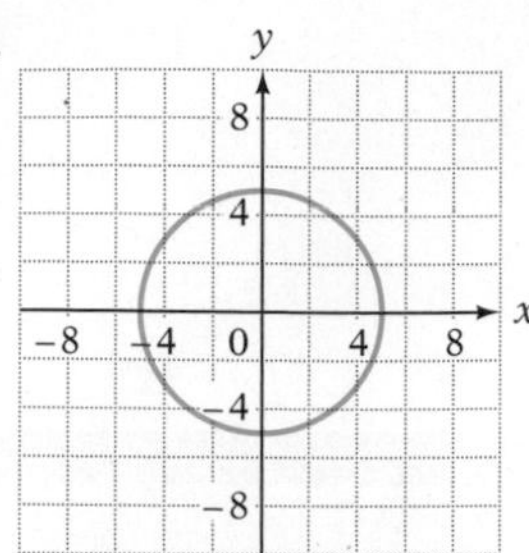

18.

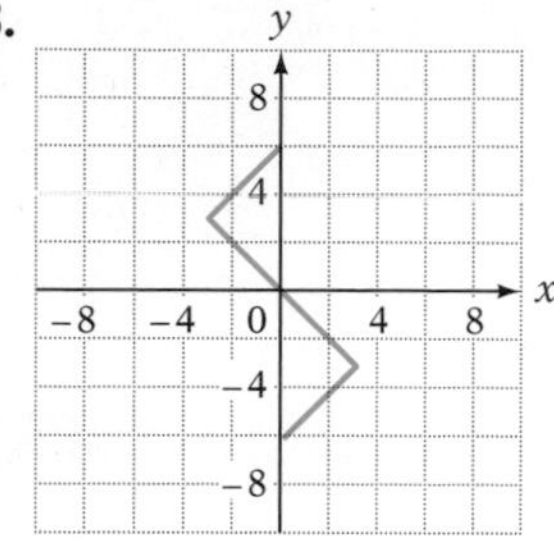

19.

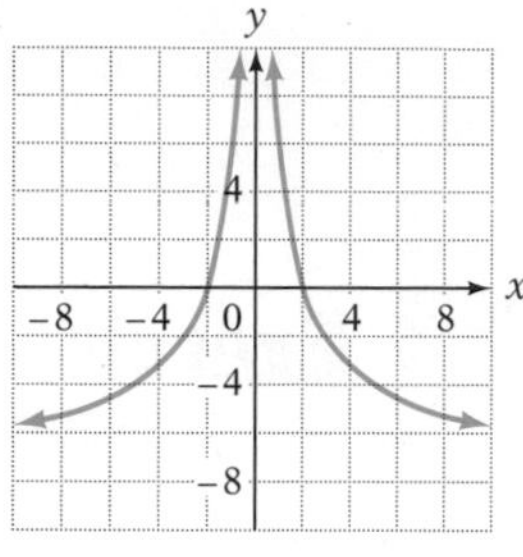

20.

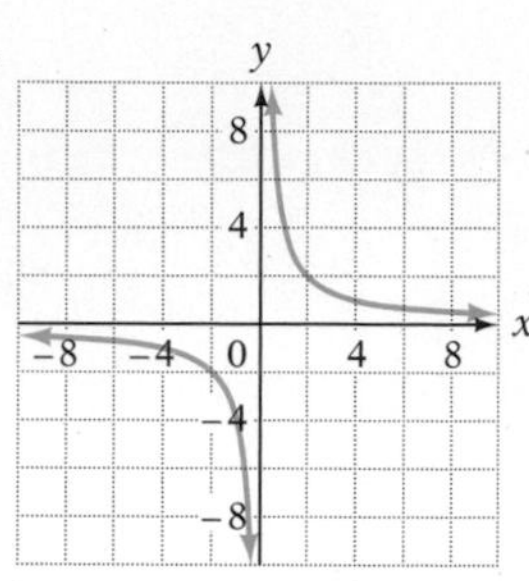

21.

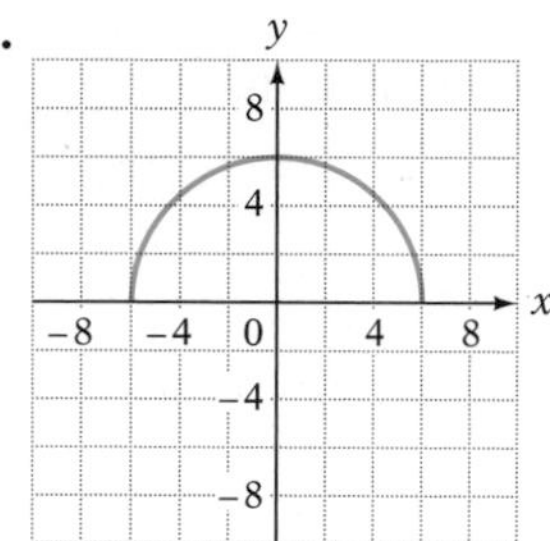

22.

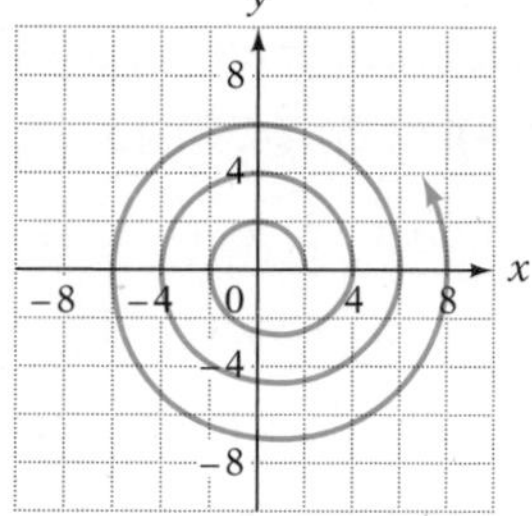

State whether the given function is a 1–1 function.

23. $f(x) = x$

24. $f(x) = a$, where a is any real number

25. $f(x) = |x| + a$, where a is any real number

26. $f(x) = x^3$

2 Find the inverse of a function (See pages 514–517.)

GETTING READY

27. If $(2, -3)$ are the coordinates of a point on the graph of the function $y = f(x)$, then (__?__, __?__) are the coordinates of a point on the graph of the inverse of $y = f(x)$.

28. a. The only functions whose inverses are also functions are __?__ functions.

b. The notation used to denote the inverse function of the 1–1 function f is __?__.

Find the inverse of the function. If the function does not have an inverse, write "No inverse."

29. $\{(1, 0), (2, 3), (3, 8), (4, 15)\}$

30. $\{(1, 0), (2, 1), (-1, 0), (-2, 1)\}$

31. $\{(3, 5), (-3, -5), (2, 5), (-2, -5)\}$

32. $\{(-5, -5), (-3, -1), (-1, 3), (1, 7)\}$

33. $f(x) = 4x - 8$

34. $f(x) = 3x + 15$

35. $f(x) = x^2 - 1$

36. $f(x) = 2x + 4$

37. $f(x) = x - 5$

38. $f(x) = \frac{1}{2}x - 1$

39. $f(x) = \frac{1}{3}x + 2$

40. $f(x) = -2x + 2$

41. $f(x) = -3x - 9$

42. $f(x) = 2x^2 + 2$

43. $f(x) = \frac{2}{3}x + 4$

44. $f(x) = \frac{3}{4}x - 4$

45. $f(x) = -\frac{1}{3}x + 1$

46. $f(x) = -\frac{1}{2}x + 2$

47. $f(x) = 2x - 5$

48. $f(x) = 3x + 4$

49. $f(x) = x^2 + 3$

50. $f(x) = 5x - 2$

Given $f(x) = 3x - 5$, find:

51. $f^{-1}(0)$

52. $f^{-1}(2)$

53. $f^{-1}(4)$

State whether the graph is the graph of a function. If it is the graph of a function, does it have an inverse?

54.

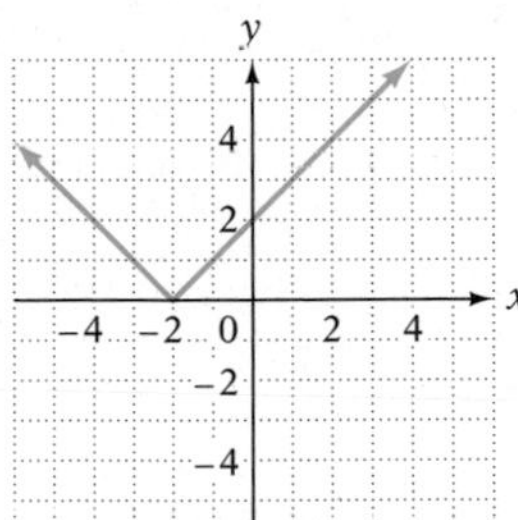

55.

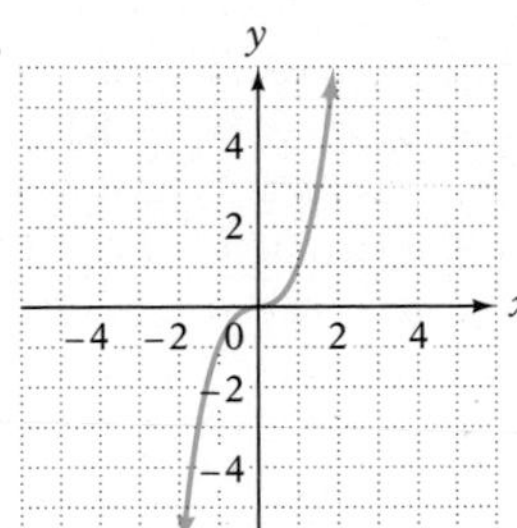

56.

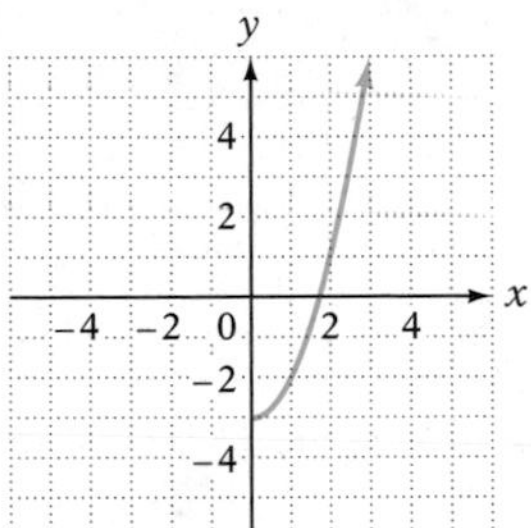

Use the Composition of Inverse Functions Property to determine whether the functions are inverses of each other.

57. $f(x) = 4x$; $g(x) = \frac{x}{4}$

58. $g(x) = x + 5$; $h(x) = x - 5$

59. $f(x) = 3x$; $h(x) = \frac{1}{3x}$

60. $h(x) = x + 2$; $g(x) = 2 - x$

61. $g(x) = 3x + 2$; $f(x) = \frac{1}{3}x - \frac{2}{3}$

62. $h(x) = 4x - 1$; $f(x) = \frac{1}{4}x + \frac{1}{4}$

63. $f(x) = \frac{1}{2}x - \frac{3}{2}$; $g(x) = 2x + 3$

64. $g(x) = -\frac{1}{2}x - \frac{1}{2}$; $h(x) = -2x + 1$

Complete.

65. The domain of the inverse function f^{-1} is the _______ of f.

66. The range of the inverse function f^{-1} is the _______ of f.

67. For any function f and its inverse f^{-1}, $f[f^{-1}(3)] =$ _______.

68. For any function f and its inverse f^{-1}, $f^{-1}[f(-4)] =$ _______.

If f is a 1–1 function and $f(0) = 5$, $f(1) = 7$, $f(2) = 9$, and $f(3) = 11$, find:

69. $f^{-1}(5)$

70. $f^{-1}(11)$

71. $f^{-1}(9)$

72. $f^{-1}(7)$

For Exercises 73 to 76, f is a 1–1 function such that $f(a) = b$, $f(b) = c$, and $f(c) = a$.

73. Find $f^{-1}(a)$.

74. Find $f^{-1}(f^{-1}(a))$.

75. Find $f(f^{-1}(a))$.

76. Find $f^{-1}(f^{-1}(b))$.

77. Given $f(x) = 3x + 2$ and $g(x) = 5x + 4$, show that $f[g(x)] = g[f(x)]$. Are f and g inverse functions?

78. Given $f(x) = 7x - 4$ and $g(x) = -2x + 2$, show that $f[g(x)] = g[f(x)]$. Are f and g inverse functions?

APPLYING CONCEPTS

Given the graph of the 1–1 function, draw the graph of its inverse function by using the technique shown in this section.

79.

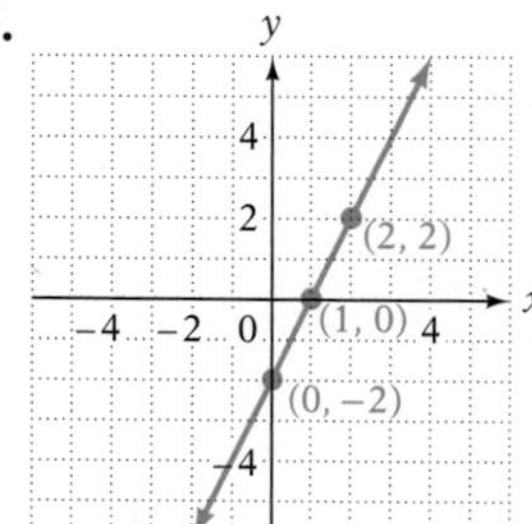

80.

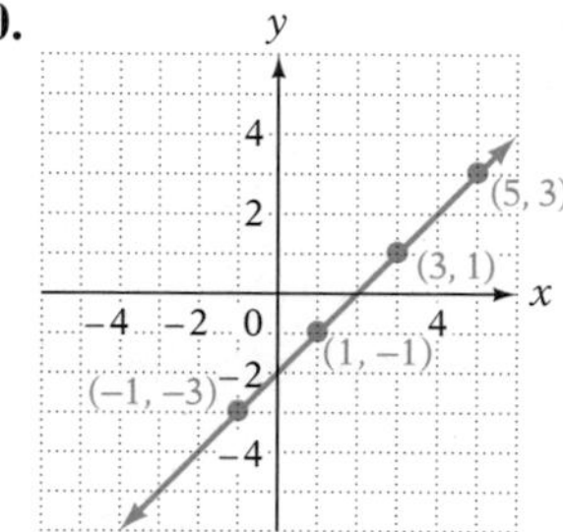

81.

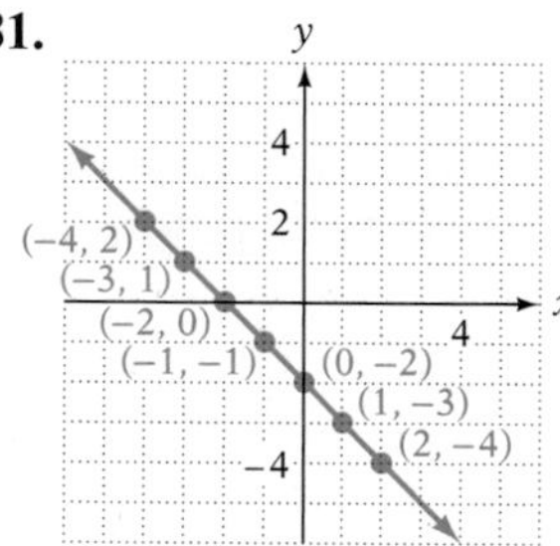

82.

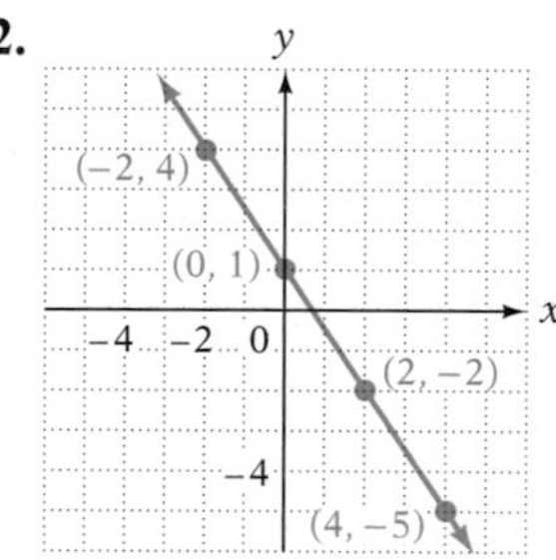

83.

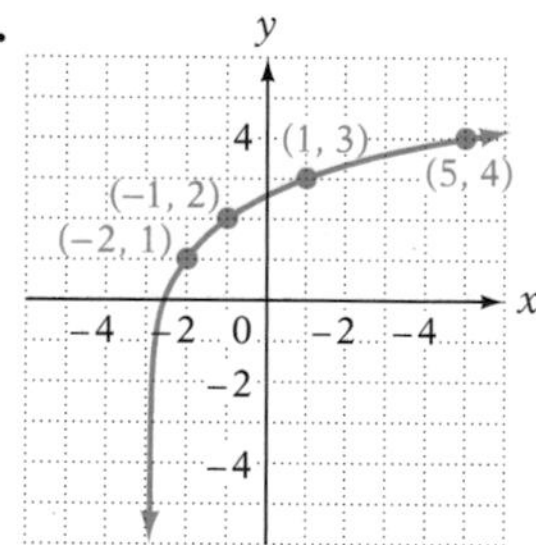

84.

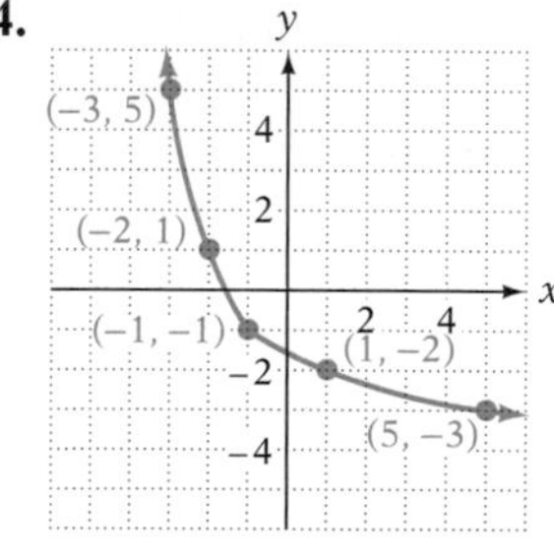

Each of the following tables defines a function. Is the inverse of the function a function? Explain your answer.

85. **Grading Scale Table**

Score	Grade
90–100	A
80–89	B
70–79	C
60–69	D
0–59	F

86. **First-Class Postage**

Weight (in oz.)	Cost
$0 < w \le 1$	\$.44
$1 < w \le 2$	\$.61
$2 < w \le 3$	\$.78
$3 < w \le 3.5$	\$.95

87. **Currency Exchange** Read the article at the right. The function $f(x) = 121.46x$ represents the exchange rate in July 2007. For this function, the U.S. dollar is the base currency, which means x is in dollars and $f(x)$ is in yen. Find an equation for f^{-1}, which represents the exchange rate with Japanese yen as the base currency. *Note:* Exchange rates less than 1 are often given to six decimal places.

88. **Currency Exchange** Read the article at the right and refer to Exercise 87. Give the exchange rate between U.S. dollars and Japanese yen in July 2010 by writing two functions that are inverses of each other. Tell which currency is the base currency for each function.

89. **Unit Conversion** The function given by $f(x) = 3x$ converts yards into feet. Find f^{-1} and explain what it does.

90. **Unit Conversion** The function given by $f(x) = 16x$ converts pounds into ounces. Find f^{-1} and explain what it does.

91. **Geometry** The formula for the perimeter of a square is $P(x) = 4x$. Find $P^{-1}(x)$ and explain what it does.

In the News

Tougher Exchange Rates for Exchange Students

The recession has made life harder for American students living in Tokyo, Japan. In July 2007, when one U.S. dollar was worth 121.460 yen, living in this very expensive city was affordable. Now, in July 2010, with an exchange rate of 87.4993 yen per dollar, American exchange students are feeling the pinch.

Source: www.temple-news.com, www.x-rates.com

PROJECTS OR GROUP ACTIVITIES

Exercises 92 and 93 demonstrate the essence of secure communication over the Internet. In practice, the coding functions are very complicated, making it extremely difficult for anyone to find the inverse and break the code.

92. **Internet Commerce** Inverse functions are used to secure business transactions over the Internet. Here is a very simple example. Let A = 10, B = 11, C = 12, ..., Z = 35. For instance, NAME would be coded as 23102214. Now choose a function that has an inverse function, say $f(x) = 2x - 4$. The function f is used to code NAME by evaluating the function at 23102214:

$$f(23102214) = 2(23102214) - 4 = 46204428 - 4 = 46204424$$

The number 46204424 would be sent over the Internet. The computer receiving the code would use the inverse of f to find the original word.

a. Find f^{-1} and show that $f^{-1}(46204424) = 23102214$.

b. Suppose the number 44205830 is received by a computer. What word was sent?

93. **Cryptography** Two friends send each other coded messages by using the scheme outlined in Exercise 92. They decide to use the coding function $f(x) = 2x + 1$. If one person receives the message 58285659, what word was sent?

CHAPTER 9 Summary

Key Words

Objective and Page Reference

Examples

The **inverse of a function** is the set of ordered pairs formed by reversing the coordinates of each ordered pair of the function.

[9.3.2, p. 514]

Function: $\{(2, 4), (3, 5), (6, 7), (8, 9)\}$

Inverse: $\{(4, 2), (5, 3), (7, 6), (9, 8)\}$

Essential Rules and Procedures

Objective and Page Reference

Examples

Vertical Translation
The graph of $y = f(x) + c, c > 0$, is the graph of $y = f(x)$ shifted up c units.

The graph of $y = f(x) - c, c > 0$, is the graph of $y = f(x)$ shifted down c units.

[9.1.1, p. 498]

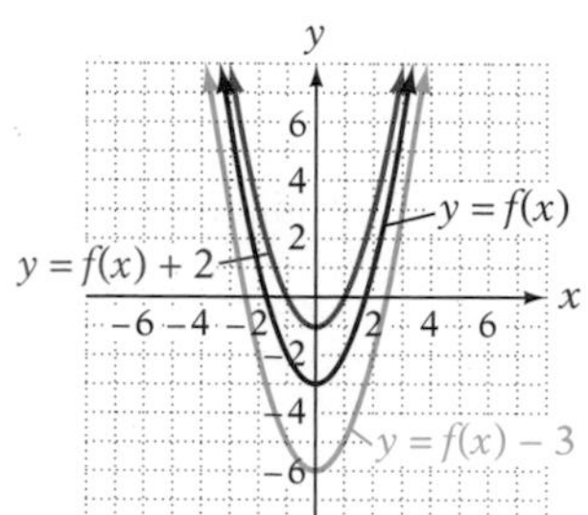

Horizontal Translation
The graph of $y = f(x - c), c > 0$, is the graph of $y = f(x)$ shifted horizontally c units to the right.

The graph of $y = f(x + c), c > 0$, is the graph of $y = f(x)$ shifted horizontally c units to the left.

[9.1.1, p. 499]

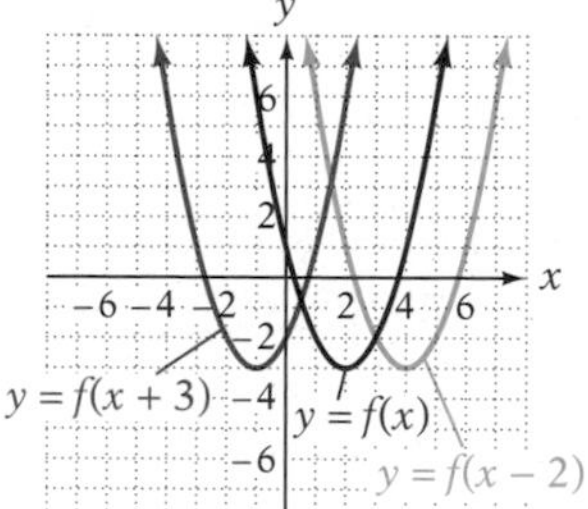

Operations on Functions
If f and g are functions and x is an element of the domain of each function, then

$(f + g)(x) = f(x) + g(x)$

$(f - g)(x) = f(x) - g(x)$

$(f \cdot g)(x) = f(x) \cdot g(x)$

$\left(\frac{f}{g}\right)(x) = \frac{f(x)}{g(x)}, g(x) \neq 0$

[9.2.1, p. 504]

Given $f(x) = x + 2$ and $g(x) = 2x$, then

$$\begin{aligned}(f + g)(4) &= f(4) + g(4)\\ &= (4 + 2) + 2(4) = 6 + 8\\ &= 14\end{aligned}$$

$$\begin{aligned}(f - g)(4) &= f(4) - g(4)\\ &= (4 + 2) - 2(4) = 6 - 8\\ &= -2\end{aligned}$$

$$\begin{aligned}(f \cdot g)(4) &= f(4) \cdot g(4)\\ &= (4 + 2) \cdot (2)(4) = 6 \cdot 8\\ &= 48\end{aligned}$$

$$\left(\frac{f}{g}\right)(4) = \frac{f(4)}{g(4)} = \frac{4 + 2}{2(4)} = \frac{6}{8} = \frac{3}{4}$$

Composition of Two Functions
$(f \circ g)(x) = f[g(x)]$

[9.2.2, p. 507]

Given $f(x) = x - 4$ and $g(x) = 4x$, then

$$\begin{aligned}(f \circ g)(2) &= f[g(2)]\\ &= f(8) \qquad \bullet\ g(2) = 8\\ &= 8 - 4 = 4\end{aligned}$$

Horizontal Line Test [9.3.1, p. 512]
A function f is a 1–1 function if any horizontal line intersects its graph at no more than one point.

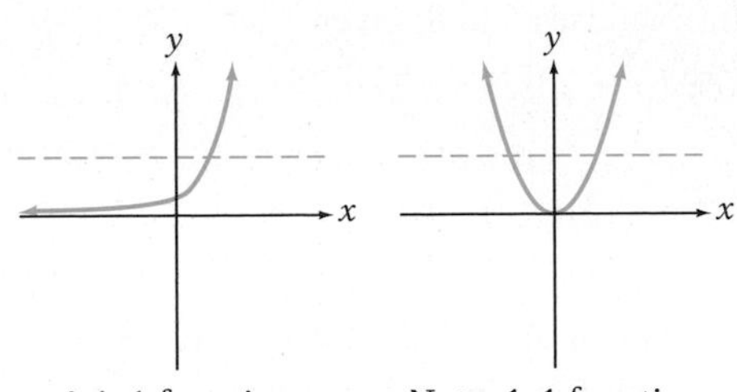

A 1–1 function Not a 1–1 function

Condition for an Inverse Function [9.3.2, p. 514]
A function f has an inverse if and only if f is a 1–1 function.

The function $f(x) = x^2$ does not have an inverse function. When $y = 4$, $x = 2$ or -2; therefore, the function $f(x)$ is not a 1–1 function.

Composition of Inverse Functions Property [9.3.2, p. 516]
$f^{-1}[f(x)] = x$ and $f[f^{-1}(x)] = x$

$f(x) = 2x - 3 \qquad f^{-1}(x) = \frac{1}{2}x + \frac{3}{2}$

$f^{-1}[f(x)] = \frac{1}{2}(2x - 3) + \frac{3}{2} = x$

$f[f^{-1}(x)] = 2\left(\frac{1}{2}x + \frac{3}{2}\right) - 3 = x$

CHAPTER 9 Review Exercises

1. Use the graph of $y = f(x)$ shown below to graph $g(x) = f(x - 1)$.

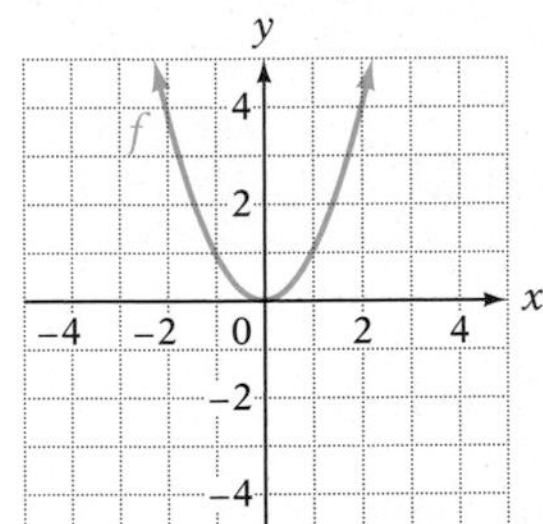

2. Use the graph of $y = f(x)$ shown below to graph $g(x) = f(x) + 3$.

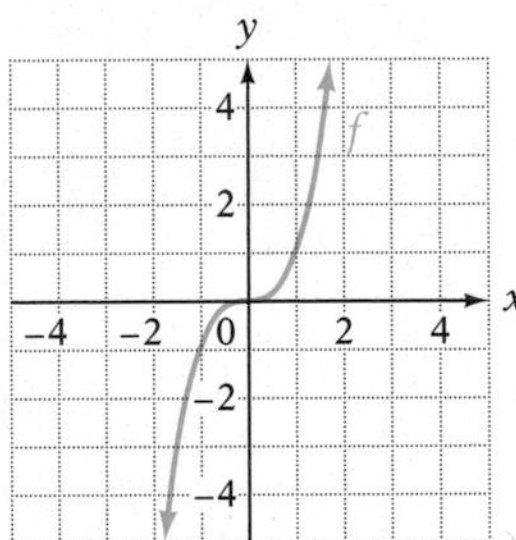

3. Use the graph of $y = f(x)$ shown below to graph $g(x) = f(x - 1) + 2$.

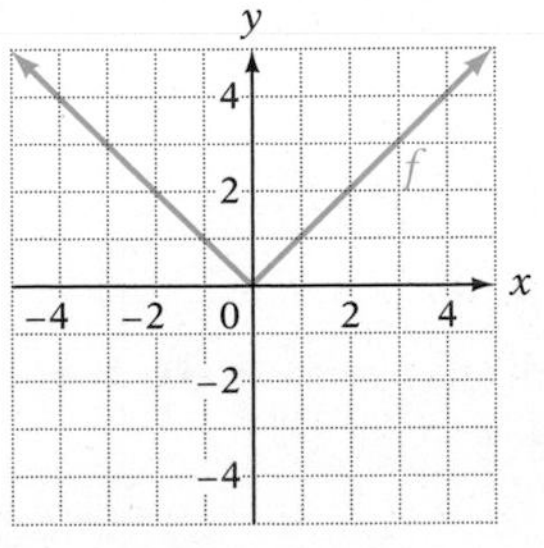

4. Use the graph of $y = f(x)$ shown below to graph $g(x) = f(x + 1) - 2$.

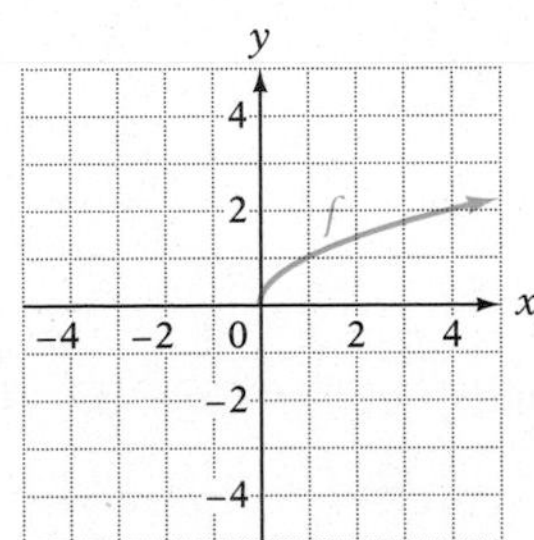

In Exercises 5 to 8, given $f(x) = x^2 + 2x - 3$ and $g(x) = x^2 - 2$, find:

5. $(f + g)(2)$

6. $(f - g)(x)$

7. $(f \cdot g)(-4)$

8. $\left(\dfrac{f}{g}\right)(3)$

9. Given $f(x) = x^2 + 4$ and $g(x) = 4x - 1$, find $f[g(0)]$.

10. Given $f(x) = 6x + 8$ and $g(x) = 4x + 2$, find $g[f(-1)]$.

11. Given $f(x) = 3x^2 - 4$ and $g(x) = 2x + 1$, find $f[g(x)]$.

12. Given $f(x) = 2x^2 + x - 5$ and $g(x) = 3x - 1$, find $g[f(x)]$.

13. Is the set of ordered pairs a 1–1 function? $\{(-3, 0), (-2, 0), (-1, 1), (0, 1)\}$

14. Find the inverse of the function $\{(-2, 1), (2, 3), (5, -4), (7, 9)\}$.

15. Determine whether the graph is the graph of a 1–1 function.

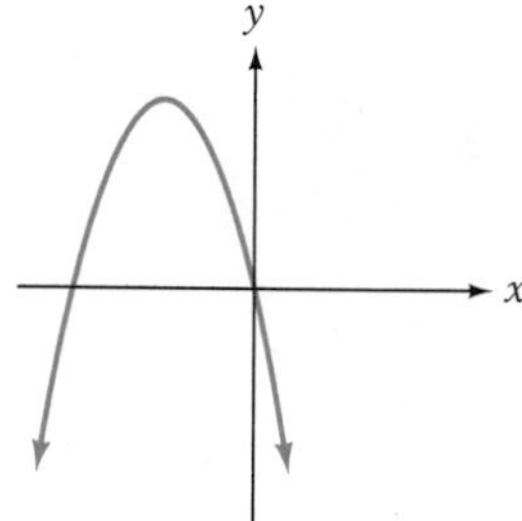

16. Determine whether the graph is the graph of a 1–1 function.

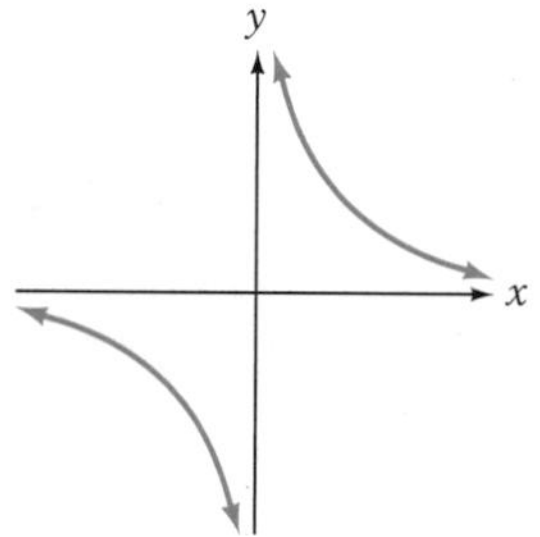

17. Find the inverse of the function $f(x) = \frac{1}{2}x + 8$.

18. Find the inverse of the function $f(x) = -6x + 4$.

19. Find the inverse of the function $f(x) = \frac{2}{3}x - 12$.

20. Are the functions $f(x) = -\frac{1}{4}x + \frac{5}{4}$ and $g(x) = -4x + 5$ inverses of each other?

CHAPTER 9 Test

1. Determine whether the graph represents the graph of a 1–1 function.

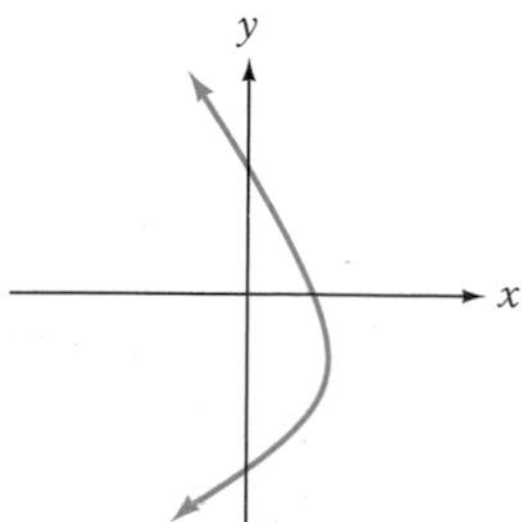

2. Given $f(x) = x^2 + 2x - 3$ and $g(x) = x^3 - 1$, find $(f - g)(2)$.

3. Given $f(x) = x^3 + 1$ and $g(x) = 2x - 3$, find $(f \cdot g)(-3)$.

4. Given $f(x) = 4x - 5$ and $g(x) = x^2 + 3x + 4$, find $\left(\frac{f}{g}\right)(-2)$.

5. Given $f(x) = x^2 + 4$ and $g(x) = 2x^2 + 2x + 1$, find $(f - g)(-4)$.

6. Given $f(x) = 4x + 2$ and $g(x) = \frac{x}{x+1}$, find $f[g(3)]$.

7. Use the graph of $y = f(x)$ shown below to graph $g(x) = f(x) - 2$.

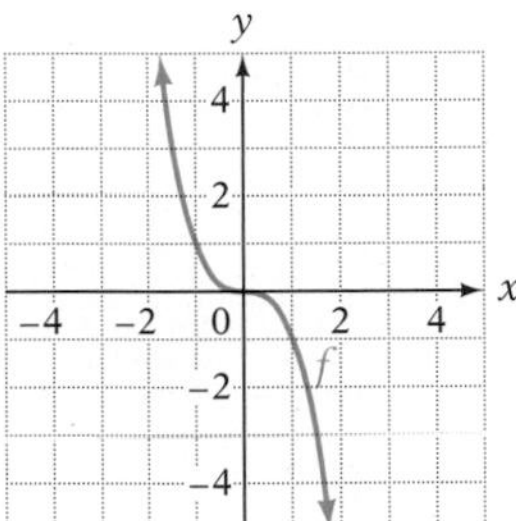

8. Use the graph of $y = f(x)$ shown below to graph $g(x) = f(x + 3)$.

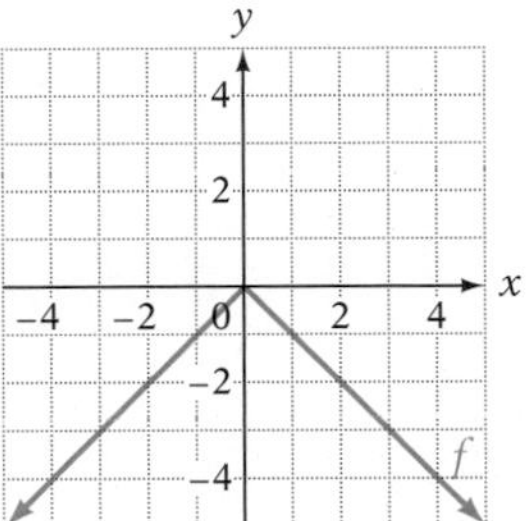

9. If $f(3) = 7$ and $g(x) = x^2 - 2$, find $g[f(3)]$.

10. Use the graph of $y = f(x)$ shown below to graph $g(x) = f(x + 2) - 3$.

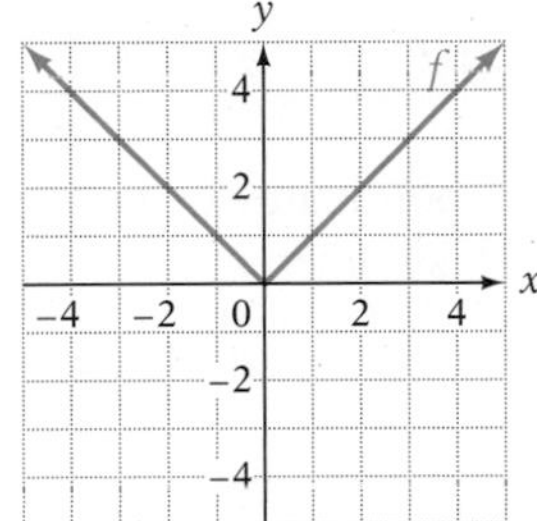

11. Use the graph of $y = f(x)$ shown below to graph $g(x) = f(x - 2) + 2$.

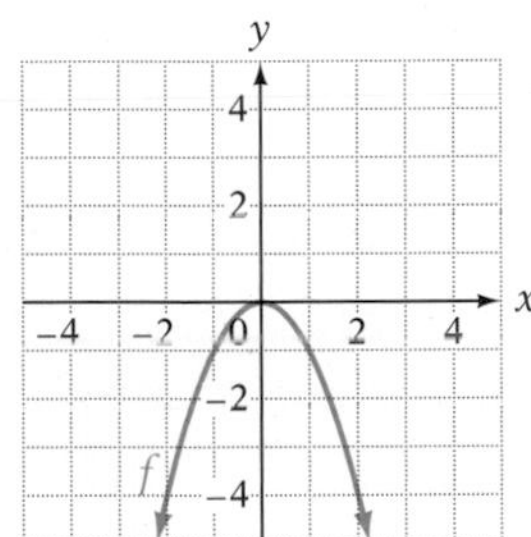

12. Find the inverse of the function $f(x) = \frac{1}{4}x - 4$.

13. Find the inverse of the function $\{(2, 6), (3, 5), (4, 4), (5, 3)\}$.

14. Are the functions $f(x) = \frac{1}{2}x + 2$ and $g(x) = 2x - 4$ inverses of each other?

15. Given $f(x) = 2x^2 - 7$ and $g(x) = x - 1$, find $f[g(x)]$.

16. Find the inverse of the function $f(x) = \frac{1}{2}x - 3$.

17. Are the functions $f(x) = \frac{2}{3}x + 3$ and $g(x) = \frac{3}{2}x - 3$ inverses of each other?

18. Determine whether the graph is the graph of a 1–1 function.

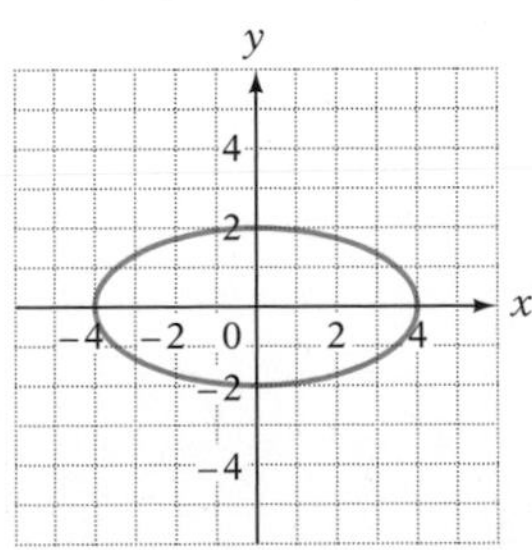

Cumulative Review Exercises

1. Evaluate $-3a + \left|\frac{3b - ab}{3b - c}\right|$ when $a = 2$, $b = 2$, and $c = -2$.

2. Graph: $\{x|x < -3\} \cap \{x|x > -4\}$

3. Solve: $\frac{3x - 1}{6} - \frac{5 - x}{4} = \frac{5}{6}$

4. Solve: $4x - 2 < -10$ or $3x - 1 > 8$

5. Graph $f(x) = \frac{1}{4}x^2$. Find the coordinates of the vertex and the equation of the axis of symmetry.

6. Graph the solution set of $3x - 4y \geq 8$.

7. Solve: $|8 - 2x| \geq 0$

8. Simplify: $\left(\frac{3a^3b}{2a}\right)^2\left(\frac{a^2}{-3b^2}\right)^3$

9. Multiply: $(x - 4)(2x^2 + 4x - 1)$

10. Factor: $a^4 - 2a^2 - 8$

11. Factor: $x^3y + x^2y^2 - 6xy^3$

12. Solve: $(b + 2)(b - 5) = 2b + 14$

13. Solve: $x^2 - 2x > 15$

14. Subtract: $\frac{x^2 + 4x - 5}{2x^2 - 3x + 1} - \frac{x}{2x - 1}$

15. Solve: $\frac{5}{x^2 + 7x + 12} = \frac{9}{x + 4} - \frac{2}{x + 3}$

16. Divide: $\frac{4 - 6i}{2i}$

17. Find the equation of the line that contains the points $P_1(-3, 4)$ and $P_2(2, -6)$.

18. Find the equation of the line that contains the point $P(-3, 1)$ and is perpendicular to the line whose equation is $2x - 3y = 6$.

19. Solve: $3x^2 = 3x - 1$

20. Solve: $\sqrt{8x + 1} = 2x - 1$

21. Find the minimum value of the function $f(x) = 2x^2 - 3$.

22. Find the range of $f(x) = |3x - 4|$ if the domain is $\{0, 1, 2, 3\}$.

23. Is this set of ordered pairs a function?
$\{(-3, 0), (-2, 0), (-1, 1), (0, 1)\}$

24. Solve: $\sqrt[3]{5x - 2} = 2$

25. Given $g(x) = 3x - 5$ and $h(x) = \frac{1}{2}x + 4$, find $g[h(2)]$.

26. Find the inverse of the function given by $f(x) = -3x + 9$.

27. **Food Mixtures** Find the cost per pound of a tea mixture made from 30 lb of tea costing \$4.50 per pound and 45 lb of tea costing \$3.60 per pound.

28. **Metallurgy** How many pounds of an 80% copper alloy must be mixed with 50 lb of a 20% copper alloy to make an alloy that is 40% copper?

29. **Mixture Problem** Six ounces of insecticide are mixed with 16 gal of water to make a spray for spraying an orange grove. How much additional insecticide is required if it is to be mixed with 28 gal of water?

30. **Work Problem** A large pipe can fill a tank in 8 min less time than it takes a smaller pipe to fill the same tank. Working together, the pipes can fill the tank in 3 min. How long would it take the larger pipe, working alone, to fill the tank?

31. **Physics** The distance (d) that a spring stretches varies directly as the force (f) used to stretch the spring. If a force of 50 lb can stretch a spring 30 in., how far can a force of 40 lb stretch the spring?

32. **Wind Power** The table at the right shows the projected use of renewable energy sources worldwide. If the projections are accurate, what will be the average rate of change in the worldwide use, in quadrillion Btu per year, of renewable energy sources between 2010 and 2016? Round to the nearest hundredth.

Year	Energy Use (in quadrillion Btu)
2010	52.05
2011	54.10
2012	56.76
2013	59.49
2014	61.80
2015	63.81
2016	66.21

Source: Energy Information Administration

33. **Music** The frequency of vibration (f) in an open pipe organ varies inversely as the length (L) of the pipe. If the air in a pipe 2 m long vibrates 60 times per minute, find the frequency in a pipe that is 1.5 m long.

Exponential and Logarithmic Functions

CHAPTER 10

Digital Vision

Focus on Success

What resources do you use when you need help in this course? You already know to read and reread the text when you are having difficulty understanding a concept. Instructors are available to help you during their office hours. Most schools have a math center where students can get help. Some schools have a tutoring program. You might also ask a student who has been successful in this class for assistance. (See Habits of Successful Students, page AIM-6.)

OBJECTIVES

10.1 ❶ Evaluate exponential functions
❷ Graph exponential functions

10.2 ❶ Write equivalent exponential and logarithmic equations
❷ The properties of logarithms

10.3 ❶ Graph logarithmic functions

10.4 ❶ Solve exponential equations
❷ Solve logarithmic equations

10.5 ❶ Application problems

PREP TEST

Are you ready to succeed in this chapter?
Take the Prep Test below to find out if you are ready to learn the new material.

1. Simplify: 3^{-2}

2. Simplify: $\left(\frac{1}{2}\right)^{-4}$

3. Complete: $\frac{1}{8} = 2^?$

4. Evaluate $f(x) = x^4 + x^3$ for $x = -1$ and $x = 3$.

5. Solve: $3x + 7 = x - 5$

6. Solve: $16 = x^2 - 6x$

7. Evaluate $A(1 + i)^n$ for $A = 5000$, $i = 0.04$, and $n = 6$. Round to the nearest hundredth.

8. Graph: $f(x) = x^2 - 1$

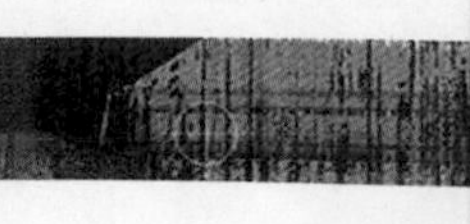

10.1 Exponential Functions

OBJECTIVE 1 Evaluate exponential functions

Data suggest that since the year 2000, the number of Internet users worldwide has been increasing at a rate of approximately 17% per year. The graph at the right shows this growth. This graph depicts an example of **exponential growth.**

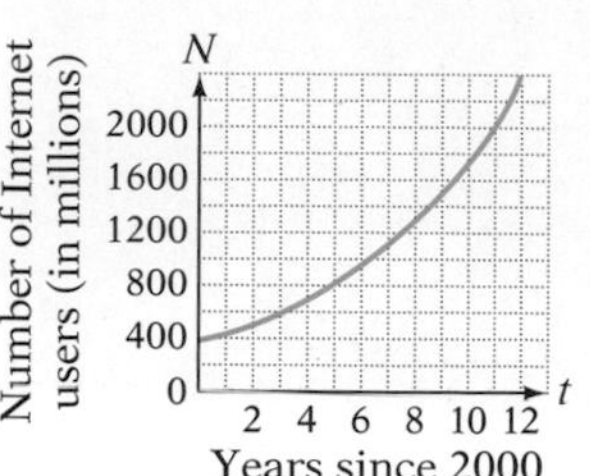

Nuclear medicine physicians use radioisotopes for the diagnosis and treatment of certain diseases. One of the most widely used isotopes is technetium-99m. One use of this isotope is in the diagnosis of cardiovascular disease. The graph at the right shows the amount of technetium-99m in a patient after its injection into the patient. This graph depicts an example of **exponential decay.**

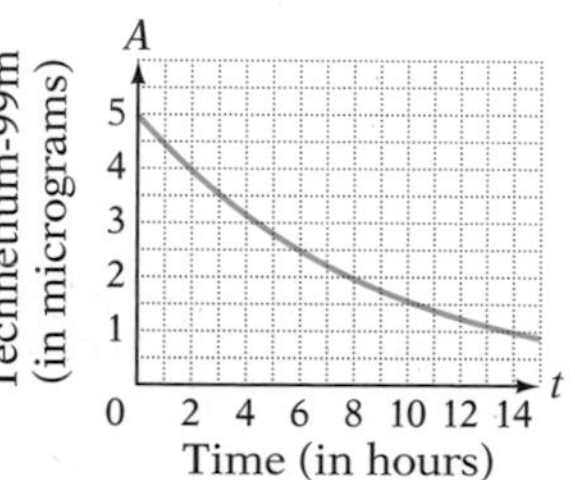

How It's Used

A photograph taken by a camera in outer space, by a surveillance camera during a robbery, or by a journalist's camera being jostled may be too blurry for the viewer to extract important information. Blurred images can be improved through the use of *Fourier transforms*, which are based on exponential functions.

DEFINITION OF AN EXPONENTIAL FUNCTION

The **exponential function** with **base *b*** is defined by

$$f(x) = b^x$$

where $b > 0$, $b \neq 1$, and x is any real number.

In the definition of an exponential function, b, the base, is required to be positive. If the base were a negative number, the value of the function would be a complex number for some values of x. For instance, the value of $f(x) = (-4)^x$ when $x = \frac{1}{2}$ is $f\left(\frac{1}{2}\right) = (-4)^{\frac{1}{2}} = \sqrt{-4} = 2i$. To avoid complex number values of a function, the base of the exponential function is a positive number.

Focus on evaluating an exponential function

Evaluate $f(x) = 2^x$ at $x = 3$ and $x = -2$.

Substitute 3 for x and simplify. $f(3) = 2^3 = 8$

Substitute -2 for x and simplify. $f(-2) = 2^{-2} = \frac{1}{2^2} = \frac{1}{4}$

To evaluate an exponential expression at an irrational number such as $\sqrt{2}$, we obtain an approximation to the value of the function by approximating the irrational number. For instance, the value of $f(x) = 4^x$ when $x = \sqrt{2}$ can be approximated by using an approximation of $\sqrt{2}$.

$$f(\sqrt{2}) = 4^{\sqrt{2}} \approx 4^{1.4142} \approx 7.1029$$

Because $f(x) = b^x$ $(b > 0, b \neq 1)$ can be evaluated at both rational and irrational numbers, the domain of f is all real numbers. And because $b^x > 0$ for all values of x, the range of f is the positive real numbers.

EXAMPLE 1 Evaluate $f(x) = \left(\frac{1}{2}\right)^x$ at $x = 2$ and $x = -3$.

Solution

$$f(x) = \left(\frac{1}{2}\right)^x$$

$$f(2) = \left(\frac{1}{2}\right)^2 = \frac{1}{4}$$

$$f(-3) = \left(\frac{1}{2}\right)^{-3} = 2^3 = 8$$

Problem 1 Evaluate $f(x) = \left(\frac{2}{3}\right)^x$ at $x = 3$ and $x = -2$.

Solution See page S31.

Try Exercise 13, page 536.

EXAMPLE 2 Evaluate $f(x) = 2^{3x-1}$ at $x = 1$ and $x = -1$.

Solution

$$f(x) = 2^{3x-1}$$

$$f(1) = 2^{3(1)-1} = 2^2 = 4$$

$$f(-1) = 2^{3(-1)-1} = 2^{-4} = \frac{1}{2^4} = \frac{1}{16}$$

Problem 2 Evaluate $f(x) = 2^{2x+1}$ at $x = 0$ and $x = -2$.

Solution See page S31.

Try Exercise 15, page 536.

Point of Interest

The natural exponential function is an extremely important function. It is used extensively in applied problems in virtually all disciplines from archeology to zoology. Leonhard Euler (1707–1783) was the first to use the letter *e* as the base of the natural exponential function.

A frequently used base in applications of exponential functions is an irrational number designated by *e*. The number *e* is approximately 2.71828183. It is an irrational number, so it has a nonterminating, nonrepeating decimal representation.

NATURAL EXPONENTIAL FUNCTION

The function defined by $f(x) = e^x$ is called the **natural exponential function.**

The $\boxed{e^x}$ key on a calculator can be used to evaluate the natural exponential function.

EXAMPLE 3 Evaluate $f(x) = e^x$ at $x = 2$, $x = -3$, and $x = \pi$. Round to the nearest ten-thousandth.

Solution

$$f(x) = e^x$$

$$f(2) = e^2 \approx 7.3891$$

$$f(-3) = e^{-3} \approx 0.0498$$

$$f(\pi) = e^{\pi} \approx 23.1407$$

Problem 3 Evaluate $f(x) = e^x$ at $x = 1.2$, $x = -2.5$, and $x = e$. Round to the nearest ten-thousandth.

Solution See page S31.

Try Exercise 21, page 536.

OBJECTIVE  Graph exponential functions

Some of the properties of an exponential function can be seen by considering its graph.

Focus on graphing an exponential function with $b > 1$

Graph: $f(x) = 2^x$

To graph $f(x) = 2^x$, think of the function as the equation $y = 2^x$. Choose values of x, and find the corresponding values of y. The results can be recorded in a table, as shown below.

Graph the ordered pairs on a rectangular coordinate system. Connect the points with a smooth curve.

x	$f(x) = 2^x$	y
-2	$f(-2) = 2^{-2} = \frac{1}{2^2} = \frac{1}{4}$	$\frac{1}{4}$
-1	$f(-1) = 2^{-1} = \frac{1}{2}$	$\frac{1}{2}$
0	$f(0) = 2^0 = 1$	1
1	$f(1) = 2^1 = 2$	2
2	$f(2) = 2^2 = 4$	4
3	$f(3) = 2^3 = 8$	8

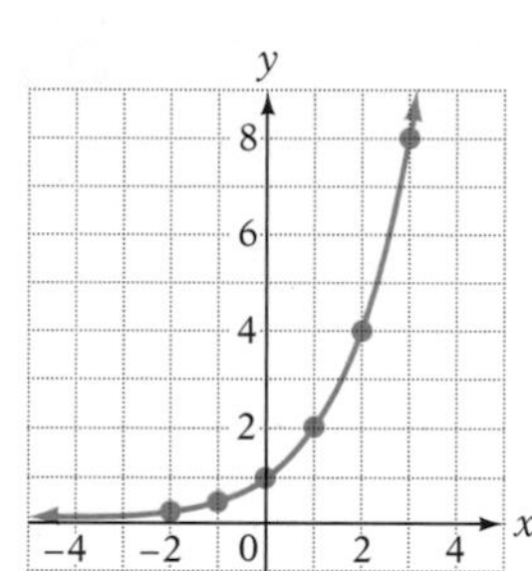

Note that a vertical line would intersect the graph in the Focus On example above at only one point. Therefore, by the vertical line test, the graph of $f(x) = 2^x$ is the graph of a function. Also note that a horizontal line would intersect the graph at only one point. Therefore, the graph of $f(x) = 2^x$ is the graph of a 1–1 function.

Focus on graphing an exponential function with $0 < b < 1$

Graph: $f(x) = \left(\frac{1}{2}\right)^x$

Think of the function as the equation $y = \left(\frac{1}{2}\right)^x$. Choose values of x, and find the corresponding values of y.

Graph the ordered pairs on a rectangular coordinate system. Connect the points with a smooth curve.

x	$f(x) = \left(\frac{1}{2}\right)^x$	y
-3	$f(-3) = \left(\frac{1}{2}\right)^{-3} = 2^3 = 8$	8
-2	$f(-2) = \left(\frac{1}{2}\right)^{-2} = 2^2 = 4$	4
-1	$f(-1) = \left(\frac{1}{2}\right)^{-1} = 2^1 = 2$	2
0	$f(0) = \left(\frac{1}{2}\right)^{0} = 1$	1
1	$f(1) = \left(\frac{1}{2}\right)^{1} = \frac{1}{2}$	$\frac{1}{2}$
2	$f(2) = \left(\frac{1}{2}\right)^{2} = \frac{1}{4}$	$\frac{1}{4}$

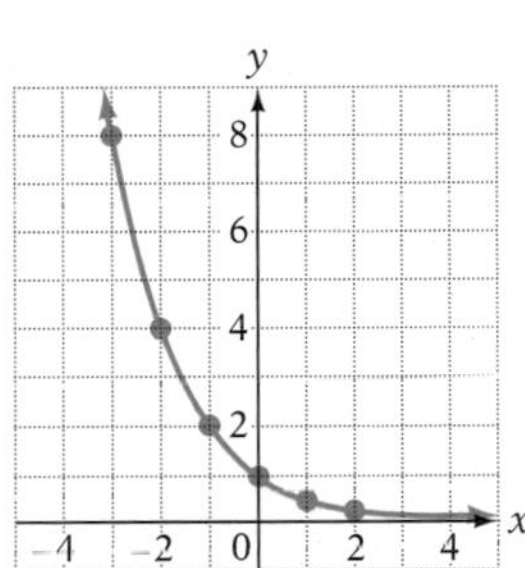

Applying the vertical and horizontal line tests to the graph in the Focus On example above reveals that the graph of $f(x) = \left(\frac{1}{2}\right)^x$ is also the graph of a 1–1 function.

EXPONENTIAL FUNCTIONS ARE 1–1

The exponential function defined by $f(x) = b^x, b > 0, b \neq 1$, is a 1–1 function.

EXAMPLE 4 Graph. **A.** $f(x) = 3^{\frac{1}{2}x-1}$ **B.** $f(x) = 2^x - 1$

Solution **A.**

x	$y = 3^{\frac{1}{2}x-1}$
-2	$\frac{1}{9}$
0	$\frac{1}{3}$
2	1
4	3

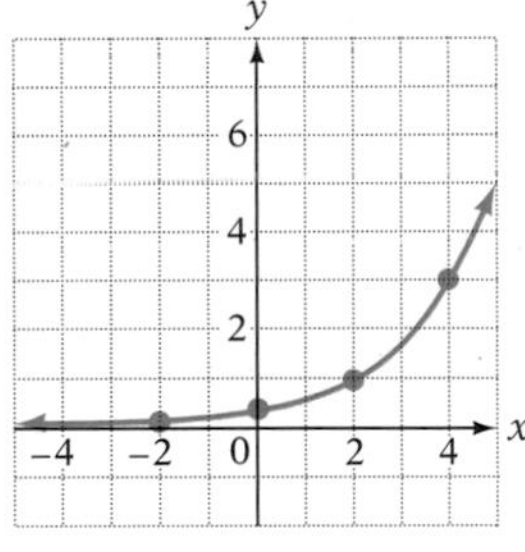

B.

x	$y = 2^x - 1$
-2	$-\frac{3}{4}$
-1	$-\frac{1}{2}$
0	0
1	1
2	3
3	7

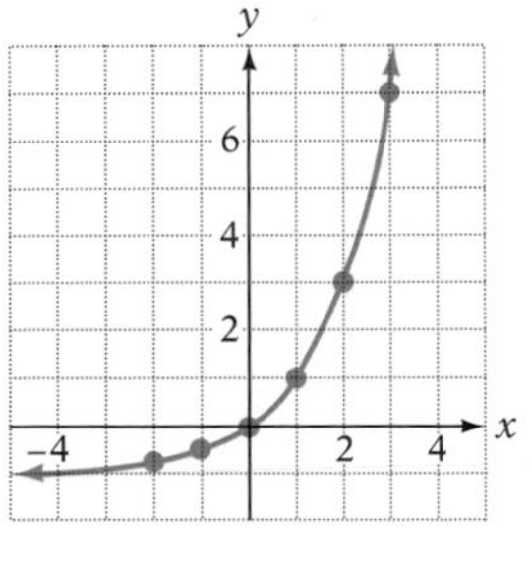

Problem 4 Graph. **A.** $f(x) = 2^{-\frac{1}{2}x}$ **B.** $f(x) = 2^x + 1$

Solution See page S31.

Try Exercise 37, page 537.

EXAMPLE 5 Graph. **A.** $f(x) = e^{x-2}$ **B.** $f(x) = e^x - 1$

Solution **A.**

x	$y = e^{x-2}$
−3	0.007
−2	0.02
−1	0.05
0	0.14
1	0.37
2	1
3	2.72
4	7.39

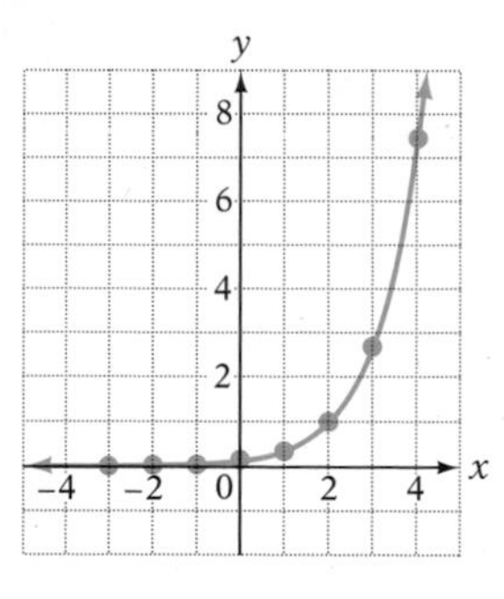

B.

x	$y = e^x - 1$
−2	−0.86
−1	−0.63
0	0
1	1.72
2	6.39

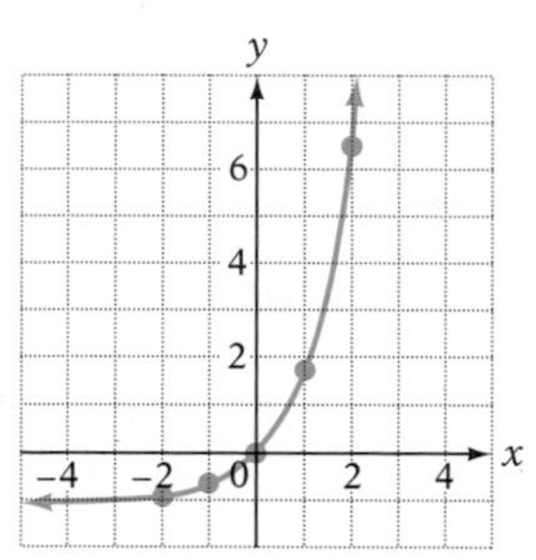

Problem 5 Graph. **A.** $f(x) = 2^{-x} + 2$ **B.** $f(x) = e^{-2x} - 4$

Solution See page S31.

Try Exercise 47, page 537.

EXAMPLE 6 Graph $f(x) = -\frac{1}{3}e^{2x} + 2$ and approximate the zero of f to the nearest tenth.

Solution Recall that a zero of f is a value of x for which $f(x) = 0$. Use the features of a graphing utility to determine the x-coordinate of the x-intercept of the graph, which is the zero of f.

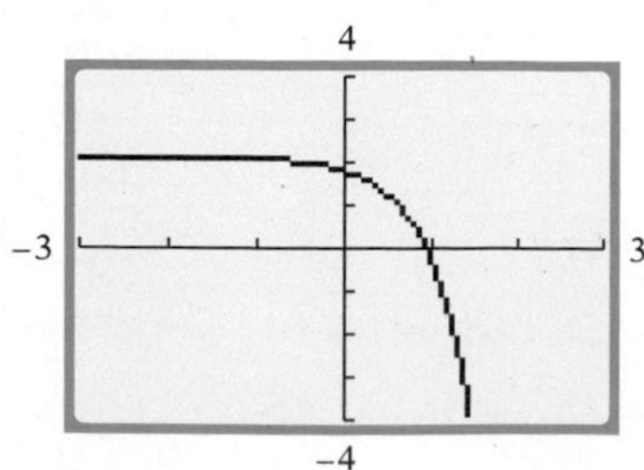

To the nearest tenth, the zero of f is 0.9.

Problem 6 Graph $f(x) = 2\left(\frac{3}{4}\right)^x - 3$ and approximate, to the nearest tenth, the value of x for which $f(x) = 1$.

Solution See page S31.

Try Exercise 61, page 538.

10.1 Exercises

CONCEPT CHECK

1. Which of the following cannot be the base of an exponential function?
(i) -2 (ii) 0 (iii) 0.3789 (iv) 478

2. Which of the following cannot be the base of an exponential function?
(i) $-\frac{1}{2}$ (ii) 1 (iii) $\frac{9}{5}$ (iv) 138.9

3. What are the coordinates of the y-intercept of the graph of $y = 3^x$?

4. True or false? $f(x) = (0.7)^x$ is a 1–1 function.

5. True or false? $f(x) = (1.7)^x$ is a 1–1 function.

6. If $b > 1$, what is the domain of $f(x) = b^x$?

7. If $b > 1$ and $y = b^x$, are there any values of x for which y is less than zero?

8. Replace the question mark with $<$ or $>$ to make a true statement. If $y = \left(\frac{2}{3}\right)^x$ and c and d are real numbers, with $c < d$, then $\left(\frac{2}{3}\right)^c \underline{\quad ? \quad} \left(\frac{2}{3}\right)^d$.

1 Evaluate exponential functions (See pages 530–531.)

9. What is an exponential function?

10. What is the natural exponential function?

GETTING READY

11. Given $f(x) = 4^{x-1}$, evaluate $f(1)$.
$f(1) = 4^{\underline{\ ?\ } - 1} = 4^{\underline{\ ?\ }} = \underline{\ ?\ }$

12. Given $g(x) = e^{3x}$, evaluate $g\left(\frac{1}{6}\right)$. Round to the nearest ten-thousandth.

$g\left(\frac{1}{6}\right) = e^{3(\underline{\ ?\ })}$ • **Replace x with $\frac{1}{6}$.**

$= e^{\underline{\ ?\ }}$ • **Simplify the exponent.**

$\approx \underline{\ ?\ }$ • **Use the e^x key on a calculator.**

13. Given $f(x) = 3^x$, evaluate:
a. $f(2)$ b. $f(0)$ c. $f(-2)$

14. Given $H(x) = 2^x$, evaluate:
a. $H(-3)$ b. $H(0)$ c. $H(2)$

15. Given $g(x) = 2^{x+1}$, evaluate:
a. $g(3)$ b. $g(1)$ c. $g(-3)$

16. Given $F(x) = 3^{x-2}$, evaluate:
a. $F(-4)$ b. $F(-1)$ c. $F(0)$

17. Given $P(x) = \left(\frac{1}{2}\right)^{2x}$, evaluate:
a. $P(0)$ b. $P\left(\frac{3}{2}\right)$ c. $P(-2)$

18. Given $R(t) = \left(\frac{1}{3}\right)^{3t}$, evaluate:
a. $R\left(-\frac{1}{3}\right)$ b. $R(1)$ c. $R(-2)$

19. Given $f(x) = 3^{-x}$, evaluate:
a. $f(2)$ b. $f(-2)$ c. $f(0)$

20. Given $f(x) = 2^{-2x}$, evaluate:
a. $f(-3)$ b. $f(2)$ c. $f(0)$

21. Given $G(x) = e^{\frac{x}{2}}$, evaluate the following. Round to the nearest ten-thousandth.
a. $G(4)$ b. $G(-2)$ c. $G\left(\frac{1}{2}\right)$

22. Given $f(x) = e^{2x}$, evaluate the following. Round to the nearest ten-thousandth.
a. $f(-2)$ b. $f\left(-\frac{2}{3}\right)$ c. $f(2)$

23. Given $H(r) = e^{-r+3}$, evaluate the following. Round to the nearest ten-thousandth.
a. $H(-1)$ b. $H(3)$ c. $H(5)$

24. Given $P(t) = e^{-\frac{1}{2}t}$, evaluate the following. Round to the nearest ten-thousandth.
a. $P(-3)$ b. $P(4)$ c. $P\left(\frac{1}{2}\right)$

25. Given $F(x) = 2^{x^2}$, evaluate:
a. $F(2)$ b. $F(-2)$ c. $F\left(\frac{3}{4}\right)$

26. Given $Q(x) = 2^{-x^2}$, evaluate:
a. $Q(3)$ b. $Q(-1)$ c. $Q(-2)$

27. Given $f(x) = e^{-\frac{x^2}{2}}$, evaluate the following. Round to the nearest ten-thousandth.
a. $f(-2)$ b. $f(2)$ c. $f(-3)$

28. Given $f(x) = e^{-2x} + 1$, evaluate the following. Round to the nearest ten-thousandth.
a. $f(-1)$ b. $f(3)$ c. $f(-2)$

Complete Exercises 29 to 32 without using a calculator.

29. Which number is a rational number, $(-3)^{\frac{1}{2}}$ or $\left(\frac{1}{2}\right)^{-3}$?

30. For which number can you give only an approximate decimal value, $(\sqrt{7})^4$ or $\sqrt[4]{7}$?

31. Is e^3 greater than or less than 27?

32. Is e^{-1} greater than or less than $\frac{1}{2}$?

2 Graph exponential functions (See pages 532–535.)

GETTING READY

33. To find the y-coordinate of the point on the graph of $f(x) = 3^x$ that has an x-coordinate of -3, evaluate $f(-3) = 3^{\underline{\ ?\ }} = \frac{1}{?} = \underline{\ ?\ }$. The coordinates of one point on the graph of $f(x) = 3^x$ are ($\underline{\ ?\ }$, $\underline{\ ?\ }$).

34. To find the y-coordinate of the point on the graph of $g(x) = 2^{x+1}$ that has an x-coordinate of -1, evaluate $g(-1) = 2^{\underline{\ ?\ }+1} = 2^{\underline{\ ?\ }} = \underline{\ ?\ }$. The coordinates of one point on the graph of $g(x) = 2^{x+1}$ are ($\underline{\ ?\ }$, $\underline{\ ?\ }$).

Graph.

35. $f(x) = 3^x$

36. $f(x) = 3^{-x}$

37. $f(x) = 2^{x+1}$

38. $f(x) = 2^{x-1}$

39. $f(x) = \left(\frac{1}{3}\right)^x$

40. $f(x) = \left(\frac{2}{3}\right)^x$

41. $f(x) = 2^{-x} + 1$

42. $f(x) = 2^x - 3$

43. $f(x) = \left(\frac{1}{3}\right)^{-x}$

44. $f(x) = \left(\frac{3}{2}\right)^{-x}$

45. $f(x) = \left(\frac{1}{2}\right)^{-x} + 2$

46. $f(x) = \left(\frac{1}{2}\right)^x - 1$

47. $f(x) = e^x + 1$

48. $f(x) = e^{x-1}$

49. $f(x) = e^{-x}$

50. $f(x) = e^{-x+2}$

51. Which of the following functions have the same graph?

(i) $f(x) = 3^x$ (ii) $f(x) = \left(\frac{1}{3}\right)^x$ (iii) $f(x) = x^3$ (iv) $f(x) = 3^{-x}$

52. Which of the following functions have the same graph?

(i) $f(x) = x^4$ (ii) $f(x) = 4^{-x}$ (iii) $f(x) = 4^x$ (iv) $f(x) = \left(\frac{1}{4}\right)^x$

53. Graph $f(x) = 3^x$ and $f(x) = 3^{-x}$ and find the coordinates of the point of intersection of the two graphs.

54. Graph $f(x) = 2^{x+1}$ and $f(x) = 2^{-x+1}$ and find the coordinates of the point of intersection of the two graphs.

55. Graph $f(x) = \left(\frac{1}{3}\right)^x$. What are the coordinates of the x- and y-intercepts of the graph of the function?

56. Graph $f(x) = \left(\frac{1}{3}\right)^{-x}$. What are the coordinates of the x- and y-intercepts of the graph of the function?

57. If the graph of $f(x) = b^x$ rises from left to right, is b greater than 1 or less than 1?

58. Which function has a graph that falls from left to right, $f(x) = 10^x$ or $g(x) = 10^{-x}$?

59. Both a and b are positive, and $b \neq 1$. Which function has an x-intercept, $f(x) = b^{x-a}$ or $g(x) = b^x - a$?

60. Is the y-intercept of the graph of $f(x) = 5^x$ above, below, or the same point as the y-intercept of the graph of $f(x) = 5^{-x}$?

Use a graphing calculator to solve Exercises 61 to 67.

61. Graph $f(x) = 2^x - 3$ and approximate the zero of f to the nearest tenth.

62. Graph $f(x) = 5 - 3^x$ and approximate the zero of f to the nearest tenth.

63. Graph $f(x) = e^x$ and approximate, to the nearest tenth, the value of x for which $f(x) = 3$.

64. Graph $f(x) = e^{-2x-3}$ and approximate, to the nearest tenth, the value of x for which $f(x) = 2$.

65. **Investments** The exponential function given by $F(n) = 500(1.00021918)^{365n}$ gives the value in n years of a \$500 investment in a certificate of deposit that earns 8% annual interest compounded daily. Graph F and determine in how many years the investment will be worth \$1000.

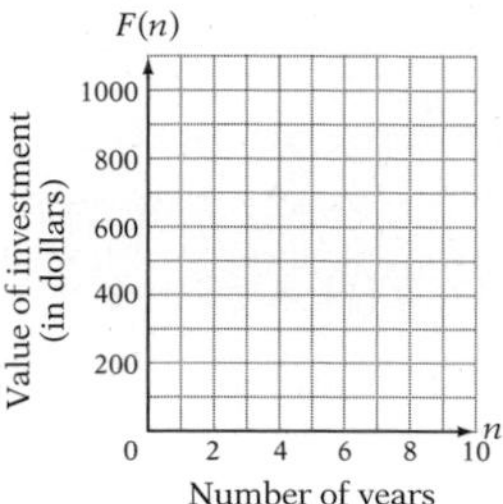

66. **Chemistry** The number of grams of radioactive cesium that remain, after t years, from an original sample of 30 g is given by $N(t) = 30(2^{-0.0322t})$. Graph N and determine in how many years there will be 20 g of cesium remaining.

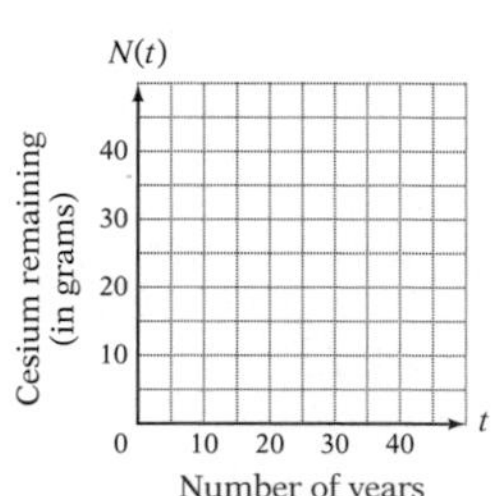

67. **Oceanography** The percent of light that reaches m meters below the surface of the ocean is given by the equation $P(m) = 100e^{-1.38m}$. Graph P and determine the depth to which 50% of the light will reach.

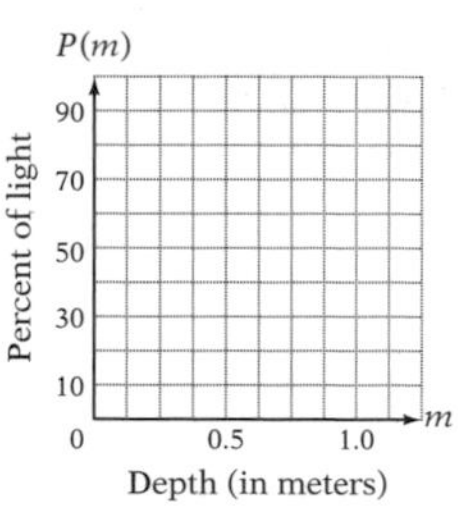

APPLYING CONCEPTS

68. Evaluate $\left(1 + \frac{1}{n}\right)^n$ for $n = 100$, 1000, 10,000, and 100,000, and compare the results with the value of e, the base of the natural exponential function. On the basis of your evaluation, complete the following sentence:

As n increases, $\left(1 + \frac{1}{n}\right)^n$ becomes closer to ______.

69. Graph $f(x) = e^x$ on a coordinate grid. Use translations to graph $g(x) = f(x - 2)$ and $h(x) = f(x + 2)$ on the same set of axes.

70. Graph $f(x) = 2^x$ on a coordinate grid. Use translations to graph $g(x) = f(x) - 2$ and $h(x) = f(x) + 2$ on the same set of axes.

PROJECTS OR GROUP ACTIVITIES

71. **Physics** If air resistance is ignored, the speed v, in feet per second, of an object t seconds after it has been dropped is given by the equation $v = 32t$. This is true regardless of the mass of the object. However, if air resistance is considered, then the speed depends on the mass (and on other things). For a certain mass, the speed t seconds after it has been dropped is given by $v = 64(1 - e^{-\frac{t}{2}})$.
    a. Graph this equation. *Suggestion:* Use Xmin $= -0.5$, Xmax $= 10$, Xscl $= 1$, Ymin $= -0.5$, Ymax $= 70$, and Yscl $= 10$.
    b. The point whose approximate coordinates are $(4, 55.3)$ is on the graph. Write a sentence that explains the meaning of this ordered pair.

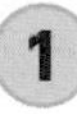

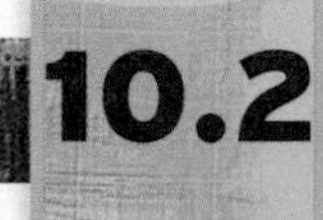

10.2 Introduction to Logarithms

OBJECTIVE 1

Write equivalent exponential and logarithmic equations

Time, t (in hours)	$1000(2^t) = A$
3 h	$1000(2^3) = 8000$
6 h	$1000(2^6) = 64{,}000$
8 h	$1000(2^8) = 256{,}000$
12 h	$1000(2^{12}) = 4{,}096{,}000$

In Michael Crichton's novel *The Andromeda Strain*, we read, "A single cell of the bacterium *E. coli* would, under ideal circumstances, divide every twenty minutes." If a colony of *E. coli* originally contained 1000 bacteria, then a modified version of Crichton's statement, where the bacteria divide once per hour, can be modeled by the exponential growth equation $A = 1000(2^t)$, where A is the number of *E. coli* present after t hours. The table at the left shows the number of *E. coli* for selected times.

To determine when there would be 32,000 bacteria in the culture, we would solve the exponential equation $32{,}000 = 1000(2^t)$ for t. Because $A = 1000(2^t)$ is a growth equation, and 32,000 is between 8000 and 64,000, t must be between 3 h and 6 h. By trial and error, we find that $t = 5$ h.

$$A = 1000(2^t)$$
$$A = 1000(2^5) \quad \text{• Replace } t \text{ by 5.}$$
$$A = 1000(32) = 32{,}000$$

Now suppose we want to find how long it would take for the colony to reach 1,000,000 bacteria. From the table, 1,000,000 is between 256,000 and 4,096,000, so this will happen sometime between 8 h and 12 h. If we try 10 h, halfway between 8 and 12, we have

$$A = 1000(2^t)$$
$$A = 1000(2^{10}) \quad \text{• Replace } t \text{ by 10.}$$
$$A = 1000(1024) = 1{,}024{,}000$$

Because 1,000,000 is less than 1,024,000, the actual value of t is less than 10 h.

We could continue to use trial and error to find t, but it would be more efficient if we could just solve the equation $1{,}000{,}000 = 1000(2^t)$ for t. Using methods described earlier in the text, we have

$$1{,}000{,}000 = 1000(2^t)$$
$$1000 = 2^t \quad \text{• Divide each side of the equation by 1000.}$$

To solve this equation, it would be helpful to have a function that would give the power of 2 that produces 1000. Around the mid-16th century, mathematicians created such a function, which we now call a *logarithmic function*. We write the solution of $1000 = 2^t$ as $t = \log_2 1000$. This is read "t equals the logarithm base 2 of 1000," and it means "t equals the power of 2 that produces 1000."

When logarithms were first introduced, tables were used to find the numerical value of t. Today, a calculator is used. Using a calculator, we can approximate the value of t as 9.97. Thus $9.97 \approx \log_2 1000$ and $1000 \approx 2^{9.97}$.

The ideas presented above are related to the concept of inverse function discussed in the last chapter. Because the exponential function given by $y = b^x$ is a 1–1 function, it has an inverse function. To find that function, we follow the same procedure that we used to find the inverses of other functions.

$$y = b^x$$

$$x = b^y \quad \text{• Interchange } x \text{ and } y.$$

Take Note

Recall, for the equation $2^t = 1000$, we said that "t equals the power of 2 that produces 1000" means "t is the logarithm base 2 of 1000." We are applying the same principle to the equation $x = b^y$.

The equation $x = b^y$ says that y equals the power of b that produces x. That is, y is the logarithm base b of x. The inverse of the exponential function $y = b^x$ is a logarithmic function.

Here is a general definition of logarithm.

DEFINITION OF LOGARITHM

If $x > 0$ and b is a positive constant not equal to 1, then $y = \log_b x$ is equivalent to $b^y = x$.

Read $\log_b x$ as "the logarithm base b of x" or "the log base b of x."

The following table shows equivalent statements written in both exponential and logarithmic form.

Exponential Form		Logarithmic Form
$2^4 = 16$	⟷	$\log_2 16 = 4$
$\left(\frac{2}{3}\right)^2 = \frac{4}{9}$	⟷	$\log_{\frac{2}{3}}\left(\frac{4}{9}\right) = 2$
$10^{-1} = 0.1$	⟷	$\log_{10}(0.1) = -1$
$b^a = x$	⟷	$\log_b x = a$

EXAMPLE 1 Write $4^5 = 1024$ in logarithmic form.

Solution $4^5 = 1024$ is equivalent to $\log_4 1024 = 5$.

Problem 1 Write $3^{-4} = \frac{1}{81}$ in logarithmic form.

Solution See page S31.

Try Exercise 13, page 548.

EXAMPLE 2 Write $\log_7 343 = 3$ in exponential form.

Solution $\log_7 343 = 3$ is equivalent to $7^3 = 343$.

Problem 2 Write $\log_{10} 0.0001 = -4$ in exponential form.

Solution See page S31.

Try Exercise 21, page 548.

In Example 2, 343 can be referred to as the *antilogarithm* base 7 of 3. If $\log_b M = N$, then N is the logarithm base b of M; M is the **antilogarithm** base b of N.

Recalling the equations $y = \log_b x$ and $x = b^y$ from the definition of a logarithm, note that because $b^y > 0$ for all values of y, x is always a positive number. Therefore, in the equation $y = \log_b x$, x is a positive number. The logarithm of a negative number is not a real number.

The 1–1 property of exponential functions can be used to evaluate some logarithms.

EQUALITY OF EXPONENTS PROPERTY

For $b > 0, b \neq 1$, if $b^u = b^v$, then $u = v$.

EXAMPLES

1. If $3^x = 3^4$, then $x = 4$.
2. If $5^{3x} = 5^6$, then $3x = 6$.

Focus on evaluating a logarithm

Evaluate: $\log_2 8$

Write an equation.	$\log_2 8 = x$
Write the equation in its equivalent exponential form.	$8 = 2^x$
Write 8 in exponential form using 2 as the base.	$2^3 = 2^x$
Use the Equality of Exponents Property.	$3 = x$
	$\log_2 8 = 3$

EXAMPLE 3 Evaluate: $\log_3\left(\frac{1}{9}\right)$

Solution

$\log_3\left(\frac{1}{9}\right) = x$	• **Write an equation.**
$\frac{1}{9} = 3^x$	• **Write the equation in its equivalent exponential form.**
$3^{-2} = 3^x$	• **Write $\frac{1}{9}$ in exponential form using 3 as the base.**
$-2 = x$	• **Solve for x using the Equality of Exponents Property.**
$\log_3\left(\frac{1}{9}\right) = -2$	

Problem 3 Evaluate: $\log_4 64$

Solution See page S31.

Try Exercise 33, page 549.

The Equality of Exponents Property can be used to solve some logarithmic equations.

Focus on solving a logarithmic equation

Solve $\log_4 x = -2$ for x. $\log_4 x = -2$

Write the equation in its equivalent exponential form. $4^{-2} = x$

Solve for x. $\frac{1}{16} = x$

The solution is $\frac{1}{16}$.

EXAMPLE 4 Solve $\log_6 x = 2$ for x.

Solution

$\log_6 x = 2$

$6^2 = x$ • **Write $\log_6 x = 2$ in its equivalent exponential form.**

$36 = x$

The solution is 36.

Problem 4 Solve $\log_2 x = -4$ for x.

Solution See page S31.

Try Exercise 43, page 549.

Take Note

The logarithms of most numbers are irrational numbers. Therefore, the value displayed on a calculator is an approximation.

Logarithms base 10 are called **common logarithms.** We usually omit the base, 10, when writing the common logarithm of a number. Therefore, $\log_{10} x$ is written $\log x$. To find the common logarithm of most numbers, a calculator is necessary. A calculator was used to find the value of log 384, shown below.

$$\log 384 \approx 2.584331224$$

When e (the base of the natural exponential function) is used as the base of a logarithm, $\log_e x$ is referred to as the **natural logarithm** and is abbreviated $\ln x$. This is read "el en x." The equivalent exponential form of $y = \ln x$ is $e^y = x$.

Using a calculator, we find that $\ln 23 \approx 3.135494216$.

EXAMPLE 5 Solve $\ln x = -1$ for x. Round to the nearest ten-thousandth.

Solution

$\ln x = -1$

$e^{-1} = x$ • **Use $\ln x = y$ is equivalent to $e^y = x$.**

$0.3679 \approx x$ • **Evaluate e^{-1}.**

The solution is 0.3679.

Problem 5 Solve $\log x = 1.5$ for x. Round to the nearest ten-thousandth.

Solution See page S31.

Try Exercise 51, page 549.

OBJECTIVE 2 The properties of logarithms

Because a logarithm is an exponent, the properties of logarithms are similar to the properties of exponents.

The table at the right shows some powers of 2 and the equivalent logarithmic form of each.

$2^0 = 1$	$\log_2 1 = 0$
$2^1 = 2$	$\log_2 2 = 1$
$2^2 = 4$	$\log_2 4 = 2$
$2^3 = 8$	$\log_2 8 = 3$
$2^4 = 16$	$\log_2 16 = 4$
$2^5 = 32$	$\log_2 32 = 5$

The table can be used to show that $\log_2 4 + \log_2 8$ equals $\log_2 32$.

$$\log_2 4 + \log_2 8 = 2 + 3 = 5$$
$$\log_2 32 = 5$$
$$\log_2 4 + \log_2 8 = \log_2 32$$

Note that $\log_2 32 = \log_2(4 \times 8) = \log_2 4 + \log_2 8$.

The property of logarithms that states that the logarithm of the product of two numbers equals the sum of the logarithms of the two numbers is similar to the property of exponents that states that to multiply two exponential expressions with the same base, we add the exponents.

Take Note

Pay close attention to this theorem. Note, for instance, that this theorem states that

$$\log_3(4p) = \log_3 4 + \log_3 p$$

It also states that

$$\log_5 9 + \log_5 z = \log_5(9z)$$

It does not state any relationship that involves $\log_b(x + y)$. **This expression cannot be simplified.**

PRODUCT PROPERTY OF LOGARITHMS

For any positive real numbers x, y, and b, $b \neq 1$,

$$\log_b(xy) = \log_b x + \log_b y$$

EXAMPLES

1. $\log_7(9z) = \log_7 9 + \log_7 z$
2. $\log[(x - 2)(x + 3)] = \log(x - 2) + \log(x + 3)$
3. $\ln(xy) = \ln x + \ln y$

The Product Property of Logarithms can be extended to more than two factors. For instance,

$$\log_b(xyz) = \log_b x + \log_b y + \log_b z$$

To prove this property, let $\log_b x = m$ and $\log_b y = n$.

Write each equation in its equivalent exponential form. $\quad x = b^m \qquad y = b^n$

Use substitution and the properties of exponents.

$$xy = b^m b^n$$
$$xy = b^{m+n}$$

Write the equation in its equivalent logarithmic form. $\quad \log_b(xy) = m + n$

Substitute $\log_b x$ for m and $\log_b y$ for n. $\quad \log_b(xy) = \log_b x + \log_b y$

A second property of logarithms involves the logarithm of the quotient of two numbers. This property of logarithms is also based on the fact that a logarithm is an exponent and that to divide two exponential expressions with the same base, we subtract the exponents.

Take Note

This theorem is used to rewrite expressions such as

$$\log_5 \frac{m}{8} = \log_5 m - \log_5 8$$

It does *not* state any relationship that involves $\frac{\log_b x}{\log_b y}$.

This expression cannot be simplified.

QUOTIENT PROPERTY OF LOGARITHMS

For any positive real numbers x, y, and b, $b \neq 1$, $\log_b \frac{x}{y} = \log_b x - \log_b y$.

EXAMPLES

1. $\log_3 \frac{y}{11} = \log_3 y - \log_3 11$
2. $\log \frac{x + 1}{x - 1} = \log(x + 1) - \log(x - 1)$
3. $\ln \frac{14}{w} = \ln 14 - \ln w$

Point of Interest

Logarithms were developed independently by Jobst Burgi (1552–1632) and John Napier (1550–1617) as a means of simplifying the calculations of astronomers. The idea was to devise a method by which two numbers could be multiplied by performing additions. Napier is usually given credit for developing logarithms because he published his results first.

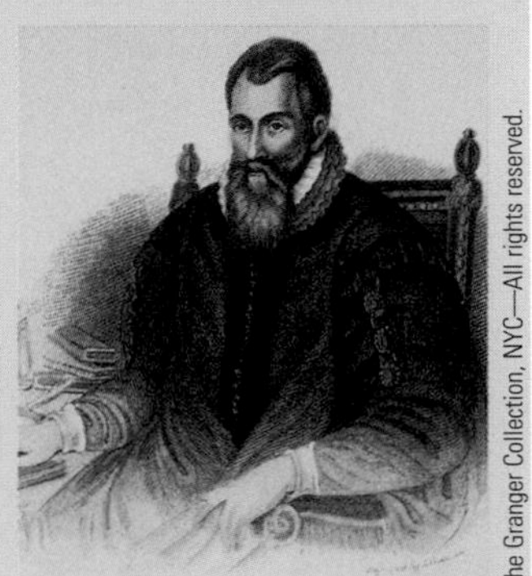

John Napier

In Napier's original work, the logarithm of 10,000,000 was 0. After this work was published, Napier, in discussions with Henry Briggs (1561–1631), decided that tables of logarithms would be easier to use if the logarithm of 1 were 0. Napier died before new tables could be determined, and Briggs took on the task. His table consisted of logarithms accurate to 30 decimal places, all accomplished without the use of a calculator!

The logarithms Briggs calculated are the common logarithms mentioned earlier.

To prove this property, let $\log_b x = m$ and $\log_b y = n$.

Write each equation in its equivalent exponential form. $\quad x = b^m \qquad y = b^n$

Use substitution and the properties of exponents. $\quad \frac{x}{y} = \frac{b^m}{b^n}$

$$\frac{x}{y} = b^{m-n}$$

Write the equation in its equivalent logarithmic form. $\quad \log_b \frac{x}{y} = m - n$

Substitute $\log_b x$ for m and $\log_b y$ for n. $\quad \log_b \frac{x}{y} = \log_b x - \log_b y$

A third property of logarithms, especially useful in computing the power of a number, is based on the fact that a logarithm is an exponent and that the power of an exponential expression is found by multiplying the exponents.

POWER PROPERTY OF LOGARITHMS

For any positive real numbers x and b, $b \neq 1$, and for any real number r, $\log_b x^r = r \log_b x$.

EXAMPLES

1. $\log_4 x^5 = 5 \log_4 x$
2. $\log 3^{2x-1} = (2x - 1)\log 3$
3. $\ln\sqrt{x} = \ln x^{\frac{1}{2}} = \frac{1}{2}\ln x$

To prove this property, let $\log_b x = m$.

Write the equation in its equivalent exponential form. $\quad x = b^m$

Raise each side to the r power. $\quad x^r = (b^m)^r$

Use the properties of exponents. $\quad x^r = b^{mr}$

Write the equation in its equivalent logarithmic form. $\quad \log_b x^r = mr$

Substitute $\log_b x$ for m. $\quad \log_b x^r = r \log_b x$

The properties of logarithms can be used in combination to write a logarithmic expression in **expanded form.**

EXAMPLE 6 Write the logarithm in expanded form.

A. $\log_b(x^2\sqrt{y})$ **B.** $\ln \frac{x}{yz^2}$ **C.** $\log_8\sqrt{x^3y}$

Solution

A. $\log_b(x^2\sqrt{y}) = \log_b x^2 + \log_b \sqrt{y}$ • **Use the Product Property of Logarithms.**

$= \log_b x^2 + \log_b y^{\frac{1}{2}}$ • **Write $\sqrt{y} = y^{\frac{1}{2}}$.**

$= 2\log_b x + \frac{1}{2}\log_b y$ • **Use the Power Property of Logarithms.**

B. $\ln \frac{x}{yz^2} = \ln x - \ln(yz^2)$ • **Use the Quotient Property of Logarithms.**

$= \ln x - (\ln y + \ln z^2)$ • **Use the Product Property of Logarithms.**

$= \ln x - (\ln y + 2\ln z)$ • **Use the Power Property of Logarithms.**

$= \ln x - \ln y - 2\ln z$ • **Use the Distributive Property.**

C. $\log_8 \sqrt{x^3 y} = \log_8 (x^3 y)^{\frac{1}{2}}$ • Write the radical expression as an exponential expression.

$= \frac{1}{2} \log_8 x^3 y$ • Use the Power Property of Logarithms.

$= \frac{1}{2}(\log_8 x^3 + \log_8 y)$ • Use the Product Property of Logarithms.

$= \frac{1}{2}(3 \log_8 x + \log_8 y)$ • Use the Power Property of Logarithms.

$= \frac{3}{2} \log_8 x + \frac{1}{2} \log_8 y$ • Use the Distributive Property.

Problem 6 Write the logarithm in expanded form.

A. $\log_b \frac{x^2}{y}$ **B.** $\ln y^{\frac{1}{3}} z^3$ **C.** $\log_8 \sqrt[3]{xy^2}$

Solution See pages S31–S32.

Try Exercise 71, page 549.

The properties of logarithms are also used to rewrite a logarithmic expression that is in expanded form as a single logarithm.

EXAMPLE 7 Express as a single logarithm with a coefficient of 1.

A. $3 \log_5 x + \log_5 y - 2 \log_5 z$

B. $2(\log_4 x + 3 \log_4 y - 2 \log_4 z)$

C. $\frac{1}{3}(2 \ln x - 4 \ln y)$

Solution **A.** $3 \log_5 x + \log_5 y - 2 \log_5 z$

$= \log_5 x^3 + \log_5 y - \log_5 z^2$ • Use the Power Property of Logarithms.

$= \log_5 x^3 y - \log_5 z^2$ • Use the Product Property of Logarithms.

$= \log_5 \frac{x^3 y}{z^2}$ • Use the Quotient Property of Logarithms.

B. $2(\log_4 x + 3 \log_4 y - 2 \log_4 z)$

$= 2(\log_4 x + \log_4 y^3 - \log_4 z^2)$ • Use the Power Property of Logarithms.

$= 2(\log_4 xy^3 - \log_4 z^2)$ • Use the Product Property of Logarithms.

$= 2 \log_4 \frac{xy^3}{z^2}$ • Use the Quotient Property of Logarithms.

$= \log_4 \left(\frac{xy^3}{z^2}\right)^2$ • Use the Power Property of Logarithms.

$= \log_4 \frac{x^2 y^6}{z^4}$ • Simplify the power of the exponential expression.

C. $\frac{1}{3}(2 \ln x - 4 \ln y)$

$= \frac{1}{3}(\ln x^2 - \ln y^4)$ • Use the Power Property of Logarithms.

$= \frac{1}{3}\left(\ln \frac{x^2}{y^4}\right)$ • Use the Quotient Property of Logarithms.

$= \ln \left(\frac{x^2}{y^4}\right)^{\frac{1}{3}} = \ln \sqrt[3]{\frac{x^2}{y^4}}$ • Use the Power Property of Logarithms. Write the exponential expression as a radical expression.

Problem 7 Express as a single logarithm with a coefficient of 1.

A. $2 \log_b x - 3 \log_b y - \log_b z$

B. $3(\log_5 x - 2 \log_5 y + 4 \log_5 z)$

C. $\frac{1}{2}(2 \ln x - 5 \ln y)$

Solution See page S32.

Try Exercise 107, page 550.

There are three other properties of logarithms that are useful in simplifying logarithmic expressions.

OTHER PROPERTIES OF LOGARITHMS

Logarithmic Property of One

For any positive real number b, $b \neq 1$, $\log_b 1 = 0$.

EXAMPLES

1. $\log_7 1 = 0$ 2. $\log 1 = 0$ 3. $\ln 1 = 0$

Inverse Property of Logarithms

For any positive real numbers x and b, $b \neq 1$, $\log_b b^x = x$ and $b^{\log_b x} = x$.

EXAMPLES

1. $\log_5 5^x = x$ 2. $\log 10^{3z-1} = 3z - 1$ 3. $\ln e^{2x+1} = 2x + 1$
4. $8^{\log_8 x} = x$ 5. $10^{\log(2y+3)} = 2y + 3$ 6. $e^{\ln(3z-7)} = 3z - 7$

1–1 Property of Logarithms

For any positive real numbers x, y, and b, $b \neq 1$, if $\log_b x = \log_b y$, then $x = y$.

EXAMPLES

1. If $\log_2(3x - 2) = \log_2(x + 4)$, then $3x - 2 = x + 4$.
2. If $\ln(x^2 + 1) = \ln(2x)$, then $x^2 + 1 = 2x$.

Although only common logarithms and natural logarithms are programmed into a calculator, the logarithms for other positive bases can be found.

Take Note

To evaluate $\frac{\log 22}{\log 5}$ on a scientific calculator, use the keystrokes

22 [log] [÷] 5 [log] [=]

The display should read 1.9205727.

Focus on finding the logarithm of a number

Evaluate $\log_5 22$. Round to the nearest ten-thousandth.

Write an equation.	$\log_5 22 = x$
Write the equation in its equivalent exponential form.	$5^x = 22$
Apply the common logarithm to each side of the equation.	$\log 5^x = \log 22$
Use the Power Property of Logarithms.	$x \log 5 = \log 22$
	$x = \frac{\log 22}{\log 5}$
This is an exact answer.	$x \approx 1.9206$
This is an approximate answer.	$\log_5 22 \approx 1.9206$

In the third step of the preceding Focus On, the natural logarithm, instead of the common logarithm, could have been applied to each side of the equation. As shown at the right, the same result would have been obtained.

$$\begin{aligned} 5^x &= 22 \\ \ln 5^x &= \ln 22 \\ x \ln 5 &= \ln 22 \\ x &= \frac{\ln 22}{\ln 5} \approx 1.9206 \end{aligned}$$

Using a procedure similar to the one used to evaluate $\log_5 22$, a formula for changing bases can be derived.

CHANGE-OF-BASE FORMULA

$$\log_a N = \frac{\log_b N}{\log_b a}$$

EXAMPLE 8 Evaluate $\log_7 32$. Round to the nearest ten-thousandth.

Solution $\log_7 32 = \dfrac{\ln 32}{\ln 7} \approx 1.7810$ • **Use the Change-of-Base Formula. $N = 32, a = 7, b = e$**

Problem 8 Evaluate $\log_4 2.4$. Round to the nearest ten-thousandth.

Solution See page S32.

Try Exercise 121, page 550.

EXAMPLE 9 Rewrite $f(x) = -3 \log_7(2x - 5)$ in terms of natural logarithms.

Solution

$$\begin{aligned} f(x) &= -3 \log_7(2x - 5) \\ &= -3 \cdot \frac{\ln(2x - 5)}{\ln 7} \\ &= -\frac{3}{\ln 7} \ln(2x - 5) \end{aligned}$$

• **Use the Change-of-Base Formula to rewrite $\log_7(2x - 5)$ as $\dfrac{\ln(2x - 5)}{\ln 7}$.**

Problem 9 Rewrite $f(x) = 4 \log_8(3x + 4)$ in terms of common logarithms.

Solution See page S32.

Try Exercise 133, page 550.

In Example 9, it is important to understand that $-\frac{3}{\ln 7} \ln(2x - 5)$ and $-3 \log_7(2x - 5)$ are *exactly* equal. If common logarithms had been used, the result would have been $f(x) = -\frac{3}{\log 7} \log(2x - 5)$. The expressions $-\frac{3}{\log 7} \log(2x - 5)$ and $-3 \log_7(2x - 5)$ are also *exactly* equal.

If you are working in a base other than base 10 or base e, the Change-of-Base Formula will enable you to calculate the value of a logarithm in that base just as though that base were programmed into the calculator.

10.2 Exercises

CONCEPT CHECK

1. What is a common logarithm?
2. What is a natural logarithm?

For each of the following, assume that x and y are positive real numbers. Determine whether the statement is true or false.

3. $\log_5(x + y) = \log_5 x + \log_5 y$
4. $\log_2 16 = 4$
5. $\log_7 12 = \log_7 4 + \log_7 3$
6. $\dfrac{\log 8}{\log 2} = \log 4$
7. $\ln \dfrac{8}{2} = \ln 4$
8. $\log_3 10 - \log_3 19 = \log_3 \dfrac{10}{19}$
9. $9^{\log_9 x} = x$
10. $\ln e^x = x$

1 Write equivalent exponential and logarithmic equations (See pages 539-542.)

GETTING READY

11. The notation "$\log_b x$" can be read as "log ___?___ of ___?___."
12. A logarithm is an exponent. For example, $\log_2 8$ is the exponent used on ___?___ to give a result of ___?___, so $\log_2 8$ is ___?___.

Write the exponential equation in logarithmic form.

13. $5^2 = 25$
14. $10^3 = 1000$
15. $4^{-2} = \dfrac{1}{16}$
16. $3^{-3} = \dfrac{1}{27}$
17. $10^y = x$
18. $e^y = x$
19. $a^x = w$
20. $b^y = c$

Write the logarithmic equation in exponential form.

21. $\log_3 9 = 2$
22. $\log_2 32 = 5$
23. $\log 0.01 = -2$
24. $\log_5 \dfrac{1}{5} = -1$
25. $\ln x = y$
26. $\log x = y$
27. $\log_b u = v$
28. $\log_c x = y$

GETTING READY

29. Complete the Equality of Exponents Property:
For $b > 0$, $b \neq 1$, if $b^u = b^v$, then ___?___.

30. Evaluate $\log_4 64$.

$\log_4 64 = x$

___?___ $= 4^{\text{?}}$ • **Write an equation.** • **Write the equation in its equivalent exponential form.**

$4^{\text{?}} = 4^x$ • **Write 64 as a power of 4.**

___?___ $= x$ • **Solve for x using the Equality of Exponents Property.**

Evaluate.

31. $\log_3 81$
32. $\log_7 49$
33. $\log_2 128$
34. $\log_5 125$
35. $\log 100$
36. $\log 0.001$
37. $\ln e^3$
38. $\ln e^2$
39. $\log_8 1$
40. $\log_3 243$
41. $\log_5 625$
42. $\log_2 64$

Solve for x.

43. $\log_3 x = 2$
44. $\log_5 x = 1$
45. $\log_4 x = 3$
46. $\log_2 x = 6$
47. $\log_7 x = -1$
48. $\log_8 x = -2$
49. $\log_6 x = 0$
50. $\log_4 x = 0$

Solve for x. Round to the nearest hundredth.

51. $\log x = 2.5$
52. $\log x = 3.2$
53. $\log x = -1.75$
54. $\log x = -2.1$
55. $\ln x = 2$
56. $\ln x = 4$
57. $\ln x = -\frac{1}{2}$
58. $\ln x = -1.7$

59. Assume $b > 1$ and m and n are positive numbers such that $m > n$. Is $\log_b m$ less than, equal to, or greater than $\log_b n$?

60. Assume $0 < b < 1$ and $m > 1$. Is $\log_b m$ less than, equal to, or greater than zero?

2 The properties of logarithms (See pages 542–547.)

61. What is the Product Property of Logarithms?

62. What is the Quotient Property of Logarithms?

GETTING READY

63. Complete the Power Property of Logarithms: For any positive real numbers x and b, $b \neq 1$, and any real number r, $\log_b x^r =$ __?__.

64. Write $\log_4 x^8y$ in expanded form.

$\log_4 (x^8y) = \log_4(\underline{\ ?\ }) + \log_4(\underline{\ ?\ })$ • **By the Product Property of Logarithms, the log of a product expands into the sum of the logs of the factors.**

$= (\underline{\ ?\ })\log_4 x + \log_4 y$ • **Use the Power Property of Logarithms.**

Write the logarithm in expanded form.

65. $\log_8(xz)$
66. $\log_7(4y)$
67. $\log_3 x^5$
68. $\log_2 y^7$
69. $\ln \frac{r}{s}$
70. $\ln \frac{z}{4}$
71. $\log_3(x^2y^6)$
72. $\log_4(t^4u^2)$
73. $\log_7 \frac{u^3}{v^4}$
74. $\log \frac{s^5}{t^2}$
75. $\log_2(rs)^2$
76. $\log_3(x^2y)^3$
77. $\log_9(x^2yz)$
78. $\log_6(xy^2z^3)$
79. $\ln\left(\frac{xy^2}{z^4}\right)$
80. $\ln\left(\frac{r^2s}{t^3}\right)$
81. $\log_8\left(\frac{x^2}{yz^2}\right)$
82. $\log_9\left(\frac{x}{y^2z^3}\right)$

83. $\log_7 \sqrt{xy}$

84. $\log_8 \sqrt[3]{xz}$

85. $\log_2 \sqrt{\frac{x}{y}}$

86. $\log_3 \sqrt[3]{\frac{r}{s}}$

87. $\ln \sqrt{x^3 y}$

88. $\ln \sqrt{x^5 y^3}$

89. $\log_7 \sqrt{\frac{x^3}{y}}$

90. $\log_b \sqrt[3]{\frac{r^2}{t}}$

Express as a single logarithm with a coefficient of 1.

91. $\log_3 x^3 - \log_3 y$

92. $\log_7 t + \log_7 v^2$

93. $\log_8 x^4 + \log_8 y^2$

94. $\log_2 r^2 + \log_2 s^3$

95. $3 \ln x$

96. $4 \ln y$

97. $3 \log_5 x + 4 \log_5 y$

98. $2 \log_6 x + 5 \log_6 y$

99. $-2 \log_4 x$

100. $-3 \log_2 y$

101. $2 \log_3 x - \log_3 y + 2 \log_3 z$

102. $4 \log_5 r - 3 \log_5 s + \log_5 t$

103. $\log_b x - (2 \log_b y + \log_b z)$

104. $2 \log_2 x - (3 \log_2 y + \log_2 z)$

105. $2(\ln x + \ln y)$

106. $3(\ln r + \ln t)$

107. $\frac{1}{2}(\log_6 x - \log_6 y)$

108. $\frac{1}{3}(\log_8 x - \log_8 y)$

109. $2(\log_4 s - 2 \log_4 t + \log_4 r)$

110. $3(\log_9 x + 2 \log_9 y - 2 \log_9 z)$

111. $\log_5 x - 2(\log_5 y + \log_5 z)$

112. $\log_4 t - 3(\log_4 u + \log_4 v)$

113. $3 \ln t - 2(\ln r - \ln v)$

114. $2 \ln x - 3(\ln y - \ln z)$

115. $\frac{1}{2}(3 \log_4 x - 2 \log_4 y + \log_4 z)$

116. $\frac{1}{3}(4 \log_5 t - 3 \log_5 u - 3 \log_5 v)$

Without using a calculator, determine if the statement is true or false.

117. $\log 10 + \log 5 = \log 15$

118. $\log 10 - \log 5 = \log 2$

119. $\log_b b^{10} - \log_b 1 = 10$

120. $\frac{\log 10}{\log 2} = \log 8$

Evaluate. Round to the nearest ten-thousandth.

121. $\log_8 6$

122. $\log_4 8$

123. $\log_5 30$

124. $\log_6 28$

125. $\log_3(0.5)$

126. $\log_5(0.6)$

127. $\log_7(1.7)$

128. $\log_6(3.2)$

129. $\log_5 15$

130. $\log_3 25$

131. $\log_{12} 120$

132. $\log_9 90$

Rewrite each function in terms of common logarithms.

133. $f(x) = \log_3(3x - 2)$

134. $f(x) = \log_5(x^2 + 4)$

135. $f(x) = 5 \log_9(6x + 7)$

136. $f(x) = 3 \log_2(2x^2 - x)$

Rewrite each function in terms of natural logarithms.

137. $f(x) = \log_2(x + 5)$

138. $f(x) = \log_4(3x + 4)$

139. $f(x) = \log_3(x^2 + 9)$

140. $f(x) = \log_7(9 - x^2)$

APPLYING CONCEPTS

141. Given $S(t) = 8 \log_5(6t + 2)$, determine $S(2)$ to the nearest hundredth.

142. Given $f(x) = 3 \log_6(2x - 1)$, determine $f(7)$ to the nearest hundredth.

143. Given $G(x) = -5 \log_7(2x + 19)$, determine $G(-3)$ to the nearest hundredth.

144. Given $P(v) = -3 \log_6(4 - 2v)$, determine $P(-4)$ to the nearest hundredth.

Solve for x.

145. $\log_2(\log_2 x) = 3$

146. $\log_3(\log_3 x) = 1$

147. $\ln(\ln x) = 1$

148. $\ln(\log x) = 2$

PROJECTS OR GROUP ACTIVITIES

149. **Biology** To discuss the variety of species that live in a certain environment, a biologist needs a precise definition of *diversity*. Let $p_1, p_2, \ldots, p_n$ be the proportions of n species that live in an environment. The biological diversity D of this system is

$$D = -(p_1 \log_2 p_1 + p_2 \log_2 p_2 + \cdots + p_n \log_2 p_n)$$

The larger the value of D, the greater the diversity of the system. Suppose an ecosystem has exactly five different varieties of grass: rye (R), Bermuda (B), blue (L), fescue (F), and St. Augustine (A).

Table 1

R	B	L	F	A
$\frac{1}{5}$	$\frac{1}{5}$	$\frac{1}{5}$	$\frac{1}{5}$	$\frac{1}{5}$

Table 2

R	B	L	F	A
$\frac{1}{8}$	$\frac{3}{8}$	$\frac{1}{16}$	$\frac{1}{8}$	$\frac{5}{16}$

Table 3

R	B	L	F	A
0	$\frac{1}{4}$	0	0	$\frac{3}{4}$

Table 4

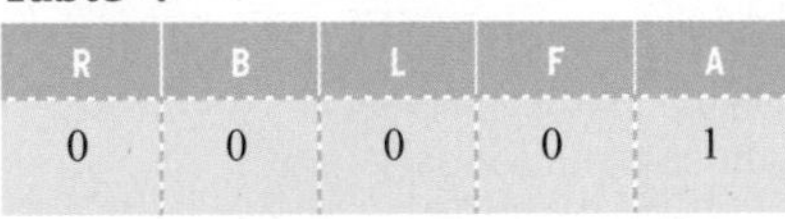

R	B	L	F	A
0	0	0	0	1

a. Calculate the diversity of this ecosystem if the proportions are as shown in Table 1.

b. Because Bermuda and St. Augustine are virulent grasses, after a time the proportions are as shown in Table 2. Does this system have more or less diversity than the one given in Table 1?

c. After an even longer period, the Bermuda and St. Augustine completely overrun the environment, and the proportions are as in Table 3. Calculate the diversity of this system. (*Note:* For purposes of the diversity definition, $0 \log_2 0 = 0$.) Does it have more or less diversity than the system given in Table 2?

d. Finally, the St. Augustine overruns the Bermuda, and the proportions are as in Table 4. Calculate the diversity of this system. Write a sentence that explains your answer.

10.3 Graphs of Logarithmic Functions

OBJECTIVE 1 Graph logarithmic functions

Point of Interest

Although logarithms were originally developed to assist with computations, logarithmic functions have a much broader use today. These functions occur in geology, acoustics, chemistry, and economics, for example.

The graph of a logarithmic function can be drawn by using the relationship between the exponential and logarithmic functions.

To graph $g(x) = \log_2 x$, think of the function as the equation $y = \log_2 x$.

$$g(x) = \log_2 x$$
$$y = \log_2 x$$

Write the equivalent exponential equation.

$$x = 2^y$$

Because the equation is solved for x in terms of y, it is easier to choose values of y and find the corresponding values of x. The results can be recorded in a table.

Graph the ordered pairs on a rectangular coordinate system.

Connect the points with a smooth curve.

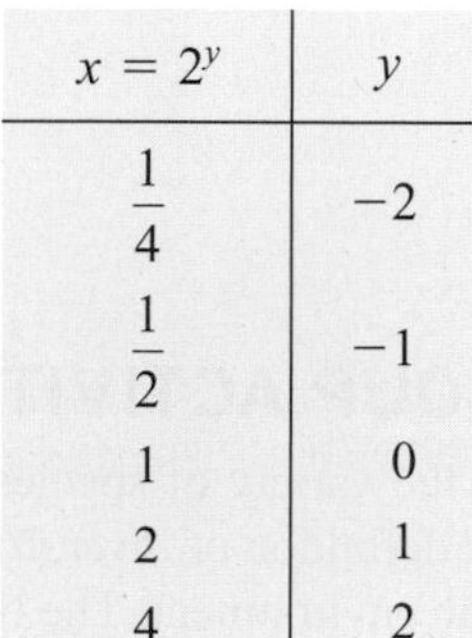

$x = 2^y$	y
$\frac{1}{4}$	-2
$\frac{1}{2}$	-1
1	0
2	1
4	2

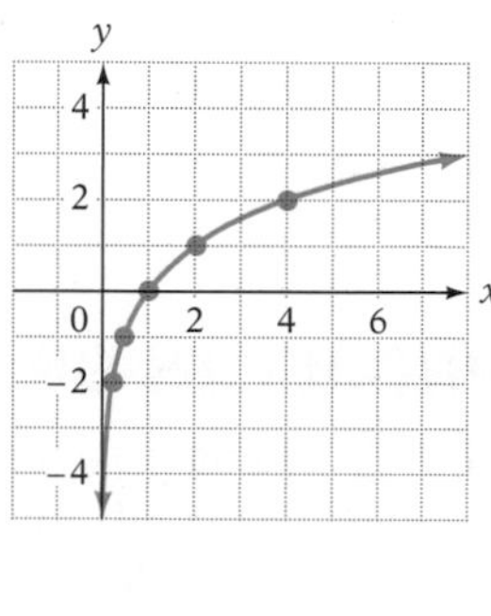

Applying the vertical line and horizontal line tests reveals that $g(x) = \log_2 x$ is the graph of a 1–1 function.

Recall that the graph of the inverse of a function f is the mirror image of the graph of f with respect to the line whose equation is $y = x$. The graph of $f(x) = 2^x$ was shown earlier. Because $g(x) = \log_2 x$ is the inverse of $f(x) = 2^x$, the graphs of these functions are mirror images of each other with respect to the line whose equation is $y = x$.

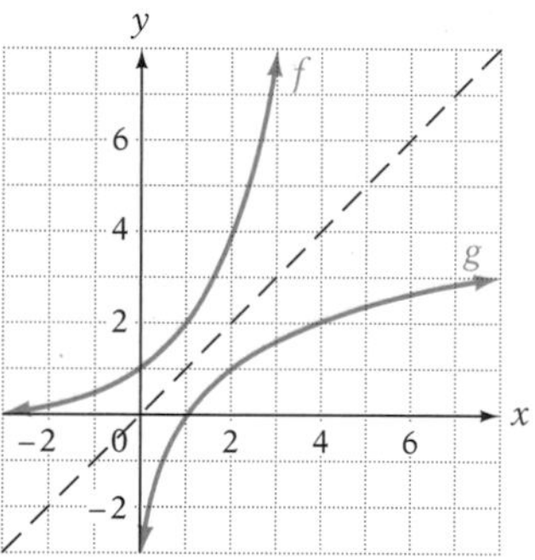

Focus on graphing a logarithmic function

Graph: $f(x) = \log_2 x + 1$

Think of the function as the equation $y = \log_2 x + 1$.

$$f(x) = \log_2 x + 1$$
$$y = \log_2 x + 1$$

Solve the equation for $\log_2 x$.

$$y - 1 = \log_2 x$$

Write the equivalent exponential equation.

$$2^{y-1} = x$$

Choose values of y, and find the corresponding values of x. Graph the ordered pairs on a rectangular coordinate system. Connect the points with a smooth curve.

$x = 2^{y-1}$	y
$\frac{1}{4}$	-1
$\frac{1}{2}$	0
1	1
2	2
4	3

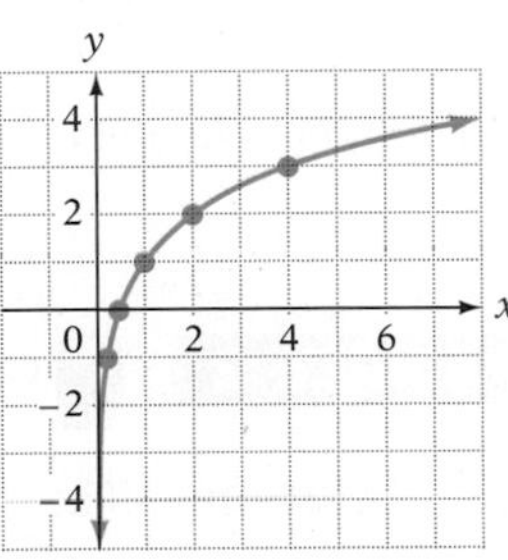

EXAMPLE 1 Graph. **A.** $f(x) = \log_3 x$ **B.** $f(x) = 2 \log_3 x$

Solution **A.** $f(x) = \log_3 x$

$y = \log_3 x$ • **Substitute y for $f(x)$.**

$x = 3^y$ • **Write the equivalent exponential equation.**

Choose values of y, and find the corresponding values of x. Graph the ordered pairs on a rectangular coordinate system. Connect the points with a smooth curve.

$x = 3^y$	y
$\frac{1}{9}$	-2
$\frac{1}{3}$	-1
1	0
3	1

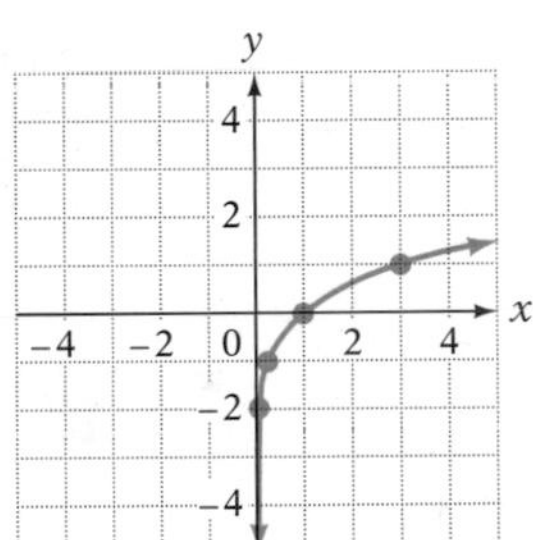

B. $f(x) = 2 \log_3 x$

$y = 2 \log_3 x$ • **Substitute y for $f(x)$.**

$\frac{y}{2} = \log_3 x$ • **Solve the equation for $\log_3 x$.**

$x = 3^{\frac{y}{2}}$ • **Write the equivalent exponential equation.**

Choose values of y, and find the corresponding values of x. Graph the ordered pairs on a rectangular coordinate system. Connect the points with a smooth curve.

$x = 3^{\frac{y}{2}}$	y
$\frac{1}{9}$	-4
$\frac{1}{3}$	-2
1	0
3	2

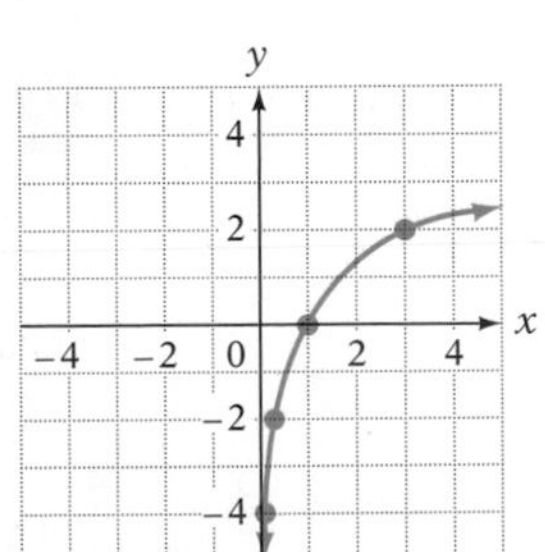

Problem 1 Graph. **A.** $f(x) = \log_2(x - 1)$ **B.** $f(x) = \log_3 2x$

Solution See page S32.

Try Exercise 15, page 556.

EXAMPLE 2  Graph: $f(x) = 2 \ln x + 3$

Solution The graph is shown below. To verify the accuracy of the graph, evaluate $f(x)$ for a few values of x, and compare the results to values found by using the TRACE feature of the graphing utility. For example, $f(1) = 2 \ln(1) + 3 = 3$, so $(1, 3)$ is a point on the graph. When $x = 3$, $f(3) = 2 \ln(3) + 3 \approx 5.2$, so $(3, 5.2)$ is a point on the graph.

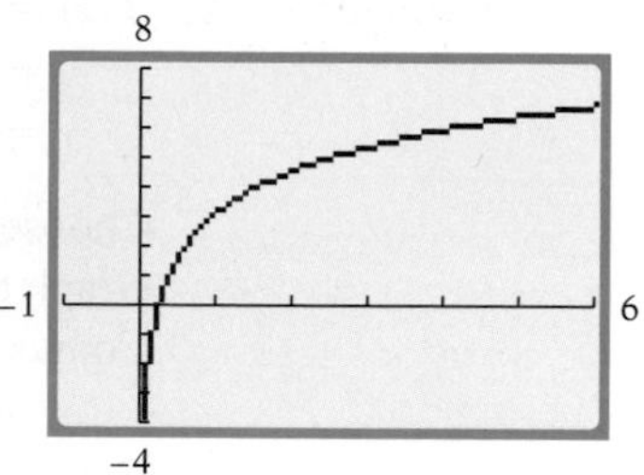

Problem 2 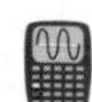 Graph: $f(x) = 10 \log(x - 2)$

Solution See page S32.

Try Exercise 19, page 556.

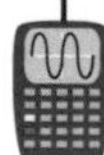

The graphs of logarithmic functions to bases other than base e or base 10 can be drawn with a graphing calculator by first using the Change-of-Base Formula $\log_a N = \dfrac{\log_b N}{\log_b a}$ to rewrite the logarithmic function in terms of base e or base 10.

Focus on using a graphing calculator to graph a logarithmic function with base other than e or 10

Graph: $f(x) = \log_3 x$

Use the Change-of-Base Formula to rewrite $\log_3 x$ in terms of $\log x$ or $\ln x$. The natural logarithmic function $\ln x$ is used here.

$$\log_3 x = \frac{\ln x}{\ln 3}$$

To graph $f(x) = \log_3 x$ using a graphing calculator, use the equivalent form $f(x) = \frac{\ln x}{\ln 3}$.

The graph is shown at the right.

The graph of $f(x) = \log_3 x$ could have been drawn by rewriting $\log_3 x$ in terms of $\log x$, as $\log_3 x = \frac{\log x}{\log 3}$. The graph of $f(x) = \log_3 x$ is identical to the graph of $f(x) = \frac{\log x}{\log 3}$.

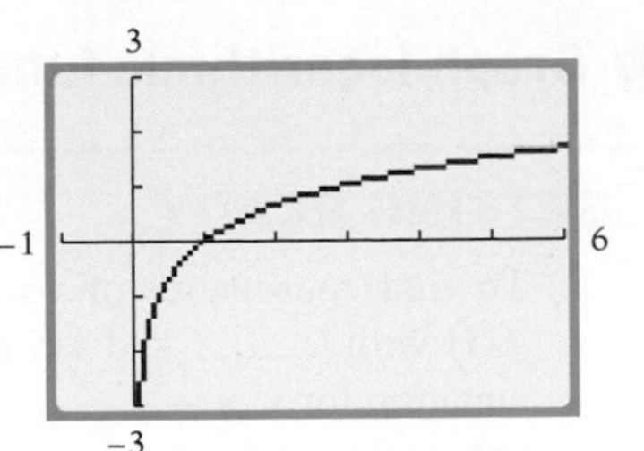

The example that follows was graphed by rewriting the logarithmic function in terms of the natural logarithmic function. The common logarithmic function could also have been used.

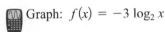

EXAMPLE 3 Graph: $f(x) = -3 \log_2 x$

Solution
$$f(x) = -3 \log_2 x$$
$$= -3 \cdot \frac{\ln x}{\ln 2} = -\frac{3}{\ln 2} \ln x$$

- **Rewrite $\log_2 x$ in terms of $\ln x$.**

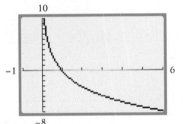

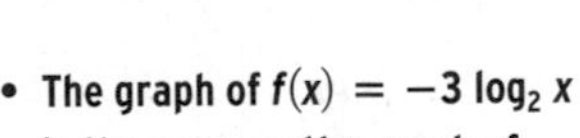

- **The graph of $f(x) = -3 \log_2 x$ is the same as the graph of $f(x) = -\frac{3}{\ln 2} \ln x$.**

Problem 3 Graph: $f(x) = 2 \log_4 x$

Solution See page S32.

Try Exercise 23, page 556.

10.3 Exercises

CONCEPT CHECK

1. Is the function $f(x) = \log x$ a 1–1 function? Why or why not?
2. Name two characteristics of the graph of $y = \log_b x$, $b > 1$.
3. What is the relationship between the graph of $x = 3^y$ and that of $y = \log_3 x$?
4. What is the relationship between the graph of $y = 3^x$ and that of $y = \log_3 x$?

1 Graph logarithmic functions (See pages 552–555.)

GETTING READY

5. To find coordinates of points on the graph of $f(x) = \log_3(x - 1)$, first replace $f(x)$ with __?__ and write the equation in __?__ form: $x - 1 = 3^y$. Solve the equation for x: $x =$ __?__. Choose values for __?__ and find the corresponding values for __?__.

6. Use the results of Exercise 5. The x-coordinate of the point on the graph of $f(x) = \log_3(x - 1)$ that has y-coordinate 2 is $3^{?} + 1 =$ __?__. The coordinates of one point on the graph of $f(x) = \log_3(x - 1)$ are (__?__, __?__).

Graph by plotting points.

7. $f(x) = \log_4 x$

8. $f(x) = \log_2(x + 1)$

9. $f(x) = \log_3(2x - 1)$

10. $f(x) = \log_2\left(\frac{1}{2}x\right)$

11. $f(x) = 3 \log_2 x$

12. $f(x) = \frac{1}{2} \log_2 x$

13. $f(x) = -\log_2 x$

14. $f(x) = -\log_3 x$

15. $f(x) = \log_2(x - 1)$

16. $f(x) = \log_3(2 - x)$

17. $f(x) = -\log_2(x - 1)$

18. $f(x) = -\log_2(1 - x)$

Use a graphing utility to graph the following.

19. $y = -\ln x + 1$

20. $y = 3 \log(x + 1)$

21. $y = \ln(x - 3)$

22. $f(x) = \log_3 x + 2$

23. $f(x) = \log_2 x - 3$

24. $f(x) = -\frac{1}{2} \log_2 x - 1$

25. $f(x) = -\log_2 x + 2$

26. $f(x) = x + \log_3(2 - x)$

27. $f(x) = x - \log_2(1 - x)$

Graph the functions on the same rectangular coordinate system.

28. $f(x) = 3^x;\ g(x) = \log_3 x$

29. $f(x) = \left(\frac{1}{2}\right)^x;\ g(x) = \log_{\frac{1}{2}} x$

30. $f(x) = 10^x;\ g(x) = \log_{10} x$

31. Which function(s) will have the same graph as the graph of the function $f(x) = \log_3 3x$?

(i) $g(x) = 3\log_3 x$ (ii) $h(x) = 1 + \log_3 x$

(iii) $F(x) = \log_3 x^3$ (iv) $G(x) = \dfrac{\log 3x}{\log 3}$

32. Which function(s) will have the same graph as the graph of the function $f(x) = -\log_b x$?

(i) $g(x) = \log_b(-x)$ (ii) $h(x) = \dfrac{1}{\log_b x}$

(iii) $F(x) = \log_b \dfrac{1}{x}$ (iv) $G(x) = \dfrac{\log b}{\log x}$

APPLYING CONCEPTS

For part (a) of Exercises 33 to 35, use a graphing utility.

33. **Employment** The proficiency of a typist decreases (without practice) over time. An equation that approximates this decrease is given by $S = 60 - 7\ln(t + 1)$, where S is the typing speed in words per minute and t is the number of months without typing.

a. Graph the equation. Use Xmin = 0, Xmax = 6, Xscl = 1, Ymin = 0, Ymax = 60, Yscl = 10.

b. The point whose approximate coordinates are (4, 49) is on the graph. Write a sentence that describes the meaning of this ordered pair.

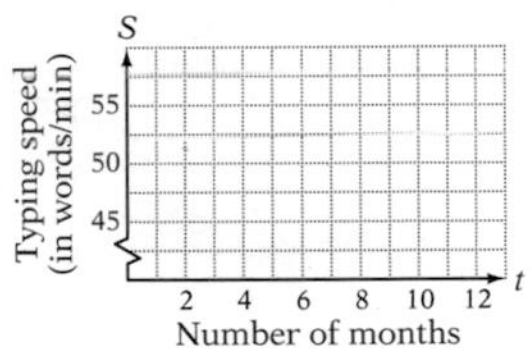

34. **Astronomy** Astronomers use the *distance modulus* of a star as a method of determining the star's distance from Earth. The formula is $M = 5\log s - 5$, where M is the distance modulus and s is the star's distance from Earth in parsecs. (One parsec $\approx 2.1 \times 10^{13}$ miles.)

a. Graph the equation. Use Xmin = 0, Xmax = 30, Xscl = 5, Ymin = −5, Ymax = 5, Yscl = 1.

b. The point whose approximate coordinates are (25.1, 2) is on the graph. Write a sentence that describes the meaning of this ordered pair.

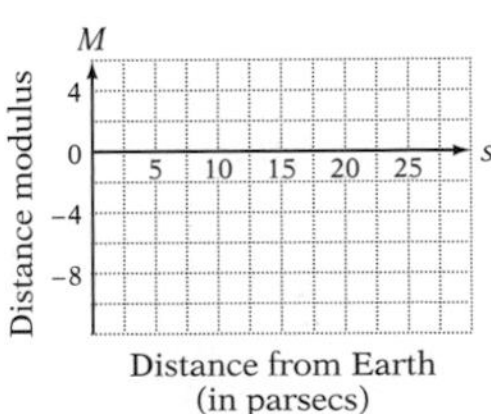

35. **Energy** The Energy Information Administration forecasts energy production for the United States. Based on data from that agency, the equation

$$y = -3.6196 + 3.4455\ln(x + 5)$$

models the estimated quadrillion Btu of energy y that will be produced in the United States by biomass fuels during year x, where x is the number of years after 2005.

a. Graph the equation. Use Xmin = −5, Xmax = 20, Ymin = −10, Ymax = 10.

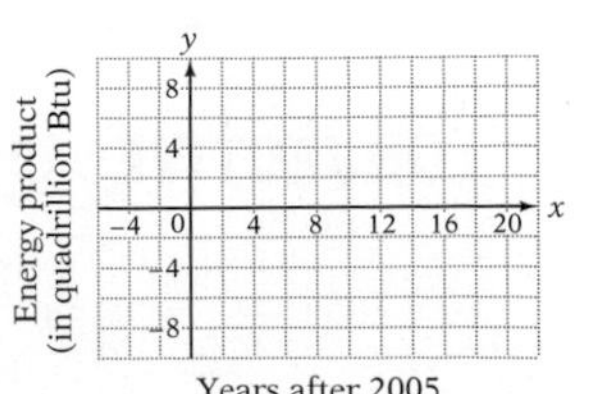

b. According to this model, what is the projected energy product, in quadrillion Btu, of biomass fuels in 2015? Round to the nearest hundredth of a quadrillion.

c. According to this model, in what year will energy produced by biomass fuels first exceed 10 quadrillion Btu?

36. By the Power Property of Logarithms, $\ln x^2 = 2 \ln x$. Graph the equations $f(x) = \ln x^2$ and $g(x) = 2 \ln x$ on the same rectangular coordinate system. Are the graphs the same? Why or why not?

37. Because $f(x) = e^x$ and $g(x) = \ln x$ are inverse functions of each other, $f[g(x)] = x$ and $g[f(x)] = x$. Graph $f[g(x)] = e^{\ln x}$ and $g[f(x)] = \ln e^x$. Explain why the graphs are different even though $f[g(x)] = g[f(x)]$.

PROJECTS OR GROUP ACTIVITIES

Plot the ordered pairs in the table. Determine whether the ordered pairs belong to a linear function, a quadratic function, an exponential function, or a logarithmic function.

38.

x	y
0	5
1	3
2	1
3	−1
4	−3

39.

x	y
−1	6
0	1
1	2
2	9
3	22

40.

x	y
0	1
1	4
2	16
3	64
4	256

41.

x	y
0	−3
2	−2
4	−1
6	0
8	1

42.

x	y
0	−3
1	−7
2	−9
3	−9
4	−7

43.

x	y
0	160
1	80
2	40
3	20
4	10

44.

x	y
$\frac{1}{2}$	−1
1	0
2	1
4	2
8	3

45.

x	y
$\frac{1}{4}$	−2
1	0
4	2
16	4
256	8

10.4 Exponential and Logarithmic Equations

OBJECTIVE 1 Solve exponential equations

An **exponential equation** is one in which a variable occurs in an exponent. The examples at the right are exponential equations.

$$6^{2x+1} = 6^{3x-2}$$
$$4^x = 3$$
$$2^{x+1} = 7$$

An exponential equation in which each side of the equation can be expressed in terms of the same base can be solved by using the Equality of Exponents Property. Recall that the Equality of Exponents Property states that

If $b^u = b^v$, then $u = v$.

Focus on solving an exponential equation in which the bases are the same

Solve: $4^{3x+2} = 4^{x-6}$

The bases are the same. $4^{3x+2} = 4^{x-6}$

Use the Equality of Exponents Property to equate the exponents. $3x + 2 = x - 6$

Solve for x. $2x = -8$

The solution is -4. $x = -4$

If the bases are not the same, as in Example 1 below, try to rewrite the equation so that both sides are written in terms of the same base.

EXAMPLE 1 Solve and check: $9^{x+1} = 27^{x-1}$

Solution

$$9^{x+1} = 27^{x-1}$$

$$(3^2)^{x+1} = (3^3)^{x-1}$$

$$3^{2x+2} = 3^{3x-3}$$

- **Rewrite each side of the equation using the same base.**

$$2x + 2 = 3x - 3$$

- **Use the Equality of Exponents Property to equate the exponents.**

$$2 = x - 3$$

$$5 = x$$

- **Solve the resulting equation.**

Check:

$$\begin{array}{r|l} 9^{x+1} & = 27^{x-1} \\ \hline 9^{5+1} & 27^{5-1} \\ 9^6 & 27^4 \\ (3^2)^6 & (3^3)^4 \\ 3^{12} & = 3^{12} \end{array}$$

The solution is 5.

Problem 1 Solve and check: $10^{3x+5} = 10^{x-3}$

Solution See page S33.

Try Exercise 27, page 563.

When both sides of an exponential equation cannot easily be expressed in terms of the same base, logarithms are used to solve the exponential equation.

EXAMPLE 2 Solve for x. Round to the nearest ten-thousandth.

A. $4^x = 7$ **B.** $3^{2x} = 4$

Solution **A.**

$$4^x = 7$$

$$\log 4^x = \log 7$$

- **Take the common logarithm of each side of the equation.**

$$x \log 4 = \log 7$$

- **Rewrite using the properties of logarithms.**

$$x = \frac{\log 7}{\log 4}$$

- **Solve for x.**

$$x \approx 1.4037$$

The solution is 1.4037.

B. $3^{2x} = 4$

$\log 3^{2x} = \log 4$ • **Take the common logarithm of each side of the equation.**

$2x \log 3 = \log 4$ • **Rewrite using the properties of logarithms.**

$x = \dfrac{\log 4}{2 \log 3}$ • **Solve for x.**

$x \approx 0.6309$

The solution is 0.6309.

Problem 2 Solve for x. Round to the nearest ten-thousandth.

A. $4^{3x} = 25$ **B.** $(1.06)^x = 1.5$

Solution See page S33.

Try Exercise 37, page 563.

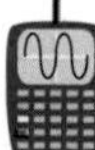

The equations in Example 2 can be solved by graphing. For Example 2(A), by subtracting 7 from each side of the equation $4^x = 7$, the equation can be written as $4^x - 7 = 0$. The graph of $f(x) = 4^x - 7$ is shown at the right. The values of x for which $f(x) = 0$ are the solutions of the equation $4^x = 7$. These values of x are the zeros of the function. By using the features of a graphing calculator, it is possible to determine a very accurate solution. The solution to the nearest tenth is shown in the graph.

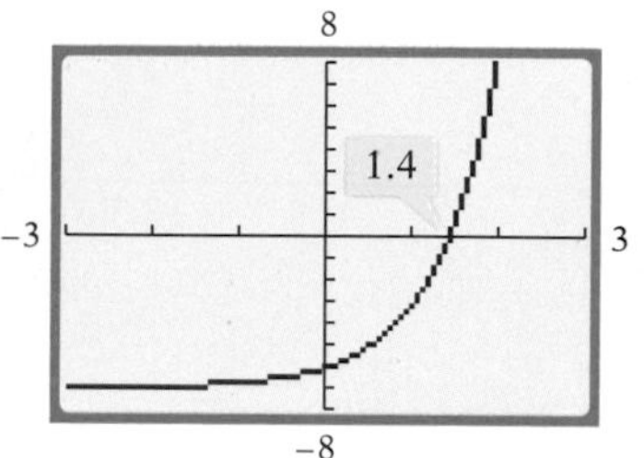

EXAMPLE 3  Solve $e^x = 2x + 1$ for x. Round to the nearest hundredth.

Solution Rewrite the equation by subtracting $2x + 1$ from each side and writing the equation as $e^x - 2x - 1 = 0$. The zeros of $f(x) = e^x - 2x - 1$ are the solutions of $e^x = 2x + 1$. Graph f and use the features of a graphing calculator to estimate the solutions to the nearest hundredth.

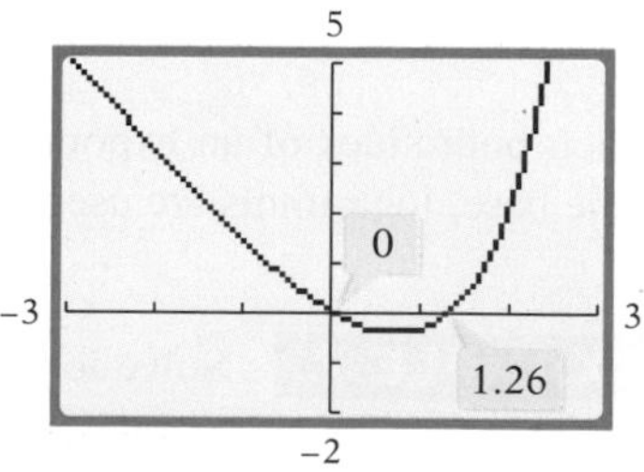

The solutions are 0 and 1.26.

Problem 3  Solve $e^x = x$ for x. Round to the nearest hundredth.

Solution See page S33.

Try Exercise 45, page 564.

OBJECTIVE 2 Solve logarithmic equations

A logarithmic equation can be solved by using the properties of logarithms.

Focus on solving a logarithmic equation

Solve: $\log_9 x + \log_9(x - 8) = 1$

Use the Product Property of Logarithms to rewrite the left side of the equation.

$$\log_9 x + \log_9(x - 8) = 1$$
$$\log_9 x(x - 8) = 1$$

Write the equation in exponential form.

$$9^1 = x(x - 8)$$

Simplify and solve for x.

$$9 = x^2 - 8x$$
$$0 = x^2 - 8x - 9$$
$$0 = (x - 9)(x + 1)$$

$$x - 9 = 0 \qquad x + 1 = 0$$
$$x = 9 \qquad x = -1$$

When x is replaced by 9 in the original equation, 9 checks as a solution. When x is replaced by -1, the original equation contains the expression $\log_9(-1)$. Because the logarithm of a negative number is not a real number, -1 does not check as a solution.

The solution of the equation is 9.

The above algebraic solution produced -1 and 9 as possible solutions, but only 9 satisfies the equation $\log_9 x + \log_9(x - 8) = 1$. The extraneous solution was introduced at the second step. The Product Property of Logarithms, $\log_b(xy) = \log_b x + \log_b y$, applies only when both x and y are positive numbers. For the equation under consideration, this occurs when $x > 8$. Therefore, a solution to this equation must be greater than 8.

EXAMPLE 4 Solve: $\log_4(x^2 - 6x) = 2$

Solution

$$\log_4(x^2 - 6x) = 2$$
$$4^2 = x^2 - 6x$$
• **Rewrite the equation in exponential form.**
$$16 = x^2 - 6x$$
• **Simplify.**
$$0 = x^2 - 6x - 16$$
• **Write the quadratic equation in standard form.**
$$0 = (x + 2)(x - 8)$$
• **Factor and use the Principle of Zero Products.**

$$x + 2 = 0 \qquad x - 8 = 0$$
$$x = -2 \qquad x = 8$$

-2 and 8 check as solutions. The solutions are -2 and 8.

Problem 4 Solve: $\log_4(x^2 - 3x) = 1$

Solution See page S33.

Try Exercise 59, page 564.

Some logarithmic equations can be solved by using the 1–1 Property of Logarithms, which states that for $x > 0$, $y > 0$, $b > 0$, $b \neq 1$,

If $\log_b x = \log_b y$, then $x = y$.

EXAMPLE 5 Solve: $\log_2 x - \log_2(x - 1) = \log_2 2$

Solution

$$\log_2 x - \log_2(x - 1) = \log_2 2$$

$$\log_2\left(\frac{x}{x - 1}\right) = \log_2 2 \qquad \text{• Use the Quotient Property of Logarithms.}$$

$$\frac{x}{x - 1} = 2 \qquad \text{• Use the 1–1 Property of Logarithms.}$$

$$(x - 1)\left(\frac{x}{x - 1}\right) = (x - 1)2 \qquad \text{• Solve for } x.$$

$$x = 2x - 2$$

$$-x = -2$$

$$x = 2$$

2 checks as a solution. The solution is 2.

Problem 5 Solve: $\log_3 x + \log_3(x + 3) = \log_3 4$

Solution See page S33.

Try Exercise 69, page 565.

Some logarithmic equations cannot be solved algebraically. In such cases, a graphical approach may be appropriate.

EXAMPLE 6  Solve $\ln(2x + 4) = x^2$ for x. Round to the nearest hundredth.

Solution Begin by subtracting x^2 from each side and rewriting the equation as $\ln(2x + 4) - x^2 = 0$. The zeros of the function defined by $f(x) = \ln(2x + 4) - x^2$ are the solutions of the equation. Graph f and then use the features of a graphing utility to estimate the solutions to the nearest hundredth.

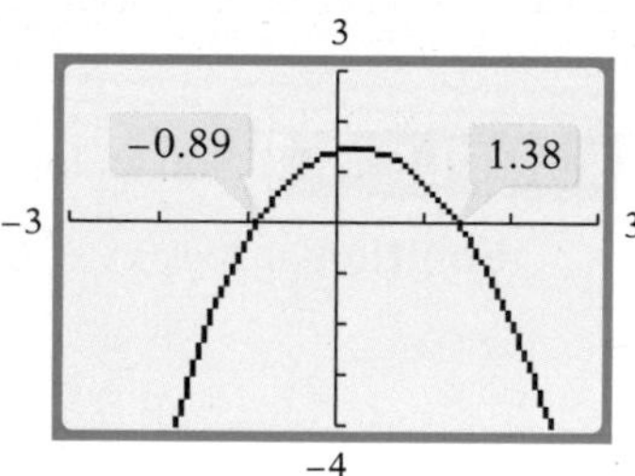

The solutions are -0.89 and 1.38.

Problem 6 Solve $\log(3x - 2) = -2x$ for x. Round to the nearest hundredth.

Solution See page S33.

Try Exercise 81, page 565.

10.4 Exercises

CONCEPT CHECK

1. What is an exponential equation?

2. a. What does the Equality of Exponents Property state?
b. Describe when you would use this property.

3. What is a logarithmic equation?

4. What does the 1–1 Property of Logarithms state?

1 Solve exponential equations (See pages 558–560.)

GETTING READY

5. Solve: $3^{3x-1} = 9^{2x}$

$3^{3x-1} = 9^{2x}$	• Both bases are powers of ___?___.
$3^{3x-1} = (3^{\underline{\ ?\ }})^{2x}$	• Write 9 as a power of 3.
$3^{3x-1} = 3^{\underline{\ ?\ }}$	• Simplify the exponent.
$\underline{\ ?\ } = \underline{\ ?\ }$	• Use the Equality of Exponents Property to equate the exponents.
$x = \underline{\ ?\ }$	• Solve for x.

6. Solve: $5^x = 18$

$5^x = 18$	• The bases are not powers of the same number.
$\log 5^x = \log 18$	• Take the ___?___ of each side of the equation.
$(\underline{\ ?\ })\log 5 = \log 18$	• Use the Power Property of Logarithms.
$x = \dfrac{?}{?}$	• Solve for x by dividing each side of the equation by ___?___.
$x \approx \underline{\ ?\ }$	• Use a calculator to evaluate x to the nearest ten-thousandth.

Solve for x. Round to the nearest ten-thousandth.

7. $5^{4x-1} = 5^{x+2}$

8. $7^{4x-3} = 7^{2x+1}$

9. $8^{x-4} = 8^{5x+8}$

10. $10^{4x-5} = 10^{x+4}$

11. $5^x = 6$

12. $7^x = 10$

13. $12^x = 6$

14. $10^x = 5$

15. $\left(\frac{1}{2}\right)^x = 3$

16. $\left(\frac{1}{3}\right)^x = 2$

17. $(1.5)^x = 2$

18. $(2.7)^x = 3$

19. $10^x = 21$

20. $10^x = 37$

21. $2^{-x} = 7$

22. $3^{-x} = 14$

23. $2^{x-1} = 6$

24. $4^{x+1} = 9$

25. $3^{2x-1} = 4$

26. $4^{-x+2} = 12$

27. $9^x = 3^{x+1}$

28. $2^{x-1} = 4^x$

29. $8^{x+2} = 16^x$

30. $9^{3x} = 81^{x-4}$

31. $5^{x^2} = 21$

32. $3^{x^2} = 40$

33. $2^{4x-2} = 20$

34. $4^{3x+8} = 12$

35. $3^{-x+2} = 18$

36. $5^{-x+1} = 15$

37. $4^{2x} = 100$

38. $3^{3x} = 1000$

39. $2.5^{-x} = 4$

40. $3.25^{x+1} = 4.2$

41. $0.25^x = 0.125$

42. $0.1^{5x} = 10^{-2}$

43. Given that a is any real number, which equations have the same solution?
(i) $9^{x-a} = 81^x$ (ii) $5^{2x-2a} = 25^{2x}$ (iii) $49^{(x-a)/4} = 7^x$ (iv) $9^x = 81^{x-a}$

44. Use the graphs shown below. Which graph(s) can be used to solve the equation $e^x - 5 = x$? How many positive solutions does the equation $e^x - 5 = x$ have?

(i)

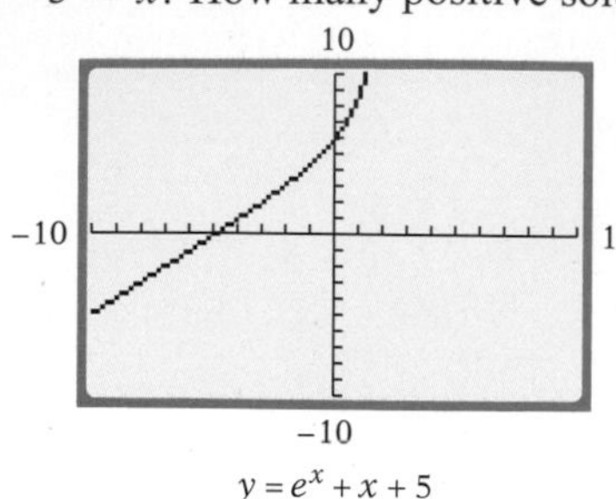

$y = e^x + x + 5$

(ii)

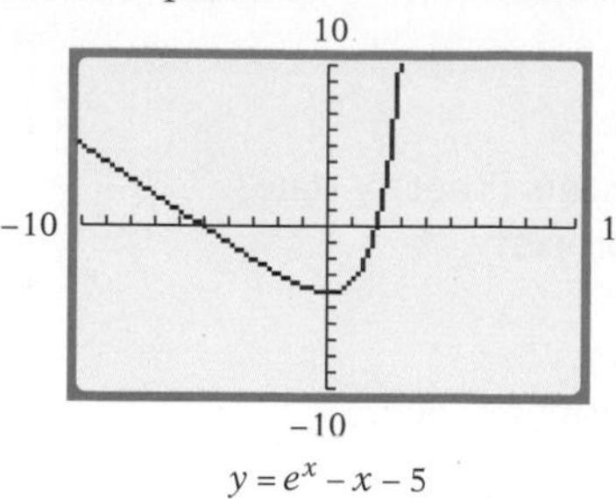

$y = e^x - x - 5$

(iii)

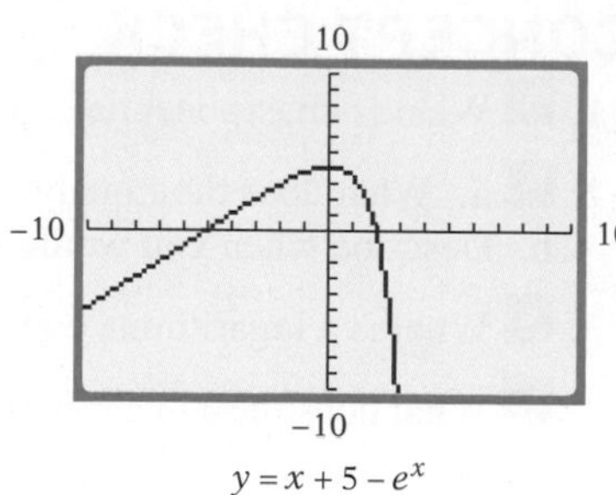

$y = x + 5 - e^x$

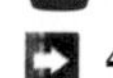 Solve for x by graphing. Round to the nearest hundredth.

45. $3^x = 2$

46. $5^x = 9$

47. $2^x = 2x + 4$

48. $3^x = -x - 1$

49. $e^x = -2x - 2$

50. $e^x = 3x + 4$

2 Solve logarithmic equations (See pages 561–562.)

GETTING READY

51. A logarithmic equation in the form $\log_b$ (expression) $= n$, where n is a real number, can be solved by rewriting the equation in __?__.

52. A logarithmic equation in the form $\log_b$(expression) $= \log_b$(expression) can be solved by using the __?__ Property of Logarithms to set the two expressions __?__ to each other.

53. Look at the equation in each exercise. State whether you will use the method described in Exercise 51 or the method described in Exercises 52 to solve the equation.
a. Exercise 63 **b.** Exercise 65 **c.** Exercise 67 **d.** Exercise 69

54. Use the graphs shown below. Which equation has no solution, $\log(x - 1) = -2x + 3$ or $\log(x - 1) = 2x - 3$?

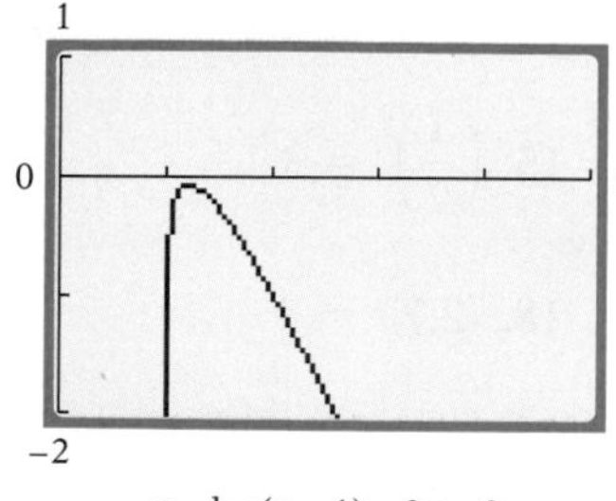

$y = \log(x - 1) - 2x + 3$

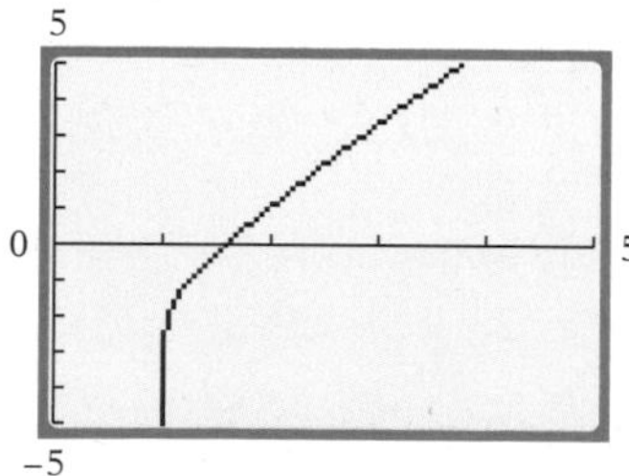

$y = \log(x - 1) + 2x - 3$

Solve for x.

55. $\log_3(x + 1) = 2$

56. $\log_5(x - 1) = 1$

57. $\log_2(2x - 3) = 3$

58. $\log_4(3x + 1) = 2$

59. $\log_2(x^2 + 2x) = 3$

60. $\log_3(x^2 + 6x) = 3$

61. $\log_5\left(\dfrac{2x}{x - 1}\right) = 1$

62. $\log_6\left(\dfrac{3x}{x + 1}\right) = 1$

63. $\log_7 x = \log_7(1 - x)$

64. $\dfrac{3}{4}\log x = 3$

65. $\dfrac{2}{3}\log x = 6$

66. $\log(x - 2) - \log x = 3$

67. $\log_2(x - 3) + \log_2(x + 4) = 3$

68. $\log x - 2 = \log(x - 4)$

69. $\log_3 x + \log_3(x - 1) = \log_3 6$

70. $\log_4 x + \log_4(x - 2) = \log_4 15$

71. $\log_2(8x) - \log_2(x^2 - 1) = \log_2 3$

72. $\log_5(3x) - \log_5(x^2 - 1) = \log_5 2$

73. $\log_9 x + \log_9(2x - 3) = \log_9 2$

74. $\log_6 x + \log_6(3x - 5) = \log_6 2$

75. $\log_8(6x) = \log_8 2 + \log_8(x - 4)$

76. $\log_7(5x) = \log_7 3 + \log_7(2x + 1)$

77. $\log_9(7x) = \log_9 2 + \log_9(x^2 - 2)$

78. $\log_3 x = \log_3 2 + \log_3(x^2 - 3)$

79. $\log(x^2 + 3) - \log(x + 1) = \log 5$

80. $\log(x + 3) + \log(2x - 4) = \log 3$

Solve for x by graphing. Round to the nearest hundredth.

81. $\log x = -x + 2$

82. $\log x = -2x$

83. $\log(2x - 1) = -x + 3$

84. $\log(x + 4) = -2x + 1$

85. $\ln(x + 2) = x^2 - 3$

86. $\ln x = -x^2 + 1$

APPLYING CONCEPTS

Solve for x. Round to the nearest ten-thousandth.

87. $8^{\frac{x}{2}} = 6$

88. $4^{\frac{x}{3}} = 2$

89. $5^{\frac{3x}{2}} = 7$

90. $9^{\frac{2x}{3}} = 8$

91. $1.2^{\frac{x}{2} - 1} = 1.4$

92. $5.6^{\frac{x}{3} + 1} = 7.8$

Solve.

93. If $4^x = 7$, find the value of $2^{(6x+3)}$.

94. The following "proof" appears to show that $0.04 < 0.008$. Explain the error.

$$2 < 3$$
$$2 \log 0.2 < 3 \log 0.2$$
$$\log(0.2)^2 < \log(0.2)^3$$
$$(0.2)^2 < (0.2)^3$$
$$0.04 < 0.008$$

PROJECTS OR GROUP ACTIVITIES

95. **Physics** A model for the distance s (in feet) that an object experiencing air resistance will fall in t seconds is given by $s = 312.5 \ln\left(\frac{e^{0.32t} + e^{-0.32t}}{2}\right)$.

a. Graph this equation. *Suggestion:* Use Xmin = 0, Xmax = 4.5, Xscl = 0.5, Ymin = 0, Ymax = 140, and Yscl = 20.

b. Determine, to the nearest hundredth of a second, the time it takes the object to fall 100 ft.

96. **Physics** A model for the distance s (in feet) that an object experiencing air resistance will fall in t seconds is given by $s = 78 \ln\left(\frac{e^{0.8t} + e^{-0.8t}}{2}\right)$.

a. Graph this equation. *Suggestion:* Use Xmin = 0, Xmax = 4.5, Xscl = 0.5, Ymin = 0, Ymax = 140, and Yscl = 20.

b. Determine, to the nearest hundredth of a second, the time it takes the object to fall 125 ft.

10.5 Applications of Exponential and Logarithmic Functions

OBJECTIVE 1 Application problems

A biologist places one single-celled bacterium in a culture, and each hour that particular species of bacterium divides into two bacteria. After 1 h, there will be two bacteria. After 2 h, each of the two bacteria will divide and there will be four bacteria. After 3 h, each of the four bacteria will divide and there will be eight bacteria.

Time, t	Number of Bacteria, N
0	1
1	2
2	4
3	8
4	16

The table at the left shows the number of bacteria in the culture after various intervals of time t, in hours. Values in this table could also be found by using the exponential equation $N = 2^t$.

The equation $N = 2^t$ is an example of an **exponential growth equation.** In general, any equation that can be written in the form $\mathbf{A = A_0 b^{kt}}$, where A is the size at time t, A_0 is the initial size, $b > 1$, and k is a positive real number, is an exponential growth equation. These equations are important not only in population growth studies but also in physics, chemistry, psychology, and economics.

Recall that interest is the amount of money one pays (or receives) when borrowing (or investing) money. **Compound interest** is interest that is computed not only on the original principal but also on the interest already earned. The compound interest formula is an exponential growth equation.

The **compound interest formula** is $\mathbf{P = A(1 + i)^n}$, where A is the original value of an investment, i is the interest rate per compounding period, n is the total number of compounding periods, and P is the value of the investment after n periods.

Focus on finding compound interest

An investment broker deposits \$1000 into an account that earns 12% annual interest compounded quarterly. What is the value of the investment after 2 years? Round to the nearest dollar.

Find i, the interest rate per quarter. The quarterly rate is the annual rate divided by 4, the number of quarters in 1 year.

$$i = \frac{12\%}{4} = \frac{0.12}{4} = 0.03$$

Find n, the number of compounding periods. The investment is compounded quarterly, 4 times a year, for 2 years.

$$n = 4 \cdot 2 = 8$$

Use the compound interest formula. $P = A(1 + i)^n$

Replace A, i, and n by their values. $P = 1000(1 + 0.03)^8$

Solve for P. $P \approx 1267$

The value of the investment after 2 years is approximately \$1267.

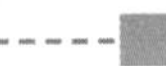

Exponential decay can also be modeled by an exponential equation. One of the most common illustrations of exponential decay is the decay of a radioactive substance.

Time, t	Amount, A
0	10
5	5
10	2.5
15	1.25
20	0.625

A radioactive isotope of cobalt has a half-life of approximately 5 years. This means that one-half of any given amount of the cobalt isotope will disintegrate in 5 years. Suppose you begin with 10 mg of a cobalt isotope. The table at the left indicates the amount of the initial 10 mg of cobalt isotope that remains after various intervals of time t, in years. Values in this table could also be found by using the exponential equation $A = 10\left(\frac{1}{2}\right)^{\frac{t}{5}}$.

The equation $A = 10\left(\frac{1}{2}\right)^{\frac{t}{5}}$ is an example of an **exponential decay equation.**

Compare this equation to the exponential growth equation, and note that for exponential growth, the base of the exponential expression is greater than 1, whereas for exponential decay, the base is between 0 and 1.

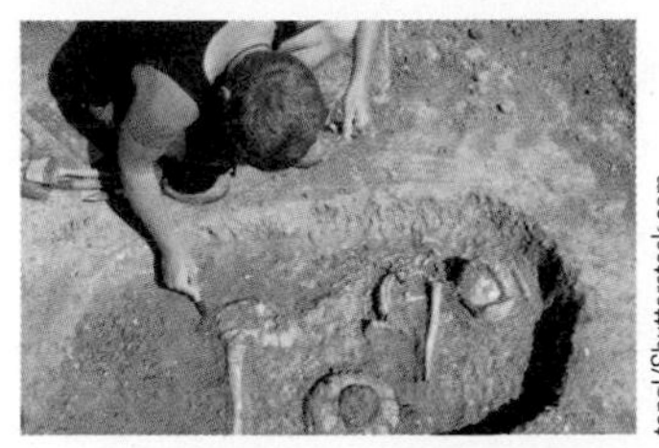
topal/Shutterstock.com

A method by which an archeologist can measure the age of a bone is based on the exponential decay half-life equation $A = A_0\left(\frac{1}{2}\right)^{\frac{t}{k}}$, where A is the amount of material remaining after time t, k is the half-life of the material, and A_0 is the original amount of material present.

Focus on using an exponential decay equation

Carbon dating is based on a radioactive isotope of carbon called carbon-14, which has a half-life of approximately 5570 years. A bone that originally contained 100 mg of carbon-14 now has 70 mg of carbon-14. What is the approximate age of the bone? Round to the nearest year.

Use the exponential decay equation.
$$A = A_0\left(\frac{1}{2}\right)^{\frac{t}{k}}$$

Replace A_0, A, and k by their given values, and solve for t.
$$70 = 100\left(\frac{1}{2}\right)^{\frac{t}{5570}}$$

Divide each side of the equation by 100.
$$\frac{70}{100} = \left(\frac{1}{2}\right)^{\frac{t}{5570}}$$

Take the common logarithm of each side of the equation.
$$\log\frac{70}{100} = \log\left(\frac{1}{2}\right)^{\frac{t}{5570}}$$

Use the Power Property of Logarithms.
$$\log\frac{70}{100} = \frac{t}{5570}\log\frac{1}{2}$$

Solve for t.
$$\frac{5570\log\frac{70}{100}}{\log\frac{1}{2}} = t$$

The age of the bone is approximately 2866 years.
$$2866 \approx t$$

EXAMPLE 1 Molybdenum-99 is a radioactive isotope used in medicine. An original 20-microgram sample of molybdenum-99 decays to 18 micrograms in 10 h. Find the half-life of molybdenum-99. Round to the nearest hour.

Strategy A_0, the original amount, is 20 micrograms. A, the final amount, is 18 micrograms. The time is 10 h. To find the half-life, solve the exponential decay equation $A = A_0\left(\frac{1}{2}\right)^{\frac{t}{k}}$ for the half-life, k.

Solution

$$A = A_0\left(\frac{1}{2}\right)^{\frac{t}{k}}$$ • Use the exponential decay equation.

$$18 = 20\left(\frac{1}{2}\right)^{\frac{10}{k}}$$ • $A_0 = 20, A = 18, t = 10$

$$\frac{18}{20} = \left(\frac{1}{2}\right)^{\frac{10}{k}}$$ • Solve for k.

$$\log\frac{18}{20} = \log\left(\frac{1}{2}\right)^{\frac{10}{k}}$$

$$\log\frac{18}{20} = \frac{10}{k}\log\frac{1}{2}$$

$$k = \frac{10\log\frac{1}{2}}{\log\frac{18}{20}}$$

$$k \approx 65.8$$

The half-life of molybdenum-99 is about 66 h.

Problem 1 The number of words per minute that a student can type will increase with practice and can be approximated by the equation $N = 100[1 - (0.9)^t]$, where N is the number of words typed per minute after t days of practice. In how many days will the student be able to type 60 words per minute? Round to the nearest whole number of days.

Solution See page S33.

Try Exercise 11, page 571.

Point of Interest

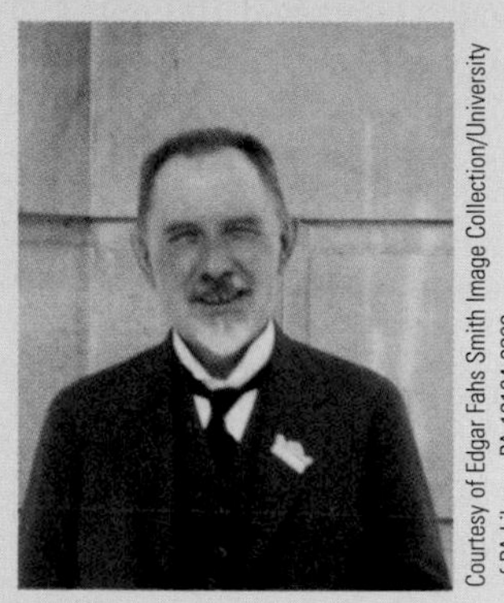

Courtesy of Edgar Fahs Smith Image Collection/University of PA Library, PA 19104-6206

Søren Sørensen

The pH scale was created by the Danish biochemist Søren Sørensen in 1909 to measure the acidity of water used in the brewing of beer. pH is an abbreviation for *pondus hydrogenii*, which translates as "potential hydrogen."

The first applications of logarithms (and the main reason why they were developed) were to reduce computational drudgery. Today, with the widespread use of calculators and computers, the computational uses of logarithms have diminished. However, a number of other applications of logarithms have emerged.

A chemist measures the acidity or alkalinity of a solution by the formula $\mathbf{pH = -\log(H^+)}$, where H^+ is the concentration of hydrogen ions in the solution. A neutral solution such as distilled water has a pH of 7, acids have a pH less than 7, and alkaline solutions (also called basic solutions) have a pH greater than 7.

Focus on finding the pH of a solution

Find the pH of vinegar for which $H^+ = 1.26 \times 10^{-3}$. Round to the nearest tenth.

Use the pH equation. $\quad \text{pH} = -\log(H^+)$

$H^+ = 1.26 \times 10^{-3}$

$$\begin{aligned} &= -\log(1.26 \times 10^{-3}) \\ &= -(\log 1.26 + \log 10^{-3}) \\ &\approx -[0.1004 + (-3)] = 2.8996 \end{aligned}$$

The pH of vinegar is approximately 2.9.

Logarithmic functions are used to scale very large or very small numbers into numbers that are easier to comprehend. For instance, the *Richter scale* magnitude of an earthquake uses a logarithmic function to convert the intensity of shock waves I into a number M, which for most earthquakes is in the range of 0 to 10. The intensity I of an earthquake is often given in terms of the constant I_0, where I_0 is the intensity of the smallest earthquake, called a **zero-level earthquake,** that can be measured on a seismograph near the earthquake's epicenter. An earthquake with an intensity I_0 has a Richter scale magnitude of $\boldsymbol{M = \log \frac{I}{I_0}}$, where I_0 is the measure of a zero-level earthquake.

Focus on finding the magnitude of an earthquake

Take Note

Note that we do not need to know the value of I_0 in order to determine the Richter scale magnitude of the quake.

The earthquake with the largest intensity ever recorded in the continental United States since 1900 occurred in San Francisco, California, in 1906. The intensity I of the quake was approximately $50{,}118{,}000I_0$. Find the Richter scale magnitude of the earthquake. Round to the nearest tenth.

$$M = \log \frac{I}{I_0}$$

$I = 50{,}118{,}000I_0$ $\qquad M = \log \dfrac{50{,}118{,}000\, I_0}{I_0}$

Divide the numerator and denominator by I_0. $\qquad M = \log 50{,}118{,}000$

Evaluate log 50,118,000. $\qquad M \approx 7.7$

The San Francisco earthquake had a Richter scale magnitude of 7.7.

If you know the Richter scale magnitude of an earthquake, you can determine the intensity of the earthquake.

Focus on finding the intensity of an earthquake

Point of Interest

© Bettmann/Corbis

The Richter scale was created by seismologist Charles F. Richter in 1935. Note that a tenfold increase in the intensity level of an earthquake increases the Richter scale magnitude of the earthquake by only 1.

In 1964, Prince William Sound, Alaska, experienced what was then the earthquake with the largest magnitude ever to have occurred in the United States. The magnitude of the earthquake was 9.2 on the Richter scale. Find the intensity of this earthquake in terms of I_0. Round to the nearest thousand.

Replace M in $M = \log \frac{I}{I_0}$ by 9.2. $\qquad 9.2 = \log \dfrac{I}{I_0}$

Write the equation in its equivalent exponential form. $\qquad 10^{9.2} = \dfrac{I}{I_0}$

Multiply each side of the equation by I_0. $\qquad 10^{9.2}I_0 = I$

Evaluate $10^{9.2}$. $\qquad 1{,}584{,}893{,}000I_0 \approx I$

The Prince William Sound earthquake had an intensity that was approximately 1,584,893,000 times the intensity of a zero-level earthquake.

The percent of light that will pass through a substance is given by the equation $\boldsymbol{\log P = -kd}$, where P is the percent of light, as a decimal, passing through the substance, k is a constant that depends on the substance, and d is the thickness of the substance in centimeters.

Focus on finding the percent of light passing through a substance

Find the percent of light that will pass through translucent glass for which $k = 0.4$ and $d = 0.5$ cm.

$$\log P = -kd$$

Replace k and d in the equation by their given values, and solve for P.

$$\log P = -(0.4)(0.5)$$
$$\log P = -0.2$$

Use the relationship between the logarithmic and exponential functions.

$$P = 10^{-0.2}$$
$$P \approx 0.6310$$

Approximately 63.1% of the light will pass through the glass.

EXAMPLE 2 Find the hydrogen ion concentration H^+ of orange juice that has a pH of 3.6.

Strategy To find the hydrogen ion concentration, replace pH by 3.6 in the equation $pH = -\log(H^+)$ and solve for H^+.

Solution

$$pH = -\log(H^+)$$
$$3.6 = -\log(H^+)$$
$$-3.6 = \log(H^+)$$
$$10^{-3.6} = H^+$$
$$0.00025 \approx H^+$$

The hydrogen ion concentration is approximately 0.00025.

Problem 2 On September 3, 2000, an earthquake measuring 5.2 on the Richter scale struck the Napa Valley, 50 mi north of San Francisco. Find the intensity of the quake in terms of I_0.

Solution See page S33.

Try Exercise 21, page 572.

10.5 Exercises

CONCEPT CHECK

1. What is meant by the phrase *exponential decay*?
2. Explain how compound interest differs from simple interest. Is compound interest an example of exponential decay or exponential growth?
3. State whether each equation is an exponential growth equation, an exponential decay equation, or neither.

   a. $y = 3\left(\frac{3}{4}\right)^x$ **b.** $y = 3x^{\frac{3}{4}}$ **c.** $y = 2e^{0.34x}$

   d. $y = 2^{x+1}$ **e.** $y = 50x^2$ **f.** $y = 2^{-3x}$
4. What is the meaning of the half-life of a radioactive isotope?

1 Application problems (See pages 566–570.)

GETTING READY

Replace the question marks in Exercises 5 and 6 with the correct number from the problem situation or with the word "unknown."

5. Problem Situation: You invest \$3000 in an account that earns 3% annual interest compounded monthly. In approximately how many years will your investment be worth \$4000?

In the formula $P = A(1 + i)^n$, $P =$ __?__, $A =$ __?__, $i =$ __?__, and $n =$ __?__.

6. Problem Situation: A sample of a radioactive isotope with a half-life of 4 h contains 40 mg. How much of the radioactive isotope is present after 24 h?

In the formula $A = A_0\left(\frac{1}{2}\right)^{\frac{t}{k}}$, $A =$ __?__, $A_0 =$ __?__, $t =$ __?__, and $k =$ __?__.

Finance For Exercises 7 to 10, use the compound interest formula $P = A(1 + i)^n$, where A is the original value of an investment, i is the interest rate per compounding period, n is the total number of compounding periods, and P is the value of the investment after n periods.

7. An investment broker deposits \$1000 into an account that earns 8% annual interest compounded quarterly. What is the value of the investment after 2 years? Round to the nearest dollar.

8. A financial advisor recommends that a client deposit \$2500 into a fund that earns 7.5% annual interest compounded monthly. What will be the value of the investment after 3 years? Round to the nearest cent.

9. To save for college tuition, the parents of a preschooler invest \$5000 in a bond fund that earns 6% annual interest compounded monthly. In approximately how many years will the investment be worth \$15,000? Round to the nearest whole number.

10. A hospital administrator deposits \$10,000 into an account that earns 9% annual interest compounded monthly. In approximately how many years will the investment be worth \$15,000? Round to the nearest whole number.

Biology For Exercises 11 to 14, use the exponential decay equation $A = A_0\left(\frac{1}{2}\right)^{\frac{t}{k}}$, where A is the amount of a radioactive material present after time t, k is the half-life of the radioactive material, and A_0 is the original amount of radioactive substance. Round to the nearest tenth.

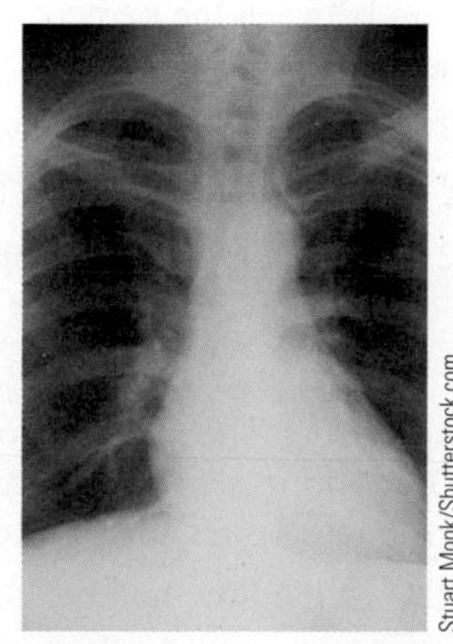
Stuart Monk/Shutterstock.com

11. An isotope of technetium is used to prepare images of internal body organs. This isotope has a half-life of approximately 6 h. A patient is injected with 30 mg of this isotope.

a. What will be the amount of technetium in the patient after 3 h?

b. How long (in hours) will it take for the amount of technetium in the patient to reach 20 mg?

12. Iodine-131 is an isotope that is used to study the functioning of the thyroid gland. This isotope has a half-life of approximately 8 days. A patient is given an injection that contains 8 micrograms of iodine-131.

a. What will be the amount of iodine in the patient after 5 days?

b. How long (in days) will it take for the amount of iodine in the patient to reach 5 micrograms?

13. A sample of promethium-147 (used in some luminous paints) contains 25 mg. One year later, the sample contains 18.95 mg. What is the half-life of promethium-147, in years? Round to the nearest tenth of a year.

14. Francium-223 is a very rare radioactive isotope discovered in 1939 by Marguerite Percy. A 3-microgram sample of francium-223 decays to 2.54 micrograms in 5 min. What is the half-life of francium-223, in minutes? Round to the nearest tenth of a minute.

15. After 6 years, $\frac{1}{8}$ of the original amount of a radioactive isotope is left. What is the half-life of the isotope?

16. After 12 h, $\frac{1}{16}$ of the original amount of a radioactive isotope is left. What is the half-life of the isotope?

17. **Atmospheric Pressure** The atmospheric pressure changes as you rise above Earth's surface. At an altitude of h kilometers, where $0 < h < 80$, the pressure P in newtons per square centimeter is approximately modeled by the equation $P(h) = 10.13e^{-0.116h}$.
   a. What is the approximate pressure at 40 km above Earth?
   b. What is the approximate pressure on Earth's surface?
   c. Does atmospheric pressure increase or decrease as you rise above Earth's surface?

18. **Wind Power** Read the article at the right. The data given in the 2009 Global Wind Report for worldwide installed wind power capacity for the years 1996 to 2009 can be modeled by the equation $y = 4.982\,(1.282^x)$, where x is the number of years since 1995 and y is the wind power capacity in gigawatts.
   a. Show that the model fits the data given in the article for the years 1996, 2009, and 2010.
   b. According to the model, during what year will the world capacity first reach 300 GW of installed power?

In the News

GWEC Issues Annual Global Wind Report

Data from the Global Wind Energy Council's 2009 Global Wind Report show that worldwide installed wind power capacity continues its rapid growth, having increased from about 6 gigawatts (GW) of power in 1996 to about 160 GW of power in 2009. The report projects that the 200-gigawatt mark will be passed in 2010.

Source: Global Wind Energy Council

Chemistry For Exercises 19 and 20, use the equation $\text{pH} = -\log(\text{H}^+)$, where H^+ is the hydrogen ion concentration of a solution. Round to the nearest hundredth.

19. Find the pH of milk, for which the hydrogen ion concentration is 3.97×10^{-7}.

20. Find the pH of a baking soda solution for which the hydrogen ion concentration is 3.98×10^{-9}.

Optics For Exercises 21 and 22, use the equation $\log P = -kd$, which gives the relationship between the percent P, as a decimal, of light passing through a substance of thickness d.

21. The value of k for a swimming pool is approximately 0.05. At what depth, in meters, will the percent of light be 75% of the light at the surface of the pool?

22. The constant k for a piece of blue stained glass is 20. What percent of light will pass through a piece of this glass that is 0.005 m thick?

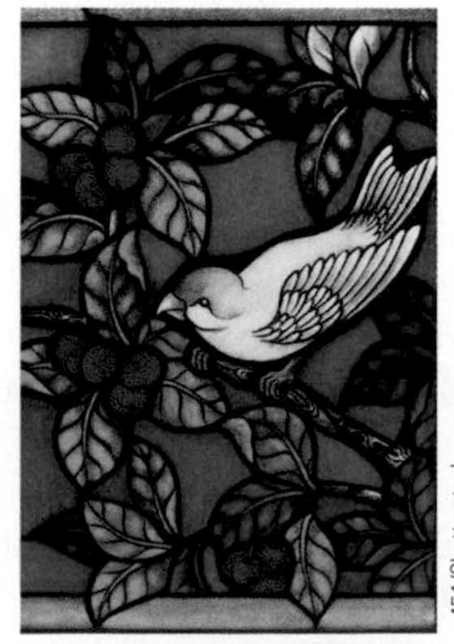
a454/Shutterstock.com

Acoustics The number of decibels D of a sound can be given by the equation $D = 10(\log I + 16)$, where I is the power of a sound measured in watts. Use this equation for Exercises 23 and 24. Round to the nearest whole number.

23. Find the number of decibels of normal conversation. The power of the sound of normal conversation is approximately 3.2×10^{-10} watts.

24. The loudest sound made by any animal is made by the blue whale and can be heard from over 500 mi away. The power of the sound is 630 watts. Find the number of decibels of sound emitted by the blue whale.

25. ***E. coli*** The actual equation for the population growth of *E. coli* bacteria mentioned in the novel *The Andromeda Strain* by Michael Crichton is given by $y = 2^{3t}$, where y is the size of the population in t hours. In the novel, Crichton says, "it can be shown that in a single day, one cell of *E. coli* could produce a super-colony equal in size and weight to the entire planet Earth." The mass of Earth is 5.98×10^{27} g, and the mass of one *E. coli* bacterium is approximately 6.7×10^{-15} g. Assuming that a colony of *E. coli* bacteria continues to grow as predicted by the equation, in how many hours would the mass of the colony be equal to the mass of Earth? Round to the nearest hour. Is this more or less time than predicted in the novel?

26. **Electricity** The current, in amperes, in an electric circuit is given by $I = 6(1 - e^{-2t})$, where t is the time in seconds. Using this model, after how many seconds is the current 4 amperes? Round to the nearest hundredth.

27. **Medicine** The intensity I of an x-ray after it has passed through a material that is x centimeters thick is given by $I = I_0 e^{-kx}$, where I_0 is the initial intensity and k is a number that depends on the material. The constant k for copper is 3.2. Find the thickness of copper that is needed so that the intensity of an x-ray after passing through the copper is 25% of the original intensity. Round to the nearest tenth.

28. **Energy** One model for the time it will take for the world's oil supply to be depleted is given by the equation $T = 14.29 \ln(0.00411r + 1)$, where r is the estimated world oil reserves in billions of barrels and T is the time before that amount of oil is depleted. Use this equation to determine how many barrels of oil are necessary to last 20 years. Round to the nearest tenth.

Geology For Exercises 29 to 32, use the Richter scale equation $M = \log \frac{I}{I_0}$, where M is the magnitude of an earthquake, I is the intensity of the shock waves, and I_0 is the measure of a zero-level earthquake.

29. The largest earthquake ever recorded occurred on May 22, 1960, off the coast of Chile. The earthquake had an intensity of approximately $I = 3{,}162{,}277{,}000I_0$. Find the Richter scale magnitude of the earthquake. Round to the nearest tenth.

30. One of the largest earthquakes ever recorded in the United States occurred on February 3, 1965, in the Rat Islands of Alaska. The earthquake had an intensity of approximately $I = 501{,}187{,}000I_0$. Find the Richter scale magnitude of the earthquake. Round to the nearest tenth.

31. Read the article at the right about the April 13, 2010, earthquake in Qinghai province, China. Find the intensity of this earthquake, in terms of I_0, for each of the magnitudes reported in the article. Round to the nearest thousand.

32. The magnitude of the Sumatra–Andaman earthquake that occurred on December 26, 2004, off the northern coast of Sumatra was 9.15 on the Richter scale. Find the intensity of the earthquake in terms of I_0.

In the News

Earthquake Strikes Remote Area of China

Rescue workers continue to look for survivors of a severe earthquake that shook a remote area of western China. The China Earthquake Commission reported the quake at magnitude 7.1 on the Richter scale, while the U.S. Geological Survey recorded the quake at magnitude 6.9.

Source: www.msnbc.msn.com

Geology Shown at the right is a **seismogram,** which is used to measure the magnitude of an earthquake. The magnitude is determined by the amplitude A of the wave and the time t between the occurrence of two types of waves, called primary waves and secondary waves. As you can see on the graph, a primary wave is abbreviated p-wave, and a secondary wave is abbreviated s-wave. The amplitude A of a wave is one-half the difference between its highest and lowest points. For this graph, A is 23 mm. The equation of the graph is $M = \log A + 3 \log 8t - 2.92$.

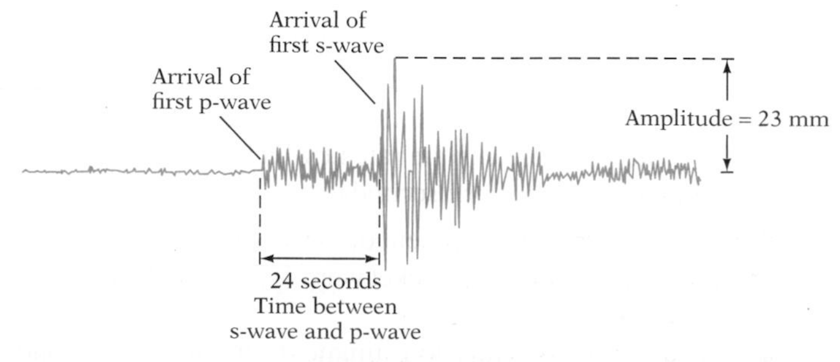

CHAPTER 10 Test

1. Evaluate $f(x) = \left(\frac{3}{4}\right)^x$ at $x = 0$.

2. Evaluate $f(x) = 4^{x-1}$ at $x = -2$.

3. Evaluate: $\log_4 64$

4. Solve for x: $\log_4 x = -2$

5. Write $\log_6 \sqrt[3]{x^2y^5}$ in expanded form.

6. Write $\frac{1}{2}(\log_5 x - \log_5 y)$ as a single logarithm with a coefficient of 1.

7. Solve for x: $\log_6 x + \log_6(x - 1) = 1$

8. Evaluate $f(x) = 3^{x+1}$ at $x = -2$.

9. Solve $3^x = 17$ for x. Round to the nearest ten-thousandth.

10. Solve for x: $\log_2 x + 3 = \log_2(x^2 - 20)$

11. Solve for x: $5^{6x-2} = 5^{3x+7}$

12. Solve for x: $4^x = 2^{3x+4}$

13. Solve for x: $\log(2x + 1) + \log x = \log 6$

14. Graph: $f(x) = 2^x - 1$

15. Graph: $f(x) = 2^x + 2$

16. Graph: $f(x) = \log_2(3x)$

17. Graph: $f(x) = \log_3(x + 1)$

18. Graph $f(x) = 3 - 2^x$ and approximate the zeros of f to the nearest tenth.

19. Investments Find the value of \$10,000 invested for 6 years at 7.5% compounded monthly. Use the compound interest formula $P = A(1 + i)^n$, where A is the original value of the investment, i is the interest rate per compounding period, and n is the number of compounding periods. Round to the nearest dollar.

20. Radioactivity Use the exponential decay equation $A = A_0\left(\frac{1}{2}\right)^{\frac{t}{k}}$, where A is the amount of a radioactive material present after time t, k is the half-life, and A_0 is the original amount of radioactive material, to find the half-life of a material that decays from 40 mg to 30 mg in 10 h. Round to the nearest whole number.

Cumulative Review Exercises

1. Solve: $4 - 2[x - 3(2 - 3x) - 4x] = 2x$

2. Solve $S = 2WH + 2WL + 2LH$ for L.

3. Solve: $|2x - 5| \leq 3$

4. Factor: $4x^2 + 7x + 3$

5. Solve: $x^2 + 4x - 5 \leq 0$

6. Simplify: $\dfrac{1 - \frac{5}{x} + \frac{6}{x^2}}{1 + \frac{1}{x} - \frac{6}{x^2}}$

7. Simplify: $\dfrac{\sqrt{xy}}{\sqrt{x} - \sqrt{y}}$

8. Simplify: $y\sqrt{18x^5y^4} - x\sqrt{98x^3y^6}$

9. Divide: $\dfrac{i}{2 - i}$

10. Find the equation of the line that contains the point $P(2, -2)$ and is parallel to the line with equation $2x - y = 5$.

11. Write a quadratic equation that has integer coefficients and has as solutions $\frac{1}{3}$ and -3.

12. Solve: $x^2 - 4x - 6 = 0$

13. Find the range of $f(x) = x^2 - 3x - 4$ if the domain is $\{-1, 0, 1, 2, 3\}$.

14. Given $f(x) = x^2 + 2x + 1$ and $g(x) = 2x - 3$, find $f[g(0)]$.

15. Solve by the addition method:

$$\begin{aligned} 3x - y + z &= 3 \\ x + y + 4z &= 7 \\ 3x - 2y + 3z &= 8 \end{aligned}$$

16. Solve: $\begin{aligned} y &= -2x - 3 \\ y &= 2x - 1 \end{aligned}$

17. Evaluate $f(x) = 3^{-x+1}$ at $x = -4$.

18. Solve for x: $\log_4 x = 3$

19. Solve for x: $2^{3x+2} = 4^{x+5}$

20. Solve for x: $\log x + \log(3x + 2) = \log 5$

21. Graph: $\{x | x < 0\} \cap \{x | x > -4\}$

22. Graph the solution set of $\dfrac{x + 2}{x - 1} \geq 0$.

23. Graph: $y = -x^2 - 2x + 3$

24. Graph: $f(x) = |x| - 2$

25. Graph: $f(x) = \left(\frac{1}{2}\right)^x - 1$

26. Graph: $f(x) = \log_2 x + 1$

27. Metallurgy An alloy containing 25% tin is mixed with an alloy containing 50% tin. How much of each were used to make 2000 lb of an alloy containing 40% tin?

28. Compensation An account executive earns \$1000 per month plus 8% commission on the amount of sales. The executive's goal is to earn a minimum of \$7000 per month. What amount of sales will enable the executive to earn \$7000 or more per month?

29. **Education** The table at the right shows the actual or projected numbers of Florida high school graduates, in thousands, for selected years. If the projections are accurate, what will be the average annual rate of change in the number of Florida high school graduates for the years 2008 to 2014? Round to the nearest whole number.

Year	High School Graduates
2008	142,102
2009	147,735
2010	145,793
2011	148,147
2012	142,695
2013	146,193
2014	146,942

Source: Florida Dept. of Education

30. Work Problem A new printer can print checks three times faster than an older printer. The old printer can print the checks in 30 min. How long would it take to print the checks with both printers operating?

31. Chemistry For a constant temperature, the pressure (P) of a gas varies inversely as the volume (V). If the pressure is 50 lb/in^2 when the volume is 250 ft^3, find the pressure when the volume is 25 ft^3.

32. Purchasing A contractor buys 45 yd of nylon carpet and 30 yd of wool carpet for \$2340. A second purchase, at the same prices, includes 25 yd of nylon carpet and 80 yd of wool carpet for \$2820. Find the cost per yard of the wool carpet.

33. Investments An investor deposits \$10,000 into an account that earns 9% interest compounded monthly. What is the value of the investment after 5 years? Use the compound interest formula $P = A(1 + i)^n$, where A is the original value of an investment, i is the interest rate per compounding period, n is the total number of compounding periods, and P is the value of the investment after n periods. Round to the nearest dollar.

Sequences and Series

CHAPTER 11

Focus on Success

Remember to prepare your brain for the material you will learn in this chapter. Read the list of objectives on this page. Look through the entire chapter, noting words that are in bold type. Read the rules and definitions that appear in boxes. By getting an overview of the new material, you will be building a foundation for learning it. (See Understand the Organization, page AIM-8.)

OBJECTIVES

11.1
1. Write the terms of a sequence
2. Evaluate a series

11.2
1. Find the *n*th term of an arithmetic sequence
2. Evaluate an arithmetic series
3. Application problems

11.3
1. Find the *n*th term of a geometric sequence
2. Finite geometric series
3. Infinite geometric series
4. Application problems

11.4
1. Expand $(a + b)^n$

PREP TEST

Are you ready to succeed in this chapter?
Take the Prep Test below to find out if you are ready to learn the new material.

1. Simplify:
   $[3(1) - 2] + [3(2) - 2] + [3(3) - 2]$

2. Evaluate $f(n) = \dfrac{n}{n + 2}$ for $n = 6$.

3. Evaluate $a_1 + (n - 1)d$ for $a_1 = 2$, $n = 5$, and $d = 4$.

4. Evaluate a_1r^{n-1} for $a_1 = -3$, $r = -2$, and $n = 6$.

5. Evaluate $\dfrac{a_1(1 - r^n)}{1 - r}$ for $a_1 = -2$, $r = -4$, and $n = 5$.

6. Simplify: $\dfrac{\frac{4}{10}}{1 - \frac{1}{10}}$

7. Simplify: $(x + y)^2$

8. Simplify: $(x + y)^3$

11.1 Introduction to Sequences and Series

OBJECTIVE 1 Write the terms of a sequence

An investor deposits $100 in an account that earns 10% interest compounded annually. The amount of interest earned each year can be determined by using the compound interest formula. The amount of interest earned in each of the first four years of the investment is shown below.

Year	1	2	3	4
Interest Earned	$10	$11	$12.10	$13.31

The list of numbers 10, 11, 12.10, 13.31 is called a sequence. A **sequence** is an ordered list of numbers. The list 10, 11, 12.10, 13.31 is ordered because the position of a number in this list indicates the year in which the amount of interest was earned. Each of the numbers of a sequence is called a **term of the sequence.**

Examples of other sequences are shown at the right. These sequences are separated into two groups. A **finite sequence** contains a finite number of terms. An **infinite sequence** contains an infinite number of terms.

$1, 1, 2, 3, 5, 8$
$1, 2, 3, 4, 5, 6, 7, 8$
$1, -1, 1, -1$
} Finite sequences

$1, 3, 5, 7, \ldots$
$1, \frac{1}{2}, \frac{1}{4}, \frac{1}{8}, \ldots$
$1, 1, 2, 3, 5, 8, \ldots$
} Infinite sequences

For the sequence at the right, the first term is 2, the second term is 4, the third term is 6, and the fourth term is 8.

$2, 4, 6, 8, \ldots$

A general sequence is shown at the right. The first term is a_1, the second term is a_2, the third term is a_3, and the nth term, also called the **general term** of the sequence, is a_n. Note that each term of the sequence is paired with a natural number.

$a_1, a_2, a_3, \ldots, a_n, \ldots$

Frequently, a sequence has a definite pattern that can be expressed by a formula.

Each term of the sequence shown at the right is paired with a natural number by the formula $a_n = 3n$. The first term, a_1, is 3. The second term, a_2, is 6. The third term, a_3, is 9. The nth term, a_n, is $3n$.

$a_n = 3n$

$$\begin{array}{lllll} a_1, & a_2, & a_3, & \ldots, & a_n, \ldots \\ 3(1), & 3(2), & 3(3), & \ldots, & 3(n), \ldots \\ 3, & 6, & 9, & \ldots, & 3n, \ldots \end{array}$$

EXAMPLE 1 Write the first three terms of the sequence whose nth term is given by the formula $a_n = 2n - 1$.

Solution

$a_n = 2n - 1$
$a_1 = 2(1) - 1 = 1$ • **Replace n by 1.**
$a_2 = 2(2) - 1 = 3$ • **Replace n by 2.**
$a_3 = 2(3) - 1 = 5$ • **Replace n by 3.**

The first term is 1, the second term is 3, and the third term is 5.

Problem 1 Write the first four terms of the sequence whose nth term is given by the formula $a_n = n(n + 1)$.

Solution See page S34.

Try Exercise 17, page 585.

EXAMPLE 2 Find the eighth and tenth terms of the sequence whose nth term is given by the formula $a_n = \frac{n}{n+1}$.

Solution

$$a_n = \frac{n}{n+1}$$

$$a_8 = \frac{8}{8+1} = \frac{8}{9}$$ • **Replace *n* by 8.**

$$a_{10} = \frac{10}{10+1} = \frac{10}{11}$$ • **Replace *n* by 10.**

The eighth term is $\frac{8}{9}$, and the tenth term is $\frac{10}{11}$.

Problem 2 Find the sixth and ninth terms of the sequence whose nth term is given by the formula $a_n = \frac{1}{n(n+2)}$.

Solution See page S34.

Try Exercise 31, page 585.

OBJECTIVE 2

Evaluate a series

At the beginning of this section, the sequence 10, 11, 12.10, 13.31 was shown to represent the amount of interest earned in each of 4 years of an investment.

10, 11, 12.10, 13.31

The sum of the terms of this sequence represents the total interest earned by the investment over the 4-year period.

10 + 11 + 12.10 + 13.31 = 46.41
The total interest earned over the 4-year period is $46.41.

The indicated sum of the terms of a sequence is called a **series.** Given the sequence 10, 11, 12.10, 13.31, the series 10 + 11 + 12.10 + 13.31 can be written.

S_n is used to indicate the sum of the first n terms of a sequence.

For the preceding example, the sums of the series S_1, S_2, S_3, and S_4 represent the total interest earned after 1, 2, 3, and 4 years, respectively.

$$S_1 = 10 = 10$$
$$S_2 = 10 + 11 = 21$$
$$S_3 = 10 + 11 + 12.10 = 33.10$$
$$S_4 = 10 + 11 + 12.10 + 13.31 = 46.41$$

How It's Used

Statisticians use math to make sense of large quantities of raw data. Many statistics formulas involve the use of sigma notation.

For the general sequence $a_1, a_2, a_3, \ldots, a_n$, the series S_1, S_2, S_3, and S_n are shown at the right.

$$S_1 = a_1$$
$$S_2 = a_1 + a_2$$
$$S_3 = a_1 + a_2 + a_3$$
$$S_n = a_1 + a_2 + a_3 + \cdots + a_n$$

It is convenient to represent a series in a compact form called **summation notation,** or **sigma notation.** The Greek letter sigma, Σ, is used to indicate a sum.

The first four terms of the sequence whose nth term is given by the formula $a_n = 2n$ are 2, 4, 6, 8. The corresponding series is shown at the right written in summation notation and is read "the summation from 1 to 4 of $2n$." The letter n is called the **index of the summation.**

$$\sum_{n=1}^{4} 2n$$

To write the terms of the series, replace n by the consecutive integers from 1 to 4.

$$\sum_{n=1}^{4} 2n = 2(1) + 2(2) + 2(3) + 2(4)$$

The series is $2 + 4 + 6 + 8$.

$$= 2 + 4 + 6 + 8$$

The sum of the series is 20.

$$= 20$$

Take Note

The placement of the parentheses in part (A) is important. Note that

$$\sum_{i=1}^{3} 2i - 1$$
$$= 2(1) + 2(2) + 2(3) - 1$$
$$= 2 + 4 + 6 - 1 = 11$$

This is *not* the same result that was obtained by evaluating $\sum_{i=1}^{3} (2i - 1)$.

EXAMPLE 3 Evaluate the series. **A.** $\sum_{i=1}^{3} (2i - 1)$ **B.** $\sum_{n=3}^{6} \frac{1}{2}n$

Solution **A.** $\sum_{i=1}^{3} (2i - 1)$

$$= [2(1) - 1] + [2(2) - 1] + [2(3) - 1]$$
- **Replace i by 1, 2, and 3.**

$$= 1 + 3 + 5$$
- **Write the series.**

$$= 9$$
- **Find the sum of the series.**

B. $\sum_{n=3}^{6} \frac{1}{2}n$

$$= \frac{1}{2}(3) + \frac{1}{2}(4) + \frac{1}{2}(5) + \frac{1}{2}(6)$$
- **Replace n by 3, 4, 5, and 6.**

$$= \frac{3}{2} + 2 + \frac{5}{2} + 3$$
- **Write the series.**

$$= 9$$
- **Find the sum of the series.**

Problem 3 Evaluate the series. **A.** $\sum_{n=1}^{4} (7 - n)$ **B.** $\sum_{i=3}^{6} (i^2 - 2)$

Solution See page S34.

Try Exercise 49, page 586.

EXAMPLE 4 Write $\sum_{i=1}^{5} x^i$ in expanded form.

Solution

$$\sum_{i=1}^{5} x^i = x + x^2 + x^3 + x^4 + x^5$$
- **This is a variable series. Replace i by 1, 2, 3, 4, and 5.**

Problem 4 Write $\sum_{n=1}^{5} nx$ in expanded form.

Solution See page S34.

Try Exercise 67, page 586.

11.1 Exercises

CONCEPT CHECK

1. What is a sequence?
2. What is the difference between a finite sequence and an infinite sequence?
3. What is a series?
4. Consider the series written in sigma notation as $\sum_{n=1}^{7} a_n$. What is the index of summation? What does a_n represent? How many terms are in the sum?

1 Write the terms of a sequence (See pages 582–583.)

GETTING READY

5. The third term of the sequence 2, 5, 8, 11, 14, . . . is ___?___.
6. The fourth term of the sequence 1, 2, 4, 8, 16, 32, . . . is ___?___.
7. The notation a_1 is used to represent the ___?___ term of a sequence. The notation a_2 is used to represent the ___?___ term of a sequence. The notation a_n is used to represent the ___?___ term of a sequence.
8. To find the first term of the sequence given by the formula $a_n = 2n - 5$, replace n in the formula by 1: $a_1 = 2(\underline{\ ?\ }) - 5 = \underline{\ ?\ }$.

   To find the second term, replace n by ___?___:

   $a_{\underline{?}} = 2(\underline{\ ?\ }) - 5 = \underline{\ ?\ }$.

Write the first four terms of the sequence whose nth term is given by the formula.

9. $a_n = n + 1$
10. $a_n = n - 1$
11. $a_n = 2n + 1$
12. $a_n = 3n - 1$
13. $a_n = 2 - 2n$
14. $a_n = 1 - 2n$
15. $a_n = 2^n$
16. $a_n = 3^n$
17. $a_n = n^2 + 1$
18. $a_n = n^2 - 1$
19. $a_n = \dfrac{n}{n^2 + 1}$
20. $a_n = \dfrac{n^2 - 1}{n}$
21. $a_n = n - \dfrac{1}{n}$
22. $a_n = n^2 - \dfrac{1}{n}$
23. $a_n = (-1)^{n+1}n$
24. $a_n = \dfrac{(-1)^{n+1}}{n + 1}$
25. $a_n = \dfrac{(-1)^{n+1}}{n^2 + 1}$
26. $a_n = (-1)^n(n^2 + 2n + 1)$
27. $a_n = (-1)^n 2^n$
28. $a_n = \dfrac{1}{3}n^3 + 1$
29. $a_n = 2\left(\dfrac{1}{3}\right)^{n+1}$

Find the indicated term of the sequence whose nth term is given by the formula.

30. $a_n = 3n + 4$; a_{12}
31. $a_n = 2n - 5$; a_{10}
32. $a_n = n(n - 1)$; a_{11}
33. $a_n = \dfrac{n}{n + 1}$; a_{12}
34. $a_n = (-1)^{n-1}n^2$; a_{15}
35. $a_n = (-1)^{n-1}(n - 1)$; a_{25}
36. $a_n = \left(\dfrac{1}{2}\right)^n$; a_8
37. $a_n = \left(\dfrac{2}{3}\right)^n$; a_5
38. $a_n = (n + 2)(n + 3)$; a_{17}

39. $a_n = (n + 4)(n + 1)$; a_7

40. $a_n = \dfrac{(-1)^{2n-1}}{n^2}$; a_6

41. $a_n = \dfrac{(-1)^{2n}}{n + 4}$; a_{16}

42. $a_n = \dfrac{3}{2}n^2 - 2$; a_8

43. $a_n = \dfrac{1}{3}n + n^2$; a_6

44. $a_n = \dfrac{3}{4}n^2 - \dfrac{1}{4}n$; a_4

45. True or false? For any real number b, all terms of the sequence given by the formula $a_n = bn$ are greater than b.

46. True or false? The 99th term of the sequence given by the formula $a_n = (-1)^{n+1}n$ is positive.

2 Evaluate a series (See pages 583-584.)

GETTING READY

47. **a.** The sigma notation $\sum_{n=1}^{4}(3n + 1)$ represents the indicated sum of the first __?__ terms of the sequence given by the formula $a_n =$ __?__.

b. To evaluate the series from part (a), first write the series in expanded form by replacing n by 1, 2, 3, and 4:

$$\sum_{n=1}^{4}(3n + 1)$$
$= [3(\_\_?\_\_) + 1] + [3(\_\_?\_\_) + 1] + [3(\_\_?\_\_) + 1] + [3(\_\_?\_\_) + 1]$
$=$ __?__ + __?__ + __?__ + __?__ • **Simplify each term.**
$=$ __?__ • **Find the sum.**

Evaluate the series.

48. $\sum_{n=1}^{5}(2n + 3)$

49. $\sum_{i=1}^{7}(i + 2)$

50. $\sum_{i=1}^{4}(2i)$

51. $\sum_{n=1}^{7}n$

52. $\sum_{i=1}^{6}i^2$

53. $\sum_{i=1}^{5}(i^2 + 1)$

54. $\sum_{n=1}^{6}(-1)^n$

55. $\sum_{n=1}^{4}\dfrac{1}{2n}$

56. $\sum_{i=3}^{6}i^3$

57. $\sum_{n=2}^{4}2^n$

58. $\sum_{n=3}^{7}\dfrac{n}{n - 1}$

59. $\sum_{i=3}^{6}\dfrac{i + 1}{i}$

60. $\sum_{i=1}^{4}\dfrac{1}{2^i}$

61. $\sum_{i=1}^{5}\dfrac{1}{2i}$

62. $\sum_{n=1}^{4}(-1)^{n-1}n^2$

63. $\sum_{i=1}^{4}(-1)^{i-1}(i + 1)$

64. $\sum_{n=3}^{5}\dfrac{(-1)^{n-1}}{n - 2}$

65. $\sum_{n=4}^{7}\dfrac{(-1)^{n-1}}{n - 3}$

Write the series in expanded form.

66. $\sum_{n=1}^{5}2x^n$

67. $\sum_{n=1}^{4}\dfrac{2n}{x}$

68. $\sum_{i=1}^{5}\dfrac{x^i}{i}$

69. $\sum_{i=1}^{4}\dfrac{x^i}{i + 1}$

70. $\sum_{i=3}^{5} \frac{x^i}{2i}$

71. $\sum_{i=2}^{4} \frac{x^i}{2i-1}$

72. $\sum_{n=1}^{5} x^{2n}$

73. $\sum_{n=1}^{4} x^{2n-1}$

Determine whether each series is equivalent to the series $\sum_{n=1}^{4} n^2$.

74. $\sum_{n=3}^{6} (n^2 + 2)$

75. $\sum_{n=3}^{6} (n^2 + 2)^2$

76. $\sum_{n=3}^{6} (n - 2)^2$

77. $\sum_{n=10}^{13} (n - 9)^2$

APPLYING CONCEPTS

Write a formula for the *n*th term of the sequence.

78. The sequence of the natural numbers

79. The sequence of the odd natural numbers

80. The sequence of the negative even integers

81. The sequence of the negative odd integers

82. The sequence of the positive multiples of 7

83. The sequence of the positive integers that are divisible by 4

Evaluate the series. Write your answer as a single logarithm.

84. $\sum_{n=1}^{5} \log n$

85. $\sum_{i=1}^{4} \log 2i$

86. **Number Problem** The first 22 numbers in the sequence 4, 44, 444, 4444, . . . are added together. What digit is in the thousands place of the sum?

87. **Number Problem** The first 31 numbers in the sequence 6, 66, 666, 6666, . . . are added together. What digit is in the hundreds place of the sum?

A **recursive sequence** is one in which each term of the sequence is defined by using preceding terms. Find the first four terms of each recursively defined sequence.

88. $a_1 = 1, a_n = na_{n-1}, n \geq 2$

89. $a_1 = 1, a_2 = 1, a_n = a_{n-1} + a_{n-2}, n \geq 3$

90. In the first box below, $\frac{1}{2}$ of the box is shaded.

In successive boxes, the sums $\frac{1}{2} + \frac{1}{4}$ and $\frac{1}{2} + \frac{1}{4} + \frac{1}{8}$ are shown.

Can you identify the sum $\frac{1}{2} + \frac{1}{4} + \frac{1}{8} + \frac{1}{16} + \cdots$? Explain your answer.

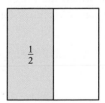

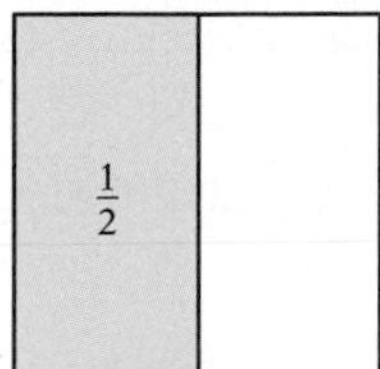

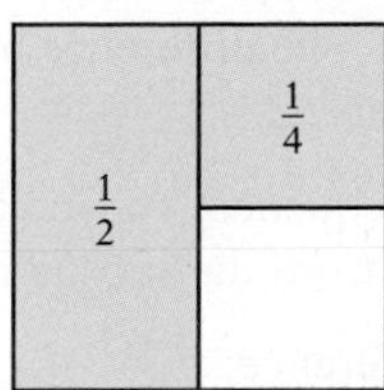

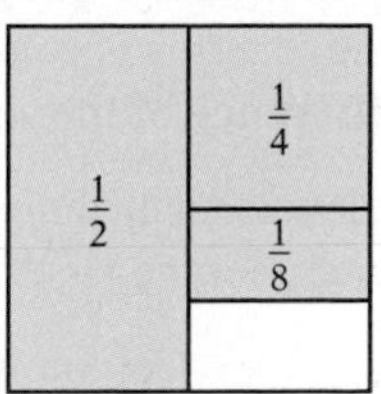

91. Rewrite $1 + \frac{1}{2} + \frac{1}{3} + \cdots + \frac{1}{n}$ using sigma notation.

PROJECTS OR GROUP ACTIVITIES

Epidemiology A model used by epidemiologists (people who study epidemics) to study the spread of a virus suggests that the number of people in a population who are newly infected on a given day is proportional to the number not yet exposed on the previous day. This relationship can be described by a recursive sequence defined by $a_n - a_{n-1} = k(P - a_{n-1})$, where P is the number of people in the original population exposed to a virus, a_n is the number of people ill with the virus n days after being exposed, a_{n-1} is the number of people ill with the virus on the previous day, and k is a constant that depends on the contagiousness of the disease. k is determined from experimental evidence.

92. Suppose that a population of 5000 people is exposed to a virus and that 150 people become ill ($a_0 = 150$). The next day, 344 people are ill ($a_1 = 344$). Determine the value of k.

93. Substitute the values of k and P into the recursive equation, and solve for a_n.

94. How many people are infected after 4 days?

11.2 Arithmetic Sequences and Series

OBJECTIVE 

Find the *n*th term of an arithmetic sequence

A company's expenses for training a new employee are quite high. To encourage employees to continue their employment with the company, a company that has a 6-month training program offers a starting salary of \$1600 per month and then a \$200-per-month pay increase each month during the training period.

The sequence below shows the employee's monthly salaries during the training period. Each term of the sequence is found by adding \$200 to the previous term.

Month	1	2	3	4	5	6
Salary	1600	1800	2000	2200	2400	2600

> **Take Note**
> An arithmetic sequence is a special type of sequence, one in which the difference between *any* two successive terms is the same constant. For instance, 5, 10, 15, 20, 25, . . . is an arithmetic sequence. The difference between any two successive terms is 5. The sequence 1, 4, 9, 16, . . . is not an arithmetic sequence because $4 - 1 \neq 9 - 4$.

The sequence 1600, 1800, 2000, 2200, 2400, 2600 is called an arithmetic sequence. An **arithmetic sequence,** or **arithmetic progression,** is a sequence in which the difference between any two consecutive terms is constant. The difference between consecutive terms is called the **common difference** of the sequence.

Each sequence shown below is an arithmetic sequence. To find the common difference of an arithmetic sequence, subtract the first term from the second term.

$2, 7, 12, 17, 22, \ldots$ Common difference: 5

$3, 1, -1, -3, -5, \ldots$ Common difference: -2

$1, \frac{3}{2}, 2, \frac{5}{2}, 3, \frac{7}{2}$ Common difference: $\frac{1}{2}$

Consider an arithmetic sequence in which the first term is a_1 and the common difference is d. Adding the common difference to each successive term of the arithmetic sequence yields a formula for the nth term.

The first term is a_1.	$a_1 = a_1$
To find the second term, add the common difference d to the first term.	$a_2 = a_1 + d$
To find the third term, add the common difference d to the second term.	$a_3 = a_2 + d = (a_1 + d) + d$ $a_3 = a_1 + 2d$
To find the fourth term, add the common difference d to the third term.	$a_4 = a_3 + d = (a_1 + 2d) + d$ $a_4 = a_1 + 3d$
Note the relationship between the term number and the number that multiplies d. The multiplier of d is 1 less than the term number.	$a_n = a_1 + (n - 1)d$

FORMULA FOR THE nTH TERM OF AN ARITHMETIC SEQUENCE

The nth term of an arithmetic sequence with a common difference of d is given by $a_n = a_1 + (n - 1)d$.

EXAMPLE 1 Find the 27th term of the arithmetic sequence $-4, -1, 2, 5, 8, \ldots$.

Solution

$d = a_2 - a_1 = -1 - (-4) = 3$ • **Find the common difference.**

$a_n = a_1 + (n - 1)d$ • **Use the Formula for the nth Term of an Arithmetic Sequence.**

$$\begin{aligned} a_{27} &= -4 + (27 - 1)3 \\ &= -4 + (26)3 \\ &= -4 + 78 \\ &= 74 \end{aligned}$$
• **$n = 27$, $a_1 = -4$, $d = 3$**

Problem 1 Find the 15th term of the arithmetic sequence $9, 3, -3, -9, \ldots$.

Solution See page S34.

Try Exercise 9, page 593.

EXAMPLE 2 Find the formula for the nth term of the arithmetic sequence $-5, -1, 3, 7, \ldots$.

Solution

$d = a_2 - a_1 = -1 - (-5) = 4$ • **Find the common difference.**

$$\begin{aligned} a_n &= a_1 + (n - 1)d \\ a_n &= -5 + (n - 1)4 \\ a_n &= -5 + 4n - 4 \\ a_n &= 4n - 9 \end{aligned}$$
• **Use the Formula for the nth Term of an Arithmetic Sequence.**
• **$a_1 = -5$, $d = 4$**

Problem 2 Find the formula for the nth term of the arithmetic sequence $-3, 1, 5, 9, \ldots$.

Solution See page S34.

Try Exercise 21, page 593.

EXAMPLE 3 Find the number of terms in the finite arithmetic sequence 7, 10, 13, . . . , 55.

Solution $d = a_2 - a_1 = 10 - 7 = 3$ • Find the common difference.

$a_n = a_1 + (n - 1)d$ • Use the Formula for the *n*th Term of an Arithmetic Sequence.

$55 = 7 + (n - 1)3$ • $a_n = 55, a_1 = 7, d = 3$

$55 = 7 + 3n - 3$ • Solve for *n*.

$55 = 3n + 4$

$51 = 3n$

$17 = n$

There are 17 terms in the sequence.

Problem 3 Find the number of terms in the finite arithmetic sequence 7, 9, 11, . . . , 59.

Solution See page S34.

Try Exercise 33, page 593.

OBJECTIVE 2 Evaluate an arithmetic series

The indicated sum of the terms of an arithmetic sequence is called an **arithmetic series.** The sum of an arithmetic series can be found by using a formula.

Point of Interest

This formula was proved in *Aryabhatiya*, which was written by Aryabhata around 499. The book is the earliest known Indian mathematical work by an identifiable author. Although the proof of the formula appears in that text, the formula was known before Aryabhata's time.

FORMULA FOR THE SUM OF *n* TERMS OF AN ARITHMETIC SERIES

Let a_1 be the first term of a finite arithmetic sequence, let n be the number of terms, and let a_n be the last term of the sequence. Then the sum of the series S_n is given by $S_n = \frac{n}{2}(a_1 + a_n)$.

Each term of the arithmetic sequence shown at the right was found by adding 3 to the previous term. 2, 5, 8, . . . , 17, 20

Each term of the reverse arithmetic sequence can be found by subtracting 3 from the previous term. 20, 17, 14, . . . , 5, 2

This idea is used in the following proof of the Formula for the Sum of n Terms of an Arithmetic Series.

Let S_n represent the sum of the series.

$$S_n = a_1 + (a_1 + d) + (a_1 + 2d) + \cdots + a_n$$

Write the terms of the sum of the series in reverse order. The sum is the same.

$$S_n = a_n + (a_n - d) + (a_n - 2d) + \cdots + a_1$$

Add the two equations.

$$2S_n = (a_1 + a_n) + (a_1 + a_n) + (a_1 + a_n) + \cdots + (a_1 + a_n)$$

Simplify the right side of the equation by using the fact that there are n terms in the sequence.

$$2S_n = n(a_1 + a_n)$$

Solve for S_n.

$$S_n = \frac{n}{2}(a_1 + a_n)$$

EXAMPLE 4 Find the sum of the first 10 terms of the arithmetic sequence 2, 4, 6, 8, ...

Solution $d = a_2 - a_1 = 4 - 2 = 2$ • **Find the common difference.**

$$a_n = a_1 + (n - 1)d$$
$$a_{10} = 2 + (10 - 1)2$$
$$= 2 + (9)2$$
$$= 2 + 18 = 20$$

• **To find the 10th term, use the Formula for the *n*th Term of an Arithmetic Sequence.**

$$S_n = \frac{n}{2}(a_1 + a_n)$$

• **Use the Formula for the Sum of *n* Terms of an Arithmetic Series.**

$$S_{10} = \frac{10}{2}(2 + 20) = 5(22) = 110$$

• $n = 10, a_1 = 2, a_n = 20$

Problem 4 Find the sum of the first 25 terms of the arithmetic sequence −4, −2, 0, 2, 4, ...

Solution See page S34.

Try Exercise 49, page 594.

EXAMPLE 5 Evaluate the arithmetic series $\sum_{n=1}^{25}(3n + 1)$.

Solution

$$a_n = 3n + 1$$
$$a_1 = 3(1) + 1 = 4$$ • **Find the first term.**
$$a_{25} = 3(25) + 1 = 76$$ • **Find the 25th term.**

$$S_n = \frac{n}{2}(a_1 + a_n)$$ • **Use the Formula for the Sum of *n* Terms of an Arithmetic Series.**

$$S_{25} = \frac{25}{2}(4 + 76)$$ • $n = 25, a_1 = 4, a_n = 76$
$$= \frac{25}{2}(80)$$
$$= 1000$$

Problem 5 Evaluate the arithmetic series $\sum_{n=1}^{18}(3n - 2)$.

Solution See page S34.

Try Exercise 57, page 594.

OBJECTIVE  Application problems

EXAMPLE 6 The distance a ball rolls down a ramp each second is given by an arithmetic sequence. The distance in feet traveled by the ball during the *n*th second is given by $2n - 1$. Find the distance the ball will travel during the first 10 s.

Point of Interest

Galileo (1564–1642) constructed inclines of various slopes similar to the one shown below, with a track in the center and equal intervals marked along the incline. He measured the distances various balls of different weights traveled in equal intervals of time. He showed that objects of different weights fall at the same rate.

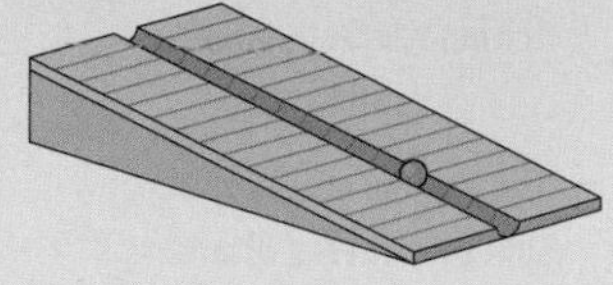

Strategy To find the distance:

- Find the first and tenth terms of the sequence.
- Use the Formula for the Sum of n Terms of an Arithmetic Series to find the sum of the first 10 terms.

Solution

$$a_n = 2n - 1$$
$$a_1 = 2(1) - 1 = 1$$
$$a_{10} = 2(10) - 1 = 19$$
$$S_n = \frac{n}{2}(a_1 + a_n)$$
$$S_{10} = \frac{10}{2}(1 + 19) = 5(20) = 100$$

The ball will roll 100 ft during the first 10 s.

Problem 6 A contest offers 20 prizes. The first prize is \$10,000, and each successive prize is \$300 less than the preceding prize. What is the value of the 20th-place prize? What is the total amount of prize money that is being awarded?

Solution See page S34.

Try Exercise 67, page 595.

11.2 Exercises

CONCEPT CHECK

1. What is an arithmetic sequence?

2. For the arithmetic sequence 5, 11, 17, 23, . . . , what is the common difference?

3. Determine whether each formula for the nth term of a sequence defines an arithmetic sequence.

a. $a_n = 2n - 7$ **b.** $a_n = 3 - 4n$ **c.** $a_n = 2n^2 + 1$

d. $a_n = 2^n - 1$ **e.** $a_n = n$ **f.** $a_n = \frac{1}{n}$

4. What is the formula for the sum of the first n terms of an arithmetic series?

1 Find the *n*th term of an arithmetic sequence (See pages 588–590.)

GETTING READY

5. a. The sequence 3, 7, 11, 15, 19, . . . is an arithmetic sequence because the difference between each pair of consecutive terms is ___?___. This number is called the ___?___ difference of the sequence.

b. Find the 50th term of the arithmetic sequence given in part (a).

$a_n = a_1 + (n - 1)d$ • Use the Formula for the ___?___ Term of an Arithmetic Sequence.

$a_{50} = 3 + (50 - 1)(4)$ • Replace n by ___?___, a_1 by ___?___, and d by ___?___.

$= $ ___?___ • Simplify.

6. Find the formula for the nth term of the arithmetic sequence 10, 8, 6, 4, ...

$d = a_2 - a_1 = $ ___?___ $-$ ___?___ • To find the common difference, subtract the first term from the second term.

$= $ ___?___

$a_n = a_1 + (n - 1)d$ • Use the Formula for the nth Term of an Arithmetic Sequence.

$a_n = $ ___?___ $+ (n - 1)($___?___$)$ • Replace a_1 by 10 and d by -2.

$a_n = 10 - $ ___?___ $+$ ___?___ • Use the Distributive Property.

$a_n = $ ___?___ • Simplify.

Find the indicated term of the arithmetic sequence.

7. 1, 11, 21, . . . ; a_{15}

8. 3, 8, 13, . . . ; a_{20}

9. $-6, -2, 2, \ldots ; a_{15}$

10. $-7, -2, 3, \ldots ; a_{14}$

11. 3, 7, 11, . . . ; a_{18}

12. $-13, -6, 1, \ldots ; a_{31}$

13. $-\frac{3}{4}, 0, \frac{3}{4}, \ldots ; a_{11}$

14. $\frac{3}{8}, 1, \frac{13}{8}, \ldots ; a_{17}$

15. $2, \frac{5}{2}, 3, \ldots ; a_{31}$

16. $1, \frac{5}{4}, \frac{3}{2}, \ldots ; a_{17}$

17. 6, 5.75, 5.50, . . . ; a_{10}

18. 4, 3.7, 3.4, . . . ; a_{12}

Find the formula for the nth term of the arithmetic sequence.

19. 1, 2, 3, . . .

20. 1, 4, 7, . . .

21. $6, 2, -2, \ldots$

22. $3, 0, -3, \ldots$

23. $2, \frac{7}{2}, 5, \ldots$

24. 7, 4.5, 2, . . .

25. $-8, -13, -18, \ldots$

26. 17, 30, 43, . . .

27. 26, 16, 6, . . .

Find the number of terms in the finite arithmetic sequence.

28. $-2, 1, 4, \ldots, 73$

29. 7, 11, 15, . . . , 171

30. $-\frac{1}{2}, \frac{3}{2}, \frac{7}{2}, \ldots, \frac{71}{2}$

31. $\frac{1}{3}, \frac{5}{3}, 3, \ldots, \frac{61}{3}$

32. 1, 5, 9, . . . , 81

33. 3, 8, 13, . . . , 98

34. $2, 0, -2, \ldots, -56$

35. $1, -3, -7, \ldots, -75$

36. $\frac{5}{2}, 3, \frac{7}{2}, \ldots, 13$

37. $\frac{7}{3}, \frac{13}{3}, \frac{19}{3}, \ldots, \frac{79}{3}$

38. $1, 0.75, 0.50, \ldots, -4$

39. $3.5, 2, 0.5, \ldots, -25$

For Exercises 40 to 43, $a_1, a_2, a_3, a_4, \ldots$ is an arithmetic sequence with common difference d. Determine whether the statement is true or false.

40. If $a_1 < 0$ and $d < 0$, then all terms of the sequence are negative.

41. $a_7 - a_3 = 4d$

42. If $d < 0$, then $a_n < a_{n+1}$.

43. If $a_1 < 0$ and $d > 0$, then the sequence has a finite number of negative terms and an infinite number of positive terms.

2 Evaluate an arithmetic series (See pages 590-591.)

GETTING READY

44. Give the meaning of each part of the Formula for the Sum of n Terms of an Arithmetic Series, $S_n = \frac{n}{2}(a_1 + a_n)$.
S_n is the __?__ of n terms of the series.
n is the __?__ of terms.
a_1 is the __?__ term.
a_n is the __?__ term.

45. Find the sum of the arithmetic series $\sum_{n=1}^{10}(n-3)$.

This series is the sum of the first __?__ terms of the sequence given by the formula $a_n =$ __?__.

$a_1 =$ __?__ $- 3 =$ __?__ • To find the first term, replace n by 1 and simplify.
$a_{10} =$ __?__ $- 3 =$ __?__ • To find the last term, replace n by 10 and simplify.
$S_n = \frac{n}{2}(a_1 + a_n)$ • Use the Formula for the Sum of n Terms of an __?__.
$S_{?} = \frac{?}{2}($ __?__ $+$ __?__ $)$ • Replace n by 10, a_1 by -2, and a_{10} by 7.
$= ($ __?__ $)($ __?__ $)$ • Simplify.
$=$ __?__ • Simplify.

Find the sum of the indicated number of terms of the arithmetic sequence.

46. $1, 3, 5, \ldots;\ n = 50$

47. $2, 4, 6, \ldots;\ n = 25$

48. $20, 18, 16, \ldots;\ n = 40$

49. $25, 20, 15, \ldots;\ n = 22$

50. $\frac{1}{2}, 1, \frac{3}{2}, \ldots;\ n = 27$

51. $2, \frac{11}{4}, \frac{7}{2}, \ldots;\ n = 10$

Evaluate the arithmetic series.

52. $\sum_{i=1}^{15}(3i - 1)$

53. $\sum_{i=1}^{15}(3i + 4)$

54. $\sum_{n=1}^{17}\left(\frac{1}{2}n + 1\right)$

55. $\sum_{n=1}^{10}(1 - 4n)$

56. $\sum_{i=1}^{15}(4 - 2i)$

57. $\sum_{n=1}^{10}(5 - n)$

58. State whether the series can be evaluated using the formula $S_n = \frac{n}{2}(a_1 + a_n)$.

a. $12 + 7 + 2 - 3 - 8$ **b.** $2 + 4 + 8 + \cdots + 64$

59. Will the sum of the series $\sum_{n=1}^{100}(5 - n)$ be positive or negative?

3 Application problems (See pages 591-592.)

GETTING READY

For Exercises 60 and 61, use this **problem situation:** A theater section has 8 rows of seats. The first row has 3 seats, the second row has 5 seats, the third row has 7 seats, and so on in an arithmetic sequence.

60. a. Give the value of each of the following for the sequence of numbers of seats per row.
$a_1 =$ __?__, $d =$ __?__, $n =$ __?__

b. Find the number of seats in the last row.

$a_n = a_1 + (n - 1)d$ • **Write the Formula for the *n*th Term of an Arithmetic Sequence.**

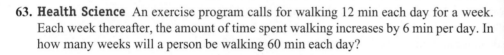

$a_{\_?\_} = \_?\_ + (\_?\_ - 1)(\_?\_)$ • **Replace *n*, a_1, and *d* by the values listed in part (a).**

$= \_?\_$ • **Simplify.**

61. Use the results of Exercise 60 to find the total number of seats in the theater section.

$S_n = \frac{n}{2}(a_1 + a_n)$ • **Write the Formula for the Sum of *n* Terms of an Arithmetic Series.**

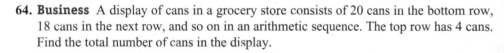

$S_{\_?\_} = \frac{?}{2}(\_?\_ + \_?\_)$ • **Replace *n*, a_1, and a_n by the values you found in Exercise 60.**

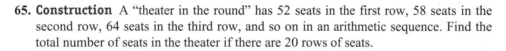

$= \_?\_$ • **Simplify.**

62. Physics The distance that an object dropped from a cliff will fall is 16 ft the first second, 48 ft the next second, 80 ft the third second, and so on in an arithmetic sequence. What is the total distance the object will fall in 6 s?

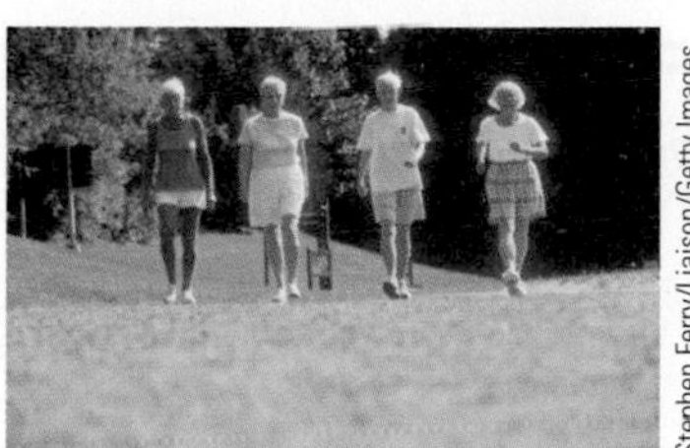
Stephen Ferry/Liaison/Getty Images

63. Health Science An exercise program calls for walking 12 min each day for a week. Each week thereafter, the amount of time spent walking increases by 6 min per day. In how many weeks will a person be walking 60 min each day?

64. Business A display of cans in a grocery store consists of 20 cans in the bottom row, 18 cans in the next row, and so on in an arithmetic sequence. The top row has 4 cans. Find the total number of cans in the display.

65. Construction A "theater in the round" has 52 seats in the first row, 58 seats in the second row, 64 seats in the third row, and so on in an arithmetic sequence. Find the total number of seats in the theater if there are 20 rows of seats.

66. Computer Viruses Read the article at the right. Find the number of minutes it will take for the number of computer viruses to reach 2,000,000. Show that the prediction given at the end of the article is correct.

In the News

Computer Viruses at Record Levels

There has been a large increase in the number of viruses attacking personal computers. In the first six months of the year, there were about 1,000,000 viruses. With new viruses arriving at the rate of 4 every minute, the number of viruses will reach 2,000,000 before the end of the year.

Source: linuxtoday.com

67. Human Resources The salary schedule for an engineering assistant is \$2200 for the first month and a \$150-per-month salary increase for the next 9 months. Find the monthly salary during the tenth month. Find the total salary for the 10-month period.

For Exercises 68 and 69, refer to the problem situations in Exercises 62 to 67. State the exercise number of the problem situation that is modeled by the given series.

68. $\sum_{n=1}^{9}(22 - 2n)$

69. $\sum_{n=1}^{6}(32n - 16)$

APPLYING CONCEPTS

70. Find the sum of the first 50 positive integers.

71. How many terms of the arithmetic sequence $-3, 2, 7, \ldots$ must be added together for the sum of the series to be 116?

72. Given an arithmetic sequence for which $a_1 = -9$, $a_n = 21$, and $S_n = 36$, find d and n.

73. The fourth term of an arithmetic sequence is 9, and the ninth term is 29. Find the first term.

74. **Business** Straight-line depreciation is used by some companies to determine the value of an asset. A model for this depreciation method is $a_n = V - dn$, where a_n is the value of the asset after n years, V is the original value of the asset, d is the annual decrease in the asset's value, and n is the number of years. Suppose that an asset has an original value of \$20,000 and that the annual decrease in value is \$3000.
   a. Substitute the values of V and d into the equation, and write an expression for a_n.

   b. Show that a_n is an arithmetic sequence.

75. **Geometry** The sum of the interior angles of a triangle is 180°. The sum is 360° for a quadrilateral and 540° for a pentagon. Assuming that this pattern continues, find the sum of the interior angles of a dodecagon (12-sided figure). Find a formula for the sum of the interior angles of an n-sided polygon.

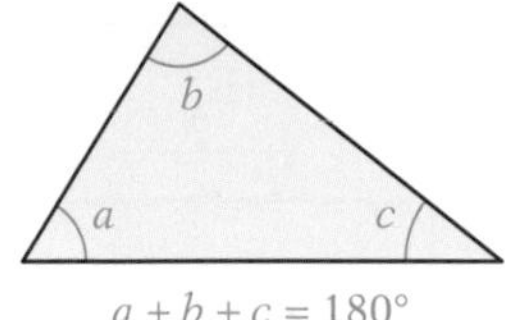

$a + b + c = 180°$

PROJECTS OR GROUP ACTIVITIES

76. **Sports** The International Amateur Athletic Federation (IAAF) specifies the designs of tracks on which world records can be set. A typical design is shaped like a rectangle with semicircles on either end, as shown below. There are eight lanes, and each lane is 1.22 m wide. The distance around the track for each lane is measured from the center of that lane. Find a formula for the sequence of radii of the circular portion of the track (the two ends). Round to the nearest hundredth.

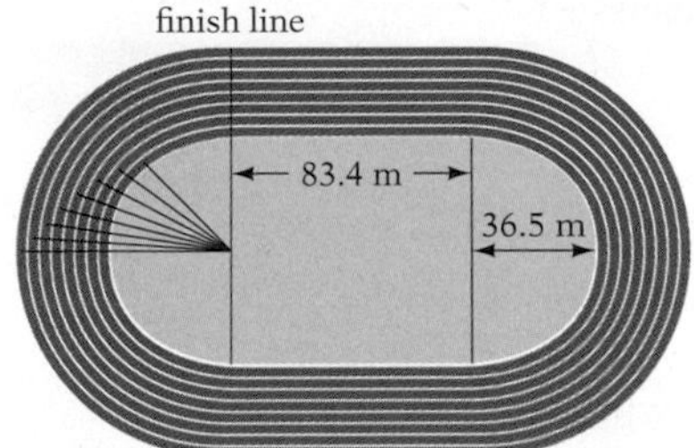

77. **a.** **Sports** Using the information in the preceding exercise, write a formula for the sequence of distances around the track for each lane. Round to the nearest hundredth, and use 3.14 as an approximation for π. Remember that the distance around the circular portion of the track is measured from the center of the lane.

   b. Explain why the IAAF staggers the starting position for runners in a 400-meter race.

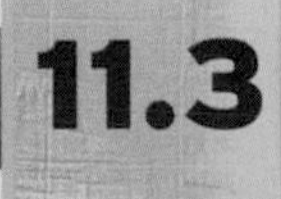

11.3 Geometric Sequences and Series

OBJECTIVE 1 Find the *n*th term of a geometric sequence

An ore sample contains 20 mg of a radioactive material with a half-life of 1 week. The amount of the radioactive material that the sample contains at the beginning of each week can be determined by using an exponential decay equation.

The following sequence represents the amount in the sample at the beginning of each week. Each term of the sequence is found by multiplying the preceding term by $\frac{1}{2}$.

Week	1	2	3	4	5
Amount	20	10	5	2.5	1.25

The sequence 20, 10, 5, 2.5, 1.25 is called a geometric sequence. A **geometric sequence,** or **geometric progression,** is a sequence in which each successive term of the sequence is the same nonzero constant multiple of the preceding term. The common multiple is called the **common ratio** of the sequence.

Take Note

Geometric sequences are different from arithmetic sequences. For a geometric sequence, every two successive terms have the same *ratio*. For an arithmetic sequence, every two successive terms have the same *difference*.

Each of the sequences shown below is a geometric sequence. To find the common ratio of a geometric sequence, divide the second term of the sequence by the first term.

$3, 6, 12, 24, 48, \ldots$ Common ratio: 2

$4, -12, 36, -108, 324, \ldots$ Common ratio: -3

$6, 4, \frac{8}{3}, \frac{16}{9}, \frac{32}{27}, \ldots$ Common ratio: $\frac{2}{3}$

Consider a geometric sequence in which the first term is a_1 and the common ratio is r. Multiplying each successive term of the geometric sequence by the common ratio yields a formula for the nth term.

The first term is a_1. $\quad a_1 = a_1$

To find the second term, multiply the first term by the common ratio r. $\quad a_2 = a_1r$

To find the third term, multiply the second term by the common ratio r. $\quad a_3 = (a_2)r = (a_1r)r$, $\quad a_3 = a_1r^2$

To find the fourth term, multiply the third term by the common ratio r. $\quad a_4 = (a_3)r = (a_1r^2)r$, $\quad a_4 = a_1r^3$

Note the relationship between the term number and the number that is the exponent on r. The exponent on r is 1 less than the term number. $\quad a_n = a_1r^{n-1}$

FORMULA FOR THE nTH TERM OF A GEOMETRIC SEQUENCE

The nth term of a geometric sequence with first term a_1 and common ratio r is given by $a_n = a_1r^{n-1}$.

Focus on finding a formula for the nth term of a geometric sequence

Find a formula for the nth term of the geometric sequence $5, 3, \frac{9}{5}, \ldots$.

Find r, the common ratio. $\quad r = \frac{a_2}{a_1} = \frac{3}{5}$

Use the Formula for the nth Term of a Geometric Sequence. $\quad a_n = a_1r^{n-1}$

$a_1 = 5, r = \frac{3}{5}$ $\quad a_n = 5\left(\frac{3}{5}\right)^{n-1}$

EXAMPLE 1 Find the sixth term of the geometric sequence 3, 6, 12, ...

Solution $r = \frac{a_2}{a_1} = \frac{6}{3} = 2$ • Find the common ratio.

$a_n = a_1 r^{n-1}$ • Use the Formula for the *n*th Term of a Geometric Sequence to find the sixth term.

$a_6 = 3(2)^{6-1}$ • $n = 6, a_1 = 3, r = 2$

$= 3(2)^5$

$= 3(32)$

$= 96$

Problem 1 Find the fifth term of the geometric sequence 5, 2, $\frac{4}{5}$, ...

Solution See page S35.

Try Exercise 13, page 603.

EXAMPLE 2 Find a_3 for the geometric sequence 8, a_2, a_3, 27, ...

Solution $a_n = a_1 r^{n-1}$

$a_4 = a_1 r^{4-1}$ • Use the Formula for the *n*th Term of a Geometric Sequence to find the common ratio. The first and fourth terms are known. Substitute 4 for *n*.

$27 = 8r^{4-1}$ • $a_4 = 27, a_1 = 8$

$\frac{27}{8} = r^3$ • Solve for *r*.

$\frac{3}{2} = r$ • Take the cube root of each side of the equation.

$a_3 = 8\left(\frac{3}{2}\right)^{3-1}$ • Use the Formula for *n*th Term of a Geometric Sequence to find a_3. $n = 3, a_1 = 8, r = \frac{3}{2}$

$= 8\left(\frac{3}{2}\right)^2$

$= 8\left(\frac{9}{4}\right) = 18$

Problem 2 Find a_3 for the geometric sequence 3, a_2, a_3, -192, ...

Solution See page S35.

Try Exercise 19, page 603.

OBJECTIVE 2 Finite geometric series

Point of Interest

Geometric series are used extensively in the mathematics of finance. Finite geometric series are used to calculate loan balances and monthly payments for amortized loans.

The indicated sum of the terms of a geometric sequence is called a **geometric series.** The sum of a geometric series can be found by a formula.

FORMULA FOR THE SUM OF *n* TERMS OF A FINITE GEOMETRIC SERIES

Let a_1 be the first term of a finite geometric sequence, let n be the number of terms, and let r be the common ratio, $r \neq 1$. Then the sum of the series S_n is given by $S_n = \frac{a_1(1 - r^n)}{1 - r}$.

Proof of the Formula for the Sum of n Terms of a Finite Geometric Series:

Let S_n represent the sum of n terms of the sequence.

$$S_n = a_1 + a_1r + a_1r^2 + \cdots + a_1r^{n-2} + a_1r^{n-1}$$

Multiply each side of the equation by r.

$$rS_n = a_1r + a_1r^2 + a_1r^3 + \cdots + a_1r^{n-1} + a_1r^n$$

Subtract the two equations.

$$S_n - rS_n = a_1 - a_1r^n$$

Factor each side of the equation.

$$(1 - r)S_n = a_1(1 - r^n)$$

Assuming that $r \neq 1$, solve for S_n.

$$S_n = \frac{a_1(1 - r^n)}{1 - r}$$

EXAMPLE 3 Find the sum of the first five terms of the geometric sequence 2, 8, 32, ...

Solution

$$r = \frac{a_2}{a_1} = \frac{8}{2} = 4$$

• **Find the common ratio.**

$$S_n = \frac{a_1(1 - r^n)}{1 - r}$$

• **Use the Formula for the Sum of n Terms of a Finite Geometric Series.**

$$S_5 = \frac{2(1 - 4^5)}{1 - 4}$$

• **$n = 5$, $a_1 = 2$, $r = 4$**

$$= \frac{2(1 - 1024)}{-3} = \frac{2(-1023)}{-3} = \frac{-2046}{-3} = 682$$

Problem 3 Find the sum of the first eight terms of the geometric sequence $1, -\frac{1}{3}, \frac{1}{9}, \ldots$.

Solution See page S35.

Try Exercise 29, page 604.

EXAMPLE 4 Evaluate the geometric series $\sum_{i=1}^{7} \left(\frac{2}{3}\right)^i$.

Solution

$$a_i = \left(\frac{2}{3}\right)^i$$

$$a_1 = \left(\frac{2}{3}\right)^1 = \frac{2}{3}$$

• **To find a_1, let $i = 1$.**

$$r = \frac{2}{3}$$

• **r is the base of the exponential expression.**

$$S_n = \frac{a_1(1 - r^n)}{1 - r}$$

• **Use the Formula for the Sum of n Terms of a Finite Geometric Series.**

$$S_7 = \frac{\frac{2}{3}\left[1 - \left(\frac{2}{3}\right)^7\right]}{1 - \frac{2}{3}} = \frac{\frac{2}{3}\left(1 - \frac{128}{2187}\right)}{\frac{1}{3}}$$

• $n = 7, a_1 = \frac{2}{3}, r = \frac{2}{3}$

$$= \frac{\frac{2}{3}\left(\frac{2059}{2187}\right)}{\frac{1}{3}} = \frac{4118}{2187}$$

Problem 4 Evaluate the geometric series $\sum_{n=1}^{5} \left(\frac{1}{2}\right)^n$.

Solution See page S35.

Try Exercise 39, page 604.

OBJECTIVE 3 Infinite geometric series

When the absolute value of the common ratio of a geometric sequence is less than 1, $|r| < 1$, then as n becomes larger, r^n becomes closer to zero.

Examples of geometric sequences for which $|r| < 1$ are shown at the right. As the number of terms increases, the absolute value of the last term listed gets closer to zero.

$$1, \frac{1}{3}, \frac{1}{9}, \frac{1}{27}, \frac{1}{81}, \frac{1}{243}, \ldots$$

$$1, -\frac{1}{2}, \frac{1}{4}, -\frac{1}{8}, \frac{1}{16}, -\frac{1}{32}, \ldots$$

The indicated sum of the terms of an infinite geometric sequence is called an **infinite geometric series.**

An example of an infinite geometric series is shown at the right. The first term is 1. The common ratio is $\frac{1}{3}$.

$$1 + \frac{1}{3} + \frac{1}{9} + \frac{1}{27} + \frac{1}{81} + \frac{1}{243} + \cdots$$

The sums of the first 5, 7, 12, and 15 terms, along with the values of r^n, are shown at the right. Note that as n increases, the sum of the terms gets closer to 1.5, and the value of r^n gets closer to zero.

n	S_n	r^n
5	1.4938272	0.0041152
7	1.4993141	0.0004572
12	1.4999972	0.0000019
15	1.4999999	0.0000001

Using the Formula for the Sum of n Terms of a Geometric Series and the fact that r^n approaches zero when $|r| < 1$ and n increases, a formula for the sum of an infinite geometric series can be found.

The sum of the first n terms of a geometric series is shown at the right. If $|r| < 1$, then r^n can be made very close to zero by using larger and larger values of n. Therefore, the sum of the first n terms is approximately $\frac{a_1}{1 - r}$.

Approximately zero

$$S_n = \frac{a_1(1 - r^n)}{1 - r}$$

$$S_n \approx \frac{a_1(1 - 0)}{1 - r} = \frac{a_1}{1 - r}$$

FORMULA FOR THE SUM OF AN INFINITE GEOMETRIC SERIES

The sum of an infinite geometric series in which $|r| < 1, r \neq 0$, and a_1 is the first term, is given by $S = \dfrac{a_1}{1 - r}$.

When $|r| \geq 1$, the infinite geometric series does not have a finite sum. For example, the sum of the infinite geometric series $1 + 2 + 4 + 8 + \cdots$ increases without bound.

EXAMPLE 5 Find the sum of the terms of the infinite geometric sequence $1, -\frac{1}{2}, \frac{1}{4}, -\frac{1}{8}, \ldots$.

Solution

$$r = \frac{a_2}{a_1} = \frac{-\frac{1}{2}}{1} = -\frac{1}{2}$$

• **Find the common ratio.**

$$S = \frac{a_1}{1 - r}$$

• **$|r| < 1$. Use the Formula for the Sum of an Infinite Geometric Series.**

$$= \frac{1}{1 - \left(-\frac{1}{2}\right)}$$

• **$a_1 = 1, r = -\frac{1}{2}$**

$$= \frac{1}{\frac{3}{2}} = \frac{2}{3}$$

Problem 5 Find the sum of the terms of the infinite geometric sequence $3, -2, \frac{4}{3}, -\frac{8}{9}, \ldots$.

Solution See page S35.

Try Exercise 49, page 605.

The sum of an infinite geometric series can be used to find a fraction that is equivalent to a nonterminating, repeating decimal.

The repeating decimal shown at the right has been rewritten as an infinite geometric series with first term $\frac{3}{10}$ and common ratio $\frac{1}{10}$.

$$0.\overline{3} = 0.3 + 0.03 + 0.003 + \cdots$$
$$= \frac{3}{10} + \frac{3}{100} + \frac{3}{1000} + \cdots$$

Use the Formula for the Sum of an Infinite Geometric Series.

$$S = \frac{a_1}{1 - r} = \frac{\frac{3}{10}}{1 - \frac{1}{10}} = \frac{\frac{3}{10}}{\frac{9}{10}} = \frac{3}{9} = \frac{1}{3}$$

$\frac{1}{3}$ is equivalent to the nonterminating, repeating decimal $0.\overline{3}$.

EXAMPLE 6 Find an equivalent fraction for $0.1\overline{2}$.

Solution Write the decimal as a sum that includes an infinite geometric series.

$$0.1\overline{2} = 0.1 + 0.02 + 0.002 + 0.0002 + \cdots$$
$$= \frac{1}{10} + \frac{2}{100} + \frac{2}{1000} + \frac{2}{10{,}000} + \cdots$$

The geometric series does not begin with the first term, $\frac{1}{10}$. The series begins with $\frac{2}{100}$. The common ratio is $\frac{1}{10}$.

$$S = \frac{a_1}{1-r} = \frac{\frac{2}{100}}{1-\frac{1}{10}}$$

• **Use the Formula for the Sum of an Infinite Geometric Series to find the sum of all the terms except the first, $\frac{1}{10}$.**

$$= \frac{\frac{2}{100}}{\frac{9}{10}} = \frac{2}{90}$$

$$0.1\overline{2} = \frac{1}{10} + \frac{2}{90} = \frac{11}{90}$$

• **Add $\frac{1}{10}$ to the sum of the series.**

An equivalent fraction for $0.1\overline{2}$ is $\frac{11}{90}$.

Problem 6 Find an equivalent fraction for $0.\overline{36}$.

Solution See page S35.

Try Exercise 57, page 605.

OBJECTIVE 

Application problems

Point of Interest

Galileo used pendulums extensively in his experiments. His investigations into their behavior early in his career led him to use them as measurement devices in later experiments.

EXAMPLE 7 On the first swing, the length of the arc through which a pendulum swings is 16 in. The length of each successive swing is $\frac{7}{8}$ that of the preceding swing. Find the length of the arc on the fifth swing. Round to the nearest tenth.

Strategy To find the length of the arc on the fifth swing, use the Formula for the *n*th Term of a Geometric Sequence.

$n = 5, a_1 = 16, r = \frac{7}{8}$

Solution $a_n = a_1 r^{n-1}$

$$a_5 = 16\left(\frac{7}{8}\right)^{5-1} = 16\left(\frac{7}{8}\right)^4 = 16\left(\frac{2401}{4096}\right) = \frac{38{,}416}{4096} \approx 9.4$$

The length of the arc on the fifth swing is approximately 9.4 in.

Problem 7 You start a chain letter and send it to three friends. Each of the three friends sends the letter to three other friends, and the sequence is repeated. If no one breaks the chain, how many letters will have been mailed from the first through the sixth mailings?

Solution See page S35.

Try Exercise 67, page 606.

11.3 Exercises

CONCEPT CHECK

1. What is a geometric sequence?

2. For the geometric sequence 2, 6, 18, 54, . . . , what is the common ratio?

3. Determine whether each formula for the nth term of a sequence defines a geometric sequence.

a. $a_n = n^2$ **b.** $a_n = 2^n$ **c.** $a_n = \left(-\frac{2}{3}\right)^n$

d. $a_n = n^{\frac{1}{2}}$ **e.** $a_n = 3^n - 1$ **f.** $a_n = \frac{1}{2^n}$

4. What is the formula for the sum of the first n terms of a geometric sequence?

1 Find the nth term of a geometric sequence (See pages 596–598.)

GETTING READY

5. a. The sequence 2, 6, 18, 54, . . . is a geometric sequence because each term can be found by multiplying the preceding term by __?__. This number is called the __?__ of the sequence.

b. Find the seventh term of the geometric sequence given in part (a).

$a_n = a_1r^{n-1}$ • Use the Formula for the __?__ Term of a Geometric Sequence.
$a_7 = 2(3)^{7-1}$ • Replace n by __?__, a_1 by __?__, and r by __?__.
$= 2(3)^{?}$ • Simplify.
$= 2(\_?\_)$ • Evaluate 3^6.
$= \_?\_$ • Multiply.

6. Find the common ratio of the geometric sequence 18, a_2, a_3, $\frac{16}{3}$, ...

$a_n = a_1r^{n-1}$ • Use the Formula for the nth Term of a Geometric Sequence.
$\frac{16}{3} = 18r^{4-1}$ • Replace a_n by __?__, a_1 by __?__, and n by __?__.
$\left(\frac{1}{?}\right)\left(\frac{16}{3}\right) = r^{?}$ • Multiply each side of the equation by $\frac{1}{18}$.
$\frac{?}{?} = r^3$ • Simplify the left side of the equation.
$\_?\_ = r$ • Take the cube root of each side of the equation.

Find a formula for the nth term of the given geometric sequence.

7. 3, 12, 48, . . . **8.** 2, 10, 50, . . . **9.** 4, −2, 1, . . .

10. $1, \frac{4}{9}, \frac{16}{81}, \ldots$ **11.** $3, \frac{9}{4}, \frac{27}{16}, \ldots$ **12.** $2, -\frac{4}{5}, \frac{8}{25}, \ldots$

Find the indicated term of the geometric sequence.

13. 2, 8, 32, . . . ; a_9 **14.** $4, 3, \frac{9}{4}, \ldots; a_8$ **15.** $6, -4, \frac{8}{3}, \ldots; a_7$

16. −5, 15, −45, . . . ; a_7 **17.** $1, \sqrt{2}, 2, \ldots; a_9$ **18.** $3, 3\sqrt{3}, 9, \ldots; a_8$

Find a_2 and a_3 for the geometric sequence.

19. $9, a_2, a_3, \frac{8}{3}, \ldots$ **20.** $8, a_2, a_3, \frac{27}{8}, \ldots$ **21.** $3, a_2, a_3, -\frac{8}{9}, \ldots$

22. $6, a_2, a_3, -48, \ldots$ **23.** $-3, a_2, a_3, 192, \ldots$ **24.** $5, a_2, a_3, 625, \ldots$

For Exercises 25 and 26, $a_1, a_2, a_3, a_4, \ldots$ is a geometric sequence with common ratio r.

25. True or false? If $r < 0$, then all terms of the sequence are negative.

26. True or false? If $0 < r < 1$, then $a_n < a_{n+1}$.

2 Finite geometric series (See pages 598–600.)

GETTING READY

27. Give the meaning of each part of the Formula for the Sum of n Terms of a Finite Geometric Series, $S_n = \dfrac{a_1(1 - r^n)}{1 - r}$.

S_n is the __?__ of n terms of the series.
n is the __?__ of terms.
a_1 is the __?__ term.
r is the __?__.

28. Evaluate the geometric series $\sum_{n=1}^{4} 10^n$.

This series is the sum of the first __?__ terms of the sequence given by the formula $a_n =$ __?__.

$a_1 = 10^{?} =$ __?__ • **To find the first term, replace *n* by 1 and simplify.**

$r =$ __?__ • ***r* is the base of the exponential expression 10^n.**

$S_n = \dfrac{a_1(1 - r^n)}{1 - r}$ • **Use the Formula for the Sum of *n* Terms of a __?__.**

$S_4 = \dfrac{(?)[1 - (?)^4]}{1 - ?}$ • **Replace a_1 and *r* with their values.**

$= \dfrac{10(?)}{?}$ • **Simplify.**

$=$ __?__

Find the sum of the indicated number of terms of the geometric sequence.

29. $2, 6, 18, \ldots; n = 7$

30. $-4, 12, -36, \ldots; n = 7$

31. $12, 9, \frac{27}{4}, \ldots; n = 5$

32. $3, 3\sqrt{2}, 6, \ldots; n = 12$

Evaluate the geometric series.

33. $\sum_{i=1}^{5} (2)^i$

34. $\sum_{n=1}^{6} \left(\frac{3}{2}\right)^n$

35. $\sum_{i=1}^{5} \left(\frac{1}{3}\right)^i$

36. $\sum_{n=1}^{6} \left(\frac{2}{3}\right)^n$

37. $\sum_{i=1}^{5} (4)^i$

38. $\sum_{n=1}^{8} (3)^n$

39. $\sum_{i=1}^{4} (7)^i$

40. $\sum_{n=1}^{5} (5)^n$

41. $\sum_{i=1}^{5} \left(\frac{3}{4}\right)^i$

42. $\sum_{n=1}^{3} \left(\frac{7}{4}\right)^n$

43. $\sum_{i=1}^{4} \left(\frac{5}{3}\right)^i$

44. $\sum_{n=1}^{6} \left(\frac{1}{2}\right)^n$

45. State whether the series can be evaluated using the formula $S_n = \dfrac{a_1(1 - r^n)}{1 - r}$.

a. $1 + 4 + 9 + \cdots + 81$

b. $\sqrt{3} - 3 + 3\sqrt{3} - 9 + 9\sqrt{3}$

46. If $b > 1$, will the sum of the series $\sum_{n=1}^{100} (-b)^n$ be positive or negative?

3 Infinite geometric series (See pages 600-602.)

GETTING READY

47. The formula $S = \dfrac{a_1}{1 - r}$ can be used to find the sum of an infinite geometric series as long as the common ratio r is a number between __?__ and __?__; that is, as long as $|r| <$ __?__.

48. Find the sum of the infinite geometric series $8 + 4 + 2 + \cdots$.

$r = \dfrac{a_2}{a_1} = \dfrac{?}{?} =$ __?__ • **Use the first two terms to find the common ratio.**

$S = \dfrac{a_1}{1 - r}$ • **$|r| < 1$. Use the Formula for the Sum of an __?__.**

$S = \dfrac{?}{1 - ?}$ • **Replace a_1 by 8 and r by $\frac{1}{2}$.**

$= \dfrac{8}{?}$ • **Simplify.**

$=$ __?__

Find the sum of the terms of the infinite geometric sequence.

49. $3, 2, \frac{4}{3}, \ldots$

50. $2, -\frac{1}{4}, \frac{1}{32}, \ldots$

51. $6, -4, \frac{8}{3}, \ldots$

52. $\frac{1}{10}, \frac{1}{100}, \frac{1}{1000}, \ldots$

53. $\frac{7}{10}, \frac{7}{100}, \frac{7}{1000}, \ldots$

54. $\frac{5}{100}, \frac{5}{10{,}000}, \frac{5}{1{,}000{,}000}, \ldots$

55. State whether the series can be evaluated using the formula $S = \dfrac{a_1}{1 - r}$.

a. $1 + \frac{\sqrt{2}}{2} + \frac{1}{2} + \cdots$

b. $\frac{4}{3} - \frac{16}{9} + \frac{64}{27} - \cdots$

56. $a_1 + a_2 + a_3 + \cdots$ is an infinite geometric series with $a_1 < 0$ and common ratio r such that $-1 < r < 0$. Is the sum of the series greater than or less than a_1?

Find an equivalent fraction for the repeating decimal.

57. $0.\overline{8}$

58. $0.\overline{5}$

59. $0.\overline{2}$

60. $0.\overline{9}$

61. $0.\overline{45}$

62. $0.\overline{18}$

63. $0.1\overline{6}$

64. $0.8\overline{3}$

4 Application problems (See page 602.)

GETTING READY

For Exercises 65 and 66, use this problem situation: A population of bacteria doubles every hour. The population starts at 350 bacteria.

65. After 1 h, the original population of 350 bacteria has doubled to __?__ bacteria. After 2 h, the population has doubled again, to __?__ bacteria. The sequence 350, 700, 1400, . . . is a __?__ sequence with $a_1 =$ __?__ and $r =$ __?__.

66. Use the results of Exercise 65 to find a formula for the population of bacteria after n hours.

$a_n = a_1 r^{n-1}$ • **Write the Formula for the nth Term of a Geometric Sequence.**

$a_n = (\;?\;)(\;?\;)^{n-1}$ • **Replace a_1 by __?__ and r by __?__.**

67. Physics A laboratory ore sample contains 500 mg of a radioactive material with a half-life of 1 day. Find the amount of radioactive material in the sample at the beginning of the seventh day.

68. Pendulum On the first swing, the length of the arc through which a pendulum swings is 18 in. The length of each successive swing is $\frac{3}{4}$ that of the preceding swing. What is the total distance the pendulum has traveled during the first five swings? Round to the nearest tenth.

69. Sports To test the bounce of a tennis ball, the ball is dropped from a height of 8 ft. The ball bounces to 80% of its previous height with each bounce. How high does the ball bounce on the fifth bounce? Round to the nearest tenth.

70. Physics The temperature of a hot-water spa is 75°F. Each hour, the temperature is 10% higher than during the previous hour. Find the temperature of the spa after 3 h. Round to the nearest tenth.

71. Real Estate A real estate broker estimates that a piece of land will increase in value at a rate of 12% each year. If the original value of the land is $15,000, what will be its value in 15 years?

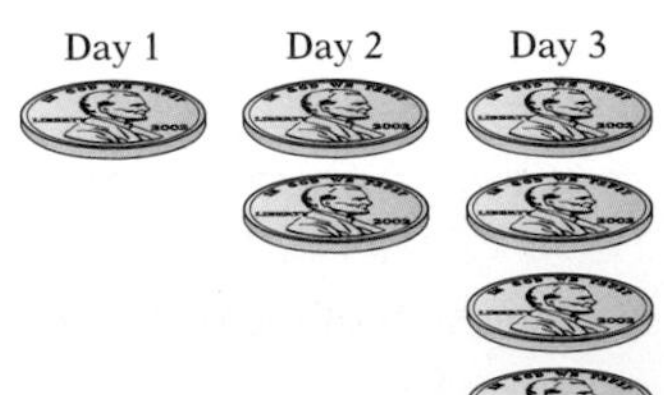

72. Business Suppose an employee receives a wage of 1¢ on the first day of work, 2¢ on the second day, 4¢ on the third day, and so on in a geometric sequence. Find the total amount of money earned for working 30 days.

73. Real Estate Assume that the average value of a home increases 5% per year. How much would a house costing $100,000 be worth in 30 years?

74. Biology A culture of bacteria doubles in size every 2 h. If there are 500 bacteria at the beginning, how many bacteria will there be after 24 h?

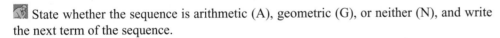

State whether the sequence is arithmetic (A), geometric (G), or neither (N), and write the next term of the sequence.

75. $4, -2, 1, \ldots$

76. $-8, 0, 8, \ldots$

77. $5, 6.5, 8, \ldots$

78. $-7, 14, -28, \ldots$

79. $1, 4, 9, 16, \ldots$

80. $\sqrt{1}, \sqrt{2}, \sqrt{3}, \sqrt{4}, \ldots$

81. $x^8, x^6, x^4, \ldots$

82. $5a^2, 3a^2, a^2, \ldots$

83. $\log x, 2 \log x, 3 \log x, \ldots$

84. $\log x, 3 \log x, 9 \log x, \ldots$

APPLYING CONCEPTS

Solve.

85. The third term of a geometric sequence is 3, and the sixth term is $\frac{1}{9}$. Find the first term.

86. Given $a_n = 162$, $r = -3$, and $S_n = 122$ for a geometric sequence, find a_1 and n.

87. For the geometric sequence given by $a_n = 2^n$, show that the sequence $b_n = \log a_n$ is an arithmetic sequence.

88. For the geometric sequence given by $a_n = e^n$, show that the sequence $b_n = \ln a_n$ is an arithmetic sequence.

89. For the arithmetic sequence given by $a_n = 3n - 2$, show that the sequence $b_n = 2^{a_n}$ is a geometric sequence.

90. For $f(n) = ab^n$, n a natural number, show that $f(n)$ is a geometric sequence.

91. Finance A car loan is normally structured so that part of each monthly payment reduces the loan amount, and the remainder of the payment pays interest on the loan. You pay interest only on the loan amount that remains to be paid (the unpaid balance). If you have a car loan of \$5000 at an annual interest rate of 9%, your monthly payment for a 5-year loan is \$103.79. The amount of the loan repaid R_n in the nth payment of the loan is a geometric sequence given by $R_n = R_1(1.0075)^{n-1}$. For the situation described above, $R_1 = 66.29$.

a. How much of the loan is repaid in the 27th payment?

b. The total amount T of the loan repaid after n payments is the sum of a geometric sequence, $T = \sum_{k=1}^{n} R_1(1.0075)^{k-1}$. Find the total amount repaid after 20 payments.

c. Determine the unpaid balance on the loan after 20 payments.

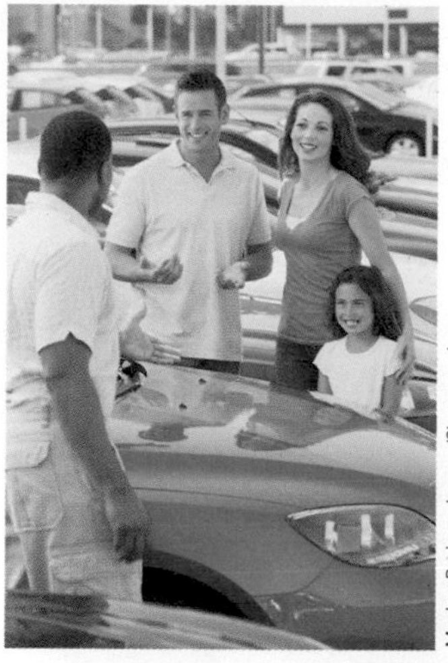

Monkey Business Images/Shutterstock.com

PROJECTS OR GROUP ACTIVITIES

92. Art The fabric designer Jhane Barnes created a fabric pattern based on the *Sierpinski triangle.* This triangle is a *fractal*, which is a geometric pattern that is repeated at ever smaller scales to produce irregular shapes. The first four stages in the construction of a Sierpinski triangle are shown below. The initial triangle is an equilateral triangle with sides 1 unit long. The cut-out triangles are formed by connecting the midpoints of the sides of the unshaded triangles. This pattern is repeated indefinitely. Find a formula for the nth term of the number of unshaded triangles.

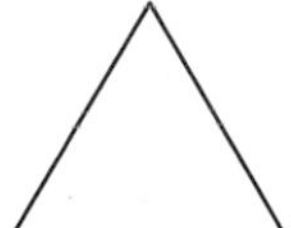 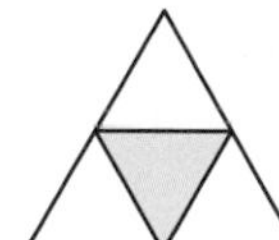 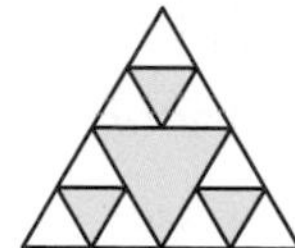 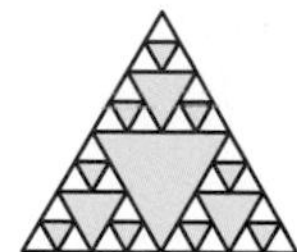

93. Art A *Sierpinski carpet* is similar to a Sierpinski triangle (see Exercise 92) except that all of the unshaded squares must be divided into nine congruent smaller squares with the one in the center shaded. The first three stages of the pattern are shown below. Find a formula for the nth term of the number of unshaded squares.

 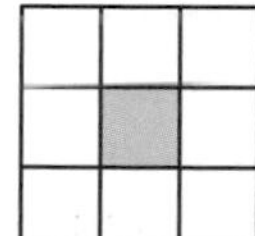 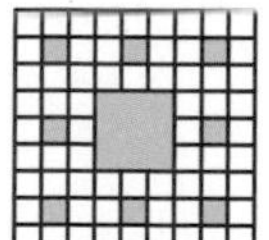

11.4 Binomial Expansions

OBJECTIVE 1

Expand $(a + b)^n$

By carefully observing the expansion of the binomial $(a + b)^n$ shown below, it is possible to identify some interesting patterns.

$$(a + b)^1 = a + b$$
$$(a + b)^2 = a^2 + 2ab + b^2$$
$$(a + b)^3 = a^3 + 3a^2b + 3ab^2 + b^3$$
$$(a + b)^4 = a^4 + 4a^3b + 6a^2b^2 + 4ab^3 + b^4$$
$$(a + b)^5 = a^5 + 5a^4b + 10a^3b^2 + 10a^2b^3 + 5ab^4 + b^5$$

PATTERNS FOR THE VARIABLE PARTS

1. The first term is a^n. The exponent on a decreases by 1 for each successive term.
2. The exponent on b increases by 1 for each successive term. The last term is b^n.
3. The degree of each term is n.

Focus on writing the variable parts of the terms of a binomial expansion

Write the variable parts of the terms in the expansion of $(a + b)^6$.

The first term is a^6. For each successive term, the exponent on a decreases by 1, and the exponent on b increases by 1. The last term is b^6.

$a^6, a^5b, a^4b^2, a^3b^3, a^2b^4, ab^5, b^6$

Point of Interest

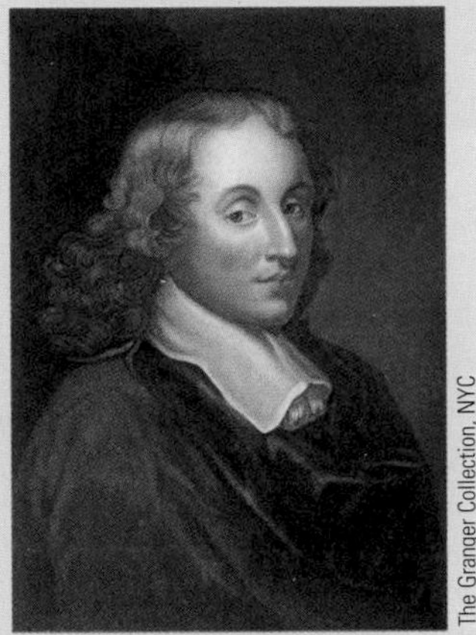

The Granger Collection, NYC

Blaise Pascal (1623–1662) is given credit for this triangle, which he first published in 1654. In that publication, the triangle looked like

```
  1 2 3 4 5...
1 1 1 1 1 1...
2 1 2 3 4...
3 1 3 6...
4 1 4...
5 1
```

Thus the triangle was rotated 45° from the way it is shown today. The first European publication of the triangle is attributed to Peter Apianus in 1527. However, there are versions of it in a Chinese text by Yang Hui that dates from 1250. In that text, Hui demonstrated how to find the third, fourth, fifth, and sixth roots of a number by using the triangle.

The variable parts of the general expansion of $(a + b)^n$ are

$$a^n, a^{n-1}b, a^{n-2}b^2, \ldots, a^{n-r}b^r, \ldots, ab^{n-1}, b^n$$

A pattern for the coefficients of the terms of the expanded binomial can be found by writing the coefficients in a triangular array known as **Pascal's Triangle.**

Each row begins and ends with the number 1. Any other number in a row is the sum of the two closest numbers above it. For example, $4 + 6 = 10$.

For $(a + b)^1$: 1 1

For $(a + b)^2$: 1 2 1

For $(a + b)^3$: 1 3 3 1

For $(a + b)^4$: 1 4 6 4 1

For $(a + b)^5$: 1 5 10 10 5 1

To write the sixth row of Pascal's Triangle, first write the numbers of the fifth row. The first and last numbers of the sixth row are 1. Each of the other numbers of the sixth row can be obtained by finding the sum of the two closest numbers above it in the fifth row.

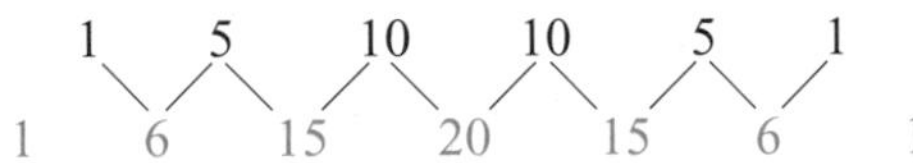

The numbers in the sixth row of Pascal's Triangle will be the coefficients of the terms in the expansion of $(a + b)^6$.

Using these numbers for the coefficients, and using the pattern for the variable part of each term, we can write the expanded form of $(a + b)^6$ as follows:

$$(a + b)^6 = a^6 + 6a^5b + 15a^4b^2 + 20a^3b^3 + 15a^2b^4 + 6ab^5 + b^6$$

Although Pascal's Triangle can be used to find the coefficients for the expanded form of the power of any binomial, this method is inconvenient when the power of the binomial is large. An alternative method for determining these coefficients is based on the concept of *factorial.*

n FACTORIAL

For a natural number n, $n!$ (which is read "n **factorial**") is the product of the first n natural numbers.

$$n! = n \cdot (n - 1) \cdot (n - 2) \cdot \cdots \cdot 3 \cdot 2 \cdot 1$$

Zero factorial is a special case and is defined as $0! = 1$.

EXAMPLES

1. $5! = 5 \cdot 4 \cdot 3 \cdot 2 \cdot 1 = 120$
2. $8! = 8 \cdot 7 \cdot 6 \cdot 5 \cdot 4 \cdot 3 \cdot 2 \cdot 1 = 40{,}320$
3. $1! = 1$
4. $0! = 1$

EXAMPLE 1 Evaluate: $\dfrac{7!}{4!\,3!}$

Solution

$$\frac{7!}{4!\,3!} = \frac{7 \cdot 6 \cdot 5 \cdot 4 \cdot 3 \cdot 2 \cdot 1}{(4 \cdot 3 \cdot 2 \cdot 1)(3 \cdot 2 \cdot 1)}$$

- **Write each factorial as a product.**

$$= 35$$

- **Simplify.**

Problem 1 Evaluate: $\dfrac{12!}{7!\,5!}$

Solution See page S36.

Try Exercise 15, page 611.

The coefficients in a binomial expansion can be given in terms of factorials. In the expansion of $(a + b)^5$ shown below, note that the coefficient of a^2b^3 can be given by $\frac{5!}{2!\,3!}$. The numerator is the factorial of the power of the binomial. The denominator is the product of the factorials of the exponents on a and b.

$$(a + b)^5 = a^5 + 5a^4b + 10a^3b^2 + 10a^2b^3 + 5ab^4 + b^5$$

$$\frac{5!}{2!\,3!} = \frac{5 \cdot 4 \cdot 3 \cdot 2 \cdot 1}{(2 \cdot 1)(3 \cdot 2 \cdot 1)} = 10$$

In general, the coefficients in the expansion of $(a + b)^n$ are given as the quotients of factorials. The coefficient of $a^{n-r}b^r$ is $\dfrac{n!}{(n - r)!\,r!}$. The symbol $\dbinom{n}{r}$ is used to express this quotient of factorials.

$$\binom{n}{r} = \frac{n!}{(n - r)!\,r!}$$

EXAMPLE 2 Evaluate: $\dbinom{8}{5}$

Solution

$$\binom{8}{5} = \frac{8!}{(8 - 5)!\,5!}$$

- **Write the quotient of the factorials.**

$$= \frac{8!}{3!5!} = \frac{8 \cdot 7 \cdot 6 \cdot 5 \cdot 4 \cdot 3 \cdot 2 \cdot 1}{(3 \cdot 2 \cdot 1)(5 \cdot 4 \cdot 3 \cdot 2 \cdot 1)} = 56$$

- **Simplify.**

Problem 2 Evaluate: $\dbinom{7}{0}$

Solution See page S36.

Try Exercise 23, page 612.

Using factorials and the pattern for the variable part of each term, we can write a formula for any natural-number power of a binomial.

BINOMIAL EXPANSION FORMULA

$$(a+b)^n = \binom{n}{0}a^n + \binom{n}{1}a^{n-1}b + \binom{n}{2}a^{n-2}b^2 + \cdots + \binom{n}{r}a^{n-r}b^r + \cdots + \binom{n}{n}b^n$$

EXAMPLES

1. $(a+b)^4 = \binom{4}{0}a^4 + \binom{4}{1}a^3b + \binom{4}{2}a^2b^2 + \binom{4}{3}ab^3 + \binom{4}{4}b^4$
   $= a^4 + 4a^3b + 6a^2b^2 + 4ab^3 + b^4$
2. $(x-2)^3 = \binom{3}{0}x^3 + \binom{3}{1}x^2(-2) + \binom{3}{2}x(-2)^2 + \binom{3}{3}(-2)^3$
   $= x^3 - 6x^2 + 12x - 8$

EXAMPLE 3 Write $(4x+3y)^3$ in expanded form.

Solution $(4x+3y)^3$

$$= \binom{3}{0}(4x)^3 + \binom{3}{1}(4x)^2(3y) + \binom{3}{2}(4x)(3y)^2 + \binom{3}{3}(3y)^3$$
$$= 1(64x^3) + 3(16x^2)(3y) + 3(4x)(9y^2) + 1(27y^3)$$
$$= 64x^3 + 144x^2y + 108xy^2 + 27y^3$$

Problem 3 Write $(3m-n)^4$ in expanded form.

Solution See page S36.

Try Exercise 39, page 612.

EXAMPLE 4 Find the first three terms in the expansion of $(x+3)^{15}$.

Solution

$$(x+3)^{15} = \binom{15}{0}x^{15} + \binom{15}{1}x^{14}(3) + \binom{15}{2}x^{13}(3)^2 + \cdots$$
$$= 1x^{15} + 15x^{14}(3) + 105x^{13}(9) + \cdots$$
$$= x^{15} + 45x^{14} + 945x^{13} + \cdots$$

Problem 4 Find the first three terms in the expansion of $(y-2)^{10}$.

Solution See page S36.

Try Exercise 43, page 612.

The Binomial Expansion Formula can also be used to write any term of a binomial expansion.

In the expansion of $(a+b)^5$ below, note that the exponent on b is 1 less than the term number.

$$(a+b)^5 = a^5 + 5a^4b + 10a^3b^2 + 10a^2b^3 + 5ab^4 + b^5$$

FORMULA FOR THE rTH TERM OF A BINOMIAL EXPANSION

The rth term in the expansion of $(a+b)^n$ is $\binom{n}{r-1}a^{n-r+1}b^{r-1}$.

EXAMPLE 5 Find the fourth term in the expansion of $(x + 3)^7$.

Solution $\binom{n}{r-1}a^{n-r+1}b^{r-1}$ • **Use the Formula for the *r*th Term of a Binomial Expansion.**

$\binom{7}{4-1}x^{7-4+1}(3)^{4-1}$ • $r = 4, n = 7, a = x, b = 3$

$= \binom{7}{3}x^4(3)^3$

$= 35x^4(27)$

$= 945x^4$

Problem 5 Find the third term in the expansion of $(t - 2s)^7$.

Solution See page S36.

Try Exercise 59, page 612.

11.4 Exercises

CONCEPT CHECK

1. What is the factorial of a natural number n?

2. What does the notation $\binom{n}{r}$ mean?

3. In the expansion of $(x + y)^n$, what is the degree of each term?

4. How many terms are in the expansion of $(x - 4)^6$?

1 Expand $(a + b)^n$ (See pages 607–611.)

GETTING READY

5. The notation $n!$ is read "n ___?___."

6. $6! =$ ___?___ · ___?___ · ___?___ · ___?___ · ___?___ · ___?___
$=$ ___?___

7. $0! =$ ___?___

8. $\binom{5}{4} = \frac{?}{(?)(?)} = \frac{(?)(?)(?)(?)(?)}{(?)(?)(?)(?)(?)}$
$=$ ___?___

Evaluate.

9. $3!$ 10. $4!$ 11. $8!$ 12. $9!$ 13. $0!$ 14. $1!$

15. $\frac{5!}{2!3!}$ 16. $\frac{8!}{5!3!}$ 17. $\frac{6!}{6!0!}$ 18. $\frac{10!}{10!0!}$ 19. $\frac{9!}{6!3!}$ 20. $\frac{10!}{2!8!}$

Evaluate.

21. $\binom{7}{2}$ **22.** $\binom{8}{6}$ **23.** $\binom{10}{2}$ **24.** $\binom{9}{6}$ **25.** $\binom{9}{0}$ **26.** $\binom{10}{10}$

27. $\binom{6}{3}$ **28.** $\binom{7}{6}$ **29.** $\binom{11}{1}$ **30.** $\binom{13}{1}$ **31.** $\binom{4}{2}$ **32.** $\binom{8}{4}$

33. True or false? $\frac{(2n)!}{n!} = 2$

34. True or false? $\frac{n!}{n} = (n - 1)!$

Write in expanded form.

35. $(x + y)^4$

36. $(r - s)^3$

37. $(x - y)^5$

38. $(y - 3)^4$

39. $(2m + 1)^4$

40. $(2x + 3y)^3$

41. $(2r - 3)^5$

42. $(x + 3y)^4$

Find the first three terms of the expansion.

43. $(a + b)^{10}$

44. $(a + b)^9$

45. $(a - b)^{11}$

46. $(a - b)^{12}$

47. $(2x + y)^8$

48. $(x + 3y)^9$

49. $(4x - 3y)^8$

50. $(2x - 5)^7$

51. $\left(x + \frac{1}{x}\right)^7$

52. $\left(x - \frac{1}{x}\right)^8$

53. $(x^2 + 3)^5$

54. $(x^2 - 2)^6$

Find the indicated term of the expansion.

55. $(2x - 1)^7$; 4th term

56. $(x + 4)^5$; 3rd term

57. $(x^2 - y^2)^6$; 2nd term

58. $(x^2 + y^2)^7$; 6th term

59. $(y - 1)^9$; 5th term

60. $(x - 2)^8$; 8th term

61. $\left(n + \frac{1}{n}\right)^5$; 2nd term

62. $\left(x + \frac{1}{2}\right)^6$; 3rd term

63. $\left(\frac{x}{2} + 2\right)^5$; 1st term

64. $\left(y - \frac{2}{3}\right)^6$; 3rd term

65. If n is odd, is the constant term in the expanded form of $(3x - 4)^n$ positive or negative?

66. Suppose $n > 7$ and b is a positive constant. Is the coefficient of the seventh term of the expanded form of $(x - b)^n$ positive or negative?

APPLYING CONCEPTS

67. Simplify: $\frac{n!}{(n - 1)!}$

68. Simplify: $\frac{n!}{(n - 2)!}$

69. For $0 \le r \le n$, show that $\binom{n}{r} = \binom{n}{n-r}$.

70. For $n \ge 1$, evaluate $\dfrac{2 \cdot 4 \cdot 6 \cdot 8 \cdot \cdots \cdot (2n)}{2^n n!}$.

71. Write a summary of the history of the development of Pascal's Triangle. Include some of the properties of the triangle.

PROJECTS OR GROUP ACTIVITIES

Powers of complex numbers can be found by using the Binomial Expansion Formula. Before doing Exercises 72–79, review the powers of i in Section 7.5, Applying Concepts.

Write each complex number in standard form.

72. $(1 + i)^5$

73. $(2 - 3i)^4$

74. $(1 - 2i)^5$

75. $(2 + i)^6$

76. $\left(\frac{1}{2} + \frac{1}{2}i\right)^4$

77. $\left(\frac{1}{2} - \frac{1}{2}i\right)^5$

78. $\left(\frac{\sqrt{3}}{2} + \frac{1}{2}i\right)^3$

79. $\left(-\frac{1}{2} + \frac{\sqrt{3}}{2}i\right)^3$

CHAPTER 11 Summary

Key Words	Objective and Page Reference	Examples
A **sequence** is an ordered list of numbers. Each of the numbers of a sequence is called a **term** of the sequence.	[11.1.1, p. 582]	$1, \frac{1}{3}, \frac{1}{9}, \frac{1}{27}$ is a sequence. The terms are $1, \frac{1}{3}, \frac{1}{9}$, and $\frac{1}{27}$.
A **finite sequence** contains a finite number of terms.	[11.1.1, p. 582]	2, 4, 6, 8 is a finite sequence.
An **infinite sequence** contains an infinite number of terms.	[11.1.1, p. 582]	1, 3, 5, 7, ... is an infinite sequence.
The indicated sum of the terms of a sequence is a **series.**	[11.1.2, p. 583]	1, 5, 9, 13 is a sequence. The series is $1 + 5 + 9 + 13$.
Summation notation is used to represent a series in compact form. The Greek letter sigma, Σ, is used to indicate the sum.	[11.1.2, p. 583]	$\sum_{n=1}^{4} 2n = 2(1) + 2(2) + 2(3) + 2(4)$
An **arithmetic sequence,** or **arithmetic progression,** is a sequence in which the difference between any two consecutive terms is constant. The difference between consecutive terms is called the **common difference** of the sequence.	[11.2.1, p. 588]	3, 9, 15, 21, ... is an arithmetic sequence. $9 - 3 = 6$; 6 is the common difference.

An **arithmetic series** is the indicated sum of the terms of an arithmetic sequence. [11.2.2, p. 590]

1, 2, 3, 4, 5 is an arithmetic sequence.

The arithmetic series is $1 + 2 + 3 + 4 + 5$.

A **geometric sequence,** or **geometric progression,** is a sequence in which each successive term is the same nonzero constant multiple of the preceding term. The common multiple is called the **common ratio** of the sequence. [11.3.1, p. 597]

$9, 3, 1, \frac{1}{3}, \ldots$ is a geometric sequence.

$\frac{3}{9} = \frac{1}{3}$; $\frac{1}{3}$ is the common ratio.

A **geometric series** is the indicated sum of the terms of a geometric sequence. [11.3.2, p. 598]

5, 10, 20, 40 is a geometric sequence.

The geometric series is $5 + 10 + 20 + 40$.

For a natural number n, **n factorial,** written $n!$, is the product of the first n natural numbers. 0! is defined to be 1. [11.4.1, p. 608]

$5! = 5 \cdot 4 \cdot 3 \cdot 2 \cdot 1 = 120$

Essential Rules and Procedures | Objective and Page Reference | Examples

Formula for the nth Term of an Arithmetic Sequence
The nth term of an arithmetic sequence with a common difference of d is given by $a_n = a_1 + (n - 1)d$. [11.2.1, p. 589]

Find the 10th term of the arithmetic sequence 4, 7, 10, ...

The common difference is $d = 7 - 4 = 3$; $a_1 = 4$.

$a_{10} = 4 + (10 - 1)3 = 31$

Formula for the Sum of n Terms of an Arithmetic Series
Let a_1 be the first term of a finite arithmetic sequence, let n be the number of terms, and let a_n be the last term of the sequence. Then the sum of the series is given by $S_n = \frac{n}{2}(a_1 + a_n)$. [11.2.2, p. 590]

Find the sum of the first 12 terms of the arithmetic sequence 5, 8, 11, ...

The common difference is $d = 8 - 5 = 3$; $a_1 = 5$.

$a_{12} = 5 + (12 - 1)3 = 38$

$S_{12} = \frac{12}{2}(5 + 38) = 258$

Formula for the nth Term of a Geometric Sequence
The nth term of a geometric sequence with first term a_1 and common ratio r is given by $a_n = a_1r^{n-1}$. [11.3.1, p. 597]

Find the 10th term of the geometric sequence 2, 6, 18, ...

The common ratio is $r = \frac{6}{2} = 3$; $a_1 = 2$.

$a_{10} = 2(3)^{10-1} = 39{,}366$

Formula for the Sum of n Terms of a Finite Geometric Series
Let a_1 be the first term of a finite geometric sequence, let n be the number of terms, and let r be the common ratio. Then the sum of the series S_n is given by $S_n = \dfrac{a_1(1 - r^n)}{1 - r}$. [11.3.2, p. 598]

Find the sum of the first 8 terms of the geometric sequence 1, 4, 16, ...

The common ratio is $r = \frac{4}{1} = 4$; $a_1 = 1$.

$S_8 = \dfrac{1(1 - 4^8)}{1 - 4} = 21{,}845$

Formula for the Sum of an Infinite Geometric Series The sum of an infinite geometric series in which $\|r\| < 1, r \neq 0$, and a_1 is the first term, is given by $S = \frac{a_1}{1-r}$.	[11.3.3, p. 601]	Find the sum of the terms of the infinite geometric sequence $2, 1, \frac{1}{2}, \ldots$. The common ratio is $r = \frac{1}{2}$; $a_1 = 2$. $S = \frac{2}{1-\frac{1}{2}} = 4$
Binomial Coefficients The coefficients of $a^{n-r}b^r$ in the expansion of $(a+b)^n$ are given by $\binom{n}{r}$, where $\binom{n}{r} = \frac{n!}{(n-r)!r!}$.	[11.4.1, p. 609]	$\binom{6}{2} = \frac{6!}{(6-2)!2!} = \frac{6!}{4!2!} = 15$
Binomial Expansion Formula $(a+b)^n =$ $\binom{n}{0}a^n + \binom{n}{1}a^{n-1}b + \binom{n}{2}a^{n-2}b^2 + \cdots$ $+ \binom{n}{r}a^{n-r}b^r + \cdots + \binom{n}{n}b^n$	[11.4.1, p. 610]	$(x+y)^4$ $= \binom{4}{0}x^4 + \binom{4}{1}x^3y + \binom{4}{2}x^2y^2 +$ $\binom{4}{3}xy^3 + \binom{4}{4}y^4$ $= x^4 + 4x^3y + 6x^2y^2 + 4xy^3 + y^4$
Formula for the *r*th Term of a Binomial Expansion The *r*th term in the expansion of $(a+b)^n$ is $\binom{n}{r-1}a^{n-r+1}b^{r-1}$.	[11.4.1, p. 610]	The sixth term in the expansion of $(2x-y)^9$ is $\binom{9}{5}(2x)^{9-6+1}(-y)^{6-1} = -2016x^4y^5$

CHAPTER 11 Review Exercises

1. Write $\sum_{i=1}^{4} 3x^i$ in expanded form.

2. Find the number of terms in the finite arithmetic sequence $-5, -8, -11, \ldots, -50$.

3. Find the seventh term of the geometric sequence $4, 4\sqrt{2}, 8, \ldots$.

4. Find the sum of the terms of the infinite geometric sequence $4, 3, \frac{9}{4}, \ldots$.

5. Evaluate: $\binom{9}{3}$

6. Write the 14th term of the sequence whose *n*th term is given by the formula $a_n = \frac{8}{n+2}$.

7. Find the 10th term of the arithmetic sequence $-10, -4, 2, \ldots$.

8. Find the sum of the first 18 terms of the arithmetic sequence $-25, -19, -13, \ldots$.

9. Find the sum of the first five terms of the geometric sequence $-6, 12, -24, \ldots$.

10. Evaluate: $\frac{8!}{4!4!}$

11. Find the seventh term in the expansion of $(3x+y)^9$.

12. Find the sum of the series $\sum_{n=1}^{4}(3n+1)$.

13. Write the sixth term of the sequence whose nth term is given by the formula $a_n = \frac{n+1}{n}$.

14. Find the formula for the nth term of the arithmetic sequence $12, 9, 6, \ldots$.

15. Find the fifth term of the geometric sequence $6, 2, \frac{2}{3}, \ldots$.

16. Find an equivalent fraction for $0.5\overline{3}$.

17. Find the 35th term of the arithmetic sequence $-13, -16, -19, \ldots$.

18. Find the sum of the first six terms of the geometric sequence $1, \frac{3}{2}, \frac{9}{4}, \ldots$.

19. Find the sum of the first 21 terms of the arithmetic sequence $5, 12, 19, \ldots$.

20. Find the fourth term in the expansion of $(x - 2y)^7$.

21. Find the number of terms in the finite arithmetic sequence $1, 7, 13, \ldots, 121$.

22. Find the eighth term of the geometric sequence $\frac{3}{8}, \frac{3}{4}, \frac{3}{2}, \ldots$.

23. Evaluate the series $\sum_{i=1}^{5} 2i$.

24. Find the sum of the first five terms of the geometric sequence $1, 4, 16, \ldots$.

25. Evaluate: $5!$

26. Find the third term in the expansion of $(x - 4)^6$.

27. Find the 30th term of the arithmetic sequence $-2, 3, 8, \ldots$.

28. Find the sum of the first 25 terms of the arithmetic sequence $25, 21, 17, \ldots$.

29. Write the fifth term of the sequence whose nth term is given by the formula $a_n = \frac{(-1)^{2n-1}n}{n^2 + 2}$.

30. Write $\sum_{i=1}^{4} 2x^{i-1}$ in expanded form.

31. Find an equivalent fraction for $0.\overline{23}$.

32. Find the sum of the infinite geometric series $4 - 1 + \frac{1}{4} - \cdots$.

33. Evaluate the geometric series $\sum_{n=1}^{5} 2(3)^n$.

34. Find the eighth term in the expansion of $(x - 2y)^{11}$.

35. Evaluate the geometric series $\sum_{n=1}^{8} \left(\frac{1}{2}\right)^n$. Round to the nearest thousandth.

36. Find the sum of the infinite geometric series $2 + \frac{4}{3} + \frac{8}{9} + \cdots$.

37. Find an equivalent fraction for $0.6\overline{3}$.

38. Write $(x - 3y^2)^5$ in expanded form.

39. Find the number of terms in the finite arithmetic sequence $8, 2, -4, \ldots, -118$.

40. Evaluate: $\frac{12!}{5!\,8!}$

41. Write $\sum_{i=1}^{5} \frac{(2x)^i}{i}$ in expanded form.

42. Evaluate the series $\sum_{n=1}^{4} \frac{(-1)^{n-1}n}{n+1}$.

43. **Compensation** The salary schedule for an apprentice electrician is \$2400 for the first month and an \$80-per-month salary increase for the next 9 months. Find the total salary for the first 9 months.

44. **Temperature** The temperature of a hot-water spa is 102°F. Each hour, the temperature is 5% lower than during the previous hour. Find the temperature of the spa after 8 h. Round to the nearest tenth.

CHAPTER 11 Test

1. Write the 14th term of the sequence whose nth term is given by the formula $a_n = \frac{6}{n+4}$.

2. Expand: $(2x + 3)^4$

3. Evaluate the series $\sum_{n=1}^{4}(2n + 3)$.

4. Write $\sum_{i=1}^{4} 2x^{2i}$ in expanded form.

5. Find the 28th term of the arithmetic sequence $-12, -16, -20, \ldots$.

6. Find the formula for the nth term of the arithmetic sequence $-3, -1, 1, \ldots$.

7. Find the number of terms in the finite arithmetic sequence $7, 3, -1, \ldots, -77$.

8. Find the sum of the first 15 terms of the arithmetic sequence $-42, -33, -24, \ldots$.

9. Find the sum of the first 24 terms of the arithmetic sequence $-4, 2, 8, \ldots$.

10. Evaluate: $\frac{10!}{5!5!}$

11. Find a formula for the nth term of the geometric sequence $4, 3, \frac{9}{4}, \ldots$.

12. Find the fifth term of the geometric sequence $5, 3, \frac{9}{5}, \ldots$.

13. Find the sum of the first five terms of the geometric sequence $1, \frac{3}{4}, \frac{9}{16}, \ldots$.

14. Find the sum of the first five terms of the geometric sequence $-5, 10, -20, \ldots$.

15. Find the sum of the terms of the infinite geometric sequence $2, 1, \frac{1}{2}, \ldots$.

16. Find an equivalent fraction for $0.2\overline{3}$.

17. Evaluate: $\binom{11}{4}$

18. Find the fifth term in the expansion of $(3x - y)^8$.

19. **Inventory** An inventory of supplies for a fabric manufacturer indicated that 7500 yd of material were in stock on January 1. On February 1, and on the first of the month for each successive month, the manufacturer sent 550 yd of material to retail outlets. How much material was in stock after the shipment on October 1?

20. **Radioactivity** An ore sample contains 320 mg of a radioactive substance with a half-life of 1 day. Find the amount of radioactive material in the sample at the beginning of the fifth day.

Cumulative Review Exercises

1. Subtract: $\frac{4x^2}{x^2 + x - 2} - \frac{3x - 2}{x + 2}$

2. Factor: $2x^6 + 16$

3. Multiply: $\sqrt{2y}(\sqrt{8xy} - \sqrt{y})$

4. Simplify: $\left(\frac{x^{-\frac{3}{4}}x^{\frac{3}{2}}}{x^{-\frac{5}{2}}}\right)^{-8}$

5. Solve: $5 - \sqrt{x} = \sqrt{x + 5}$

6. Solve: $2x^2 - x + 7 = 0$

7. Solve by the addition method: $\begin{aligned} 3x - 3y &= 2 \\ 6x - 4y &= 5 \end{aligned}$

8. Solve: $2x - 1 > 3$ or $1 - 3x > 7$

9. Evaluate the determinant: $\begin{vmatrix} -3 & 1 \\ 4 & 2 \end{vmatrix}$

10. Write $\log_5 \sqrt{\frac{x}{y}}$ in expanded form.

11. Solve for x: $4^x = 8^{x-1}$

12. Write the fifth and sixth terms of the sequence whose nth term is given by the formula $a_n = n(n - 1)$.

13. Evaluate the series $\sum_{n=1}^{7} (-1)^{n-1}(n + 2)$.

14. Solve by the addition method:
$$\begin{aligned} x + 2y + z &= 3 \\ 2x - y + 2z &= 6 \\ 3x + y - z &= 5 \end{aligned}$$

15. Solve for x: $\log_6 x = 3$

16. Divide: $(4x^3 - 3x + 5) \div (2x + 1)$

17. For $g(x) = -3x + 4$, find $g(1 + h)$.

18. Find the range of $f(a) = \dfrac{a^3 - 1}{2a + 1}$ if the domain is $\{0, 1, 2\}$.

19. Graph: $3x - 2y = -4$

20. Graph the solution set of $2x - 3y < 9$.

21. Work Problem A new computer can complete a payroll in 16 min less time than it takes an older computer to complete the same payroll. Working together, both computers can complete the payroll in 15 min. How long would it take each computer, working alone, to complete the payroll?

22. Uniform Motion A boat traveling with the current went 15 mi in 2 h. Against the current, it took 3 h to travel the same distance. Find the rate of the boat in calm water and the rate of the current.

23. Radioactivity An 80-milligram sample of a radioactive material decays to 55 mg in 30 days. Use the exponential decay equation $A = A_0\left(\frac{1}{2}\right)^{\frac{t}{k}}$, where A is the amount of radioactive material present after time t, k is the half-life, and A_0 is the original amount of radioactive material, to find the half-life of the 80-milligram sample. Round to the nearest whole number.

24. Theaters A "theater in the round" has 62 seats in the first row, 74 seats in the second row, 86 seats in the third row, and so on in an arithmetic sequence. Find the total number of seats in the theater if there are 12 rows of seats.

25. Sports To test the "bounce" of a ball, the ball is dropped from a height of 10 ft. The ball bounces to 80% of its previous height with each bounce. How high does the ball bounce on the fifth bounce? Round to the nearest tenth.

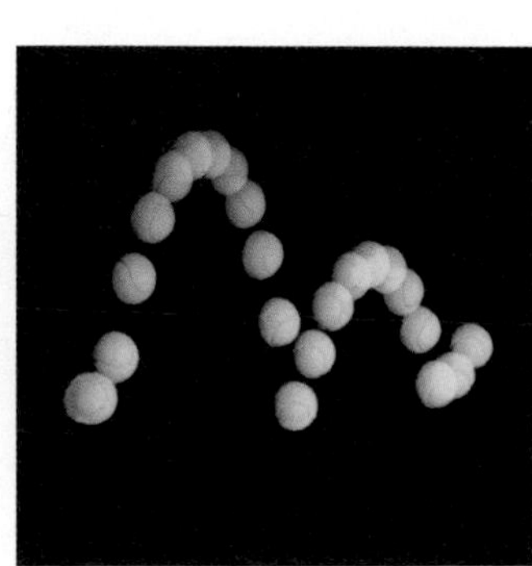

Conic Sections

CHAPTER

12

Focus on Success

The end of the semester is generally a very busy and stressful time. You may have papers or projects due, assignments to complete, and the anxiety of taking final exams. You have covered a great deal of material in this course, and reviewing all of it may be daunting. You might begin by reviewing the Chapter Summary for each chapter that you were assigned during the term. Then take the final exam on page 657. The answer to each exercise is given at the back of the book, along with the objective the exercise relates to. If you have trouble with any of the questions, restudy the objectives the questions are taken from and retry some of the exercises in those objectives. (See Ace the Test, page AIM-11.)

OBJECTIVES

12.1
1. Graph parabolas

12.2
1. Find the equation of a circle and then graph the circle
2. Write the equation of a circle in standard form and then graph the circle

12.3
1. Graph an ellipse with center at the origin
2. Graph a hyperbola with center at the origin

12.4
1. Solve nonlinear systems of equations

12.5
1. Graph the solution set of a quadratic inequality in two variables
2. Graph the solution set of a nonlinear system of inequalities

PREP TEST

Are you ready to succeed in this chapter?
Take the Prep Test below to find out if you are ready to learn the new material.

1. Find the distance between the points whose coordinates are $(-2, 3)$ and $(4, -1)$. Round to the nearest hundredth.
2. Complete the square on $x^2 - 8x$. Write the resulting perfect-square trinomial as the square of a binomial.
3. Solve $\dfrac{x^2}{16} + \dfrac{y^2}{9} = 1$ for x when $y = 3$ and when $y = 0$.
4. Solve by the substitution method:
$$7x + 4y = 3$$
$$y = x - 2$$
5. Solve by the addition method:
$$4x - y = 9$$
$$2x + 3y = -13$$
6. Find the equation of the axis of symmetry and the coordinates of the vertex of the graph of $y = x^2 - 4x + 2$.
7. Graph: $f(x) = -2x^2 + 4x$
8. Graph the solution set:
$$x + 2y \le 4$$
$$x - y \le 2$$

12.1 The Parabola

OBJECTIVE  Graph parabolas

The **conic sections** are curves that can be constructed from the intersection of a plane and a right circular cone. The four conic sections are the parabola, circle, ellipse, and hyperbola. The parabola was introduced earlier. Here we will review some of that previous discussion and look at equations of parabolas that were not discussed before.

Point of Interest

The four conic sections (parabola, circle, ellipse, and hyperbola) are obtained by slicing a cone with planes of various orientations.

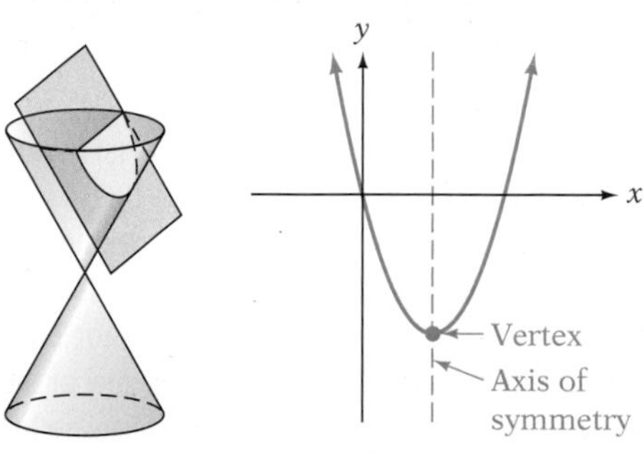

A **parabola** is a conic section formed by the intersection of a right circular cone and a plane parallel to the side of the cone. Every parabola has an **axis of symmetry** and a **vertex** that is on the axis of symmetry. To understand the axis of symmetry, think of folding the paper along that axis. The two halves of the curve will match up.

The graph of the equation $y = ax^2 + bx + c$, $a \neq 0$, is a parabola with the axis of symmetry parallel to the y-axis. The parabola opens up when $a > 0$ and opens down when $a < 0$. When the parabola opens up, the vertex is the lowest point on the parabola. When the parabola opens down, the vertex is the highest point on the parabola.

The coordinates of the vertex can be found by completing the square.

Focus on finding the vertex of a parabola by completing the square

Find the coordinates of the vertex of the parabola with equation $y = x^2 - 4x + 5$.

$$y = x^2 - 4x + 5$$

Group the terms involving x.

$$y = (x^2 - 4x) + 5$$

Complete the square on $x^2 - 4x$. Note that 4 is added and subtracted. Because $4 - 4 = 0$, the equation is not changed.

$$y = (x^2 - 4x + 4) - 4 + 5$$

Factor the trinomial and combine like terms.

$$y = (x - 2)^2 + 1$$

The coefficient of x^2 is positive, so the parabola opens up. The vertex is the lowest point on the parabola, or the point that has the least y-coordinate.

Because $(x - 2)^2 \geq 0$ for all x, the least y-coordinate occurs when $(x - 2)^2 = 0$, which occurs when $x = 2$. This means that the x-coordinate of the vertex is 2.

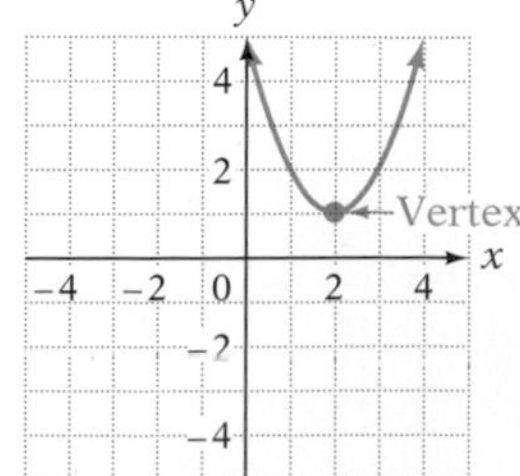

To find the y-coordinate of the vertex, replace x in $y = (x - 2)^2 + 1$ by 2 and solve for y.

$$y = (x - 2)^2 + 1$$
$$= (2 - 2)^2 + 1 = 1$$

The coordinates of the vertex are (2, 1).

Point of Interest

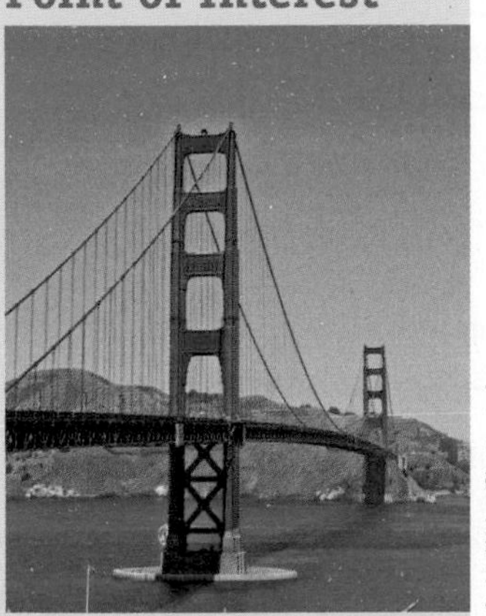

Cedr c Weber/Shutterstock.com

Golden Gate Bridge

The suspension cables for some bridges, such as the Golden Gate Bridge, hang in the shape of a parabola. Parabolic shapes are also used for mirrors in telescopes and in certain antenna designs.

By following the procedure of the last example and completing the square on the equation $y = ax^2 + bx + c$, we find that the **x-coordinate of the vertex** is $-\frac{b}{2a}$. The y-coordinate of the vertex can be determined by substituting this value of x into $y = ax^2 + bx + c$ and solving for y.

Because the axis of symmetry is parallel to the y-axis and passes through the vertex, the equation of the **axis of symmetry** is $x = -\frac{b}{2a}$.

EXAMPLE 1 Find the coordinates of the vertex and the equation of the axis of symmetry of the parabola given by the equation $y = x^2 + 2x - 3$. Then sketch the graph of the parabola.

Solution $-\dfrac{b}{2a} = -\dfrac{2}{2(1)} = -1$

- **The x-coordinate of the vertex is $-\dfrac{b}{2a}$. $a = 1, b = 2$**

$y = x^2 + 2x - 3$
$y = (-1)^2 + 2(-1) - 3$
$y = -4$

- **Find the y-coordinate of the vertex by replacing x by -1 and solving for y.**

The coordinates of the vertex are $(-1, -4)$.

The equation of the axis of symmetry is $x = -1$.

- **The equation of the axis of symmetry is $x = -\dfrac{b}{2a}$.**

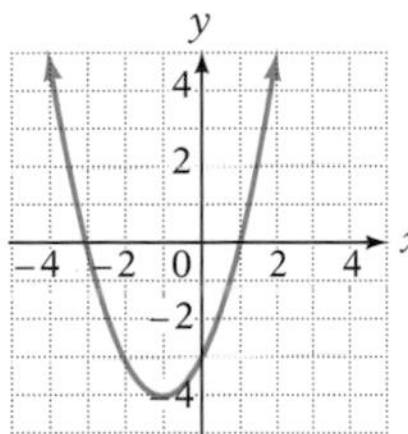

- **Because a is positive, the parabola opens up. Use the vertex and axis of symmetry to sketch the graph.**

Problem 1 Find the coordinates of the vertex and the equation of the axis of symmetry of the parabola given by the equation $y = -x^2 + x + 3$. Then sketch the graph of the parabola.

Solution See page S36.

Try Exercise 21, page 623.

The graph of an equation of the form $x = ay^2 + by + c$, $a \neq 0$, is also a parabola. In this case, the parabola opens to the right when a is positive and opens to the left when a is negative.

For a parabola of this form, the **y-coordinate of the vertex** is $-\frac{b}{2a}$. The **axis of symmetry** is the line whose equation is $y = -\frac{b}{2a}$.

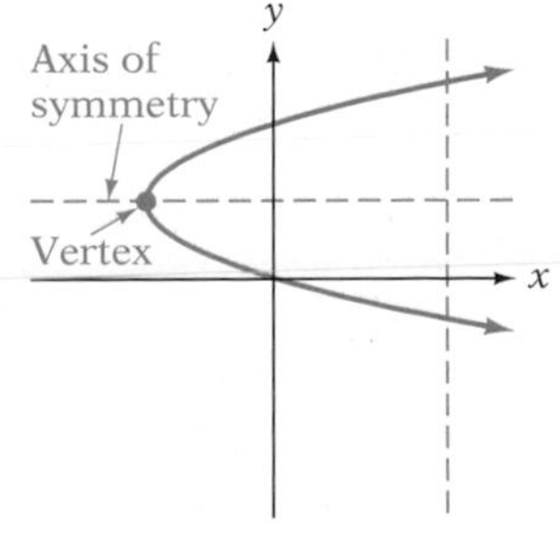

The vertical line test reveals that the graph of a parabola of this form is not the graph of a function. The graph of $x = ay^2 + by + c$ is the graph of a relation.

EXAMPLE 2 Find the coordinates of the vertex and the equation of the axis of symmetry of the parabola with equation $x = 2y^2 - 8y + 5$. Then sketch its graph.

Solution $-\frac{b}{2a} = -\frac{-8}{2(2)} = 2$
- **Find the y-coordinate of the vertex. $a = 2, b = -8$**

$x = 2y^2 - 8y + 5$
$x = 2(2)^2 - 8(2) + 5$
$x = -3$
- **Find the x-coordinate of the vertex by replacing y by 2 and solving for x.**

The coordinates of the vertex are $(-3, 2)$.

The equation of the axis of symmetry is $y = 2$.
- **The equation of the axis of symmetry is $y = -\frac{b}{2a}$.**

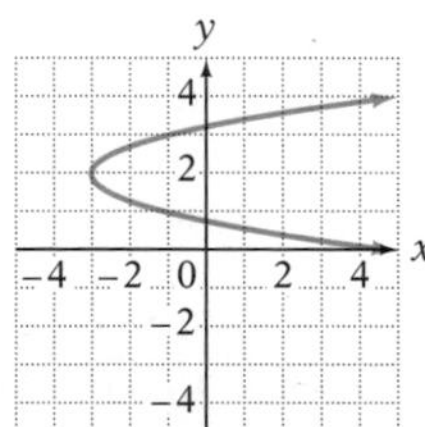

- **Because *a* is positive, the parabola opens to the right. Use the vertex and axis of symmetry to sketch the graph.**

Problem 2 Find the coordinates of the vertex and the equation of the axis of symmetry of the parabola with equation $x = -2y^2 - 4y - 3$. Then sketch its graph.

Solution See page S36.

Try Exercise 25, page 623.

12.1 Exercises

CONCEPT CHECK

1. Which of the following equations have a graph that is *not* a parabola?
(i) $y^2 = x$ (ii) $x = 3y - 5$ (iii) $x^2 - 3x + 4 - y = 0$ (iv) $y = -x + 5$

2. How many x-intercepts are possible for the graph of $y = ax^2 + bx + c, a \neq 0$?

State **a.** whether the axis of symmetry is a vertical or a horizontal line, and **b.** the direction in which the parabola opens.

3. $y = 3x^2 - 4x + 7$
4. $y = -x^2 + 5x - 2$
5. $x = y^2 + 2y - 8$
6. $x = -3y^2 - y + 9$
7. $x = -\frac{1}{2}y^2 - 4y - 7$
8. $y = \frac{1}{4}x^2 + 6x - 1$

1 Graph parabolas (See pages 620–622.)

(See pages 620–622.)

GETTING READY

9. For a parabola with an equation in the form $x = ay^2 + by + c$, $-\frac{b}{2a}$ is the __?__-coordinate of the vertex. The parabola opens to the right when a is __?__. The parabola opens to the left when a is __?__.

10. a. For the equation $y = 3x^2 - 4y + 6$, the value of $-\frac{b}{2a}$ is $-\frac{?}{2(?)} =$ __?__.

b. The equation of the axis of symmetry of the parabola with equation $y = 3x^2 - 4y + 6$ is __?__ $= \frac{2}{3}$.

Find the coordinates of the vertex and the equation of the axis of symmetry of the parabola with the given equation.

11. $y = 2x^2 + 8x - 1$ **12.** $x = -y^2 - 4y + 1$ **13.** $x = y^2 + 3$ **14.** $y = 2x^2 + 5$

15. The equation of the axis of symmetry of a parabola is $x = 2$, and $(3, 4)$ are the coordinates of a point on the parabola. Use this information to find the coordinates of a second point on the parabola.

16. The equation of the axis of symmetry of a parabola is $y = -1$, and $(-2, 5)$ are the coordinates of a point on the parabola. Use this information to find the coordinates of a second point on the parabola.

17. The equation of the axis of symmetry of a parabola is $y = 3$, and $(7, 0)$ are the coordinates of a point on the parabola. Use this information to find the coordinates of a second point on the parabola.

18. The equation of the axis of symmetry of a parabola is $x = -3$, and $(0, -2)$ are the coordinates of a point on the parabola. Use this information to find the coordinates of a second point on the parabola.

Find the coordinates of the vertex and the equation of the axis of symmetry of the parabola given by the equation. Then sketch its graph.

19. $y = x^2 - 2x - 4$ **20.** $y = x^2 + 4x - 4$

21. $y = -x^2 + 2x - 3$ **22.** $y = -x^2 + 4x - 5$

23. $x = y^2 + 6y + 5$ **24.** $x = y^2 - y - 6$

25. $x = -y^2 + 3y - 4$ **26.** $x = -y^2 - 4y + 1$

27. $y = 2x^2 - 4x + 1$

28. $y = 2x^2 + 4x - 5$

29. $y = x^2 - 5x + 4$

30. $y = x^2 + 5x + 6$

31. $x = y^2 - 2y - 5$

32. $x = y^2 - 3y - 4$

33. $y = -3x^2 - 9x$

34. $y = -2x^2 + 6x$

35. $x = -\frac{1}{2}y^2 + 4$

36. $x = -\frac{1}{4}y^2 - 1$

37. $x = \frac{1}{2}y^2 - y + 1$

38. $x = -\frac{1}{2}y^2 + 2y - 3$

39. $y = \frac{1}{2}x^2 + 2x - 6$

40. $y = -\frac{1}{2}x^2 + x - 3$

 State whether the equation of the parabola shown in the graph is of the form $y = ax^2 + bx + c$ or of the form $x = ay^2 + by + c$. Then state whether a is positive or negative and whether c is positive, zero, or negative.

41.

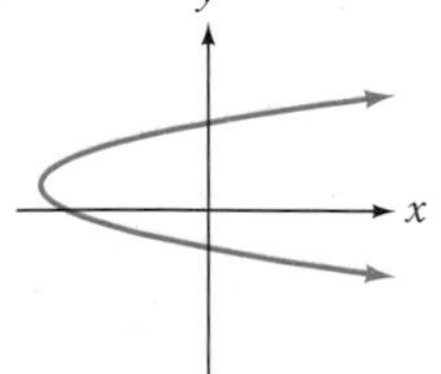

42.

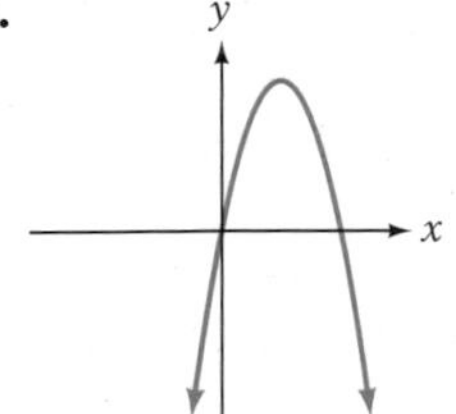

43.

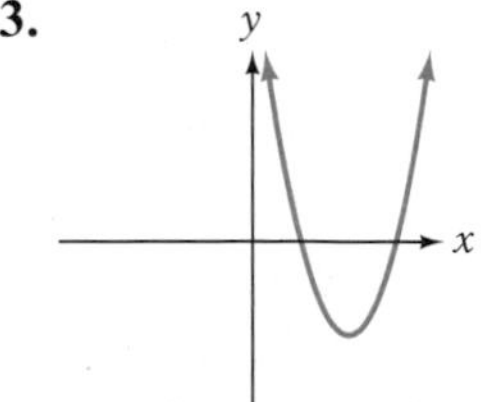

44.

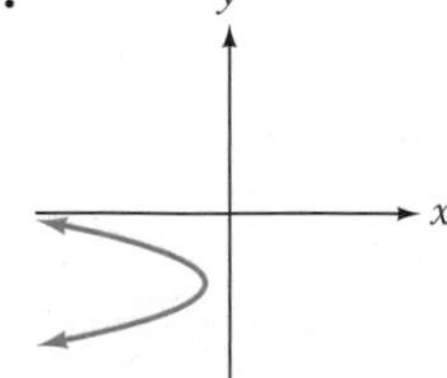

APPLYING CONCEPTS

Parabolas have a unique property that is important in the design of telescopes and antennas. If a parabola has a mirrored surface, then all light rays parallel to the axis of symmetry of the parabola are reflected to a single point called the **focus** of the parabola. The location of this point is p units from the vertex on the axis of symmetry. The value of p is given by $p = \frac{1}{4a}$, where $y = ax^2$ is the equation of a parabola with vertex at the origin. For the graph of $y = \frac{1}{4}x^2$ shown at the right, the coordinates of the focus are $(0, 1)$. For Exercises 45 and 46, find the coordinates of the focus of the parabola given by the equation.

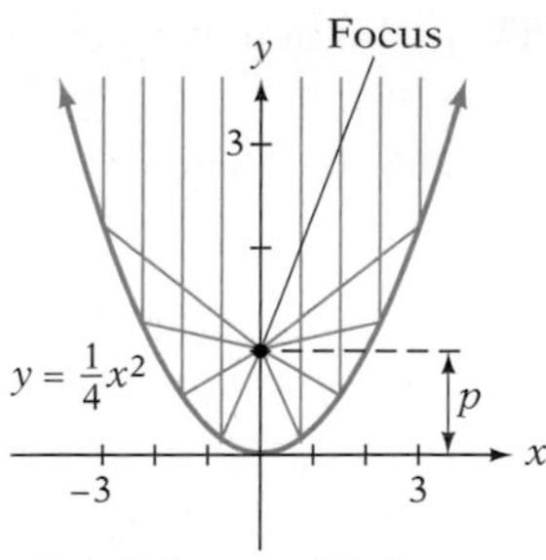

Parallel rays of light are reflected to the focus

$a = \frac{1}{4}$

$p = \frac{1}{4a} = \frac{1}{4(1/4)} = 1$

45. $y = 2x^2$

46. $y = \frac{1}{10}x^2$

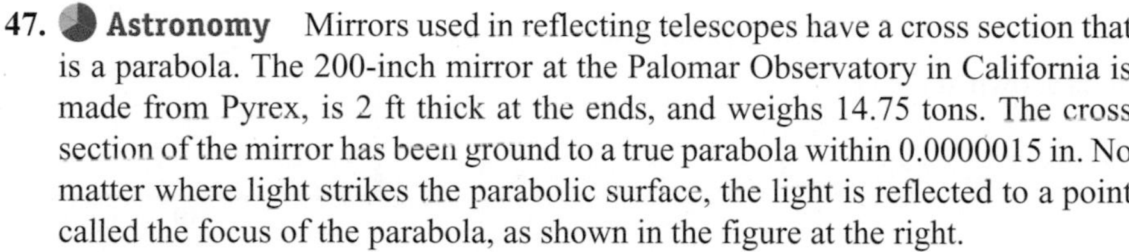

47. **Astronomy** Mirrors used in reflecting telescopes have a cross section that is a parabola. The 200-inch mirror at the Palomar Observatory in California is made from Pyrex, is 2 ft thick at the ends, and weighs 14.75 tons. The cross section of the mirror has been ground to a true parabola within 0.0000015 in. No matter where light strikes the parabolic surface, the light is reflected to a point called the focus of the parabola, as shown in the figure at the right.

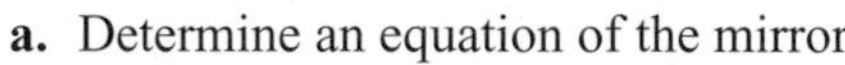

a. Determine an equation of the mirror.

b. Over what interval for x is the equation valid?

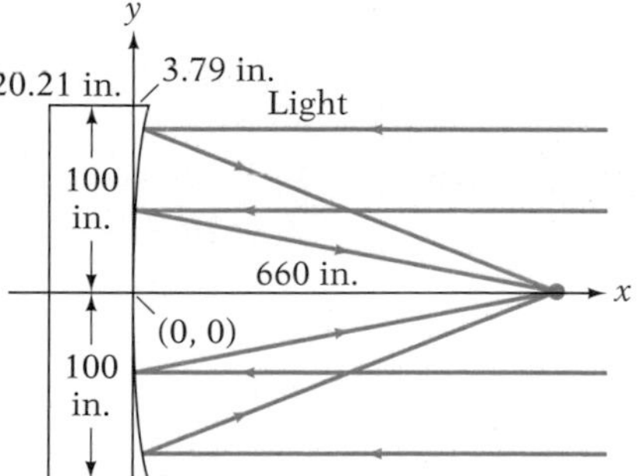

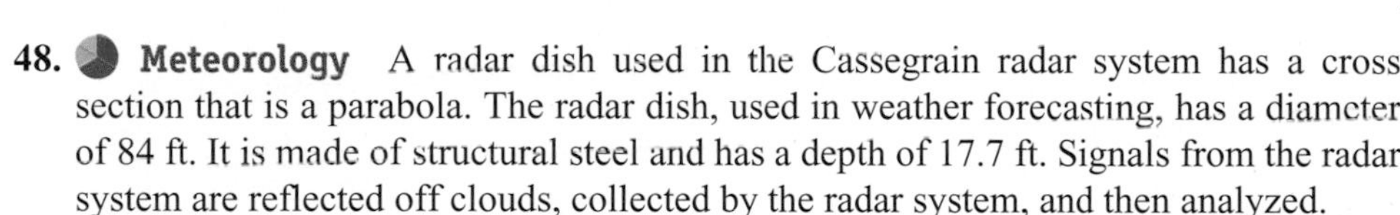

48. **Meteorology** A radar dish used in the Cassegrain radar system has a cross section that is a parabola. The radar dish, used in weather forecasting, has a diameter of 84 ft. It is made of structural steel and has a depth of 17.7 ft. Signals from the radar system are reflected off clouds, collected by the radar system, and then analyzed.

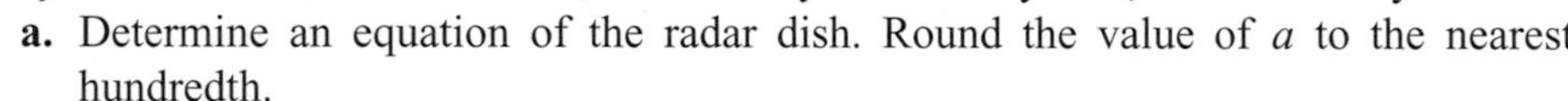

a. Determine an equation of the radar dish. Round the value of a to the nearest hundredth.

b. Over what interval for x is the equation valid?

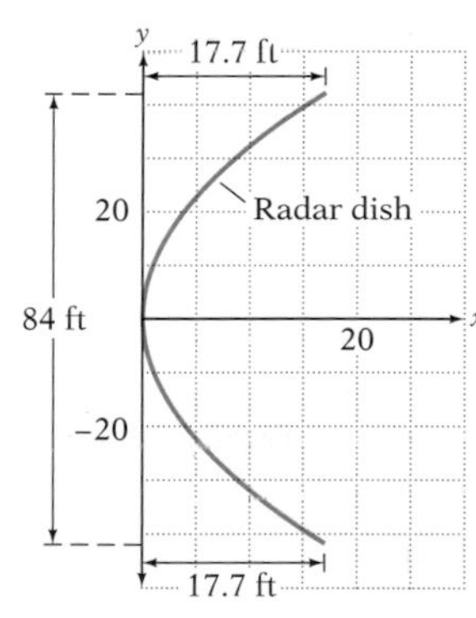

Cassegrain Radar Dish

49. Explain how the graph of $f(x) = ax^2$ changes depending on the value of a.

50. Explain why the equation $x = y^2 + 4$ does not define y as a function of x.

51. When $y = 0$, the x values of the equation $y = x^2 + 2x + 3$ are imaginary numbers. What does this mean for the graph of the equation?

PROJECTS OR GROUP ACTIVITIES

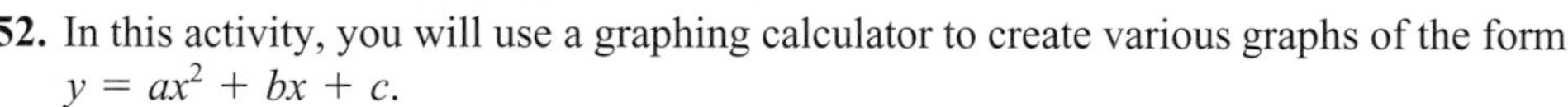

52. In this activity, you will use a graphing calculator to create various graphs of the form $y = ax^2 + bx + c$.

a. Graph $y = ax^2 + 3x - 2$ for $a = 0.25, 0.5, 1, 2,$ and 3. How does the graph change as the value of a changes?

b. Graph $y = x^2 + bx - 2$ for $b = -2, -1, 0, 1,$ and 2. How does the graph change as the value of b changes?

c. Graph $y = x^2 + 2x + c$ for $c = -2, -1, 0, 1,$ and 2. How does the graph change as the value of c changes?

d. Write an expression involving a, b, and c that you can use to determine whether the graph of $y = ax^2 + bx + c$ has zero, one, or two x-intercepts. (*Note:* The expression was discussed earlier in the text.)

53. In this activity, you will use a graphing calculator to create various graphs of the form $x = ay^2 + by + c$. Recall that the graph of an equation of this form is not the graph of a function. To create the graph, first solve the equation for y. For instance, to graph $x = y^2 - 5y + 3$, first solve for y.

$$x = y^2 - 5y + 3$$

$$0 = y^2 - 5y + 3 - x$$ • **Write the equation in standard form.**

$$y = \frac{-(-5) \pm \sqrt{(-5)^2 - 4(1)(3 - x)}}{2(1)}$$ • **Use the quadratic formula.** $a = 1, b = -5, c = 3 - x$

$$y = \frac{5 \pm \sqrt{25 - 4(3 - x)}}{2}$$

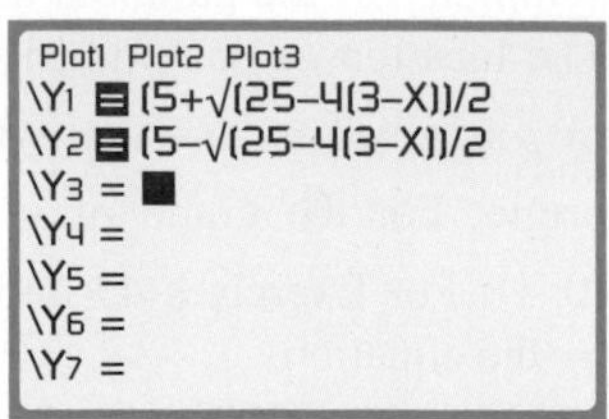

This yields two equations, $y = \frac{5 + \sqrt{25 - 4(3 - x)}}{2}$ and $y = \frac{5 - \sqrt{25 - 4(3 - x)}}{2}$, that represent the top and bottom halves of the graph. Enter these equations into Y_1 and Y_2. Then graph the equations. The result is shown at the right.

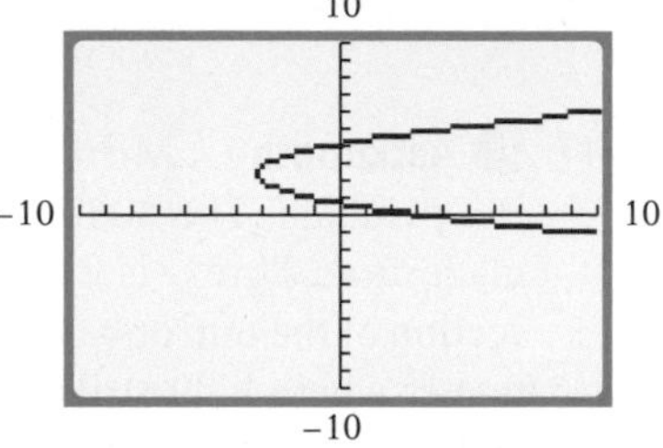

a. Graph $x = y^2 - 4y + 2$.
b. Graph $x = 2y^2 - 8y + 1$.
c. Graph $x = y^2 + 2y - 3$.

12.2 The Circle

OBJECTIVE 1 Find the equation of a circle and then graph the circle

A *circle* is a conic section formed by the intersection of a cone and a plane that is parallel to the base of the cone.

Take Note

As the angle of the plane that intersects the cone changes, different conic sections are formed. For a parabola, the plane is *parallel to the side* of the cone. For a circle, the plane is *parallel to the base* of the cone.

A **circle** can be defined as all the points $P(x, y)$ in the plane that are a fixed distance from a given point $C(h, k)$ called the **center.** The fixed distance is the **radius** of the circle.

The equation of a circle can be determined by using the distance formula.

Let (h, k) be the coordinates of the center of the circle, let r be the radius, and let (x, y) be the coordinates of any point on the circle. Then, by the distance formula,

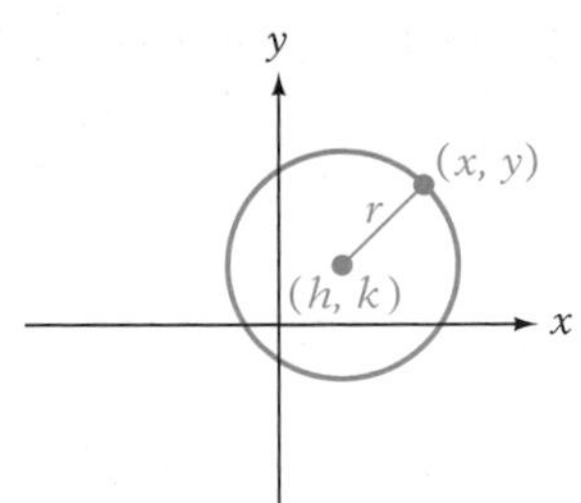

$$r = \sqrt{(x - h)^2 + (y - k)^2}$$

Squaring each side of the equation gives the standard form of the equation of a circle.

$$r^2 = \left[\sqrt{(x - h)^2 + (y - k)^2}\right]^2$$
$$r^2 = (x - h)^2 + (y - k)^2$$

Take Note

In the standard form of the equation of a circle, $(x - h)^2$ and $(y - k)^2$ are written using subtraction. Because $(y + 3)^2$ is written using addition, the expression is rewritten using subtraction as $[y - (-3)]^2$. Note that $y - (-3) = y + 3$. Also,

$$r^2 = 16$$
$$\sqrt{r^2} = \sqrt{16}$$
$$r = 4$$

STANDARD FORM OF THE EQUATION OF A CIRCLE

The standard form of the equation of a circle with center $C(h, k)$ and radius r is

$$(x - h)^2 + (y - k)^2 = r^2$$

EXAMPLES

1. The equation $(x - 3)^2 + (y - 1)^2 = 6^2$ is the equation of a circle in standard form, with $h = 3$ and $k = 1$. Therefore, the coordinates of the center are (3, 1). Because $r = 6$, the radius of the circle is 6.
2. The equation $(x - 2)^2 + (y + 3)^2 = 16$ is not in standard form. In standard form, the equation is written as $(x - 2)^2 + [y - (-3)]^2 = 4^2$, with $h = 2$ and $k = -3$. Therefore, the coordinates of the center are $(2, -3)$. Because $r = 4$, the radius of the circle is 4. See the Take Note at the left.

Focus on using the equation of a circle in standard form

A. Find the coordinates of the center and the radius of the circle given by the equation $(x + 3)^2 + y^2 = 25$.

$$(x + 3)^2 + y^2 = 25$$

Write the equation in standard form. $\quad [x - (-3)]^2 + (y - 0)^2 = 5^2$

$(x + 3)^2 = [x - (-3)]^2$;
$y^2 = (y - 0)^2$; $r = \sqrt{25} = 5$

The coordinates of the center are $(-3, 0)$. The radius is 5.

B. Find the equation of the circle with radius 4 and center $C(-1, 2)$.

Use the standard form of the equation of a circle. $\quad (x - h)^2 + (y - k)^2 = r^2$

Replace r by **4**, h by -1, and k by 2. $\quad [x - (-1)]^2 + (y - 2)^2 = 4^2$

Simplify. $\quad (x + 1)^2 + (y - 2)^2 = 16$

The equation of the circle is $(x + 1)^2 + (y - 2)^2 = 16$.

Point of Interest

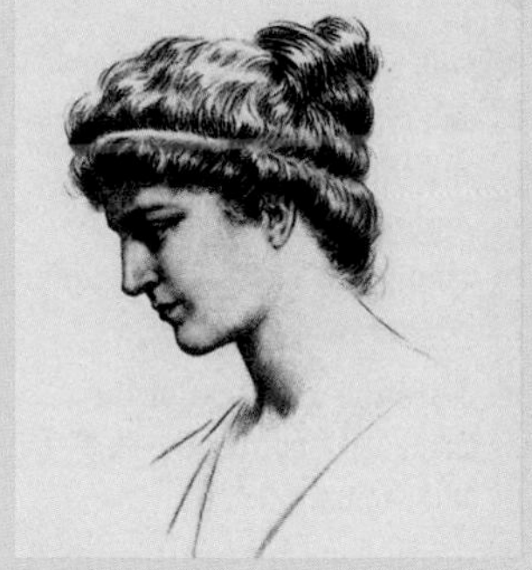

Hypatia

Hypatia (c. 340–415) is considered the first prominent woman mathematician. She lectured in mathematics and philosophy at the Museum in Alexandria, at that time the most distinguished place of learning in the world. One of the topics on which Hypatia lectured was conic sections. One historian has claimed that with the death (actually the murder) of Hypatia, "the long and glorious history of Greek mathematics was at an end."

EXAMPLE 1 Find the equation of the circle that passes through the point $P(-1, 4)$ and whose center is the point $C(2, -3)$.

Solution The radius of the circle is the distance from the center C to the point P. Use the distance formula to find this distance.

$$r = \sqrt{(x_2 - x_1)^2 + (y_2 - y_1)^2}$$

$$r = \sqrt{[2 - (-1)]^2 + (-3 - 4)^2}$$ • $(x_1, y_1) = (-1, 4)$, $(x_2, y_2) = (2, -3)$

$$r = \sqrt{3^2 + (-7)^2} = \sqrt{9 + 49}$$

$$r = \sqrt{58}$$

$$(x - 2)^2 + [y - (-3)]^2 = (\sqrt{58})^2$$ • **The radius of the circle is $\sqrt{58}$. Use the coordinates of the center $C(2, -3)$ and the radius to write the equation.**

$$(x - 2)^2 + (y + 3)^2 = 58$$

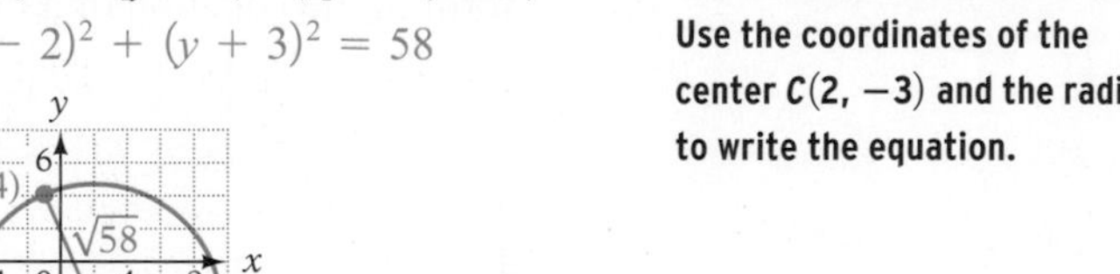

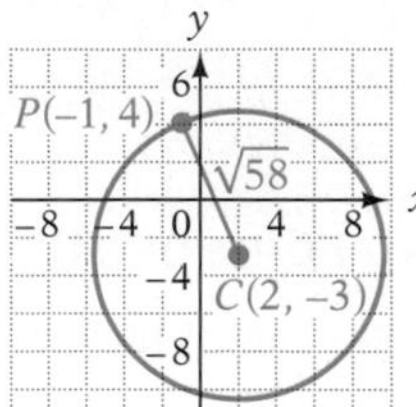

Problem 1 Find the equation of the circle that passes through the point $P(-2, 3)$ and whose center is $C(3, 5)$.

Solution See page S36.

Try Exercise 19, page 631.

EXAMPLE 2 Find the equation of the circle for which a diameter has endpoints $P_1(-4, -1)$ and $P_2(2, 3)$.

Solution

$$x_m = \frac{x_1 + x_2}{2} \qquad y_m = \frac{y_1 + y_2}{2}$$

$$x_m = \frac{-4 + 2}{2} \qquad y_m = \frac{-1 + 3}{2}$$

$$x_m = -1 \qquad y_m = 1$$

• **Let $(x_1, y_1) = (-4, -1)$ and $(x_2, y_2) = (2, 3)$. Find the center of the circle by finding the midpoint of the diameter.**

Center: $(x_m, y_m) = (-1, 1)$

$$r = \sqrt{(x_1 - x_m)^2 + (y_1 - y_m)^2}$$

$$r = \sqrt{[-4 - (-1)]^2 + (-1 - 1)^2}$$

$$r = \sqrt{9 + 4}$$

$$r = \sqrt{13}$$

• **Find the radius of the circle. Use either point on the circle and the coordinates of the center of the circle. P_1 is used here.**

$$(x + 1)^2 + (y - 1)^2 = 13$$

• **Write the equation of the circle with center $C(-1, 1)$ and radius $\sqrt{13}$.**

Problem 2 Find the equation of the circle for which a diameter has endpoints $P_1(-2, 1)$ and $P_2(4, -1)$.

Solution See page S36.

Try Exercise 21, page 631.

OBJECTIVE 2 Write the equation of a circle in standard form and then graph the circle

The equation of a circle can also be expressed in **general form** as

$$x^2 + y^2 + ax + by + c = 0$$

To rewrite this equation in standard form, it is necessary to complete the square on the x and y terms.

Focus on writing the equation of a circle in standard form

Write the equation of the circle $x^2 + y^2 + 4x + 2y + 1 = 0$ in standard form.

$$x^2 + y^2 + 4x + 2y + 1 = 0$$

Subtract the constant term from each side of the equation.

$$x^2 + y^2 + 4x + 2y = -1$$

Rewrite the equation by grouping terms involving x and terms involving y.

$$(x^2 + 4x) + (y^2 + 2y) = -1$$

Complete the square on $x^2 + 4x$ and $y^2 + 2y$.

$$(x^2 + 4x + 4) + (y^2 + 2y + 1) = -1 + 4 + 1$$
$$(x^2 + 4x + 4) + (y^2 + 2y + 1) = 4$$

Factor each trinomial.

$$(x + 2)^2 + (y + 1)^2 = 4$$

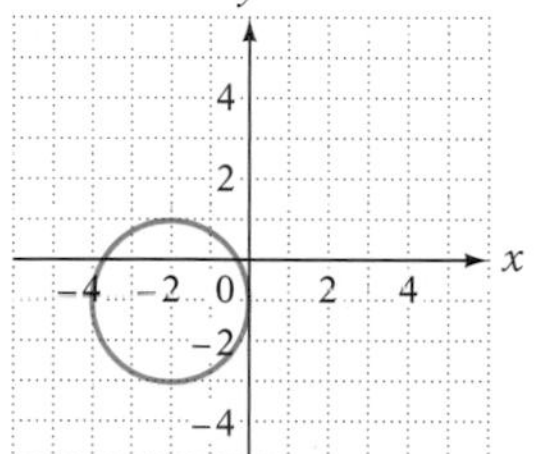

The graph of the equation $(x + 2)^2 + (y + 1)^2 = 4$ is shown at the left.

EXAMPLE 3 Write the equation of the circle $x^2 + y^2 + 3x - 2y = 1$ in standard form. Then sketch its graph.

Solution

$$x^2 + y^2 + 3x - 2y = 1$$
$$(x^2 + 3x) + (y^2 - 2y) = 1$$

- Group terms involving x and terms involving y.

$$\left(x^2 + 3x + \frac{9}{4}\right) + (y^2 - 2y + 1) = 1 + \frac{9}{4} + 1$$

- Complete the square on $x^2 + 3x$ and $y^2 - 2y$.

$$\left(x + \frac{3}{2}\right)^2 + (y - 1)^2 = \frac{17}{4}$$

- Factor each trinomial.

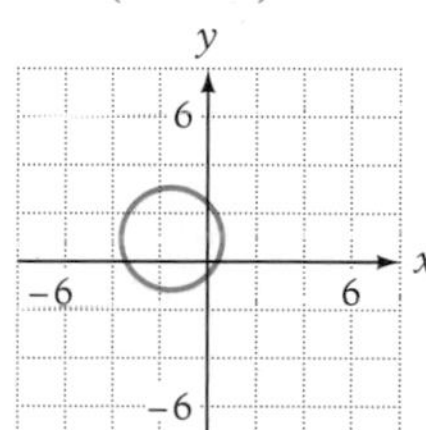

- Draw a circle with center $\left(-\frac{3}{2}, 1\right)$ and radius $\sqrt{\frac{17}{4}} = \frac{\sqrt{17}}{2} \approx 2.1$.

Problem 3 Write the equation of the circle $x^2 + y^2 - 4x + 8y + 15 = 0$ in standard form. Then sketch its graph.

Solution See page S36.

Try Exercise 27, page 632.

12.2 Exercises

CONCEPT CHECK

1. Describe how the points on the circumference of a circle are related to the center of the circle.

2. What do the values of h, k, and r represent in the equation of a circle in standard form?

3. Is the graph of a circle the graph of a function? Why or why not?

4. Which of the following are *not* the equation of a circle?
(i) $(x - 3) + (y - 4)^2 = 16$ (ii) $x^2 + y^2 = 1$ (iii) $(x + 1)^2 - (y - 3)^2 = 36$
(iv) $(x + 5)^2 + (y + 5)^2 = 5$ (v) $(x - 1) + (y - 4) = 25$ (vi) $(x + 3)^2 + (y - 2)^2 = -25$

1 Find the equation of a circle and then graph the circle (See pages 626–628.)

GETTING READY

5. The equation $(x - 1)^2 + (y + 4)^2 = 36$ is in the form

$$(x - h)^2 + (y - k)^2 = r^2$$

where $h =$ __?__, $k =$ __?__, and $r =$ __?__. This means that $(x - 1)^2 + (y + 4)^2 = 36$ is the equation of the circle with center $C($__?__, __?__$)$ and radius __?__.

6. One way to sketch the graph of the circle in Exercise 5 is to locate four points on the circle:
   a. The coordinates of the point that is 6 units above the center $C(1, -4)$ are $(1, -4 + 6)$, or $(1,$ __?__$)$.
   b. The coordinates of the point that is 6 units below the center $C(1, -4)$ are $(1, -4 -$ __?__$)$, or $(1,$ __?__$)$.
   c. The coordinates of the point that is 6 units to the right of the center $C(1, -4)$ are $(1 + 6,$ __?__$)$, or $($__?__, __?__$)$.
   d. The coordinates of the point that is 6 units to the left of the center $C(1, -4)$ are $($__?__, __?__$)$, or $($__?__, __?__$)$.

Sketch a graph of the circle given by the equation.

7. $(x - 2)^2 + (y + 2)^2 = 9$

8. $(x + 2)^2 + (y - 3)^2 = 16$

9. $(x + 3)^2 + (y - 1)^2 = 25$

10. $(x - 2)^2 + (y + 3)^2 = 4$

11. $(x - 4)^2 + (y + 2)^2 = 1$

12. $(x - 3)^2 + (y - 2)^2 = 16$

13. $(x + 5)^2 + (y + 2)^2 = 4$

14. $(x + 1)^2 + (y - 1)^2 = 9$

15. If $h > 0$ and $k < 0$, in which quadrant does the center of the circle with equation $(x - h)^2 + (y - k)^2 = r^2$ lie?

16. Does the circle with equation $(x + 2)^2 + (y - 3)^2 = 4$ pass through quadrant III?

17. Find the equation of the circle with radius 2 and center $C(2, -1)$. Then sketch its graph.

18. Find the equation of the circle with radius 3 and center $C(-1, -2)$. Then sketch its graph.

19. Find the equation of the circle that passes through the point $P(1, 2)$ and whose center is the point $C(-1, 1)$. Then sketch its graph.

20. Find the equation of the circle that passes through the point $P(-1, 3)$ and whose center is the point $C(-2, 1)$. Then sketch its graph.

21. Find the equation of the circle for which a diameter has endpoints $P_1(-1, 4)$ and $P_2(-5, 8)$.

22. Find the equation of the circle for which a diameter has endpoints $P_1(2, 3)$ and $P_2(5, -2)$.

23. Find the equation of the circle for which a diameter has endpoints $P_1(-4, 2)$ and $P_2(0, 0)$.

24. Find the equation of the circle for which a diameter has endpoints $P_1(-8, -3)$ and $P_2(0, -4)$.

2 Write the equation of a circle in standard form and then graph the circle (See pages 628–629.)

GETTING READY

25. a. The first two steps in writing the equation of the circle

$$x^2 + y^2 + 8x - 4y - 5 = 0$$

in standard form are (1) to add 5 to each side of the equation and (2) to group the terms involving x and group the terms involving y. The result is

$(x^2 + \underline{\quad ? \quad}) + (y^2 - \underline{\quad ? \quad}) = \underline{\quad ? \quad}$.

b. The next step is to complete the square on $x^2 + 8x$ by adding $\underline{\quad ? \quad}$ to each side of the equation, and to complete the square on $y^2 - 4y$ by adding $\underline{\quad ? \quad}$ to each side of the equation. The result is

$(x^2 + 8x + \underline{\quad ? \quad}) + (y^2 - 4y + \underline{\quad ? \quad}) = 5 + \underline{\quad ? \quad} + \underline{\quad ? \quad}$.

c. Write the equation from part (b) in standard form by factoring each perfect-square trinomial on the left side of the equation and simplifying the right side of the equation:

$(\underline{\quad ? \quad})^2 + (\underline{\quad ? \quad})^2 = \underline{\quad ? \quad}$.

26. Use the result of part (c) of Exercise 25 to find the coordinates of the center and the radius of the circle with equation $x^2 + y^2 + 8x - 4y - 5 = 0$. The coordinates of the center are $\underline{\quad ? \quad}$ and the radius is $\underline{\quad ? \quad}$.

Write the equation of the circle in standard form. Then sketch its graph.

27. $x^2 + y^2 - 2x + 4y - 20 = 0$

28. $x^2 + y^2 - 4x + 8y + 4 = 0$

29. $x^2 + y^2 + 6x + 8y + 9 = 0$

30. $x^2 + y^2 - 6x + 10y + 25 = 0$

31. $x^2 + y^2 - x + 4y + \frac{13}{4} = 0$

32. $x^2 + y^2 + 4x + y + \frac{1}{4} = 0$

33. $x^2 + y^2 - 6x + 4y + 4 = 0$

34. $x^2 + y^2 - 10x + 8y + 40 = 0$

Use the general form of the equation of a circle, $x^2 + y^2 + ax + by + c = 0$. Match the conditions given for a and b with circle A, B, C, or D on the graph.

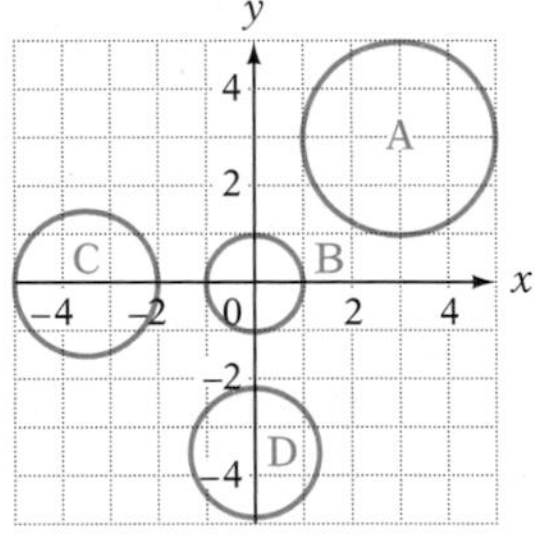

35. $a = 0$ and $b = 0$

36. $a < 0$ and $b < 0$

37. $a = 0$ and $b > 0$

38. $a > 0$ and $b = 0$

APPLYING CONCEPTS

In Exercises 39 to 41, write the equation of the circle in standard form.

39. The circle has its center at the point $C(3, 0)$ and passes through the origin.

40. The circle has radius 5 and center $C(6, -3)$.

41. The circle has radius 1, is tangent to both the x- and y-axes, and lies in quadrant II.

42. Is $x^2 + y^2 + 4x + 8y + 24 = 0$ the equation of a circle? If not, explain why not. If so, find the radius and the coordinates of the center.

PROJECTS OR GROUP ACTIVITIES

43. **Geometry** The radius of a sphere is 12 in. What is the radius of the circle that is formed by the intersection of the sphere and a plane that is 6 in. from the center of the sphere?

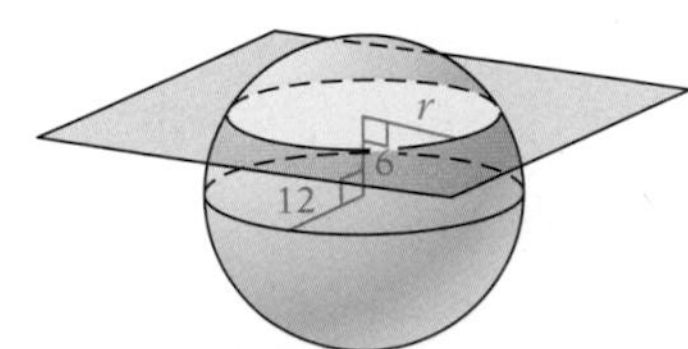

44. Find the area of the smallest region bounded by the graphs of $y = |x|$ and $x^2 + y^2 = 4$.

45. The line with equation $x = 3$ crosses the circle with equation $x^2 + y^2 = 34$ at points A and B. Find the length of AB.

12.3 The Ellipse and the Hyperbola

OBJECTIVE 1 Graph an ellipse with center at the origin

The orbits of the planets around the sun are "oval" shaped. This oval shape can be described as an **ellipse,** which is another of the conic sections.

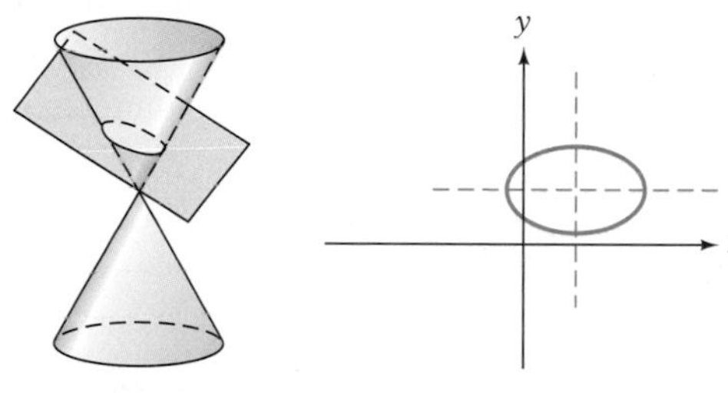

An ellipse has two **axes of symmetry.** The intersection of these two axes is the **center** of the ellipse.

An ellipse with center at the origin is shown at the right. Note that there are two x-intercepts and two y-intercepts.

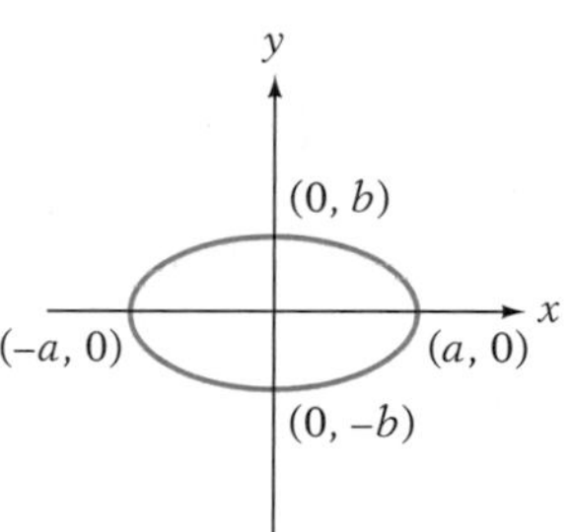

Point of Interest

The word *ellipse* comes from the Greek word *ellipsis*, which means "deficient." The method by which the early Greeks analyzed the conics caused a certain area in the construction of the ellipse to be less than another area (deficient). The word *ellipsis* in English, which means "omission," has the same Greek root as the word *ellipse*.

STANDARD FORM OF THE EQUATION OF AN ELLIPSE WITH CENTER AT THE ORIGIN

The equation of an ellipse with center at the origin is $\frac{x^2}{a^2} + \frac{y^2}{b^2} = 1$. The coordinates of the x-intercepts are $(a, 0)$ and $(-a, 0)$. The coordinates of the y-intercepts are $(0, b)$ and $(0, -b)$.

By finding the coordinates of the x- and y-intercepts of an ellipse and using the fact that the ellipse is "oval" shaped, we can sketch a graph of an ellipse.

EXAMPLE 1 Sketch a graph of the ellipse given by the equation.

A. $\frac{x^2}{9} + \frac{y^2}{4} = 1$ **B.** $\frac{x^2}{16} + \frac{y^2}{12} = 1$

Solution **A.** $\frac{x^2}{9} + \frac{y^2}{4} = 1$ • $a^2 = 9, b^2 = 4$

x-intercepts: $(3, 0)$ and $(-3, 0)$ • **The coordinates of the x-intercepts are $(a, 0)$ and $(-a, 0)$.**

y-intercepts: $(0, 2)$ and $(0, -2)$ • **The coordinates of the y-intercepts are $(0, b)$ and $(0, -b)$.**

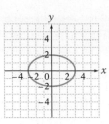

Take Note

Applying the vertical line test reveals that the graph of an ellipse is not the graph of a function. The graph of an ellipse is the graph of a relation.

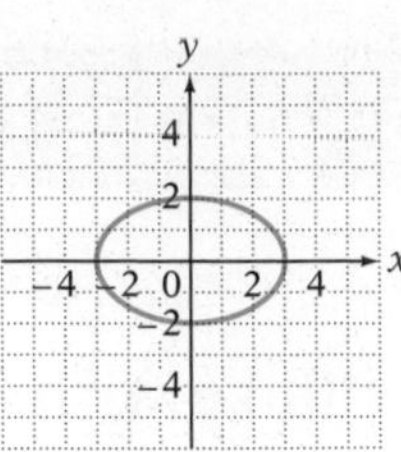

- **Use the intercepts and symmetry to sketch the graph of the ellipse.**

B. $\frac{x^2}{16} + \frac{y^2}{12} = 1$

- **$a^2 = 16, b^2 = 12$**

x-intercepts:
(4, 0) and (−4, 0)

- **The coordinates of the x-intercepts are $(a, 0)$ and $(-a, 0)$.**

y-intercepts:
$(0, 2\sqrt{3})$ and $(0, -2\sqrt{3})$

- **The coordinates of the y-intercepts are $(0, b)$ and $(0, -b)$.**

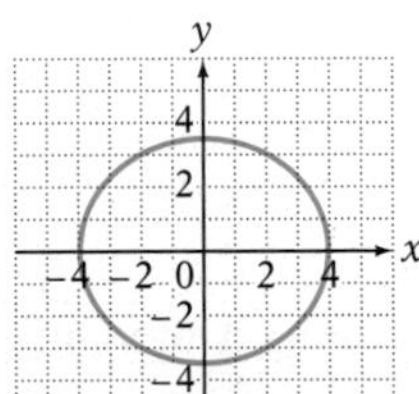

- **Use the intercepts and symmetry to sketch the graph of the ellipse. $2\sqrt{3} \approx 3.5$**

Point of Interest

Early Greek astronomers thought that each planet had a circular orbit. Today we know that the planets have elliptical orbits. However, in most cases, the ellipse is very nearly a circle. Note that when $a = b$ in the equation $\frac{x^2}{a^2} + \frac{y^2}{b^2} = 1$, the graph is the graph of a circle.

Problem 1 Sketch a graph of the ellipse given by the equation.

A. $\frac{x^2}{4} + \frac{y^2}{25} = 1$ **B.** $\frac{x^2}{18} + \frac{y^2}{9} = 1$

Solution See pages S36–S37.

Try Exercise 7, page 637.

OBJECTIVE 2 Graph a hyperbola with center at the origin

A **hyperbola** is a conic section that is formed by the intersection of a right circular cone and a plane perpendicular to the base of the cone.

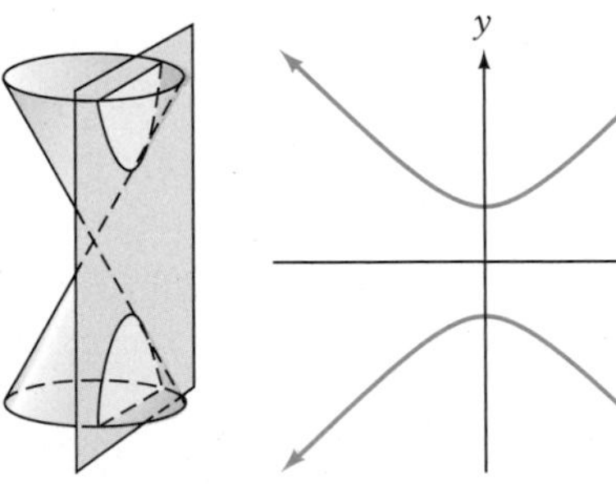

The hyperbola has two **vertices** and an **axis of symmetry** that passes through the vertices. The **center** of a hyperbola is the point halfway between the vertices.

The graphs below show two graphs of a hyperbola with center at the origin.

In the first graph, the axis of symmetry that contains the vertices is the x-axis.

In the second graph, the axis of symmetry that contains the vertices is the y-axis.

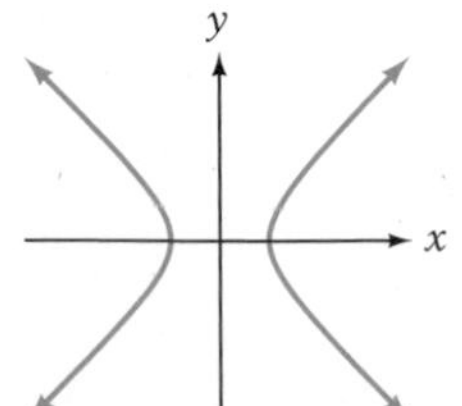

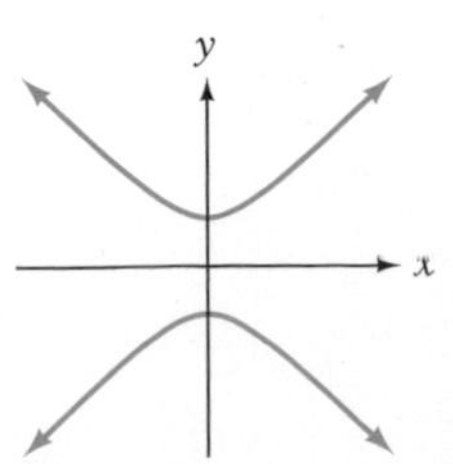

Note that in either case, the graph of a hyperbola is not the graph of a function. The graph of a hyperbola is the graph of a relation.

STANDARD FORM OF THE EQUATION OF A HYPERBOLA WITH CENTER AT THE ORIGIN

The equation of a hyperbola for which the axis of symmetry that contains the vertices is the x-axis is $\frac{x^2}{a^2} - \frac{y^2}{b^2} = 1$. The coordinates of the vertices are $(a, 0)$ and $(-a, 0)$.

The equation of a hyperbola for which the axis of symmetry that contains the vertices is the y-axis is $\frac{y^2}{b^2} - \frac{x^2}{a^2} = 1$. The coordinates of the vertices are $(0, b)$ and $(0, -b)$.

Point of Interest

The word *hyperbola* comes from the Greek word *yperboli*, which means "exceeding." The method by which the early Greeks analyzed the conics caused a certain area in the construction of the hyperbola to be greater than (exceed) another area. The word *hyperbole* in English, which means "exaggeration," has the same Greek root as the word *hyperbola*.

The word *asymptote* comes from the Greek word *asymptotos*, which means "not capable of meeting."

To sketch a hyperbola, it is helpful to draw two lines that are "approached" by the hyperbola. These two lines are called **asymptotes.** As a point on the hyperbola gets farther from the origin, the hyperbola "gets closer to" the asymptotes.

Because the asymptotes are straight lines, their equations are linear equations. The equations of the asymptotes of a hyperbola with center at the origin are $y = \frac{b}{a}x$ and $y = -\frac{b}{a}x$.

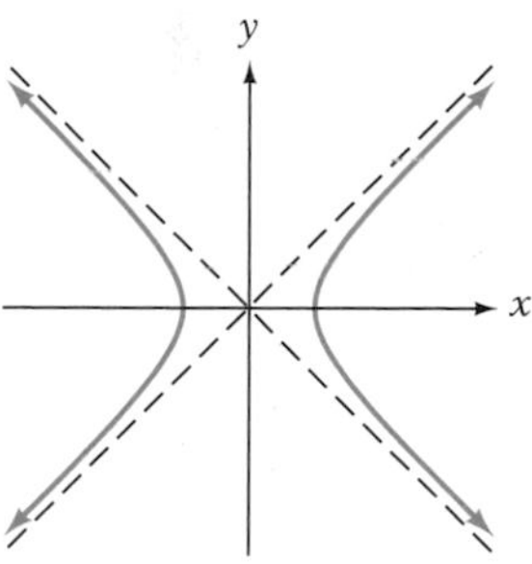

EXAMPLE 2 Sketch a graph of the hyperbola given by the equation.

A. $\frac{x^2}{16} - \frac{y^2}{4} = 1$ **B.** $\frac{y^2}{16} - \frac{x^2}{25} = 1$

Solution **A.** $\frac{x^2}{16} - \frac{y^2}{4} = 1$

- **$a^2 = 16, b^2 = 4$**

Axis of symmetry:
x-axis

Vertices:
$(4, 0)$ and $(-4, 0)$

- **The coordinates of the vertices are $(a, 0)$ and $(-a, 0)$.**

Asymptotes:
$y = \frac{1}{2}x$ and $y = -\frac{1}{2}x$

- **The equations of the asymptotes are $y = \frac{b}{a}x$ and $y = -\frac{b}{a}x$.**

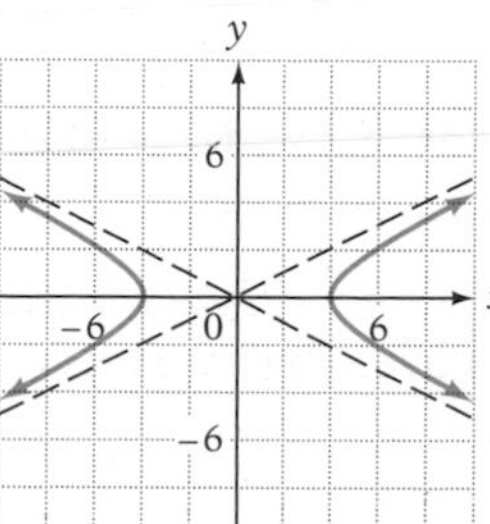

- **Sketch the asymptotes. Use symmetry and the fact that the hyperbola will approach the asymptotes to sketch its graph.**

How It's Used

Hyperbolas are used in "loran" (LOng RAnge Navigation) as a method by which a ship's navigator can determine the position of the ship relative to three transmitters, T_1, T_2, and T_3, as shown in the figure below.

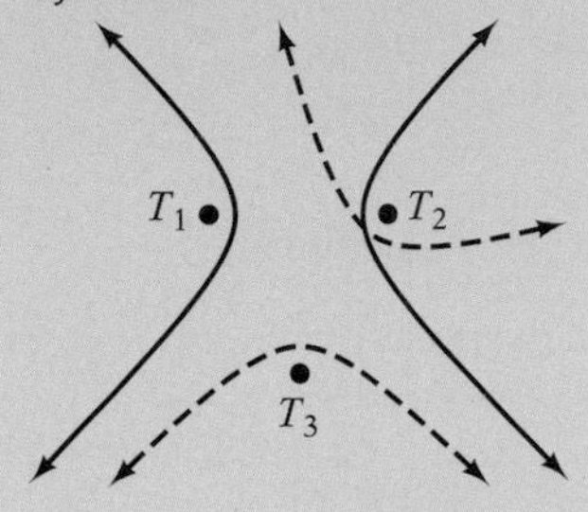

B. $\dfrac{y^2}{16} - \dfrac{x^2}{25} = 1$ • $a^2 = 25, b^2 = 16$

Axis of symmetry: y-axis

Vertices: $(0, 4)$ and $(0, -4)$ • The coordinates of the vertices are $(0, b)$ and $(0, -b)$.

Asymptotes: $y = \dfrac{4}{5}x$ and $y = -\dfrac{4}{5}x$ • The equations of the asymptotes are $y = \dfrac{b}{a}x$ and $y = -\dfrac{b}{a}x$.

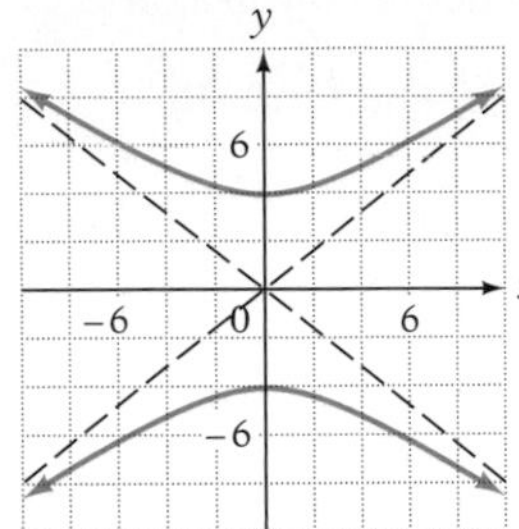

• Sketch the asymptotes. Use symmetry and the fact that the hyperbola will approach the asymptotes to sketch its graph.

Problem 2 Sketch a graph of the hyperbola given by the equation.

A. $\dfrac{x^2}{9} - \dfrac{y^2}{25} = 1$ **B.** $\dfrac{y^2}{9} - \dfrac{x^2}{9} = 1$

Solution See page S37.

Try Exercise 27, page 638.

12.3 Exercises

CONCEPT CHECK

1. Is the graph of an ellipse the graph of a function? Why or why not?

2. Is the graph of a hyperbola the graph of a function? Why or why not?

3. Identify each of the following as the equation of an ellipse, the equation of a hyperbola, or neither of these.

a. $\dfrac{x}{16} + \dfrac{y}{9} = 1$ **b.** $\dfrac{x^2}{5} - \dfrac{y^2}{3} = 1$ **c.** $\dfrac{y^2}{9} + \dfrac{x^2}{16} = 1$

d. $\dfrac{x^2}{16} - \dfrac{y}{9} = 1$ **e.** $x^2 - y^2 = 1$ **f.** $\dfrac{x^2}{7} + \dfrac{y^2}{3} = 1$

4. What are the axes of symmetry for $\dfrac{x^2}{a^2} + \dfrac{y^2}{b^2} = 1$?

1 Graph an ellipse with center at the origin (See pages 633–634.)

GETTING READY

5. The equation $\frac{x^2}{a^2} + \frac{y^2}{b^2} = 1$ is the equation of an __?__ with center at $C($__?__, __?__$)$.

The coordinates of the x-intercepts are __?__ and __?__, and the coordinates of the y-intercepts are __?__ and __?__.

6. The equation $\frac{x^2}{36} + \frac{y^2}{49} = 1$ is in the form $\frac{x^2}{a^2} + \frac{y^2}{b^2} = 1$, with $a^2 =$ __?__ and $b^2 =$ __?__. The coordinates of the x-intercepts of the ellipse are __?__ and __?__. The coordinates of the y-intercepts of the ellipse are __?__ and __?__.

Sketch the graph of the ellipse given by the equation.

7. $\frac{x^2}{4} + \frac{y^2}{9} = 1$

8. $\frac{x^2}{25} + \frac{y^2}{16} = 1$

9. $\frac{x^2}{25} + \frac{y^2}{9} = 1$

10. $\frac{x^2}{16} + \frac{y^2}{9} = 1$

11. $\frac{x^2}{36} + \frac{y^2}{16} = 1$

12. $\frac{x^2}{49} + \frac{y^2}{64} = 1$

13. $\frac{x^2}{9} + \frac{y^2}{25} = 1$

14. $\frac{x^2}{16} + \frac{y^2}{36} = 1$

15. $\frac{x^2}{36} + \frac{y^2}{9} = 1$

16. $\frac{x^2}{4} + \frac{y^2}{16} = 1$

17. $\frac{x^2}{12} + \frac{y^2}{4} = 1$

18. $\frac{x^2}{8} + \frac{y^2}{25} = 1$

19. An ellipse has the equation $\frac{x^2}{a^2} + \frac{y^2}{b^2} = 1$, where $a > b$. Is the distance between the y-intercepts of the ellipse less than, equal to, or greater than the distance between the x-intercepts of the ellipse?

20. Suppose $\frac{x^2}{a^2} + \frac{y^2}{4} = 1$ is the equation of an ellipse with $a > 2$. If you were to draw ellipses with ever larger values of a, do the ellipses become rounder or flatter?

2 Graph a hyperbola with center at the origin (See pages 634–636.)

21. How can you tell from an equation whether the graph will be that of an ellipse or that of a hyperbola?

22. What are the asymptotes of a hyperbola?

GETTING READY

For Exercises 23 to 26, use the hyperbola with equation $\frac{y^2}{49} - \frac{x^2}{25} = 1$.

23. For this hyperbola, $a^2 =$ ___?___ and $b^2 =$ ___?___.

24. The axis of symmetry that contains the vertices is the ___?___-axis.

25. The vertices of the hyperbola are on the ___?___-axis. Their coordinates are ___?___ and ___?___.

26. The asymptotes of the hyperbola are the lines with equations ___?___ and ___?___.

Sketch a graph of the hyperbola given by the equation.

27. $\frac{x^2}{9} - \frac{y^2}{16} = 1$

28. $\frac{x^2}{25} - \frac{y^2}{4} = 1$

29. $\frac{y^2}{16} - \frac{x^2}{9} = 1$

30. $\frac{y^2}{4} - \frac{x^2}{9} = 1$

31. $\frac{x^2}{4} - \frac{y^2}{25} = 1$

32. $\frac{x^2}{9} - \frac{y^2}{49} = 1$

33. $\frac{y^2}{25} - \frac{x^2}{9} = 1$

34. $\frac{y^2}{4} - \frac{x^2}{16} = 1$

35. $\frac{x^2}{25} - \frac{y^2}{16} = 1$

36. $\frac{x^2}{9} - \frac{y^2}{9} = 1$

37. $\frac{y^2}{16} - \frac{x^2}{4} = 1$

38. $\frac{y^2}{9} - \frac{x^2}{36} = 1$

39. $\frac{x^2}{25} - \frac{y^2}{9} = 1$

40. $\frac{x^2}{16} - \frac{y^2}{25} = 1$

41. $\frac{y^2}{16} - \frac{x^2}{16} = 1$

42. The equation $25x^2 - 4y^2 = 100$ is the equation of a hyperbola. What step would you take to rewrite the equation in the standard form $\frac{x^2}{a^2} - \frac{y^2}{b^2} = 1$?

APPLYING CONCEPTS

Sketch a graph of the conic section given by the equation. (*Hint:* Divide each term by the number on the right-hand side of the equation.)

43. $4x^2 + y^2 = 16$

44. $x^2 - y^2 = 9$

45. $y^2 - 4x^2 = 16$

46. $9x^2 + 4y^2 = 144$

47. $9x^2 - 25y^2 = 225$

48. $4y^2 - x^2 = 36$

49. Prepare a report on the system of navigation called loran (long-range navigation).

PROJECTS OR GROUP ACTIVITIES

50. **Astronomy** The orbit of comet Halley is an ellipse with a major axis of approximately 36 AU and a minor axis of approximately 9 AU. (One AU is 1 astronomical unit and is approximately 92,960,000 mi, the average distance of Earth from the sun.)

a. Determine an equation for the orbit of comet Halley in terms of astronomical units. See the diagram at the right.

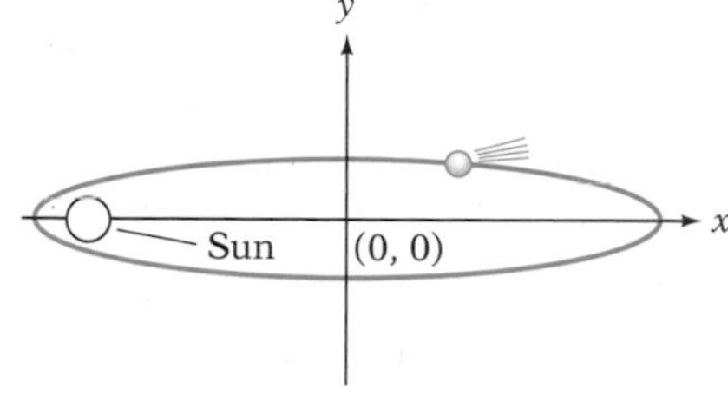

b. The distance of the sun from the center of comet Halley's elliptical orbit is $\sqrt{a^2 - b^2}$. The aphelion of the orbit (the point at which the comet is farthest from the sun) is a vertex on the major axis. Determine the distance, to the nearest hundred-thousand miles, from the sun to the point at the aphelion of comet Halley.

c. The perihelion of the orbit (the point at which the comet is closest to the sun) is a vertex on the major axis. Determine the distance, to the nearest hundred-thousand miles, from the sun to the point at the perihelion of comet Halley.

51. **Astronomy** The orbit of comet Hale–Bopp is an ellipse as shown at the right. The units are astronomical units (abbreviated AU). 1 AU ≈ 92,960,000 mi.

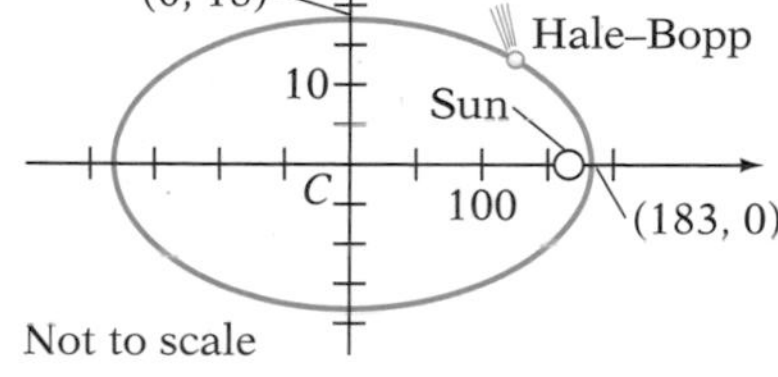

a. Find the equation of the orbit of the comet.

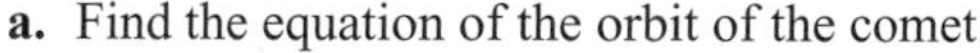

b. The distance from the center C of the orbit to the sun is approximately 182.085 AU. Find the aphelion (the point at which the comet is farthest from the sun) in miles. Round to the nearest million miles.

c. Find the perihelion (the point at which the comet is closest to the sun) in miles. Round to the nearest hundred-thousand miles.

52. **Astronomy** As mentioned in a Point of Interest in this section, the orbits of the planets are ellipses. The length of the major axis of Mars's orbit is 3.04 AU (see Exercise 50), and the length of the minor axis is 2.99 AU.

a. Determine an equation for the orbit of Mars.

b. Determine the aphelion to the nearest hundred-thousand miles.

c. Determine the perihelion to the nearest hundred-thousand miles.

12.4 Solving Nonlinear Systems of Equations

OBJECTIVE 1 Solve nonlinear systems of equations

A **nonlinear system of equations** is one in which one or more equations of the system is not a linear equation. Some examples of nonlinear systems and their graphs are shown on the next page.

$$x^2 + y^2 = 4$$
$$y = x^2 + 2$$

The graphs intersect at one point.
The system of equations has one solution.

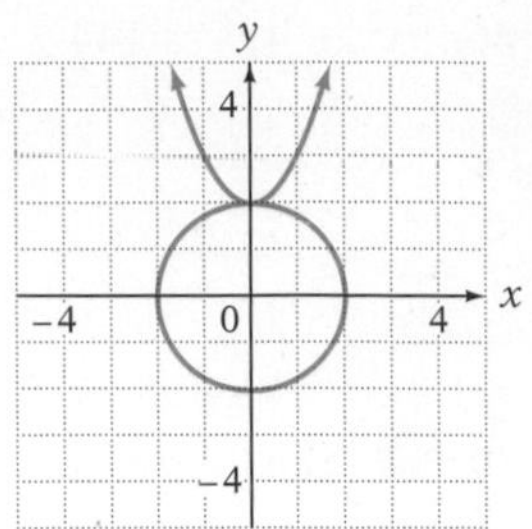

$$y = x^2$$
$$y = -x + 2$$

The graphs intersect at two points.
The system of equations has two solutions.

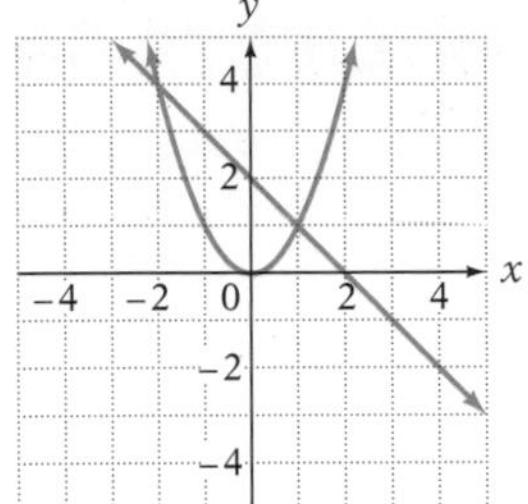

$$(x + 2)^2 + (y - 2)^2 = 4$$
$$x = y^2$$

The graphs do not intersect.
The system of equations has no solutions.

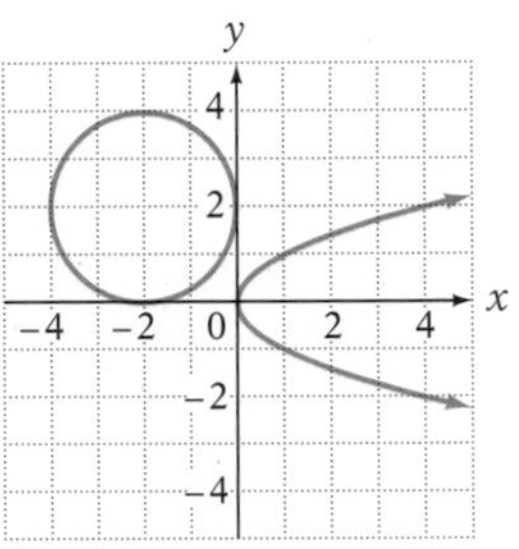

Nonlinear systems of equations can be solved by using either a substitution method or an addition method.

Focus on solving a nonlinear system of equations using the substitution method

Solve: $2x - y = 4$ (1)
$y^2 = 4x$ (2)

When a system contains both a linear and a quadratic equation, the substitution method is used.

Solve equation (1) for y.

$$2x - y = 4$$
$$-y = -2x + 4$$
$$y = 2x - 4$$

Substitute $2x - 4$ for y into equation (2).

$$y^2 = 4x$$
$$(2x - 4)^2 = 4x$$

Write the quadratic equation in standard form.

$$4x^2 - 16x + 16 = 4x$$
$$4x^2 - 20x + 16 = 0$$

Solve for x by factoring.

$$4(x^2 - 5x + 4) = 0$$
$$4(x - 4)(x - 1) = 0$$

$$x - 4 = 0 \qquad x - 1 = 0$$
$$x = 4 \qquad x = 1$$

Substitute the values of x into the equation $y = 2x - 4$, and solve for y.

$$y = 2x - 4 \qquad y = 2x - 4$$
$$y = 2(4) - 4 \qquad y = 2(1) - 4$$
$$y = 4 \qquad y = -2$$

The solutions are (4, 4) and (1, −2).

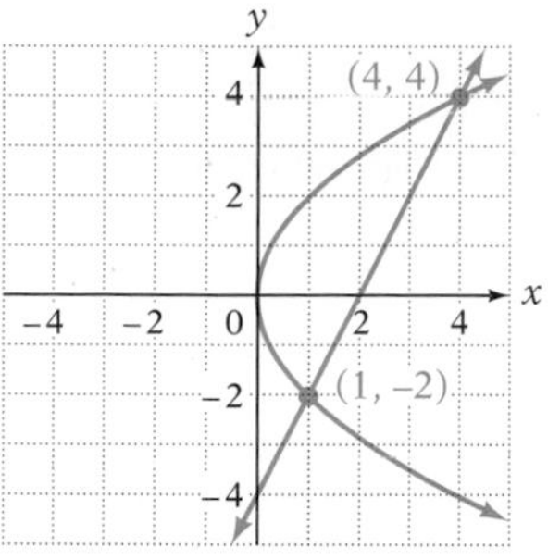

The graph of the system that was just solved is shown at the left. Note that the line intersects the parabola at two points. These points correspond to the solutions.

Solve: $x^2 + y^2 = 4$ (1)
$y = x + 4$ (2)

The system of equations contains a linear equation. The substitution method is used to solve the system.

Substitute the expression for y into equation (1).

$$x^2 + y^2 = 4$$
$$x^2 + (x + 4)^2 = 4$$

Write the equation in standard form.

$$x^2 + x^2 + 8x + 16 = 4$$
$$2x^2 + 8x + 16 = 4$$
$$2x^2 + 8x + 12 = 0$$

Use the quadratic formula to solve for x. Because the solutions are complex numbers, the graphs do not intersect.

$$x = \frac{-8 \pm \sqrt{8^2 - 4(2)(12)}}{2(2)}$$
$$x = \frac{-8 \pm \sqrt{64 - 96}}{4}$$
$$x = \frac{-8 \pm \sqrt{-32}}{4} = \frac{-8 \pm 4i\sqrt{2}}{4}$$
$$x = -2 \pm i\sqrt{2}$$

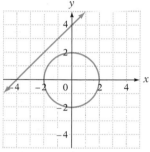

The system of equations has no real number solutions.

The graph of the system of equations that was solved above is shown at the left. Note that the two graphs do not intersect.

Focus on solving a nonlinear system of equations using the addition method

Solve: $4x^2 + y^2 = 16$ (1)
$x^2 + y^2 = 4$ (2)

Use the addition method to solve this system of equations.

Multiply equation (2) by -1 and add it to equation (1).

$$4x^2 + y^2 = 16$$
$$-x^2 - y^2 = -4$$
$$3x^2 = 12$$

Solve for x.

$$x^2 = 4$$
$$x = \pm 2$$

Substitute the values of x into equation (2), and solve for y.

$$x^2 + y^2 = 4 \qquad x^2 + y^2 = 4$$
$$2^2 + y^2 = 4 \qquad (-2)^2 + y^2 = 4$$
$$y^2 = 0 \qquad y^2 = 0$$
$$y = 0 \qquad y = 0$$

The solutions are (2, 0) and (−2, 0).

Take Note

Note from the examples in this section that the number of points at which the graphs of the equations of the system intersect is the same as the number of real number solutions of the system of equations.

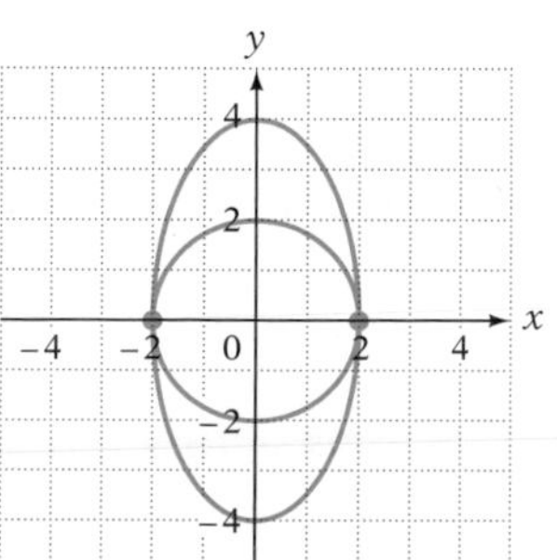

The graph of the system that was solved above is shown at the left. Note that the graphs intersect at two points.

EXAMPLE 1 Solve: $y = 2x^2 - 3x - 1$
$y = x^2 - 2x + 5$

Solution

$$(1) \qquad y = 2x^2 - 3x - 1$$
$$(2) \qquad y = x^2 - 2x + 5$$

$$2x^2 - 3x - 1 = x^2 - 2x + 5$$
$$x^2 - x - 6 = 0$$
$$(x + 2)(x - 3) = 0$$

• **Use the substitution method to solve for x.**

$$x + 2 = 0 \qquad x - 3 = 0$$
$$x = -2 \qquad x = 3$$

$$y = 2x^2 - 3x - 1$$
$$y = 2(-2)^2 - 3(-2) - 1$$
$$y = 8 + 6 - 1$$
$$y = 13$$

• **Substitute each value of x into equation (1) or equation (2) and solve for y. We will use equation (1).**

• **When $x = -2$, $y = 13$.**

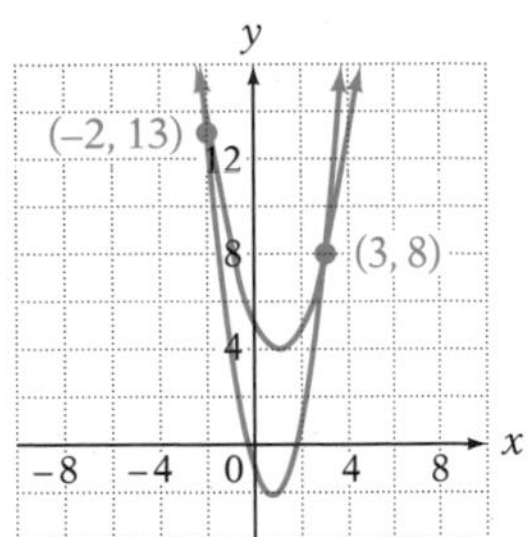

$$y = 2x^2 - 3x - 1$$
$$y = 2(3)^2 - 3(3) - 1$$
$$y = 18 - 9 - 1$$
$$y = 8$$

• **When $x = 3$, $y = 8$.**

The solutions are $(-2, 13)$ and $(3, 8)$.

Problem 1 Solve: $y = 2x^2 + x - 3$
$y = 2x^2 - 2x + 9$

Solution See page S37.

Try Exercise 17, page 644.

EXAMPLE 2 Solve: $3x^2 - 2y^2 = 26$
$x^2 - y^2 = 5$

Solution

$$(1) \quad 3x^2 - 2y^2 = 26$$
$$(2) \quad x^2 - y^2 = 5$$

$$3x^2 - 2y^2 = 26$$
$$\underline{-2x^2 + 2y^2 = -10}$$
$$x^2 = 16$$
$$x = \pm 4$$

• **Use the addition method. We will eliminate y. Multiply equation (2) by -2 and solve for x.**

$$x^2 - y^2 = 5$$
$$(-4)^2 - y^2 = 5$$
$$16 - y^2 = 5$$
$$-y^2 = -11$$
$$y^2 = 11$$
$$y = \pm\sqrt{11}$$

• **Substitute each value of x into equation (1) or equation (2) and solve for y. We will use equation (2).**

• **When $x = -4$, $y = -\sqrt{11}$ or $y = \sqrt{11}$.**

$$x^2 - y^2 = 5$$
$$4^2 - y^2 = 5$$
$$16 - y^2 = 5$$
$$-y^2 = -11$$
$$y^2 = 11$$
$$y = \pm\sqrt{11}$$

• **When $x = 4$, $y = -\sqrt{11}$ or $y = \sqrt{11}$.**

The solutions are $(-4, -\sqrt{11})$, $(-4, \sqrt{11})$, $(4, -\sqrt{11})$, and $(4, \sqrt{11})$.

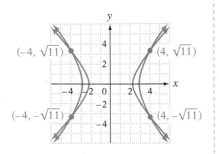

Problem 2 Solve: $x^2 - y^2 = 10$
$x^2 + y^2 = 8$

Solution See page S37.

Try Exercise 21, page 644.

12.4 Exercises

CONCEPT CHECK

1. A system of equations contains the equation of a line and the equation of a hyperbola. How many solutions can the system of equations have?

2. A system of equations contains the equation of a line and the equation of a parabola. How many solutions can the system of equations have?

3. A system of equations contains the equation of a parabola and the equation of an ellipse. How many solutions can the system of equations have?

4. A system of equations contains the equation of an ellipse and the equation of a hyperbola. How many solutions can the system of equations have?

1 Solve nonlinear systems of equations (See pages 639–642.)

5. How do nonlinear systems of equations differ from linear systems of equations?

GETTING READY

6. Solve the nonlinear system of equations: $y = x^2 - 5x + 6$
$y = x - 2$

$y = x^2 - 5x + 6$ $y = x - 2$	• **The second equation has variables raised to the first power. Use the substitution method.**
$\underline{\quad ? \quad} = x^2 - 5x + 6$	• **Substitute $x - 2$ for y in the first equation.**
$0 = x^2 - \underline{\quad ? \quad}x + \underline{\quad ? \quad}$	• **Write the quadratic equation in standard form by subtracting x from each side and adding 2 to each side.**
$0 = (x - 4)(x - \underline{\quad ? \quad})$	• **Factor the trinomial.**
$0 = \underline{\quad ? \quad}$ $\quad 0 = \underline{\quad ? \quad}$	• **Use the Principle of Zero Products.**
$\underline{\quad ? \quad} = x$ $\quad \underline{\quad ? \quad} = x$	• **Solve for x. These are the x-coordinates of the solutions of the system.**
$y = 4 - 2$ $\quad y = 2 - 2$	• **Substitute the x-values into the equation $y = \underline{\quad ? \quad}$.**
$y = 2$ $\quad y = 0$	• **Simplify. These are the $\underline{\quad ? \quad}$-coordinates of the solutions of the system.**

The solutions of the system are $(\underline{\quad ? \quad}, \underline{\quad ? \quad})$ and $(\underline{\quad ? \quad}, \underline{\quad ? \quad})$.

Solve.

7. $y = x^2 - x - 1$
$y = 2x + 9$

8. $y = x^2 - 3x + 1$
$y = x + 6$

9. $y^2 = -x + 3$
$x - y = 1$

10. $y^2 = 4x$
$x - y = -1$

11. $y^2 = 2x$
$x + 2y = -2$

12. $y^2 = 2x$
$x - y = 4$

13. $x^2 + 2y^2 = 12$
$2x - y = 2$

14. $x^2 + 4y^2 = 37$
$x - y = -4$

15. $x^2 + y^2 = 13$
$x + y = 5$

16. $x^2 + y^2 = 16$
$x - 2y = -4$

17. $4x^2 + y^2 = 12$
$y = 4x^2$

18. $2x^2 + y^2 = 6$
$y = 2x^2$

19. $y = x^2 - 2x - 3$
$y = x - 6$

20. $y = x^2 + 4x + 5$
$y = -x - 3$

21. $3x^2 - y^2 = -1$
$x^2 + 4y^2 = 17$

22. $x^2 + y^2 = 10$
$x^2 + 9y^2 = 18$

23. $2x^2 + 3y^2 = 30$
$x^2 + y^2 = 13$

24. $x^2 + y^2 = 61$
$x^2 - y^2 = 11$

25. $y = 2x^2 - x + 1$
$y = x^2 - x + 5$

26. $y = -x^2 + x - 1$
$y = x^2 + 2x - 2$

27. $2x^2 + 3y^2 = 24$
$x^2 - y^2 = 7$

28. $2x^2 + 3y^2 = 21$
$x^2 + 2y^2 = 12$

29. $x^2 + y^2 = 36$
$4x^2 + 9y^2 = 36$

30. $2x^2 + 3y^2 = 12$
$x^2 - y^2 = 25$

31. $11x^2 - 2y^2 = 4$
$3x^2 + y^2 = 15$

32. $x^2 + 4y^2 = 25$
$x^2 - y^2 = 5$

33. $2x^2 - y^2 = 7$
$2x - y = 5$

34. $3x^2 + 4y^2 = 7$
$x - 2y = -3$

35. $y = 3x^2 + x - 4$
$y = 3x^2 - 8x + 5$

36. $y = 2x^2 + 3x + 1$
$y = 2x^2 + 9x + 7$

For Exercises 37 to 40, a nonlinear system of equations is given, along with its solutions. Describe the graph of the system by stating the names of the two figures and the number of points at which they intersect.

37. $x^2 + 4y^2 = 16$
$x^2 + y^2 = 4$
Solutions: $(0, 2), (0, -2)$

38. $2x^2 + y^2 = 4$
$x^2 - 2y^2 = 12$
No solutions

39. $y = x^2 - 3x - 4$
$6x - 2y = 26$
Solution: $(3, -4)$

40. $x^2 - y^2 = 8$
$x = y^2 - 4$
Solutions: $(4, 2\sqrt{2}), (4, -2\sqrt{2}),$
$(-3, 1), (-3, -1)$

APPLYING CONCEPTS

Solve the system by graphing. Approximate the solutions to the nearest thousandth.

41. $y = 2^x$
$x + y = 3$

42. $y = 3^{-x}$
$x^2 + y^2 = 9$

43. $y = \log_2 x$
$\frac{x^2}{9} + \frac{y^2}{1} = 1$

44. $y = \log_3 x$
$x^2 + y^2 = 4$

45. $y = -\log_3 x$
$x + y = 4$

46. $y = \left(\dfrac{1}{2}\right)^x$
$\dfrac{x^2}{9} + \dfrac{y^2}{4} = 1$

47. Is it possible for two circles with centers at the origin to intersect at exactly two points?

PROJECTS OR GROUP ACTIVITIES

48. Astronomy Suppose you have been hired to track incoming meteorites and determine whether they will strike Earth. The equation of Earth's surface is $x^2 + y^2 = 40$. You observe a meteorite moving along a path whose equation is $18x - y^2 = -144$. Will the meteorite hit Earth?

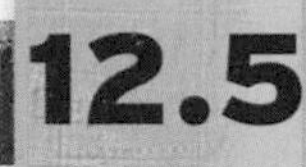

12.5 Quadratic Inequalities and Systems of Inequalities

OBJECTIVE 

Graph the solution set of a quadratic inequality in two variables

The **graph of a quadratic inequality in two variables** is a region of the plane that is bounded by one of the conic sections (parabola, circle, ellipse, or hyperbola). When graphing an inequality of this type, first replace the inequality symbol with an equals sign. Graph the resulting conic using a dashed curve when the original inequality is less than ($<$) or greater than ($>$). Use a solid curve when the original inequality is $\le$ or $\ge$. Use the point with coordinates $(0, 0)$ to determine which region of the plane to shade. If $(0, 0)$ is a solution of the inequality, then shade the region of the plane containing the point with coordinates $(0, 0)$. If not, shade the other portion of the plane.

Focus on graphing the solution set of a quadratic inequality

Graph the solution set of $x^2 + y^2 > 9$.

$x^2 + y^2 > 9$
$x^2 + y^2 = 9$

Change the inequality to an equality. This is the equation of a circle with center $C(0, 0)$ and radius 3.

Because the inequality is $>$, the graph of $x^2 + y^2 = 9$ is drawn as a dashed circle.

Substitute $(0, 0)$ into the inequality. Because $0^2 + 0^2 > 9$ is not true, the point with coordinates $(0, 0)$ should not be in the shaded region.

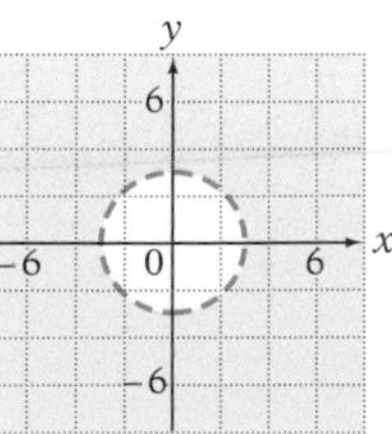

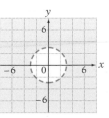

EXAMPLE 1 Graph the solution set.

A. $y \le x^2 + 2x + 2$ **B.** $\dfrac{y^2}{9} - \dfrac{x^2}{4} \ge 1$

Solution **A.** $y \le x^2 + 2x + 2$

$y = x^2 + 2x + 2$

- **Change the inequality to an equality. This is the equation of a parabola that opens up. The coordinates of the vertex are $(-1, 1)$. The equation of the axis of symmetry is $x = -1$.**
- **Because the inequality is $\le$, the graph is drawn as a solid curve.**
- **Substitute (0, 0) into the inequality. Because the inequality $0 < 0^2 + 2(0) + 2$ is true, the point with coordinates (0, 0) should be in the shaded region.**

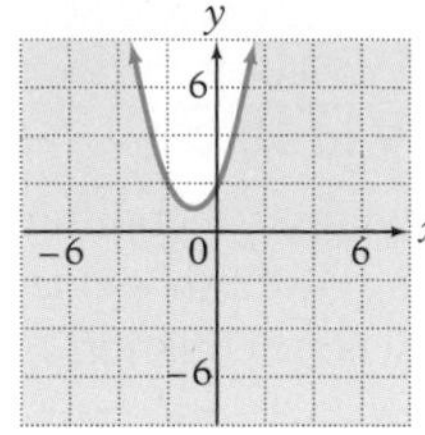

B. $\dfrac{y^2}{9} - \dfrac{x^2}{4} \ge 1$

$\dfrac{y^2}{9} - \dfrac{x^2}{4} = 1$

- **Change the inequality to an equality. This is the equation of a hyperbola. The coordinates of the vertices are $(0, -3)$ and $(0, 3)$. The equations of the asymptotes are**

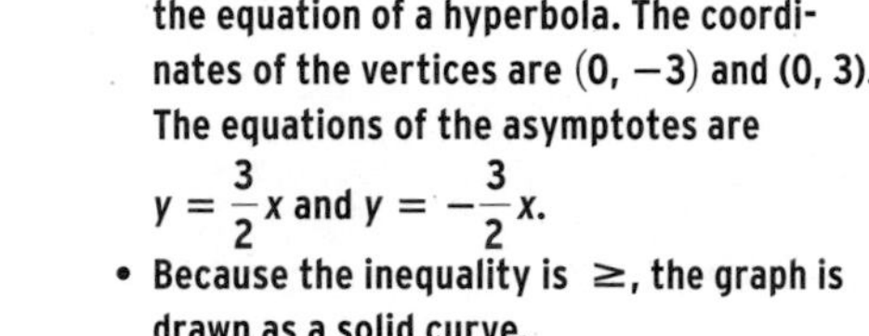

$y = \dfrac{3}{2}x$ and $y = -\dfrac{3}{2}x$.

- **Because the inequality is $\ge$, the graph is drawn as a solid curve.**
- **Substitute (0, 0) into the inequality. Because the inequality $\dfrac{0^2}{9} - \dfrac{0^2}{4} \ge 1$ is not true, the point with coordinates (0, 0) should not be in the shaded region.**

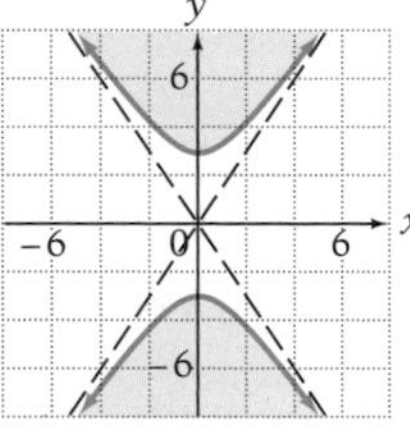

Problem 1 Graph the solution set.

A. $\dfrac{x^2}{9} + \dfrac{y^2}{16} \le 1$ **B.** $\dfrac{x^2}{9} - \dfrac{y^2}{4} \le 1$

Solution See page S37.

Try Exercise 7, page 648.

OBJECTIVE 2 Graph the solution set of a nonlinear system of inequalities

A **nonlinear system of inequalities** is a system in which one or more of the inequalities is a nonlinear inequality. The **solution set of a nonlinear system of inequalities** is the intersection of the solution sets of the individual inequalities. To graph the solution set of a system of inequalities, first graph the solution set for each inequality. The graph of the solution set of the system of inequalities is the region of the plane represented by the intersection of the two shaded regions.

EXAMPLE 2 Graph the solution set. **A.** $y > x^2$
$y < x + 2$

B. $\frac{x^2}{9} - \frac{y^2}{16} \geq 1$
$x^2 + y^2 \leq 4$

Solution **A.**

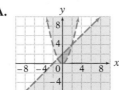

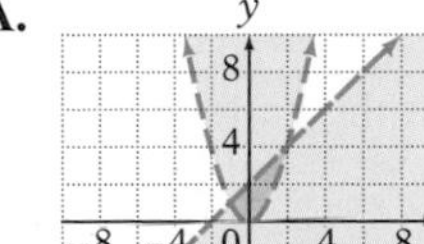

- **Graph the solution set of each inequality.**
- **$y = x^2$ is the equation of a parabola. Use a dashed curve. Shade inside the parabola.**
- **$y = x + 2$ is the equation of a line. Use a dashed line. Shade below the line.**

The solution set is the region of the plane represented by the intersection of the solution sets of the individual inequalities.

B.

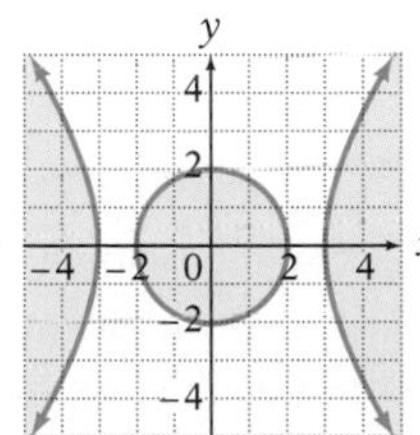

- **Graph the solution set of each inequality.**
- **$\frac{x^2}{9} - \frac{y^2}{16} = 1$ is the equation of a hyperbola. Use a solid curve. The point with coordinates (0, 0) should not be in the shaded region.**
- **$x^2 + y^2 = 4$ is the equation of a circle. Use a solid curve. Shade inside the circle.**

The solution sets of the two inequalities do not intersect. The system of inequalities has no real number solutions.

Problem 2 Graph the solution set.

A. $\frac{x^2}{4} + \frac{y^2}{9} \leq 1$
$x > y^2 - 2$

B. $\frac{x^2}{16} + \frac{y^2}{25} \geq 1$
$x^2 + y^2 < 9$

Solution See page S37.

Try Exercise 37, page 650.

12.5 Exercises

CONCEPT CHECK

1. Explain how you can check that you have shaded the correct region of the plane when graphing the solution set of a quadratic inequality.

2. If the graphs of the inequalities in a system of inequalities do not intersect, what is the solution set of the system of inequalities?

1 Graph the solution set of a quadratic inequality in two variables (See pages 645–646.)

GETTING READY

3. Complete the following statements about the graph of the inequality $y < x^2 - 8x + 16$.

a. The boundary curve is a __?__.

b. Because the inequality is $<$, the boundary will be shown as a __?__ curve.

c. (0, 0) is a solution of the inequality, so the point with coordinates (0, 0) __?__ (*is* or *is not*) in the shaded region of the graph.

4. Complete the following statements about the graph of the inequality $\frac{x^2}{4} - \frac{y^2}{25} \geq 1$.

a. The boundary curve is a __?__.

b. Because the inequality is $\geq$, the boundary will be shown as a __?__ curve.

c. (0, 0) is not a solution of the inequality, so the point with coordinates (0, 0) __?__ (*is* or *is not*) in the shaded region of the graph.

Graph the solution set.

5. $y \leq x^2 - 4x + 3$

6. $y < x^2 - 2x - 3$

7. $(x - 1)^2 + (y + 2)^2 \leq 9$

8. $(x + 2)^2 + (y - 3)^2 > 4$

9. $(x + 3)^2 + (y - 2)^2 \geq 9$

10. $(x - 2)^2 + (y + 1)^2 \leq 16$

11. $\frac{x^2}{16} + \frac{y^2}{25} < 1$

12. $\frac{x^2}{9} + \frac{y^2}{4} \geq 1$

13. $\frac{x^2}{25} - \frac{y^2}{9} \leq 1$

14. $\frac{y^2}{25} - \frac{x^2}{36} > 1$

15. $\frac{x^2}{4} + \frac{y^2}{16} \geq 1$

16. $\frac{x^2}{4} - \frac{y^2}{16} \leq 1$

17. $y \leq x^2 - 2x + 3$

18. $x \leq y^2 + 2y + 1$

19. $\frac{y^2}{9} - \frac{x^2}{16} \leq 1$

20. $\frac{x^2}{16} - \frac{y^2}{4} < 1$

21. $\frac{x^2}{9} + \frac{y^2}{1} \leq 1$

22. $\frac{x^2}{16} + \frac{y^2}{4} > 1$

23. $(x - 1)^2 + (y + 3)^2 \leq 25$

24. $(x + 1)^2 + (y - 2)^2 \geq 16$

25. $\frac{y^2}{25} - \frac{x^2}{4} \leq 1$

26. $\frac{x^2}{9} - \frac{y^2}{25} \geq 1$

27. $\frac{x^2}{25} + \frac{y^2}{9} \leq 1$

28. $\frac{x^2}{36} + \frac{y^2}{4} \leq 1$

29. True or false? For all real numbers h, k, and r, the graph of the inequality $(x - h)^2 + (y - k)^2 < r^2$ will be the region inside the circle with equation $(x - h)^2 + (y - k)^2 = r^2$.

30. True or false? For all real numbers a, b, and c, $a \neq 0$, the graph of the inequality $y > ax^2 + bx + c$ will be the region inside the parabola with equation $y = ax^2 + bx + c$.

2 Graph the solution set of a nonlinear system of inequalities (See pages 646–647.)

GETTING READY

31. To graph the solution set of a system of inequalities, find the __?__ of the graphs of the individual inequalities.

32. If there is no overlap between the graphs of the individual inequalities of a system of inequalities, then the system of inequalities has __?__ solutions.

Graph the solution set.

33. $y \leq x^2 - 4x + 4$
$y + x > 4$

34. $x^2 + y^2 < 1$
$x + y \geq 4$

35. $x^2 + y^2 < 16$
$y > x + 1$

36. $y > x^2 - 4$
$y < x - 2$

37. $\frac{x^2}{4} + \frac{y^2}{16} \le 1$
$y \le -\frac{1}{2}x + 2$

38. $\frac{y^2}{4} - \frac{x^2}{25} \ge 1$
$y \le \frac{2}{3}x + 4$

39. $x \ge y^2 - 3y + 2$
$y \ge 2x - 2$

40. $x^2 + y^2 \le 25$
$y \le -\frac{1}{3}x + 2$

41. $x^2 + y^2 < 25$
$\frac{x^2}{9} + \frac{y^2}{36} < 1$

42. $\frac{x^2}{9} - \frac{y^2}{4} < 1$
$\frac{x^2}{25} + \frac{y^2}{9} < 1$

43. $x^2 + y^2 > 4$
$x^2 + y^2 < 25$

44. $\frac{x^2}{25} + \frac{y^2}{16} \le 1$
$\frac{x^2}{4} + \frac{y^2}{4} \ge 1$

For Exercises 45 to 48, a and b are positive numbers such that $a < b$. Describe the solution set of each system of inequalities.

45. $x^2 + y^2 < a$
$x^2 + y^2 < b$

46. $x^2 + y^2 < a$
$x^2 + y^2 > b$

47. $x^2 + y^2 > a$
$x^2 + y^2 < b$

48. $x^2 + y^2 > a$
$x^2 + y^2 > b$

APPLYING CONCEPTS

Graph the solution set.

49. $y > x^2 - 3$
$y < x + 3$
$x \le 0$

50. $x^2 + y^2 \le 25$
$y > x + 1$
$x \ge 0$

51. $x^2 + y^2 < 3$
$x > y^2 - 1$
$y \ge 0$

52. $\frac{x^2}{4} - \frac{y^2}{25} \le 1$
$\frac{x^2}{4} + \frac{y^2}{4} \le 1$
$y \ge 0$

53. $\frac{x^2}{4} + \frac{y^2}{1} \le 4$
$x^2 + y^2 \le 4$
$x \ge 0$
$y \le 0$

54. $\frac{x^2}{4} + \frac{y^2}{25} \le 1$
$x > y^2 - 4$
$x \le 0$
$y \ge 0$

55. $y > 2^x$
$x + y < 4$

56. $y < \left(\frac{1}{2}\right)^x$
$2x - y \geq 2$

57. $y \geq \log_2 x$
$x^2 + y^2 < 9$

58. $y \leq -\log_3 x$
$\frac{x^2}{9} + \frac{y^2}{4} < 1$

59. $y < 3^{-x}$
$\frac{x^2}{4} - \frac{y^2}{1} \geq 1$

60. $y \geq 2^{x-1}$
$2x + 3y > 6$

PROJECTS OR GROUP ACTIVITIES

61. a. Graph $xy > 1$ and $y > \frac{1}{x}$ on different coordinate grids.

b. Note that dividing each side of $xy > 1$ by x yields $y > \frac{1}{x}$, but the graphs you drew in part (a) are not the same. Explain.

CHAPTER 12 Summary

Key Words	Objective and Page Reference	Examples
Conic sections are curves that can be constructed from the intersection of a plane and a cone. The four conic sections are the **parabola, circle, ellipse,** and **hyperbola.**	[12.1.1, p. 620]	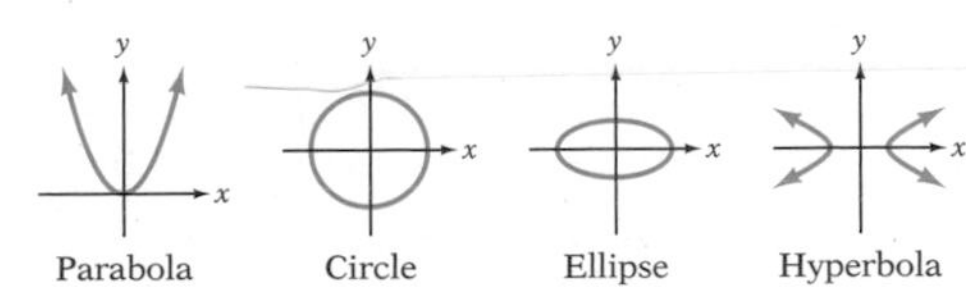

The **asymptotes** of a hyperbola are the two lines that are "approached" by the hyperbola. As the graph of the hyperbola gets farther from the origin, the hyperbola "gets closer to" the asymptotes.

[12.3.2, p. 635]

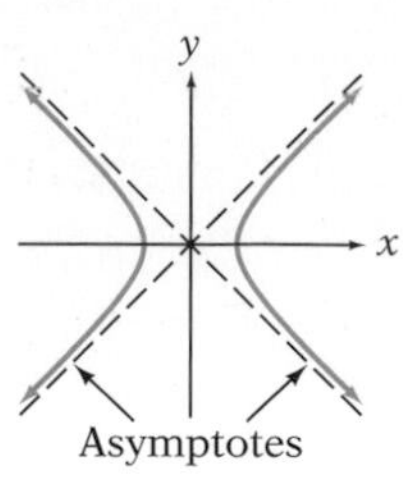

A **nonlinear system of equations** is a system in which one or more of the equations is not a linear equation.

[12.4.1, p. 639]

$$x^2 + y^2 = 16$$
$$y = x^2 - 2$$

The **graph of a quadratic inequality in two variables** is a region of the plane that is bounded by one of the conic sections.

[12.5.1, p. 645]

$y \geq x^2 - 2x - 4$

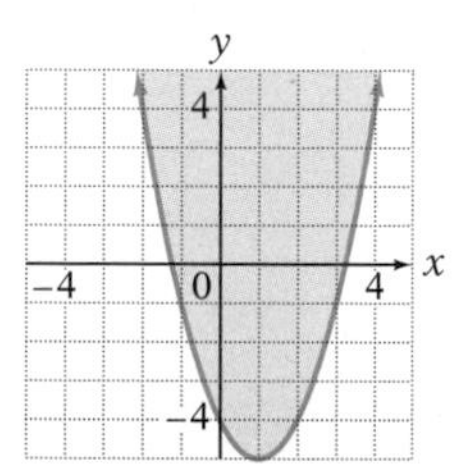

A **nonlinear system of inequalities** is a system in which one or more of the inequalities is not a linear inequality. The **solution set of a nonlinear system of inequalities** is the intersection of the solution sets of the individual inequalities.

[12.5.2, p. 646]

Graph the solution set of

$y \leq -x^2 + 3$

$y \geq x + 1$

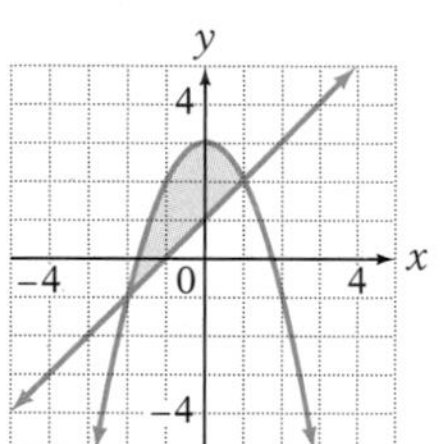

Essential Rules and Procedures | Objective and Page Reference | Examples

Equation of a parabola

$y = ax^2 + bx + c$

When $a > 0$, the parabola opens up.
When $a < 0$, the parabola opens down.
The x-coordinate of the vertex is $-\frac{b}{2a}$.
The equation of the axis of symmetry is $x = -\frac{b}{2a}$.

[12.1.1, pp. 620–621]

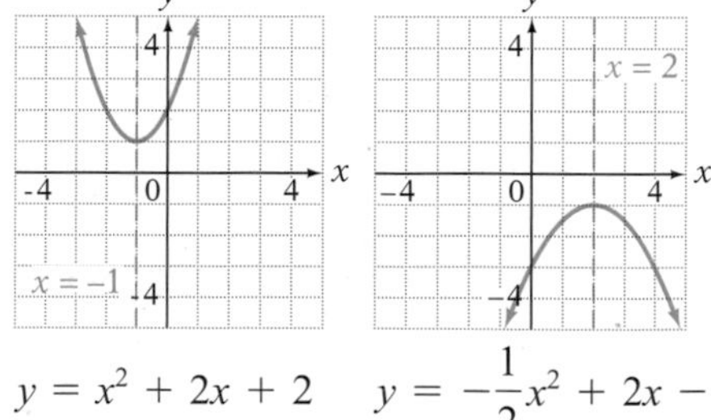

$y = x^2 + 2x + 2 \qquad y = -\frac{1}{2}x^2 + 2x - 3$

$x = ay^2 + by + c$

When $a > 0$, the parabola opens to the right.
When $a < 0$, the parabola opens to the left.
The y-coordinate of the vertex is $-\frac{b}{2a}$.
The equation of the axis of symmetry is $y = -\frac{b}{2a}$.

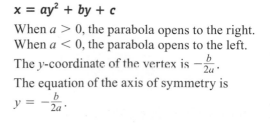

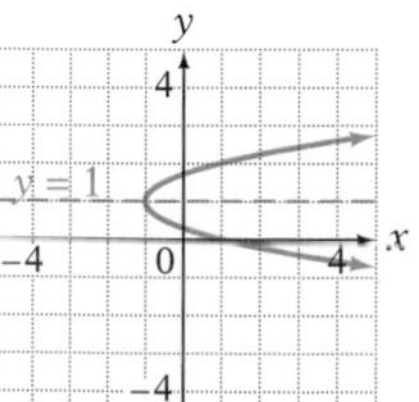

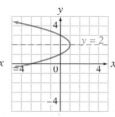

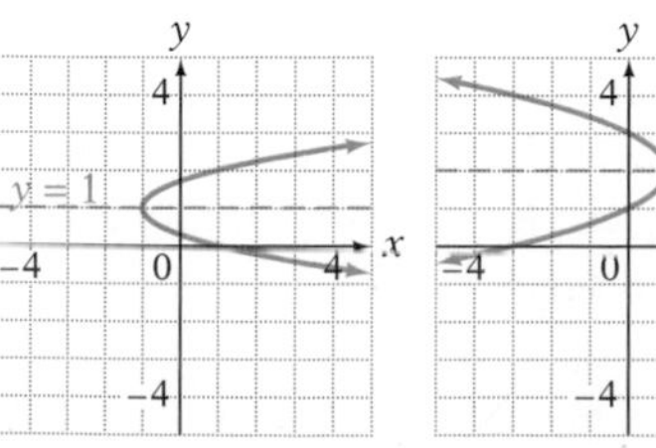

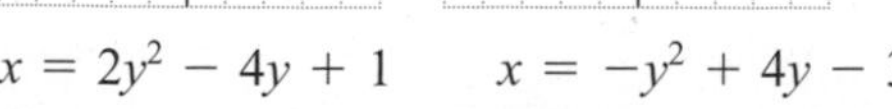

$x = 2y^2 - 4y + 1 \qquad x = -y^2 + 4y - 3$

Equation of a circle
$(x - h)^2 + (y - k)^2 = r^2$
The center of the circle is $C(h, k)$, and the radius is r.

[12.2.1, p. 627]

$(x - 2)^2 + (y + 1)^2 = 9$
$(h, k) = (2, -1)$
$r = 3$

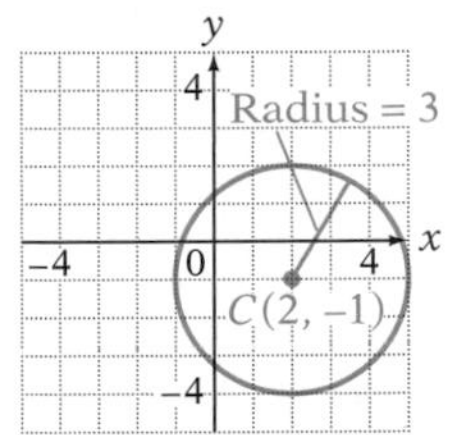

Equation of an ellipse
$\dfrac{x^2}{a^2} + \dfrac{y^2}{b^2} = 1$
The coordinates of the x-intercepts are $(a, 0)$ and $(-a, 0)$.
The coordinates of the y-intercepts are $(0, b)$ and $(0, -b)$.

[12.3.1, p. 633]

$\dfrac{x^2}{9} + \dfrac{y^2}{1} = 1$
x-intercepts:
$(3, 0)$ and $(-3, 0)$
y-intercepts:
$(0, 1)$ and $(0, -1)$

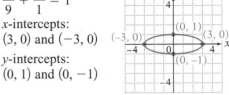

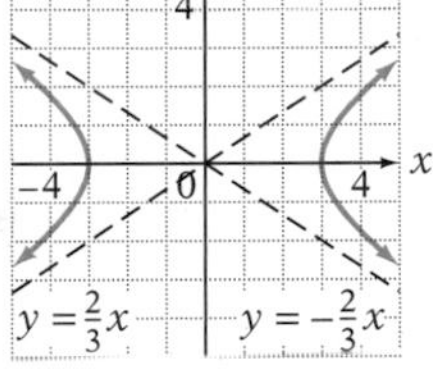

Equation of a hyperbola
$\dfrac{x^2}{a^2} - \dfrac{y^2}{b^2} = 1$
The axis of symmetry that contains the vertices is the x-axis.
The coordinates of the vertices are $(a, 0)$ and $(-a, 0)$.
The equations of the asymptotes are $y = \frac{b}{a}x$ and $y = -\frac{b}{a}x$.

[12.3.2, p. 635]

$\dfrac{x^2}{9} - \dfrac{y^2}{4} = 1$
Vertices:
$(3, 0)$ and $(-3, 0)$
Asymptotes:
$y = \frac{2}{3}x$ and $y = -\frac{2}{3}x$

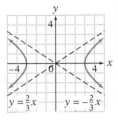

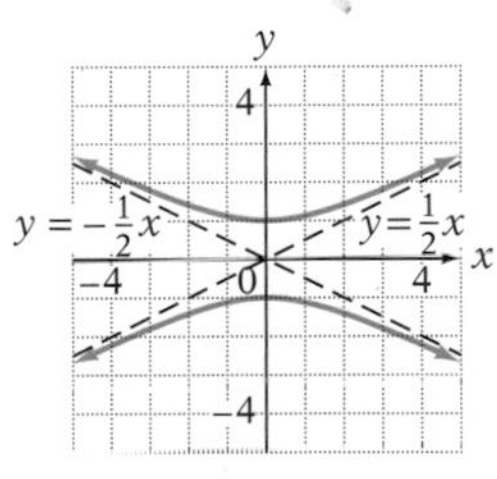

$\dfrac{y^2}{b^2} - \dfrac{x^2}{a^2} = 1$
The axis of symmetry that contains the vertices is the y-axis.
The coordinates of the vertices are $(0, b)$ and $(0, -b)$.
The equations of the asymptotes are $y = \frac{b}{a}x$ and $y = -\frac{b}{a}x$.

$\dfrac{y^2}{1} - \dfrac{x^2}{4} = 1$
Vertices:
$(0, 1)$ and $(0, -1)$
Asymptotes:
$y = \frac{1}{2}x$ and
$y = -\frac{1}{2}x$

CHAPTER 12 Review Exercises

1. Find the coordinates of the vertex and the equation of the axis of symmetry of the parabola with equation $y = x^2 - 4x + 8$.

2. Find the coordinates of the vertex and the equation of the axis of symmetry of the parabola with equation $y = -x^2 + 7x - 8$.

3. Sketch a graph of $y = -2x^2 + x - 2$.

4. Sketch a graph of $x = 2y^2 - 6y + 5$.

5. Find the equation of the circle that passes through the point $P(2, -1)$ and whose center is the point $C(-1, 2)$.

6. Find the equation of the circle with radius 6 and center $C(-1, 5)$.

7. Sketch a graph of $(x + 3)^2 + (y + 1)^2 = 1$.

8. Sketch a graph of $x^2 + (y - 2)^2 = 9$.

9. Find the equation of the circle that passes through the point $P(4, 6)$ and whose center is the point $C(0, -3)$.

10. Write the equation $x^2 + y^2 + 4x - 2y = 4$ in standard form.

11. Sketch a graph of $\frac{x^2}{1} + \frac{y^2}{9} = 1$.

12. Sketch a graph of $\frac{x^2}{25} + \frac{y^2}{9} = 1$.

13. Sketch a graph of $\frac{x^2}{25} - \frac{y^2}{1} = 1$.

14. Sketch a graph of $\frac{y^2}{16} - \frac{x^2}{9} = 1$.

15. Solve: $y = x^2 + 5x - 6$
$y = x - 10$

16. Solve: $2x^2 + y^2 = 19$
$3x^2 - y^2 = 6$

17. Solve: $x = 2y^2 - 3y + 1$
$3x - 2y = 0$

18. Solve: $y^2 = 2x^2 - 3x + 6$
$y^2 = 2x^2 + 5x - 2$

19. Graph the solution set: $(x - 2)^2 + (y + 1)^2 \leq 16$

20. Graph the solution set: $\frac{x^2}{9} - \frac{y^2}{16} < 1$

21. Graph the solution set: $y \geq -x^2 - 2x + 3$

22. Graph the solution set: $\frac{x^2}{16} + \frac{y^2}{4} > 1$

23. Graph the solution set: $y \geq x^2 - 4x + 2$
$y \leq \frac{1}{3}x - 1$

24. Graph the solution set: $\frac{x^2}{25} + \frac{y^2}{16} \leq 1$
$\frac{y^2}{4} - \frac{x^2}{4} \geq 1$

25. Graph the solution set: $\frac{x^2}{9} + \frac{y^2}{1} \geq 1$
$\frac{x^2}{4} - \frac{y^2}{1} \leq 1$

26. Graph the solution set: $\frac{x^2}{16} + \frac{y^2}{4} < 1$
$x^2 + y^2 > 9$

CHAPTER 12 Test

1. Find the equation of the axis of symmetry of the parabola with equation $y = -x^2 + 6x - 5$.

2. Find the coordinates of the vertex of the parabola with equation $y = -x^2 + 3x - 2$.

3. Sketch a graph of $y = \frac{1}{2}x^2 + x - 4$.

4. Sketch a graph of $x = y^2 - y - 2$.

5. Find the equation of the circle with radius 4 and center $C(-3, -3)$.

6. Solve: $x^2 + 2y^2 = 4$
$x + y = 2$

7. Solve: $x = 3y^2 + 2y - 4$
$x = y^2 - 5y$

8. Solve: $x^2 - y^2 = 24$
$2x^2 + 5y^2 = 55$

9. Find the equation of the circle that passes through the point $P(2, 4)$ and whose center is the point $C(-1, -3)$.

10. Sketch a graph of $(x - 2)^2 + (y + 1)^2 = 9$.

11. Find the equation of the circle with radius 3 and center $C(-2, 4)$.

12. Find the equation of the circle that passes through the point $P(2, 5)$ and whose center is the point $C(-2, 1)$.

13. Write the equation $x^2 + y^2 - 4x + 2y + 1 = 0$ in standard form, and then sketch its graph.

14. Sketch a graph of $\frac{y^2}{25} - \frac{x^2}{16} = 1$.

15. Sketch a graph of $\frac{x^2}{9} - \frac{y^2}{4} = 1$.

16. Sketch a graph of $\frac{x^2}{16} + \frac{y^2}{4} = 1$.

17. Graph the solution set: $\frac{x^2}{16} - \frac{y^2}{25} < 1$

18. Graph the solution set: $x^2 + y^2 < 36$
$x + y > 4$

19. Graph the solution set: $\frac{x^2}{25} + \frac{y^2}{4} \leq 1$

20. Graph the solution set: $\frac{x^2}{25} - \frac{y^2}{16} \geq 1$
$x^2 + y^2 \leq 9$

Final Exam

1. Simplify: $12 - 8[3 - (-2)]^2 \div 5 - 3$

2. Evaluate $\dfrac{a^2 - b^2}{a - b}$ when $a = 3$ and $b = -4$.

3. Simplify: $5 - 2[3x - 7(2 - x) - 5x]$

4. Solve: $\dfrac{3}{4}x - 2 = 4$

5. Solve: $\dfrac{2 - 4x}{3} - \dfrac{x - 6}{12} = \dfrac{5x - 2}{6}$

6. Solve: $8 - |5 - 3x| = 1$

7. Solve: $|2x + 5| < 3$

8. Solve: $2 - 3x < 6$ and $2x + 1 > 4$

9. Find the equation of the line that contains the point $P(-2, 1)$ and is perpendicular to the line with equation $3x - 2y = 6$.

10. Simplify: $2a[5 - a(2 - 3a) - 2a] + 3a^2$

11. Divide: $\dfrac{3}{2 + i}$

12. Write a quadratic equation that has integer coefficients and has solutions $-\frac{1}{2}$ and 2.

13. Factor: $8 - x^3y^3$

14. Factor: $x - y - x^3 + x^2y$

15. Divide: $(2x^3 - 7x^2 + 4) \div (2x - 3)$

16. Divide: $\dfrac{x^2 - 3x}{2x^2 - 3x - 5} \div \dfrac{4x - 12}{4x^2 - 4}$

17. Subtract: $\dfrac{x - 2}{x + 2} - \dfrac{x + 3}{x - 3}$

18. Simplify: $\dfrac{\dfrac{3}{x} + \dfrac{1}{x + 4}}{\dfrac{1}{x} + \dfrac{3}{x + 4}}$

19. Solve: $\dfrac{5}{x - 2} - \dfrac{5}{x^2 - 4} = \dfrac{1}{x + 2}$

20. Solve $a_n = a_1 + (n - 1)d$ for d.

21. Simplify: $\left(\dfrac{4x^2y^{-1}}{3x^{-1}y}\right)^{-2}\left(\dfrac{2x^{-1}y^2}{9x^{-2}y^2}\right)^3$

22. Simplify: $\left(\dfrac{3x^{\frac{2}{3}}y^{\frac{1}{2}}}{6x^2y^{\frac{4}{3}}}\right)^6$

23. Subtract: $x\sqrt{18x^2y^3} - y\sqrt{50x^4y}$

24. Simplify: $\dfrac{\sqrt{16x^5y^4}}{\sqrt{32xy^7}}$

25. Solve by using the quadratic formula: $2x^2 - 3x - 1 = 0$

26. Solve: $x^{\frac{2}{3}} - x^{\frac{1}{3}} - 6 = 0$

27. Find the equation of the line containing the points $P_1(3, -2)$ and $P_2(1, 4)$.

28. Solve: $\frac{2}{x} - \frac{2}{2x + 3} = 1$

29. Solve by the addition method: $3x - 2y = 1$, $5x - 3y = 3$

30. Evaluate the determinant: $\begin{vmatrix} 3 & 4 \\ -1 & 2 \end{vmatrix}$

31. Solve for x: $\log_3 x - \log_3(x - 3) = \log_3 2$

32. Write $\sum_{i=1}^{5} 2y^i$ in expanded form.

33. Find an equivalent fraction for $0.5\overline{1}$.

34. Find the third term in the expansion of $(x - 2y)^9$.

35. Solve: $x^2 - y^2 = 4$, $x + y = 1$

36. Find the inverse function of $f(x) = \frac{2}{3}x - 4$.

37. Write $2(\log_2 a - \log_2 b)$ as a single logarithm with a coefficient of 1.

38. Graph $2x - 3y = 9$ by using the x- and y-intercepts.

39. Graph the solution set of $3x + 2y > 6$.

40. Graph: $f(x) = -x^2 + 4$

41. Graph: $\frac{x^2}{16} + \frac{y^2}{4} = 1$

42. Graph: $f(x) = \log_2(x + 1)$

43. Graph $f(x) = x + 2^{-x}$ and approximate, to the nearest tenth, the values of x for which $f(x) = 2$.

44. Given the graph of $y = f(x)$ shown below, graph $g(x) = f(x + 3)$.

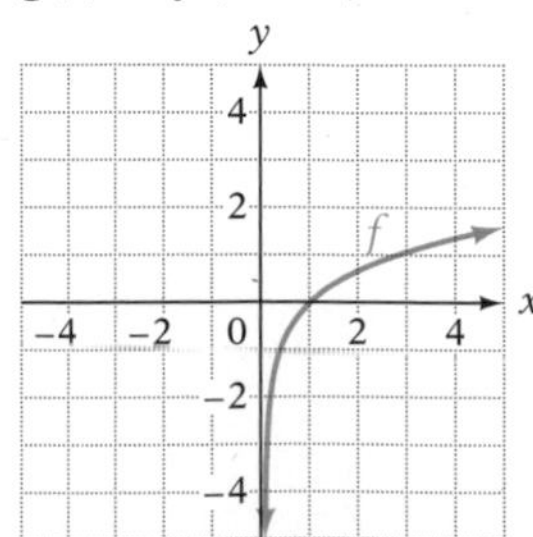

45. **Test Scores** An average score of 70 to 79 in a history class receives a C grade. A student has grades of 64, 58, 82, and 77 on four history tests. Find the range of scores on the fifth test that will give the student a C grade for the course.

46. **Uniform Motion** A jogger and a cyclist set out at 8 A.M. from the same point headed in the same direction. The average speed of the cyclist is two and a half times the average speed of the jogger. In 2 h, the cyclist is 24 mi ahead of the jogger. How far did the cyclist ride?

47. **Investments** You have a total of $12,000 invested in two simple interest accounts. On one account, a money market fund, the annual simple interest rate is 8.5%. On the other account, a tax-free bond fund, the annual simple interest rate is 6.4%. The total annual interest earned by the two accounts is $936. How much do you have invested in each account?

48. **Geometry** The length of a rectangle is 1 ft less than three times the width. The area of the rectangle is 140 ft^2. Find the length and width of the rectangle.

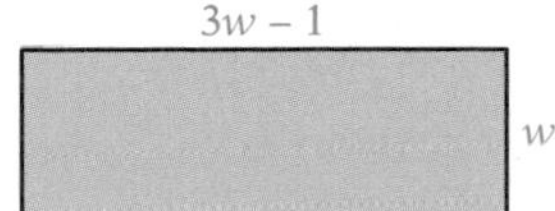

49. **Investments** Three hundred shares of a utility stock earn a yearly dividend of $486. How many additional shares of the utility stock would give a total dividend income of $810?

50. **Uniform Motion** An account executive traveled 45 mi by car and then an additional 1050 mi by plane. The rate of the plane was seven times the rate of the car. The total time of the trip was $3\frac{1}{4}$ h. Find the rate of the plane.

51. **Physics** An object is dropped from the top of a building. Find the distance the object has fallen when its speed reaches 75 ft/s. Use the equation $v = \sqrt{64d}$, where v is the speed of the object and d is the distance. Round to the nearest whole number.

52. **Uniform Motion** A small plane made a trip of 660 mi in 5 h. The plane traveled the first 360 mi at a constant rate before increasing its speed by 30 mph. Another 300 mi was traveled at the increased speed. Find the rate of the plane for the first 360 mi.

53. **Light** The intensity (L) of a light source is inversely proportional to the square of the distance (d) from the source. If the intensity is 8 lumens at a distance of 20 ft, what is the intensity when the distance is 4 ft?

54. **Uniform Motion** A motorboat traveling with the current can go 30 mi in 2 h. Against the current, it takes 3 h to go the same distance. Find the rate of the motorboat in calm water and the rate of the current.

55. **Investments** An investor deposits $4000 into an account that earns 9% annual interest compounded monthly. Use the compound interest formula $P = A(1 + i)^n$, where A is the original value of the investment, i is the interest rate per compounding period, n is the total number of compounding periods, and P is the value of the investment after n periods, to find the value of the investment after 2 years. Round to the nearest cent.

56. **Appreciation** Assume the average value of a home increases 6% per year. How much would a house costing $180,000 be worth in 20 years? Round to the nearest dollar.

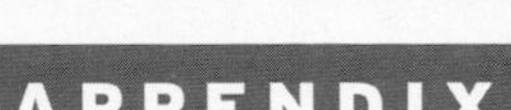

Table of Properties

APPENDIX

Properties of Real Numbers

The Associative Property of Addition
If a, b, and c are real numbers, then $(a + b) + c = a + (b + c)$.

The Commutative Property of Addition
If a and b are real numbers, then $a + b = b + a$.

The Addition Property of Zero
If a is a real number, then $a + 0 = 0 + a = a$.

The Multiplication Property of Zero
If a is a real number, then $a \cdot 0 = 0 \cdot a = 0$.

The Inverse Property of Addition
If a is a real number, then $a + (-a) = (-a) + a = 0$.

The Associative Property of Multiplication
If a, b, and c are real numbers, then $(a \cdot b) \cdot c = a \cdot (b \cdot c)$.

The Commutative Property of Multiplication
If a and b are real numbers, then $a \cdot b = b \cdot a$.

The Multiplication Property of One
If a is a real number, then $a \cdot 1 = 1 \cdot a = a$.

The Inverse Property of Multiplication
If a is a real number and $a \neq 0$, then $a \cdot \frac{1}{a} = \frac{1}{a} \cdot a = 1$.

Distributive Property
If a, b, and c are real numbers, then $a(b + c) = ab + ac$ or $(b + c)a = ba + ca$.

Properties of Equations

Addition Property of Equations
The same number or variable term can be added to each side of an equation without changing the solution of the equation.

Multiplication Property of Equations
Each side of an equation can be multiplied by the same nonzero number without changing the solution of the equation.

Properties of Inequalities

Addition Property of Inequalities
If $a > b$, then $a + c > b + c$.

If $a < b$, then $a + c < b + c$.

Multiplication Property of Inequalities
If $a > b$ and $c > 0$, then $ac > bc$.

If $a < b$ and $c > 0$, then $ac < bc$.

If $a > b$ and $c < 0$, then $ac < bc$.

If $a < b$ and $c < 0$, then $ac > bc$.

Properties of Exponents

If m and n are integers, then $x^m \cdot x^n = x^{m+n}$.

If m and n are integers, then $(x^m)^n = x^{mn}$.

If $x \neq 0$, then $x^0 = 1$.

If m and n are integers and $x \neq 0$, then
$\dfrac{x^m}{x^n} = x^{m-n}$.

If m, n, and p are integers, then $(x^m \cdot y^n)^p = x^{mp}y^{np}$.

If n is a positive integer and $x \neq 0$, then
$x^{-n} = \dfrac{1}{x^n}$ and $\dfrac{1}{x^{-n}} = x^n$.

If m, n, and p are integers and $y \neq 0$, then
$\left(\dfrac{x^m}{y^n}\right)^p = \dfrac{x^{mp}}{y^{np}}$.

Principle of Zero Products

If $a \cdot b = 0$, then $a = 0$ or $b = 0$.

Properties of Radical Expressions

If a and b are positive real numbers, then
$\sqrt[n]{ab} = \sqrt[n]{a}\sqrt[n]{b}$.

If a and b are positive real numbers, then
$\sqrt[n]{\dfrac{a}{b}} = \dfrac{\sqrt[n]{a}}{\sqrt[n]{b}}$.

Property of Squaring Both Sides of an Equation

If a and b are real numbers and $a = b$, then $a^2 = b^2$.

Properties of Logarithms

If x, y, and b are positive real numbers and $b \neq 1$, then
$\log_b(xy) = \log_b x + \log_b y$.

If x, y, and b are positive real numbers and $b \neq 1$, then
$\log_b \dfrac{x}{y} = \log_b x - \log_b y$.

If x and b are positive real numbers, $b \neq 1$, and r is any real number, then $\log_b x^r = r \log_b x$.

If x and b are positive real numbers and $b \neq 1$, then $\log_b b^x = x$.

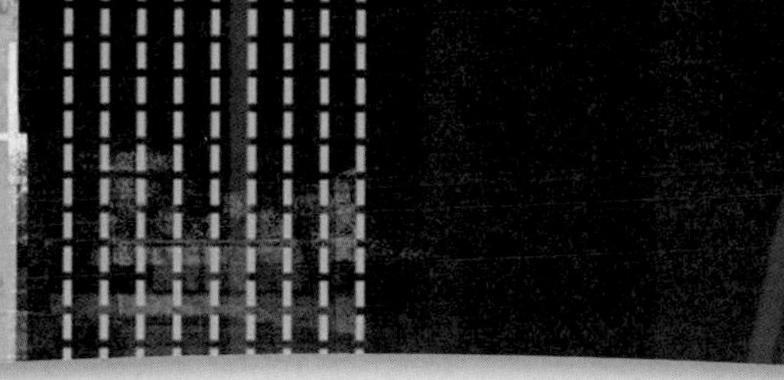

Keystroke Guide for the TI-83 Plus and TI-84 Plus

APPENDIX

Basic Operations

Numerical calculations are performed on the **home screen.** You can always return to the home screen by pressing 2ND QUIT. Pressing CLEAR erases the home screen.

> **Take Note**
>
> The descriptions in the margins (for example, Basic Operations and Complex Numbers) are the same as those used in the text and are arranged alphabetically.

To evaluate the expression $-2(3 + 5) - 8 \div 4$, use the following keystrokes.

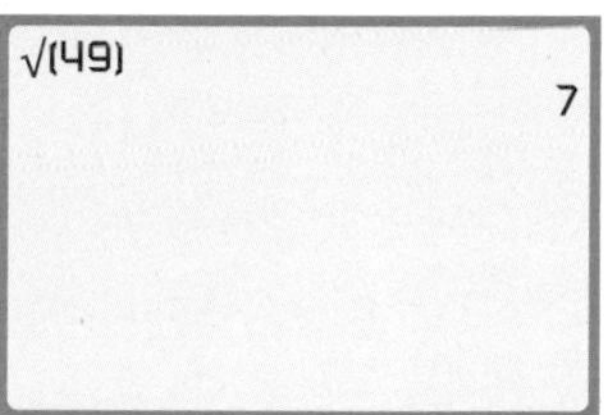

(-) 2 (3 + 5) − 8 ÷ 4 ENTER

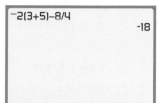

Note: There is a difference between the key to enter a negative number, (-), and the key for subtraction, −. You cannot use these keys interchangeably.

The 2ND key is used to access the commands in blue writing above a key. For instance, to evaluate the $\sqrt{49}$, press 2ND √ 49) ENTER.

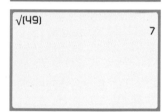

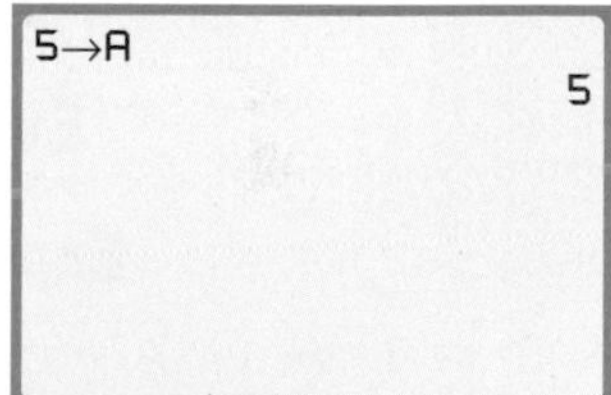

The ALPHA key is used to place a letter on the screen. One reason to do this is to store a value of a variable. The following keystrokes give A the value of 5.

5 STO▸ ALPHA A ENTER

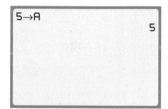

This value is now available in calculations. For instance, we can find the value of $3a^2$ by using the following keystrokes: 3 ALPHA A x^2. To display the value of the variable on the screen, press 2ND RCL ALPHA A.

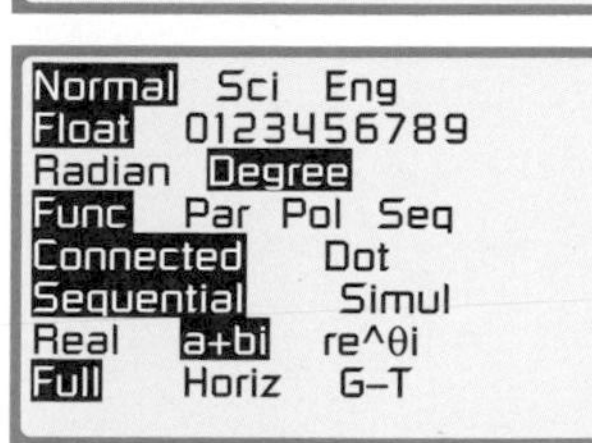

Note: When you use the ALPHA key, only capital letters are available on the TI-83 calculator.

Complex Numbers

To perform operations on complex numbers, first press MODE and then use the arrow keys to select a + bi. Then press ENTER 2ND QUIT.

Normal Sci Eng
Float 0123456789
Radian Degree
Func Par Pol Seq
Connected Dot
Sequential Simul
Real a+bi re^θi
Full Horiz G-T

Addition of complex numbers To add $(3 + 4i) + (2 - 7i)$, use the keystrokes

(3 + 4 2ND *i*) +

(2 − 7 2ND *i*) ENTER.

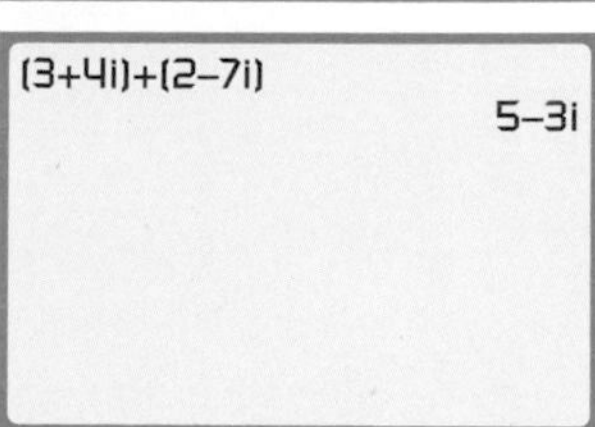

Division of complex numbers To divide $\frac{26 + 2i}{2 + 4i}$, use the keystrokes

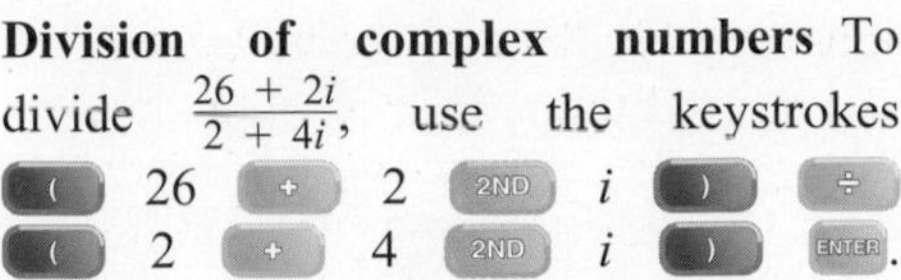

Note: Operations for subtraction and multiplication are similar.

Additional operations on complex numbers can be found by selecting CPX under the MATH key.

To find the absolute value of $2 - 5i$, press MATH (scroll to CPX) (scroll to abs) ENTER (2 − 5 2ND i) ENTER.

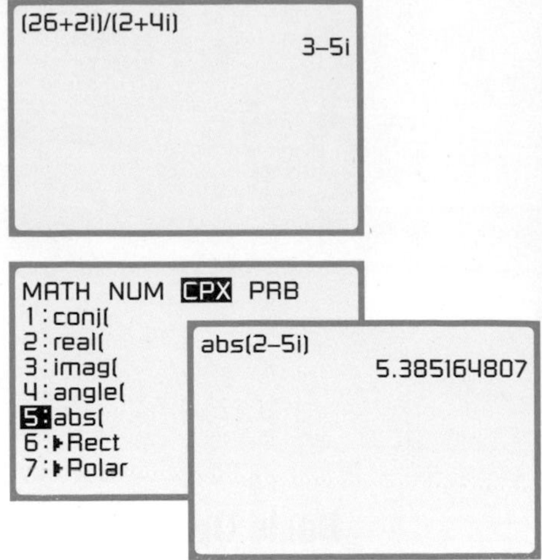

Evaluating Functions

There are various methods of evaluating a function, but all methods require that the expression be entered as one of the ten functions Y1 to Y0. To evaluate $f(x) = \frac{x^2}{x - 1}$ when $x = -3$, enter the expression into, for instance, Y1, and then press VARS ▸ 1 1 ((−) 3) ENTER.

> **Take Note**
>
> Use the down arrow key to scroll past Y_7 to see Y_8, Y_9, and Y_0.

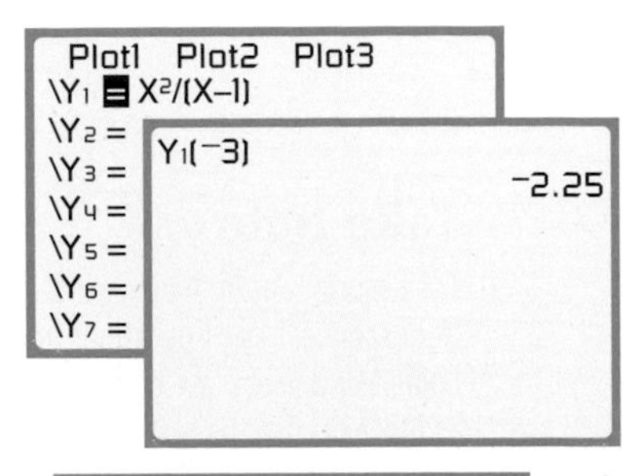

Note: If you try to evaluate a function at a number that is not in the domain of the function, you will get an error message. For instance, 1 is not in the domain of $f(x) = \frac{x^2}{x - 1}$. If we try to evaluate the function at 1, the error screen at the right appears.

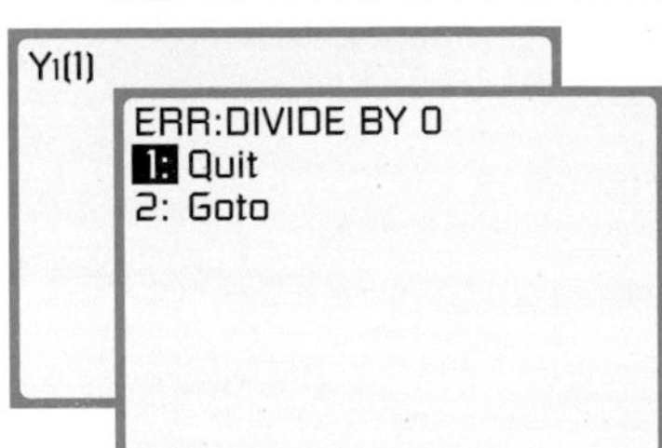

Evaluating Variable Expressions

To evaluate a variable expression, first store the values of each variable. Then enter the variable expression. For instance, to evaluate $s^2 + 2sl$ when $s = 4$ and $l = 5$, use the following keystrokes.

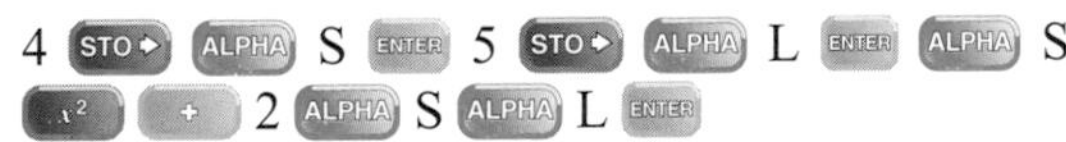

4 STO▸ ALPHA S ENTER 5 STO▸ ALPHA L ENTER ALPHA S
x² + 2 ALPHA S ALPHA L ENTER

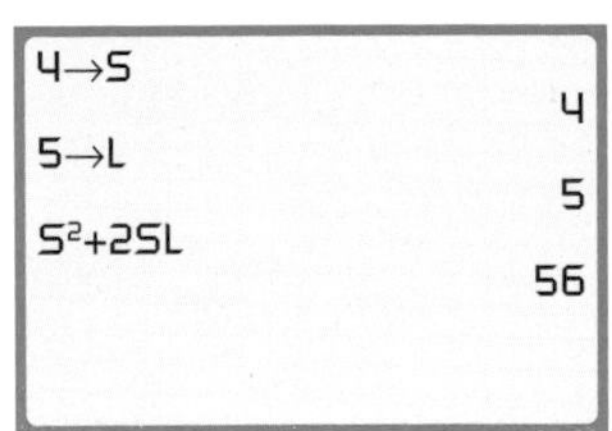

Graph

To graph a function, use the Y= key to enter the expression for the function, select a suitable viewing window, and then press GRAPH. For instance, to graph $f(x) = 0.1x^3 - 2x - 1$ in the standard viewing window, use the following keystrokes.

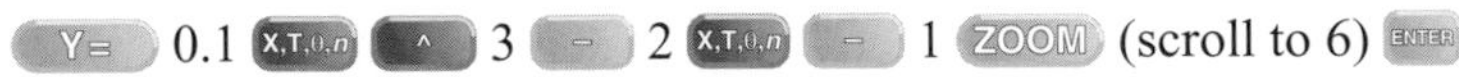

Y= 0.1 X,T,θ,n ^ 3 − 2 X,T,θ,n − 1 ZOOM (scroll to 6) ENTER

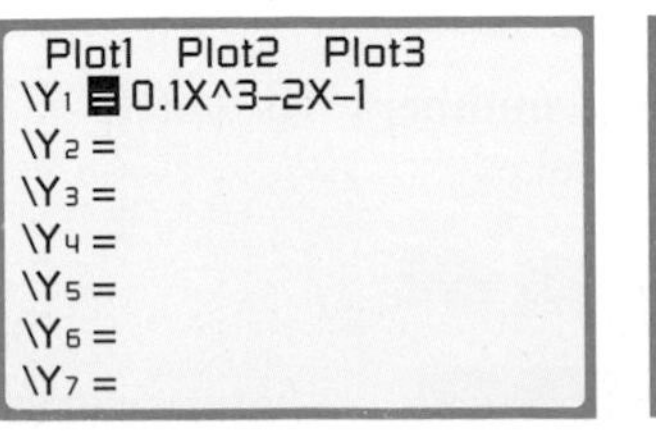

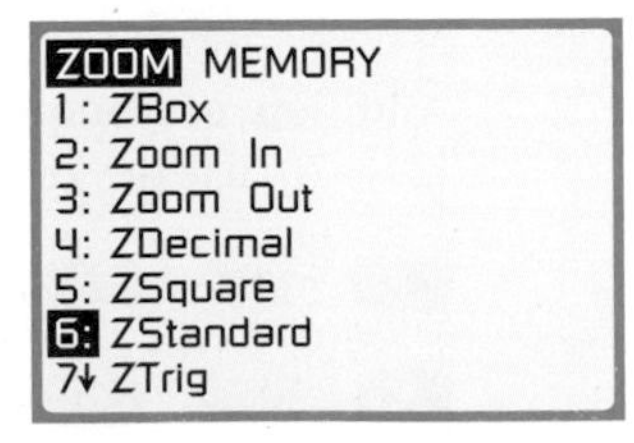

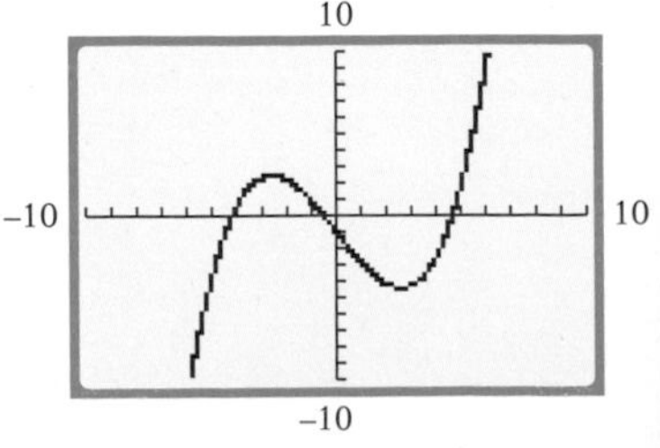

Note: For the keystrokes above, you do not have to scroll to 6. Alternatively, use ZOOM 6. This will select the standard viewing window and automatically start the graph. Use the WINDOW key to create a custom window for a graph.

Graphing Inequalities

To illustrate this feature, we will graph $y \leq 2x - 1$. Enter $2x - 1$ into Y1. Because $y \leq 2x - 1$, we want to shade below the graph. Move the cursor to the left of Y1 and press ENTER three times. Press GRAPH.

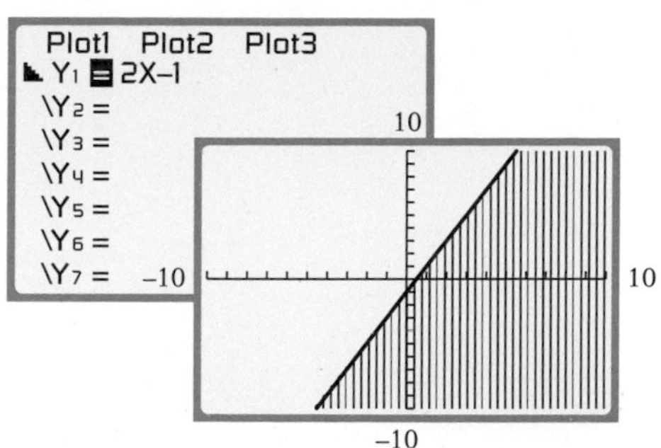

Note: To shade above the graph, move the cursor to the left of Y1 and press ENTER two times. An inequality with the symbol $\leq$ or $\geq$ should be graphed with a solid line, and an inequality with the symbol $<$ or $>$ should be graphed with a dashed line. However, the graph of a linear inequality on a graphing calculator does not distinguish between a solid line and a dashed line.

To graph the solution set of a system of inequalities, solve each inequality for y and graph each inequality. The solution set is the intersection of the two inequalities. The solution set of $\begin{array}{l} 3x + 2y > 10 \\ 4x - 3y \leq 5 \end{array}$ is shown at the right.

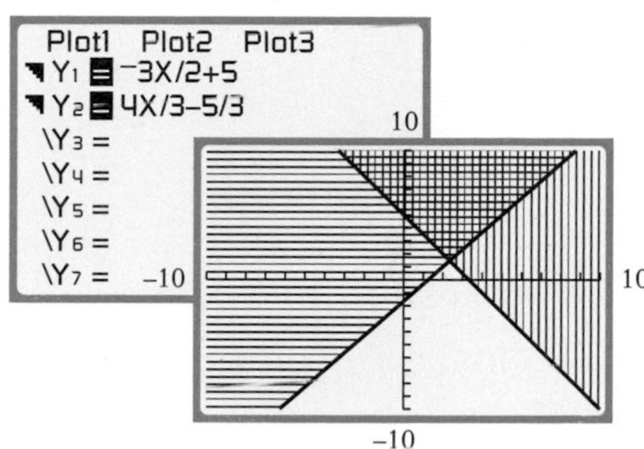

Intersect

The INTERSECT feature is used to solve a system of equations. To illustrate this feature, we will use the system of equations $\begin{array}{l} 2x - 3y = 13 \\ 3x + 4y = -6 \end{array}$.

Note: Some equations can be solved by this method. See the section "Solve an equation" below. Also, this method is used to find a number in the domain of a function for a given number in the range. See the section "Find a domain element."

Solve each of the equations in the system of equations for y. In this case, we have $y = \frac{2}{3}x - \frac{13}{3}$ and $y = -\frac{3}{4}x - \frac{3}{2}$.

Use the Y-editor to enter $\frac{2}{3}x - \frac{13}{3}$ into Y1 and $-\frac{3}{4}x - \frac{3}{2}$ into Y2. Graph the two functions in the standard viewing window. (If the window does not show the point of intersection of the two graphs, adjust the window until you can see the point of intersection.)

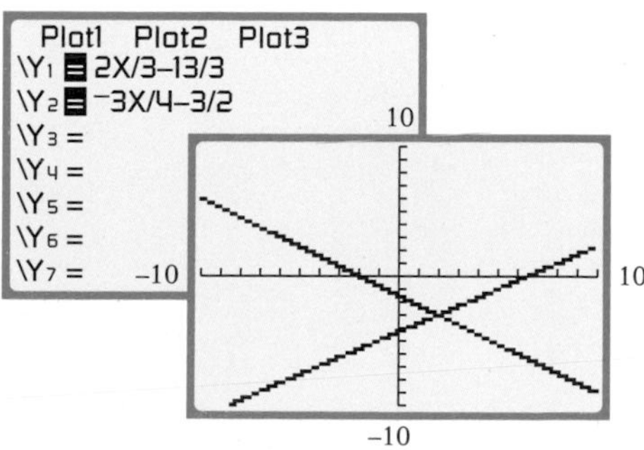

Press 2ND CALC (scroll to 5, intersect) ENTER.

Alternatively, you can just press 2ND CALC 5.

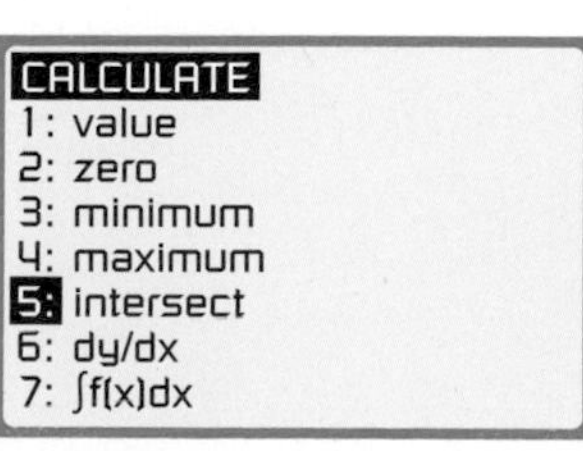

First curve? is shown at the bottom of the screen and identifies one of the two graphs on the screen. Press ENTER.

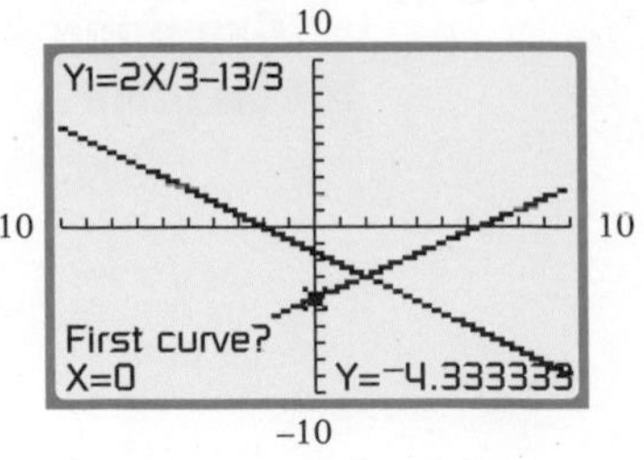

Second curve? is shown at the bottom of the screen and identifies the second of the two graphs on the screen. Press ENTER.

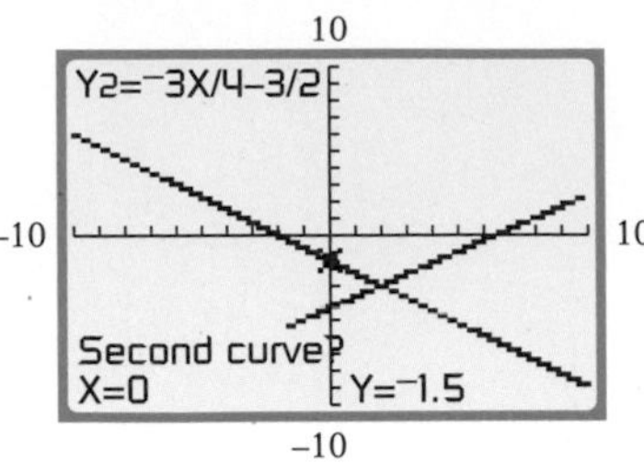

Guess? shown at the bottom of the screen asks you to use the left or right arrow key to move the cursor to the *approximate* location of the point of intersection. (If there are two or more points of intersection, it does not matter which one you choose first.) Press ENTER.

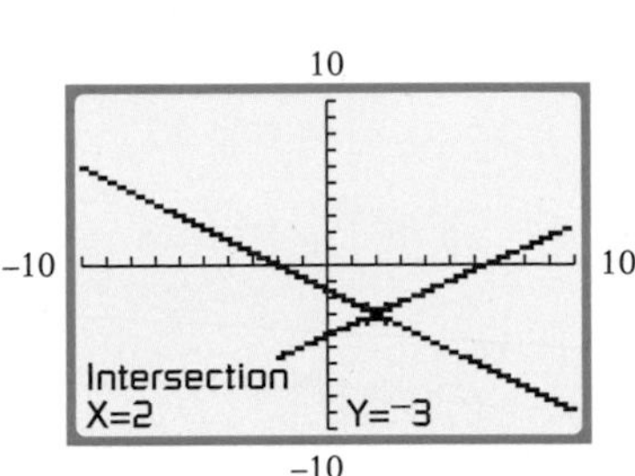

The solution of the system of equations is $(2, -3)$.

Solve an equation To illustrate the steps involved, we will solve the equation $2x + 4 = -3x - 1$. The idea is to write the equation as the system of equations $\begin{matrix} y = 2x + 4 \\ y = -3x - 1 \end{matrix}$ and then use the steps for solving a system of equations.

Use the Y-editor to enter the left and right sides of the equation into Y1 and Y2. Graph the two functions and then follow the steps for **Intersect**.

The solution is -1, the x-coordinate of the point of intersection.

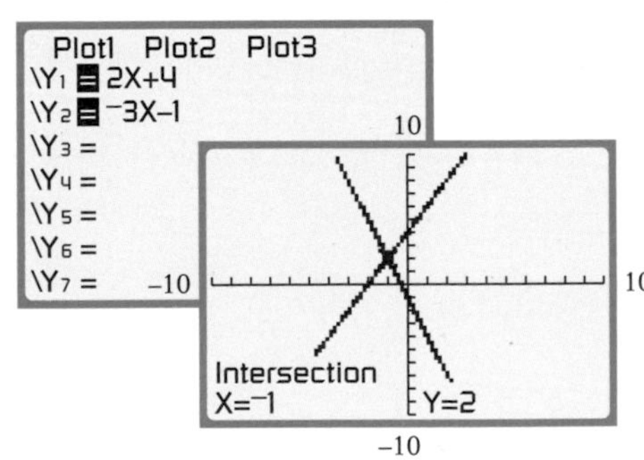

Find a domain element For this example, we will find a number in the domain of $f(x) = -\frac{2}{3}x + 2$ that corresponds to 4 in the range of the function. This is like solving the system of equations $y = -\frac{2}{3}x + 2$ and $y = 4$.

Use the Y-editor to enter the expression for the function in Y1 and the desired output, 4, in Y2. Graph the two functions and then follow the steps for **Intersect**.

The point of intersection is $(-3, 4)$. The number -3 in the domain of f produces an output of 4 in the range of f.

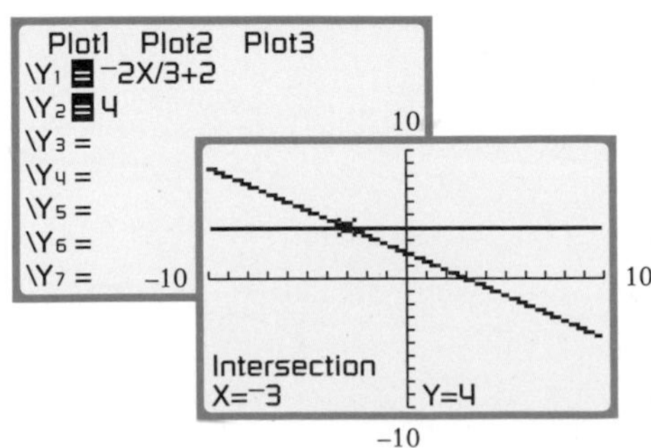

Math

Pressing MATH gives you access to many built-in functions. The following keystrokes will convert 0.125 into a fraction: .125 MATH 1 ENTER.

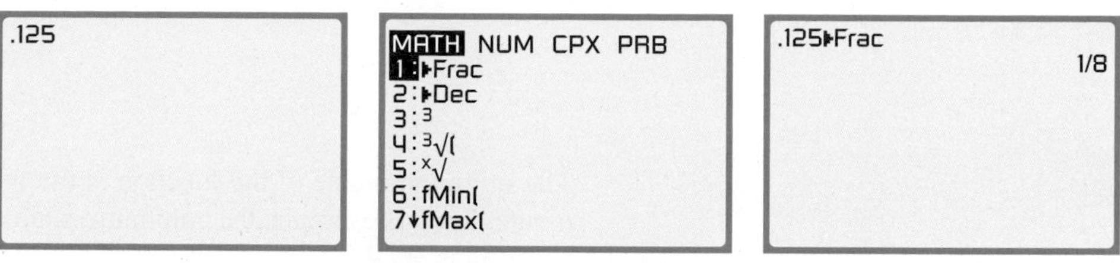

Additional built-in functions under MATH can be found by pressing MATH ▸. For instance, to evaluate $-|-25|$, press (–) MATH ▸ 1 (–) 25) ENTER.

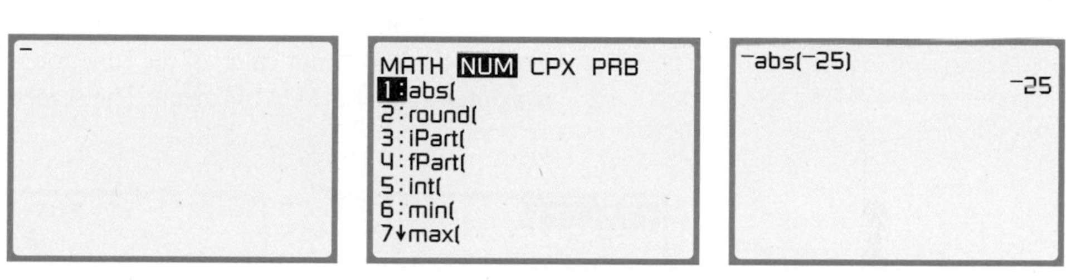

See your owner's manual for assistance with other functions under the MATH key.

Min and Max

The local minimum and the local maximum values of a function are calculated by accessing the CALC menu. For this demonstration, we will find the minimum value and the maximum value of $f(x) = 0.2x^3 + 0.3x^2 - 3.6x + 2$.

Enter the function into Y1. Press 2ND CALC (scroll to 3 for minimum of the function) ENTER.

Alternatively, you can just press 2ND CALC 3.

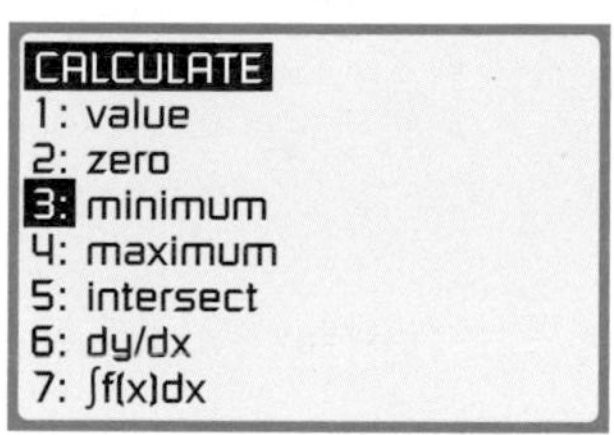

Left Bound? shown at the bottom of the screen asks you to use the left or right arrow key to move the cursor to the *left* of the minimum. Press ENTER.

Right Bound? shown at the bottom of the screen asks you to use the left or right arrow key to move the cursor to the *right* of the minimum. Press ENTER.

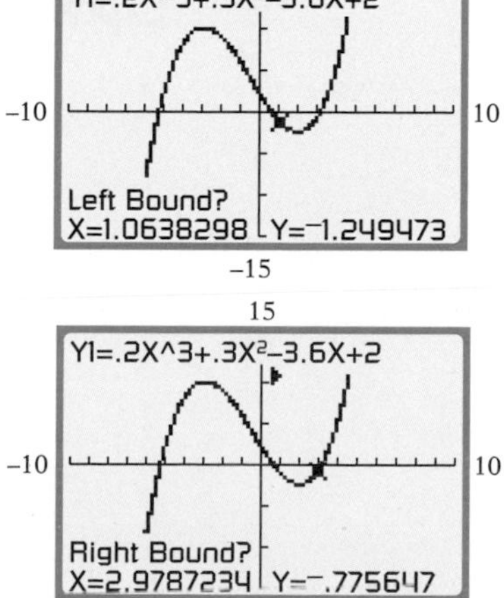

Guess? shown at the bottom of the screen asks you to use the left or right arrow key to move the cursor to the *approximate* location of the minimum. Press ENTER.

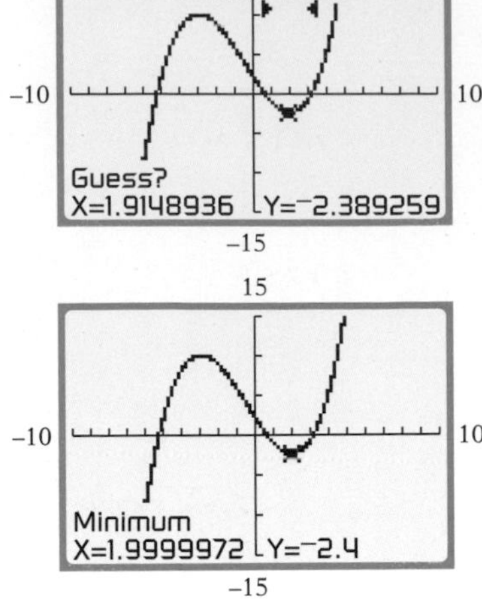

The minimum value of the function is the *y*-coordinate. For this example, the minimum value of the function is -2.4.

The *x*-coordinate for the minimum is 2. However, because of rounding errors in the calculation, it is shown as a number close to 2.

To find the maximum value of the function, follow the same steps as above except select maximum under the CALC menu. The screens for this calculation are shown below.

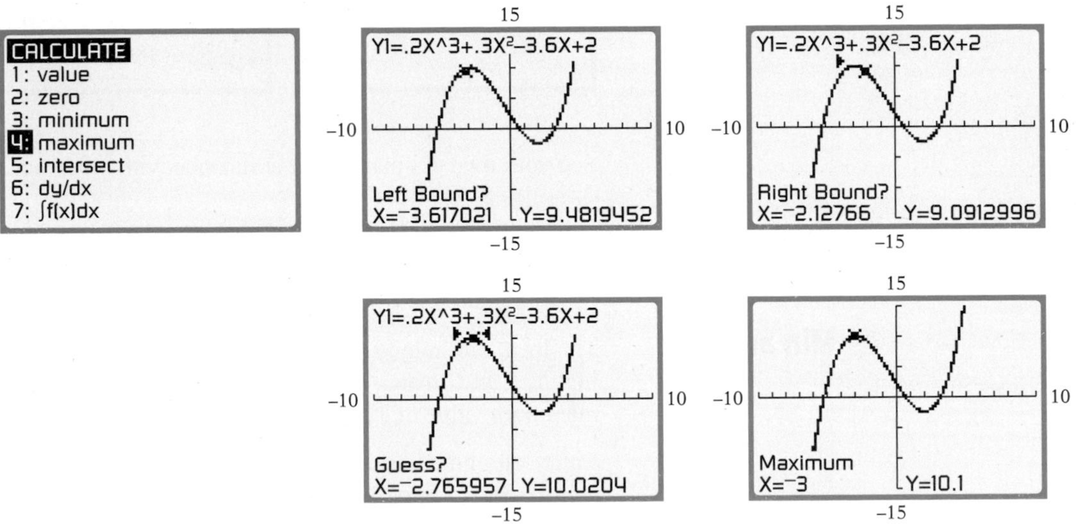

The maximum value of the function is 10.1.

Radical Expressions

To evaluate a square-root expression, press 2ND $\sqrt{}$.

For instance, to evaluate $0.15\sqrt{p^2 + 4p + 10}$ when $p = 100{,}000$, first store 100,000 in P. Then press 0.15 2ND $\sqrt{}$ ALPHA P x^2 + 4 ALPHA P + 10) ENTER.

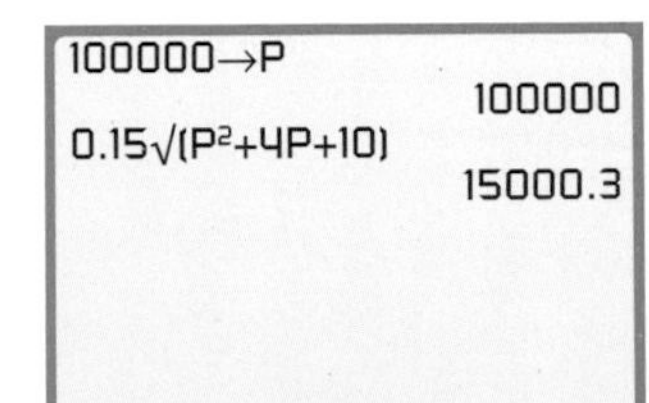

To evaluate a radical expression other than a square root, access $\sqrt[x]{}$ by pressing MATH. For instance, to evaluate $\sqrt[4]{67}$, press 4 (the index of the radical) MATH (scroll to 5) ENTER 67 ENTER.

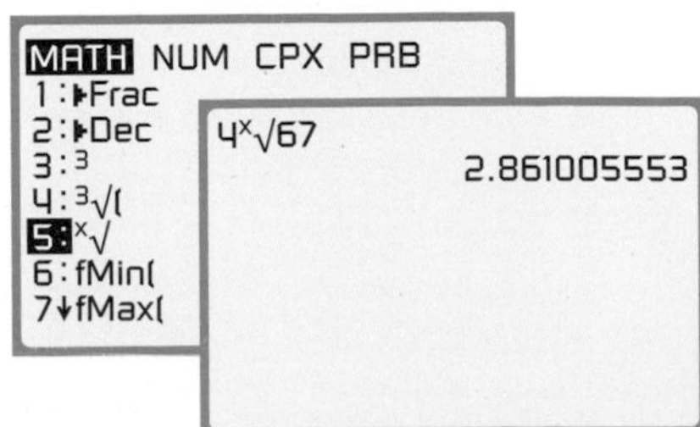

Scientific Notation

To enter a number in scientific notation, use 2ND EE. For instance, to find $\frac{3.45 \times 10^{-12}}{1.5 \times 10^{25}}$, press 3.45 2ND EE (-) 12 ÷ 1.5 2ND EE 25 ENTER. The answer is 2.3×10^{-37}.

3.45E-12/1.5E25
2.3E-37

Sequences and Series

The terms of a sequence and the sum of a series can be calculated by using the 2ND LIST feature.

Store a sequence A sequence is stored in one of the lists L1 through L6. For instance, to store the sequence 1, 3, 5, 7, 9 in L1, use the following keystrokes.

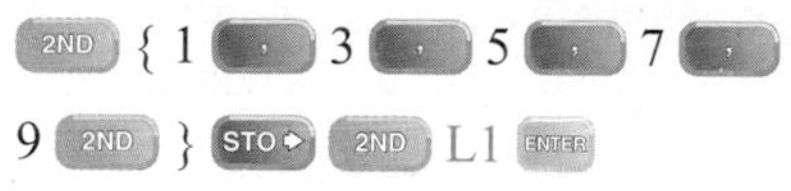

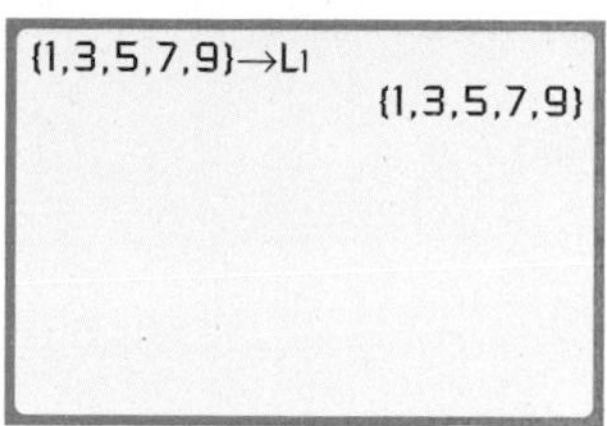

Display the terms of a sequence The terms of a sequence are displayed by using the function seq(expression, variable, begin, end, increment). For instance, to display the third through eighth terms of the sequence given by $a_n = n^2 + 6$, enter the following keystrokes.

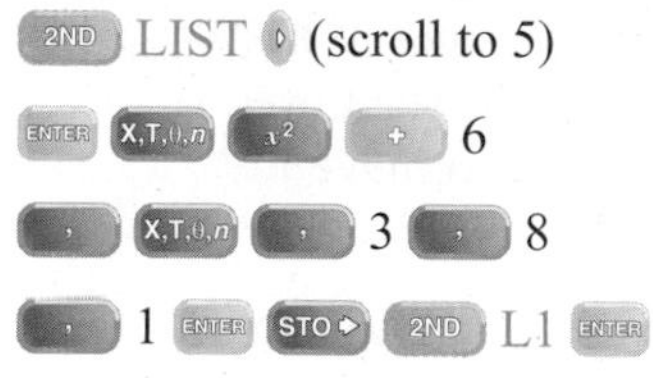

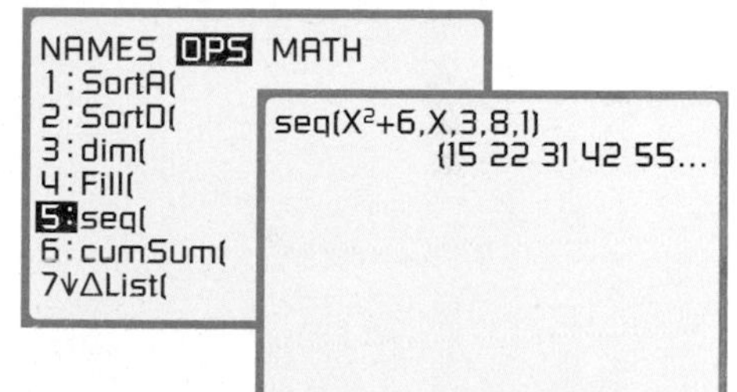

The keystrokes STO▸ 2ND L1 ENTER store the terms of the sequence in L1. This is not necessary but is sometimes helpful if additional work will be done with that sequence.

Find a sequence of partial sums To find a sequence of partial sums, use the cumSum(function. For instance, to find the sequence of partial sums for 2, 4, 6, 8, 10, use the following keystrokes.

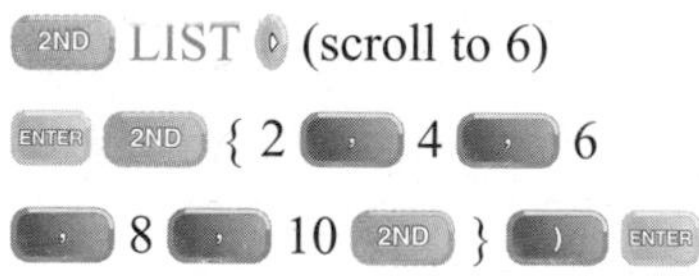

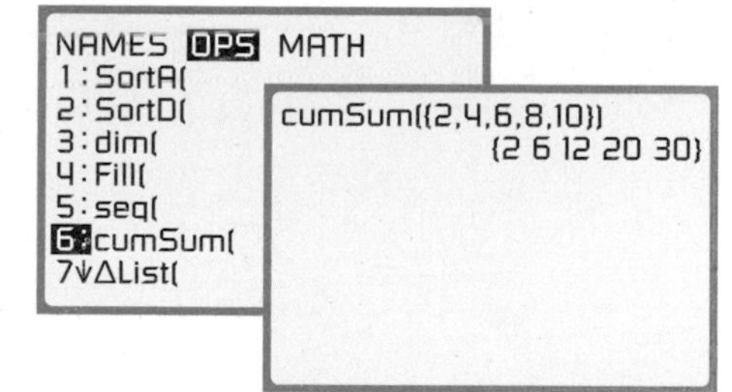

If a sequence is stored as a list in L1, then the sequence of partial sums can be calculated by pressing 2ND LIST ▸ (scroll to 6 [or press 6]) ENTER 2ND L1) ENTER.

Find the sum of a series The sum of a series is calculated using sum<list, start, end>. For instance, to find $\sum_{n=3}^{6} (n^2 + 2)$, enter the following keystrokes.

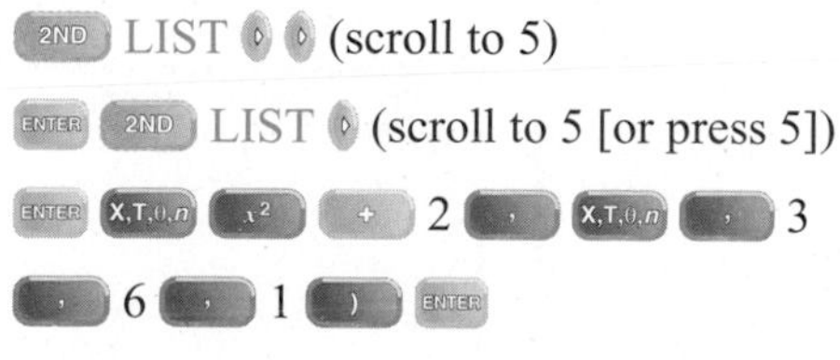

2ND LIST ▸ ▸ (scroll to 5)
ENTER 2ND LIST ▸ (scroll to 5 [or press 5])
ENTER X,T,θ,n x² + 2 , X,T,θ,n , 3
, 6 , 1) ENTER

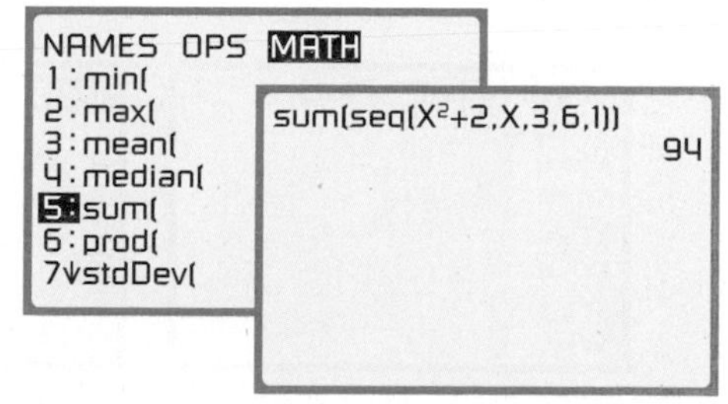

Table

There are three steps in creating an input/output table for a function. First use the Y= editor to input the function. The second step is setting up the table, and the third step is displaying the table.

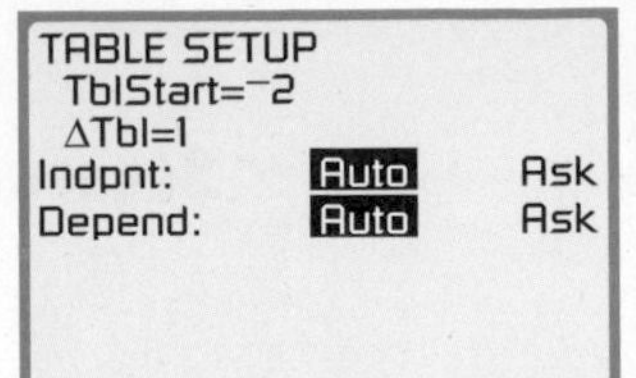

To set up the table, press 2ND TBLSET. TblStart is the first value of the independent variable in the input/output table. ΔTbl is the difference between successive values. Setting this to 1 means that, for this table, the input values are $-2, -1, 0, 1, 2, \ldots$. If ΔTbl $= 0.5$, then the input values are $-2, -1.5, -1, -0.5, 0, 0.5, \ldots$.

Indpnt is the independent variable. When this is set to Auto, values of the independent variable are automatically entered into the table. Depend is the dependent variable. When this is set to Auto, values of the dependent variable are automatically entered into the table.

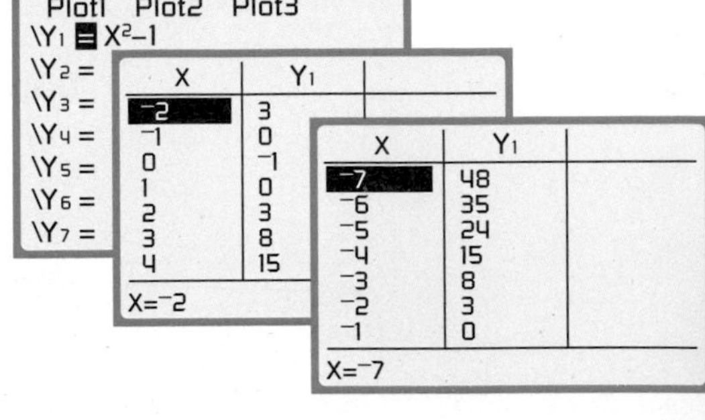

To display the table, press 2ND TABLE. An input/output table for $f(x) = x^2 - 1$ is shown at the right.

Once the table is on the screen, the up and down arrow keys can be used to display more values in the table. For the table at the right, we used the up arrow key to move to $x = -7$.

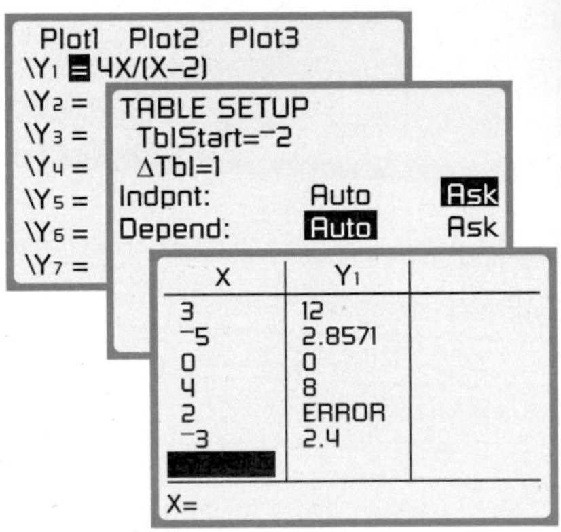

An input/output table for any given input can be created by selecting Ask for the independent variable. The table at the right shows an input/output table for $f(x) = \frac{4x}{x-2}$ for selected values of x. Note that the word ERROR when 2 was entered. This occurred because f is not defined when $x = 2$.

Note: Using the table feature in Ask mode is the same as evaluating a function for given values of the independent variable. For instance, from the table above, we have $f(4) = 8$.

Test

The TEST feature has many uses, one of which is to graph the solution set of a linear inequality in one variable. To illustrate this feature, we will graph the solution set of $x - 1 < 4$. Press Y= X,T,θ,n − 1 2ND TEST (scroll to 5) ENTER 4 GRAPH.

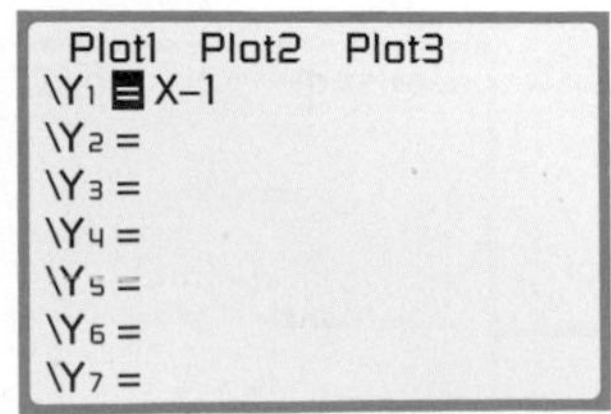

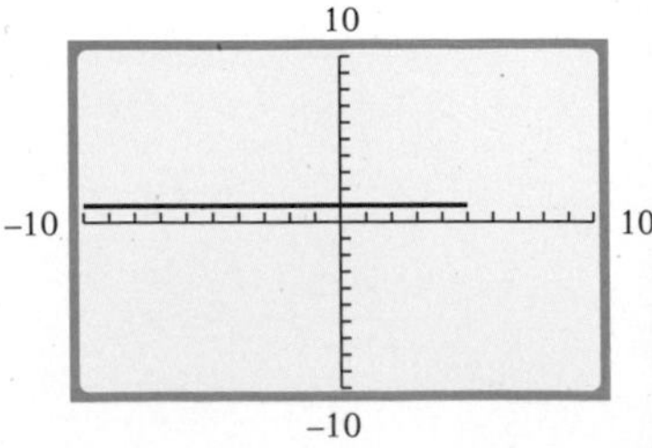

Trace

Once a graph is drawn, pressing TRACE will place a cursor on the screen, and the coordinates of the point below the cursor are shown at the bottom of the screen. Use the left and right arrow keys to move the cursor along the graph. For the graph at the right, we have $f(4.8) = 3.4592$, where $f(x) = 0.1x^3 - 2x + 2$ is shown at the top left of the screen.

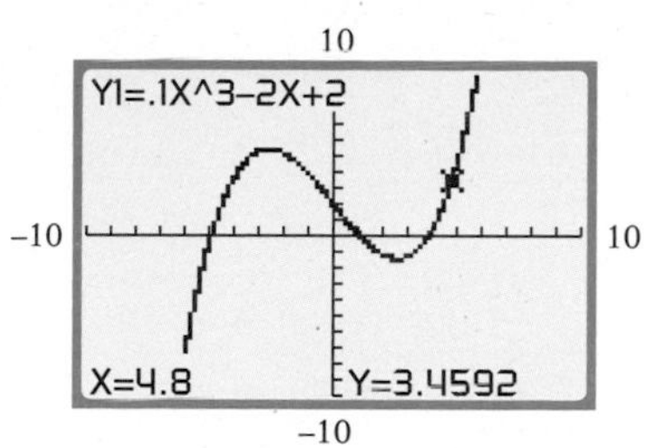

In TRACE mode, you can evaluate a function at any value of the independent variable that is within Xmin and Xmax. To do this, first graph the function. Now press TRACE (the value of x) ENTER. For the graph at the left below, we used $x = -3.5$. If a value of x is chosen outside the window, an error message is displayed.

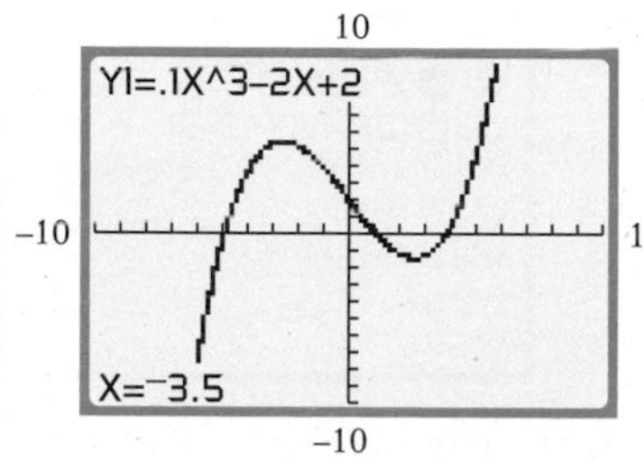

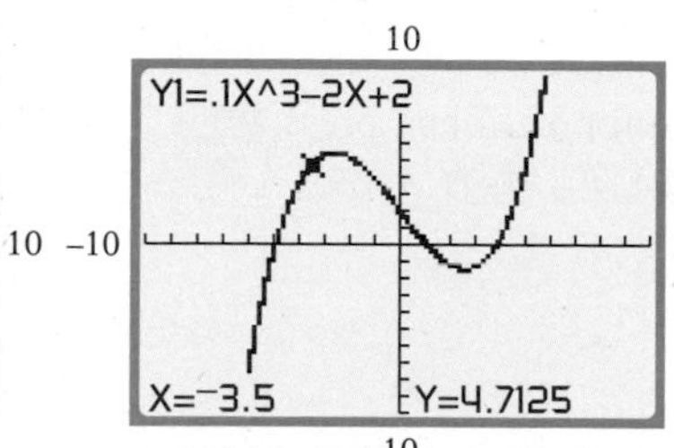

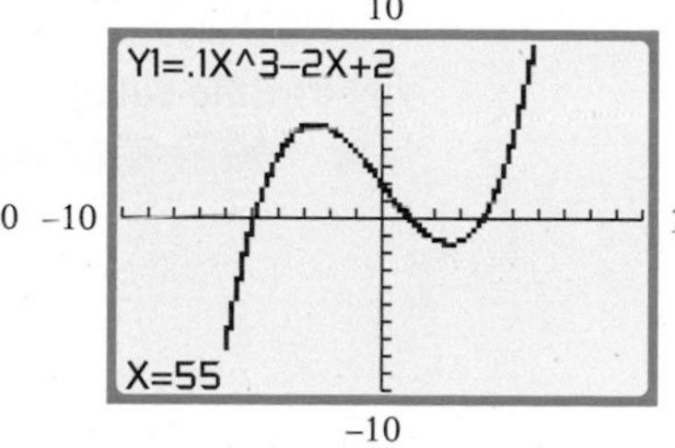

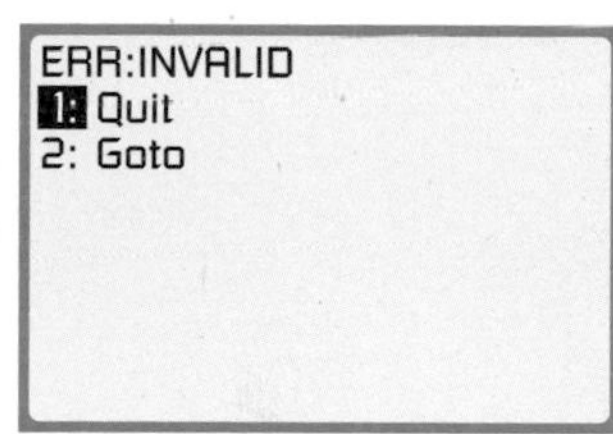

In the example above where we entered -3.5 for x, the value of the function was calculated as 4.7125. This means that $f(-3.5) = 4.7125$. The keystrokes 2ND QUIT VARS ▸ 11 MATH 1 ENTER will convert the decimal value into a fraction.

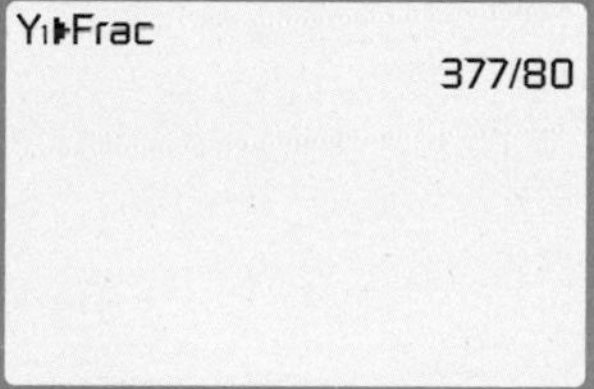

When the TRACE feature is used with two or more graphs, the up and down arrow keys are used to move between the graphs. The graphs below are for the functions $f(x) = 0.1x^3 - 2x + 2$ and $g(x) = 2x - 3$. By using the up and down arrows, we can place the cursor on either graph. The right and left arrows are used to move along the graph.

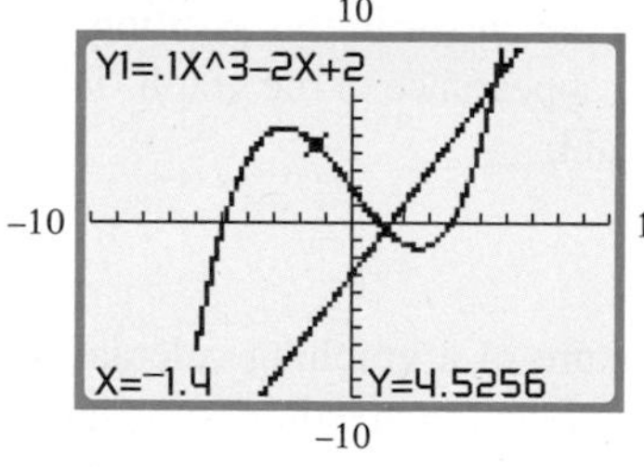

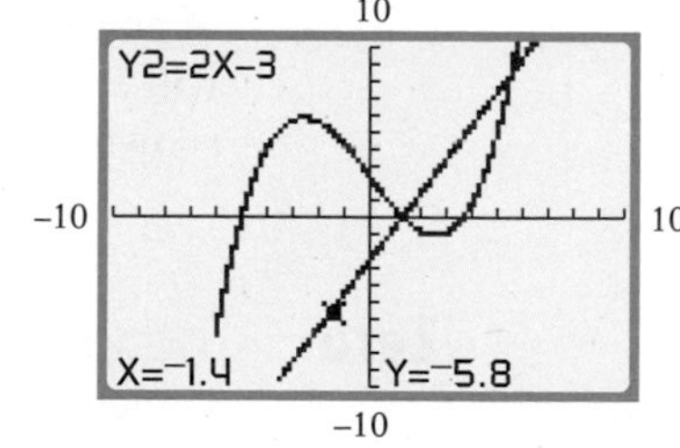

Window

The viewing window for a graph is controlled by pressing WINDOW. Xmin and Xmax are the minimum value and maximum value, respectively, of the independent variable shown on the graph. Xscl is the distance between tick marks on the x-axis. Ymin and Ymax are the minimum value and maximum value, respectively, of the dependent variable shown on the graph. Yscl is the distance between tick marks on the y-axis. Leave Xres as 1.

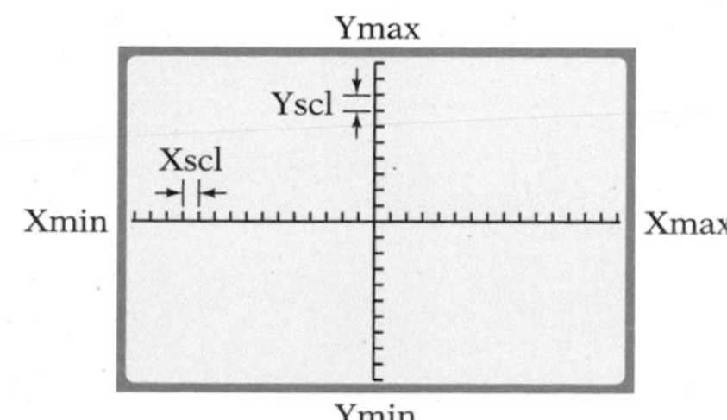

Note: In the standard viewing window, the distance between tick marks on the x-axis is different from the distance between tick marks on the y-axis. This will distort a graph. A more accurate picture of a graph can be created by using a square viewing window. See ZOOM.

Y= The Y= editor is used to enter the expression for a function. There are ten possible functions, labeled Y_1 to Y_0, that can be active at any one time. For instance, to enter $f(x) = x^2 + 3x - 2$ as Y_1, use the following keystrokes.

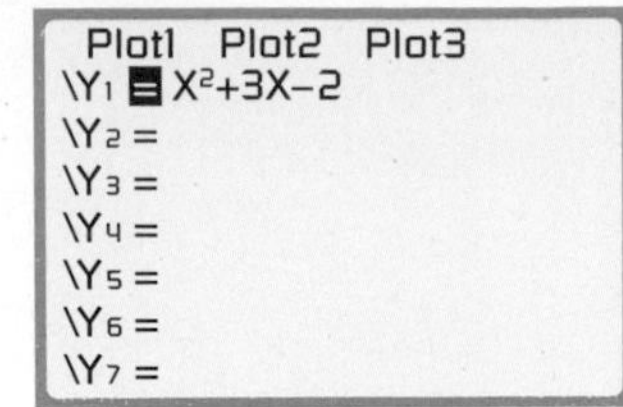

Y= X,T,θ,n x^2 + 3 X,T,θ,n − 2

Note: If an expression is already entered for Y_1, place the cursor anywhere on that expression and press CLEAR.

To enter $s = \frac{2v - 1}{v^3 - 3}$ into Y_2, place the cursor to the right of the equals sign for Y_2. Then press (2 X,T,θ,n − 1) ÷ (X,T,θ,n ^ 3 − 3).

Plot1 Plot2 Plot3
\Y1 ■ X²+3X−2
\Y2 ■ (2X−1)/(X^3−3)
\Y3 =
\Y4 =
\Y5 =
\Y6 =
\Y7 =

Note: When we enter an equation, the independent variable, v in the expression above, is entered using X,T,θ,n. The dependent variable, s in the expression above, is one of Y_1 to Y_0. Also note the use of parentheses to ensure the correct order of operations.

Observe the black rectangle that covers the equals sign for the two examples we have shown. This rectangle means that the function is "active." If we were to press GRAPH, then the graph of both functions would appear. You can make a function inactive by using the arrow keys to move the cursor over the equals sign of that function and then pressing ENTER. This will remove the black rectangle. We have done that for Y_2, as shown here. Now if GRAPH is pressed, only Y_1 will be graphed.

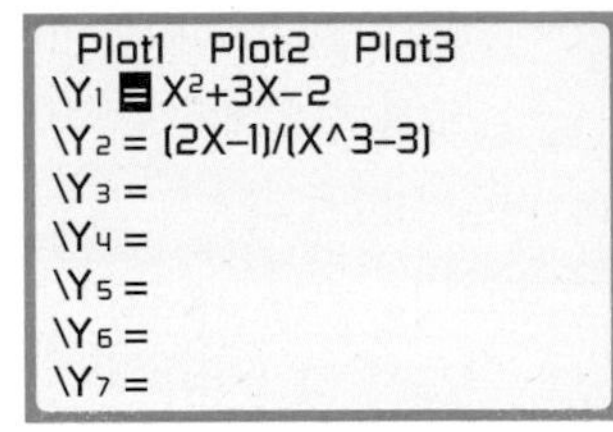

It is also possible to control the appearance of the graph by moving the cursor on the Y= screen to the left of any Y. With the cursor in this position, pressing ENTER will change the appearance of the graph. The options are shown at the right.

Plot1 Plot2 Plot3
\Y1 = Default graph line
\Y2 = Bold graph line
◥Y3 = Shade above graph
◣Y4 = Shade below graph
-0Y5 = Draw path of graph
0Y6 = Travel path of graph
⋱Y7 = Dashed graph line

Zero

The ZERO feature of a graphing calculator is used for various calculations: to find the x-intercepts of a function, to solve some equations, and to find the zero of a function.

x-intercepts To illustrate the procedure for finding x-intercepts, we will use $f(x) = x^2 + x - 2$.

First, use the Y-editor to enter the expression for the function and then graph the function in the standard viewing window. (It may be necessary to adjust this window so that the intercepts are visible.) Once the graph is displayed, use the following keystrokes to find the x-intercepts of the graph of the function.

Press 2ND CALC (scroll to 2 for zero of the function) ENTER.

Alternatively, you can just press 2ND CALC 2.

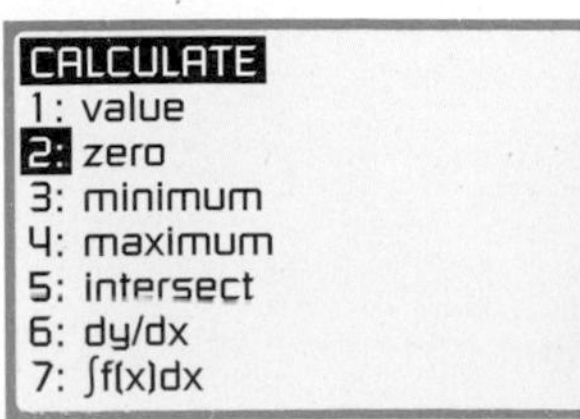

Left Bound? shown at the bottom of the screen asks you to use the left or right arrow key to move the cursor to the *left* of the desired x-intercept. Press ENTER.

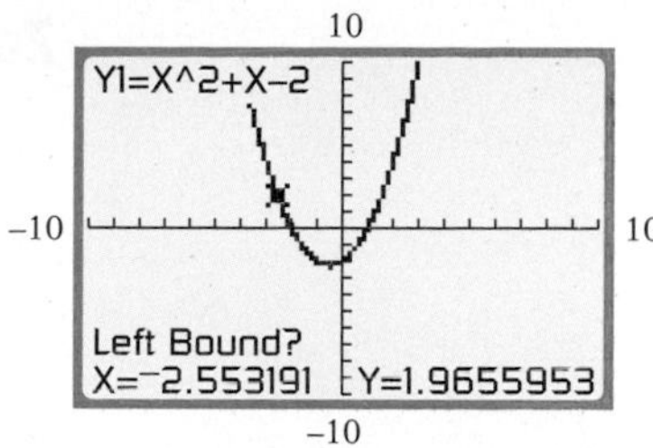

Right Bound? shown at the bottom of the screen asks you to use the left or right arrow key to move the cursor to the *right* of the desired x-intercept. Press ENTER.

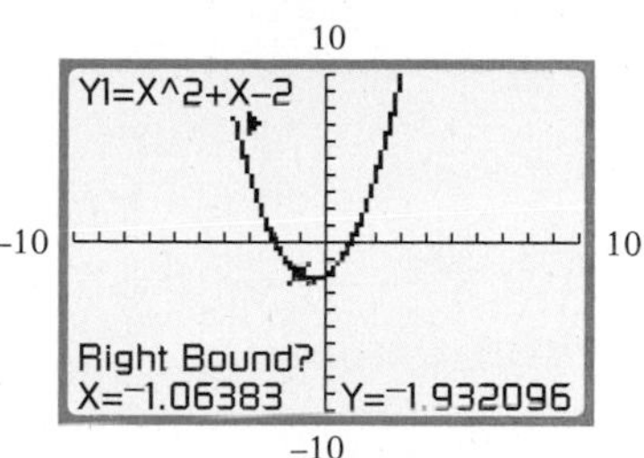

Guess? shown at the bottom of the screen asks you to use the left or right arrow key to move the cursor to the *approximate* location of the desired x-intercept. Press ENTER.

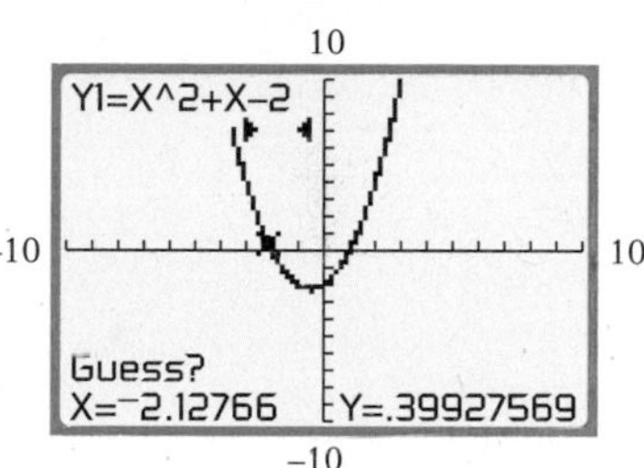

The x-coordinate of an x-intercept is -2. Therefore, an x-intercept is $(-2, 0)$.

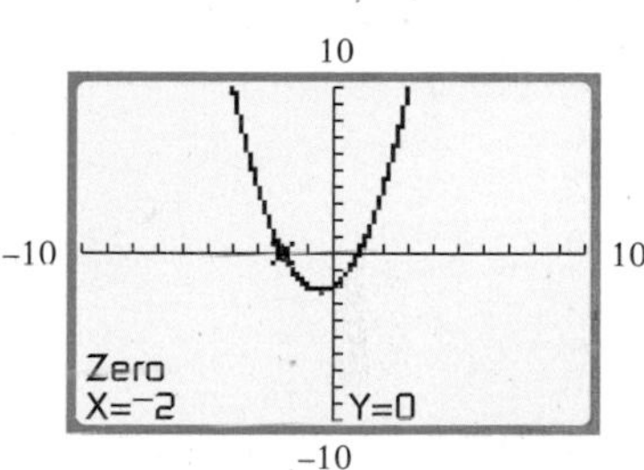

To find the other x-intercept, follow the same steps as above. The screens for this calculation are shown below.

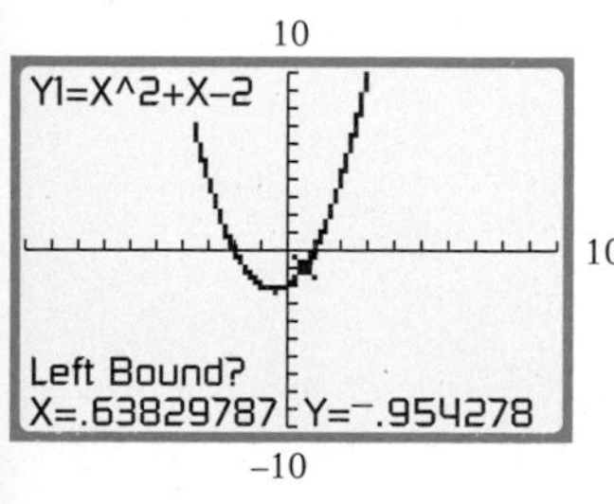

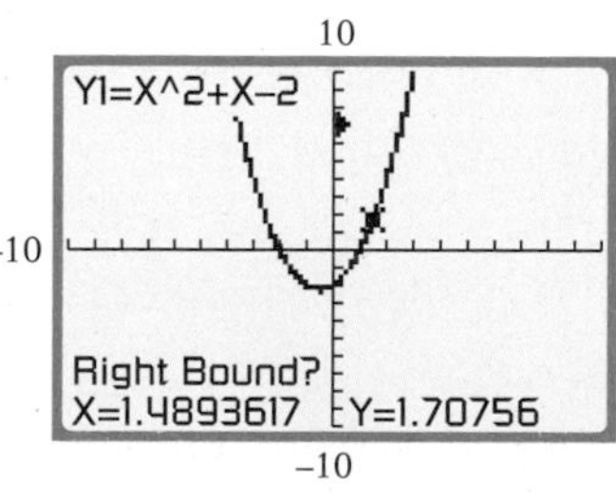

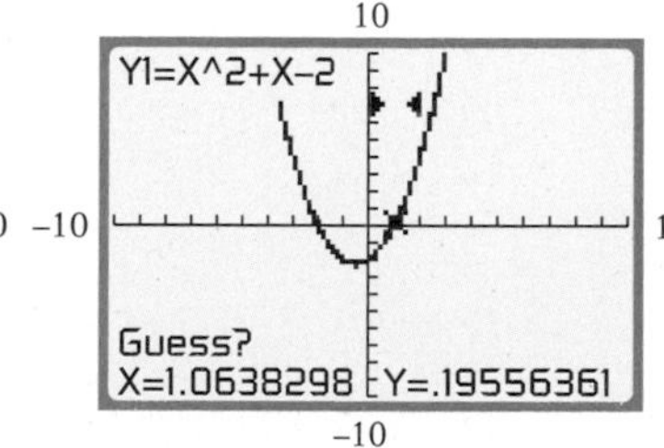

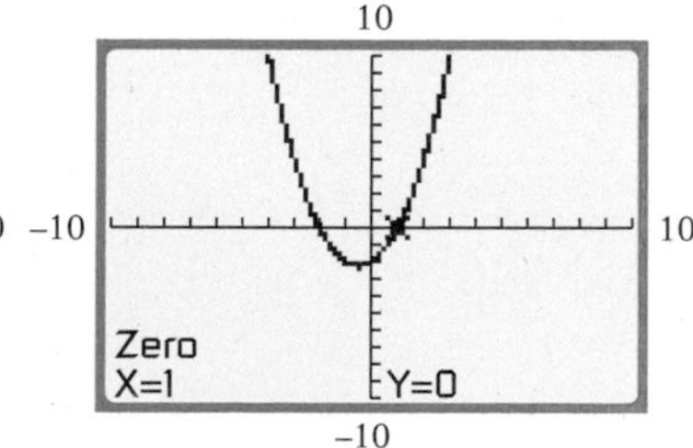

A second x-intercept is $(1, 0)$.

Solve an equation To use the ZERO feature to solve an equation, first rewrite the equation with all terms on one side. For instance, one way to solve the equation $x^3 - x + 1 = -2x + 3$ is first to rewrite it as $x^3 + x - 2 = 0$. Enter $x^3 + x - 2$ into Y1 and then follow the steps for finding x-intercepts.

Find the real zeros of a function To find the real zeros of a function, follow the steps for finding x-intercepts.

Zoom

Pressing ZOOM enables you to select some preset viewing windows. This key also gives you access to ZBox, Zoom In, and Zoom Out. These functions enable you to redraw a selected portion of a graph in a new window. Some windows used frequently in this text are shown below.

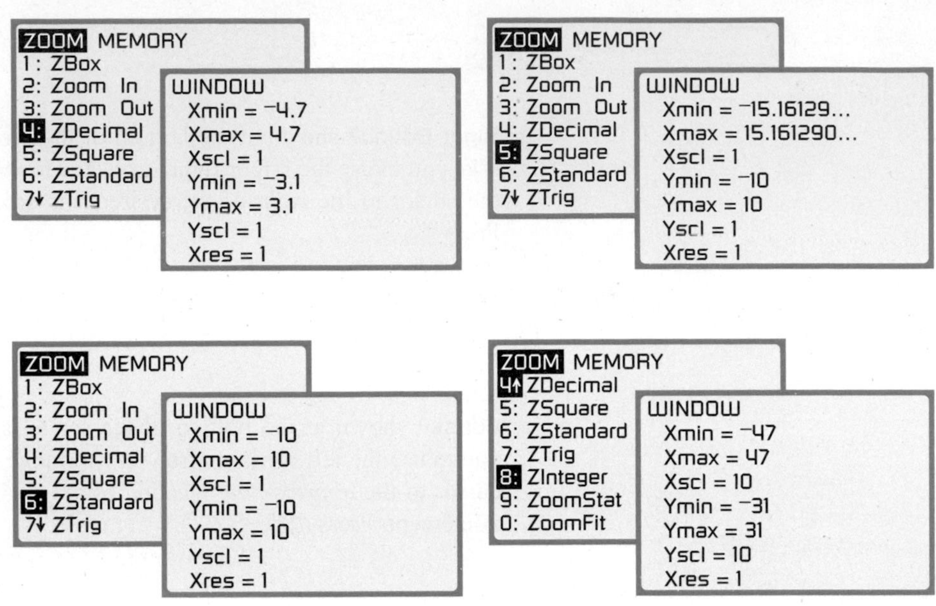

Solutions to Chapter Problems

Solutions to Chapter 1 Problems

SECTION 1.1

Problem 1

Replace z by each element of the set and determine whether the statement is true.

$z \le 0$
$-2 \le 0$ A true statement.
$-1 \le 0$ A true statement.
$0 \le 0$ A true statement.
$1 \le 0$ A false statement.
$2 \le 0$ A false statement.

The inequality is true for -2, -1, and 0.

Problem 2

$-v$ • **Write the expression for the additive inverse of v.**

$-(-8) = 8$ • **Replace v by each element of the set and determine the value of the expression.**
$-(0) = 0$
$-(9) = -9$

Problem 3

From the definition of absolute value, $|-23| = 23$.

Problem 4

$\{-7, -5 -3, -1\}$

Problem 5

$\{x | x \le 7, x \in \text{integers}\}$

Problem 6

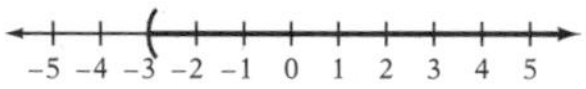

(number line: −5 to 5)

Problem 7

	Set-builder notation	Interval notation	Graph
A.	$\{x \mid -2 < x < 0\}$	$(-2, 0)$	(number line)
B.	$\{x \mid -1 < x \le 2\}$	$(-1, 2]$	(number line)
C.	$\{x \mid x \ge -2\}$	$[-2, \infty)$	(number line)

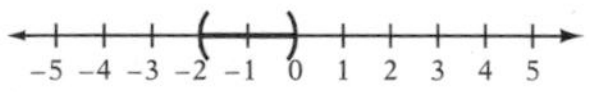

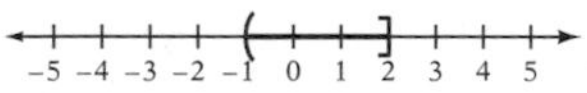

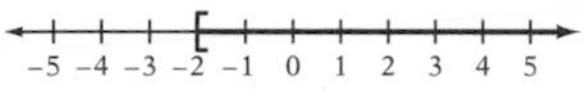

Problem 8

A. $(-\infty, -1] \cup [2, 4)$ is the set of real numbers less than or equal to -1 *or* greater than or equal to 2 and less than 4.

The graph of $(-\infty, -1] \cup [2, 4)$ contains all the points on the graphs of $(-\infty, -1]$ and $[2, 4)$.

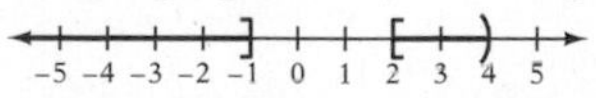

(number line: −5 to 5)

B. $\{x | x \le 3\} \cap \{x | -3 < x < 5\}$ is the set of real numbers less than or equal to 3 *and* greater than -3 and less than 5.

The graph of $x \le 3$ is shown by the bracket at 3 and the arrow that points to the left, and the graph of $-3 < x < 5$ is the segment between -3 and 5.

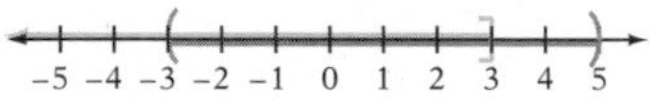

(number line: −5 to 5)

Real numbers that satisfy both $x \le 3$ *and* $-3 < x < 5$ correspond to the points in the section of overlap; thus, $\{x | x \le 3\} \cap \{x | -3 < x < 5\}$ is the interval $(-3, 3]$.

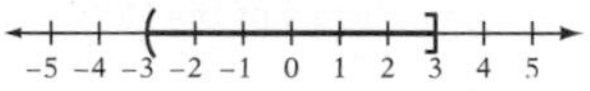

(number line: −5 to 5)

SECTION 1.2

Problem 1

A. $-21 + 32 = 11$
B. $7 - (-12) = 7 + 12 = 19$

Problem 2

A. $(-14)(-5) = 70$
B. $(-36) \div 9 = -4$

Problem 3

A. $-5^3 = -(5 \cdot 5 \cdot 5) = -125$
B. $(-2)^7 = (-2)(-2)(-2)(-2)(-2)(-2)(-2) = -128$
C. $3^4 \cdot (-2)^2 = 3 \cdot 3 \cdot 3 \cdot 3 \cdot (-2) \cdot (-2) = 324$

Problem 4

$24 - 18 \div 6(3 - 6)^3$
$= 24 - 18 \div 6(-3)^3$ • **Perform operations inside grouping symbols.**
$= 24 - 18 \div 6(-27)$ • **Simplify exponential expressions.**
$= 24 - 3(-27)$ • **Do multiplications and divisions from left to right.**
$= 24 + 81$
$= 105$ • **Do additions and subtractions from left to right.**

Problem 5

$4 - 2[(25 - 9) \div 2^3]^2$
$= 4 - 2[16 \div 2^3]^2$ • **Perform operations inside grouping symbols. First, subtract.**
$= 4 - 2[16 \div 8]^2$ • **Simplify exponential expressions.**
$= 4 - 2[2]^2$ • **Divide.**
$= 4 - 2 \cdot 4$ • **Simplify exponential expressions.**
$= 4 - 8$ • **Multiply.**
$= -4$ • **Subtract.**

SECTION 1.3

Problem 1

The signs are different. The product is negative.

$$\frac{10}{21}\left(-\frac{14}{25}\right) = -\frac{10 \cdot 14}{21 \cdot 25} = -\frac{4}{15}$$

Problem 2

$$\left(-\frac{2}{5}\right)^3 = \left(-\frac{2}{5}\right)\left(-\frac{2}{5}\right)\left(-\frac{2}{5}\right) = -\frac{8}{125}$$

Problem 3

$$\left(-\frac{5}{6}\right) \div \left(-\frac{25}{12}\right)$$

$$= \left(-\frac{5}{6}\right) \cdot \left(-\frac{12}{25}\right)$$ • **Multiply by the reciprocal of the divisor.**

$$= \frac{5 \cdot 12}{6 \cdot 25} = \frac{2}{5}$$ • **The signs are the same. The product is positive.**

Problem 4

$$\frac{5}{12} - \left(-\frac{3}{8}\right)$$

$$= \frac{5}{12} + \frac{3}{8}$$ • **Rewrite subtraction as addition of the opposite.**

$$= \frac{10}{24} + \frac{9}{24}$$ • **Write each fraction in terms of the LCD.**

$$= \frac{19}{24}$$ • **Add the numerators.**

Problem 5

$$\frac{5}{16} - \left(\frac{3}{4}\right)^2 + \frac{5}{8}$$

$$= \frac{5}{16} - \frac{9}{16} + \frac{5}{8}$$ • **Simplify exponential expressions.**

$$= \frac{5}{16} - \frac{9}{16} + \frac{10}{16} = \frac{6}{16} = \frac{3}{8}$$ • **Write each fraction in terms of the LCD. Then simplify.**

Problem 6

$$\left(\frac{4}{3} - \frac{5}{6}\right)^2 + \frac{7}{8} \div \frac{3}{4}$$

$$= \left(\frac{8}{6} - \frac{5}{6}\right)^2 + \frac{7}{8} \div \frac{3}{4}$$ • **Subtract the fractions inside the parentheses. The LCD is 6.**

$$= \left(\frac{1}{2}\right)^2 + \frac{7}{8} \div \frac{3}{4}$$

$$= \frac{1}{4} + \frac{7}{8} \div \frac{3}{4}$$ • **Simplify exponential expressions.**

$$= \frac{1}{4} + \frac{7}{6}$$ • **Divide.**

$$= \frac{17}{12}$$ • **Add.**

Problem 7

$$\frac{\frac{2}{3} - \frac{3}{4}}{\frac{3}{10} - \frac{1}{5}} + \left(\frac{2}{3}\right)^2$$

$$= \frac{-\frac{1}{12}}{\frac{1}{10}} + \left(\frac{2}{3}\right)^2$$ • **Simplify the numerator and denominator of the complex fraction.**

$$= -\frac{1}{12} \cdot \frac{10}{1} + \frac{4}{9}$$ • **Rewrite the complex fraction and simplify the exponential expression.**

$$= -\frac{5}{6} + \frac{4}{9}$$ • **Multiply first. Then add.**

$$= -\frac{7}{18}$$

Problem 8

$$\begin{array}{r} 0.7142857 \\ 7\overline{)5.0000000} \\ \underline{-4\,9} \\ 10 \\ \underline{-7} \\ 30 \\ \underline{-28} \\ 20 \\ \underline{-14} \\ 60 \\ \underline{-56} \\ 40 \\ \underline{-35} \\ 50 \\ \underline{-49} \\ 1 \end{array}$$

• **The difference begins to repeat.**

$$\frac{5}{7} = 0.\overline{714285}$$

Problem 9

$$6.4 \div (-0.8) + 1.2(0.3^2 - 0.2)$$

$$= 6.4 \div (-0.8) + 1.2(0.09 - 0.2)$$ • **Perform operations within grouping symbols.**

$$= 6.4 \div (-0.8) + 1.2(-0.11)$$

$$= -8 + (-0.132)$$ • **Multiply and divide from left to right.**

$$= -8.132$$ • **Add.**

SECTION 1.4

Problem 1

$$(x)\left(\frac{1}{4}\right) = \left(\frac{1}{4}\right)(x)$$

Problem 2

The Associative Property of Addition

Problem 3

$(b - c)^2 \div ab$

$[2 - (-4)]^2 \div (-3)(2)$ • Replace each variable with its value.

$= [6]^2 \div (-3)(2)$ • Use the Order of Operations Agreement.

$= 36 \div (-3)(2)$

$= -12(2)$

$= -24$

Problem 4

$SA = \pi r^2 + \pi rl$ • The formula for surface area of a right circular cone.

$SA = \pi(5)^2 + \pi(5)(12)$

$SA = 25\pi + 60\pi$

$SA = 85\pi$

$SA \approx 267.04$ • Use the π key on a calculator.

The exact area is 85π cm^2.
The approximate area is 267.04 cm^2.

Problem 5

$(2x + xy - y) - (5x - 7xy + y)$

$= 2x + xy - y - 5x + 7xy - y$ • Distributive Property

$= -3x + 8xy - 2y$ • Combine like terms.

Problem 6

$2x - 3[y - 3(x - 2y + 4)]$

$= 2x - 3[y - 3x + 6y - 12]$ • Distributive Property

$= 2x - 3[7y - 3x - 12]$ • Combine like terms.

$= 2x - 21y + 9x + 36$ • Distributive Property

$= 11x - 21y + 36$ • Combine like terms.

SECTION 1.5

Problem 1

the unknown number: n • Assign a variable to the unknown number.

twice the unknown number: $2n$ • Use the variable to write an expression for any other unknown quantity.

the difference between 8 and twice the unknown number: $8 - 2n$

$n - (8 - 2n)$ • Write the variable expression.

$= n - 8 + 2n$ • Simplify.

$= 3n - 8$

Problem 2

the unknown number: n • Assign a variable to the unknown number.

three-eighths of the number: $\frac{3}{8}n$ • Use the variable to write expressions for the numbers in the sum.

five-twelfths of the number: $\frac{5}{12}n$

$\frac{3}{8}n + \frac{5}{12}n$ • Write the variable expression.

$= \frac{9}{24}n + \frac{10}{24}n = \frac{19}{24}n$ • Simplify.

Problem 3

The pounds of caramel: c
The pounds of milk chocolate: $c + 3$

Problem 4

The depth of the shallow end: D
The depth of the deep end: $2D + 2$

Solutions to Chapter 2 Problems

SECTION 2.1

Problem 1

$\frac{6x}{5} - 3 = -7$

$\frac{6x}{5} - 3 + 3 = -7 + 3$ • Add 3 to each side.

$\frac{6x}{5} = -4$ • Simplify.

$\frac{5}{6}\left(\frac{6}{5}x\right) = \frac{5}{6}(-4)$ • Multiply each side by the reciprocal of $\frac{6}{5}$.

$x = -\frac{10}{3}$ • Simplify.

The solution is $-\frac{10}{3}$.

Problem 2

$3x - 5 = 14 - 5x$

$3x + 5x - 5 = 14 - 5x + 5x$ • Add $5x$ to each side.

$8x - 5 = 14$ • Simplify.

$8x - 5 + 5 = 14 + 5$ • Add 5 to each side.

$8x = 19$ • Simplify.

$\frac{8x}{8} = \frac{19}{8}$ • Divide each side by 8.

$x = \frac{19}{8}$ • Simplify.

The solution is $\frac{19}{8}$.

Problem 3

$6(5 - x) - 12 = 2x - 3(4 + x)$

$30 - 6x - 12 = 2x - 12 - 3x$ • Distributive Property

$18 - 6x = -x - 12$ • Simplify.

$18 - 5x = -12$ • Add x to each side.

$-5x = -30$ • Subtract 18 from each side.

$x = 6$ • Divide each side by -5.

The solution is 6.

Problem 4

The LCM of 3, 5, and 30 is 30.

$$\frac{2x - 7}{3} - \frac{5x + 4}{5} = \frac{-x - 4}{30}$$

$$30\left(\frac{2x - 7}{3} - \frac{5x + 4}{5}\right) = 30\left(\frac{-x - 4}{30}\right)$$

$$\frac{30(2x - 7)}{3} - \frac{30(5x + 4)}{5} = \frac{30(-x - 4)}{30}$$

$$10(2x - 7) - 6(5x + 4) = -x - 4$$

$$20x - 70 - 30x - 24 = -x - 4$$

$$-10x - 94 = -x - 4$$

$$-9x - 94 = -4$$

$$-9x = 90$$

$$x = -10$$

The solution is -10.

Problem 5

Strategy Next year's salary: s
Next year's salary is the sum of this year's salary and the raise.

Solution
$$\begin{aligned} s &= 34{,}500 + 0.04(34{,}500) \\ &= 34{,}500 + 1380 \\ &= 35{,}880 \end{aligned}$$

Next year's salary is \$35,880.

SECTION 2.2

Problem 1

Strategy ▶ Pounds of \$4.00 hamburger: x
Pounds of \$2.80 hamburger: $75 - x$

	Amount	Cost	Value
\$4.00 hamburger	x	4.00	$4.00x$
\$2.80 hamburger	$75 - x$	2.80	$2.80(75 - x)$
Mixture	75	3.20	$75(3.20)$

▶ The sum of the values before mixing equals the value after mixing.

Solution
$$\begin{aligned} 4.00x + 2.80(75 - x) &= 75(3.20) \\ 4x + 210 - 2.80x &= 240 \\ 1.2x + 210 &= 240 \\ 1.2x &= 30 \\ x &= 25 \end{aligned}$$

The value of x is the amount of \$4.00 hamburger. To find the amount of \$2.80 hamburger, substitute the value of x into the expression for the amount of \$2.80 hamburger.

$75 - x = 75 - 25 = 50$

The mixture must contain 25 lb of the \$4.00 hamburger and 50 lb of the \$2.80 hamburger.

Problem 2

Strategy The distance is 350 mi. Therefore, $d = 350$. The plane is flying into a headwind, so the rate of the plane is the difference between its rate in calm air (175 mph) and the rate of the wind (35 mph): 175 mph − 35 mph = 140 mph. Therefore, $r = 140$. To find the time, solve the equation $d = rt$ for t.

Solution
$$\begin{aligned} d &= rt \\ 350 &= 140t \qquad \bullet\ d = 350, r = 140 \\ \frac{350}{140} &= \frac{140t}{140} \\ 2.5 &= t \end{aligned}$$

It takes the plane 2.5 h to travel 350 mi.

Problem 3

Strategy ▶ Rate of the second plane: r
Rate of the first plane: $r + 30$

	Rate	Time	Distance
First plane	$r + 30$	4	$4(r + 30)$
Second plane	r	4	$4r$

▶ The total distance traveled by the two planes is 1160 mi.

Solution
$$\begin{aligned} 4(r + 30) + 4r &= 1160 \\ 4r + 120 + 4r &= 1160 \\ 8r + 120 &= 1160 \\ 8r &= 1040 \\ r &= 130 \end{aligned}$$

The value of r is the rate of the second plane. To find the rate of the first plane, substitute the value of r into the expression for the rate of the first plane.

$r + 30 = 130 + 30 = 160$

The first plane is traveling 160 mph.
The second plane is traveling 130 mph.

SECTION 2.3

Problem 1

Strategy ▶ Amount invested at 7.5%: x

	Principal	Rate	Interest
Amount at 5.2%	3500	0.052	$0.052(3500)$
Amount at 7.5%	x	0.075	$0.075x$

▶ The sum of the amounts of interest earned by the two investments equals the total annual interest earned (\$575).

Solution
$$\begin{aligned} 0.052(3500) + 0.075x &= 575 \\ 182 + 0.075x &= 575 \\ 0.075x &= 393 \\ x &= 5240 \end{aligned}$$

The amount invested at 7.5% is \$5240.

Problem 2

Strategy ▶ Pounds of 22% fat hamburger: x
Pounds of 12% fat hamburger: $80 - x$

	Amount	Percent	Quantity
22%	x	0.22	$0.22x$
12%	$80 - x$	0.12	$0.12(80 - x)$
18%	80	0.18	$0.18(80)$

▶ The sum of the quantities before mixing is equal to the quantity after mixing.

Solution

$$0.22x + 0.12(80 - x) = 0.18(80)$$
$$0.22x + 9.6 - 0.12x = 14.4$$
$$0.10x + 9.6 = 14.4$$
$$0.10x = 4.8$$
$$x = 48$$

The value of x is the amount of 22% fat hamburger. To find the amount of 12% fat hamburger, substitute the value of x into the expression for the amount of 12% fat hamburger.

$80 - x = 80 - 48 = 32$

The butcher needs 48 lb of the hamburger that is 22% fat and 32 lb of the hamburger that is 12% fat.

SECTION 2.4

Problem 1

$2x - 1 < 6x + 7$
$-4x - 1 < 7$ • **Subtract 6*x* from each side.**
$-4x < 8$ • **Add 1 to each side.**
$\frac{-4x}{-4} > \frac{8}{-4}$ • **Divide each side by −4 and reverse the inequality symbol.**
$x > -2$ • **Simplify.**

The solution set is $\{x|x > -2\}$.

Problem 2

$5x - 2 \leq 4 - 3(x - 2)$
$5x - 2 \leq 4 - 3x + 6$ • **Distributive Property**
$5x - 2 \leq 10 - 3x$ • **Simplify.**
$8x - 2 \leq 10$ • **Add 3*x* to each side.**
$8x \leq 12$ • **Add 2 to each side.**
$\frac{8x}{8} \leq \frac{12}{8}$ • **Divide each side by 8.**
$x \leq \frac{3}{2}$ • **Simplify.**

The solution set is $\left(-\infty, \frac{3}{2}\right]$.

Problem 3

$2 - 3x > 11$ or $5 + 2x > 7$
$-3x > 9$ $\quad 2x > 2$
$x < -3$ $\quad x > 1$
$\{x|x < -3\}$ $\quad \{x|x > 1\}$

The solution set is $\{x|x < -3\} \cup \{x|x > 1\}$.

Problem 4

$5 - 4x > 1$ and $6 - 5x < 11$
$-4x > -4$ $\quad -5x < 5$
$x < 1$ $\quad x > -1$
$\{x|x < 1\}$ $\quad \{x|x > -1\}$
$\{x|x < 1\} \cap \{x|x > -1\} = (-1, 1)$

The solution set is $(-1, 1)$.

Problem 5

$-2 \leq 5x + 3 \leq 13$
$-2 - 3 \leq 5x + 3 - 3 \leq 13 - 3$ • **Subtract 3 from each part.**
$-5 \leq 5x \leq 10$ • **Simplify.**
$\frac{-5}{5} \leq \frac{5x}{5} \leq \frac{10}{5}$ • **Divide each part by 5.**
$-1 \leq x \leq 2$ • **Simplify.**

The solution set is $[-1, 2]$.

Problem 6

Strategy To find the number of miles, write and solve an inequality using N to represent the number of miles.

Solution

Cost of Company A car < Cost of Company B car

$$6(5) + 0.14N < 12(5) + 0.08N$$
$$30 + 0.14N < 60 + 0.08N$$
$$30 + 0.06N < 60$$
$$0.06N < 30$$
$$N < 500$$

It is less expensive to rent from Company A if the car is driven less than 500 mi.

Problem 7

Strategy Number of pounds of pecans: n

The sum of the values of the pecans and walnuts: $7n + 4(8) = 7n + 32$

Number of pounds of the mixed nuts: $n + 8$
Value of the mixture at \$5 per pound:
$5(n + 8) = 5n + 40$

Value of the mixture at \$6 per pound:
$6(n + 8) = 6n + 48$

The sum of the values of the pecans and walnuts is greater than the value of the mixture at \$5 per pound and less than the value of the mixture at \$6 per pound.

Solution

$7n + 32 > 5n + 40$ and $7n + 32 < 6n + 48$
$2n + 32 > 40$ $\quad n + 32 < 48$
$2n > 8$ $\quad n < 16$
$n > 4$

The mixture should contain between 4 lb and 16 lb of pecans if it is to cost between \$5 and \$6 per pound.

SECTION 2.5

Problem 1

A. $|x| = 25$
$x = 25$ $\quad x = -25$

The solutions are 25 and −25.

B. $|2x - 3| = 5$
$2x - 3 = 5$ $\quad 2x - 3 = -5$ • **Write two equations.**
$2x = 8$ $\quad 2x = -2$ • **Solve each equation.**
$x = 4$ $\quad x = -1$

The solutions are 4 and −1.

C. $5 - |3x + 5| = 3$

$$-|3x + 5| = -2$$ • Solve for the absolute value.

$$|3x + 5| = 2$$

$3x + 5 = 2 \qquad 3x + 5 = -2$ • Write two equations.

$3x = -3 \qquad 3x = -7$ • Solve each equation.

$x = -1 \qquad x = -\frac{7}{3}$

The solutions are -1 and $-\frac{7}{3}$.

Problem 2

$$|3x + 2| < 8$$
$$-8 < 3x + 2 < 8$$ • Solve the equivalent compound inequality.
$$-8 - 2 < 3x + 2 - 2 < 8 - 2$$
$$-10 < 3x < 6$$
$$\frac{-10}{3} < \frac{3x}{3} < \frac{6}{3}$$
$$-\frac{10}{3} < x < 2$$

The solution set is $\left\{x \middle| -\frac{10}{3} < x < 2\right\}$.

Problem 3

$$|5x + 3| > 8$$

$5x + 3 < -8$ or $5x + 3 > 8$ • Solve the equivalent compound inequality.

$5x < -11 \qquad 5x > 5$

$x < -\frac{11}{5} \qquad x > 1$

$\left\{x \middle| x < -\frac{11}{5}\right\} \qquad \{x|x > 1\}$

The solution set is $\left\{x \middle| x < -\frac{11}{5}\right\} \cup \{x|x > 1\}$.

Problem 4

Strategy Let b represent the desired diameter of the bushing, T the tolerance, and d the actual diameter of the bushing. Solve the absolute value inequality $|d - b| \le T$ for d.

Solution

$$|d - b| \le T$$
$$|d - 2.55| \le 0.003$$ • $b = 2.55, T = 0.003$
$$-0.003 \le d - 2.55 \le 0.003$$
$$-0.003 + 2.55 \le d - 2.55 + 2.55 \le 0.003 + 2.55$$
$$2.547 \le d \le 2.553$$

The lower and upper limits of the diameter of the bushing are 2.547 in. and 2.553 in.

Solutions to Chapter 3 Problems

SECTION 3.1

Problem 1

$(x_1, y_1) = (5, -2) \qquad (x_2, y_2) = (-4, 3)$

$$d = \sqrt{(x_2 - x_1)^2 + (y_2 - y_1)^2}$$
$$= \sqrt{(-4 - 5)^2 + (3 - (-2))^2}$$
$$= \sqrt{(-9)^2 + (5)^2}$$
$$= \sqrt{81 + 25}$$
$$= \sqrt{106}$$

The distance between the points is $\sqrt{106}$.

Problem 2

$(x_1, y_1) = (-3, -5) \qquad (x_2, y_2) = (-2, 3)$

$$x_m = \frac{x_1 + x_2}{2} = \frac{-3 + (-2)}{2} = -\frac{5}{2} \qquad y_m = \frac{y_1 + y_2}{2} = \frac{-5 + 3}{2} = -1$$

The coordinates of the midpoint are $(-\frac{5}{2}, -1)$.

Problem 3

Replace x by -2 and solve for y.

$$y = \frac{3x}{x + 1} = \frac{3(-2)}{-2 + 1} = \frac{-6}{-1} = 6$$

The ordered-pair solution is $(-2, 6)$.

Problem 4

Determine the ordered-pair solutions for the given values of x.

x	$y = -\frac{1}{2}x + 3$	y	(x, y)
-4	$y = -\frac{1}{2}(-4) + 3 = 5$	5	$(-4, 5)$
-2	$y = -\frac{1}{2}(-2) + 3 = 4$	4	$(-2, 4)$
0	$y = -\frac{1}{2}(0) + 3 = 3$	3	$(0, 3)$
2	$y = -\frac{1}{2}(2) + 3 = 2$	2	$(2, 2)$
4	$y = -\frac{1}{2}(4) + 3 = 1$	1	$(4, 1)$

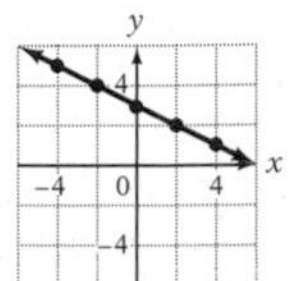

Problem 5

Determine the ordered-pair solutions for the given values of x.

x	$y = -x^2 + 4$	y	(x, y)
-3	$y = -(-3)^2 + 4 = -5$	-5	$(-3, -5)$
-2	$y = -(-2)^2 + 4 = 0$	0	$(-2, 0)$
-1	$y = -(-1)^2 + 4 = 3$	3	$(-1, 3)$
0	$y = -(0)^2 + 4 = 4$	4	$(0, 4)$
1	$y = -(1)^2 + 4 = 3$	3	$(1, 3)$
2	$y = -(2)^2 + 4 = 0$	0	$(2, 0)$
3	$y = -(3)^2 + 4 = -5$	-5	$(3, -5)$

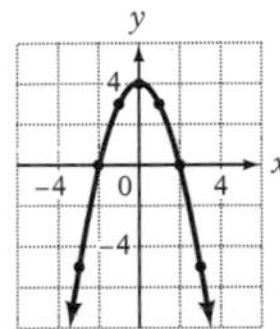

Problem 6

Determine the ordered-pair solutions for the given values of x.

x	$y = 3 - \|x\|$	y	(x, y)
-3	$y = 3 - \|-3\| = 0$	0	$(-3, 0)$
-2	$y = 3 - \|-2\| = 1$	1	$(-2, 1)$
-1	$y = 3 - \|-1\| = 2$	2	$(-1, 2)$
0	$y = 3 - \|0\| = 3$	3	$(0, 3)$
1	$y = 3 - \|1\| = 2$	2	$(1, 2)$
2	$y = 3 - \|2\| = 1$	1	$(2, 1)$
3	$y = 3 - \|3\| = 0$	0	$(3, 0)$

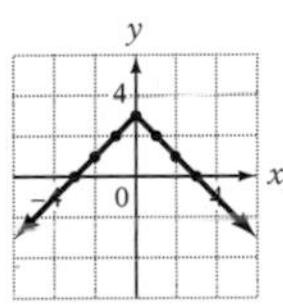

SECTION 3.2

Problem 1

$\{(-2, 6), (-1, 3), (0, 0), (1, -3), (2, -6), (3, -9)\}$

The domain is $\{-2, -1, 0, 1, 2, 3\}$. • **The domain of the relation is the set of first coordinates of the ordered pairs.**

The range is $\{-9, -6, -3, 0, 3, 6\}$. • **The range of the relation is the set of second coordinates of the ordered pairs.**

The relation is a function. • **There are no two ordered pairs with the same first coordinate and different second coordinates.**

Problem 2

A. $f(z) = z + |z|$

$f(-4) = -4 + |-4| = -4 + 4$ • **Replace z by -4. Then simplify.**

$f(-4) = 0$

B. $f(z) = z + |z|$

$f(3) = 3 + |3| = 3 + 3$ • **Replace z by 3. Then simplify.**

$f(3) = 6$

Problem 3

$r(s) = 3s - 6$

$r(2a + 3) = 3(2a + 3) - 6$ • **Replace s by $2a + 3$.**

$= 6a + 9 - 6$ • **Simplify.**

$= 6a + 3$

Problem 4

Evaluate the function for each element of the domain. The range includes the values of $f(-3)$, $f(-2)$, $f(-1)$, $f(0)$, $f(1)$, and $f(2)$.

$$f(x) = x^3 + x$$
$$f(-3) = (-3)^3 + (-3) = -30$$
$$f(-2) = (-2)^3 + (-2) = -10$$
$$f(-1) = (-1)^3 + (-1) = -2$$
$$f(0) = (0)^3 + 0 = 0$$
$$f(1) = (1)^3 + 1 = 2$$
$$f(2) = (2)^3 + 2 = 10$$

The range is $\{-30, -10, -2, 0, 2, 10\}$.

Problem 5

$f(x) = 2x - 5$

$f(c) = 2c - 5$ • **Replace x by c.**

$-3 = 2c - 5$ • **$f(c) = -3$**

$2 = 2c$ • **Solve for c.**

$1 = c$

The value of c is 1. The corresponding ordered pair is $(1, -3)$.

Problem 6

x	$y = f(x) = -x^2 - 4x + 2$	(x, y)
-5	$f(-5) = -(-5)^2 - 4(-5) + 2 = -3$	$(-5, -3)$
-4	$f(-4) = -(-4)^2 - 4(-4) + 2 = 2$	$(-4, 2)$
-3	$f(-3) = -(-3)^2 - 4(-3) + 2 = 5$	$(-3, 5)$
-2	$f(-2) = -(-2)^2 - 4(-2) + 2 = 6$	$(-2, 6)$
-1	$f(-1) = -(-1)^2 - 4(-1) + 2 = 5$	$(-1, 5)$
0	$f(0) = -(0)^2 - 4(0) + 2 = 2$	$(0, 2)$
1	$f(1) = -(1)^2 - 4(1) + 2 = -5$	$(1, -5)$

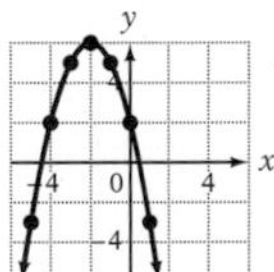

Problem 7

x	$y = f(x) = \|x - 2\|$	(x, y)
-1	$f(-1) = \|-1 - 2\| = 3$	$(-1, 3)$
0	$f(0) = \|0 - 2\| = 2$	$(0, 2)$
1	$f(1) = \|1 - 2\| = 1$	$(1, 1)$
2	$f(2) = \|2 - 2\| = 0$	$(2, 0)$
3	$f(3) = \|3 - 2\| = 1$	$(3, 1)$
4	$f(4) = \|4 - 2\| = 2$	$(4, 2)$
5	$f(5) = \|5 - 2\| = 3$	$(5, 3)$

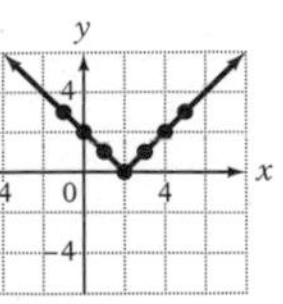

Problem 8

A. Any vertical line intersects the graph at most once. The graph is the graph of a function.

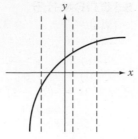

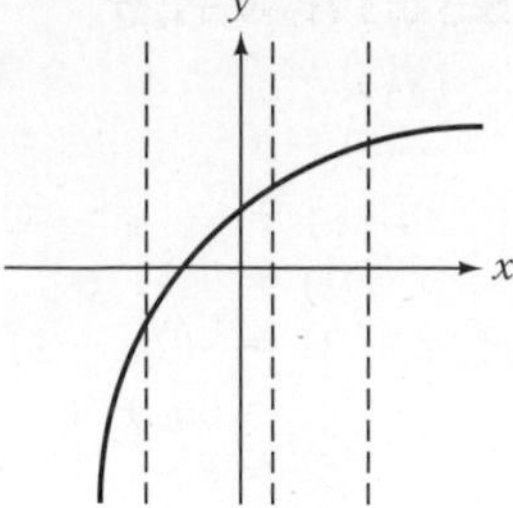

B. There are vertical lines that intersect the graph at more than one point. The graph is not the graph of a function.

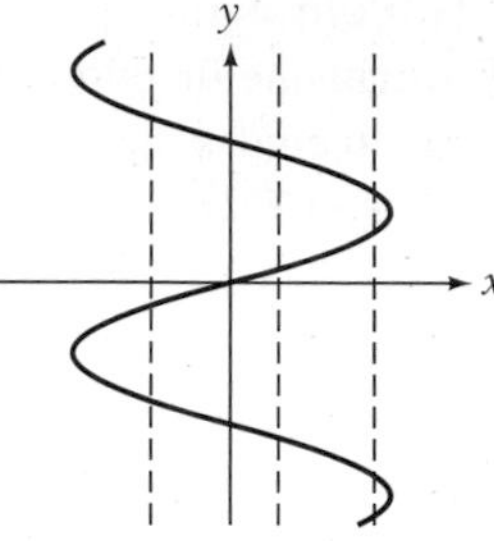

SECTION 3.3

Problem 1

x	$y = \frac{3}{5}x - 4$
-5	-7
0	-4
5	-1

• **Find at least three ordered pairs. Choose values of *x* that are divisible by 5.**

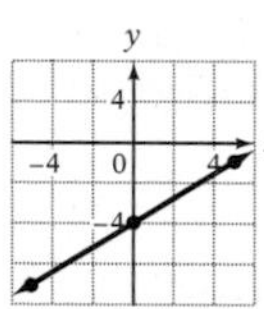

• **Graph the ordered pairs and draw a line through the points.**

Problem 2

$$-3x + 2y = 4$$
$$2y = 3x + 4$$
$$y = \frac{3}{2}x + 2$$

• **Solve the equation for *y*.**

x	$y = \frac{3}{2}x + 2$
-2	-1
0	2
2	5

• **Find at least three solutions.**

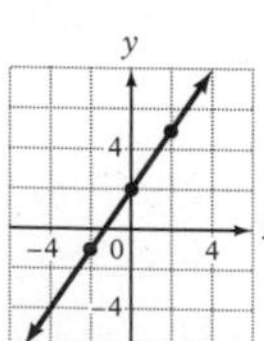

• **Graph the ordered pairs and draw a line through the points.**

Problem 3

$$y - 3 = 0$$
$$y = 3$$

• **Solve for *y*.**

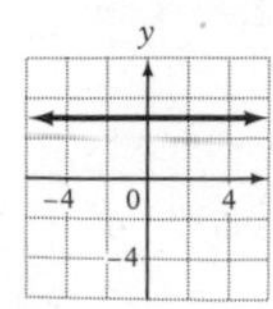

• **The graph of $y = 3$ is a horizontal line.**

Problem 4

x-intercept:

$$3x - y = 2$$
$$3x - 0 = 2$$
$$3x = 2$$
$$x = \frac{2}{3}$$
$$\left(\frac{2}{3}, 0\right)$$

y-intercept:

$$3x - y = 2$$
$$3(0) - y = 2$$
$$-y = 2$$
$$y = -2$$
$$(0, -2)$$

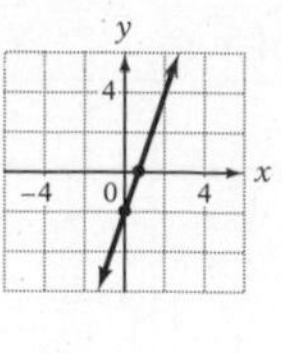

Problem 5

$$f(x) = \frac{2}{3}x + 4$$
$$0 = \frac{2}{3}x + 4$$ • **Let $f(x) = 0$.**
$$-4 = \frac{2}{3}x$$ • **Solve for *x*.**
$$-6 = x$$

The zero is -6.

Problem 6

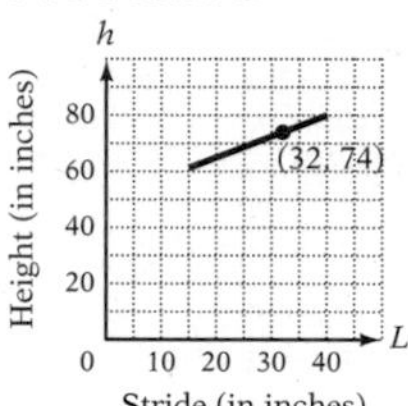

The ordered pair (32, 74) means that a person with a stride of 32 in. is 74 in. tall.

SECTION 3.4

Problem 1

Let $P_1 = (4, -3)$ and $P_2 = (2, 7)$.

$$m = \frac{y_2 - y_1}{x_2 - x_1} = \frac{7 - (-3)}{2 - 4} = \frac{10}{-2} = -5$$

The slope is -5.

Problem 2

$$m = \frac{55{,}000 - 25{,}000}{2 - 5} = \frac{30{,}000}{-3} = -10{,}000$$

A slope of $-10{,}000$ means that the value of the printing press is decreasing by \$10,000 per year.

Problem 3

$$2x + 3y = 6$$
$$3y = -2x + 6$$
$$y = -\frac{2}{3}x + 2$$

• **Solve the equation for *y*.**

$$m = -\frac{2}{3} = \frac{-2}{3}$$
y-intercept $= (0, 2)$

• **Determine the slope and *y*-intercept from the equation.**

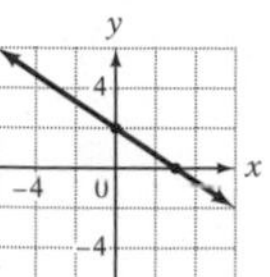

• **Begin at the *y*-intercept and use the slope to find a second point.**

Problem 4

$m = 3 = \frac{3}{1} = \frac{\text{change in } y}{\text{change in } x}$

- Write the slope with a denominator of 1.

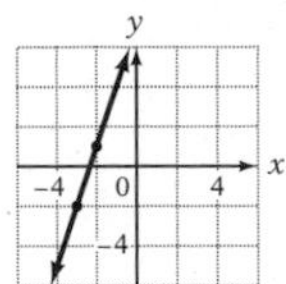

- Graph $(-3, -2)$. From that point, use the slope to move 3 units up and 1 unit right.

Problem 5

Find the y-coordinates for the given x-coordinates.

$$y_1 = f(x_1)$$
$$= (-4)^2 + (-4) - 1 = 11$$
$$y_2 = f(x_2)$$
$$= (-1)^2 + (-1) - 1 = -1$$

Find the slope of the line between $P_1(-4, 11)$ and $P_2(-1, -1)$.

$$m = \frac{y_2 - y_1}{x_2 - x_1}$$
$$= \frac{-1 - 11}{-1 - (-4)} = \frac{-12}{3} = -4$$

The average rate of change between the two points is -4.

Problem 6

In 1900, the population was 1.5 million: (1900, 1.5)
In 2000, the population was 33.9 million: (2000, 33.9)

$$m = \frac{33.9 - 1.5}{2000 - 1900}$$
$$= \frac{32.4}{100} = 0.324$$

The average rate of change in the population of California from 1900 to 2000 was 324,000 people per year.

Problem 7

A. In 1995, the median salary was 531,000.
(1995, 531,000)
In 2010, the median salary was 5,500,000.
(2010, 5,500,000)

$$m = \frac{5{,}500{,}000 - 531{,}000}{2010 - 1995}$$
$$= \frac{4{,}969{,}000}{15} \approx 331{,}000$$

The average annual rate of change in median salary was approximately \$331,000 per year.

B. \$331,000 > \$231,000
The average annual rate of change in the median salary of New York Yankees players between 1995 and 2010 is greater than the average annual rate of change in the median salary of Boston Red Sox players between 1995 and 2010.

SECTION 3.5

Problem 1

$$y - y_1 = m(x - x_1)$$
- Use the point-slope formula.

$$y - (-3) = -3(x - 4)$$
- $m = -3$ and $(x_1, y_1) = (4, -3)$.

$$y + 3 = -3x + 12$$
$$y = -3x + 9$$
- Solve for y.

The equation of the line is $y = -3x + 9$.

Problem 2

$(x_1, y_1) = (2, 0)$ and $(x_2, y_2) = (5, 3)$.

$$m = \frac{y_2 - y_1}{x_2 - x_1} = \frac{3 - 0}{5 - 2} = \frac{3}{3} = 1$$
- Find the slope.

$$y - y_1 = m(x - x_1)$$
- Point-slope formula

$$y - 0 = 1(x - 2)$$
- Substitute the slope and the coordinates of P_1.

$$y = x - 2$$
- Solve for y.

The equation of the line is $y = x - 2$.

Problem 3

Strategy ▶ Select the independent and dependent variables. Because the function will predict the Celsius temperature, that quantity is the dependent variable, y. The Fahrenheit temperature is the independent variable.

▶ From the given data, two ordered pairs are (212, 100) and (32, 0). Use these ordered pairs to determine the linear function.

Solution Let $(x_1, y_1) = (32, 0)$ and $(x_2, y_2) = (212, 100)$.

$$m = \frac{y_2 - y_1}{x_2 - x_1} = \frac{100 - 0}{212 - 32} = \frac{100}{180} = \frac{5}{9}$$
- Find the slope.

$$y - y_1 = m(x - x_1)$$
- Point-slope formula

$$y - 0 = \frac{5}{9}(x - 32)$$
- Substitute for m, x_1 and y_1.

$$y = \frac{5}{9}(x - 32), \text{ or } C = \frac{5}{9}(F - 32)$$
- Solve for y.

The linear function is $f(F) = \frac{5}{9}(F - 32)$.

SECTION 3.6

Problem 1

$$m_1 = \frac{-1 - 1}{-5 - (-2)} = \frac{-2}{-3} = \frac{2}{3}$$
- Find the slope of the line that contains the points with coordinates $(-2, 1)$ and $(-5, -1)$.

$$m_2 = \frac{2 - 0}{4 - 1} = \frac{2}{3}$$
- Find the slope of the line that contains the points with coordinates $(1, 0)$ and $(4, 2)$.

$$m_1 = m_2$$
- Compare the slopes.

Because the slopes are the same, the lines are parallel.

Problem 2

$y - y_1 = m(x - x_1)$ • Use the point-slope formula.

$y - 1 = \frac{3}{4}(x - 4)$ • $m = \frac{3}{4}$, the slope of the line parallel to the graph of $y = \frac{3}{4}x - 1$. $(x_1, y_1) = (4, 1)$, the given point.

$y - 1 = \frac{3}{4}x - 3$ • Solve for y.

$y = \frac{3}{4}x - 2$

The equation of the line is $y = \frac{3}{4}x - 2$.

Problem 3

$4x - y = -2$ $\quad x + 4y = -12$ • Solve each equation for y.

$-y = -4x - 2$ $\quad 4y = -x - 12$

$y = 4x + 2$ $\quad y = -\frac{1}{4}x - 3$

$m_1 = 4$ $\quad m_2 = -\frac{1}{4}$ • Identify the slope of the graph of each equation.

$m_1 \cdot m_2 = 4\left(-\frac{1}{4}\right) = -1$ • Find the product of the slopes.

Because the product of the slopes is -1, the graphs are perpendicular.

Problem 4

$x - 4y = 3$ • Solve the equation of the given line for y.

$-4y = -x + 3$

$y = \frac{1}{4}x - \frac{3}{4}$

$m_1 = \frac{1}{4}$ • The slope of the given line

$m_2 = -4$ • The slope of the perpendicular line is the negative reciprocal of m_1.

$y - y_1 = m(x - x_1)$ • Point-slope formula

$y - 3 = -4[x - (-2)]$ • Substitute the coordinates of the given point and the slope of the perpendicular line, m_2.

$y - 3 = -4(x + 2)$

$y - 3 = -4x - 8$ • Solve for y.

$y = -4x - 5$

The equation of the line is $y = -4x - 5$.

SECTION 3.7

Problem 1

$x + 3y > 6$ • Solve the inequality for y.

$3y > -x + 6$

$y > -\frac{1}{3}x + 2$

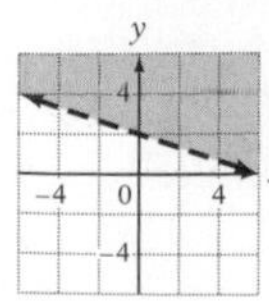

• The inequality symbol is $>$. Graph $y = -\frac{1}{3}x + 2$ as a dashed line and shade above the line.

Problem 2

$y < 2$ • The inequality symbol is $<$. Graph $y = 2$ as a dashed line and shade below the line.

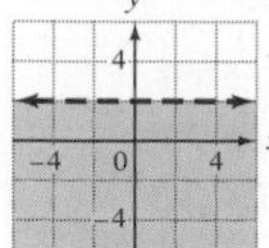

Solutions to Chapter 4 Problems

SECTION 4.1

Problem 1

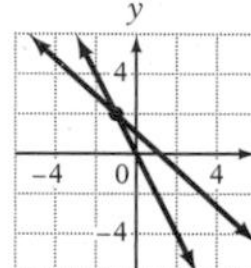

• Graph each line. The coordinates of the point of intersection give the ordered pair solution of the system.

The solution is $(-1, 2)$.

Problem 2

A. • Graph each line. The lines are parallel. The system of equations is inconsistent.

The system of equations has no solution.

B.

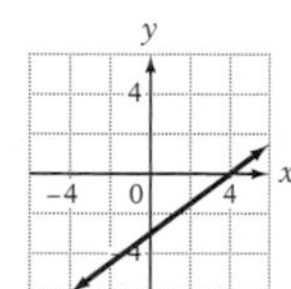

• The graphs are the same line. The system of equations is dependent.

The solutions are the ordered pairs $(x, \frac{3}{4}x - 3)$.

Problem 3

A. (1) $3x - y = 3$

(2) $6x + 3y = -4$

$3x - y = 3$ • Solve equation (1) for y. The result is equation (3).

$-y = -3x + 3$

(3) $y = 3x - 3$

$6x + 3y = -4$ • Use equation (2).

$6x + 3(3x - 3) = -4$ • Substitute for y.

$6x + 9x - 9 = -4$ • Solve for x.

$15x - 9 = -4$

$15x = 5$

$x = \frac{5}{15} = \frac{1}{3}$

$y = 3x - 3$

$y = 3\left(\frac{1}{3}\right) - 3$ • Substitute into equation (3).

$= 1 - 3$

$= -2$

The solution is $\left(\frac{1}{3}, -2\right)$.

B. (1) $6x - 3y = 6$
(2) $2x - y = 2$

$2x - y = 2$
$-y = -2x + 2$
$y = 2x - 2$
• Solve equation (2) for y.

$6x - 3y = 6$
• Use equation (1).

$6x - 3(2x - 2) = 6$
• Substitute for y.

$6x - 6x + 6 = 6$
$6 = 6$
• Solve for x.

$6 = 6$ is a true equation. The system of equations is dependent. The solutions are the ordered pairs $(x, 2x - 2)$.

SECTION 4.2

Problem 1

A. (1) $2x + 5y = 6$
(2) $3x - 2y = 6x + 2$

$2x + 5y = 6$
$-3x - 2y = 2$
• Write equation (2) in the form $Ax + By = C$.

$2(2x + 5y) = 2(6)$
$5(-3x - 2y) = 5(2)$
• Eliminate y. Multiply equation (1) by 2 and multiply equation (2) by 5.

$4x + 10y = 12$
$-15x - 10y = 10$
• Simplify.

$-11x = 22$
$x = -2$
• Add the equations.

(1) $2x + 5y = 6$
$2(-2) + 5y = 6$
$-4 + 5y = 6$
$5y = 10$
$y = 2$
• Replace x in equation (1). Solve for y.

The solution is $(-2, 2)$.

B. (1) $2x + y = 5$
(2) $4x + 2y = 6$

$-2(2x + y) = -2(5)$
$4x + 2y = 6$
• Eliminate y. Multiply equation (1) by -2.

$-4x - 2y = -10$
$4x + 2y = 6$
• Simplify.

$0x + 0y = -4$
$0 = -4$
• Add the equations.

$0 = -4$ is not a true equation. The system is inconsistent and therefore has no solution.

Problem 2

(1) $x - y + z = 6$
(2) $2x + 3y - z = 1$
(3) $x + 2y + 2z = 5$

$x - y + z = 6$
$2x + 3y - z = 1$
(4) $3x + 2y = 7$
• Eliminate z. Add equations (1) and (2).

$4x + 6y - 2z = 2$
$x + 2y + 2z = 5$
(5) $5x + 8y = 7$
• Eliminate z. Multiply equation (2) by 2 and add to equation (3).

(4) $3x + 2y = 7$
(5) $5x + 8y = 7$
• Solve the system of two equations, equations (4) and (5).

$-12x - 8y = -28$
$5x + 8y = 7$
$-7x = -21$
$x = 3$
• Multiply equation (4) by -4 and add to equation (5). Solve for x.

(4) $3x + 2y = 7$
$3(3) + 2y = 7$
$9 + 2y = 7$
$2y = -2$
$y = -1$
• Replace x by 3 in equation (4). Solve for y.

$x - y + z = 6$
$3 - (-1) + z = 6$
$4 + z = 6$
$z = 2$
• Replace x by 3 and y by -1 in equation (1). Solve for z.

The solution is $(3, -1, 2)$.

SECTION 4.3

Problem 1

A. $\begin{vmatrix} -1 & -4 \\ 3 & -5 \end{vmatrix} = -1(-5) - 3(-4)$
$= 5 + 12 = 17$

B. There are no zeros in any row or column. Expand by cofactors of any row or column. We will use row 1.

$$\begin{vmatrix} 2 & -1 & 3 \\ 4 & -1 & 1 \\ 2 & 2 & 1 \end{vmatrix}$$

$$= 2(-1)^{1+1}\begin{vmatrix} -1 & 1 \\ 2 & 1 \end{vmatrix} + (-1)(-1)^{1+2}\begin{vmatrix} 4 & 1 \\ 2 & 1 \end{vmatrix} + 3(-1)^{1+3}\begin{vmatrix} 4 & -1 \\ 2 & 2 \end{vmatrix}$$

$$= 2(1)\begin{vmatrix} -1 & 1 \\ 2 & 1 \end{vmatrix} + (-1)(-1)\begin{vmatrix} 4 & 1 \\ 2 & 1 \end{vmatrix} + 3(1)\begin{vmatrix} 4 & -1 \\ 2 & 2 \end{vmatrix}$$

$= 2(-1 - 2) + 1(4 - 2) + 3(8 - (-2))$
$= 2(-3) + 1(2) + 3(10) = -6 + 2 + 30$
$= 26$

Problem 2

$$D = \begin{vmatrix} 6 & -6 \\ 2 & -10 \end{vmatrix} = -48$$

- **Evaluate the coefficient determinant.**

$$D_x = \begin{vmatrix} 5 & -6 \\ -1 & -10 \end{vmatrix} = -56$$

- **Evaluate each numerator determinant.**

$$D_y = \begin{vmatrix} 6 & 5 \\ 2 & -1 \end{vmatrix} = -16$$

$$x = \frac{D_x}{D} = \frac{-56}{-48} = \frac{7}{6}$$

$$y = \frac{D_y}{D} = \frac{-16}{-48} = \frac{1}{3}$$

- **Use Cramer's Rule to find the coordinates of the solution.**

The solution is $\left(\frac{7}{6}, \frac{1}{3}\right)$.

Problem 3

$$D = \begin{vmatrix} 2 & -1 & 1 \\ 3 & 2 & -1 \\ 1 & 3 & 1 \end{vmatrix} = 21$$

- **Evaluate the coefficient determinant.**

$$D_x = \begin{vmatrix} -1 & -1 & 1 \\ 3 & 2 & -1 \\ -2 & 3 & 1 \end{vmatrix} = 9$$

- **Evaluate each numerator determinant.**

$$D_y = \begin{vmatrix} 2 & -1 & 1 \\ 3 & 3 & -1 \\ 1 & -2 & 1 \end{vmatrix} = -3$$

$$D_z = \begin{vmatrix} 2 & -1 & -1 \\ 3 & 2 & 3 \\ 1 & 3 & -2 \end{vmatrix} = -42$$

$$x = \frac{D_x}{D} = \frac{9}{21} = \frac{3}{7}$$

$$y = \frac{D_y}{D} = \frac{-3}{21} = -\frac{1}{7}$$

$$z = \frac{D_z}{D} = \frac{-42}{21} = -2$$

- **Use Cramer's Rule to find the coordinates of the solution.**

The solution is $\left(\frac{3}{7}, -\frac{1}{7}, -2\right)$.

Problem 4

Each row of the matrix represents one equation of the system. The numbers to the left of the vertical line are the coefficients of x, y, and z. The numbers to the right of the vertical line are the constant terms. The system of equations for the augmented matrix $\left[\begin{array}{ccc|c} 2 & -3 & 1 & 4 \\ 1 & 0 & -2 & 3 \\ 0 & 1 & 2 & -3 \end{array}\right]$ is

$$\begin{aligned} 2x - 3y + z &= 4 \\ x - 2z &= 3 \\ y + 2z &= -3 \end{aligned}$$

Problem 5

A. $R_2 \leftrightarrow R_3$ means to interchange row 2 and row 3.

$$\left[\begin{array}{ccc|c} 1 & 8 & -2 & 3 \\ 2 & -3 & 4 & 1 \\ 3 & 5 & -7 & 3 \end{array}\right] \; R_2 \leftrightarrow R_3 \; \left[\begin{array}{ccc|c} 1 & 8 & -2 & 3 \\ 3 & 5 & -7 & 3 \\ 2 & -3 & 4 & 1 \end{array}\right]$$

B. $3R_2$ means to multiply row 2 by 3.

$$\left[\begin{array}{ccc|c} 1 & 8 & -2 & 3 \\ 2 & -3 & 4 & 1 \\ 3 & 5 & -7 & 3 \end{array}\right] \; 3R_2 \rightarrow \; \left[\begin{array}{ccc|c} 1 & 8 & -2 & 3 \\ 6 & -9 & 12 & 3 \\ 3 & 5 & -7 & 3 \end{array}\right]$$

C. $-3R_1 + R_3$ means to multiply row 1 by -3 and then add the result to row 3. The result replaces row 3.

$$\left[\begin{array}{ccc|c} 1 & 8 & -2 & 3 \\ 2 & -3 & 4 & 1 \\ 3 & 5 & -7 & 3 \end{array}\right] \; -3R_1 + R_3 \rightarrow \; \left[\begin{array}{ccc|c} 1 & 8 & -2 & 3 \\ 2 & -3 & 4 & 1 \\ 0 & -19 & -1 & -6 \end{array}\right]$$

Problem 6

Note: Row echelon form is not unique. One set of steps to reach a row echelon form of the matrix is shown.

a_{11} is 1. To change a_{21} to 0, multiply row 1 by the opposite of a_{21} and add the result to row 2.

$$\left[\begin{array}{ccc|c} 1 & -3 & 2 & 1 \\ -4 & 14 & 0 & -2 \\ 2 & -5 & -3 & 16 \end{array}\right] \; 4R_1 + R_2 \rightarrow \; \left[\begin{array}{ccc|c} 1 & -3 & 2 & 1 \\ 0 & 2 & 8 & 2 \\ 2 & -5 & -3 & 16 \end{array}\right]$$

To change a_{31} to 0, multiply row 1 by the opposite of a_{31} and add the result to row 3.

$$\left[\begin{array}{ccc|c} 1 & -3 & 2 & 1 \\ 0 & 2 & 8 & 2 \\ 2 & -5 & -3 & 16 \end{array}\right] \; -2R_1 + R_3 \rightarrow \; \left[\begin{array}{ccc|c} 1 & -3 & 2 & 1 \\ 0 & 2 & 8 & 2 \\ 0 & 1 & -7 & 14 \end{array}\right]$$

To change a_{22} to 1, multiply row 2 by the reciprocal of a_{22}.

$$\left[\begin{array}{ccc|c} 1 & -3 & 2 & 1 \\ 0 & 2 & 8 & 2 \\ 0 & 1 & -7 & 14 \end{array}\right] \; \frac{1}{2}R_2 \rightarrow \; \left[\begin{array}{ccc|c} 1 & -3 & 2 & 1 \\ 0 & 1 & 4 & 1 \\ 0 & 1 & -7 & 14 \end{array}\right]$$

To change a_{32} to 0, multiply row 2 by the opposite of a_{32} and add the result to row 3.

$$\left[\begin{array}{ccc|c} 1 & -3 & 2 & 1 \\ 0 & 1 & 4 & 1 \\ 0 & 1 & -7 & 14 \end{array}\right] \; -1R_2 + R_3 \rightarrow \; \left[\begin{array}{ccc|c} 1 & -3 & 2 & 1 \\ 0 & 1 & 4 & 1 \\ 0 & 0 & -11 & 13 \end{array}\right]$$

To change a_{33} to 1, multiply row 3 by the reciprocal of a_{33}.

$$\left[\begin{array}{rrr|r} 1 & -3 & 2 & 1 \\ 0 & 1 & 4 & 1 \\ 0 & 0 & -11 & 13 \end{array}\right] -\frac{1}{11}R_3 \rightarrow \left[\begin{array}{rrr|r} 1 & -3 & 2 & 1 \\ 0 & 1 & 4 & 1 \\ 0 & 0 & 1 & -\frac{13}{11} \end{array}\right]$$

A row echelon form of the matrix is $\left[\begin{array}{rrr|r} 1 & -3 & 2 & 1 \\ 0 & 1 & 4 & 1 \\ 0 & 0 & 1 & -\frac{13}{11} \end{array}\right]$.

Problem 7

$4x - 5y = 17$
$3x + 2y = 7$

Write the augmented matrix for the system of equations and then use elementary row operations to rewrite the matrix in row echelon form. One set of steps to reach a row echelon form of the matrix is shown.

$$\left[\begin{array}{rr|r} 4 & -5 & 17 \\ 3 & 2 & 7 \end{array}\right] \xrightarrow[\frac{1}{4}R_1]{\text{Change } a_{11} \text{ to } 1.} \left[\begin{array}{rr|r} 1 & -\frac{5}{4} & \frac{17}{4} \\ 3 & 2 & 7 \end{array}\right]$$

$$\left[\begin{array}{rr|r} 1 & -\frac{5}{4} & \frac{17}{4} \\ 3 & 2 & 7 \end{array}\right] \xrightarrow[-3R_1 + R_2]{\text{Change } a_{21} \text{ to } 0.} \left[\begin{array}{rr|r} 1 & -\frac{5}{4} & \frac{17}{4} \\ 0 & \frac{23}{4} & -\frac{23}{4} \end{array}\right]$$

$$\left[\begin{array}{rr|r} 1 & -\frac{5}{4} & \frac{17}{4} \\ 0 & \frac{23}{4} & -\frac{23}{4} \end{array}\right] \xrightarrow[\frac{4}{23}R_2]{\text{Change } a_{22} \text{ to } 1.} \left[\begin{array}{rr|r} 1 & -\frac{5}{4} & \frac{17}{4} \\ 0 & 1 & -1 \end{array}\right]$$

(1) $x - \frac{5}{4}y = \frac{17}{4}$
(2) $y = -1$

- **Write the system of equations corresponding to the row echelon form of the matrix.**

$x - \frac{5}{4}(-1) = \frac{17}{4}$
$x + \frac{5}{4} = \frac{17}{4}$
$x = 3$

- **Substitute −1 for *y* in equation (1) and solve for *x*.**

The solution is $(3, -1)$.

Problem 8

$2x + 3y + 3z = -2$
$x + 2y - 3z = 9$
$3x - 2y - 4z = 1$

Write the augmented matrix for the system of equations and then use elementary row operations to rewrite the matrix in row echelon form. One set of steps to reach a row echelon form of the matrix is shown.

$$\left[\begin{array}{rrr|r} 2 & 3 & 3 & -2 \\ 1 & 2 & -3 & 9 \\ 3 & -2 & -4 & 1 \end{array}\right] R_1 \leftrightarrow R_2 \left[\begin{array}{rrr|r} 1 & 2 & -3 & 9 \\ 2 & 3 & 3 & -2 \\ 3 & -2 & -4 & 1 \end{array}\right]$$

$$\left[\begin{array}{rrr|r} 1 & 2 & -3 & 9 \\ 2 & 3 & 3 & -2 \\ 3 & -2 & -4 & 1 \end{array}\right] -2R_1 + R_2 \rightarrow \left[\begin{array}{rrr|r} 1 & 2 & -3 & 9 \\ 0 & -1 & 9 & -20 \\ 3 & -2 & -4 & 1 \end{array}\right]$$

$$\left[\begin{array}{rrr|r} 1 & 2 & -3 & 9 \\ 0 & -1 & 9 & -20 \\ 3 & -2 & -4 & 1 \end{array}\right] -3R_1 + R_3 \rightarrow \left[\begin{array}{rrr|r} 1 & 2 & -3 & 9 \\ 0 & -1 & 9 & -20 \\ 0 & -8 & 5 & -26 \end{array}\right]$$

$$\left[\begin{array}{rrr|r} 1 & 2 & -3 & 9 \\ 0 & -1 & 9 & -20 \\ 0 & -8 & 5 & -26 \end{array}\right] -1R_2 \rightarrow \left[\begin{array}{rrr|r} 1 & 2 & -3 & 9 \\ 0 & 1 & -9 & 20 \\ 0 & -8 & 5 & -26 \end{array}\right]$$

$$\left[\begin{array}{rrr|r} 1 & 2 & -3 & 9 \\ 0 & 1 & -9 & 20 \\ 0 & -8 & 5 & -26 \end{array}\right] 8R_2 + R_3 \rightarrow \left[\begin{array}{rrr|r} 1 & 2 & -3 & 9 \\ 0 & 1 & -9 & 20 \\ 0 & 0 & -67 & 134 \end{array}\right]$$

$$\left[\begin{array}{rrr|r} 1 & 2 & -3 & 9 \\ 0 & 1 & -9 & 20 \\ 0 & 0 & -67 & 134 \end{array}\right] -\frac{1}{67}R_2 \rightarrow \left[\begin{array}{rrr|r} 1 & 2 & -3 & 9 \\ 0 & 1 & -9 & 20 \\ 0 & 0 & 1 & -2 \end{array}\right]$$

(1) $x + 2y - 3z = 9$
(2) $y - 9z = 20$
(3) $z = -2$

- **Write the system of equations corresponding to the row echelon form of the matrix.**

$y - 9(-2) = 20$
$y + 18 = 20$
$y = 2$

- **Substitute −2 for *z* in equation (2) and solve for *y*.**

$x + 2(2) - 3(-2) = 9$
$x + 10 = 9$
$x = -1$

- **Substitute −2 for *z* and 2 for *y* in equation (1) and solve for *x*.**

The solution is $(-1, 2, -2)$.

SECTION 4.4

Problem 1

Strategy ▶ Rate of the rowing team in calm water: t
Rate of the current: c

	Rate	Time	Distance
With current	$t + c$	2	$2(t + c)$
Against current	$t - c$	2	$2(t - c)$

▶ The distance traveled with the current is 18 mi. The distance traveled against the current is 10 mi.

Solution

$2(t + c) = 18$
$2(t - c) = 10$

- **Write a system of equations.**

$t + c = 9$
$t - c = 5$

- **Divide each side of both equations by 2.**

$2t = 14$

- **Add the equations.**

$t = 7$

- **Solve for *t*.**

$t + c = 9$
$7 + c = 9$
$c = 2$

- **Substitute the value of *t* into one equation and solve for *c*.**

The rate of the rowing team in calm water is 7 mph.
The rate of the current is 2 mph.

Problem 2

Strategy ▶ Cost of an orange tree: x
Cost of a grapefruit tree: y

First purchase:

	Amount	Unit cost	Value
Orange trees	25	x	$25x$
Grapefruit trees	20	y	$20y$

Second purchase:

	Amount	Unit cost	Value
Orange trees	20	x	$20x$
Grapefruit trees	30	y	$30y$

▶ The total of the first purchase was \$290.
The total of the second purchase was \$330.

Solution

$$\begin{aligned} 25x + 20y &= 290 \\ 20x + 30y &= 330 \end{aligned}$$

- **Write a system of equations.**

$$\begin{aligned} 4(25x + 20y) &= 4 \cdot 290 \\ -5(20x + 30y) &= -5 \cdot 330 \end{aligned}$$

- **Change the coefficients of x into additive inverses.**

$$\begin{aligned} 100x + 80y &= 1160 \\ -100x - 150y &= -1650 \end{aligned}$$

- **Simplify.**

$$\begin{aligned} -70y &= -490 \\ y &= 7 \end{aligned}$$

- **Add the equations.**
- **Solve for y.**

$$\begin{aligned} 25x + 20y &= 290 \\ 25x + 20(7) &= 290 \\ 25x + 140 &= 290 \\ 25x &= 150 \\ x &= 6 \end{aligned}$$

- **Substitute the value of y into one equation and solve for x.**

The cost of an orange tree is \$6.
The cost of a grapefruit tree is \$7.

Problem 3

Strategy Number of full-price tickets sold: x
Number of member tickets sold: y
Number of student tickets sold: z
There were 750 tickets sold: $x + y + z = 750$
The income was \$5400:
$10x + 7y + 5z = 5400$
20 more student tickets were sold than full-price tickets: $z = x + 20$

Solution

(1) $x + y + z = 750$
(2) $10x + 7y + 5z = 5400$
(3) $z = x + 20$

- **Write a system of equations.**

$$\begin{aligned} x + y + (x + 20) &= 750 \\ 10x + 7y + 5(x + 20) &= 5400 \end{aligned}$$

- **Replace z in equations (1) and (2) by $x + 20$.**

(4) $2x + y = 730$
(5) $15x + 7y = 5300$

$$\begin{aligned} -14x - 7y &= -5110 \\ 15x + 7y &= 5300 \\ \hline x &= 190 \end{aligned}$$

- **Multiply equation (4) by -7 and add to equation (5).**

$$\begin{aligned} 2x + y &= 730 \\ 2(190) + y &= 730 \\ 380 + y &= 730 \\ y &= 350 \end{aligned}$$

- **Substitute the value of x into equation (4) and solve for y.**

$$\begin{aligned} x + y + z &= 750 \\ 190 + 350 + z &= 750 \\ 540 + z &= 750 \\ z &= 210 \end{aligned}$$

- **Substitute for x and y in equation (1) and solve for z.**

The museum sold 190 full-price tickets, 350 member tickets, and 210 student tickets.

SECTION 4.5

Problem 1

A.

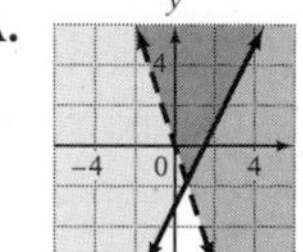

- **Shade above the solid line graph of $y = 2x - 3$.**
- **Shade above the dashed line graph of $y = -3x$.**
- **The solution set of the system is the intersection of the solution sets of the individual inequalities.**

B.

$$\begin{aligned} 3x + 4y &> 12 \\ 4y &> -3x + 12 \\ y &> -\frac{3}{4}x + 3 \end{aligned}$$

- **Solve $3x + 4y > 12$ for y.**

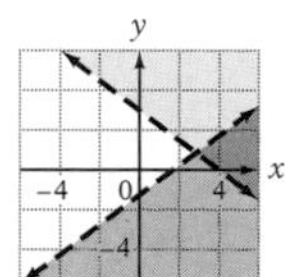

- **Shade above the dashed line graph of $y = -\frac{3}{4}x + 3$.**
- **Shade below the dashed line graph of $y = \frac{3}{4}x - 1$.**

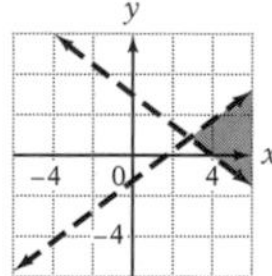

- **The solution set of the system is the intersection of the solution sets of the individual inequalities.**

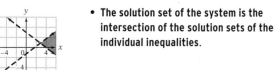

Solutions to Chapter 5 Problems

SECTION 5.1

Problem 1

$$\begin{aligned} &(7xy^3)(-5x^2y^2)(-xy^2) \\ &= [7(-5)(-1)](x \cdot x^2 \cdot x)(y^3 \cdot y^2 \cdot y^2) \\ &= 35x^4y^7 \end{aligned}$$

- **Use the Commutative and Associative Properties.**
- **Add exponents on like bases.**

Problem 2

A. $(-x^2y^4)^5 = (-1)^{1 \cdot 5}x^{2 \cdot 5}y^{4 \cdot 5}$
$= (-1)^5x^{10}y^{20} = -x^{10}y^{20}$

- **Rule for Simplifying Powers of Products**

B. $(2a^3b^2)(-3a^4b)^3$
$= (2a^3b^2)[(-3)^{1\cdot3}a^{4\cdot3}b^{1\cdot3}]$ • **Rule for Simplifying Powers of Products**
$= (2a^3b^2)[(-3)^3a^{12}b^3]$
$= (2a^3b^2)[-27a^{12}b^3]$
$= -54a^{15}b^5$ • **Rule for Multiplying Exponential Expressions**

Problem 3

A. $\dfrac{-8x^7y^5}{16xy^4} = -\dfrac{x^{7-1}y^{5-4}}{2}$ • **Write $\frac{-8}{16}$ in simplest form. Subtract the exponents on like bases.**

$= -\dfrac{x^6y}{2}$

B. $\dfrac{5a^5b^3}{8a^2b} = \dfrac{5a^{5-2}b^{3-1}}{8}$ • **$\frac{5}{8}$ is in simplest form. Subtract the exponents on like bases.**

$= \dfrac{5a^3b^2}{8}$

Problem 4

A. $\dfrac{18x^{-6}y}{9x^{-6}y^7} = 2x^{-6-(-6)}y^{1-7}$ • **Rule for Dividing Exponential Expressions**

$= 2x^0y^{-6}$

$= \dfrac{2}{y^6}$ • **Definitions of Zero and Negative Exponents**

B. $\dfrac{6x^8y^{-3}}{4x^{-2}y^4} = \dfrac{3x^{8-(-2)}y^{-3-4}}{2}$ • **Rule for Dividing Exponential Expressions**

$= \dfrac{3x^{10}y^{-7}}{2}$

$= \dfrac{3x^{10}}{2y^7}$ • **Definition of a Negative Exponent**

Problem 5

A. $\left(\dfrac{x^2}{y^{-3}}\right)^4 = \dfrac{x^{2\cdot4}}{y^{-3\cdot4}} = \dfrac{x^8}{y^{-12}}$ • **Rule for Simplifying Powers of Quotients**

$= x^8y^{12}$ • **Definition of a Negative Exponent**

B. $\left(\dfrac{x^5}{y^4}\right)^{-3} = \dfrac{x^{-15}}{y^{-12}}$ • **Rule for Simplifying Powers of Quotients**

$= \dfrac{y^{12}}{x^{15}}$ • **Definition of a Negative Exponent**

Problem 6

A. $(2x^{-5}y)(5x^4y^{-3}) = 10x^{-5+4}y^{1+(-3)}$ • **Rule for Multiplying Exponential Expressions**
$= 10x^{-1}y^{-2}$

$= \dfrac{10}{xy^2}$ • **Definition of a Negative Exponent**

B. $\left(\dfrac{2^{-1}x^2y^{-3}}{4x^{-2}y^{-5}}\right)^{-2} = \left(\dfrac{x^4y^2}{8}\right)^{-2} = \dfrac{x^{-8}y^{-4}}{8^{-2}} = \dfrac{8^2}{x^8y^4}$

$= \dfrac{64}{x^8y^4}$

Problem 7

Move the decimal point 8 places to the left.
$942{,}000{,}000 = 9.42 \times 10^8$

Problem 8

Move the decimal point 5 places to the left.
$2.7 \times 10^{-5} = 0.000027$

Problem 9

$\dfrac{5{,}600{,}000 \times 0.000000081}{900 \times 0.000000028}$

$= \dfrac{5.6 \times 10^6 \times 8.1 \times 10^{-8}}{9 \times 10^2 \times 2.8 \times 10^{-8}}$ • **Write the numbers in scientific notation.**

$= \dfrac{(5.6)(8.1) \times 10^{6+(-8)-2-(-8)}}{(9)(2.8)}$ • **Use the Rules for Multiplying and Dividing Exponential Expressions.**

$= 1.8 \times 10^4$

Problem 10

Strategy To find the number of arithmetic operations:
- ▶ Find the reciprocal of 1×10^{-7}, which is the number of operations performed in 1 s.
- ▶ Write the number of seconds in 1 min (60) in scientific notation.
- ▶ Multiply the number of arithmetic operations per second by the number of seconds in 1 min.

Solution $\dfrac{1}{1 \times 10^{-7}} = 10^7$

$60 = 6 \times 10$

$6 \times 10 \times 10^7 = 6 \times 10^8$

The computer can perform 6×10^8 operations in 1 min.

SECTION 5.2

Problem 1

The term with the greatest exponent is $-3x^4$. The term without a variable is -12.

The leading coefficient is -3, the constant term is -12, and the degree is 4.

Problem 2

Strategy Evaluate the function when $r = 0.4$.

Solution $v(r) = 600r^2 - 1000r^3$
$V(0.4) = 600(0.4)^2 - 1000(0.4)^3$
$= 32$

The velocity is 32 cm/s.

Problem 3

x	$y = x^2 - 2x$
-1	3
0	0
1	-1
2	0
3	3

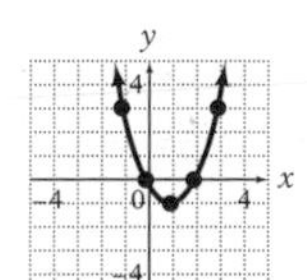

Problem 4

x	$y = -x^3 + 1$
-3	28
-2	9
-1	2
0	1
1	0
2	-7
3	-26

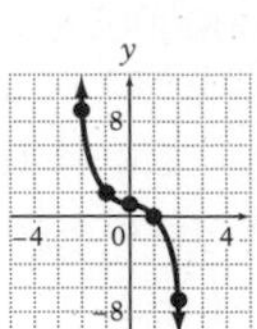

Problem 5

$(x^3 - x + 2) + (x^2 + x - 6)$

$= x^3 + x^2 + (-x + x) + (2 - 6)$ • Use the Commutative and Associate Properties to group like terms.

$= x^3 + x^2 - 4$ • Combine like terms.

Problem 6

$(-5x^2 + 2x - 3) + (-6x^2 - 3x + 7)$ • Rewrite the subtraction as addition of the opposite.

$$\begin{array}{r} -5x^2 + 2x - 3 \\ -6x^2 - 3x + 7 \\ \hline -11x^2 - x + 4 \end{array}$$

• Write in a vertical format.

• Combine like terms.

Problem 7

$D(x) = P(x) - R(x)$

$= (4x^3 - 3x^2 + 2) - (-2x^2 + 2x - 3)$

$= (4x^3 - 3x^2 + 2) + (2x^2 - 2x + 3)$

$= 4x^3 - x^2 - 2x + 5$

SECTION 5.3

Problem 1

A. $-4y(y^2 - 3y + 2)$

$= -4y(y^2) - (-4y)(3y) + (-4y)(2)$ • Distributive Property

$= -4y^3 + 12y^2 - 8y$ • Rule for Multiplying Exponential Expressions

B. $x^2 - 2x[x - x(4x - 5) + x^2]$

$= x^2 - 2x[x - 4x^2 + 5x + x^2]$ • Distributive Property

$= x^2 - 2x[6x - 3x^2]$ • Combine like terms.

$= x^2 - 12x^2 + 6x^3$ • Distributive Property

$= 6x^3 - 11x^2$ • Combine like terms.

Problem 2

$$\begin{array}{r} -2b^2 + 5b - 4 \\ -3b + 2 \\ \hline -4b^2 + 10b - 8 \\ 6b^3 - 15b^2 + 12b \phantom{{}- 8} \\ \hline 6b^3 - 19b^2 + 22b - 8 \end{array}$$

Problem 3

A. $(3x + 4)(5x - 2)$

$= 3x(5x) + 3x(-2) + 4(5x) + 4(-2)$ • FOIL

$= 15x^2 - 6x + 20x - 8$ • Simplify.

$= 15x^2 + 14x - 8$

B. $(x - 3y)(2x - 7y)$

$= x(2x) + x(-7y) + (-3y)2x + (-3y)(-7y)$ • FOIL

$= 2x^2 - 7xy - 6xy + 21y^2$ • Simplify.

$= 2x^2 - 13xy + 21y^2$

C. $(x + 3)(x - 4)(2x - 3)$

$= [x(x) - 4x + 3x - 12](2x - 3)$ • Multiply $(x + 3)(x - 4)$.

$= [x^2 - x - 12](2x - 3)$ • Simplify.

$= (x^2 - x - 12)2x - (x^2 - x - 12)3$ • Distributive Property

$= (2x^3 - 2x^2 - 24x) - (3x^2 - 3x - 36)$ • Distributive Property

$= 2x^3 - 5x^2 - 21x + 36$ • Simplify.

Problem 4

A. $(4x^2 - y)(4x^2 + y)$

$= (4x^2)^2 - y^2 = 16x^4 - y^2$ • $(a - b)(a + b) = a^2 - b^2$

B. $(3x^2 + 5y^2)^2$

$= (3x^2)^2 + 2(3x^2)(5y^2) + (5y^2)^2$ • $(a + b)^2 = a^2 + 2ab + b^2$

$= 9x^4 + 30x^2y^2 + 25y^4$

Problem 5

Strategy To find the area, replace the variables b and h in the equation $A = \frac{1}{2}bh$ with the given values, and solve for A.

Solution

$A = \dfrac{1}{2}bh$

$A = \dfrac{1}{2}(2x + 6)(x - 4)$

$A = (x + 3)(x - 4)$

$A = x^2 - 4x + 3x - 12$

$A = x^2 - x - 12$

The area is $(x^2 - x - 12)$ ft^2.

Problem 6

Strategy To find the volume, subtract the volume of the small rectangular solid from the volume of the large rectangular solid.

Large rectangular solid: Length $= L_1 = 12x$
Width $= W_1 = 7x + 2$
Height $= H_1 = 5x - 4$

Small rectangular solid: Length $= L_2 = 12x$
Width $= W_2 = x$
Height $= H_2 = 2x$

Solution

$V =$ Volume of large rectangular solid − volume of small rectangular solid

$V = (L_1 \cdot W_1 \cdot H_1) - (L_2 \cdot W_2 \cdot H_2)$

$V = (12x)(7x + 2)(5x - 4) - (12x)(x)(2x)$

$V = (84x^2 + 24x)(5x - 4) - (12x^2)(2x)$

$V = (420x^3 - 336x^2 + 120x^2 - 96x) - (24x^3)$

$V = 396x^3 - 216x^2 - 96x$

The volume is $(396x^3 - 216x^2 - 96x)$ ft^3.

SECTION 5.4

Problem 1

$$\frac{15y^3 + 10y^2 - 25y}{-5y}$$
$$= \frac{15y^3}{-5y} + \frac{10y^2}{-5y} - \frac{25y}{-5y}$$ • Divide each term by $-5y$.
$$= -3y^2 - 2y + 5$$ • Simplify.

Problem 2

$$\frac{12x^2y^2 - 16xy^2 - 8x}{4x^2y}$$
$$= \frac{12x^2y^2}{4x^2y} - \frac{16xy^2}{4x^2y} - \frac{8x}{4x^2y}$$ • Divide each term by $4x^2y$.
$$= 3y - \frac{4y}{x} - \frac{2}{xy}$$ • Simplify.

Problem 3

A.
$$\begin{array}{r} 5x - 1 \\ 3x + 4\overline{)15x^2 + 17x - 20} \\ \underline{15x^2 + 20x} \\ -3x - 20 \\ \underline{-3x - 4} \\ -16 \end{array}$$

$$\frac{15x^2 + 17x - 20}{3x + 4} = 5x - 1 - \frac{16}{3x + 4}$$

B.
$$\begin{array}{r} x^2 + 3x - 1 \\ 3x - 1\overline{)3x^3 + 8x^2 - 6x + 2} \\ \underline{3x^3 - x^2} \\ 9x^2 - 6x \\ \underline{9x^2 - 3x} \\ -3x + 2 \\ \underline{-3x + 1} \\ 1 \end{array}$$

$$\frac{3x^3 + 8x^2 - 6x + 2}{3x - 1} = x^2 + 3x - 1 + \frac{1}{3x - 1}$$

Problem 4

A. • $x - a = x + 2$; $a = -2$

−2	6	8	−5
		−12	8
	6	−4	3

$$(6x^2 + 8x - 5) \div (x + 2) = 6x - 4 + \frac{3}{x + 2}$$

B. • Insert a 0 for the missing term.

3	2	−3	−8	0	−2
		6	9	3	9
	2	3	1	3	7

$$(2x^4 - 3x^3 - 8x^2 - 2) \div (x - 3)$$
$$= 2x^3 + 3x^2 + x + 3 + \frac{7}{x - 3}$$

Problem 5

• Use synthetic division with $a = 3$.

3	2	0	−4	−5
		6	18	42
	2	6	14	37

By the Remainder Theorem, $P(3) = 37$.

SECTION 5.5

Problem 1

A. $16x^6yz^4 = 2^4 \cdot x^6 \cdot y \cdot z^4$
$9y^2z^5 = 3^2 \cdot y^2 \cdot z^5$

There is no common numerical factor. The common variable factors are y, with smallest exponent 1, and z, with smallest exponent 4.

The GCF of $16x^6yz^4$ and $9y^2z^5$ is yz^4.

B. $36x^5y^3 = 2^2 \cdot 3^2 \cdot x^5 \cdot y^3$
$12x^2y^4z^5 = 2^2 \cdot 3 \cdot x^2 \cdot y^4 \cdot z^5$
$15x^2y^4 = 3 \cdot 5 \cdot x^2 \cdot y^4$

The common numerical factor is 3, with smallest exponent 1. The common variable factors are x, with smallest exponent 2, and y, with smallest exponent 3.

The GCF of $36x^5y^3$, $12x^2y^4z^5$, and $15x^2y^4$ is $3x^2y^3$.

Problem 2

A. The GCF of $6x^6z^3$ and $8x^5z^2$ is $2x^5z^2$.
$$6x^6z^3 + 8x^5z^2 = 2x^5z^2(3xz) + 2x^5z^2(4)$$
$$= 2x^5z^2(3xz + 4)$$

B. The GCF of $16x^5y^4z^3$, $30xy^6z^3$, and $30xy^6z$ is $2xy^4z$.
$$16x^5y^4z^3 + 30xy^6z^3 - 30xy^6z$$
$$= 2xy^4z(8x^4z^2) + 2xy^4z(15y^2z^2) - 2xy^4z(15y^2)$$
$$= 2xy^4z(8x^4z^2 + 15y^2z^2 - 15y^2)$$

Problem 3

A. $6a(2b + 5) - 7(5 + 2b)$
$= 6a(2b + 5) - 7(2b + 5)$ • $5 + 2b = 2b + 5$
$= (2b + 5)(6a - 7)$ • Factor out $2b + 5$.

B. $3rs - 2r - 3s + 2$
$= (3rs - 2r) - (3s - 2)$ • $-3s + 2 = -(3s - 2)$
$= r(3s - 2) - (3s - 2)$ • Factor out r.
$= (3s - 2)(r - 1)$ • Factor out $3s - 2$.

Problem 4

$2x^3 - 4x^2 + 3x - 6$
$= (2x^3 - 4x^2) + (3x - 6)$ • Group the first two terms and the last two terms.
$= 2x^2(x - 2) + 3(x - 2)$ • Factor out the GCF from each group.
$= (x - 2)(2x^2 + 3)$ • Factor out $x - 2$.

SECTION 5.6

Problem 1

A. Find two factors of 42 whose sum is 13.
$x^2 + 13x + 42 = (x + 6)(x + 7)$

B. Find two factors of -20 whose sum is -1.
$x^2 - x - 20 = (x + 4)(x - 5)$

Problem 2

$6x^3y + 6x^2y - 36xy$
$= 6xy(x^2 + x - 6)$ • Factor out the GCF.
$= 6xy(x + 3)(x - 2)$ • Factor the trinomial.

Problem 3

$x^2 + 5x - 1$

There are no factors of -1 whose sum is 5.

The trinomial is nonfactorable over the integers.

Problem 4

A. $4x^2 + 15x - 4$

Factors of 4	Factors of -4
1, 4	1, -4
2, 2	-1, 4
	2, -2

• Find the factors of *a* and *c*.

Trial Factors	Middle Term
$(4x + 1)(x - 4)$	$-16x + x = -15x$
$(4x - 1)(x + 4)$	$16x - x = 15x$

• Write trial factors and check the middle term.

$4x^2 + 15x - 4 = (4x - 1)(x + 4)$

B. $10x^2 + 39x + 14$

Factors of 10	Factors of 14
1, 10	1, 14
2, 5	2, 7

• Find the factors of *a* and *c*.

Trial Factors	Middle Term
$(x + 2)(10x + 7)$	$7x + 20x = 27x$
$(2x + 1)(5x + 14)$	$28x + 5x = 33x$
$(10x + 1)(x + 14)$	$140x + x = 141x$
$(5x + 2)(2x + 7)$	$35x + 4x = 39x$

• Write trial factors and check the middle term.

$10x^2 + 39x + 14 = (5x + 2)(2x + 7)$

Problem 5

A. $6x^2 + 7x - 20$

$a \cdot c = -120$

Find two factors of -120 whose sum is 7. The factors are 15 and -8.

$6x^2 + 7x - 20$
$= 6x^2 + 15x - 8x - 20$ • Rewrite $7x$ as $15x - 8x$.
$= (6x^2 + 15x) - (8x + 20)$ • Factor by grouping.
$= 3x(2x + 5) - 4(2x + 5)$
$= (2x + 5)(3x - 4)$

B. $2 - x - 6x^2$

$a \cdot c = -12$

Find two factors of -12 whose sum is -1. The factors are -4 and 3.

$2 - x - 6x^2$
$= 2 - 4x + 3x - 6x^2$ • Rewrite $-x$ as $-4x + 3x$.
$= (2 - 4x) + (3x - 6x^2)$ • Factor by grouping.
$= 2(1 - 2x) + 3x(1 - 2x)$
$= (1 - 2x)(2 + 3x)$

Problem 6

A. $3a^3b^3 + 3a^2b^2 - 60ab$ • The GCF is $3ab$.
$= 3ab(a^2b^2 + ab - 20)$ • Factor out the GCF.
$= 3ab(ab + 5)(ab - 4)$ • Factor the trinomial.

B. $40a - 10a^2 - 15a^3$ • The GCF is $5a$.
$= 5a(8 - 2a - 3a^2)$ • Factor out the GCF.
$= 5a(2 + a)(4 - 3a)$ • Factor the trinomial.

SECTION 5.7

Problem 1

A. $81x^2 - 4y^2 = (9x)^2 - (2y)^2$ • $81x^2 - 4y^2$ is a difference of two squares.
$= (9x + 2y)(9x - 2y)$ • $a^2 - b^2 = (a + b)(a - b)$

B. $20x^3y^2 - 45xy^2$
$= 5xy^2(4x^2) - 5xy^2(9)$ • The GCF is $5xy^2$.
$= 5xy^2(4x^2 - 9)$ • Factor out the GCF.
$= 5xy^2[(2x)^2 - 3^2]$ • $4x^2 - 9$ is a difference of two squares.
$= 5xy^2(2x + 3)(2x - 3)$ • $a^2 - b^2 = (a + b)(a - b)$

Problem 2

$9x^2 + 12x + 4$ • $9x^2 = (3x)^2$ and $4 = 2^2$.
$= (3x + 2)^2$ • Check that $(3x + 2)^2 = 9x^2 + 12x + 4$.

Problem 3

A. $a^3 + 64b^3$ • $a^3 + 64b^3$ is the sum of two cubes.
$= a^3 + (4b)^3$
$= (a + 4b)[a^2 - a(4b) + (4b)^2]$ • Use the Sum of Two Cubes formula.
$= (a + 4b)(a^2 - 4ab + 16b^2)$

B. $32x^4y^3 - 108xy^3$ • The GCF is $4xy^3$.
$= 4xy^3(8x^3 - 27)$ • Factor out the GCF.
$= 4xy^3[(2x)^3 - 3^3]$ • $8x^3 - 27$ is the sum of two cubes.
$= 4xy^3(2x - 3)[(2x)^2 + 2x(3) + 3^2]$ • Use the Sum of Two Cubes formula.
$= 4xy^3(2x - 3)(4x^2 + 6x + 9)$

Problem 4

A. $6x^2y^2 - 19xy + 10$ • Let $u = xy$.
$= (3xy - 2)(2xy - 5)$ • Factor $6u^2 - 19u + 10$.

B. $3x^4 + 4x^2 - 4$ • Let $u = x^2$.
$= (x^2 + 2)(3x^2 - 2)$ • Factor $3u^2 + 4u - 4$.

Problem 5

A. $4x - 4y - x^3 + x^2y$
$= (4x - 4y) - (x^3 - x^2y)$ • Factor by grouping.
$= 4(x - y) - x^2(x - y)$
$= (x - y)(4 - x^2)$
$= (x - y)(2 + x)(2 - x)$ • Factor the difference of squares.

B. $x^4 - 8x^2 - 9$ • Let $u = x^2$.
$= (x^2 - 9)(x^2 + 1)$ • Factor $u^2 - 8u - 9$.
$= (x + 3)(x - 3)(x^2 + 1)$ • Factor the difference of squares.

SECTION 5.8

Problem 1

A. $4x^2 + 11x = 3$
$4x^2 + 11x - 3 = 0$ • Write in standard form.
$(4x - 1)(x + 3) = 0$ • Factor.
$4x - 1 = 0 \quad x + 3 = 0$ • Principle of Zero Products
$4x = 1 \quad x = -3$
$x = \frac{1}{4}$

The solutions are $\frac{1}{4}$ and -3.

B. $(x-2)(x+5)=8$
$x^2+3x-10=8$ • Multiply.
$x^2+3x-18=0$ • Write in standard form.
$(x+6)(x-3)=0$ • Factor.
$x+6=0 \quad x-3=0$ • Principle of Zero Products
$x=-6 \quad x=3$

The solutions are -6 and 3.

Problem 2

$s(c)=4$
$c^2-c-2=4$ • $s(c)=c^2-c-2$
$c^2-c-6=0$ • Write in standard form.
$(c+2)(c-3)=0$ • Factor.
$c+2=0 \quad c-3=0$ • Principle of Zero Products
$c=-2 \quad c=3$

The values of c are -2 and 3.

Problem 3

We must find the values of c for which $s(c)=0$.

$s(c)=0$
$c^2+5c-14=0$ • $s(t)=t^2+5t-14$; thus $s(c)=c^2+5c-14$.
$(c+7)(c-2)=0$ • Factor.
$c+7=0 \quad c-2=0$ • Principle of Zero Products
$c=-7 \quad c=2$

The zeros of s are -7 and 2.

Problem 4

Strategy Replace D by 20 in the equation $D=\frac{n(n-3)}{2}$, and then solve for n.

Solution $D=\dfrac{n(n-3)}{2}$
$20=\dfrac{n(n-3)}{2}$ • $D=20$
$40=n^2-3n$ • Multiply each side by 2.
$0=n^2-3n-40$ • Write in standard form.
$0=(n-8)(n+5)$ • Factor.
$n-8=0 \quad n+5=0$ • Principle of Zero Products
$n=8 \quad n=-5$

The answer -5 does not make sense in the context of this problem.

The polygon has 8 sides.

Solutions to Chapter 6 Problems

SECTION 6.1

Problem 1

$f(x)=\dfrac{3-5x}{x^2+5x+6}$
$f(2)=\dfrac{3-5(2)}{2^2+5(2)+6}$ • Substitute 2 for x.
$f(2)=\dfrac{3-10}{4+10+6}$
$f(2)=\dfrac{-7}{20}$
$f(2)=-\dfrac{7}{20}$

Problem 2

The domain must exclude values of x for which $x^2-4=0$.

$x^2-4=0$
$(x-2)(x+2)=0$ • Factor.
$x-2=0 \quad x+2=0$ • Principle of Zero Products
$x=2 \quad x=-2$

The domain is $\{x \mid x \neq -2, x \neq 2\}$.

Problem 3

A. $\dfrac{6x^4-24x^3}{12x^3-48x^2}=\dfrac{6x^3(x-4)}{12x^2(x-4)}$ • Factor.

$=\dfrac{6x^3\overset{1}{\cancel{(x-4)}}}{12x^2\underset{1}{\cancel{(x-4)}}}=\dfrac{x}{2}$ • Divide by common factors.

B. $\dfrac{20x-15x^2}{15x^3-5x^2-20x}=\dfrac{5x(4-3x)}{5x(3x^2-x-4)}$ • Factor out each GCF.

$=\dfrac{5x(4-3x)}{5x(3x-4)(x+1)}$ • Factor the trinomial.

$=\dfrac{5x\overset{-1}{\cancel{(4-3x)}}}{5x\underset{1}{\cancel{(3x-4)}}(x+1)}$ • Divide by common factors.

$=-\dfrac{1}{x+1}$

SECTION 6.2

Problem 1

$$\frac{12 + 5x - 3x^2}{x^2 + 2x - 15} \cdot \frac{2x^2 + x - 45}{3x^2 + 4x}$$

$$= \frac{(4 + 3x)(3 - x)}{(x + 5)(x - 3)} \cdot \frac{(2x - 9)(x + 5)}{x(3x + 4)}$$ • Factor.

$$= \frac{(4 + 3x)(3 - x)(2x - 9)(x + 5)}{(x + 5)(x - 3) \cdot x(3x + 4)}$$ • Multiply.

$$= \frac{\overset{1}{\cancel{(4 + 3x)}}\overset{-1}{\cancel{(3 - x)}}(2x - 9)\overset{1}{\cancel{(x + 5)}}}{\underset{1}{\cancel{(x + 5)}}\underset{1}{\cancel{(x - 3)}} \cdot x\underset{1}{\cancel{(3x + 4)}}}$$ • Divide by common factors.

$$= -\frac{2x - 9}{x}$$

Problem 2

A. $$\frac{6x^2 - 3xy}{10ab^4} \div \frac{16x^2y^2 - 8xy^3}{15a^2b^2}$$

$$= \frac{6x^2 - 3xy}{10ab^4} \cdot \frac{15a^2b^2}{16x^2y^2 - 8xy^3}$$ • Multiply by the reciprocal.

$$= \frac{3x(2x - y)}{10ab^4} \cdot \frac{15a^2b^2}{8xy^2(2x - y)}$$ • Factor.

$$= \frac{45a^2b^2x(2x - y)}{80ab^4xy^2(2x - y)}$$ • Multiply.

$$= \frac{45a^2b^2x\overset{1}{\cancel{(2x - y)}}}{80ab^4xy^2\underset{1}{\cancel{(2x - y)}}} = \frac{9a}{16b^2y^2}$$ • Divide by common factors.

B. $$\frac{6x^2 - 7x + 2}{3x^2 + x - 2} \div \frac{4x^2 - 8x + 3}{5x^2 + x - 4}$$

$$= \frac{6x^2 - 7x + 2}{3x^2 + x - 2} \cdot \frac{5x^2 + x - 4}{4x^2 - 8x + 3}$$ • Multiply by the reciprocal.

$$= \frac{(2x - 1)(3x - 2)}{(x + 1)(3x - 2)} \cdot \frac{(x + 1)(5x - 4)}{(2x - 1)(2x - 3)}$$ • Factor.

$$= \frac{(2x - 1)(3x - 2)(x + 1)(5x - 4)}{(x + 1)(3x - 2)(2x - 1)(2x - 3)}$$ • Multiply.

$$= \frac{\overset{1}{\cancel{(2x - 1)}}\overset{1}{\cancel{(3x - 2)}}\overset{1}{\cancel{(x + 1)}}(5x - 4)}{\underset{1}{\cancel{(x + 1)}}\underset{1}{\cancel{(3x - 2)}}\underset{1}{\cancel{(2x - 1)}}(2x - 3)}$$ • Simplify.

$$= \frac{5x - 4}{2x - 3}$$

Problem 3

The LCD is xy^2.

$$\frac{5y}{x} - \frac{9}{xy} - \frac{4}{y^2}$$

$$= \frac{5y}{x} \cdot \frac{y^2}{y^2} - \frac{9}{xy} \cdot \frac{y}{y} - \frac{4}{y^2} \cdot \frac{x}{x}$$ • Write each fraction in terms of the LCD.

$$= \frac{5y^3}{xy^2} - \frac{9y}{xy^2} - \frac{4x}{xy^2}$$ • Simplify.

$$= \frac{5y^3 - 9y - 4x}{xy^2}$$ • Subtract the numerators. Place the difference over the common denominator.

Problem 4

The LCD is $a(a - 5)(a + 5)$.

$$\frac{a - 3}{a^2 - 5a} + \frac{a - 9}{a^2 - 25}$$

$$= \frac{a - 3}{a(a - 5)} \cdot \frac{a + 5}{a + 5} + \frac{a - 9}{(a - 5)(a + 5)} \cdot \frac{a}{a}$$

$$= \frac{a^2 + 2a - 15}{a(a - 5)(a + 5)} + \frac{a^2 - 9a}{a(a - 5)(a + 5)}$$

$$= \frac{(a^2 + 2a - 15) + (a^2 - 9a)}{a(a - 5)(a + 5)}$$

$$= \frac{a^2 + 2a - 15 + a^2 - 9a}{a(a - 5)(a + 5)}$$

$$= \frac{2a^2 - 7a - 15}{a(a - 5)(a + 5)} = \frac{(2a + 3)(a - 5)}{a(a - 5)(a + 5)}$$

$$= \frac{(2a + 3)\overset{1}{\cancel{(a - 5)}}}{a\underset{1}{\cancel{(a - 5)}}(a + 5)} = \frac{2a + 3}{a(a + 5)}$$

Problem 5

A. The LCD is $(x + 2)(x - 3)$.

$$\frac{1}{x + 2} - \frac{x}{x - 3} + 2$$

$$= \frac{1}{x + 2} \cdot \frac{x - 3}{x - 3} - \frac{x}{x - 3} \cdot \frac{x + 2}{x + 2} + \frac{2}{1} \cdot \frac{(x + 2)(x - 3)}{(x + 2)(x - 3)}$$

$$= \frac{x - 3}{(x + 2)(x - 3)} - \frac{x^2 + 2x}{(x + 2)(x - 3)} + \frac{2x^2 - 2x - 12}{(x + 2)(x - 3)}$$

$$= \frac{x - 3 - x^2 - 2x + 2x^2 - 2x - 12}{(x + 2)(x - 3)}$$

$$= \frac{x^2 - 3x - 15}{(x + 2)(x - 3)}$$

B. The LCD is $(x - 2)(2x - 3)$.

$$\frac{x - 1}{x - 2} - \frac{7 - 6x}{2x^2 - 7x + 6} + \frac{4}{2x - 3}$$

$$= \frac{x - 1}{x - 2} \cdot \frac{2x - 3}{2x - 3} - \frac{7 - 6x}{(x - 2)(2x - 3)} + \frac{4}{2x - 3} \cdot \frac{x - 2}{x - 2}$$

$$= \frac{2x^2 - 5x + 3}{(x - 2)(2x - 3)} - \frac{7 - 6x}{(x - 2)(2x - 3)} + \frac{4x - 8}{(x - 2)(2x - 3)}$$

$$= \frac{(2x^2 - 5x + 3) - (7 - 6x) + (4x - 8)}{(x - 2)(2x - 3)}$$

$$= \frac{2x^2 - 5x + 3 - 7 + 6x + 4x - 8}{(x - 2)(2x - 3)}$$

$$= \frac{2x^2 + 5x - 12}{(x - 2)(2x - 3)} = \frac{(x + 4)(2x - 3)}{(x - 2)(2x - 3)}$$

$$= \frac{(x + 4)\overset{1}{\cancel{(2x - 3)}}}{(x - 2)\underset{1}{\cancel{(2x - 3)}}} = \frac{x + 4}{x - 2}$$

SECTION 6.3

Problem 1

A. Multiply the numerator and denominator of the complex fraction by the LCD. The LCD is x^2.

$$\frac{3+\frac{16}{x}+\frac{16}{x^2}}{6+\frac{5}{x}-\frac{4}{x^2}}=\frac{3+\frac{16}{x}+\frac{16}{x^2}}{6+\frac{5}{x}-\frac{4}{x^2}}\cdot\frac{x^2}{x^2}$$

$$=\frac{3\cdot x^2+\frac{16}{x}\cdot x^2+\frac{16}{x^2}\cdot x^2}{6\cdot x^2+\frac{5}{x}\cdot x^2-\frac{4}{x^2}\cdot x^2}$$ • **Distributive Property**

$$=\frac{3x^2+16x+16}{6x^2+5x-4}$$ • **Simplify.**

$$=\frac{(3x+4)(x+4)}{(2x-1)(3x+4)}$$ • **Factor.**

$$=\frac{\overset{1}{\cancel{(3x+4)}}(x+4)}{(2x-1)\underset{1}{\cancel{(3x+4)}}}=\frac{x+4}{2x-1}$$

B. Multiply the numerator and denominator of the complex fraction by the LCD. The LCD is $x-3$.

$$\frac{2x+5+\frac{14}{x-3}}{4x+16+\frac{49}{x-3}}=\frac{2x+5+\frac{14}{x-3}}{4x+16+\frac{49}{x-3}}\cdot\frac{x-3}{x-3}$$

$$=\frac{(2x+5)(x-3)+\frac{14}{x-3}(x-3)}{(4x+16)(x-3)+\frac{49}{x-3}(x-3)}$$ • **Distributive Property**

$$=\frac{2x^2-x-15+14}{4x^2+4x-48+49}=\frac{2x^2-x-1}{4x^2+4x+1}$$ • **Simplify.**

$$=\frac{(2x+1)(x-1)}{(2x+1)(2x+1)}=\frac{\overset{1}{\cancel{(2x+1)}}(x-1)}{\underset{1}{\cancel{(2x+1)}}(2x+1)}$$ • **Factor.**

$$=\frac{x-1}{2x+1}$$

SECTION 6.4

Problem 1

A.

$$\frac{5}{2x-3}=\frac{-2}{x+1}$$
$$(x+1)(2x-3)\frac{5}{(2x-3)}=(x+1)(2x-3)\frac{-2}{(x+1)}$$
$$5(x+1)=-2(2x-3)$$
$$5x+5=-4x+6$$
$$9x+5=6$$
$$9x=1$$
$$x=\frac{1}{9}$$

The solution is $\frac{1}{9}$.

B.

$$\frac{4x+1}{2x-1}=2+\frac{3}{x-3}$$
$$(2x-1)(x-3)\frac{4x+1}{(2x-1)}=(2x-1)(x-3)\left(2+\frac{3}{x-3}\right)$$
$$(x-3)(4x+1)=(2x-1)(x-3)2+(2x-1)3$$
$$4x^2-11x-3=4x^2-14x+6+6x-3$$
$$-11x-3=-8x+3$$
$$-3x=6$$
$$x=-2$$

The solution is -2.

Problem 2

Strategy ► Time required for the small pipe to fill the tank: x

	Rate	·	Time	=	Part
Large pipe	$\frac{1}{9}$	·	6	=	$\frac{6}{9}$
Small pipe	$\frac{1}{x}$	·	6	=	$\frac{6}{x}$

► The sum of the part of the task completed by the large pipe and the part of the task completed by the small pipe is 1.

Solution

$$\frac{6}{9}+\frac{6}{x}=1$$
$$\frac{2}{3}+\frac{6}{x}=1$$
$$3x\left(\frac{2}{3}+\frac{6}{x}\right)=3x\cdot 1$$
$$2x+18=3x$$
$$18=x$$

The small pipe, working alone, will fill the tank in 18 h.

Problem 3

Strategy ► Rate of the wind: r

	Distance	÷	Rate	=	Time
With wind	700	÷	$150+r$	=	$\frac{700}{150+r}$
Against wind	500	÷	$150-r$	=	$\frac{500}{150-r}$

► The time flying with the wind equals the time flying against the wind.

Solution

$$\frac{700}{150+r} = \frac{500}{150-r}$$
$$(150+r)(150-r)\left(\frac{700}{150+r}\right) = (150+r)(150-r)\left(\frac{500}{150-r}\right)$$
$$(150-r)700 = (150+r)500$$
$$105{,}000 - 700r = 75{,}000 + 500r$$
$$30{,}000 = 1200r$$
$$25 = r$$

The rate of the wind is 25 mph.

SECTION 6.5

Problem 1

Strategy To find the cost, write and solve a proportion using x to represent the cost.

Solution

$$\frac{2}{5.80} = \frac{15}{x}$$
$$x(5.80)\frac{2}{5.80} = x(5.80)\frac{15}{x}$$
$$2x = 15(5.80)$$
$$2x = 87$$
$$x = 43.5$$

The cost of 15 lb of cashews is \$43.50.

Problem 2

Strategy To find the distance:

- Write the basic direct variation equation, replace the variables by the given values, and solve for k.
- Write the direct variation equation, replacing k by its value. Substitute 5 for t, and solve for s.

Solution

$$s = kt^2$$
$$64 = k(2)^2$$
$$64 = k \cdot 4$$
$$16 = k$$
$$s = 16t^2$$
$$s = 16(5)^2 = 400$$

The object will fall 400 ft in 5 s.

Problem 3

Strategy To find the resistance:

- Write the basic inverse variation equation, replace the variables by the given values, and solve for k.
- Write the inverse variation equation, replacing k by its value. Substitute 0.02 for d, and solve for R.

Solution

$$R = \frac{k}{d^2}$$
$$0.5 = \frac{k}{(0.01)^2}$$
$$0.5 = \frac{k}{0.0001}$$
$$0.00005 = k$$
$$R = \frac{0.00005}{d^2}$$
$$R = \frac{0.00005}{(0.02)^2} = 0.125$$

The resistance is 0.125 ohm.

Problem 4

Strategy To find the strength of the beam:

- Write the basic combined variation equation, replace the variables by the given values, and solve for k.
- Write the combined variation equation, replacing k by its value and substituting 4 for W, 8 for d, and 16 for L. Solve for s.

Solution

$$s = \frac{kWd^2}{L}$$
$$1200 = \frac{k(2)(12)^2}{12}$$
$$1200 = \frac{k(288)}{12}$$
$$14{,}400 = 288k$$
$$50 = k$$

$$s = \frac{50Wd^2}{L}$$
$$s = \frac{50(4)8^2}{16}$$
$$s = \frac{12{,}800}{16}$$
$$s = 800$$

The strength of the beam is 800 lb.

SECTION 6.6

Problem 1

A.

$$\frac{1}{R_1} + \frac{1}{R_2} = \frac{1}{R}$$
$$RR_1R_2\left(\frac{1}{R_1} + \frac{1}{R_2}\right) = RR_1R_2\left(\frac{1}{R}\right)$$
$$RR_1R_2\left(\frac{1}{R_1}\right) + RR_1R_2\left(\frac{1}{R_2}\right) = R_1R_2$$
$$RR_2 + RR_1 = R_1R_2$$
$$R(R_2 + R_1) = R_1R_2$$
$$R = \frac{R_1R_2}{R_2 + R_1}$$

B.

$$t = \frac{r}{r+1}$$

$(r+1)t = (r+1)\left(\frac{r}{r+1}\right)$ • **Multiply each side by $r + 1$.**

$tr + t = r$

$t = r - tr$ • **Subtract tr from each side.**

$t = r(1 - t)$ • **Factor.**

$\frac{t}{1-t} = r$ • **Divide each side by $1 - t$.**

Solutions to Chapter 7 Problems

SECTION 7.1

Problem 1

A. $64^{\frac{2}{3}} = (2^6)^{\frac{2}{3}}$ • Rewrite 64 as 2^6.
$= 2^4 = 16$ • Multiply the exponents.

B. $16^{-\frac{3}{4}} = (2^4)^{-\frac{3}{4}}$ • Rewrite 16 as 2^4.
$= 2^{-3}$ • Multiply the exponents.
$= \frac{1}{2^3} = \frac{1}{8}$ • Definition of a Negative Exponent

Problem 2

A. $\frac{x^{\frac{1}{2}}y^{-\frac{5}{4}}}{x^{-\frac{4}{3}}y^{\frac{1}{3}}} = \frac{x^{\frac{1}{2}+\frac{4}{3}}}{y^{\frac{1}{3}+\frac{5}{4}}} = \frac{x^{\frac{11}{6}}}{y^{\frac{19}{12}}}$ • Rule for Dividing Exponential Expressions

B. $(x^{\frac{3}{4}}y^{\frac{1}{2}}z^{-\frac{2}{3}})^{-\frac{4}{3}} = x^{-1}y^{-\frac{2}{3}}z^{\frac{8}{9}}$ • Multiply the exponents.
$= \frac{z^{\frac{8}{9}}}{xy^{\frac{2}{3}}}$ • Definition of a Negative Exponent

C. $\frac{(3x^{\frac{3}{4}}y^{-\frac{1}{4}})^4}{(9x^2y^4)^{\frac{1}{2}}} = \frac{3^4x^3y^{-1}}{9^{\frac{1}{2}}xy^2}$ • Rule for Simplifying Powers of Products
$= \frac{3^4x^3y^{-1}}{3xy^2}$ • Simplify $9^{\frac{1}{2}}$.
$= 3^{4-1}x^{3-1}y^{-1-2}$ • Rule for Dividing Exponential Expressions
$= 3^3x^2y^{-3} = \frac{27x^2}{y^3}$

Problem 3

A. The denominator of the rational exponent is the index of the radical. The numerator is the power of the radicand.
$(2x^3)^{\frac{3}{4}} = \sqrt[4]{(2x^3)^3} = \sqrt[4]{8x^9}$

B. Note that -5 is not raised to the power.
$-5a^{\frac{5}{6}} = -5(a^5)^{\frac{1}{6}} = -5\sqrt[6]{a^5}$

Problem 4

A. $\sqrt[3]{3ab} = (3ab)^{\frac{1}{3}}$ • The index of the radical is the denominator of the exponent.

B. $\sqrt[4]{x^4 + y^4} = (x^4 + y^4)^{\frac{1}{4}}$ • The index of the radical is the denominator of the exponent.

Problem 5

A. $\sqrt{121x^{10}} = 11x^5$ • The radicand is a perfect square. Divide the exponent by 2.

B. $\sqrt[3]{-8x^{12}y^3} = -2x^4y$ • The radicand is a perfect cube. Divide each exponent by 3.

C. $-\sqrt[4]{81x^{12}y^8} = -3x^3y^2$ • The radicand is a perfect 4th power. Divide each exponent by 4.

SECTION 7.2

Problem 1

$\sqrt[5]{128x^7} = \sqrt[5]{32x^5(4x^2)}$ • Rewrite $128x^7$ using a perfect fifth power.
$= \sqrt[5]{32x^5}\sqrt[5]{4x^2}$ • Product Property of Radicals
$= 2x\sqrt[5]{4x^2}$ • Simplify.

Problem 2

A. $4\sqrt{45x^4y^3} + 2xy\sqrt{20x^2y}$
$= 4\sqrt{9x^4y^2(5y)} + 2xy\sqrt{4x^2(5y)}$
$= 4\sqrt{9x^4y^2} \cdot \sqrt{5y} + 2xy\sqrt{4x^2} \cdot \sqrt{5y}$ • Use the Product Property of Radicals to simplify each radical.
$= 4(3x^2y) \cdot \sqrt{5y} + 2xy(2x) \cdot \sqrt{5y}$ • Simplify $\sqrt{9x^4y^2}$ and $\sqrt{4x^2}$.
$= 12x^2y\sqrt{5y} + 4x^2y\sqrt{5y}$ • Simplify.
$= 16x^2y\sqrt{5y}$ • Combine like terms.

B. $5a\sqrt[4]{32b^5} - 7b\sqrt[4]{162a^4b}$
$= 5a\sqrt[4]{16b^4(2b)} - 7b\sqrt[4]{81a^4(2b)}$
$= 5a\sqrt[4]{16b^4} \cdot \sqrt[4]{2b} - 7b\sqrt[4]{81a^4} \cdot \sqrt[4]{2b}$ • Use the Product Property of Radicals to simplify each radical.
$= 5a(2b) \cdot \sqrt[4]{2b} - 7b(3a) \cdot \sqrt[4]{2b}$ • Simplify $\sqrt[4]{16b^4}$ and $\sqrt[4]{81a^4}$.
$= 10ab\sqrt[4]{2b} - 21ab\sqrt[4]{2b}$ • Simplify.
$= -11ab\sqrt[4]{2b}$ • Combine like terms.

Problem 3

A. $(-3\sqrt{7})^2 = (-3\sqrt{7})(-3\sqrt{7})$ • Product Property of Radicals
$= 9\sqrt{49} = 9(7) = 63$

B. $\sqrt{15xy^3}\sqrt{5xy} = \sqrt{75x^2y^4}$ • Product Property of Radicals
$= \sqrt{25x^2y^4}\sqrt{3}$
$= 5xy^2\sqrt{3}$

C. $\sqrt[3]{20ab^4}\sqrt[3]{2a^4b^2} = \sqrt[3]{40a^5b^6}$ • Product Property of Radicals
$= \sqrt[3]{8a^3b^6}\sqrt[3]{5a^2}$
$= 2ab^2\sqrt[3]{5a^2}$

Problem 4

$\sqrt{5b}(\sqrt{3b} - \sqrt{10})$
$= \sqrt{15b^2} - \sqrt{50b}$ • Distributive Property
$= \sqrt{b^2}\sqrt{15} - \sqrt{25}\sqrt{2b}$ • Simplify.
$= b\sqrt{15} - 5\sqrt{2b}$

Problem 5

A. $(2\sqrt{5} + 3)(4\sqrt{5} - 7)$
$= 8\sqrt{25} - 14\sqrt{5} + 12\sqrt{5} - 21$ • Use FOIL.
$= 8(5) - 2\sqrt{5} - 21$ • Simplify $\sqrt{25}$ and combine like terms.
$= 40 - 2\sqrt{5} - 21$
$= 19 - 2\sqrt{5}$

B. $(2\sqrt{a} + 3\sqrt{b})(\sqrt{a} - 2\sqrt{b})$
$= 2\sqrt{a^2} - 4\sqrt{ab} + 3\sqrt{ab} - 6\sqrt{b^2}$ • Use FOIL.
$= 2a - \sqrt{ab} - 6b$ • Simplify $\sqrt{a^2}$ and $\sqrt{b^2}$. Combine like terms.

Problem 6

A. $\sqrt{\frac{3}{7}} = \frac{\sqrt{3}}{\sqrt{7}}$ • Quotient Property of Radicals
$= \frac{\sqrt{3}}{\sqrt{7}} \cdot \frac{\sqrt{7}}{\sqrt{7}} = \frac{\sqrt{21}}{\sqrt{49}}$ • Multiply numerator and denominator by $\sqrt{7}$.
$= \frac{\sqrt{21}}{7}$ • Simplify.

B. $\dfrac{y}{\sqrt{3y}} = \dfrac{y}{\sqrt{3y}} \cdot \dfrac{\sqrt{3y}}{\sqrt{3y}}$ • **Multiply numerator and denominator by $\sqrt{3y}$.**

$= \dfrac{y\sqrt{3y}}{\sqrt{9y^2}}$

$= \dfrac{y\sqrt{3y}}{3y} = \dfrac{\sqrt{3y}}{3}$ • **Simplify.**

C. $\dfrac{3}{\sqrt[3]{3x^2}} = \dfrac{3}{\sqrt[3]{3x^2}} \cdot \dfrac{\sqrt[3]{9x}}{\sqrt[3]{9x}}$ • **Multiply numerator and denominator by $\sqrt[3]{9x}$.**

$= \dfrac{3\sqrt[3]{9x}}{\sqrt[3]{27x^3}}$

$= \dfrac{3\sqrt[3]{9x}}{3x} = \dfrac{\sqrt[3]{9x}}{x}$ • **Simplify.**

Problem 7

A. $\dfrac{4+\sqrt{2}}{3-\sqrt{3}} = \dfrac{4+\sqrt{2}}{3-\sqrt{3}} \cdot \dfrac{3+\sqrt{3}}{3+\sqrt{3}}$ • **Multiply numerator and denominator by the conjugate of $3 - \sqrt{3}$.**

$= \dfrac{12 + 4\sqrt{3} + 3\sqrt{2} + \sqrt{6}}{9 - (\sqrt{3})^2}$ • **Use FOIL.**

$= \dfrac{12 + 4\sqrt{3} + 3\sqrt{2} + \sqrt{6}}{6}$

B. $\dfrac{\sqrt{2}+\sqrt{x}}{\sqrt{2}-\sqrt{x}} = \dfrac{\sqrt{2}+\sqrt{x}}{\sqrt{2}-\sqrt{x}} \cdot \dfrac{\sqrt{2}+\sqrt{x}}{\sqrt{2}+\sqrt{x}}$ • **Multiply numerator and denominator by the conjugate of $\sqrt{2} - \sqrt{x}$.**

$= \dfrac{\sqrt{4} + \sqrt{2x} + \sqrt{2x} + \sqrt{x^2}}{(\sqrt{2})^2 - (\sqrt{x})^2}$ • **Use FOIL.**

$= \dfrac{2 + 2\sqrt{2x} + x}{2 - x}$

SECTION 7.3

Problem 1

A. Because Q contains an odd root, there are no restrictions on the radicand.
The domain is $(-\infty, \infty)$.

B. T contains an even root. The radicand must be positive or zero.

$3x + 9 \geq 0$

$3x \geq -9$

$x \geq -3$

The domain is $[-3, \infty)$.

Problem 2

$x - 2 \geq 0$

$x \geq 2$

• **Because F contains an even root, the radicand must be positive or zero.**

• **Find ordered pairs by evaluating $F(x)$ for integer values of x in the domain $\{x \mid x \geq 2\}$.**

Problem 3

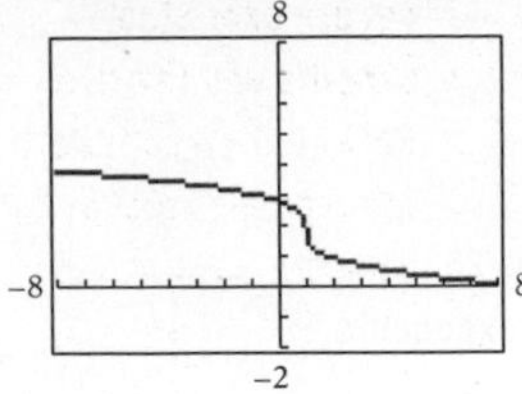

•**Because y contains an odd root, the domain is all real numbers.**

SECTION 7.4

Problem 1

A. $\sqrt{4x+5} - 12 = -5$ • **Add 12 to each side so that the radical is alone.**

$\sqrt{4x+5} = 7$

$(\sqrt{4x+5})^2 = 7^2$ • **Square each side.**

$4x + 5 = 49$

$4x = 44$

$x = 11$

Check: $\sqrt{4x+5} - 12 = -5$

$\sqrt{4 \cdot 11 + 5} - 12$	-5
$\sqrt{44 + 5} - 12$	-5
$\sqrt{49} - 12$	-5
$7 - 12$	-5

$-5 = -5$

The solution is 11.

B. $\sqrt[4]{x-8} = 3$ • **The radical is alone.**

$(\sqrt[4]{x-8})^4 = 3^4$ • **Raise each side to the 4th power.**

$x - 8 = 81$

$x = 89$

Check: $\sqrt[4]{x-8} = 3$

$\sqrt[4]{89-8}$	3
$\sqrt[4]{81}$	3

$3 = 3$

The solution is 89.

Problem 2

A. $x + 3\sqrt{x+2} = 8$ • **Subtract x from each side so that the radical is alone.**

$3\sqrt{x+2} = 8 - x$

$(3\sqrt{x+2})^2 = (8-x)^2$ • **Square each side.**

$9(x+2) = 64 - 16x + x^2$ • **This is a quadratic equation.**

$9x + 18 = 64 - 16x + x^2$

$0 = x^2 - 25x + 46$ • **Write in standard form.**

$0 = (x-2)(x-23)$ • **Factor.**

$x - 2 = 0 \quad x - 23 = 0$ • **Principle of Zero Products**

$x = 2 \quad x = 23$

Check:

$x + 3\sqrt{x+2} = 8$

$2 + 3\sqrt{2+2}$	8
$2 + 3\sqrt{4}$	8
$2 + 3 \cdot 2$	8
$2 + 6$	8

$8 = 8$

$x + 3\sqrt{x+2} = 8$

$23 + 3\sqrt{23+2}$	8
$23 + 3\sqrt{25}$	8
$23 + 3 \cdot 5$	8
$23 + 15$	8

$38 \neq 8$

23 does not check as a solution. The solution is 2.

B.

$$\begin{aligned}\sqrt{x+5} &= 5-\sqrt{x} && \bullet\ \text{There are two radicals.}\\ (\sqrt{x+5})^2 &= (5-\sqrt{x})^2 && \bullet\ \text{Square each side.}\\ x+5 &= 25-10\sqrt{x}+x && \bullet\ \text{Simplify and rewrite the equation so that the radical is alone.}\\ -20 &= -10\sqrt{x}\\ 2 &= \sqrt{x}\\ 2^2 &= (\sqrt{x})^2 && \bullet\ \text{Square each side.}\\ 4 &= x\end{aligned}$$

Check: $\sqrt{x+5} = 5-\sqrt{x}$

$$\begin{array}{c|c} \sqrt{4+5} & 5-\sqrt{4}\\ \sqrt{9} & 5-2\\ 3 & 3 \end{array}$$

The solution is 4.

Problem 3

Strategy To find the diagonal, use the Pythagorean Theorem. One leg is the length of the rectangle. The second leg is the width of the rectangle. The hypotenuse is the diagonal of the rectangle.

Solution

$$\begin{aligned}c^2 &= a^2+b^2\\ c^2 &= (6)^2+(3)^2\\ c^2 &= 36+9\\ c^2 &= 45\\ \sqrt{c^2} &= \sqrt{45}\\ c &= \sqrt{45}\\ c &\approx 6.7\end{aligned}$$

The diagonal is approximately 6.7 cm.

SECTION 7.5

Problem 1

$-5\sqrt{-80} = -5i\sqrt{80} = -5i(4\sqrt{5}) = -20i\sqrt{5}$

Problem 2

$$\begin{aligned}\frac{12-\sqrt{-72}}{3} &= \frac{12-i\sqrt{72}}{3} && \bullet\ \sqrt{-72} = i\sqrt{72}\\ &= \frac{12-6i\sqrt{2}}{3} && \bullet\ \text{Simplify the radical.}\\ &= \frac{3(4-2i\sqrt{2})}{3} && \bullet\ \text{Factor 3 from the numerator.}\\ &= 4-2i\sqrt{2} && \bullet\ \text{Simplify.}\end{aligned}$$

Problem 3

A. $(5-7i)+(2-i)$

$$\begin{aligned}&= (5+2)+[-7+(-1)]i && \bullet\ \text{Add the real parts and add the imaginary parts.}\\ &= 7-8i\end{aligned}$$

B. $(-4+2i)-(6-8i)$

$$\begin{aligned}&= (-4-6)+[2-(-8)]i && \bullet\ \text{Subtract the real parts and subtract the imaginary parts.}\\ &= -10+10i\end{aligned}$$

Problem 4

A. $(4-3i)(2-i)$

$$\begin{aligned}&= 8-4i-6i+3i^2 = 8-10i+3i^2 && \bullet\ \text{FOIL}\\ &= 8-10i+3(-1) = 5-10i && \bullet\ i^2=-1\end{aligned}$$

B. $(3-i)\left(\frac{3}{10}+\frac{1}{10}i\right)$

$$\begin{aligned}&= \frac{9}{10}+\frac{3}{10}i-\frac{3}{10}i-\frac{1}{10}i^2 = \frac{9}{10}-\frac{1}{10}i^2 && \bullet\ \text{FOIL}\\ &= \frac{9}{10}-\frac{1}{10}(-1) = \frac{9}{10}+\frac{1}{10} = 1 && \bullet\ i^2=-1\end{aligned}$$

C. $(3+6i)(3-6i)$

$$\begin{aligned}&= 9-18i+18i-36i^2 = 9-36i^2 && \bullet\ \text{FOIL}\\ &= 9-36(-1) = 9+36 = 45 && \bullet\ i^2=-1\end{aligned}$$

D.

$$\begin{aligned}(1+5i)^2 &= 1+10i+25i^2 && \bullet\ (a+b)^2 = a^2+2ab+b^2\\ &= 1+10i+25(-1) && \bullet\ i^2=-1\\ &= 1+10i-25\\ &= -24+10i\end{aligned}$$

Problem 5

A.

$$\begin{aligned}\frac{-3}{6i} &= \frac{-3}{6i}\cdot\frac{i}{i} = \frac{-3i}{6i^2} && \bullet\ \text{Multiply by } \frac{i}{i}=1.\\ &= \frac{-3i}{6(-1)} = \frac{-3i}{-6} = \frac{i}{2} = \frac{1}{2}i && \bullet\ i^2=-1\end{aligned}$$

B.

$$\begin{aligned}\frac{2-3i}{4i} &= \frac{2-3i}{4i}\cdot\frac{i}{i} = \frac{2i-3i^2}{4i^2} && \bullet\ \text{Multiply by } \frac{i}{i}=1.\\ &= \frac{2i-3(-1)}{4(-1)} = \frac{3+2i}{-4} = -\frac{3}{4}-\frac{1}{2}i && \bullet\ i^2=-1\end{aligned}$$

Problem 6

A.

$$\begin{aligned}\frac{5}{1+3i} &= \frac{5}{1+3i}\cdot\frac{1-3i}{1-3i} && \bullet\ \text{Multiply numerator and denominator by the conjugate of the denominator.}\\ &= \frac{5(1-3i)}{1^2+3^2} = \frac{5-15i}{1+9}\\ &= \frac{5-15i}{10} = \frac{1}{2}-\frac{3}{2}i && \bullet\ \text{Write in } a+bi \text{ form.}\end{aligned}$$

B.

$$\begin{aligned}\frac{2+5i}{3-2i} &= \frac{(2+5i)}{(3-2i)}\cdot\frac{(3+2i)}{(3+2i)} && \bullet\ \text{Multiply numerator and denominator by the conjugate of the denominator.}\\ &= \frac{6+4i+15i+10i^2}{3^2+2^2}\\ &= \frac{6+19i+10(-1)}{13} && \bullet\ i^2=-1\\ &= \frac{-4+19i}{13} = -\frac{4}{13}+\frac{19}{13}i && \bullet\ \text{Write in } a+bi \text{ form.}\end{aligned}$$

Solutions to Chapter 8 Problems

SECTION 8.1

Problem 1

$3x^2+14x+8 = 0$ • The equation is in standard form.

$(3x+2)(x+4) = 0$ • Factor.

$$\begin{aligned}3x+2 &= 0 \qquad & x+4 &= 0 && \bullet\ \text{Principle of Zero Products}\\ 3x &= -2 & x &= -4\\ x &= -\frac{2}{3}\end{aligned}$$

The solutions are -4 and $-\frac{2}{3}$.

Problem 2

$(x + 5)(x - 1) = 7$

$x^2 + 4x - 5 = 7$ • Multiply the binomials.

$x^2 + 4x - 12 = 0$ • Write in standard form.

$(x + 6)(x - 2) = 0$ • Factor.

$x + 6 = 0 \qquad x - 2 = 0$ • Principle of Zero Products

$x = -6 \qquad x = 2$

The solutions are -6 and 2.

Problem 3

$(x - r)(x - s) = 0$

$\left[x - \left(-\frac{2}{3}\right)\right]\left(x - \frac{1}{6}\right) = 0$ • $r = -\frac{2}{3}, s = \frac{1}{6}$

$\left(x + \frac{2}{3}\right)\left(x - \frac{1}{6}\right) = 0$ • Simplify.

$x^2 + \frac{3}{6}x - \frac{2}{18} = 0$ • Multiply the binomials.

$18\left(x^2 + \frac{3}{6}x - \frac{2}{18}\right) = 18 \cdot 0$ • Multiply each side by the LCD.

$18x^2 + 9x - 2 = 0$

Problem 4

$5(x + 4)^2 + 7 = 17$

$5(x + 4)^2 = 10$ • Solve for $(x + 4)^2$.

$(x + 4)^2 = 2$

$\sqrt{(x + 4)^2} = \sqrt{2}$ • Take the square root of each side.

$|x + 4| = \sqrt{2}$ • Simplify.

$x + 4 = \pm\sqrt{2}$

$x + 4 = -\sqrt{2} \qquad x + 4 = \sqrt{2}$ • Solve for *x*.

$x = -4 - \sqrt{2} \qquad x = -4 + \sqrt{2}$

The solutions are $-4 - \sqrt{2}$ and $-4 + \sqrt{2}$.

SECTION 8.2

Problem 1

A. $4x^2 - 4x - 1 = 0$

$4x^2 - 4x = 1$ • Add 1 to each side.

$\frac{4x^2 - 4x}{4} = \frac{1}{4}$ • Divide each side by 4.

$x^2 - x = \frac{1}{4}$ • The coefficient of x^2 is 1.

Complete the square.

$x^2 - x + \frac{1}{4} = \frac{1}{4} + \frac{1}{4}$ • $\left[\frac{1}{2}(-1)\right]^2 = \frac{1}{4}$

$\left(x - \frac{1}{2}\right)^2 = \frac{2}{4}$ • Factor the perfect-square trinomial.

$\sqrt{\left(x - \frac{1}{2}\right)^2} = \sqrt{\frac{2}{4}}$ • Take the square root of each side.

$\left|x - \frac{1}{2}\right| = \frac{\sqrt{2}}{2}$ • Simplify.

$x - \frac{1}{2} = \pm\frac{\sqrt{2}}{2}$

$x - \frac{1}{2} = \frac{\sqrt{2}}{2} \qquad x - \frac{1}{2} = -\frac{\sqrt{2}}{2}$ • Solve for *x*.

$x = \frac{1}{2} + \frac{\sqrt{2}}{2} \qquad x = \frac{1}{2} - \frac{\sqrt{2}}{2}$

The solutions are $\frac{1 + \sqrt{2}}{2}$ and $\frac{1 - \sqrt{2}}{2}$.

B. $2x^2 + x - 5 = 0$

$2x^2 + x = 5$ • Add 5 to each side.

$\frac{2x^2 + x}{2} = \frac{5}{2}$ • Divide each side by 2.

$x^2 + \frac{1}{2}x = \frac{5}{2}$ • The coefficient of x^2 is 1.

Complete the square.

$x^2 + \frac{1}{2}x + \frac{1}{16} = \frac{5}{2} + \frac{1}{16}$ • $\left[\frac{1}{2}\left(\frac{1}{2}\right)\right]^2 = \frac{1}{16}$

$\left(x + \frac{1}{4}\right)^2 = \frac{41}{16}$ • Factor the perfect-square trinomial.

$\sqrt{\left(x + \frac{1}{4}\right)^2} = \sqrt{\frac{41}{16}}$ • Take the square root of each side.

$\left|x + \frac{1}{4}\right| = \frac{\sqrt{41}}{4}$ • Simplify.

$x + \frac{1}{4} = \pm\frac{\sqrt{41}}{4}$

$x + \frac{1}{4} = \frac{\sqrt{41}}{4} \qquad x + \frac{1}{4} = -\frac{\sqrt{41}}{4}$ • Solve for *x*.

$x = -\frac{1}{4} + \frac{\sqrt{41}}{4} \qquad x = -\frac{1}{4} - \frac{\sqrt{41}}{4}$

The solutions are $\frac{-1 + \sqrt{41}}{4}$ and $\frac{-1 - \sqrt{41}}{4}$.

Problem 2

A. $x^2 + 6x - 9 = 0$ • The equation is in standard form.

$x = \frac{-b \pm \sqrt{b^2 - 4ac}}{2a}$

$= \frac{-6 \pm \sqrt{6^2 - 4(1)(-9)}}{2 \cdot 1}$ • $a = 1, b = 6, c = -9$

$= \frac{-6 \pm \sqrt{36 + 36}}{2}$

$= \frac{-6 \pm \sqrt{72}}{2}$

$= \frac{-6 \pm 6\sqrt{2}}{2} = -3 \pm 3\sqrt{2}$

The solutions are $-3 + 3\sqrt{2}$ and $-3 - 3\sqrt{2}$.

B.

$$4x^2 = 4x - 1$$
$$4x^2 - 4x + 1 = 0$$ • Write in standard form.

$$x = \frac{-b \pm \sqrt{b^2 - 4ac}}{2a}$$
$$= \frac{-(-4) \pm \sqrt{(-4)^2 - 4(4)(1)}}{2 \cdot 4}$$ • $a = 4, b = -4, c = 1$
$$= \frac{4 \pm \sqrt{16 - 16}}{8}$$
$$= \frac{4 \pm \sqrt{0}}{8} = \frac{4}{8} = \frac{1}{2}$$

The solution is $\frac{1}{2}$.

Problem 3

$$3x^2 - x - 1 = 0$$
$$b^2 - 4ac$$
$$= (-1)^2 - 4(3)(-1)$$ • $a = 3, b = -1, c = -1$
$$= 1 + 12 = 13$$
$$13 > 0$$ • The discriminant is greater than zero.

The equation has two real number solutions.

SECTION 8.3

Problem 1

A.

$$x - 5x^{\frac{1}{2}} + 6 = 0$$
$$(x^{\frac{1}{2}})^2 - 5(x^{\frac{1}{2}}) + 6 = 0$$
$$u^2 - 5u + 6 = 0$$ • Let $x^{\frac{1}{2}} = u$.
$$(u - 2)(u - 3) = 0$$ • Solve for u by factoring.

$u - 2 = 0$ $\quad$ $u - 3 = 0$
$u = 2$ $\quad$ $u = 3$
$x^{\frac{1}{2}} = 2$ $\quad$ $x^{\frac{1}{2}} = 3$ • Replace u with $x^{\frac{1}{2}}$.
$(x^{\frac{1}{2}})^2 = 2^2$ $\quad$ $(x^{\frac{1}{2}})^2 = 3^2$ • Solve for x.
$x = 4$ $\quad$ $x = 9$

4 and 9 check as solutions.
The solutions are 4 and 9.

B.

$$4x^4 + 35x^2 - 9 = 0$$
$$4(x^2)^2 + 35(x^2) - 9 = 0$$
$$4u^2 + 35u - 9 = 0$$ • Let $x^2 = u$.
$$(4u - 1)(u + 9) = 0$$ • Solve for u by factoring.

$4u - 1 = 0$ $\quad$ $u + 9 = 0$
$4u = 1$ $\quad$ $u = -9$
$u = \frac{1}{4}$

$x^2 = \frac{1}{4}$ $\quad$ $x^2 = -9$ • Replace u with x^2.
$\sqrt{x^2} = \sqrt{\frac{1}{4}}$ $\quad$ $\sqrt{x^2} = \sqrt{-9}$ • Solve for x.
$|x| = \frac{1}{2}$ $\quad$ $|x| = 3i$
$x = \pm\frac{1}{2}$ $\quad$ $x = \pm 3i$

The solutions are $\frac{1}{2}$, $-\frac{1}{2}$, $3i$, and $-3i$.

Problem 2

$$\sqrt{2x + 1} + x = 7$$
$$\sqrt{2x + 1} = 7 - x$$ • Solve for the radical.
$$(\sqrt{2x + 1})^2 = (7 - x)^2$$ • Square each side.
$$2x + 1 = 49 - 14x + x^2$$
$$0 = x^2 - 16x + 48$$ • Write in standard form.
$$0 = (x - 4)(x - 12)$$ • Solve for x by factoring.

$x - 4 = 0$ $\quad$ $x - 12 = 0$
$x = 4$ $\quad$ $x = 12$

4 checks as a solution, but 12 does not check as a solution. The solution is 4.

Problem 3

$$\sqrt{2x - 1} + \sqrt{x} = 2$$
$$\sqrt{2x - 1} = 2 - \sqrt{x}$$ • Solve for one radical.
$$(\sqrt{2x - 1})^2 = (2 - \sqrt{x})^2$$ • Square each side.
$$2x - 1 = 4 - 4\sqrt{x} + x$$ • Solve for the other radical.
$$x - 5 = -4\sqrt{x}$$
$$(x - 5)^2 = (-4\sqrt{x})^2$$ • Square each side.
$$x^2 - 10x + 25 = 16x$$
$$x^2 - 26x + 25 = 0$$ • Write in standard form.
$$(x - 1)(x - 25) = 0$$ • Solve for x by factoring.

$x - 1 = 0$ $\quad$ $x - 25 = 0$
$x = 1$ $\quad$ $x = 25$

1 checks as a solution, but 25 does not check as a solution. The solution is 1.

Problem 4

$$3y + \frac{25}{3y - 2} = -8$$
$$(3y - 2)\left(3y + \frac{25}{3y - 2}\right) = (3y - 2)(-8)$$
$$(3y - 2)(3y) + (3y - 2)\left(\frac{25}{3y - 2}\right) = (3y - 2)(-8)$$
$$9y^2 - 6y + 25 = -24y + 16$$
$$9y^2 + 18y + 9 = 0$$
$$9(y^2 + 2y + 1) = 0$$
$$9(y + 1)(y + 1) = 0$$

$y + 1 = 0$ $\quad$ $y + 1 = 0$
$y = -1$ $\quad$ $y = -1$

-1 checks as a solution.
The solution is -1.

SECTION 8.4

Problem 1

Strategy To find the hang time, replace h by 0 in the equation $h = -16t^2 + 60.6t + 4$ and solve for t.

Solution

$h = -16t^2 + 60.6t + 4$

$0 = -16t^2 + 60.6t + 4$ • **Replace *h* by 0.**

$t = \dfrac{-60.6 \pm \sqrt{(60.6)^2 - 4(-16)(4)}}{2(-16)}$ • **Use the quadratic formula.**

$t = \dfrac{-60.6 \pm \sqrt{3928.36}}{2(-16)} \approx \dfrac{-60.6 \pm 62.68}{-32}$

$t \approx \dfrac{-60.6 + 62.68}{-32}$ $\quad t \approx \dfrac{-60.6 - 62.68}{-32}$

$t = -0.065$ $\quad t = 3.8525$

The time cannot be negative. The hang time is approximately 3.9 s.

Problem 2

Strategy
- This is a work problem.
- The unknown time for Tessa: t
- The unknown time for Seth: $t + 3$
- They work together for 2 h.

	Rate	·	Time	=	Part
Tessa	$\frac{1}{t}$	·	2	=	$\frac{2}{t}$
Seth	$\frac{1}{t+3}$	·	2	=	$\frac{2}{t+3}$

- The sum of the part of the task completed by Tessa and the part completed by Seth is 1.

Solution

$\dfrac{2}{t} + \dfrac{2}{t+3} = 1$

$t(t+3)\left(\dfrac{2}{t} + \dfrac{2}{t+3}\right) = t(t+3) \cdot 1$ • **Multiply each side by the LCD.**

$2(t+3) + 2t = t^2 + 3t$ • **Simplify.**

$4t + 6 = t^2 + 3t$

$0 = t^2 - t - 6$ • **Standard form**

$0 = (t+2)(t-3)$ • **Factor.**

$t + 2 = 0 \quad t - 3 = 0$ • **Solve for *t*.**

$t = -2 \quad t = 3$

The time cannot be negative. It would take Tessa 3 h to wash the windows.

Problem 3

Strategy
- This is a uniform motion problem.
- The unknown rate of the wind: r

	Distance	÷	Rate	=	Time
With wind	1200	÷	$250 + r$	=	$\frac{1200}{250+r}$
Against wind	1200	÷	$250 - r$	=	$\frac{1200}{250-r}$

- The time flying with the wind is 2 h less than the time flying against the wind.

Solution

$\dfrac{1200}{250+r} = \dfrac{1200}{250-r} - 2$

$(250+r)(250-r)\left(\dfrac{1200}{250+r}\right) = (250+r)(250-r)\left(\dfrac{1200}{250-r} - 2\right)$

$1200(250-r) = 1200(250+r) - 2(62{,}500 - r^2)$

$-2400r = -125{,}000 + 2r^2$

$0 = 2r^2 + 2400r - 125{,}000$

$0 = 2(r + 1250)(r - 50)$

$0 = (r + 1250)(r - 50)$

$r + 1250 = 0 \quad r - 50 = 0$

$r = -1250 \quad r = 50$

The rate cannot be negative. The rate of the wind is 50 mph.

SECTION 8.5

Problem 1

x-coordinate: $-\dfrac{b}{2a} = -\dfrac{0}{2(1)} = 0$ • **$a = 1, b = 0$**

$y = x^2 - 2$ • **Find *y* when *x* = 0.**

$= 0^2 - 2$

$= -2$

The coordinates of the vertex are $(0, -2)$.

The equation of the axis of symmetry is $x = 0$.

x	$y = x^2 - 2$
1	-1
2	2

• **Find ordered-pair solutions using values of *x* greater than 0.**

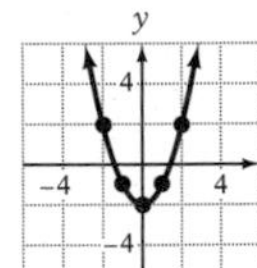

• **Use symmetry to find two more points on the other side of the axis of symmetry.**

Problem 2

Because $a > 0$, the graph of g will open up.

The coordinates of the vertex are

$x = -\dfrac{b}{2a} = \dfrac{-4}{2(1)} = -2$

$y = g(-2) = (-2)^2 + 4(-2) - 2 = -6$

Find several ordered-pair solutions and then use symmetry to draw the graph.

The domain is $\{x | x \in \text{real numbers}\}$.

The range is $\{y | y \geq -6\}$.

Problem 3

$f(x) = x^2 + 2x - 8$

$0 = x^2 + 2x - 8$ • **Replace *f*(*x*) by 0.**

$0 = (x + 4)(x - 2)$ • **Factor.**

$x + 4 = 0 \quad x - 2 = 0$ • **Solve for *x*.**

$x = -4 \quad x = 2$

The coordinates of the x-intercepts are $(-4, 0)$ and $(2, 0)$.

Problem 4

$$f(x) = 2x^2 + 3x - 1$$
$$0 = 2x^2 + 3x - 1 \quad \text{• Replace } f(x) \text{ by 0.}$$
$$x = \frac{-3 \pm \sqrt{3^2 - 4(2)(-1)}}{2(2)} \quad \text{• Use the quadratic formula to solve for } x.$$
$$= \frac{-3 \pm \sqrt{9+8}}{4} = \frac{-3 \pm \sqrt{17}}{4}$$

The zeros are $\frac{-3 - \sqrt{17}}{4}$ and $\frac{-3 + \sqrt{17}}{4}$.

Problem 5

$$f(x) = 4x^2 - 8x + 5$$
$$0 = 4x^2 - 8x + 5 \quad \text{• Replace } f(x) \text{ by 0.}$$
$$x = \frac{-(-8) \pm \sqrt{(-8)^2 - 4(4)(5)}}{2(4)} \quad \text{• Use the quadratic formula to solve for } x.$$
$$= \frac{8 \pm \sqrt{64 - 80}}{8}$$
$$= \frac{8 \pm \sqrt{-16}}{8} = \frac{8 \pm 4i}{8} = 1 \pm \frac{1}{2}i$$

The zeros are $1 - \frac{1}{2}i$ and $1 + \frac{1}{2}i$.

Problem 6

$$y = x^2 - x - 6$$
$$a = 1, b = -1, c = -6$$
$$b^2 - 4ac = (-1)^2 - 4(1)(-6) \quad \text{• Evaluate the discriminant.}$$
$$= 1 + 24 = 25$$

The discriminant is positive. The parabola has two x-intercepts.

SECTION 8.6

Problem 1

$$x = -\frac{b}{2a} = -\frac{-3}{2(2)} = \frac{3}{4} \quad \text{• Find the } x\text{-coordinate of the vertex.}$$
$$f(x) = 2x^2 - 3x + 1$$
$$f\left(\frac{3}{4}\right) = 2\left(\frac{3}{4}\right)^2 - 3\left(\frac{3}{4}\right) + 1 \quad \text{• Evaluate } f \text{ at } x = \frac{3}{4}.$$
$$= \frac{9}{8} - \frac{9}{4} + 1$$
$$= -\frac{1}{8}$$

Because a is positive, the function has a minimum value. The minimum value of the function is $-\frac{1}{8}$.

Problem 2

Strategy
- ▶ To find the time it takes the ball to reach its maximum height, find the t-coordinate of the vertex.
- ▶ To find the maximum height, evaluate the function at the t-coordinate of the vertex.

Solution $t = -\frac{b}{2a} = -\frac{64}{2(-16)} = 2$

The ball reaches its maximum height in 2 s.

$$s(t) = -16t^2 + 64t$$
$$s(2) = -16(2)^2 + 64(2) = -64 + 128 = 64$$

The maximum height is 64 ft.

Problem 3

Strategy
- ▶ Width of rectangle: x
  Length of rectangle: y
  Amount of fencing, F: 100 ft
- ▶ Express the length of the rectangle in terms of x.

$$F = 2x + y$$
$$100 = 2x + y \quad \text{• } F = 100$$
$$100 - 2x = y \quad \text{• Solve for } y.$$

  Express the area of the rectangle in terms of x.

$$A = xy$$
$$A = x(100 - 2x) \quad \text{• } y = 100 - 2x$$
$$A = -2x^2 + 100x$$

- ▶ To find the width, find the x-coordinate of the vertex of $f(x) = -2x^2 + 100x$.
- ▶ To find the length, replace x in $y = 100 - 2x$ by the x-coordinate of the vertex.

Solution $x = -\frac{b}{2a} = -\frac{100}{2(-2)} = 25$

The width is 25 ft.

$$100 - 2x = 100 - 2(25)$$
$$= 100 - 50 = 50$$

The length is 50 ft.

SECTION 8.7

Problem 1

$$2x^2 - x - 10 \le 0$$
$$(2x - 5)(x + 2) \le 0$$

2x − 5	− − −	− − − − − − − −	+ + +
x + 2	− − −	+ + + + + + + +	+ + +

−3 −2 −1 0 1 2 3

The solution set is $\left\{x \middle| -2 \le x \le \frac{5}{2}\right\}$.

−5 −4 −3 −2 −1 0 1 2 3 4 5

Problem 2

$$\frac{x}{x-2} \le 0$$

x	− − − −	+ + +	+ + + + + +
x − 2	− − − −	− − −	+ + + + + +

−1 0 1 2 3 4

The solution set is $[0, 2)$.

−5 −4 −3 −2 −1 0 1 2 3 4 5

Solutions to Chapter 9 Problems

SECTION 9.1

Problem 1

The graph of $g(x) = f(x) - 3$ is the graph of $y = f(x)$ shifted 3 units down.

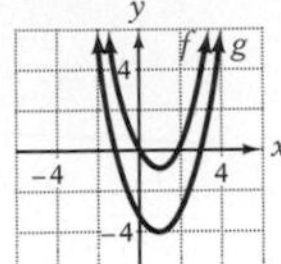

Problem 2

The graph of $g(x) = f(x - 3)$ is the graph of $y = f(x)$ shifted 3 units to the right.

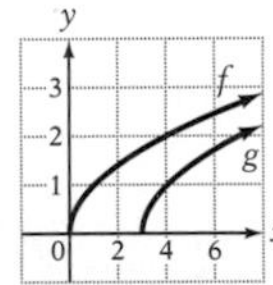

Problem 3

The graph of $V(x) = g(x - 2) + 1$ is the graph of $y = g(x)$ shifted 1 unit up and 2 units to the right.

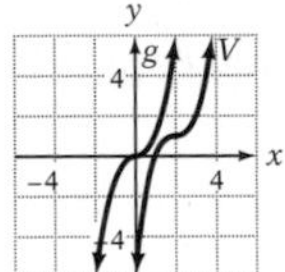

SECTION 9.2

Problem 1

Given $f(x) = x^2 + 2x$ and $g(x) = 5x - 2$,

$$\begin{aligned}(f+g)(-2) &= f(-2) + g(-2)\\ &= [(-2)^2 + 2(-2)] + [5(-2) - 2]\\ &= (4 - 4) + (-10 - 2)\\ (f+g)(-2) &= -12\end{aligned}$$

Problem 2

Given $f(x) = x^2 + 3$ and $g(x) = 3x - 5$,

$$\begin{aligned}(f\cdot g)(x) &= f(x)\cdot g(x)\\ &= (x^2+3)(3x-5)\\ (f\cdot g)(x) &= 3x^3 - 5x^2 + 9x - 15\end{aligned}$$

Problem 3

Given $f(x) = x^2 - 4$ and $g(x) = x^2 + 2x + 1$,

$$\begin{aligned}\left(\frac{f}{g}\right)(4) &= \frac{f(4)}{g(4)}\\ &= \frac{4^2 - 4}{4^2 + 2\cdot 4 + 1}\\ &= \frac{16 - 4}{16 + 8 + 1}\\ \left(\frac{f}{g}\right)(4) &= \frac{12}{25}\end{aligned}$$

Problem 4

A.

$$\begin{aligned}h(x) &= x^2 + 1\\ h(0) &= 0^2 + 1 = 1 && \text{• Find } h(0).\\ g(x) &= 3x - 2\\ g[h(0)] &= 3[h(0)] - 2 && \text{• Find } g[h(0)].\\ &= 3(1) - 2 && \text{• Replace } h(0) \text{ by 1.}\\ g[h(0)] &= 1\end{aligned}$$

B.

$$\begin{aligned}h(x) &= x^2 + 1\\ h[g(x)] &= [g(x)]^2 + 1 && \text{• Replace } x \text{ by } g(x).\\ &= (3x - 2)^2 + 1 && \text{• Replace } g(x) \text{ by } 3x - 2.\\ &= 9x^2 - 12x + 4 + 1\\ h[g(x)] &= 9x^2 - 12x + 5\end{aligned}$$

SECTION 9.3

Problem 1

A. Because any vertical line will intersect the graph at no more than one point, and any horizontal line will intersect the graph at no more than one point, the graph is the graph of a 1–1 function.

B. Because any vertical line will intersect the graph at no more than one point, and any horizontal line will intersect the graph at no more than one point, the graph is the graph of a 1–1 function.

Problem 2

$$\begin{aligned}f(x) &= 4x + 2\\ y &= 4x + 2 && \text{• Replace } f(x) \text{ by } y.\\ x &= 4y + 2 && \text{• Interchange } x \text{ and } y.\\ 4y &= x - 2 && \text{• Solve for } y.\\ y &= \frac{1}{4}x - \frac{1}{2}\\ f^{-1}(x) &= \frac{1}{4}x - \frac{1}{2}\end{aligned}$$

Problem 3

Check that the functions h and g satisfy the Composition of Inverse Functions Property.

$$\begin{aligned}h[g(x)] &= h\left(\frac{1}{4}x - \frac{1}{2}\right)\\ &= 4\left(\frac{1}{4}x - \frac{1}{2}\right) + 2\\ &= x - 2 + 2 = x && \text{• } h[g(x)] = x\end{aligned}$$

$$\begin{aligned}g[h(x)] &= g(4x + 2)\\ &= \frac{1}{4}(4x + 2) - \frac{1}{2}\\ &= x + \frac{1}{2} - \frac{1}{2} = x && \text{• } g[h(x)] = x\end{aligned}$$

The functions are inverses of each other.

Solutions to Chapter 10 Problems

SECTION 10.1

Problem 1

$$f(x) = \left(\frac{2}{3}\right)^x$$
$$f(3) = \left(\frac{2}{3}\right)^3 = \frac{8}{27}$$
$$f(-2) = \left(\frac{2}{3}\right)^{-2} = \left(\frac{3}{2}\right)^2 = \frac{9}{4}$$

Problem 2

$$f(x) = 2^{2x+1}$$
$$f(0) = 2^{2(0)+1} = 2^1 = 2$$
$$f(-2) = 2^{2(-2)+1} = 2^{-3} = \frac{1}{2^3} = \frac{1}{8}$$

Problem 3

$$f(x) = e^x$$
$$f(1.2) = e^{1.2} \approx 3.3201$$
$$f(-2.5) = e^{-2.5} \approx 0.0821$$
$$f(e) = e^e \approx 15.1543$$

Problem 4

A.

x	$y = 2^{-\frac{1}{2}x}$
−4	4
−2	2
0	1
2	$\frac{1}{2}$

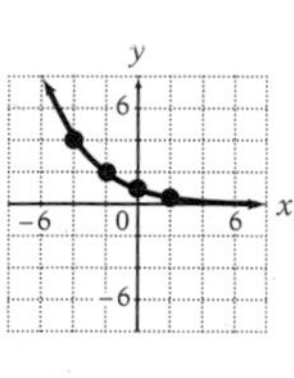

B.

x	$y = 2^x + 1$
−4	$1\frac{1}{16}$
−2	$1\frac{1}{4}$
0	2
2	5

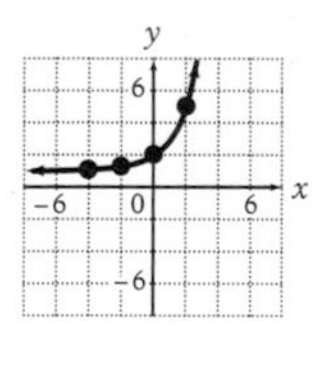

Problem 5

A.

x	$y = 2^{-x} + 2$
−2	6
−1	4
0	3
1	$2\frac{1}{2}$
2	$2\frac{1}{4}$

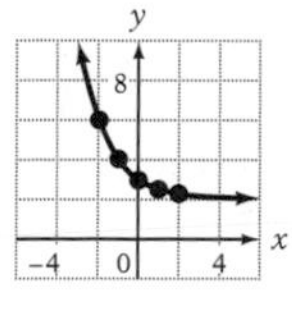

B.

x	$y = e^{-2x} - 4$
−2	50.60
−1	3.39
0	−3
2	−3.98
4	−3.99

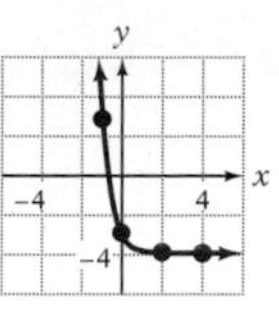

Problem 6

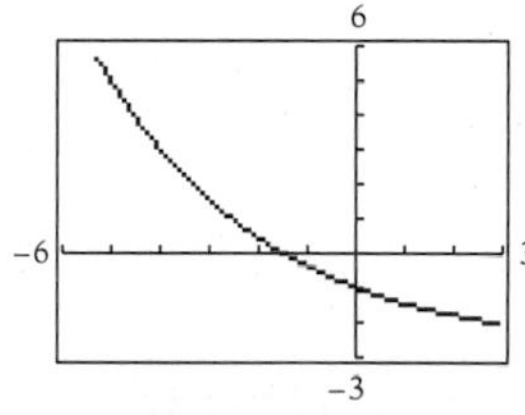

To the nearest tenth, the value of x for which $f(x) = 1$ is -2.4.

SECTION 10.2

Problem 1

$3^{-4} = \frac{1}{81}$ is equivalent to $\log_3 \frac{1}{81} = -4$.

Problem 2

$\log_{10} 0.0001 = -4$ is equivalent to $10^{-4} = 0.0001$.

Problem 3

$\log_4 64 = x$ • **Write an equation.**
$64 = 4^x$ • **Write in exponential form.**
$4^3 = 4^x$ • **Write 64 using 4 as a base.**
$3 = x$ • **Equality of Exponents Property**
$\log_4 64 = 3$

Problem 4

$\log_2 x = -4$
$2^{-4} = x$ • **Write in exponential form.**
$\frac{1}{2^4} = x$
$\frac{1}{16} = x$

The solution is $\frac{1}{16}$.

Problem 5

$\log x = 1.5$
$10^{1.5} = x$ • **Write in exponential form.**
$31.6228 \approx x$ • **Evaluate.**

The solution is approximately 31.6228.

Problem 6

A. $\log_b \frac{x^2}{y} = \log_b x^2 - \log_b y$ • **Quotient Property**
$= 2 \log_b x - \log_b y$ • **Power Property**

B. $\ln y^{\frac{1}{3}}z^3 = \ln y^{\frac{1}{3}} + \ln z^3$ • **Product Property**
$= \frac{1}{3} \ln y + 3 \ln z$ • **Power Property**

C. $\log_8 \sqrt[3]{xy^2} = \log_8(xy^2)^{\frac{1}{3}}$ • Write the radical as an exponential expression.

$= \frac{1}{3}\log_8 (xy^2)$ • Power Property

$= \frac{1}{3}(\log_8 x + \log_8 y^2)$ • Product Property

$= \frac{1}{3}(\log_8 x + 2\log_8 y)$ • Power Property

$= \frac{1}{3}\log_8 x + \frac{2}{3}\log_8 y$ • Distributive Property

Problem 7

A. $2\log_b x - 3\log_b y - \log_b z$

$= \log_b x^2 - \log_b y^3 - \log_b z$ • Power Property

$= \log_b \frac{x^2}{y^3} - \log_b z$ • Quotient Property

$= \log_b \frac{x^2}{y^3 z}$ • Quotient Property

B. $3(\log_5 x - 2\log_5 y + 4\log_5 z)$

$= 3(\log_5 x - \log_5 y^2 + \log_5 z^4)$ • Power Property

$= 3\left(\log_5 \frac{x}{y^2} + \log_5 z^4\right)$ • Quotient Property

$= 3\log_5 \frac{xz^4}{y^2}$ • Product Property

$= \log_5 \left(\frac{xz^4}{y^2}\right)^3$ • Power Property

$= \log_5 \frac{x^3z^{12}}{y^6}$ • Simplify.

C. $\frac{1}{2}(2\ln x - 5\ln y) = \frac{1}{2}(\ln x^2 - \ln y^5)$ • Power Property

$= \frac{1}{2}\left(\ln \frac{x^2}{y^5}\right)$ • Quotient Property

$= \ln \left(\frac{x^2}{y^5}\right)^{\frac{1}{2}}$ • Power Property

$= \ln \sqrt{\frac{x^2}{y^5}}$ • Write the exponential expression as a radical.

Problem 8

$\log_4 2.4 = \frac{\ln 2.4}{\ln 4} \approx 0.6315$

Problem 9

$f(x) = 4\log_8(3x + 4) = 4 \cdot \frac{\log(3x + 4)}{\log 8}$ • Change-of-Base Formula

$= \frac{4}{\log 8}\log(3x + 4)$

SECTION 10.3

Problem 1

A. $f(x) = \log_2(x - 1)$

$y = \log_2(x - 1)$ • Substitute y for $f(x)$.

$2^y = x - 1$ • Write the equivalent exponential equation.

$2^y + 1 = x$ • Solve for x.

$x = 2^y + 1$	y
$1\frac{1}{4}$	-2
$1\frac{1}{2}$	-1
2	0
3	1
5	2

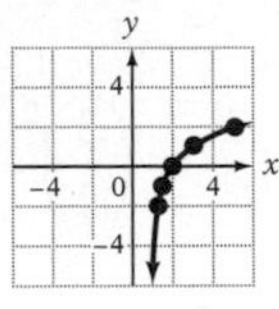

B. $f(x) = \log_3 2x$

$y = \log_3 2x$ • Substitute y for $f(x)$.

$3^y = 2x$ • Write the equivalent exponential equation.

$\frac{3^y}{2} = x$ • Solve for x.

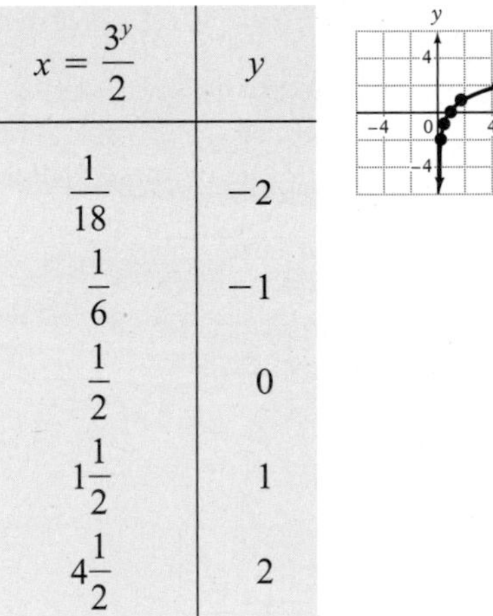

$x = \frac{3^y}{2}$	y
$\frac{1}{18}$	-2
$\frac{1}{6}$	-1
$\frac{1}{2}$	0
$1\frac{1}{2}$	1
$4\frac{1}{2}$	2

Problem 2

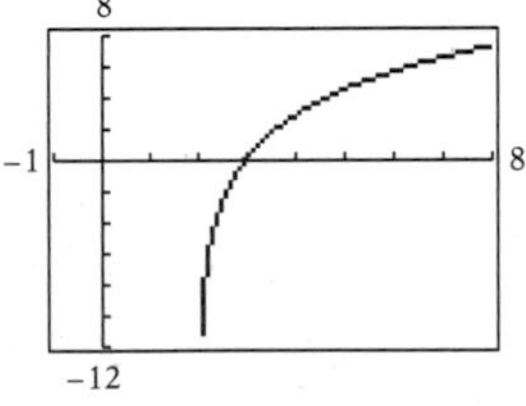

Problem 3

$f(x) = 2\log_4 x$

$= 2 \cdot \frac{\ln x}{\ln 4}$

$= \frac{2}{\ln 4}\ln x$

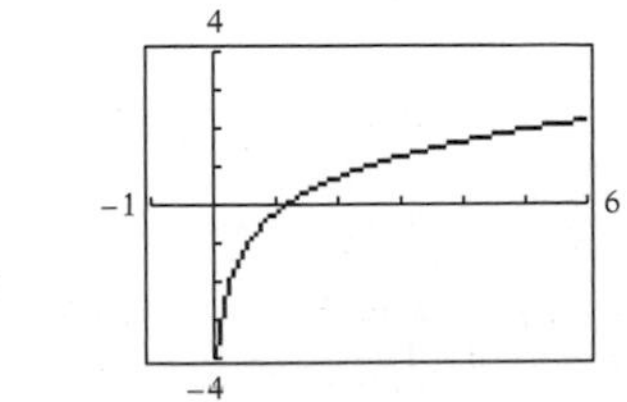

SECTION 10.4

Problem 1

$10^{3x+5} = 10^{x-3}$

$3x + 5 = x - 3$ • **Equality of Exponents Property**

$2x + 5 = -3$ • **Solve for *x*.**

$2x = -8$

$x = -4$

Check: $10^{3x+5} = 10^{x-3}$

$10^{3(-4)+5} \mid 10^{-4-3}$

$10^{-12+5} \mid 10^{-7}$

$10^{-7} = 10^{-7}$

The solution is −4.

Problem 2

A. $4^{3x} = 25$

$\log 4^{3x} = \log 25$ • **Take the common logarithm of each side.**

$3x \log 4 = \log 25$ • **Power Property of Logarithms**

$3x = \dfrac{\log 25}{\log 4}$ • **Solve for *x*.**

$x = \dfrac{\log 25}{3 \log 4}$

$x \approx 0.7740$

The solution is 0.7740.

B. $(1.06)^x = 1.5$

$\log(1.06)^x = \log 1.5$ • **Take the common logarithm of each side.**

$x \log 1.06 = \log 1.5$ • **Power Property of Logarithms**

$x = \dfrac{\log 1.5}{\log 1.06}$ • **Solve for *x*.**

$x \approx 6.9585$

The solution is 6.9585.

Problem 3

$e^x = x$

$e^x - x = 0$

Graph $f(x) = e^x - x$ and find any zeros.

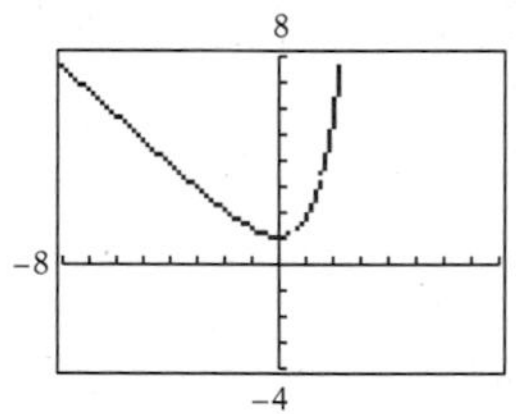

The equation has no real number solutions.

Problem 4

$\log_4(x^2 - 3x) = 1$

$4^1 = x^2 - 3x$ • **Write in exponential form.**

$4 = x^2 - 3x$

$0 = x^2 - 3x - 4$ • **Write in standard form.**

$0 = (x + 1)(x - 4)$ • **Factor.**

$x + 1 = 0 \qquad x - 4 = 0$ • **Principle of Zero Products**

$x = -1 \qquad x = 4$

The solutions are −1 and 4.

Problem 5

$\log_3 x + \log_3(x + 3) = \log_3 4$

$\log_3[x(x + 3)] = \log_3 4$ • **Product Property of Logarithms**

$x(x + 3) = 4$ • **1–1 Property of Logarithms**

$x^2 + 3x = 4$

$x^2 + 3x - 4 = 0$ • **Write in standard form.**

$(x + 4)(x - 1) = 0$ • **Factor.**

$x + 4 = 0 \qquad x - 1 = 0$ • **Principle of Zero Products**

$x = -4 \qquad x = 1$

−4 does not check as a solution. The solution is 1.

Problem 6

$\log(3x - 2) = -2x$

$\log(3x - 2) + 2x = 0$

Graph $f(x) = \log(3x - 2) + 2x$ and find any zeros.

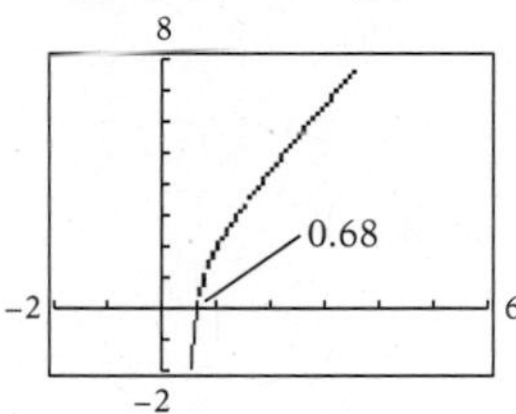

The solution is 0.68.

SECTION 10.5

Problem 1

Strategy To find the number of days, replace *N* by 60 and solve for *t*.

Solution

$N = 100[1 - (0.9)^t]$

$60 = 100[1 - (0.9)^t]$

$0.6 = 1 - (0.9)^t$

$-0.4 = -(0.9)^t$

$0.4 = (0.9)^t$

$\log 0.4 = \log(0.9)^t$

$\log 0.4 = t \log 0.9$

$t = \dfrac{\log 0.4}{\log 0.9} \approx 8.6967184$

After approximately 9 days, the student will type 60 words per minute.

Problem 2

Strategy To find the intensity, use the equation for the Richter scale magnitude of an earthquake, $M = \log \dfrac{I}{I_0}$. Replace *M* by 5.2 and solve for *I*.

Solution

$M = \log \dfrac{I}{I_0}$

$5.2 = \log \dfrac{I}{I_0}$ • **Replace *M* by 5.2.**

$10^{5.2} = \dfrac{I}{I_0}$ • **Write in exponential form.**

$10^{5.2} I_0 = I$ • **Solve for *I*.**

$158{,}489 I_0 \approx I$

The earthquake had an intensity that was approximately 158,489 times the intensity of a zero-level earthquake.

Solutions to Chapter 11 Problems

SECTION 11.1

Problem 1

$a_n = n(n + 1)$
$a_1 = 1(1 + 1) = 2$ The first term is 2.
$a_2 = 2(2 + 1) = 6$ The second term is 6.
$a_3 = 3(3 + 1) = 12$ The third term is 12.
$a_4 = 4(4 + 1) = 20$ The fourth term is 20.

Problem 2

$a_n = \dfrac{1}{n(n+2)}$

$a_6 = \dfrac{1}{6(6+2)} = \dfrac{1}{48}$ The sixth term is $\frac{1}{48}$.

$a_9 = \dfrac{1}{9(9+2)} = \dfrac{1}{99}$ The ninth term is $\frac{1}{99}$.

Problem 3

A. $\sum_{n=1}^{4}(7 - n)$ • **Replace *n* by 1, 2, 3, and 4.**
$= (7 - 1) + (7 - 2) + (7 - 3) + (7 - 4)$
$= 6 + 5 + 4 + 3 = 18$

B. $\sum_{i=3}^{6}(i^2 - 2)$ • **Replace *i* by 3, 4, 5, and 6.**
$= (3^2 - 2) + (4^2 - 2) + (5^2 - 2) + (6^2 - 2)$
$= 7 + 14 + 23 + 34 = 78$

Problem 4

$\sum_{n=1}^{5} nx = x + 2x + 3x + 4x + 5x$

SECTION 11.2

Problem 1

$9, 3, -3, -9, \ldots$
$d = a_2 - a_1 = 3 - 9 = -6$ • **Find the common difference.**

$a_n = a_1 + (n - 1)d$
$a_{15} = 9 + (15 - 1)(-6)$ • **$n = 15, a_1 = 9, d = -6$**
$= 9 + (14)(-6)$
$= 9 - 84 = -75$

Problem 2

$-3, 1, 5, 9, \ldots$
$d = a_2 - a_1 = 1 - (-3) = 4$ • **Find the common difference.**

$a_n = a_1 + (n - 1)d$
$a_n = -3 + (n - 1)4$ • **$a_1 = -3, d = 4$**
$a_n = -3 + 4n - 4$
$a_n = 4n - 7$

Problem 3

$7, 9, 11, \ldots, 59$
$d = a_2 - a_1 = 9 - 7 = 2$ • **Find the common difference.**

$a_n = a_1 + (n - 1)d$
$59 = 7 + (n - 1)2$ • **$a_n = 59, a_1 = 7, d = 2$**
$59 = 7 + 2n - 2$ • **Solve for *n*.**
$59 = 5 + 2n$
$54 = 2n$
$27 = n$

There are 27 terms in the sequence.

Problem 4

$-4, -2, 0, 2, 4, \ldots$
$d = a_2 - a_1 = -2 - (-4) = 2$ • **Find the common difference.**

$a_n = a_1 + (n - 1)d$ • **Find the 25th term.**
$a_{25} = -4 + (25 - 1)2$ • **$n = 25, a_1 = -4, d = 2$**
$= -4 + (24)2$
$= -4 + 48 = 44$

$S_n = \dfrac{n}{2}(a_1 + a_n)$ • **Formula for the Sum of *n* Terms of an Arithmetic Series**

$S_{25} = \dfrac{25}{2}(-4 + 44)$ • **$n = 25, a_1 = -4, a_n = a_{25} = 44$**

$= \dfrac{25}{2}(40) = 25(20) = 500$

Problem 5

$\sum_{n=1}^{18}(3n - 2)$

$a_n = 3n - 2$
$a_1 = 3(1) - 2 = 1$ • **Find the first term.**
$a_{18} = 3(18) - 2 = 52$ • **Find the 18th term.**

$S_n = \dfrac{n}{2}(a_1 + a_n)$ • **Formula for the Sum of *n* Terms of an Arithmetic Series**

$S_{18} = \dfrac{18}{2}(1 + 52)$ • **$n = 18, a_1 = 1, a_n = a_{18} = 52$**
$= 9(53) = 477$

Problem 6

Strategy To find the value of the 20th-place prize:

► Write the equation for the *n*th-place prize.

► Find the 20th term of the sequence.

To find the total amount of prize money being awarded, use the Formula for the Sum of *n* Terms of an Arithmetic Series.

Solution $10{,}000, 9700, \ldots$

$d = a_2 - a_1 = 9700 - 10{,}000 = -300$

$a_n = a_1 + (n - 1)d$
$= 10{,}000 + (n - 1)(-300)$
$= 10{,}000 - 300n + 300$
$= -300n + 10{,}300$
$a_{20} = -300(20) + 10{,}300$
$= -6000 + 10{,}300 = 4300$

$S_n = \dfrac{n}{2}(a_1 + a_n)$

$S_{20} = \dfrac{20}{2}(10{,}000 + 4300)$
$= 10(14{,}300) = 143{,}000$

The value of the 20th-place prize is \$4300.

The total amount of prize money being awarded is \$143,000.

SECTION 11.3

Problem 1

$5, 2, \frac{4}{5}, \ldots$

$r = \frac{a_2}{a_1} = \frac{2}{5}$ • Find the common ratio.

$a_n = a_1 r^{n-1}$

$a_5 = 5\left(\frac{2}{5}\right)^{5-1}$ • $n = 5, a_1 = 5, r = \frac{2}{5}$

$= 5\left(\frac{2}{5}\right)^4 = 5\left(\frac{16}{625}\right) = \frac{16}{125}$

Problem 2

$3, a_2, a_3, -192, \ldots$

$a_n = a_1 r^{n-1}$

$a_4 = 3r^{4-1}$ • $n = 4$

$-192 = 3r^{4-1}$ • $a_4 = -192, a_1 = 3$

$-192 = 3r^3$ • Solve for r.

$-64 = r^3$

$-4 = r$

$a_n = a_1 r^{n-1}$

$a_3 = 3(-4)^{3-1}$ • $n = 3, a_1 = 3, r = -4$

$= 3(-4)^2 = 3(16) = 48$

Problem 3

$1, -\frac{1}{3}, \frac{1}{9}, \ldots$

$r = \frac{a_2}{a_1} = \frac{-\frac{1}{3}}{1} = -\frac{1}{3}$ • Find the common ratio.

$S_n = \frac{a_1(1 - r^n)}{1 - r}$ • Formula for the Sum of n Terms of a Finite Geometric Series

$S_8 = \frac{1\left[1 - \left(-\frac{1}{3}\right)^8\right]}{1 - \left(-\frac{1}{3}\right)}$ • $n = 8, a_1 = 1, r = -\frac{1}{3}$

$= \frac{1 - \frac{1}{6561}}{\frac{4}{3}} = \frac{\frac{6560}{6561}}{\frac{4}{3}} = \frac{6560}{6561} \cdot \frac{3}{4} = \frac{1640}{2187}$

Problem 4

$\sum_{n=1}^{5} \left(\frac{1}{2}\right)^n$

$a_n = \left(\frac{1}{2}\right)^n$

$a_1 = \left(\frac{1}{2}\right)^1 = \frac{1}{2}$ • To find a_1, let $n = 1$.

$r = \frac{1}{2}$ • r is the base of the exponential expression.

$S_n = \frac{a_1(1 - r^n)}{1 - r}$ • Formula for the Sum of n Terms of a Finite Geometric Series

$S_5 = \frac{\frac{1}{2}\left[1 - \left(\frac{1}{2}\right)^5\right]}{1 - \frac{1}{2}}$ • $n = 5, a_1 = \frac{1}{2}, r = \frac{1}{2}$

$= \frac{\frac{1}{2}\left(1 - \frac{1}{32}\right)}{\frac{1}{2}} = \frac{\frac{1}{2}\left(\frac{31}{32}\right)}{\frac{1}{2}} = \frac{\frac{31}{64}}{\frac{1}{2}}$

$= \frac{31}{64} \cdot \frac{2}{1} = \frac{31}{32}$

Problem 5

$3, -2, \frac{4}{3}, -\frac{8}{9}, \ldots$

$r = \frac{a_2}{a_1} = \frac{-2}{3} = -\frac{2}{3}$ • Find the common ratio.

$S = \frac{a_1}{1 - r}$ • $|r| < 1$. Use the Formula for the Sum of an Infinite Geometric Series.

$= \frac{3}{1 - \left(-\frac{2}{3}\right)}$ • $a_1 = 3, r = -\frac{2}{3}$

$= \frac{3}{1 + \frac{2}{3}} = \frac{3}{\frac{5}{3}} = \frac{9}{5}$

Problem 6

$0.\overline{36} = 0.36 + 0.0036 + 0.000036 + \cdots$

$S = \frac{a_1}{1 - r} = \frac{\frac{36}{100}}{1 - \frac{1}{100}}$ • $a_1 = \frac{36}{100}, r = \frac{1}{100}$

$= \frac{\frac{36}{100}}{\frac{99}{100}} = \frac{36}{99} = \frac{4}{11}$

An equivalent fraction for $0.\overline{36}$ is $\frac{4}{11}$.

Problem 7

Strategy To find the total number of letters mailed, use the Formula for the Sum of n Terms of a Finite Geometric Series.

$n = 6, a_1 = 3, r = 3$

Solution $S_n = \frac{a_1(1 - r^n)}{1 - r}$

$S_6 = \frac{3(1 - 3^6)}{1 - 3} = \frac{3(1 - 729)}{1 - 3} = \frac{3(-728)}{-2}$

$= \frac{-2184}{-2} = 1092$

From the first through the sixth mailings, 1092 letters will have been mailed.

SECTION 11.4

Problem 1

$$\frac{12!}{7!5!} = \frac{12 \cdot 11 \cdot 10 \cdot 9 \cdot 8 \cdot 7 \cdot 6 \cdot 5 \cdot 4 \cdot 3 \cdot 2 \cdot 1}{(7 \cdot 6 \cdot 5 \cdot 4 \cdot 3 \cdot 2 \cdot 1)(5 \cdot 4 \cdot 3 \cdot 2 \cdot 1)}$$
$$= 792$$

Problem 2

$$\binom{7}{0} = \frac{7!}{(7-0)!0!}$$
$$= \frac{7!}{7!0!}$$
$$= \frac{7 \cdot 6 \cdot 5 \cdot 4 \cdot 3 \cdot 2 \cdot 1}{(7 \cdot 6 \cdot 5 \cdot 4 \cdot 3 \cdot 2 \cdot 1)(1)} = 1$$

• $\binom{n}{r} = \frac{n!}{(n-r)!r!}$

Problem 3

$(3m - n)^4$

$$= \binom{4}{0}(3m)^4 + \binom{4}{1}(3m)^3(-n) + \binom{4}{2}(3m)^2(-n)^2 + \binom{4}{3}(3m)(-n)^3 + \binom{4}{4}(-n)^4$$
$$= 1(81m^4) + 4(27m^3)(-n) + 6(9m^2)(n^2) + 4(3m)(-n^3) + 1(n^4)$$
$$= 81m^4 - 108m^3n + 54m^2n^2 - 12mn^3 + n^4$$

Problem 4

$$(y-2)^{10} = \binom{10}{0}y^{10} + \binom{10}{1}y^9(-2) + \binom{10}{2}y^8(-2)^2 + \cdots$$
$$= 1(y^{10}) + 10y^9(-2) + 45y^8(4) + \cdots$$
$$= y^{10} - 20y^9 + 180y^8 - \cdots$$

Problem 5

$(t - 2s)^7$

$$\binom{n}{r-1} = a^{n-r+1}b^{r-1}$$

• **Formula for the *r*th Term of a Binomial Expansion**

$$\binom{7}{3-1}(t)^{7-3+1}(-2s)^{3-1}$$

• $r = 3, n = 7, a = t, b = -2s$

$$= \binom{7}{2}(t)^5(-2s)^2$$
$$= 21t^5(4s^2) = 84t^5s^2$$

Solutions to Chapter 12 Problems

SECTION 12.1

Problem 1

$$-\frac{b}{2a} = -\frac{1}{2(-1)} = \frac{1}{2}$$

• **Find the *x*-coordinate of the vertex.**

$$y = -x^2 + x + 3$$
$$y = -\left(\frac{1}{2}\right)^2 + \frac{1}{2} + 3 = \frac{13}{4}$$

• **Find the *y*-coordinate of the vertex.**

The coordinates of the vertex are $\left(\frac{1}{2}, \frac{13}{4}\right)$.

The equation of the axis of symmetry is $x = \frac{1}{2}$.

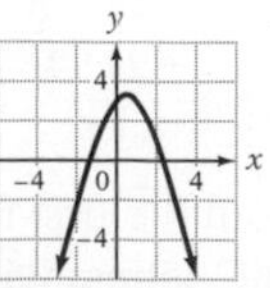

Problem 2

$$-\frac{b}{2a} = -\frac{-4}{2(-2)} = -1$$

• **Find the *y*-coordinate of the vertex.**

$$x = -2y^2 - 4y - 3$$
$$x = -2(-1)^2 - 4(-1) - 3$$
$$x = -1$$

• **Find the *x*-coordinate of the vertex.**

The coordinates of the vertex are $(-1, -1)$.

The equation of the axis of symmetry is $y = -1$.

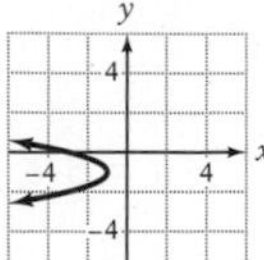

SECTION 12.2

Problem 1

Use the distance formula to find the radius.

$$r = \sqrt{[3 - (-2)]^2 + (5 - 3)^2}$$

• $(x_1, y_1) = (-2, 3)$, $(x_2, y_2) = (3, 5)$

$$r = \sqrt{5^2 + 2^2}$$
$$r = \sqrt{25 + 4}$$
$$r = \sqrt{29}$$

• **The radius is $\sqrt{29}$.**

$$(x - h)^2 + (y - k)^2 = r^2$$
$$(x - 3)^2 + (y - 5)^2 = (\sqrt{29})^2$$

• $(h, k) = (3, 5), r = \sqrt{29}$

$$(x - 3)^2 + (y - 5)^2 = 29$$

Problem 2

$$x_m = \frac{x_1 + x_2}{2} \qquad y_m = \frac{y_1 + y_2}{2}$$
$$x_m = \frac{-2 + 4}{2} = 1 \qquad y_m = \frac{1 + (-1)}{2} = 0$$

Center: $(x_m, y_m) = (1, 0)$

$$r = \sqrt{(x_1 - x_m)^2 + (y_1 - y_m)^2}$$
$$r = \sqrt{(-2 - 1)^2 + (1 - 0)^2} = \sqrt{9 + 1}$$
$$= \sqrt{10}$$

Radius: $\sqrt{10}$

$$(x - h)^2 + (y - k)^2 = r^2$$
$$(x - 1)^2 + (y - 0)^2 = (\sqrt{10})^2$$
$$(x - 1)^2 + y^2 = 10$$

Problem 3

$$x^2 + y^2 - 4x + 8y + 15 = 0$$
$$(x^2 - 4x) + (y^2 + 8y) = -15$$
$$(x^2 - 4x + 4) + (y^2 + 8y + 16) = -15 + 4 + 16$$
$$(x - 2)^2 + (y + 4)^2 = 5$$

Center: $(2, -4)$

Radius: $\sqrt{5}$

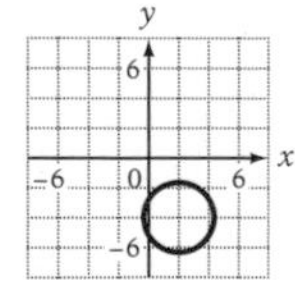

SECTION 12.3

Problem 1

A. x-intercepts:
$(2, 0)$ and $(-2, 0)$

y-intercepts:
$(0, 5)$ and $(0, -5)$

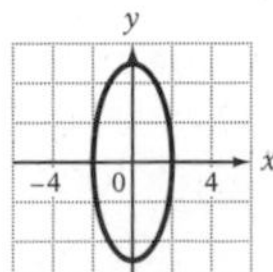

B. x-intercepts:
$(3\sqrt{2}, 0)$ and $(-3\sqrt{2}, 0)$
y-intercepts:
$(0, 3)$ and $(0, -3)$
$\left(3\sqrt{2} \approx 4\frac{1}{4}\right)$

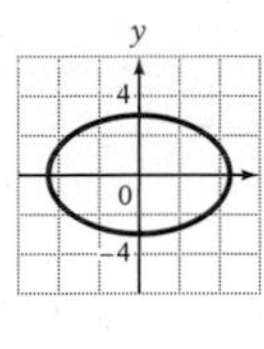

Problem 2

A. Axis of symmetry:
x-axis
Vertices:
$(3, 0)$ and $(-3, 0)$
Asymptotes:
$y = \frac{5}{3}x$ and $y = -\frac{5}{3}x$

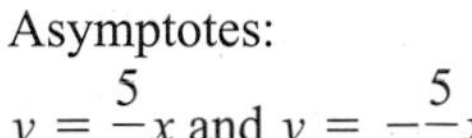
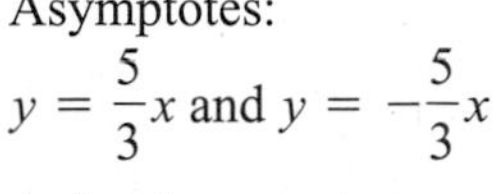
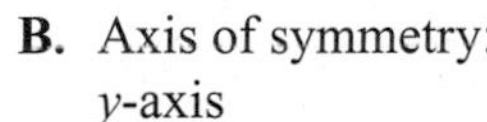

B. Axis of symmetry:
y-axis
Vertices:
$(0, 3)$ and $(0, -3)$
Asymptotes:
$y = x$ and $y = -x$

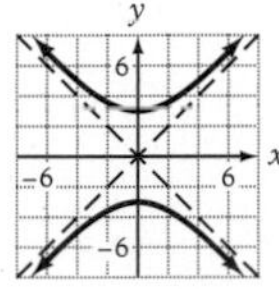

SECTION 12.4

Problem 1

(1) $y = 2x^2 + x - 3$
(2) $y = 2x^2 - 2x + 9$

Use the substitution method.

$$2x^2 - 2x + 9 = 2x^2 + x - 3$$
$$-3x + 9 = -3$$
$$-3x = -12$$
$$x = 4$$

Substitute the value of x into equation (1).

$$y = 2x^2 + x - 3$$
$$y = 2(4)^2 + 4 - 3$$
$$y = 32 + 4 - 3$$
$$y = 33$$

The solution is $(4, 33)$.

Problem 2

(1) $x^2 - y^2 = 10$
(2) $x^2 + y^2 = 8$

Use the addition method.

$$2x^2 = 18$$
$$x^2 = 9$$
$$x = \pm\sqrt{9} = \pm 3$$

Substitute the values of x into equation (2).

$$x^2 + y^2 = 8 \qquad x^2 + y^2 = 8$$
$$3^2 + y^2 = 8 \qquad (-3)^2 + y^2 = 8$$
$$9 + y^2 = 8 \qquad 9 + y^2 = 8$$
$$y^2 = -1 \qquad y^2 = -1$$
$$y = \pm\sqrt{-1} \qquad y = \pm\sqrt{-1}$$

y is not a real number. Therefore, the system of equations has no real number solutions.

SECTION 12.5

Problem 1

A. Graph the ellipse $\frac{x^2}{9} + \frac{y^2}{16} = 1$ as a solid curve.
Shade the region of the plane that includes $(0, 0)$.

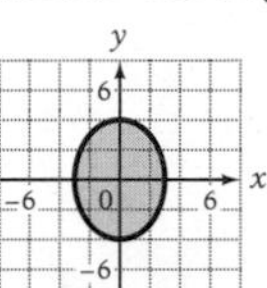

B. Graph the hyperbola $\frac{x^2}{9} - \frac{y^2}{4} = 1$ as a solid curve.
Shade the region that includes $(0, 0)$.

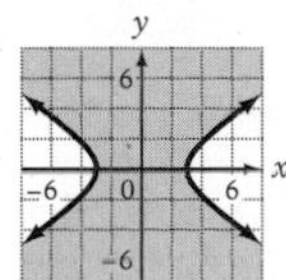

Problem 2

A. Graph the ellipse $\frac{x^2}{4} + \frac{y^2}{9} = 1$ as a solid curve.
Shade inside the ellipse.
Graph the parabola $x = y^2 - 2$ as a dashed curve.
Shade inside the parabola.
The solution set is the region of the plane represented by the intersection of the graphs of the solution sets of the two inequalities.

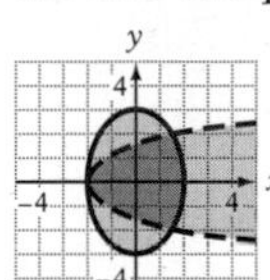

B. Graph the ellipse $\frac{x^2}{16} + \frac{y^2}{25} = 1$ as a solid curve.
Shade outside the ellipse.
Graph the circle $x^2 + y^2 = 9$ as a dashed curve.
Shade inside the circle.
The graphs of the solution sets of the two inequalities do not intersect. The system of inequalities has no real number solutions.

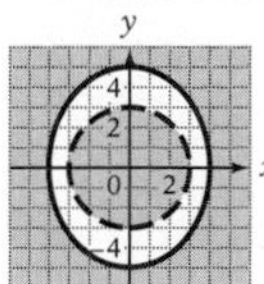

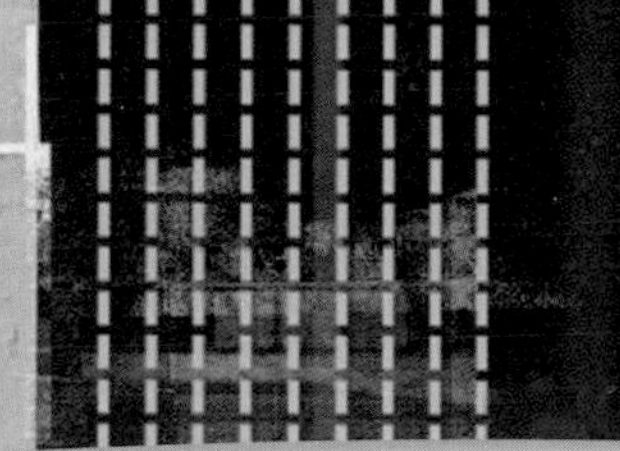

Answers to Selected Exercises

Answers to Chapter 1 Selected Exercises

PREP TEST

1. $\frac{13}{20}$ **2.** $\frac{11}{60}$ **3.** $\frac{2}{9}$ **4.** $\frac{2}{3}$ **5.** 44.405 **6.** 73.63 **7.** 7.446 **8.** 54.06 **9.** i, iii, iv **10.** a and C; b and D; c and A; d and B

SECTION 1.1

1. a. Natural numbers: 9, 53 **b.** Whole numbers: 0, 9, 53 **c.** Integers: $-14, 9, 0, 53, -626$ **d.** Positive integers: 9, 53 **e.** Negative integers: $-14, -626$ **3. a.** Integers: $0, -3$ **b.** Rational: $-\frac{15}{2}, 0, -3, 2.\overline{33}$ **c.** Irrational: π, 4.232232223 . . . , $\frac{\sqrt{5}}{4}$, $\sqrt{7}$ **d.** Real: all **11.** irrational **13.** is an element of **15.** -27 **17.** $-\frac{3}{4}$ **19.** 0 **21.** $\sqrt{33}$ **23.** 91 **25.** $-3, 0$ **27.** 7 **29.** $-2, -1$ **31.** $9, 0, -9$ **33.** 4, 0, 4 **35.** $-6, -2, 0, -1, -4$ **37.** Yes; negative real numbers **39.** roster; set-builder **41.** infinity **43.** $\{-2, -1, 0, 1, 2, 3, 4\}$ **45.** $\{2, 4, 6, 8, 10, 12\}$ **47.** $\{3, 6, 9, 12, 15, 18, 21, 24, 27, 30\}$ **49.** $\{-35, -30, -25, -20, -15, -10, -5\}$ **51.** $\{x|x > 4, x \in \text{integers}\}$ **53.** $\{x|x \geq -2\}$ **55.** $\{x|0 < x < 1\}$ **57.** $\{x|1 \leq x \leq 4\}$ **59.** **61.** **63.** **65.** **67.** $\{x|0 < x < 8\}$ **69.** $\{x|-5 \leq x \leq 7\}$ **71.** $\{x|-3 \leq x < 6\}$ **73.** $\{x|x \leq 4\}$ **75.** $\{x|x > 5\}$ **77.** $(-2, 4)$ **79.** $[-1, 5]$ **81.** $(-\infty, 1)$ **83.** $[-2, 6)$ **85.** $(-\infty, \infty)$ **87.** **89.** **91.** **93.** **95.** $\{1, 2, 4, 6, 9\}; \{4\}$ **97.** $\{2, 3, 5, 8, 9, 10\}; \varnothing$ **99.** $\{-4, -2, 0, 2, 4, 8\}; \{0, 4\}$ **101.** $\{1, 2, 3, 4, 5\}; \{3, 4, 5\}$ **103.** **105.** **107.** **109.** **111.** **113.** **115.** **117.** iii **119.** A **121.** B **123.** A **125.** R **127.** R **129.** 0 **131.** **133.** **135.** ii and iii **137.** $\{1, 4, 6, 8, 9\}$ **139.** $\{x|x \in \text{irrational numbers}\}$

SECTION 1.2

3. No. For instance, $-5 + (-3) = -8$. **5.** One number is positive and the other is negative. **7.** At least one of the numbers is zero. **9.** $(-5)^6$ **11.** $3; -5$ **13.** Negative **15.** 9 **17.** -12 **19.** 27 **21.** -232 **23.** -40 **25.** 17 **27.** -16 **29.** -27 **31.** 90 **33.** -12 **35.** -4 **37.** -64 **39.** -4 **41.** -26 **43.** -54 **45.** 125 **47.** -8 **49.** -125 **51.** 324 **53.** -36 **55.** -72 **57. a.** Negative **b.** Positive **c.** Negative **d.** Positive **59.** 5; 25; 125; 25; 0 **61.** -2 **63.** 18 **65.** -11 **67.** -40 **69.** 20 **71.** 24 **73.** 2 **75.** -8 **77.** 9413 **79.** 8 **81.** 2 **83.** ii **85.** 9 **87.** 625 **89.** Find b^c; then find $a^{(b^c)}$. **91.** Perfect

SECTION 1.3

3. Yes **5.** Yes; no **7.** 56; 35; 2; 16 **9.** $-\frac{8}{25}$ **11.** $-\frac{166}{15}$ **13.** $\frac{45}{112}$ **15.** $-\frac{307}{38}$ **17.** $\frac{81}{16}$ **19.** $-\frac{729}{8}$ **21.** $-\frac{289}{16}$ **23.** $-\frac{19}{3}$ **25.** $-\frac{64}{21}$ **27.** $\frac{39}{10}$ **29.** $-\frac{330}{31}$ **31.** $\frac{1}{100}$ **33.** $-\frac{30}{49}$ **35.** $-\frac{135}{22}$ **37.** $\frac{1}{12}$ **39.** $-\frac{59}{36}$ **41.** $-\frac{22}{15}$ **43.** $-\frac{25}{56}$ **45.** $\frac{11}{24}$ **47.** No. For instance, there is no integer between 2 and 3. **49.** fractions **51.** $\frac{12}{11}$ **53.** $-\frac{7}{18}$ **55.** $-\frac{25}{18}$ **57.** $\frac{2}{15}$ **59.** $\frac{27}{32}$ **61.** $-\frac{25}{42}$ **63.** $\frac{5}{6}$ **65.** $-\frac{2}{3}$ **67.** $-\frac{4}{3}$ **69.** $\frac{181}{8}$ **71.** $\frac{1}{2}$ **73.** $\frac{5}{2}$ **75.** i, ii, and iv **77.** 0.625 **79.** $0.1\overline{6}$ **81.** $0.1\overline{45}$ **83.** $0.\overline{230769}$ **85.** 3 **87.** −0.0012 **89.** 3.4992 **91.** −0.3897 **93.** 0.000456 **95.** −0.018 **97.** −0.013 **99.** −14.82 **101.** 0.0254 **103.** −0.012 **105.** 6.284 **107.** 5.4375 **109.** 6.44 **111.** No. $\frac{5}{23}$ is a rational number, so its decimal representation either terminates or repeats. **113.** $\frac{7}{2}$ **115.** $\frac{55}{21}$ **117.** True **119.** True **121.** False; for example, $\pi + (-\pi) = 0$. **123.** For example, $\frac{5}{12}$

SECTION 1.4

1. Addition, multiplication **3.** a **5.** Distributive Property **7.** No. The variable parts are not the same. **9.** Commutative **11.** zero **13.** 3 **15.** 3 **17.** 0 **19.** 6 **21.** 0 **23.** 1 **25.** $(2 \cdot 3)$ **27.** Division Property of Zero **29.** Inverse Property of Multiplication **31.** Addition Property of Zero **33.** Division Property of Zero **35.** Distributive Property **37.** Associative Property of Multiplication **39.** Zero **43.** −5; −3; −2; 2; 25; 50 **45.** 10 **47.** 15 **49.** 0 **51.** $-\frac{1}{7}$ **53.** 17 **55.** $\frac{1}{2}$ **57.** 2 **59.** 3 **61.** −12 **63.** 2 **65.** 6 **67.** −2 **69.** −14 **71.** 56 **73.** 256 **75.** The volume of the rectangular solid is 840 in^3. **77.** The volume of the pyramid is 15 ft^3. **79.** The volume of the sphere is exactly 4.5π cm^3. The volume of the sphere is approximately 14.14 cm^3. **81.** The surface area of the rectangular solid is 94 m^2. **83.** The surface area of the pyramid is 56 m^2. **85.** The surface area of the cylinder is exactly 96π in^2. The surface area of the cylinder is approximately 301.59 in^2. **87.** Negative **91.** −5; 25; Commutative; Associative; $4y$ **93.** $13x$ **95.** x **97.** $-3x + 6$ **99.** $5x + 10$ **101.** $x + y$ **103.** $-12a + 22$ **105.** $-11m - 6$ **107.** $-12y + 13$ **109.** $-11a + 21$ **111.** $-x + 6y$ **113.** $140 - 30a$ **115.** $-10y + 30x$ **117.** $-10a + 2b$ **119.** $-12a + b$ **121.** $-2x - 144y - 96$ **123.** $5x - 32 + 3y$ **125.** $x + 6$ **127. a.** No **b.** Yes **129.** Distributive Property **131.** Incorrect use of the Distributive Property; $2 + 3x = 2 + 3x$ **133.** Incorrect use of the Associative Property of Multiplication; $2(3y) = (2 \cdot 3)y = 6y$ **135.** Commutative Property of Addition **137.** Yes **139.** Yes

SECTION 1.5

1. For instance, "the sum of y and 6" and "6 more than y" **3.** No. The first expression translates to $x - 2$, and the second expression translates to $2 - x$. **5.** $14 - x$ **7.** more than; product **9.** difference between; times; times **11.** $n - (n + 2)$; −2 **13.** $\frac{1}{3}n + \frac{4}{5}n$; $\frac{17}{15}n$ **15.** $5(8n)$; $40n$ **17.** $17n - 2n$; $15n$ **19.** $n^2 - (12 + n^2)$; −12 **21.** $15n + (5n + 12)$; $20n + 12$ **23.** $2x + (15 - x + 2)$; $x + 17$ **25.** $(34 - x + 2) - 2x$; $-3x + 36$ **27.** iii **29.** width; $W + 8$ **31.** Money to be spent on high-speed rails: A; money spent on highways: $8A$ **33.** The distance from Earth to the moon: d; the distance from Earth to the sun: $390d$ **35.** $10{,}000 - x$ **37.** The flying time between San Diego and New York: t; the flying time between New York and San Diego: $13 - t$ **39.** The measure of angle B: x; the measure of angle A: $2x$; the measure of angle C: $4x$ **41.** $\frac{1}{2}gt^2$ **43.** Av^2 **45.** The sum of twice a number and three **47.** Twice the sum of a number and three

CHAPTER 1 REVIEW EXERCISES[1]

1. $\frac{3}{4}$ [1.4.1] **2.** 0, 2 [1.1.1] **3.** −4, 0, −7 [1.1.1] **4.** {−2, −1, 0, 1, 2, 3} [1.1.2] **5.** $\{x \mid x < -3\}$ [1.1.2] **6.** $\{x \mid -2 \le x \le 3\}$ [1.1.2] **7.** {1, 2, 3, 4, 5, 6, 7, 8} [1.1.2] **8.** {2, 3} [1.1.2] **9.** −5 −4 −3 −2 −1 0 1 2 3 4 5 [1.1.2] **10.** −5 −4 −3 −2 −1 0 1 2 3 4 5 [1.1.2] **11.** −5 −4 −3 −2 −1 0 1 2 3 4 5 [1.1.2] **12.** −5 −4 −3 −2 −1 0 1 2 3 4 5 [1.1.2] **13.** −5 −4 −3 −2 −1 0 1 2 3 4 5 [1.1.2] **14.** −5 −4 −3 −2 −1 0 1 2 3 4 5 [1.1.2] **15.** −7 [1.2.1] **16.** 12 [1.2.1] **17.** $-\frac{13}{24}$ [1.3.1] **18.** $-\frac{5}{8}$ [1.3.1] **19.** −1.77 [1.3.3] **20.** −25 [1.2.2] **21.** 31 [1.2.2] **22.** $\frac{5}{8}$ [1.3.2] **23.** 3 [1.3.2] **24.** −3.22 [1.3.3]

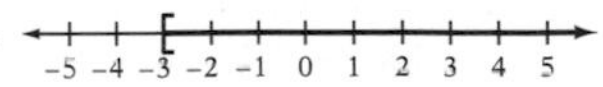

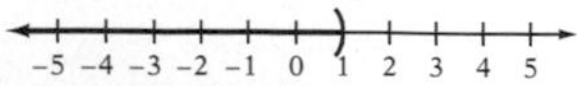

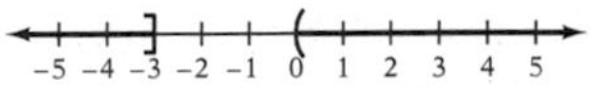

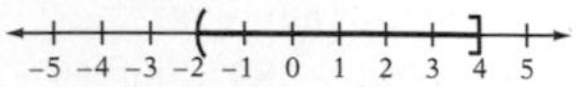

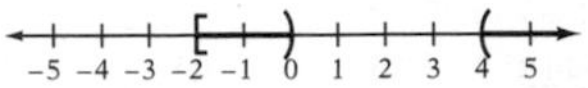

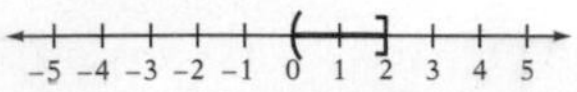

[1]The numbers in brackets following the answers in the Chapter Review Exercises are a reference to the objective that corresponds to that problem. For example, the reference [1.2.1] stands for Section 1.2, Objective 1. This notation will be used for all Prep Tests, Chapter Reviews, Chapter Tests, and Cumulative Reviews throughout the text.

25. $21y$ [1.4.1] **26.** (ab) [1.4.1] **27.** The Inverse Property of Addition [1.4.1] **28.** The Associative Property of Multiplication [1.4.1] **29.** 41 [1.4.2] **30.** -257 [1.4.2] **31.** $-8a + 10$ [1.4.3] **32.** $-6x + 14$ [1.4.3] **33.** $16y - 15x + 18$ [1.4.3] **34.** $14x - 12y - 7$ [1.4.3] **35.** $4(x + 4)$; $4x + 16$ [1.5.1] **36.** $2(x - 2) + 8$; $2x + 4$ [1.5.1] **37.** $2x + [(40 - x) + 5]$; $x + 45$ [1.5.1] **38.** $[2(9 - x) + 3] - (x + 1)$; $-3x + 20$ [1.5.1] **39.** The width of the rectangle: W; the length of the rectangle: $3W - 3$ [1.5.2] **40.** The first integer: x; the second integer: $4x + 5$ [1.5.2]

CHAPTER 1 TEST

1. 12 [1.1.1, Example 2] **2.** -5 [1.1.1, Example 1] **3.** 12 [1.2.1, 1st and 2nd Focus On, page 14] **4.** -30 [1.2.1, Multiplication or Division of Real Numbers, Examples 1 and 2] **5.** -15 [1.2.1, Multiplication or Division of Real Numbers, Examples 3 and 4] **6.** 2 [1.2.1, 2nd Focus On, page 14] **7.** -100 [1.2.1, Example 3] **8.** -72 [1.2.1, Example 3] **9.** $\frac{25}{36}$ [1.3.1, Example 4] **10.** $-\frac{4}{27}$ [1.3.1, Example 1] **11.** -1.41 [1.3.3, Focus On, page 29] **12.** -4.9 [1.3.3, Focus On, page 29] **13.** 10 [1.2.2, Example 5] **14.** 6 [1.2.2, Example 4] **15.** -5 [1.4.2, Example 3] **16.** 2 [1.4.2, Example 3] **17.** 4 [1.4.1, Example 1] **18.** The Distributive Property [1.4.1, Example 2] **19.** $13x - y$ [1.4.3, Example 5] **20.** $14x + 48y$ [1.4.3, Example 6] **21.** $13 - (n - 3)(9)$; $40 - 9n$ [1.5.1, Example 1] **22.** $\frac{1}{3}(12n + 27)$; $4n + 9$ [1.5.1, Example 2] **23.** $\{1, 2, 3, 4, 5, 7\}$ [1.1.2, Union of Two Sets, Examples 1–3] **24.** $\{-2, -1, 0, 1, 2, 3\}$ [1.1.2, Union of Two Sets, Examples 1–3] **25.** $\{5, 7\}$ [1.1.2, Intersection of Two Sets, Examples 1–3] **26.** $\{-1, 0, 1\}$ [1.1.2, Intersection of Two Sets, Examples 1–3] **27.** -5 -4 -3 -2 -1 0 1 2 3 4 5 [1.1.2, Example 7B]

28. -5 -4 -3 -2 -1 0 1 2 3 4 5 [1.1.2, Example 7B] **29.** -5 -4 -3 -2 -1 0 1 2 3 4 5 [1.1.2, Example 8]

30. -5 -4 -3 -2 -1 0 1 2 3 4 5 [1.1.2, Example 8]

Answers to Chapter 2 Selected Exercises

PREP TEST

1. -4 [1.2.1] **2.** -6 [1.2.1] **3.** 3 [1.2.1] **4.** 1 [1.3.1] **5.** $-\frac{1}{2}$ [1.3.1] **6.** $10x - 5$ [1.4.3] **7.** $6x - 9$ [1.4.3] **8.** $3n + 6$ [1.4.3] **9.** $0.03x + 20$ [1.4.3] **10.** $20 - n$ [1.5.2]

SECTION 2.1

5. No **7.** Yes **9.** No **11.** Yes **13.** Yes **15.** 42; 55 **17.** $-\frac{5}{2}$; -20 **19.** The solution is 9. **21.** The solution is -10. **23.** The solution is 4. **25.** The solution is -7. **27.** The solution is $\frac{11}{21}$. **29.** The solution is -49. **31.** The solution is $-\frac{21}{20}$. **33.** The solution is -3.73. **35.** The solution is $\frac{3}{2}$. **37.** The solution is 8. **39.** The solution is $-\frac{5}{2}$. **41.** The solution is -3. **43.** The solution is 1. **45.** The solution is $-\frac{3}{2}$. **47.** The solution is $\frac{11}{2}$. **49.** Greater than **51.** 42 **53.** The solution is 6. **55.** The solution is $\frac{1}{2}$. **57.** The solution is -6. **59.** The solution is $-\frac{4}{3}$. **61.** The solution is $\frac{6}{7}$. **63.** The solution is $\frac{35}{12}$. **65.** The solution is -1. **67.** The solution is -33. **69.** The solution is 6. **71.** The solution is 1. **73.** The solution is $\frac{25}{14}$. **75.** The solution is $\frac{3}{4}$. **77.** The solution is $-\frac{4}{29}$. **79.** The solution is 3. **81.** The solution is -10. **83.** The solution is 11. **85.** The solution is 9.4. **87.** The solution is 0.5. **89.** i **91.** 8; 3; $10 - n$ **93.** The temperature was 41°F. **95.** The customer bought 8 bags of feed. **97.** The teachers will be paid $21.58 per hour. **99.** Charlotte's annual income was $48,000. **101.** The solution is -9. **103.** The solution is $-\frac{15}{2}$. **105.** No solution **107.** The solution is -1. **109.** All real numbers are solutions. **111.** 10, 11, 12 **113.** 20, 22, 24, 26

SECTION 2.2

1. ii, iii **3.** i **5.** Less than **7. a.** $8.40 **b.** $.70 **9.** iv and vi **11.** The vegetable medley costs $1.66 per pound. **13.** There were 320 adult tickets sold. **15.** The mixture must contain 225 L of imitation maple syrup. **17.** The instructor used 8 lb of nuts. **19.** The cost of the mixture is $4.04 per pound. **21.** The mixture contained 37.5 gal of cranberry juice. **23. a.** $12t$; $15t$ **b.** 400; 360 **25. a.** Less than **b.** Equal to **c.** 3 mi **27.** The student drives at an average rate of 40 mph. **29.** It would take a Boeing 737-800 jet about 3.2 h to make the same trip. **31.** The two cyclists will meet at 2:30 P.M.

33. The bicyclist overtakes the in-line skater 11.25 mi from the starting point. **35.** The speed of the first plane is 420 mph. The speed of the second plane is 500 mph. **37.** The rate of the first plane is 255 mph. The rate of the second plane is 305 mph. **39.** The distance to the island is 43.2 mi. **41.** The distance between Marcella's home and the bicycle shop is 2.8 mi. **43.** The two trains will pass each other in 2 h. **45.** It would take a spacecraft about 1600 years to reach this star. **47.** The cars are $3\frac{1}{3}$ mi apart 2 min before impact. **49.** No **51. a.** The cost of the mixture is \$15/lb. **b.** The cost is greater than the cost found in part (a).

SECTION 2.3

1. a. The principal is \$1500. **b.** The interest rate is 4%. **c.** The interest earned is \$60. **3.** i, iv, v **7.** $0.0625x$; $0.06(x - 500)$; 115 **9.** Joseph will earn \$327.50 from the two accounts. **11.** Deon must invest \$1600 in the account that earns 8% interest. **13.** The amount invested in the CD is \$15,000. **15.** An additional \$2000 must be invested at an annual simple interest rate of 10.5%. **17.** The amount invested at 8.5% is \$3000. The amount invested at 6.4% is \$5000. **19.** She must invest an additional \$3000 at an annual simple interest rate of 10%. **21.** The amount that should be invested at 4.2% is \$8000. The amount that should be invested at 6% is \$5600. **23. a.** The interest rates were 5.5% and 7.2%. **b.** Will had invested \$6000. **25.** 0.10; 3.2 **27.** A 40-ounce bottle of Orange Ade contains 10 oz of orange juice. **29.** The 850-milliliter solution contains 12.5 more milliliters of hydrogen peroxide. **31.** The resulting alloy is 30% silver. **33.** The resulting alloy is $33\frac{1}{3}$% silver. **35.** 500 lb of the 12% aluminum alloy are needed. **37.** 25 L of the 65% solution and 25 L of the 15% solution were used. **39.** 3 qt of water must be added. **41.** 30 oz of pure water must be added. **43.** The result is 6% real fruit juice. **45.** The resulting alloy is 31.1% copper. **47.** ii **49.** The amount invested at 9% was \$5000. The amount invested at 8% was \$6000. The amount invested at 9.5% was \$9000. **51.** The cost per pound of the tea mixture is \$4.84. **53.** 60 g of pure water were in the beaker before the acid was added. **55. a.** 2002 **b.** 2000 **57.** Equal to

SECTION 2.4

5. Yes **7.** i, iii **9.** remains the same **11.** is reversed **13.** The solution set is $\{x|x < 5\}$. **15.** The solution set is $\{x|x \le 2\}$. **17.** The solution set is $\{x|x < -4\}$. **19.** The solution set is $\{x|x > 3\}$. **21.** The solution set is $\{x|x > 4\}$. **23.** The solution set is $\{x|x > -2\}$. **25.** The solution set is $\{x|x \ge 2\}$. **27.** The solution set is $\{x|x \le 3\}$. **29.** The solution set is $\{x|x < -3\}$. **31.** The solution set is $(-\infty, 5]$. **33.** The solution set is $[1, \infty)$. **35.** The solution set is $(-\infty, -5)$. **37.** The solution set is $\left(-\infty, \frac{23}{16}\right)$. **39.** The solution set is $\left[\frac{8}{3}, \infty\right)$. **41.** The solution set is $(-\infty, 1)$. **43.** The solution set is $(-\infty, 3)$. **45.** Only positive numbers **47.** Only negative numbers **49. a.** Union **b.** Intersection **51.** 4; 2; 1; 6 **53.** One interval of real numbers **55.** Empty set **57.** The solution set is $(-1, 2)$. **59.** The solution set is $(-\infty, 1] \cup [3, \infty)$. **61.** The solution set is $(-2, 4)$. **63.** The solution set is $(-\infty, -3) \cup (0, \infty)$. **65.** The solution set is $[3, \infty)$. **67.** The solution set is $\varnothing$. **69.** The solution set is $\varnothing$. **71.** The solution set is $(-\infty, 1) \cup (3, \infty)$. **73.** The solution set is $\{x|-3 < x < 4\}$. **75.** The solution set is $\{x|3 \le x \le 5\}$. **77.** The solution set is $\{x|x > 3\} \cup \left\{x\middle|x \le -\frac{5}{2}\right\}$. **79.** The solution set is $\{x|x > 4\}$. **81.** The solution set is $\varnothing$. **83.** The solution set is $\{x|x \in \text{real numbers}\}$. **85.** The solution set is $\{x|-4 < x < -2\}$. **87.** The solution set is $\{x|x < -4\} \cup \{x|x > 3\}$. **89.** The solution set is $\{x|-4 \le x < 1\}$. **91.** The solution set is $\left\{x\middle|x > \frac{27}{2}\right\} \cup \left\{x\middle|x < \frac{5}{2}\right\}$. **93.** The solution set is $\left\{x\middle|-10 \le x \le \frac{11}{2}\right\}$. **95.** $n \ge 40$ **97.** A customer can use a cellular phone for 60 min before the charges exceed those of the first option. **99.** The AirTouch Plan is less expensive for more than 460 messages per month. **101.** If a business chooses the Glendale Federal Bank, then the business writes more than 200 checks per month. **103.** The buses can travel between 392 mi and 560 mi on a full tank of fuel. **105.** The range of scores that will give the student a B for the course is $77 \le S \le 100$. **107.** The mixture should contain between 20 lb and 37.5 lb of peanuts. **109.** The miller should use between 10 bushels and 20 bushels of soybeans. **111.** The maximum width of the rectangle is 2 ft. **113.** The length of the second side could be 5 in., 6 in., or 7 in. **115.** The solution set is $\{1, 2\}$. **117.** The solution set is $\{1, 2, 3, 4, 5, 6\}$. **119.** The solution set is $\{1, 2, 3, 4\}$. **121.** The solution set is $\{1, 2\}$. **123.** The largest whole number of minutes a call can last is 10 min. **125. a.** $3.255 \le b < 3.265$ **b.** $5.125 \le h < 5.135$ **c.** $8.3409375 \le A < 8.3828875$

SECTION 2.5

3. Union **5.** Yes **7.** Yes **9.** 8, −8 **11.** The solutions are −7 and 7. **13.** The solutions are −3 and 3. **15.** The solutions are −4 and 4. **17.** The solutions are $-\frac{4}{3}$ and $\frac{4}{3}$. **19.** The solutions are −4 and 4. **21.** The solutions are −5 and 1. **23.** The solutions are 8 and 2. **25.** The solution is 2. **27.** The equation has no solution. **29.** The solutions are $\frac{1}{2}$ and $\frac{9}{2}$. **31.** The solutions are 0 and $\frac{4}{5}$. **33.** The solution is −1. **35.** The solutions are 4 and 14. **37.** The solutions are 4 and 12. **39.** The solution is $\frac{7}{4}$. **41.** The solutions are −3 and $\frac{3}{2}$. **43.** The solutions are $\frac{4}{5}$ and 0.

45. The solutions are -2 and 1. **47.** The solutions are $-\frac{11}{5}$ and 1. **49.** Two positive solutions **51.** Two negative solutions **53.** $<$; $>$ **55.** The solution set is $\{x|x > 3\} \cup \{x|x < -3\}$. **57.** The solution set is $\{x|x > 1\} \cup \{x|x < -3\}$. **59.** The solution set is $\{x|4 \le x \le 6\}$. **61.** The solution set is $\{x|x \le -1\} \cup \{x|x \ge 5\}$. **63.** The solution set is $\{x|-3 < x < 2\}$. **65.** The solution set is $\{x|x > 2\} \cup \left\{x \middle| x < -\frac{14}{5}\right\}$. **67.** The solution set is $\varnothing$. **69.** The solution set is $\{x|x \in \text{real numbers}\}$. **71.** The solution set is $\left\{x \middle| x \le -\frac{1}{3}\right\} \cup \{x|x \ge 3\}$. **73.** The solution set is $\left\{x \middle| -2 \le x \le \frac{9}{2}\right\}$. **75.** The solution set is $\{x|x = 2\}$. **77.** The solution set is $\{x|x < -2\} \cup \left\{x \middle| x > \frac{22}{9}\right\}$. **79.** All negative solutions **81.** tolerance; 50.5; upper; 49.5; lower **83.** The lower and upper limits of the diameter of the bushing are 1.742 in. and 1.758 in. **85.** The lower and upper limits of the amount of medicine to be given to the patient are 2.3 cc and 2.7 cc. **87.** The lower and upper limits for the percent of voters who felt the economy was the most important election issue are 38% and 44%. **89.** The lower and upper limits of voltage to the computer are 93.5 volts and 126.5 volts. **91. a.** The lower and upper limits for the girth of an NCAA football are $20\frac{3}{4}$ in. and $21\frac{1}{4}$ in. **b.** The lower and upper limits for the circumference of an NCAA football are $27\frac{3}{4}$ in. and $28\frac{1}{2}$ in. **c.** The lower and upper limits for the length of an NCAA football are $10\frac{7}{8}$ in. and $11\frac{3}{16}$ in. **93.** The desired diameter is 5 in. The tolerance is 0.01 in. **95.** The lower and upper limits of the resistor are 13,500 ohms and 16,500 ohms. **97.** The solutions are 2 and $-\frac{2}{3}$. **99.** The solution set is $\{x|-7 \le x \le 8\}$. **101.** The solutions are -2, $-\frac{6}{5}$, $\frac{6}{5}$, and 2. **103.** The solution set is $\{y|y \ge -6\}$. **105.** The solution set is $\{b|b \le 7\}$. **107.** $|x - 2| < 5$ **109.** i

CHAPTER 2 REVIEW EXERCISES

1. The solution is -9. [2.1.1] **2.** The solution is $\frac{2}{3}$. [2.1.1] **3.** The solution is -1. [2.1.1] **4.** The solution is $-\frac{2}{7}$. [2.1.1] **5.** The solution is -3. [2.1.1] **6.** The solution is -15. [2.1.1] **7.** The solution is $\frac{8}{5}$. [2.1.1] **8.** The solution is 6. [2.1.1] **9.** The solution is $\frac{26}{17}$. [2.1.2] **10.** The solution is $\frac{5}{2}$. [2.1.2] **11.** The solution is $-\frac{17}{2}$. [2.1.2] **12.** The solution is $-\frac{9}{19}$. [2.1.2] **13.** The solution is $\left(\frac{5}{3}, \infty\right)$. [2.4.1] **14.** The solution is $(-4, \infty)$. [2.4.1] **15.** The solution is $\left\{x \middle| x \le -\frac{87}{14}\right\}$. [2.4.1] **16.** The solution is $\left\{x \middle| x \ge \frac{16}{13}\right\}$. [2.4.1] **17.** The solution is $(-1, 2)$. [2.4.2] **18.** The solution set is $(-\infty, -2) \cup (2, \infty)$. [2.4.2] **19.** The solution set is $\left\{x \middle| -3 < x < \frac{4}{3}\right\}$. [2.4.2] **20.** The solution set is $\{x|x \in \text{real numbers}\}$. [2.4.2] **21.** The solutions are -1 and $\frac{9}{2}$. [2.5.1] **22.** The solution is $-\frac{8}{5}$. [2.5.1] **23.** The equation has no solution. [2.5.1] **24.** The solution set is $\{x|1 \le x \le 4\}$. [2.5.2] **25.** The solution set is $\left\{x \middle| x \le \frac{1}{2}\right\} \cup \{x|x \ge 2\}$. [2.5.2] **26.** The solution set is $\varnothing$. [2.5.2] **27.** The lower and upper limits of the diameter of the bushing are 2.747 in. and 2.753 in. [2.5.3] **28.** The lower and upper limits of the amount of medicine to be given are 1.75 cc and 2.25 cc. [2.5.3] **29.** The mixture costs \$9.75 per ounce. [2.2.1] **30.** The mixture must contain 52 gal of apple juice. [2.2.1] **31.** The cyclist reached the jogger in 1 h. [2.2.2] **32.** The speed of the first plane is 440 mph. The speed of the second plane is 520 mph. [2.2.2] **33.** The amount invested at 10.5% was \$3000. The amount invested at 6.4% was \$5000. [2.3.1] **34.** 375 lb of the 30% tin alloy and 125 lb of the 70% tin alloy were used. [2.3.2] **35.** The executive's amount of sales must be \$55,000 or more. [2.4.3] **36.** The range of scores to receive a B grade is $82 \le S \le 100$. [2.4.3] **37.** The amount of pure silver should be between 10 oz and 25 oz. [2.4.3]

CHAPTER 2 TEST

1. The solution is -2. [2.1.1, Focus On A, pages 56–57] **2.** The solution is $-\frac{1}{8}$. [2.1.1, Focus On B, pages 56–57] **3.** The solution is $\frac{5}{6}$. [2.1.1, Focus On A, page 57] **4.** The solution is 4. [2.1.1, Example 1] **5.** The solution is $\frac{32}{3}$. [2.1.1, Problem 1] **6.** The solution is $-\frac{1}{5}$. [2.1.1, Example 2] **7.** The solution is 1. [2.1.2, Example 3] **8.** The solution is -24. [2.1.2, Focus On, page 59] **9.** The solution is $\frac{12}{7}$. [2.1.2, Example 4] **10.** The solution set is $(-\infty, -3]$. [2.4.1, Example 1] **11.** The solution set is $(-1, \infty)$. [2.4.1, Problem 2] **12.** The solution set is $\{x|x > -2\}$. [2.4.2, Problem 3] **13.** The solution set is $\varnothing$. [2.4.2, Example 4] **14.** The solutions are 3 and $-\frac{9}{5}$. [2.5.1, Example 1] **15.** The solutions are 7 and -2. [2.5.1, Example 1] **16.** The solution set is $\left\{x \middle| -\frac{1}{3} \le x \le 1\right\}$. [2.5.2, Example 2] **17.** The solution set is $\{x|x > 2\} \cup \{x|x < -1\}$.

[2.5.2, Example 3] **18.** The equation has no solution. [2.5.1, Example 1] **19.** It costs less to rent from Agency A if the car is driven less than 120 mi. [2.4.3, Problem 6] **20.** The lower and upper limits of the amount of medication to be given are 2.9 cc and 3.1 cc. [2.5.3, Example 4] **21.** The cost of the hamburger mixture is \$3.20/lb. [2.2.1, Example 1] **22.** The jogger ran a total distance of 12 mi. [2.2.2, Example 3] **23.** The amount invested at 7.8% was \$5000. The amount invested at 9% was \$7000. [2.3.1, Example 1] **24.** 100 oz of pure water must be added. [2.3.2, Example 2]

CUMULATIVE REVIEW EXERCISES

1. -108 [1.2.1] **2.** 3 [1.2.2] **3.** -64 [1.3.2] **4.** -8 [1.4.2] **5.** The Commutative Property of Addition [1.4.1] **6.** $\{3, 9\}$ [1.1.2] **7.** $-17x + 2$ [1.4.3] **8.** $25y$ [1.4.3] **9.** The solution is 2. [2.1.1] **10.** The solution is $\frac{1}{2}$. [2.1.1] **11.** The solution is 1. [2.1.1] **12.** The solution is 24. [2.1.1] **13.** The solution is 2. [2.1.2] **14.** The solution is 2. [2.1.2] **15.** The solution is $-\frac{13}{5}$. [2.1.2] **16.** The solution set is $\{x|x \le -3\}$. [2.4.1] **17.** The solution set is $\varnothing$. [2.4.2] **18.** The solution set is $\{x|x > -2\}$. [2.4.2] **19.** The solutions are -1 and 4. [2.5.1] **20.** The solutions are -4 and 7. [2.5.1] **21.** The solution set is $\left\{x\middle|\frac{1}{3} \le x \le 3\right\}$. [2.5.2] **22.** The solution set is $\{x|x > 2\} \cup \left\{x\middle|x < -\frac{1}{2}\right\}$. [2.5.2]

23. −5 −4 −3 −2 −1 0 1 2 3 4 5 [1.1.2] **24.** −5 −4 −3 −2 −1 0 1 2 3 4 5 [1.1.2] **25.** $3n + (3n + 6)$; $6n + 6$ [1.5.1]

26. 48 adult tickets were sold. [2.2.1] **27.** The speed of the faster plane is 340 mph. [2.2.2] **28.** 3 L of 12% acid solution must be added to the mixture. **[2.3.2]** **29.** \$6500 was invested at 9.8%. [2.3.1] **30.** The amount of mixed nuts should be between 3.5 lb and 6 lb. [2.4.3]

Answers to Chapter 3 Selected Exercises

PREP TEST

1. $-4x + 12$ [1.4.3] **2.** 10 [1.2.2] **3.** -2 [1.2.2] **4.** 11 [1.4.2] **5.** 2.5 [1.4.2] **6.** 5 [1.4.2] **7.** 1 [1.4.2] **8.** 4 [2.1.1]

SECTION 3.1

1. 0 **3. a.** II **b.** I **c.** III **d.** IV **5.** I and IV **7. a.** Yes **b.** No **c.** No **d.** Yes **9.** 3; -1; -3; 4 **11. a.** The distance between the points is 5. **b.** The coordinates of the midpoint are $\left(\frac{7}{2}, 7\right)$. **13. a.** The distance between the points is 17. **b.** The coordinates of the midpoint are $\left(2, -\frac{3}{2}\right)$. **15. a.** The distance between the points is $\sqrt{17}$. **b.** The coordinates of the midpoint are $\left(\frac{7}{2}, 3\right)$. **17. a.** The distance between the points is $\sqrt{5}$. **b.** The coordinates of the midpoint are $\left(-1, \frac{7}{2}\right)$. **19. a.** The distance between the points is $\sqrt{26}$. **b.** The coordinates of the midpoint are $\left(-\frac{1}{2}, -\frac{9}{2}\right)$. **21. a.** The distance between the points is $\sqrt{85}$. **b.** The coordinates of the midpoint are $\left(2, \frac{3}{2}\right)$. **23.** IV

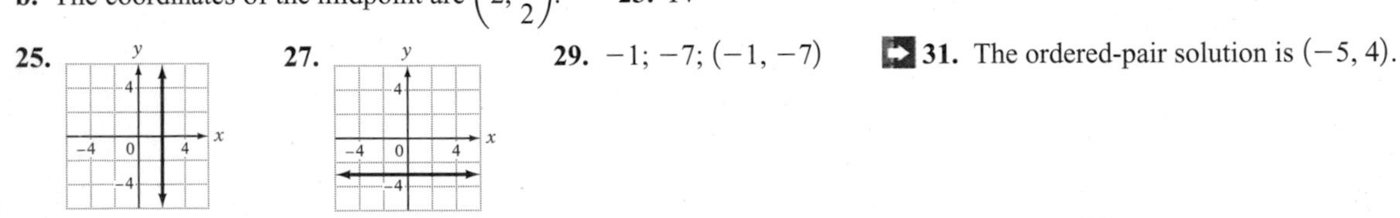

33. The ordered-pair solution is $(-1, 1)$. **35.** The ordered-pair solution is $(0, -2)$.

37. **39.**

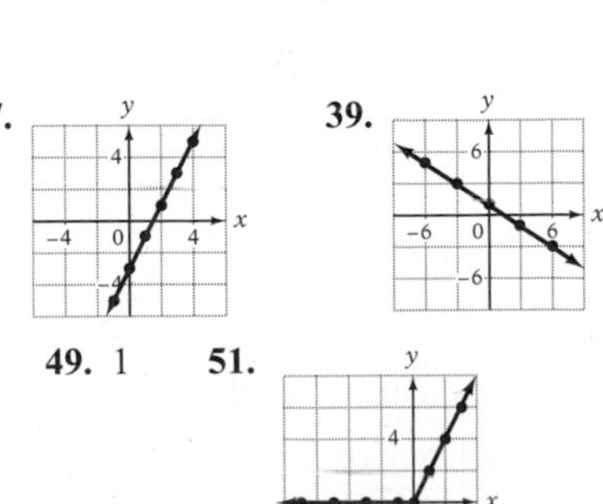

41.

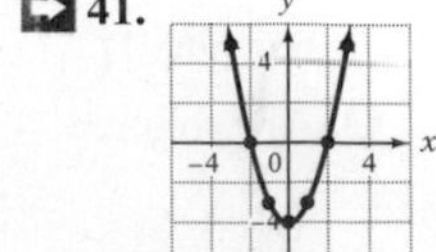

43.

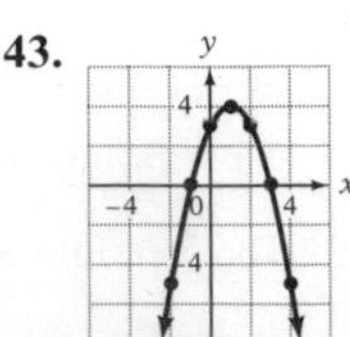

45.

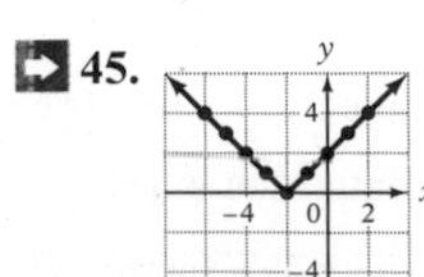

47.

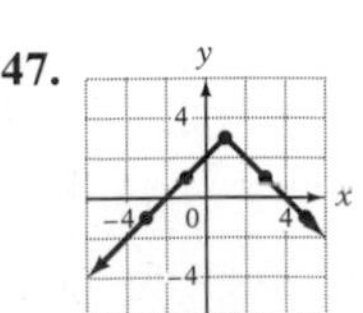

49. 1 **51.**

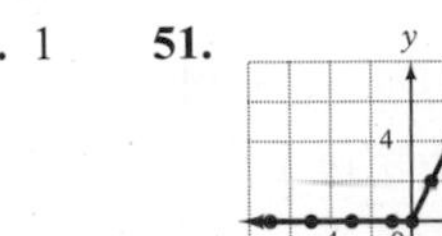

53. 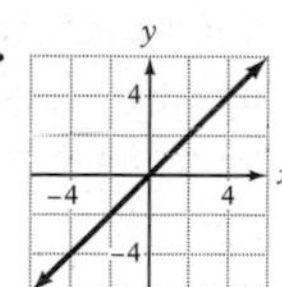 **57. a.** (3, −5) **b.** (5, 3) **c.** (−3, −4) **d.** (0, 5)

SECTION 3.2

3. exactly one **5.** 4 **7.** No; Yes **9.** The relation is a function. The domain is $\{-3, -2, 1, 4\}$. The range is $\{-7, 1, 2, 5\}$. **11.** The relation is a function. The domain is $\{1, 2, 3, 4, 5\}$. The range is $\{5\}$. **13.** The relation is a function. The domain is $\{-2, -1, 1, 2, 3\}$. The range is $\left\{-1, -\frac{1}{2}, \frac{1}{3}, \frac{1}{2}, 1\right\}$. **15.** The relation is not a function. The domain is $\{2, 4, 6, 8\}$. The range is $\{3, 5, 7, 8, 9\}$. **17.** Yes, each element of the domain is paired with exactly one element of the range. **19.** No, 1 and 4 in the domain are paired with more than one element in the range. **21.** Yes, each element of the domain is paired with exactly one element of the range. **23. a.** Yes **b.** \$24.70 **c.** \$23.20 **d.** \$20.70 **27.** True **29.** 3; 3; 8 **31. a.** 13 **b.** −3 **c.** 5 **33. a.** −3 **b.** 12 **c.** 8 **35. a.** −8 **b.** 1 **c.** −4 **37. a.** 1 **b.** −1 **c.** $-\frac{1}{2}$ **39. a.** $4a + 11$ **b.** $8a - 1$ **41. a.** 17 **b.** 7 **c.** −1 **43. a.** 15 **b.** 2 **c.** $3w^2 - 4w - 5$ **45. a.** 2 **b.** 2 **c.** $a^2 + 2a - 1$ **47. a.** 35 **b.** 0 **c.** $4a^2 + 8a + 3$ **49. a.** 4 **b.** 2 **c.** 0 **51. a.** 7 **b.** 10 **c.** −2 **53. a.** 5 **b.** −10 **c.** −10 **55.** 0 **57.** 13 **59.** 1 **61.** The range is $\{-14, -11, -8, -5, -2, 1\}$. **63.** The range is $\left\{-\frac{3}{2}, -1, -\frac{1}{2}, 0, \frac{1}{2}, 1\right\}$. **65.** The range is $\{3, 4, 7, 12\}$. **67.** The range is $\{-26, -8, -2, 4\}$. **69.** The value of c is 4. The ordered pair of the function is (4, 5). **71.** The value of c is 4. The ordered pair of the function is (4, −7). **73.** The value of c is 3. The ordered pair of the function is (3, 0). **75.** −5; −15; −19; (−5, −19) **77.** No **79.** No **81.** Yes

83. **85.** **87.** **89.** **91.**

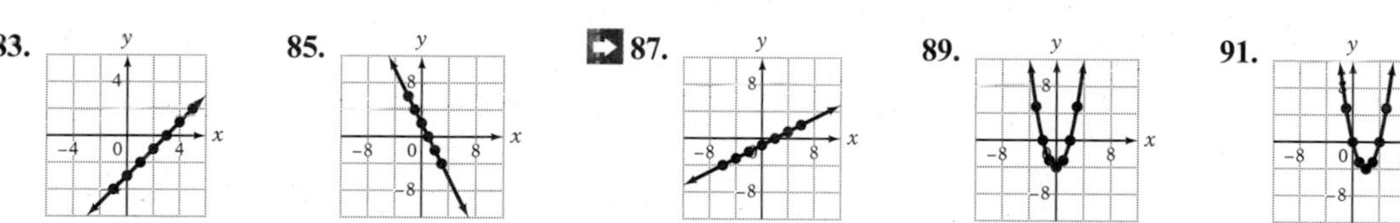

93. 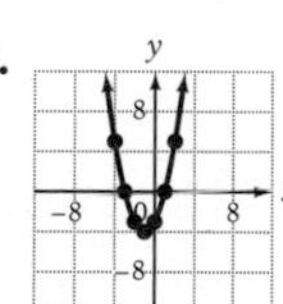 **95.** 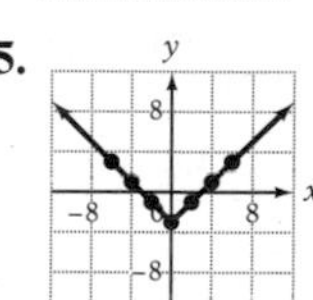 **97.** 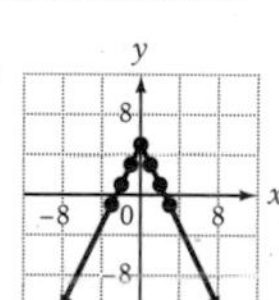 **99.** 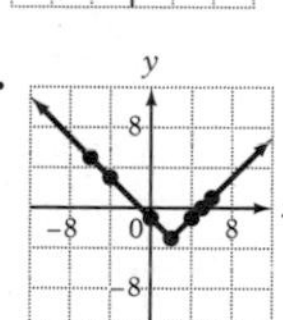 **101.** every **103.** Function

105. Function **107.** Function **109.** Not a function **111.** Function **113.** −2, 6 **115.** $4h$ **119.** The car will skid 61.2 ft. **121. a.** The temperature of the cola is 60°F. **b.** The temperature of the cola is 52°F. **123. a.** The employee would receive a score of 90%. **b.** The employee would receive a score of 100%. **125.** Answers will vary. **127.** Answers will vary.

SECTION 3.3

1. Yes **3.** Yes **5.** No **9. a.** −7; −1; −7 **b.** −5; 0; −5 **c.** −3; 1; −3 **11.** 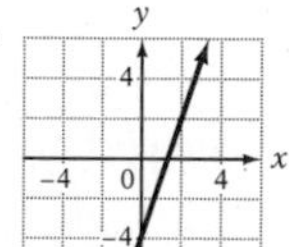 **13.** 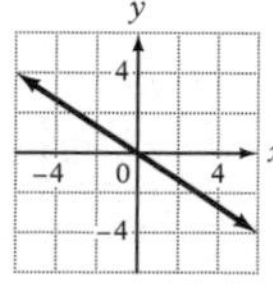

15. 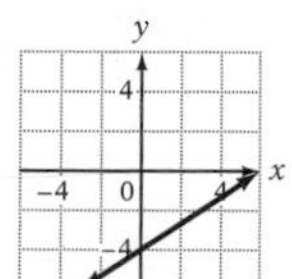 **17.** 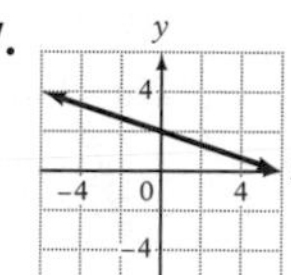 **19.** 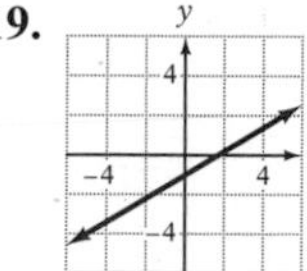 **21.** 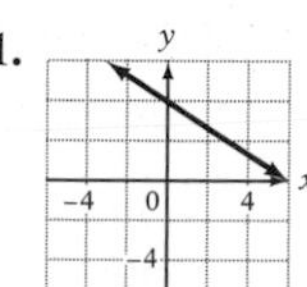 **23.** 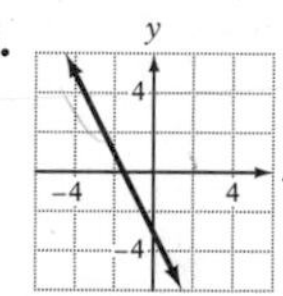 **25.** 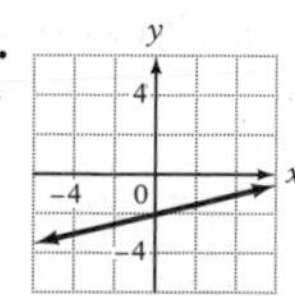

27. 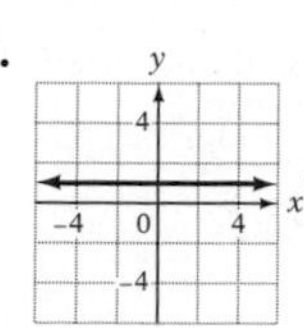 **29.** 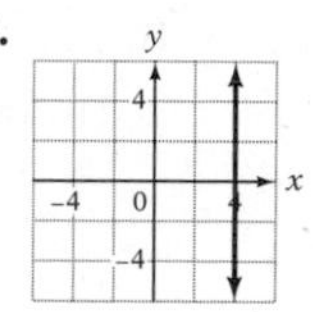 **31.** 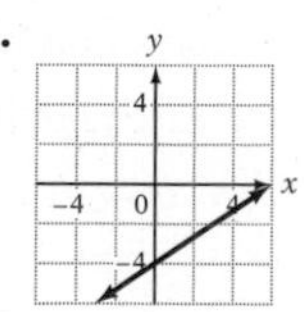 **33.** 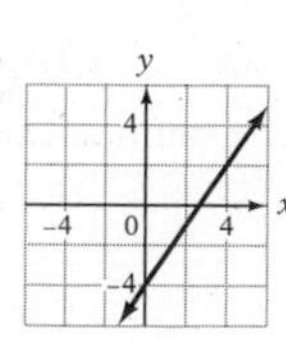 **35.** x-axis; y

37. x-intercept: $(-4, 0)$
y-intercept: $(0, 2)$

39. x-intercept: $\left(\frac{9}{2}, 0\right)$
y-intercept: $(0, -3)$

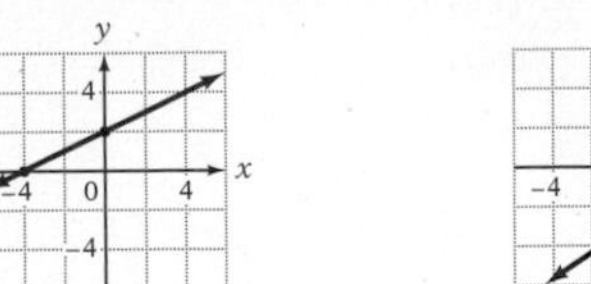

41. x-intercept: $(2, 0)$
y-intercept: $(0, -4)$

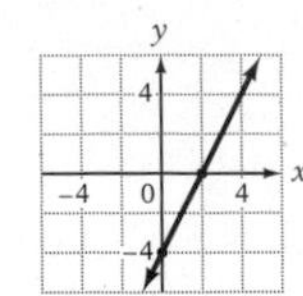

43. x-intercept: $\left(\frac{5}{3}, 0\right)$
y-intercept: $\left(0, \frac{5}{2}\right)$

45. x-intercept: $\left(\frac{4}{3}, 0\right)$
y-intercept: $(0, 2)$

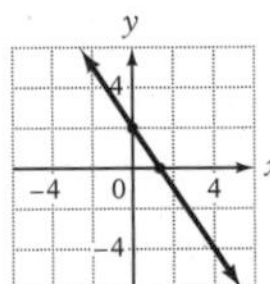

47. x-intercept: $(3, 0)$
y-intercept: $\left(0, -\frac{9}{5}\right)$

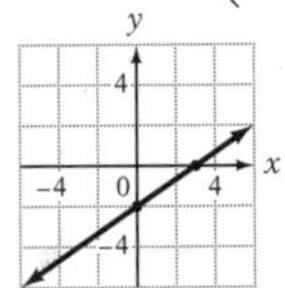

49. Opposite

51. The zero is -3. **53.** The zero is -1. **55.** The zero is 3. **57.** The zero is 3. **59.** The zero is $\frac{5}{2}$. **61.** The zero is 8. **63.** The zero is 6. **65.** The zero is $-\frac{8}{3}$. **67.** The zero is $\frac{8}{3}$.

69. The roller coaster travels 940 ft in 5 s.

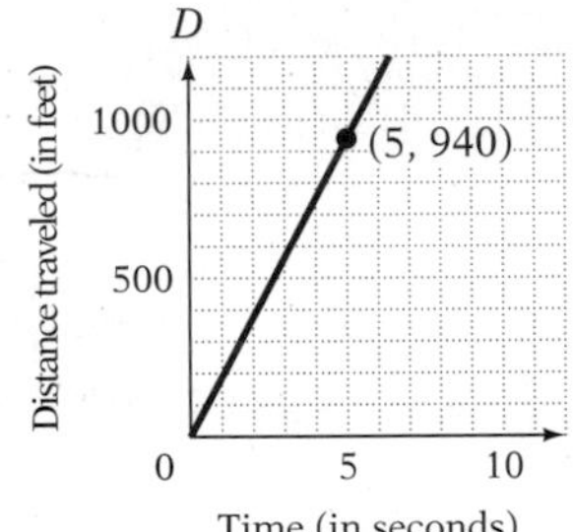

71. The realtor will earn a monthly income of \$4000 for selling \$60,000 worth of property.

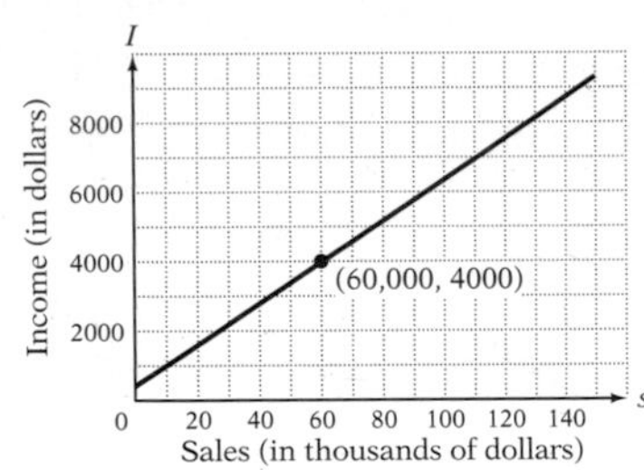

73. The caterer will charge \$614 for 120 hot appetizers.

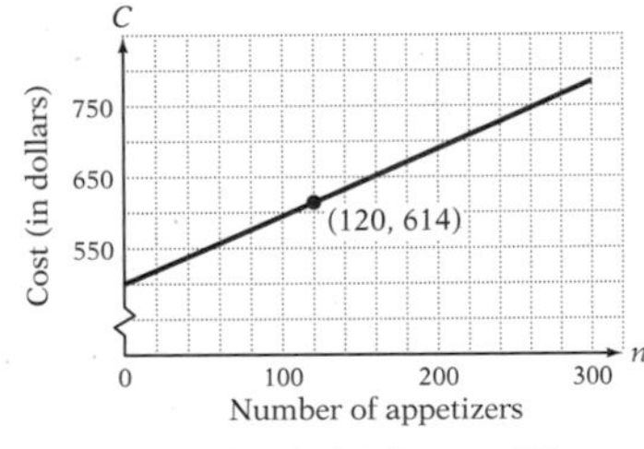

75. The technician charges \$60 to work for 15 min.

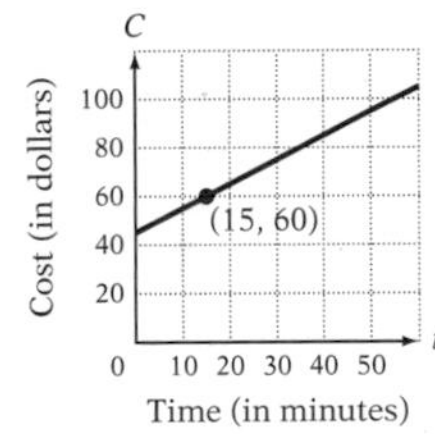

77. The sale price is \$160. **79.** **81.** **83.**

89. x-intercept: $(2, 0)$; y-intercept: $(0, -3)$ **91.** The equation is $y = -3x + 6$.

SECTION 3.4

1. increases **3.** The slope is $-\frac{3}{4}$. The coordinates of the y-intercept are $(0, 2)$. **5.** The slope is -1. The coordinates of the y-intercept are $(0, 3)$. **7.** -5; -4; 3; 2 **9.** The slope is -1. **11.** The slope is $\frac{1}{3}$. **13.** The slope is $-\frac{2}{3}$. **15.** The slope is $-\frac{3}{4}$. **17.** The slope is undefined. **19.** The slope is $\frac{7}{5}$. **21.** The slope is 0. **23.** The slope is $-\frac{1}{2}$. **25.** The slope is undefined. **27.** The slope is 40. The motorist is traveling 40 mph. **29.** The slope is -0.05. For each mile the car is driven, approximately 0.05 gal of fuel is used. **31.** Line A represents Lois's distance; line B, Tanya's distance; and line C, the distance between Lois and Tanya. **33. a.** The ramp does not meet the ANSI requirements. **b.** The ramp does meet the ANSI requirements. **35.** 5; $(0, -3)$

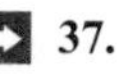 **37.** 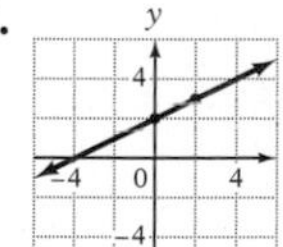 **39.** 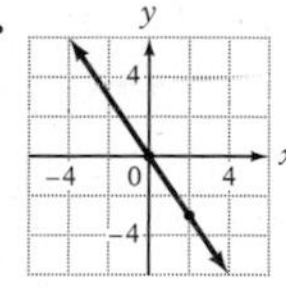 **41.** 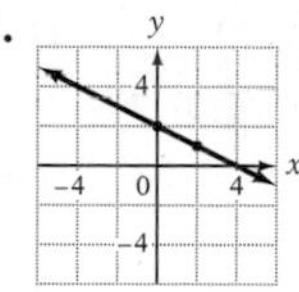 **43.** 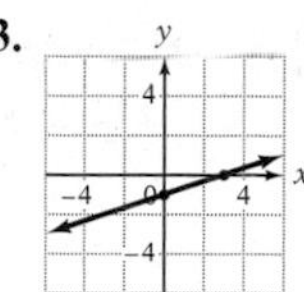 **45.** 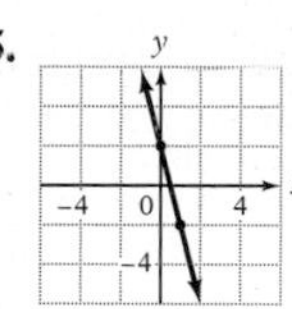 **47.** Below; positive

 **49.** 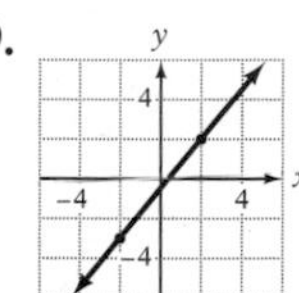 **51.** 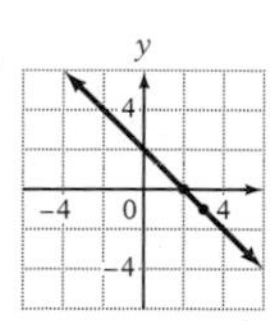 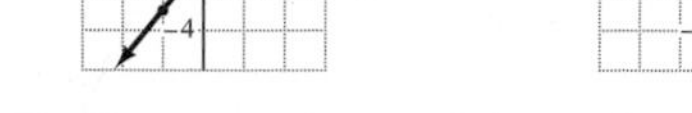

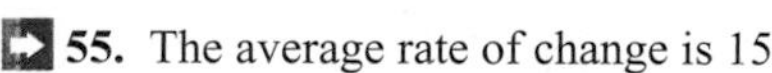 **53.** y-coordinate **55.** The average rate of change is 15.

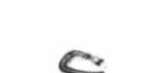

57. The average rate of change is -14. **59.** The average rate of change is 38. **61.** The average rate of change is 27. **63. a.** The average rate of change in the population was 179,000 people per year. **b.** The average annual rate of change in the population from 1900 to 1950 was less than the average annual rate of change from 1950 to 2000. **c.** The average annual rate of change in the population of Texas from 1980 to 2000 was less than the average annual rate of change in the population of California from 1980 to 2000. **65. a.** The goal for the average rate of change in green-house gas emissions from 2020 to 2030 is -0.18 billion metric tons per year. **b.** The goal for the average rate of change in green-house gas emissions from 2030 to 2050 is -0.15 billion metric tons per year. **67. a.** The average annual rate of change in the number of pedestrian fatalities from 1998 to 2008 was -85 pedestrians per year. **b.** The answer to part (a) can be considered encouraging because it reflects the fact that the number of fatalities is decreasing. **69. a.** The average rate of change in the number of applications from 1990 to 2000 was 25,500 applications per year. **b.** The average annual rate of change in the number of applications from 1995 to 2000 was 29,000 applications per year greater than the average rate of change from 1990 to 1995. **71.** 10 **73.** i and D; ii and C; iii and B; iv and F; v and E; vi and A **75.** increases by 2 **77.** increases by $\frac{1}{2}$

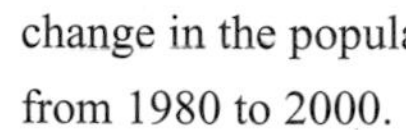

SECTION 3.5

1. One **3.** Yes **5.** $-\frac{6}{5}$; 2; $-\frac{6}{5}x + 2$ **7.** The equation of the line is $y = 2x + 5$. **9.** The equation of the line is $y = \frac{1}{2}x + 2$. **11.** The equation of the line is $y = -\frac{5}{3}x + 5$. **13.** The equation of the line is $y = -3x + 4$. **15.** The equation of the line is $y = \frac{1}{2}x$. **17.** The equation of the line is $y = 3x - 9$. **19.** The equation of the line is $y = -\frac{2}{3}x + 7$. **21.** The equation of the line is $y = -x - 3$. **23.** The equation of the line is $x = 3$. **25.** The equation of the line is $y = -3$. **27.** The equation of the line is $y = -2x + 3$. **29.** The equation of the line is $x = -5$. **33.** slope **35.** The equation of the line is $y = x + 2$. **37.** The equation of the line is $y = -2x - 3$. **39.** The equation of the line is $y = \frac{1}{3}x + \frac{10}{3}$. **41.** The equation of the line is $y = -\frac{3}{2}x + 3$. **43.** The equation of the line is $y = x - 1$. **45.** The equation of the line is $y = \frac{2}{3}x + \frac{5}{3}$. **47.** The equation of the line is $y = \frac{1}{2}x - 1$. **49.** The equation of the line is $y = -4$. **51.** The equation of the line is $y = \frac{3}{4}x$. **53.** The equation of the line is $x = -2$. **55.** The equation of the line is $y = x - 1$. **57.** The equation of the line is $y = -x + 3$. **59.** 1000; 500 **61.** $f(x) = 1200x$; the function predicts the plane will be 13,200 ft above sea level in 11 min. **63.** $f(x) = 0.59x + 4.95$; the function gives a cost of $12.62 for using a cell phone for 13 min in one month. **65.** $f(x) = 63x$; the function gives an estimate of 315 calories in a 5-ounce serving of lean hamburger. **67.** $f(x) = 0.625x - 1170.5$; the function predicts that 92% of trees at 2600 ft will be hardwoods in 2020. **69.** $f(x) = -0.032x + 16$; the function predicts that this car would contain 11.2 gal of gas after driving 150 mi. **71.** $\left(-\frac{b}{m}, 0\right)$ **73.** 0 **75.** Changing b moves the graph of the line up or down. **79.** -2

SECTION 3.6

3. slope **5.** The slope is -5. **7.** The slope is $-\frac{1}{4}$. **9.** The slope is $-\frac{1}{3}$. **11.** Yes **13.** No **15.** No **17.** Yes **19.** Yes **21.** Yes **23.** No **25.** Yes **27.** Yes **29.** The equation of the line is $y = \frac{2}{3}x - \frac{8}{3}$. **31.** The equation of the line is $y = -\frac{5}{4}x - 6$. **33.** The equation of the line is $y = \frac{1}{3}x - \frac{1}{3}$. **35.** The equation of the line is $y = -\frac{5}{3}x - \frac{14}{3}$. **37.** Downward to the right **39.** $\frac{A_1}{B_1} = \frac{A_2}{B_2}$ **41.** The equation of the line is $y = -2x + 15$. **43.** Any equation of the form $y = 2x + b$, where $b \neq -13$, or of the form $y = -\frac{3}{2}x + c$, where $c \neq 8$

SECTION 3.7

3. Yes **5.** No **7.** above **9.**

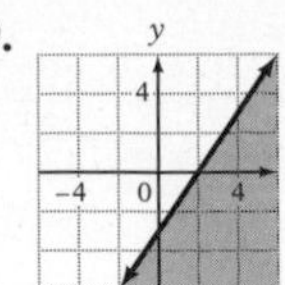

11.

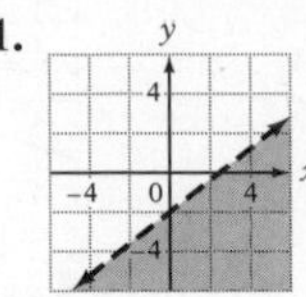

13.

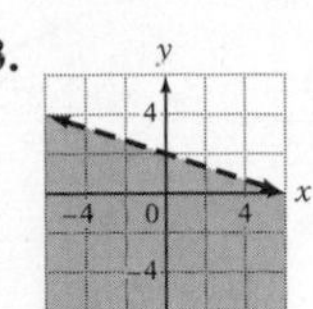

15.

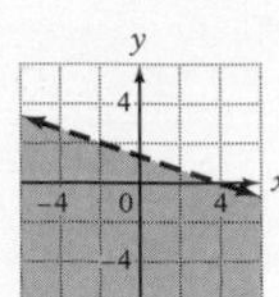

17.

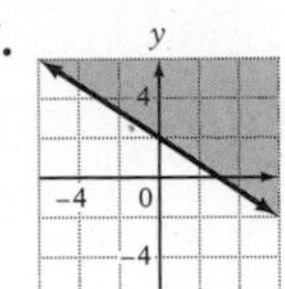

19.

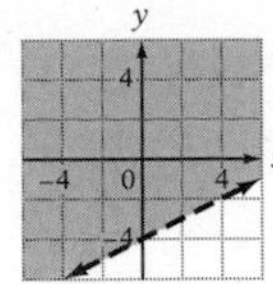

21.

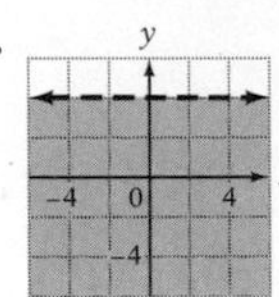

23.

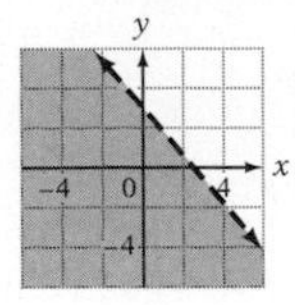

25.

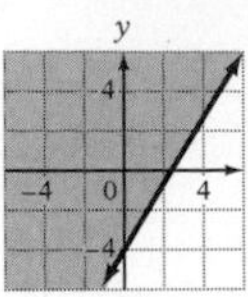

27. Negative

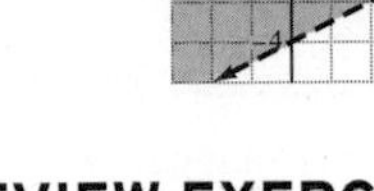

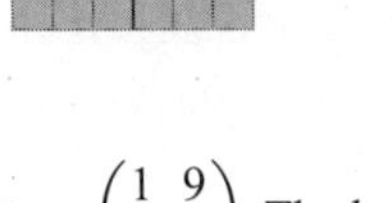

CHAPTER 3 REVIEW EXERCISES

1. (4, 2) [3.2.1] **2.** The coordinates of the midpoint are $\left(\frac{1}{2}, \frac{9}{2}\right)$. The length is $\sqrt{26}$. [3.1.1] **3.**

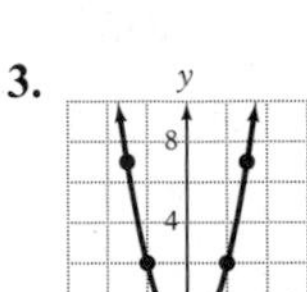

[3.2.2]

4.

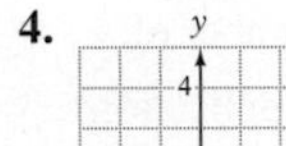

[3.3.2] **5.** $P(-2) = -2$; $P(a) = 3a + 4$ [3.2.1] **6.** The domain is $\{-1, 0, 1, 2, 5\}$. The range is $\{0, 2, 3\}$. Yes, the relation is a function. [3.2.1]

7. The range is $\{-2, -1, 2\}$. [3.2.1] **8.** The value of c is 2. The ordered pair is (2, 5). [3.2.1]

9. x-intercept: (3, 0)
y-intercept: (0, −2)

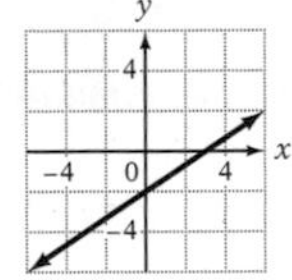

[3.3.2] **10.**

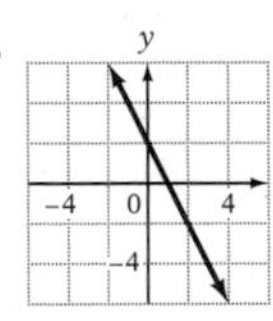

[3.3.1] **11.**

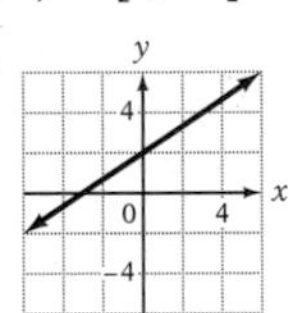

[3.3.2]

12. The slope is −1. [3.4.1] **13.** x-intercept: $\left(-\frac{4}{3}, 0\right)$
y-intercept: (0, −2)

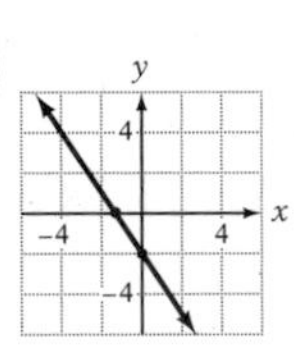

[3.3.2] **14.**

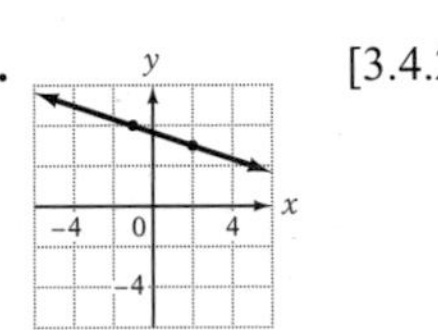

[3.4.2]

15. The equation of the line is $y = \frac{5}{2}x + \frac{23}{2}$. [3.5.1] **16.** The equation of the line is $y = -\frac{7}{6}x + \frac{5}{3}$. [3.5.2]

17. The equation of the line is $y = -3x + 7$. [3.6.1] **18.** The equation of the line is $y = \frac{2}{3}x - \frac{8}{3}$. [3.6.1]

19. The equation of the line is $y = \frac{3}{2}x + 2$. [3.6.1] **20.** The equation of the line is $y = -\frac{1}{2}x - \frac{5}{2}$. [3.6.1]

21. The zero is 4. [3.3.2] **22.** Yes [3.2.3] **23.** [3.7.1] **24.**

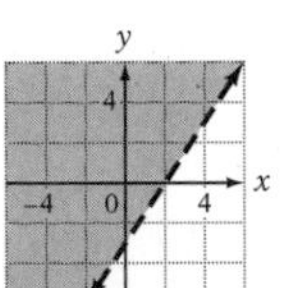

[3.7.1]

25. a. The average annual rate of change in the number of foreign-born people living in the United States from 1980 to 2000 was 715,000 people per year. **b.** The average annual rate of change in the number of foreign-born people living in the United States from 1970 to 1980 was 450,000 people per year. This is less than the average annual rate of change from 1980 to 1990. [3.4.3]

26.

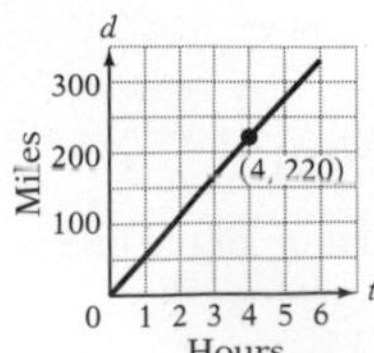

After 4 h, the car has traveled 220 mi. [3.3.3]

27. The slope is 20. The cost of manufacturing one calculator is \$20. [3.4.1] **28.** $f(x) = 80x + 25{,}000$; this function gives a cost of \$185,000 to build a house that contains 2000 ft^2. [3.5.3]

CHAPTER 3 TEST

1. [3.2.2, Example 6] **2.** The ordered-pair solution is $(-3, 0)$. [3.1.2, Example 3]

3. [3.3.1, Example 1] **4.** 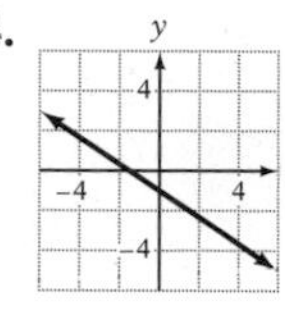 [3.3.2, Example 2] **5.** The equation of the line is $x = -2$. [3.5.1, Focus On part (B), page 162] **6.** The coordinates of the midpoint are $\left(-\frac{1}{2}, 5\right)$. The length is $\sqrt{117}$. [3.1.1, Examples 1 and 2] **7.** The slope is $-\frac{1}{6}$. [3.4.1, Example 1] **8.** 9 [3.2.1, Example 2] **9.** 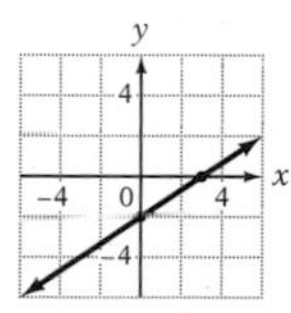 [3.3.2, Example 4]

10. [3.4.2, Example 4] **11.** The equation of the line is $y = \frac{2}{5}x + 4$. [3.5.1, Example 1]

12. The value of c is 1. The ordered pair is $(1, 3)$. [3.2.1, Example 5] **13.** The equation of the line is $y = -\frac{7}{5}x + \frac{1}{5}$. [3.5.2, Example 2] **14.** The zero is 6. [3.3.2, Example 5] **15.** The domain is $\{-4, -2, 0, 3\}$. The range is $\{0, 2, 5\}$. Yes, the relation is a function. [3.2.1, Example 1] **16.** The equation of the line is $y = -\frac{3}{2}x + \frac{7}{2}$. [3.6.1, Example 2] **17.** The equation of the line is $y = 2x + 1$. [3.6.1, Example 4] **18.** [3.7.1, Example 1] **19.** No [3.2.3, Example 8]

20. $f(x) = -\frac{3}{10}x + 175$; this function predicts that 85 students will enroll when the tuition is \$300. [3.5.3, Example 3]

21. The slope is $-\frac{10,000}{3}$. The value of the house decreases by \$3333.33 each year. [3.4.1, Example 2]

CUMULATIVE REVIEW EXERCISES

1. Commutative Property of Multiplication [1.3.1] **2.** The solution is $\frac{9}{2}$. [2.1.2] **3.** The solution is $\frac{8}{9}$. [2.1.2] **4.** The solution is $-\frac{1}{14}$. [2.1.2] **5.** The solution set is $\{x|x < -1\} \cup \left\{x|x > \frac{1}{2}\right\}$. [2.5.2] **6.** The solutions are $\frac{5}{2}$ and $-\frac{3}{2}$. [2.6.1] **7.** The solution set is $\left\{x\middle|0 < x < \frac{10}{3}\right\}$. [2.6.2] **8.** 8 [1.2.2] **9.** -18 [1.4.2] **10.** [1.1.2] **11.** The solution set is $\varnothing$. [2.5.2] **12.** 14 [3.2.1] **13.** The ordered-pair solution is $(-8, 13)$. [3.1.1] **14.** The slope is $-\frac{7}{4}$. [3.4.1] **15.** The equation of the line is $y = \frac{3}{2}x + \frac{13}{2}$. [3.5.1] **16.** The equation of the line is $y = -\frac{5}{4}x + 3$. [3.5.2] **17.** The equation of the line is $y = -\frac{3}{2}x + 7$. [3.6.1] **18.** The equation of the line is $y = -\frac{2}{3}x + \frac{8}{3}$. [3.6.1] **19.** [3.3.2] **20.** [3.4.2]

21. 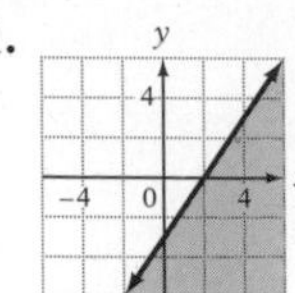

[3.7.1] **22.** The speed of the first plane is 200 mph, and the speed of the second plane is 400 mph. [2.3.2]

23. The mixture consists of 32 lb of $8 coffee and 48 lb of $3 coffee. [2.3.1] **24.** The slope is −2500. The value of the truck decreases by $2500 each year. [3.4.1]

Answers to Chapter 4 Selected Exercises

PREP TEST

1. $6x + 5y$ [1.4.3] **2.** 7 [1.4.2] **3.** 0 [2.1.1] **4.** The solution is −3. [2.1.2] **5.** The solution is 1000. [2.1.2]

6. 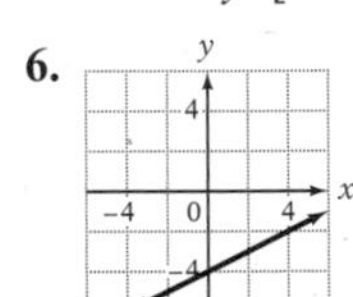

[3.3.2] **7.** 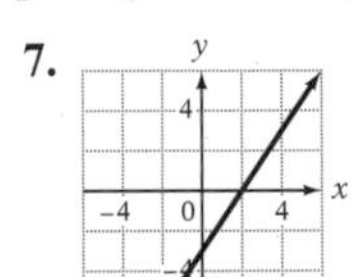

[3.3.2] **8.** 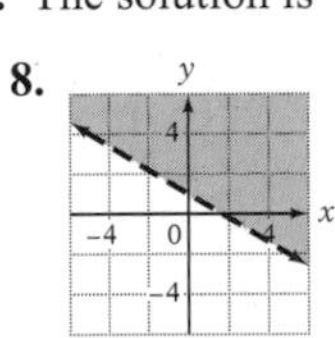

[3.7.1]

SECTION 4.1

1. exactly one solution; no solution; infinitely many solutions **3.** Yes **5.** Independent **7.** Inconsistent **9.** (−2, 3)

11. 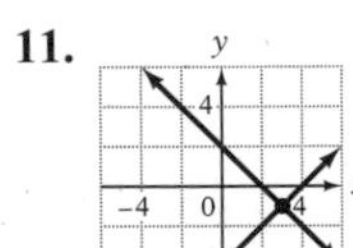

The solution is (3, −1).

13. 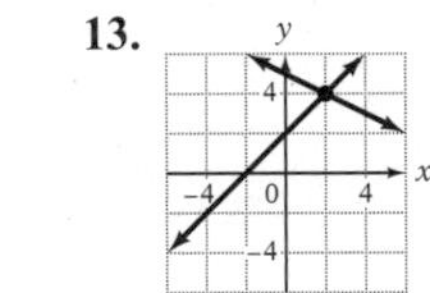

The solution is (2, 4).

15. 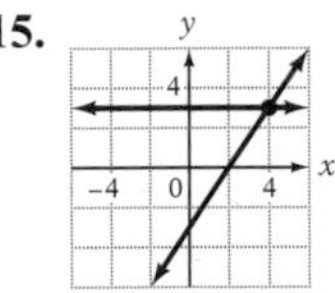

The solution is (4, 3).

17. 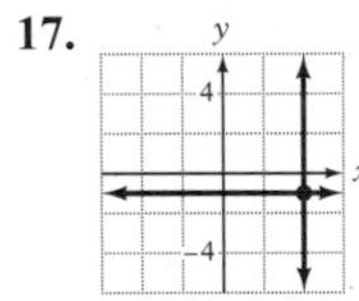

The solution is (4, −1).

19. 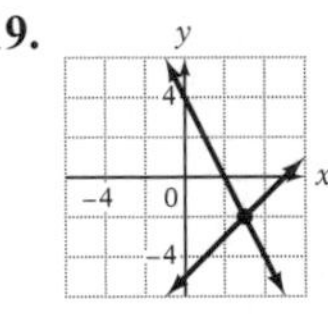

The solution is (3, −2).

21. 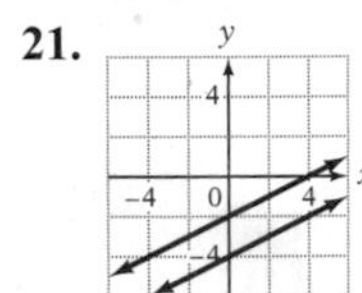

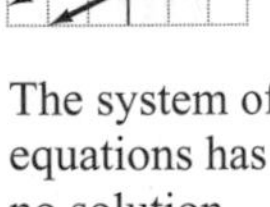
The system of equations has no solution.

23. 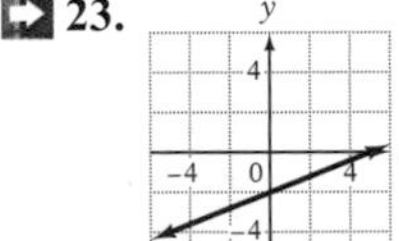

The solutions are the ordered pairs $\left(x, \frac{2}{5}x - 2\right)$.

25. 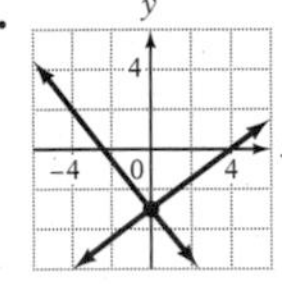

The solution is (0, −3).

27. $a_1 \neq a_2$ **31.** 3; −2; 3; −2

33. The solution is (2, 1). **35.** The solution is (1, 1). **37.** The solution is (2, 1). **39.** The solution is (−2, −3).

41. The solution is (3, −4). **43.** The solution is (0, −1). **45.** The solution is $\left(\frac{1}{2}, 3\right)$. **47.** The solution is (1, 2).

49. The solution is (−1, 2). **51.** The solution is (−2, 5). **53.** The solution is (0, 0). **55.** The solution is (−4, −6).

57. The system of equations has no solution. **59.** The solutions are the ordered pairs $\left(x, \frac{1}{2}x - 4\right)$.

61. The system of equations has no solution. **63.** The solution is (1, 5). **65.** The solution is (5, 2).

67. The solution is (1, 4). **69.** Any real number except 3 **71.** $\frac{a_1}{b_1} \neq \frac{a_2}{b_2}$ **73.** The solution is (1.20, 1.40).

75. The solution is (0.64, −0.10).

SECTION 4.2

1. No **5.** $A = -2, B = 12, C = -4$ **7.** The solution is (6, 1). **9.** The solution is (1, 1).

11. The solution is (2, 1). **13.** The solutions are the ordered pairs $(x, 3x - 4)$. **15.** The solution is $\left(-\frac{1}{2}, 2\right)$.

17. The system of equations has no solution. **19.** The solution is (−1, −2). **21.** The solution is (2, 5).

23. The solution is $\left(\frac{1}{2}, \frac{3}{4}\right)$. **25.** The solution is (0, 0). **27.** The solution is $\left(\frac{2}{3}, -\frac{2}{3}\right)$. **29.** The solution is (1, −1).

31. The solutions are the ordered pairs $\left(x, \frac{4}{3}x - 6\right)$. **33.** The solution is (5, 3). **35.** The solution is $\left(\frac{1}{3}, -1\right)$.

37. The solution is $\left(\frac{5}{3}, \frac{1}{3}\right)$. **39.** The system of equations has no solution. **43.** Dependent **45.** -2
47. The solution is $(-1, 2, 1)$. **49.** The solution is $(6, 2, 4)$. **51.** The solution is $(4, 1, 5)$. **53.** The solution is $(3, 1, 0)$.
55. The solution is $(-1, -2, 2)$. **57.** The system of equations has no solution. **59.** The solution is $(2, 1, -3)$.
61. The solution is $(2, -1, 3)$. **63.** The solution is $(6, -2, 2)$. **65.** The solution is $(0, -2, 0)$. **67.** The solution is $(2, 3, 1)$.
69. The solution is $(1, 1, 3)$. **71.** The solution is $(2, 1)$. **73.** The solution is $(2, -1)$. **75.** The solution is $(2, 0, -1)$.
77. $A = 2, B = 3, C = -3$ **79.** 3

SECTION 4.3

3. 3×4 **5.** 2, 0, 3 **11.** 8; -1; -3 **13.** The minor of 7 is $\begin{vmatrix} -3 & 4 \\ 0 & 8 \end{vmatrix}$. The minor of 8 is $\begin{vmatrix} 2 & 7 \\ -3 & -1 \end{vmatrix}$.

15. 11 **17.** 18 **19.** 0 **21.** 15 **23.** -30 **25.** 0 **27. a.** 2; -3; -5; 4 **b.** -7 **29. a.** 2; 5; -5; -2
b. 21 **31.** The solution is $(3, -4)$. **33.** The solution is $(4, -1)$. **35.** The solution is $\left(\frac{11}{14}, \frac{17}{21}\right)$.

37. The solution is $\left(\frac{1}{2}, 1\right)$. **39.** Not possible by Cramer's Rule **41.** The solution is $(-1, 0)$.

43. The solution is $(1, -1, 2)$. **45.** The solution is $(2, -2, 3)$. **47.** Not possible by Cramer's Rule

49. The solution is $\left(\frac{68}{25}, \frac{56}{25}, -\frac{8}{25}\right)$. **51.** -3 **53.** 1; 0 **55.** The augmented matrix is $\left[\begin{array}{cc|c} 2 & 1 & 4 \\ 1 & -3 & -10 \end{array}\right]$.

57. The augmented matrix is $\left[\begin{array}{ccc|c} 3 & 1 & -3 & -5 \\ 1 & -1 & 2 & 6 \\ -1 & 4 & -1 & 0 \end{array}\right]$. **59.** $8x - y = -15$, $2x + y = -3$ **61.** $x - 4y - z = 0$, $y + 6z = 15$, $z = 2$

63. $\left[\begin{array}{ccc|c} 5 & -1 & 3 & -2 \\ 2 & -3 & 4 & 1 \\ 1 & 2 & -3 & 7 \end{array}\right]$ **65.** $\left[\begin{array}{cc|c} 1 & \frac{4}{3} & -\frac{2}{3} \\ 2 & -3 & 5 \end{array}\right]$ **67.** $\left[\begin{array}{ccc|c} 1 & 3 & -2 & 1 \\ 0 & 4 & -3 & 2 \\ 0 & 7 & -6 & 4 \end{array}\right]$

69. A row echelon form of the matrix is $\left[\begin{array}{cc|c} 1 & -2 & \frac{1}{2} \\ 0 & 1 & \frac{5}{2} \end{array}\right]$. **71.** A row echelon form of the matrix is $\left[\begin{array}{cc|c} 1 & -\frac{2}{5} & \frac{3}{5} \\ 0 & 1 & 26 \end{array}\right]$.

73. A row echelon form of the matrix is $\left[\begin{array}{ccc|c} 1 & 4 & 1 & -2 \\ 0 & 1 & 4 & -8 \\ 0 & 0 & 1 & -\frac{32}{19} \end{array}\right]$. **75.** A row echelon form of the matrix is $\left[\begin{array}{ccc|c} 1 & -3 & \frac{1}{2} & -\frac{3}{2} \\ 0 & 1 & \frac{3}{2} & \frac{5}{2} \\ 0 & 0 & 1 & 3 \end{array}\right]$.

77. The solution is $(-7, 4, 3)$. **79.** The solution is $(1, 3)$. **81.** The solution is $(-1, -3)$. **83.** The solution is $(3, 2)$.
85. The system of equations has no solution. **87.** The solution is $(-2, 2)$. **89.** The solution is $(0, 0, -3)$.
91. The solution is $(1, -1, -1)$. **93.** The system of equations has no solution. **95.** The solution is $\left(\frac{1}{3}, \frac{1}{2}, 0\right)$.
97. The solution is $\left(\frac{1}{5}, \frac{2}{5}, -\frac{3}{5}\right)$. **99.** The solution is $\left(\frac{1}{4}, 0, -\frac{2}{3}\right)$. **101.** -14 **103.** 0 **105. a.** 0 **b.** 0
107. The area of the polygon is 239 ft^2.

SECTION 4.4

1. The speed of the plane is 450 mph. **3.** $50x + 100y$ **5.** Row 1: $b + c$; 3; $3(b + c)$; Row 2: $b - c$; 5; $5(b - c)$
7. Less than **9.** The rate of the plane in calm air is 150 mph. The rate of the wind is 10 mph. **11.** The rate of the cabin cruiser in calm water is 14 mph. The rate of the current is 2 mph. **13.** The rate of the plane in calm air is 165 mph. The rate of the wind is 15 mph. **15.** The rate of the boat in calm water is 19 km/h. The rate of the current is 3 km/h. **17.** The rate of the plane in calm air is 105 mph. The rate of the wind is 15 mph. **19.** The rate of the boat in calm water is 16.5 mph. The rate of the current is 1.5 mph. **21.** Table 1, Row 1: 200; h; $200h$; Row 2: 300; w; $300w$; Table 2, Row 1: 350; h; $350h$; Row 2: 100; w; $100w$
23. The cost of the pine is \$.18 per foot. The cost of the redwood is \$.30 per foot. **25.** The cost per unit of gas is \$.16.
27. The company plans to manufacture 25 mountain bikes during the week. **29.** The owner drove 322 mi in the city and 72 mi on the highway. **31.** The pharmacist should use 200 mg of the first powder and 450 mg of the second powder. **33.** The manufacturer charges \$4000 for a Model VI computer. **35.** \$8000 was invested at 9%, \$6000 at 7%, and \$4000 at 5%. **37.** The distances are $d_1 = 6$ in., $d_2 = 3$ in., and $d_3 = 9$ in. **39.** The measures of the two angles are 35° and 145°. **41.** The point of no return is approximately 1517 mi from New York.

SECTION 4.5

1. ii **3.** Intersection **5. a.** A; D **b.** A; B **c.** A **7.**

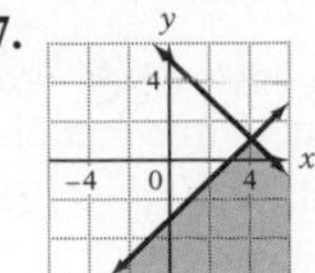

9.

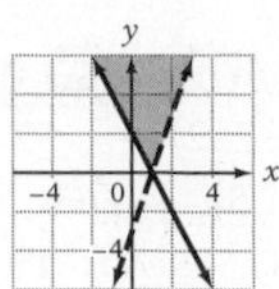

11.

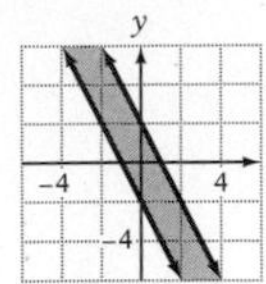

13.

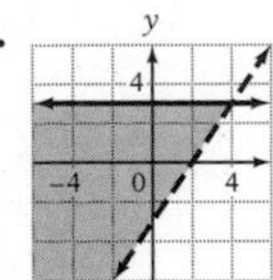

15.

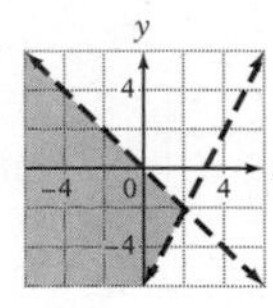

17.

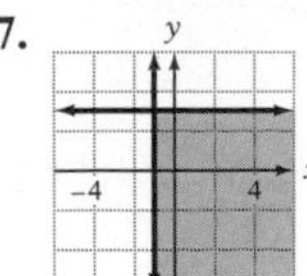

19.

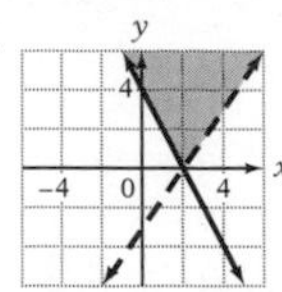

21.

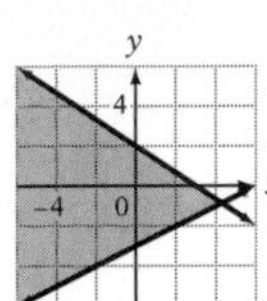

23.

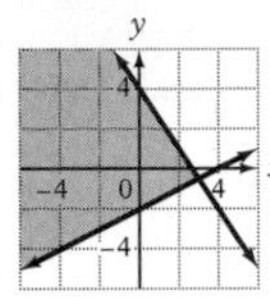

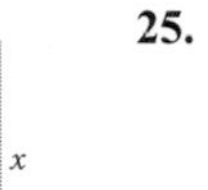

25. Points below the graph of the line $x + y = b$ **27.** Region between the graphs of the parallel lines $x + y = a$ and $x + y = b$

29.

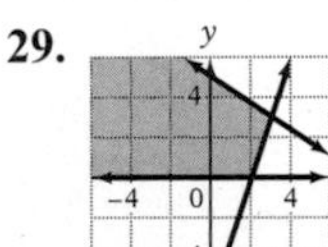

31.

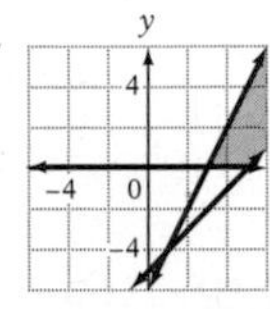

33.

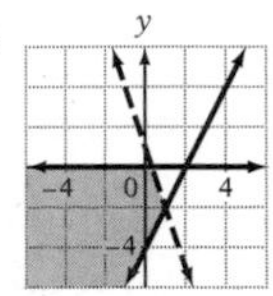

35. ii, iii

CHAPTER 4 REVIEW EXERCISES

1. The system of equations has no solution. [4.1.2] **2.** The solutions are the ordered pairs $\left(x, -\frac{1}{4}x + \frac{3}{2}\right)$. [4.1.2]

3. The solution is $(-4, 7)$. [4.2.1] **4.** The solutions are the ordered pairs $\left(x, \frac{1}{3}x - 2\right)$. [4.2.1]

5. The solution is $(5, -2, 3)$. [4.2.2] **6.** The solution is $(3, -1, -2)$. [4.2.2] **7.** 28 [4.3.1] **8.** 0 [4.3.1]

9. The solution is $(3, -1)$. [4.3.2] **10.** The solution is $\left(\frac{110}{23}, \frac{25}{23}\right)$. [4.3.2] **11.** The solution is $(-1, -3, 4)$. [4.3.2]

12. The solution is $(2, 3, -5)$. [4.3.2] **13.** The solution is $(1, -1, 4)$. [4.2.2] **14.** The solution is $\left(\frac{8}{5}, \frac{7}{5}\right)$. [4.3.2]

15. The solution is $\left(\frac{1}{2}, -1, \frac{1}{3}\right)$. [4.3.3] **16.** 12 [4.3.1] **17.** The solution is $(2, -3)$. [4.3.2]

18. The solution is $(2, -3, 1)$. [4.3.3] **19.** [4.1.1]

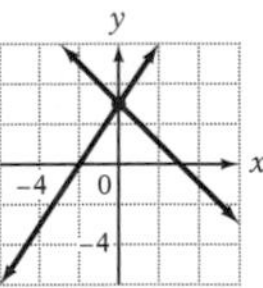

The solution is $(0, 3)$.

20. [4.1.1]

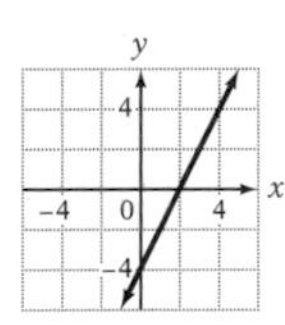

The solutions are the ordered pairs $(x, 2x - 4)$.

21. [4.5.1] **22.** [4.5.1]

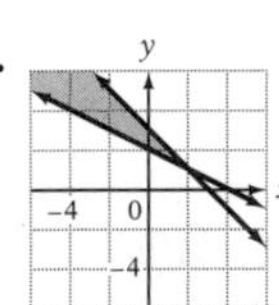

23. The rate of the cabin cruiser in calm water is 16 mph. The rate of the current is 4 mph. [4.4.1]

24. The rate of the plane in calm air is 175 mph. The rate of the wind is 25 mph. [4.4.1] **25.** The number of children attending on Friday was 100. [4.4.2] **26.** There is \$10,400 invested at 8%, \$5200 invested at 6%, and \$9400 invested at 4%. [4.2.2]

CHAPTER 4 TEST

1. The solution is $\left(\frac{3}{4}, \frac{7}{8}\right)$. [4.1.2, Example 3A] **2.** The solution is $(-3, -4)$. [4.1.2, Focus On, page 191]

3. The solution is $(2, -1)$. [4.1.2, Example 3A] **4.** The solution is $(-2, 1)$. [4.3.3, Example 7]
5. The system of equations has no solution. [4.2.1, Problem 1B] **6.** The solution is $(1, 1)$. [4.2.1, Examples 1 and 2]
7. The system of equations has no solution. [4.2.2, Focus On, part B, page 200] **8.** The solution is $(2, -1, -2)$. [4.3.3, Example 8]

9. 10 [4.3.1, Example 1A] **10.** −32 [4.3.1, Example 1B] **11.** The solution is $\left(-\frac{1}{3}, -\frac{10}{3}\right)$. [4.3.2, Example 2]

12. The solution is $\left(\frac{1}{5}, -\frac{6}{5}, \frac{3}{5}\right)$. [4.3.2, Example 3] **13.** The solution is $(0, -2, 3)$. [4.3.2, Example 3]

14. 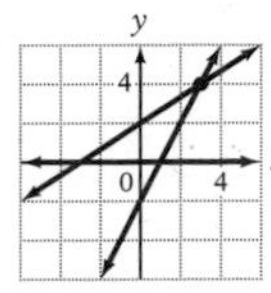 [4.1.1, Example 1]

The solution is $(3, 4)$.

15. 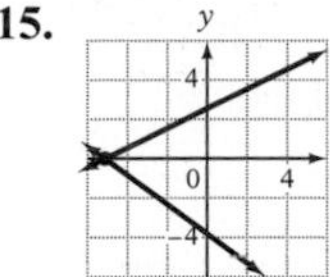 [4.1.1, Example 1]

The solution is $(-5, 0)$.

16. 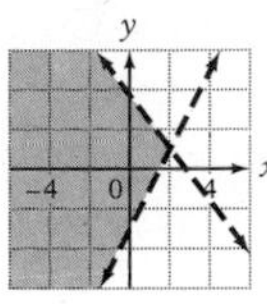 [4.5.1, Example 1] **17.** [4.5.1, Example 1]

18. The rate of the plane in calm air is 150 mph. The rate of the wind is 25 mph. [4.4.1, Problem 1] **19.** The cost per yard of cotton is \$9. The cost per yard of wool is \$14. [4.4.2, Problem 2]

CUMULATIVE REVIEW EXERCISES

1. The solution is $-\frac{11}{28}$. [2.1.1] **2.** The equation is $y = 5x - 11$. [3.5.2] **3.** $3x - 24$ [1.4.3] **4.** −4 [1.4.2]
5. The solution set is $\{x|x < 6\}$. [2.4.2] **6.** The solution set is $\{x| -4 < x < 8\}$. [2.5.2]
7. The solution set is $\{x|x > 4\} \cup \{x|x < -1\}$. [2.5.2] **8.** −98 [3.2.1]
9. The range is $\{0, 1, 5, 8, 16\}$. [3.2.1] **10.** 1 [3.2.1] **11.** $3h$ [3.2.1] **12.** −5 −4 −3 −2 −1 0 1 2 3 4 5 [1.1.2]

13. The equation is $y = -\frac{2}{3}x + \frac{5}{3}$. [3.5.1] **14.** The equation is $y = -\frac{3}{2}x + \frac{1}{2}$. [3.6.1] **15.** The distance is $2\sqrt{10}$. [3.1.1]

16. The coordinates of the midpoint are $\left(-\frac{1}{2}, 4\right)$. [3.1.1] **17.** 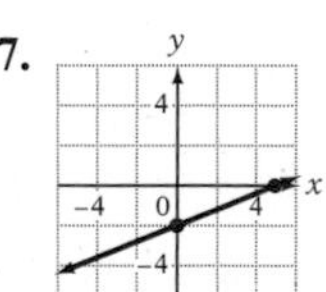 [3.4.2] **18.** 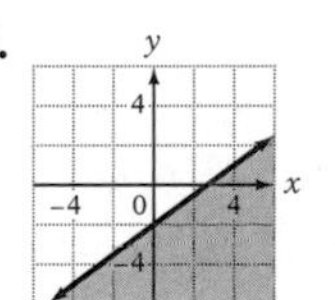 [3.7.1]

19. The solution is $(-5, -11)$. [4.1.2] **20.** The solution is $(1, 0, -1)$. [4.2.2] **21.** 3 [4.3.1] **22.** [4.1.1]

The solution is $(2, 0)$.

23. The solution is $(2, -3)$. [4.3.2] **24.** 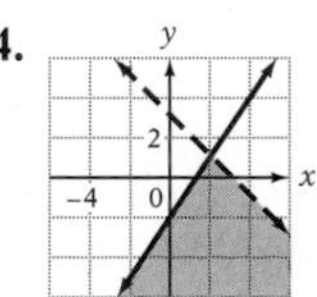 [4.5.1] **25.** The amount of water that should be added is 60 ml. [2.3.2]

26. The rate of the wind is 12.5 mph. [4.4.1] **27.** The cost per pound of steak is \$5. [4.4.2] **28.** The lower and upper limits of the resistor are 10,200 ohms and 13,800 ohms. [2.5.3] **29.** The slope is 40. The account executive earns \$40 for each \$1000 of sales. [3.4.1]

Answers to Chapter 5 Selected Exercises

PREP TEST

1. $-12y$ [1.4.3] **2.** -8 [1.2.1] **3.** $3a - 8b$ [1.4.3] **4.** $11x - 2y - 2$ [1.4.3] **5.** $-x + y$ [1.4.3]
6. $2 \cdot 2 \cdot 2 \cdot 5$ [1.3.1] **7.** 4 [1.3.1] **8.** -13 [1.4.2] **9.** The solution is $-\frac{1}{3}$. [2.1.1]

SECTION 5.1

1. i, ii, iv **5.** 32 **7.** x^4y^5; 9 **9. a.** Yes **b.** No **c.** No **d.** Yes **11.** $24r^3t^6$ **13.** $a^2b^3c^9$ **15.** b^{10}
17. x^4y^{24} **19.** x^8y^{16} **21.** $81a^4$ **23.** $256c^{20}$ **25.** $64x^{30}y^6$ **27.** $-8a^3b^6$ **29.** $-32x^{10}y^5z^{15}$ **31.** $-6x^5y^5z^4$
33. $-12a^2b^9c^2$ **35.** $-6x^4y^4z^5$ **37.** $-432a^7b^{11}$ **39.** $54a^{13}b^{17}$ **43.** a^{-8}; a^{-16}; $1/a^{16}$ **45.** $3a^3$ **47.** $\frac{a^5}{b^3}$
49. $\frac{2ab^2c}{3}$ **51.** y^3 **53.** 243 **55.** y^3 **57.** $\frac{a^3b^2}{4}$ **59.** $\frac{1}{x^8}$ **61.** $\frac{1}{125x^6}$ **63.** x^9 **65.** a^2 **67.** $\frac{1}{x^6y^{10}}$
69. $\frac{1}{2187a}$ **71.** $\frac{y^6}{x^3}$ **73.** $\frac{y^4}{x^8}$ **75.** $\frac{b^2c^2}{a^4}$ **77.** $-\frac{9a}{8b^6}$ **79.** $\frac{b^{10}}{a^{10}}$ **81.** $\frac{4x^5}{9y^4}$ **83.** $\frac{8b^{15}}{3a^{18}}$ **85.** $x^{12}y^6$
87. $\frac{b^2}{a^4}$ **89. a.** False **b.** True **91.** 7; right; -7 **93.** 5×10^{-8} **95.** 4.3×10^6 **97.** 9.8×10^9
99. 0.0000000000062 **101.** 634,000 **103.** 4,350,000,000 **105.** 2.848×10^{-10} **107.** 4.608×10^{-3}
109. 9×10^9 **111.** 2×10^{10} **113.** 6.92×10^{-1} **115.** 1.38 **117.** 6×10^{-15} **119.** Less than **121.** 5.4; 9
123. Light travels 2.592×10^{13} m in 1 day. **125.** The proton is 1.83664508×10^3 times heavier than the electron.
127. The *Phoenix* traveled an average of 1.43×10^6 mi/day. **129.** 1.44×10^{11} new seedlings would be growing.
131. It takes light 5×10^2 s to travel from the sun to Earth. **133.** The volume of the cell is $1.41371669 \times 10^{-11}$ mm^3.
135. The spaceship travels across the galaxy in 2.24×10^{15} h. **137. a.** $\frac{1}{16}$ **b.** $\frac{1}{512}$ **c.** $\frac{1}{2}$ **d.** $\frac{1}{16}$ **139.** $-6x^{2n}y^{3m}$
141. x^{3n} **143.** $\frac{x^{6n}}{9}$ **145.** $6x^3y^6$ **147.** $\frac{4}{m^4}$ **149.** $a^{-1} > b^{-1}$ **151.** 15; $2^4 - 1$

SECTION 5.2

1. a. binomial **b.** binomial **c.** monomial **d.** trinomial **e.** none **f.** monomial **3.** All real numbers
7. 2; leading; constant **9. a.** The leading coefficient is -1. **b.** The constant term is 8. **c.** The degree is 2.
11. The expression is not a polynomial. **13.** The expression is not a polynomial. **15. a.** The leading coefficient is 3.
b. The constant term is π. **c.** The degree is 5. **17. a.** The leading coefficient is -5. **b.** The constant term is 2.
c. The degree is 3. **19. a.** The leading coefficient is 14. **b.** The constant term is 14. **c.** The degree is 0.
21. 13 **23.** 10 **25.** -11 **27.** **29.**

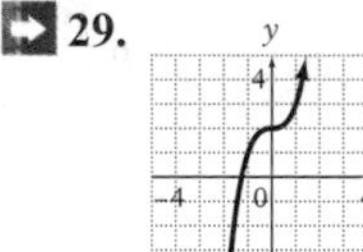

31.

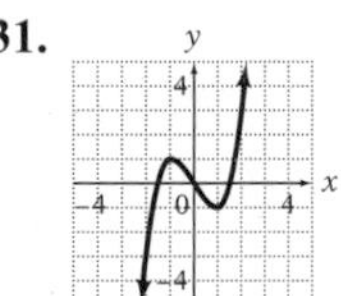

33. The length of the wave is 23.1 m. **35.** There must be 56 games scheduled. **37.** The Baked Alaska needs 260 in^3 of meringue. **39.** A **41.** $-8x^2 - 3x + 5$ **43.** $6x^2 - 6x + 5$ **45.** $-x^2 + 1$ **47.** $17x^3 + x^2 - 2x - 6$
49. $-8x^2 + 6x + 9$ **51.** $13x^2 - 8x + 7$ **53.** $5y^2 - 15y + 2$ **55.** $7a^2 - a + 2$ **57.** $6x^2 + 13x$
59. $9x^3 - 4x^2 + 11x - 8$ **61.** $-3x^3 - 8x^2 - 2x + 6$ **63.** $5x^3 - 4x^2 + 4x - 7$ **65.** $3x^4 - 8x^2 + 2x$
67. $-2x^3 + 4x^2 + 8$ **69.** $P(x) + Q(x)$ is a fourth-degree polynomial. **71.** -2 **73.** 2 **75.**

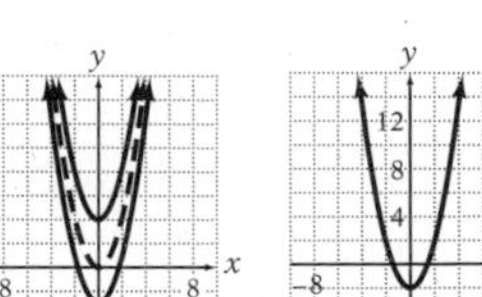

The graph of k is the graph of f moved 2 units down. **77.** The maximum deflection of the beam is 3.125 in. **81.** It would take the shamans about 5.85×10^{11} years to move all the golden disks.

SECTION 5.3

1. 5 **3.** True **7.** $-3y^3$; $-3y^3$; Distributive; $-3y^5$; $30y^3$ **9.** $2x^2 - 6x$ **11.** $6x^4 - 3x^3$ **13.** $6x^2y - 9xy^2$
15. $-4b^2 + 10b$ **17.** $5a^2 - 6a$ **19.** $-6a^4 + 4a^3 - 6a^2$ **21.** $9b^5 - 9b^3 + 24b$ **23.** $-20x^5 - 15x^4 + 15x^3 - 20x^2$
25. $-2x^4y + 6x^3y^2 - 4x^2y^3$ **27.** $5y^2 - 11y$ **29.** $6y^2 - 31y$ **31.** No **33.** No **35.** $3x$; x; $3x$; 5; -2; x; -2; 5

37. $24x^3 - 61x^2 - 5x - 8$ **39.** $-6x^3 + 25x^2 - 16$ **41.** $-12x^3 + 24x^2 - 6x - 6$ **43.** $-35x^4 - 22x^3 + 53x^2 - 18x - 18$ **45.** $4x^4 + 28x^3 - 37x^2 - 35x + 40$ **47.** $4x^4 - 4x^3 + 5x^2 - 9x + 4$ **49.** $-5x^4 + 7x^3 + 24x^2 + 20x + 32$ **51.** $24x^4 + 55x^3 + 19x^2 + 15x - 18$ **53.** $-2x^4 - 12x^3 - 3x^2 + 32x - 15$ **55.** $2x^3 + 3x^2 - 23x - 12$ **57.** $3x^3 + x^2 - 27x - 9$ **59.** $4x^3 - 57x + 28$ **61.** $x^2 + 28x + 195$ **63.** $x^2 - 21x + 104$ **65.** $x^2 + 8x - 48$ **67.** $x^2 + 12x + 27$ **69.** $x^2 + 4x - 32$ **71.** $12x^2 + 24x + 12$ **73.** $-9x^2 - 21x - 12$ **75.** $-5x^2 - 31x - 6$ **77.** $9x^2 - 15x - 24$ **79.** $2x^2 + 5x + 2$ **81.** $-24x^2 - 46x - 10$ **83.** $x^2 - 2xy - 8y^2$ **85.** $3a^2 - 8ab - 3b^2$ **87.** $7x^2 + 34xy - 5y^2$ **89.** $x^2y^2 - xy - 20$ **91.** $3a^2b^2 + 4ab - 32$ **93.** $10a^4 + 3a^2 - 18$ **95.** No **97.** $4x$; $4x$; 1; 1; $16x^2$; $8x$; 1 **99.** $b^2 - 49$ **101.** $b^2 - 121$ **103.** $25x^2 - 40xy + 16y^2$ **105.** $x^4 + 2x^2y^2 + y^4$ **107.** $4a^2 - 9b^2$ **109.** $x^4 - 1$ **111.** $25a^2 - 81b^2$ **113.** $9x^2 - 49y^2$ **115.** $36 - x^2$ **117.** $9a^2 - 24ab + 16b^2$ **119.** $9x^6 + 12x^3 + 4$ **121.** Zero **123.** Positive **125. a.** $3x$; 3 **b.** 3; $9x$; x; $x^2 + 3x$; $9x$; $x^2 + 3x$; $x^2 + 12x$ **127.** The area is $\left(x^2 + \frac{x}{2} - 3\right)$ ft². **129.** The area is $(x^2 + 12x + 16)$ ft². **131.** The volume is $(4x^3 - 72x^2 + 324x)$ in³. **133.** The volume is $(2x^3 - 7x^2 - 15x)$ cm³. **135.** The volume is $(4x^3 + 32x^2 + 48x)$ cm³. **137.** $2x^2 + 4xy$ **139.** $6x^5 - 5x^4 + 8x^3 + 13x^2$ **141.** $5x^2 - 5y^2$ **143.** $2x^{2n} + 5x^n - 12$ **145.** $x^{2n} - 49$ **147.** $x^{2n} - y^{2n}$ **149.** $9x^{2n} + 30x^n + 25$ **151.** 2 **153.** $2x^2 + 11x - 21$ **155.** For example, $(x + 3)(2x^2 - 1)$ **157.** 1 and 4, 2 and 3

SECTION 5.4

1. 4 **3.** -5 **5.** $3y$; $3y$; $3y$; 2 **7.** $12x^2 + 6x = 3x(4x + 2)$ **9.** $a - 2$ **11.** $3w + 2$ **13.** $t^2 - 3t + 4$ **15.** $-y + 2x$ **17.** $2v^2 - \frac{3}{2}v + 3$ **19.** $-2x + 3 - \frac{6}{x}$ **21.** $3y - 4 + \frac{5y}{x}$ **25.** $x + 8$ **27.** $2x + 3$ **29.** $5x + 7 + \frac{2}{2x - 1}$ **31.** $2x - 1 - \frac{3}{x + 2}$ **33.** $x^2 - 3x - 4$ **35.** $x^2 - 4x - 2$ **37.** $x^2 - x + 7 - \frac{33}{x + 4}$ **39.** $x^2 + \frac{2}{x - 3}$ **41.** $x^2 - 2x - 3 + \frac{1}{x + 2}$ **43.** $2x^2 - x + 3 - \frac{15}{x + 3}$ **45.** $4x^2 + 6x + 9 + \frac{18}{2x - 3}$ **47.** $x^2 - 3x - 10$ **49.** $-x^2 + 2x - 1 + \frac{1}{x - 3}$ **51.** $2x^3 - 3x^2 + x - 4$ **53.** $3x + 1 + \frac{10x + 7}{2x^2 - 3}$ **55.** False **57.** binomial; $x - a$ **59.** $2x - 8$ **61.** $x + 1 - \frac{13}{x + 4}$ **63.** $3x + 3 - \frac{1}{x - 1}$ **65.** $x - 2 + \frac{16}{x + 2}$ **67.** $2x^2 - 3x + 9$ **69.** $x^2 - 5x + 16 - \frac{41}{x + 2}$ **71.** $x^2 - x + 2 - \frac{4}{x + 1}$ **73.** $3x^3 - x + 4 - \frac{2}{x + 1}$ **75.** $2x^3 - 3x^2 + x - 4$ **77.** $2x^3 + 6x^2 + 17x + 51 + \frac{155}{x - 3}$ **79.** $x^2 - 5x + 25$ **81.** Yes **83.** 8; 9; 18; 9; 9 **85.** $P(3) = 8$ **87.** $R(4) = 43$ **89.** $P(-2) = -39$ **91.** $Q(2) = 31$ **93.** $F(-3) = 190$ **95.** $P(5) = 122$ **97.** $R(-3) = 302$ **99.** $Q(2) = 0$ **101.** $x^2 - 2x + 4$ **103.** $x^2 + x + 5$ **105.** 3 **107.** -1 **109.** $x - 3$ **111.** Yes

SECTION 5.5

3. $5x(3x - 2)$ **5.** $6x^4y^3$ **7.** 2 **9.** $6x^2$ **11.** $6x$ **13.** $4x^3yz^2$ **15.** xz **17.** $4x^3y$ **19.** 1 **21.** x^2y **23.** $5x - 3y$ **25.** $6xy^2(5xy^2 + 3)$ **27.** $4(3x^2y^6 + 5)$ **29.** $2xy^3(9x^5 - 5)$ **31.** $2y(4x^2 + 3)$ **33.** $2x^3y^4z^3(5x^3y - 5x + 3z^3)$ **35.** $x^4z^4(20x + 15y^3z^2 + 8z^2)$ **37.** $2x^2y^3(9x^2z^2 - 5y + 4z^3)$ **39.** $6y(4x^5z + 6xy^2 - 5y^4)$ **41.** $5(6x^5 + 4y^4z^3 - 5y^3)$ **43.** x^{2n} **45.** $(x - 5)$ **47.** $(a + 2)(x - 2)$ **49.** $(x - 2)(a + b)$ **51.** $(x + 3)(x + 2)$ **53.** $(x + 4)(y - 2)$ **55.** $(a + b)(x - y)$ **57.** $(y - 3)(x^2 - 2)$ **59.** $(3 + y)(2 + x^2)$ **61.** $(2a + b)(x^2 - 2y)$ **63.** $(x + 1)(x^2 + 2)$ **65.** $(2x - 1)(x^2 + 2)$ **67.** $(x + 6)(x^2 - 6)$ **69.** $(3x + 7)(x^2 - 7)$ **71.** $2x^{-4}(2x^2 + 3)$ **73.** $6x^{-2}y^{-2}(2x - 3y)$ **75.** $(a + b)(c + d + 2)$ **77.** For example, $6x^3y^2 + 15x^2y$ **79.** For example, $3a(2a + b) + b(2a + b)$

SECTION 5.6

1. ii **3. a.** -2 and -9 **b.** -2 and 6 **c.** -5 and 4 **d.** -3 and 8 **5. a.** -3 and 7 **b.** -4 and -5 **7.** $(x - 5)(x - 3)$ **9.** $(a + 11)(a + 1)$ **11.** $(b + 7)(b - 5)$ **13.** $(y - 3)(y - 13)$ **15.** $(a - 7)(a - 8)$ **17.** $(x + 5)(x - 1)$ **19.** $(a + 6b)(a + 5b)$ **21.** $(x - 12y)(x - 2y)$ **23.** $(y + 9x)(y - 7x)$ **25.** Nonfactorable **27.** $x^2y^4(x - 7)(x + 3)$ **29.** $6x^4(x - 9)(x - 1)$ **31.** $3x^3y^4(x - 5)(x - 3)$ **33.** $3x^4y(x - 8)(x - 7)$ **35.** $2x^3y^4(x - 3)(x + 5)$ **37.** Nonfactorable **39. a.** Positive **b.** Negative **c.** Less than **41.** Two negative **43.** Two positive **45.** $(2x + 5)(x - 8)$ **47.** $(4y - 3)(y - 3)$ **49.** $(2a + 1)(a + 6)$ **51.** Nonfactorable **53.** $(5x + 1)(x + 5)$ **55.** $(11x - 1)(x - 11)$ **57.** $(12x - 5)(x - 1)$ **59.** $(4y - 3)(2y - 3)$ **61.** Nonfactorable

63. Nonfactorable **65.** $(5 - x)(8 + x)$ **67.** $(8 - x)(3 + 2x)$ **69.** $6x^2(4x - 3)(5x - 9)$ **71.** $4x^2y^3(x + 2)(8x - 7)$ **73.** $6x(2x - 7)(2x - 1)$ **75.** $4xy^4(9x - 5y)(4x + 3y)$ **77.** 5, −5, 1, −1 **79.** 3, −3, 9, −9 **81.** 1, −1, 5, −5 **83.** 5, 8, and 9 **85.** $m + n$; mn

SECTION 5.7

1. a. No **b.** Yes **c.** Yes **d.** No **3. a.** Yes **b.** No **c.** Yes **d.** No **5. a.** Yes **b.** No **c.** Yes **d.** Yes **7.** 4; $25x^6$; $100x^4y^4$ **9.** $4z^4$ **11.** $9a^2b^3$ **13. a.** $4x$; 1 **b.** $4x + 1$; $4x - 1$ **17.** $(x + 4)(x - 4)$ **19.** $(2x + 1)(2x - 1)$ **21.** $(b - 1)^2$ **23.** $(4x - 5)^2$ **25.** Nonfactorable **27.** $(x + 3y)^2$ **29.** $(2x + y)(2x - y)$ **31.** $(4x + 11)(4x - 11)$ **33.** $(1 + 3a)(1 - 3a)$ **35.** Nonfactorable **37.** Nonfactorable **39.** $(5a - 4b)^2$ **41.** $-5xy(3x - 4)^2$ **43.** $6x^3y^4(3x - 4)^2$ **45.** $5x^2y^3(3x + 1)^2$ **47.** $-x^4(2x + 3)(2x - 3)$ **49.** $-5x^3y(3x + 1)(3x - 1)$ **51.** ii and iv **53.** 8; x^9; $27c^{15}d^{18}$ **55.** $2x^3$ **57.** $4a^2b^6$ **59. a.** $5x$; 2 **b.** $5x + 2$; $25x^2 - 10x + 4$ **61.** $(x - 3)(x^2 + 3x + 9)$ **63.** $(2x - 1)(4x^2 + 2x + 1)$ **65.** $(x - 2y)(x^2 + 2xy + 4y^2)$ **67.** $(4x + 1)(16x^2 - 4x + 1)$ **69.** $(3x - 2y)(9x^2 + 6xy + 4y^2)$ **71.** $(xy + 4)(x^2y^2 - 4xy + 16)$ **73.** $3x^3y(2x - 1)(4x^2 + 2x + 1)$ **75.** $2x^3y(3x + 1)(9x^2 - 3x + 1)$ **77.** $3y(3x + 1)(9x^2 - 3x + 1)$ **79.** x^2 **83.** $(xy - 3)(xy - 5)$ **85.** $(x^2 - 6)(x^2 - 3)$ **87.** $(b^2 - 18)(b^2 + 5)$ **89.** $(x^2y^2 - 6)(x^2y^2 - 2)$ **91.** $(3xy - 5)(xy - 3)$ **93.** $(2ab - 3)(3ab - 7)$ **95.** $(2x^2 - 15)(x^2 + 1)$ **97.** $xy(2x^2 + 3)(3x^2 - 4)$ **99.** $4x^2y^2(2x^2 - 5)^2$ **101.** $2x^2y^3(2x^2 - 1)(3x^2 + 2)$ **103.** Common **105.** $3(2x - 3)^2$ **107.** $a(3a - 1)(9a^2 + 3a + 1)$ **109.** $5(2x + 1)(2x - 1)$ **111.** $y^3(y + 11)(y - 5)$ **113.** $(4x^2 + 9)(2x + 3)(2x - 3)$ **115.** $2a(2 - a)(4 + 2a + a^2)$ **117.** $(x + 2)(x + 1)(x - 1)$ **119.** $(x + 2)(2x^2 - 3)$ **121.** $(x + 1)(x + 4)(x - 4)$ **123.** $x^2(x - 7)(x + 5)$ **125.** Nonfactorable **127.** $2x^3(3x + 1)(x + 12)$ **129.** $(4a^2 + b^2)(2a + b)(2a - b)$ **131.** Nonfactorable **133.** $3b^2(b - 2)(b^2 + 2b + 4)$ **135.** $x^2y^2(x - 3y)(x - 2y)$ **137.** $2xy(4x + 7y)(2x - 3y)$ **139.** $(x - 2)(x + 1)(x - 1)$ **141.** $n + 2$ **143.** 20, −20 **145.** 8, −8 **147.** $(a - 2b)(a^2 - ab + b^2)$ **149.** $(x^n + 7)(x^n - 7)$ **151.** $(x^n + 2)(x^{2n} - 2x^n + 4)$ **153.** $(x^n + 2)(x^n - 2)(x^{2n} + 4)$ **155.** $(x^2 - 4x + 8)(x^2 + 4x + 8)$ **159.** $(x - 3)(x^2 + 2x + 3)$

SECTION 5.8

3. a. Yes **b.** No **c.** Yes **d.** No **e.** Yes **f.** No **7.** 5 **9.** The solutions are −4 and −6. **11.** The solutions are 0 and 7. **13.** The solutions are $-\frac{3}{2}$ and 7. **15.** The solutions are 7 and −7. **17.** The solutions are −5 and 1. **19.** The solutions are $-\frac{3}{2}$ and 4. **21.** The solutions are 0 and 9. **23.** The solutions are 7 and −4. **25.** The solutions are $\frac{2}{3}$ and −5. **27.** The solutions are $\frac{2}{5}$ and 3. **29.** The solutions are $\frac{3}{2}$ and $-\frac{1}{4}$. **31.** The solutions are 7 and −5. **33.** The solutions are $-\frac{4}{3}$ and 2. **35.** The solutions are $\frac{4}{3}$ and −5. **37.** The solutions are 5 and −3. **39.** The solutions are −11 and −4. **41.** The solutions are 2 and 3. **43.** The solutions are −4, −1, and 1. **45.** The solutions are −5, 4, and 5. **47.** The solutions are −4, $-\frac{2}{3}$, and 4. **49.** The solutions are $-\frac{5}{3}$, 1, and $\frac{5}{3}$. **51.** The values of c are 1 and 2. **53.** The values of c are $-\frac{1}{2}$ and 1. **55.** The zeros are −4 and 1. **57.** The zeros are −2 and 6. **59.** The zeros are 1 and $\frac{3}{2}$. **61.** The zeros are −3.30 and 0.30. **63.** The zeros are −0.28 and 1.78. **65.** A polygon with 54 diagonals has 12 sides. **67.** The length of the e-reader is 15 cm. The width is 10 cm. **69.** The height is 4 cm. The base is 12 cm. **71.** The length of the larger rectangle is 11 cm. The width is 8 cm. **73.** The value of x is 2 m or 3 m. **75.** The brick walkway should be 3 ft wide. **77.** The ball will be 63 ft above the ground after $\frac{3}{2}$ s (on the way up) and after $\frac{5}{2}$ s (on the way down). **79.** 175 or 15 **81.** −16 or 4 **83.** The values of c are −3, −2, and 2. **85.** The volume of the box is 160 cm³. **89.** The solutions are $-12y$ and $3y$. **91.** The solutions are $-\frac{b}{2}$ and $-b$.

CHAPTER 5 REVIEW EXERCISES

1. $2x^2 - 5x - 2$ [5.2.2] **2.** $4x^2 - 8xy + 5y^2$ [5.2.2] **3.** $70xy^2z^6$ [5.1.2] **4.** $-\frac{48y^9z^{17}}{x}$ [5.1.2] **5.** $-\frac{x^3}{4y^2z^3}$ [5.1.2] **6.** $\frac{c^{10}}{2b^{17}}$ [5.1.2] **7.** 9.3×10^7 [5.1.3] **8.** 0.00254 [5.1.3] **9.** 2×10^{-6} [5.1.3] **10.** $P(-2) = -7$ [5.2.1]

11. 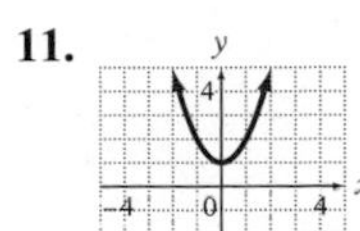 [5.2.1] **12. a.** The leading coefficient is 3. **b.** The constant term is 8. **c.** The degree is 5. [5.2.1]

13. $P(-3) = 68$ [5.4.3] **14.** $2y^2 - 6xy + 4x^2$ [5.4.1] **15.** $5x + 4 + \frac{6}{3x - 2}$ [5.4.1] **16.** $2x - 3 - \frac{4}{6x + 1}$ [5.4.1]
17. $4x^2 + 3x - 8 + \frac{50}{x + 6}$ [5.4.1/5.4.2] **18.** $x^2 + 2x + 2 + \frac{6}{x - z}$ [5.4.1/5.4.2] **19.** $12x^5y^3 + 8x^3y^2 - 28x^2y^4$ [5.3.1]
20. $9x^2 + 8x$ [5.3.1] **21.** $x^4 + 3x^3 - 23x^2 - 29x + 6$ [5.3.2] **22.** $6x^3 - 29x^2 + 14x + 24$ [5.3.2]
23. $25a^2 - 4b^2$ [5.3.3] **24.** $16x^2 - 24xy + 9y^2$ [5.3.3] **25.** $6a^2b(3a^3b - 2ab^2 + 5)$ [5.5.1]
26. $(y - 3)(x - 4)$ [5.5.2] **27.** $(a + 2b)(2x - 3y)$ [5.5.2] **28.** $(x + 5)(x + 7)$ [5.6.1] **29.** $(3 + x)(4 - x)$ [5.6.1]
30. $(x - 7)(x - 9)$ [5.6.1] **31.** $(3x - 2)(2x - 9)$ [5.6.2] **32.** $(8x - 1)(3x + 8)$ [5.6.2]
33. $(xy + 3)(xy - 3)$ [5.7.1] **34.** $(2x + 3y)^2$ [5.7.1] **35.** $(3x + 2)(9x^2 - 6x + 4)$ [5.7.2]
36. $(4a - 3b)(16a^2 + 12ab + 9b^2)$ [5.7.2] **37.** $(x + 3)(x - 3)(x + 2)(x - 2)$ [5.7.3] **38.** $(3x^2 + 2)(5x^2 - 3)$ [5.7.3]
39. $(6x^4 - 5)(6x^4 - 1)$ [5.7.3] **40.** $(3x^2y^2 + 2)(7x^2y^2 + 3)$ [5.7.3] **41.** $3a^2(a^2 - 6)(a^2 + 1)$ [5.7.4]
42. $3ab(a - b)(a^2 + ab + b^2)$ [5.7.4] **43.** The solutions are -2, 0, and 3. [5.8.1] **44.** The solutions are $\frac{5}{2}$ and 4. [5.8.1]
45. The solutions are -5, -2, and 2. [5.8.1] **46.** The solutions are -1, -6, and 6. [5.8.1] **47.** The zeros are -1 and 6. [5.8.1]
48. The values of c are -6 and 2. [5.8.1] **49.** The distance from Earth to the Great Galaxy of Andromeda is 1.137048×10^{23} mi. [5.1.4] **50.** The sun generates 1.09×10^{21} horsepower. [5.1.4] **51.** The area is $(10x^2 - 29x - 21)$ cm^2. [5.3.4] **52.** The area is $(5x^2 + 8x - 8)$ in^2. [5.3.4] **53.** The length of the rectangle is 12 m. [5.8.2]
54. The ball will be 143 ft above the ground after $\frac{5}{2}$ s and after $\frac{7}{2}$ s. [5.8.2]

CHAPTER 5 TEST

1. $2x^3 - 4x^2 + 6x - 14$ [5.2.2, Focus On, page 260] **2.** $64a^7b^7$ [5.1.1, Example 1] **3.** $\frac{2b^7}{a^{10}}$ [5.1.2, Example 6]
4. 5.01×10^{-7} [5.1.3, Example 7] **5.** 6.048×10^5 s [5.1.4, Example 10] **6.** $\frac{x^{12}}{16y^4}$ [5.1.2, Example 5]
7. $35x^2 - 55x$ [5.3.1, Example 1B] **8.** $6a^2 - 13ab - 28b^2$ [5.3.2, Example 3B] **9.** $2x^3 + 9x^2 - x - 15$ [5.3.2, Example 2]
10. $9z^2 - 30z + 25$ [5.3.3, Example 4C] **11.** $5y - 3 + \frac{4y}{x}$ [5.4.1, Example 2] **12.** $2x^2 + 3x + 5$ [5.4.2, Focus On, pages 276–277] **13.** $x^2 - 2x - 1 + \frac{2}{x - 3}$ [5.4.2, Example 3/5.4.3, Focus On, page 279]
14. $P(2) = -3$ [5.2.1, Focus On, page 257/5.4.4, Focus On, page 280] **15.** $P(-2) = -8$ [5.4.4, Focus On, page 280]
16. $(2a^2 - 5)(3a^2 + 1)$ [5.7.3, Focus On, page 302] **17.** $3x(2x - 3)(2x + 5)$ [5.6.1, Example 2]
18. $(4x - 5)(4x + 5)$ [5.7.1, Example 1A] **19.** $(4t + 3)^2$ [5.7.1, Focus On, page 300]
20. $(3x - 2)(9x^2 + 6x + 4)$ [5.7.2, Example 3] **21.** $(3x - 2)(2x - a)$ [5.5.2, Focus On, page 287]
22. The zeros are $-\frac{3}{2}$ and 4. [5.8.1, Example 3] **23.** The solutions are $-\frac{1}{3}$ and $\frac{1}{2}$. [5.8.1, Problem 1]
24. The solutions are -1, $-\frac{1}{6}$, and 1. [5.8.1, Focus On, page 309] **25.** The area of the rectangle is $(10x^2 - 3x - 1)$ ft^2. [5.3.4, Example 5] **26.** It takes 12 h for the space vehicle to reach the moon. [5.1.4, Example 10]

CUMULATIVE REVIEW EXERCISES

1. 4 [1.2.2] **2.** $-\frac{5}{4}$ [1.4.2] **3.** Inverse Property of Addition [1.4.1] **4.** $-18x + 8$ [1.4.3]
5. The solution is $-\frac{1}{6}$. [2.1.1] **6.** The solution is $-\frac{11}{4}$. [2.1.1] **7.** $x^2 + 3x + 9 + \frac{24}{x - 3}$ [5.4.2/5.4.3]
8. The solutions are -1 and $\frac{7}{3}$. [2.5.1] **9.** $P(-2) = 18$ [3.2.1/5.2.1/5.4.4] **10.** 2 [3.2.1]
11. The range is $\{-4, -1, 8\}$. [3.2.1] **12.** The slope is $-\frac{1}{6}$. [3.4.1] **13.** The equation of the line is $y = -\frac{3}{2}x + \frac{1}{2}$. [3.5.1]
14. The equation of the line is $y = \frac{2}{3}x + \frac{16}{3}$. [3.6.1] **15.** The solution is $\left(-\frac{7}{5}, -\frac{8}{5}\right)$. [4.3.2]
16. The solution is $\left(-\frac{9}{7}, \frac{2}{7}, \frac{11}{7}\right)$. [4.2.2] **17.** 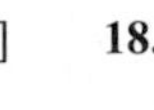 [3.3.2] **18.** 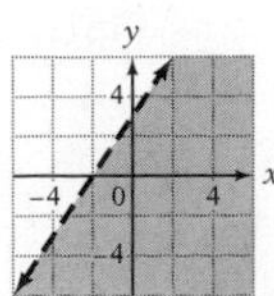 [3.7.1]

19. 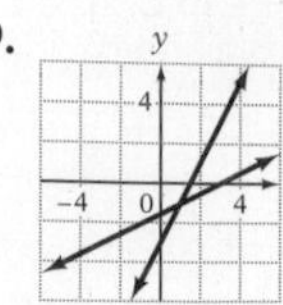 The solution is $(1, -1)$. [4.1.1] **20.** 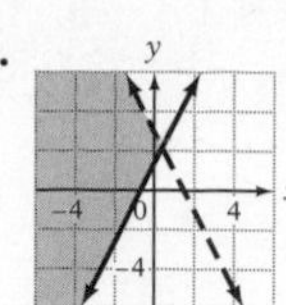 [4.5.1] **21.** $\frac{b^5}{a^8}$ [5.1.2] **22.** $\frac{y^2}{25x^6}$ [5.1.2]

23. $x^2 - x - 1 + \frac{4}{x + 3}$ [5.4.2/5.4.3] **24.** $4x^3 - 7x + 3$ [5.3.2] **25.** $-2x(2x - 3)(x - 2)$ [5.6.2]
26. $(x - y)(a + b)$ [5.5.2] **27.** $(x - 2)(x + 2)(x^2 + 4)$ [5.7.4] **28.** $2(x - 2)(x^2 + 2x + 4)$ [5.7.4]
29. The biologist uses 7.5 L of the 4% solution and 2.5 L of the 8% solution. [2.3.2] **30.** 40 oz of pure gold must be mixed with the alloy. [2.2.1] **31.** The slower cyclist travels at 5 mph, and the faster cyclist travels at 7.5 mph. [2.2.2]
32. The additional investment is \$4500. [2.3.1] **33.** $m = 50$; a slope of 50 means that the average speed was 50 mph. [3.4.1]

Answers to Chapter 6 Selected Exercises

PREP TEST

1. 50 [1.3.1] **2.** $-\frac{1}{6}$ [1.3.1] **3.** $-\frac{3}{2}$ [1.3.1] **4.** $\frac{1}{24}$ [1.3.1] **5.** $\frac{5}{24}$ [1.3.1] **6.** $-\frac{2}{9}$ [1.3.2] **7.** $\frac{1}{3}$ [1.4.2]
8. -2 [2.1.2] **9.** $\frac{10}{7}$ [2.1.2] **10.** The rate of the first plane is 110 mph. The rate of the second plane is 130 mph. [2.2.2]

SECTION 6.1

1. a. Yes **b.** No **c.** Yes **3.** $x + 6$ **7.** $-3; -3; -1; 6; -\frac{1}{7}$ **9.** $f(4) = 2$ **11.** $f(-2) = -2$ **13.** $f(3) = \frac{1}{35}$
15. $f(-1) = \frac{3}{4}$ **17.** The domain is $\{x|x \neq 3\}$. **19.** The domain is $\{x|x \neq -4\}$.
21. The domain is $\{x|x \neq -3\}$. **23.** The domain is $\{x|x \neq -2, x \neq 4\}$. **25.** The domain is $\left\{x\middle|x \neq -\frac{5}{2}, x \neq 2\right\}$.
27. The domain is $\{x|x \neq 0\}$. **29.** The domain is $\{x|x \in \text{real numbers}\}$. **31.** The domain is $\{x|x \neq -3, x \neq 2\}$.
33. The domain is $\{x|x \neq -6, x \neq 4\}$. **35.** The domain is $\{x|x \in \text{real numbers}\}$. **37.** The domain is $\left\{x\middle|x \neq \frac{2}{3}, x \neq \frac{3}{2}\right\}$.
39. The domain is $\left\{x\middle|x \neq 0, x \neq \frac{1}{2}, x \neq -5\right\}$. **41.** Negative **45.** factored; common **47.** $1 - 2x$ **49.** $3x - 1$
51. $2x$ **53.** $-\frac{2}{x}$ **55.** $\frac{x}{2}$ **57.** The expression is in simplest form. **59.** $4x^2 - 2x + 3$ **61.** $a^2 + 2a - 3$
63. $\frac{x - 3}{x - 5}$ **65.** $\frac{x + y}{x - y}$ **67.** $-\frac{x + 3}{3x - 4}$ **69.** $-\frac{x + 7}{x - 7}$ **71.** The expression is in simplest form. **73.** $\frac{x + y}{3x}$
75. $\frac{x + 2y}{3x + 4y}$ **77.** $-\frac{2x - 3}{2(2x + 3)}$ **79.** $\frac{x^2 + 2}{(x + 1)(x - 1)}$ **81.** $\frac{xy + 7}{xy - 7}$ **83.** False **85.** $\frac{a^n - 2}{a^n + 2}$ **87.** decrease
91. a.

f(x)
Distance between lens and film (in millimeters)
100 80 60 40 20
0 2000 4000 x
Distance between lens and object (in meters)

b. The ordered pair (2000, 51) means that when the distance between the object and the lens is 2000 m, the distance between the lens and the film is 51 mm. **c.** For $x = 50$, the expression $\frac{50x}{x - 50}$ is undefined. For $0 < x < 50$, $f(x)$ is negative, and distance cannot be negative. Therefore, the domain is $x > 50$. **d.** For $x > 1000$, $f(x)$ changes very little for large changes in x.

SECTION 6.2

1. No **3.** $\frac{x - 1}{x^2}$ **7.** factored **9.** $\frac{15b^4xy}{4}$ **11.** 1 **13.** $\frac{2x - 3}{x^2(x + 1)}$ **15.** $-\frac{x - 1}{x - 5}$ **17.** $-\frac{x - 3}{2x + 3}$ **19.** $x + y$
21. $\frac{a^2y}{10x}$ **23.** $\frac{3}{2}$ **25.** $\frac{5(x - y)}{xy(x + y)}$ **27.** $\frac{(x - 3)^2}{(x + 3)(x - 4)}$ **29.** $-\frac{x + 2}{x + 5}$ **31.** $\frac{(x - 3)(3x - 5)}{(2x + 5)(x + 4)}$ **33.** -1

35. $-\frac{2(x^2 + x + 1)}{(x + 1)^2}$ **37.** iii **39.** $(x - 3)^2$; $(x + 3)(x - 3)$; two; one; $(x - 3)^2(x + 3)$ **41. a.** Two **b.** Zero **c.** One

43. $\frac{9y^3}{12x^2y^4}, \frac{17x}{12x^2y^4}$ **45.** $\frac{6x^2 + 9x}{(2x - 3)(2x + 3)}, \frac{10x^2 - 15x}{(2x - 3)(2x + 3)}$ **47.** $\frac{2x}{(x + 3)(x - 3)}, \frac{x^2 + 4x + 3}{(x + 3)(x - 3)}$

49. $\frac{3x^2 - 3x}{(x + 1)(x - 1)^2}, \frac{5x^2 + 5x}{(x + 1)(x - 1)^2}$ **51.** $(x - 4)(x + 3)$; $5x + 15$; $2x - 8$; $5x + 15$; $2x - 8$; $2x$; 8; $3x$; 23

53. $-\frac{13}{2xy}$ **55.** $\frac{1}{x - 1}$ **57.** $\frac{15 - 16xy - 9x}{10x^2y}$ **59.** $-\frac{6x + 19}{36x}$ **61.** $\frac{10x + 6y + 7xy}{12x^2y^2}$ **63.** $\frac{-x - 18}{(x + 3)(x - 2)}$

65. $-\frac{x^2 + x}{(x - 3)(x - 5)}$ **67.** $\frac{2x^2 - 3x - 11}{(x - 3)(x - 5)}$ **69.** $\frac{4x^2 + 20x - 13}{(2x - 3)(3x + 2)}$ **71.** $\frac{5x - 8}{(x + 5)(x - 5)}$ **73.** $-\frac{3x^2 - 4x + 8}{x(x - 4)}$

75. $\frac{4x^2 - 14x + 15}{2x(2x - 3)}$ **77.** $\frac{2x^2 + x + 3}{(x + 1)^2(x - 1)}$ **79.** $\frac{x + 1}{x + 3}$ **81.** $\frac{x + 2}{x + 3}$ **83.** $\frac{12x + 13}{(2x + 3)(2x - 3)}$ **85.** $\frac{2x + 3}{2x - 3}$

87. $\frac{2x + 5}{(x + 3)(x - 2)(x + 1)}$ **89.** $\frac{3x^2 + 4x + 3}{(2x + 3)(x + 4)(x - 3)}$ **91.** $\frac{x - 2}{x - 3}$ **93.** $-\frac{2x^2 - 9x - 9}{(x + 6)(x - 3)}$

95. $\frac{1}{x^2 + 1}$ **97.** $\frac{6(x - 1)}{x(2x + 1)}$ **99.** $\frac{2x - 1}{x(x - 3)}$ **103. a.** $f(x) = \frac{2x^3 + 528}{x}$

b.

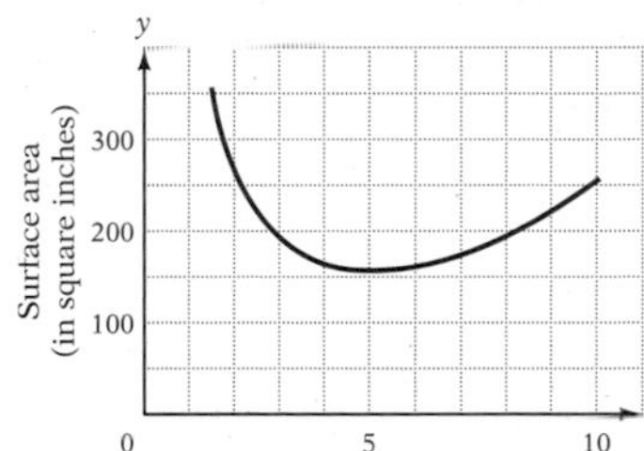

c. When the height of the box is 4 in., 164 in^2 of cardboard are needed. **d.** The height of the box is 5.1 in. **e.** The minimum amount of cardboard is 155.5 in^2.

SECTION 6.3

3. $x(x - 3)$ **5.** $\frac{2}{a}; \frac{1}{a^2}; a^2$ **7.** $a^2 + 1$ **9.** $\frac{5}{23}$ **11.** $\frac{2}{5}$ **13.** $\frac{1 - y}{y}$ **15.** $\frac{5 - a}{a}$ **17.** $\frac{b}{2(2 - b)}$

19. $\frac{4}{5}$ **21.** $\frac{2a^2 - 3a + 3}{3a - 2}$ **23.** $\frac{x + 2}{x - 1}$ **25.** $-\frac{(x - 3)(x + 5)}{4x^2 - 5x + 4}$ **27.** $\frac{1}{x + 4}$ **29.** $\frac{x - 7}{x + 5}$ **31.** $-\frac{2(a + 1)}{7a - 4}$

33. $\frac{x + y}{x - y}$ **35.** $-\frac{2x}{x^2 + 1}$ **37.** iii **39.** $\frac{2y}{x}$ **41.** $x - 1$ **43.** $\frac{c - 3}{c - 2}$ **45.** $-\frac{3x + 2}{x - 2}$ **47.** $-\frac{2x + h}{x^2(x + h)^2}$

49. a. $P(x) = \frac{Cx(x + 1)^{60}}{(x + 1)^{60} - 1}$ **b.** The ordered pair (0.005, 386.66) means that when the monthly interest rate on a car loan is 0.5%, the monthly payment on the loan is $386.66. **c.** The monthly payment for a loan amount of $20,000 at an annual interest rate of 8% is $406.

SECTION 6.4

1. -1 and 0 **3.** Hana **5.** x; $x + 3$; $x(x + 3)$ **7.** $8(x + 3)$; $5x$; -8 **9.** The solution is -5. **11.** The solution is -1. **13.** The solution is $\frac{5}{2}$. **15.** The solution is 0. **17.** The solutions are -3 and 3. **19.** The solution is 8. **21.** The solution is 5. **23.** The solution is -3. **25.** The equation has no solution. **27.** The solutions are 4 and 10. **29.** The solution is -1. **31.** The solutions are -3 and 4. **33.** The solution is 2. **35.** The solution is $\frac{7}{2}$. **37.** The equation has no solution. **39.** 3 and -4 **43.** Row 1: $\frac{1}{6}$; t; $\frac{t}{6}$; Row 2: $\frac{1}{4}$; t; $\frac{t}{4}$ **45.** It would take 1.2 h to process the data with both computers working. **47.** With both panels operating, it would take 18 min to raise the temperature 1 degree. **49.** With both members working together, it would take 3 h to wire the telephone lines. **51.** Working alone, the new machine would take 10 h to package the transistors. **53.** Working alone, the smaller printer would take 80 min to print the payroll. **55.** It would take the apprentice 20 h to shingle the roof working alone. **57.** It will take them 40 min to address 140 envelopes. **59.** In 30 min, the bathtub will start to overflow. **61.** With all three pipes open, it would take 30 h to fill the tank. **63.** Less than n **65. a.** Row 1: 1320; $470 - w$; $\frac{1320}{470 - w}$; Row 2: 1500; $470 + w$; $\frac{1500}{470 + w}$ **b.** $\frac{1320}{470 - w}; \frac{1500}{470 + w}$ **67.** The rate of the first skater is 15 mph. The rate of the second skater is 12 mph. **69.** The rate of the freight train is 45 mph. The rate of the passenger train is 59 mph. **71.** The cyclist was riding at the rate of 12 mph. **73.** The motorist walked at the rate of 4 mph.

75. The business executive can walk at the rate of 6 ft/s. **77.** The rate of the single-engine plane is 180 mph. The rate of the jet is 720 mph. **79.** The rate of the Gulf Stream is 6 mph. **81.** The rate of the current is 3 mph. **83.** The bus usually travels at 60 mph. **85.** The total time is $\frac{5}{8}$ h. **87. a.** Unit fractions are fractions for which the numerator is 1. **b.** $\frac{1}{4}+\frac{1}{8}$ **c.** For example, $\frac{1}{2}+\frac{1}{10}$ **d.** For example, $\frac{1}{2}+\frac{1}{12}$ or $\frac{1}{3}+\frac{1}{4}$

SECTION 6.5

1. iv **3.** $y = kxz$ **7.** 5 mi; 2 cm; 2 cm; 5 mi **9.** The solution is 3. **11.** The solution is −2. **13.** The solution is −4. **15.** The solution is $-\frac{6}{5}$. **17.** True **19.** There are 4000 ducks in the preserve. **21.** The dimensions of the room are 17 ft by 22 ft. **23.** To lose 1 lb, a person would need to walk 21.54 mi. **25.** An additional 0.5 oz of medication is required. **27.** An additional 34 oz of insecticide are required. **29.** It would take 2.5 s to download a 5-megabyte file. **31.** The length of the whale is approximately 96 ft. **35.** Direct variation **37.** When the company sells 300 products, the profit is $37,500. **39.** About 4.8 million barrels of oil leaked during the 86 days. **41.** The length of the person's face is 10.2 in. **43.** A force of 12.5 lb is required to stretch the spring 5 in. **45.** The horizon is approximately 24.03 mi from a point that is 800 ft high. **47.** The ball will roll 80 ft in 4 s. **49.** When the volume is 150 ft^3, the pressure is $66\frac{2}{3}$ lb/in^2. **51.** A gear that has 30 teeth will make 32 revolutions per minute. **53.** The repulsive force is 112.5 lb when the distance is 1.2 in. **55.** The resistance is 56.25 ohms. **57.** The wind force is 180 lb. **59.** x is halved. **61.** directly; inversely **63.** directly

SECTION 6.6

3. Divide each side by $b + 1$. **5.** n; n; n; $R - C$; R; $R - nP$ **7.** $W = \frac{P - 2L}{2}$ **9.** $C = \frac{S}{1 - r}$ **11.** $R = \frac{PV}{nT}$ **13.** $m_2 = \frac{Fr^2}{Gm_1}$ **15.** $R = \frac{E - Ir}{I}$ **17.** $b_2 = \frac{2A - hb_1}{h}$ **19.** $R_2 = \frac{RR_1}{R_1 - R}$ **21.** $d = \frac{a_n - a_1}{n - 1}$ **23.** $H = \frac{S - 2WL}{2W + 2L}$ **25.** $x = \frac{-by - c}{a}$ **27.** $x = \frac{d - b}{a - c}$ **29.** $x = \frac{ac}{b}$ **31.** $x = \frac{ab}{a + b}$ **33.** Yes **35.** $0, y$ **37.** $\frac{y}{2}, 5y$ **39.** You can afford to borrow $12,300.

CHAPTER 6 REVIEW EXERCISES

1. $P(4) = 4$ [6.1.1] **2.** $P(-2) = \frac{2}{21}$ [6.1.1] **3.** The domain is $\{x \mid x \neq 3\}$. [6.1.1] **4.** The domain is $\{x \mid x \neq -3, x \neq 2\}$. [6.1.1] **5.** The domain is $\{x \mid x \in \text{real numbers}\}$. [6.1.1] **6.** $3a^2 + 2a - 1$ [6.1.2] **7.** $-\frac{x + 4}{x(x + 2)}$ [6.1.2] **8.** $\frac{x^2 + 3x + 9}{x + 3}$ [6.1.2] **9.** $\frac{1}{a - 1}$ [6.2.1] **10.** x [6.2.1] **11.** $-\frac{(x + 4)(x + 3)(x + 2)}{6(x - 1)(x - 4)}$ [6.2.1] **12.** $\frac{x + 6}{x - 3}$ [6.2.1] **13.** $\frac{3x + 2}{x}$ [6.2.1] **14.** $-\frac{x - 3}{x + 3}$ [6.2.1] **15.** $\frac{21a + 40b}{24a^2b^4}$ [6.2.2] **16.** $\frac{4x - 1}{x + 2}$ [6.2.2] **17.** $\frac{3x + 26}{(3x - 2)(3x + 2)}$ [6.2.2] **18.** $\frac{9x^2 + x + 4}{(3x - 1)(x - 2)}$ [6.2.2] **19.** $-\frac{5x^2 - 17x - 9}{(x - 3)(x + 2)}$ [6.2.2] **20.** $\frac{x - 4}{x + 5}$ [6.3.1] **21.** $\frac{x - 1}{4}$ [6.3.1] **22.** The solution is $\frac{15}{13}$. [6.4.1] **23.** The solutions are −3 and 5. [6.4.1] **24.** The solution is 6. [6.4.1] **25.** The equation has no solution. [6.4.1] **26.** $R = \frac{V}{I}$ [6.6.1] **27.** $N = \frac{S}{1 - Q}$ [6.6.1] **28.** $r = \frac{S - a}{S}$ [6.6.1] **29.** The number of tanks of fuel required is $6\frac{2}{3}$. [6.5.1] **30.** The number of miles is 48. [6.5.1] **31.** Working alone, the apprentice would take 104 min to install the ceiling fan. [6.4.2] **32.** With both pipes open, it would take 40 min to empty the tub. [6.4.2] **33.** It would take 4 h for the three students to paint the dormitory room together. [6.4.2] **34.** The rate of the current is 2 mph. [6.4.3] **35.** The rate of the bus is 45 mph. [6.4.3] **36.** The rate of the tractor is 30 mph. [6.4.3] **37.** The pressure is 40 lb. [6.5.2] **38.** The illumination is 300 lumens 2 ft from the light. [6.5.2] **39.** The resistance of the cable is 0.4 ohm. [6.5.2]

CHAPTER 6 TEST

1. $\frac{v(v + 2)}{2v - 1}$ [6.1.2, Example 3A] **2.** $-\frac{2(a - 2)}{3a + 2}$ [6.1.2, Example 3B] **3.** $\frac{6(x - 2)}{5}$ [6.2.1, Focus On, page 331] **4.** $P(-1) = 2$ [6.1.1, Example 1] **5.** $\frac{x + 1}{3x - 4}$ [6.2.1, Example 2B] **6.** $\frac{x - 1}{x + 2}$ [6.2.1, Focus On, page 331]

7. $\frac{4y^2 + 6x^2 - 5xy}{2x^2y^2}$ [6.2.2, Example 3] **8.** $\frac{2(5x + 2)}{(x - 2)(x + 2)}$ [6.2.2, Example 5A] **9.** The domain is $\{x \mid x \neq -3, x \neq 3\}$. [6.1.1, Example 2] **10.** $-\frac{x^2 + 5x - 2}{(x + 1)(x + 4)(x - 1)}$ [6.2.2, Example 4] **11.** $\frac{x - 4}{x + 3}$ [6.3.1, Example 1A] **12.** $\frac{x + 4}{x + 2}$ [6.3.1, Example 1B] **13.** The solution is 2. [6.4.1, Example 1A] **14.** The equation has no solution. [6.4.1, Example 1B] **15.** $x = \frac{c}{a - b}$ [6.6.1, Example 1] **16.** It would take 80 min to empty the full tank with both pipes open. [6.4.2, Example 2] **17.** The office requires 14 rolls of wallpaper. [6.5.1, Focus On, pages 358–359] **18.** Working together, the landscapers would complete the task in 10 min. [6.4.2, Example 2] **19.** The rate of the cyclist is 10 mph. [6.4.3, Example 3] **20.** The stopping distance of the car is 61.2 ft. [6.5.2, Problem 2]

CUMULATIVE REVIEW EXERCISES

1. $\frac{36}{5}$ [1.1.2] **2.** The solution is $\frac{15}{2}$. [2.1.2] **3.** The solutions are 7 and 1. [2.6.1] **4.** The domain is $\{x \mid x \neq 3\}$. [6.1.1] **5.** $P(-2) = \frac{3}{7}$ [6.1.1] **6.** 3.5×10^{-8} [5.1.3] **7.** The equation has no solution. [6.4.1] **8.** The solutions are $\frac{1}{9}$ and 4. [5.8.1] **9.** $\frac{8b^3}{a}$ [5.1.2] **10.** The solution set is $\{x \mid x \leq 4\}$. [2.5.1] **11.** $-4a^4 + 6a^3 - 2a^2$ [5.3.1] **12.** $(2x - 1)(x + 2)$ [5.6.2] **13.** $(xy - 3)(x^2y^2 + 3xy + 9)$ [5.7.2] **14.** $x + 3y$ [6.1.2] **15.** The equation is $y = \frac{3}{2}x + 2$. [3.6.1] **16.** $\frac{x}{x^2 - 1}$ [5.1.2] **17.** [3.3.2] **18.** [4.5.1] **19.** $4x^2 + 10x + 10 + \frac{21}{x - 2}$ [5.4.2]

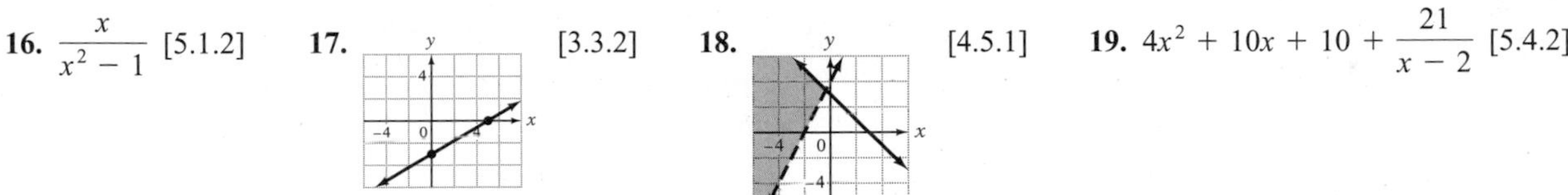

20. $\frac{3xy}{x - y}$ [6.2.1] **21.** The domain is $\{x \mid x \in \text{real numbers}\}$. [6.1.1] **22.** $-\frac{x}{(3x + 2)(x + 1)}$ [6.2.2] **23.** -28 [4.3.1]

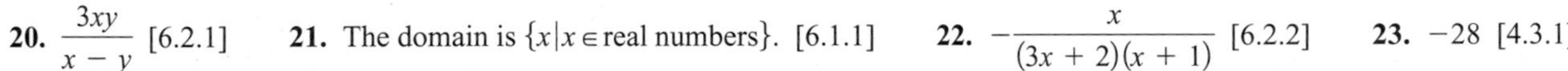

24. $\frac{x - 3}{x + 3}$ [6.3.1] **25.** (1, 1, 1) [4.2.2 or 4.3.2] **26.** The values of c are 1 and 2. [5.8.1] **27.** The solution is 9. [6.5.1] **28.** The solution is 13. [6.4.1] **29.** $a^4 + 5a^3 - 3a^2 - 11a + 20$ [5.3.2] **30.** $r = \frac{E - IR}{I}$ [6.6.1] **31.** No [3.6.1] **32.** [5.2.1] **33.** The number of pounds of almonds is 50. [2.3.1]

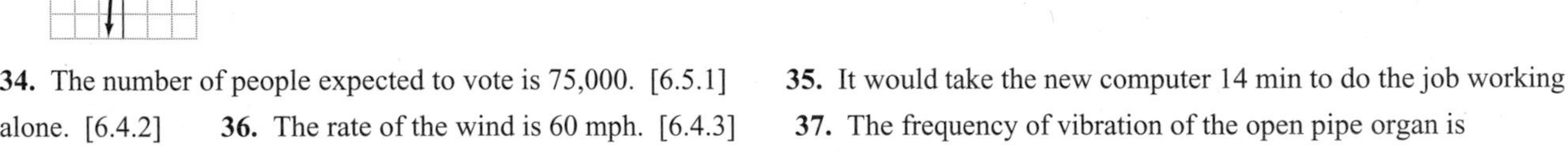

34. The number of people expected to vote is 75,000. [6.5.1] **35.** It would take the new computer 14 min to do the job working alone. [6.4.2] **36.** The rate of the wind is 60 mph. [6.4.3] **37.** The frequency of vibration of the open pipe organ is 80 vibrations/min. [6.5.2]

Answers to Chapter 7 Selected Exercises

PREP TEST

1. 16 [1.2.1] **2.** 32 [1.2.1] **3.** 9 [1.3.1] **4.** $\frac{1}{12}$ [1.3.1] **5.** $-5x - 1$ [1.4.3] **6.** $\frac{xy^5}{4}$ [5.1.2] **7.** $9x^2 - 12x + 4$ [5.3.3] **8.** $-12x^2 + 14x + 10$ [5.3.2] **9.** $36x^2 - 1$ [5.3.3] **10.** The solutions are -1 and 15. [5.8.1]

SECTION 7.1

1. 64 **3.** -32 **5.** Rational **7.** Irrational **9.** 10 **11.** 3 **13.** 8; 8; -3; 8^3; $\frac{1}{512}$ **15.** iv **17.** 2 **19.** 27 **21.** $\frac{1}{9}$ **23.** 4 **25.** $\frac{343}{125}$ **27.** x **29.** $y^{\frac{1}{2}}$ **31.** $x^{\frac{1}{12}}$ **33.** $a^{\frac{7}{12}}$ **35.** $\frac{1}{a}$ **37.** $\frac{1}{y}$ **39.** $y^{\frac{3}{2}}$ **41.** $\frac{1}{x}$ **43.** $\frac{1}{x^4}$

45. a **47.** $x^{\frac{3}{10}}$ **49.** x^4y **51.** $x^6y^3z^9$ **53.** $\dfrac{x}{y^2}$ **55.** $\dfrac{x^{\frac{3}{2}}}{y^{\frac{1}{4}}}$ **57.** $18x^3y^{\frac{7}{5}}$ **59.** $-8x^{\frac{7}{4}}y^{\frac{12}{5}}$ **61.** $30x^{\frac{9}{5}}y^{\frac{27}{20}}$ **63.** $\dfrac{3x^{\frac{6}{5}}}{y^{\frac{4}{3}}}$
65. $-\dfrac{2x^{\frac{2}{3}}}{3}$ **67.** $\dfrac{5x^{\frac{5}{2}}}{3y^{\frac{3}{10}}}$ **69.** a^5b^{13} **71.** $\dfrac{m^2}{4n^{\frac{3}{2}}}$ **73.** $\dfrac{y^{\frac{17}{2}}}{x^3}$ **75.** $\dfrac{16b^2}{a^{\frac{1}{3}}}$ **77.** $\dfrac{y^2}{x^3}$ **79.** $a - a^2$ **81.** $y + 1$
83. $a - \dfrac{1}{a}$ **85.** 4; 5; $\dfrac{4}{5}$ **87.** True **89.** $\sqrt{5}$ **91.** $\sqrt[3]{b^4}$ **93.** $\sqrt[3]{9x^2}$ **95.** $-3\sqrt[5]{a^2}$ **97.** $\sqrt{a^9b^{21}}$
99. $\sqrt[3]{3x - 2}$ **101.** $14^{\frac{1}{2}}$ **103.** $x^{\frac{1}{3}}$ **105.** $x^{\frac{4}{3}}$ **107.** $b^{\frac{3}{5}}$ **109.** $(2x^2)^{\frac{1}{3}}$ **111.** $-(3x^5)^{\frac{1}{2}}$ **113.** $3xy^{\frac{2}{3}}$ **115.** 2; a^2; b^3
117. Negative **119.** Positive **121.** y^7 **123.** $-a^3$ **125.** a^7b^3 **127.** $11y^6$ **129.** a^2b^4 **131.** $-a^3b^3$
133. $5b^5$ **135.** $-a^2b^3$ **137.** $5x^4y$ **139.** $2a^7b^2$ **141.** $-3ab^5$ **143.** y^3 **145.** $3a^5$ **147.** $-a^4b$ **149.** ab^5
151. $2a^2b^5$ **153.** $-2x^3y^4$ **155.** $\dfrac{3b}{a^3}$ **157.** $2x + 3$ **159.** $x + 1$ **161.** x **163.** b **165.** ab^2
167. $\sqrt{3} \approx 1.7321$, and the value of the continued fraction is approximately 1.7333.

SECTION 7.2

1. No; 25 is a perfect square factor. **3.** Yes **5.** No; there is a radical in the denominator. **7.** $3 + \sqrt{5}$ **9.** $-\sqrt{3} - \sqrt{5}$
11. $\sqrt{7}$ **13.** $\sqrt[4]{x}$ **15.** 25; 25; 2; 5 **17. a.** No **b.** Yes **21.** $3\sqrt{2}$ **23.** $7\sqrt{2}$ **25.** $2\sqrt[3]{9}$ **27.** $2\sqrt[3]{2}$
29. $2\sqrt[4]{3}$ **31.** $2\sqrt[5]{3}$ **33.** $x^2yz^2\sqrt{yz}$ **35.** $2ab^4\sqrt{2a}$ **37.** $3xyz^2\sqrt{5yz}$ **39.** $5y\sqrt[3]{x^2y}$ **41.** $-6xy^3\sqrt[3]{x^2}$
43. $ab^2\sqrt[3]{a^2b^2}$ **45.** radicand; index **47. a.** No **b.** No **c.** Yes **d.** Yes **49.** $2\sqrt{2}$ **51.** $3\sqrt[3]{7}$ **53.** $-2\sqrt{2}$
55. $6\sqrt{3} - 6\sqrt{2}$ **57.** $9\sqrt[3]{2}$ **59.** $-7\sqrt[3]{2}$ **61.** -19 **63.** $\sqrt{2x}$ **65.** $10x\sqrt{2x}$ **67.** $2x\sqrt{3xy}$ **69.** $3\sqrt{3a} - 2\sqrt{2a}$
71. $xy\sqrt{2y}$ **73.** $5a^2b^2\sqrt{ab}$ **75.** $-7a\sqrt[3]{3a}$ **77.** $-5xy^2\sqrt[3]{x^2y}$ **79.** $3a\sqrt[4]{2a}$ **81.** $16ab\sqrt[4]{ab}$ **83.** $6x^3y^2\sqrt{xy}$
85. $-17xy\sqrt[3]{xy}$ **87.** 18; $3x$; 3 **89.** 16 **91.** $45\sqrt{2}$ **93.** $2\sqrt[3]{4}$ **95.** $-20\sqrt[3]{15}$ **97.** $xy^3\sqrt{x}$ **99.** $8xy\sqrt{x}$
101. $2x^2y\sqrt[3]{2}$ **103.** $2ab\sqrt[4]{3a^2b}$ **105.** $672x^2y^2$ **107.** $4a^3b^3\sqrt[3]{a}$ **109.** 6 **111.** $x - \sqrt{2x}$ **113.** $4x - 8\sqrt{x}$
115. $-8\sqrt{5}$ **117.** $-78 + \sqrt{7}$ **119.** $-12 - 21\sqrt{2}$ **121.** $84 + 16\sqrt{5}$ **123.** $x - 3\sqrt{x} - 28$ **125.** $6z + \sqrt{z} - 12$
127. $x - y^2$ **129.** $12x - y$ **131.** $x - 6\sqrt{x} + 9$ **133.** $1 - 10\sqrt{x} + 25x$ **135.** $\sqrt[3]{x^2} + \sqrt[3]{x} - 20$ **137.** False
139. Quotient; 9; 8; $2x^2$ **141. a.** Yes **b.** No **c.** No **d.** No **143.** $4\sqrt{x}$ **145.** $b^2\sqrt{3a}$ **147.** $\dfrac{\sqrt{5}}{5}$
149. $\dfrac{2\sqrt{7}}{3}$ **151.** $\dfrac{\sqrt{10}}{4}$ **153.** $\dfrac{\sqrt{2x}}{2x}$ **155.** $\dfrac{\sqrt{5x}}{x}$ **157.** $\dfrac{\sqrt{5x}}{5}$ **159.** $\dfrac{3\sqrt[3]{4}}{2}$ **161.** $\dfrac{3\sqrt[3]{2x}}{2x}$ **163.** $\dfrac{\sqrt{2xy}}{2y}$
165. $\dfrac{2\sqrt{3ab}}{3b^2}$ **167.** $\dfrac{3\sqrt[4]{2x}}{2x}$ **169.** $\dfrac{2\sqrt[5]{2a^3}}{a}$ **171.** $\dfrac{\sqrt[5]{16x^2}}{2}$ **173.** $2\sqrt{5} - 4$ **175.** $\dfrac{16 - 4\sqrt{6}}{5}$ **177.** $\dfrac{5\sqrt{3} + 3\sqrt{5}}{10}$
179. $\dfrac{3\sqrt{y} + 6}{y - 4}$ **181.** $-5 + 2\sqrt{6}$ **183.** $8 + 4\sqrt{3} - 2\sqrt{2} - \sqrt{6}$ **185.** $\dfrac{\sqrt{15} + \sqrt{10} - \sqrt{6} - 5}{3}$
187. $\dfrac{1 + 2\sqrt{ab} + ab}{1 - ab}$ **189.** $\dfrac{5x\sqrt{y} + 5y\sqrt{x}}{x - y}$ **191.** $2\sqrt{2}$ **193.** $29\sqrt{2} - 45$ **195.** $\dfrac{3\sqrt{y + 1} - 3}{y}$
197. $\dfrac{1}{\sqrt{4 + h} + 2}$ **203.** $\sqrt[3]{25} + \sqrt[3]{5} + 1$

SECTION 7.3

1. Yes **3.** Yes **5.** Yes **9. a.** ≥ 0 **b.** $\in$ real numbers **11. a.** $(-\infty, 0]$ **b.** Empty set **c.** $(-\infty, \infty)$ **d.** $(-\infty, \infty)$
13. $\{x \mid x \in \text{real numbers}\}$ **15.** $\{x \mid x \geq -1\}$ **17.** $\{x \mid x \geq 0\}$ **19.** $\{x \mid x \geq 0\}$ **21.** $\{x \mid x \geq 2\}$ **23.** $(-\infty, \infty)$
25. $(-\infty, 3]$ **27.** $[2, \infty)$ **29.** $\left(-\infty, \dfrac{2}{3}\right]$ **31.** ≥ 5 **33.** 14; 9; -3 **35.** **37.**

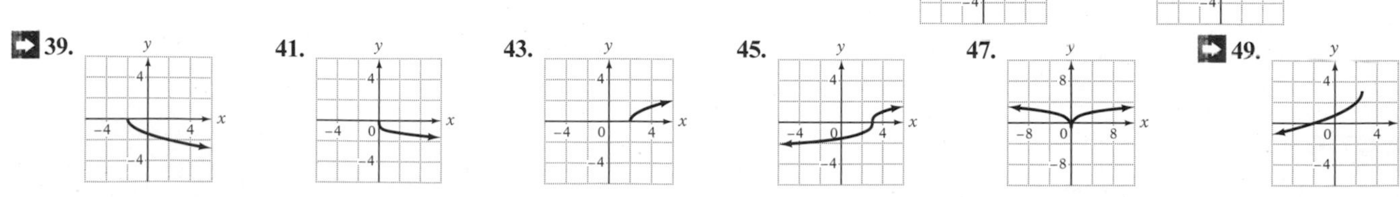

51. 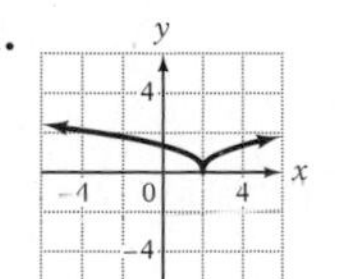

53. 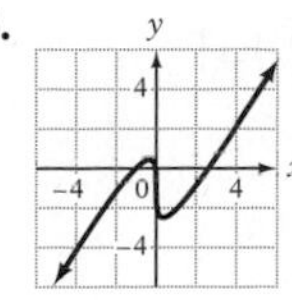

55. 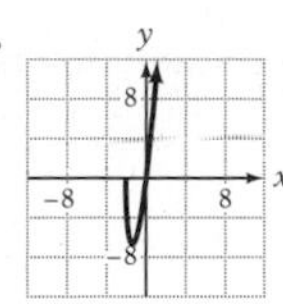

57. a. Both **b.** Below **c.** Above **d.** Above

SECTION 7.4

1. Sometimes true **3.** Sometimes true **5.** No **7.** $\sqrt{2x+1}=7$ **11.** The solution is 48. **13.** The solution is −2. **15.** The equation has no solution. **17.** The solution is 9. **19.** The solution is −23. **21.** The solution is −16. **23.** The solution is $\frac{9}{4}$. **25.** The solution is 14. **27.** The solution is 7. **29.** The solution is −122. **31.** The solution is 45. **33.** The solution is 23. **35.** The solution is −4. **37.** The solution is $\frac{25}{72}$. **39.** The solution is 2. **41.** The equation has no solution. **43.** The solution is 5. **45.** The solution is 5. **47.** The solutions are −2 and 4. **49.** The solution is 6. **51.** The solutions are 0 and 1. **53.** The solutions are 2 and 11. **55.** The solution is 3. **57. a.** Once **b.** Twice **59.** iv **61.** The elephant weighs approximately 3700 lb. **63.** The mean distance is approximately 295,000 km. **65.** The air pressure was 892 mb. The wind speed increases. **67.** The distance is 10 ft. **69.** The distance between the joggers after 3 min is approximately 829.76 m. **71.** The periscope must be approximately 8.17 ft above the water. **73.** The distance is 225 ft. **75.** The distance is 96 ft. **77.** The A-Train is in orbit about 700,000 m above Earth's surface. **79.** The distance between the corners is approximately 7.81 in.

SECTION 7.5

3. The real part is 5. The imaginary part is −7. **5.** The real part is 9. The imaginary part is 0. **7.** −1 **9.** i; 9; 6; $3i$ **11.** $10-\sqrt{-9}$ **13.** $2i$ **15.** $50i$ **17.** $-12i$ **19.** $7i\sqrt{2}$ **21.** $3i\sqrt{3}$ **23.** $20i\sqrt{2}$ **25.** $-28i\sqrt{2}$ **27.** $5+7i$ **29.** $-7-4i\sqrt{5}$ **31.** $11-12i\sqrt{6}$ **33.** $2-3i$ **35.** $-\frac{3}{4}-\frac{5}{4}i$ **37.** $2+4i\sqrt{5}$ **39.** $\frac{2}{3}-\frac{5\sqrt{3}}{6}i$ **41.** 6; $-8i$; 6; $8i$ **43.** $10-7i$ **45.** $-5-3i$ **47.** $-3-16i$ **49.** $-14+7i$ **51.** 3 **53.** $6-4i$ **55.** 0 **57.** $-3+6i$ **59.** True **61.** Negative **63.** $8i$; $12i^2$; −1; −1; $14+5i$ **65.** −24 **67.** −15 **69.** $-5\sqrt{2}$ **71.** $-15-12i$ **73.** $44-12i$ **75.** $27-21i$ **77.** $-10i$ **79.** $29-2i$ **81.** 1 **83.** 1 **85.** −26 **87.** $17i$ **89.** $7-24i$ **91.** $-2i$ **93.** False **95.** $2+3i$ **97.** $-\frac{4}{5}i$ **99.** $-\frac{5}{3}+\frac{16}{3}i$ **101.** $\frac{30}{29}-\frac{12}{29}i$ **103.** $\frac{20}{17}+\frac{5}{17}i$ **105.** $-\frac{27}{53}+\frac{15}{53}i$ **107.** $-\frac{5}{17}-\frac{3}{17}i$ **109.** $\frac{11}{13}+\frac{29}{13}i$ **111.** $3-8i$ **113.** $1-8i$ **115.** $-1+4i$ **117.** True **119.** −1 **121.** i **123.** −1 **125.** −1 **127. a.** Yes **b.** Yes

129. 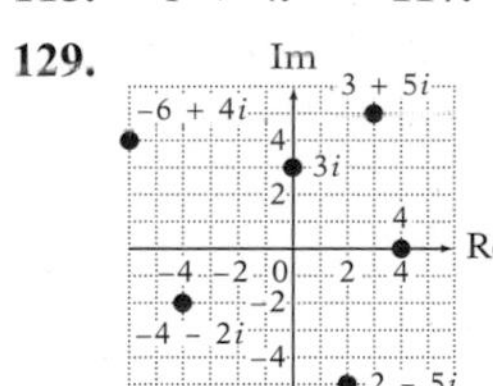

CHAPTER 7 REVIEW EXERCISES

1. $\frac{1}{3}$ [7.1.1] **2.** $\frac{1}{x^5}$ [7.1.1] **3.** $\frac{1}{a^{10}}$ [7.1.1] **4.** $\frac{160y^{11}}{x}$ [7.1.1] **5.** $81x^4$ [7.1.1] **6.** $3\sqrt[4]{x^3}$ [7.1.2] **7.** $7x^{\frac{2}{3}}y$ [7.1.2] **8.** $3a^2b^3$ [7.1.3] **9.** $-7x^3y^8$ [7.1.3] **10.** $-2a^2b^4$ [7.1.3] **11.** $3ab^3\sqrt{2a}$ [7.2.1] **12.** $-2ab^2\sqrt[5]{2a^3b^2}$ [7.2.1] **13.** $xy^2z^2\sqrt[4]{x^2z^2}$ [7.2.1] **14.** $5\sqrt{6}$ [7.2.2] **15.** $-x^2y\sqrt[3]{y^2}$ [7.2.2] **16.** $2a^2b\sqrt{2b}$ [7.2.2] **17.** $6x^2\sqrt{3y}$ [7.2.2] **18.** 40 [7.2.3] **19.** $4xy^2\sqrt[3]{x^2}$ [7.2.3] **20.** $3x+3\sqrt{3x}$ [7.2.3] **21.** $31-10\sqrt{6}$ [7.2.3] **22.** $-13+6\sqrt{3}$ [7.2.3] **23.** $6a-8\sqrt{ab}-30b$ [7.2.3] **24.** $5x\sqrt{x}$ [7.2.4] **25.** $\frac{8\sqrt{3y}}{3y}$ [7.2.4] **26.** $\frac{2\sqrt[3]{3x^2}}{x}$ [7.2.4] **27.** $\frac{21+9\sqrt{5}}{2}$ [7.2.4] **28.** $-\frac{32+9\sqrt{3}}{71}$ [7.2.4] **29.** $\frac{x+2\sqrt{xy}+y}{x-y}$ [7.2.4] **30.** The domain is $\left\{x \mid x \geq \frac{2}{3}\right\}$. [7.3.1] **31.** The domain is $\{x \mid x \in \text{real numbers}\}$. [7.3.1] **32.** [7.3.2]

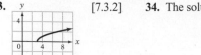

33. 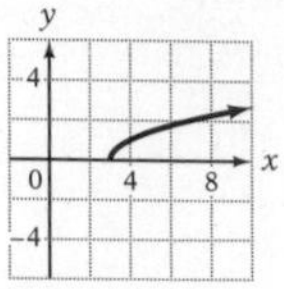

[7.3.2] **34.** The solution is −24. [7.4.1] **35.** The solution is $\frac{13}{3}$. [7.4.1]

36. The solution is −2. [7.4.1] **37.** The solution is 5. [7.4.1] **38.** $6i$ [7.5.1] **39.** $5i\sqrt{2}$ [7.5.1]
40. $\frac{2}{3} - \frac{6\sqrt{2}}{7}i$ [7.5.1] **41.** $9 - i$ [7.5.2] **42.** $-12 + 10i$ [7.5.2] **43.** $10 - 9i$ [7.5.2] **44.** -16 [7.5.3]
45. $7 + 3i$ [7.5.3] **46.** $39 - 2i$ [7.5.3] **47.** $6i$ [7.5.4] **48.** $\frac{2}{3} - \frac{5}{3}i$ [7.5.4] **49.** $\frac{14}{5} + \frac{7}{5}i$ [7.5.4]
50. $-2 + 7i$ [7.5.4] **51.** The width is 5 in. [7.4.2] **52.** The amount of power is approximately 120 watts. [7.4.2]
53. The distance required is 242 ft. [7.4.2] **54.** The distance is approximately 6.63 ft. [7.4.2]

CHAPTER 7 TEST

1. $r^{\frac{1}{6}}$ [7.1.1, Problem 2A] **2.** $\frac{64x^3}{y^6}$ [7.1.1, Problem 2B] **3.** $\frac{b^3}{8a^6}$ [7.1.1, Example 2C] **4.** $3\sqrt[5]{y^2}$ [7.1.2, Example 3]
5. $\frac{1}{2}x^{\frac{3}{4}}$ [7.1.2, Example 4] **6.** The domain is $\{x \mid x \leq 4\}$. [7.3.1, Example 1A] **7.** The domain is $(-\infty, \infty)$. [7.3.1, Example 1B]
8. $3abc^2\sqrt[3]{ac}$ [7.2.1, Example 1] **9.** $8a\sqrt{2a}$ [7.2.2, Focus On, part (A), page 390] **10.** $-2x^2y\sqrt[3]{2x}$ [7.2.2, Example 2]
11. $-4x\sqrt{3}$ [7.2.3, Example 4] **12.** $14 + 10\sqrt{3}$ [7.2.3, Example 5A] **13.** $2a - \sqrt{ab} - 15b$ [7.2.3, Example 5]
14. $4x + 4\sqrt{xy} + y$ [7.2.3, Example 5B] **15.** $\frac{4x^2}{y}$ [7.2.4, Focus On, part (B), page 392] **16.** 2 [7.2.4, Example 7A]
17. $\frac{x + \sqrt{xy}}{x - y}$ [7.2.4, Example 7B] **18.** −4 [7.5.3, Focus On, part (B), page 417] **19.** $-3 + 2i$ [7.5.2, Example 3B]
20. $18 + 16i$ [7.5.3, Example 4] **21.** $-\frac{4}{5} + \frac{7}{5}i$ [7.5.4, Example 6] **22.** 4 [7.5.2, Example 3A]
23. The solution is 4. [7.4.1, Example 1B] **24.** The solution is −3. [7.4.1, Example 2B]
25. The length of the guy wire is 30.6 ft. [7.4.2, Example 3]

CUMULATIVE REVIEW EXERCISES

1. The Distributive Property [1.4.1] **2.** $-x - 24$ [1.4.3] **3.** $\varnothing$ [1.1.2] **4.** The solution is 16. [7.4.1]
5. The solution is $\frac{3}{2}$. [2.1.1] **6.** The solution is $\frac{2}{3}$. [2.1.2] **7.** The solution set is $\{x \mid x \geq -1\}$. [2.4.1]
8. The solution set is $\{x \mid 4 < x < 5\}$. [2.4.2] **9.** The solution set is $\left\{x \,\middle|\, x < 2\right\} \cup \left\{x \,\middle|\, x > \frac{8}{3}\right\}$. [2.5.2]
10. $(8a + b)(8a - b)$ [5.7.1] **11.** $x(x^2 + 3)(x + 1)(x - 1)$ [5.7.4] **12.** The solutions are $\frac{2}{3}$ and −5. [5.8.1]

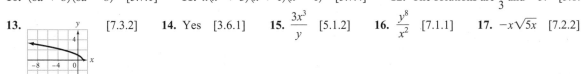

13. 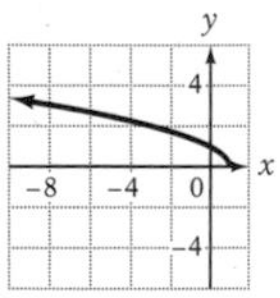

[7.3.2] **14.** Yes [3.6.1] **15.** $\frac{3x^3}{y}$ [5.1.2] **16.** $\frac{y^8}{x^2}$ [7.1.1] **17.** $-x\sqrt{5x}$ [7.2.2]

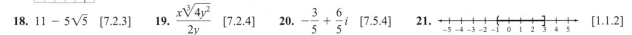

18. $11 - 5\sqrt{5}$ [7.2.3] **19.** $\frac{x\sqrt[3]{4y^2}}{2y}$ [7.2.4] **20.** $-\frac{3}{5} + \frac{6}{5}i$ [7.5.4] **21.** 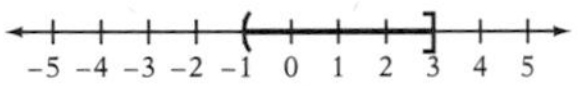

[1.1.2]
22. The domain is $\{x \mid x \in \text{real numbers}\}$. [7.3.1] **23.** 34 [3.2.1] **24.** An equation is $y = \frac{1}{3}x + \frac{7}{3}$. [3.5.2] **25.** 3 [4.3.1]

26. The solution is $\left(\frac{17}{4}, \frac{9}{2}\right)$. [4.3.2] **27.** $m = \frac{3}{2}, b = 3$ 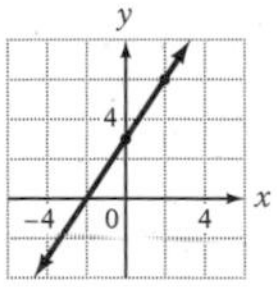

[3.4.2] **28.** 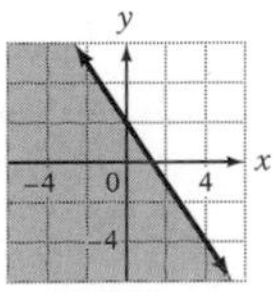

[3.7.1]

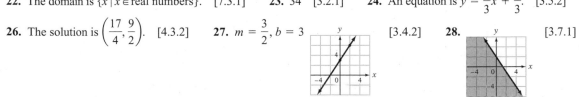

29. The additional investment must be \$4000. [2.3.1] **30.** The length is 12 ft. The width is 6 ft. [5.8.2]
31. The rate of the plane is 250 mph. [6.4.3] **32.** The time is 1.25 s. [5.1.4] **33.** The height of the periscope above the water is approximately 32.7 ft. [7.4.2] **34.** The slope is 0.08. A slope of 0.08 means that the annual interest income is 8% of the investment. [3.4.1]

Answers to Chapter 8 Selected Exercises

PREP TEST

1. $3\sqrt{2}$ [7.2.1] **2.** $3i$ [7.5.1] **3.** $\frac{2x-1}{x-1}$ [6.2.2] **4.** 8 [1.4.2] **5.** Yes [5.7.1] **6.** $(2x-1)^2$ [5.7.1] **7.** $(3x+2)(3x-2)$ [5.7.1] **8.** 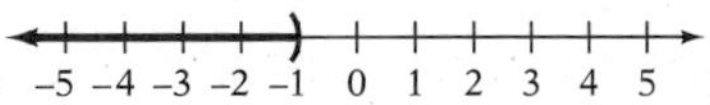 [1.1.2] **9.** The solutions are -3 and 5. [5.8.1] **10.** The solution is 4. [6.4.1]

SECTION 8.1

1. i and iii **3.** $x^2 + 3x - 7 = 0$ **5.** $4x^2 - 15 = 0$ **7.** Never true **9.** The solution is -3. **11.** The solutions are 0 and 5. **15.** standard; 3; 9; 0; 0; -3; 9 **17.** The solutions are 0 and 4. **19.** The solutions are 0 and 2. **21.** The solutions are -2 and 3. **23.** The solutions are -5 and 5. **25.** The solution is 3. **27.** The solutions are -2 and 5. **29.** The solutions are 2 and 5. **31.** The solutions are 6 and $-\frac{3}{2}$. **33.** The solutions are $\frac{1}{4}$ and 2. **35.** The solutions are -4 and $\frac{1}{3}$. **37.** The solutions are $-\frac{2}{3}$ and $\frac{9}{2}$. **39.** The solutions are -4 and $\frac{1}{4}$. **41.** The solutions are -2 and $-\frac{3}{4}$. **43.** The solutions are $-\frac{5}{2}$ and 2. **45.** The solutions are $-\frac{3}{2}$ and -4. **47. a.** One positive and one negative **b.** Two negative **c.** Two positive **49.** $x^2 - 7x + 10 = 0$ **51.** $x^2 - 5x - 6 = 0$ **53.** $x^2 - 9 = 0$ **55.** $x^2 - 8x + 16 = 0$ **57.** $x^2 - 5x = 0$ **59.** $2x^2 - 7x + 3 = 0$ **61.** $3x^2 + 11x + 10 = 0$ **63.** $9x^2 - 4 = 0$ **65.** $10x^2 - 7x - 6 = 0$ **67.** $8x^2 + 6x + 1 = 0$ **69.** $50x^2 - 25x - 3 = 0$ **71.** b is positive and c is positive. **73.** b is positive and c is negative. **75.** -100; 100; -25; 4; square root; $5i$; $5i$ **77.** The solutions are 7 and -7. **79.** The solutions are 2 and -2. **81.** The solutions are $\frac{9}{2}$ and $-\frac{9}{2}$. **83.** The solutions are $4\sqrt{3}$ and $-4\sqrt{3}$. **85.** The solutions are $4\sqrt{2}$ and $-4\sqrt{2}$. **87.** The solutions are 7 and -5. **89.** The solutions are $-4 + 5\sqrt{2}$ and $-4 - 5\sqrt{2}$. **91.** The solutions are $3 + 2\sqrt{2}$ and $3 - 2\sqrt{2}$. **93.** The solutions are $-\frac{1}{5}$ and 1. **95.** The solutions are $\frac{7}{3}$ and 1. **97.** The solutions are $7i$ and $-7i$. **99.** The solutions are $3i\sqrt{2}$ and $-3i\sqrt{2}$. **101.** The solutions are $-5 + 5i$ and $-5 - 5i$. **103.** The solutions are $-1 + 2i\sqrt{3}$ and $-1 - 2i\sqrt{3}$. **105.** The solutions are $\frac{2}{3} + \frac{2i\sqrt{3}}{3}$ and $\frac{2}{3} - \frac{2i\sqrt{3}}{3}$. **107.** The solutions are $-\frac{4}{5} + \frac{3i\sqrt{5}}{5}$ and $-\frac{4}{5} - \frac{3i\sqrt{5}}{5}$. **109.** Two complex solutions **111.** Two real solutions **113.** $x^2 - 8x + 4 = 0$ **115.** $x^2 + 3 = 0$ **117.** $x^2 - 2x + 26 = 0$ **119.** Five

SECTION 8.2

1. a. No **b.** Yes **c.** Yes **d.** No **3.** Yes **5.** i, iii, and iv **7. a.** perfect-square; x; 14; 49 **b.** 49; $x + 7$ **11.** The solutions are 5 and -1. **13.** The solution is 3. **15.** The solutions are $-2 + \sqrt{11}$ and $-2 - \sqrt{11}$. **17.** The solutions are $3 + \sqrt{2}$ and $3 - \sqrt{2}$. **19.** The solutions are $1 + i$ and $1 - i$. **21.** The solutions are $\frac{1 + \sqrt{5}}{2}$ and $\frac{1 - \sqrt{5}}{2}$. **23.** The solutions are $3 + \sqrt{13}$ and $3 - \sqrt{13}$. **25.** The solutions are 3 and 5. **27.** The solutions are $2 + 3i$ and $2 - 3i$. **29.** The solutions are $-3 + 2i$ and $-3 - 2i$. **31.** The solutions are $1 + 3\sqrt{2}$ and $1 - 3\sqrt{2}$. **33.** The solutions are $\frac{1 + \sqrt{17}}{2}$ and $\frac{1 - \sqrt{17}}{2}$. **35.** The solutions are $1 + 2i\sqrt{3}$ and $1 - 2i\sqrt{3}$. **37.** The solutions are $-\frac{1}{2}$ and -1. **39.** The solutions are $\frac{1}{2}$ and $\frac{3}{2}$. **41.** The solutions are $\frac{1}{2} + i$ and $\frac{1}{2} - i$. **43.** The solutions are $\frac{1}{3} + \frac{1}{3}i$ and $\frac{1}{3} - \frac{1}{3}i$. **45.** The solutions are $\frac{2 + \sqrt{14}}{2}$ and $\frac{2 - \sqrt{14}}{2}$. **47.** The solutions are $1 + \sqrt{5}$ and $1 - \sqrt{5}$. **49.** The solutions are $\frac{1}{2}$ and 5. **51.** The solutions are $2 + \sqrt{5}$ and $2 - \sqrt{5}$. **53.** The solutions are approximately -3.236 and 1.236. **55.** The solutions are approximately 1.707 and 0.293. **57.** The solutions are approximately 0.309 and -0.809. **59.** False **65.** $6x$; 10; $x^2 - 6x + 10$; 1; -6; 10 **67.** complex **69.** The solutions are 5 and -2. **71.** The solutions are 3 and -8. **73.** The solutions are $-\frac{1}{4}$ and $\frac{3}{2}$. **75.** The solutions are $1 + 2\sqrt{2}$ and $1 - 2\sqrt{2}$. **77.** The solutions are 10 and -2. **79.** The solutions are $6 + 2\sqrt{3}$ and $6 - 2\sqrt{3}$. **81.** The solutions are $\frac{1 + 2\sqrt{2}}{2}$ and $\frac{1 - 2\sqrt{2}}{2}$. **83.** The solutions are $\frac{1}{2}$ and 1. **85.** The solutions are $\frac{-5 + \sqrt{7}}{3}$ and $\frac{-5 - \sqrt{7}}{3}$.

87. The solutions are $\frac{2}{3}$ and $\frac{5}{2}$. **89.** The solutions are $2 + i$ and $2 - i$. **91.** The solutions are $-3 + 2i$ and $-3 - 2i$. **93.** The solutions are $-\frac{3}{2}$ and $-\frac{1}{2}$. **95.** The solutions are $-\frac{1}{2} + \frac{5}{2}i$ and $-\frac{1}{2} - \frac{5}{2}i$. **97.** The solutions are $\frac{-3 + \sqrt{6}}{3}$ and $\frac{-3 - \sqrt{6}}{3}$. **99.** The solutions are $1 + \frac{\sqrt{6}}{3}i$ and $1 - \frac{\sqrt{6}}{3}i$. **101.** The solutions are $-\frac{3}{2}$ and -1. **103.** The solutions are $-\frac{3}{2}$ and 4. **105.** The solutions are approximately 7.606 and 0.394. **107.** The solutions are approximately 0.236 and -4.236. **109.** The solutions are approximately 1.851 and -1.351. **111.** The equation has two complex solutions. **113.** The equation has one real solution. **115.** The equation has two real solutions. **117.** The equation has two real solutions. **119.** True **121.** The solutions are $-3\sqrt{2}$ and $\frac{\sqrt{2}}{2}$. **123.** The solutions are $\frac{\sqrt{3}}{2} + \frac{1}{2}i$ and $\frac{\sqrt{3}}{2} - \frac{1}{2}i$. **125.** The solutions are $2a$ and $-a$. **127.** The solutions are $1 + \sqrt{y+1}$ and $1 - \sqrt{y+1}$. **129.** The solution set is $\{p \mid p < 9, p \in \text{real numbers}\}$. **131.** The solution set is $\{p \mid p > 1, p \in \text{real numbers}\}$. **133.** The ball takes approximately 4.43 s to hit the ground. **135. a.** 1.05; The model predicts that the population will increase. **b.** 0.96; The model predicts that the population will decrease. **137.** The discriminant is 81. **139.** The discriminant is 17.

SECTION 8.3

1. a. No **b.** Yes **c.** Yes **d.** Yes **5.** x^4 **9.** The solutions are 3, -3, 2, and -2. **11.** The solutions are 2, -2, $\sqrt{2}$, and $-\sqrt{2}$. **13.** The solutions are 1 and 4. **15.** The solution is 16. **17.** The solutions are $2i$, $-2i$, 1, and -1. **19.** The solutions are $4i$, $-4i$, 2, and -2. **21.** The solution is 16. **23.** The solutions are 512 and 1. **25.** The solutions are 2, 1, $-1 + i\sqrt{3}$, $-1 - i\sqrt{3}$, $-\frac{1}{2} + \frac{\sqrt{3}}{2}i$, and $-\frac{1}{2} - \frac{\sqrt{3}}{2}i$. **27.** The solutions are 1, -1, 2, -2, i, $-i$, $2i$, and $-2i$. **29.** The solutions are -64 and 8. **31.** The solutions are $\frac{1}{4}$ and 1. **33.** i and iv **35.** $\sqrt{a+1}$; 7; square **37.** 39, 40, 41, 42, 43, 44, 51, 52, 54 **39.** The solutions are 2 and -1. **41.** The solution is -2. **43.** The solution is 5. **45.** The solution is 3. **47.** The solution is 9. **49.** The solutions are $-\frac{1}{2}$ and 2. **51.** The solution is 1. **53.** The solution is 1. **55.** The solution is -3. **57. a.** y; $y - 4$; $y(y - 4)$ **b.** $4y - 16$ **c.** $y^2 - 6y$ **d.** $4y - 16$; $y^2 - 6y$ **59.** The solutions are 10 and -1. **61.** The solutions are 1 and -3. **63.** The solutions are 0 and -1. **65.** The solutions are $\frac{1}{2}$ and $-\frac{1}{3}$. **67.** The solutions are 6 and $-\frac{2}{3}$. **69.** The solutions are $\frac{4}{3}$ and 3. **71.** The solutions are $-\frac{1}{4}$ and 3. **73.** The solutions are -2 and 5. **75.** The solutions are 5 and 4. **77.** The solutions are 3, -3, i, and $-i$. **79.** The solution is $\sqrt{2}$. **81.** The solution is 4.

SECTION 8.4

1. $\frac{1}{t}$ **3.** $\frac{8}{15}$ **5.** $\frac{500}{r - 50}$ **7.** The projectile takes 12.5 s to return to Earth. **9.** The maximum speed is approximately 72.5 km/h. **11.** The rocket will be 300 ft above the ground after approximately 1.74 s and after approximately 10.76 s. **13.** Working alone, it would take the smaller pipe 12 min to fill the tank. It would take the larger pipe 6 min. **15.** Working alone, it would take the faster computer 2 h to complete the program. **17.** Working alone, it would take the small sprinkler 24 min to soak the lawn. It would take the large sprinkler 8 min. **19.** The rate of the cruise ship for the first 40 mi was 10 mph. **21.** The rate of the wind is 20 mph. **23.** The rate of the jet stream is approximately 92.8 mph. **25.** The width is 8 cm. The length is 16 cm. The height is 2 cm. **27. a.** The length of the east-end-zone screen is 137 ft. The width is 50 ft. **b.** The length of the west-end-zone screen is 99 ft. The width is 50 ft. **29.** The width of the border is 4 ft. **31.** The maximum safe speed of the car is approximately 33 mph. **33.** The bottom of the scoop of ice cream is approximately 2.3 in. from the bottom of the cone.

SECTION 8.5

1. a. Yes **b.** No **c.** No **d.** Yes **3.** 6 **5.** None **7.** (3, 1) **9.** (3, 0) **11.** Two **15.** -5 **17.** $x = 7$ **19. a.** 3; 0; -1 **b.** parabola; up **c.** 0; 3; 0 **d.** 0; -1; 0; -1

21. The coordinates of the vertex are $(0, -2)$. The equation of the axis of symmetry is $x = 0$.

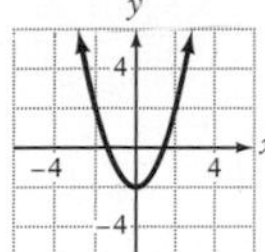

23. The coordinates of the vertex are $(0, 3)$. The equation of the axis of symmetry is $x = 0$.

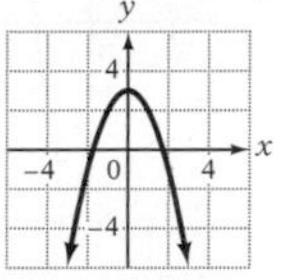

25. The coordinates of the vertex are $(0, 0)$. The equation of the axis of symmetry is $x = 0$.

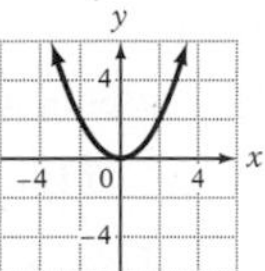

27. The coordinates of the vertex are $(0, -1)$. The equation of the axis of symmetry is $x = 0$.

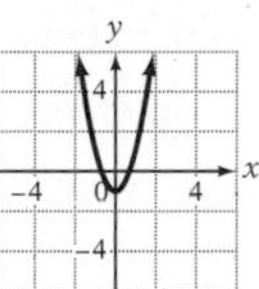

29. The coordinates of the vertex are $(1, -1)$. The equation of the axis of symmetry is $x = 1$.

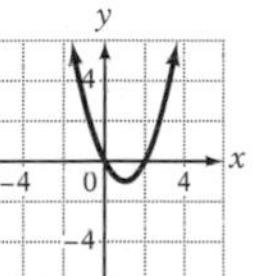

31. The coordinates of the vertex are $(1, 2)$. The equation of the axis of symmetry is $x = 1$.

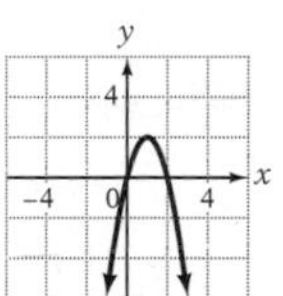

33. The coordinates of the vertex are $\left(\frac{1}{2}, -\frac{9}{4}\right)$. The equation of the axis of symmetry is $x = \frac{1}{2}$.

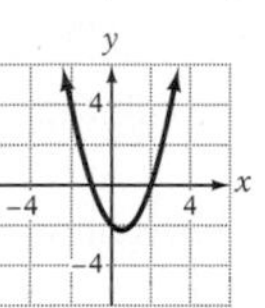

35. The coordinates of the vertex are $\left(\frac{1}{4}, -\frac{41}{8}\right)$. The equation of the axis of symmetry is $x = \frac{1}{4}$.

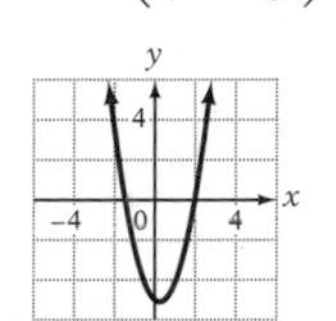

37. The coordinates of the vertex are $(1, 3)$. The equation of the axis of symmetry is $x = 1$.

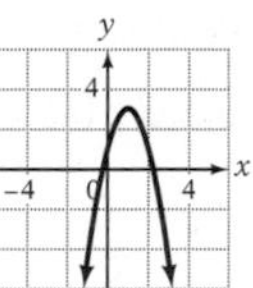

39. The domain is $\{x|x \in \text{real numbers}\}$. The range is $\{y|y \geq -7\}$.

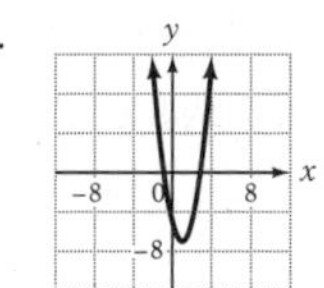

41. The domain is $\{x|x \in \text{real numbers}\}$. The range is $\left\{y \middle| y \leq \frac{25}{8}\right\}$.

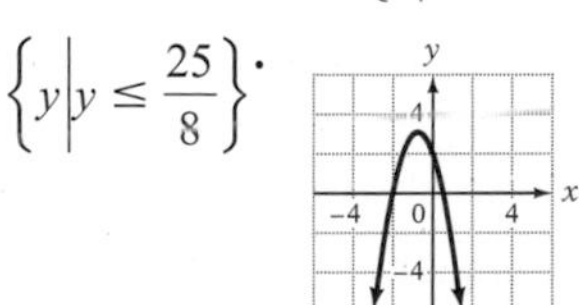

43. The domain is $\{x|x \in \text{real numbers}\}$. The range is $\{y|y \geq 0\}$.

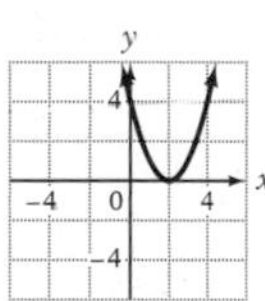

45. The domain is $\{x|x \in \text{real numbers}\}$. The range is $\{y|y \geq -7\}$.

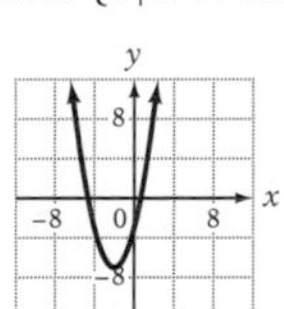

47. The domain is $\{x|x \in \text{real numbers}\}$. The range is $\{y|y \leq -1\}$.

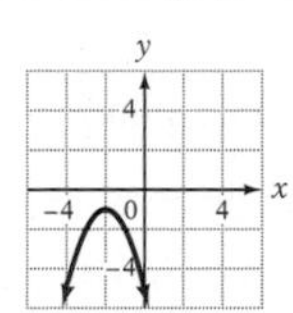

49. False **51.** Right **53.** 3; 0; 3; 0; x **57.** The coordinates of the x-intercepts are $(2, 0)$ and $(-2, 0)$. **59.** The coordinates of the x-intercepts are $(0, 0)$ and $(2, 0)$. **61.** The coordinates of the x-intercepts are $(2, 0)$ and $(-1, 0)$. **63.** The coordinates of the x-intercepts are $(3, 0)$ and $\left(-\frac{1}{2}, 0\right)$. **65.** The coordinates of the x-intercepts are $\left(-\frac{2}{3}, 0\right)$ and $(7, 0)$. **67.** The coordinates of the x-intercept are $\left(\frac{2}{3}, 0\right)$. **69.** The coordinates of the x-intercepts are $\left(\frac{\sqrt{2}}{3}, 0\right)$ and $\left(-\frac{\sqrt{2}}{3}, 0\right)$. **71.** The coordinates of the x-intercepts are $\left(-\frac{3}{2}, 0\right)$ and $\left(\frac{5}{2}, 0\right)$. **73.** The coordinates of the x-intercepts are $(-2 + \sqrt{7}, 0)$ and $(-2 - \sqrt{7}, 0)$. **75.** The parabola has no x-intercepts. **77.** The coordinates of the x-intercepts are $(-1 + \sqrt{2}, 0)$ and $(-1 - \sqrt{2}, 0)$. **79.** The zeros of the function are 0 and $\frac{3}{2}$. **81.** The zeros of the function are -2 and -1. **83.** The zero of the function is $-\frac{2}{3}$. **85.** The zeros of the function are 4 and -4. **87.** The zeros of the function are $\frac{3 + \sqrt{41}}{2}$ and $\frac{3 - \sqrt{41}}{2}$. **89.** The zeros of the function are $-\frac{3}{4} + \frac{\sqrt{7}}{4}i$ and $-\frac{3}{4} - \frac{\sqrt{7}}{4}i$.

91.

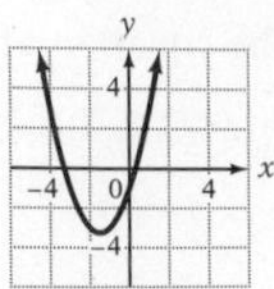

The zeros are approximately −3.3 and 0.3.

93.

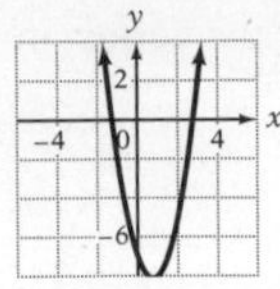

The zeros are approximately −1.3 and 2.8.

95.

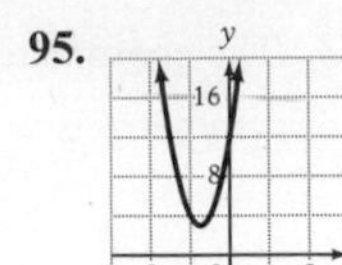

There are no real zeros.

97. The parabola has no x-intercepts. **99.** The parabola has two x-intercepts. **101.** The parabola has one x-intercept. **103.** The parabola has no x-intercepts. **105.** The parabola has two x-intercepts. **107.** The parabola has no x-intercepts. **109.** The parabola has one x-intercept. **111.** The parabola has two x-intercepts. **113.** True **115.** True **117.** The value of k is −2. **119.** The value of k is 1. **121.** The zeros of g are −2 and 3. **123.** The values of x for which $f(x) = 5$ are 2 and 4. **125.** The values of x for which $f(x) = -3$ are −2 and $-\frac{1}{2}$. **127.** $y = (x - 2)^2 + 3$; the coordinates of the vertex are (2, 3). **129.** $y = \left(x + \frac{1}{2}\right)^2 + \frac{7}{4}$; the coordinates of the vertex are $\left(-\frac{1}{2}, \frac{7}{4}\right)$. **131.** $y = 3x^2 - 6x + 5$

SECTION 8.6

1. a. Maximum value **b.** Neither **c.** Minimum value **3.** 3 **7.** −8; down; y **9. a.** Maximum value **b.** Minimum value **c.** Maximum value **11.** The minimum value is 2. **13.** The maximum value is −1. **15.** The maximum value is $\frac{41}{8}$. **17.** The minimum value is $-\frac{73}{8}$. **19.** The maximum value is $-\frac{71}{12}$. **21.** The minimum value is $-\frac{13}{4}$. **23.** i **25.** The function has a minimum value at 7. **27.** True **29.** t; 30; −5; 3; 3 **31.** The rock will attain a maximum height above the beach of 114 ft. **33.** The maximum height of the NASA airplane is approximately 32,000 ft. **35.** The maximum profit the tour operator can expect is $1600. **37.** The minimum height of the cable is 24.36 ft. **39.** The water will reach a height of $31\frac{2}{3}$ ft on the building. **41.** The two numbers are 10 and 10. **43.** The dimensions that will maximize the enclosed area are 100 ft by 50 ft. **45.** The minimum value of S is 4.5. The average of 2, 5, 4, and 7 is 4.5. **47.** The minimum value is 3.0. **49.** The maximum value is approximately 0.5. **51.** The maximum height of the football field is approximately 1.5 ft.

SECTION 8.7

5. −3; 8 **7.** The solution set is $\{x|x < -2\} \cup \{x|x > 4\}$. **9.** The solution set is $\{x|x \le 1\} \cup \{x|x \ge 2\}$. **11.** The solution set is $\{x|-3 < x < 4\}$. **13.** The solution set is $\{x|x < -2\} \cup \{x|1 < x < 3\}$. **15.** The solution set is $\{x|x < -2\} \cup \{x|x > 4\}$. **17.** The solution set is $\{x|-1 < x \le 3\}$. **19.** The solution set is $\{x|x \le -2\} \cup \{x|1 \le x < 3\}$. **21.** i and iv **23.** The solution set is $(-\infty, -4) \cup (4, \infty)$. **25.** The solution set is $(-\infty, 2) \cup (2, \infty)$. **27.** The solution set is $[-3, 12]$. **29.** The solution set is $\left(-\infty, \frac{1}{2}\right] \cup [2, \infty)$. **31.** The solution set is $\left(\frac{1}{2}, \frac{3}{2}\right)$. **33.** The solution set is $(-\infty, -3] \cup [2, 6]$. **35.** The solution set is $\left(-\frac{3}{2}, \frac{1}{2}\right) \cup (4, \infty)$. **37.** The solution set is $(-\infty, -3] \cup [-1, 1]$. **39.** The solution set is $[-2, 1] \cup [2, \infty)$. **41.** The solution set is $(-\infty, -1) \cup (2, \infty)$. **43.** The solution set is $(-1, 0]$. **45.** The solution set is $(-2, 0] \cup (1, \infty)$. **47.** The solution set is $(-\infty, 0) \cup \left(\frac{1}{2}, \infty\right)$.

49. **51.** **53.** **55.** **57.**

59. The arrow will be more than 200 m high between 4 s and 10 s after it is shot.

CHAPTER 8 REVIEW EXERCISES

1. The solutions are 0 and $\frac{3}{2}$. [8.1.1] **2.** The solutions are -2 and $\frac{1}{2}$. [8.1.1] **3.** The solutions are $-4\sqrt{3}$ and $4\sqrt{3}$. [8.1.2] **4.** The solutions are $-\frac{1}{2} - 2i$ and $-\frac{1}{2} + 2i$. [8.1.2] **5.** The minimum value is $-\frac{17}{4}$. [8.6.1] **6.** The maximum value is 3. [8.6.1] **7.** $3x^2 + 8x - 3 = 0$ [8.1.1] **8.** The solutions are -5 and $\frac{1}{2}$. [8.1.1] **9.** The solutions are $-1 - 3\sqrt{2}$ and $-1 + 3\sqrt{2}$. [8.1.2] **10.** The solutions are $-3 - i$ and $-3 + i$. [8.2.1] **11.** The solutions are 2 and 3. [8.3.3] **12.** The solutions are $-\sqrt{2}, \sqrt{2}, -2$, and 2. [8.3.1] **13.** The solution is $\frac{25}{18}$. [8.3.2] **14.** The solutions are -8 and $\frac{1}{8}$. [8.3.1] **15.** The solution is 2. [8.3.2] **16.** The solutions are $3 - \sqrt{11}$ and $3 + \sqrt{11}$. [8.2.2] **17.** The solutions are 5 and $-\frac{4}{9}$. [8.3.3] **18.** The solutions are $\frac{1 + \sqrt{3}}{2}$ and $\frac{1 - \sqrt{3}}{2}$. [8.2.2] **19.** The solution is $\frac{5}{4}$. [8.3.2] **20.** The solutions are 3 and -1. [8.3.3] **21.** The solution is -1. [8.3.3] **22.** The solutions are $\frac{-3 + \sqrt{249}}{10}$ and $\frac{-3 - \sqrt{249}}{10}$. [8.3.3] **23.** The solutions are $\frac{-11 - \sqrt{129}}{2}$ and $\frac{-11 + \sqrt{129}}{2}$. [8.3.3] **24.** The equation is $x = 3$. [8.5.1] **25.** The coordinates of the vertex are $\left(\frac{3}{2}, \frac{1}{4}\right)$. [8.5.1] **26.** The zeros are $4 + 2\sqrt{5}$ and $4 - 2\sqrt{5}$. [8.5.2] **27.** The parabola has two x-intercepts. [8.5.2] **28.** The coordinates of the x-intercepts are $\left(\frac{-3 - \sqrt{5}}{2}, 0\right)$ and $\left(\frac{-3 + \sqrt{5}}{2}, 0\right)$. [8.5.2] **29.** The coordinates of the x-intercepts are $(-2, 0)$ and $\left(\frac{1}{2}, 0\right)$. [8.5.2] **30.** The zeros are $-\frac{1}{3} + \frac{\sqrt{5}}{3}i$ and $-\frac{1}{3} - \frac{\sqrt{5}}{3}i$. [8.5.2] **31.** The solution set is $\left(-3, \frac{5}{2}\right)$. [8.7.1] **32.** The solution set is $\{x|x \le -4\} \cup \left\{x\middle|-\frac{3}{2} \le x \le 2\right\}$. [8.7.1]

33. The solution set is $\left(-\infty, \frac{3}{2}\right) \cup [2, \infty)$. –5 –4 –3 –2 –1 0 1 2 3 4 5 [8.7.1]

34. The solution set is $\{x|x \le -3\} \cup \left\{x\middle|\frac{1}{2} \le x < 4\right\}$. –5 –4 –3 –2 –1 0 1 2 3 4 5 [8.7.1]

35. 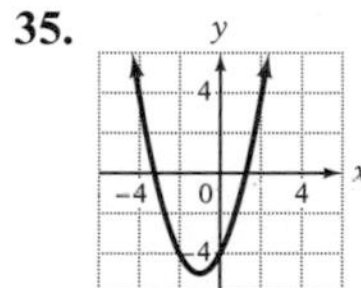 [8.5.1]

The domain is $\{x|x \in \text{real numbers}\}$.
The range is $\{y|y \ge -5\}$.

36. [8.5.1]

The coordinates of the vertex are $(1, 2)$.
The equation of the axis of symmetry is $x = 1$.

37. The width of the rectangle is 5 cm. The length of the rectangle is 12 cm. [8.4.1] **38.** Working alone, the new computer can print the payroll in 12 min. [8.4.1] **39.** The rate of the first car is 40 mph. The rate of the second car is 50 mph. [8.4.1]

CHAPTER 8 TEST

1. The solutions are $\frac{3}{2}$ and -2. [8.1.1, Example 1] **2.** The solutions are $\frac{3}{4}$ and $-\frac{4}{3}$. [8.1.1, Problem 1] **3.** The maximum value is 9. [8.6.1, Example 1] **4.** $3x^2 - 8x - 3 = 0$ [8.1.1, Example 3] **5.** The solutions are $-3 + 3\sqrt{2}$ and $-3 - 3\sqrt{2}$. [8.1.2, Example 4] **6.** The solutions are $-2 + \sqrt{5}$ and $-2 - \sqrt{5}$. [8.2.2, first Focus On, page 444] **7.** The zeros are $\frac{-3 + \sqrt{41}}{2}$ and $\frac{-3 - \sqrt{41}}{2}$. [8.5.2, Example 4] **8.** The solutions are $\frac{1}{6} - \frac{\sqrt{95}}{6}i$ and $\frac{1}{6} + \frac{\sqrt{95}}{6}i$. [8.2.2, Example 2B] **9.** The solutions are $-3 - \sqrt{10}$ and $-3 + \sqrt{10}$. [8.3.3, Focus On, page 454] **10.** The solution is $\frac{1}{4}$. [8.3.1, Focus On, page 451] **11.** The solutions are $-3, 3, -\sqrt{2}$, and $\sqrt{2}$. [8.3.1, Example 1A] **12.** The solution is 4. [8.3.2, Example 2] **13.** The zeros are $2 + i$ and $2 - i$. [8.5.2, Example 5] **14.** The parabola has two x-intercepts. [8.5.2, Example 1B] **15.** The coordinates of the x-intercepts are $(-4, 0)$ and $\left(\frac{3}{2}, 0\right)$. [8.5.2, Problem 6] **16.** The equation is $x = -\frac{3}{2}$. [8.5.1, Example 3]

17. [8.5.1, Example 2] **18.** The solution set is $\left\{x \middle| -4 < x \le \frac{3}{2}\right\}$. [8.7.1, Focus On, page 486]

The domain is $\{x|x \in \text{real numbers}\}$.
The range is $\{y|y \ge -4.5\}$.

19. Working alone, it would take the small pipe 12 min to fill the tank. It would take the large pipe 6 min. [8.4.1, Problem 2]

20. The rowing rate in calm water is 4 mph. [8.4.1, Example 3]

CUMULATIVE REVIEW EXERCISES

1. 14 [1.4.2] **2.** The solution is -28. [2.1.2] **3.** The slope is $-\frac{3}{2}$. [3.4.1] **4.** The equation of the line is $y = x + 1$. [3.6.1] **5.** $-3xy(x^2 - 2xy + 3y^2)$ [5.5.1] **6.** $(2x - 5)(3x + 4)$ [5.6.2] **7.** $(x + y)(a - 2)$ [5.5.2] **8.** $x^2 - 3x - 4 - \frac{6}{3x - 4}$ [5.4.2] **9.** $\frac{x}{2}$ [6.2.1] **10.** The distance is $2\sqrt{5}$. [3.1.1] **11.** $b = \frac{2S - an}{n}$ [6.6.1] **12.** $-8 - 14i$ [7.5.3] **13.** $1 - a$ [7.1.1] **14.** $\frac{x\sqrt[3]{4y^2}}{2y}$ [7.2.4] **15.** The solutions are $-\frac{3}{2}$ and -1. [8.3.3] **16.** The solution is $-\frac{1}{2}$. [8.3.3] **17.** The solutions are 2, -2, $\sqrt{2}$, and $-\sqrt{2}$. [8.3.1] **18.** The solutions are 0 and 1. [8.3.2] **19.** The solution set is $\left\{x \middle| -2 < x < \frac{10}{3}\right\}$. [2.5.2] **20.** The coordinates of the x-intercept are $\left(\frac{5}{2}, 0\right)$. The coordinates of the y-intercept are $(0, -3)$. [3.3.2] **21.**

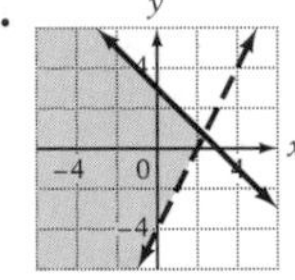

[4.5.1] **22.** The solution is $(1, -1, 2)$. [4.2.2]

23. $-\frac{7}{3}$ [6.1.1] **24.** The domain is $\{x|x \ne 5, x \ne -3\}$. [6.1.1]

25. The solution set is $\{x|x < -3\} \cup \{x|0 < x < 2\}$. [8.7.1]

26. The solution set is $(-3, 1] \cup [5, \infty)$. [8.7.1]

27. The average annual rate of change in height is approximately 5.8 ft. [3.4.3] **28.** The lower and upper limits of the length of the piston rod are $9\frac{23}{64}$ in. and $9\frac{25}{64}$ in. [2.5.3] **29.** The area of the triangle is $(x^2 + 6x - 16)$ ft^2. [5.3.4]

30. $m = -\frac{25{,}000}{3}$; a slope of $-\frac{25{,}000}{3}$ means that the value of the building decreases \$8333.33 each year. [3.4.1]

Answers to Chapter 9 Selected Exercises

PREP TEST

1. 2 [1.4.2] **2.** -7 [1.4.2] **3.** 30 [3.2.1] **4.** $h^2 + 4h - 1$ [3.2.1] **5.** $y = \frac{1}{2}x - 2$ [3.3.2]

6. The domain is $\{-2, 3, 4, 6\}$; the range is $\{4, 5, 6\}$; yes. [3.2.1] **7.** 8 [6.1.1]

8.

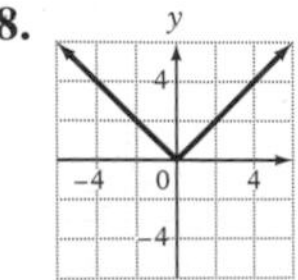

[3.2.2] **9.** [8.5.1]

SECTION 9.1

1. left **3.** up **5.** 5 units down **7.** 7 units left **9.**

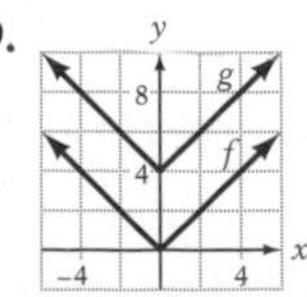

11.

13.

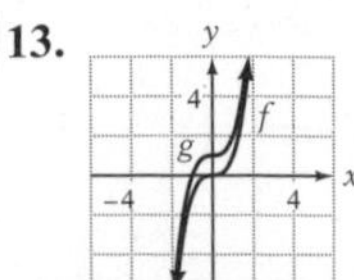

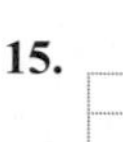
15.
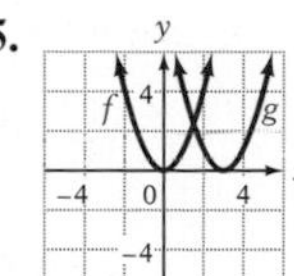
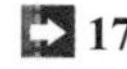
17.
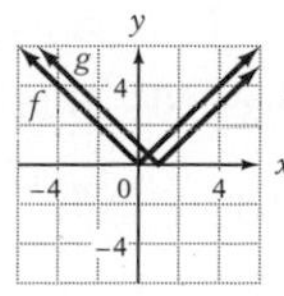
19.
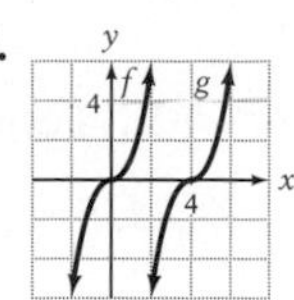
21.
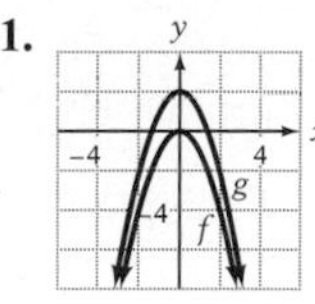
23.
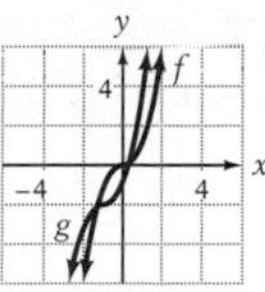

25.
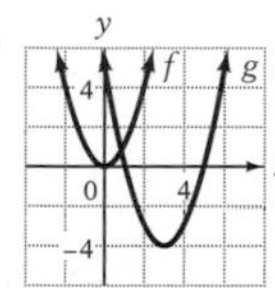
27.
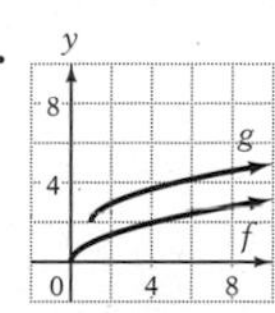
29.
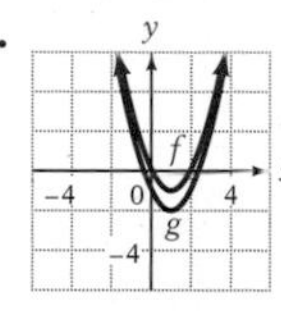
31.
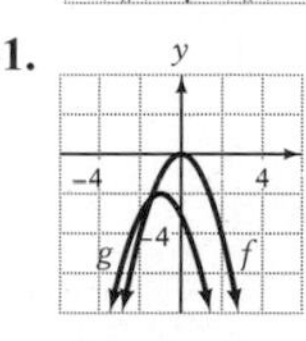
33.
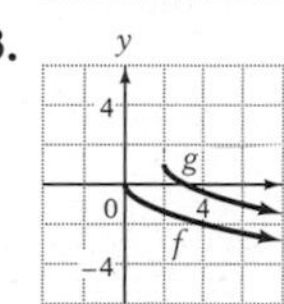

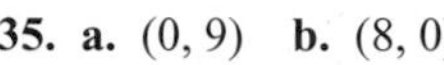
35. a. (0, 9) **b.** (8, 0) **37.**
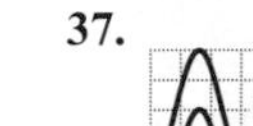
39.
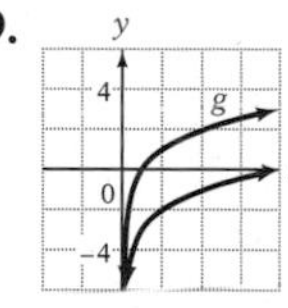
41.
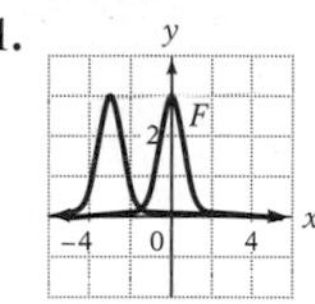
43. The equation of the graph is $y = |x - 1| + 1$.

SECTION 9.2

3. $g; f$ **5.** $-2; -2; -2; 4; 4; -6; 0; -7; 7$ **7.** 5 **9.** $2x^2 - 2x + 1$ **11.** 0 **13.** $-4x^3 + 8x^2 + 6x - 12$ **15.** $4x^2 - 16x + 16$ **17.** 39 **19.** Undefined **21.** $2x^2 + 5x - 5$ **23.** $4x^3 - 2x^2 - 14x + 4$ **25.** -2 **27.** 7 **29.** Undefined **31.** Defined **33.** $4x; 4x; 4x; 16x^2; 8x$ **37.** -13 **39.** -29 **41.** $8x - 13$ **43.** 1 **45.** 5 **47.** $x^2 + 1$ **49.** 5 **51.** 11 **53.** $3x^2 + 3x + 5$ **55.** -3 **57.** -27 **59.** $x^3 - 6x^2 + 12x - 8$ **61.** No **63.** Yes **65.** $6h + h^2$ **67.** $2 + h$ **69.** $h + 2a$ **71.** -1 **73.** -6 **75.** $6x - 13$ **77.** -2 **79.** -2 **81.** 0 **83.** 6

SECTION 9.3

1. Yes **3.** No **5.** Yes **9.** $-2; 4$ **11.** Yes **13.** No **15.** Yes **17.** No **19.** No **21.** No **23.** Yes **25.** No **27.** $-3; 2$ **29.** $\{(0, 1), (3, 2), (8, 3), (15, 4)\}$ **31.** No inverse **33.** $f^{-1}(x) = \frac{1}{4}x + 2$ **35.** No inverse **37.** $f^{-1}(x) = x + 5$ **39.** $f^{-1}(x) = 3x - 6$ **41.** $f^{-1}(x) = -\frac{1}{3}x - 3$ **43.** $f^{-1}(x) = \frac{3}{2}x - 6$ **45.** $f^{-1}(x) = -3x + 3$ **47.** $f^{-1}(x) = \frac{1}{2}x + \frac{5}{2}$ **49.** No inverse **51.** $\frac{5}{3}$ **53.** 3 **55.** Yes; yes **57.** Yes **59.** No **61.** Yes **63.** Yes **65.** range **67.** 3 **69.** 0 **71.** 2 **73.** c **75.** a

77. $f[g(x)] = f(5x + 4) = 3(5x + 4) + 2 = 15x + 14$; $g[f(x)] = g(3x + 2) = 5(3x + 2) + 4 = 15x + 14$; No

79.
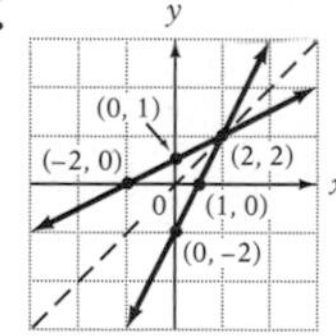

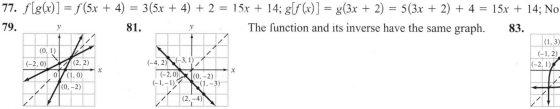

81. The function and its inverse have the same graph. **83.**
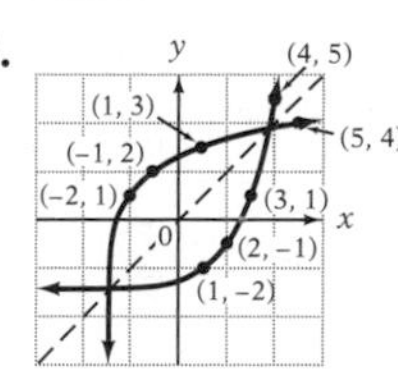

87. $f^{-1}(x) = 0.008233x$ **93.** TEST

CHAPTER 9 REVIEW EXERCISES

1.
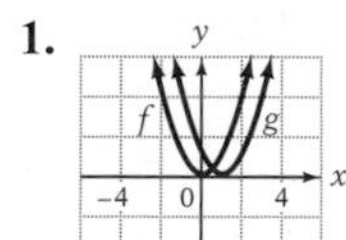
[9.1.1] **2.**
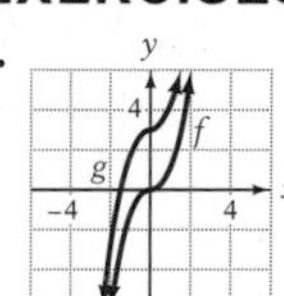
[9.1.1] **3.**
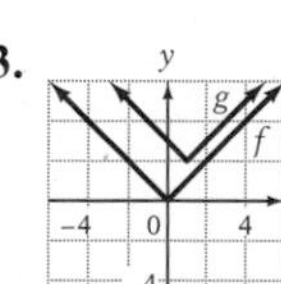
[9.1.1] **4.**
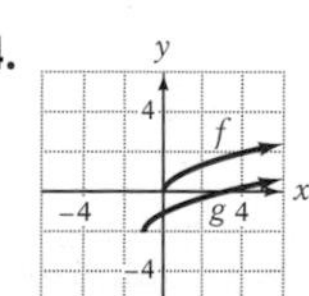
[9.1.1] **5.** 7 [9.2.1]

6. $2x - 1$ [9.2.1] **7.** 70 [9.2.1] **8.** $\frac{12}{7}$ [9.2.1] **9.** 5 [9.2.2] **10.** 10 [9.2.2] **11.** $12x^2 + 12x - 1$ [9.2.2] **12.** $6x^2 + 3x - 16$ [9.2.2] **13.** No [9.3.1] **14.** $\{(1, -2), (3, 2), (-4, 5), (9, 7)\}$ [9.3.2] **15.** No [9.3.1] **16.** Yes [9.3.1] **17.** $f^{-1}(x) = 2x - 16$ [9.3.2] **18.** $f^{-1}(x) = -\frac{1}{6}x + \frac{2}{3}$ [9.3.2] **19.** $f^{-1}(x) = \frac{3}{2}x + 18$ [9.3.2] **20.** Yes [9.3.2]

CHAPTER 9 TEST

1. No [9.3.1, Example 1] **2.** -2 [9.2.1, Example 1] **3.** 234 [9.2.1, Focus On, pages 504–505]
4. $-\frac{13}{2}$ [9.2.1, Example 3] **5.** -5 [9.2.1, Example 1] **6.** 5 [9.2.2, first Focus On, page 508]
7. [9.1.1, Problem 1] **8.** [9.1.1, Example 1] **9.** 47 [9.2.2, Example 4A]

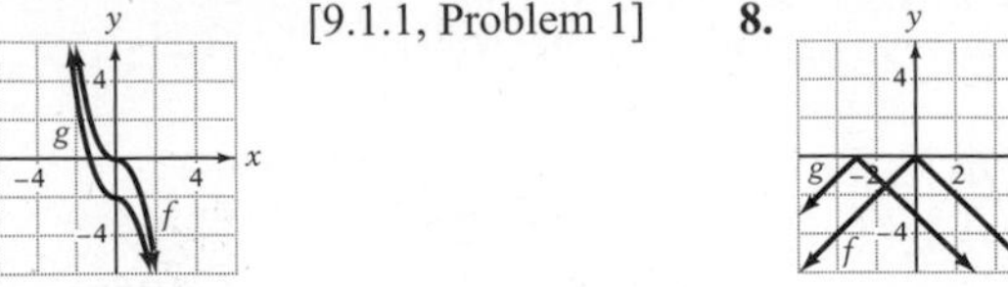

10. [9.1.1, Example 2] **11.** [9.1.1, Example 1]

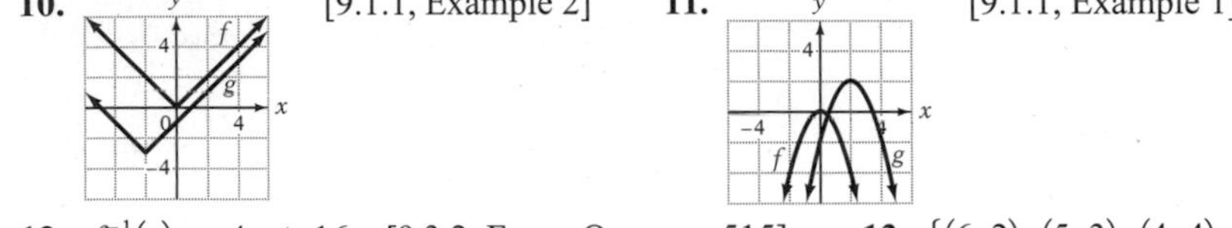

12. $f^{-1}(x) = 4x + 16$ [9.3.2, Focus On, page 515] **13.** $\{(6, 2), (5, 3), (4, 4), (3, 5)\}$ [9.3.2, Focus On, page 514]
14. Yes [9.3.2, Focus On, page 516] **15.** $2x^2 - 4x - 5$ [9.2.2, Example 4B] **16.** $f^{-1}(x) = 2x + 6$ [9.3.2, Example 2]
17. No [9.3.2, Example 3] **18.** No [9.3.1, Example 1]

CUMULATIVE REVIEW EXERCISES

1. $-\frac{23}{4}$ [1.4.2] **2.** –5 –4 –3 –2 –1 0 1 2 3 4 5 [1.1.2] **3.** The solution is 3. [2.1.2]
4. The solution set is $\{x|x < -2\} \cup \{x|x > 3\}$. [2.4.2] **5.**

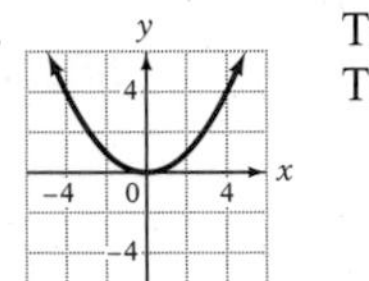

The coordinates of the vertex are $(0, 0)$. The equation of the axis of symmetry is $x = 0$. [8.5.1]

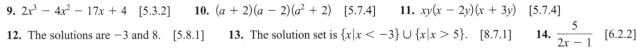

6. [3.7.1] **7.** The solution set is $\{x|x \in \text{real numbers}\}$. [2.5.2] **8.** $-\frac{a^{10}}{12b^4}$ [5.1.2]

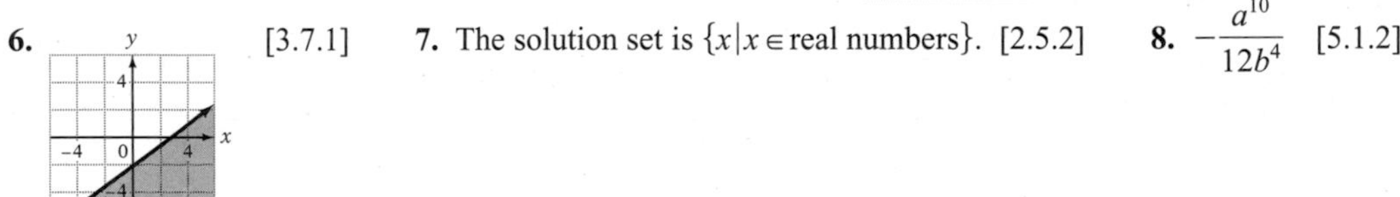

9. $2x^3 - 4x^2 - 17x + 4$ [5.3.2] **10.** $(a + 2)(a - 2)(a^2 + 2)$ [5.7.4] **11.** $xy(x - 2y)(x + 3y)$ [5.7.4]
12. The solutions are -3 and 8. [5.8.1] **13.** The solution set is $\{x|x < -3\} \cup \{x|x > 5\}$. [8.7.1] **14.** $\frac{5}{2x - 1}$ [6.2.2]
15. The solution is -2. [6.4.1] **16.** $-3 - 2i$ [7.5.4] **17.** The equation of the line is $y = -2x - 2$. [3.5.2]
18. The equation of the line is $y = -\frac{3}{2}x - \frac{7}{2}$. [3.6.1] **19.** The solutions are $\frac{1}{2} + \frac{\sqrt{3}}{6}i$ and $\frac{1}{2} - \frac{\sqrt{3}}{6}i$. [8.2.2]
20. The solution is 3. [7.4.1] **21.** The minimum value is -3. [8.6.1] **22.** The range is $\{1, 2, 4, 5\}$. [3.2.1]
23. Yes [3.2.1] **24.** The solution is 2. [7.4.1] **25.** 10 [9.2.2] **26.** $f^{-1}(x) = -\frac{1}{3}x + 3$ [9.3.2]
27. The cost per pound of the mixture is \$3.96. [2.2.1] **28.** 25 lb of the 80% copper alloy must be used. [2.3.2]
29. An additional 4.5 oz of insecticide are required. [6.5.1] **30.** It would take the larger pipe 4 min to fill the tank. [6.4.2]
31. A force of 40 lb will stretch the spring 24 in. [6.5.2] **32.** The average rate of change in the worldwide use of renewable energy sources is 2.36 quadrillion Btu per year. [3.4.3] **33.** The frequency is 80 vibrations per minute. [6.5.2]

Answers to Chapter 10 Selected Exercises

PREP TEST

1. $\frac{1}{9}$ [5.1.2] **2.** 16 [5.1.2] **3.** -3 [5.1.2] **4.** $f(-1) = 0$; $f(3) = 108$ [3.2.1] **5.** The solution is -6. [2.1.1]
6. The solutions are -2 and 8. [8.1.1] **7.** 6326.60 [1.4.2] **8.** [3.2.2]

SECTION 10.1

1. i and ii **3.** (0, 1) **5.** True **7.** No **11.** 1; 0; 1 **13. a.** 9 **b.** 1 **c.** $\frac{1}{9}$ **15. a.** 16 **b.** 4 **c.** $\frac{1}{4}$ **17. a.** 1 **b.** $\frac{1}{8}$ **c.** 16 **19. a.** $\frac{1}{9}$ **b.** 9 **c.** 1 **21. a.** 7.3891 **b.** 0.3679 **c.** 1.2840 **23. a.** 54,5982 **b.** 1 **c.** 0.1353 **25. a.** 16 **b.** 16 **c.** 1.4768 **27. a.** 0.1353 **b.** 0.1353 **c.** 0.0111 **29.** $\left(\frac{1}{2}\right)^{-3}$ **31.** Less than **33.** $-3; 3^3; \frac{1}{27}; -3; \frac{1}{27}$ **35.** **37.** **39.**

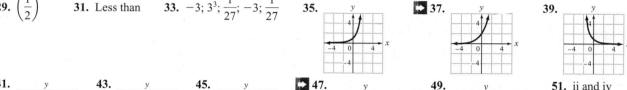

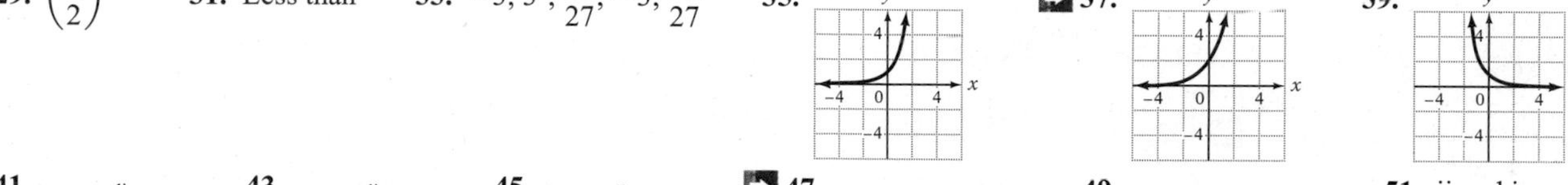

41. **43.** **45.** **47.** **49.** **51.** ii and iv

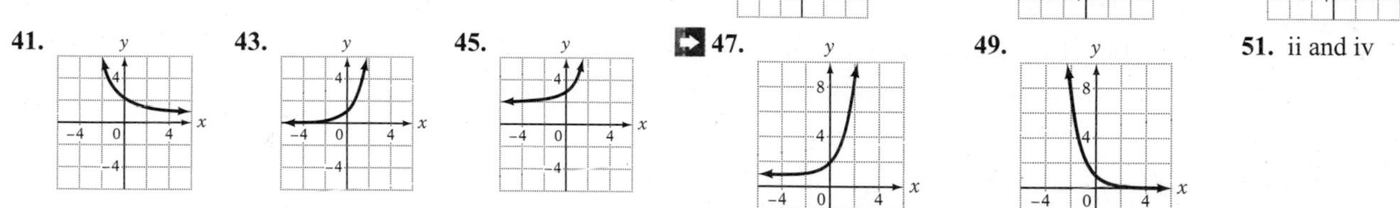

53. The coordinates of the point of intersection are (0, 1). **55.** The graph has no x-intercept. The coordinates of the y-intercept are (0, 1). **57.** Greater than **59.** $g(x) = b^x - a$ **61.** The zero is approximately 1.6.

63. The value of x for which $f(x) = 3$ is 1.1.

65. The investment will be worth $1000 in about 9 years.

67. 50% of the light will reach to a depth of approximately 0.5 m.

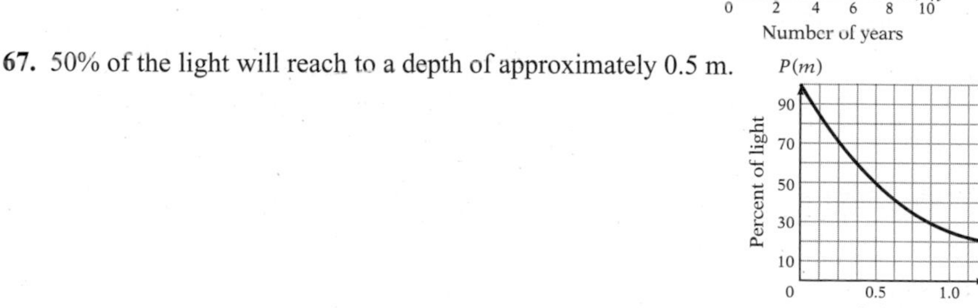

69. **71. a.**

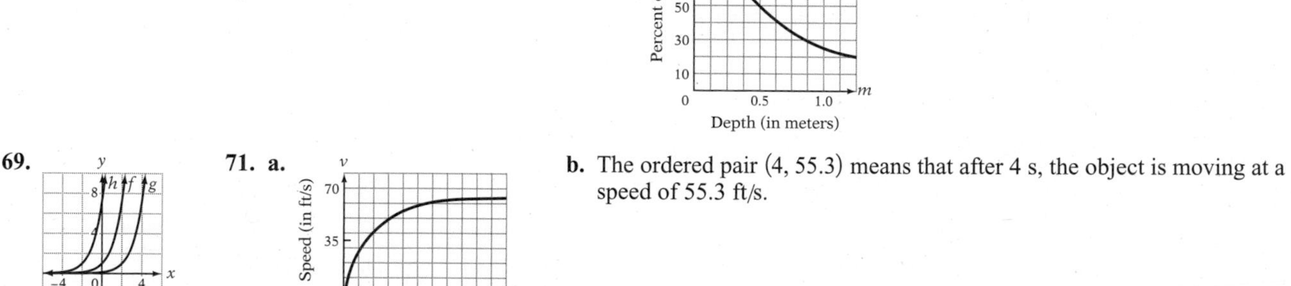

b. The ordered pair (4, 55.3) means that after 4 s, the object is moving at a speed of 55.3 ft/s.

SECTION 10.2

3. False **5.** True **7.** True **9.** True **11.** base b; x **13.** $\log_5 25 = 2$ **15.** $\log_4 \frac{1}{16} = -2$ **17.** $\log x = y$ **19.** $\log_a w = x$ **21.** $3^2 = 9$ **23.** $10^{-2} = 0.01$ **25.** $e^y = x$ **27.** $b^v = u$ **29.** $u = v$ **31.** 4 **33.** 7 **35.** 2 **37.** 3 **39.** 0 **41.** 4 **43.** 9 **45.** 64 **47.** $\frac{1}{7}$ **49.** 1 **51.** 316.23 **53.** 0.02 **55.** 7.39 **57.** 0.61 **59.** Greater than **63.** $r \log_b x$ **65.** $\log_8 x + \log_8 z$ **67.** $5 \log_3 x$ **69.** $\ln r - \ln s$

71. $2\log_3 x + 6\log_3 y$ **73.** $3\log_7 u - 4\log_7 v$ **75.** $2\log_2 r + 2\log_2 s$ **77.** $2\log_9 x + \log_9 y + \log_9 z$
79. $\ln x + 2\ln y - 4\ln z$ **81.** $2\log_8 x - \log_8 y - 2\log_8 z$ **83.** $\frac{1}{2}\log_7 x + \frac{1}{2}\log_7 y$ **85.** $\frac{1}{2}\log_2 x - \frac{1}{2}\log_2 y$
87. $\frac{3}{2}\ln x + \frac{1}{2}\ln y$ **89.** $\frac{3}{2}\log_7 x - \frac{1}{2}\log_7 y$ **91.** $\log_3 \frac{x^3}{y}$ **93.** $\log_8 (x^4y^2)$ **95.** $\ln x^3$ **97.** $\log_5 (x^3y^4)$ **99.** $\log_4 \frac{1}{x^2}$
101. $\log_3 \frac{x^2z^2}{y}$ **103.** $\log_b \frac{x}{y^2z}$ **105.** $\ln (x^2y^2)$ **107.** $\log_6 \sqrt{\frac{x}{y}}$ **109.** $\log_4 \frac{s^2r^2}{t^4}$ **111.** $\log_5 \frac{x}{y^2z^2}$ **113.** $\ln \frac{t^3v^2}{r^2}$
115. $\log_4 \sqrt{\frac{x^3z}{y^2}}$ **117.** False **119.** True **121.** 0.8617 **123.** 2.1133 **125.** −0.6309 **127.** 0.2727
129. 1.6826 **131.** 1.9266 **133.** $\frac{\log(3x-2)}{\log 3}$ **135.** $\frac{5}{\log 9}\log(6x+7)$ **137.** $\frac{\ln(x+5)}{\ln 2}$ **139.** $\frac{\ln(x^2+9)}{\ln 3}$
141. 13.12 **143.** −6.59 **145.** The solution is 256. **147.** The solution is e^e. **149. a.** 2.3219281 **b.** Less
c. Less **d.** 0; Because this system has only one species, there is no diversity in the system.

SECTION 10.3

5. y; exponential; $3^y + 1$; y; x **7.** **9.** **11.** **13.**

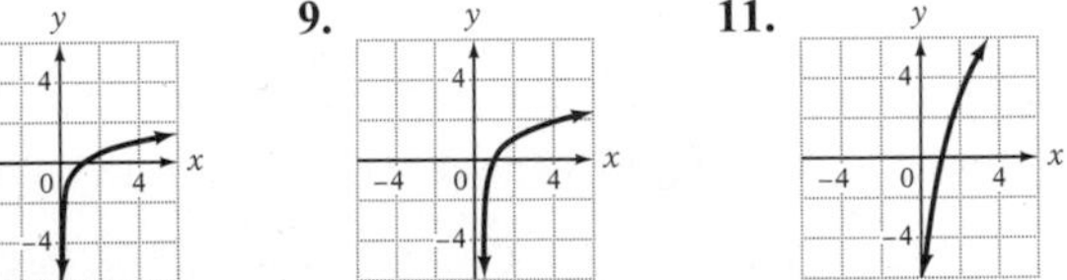

15. **17.** **19.** **21.** **23.**

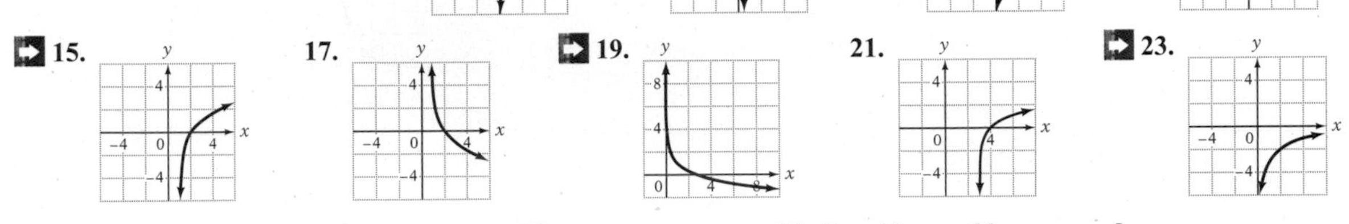

25. **27.** **29.** **31.** ii and iv **33. a.**

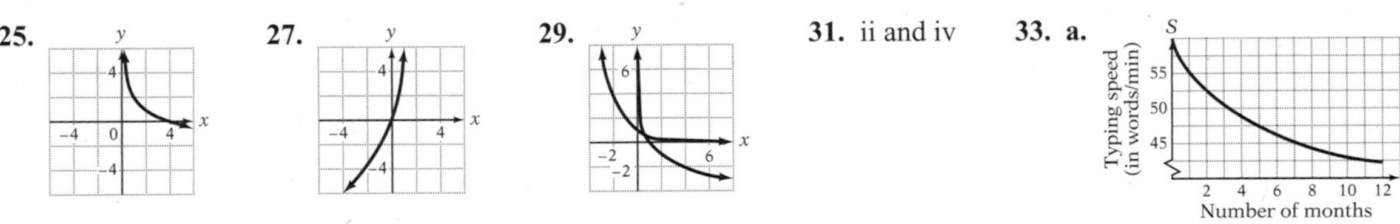

b. The ordered pair (4, 49) means that after 4 months, the typist's proficiency has dropped to 49 words per minute.

35. a.

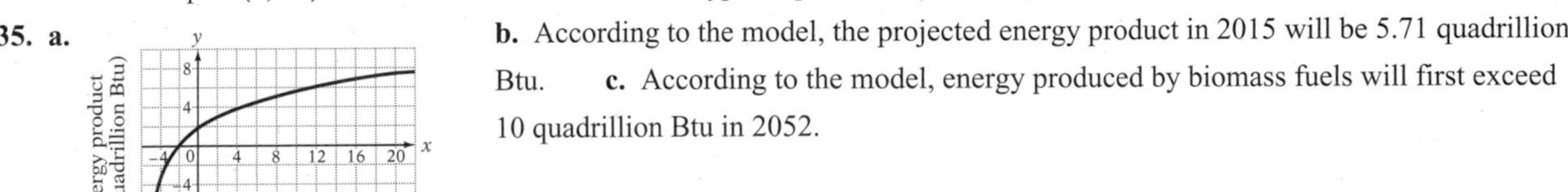

b. According to the model, the projected energy product in 2015 will be 5.71 quadrillion Btu. **c.** According to the model, energy produced by biomass fuels will first exceed 10 quadrillion Btu in 2052.

39. A quadratic function **41.** A linear function **43.** An exponential function **45.** A logarithmic function

SECTION 10.4

5. 3; 2; $4x$; $3x - 1$; $4x$; −1 **7.** The solution is 1. **9.** The solution is −3. **11.** The solution is 1.1133.
13. The solution is 0.7211. **15.** The solution is −1.5850. **17.** The solution is 1.7095. **19.** The solution is 1.3222.
21. The solution is −2.8074. **23.** The solution is 3.5850. **25.** The solution is 1.1309. **27.** The solution is 1.
29. The solution is 6. **31.** The solutions are 1.3754 and −1.3754. **33.** The solution is 1.5805. **35.** The solution is −0.6309.
37. The solution is 1.6610. **39.** The solution is −1.5129. **41.** The solution is 1.5. **43.** i, ii, and iii
45. The solution is 0.63. **47.** The solutions are −1.86 and 3.44. **49.** The solution is −1.16. **51.** exponential form
53. a. Exercise 52 **b.** Exercise 51 **c.** Exercise 51 **d.** Exercise 52 **55.** The solution is 8. **57.** The solution is $\frac{11}{2}$.
59. The solutions are 2 and −4. **61.** The solution is $\frac{5}{3}$. **63.** The solution is $\frac{1}{2}$. **65.** The solution is 1,000,000,000.
67. The solution is 4. **69.** The solution is 3. **71.** The solution is 3. **73.** The solution is 2. **75.** The equation has no solution.
77. The solution is 4. **79.** The solutions are $\frac{5+\sqrt{33}}{2}$ and $\frac{5-\sqrt{33}}{2}$. **81.** The solution is 1.76.
83. The solution is 2.42. **85.** The solutions are −1.51 and 2.10. **87.** The solution is 1.7233. **89.** The solution is 0.8060.

91. The solution is 5.6910. **93.** 2744 **95. a.**

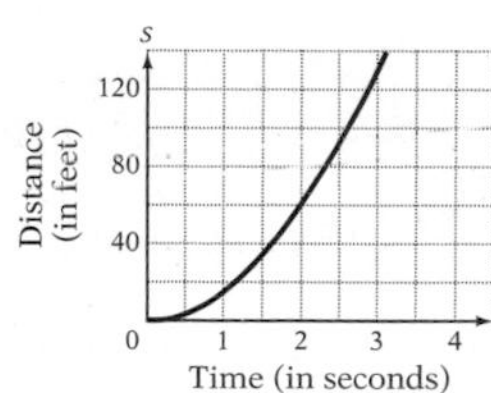

b. It will take the object approximately 2.64 s to fall 100 ft.

SECTION 10.5

3. a. Exponential decay equation **b.** Neither **c.** Exponential growth equation **d.** Exponential growth equation **e.** Neither **f.** Exponential decay equation **5.** 4000; 3000; 0.0025; unknown **7.** The value of the investment after 2 years is \$1172. **9.** The investment will be worth \$15,000 in 18 years. **11. a.** The amount of technetium will be approximately 21.2 mg after 3 h. **b.** It will take about 3.5 h for the amount of technetium to reach 20 mg. **13.** The half-life of promethium-147 is approximately 2.5 years. **15.** The half-life is 2 years. **17. a.** The approximate pressure at 40 km above Earth is 0.098 newton per square centimeter. **b.** The approximate pressure on Earth's surface is 10.13 newtons per square centimeter. **c.** The atmospheric pressure decreases as you rise above Earth's surface. **19.** The pH of milk is 6.4. **21.** At a depth of approximately 2.5 m, the percent of light will be 75% of the light at the surface of the pool. **23.** Normal conversation is 65 decibels. **25.** The mass of the colony would be equal to the mass of Earth in approximately 46 h. This is more time than predicted in the novel. **27.** A thickness of approximately 0.4 cm is needed. **29.** The Richter scale magnitude of the earthquake was 9.5. **31.** At magnitude 7.1, the intensity of the earthquake would be 12,598,000 I_0. At magnitude 6.9, the intensity of the earthquake would be 7,943,000 I_0. **33.** The magnitude of the earthquake is 5.3. **35.** The magnitude of the earthquake is 5.6. **37.** Prices will double in about 14 years. **39.** The value of the investment after 5 years is \$3210.06. **43. a.** About 104 months are required to reduce the loan amount to \$90,000. **b.** Approximately 269 months are required to reduce the loan amount to one-half the original amount. **c.** The total interest paid exceeds \$100,000 in month 163.

CHAPTER 10 REVIEW EXERCISES

1. 2 [10.2.1] **2.** $\log_3 \sqrt{\frac{x}{y}}$ [10.2.2] **3.** 1 [10.1.1] **4.** The solution is -3. [10.4.1] **5.** 1 [10.1.1] **6.** The solution is $\frac{1}{9}$. [10.2.1] **7.** $\log_2 32 = 5$ [10.2.1] **8.** The solution is 6. [10.4.2] **9.** $\frac{1}{2}\log_6 x + \frac{3}{2}\log_6 y$ [10.2.2] **10.** The solution is 2. [10.4.1] **11.** The solution is -2. [10.4.1] **12.** $\frac{1}{3}$ [10.1.1] **13.** 4 [10.2.1] **14.** The solution is 2. [10.4.2] **15.** 2.3219 [10.2.2] **16.** 1.7251 [10.2.2] **17.** The solution is 3. [10.4.1] **18.** 4 [10.1.1] **19.** $\frac{1}{2}\log_5 x - \frac{1}{2}\log_5 y$ [10.2.2] **20.** The solution is $\frac{1}{4}$. [10.4.2] **21.** The solution is 125. [10.2.1] **22.** The solution is $\frac{1}{2}$. [10.4.2] **23.** $\log_b \frac{x^3}{y^5}$ [10.2.2] **24.** 4 [10.1.1] **25.** 2.6801 [10.2.2] **26.** The solution is -0.5350. [10.4.1] **27.**

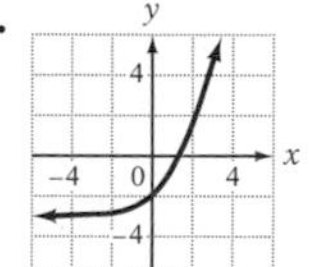

[10.1.2]

28.

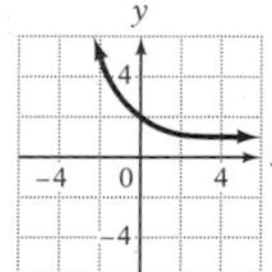

[10.1.2] **29.** [10.3.1] **30.** [10.3.1]

31.

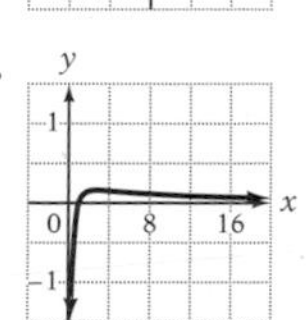

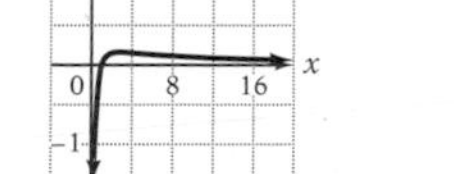

The zero is 1.0. [10.3.1] **32.** The half-life is approximately 33 h. [10.5.1]

33. The material must be approximately 0.602 cm thick. [10.5.1]

CHAPTER 10 TEST

1. 1 [10.1.1, Example 1] **2.** $\frac{1}{64}$ [10.1.1, Example 2] **3.** 3 [10.2.1, Example 3] **4.** The solution is $\frac{1}{16}$. [10.2.1, Example 4] **5.** $\frac{2}{3}\log_6 x + \frac{5}{3}\log_6 y$ [10.2.2, Example 6C] **6.** $\log_5\sqrt{\frac{x}{y}}$ [10.2.2, Example 7C] **7.** The solution is 3. [10.4.2, Focus On, page 561] **8.** $\frac{1}{3}$ [10.1.1, Example 2] **9.** The solution is 2.5789. [10.4.1, Example 2] **10.** The solution is 10. [10.4.2, Focus On, page 561] **11.** The solution is 3. [10.4.1, Problem 1] **12.** The solution is -4. [10.4.1, Example 1] **13.** The solution is $\frac{3}{2}$. [10.4.2, Example 5] **14.**

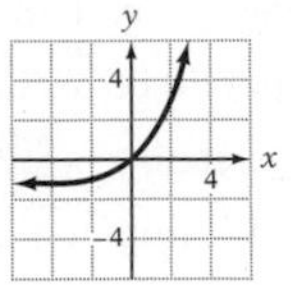

[10.1.2, Example 4]

15.

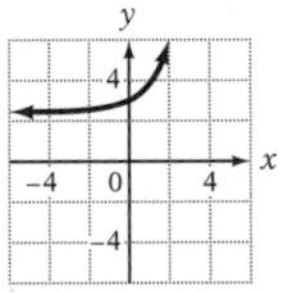

[10.1.2, Problem 4] **16.**

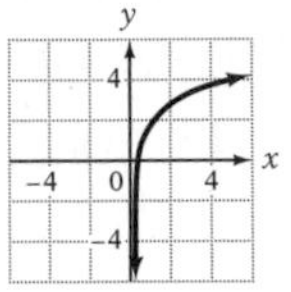

[10.3.1, Problem 1B]

17.

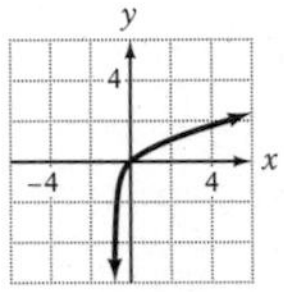

[10.3.1, Problem 1A] **18.**

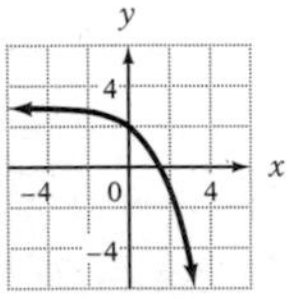

1.6 [10.1.2, Example 6]

19. The value of the investment after 6 years is about \$15,661. [10.5.1, Focus On, page 566]
20. The half-life is 24 h. [10.5.1, Example 1]

CUMULATIVE REVIEW EXERCISES

1. The solution is $\frac{8}{7}$. [2.1.2] **2.** $L = \frac{S - 2WH}{2W + 2H}$ [6.6.1] **3.** The solution set is $\{x \mid 1 \le x \le 4\}$. [2.5.2] **4.** $(4x + 3)(x + 1)$ [5.6.2] **5.** The solution set is $\{x \mid -5 \le x \le 1\}$. [8.7.1] **6.** $\frac{x - 3}{x + 3}$ [6.3.1] **7.** $\frac{x\sqrt{y} + y\sqrt{x}}{x - y}$ [7.2.4] **8.** $-4x^2y^3\sqrt{2x}$ [7.2.2] **9.** $-\frac{1}{5} + \frac{2}{5}i$ [7.5.4] **10.** The equation of the line is $y = 2x - 6$. [3.6.1] **11.** An equation is $3x^2 + 8x - 3 = 0$. [8.1.1] **12.** The solutions are $2 + \sqrt{10}$ and $2 - \sqrt{10}$. [8.2.1, 8.2.2] **13.** The range is $\{-6, -4, 0\}$. [3.2.1] **14.** 4 [9.2.2] **15.** The solution is $(0, -1, 2)$. [4.2.2] **16.** The solution is $\left(-\frac{1}{2}, -2\right)$. [4.1.2] **17.** 243 [10.1.1] **18.** The solution is 64. [10.2.1] **19.** The solution is 8. [10.4.1] **20.** The solution is 1. [10.4.2] **21.**

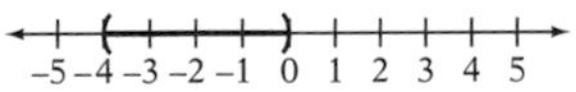

[1.1.2] **22.**

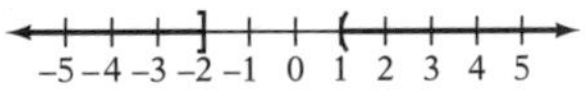

[8.7.1] **23.**

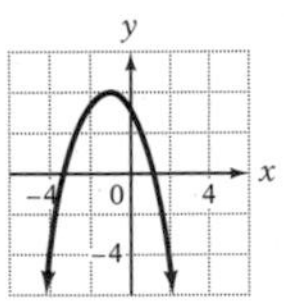

[8.5.1] **24.**

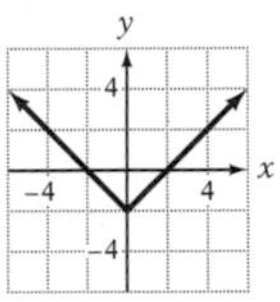

[3.2.2] **25.**

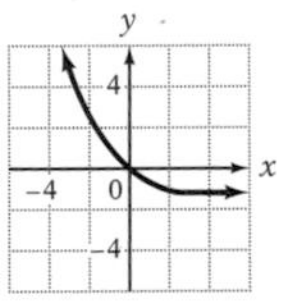

[10.1.2] **26.**

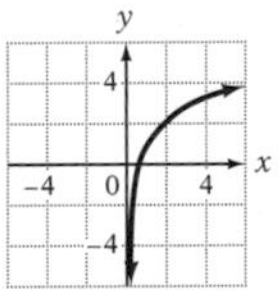

[10.3.1]

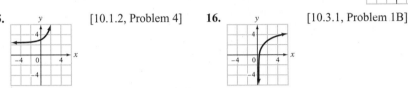

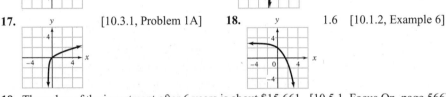

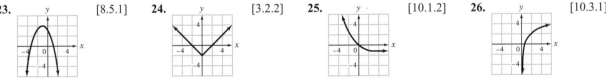

27. 800 lb of the alloy containing 25% tin and 1200 lb of the alloy containing 50% tin were used. [2.3.2] **28.** To earn \$7000 or more a month, the sales executive must make sales amounting to \$75,000 or more. [2.4.3] **29.** The rate of change will be 807 graduates per year. [3.4.3] **30.** With both printers operating, it would take 7.5 min to print the checks. [6.4.2] **31.** When the volume is 25 ft^3, the pressure is 500 lb/in^2. [6.5.2] **32.** The cost per yard of the wool carpet is \$24. [4.4.2] **33.** The value of the investment after 5 years is approximately \$15,657. [10.5.1]

Answers to Chapter 11 Selected Exercises

PREP TEST

1. 12 [1.2.2] **2.** $\frac{3}{4}$ [3.2.1] **3.** 18 [1.3.2] **4.** 96 [1.3.2] **5.** -410 [1.3.2] **6.** $\frac{4}{9}$ [1.3.2] **7.** $x^2 + 2xy + y^2$ [5.3.3] **8.** $x^3 + 3x^2y + 3xy^2 + y^3$ [5.3.2, 5.3.3]

SECTION 11.1

5. 8 **7.** first; second; nth **9.** 2, 3, 4, 5 **11.** 3, 5, 7, 9 **13.** 0, −2, −4, −6 **15.** 2, 4, 8, 16 **17.** 2, 5, 10, 17 **19.** $\frac{1}{2}, \frac{2}{5}, \frac{3}{10}, \frac{4}{17}$ **21.** $0, \frac{3}{2}, \frac{8}{3}, \frac{15}{4}$ **23.** 1, −2, 3, −4 **25.** $\frac{1}{2}, -\frac{1}{5}, \frac{1}{10}, -\frac{1}{17}$ **27.** −2, 4, −8, 16 **29.** $\frac{2}{9}, \frac{2}{27}, \frac{2}{81}, \frac{2}{243}$ **31.** 15 **33.** $\frac{12}{13}$ **35.** 24 **37.** $\frac{32}{243}$ **39.** 88 **41.** $\frac{1}{20}$ **43.** 38 **45.** False **47. a.** 4; $3n + 1$ **b.** 1; 2; 3; 4; 4; 7; 10; 13; 34 **49.** 42 **51.** 28 **53.** 60 **55.** $\frac{25}{24}$ **57.** 28 **59.** $\frac{99}{20}$ **61.** $\frac{137}{120}$ **63.** −2 **65.** $-\frac{7}{12}$ **67.** $\frac{2}{x} + \frac{4}{x} + \frac{6}{x} + \frac{8}{x}$ **69.** $\frac{x}{2} + \frac{x^2}{3} + \frac{x^3}{4} + \frac{x^4}{5}$ **71.** $\frac{x^2}{3} + \frac{x^3}{5} + \frac{x^4}{7}$ **73.** $x + x^3 + x^5 + x^7$ **75.** No **77.** Yes **79.** $a_n = 2n - 1$ **81.** $a_n = -2n + 1$ **83.** $a_n = 4n$ **85.** log 384 **87.** 3 **89.** 1, 2, 3, 5 **91.** $\sum_{i=1}^{n} \frac{1}{i}$ **93.** $a_n = 200 + 0.96a_{n-1}$

SECTION 11.2

3. a. Yes **b.** Yes **c.** No **d.** No **e.** Yes **f.** No **5. a.** 4; common **b.** nth; 50; 3; 4; 199 **7.** 141 **9.** 50 **11.** 71 **13.** $\frac{27}{4}$ **15.** 17 **17.** 3.75 **19.** $a_n = n$ **21.** $a_n = -4n + 10$ **23.** $a_n = \frac{3n + 1}{2}$ **25.** $a_n = -5n - 3$ **27.** $a_n = -10n + 36$ **29.** There are 42 terms in the sequence. **31.** There are 16 terms in the sequence. **33.** There are 20 terms in the sequence. **35.** There are 20 terms in the sequence. **37.** There are 13 terms in the sequence. **39.** There are 20 terms in the sequence. **41.** True **43.** True **45.** 10; $n - 3$; 1; −2; 10; 7; Arithmetic Series; 10; 10; −2; 7; 5; 5; 25 **47.** 650 **49.** −605 **51.** $\frac{215}{4}$ **53.** 420 **55.** −210 **57.** −5 **59.** Negative **61.** 8; 8; 3; 17; 80 **63.** In 9 weeks the person will walk 60 min per day. **65.** There are 2180 seats in the theater. **67.** The salary for the tenth month is \$3550. The total salary for the 10-month period is \$28,750. **69.** Exercise 62 **71.** 8 **73.** −3 **75.** 1800°; $180(n - 2)$ **77. a.** $a_n = 7.66n + 392.19$

SECTION 11.3

3. a. No **b.** Yes **c.** Yes **d.** No **e.** No **f.** Yes **5. a.** 3; common ratio **b.** nth; 7; 2; 3; 6; 729; 1458 **7.** $a_n = 3(4)^{n-1}$ **9.** $a_n = 4\left(-\frac{1}{2}\right)^{n-1}$ **11.** $a_n = 3\left(\frac{3}{4}\right)^{n-1}$ **13.** 131,072 **15.** $\frac{128}{243}$ **17.** 16 **19.** $a_2 = 6, a_3 = 4$ **21.** $a_2 = -2, a_3 = \frac{4}{3}$ **23.** $a_2 = 12, a_3 = -48$ **25.** False **27.** sum; number; first; common ratio **29.** 2186 **31.** $\frac{2343}{64}$ **33.** 62 **35.** $\frac{121}{243}$ **37.** 1364 **39.** 2800 **41.** $\frac{2343}{1024}$ **43.** $\frac{1360}{81}$ **45. a.** No **b.** Yes **47.** −1; 1; 1 **49.** 9 **51.** $\frac{18}{5}$ **53.** $\frac{7}{9}$ **55. a.** Yes **b.** No **57.** $\frac{8}{9}$ **59.** $\frac{2}{9}$ **61.** $\frac{5}{11}$ **63.** $\frac{1}{6}$ **65.** 700; 1400; geometric; 350; 2 **67.** There will be 7.8125 mg of radioactive material in the sample at the beginning of the seventh day. **69.** The ball bounces to a height of approximately 2.6 ft on the fifth bounce. **71.** The value of the land in 15 years will be \$82,103.49. **73.** The value of the house in 30 years will be \$432,194.24. **75.** G; $-\frac{1}{2}$ **77.** A; 9.5 **79.** N; 25 **81.** G; x^2 **83.** A; 4 log x **85.** 27 **87.** The common difference is log 2. **89.** The common ratio is 8. **91. a.** \$80.50 of the loan is repaid in the 27th payment. **b.** The total amount repaid after 20 payments is \$1424.65. **c.** The unpaid balance on the loan after 20 payments is \$3575.35. **93.** A formula for the nth term is $a_n = 8^{n-1}$.

SECTION 11.4

3. n **5.** factorial **7.** 1 **9.** 6 **11.** 40,320 **13.** 1 **15.** 10 **17.** 1 **19.** 84 **21.** 21 **23.** 45 **25.** 1 **27.** 20 **29.** 11 **31.** 6 **33.** False **35.** $x^4 + 4x^3y + 6x^2y^2 + 4xy^3 + y^4$ **37.** $x^5 - 5x^4y + 10x^3y^2 - 10x^2y^3 + 5xy^4 - y^5$ **39.** $16m^4 + 32m^3 + 24m^2 + 8m + 1$ **41.** $32r^5 - 240r^4 + 720r^3 - 1080r^2 + 810r - 243$ **43.** $a^{10} + 10a^9b + 45a^8b^2$ **45.** $a^{11} - 11a^{10}b + 55a^9b^2$ **47.** $256x^8 + 1024x^7y + 1792x^6y^2$ **49.** $65{,}536x^8 - 393{,}216x^7y + 1{,}032{,}192x^6y^2$ **51.** $x^7 + 7x^5 + 21x^3$ **53.** $x^{10} + 15x^8 + 90x^6$ **55.** $-560x^4$ **57.** $-6x^{10}y^2$ **59.** $126y^5$ **61.** $5n^3$ **63.** $\frac{x^5}{32}$ **65.** Negative **67.** n **73.** $-119 + 120i$ **75.** $-117 + 44i$ **77.** $-\frac{1}{8} + \frac{1}{8}i$ **79.** 1

CHAPTER 11 REVIEW EXERCISES

1. $3x + 3x^2 + 3x^3 + 3x^4$ [11.1.2] **2.** 16 [11.2.1] **3.** 32 [11.3.1] **4.** 16 [11.3.3] **5.** 84 [11.4.1] **6.** $\frac{1}{2}$ [11.1.1] **7.** 44 [11.2.1] **8.** 468 [11.2.2] **9.** -66 [11.3.2] **10.** 70 [11.4.1] **11.** $2268x^3y^6$ [11.4.1] **12.** 34 [11.2.2] **13.** $\frac{7}{6}$ [11.1.1] **14.** $a_n = -3n + 15$ [11.2.1] **15.** $\frac{2}{27}$ [11.3.1] **16.** $\frac{8}{15}$ [11.3.3] **17.** -115 [11.2.1] **18.** $\frac{665}{32}$ [11.3.2] **19.** 1575 [11.2.2] **20.** $-280x^4y^3$ [11.4.1] **21.** 21 [11.2.1] **22.** 48 [11.3.1] **23.** 30 [11.2.2] **24.** 341 [11.3.2] **25.** 120 [11.4.1] **26.** $240x^4$ [11.4.1] **27.** 143 [11.2.1] **28.** -575 [11.2.2] **29.** $-\frac{5}{27}$ [11.1.1] **30.** $2 + 2x + 2x^2 + 2x^3$ [11.1.2] **31.** $\frac{23}{99}$ [11.3.3] **32.** $\frac{16}{5}$ [11.3.3] **33.** 726 [11.3.2] **34.** $-42{,}240x^4y^7$ [11.4.1] **35.** 0.996 [11.3.2] **36.** 6 [11.3.3] **37.** $\frac{19}{30}$ [11.3.3] **38.** $x^5 - 15x^4y^2 + 90x^3y^4 - 270x^2y^6 + 405xy^8 - 243y^{10}$ [11.4.1] **39.** 22 [11.2.1] **40.** 99 [11.4.1] **41.** $2x + 2x^2 + \frac{8x^3}{3} + 4x^4 + \frac{32x^5}{5}$ [11.1.2] **42.** $-\frac{13}{60}$ [11.1.2] **43.** The total salary for the 9-month period is \$24,480. [11.2.3] **44.** The temperature is approximately 67.7°F. [11.3.4]

CHAPTER 11 TEST

1. $\frac{1}{3}$ [11.1.1, Example 2] **2.** $16x^4 + 96x^3 + 216x^2 + 216x + 81$ [11.4.1, Example 3] **3.** 32 [11.1.2, Example 3] **4.** $2x^2 + 2x^4 + 2x^6 + 2x^8$ [11.1.2, Example 4] **5.** -120 [11.2.1, Example 1] **6.** $a_n = 2n - 5$ [11.2.1, Example 2] **7.** 22 [11.2.1, Example 3] **8.** 315 [11.2.2, Example 4] **9.** 1560 [11.2.2, Problem 4] **10.** 252 [11.4.1, Example 1] **11.** $a_n = 4\left(\frac{3}{4}\right)^{n-1}$ [11.3.1, Focus On, page 597] **12.** $\frac{81}{125}$ [11.3.1, Problem 1] **13.** $\frac{781}{256}$ [11.3.2, Problem 3] **14.** -55 [11.3.2, Example 3] **15.** 4 [11.3.3, Example 5] **16.** $\frac{7}{30}$ [11.3.3, Example 6] **17.** 330 [11.4.1, Example 2] **18.** $5670x^4y^4$ [11.4.1, Example 6] **19.** The inventory after the October 1 shipment was 2550 yd. [11.2.3, Problem 6] **20.** There will be 20 mg of radioactive material in the sample at the beginning of the fifth day. [11.3.4, Example 7]

CUMULATIVE REVIEW EXERCISES

1. $\frac{x^2 + 5x - 2}{(x + 2)(x - 1)}$ [6.2.2] **2.** $2(x^2 + 2)(x^4 - 2x^2 + 4)$ [5.7.4] **3.** $4y\sqrt{x} - y\sqrt{2}$ [7.2.3] **4.** $\frac{1}{x^{26}}$ [7.1.1] **5.** The solution is 4. [7.4.1] **6.** The solutions are $\frac{1}{4} - \frac{\sqrt{55}}{4}i$ and $\frac{1}{4} + \frac{\sqrt{55}}{4}i$. [8.2.2] **7.** The solution is $\left(\frac{7}{6}, \frac{1}{2}\right)$. [4.2.1] **8.** The solution set is $\{x|x < -2\} \cup \{x|x > 2\}$. [2.4.2] **9.** -10 [4.3.1] **10.** $\frac{1}{2}\log_5 x - \frac{1}{2}\log_5 y$ [10.2.2] **11.** The solution is 3. [10.4.1] **12.** $a_5 = 20; a_6 = 30$ [11.1.1] **13.** 6 [11.1.2] **14.** The solution is (2, 0, 1). [4.2.2] **15.** The solution is 216. [10.4.2] **16.** $2x^2 - x - 1 + \frac{6}{2x + 1}$ [5.4.2] **17.** $-3h + 1$ [3.2.1] **18.** The range is $\left\{-1, 0, \frac{7}{5}\right\}$. [3.2.1] **19.** [3.3.2] **20.** [3.7.1]

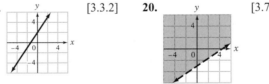

21. The new computer would take 24 min to complete the payroll. The older computer would take 40 min to complete the payroll. [6.4.2] **22.** The rate of the boat in calm water is 6.25 mph. The rate of the current is 1.25 mph. [4.4.1] **23.** The half-life is about 55 days. [10.5.1] **24.** The total number of seats in the theater is 1536. [11.2.3] **25.** The height the ball reaches on the fifth bounce is approximately 3.3 ft. [11.3.4]

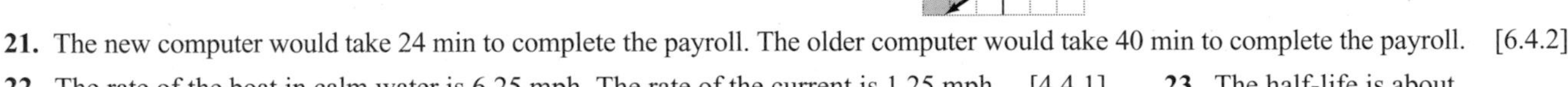

PREP TEST

1. The distance is 7.21. [3.1.1] **2.** $x^2 - 8x + 16; (x - 4)^2$ [8.2.1] **3.** The solution is 0; the solutions are 4 and -4. [8.1.2] **4.** The solution is $(1, -1)$. [4.1.2] **5.** The solution is $(1, -5)$. [4.2.1] **6.** The equation of the axis of symmetry is $x = 2$.

The coordinates of the vertex are $(2, -2)$. [8.5.1] **7.** 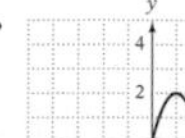 [8.6.1] **8.** 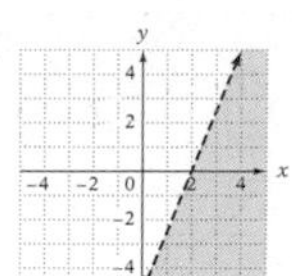 [3.7.1]

9. 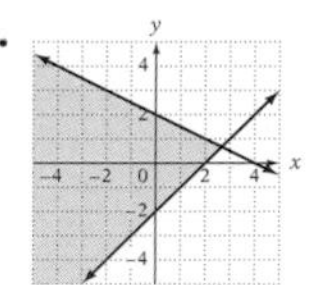 [4.5.1]

SECTION 12.1

1. ii and iv **3. a.** The axis of symmetry is a vertical line. **b.** The parabola opens up. **5. a.** The axis of symmetry is a horizontal line. **b.** The parabola opens right. **7. a.** The axis of symmetry is a horizontal line. **b.** The parabola opens left. **9.** y; positive; negative **11.** The coordinates of the vertex are $(-2, -9)$. The equation of the axis of symmetry is $x = -2$. **13.** The coordinates of the vertex are $(3, 0)$. The equation of the axis of symmetry is $y = 0$. **15.** $(1, 4)$ **17.** $(7, 6)$

19. The coordinates of the vertex are $(1, -5)$. The equation of the axis of symmetry is $x = 1$.

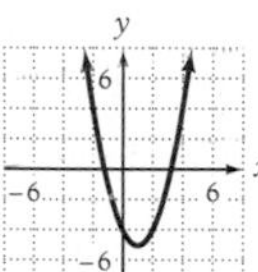

21. The coordinates of the vertex are $(1, -2)$. The equation of the axis of symmetry is $x = 1$.

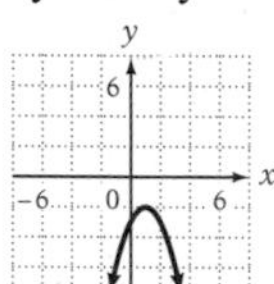

23. The coordinates of the vertex are $(-4, -3)$. The equation of the axis of symmetry is $y = -3$.

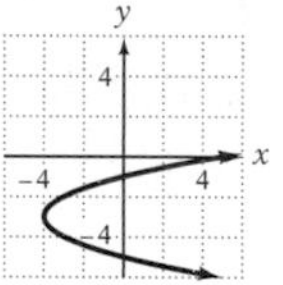

25. The coordinates of the vertex are $\left(-\frac{7}{4}, \frac{3}{2}\right)$. The equation of the axis of symmetry is $y = \frac{3}{2}$.

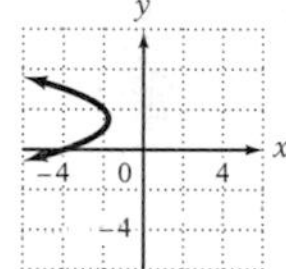

27. The coordinates of the vertex are $(1, -1)$. The equation of the axis of symmetry is $x = 1$.

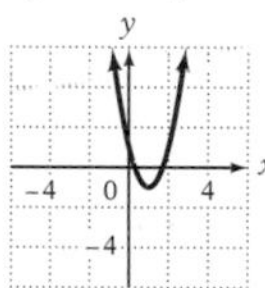

29. The coordinates of the vertex are $\left(\frac{5}{2}, -\frac{9}{4}\right)$. The equation of the axis of symmetry is $x = \frac{5}{2}$.

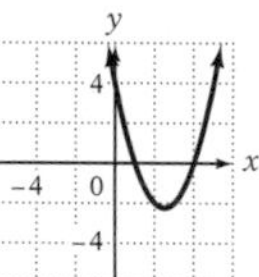

31. The coordinates of the vertex are $(-6, 1)$. The equation of the axis of symmetry is $y = 1$.

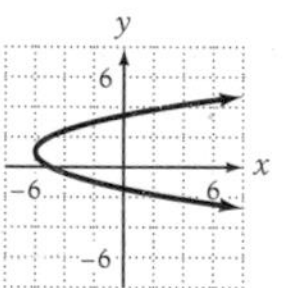

33. The coordinates of the vertex are $\left(-\frac{3}{2}, \frac{27}{4}\right)$. The equation of the axis of symmetry is $x = -\frac{3}{2}$.

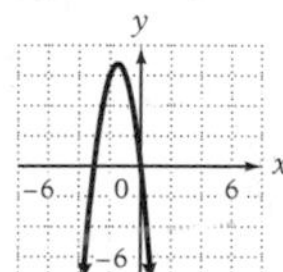

35. The coordinates of the vertex are $(4, 0)$. The equation of the axis of symmetry is $y = 0$.

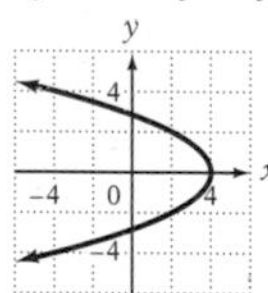

37. The coordinates of the vertex are $\left(\frac{1}{2}, 1\right)$. The equation of the axis of symmetry is $y = 1$.

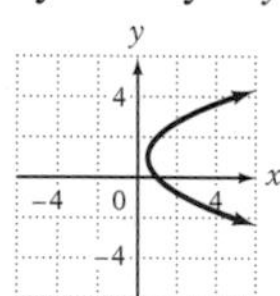

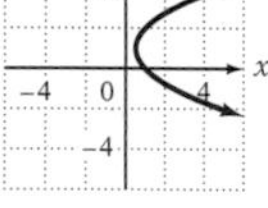

39. The coordinates of the vertex are $(-2, -8)$. The equation of the axis of symmetry is $x = -2$.

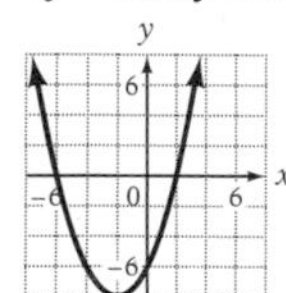

41. $x = ay^2 + by + c$; a is positive; c is negative.

43. $y = ax^2 + bx + c$; a is positive; c is positive. **45.** The coordinates of the focus are $\left(0, \frac{1}{8}\right)$. **47. a.** An equation of the mirror is $x = \frac{1}{2640}y^2$. **b.** The equation is valid over the interval $[0, 3.79]$. **53. a.**

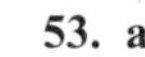

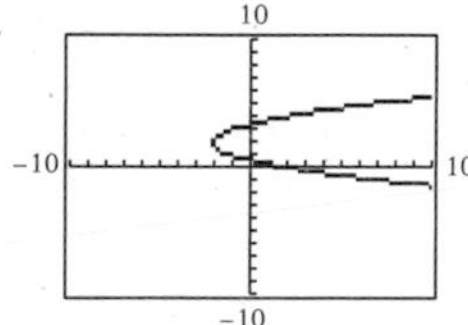

b.

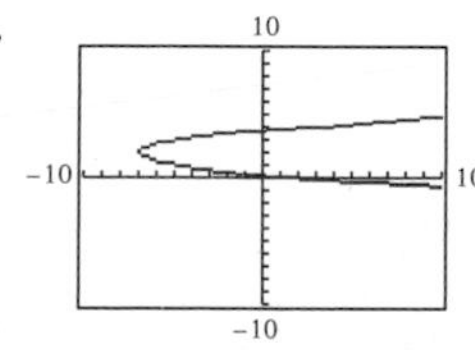

c.

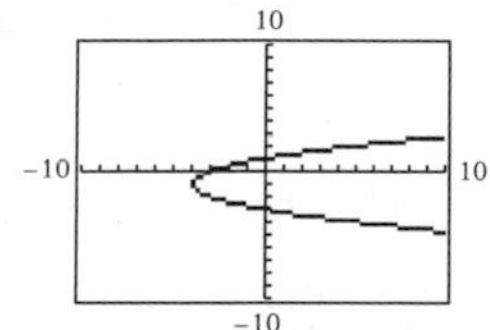

SECTION 12.2

5. 1; −4; 6; 1; −4; 6 **7.** **9.** 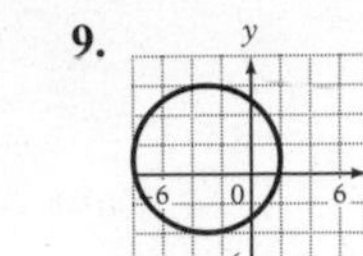 **11.** 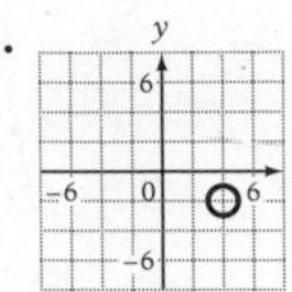 **13.** 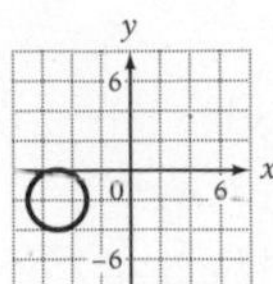 **15.** IV

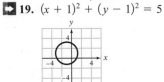

17. $(x - 2)^2 + (y + 1)^2 = 4$

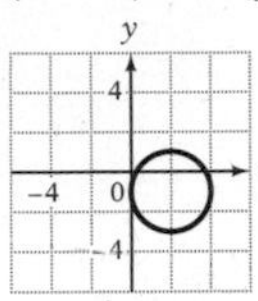

19. $(x + 1)^2 + (y - 1)^2 = 5$

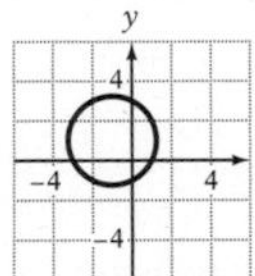

21. $(x + 3)^2 + (y - 6)^2 = 8$

23. $(x + 2)^2 + (y - 1)^2 = 5$ **25. a.** $8x$; $4y$; 5 **b.** 16; 4; 16; 4; 16; 4 **c.** $x + 4$; $y - 2$; 25

27. $(x - 1)^2 + (y + 2)^2 = 25$

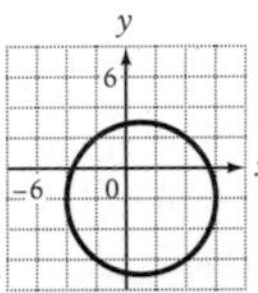

29. $(x + 3)^2 + (y + 4)^2 = 16$

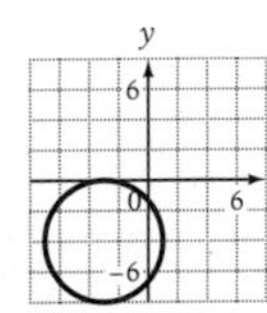

31. $\left(x - \frac{1}{2}\right)^2 + (y + 2)^2 = 1$

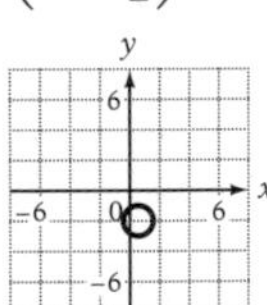

33. $(x - 3)^2 + (y + 2)^2 = 9$

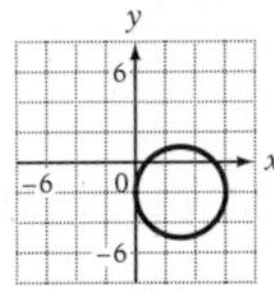

35. Circle B **37.** Circle D **39.** $(x - 3)^2 + y^2 = 9$ **41.** $(x + 1)^2 + (y - 1)^2 = 1$

43. The radius is $6\sqrt{3}$ in. **45.** 10 units

SECTION 12.3

3. a. Neither **b.** Hyperbola **c.** Ellipse **d.** Neither **e.** Hyperbola **f.** Ellipse **5.** ellipse; 0; 0; $(a, 0)$; $(-a, 0)$; $(0, b)$; $(0, -b)$

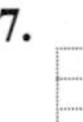

7. 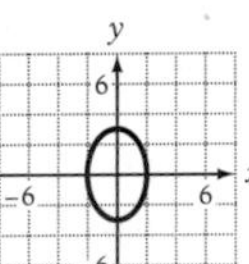 **9.** 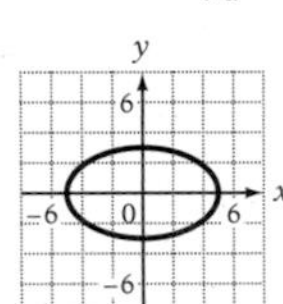 **11.** 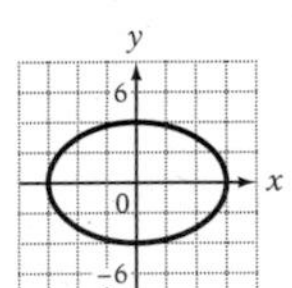 **13.** 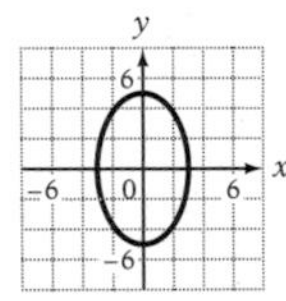

15. 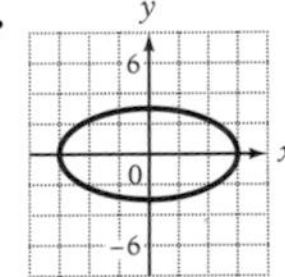 **17.** 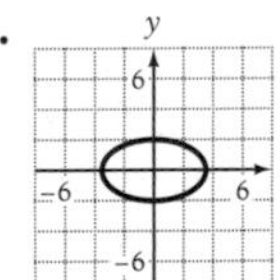 **19.** Less than **23.** 25; 49 **25.** y; $(0, 7)$; $(0, -7)$ **27.** 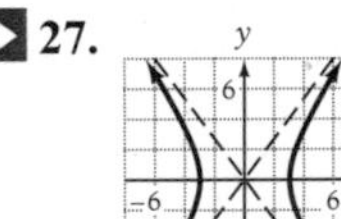

29. 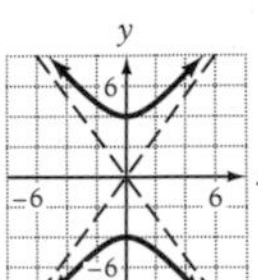 **31.** 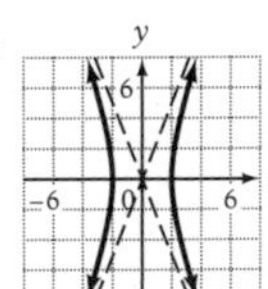 **33.** 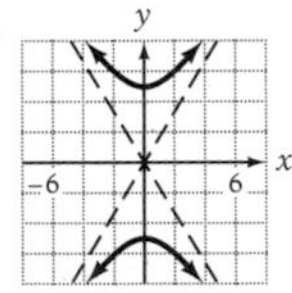 **35.** 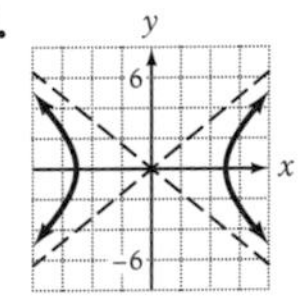 **37.** 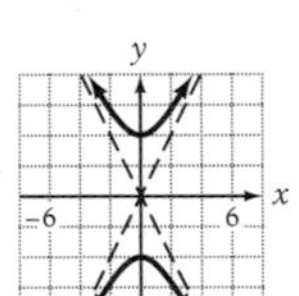 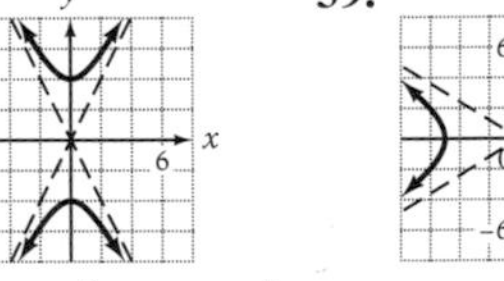 **39.** 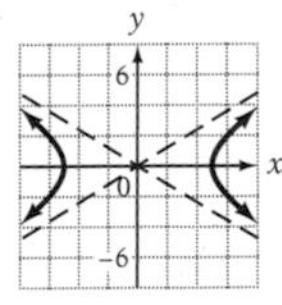

41. 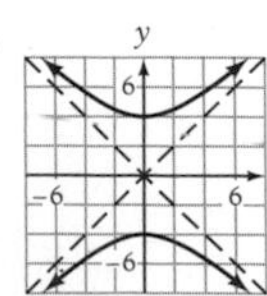 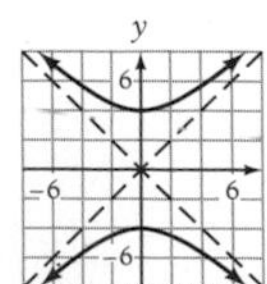 **43.** 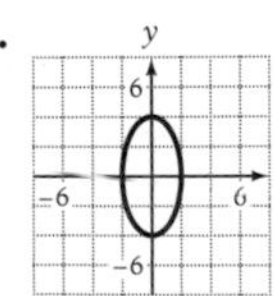 **45.** 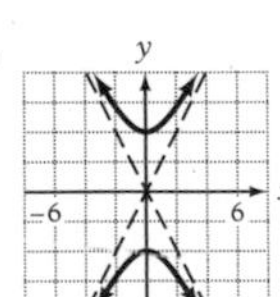 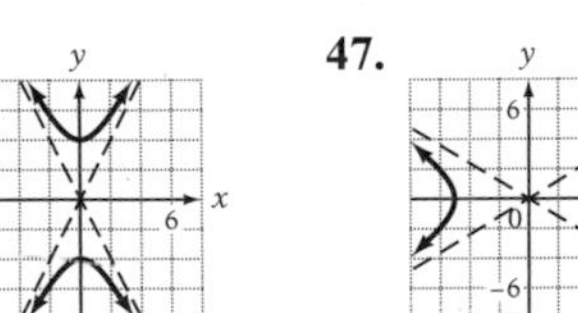 **47.** 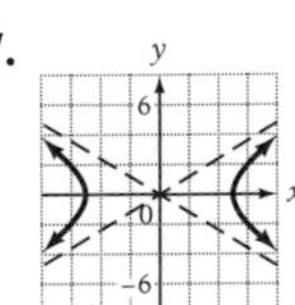 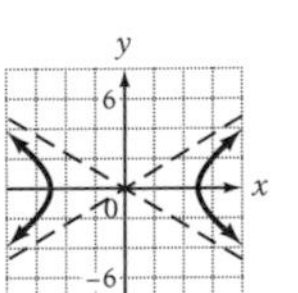

51. a. $\frac{x^2}{33{,}489} + \frac{y^2}{324} = 1$

b. The aphelion is 33,938,000,000 mi from the sun.

c. The perihelion is 85,100,000 mi from the sun.

SECTION 12.4

1. 0, 1, or 2 **3.** 0, 1, 2, 3, or 4 **7.** The solutions are $(-2, 5)$ and $(5, 19)$. **9.** The solutions are $(-1, -2)$ and $(2, 1)$.

11. The solution is $(2, -2)$. **13.** The solutions are $(2, 2)$ and $\left(-\frac{2}{9}, -\frac{22}{9}\right)$. **15.** The solutions are $(3, 2)$ and $(2, 3)$.

17. The solutions are $\left(\frac{\sqrt{3}}{2}, 3\right)$ and $\left(-\frac{\sqrt{3}}{2}, 3\right)$. **19.** The system of equations has no solution.

21. The solutions are (1, 2), (1, −2), (−1, 2), and (−1, −2). **23.** The solutions are (3, 2), (3, −2), (−3, 2), and (−3, −2). **25.** The solutions are (2, 7) and (−2, 11). **27.** The solutions are $(3, \sqrt{2})$, $(3, -\sqrt{2})$, $(-3, \sqrt{2})$, and $(-3, -\sqrt{2})$. **29.** The system of equations has no solution. **31.** The solutions are $(\sqrt{2}, 3)$, $(\sqrt{2}, -3)$, $(-\sqrt{2}, 3)$, and $(-\sqrt{2}, -3)$. **33.** The solutions are (2, −1) and (8, 11). **35.** The solution is (1, 0). **37.** An ellipse and a circle that intersect at two points **39.** A parabola and a straight line that intersect at one point **41.** The solution is (1.000, 2.000). **43.** The solutions are (1.755, 0.811) and (0.505, −0.986). **45.** The solutions are (5.562, −1.562) and (0.013, 3.987).

SECTION 12.5

3. a. parabola **b.** dashed **c.** is **5.**

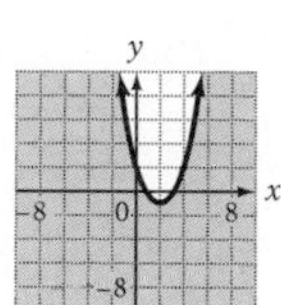

7.

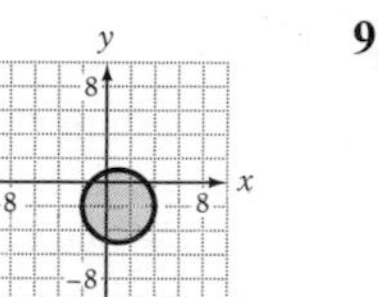

9.

11.

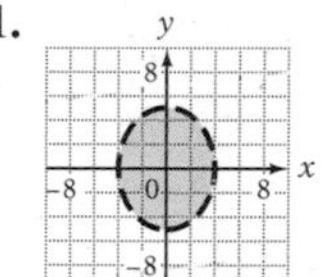

13.

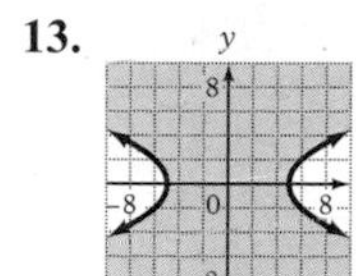

15.

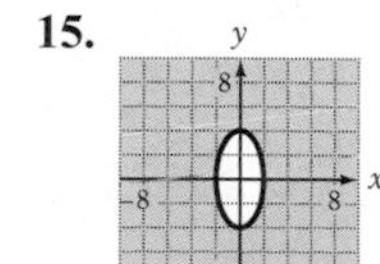

17.

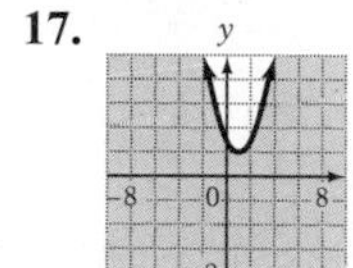

19.

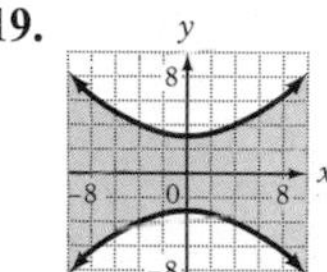

21.

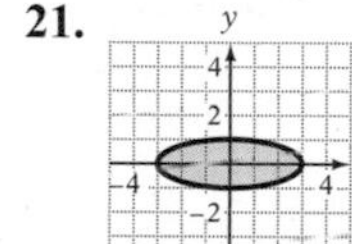

23.

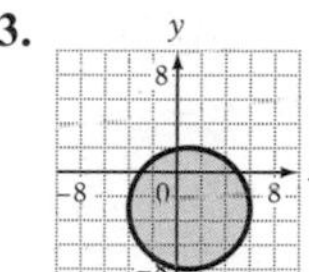

25.

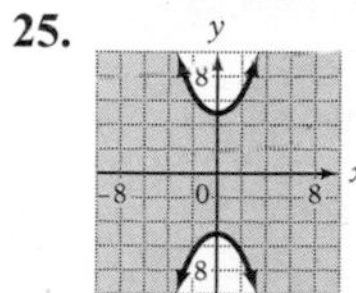

27.

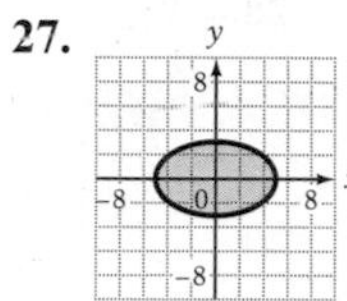

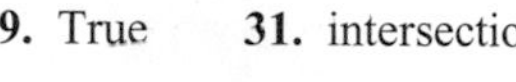

29. True **31.** intersection **33.**

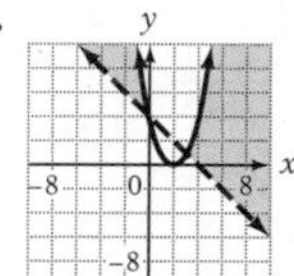

35.

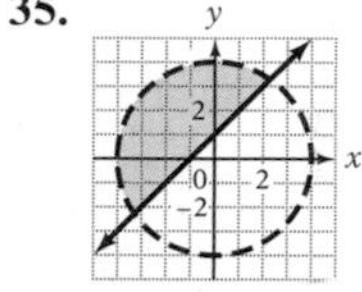

37.

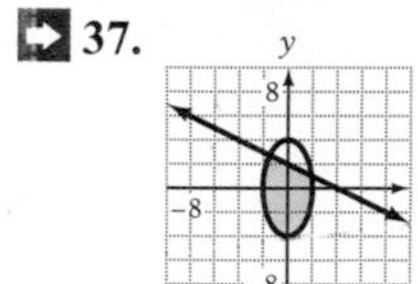

39.

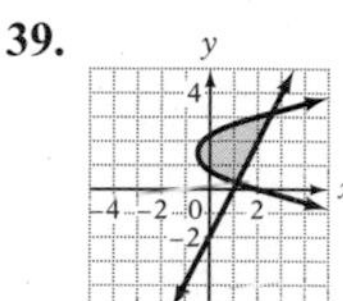

41.

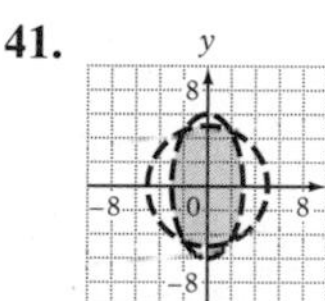

43.

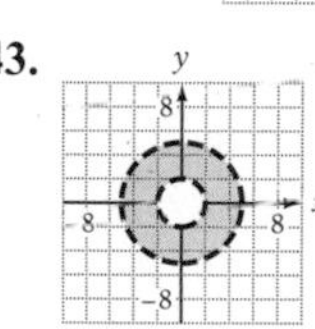

45. The solution set is the region inside the circle $x^2 + y^2 = a$.

47. The solution set is the region between the circles $x^2 + y^2 = a$ and $x^2 + y^2 = b$. **49.**

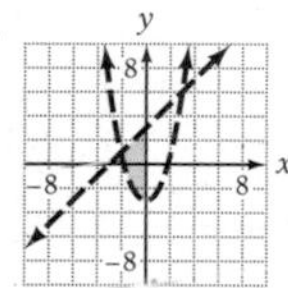

51.

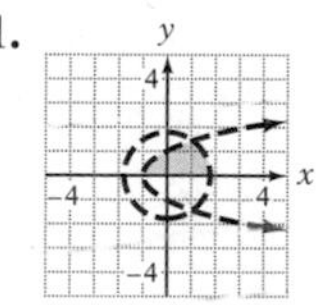

53.

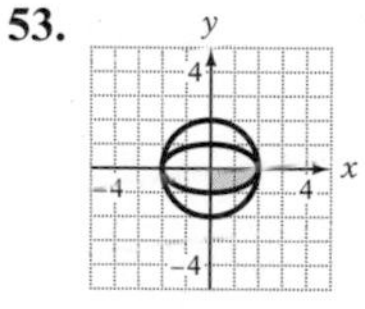

55.

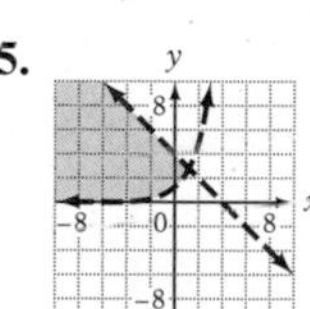

57.

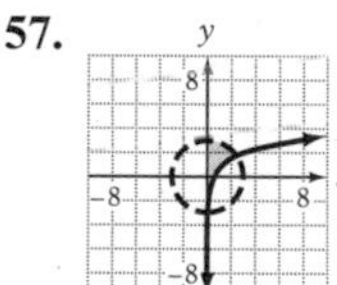

59.

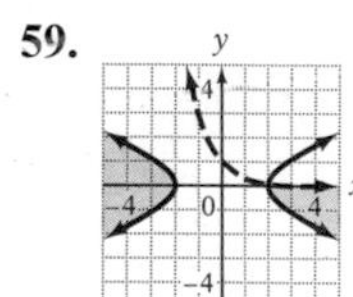

61. a.

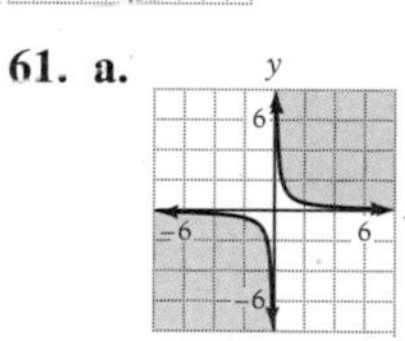

$xy > 1$

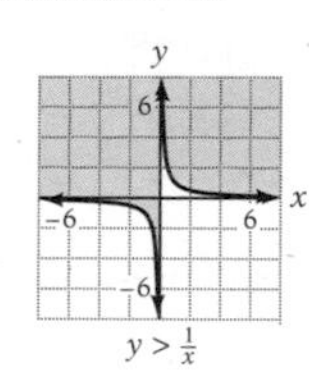

$y > \frac{1}{x}$

CHAPTER 12 REVIEW EXERCISES

1. The coordinates of the vertex are (2, 4). The equation of the axis of symmetry is $x = 2$. [12.1.1]

2. The coordinates of the vertex are $\left(\frac{7}{2}, \frac{17}{4}\right)$. The equation of the axis of symmetry is $x = \frac{7}{2}$. [12.1.1]

3.

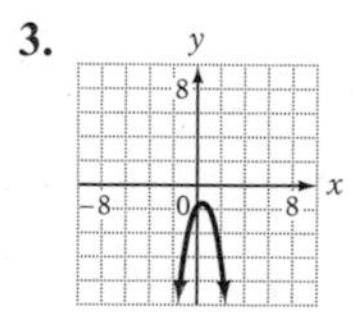

[12.1.1] **4.**

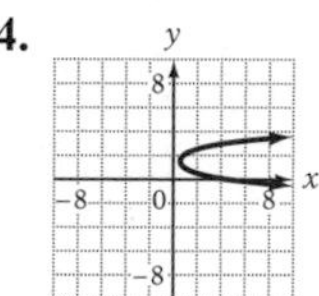

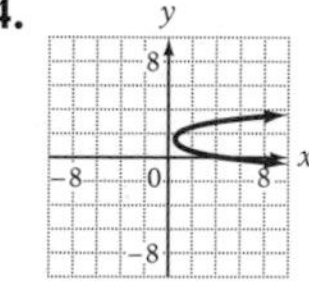

[12.1.1] **5.** $(x + 1)^2 + (y - 2)^2 = 18$ [12.2.1]

6. $(x + 1)^2 + (y - 5)^2 = 36$ [12.2.1] **7.** [12.2.1] **8.**

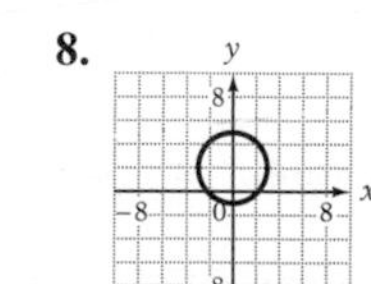

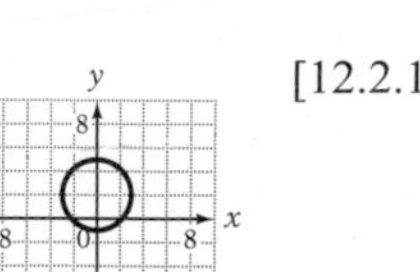

[12.2.1]

9. $x^2 + (y + 3)^2 = 97$ [12.2.1] **10.** $(x + 2)^2 + (y - 1)^2 = 9$ [12.2.2] **11.** [12.3.1]

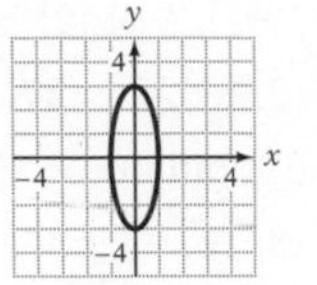

12. [12.3.1] **13.** [12.3.2]

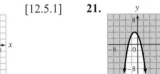

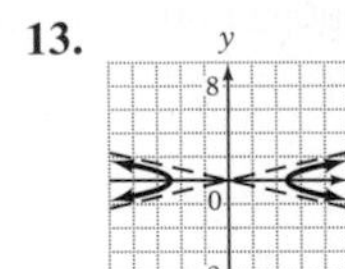

14. [12.3.2]

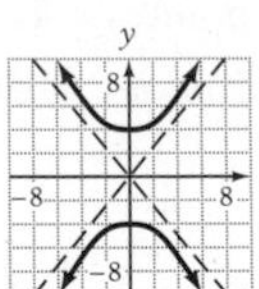

15. The solution is $(-2, -12)$. [12.4.1] **16.** The solutions are $(\sqrt{5}, 3)$, $(-\sqrt{5}, 3)$, $(\sqrt{5}, -3)$, and $(-\sqrt{5}, -3)$. [12.4.1]

17. The solutions are $\left(\frac{2}{9}, \frac{1}{3}\right)$ and $\left(1, \frac{3}{2}\right)$. [12.4.1] **18.** The solutions are $(1, \sqrt{5})$ and $(1, -\sqrt{5})$. [12.4.1]

19. [12.5.1]

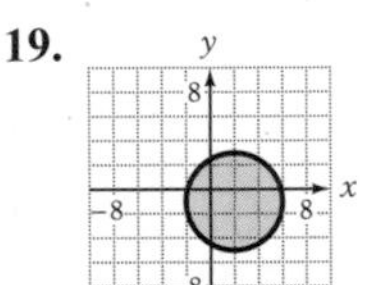

20. [12.5.1]

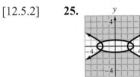

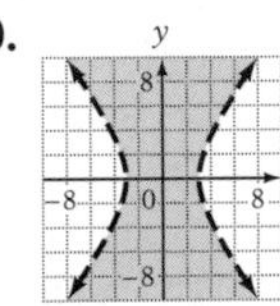

21. [12.5.1]

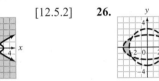

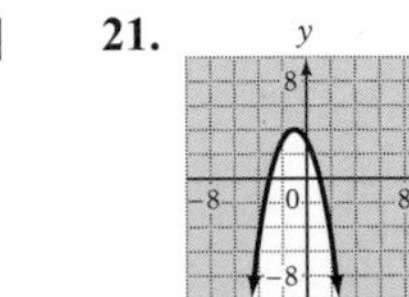

22. [12.5.1]

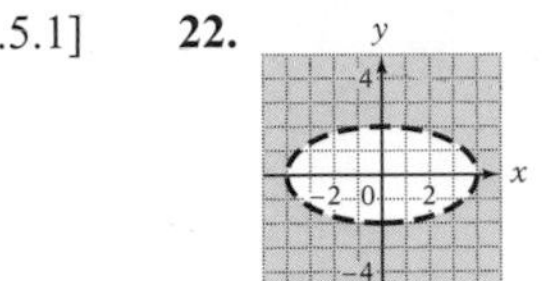

23. [12.5.2]

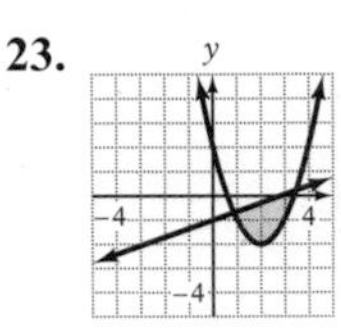

24. [12.5.2]

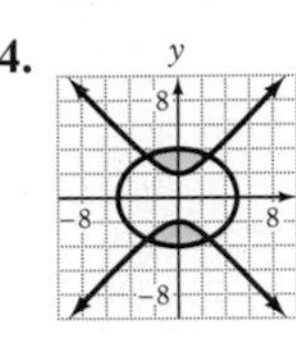

25. [12.5.2]

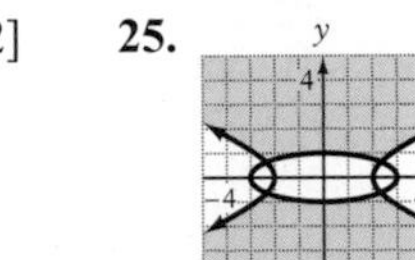

26. [12.5.2]

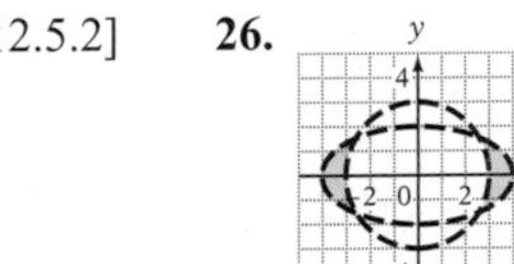

CHAPTER 12 TEST

1. The equation of the axis of symmetry is $x = 3$. [12.1.1, Example 1]

2. The coordinates of the vertex are $\left(\frac{3}{2}, \frac{1}{4}\right)$. [12.1.1, Problem 1] **3.** [12.1.1, Example 1]

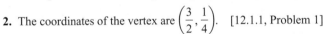

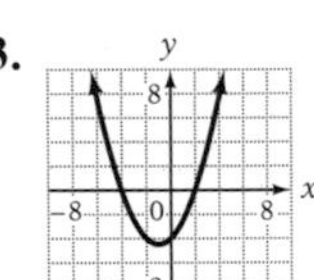

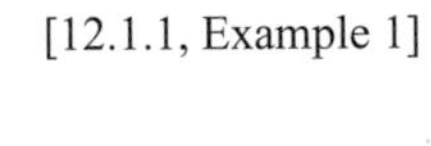

4. [12.1.1, Example 2]

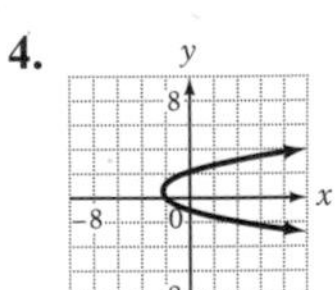

5. $(x + 3)^2 + (y + 3)^2 = 16$ [12.2.1, Focus On, Part B, page 627]

6. The solutions are $(2, 0)$ and $\left(\frac{2}{3}, \frac{4}{3}\right)$. [12.4.1, Focus On, page 640] **7.** The solutions are $(36, -4)$ and $\left(-\frac{9}{4}, \frac{1}{2}\right)$. [12.4.1, Example 1]

8. The solutions are $(5, 1)$, $(-5, 1)$, $(5, -1)$, and $(-5, -1)$. [12.4.1, Focus On, page 641]

9. $(x + 1)^2 + (y + 3)^2 = 58$ [12.2.1, Example 1] **10.** [12.2.1, Example 3]

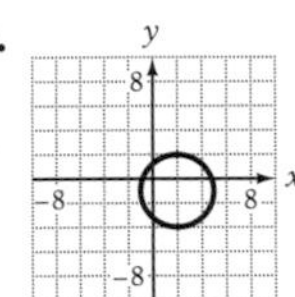

11. $(x + 2)^2 + (y - 4)^2 = 9$ [12.2.1, Focus On, Part B, page 627] **12.** $(x + 2)^2 + (y - 1)^2 = 32$ [12.2.1, Problem 1]

13. $(x - 2)^2 + (y + 1)^2 = 4$ [12.2.2, Problem 3] **14.** [12.3.2, Example 2B]

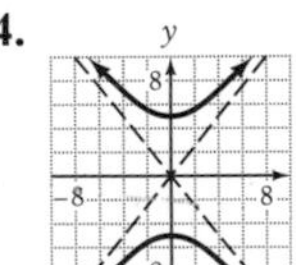

15. [12.3.2, Example 2A] **16.** [12.3.1, Example 1] **17.** [12.5.1, Example 1B]

18. 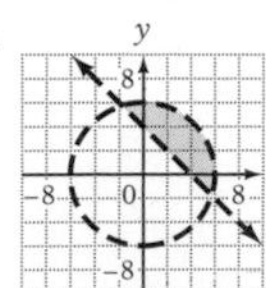 [12.5.2, Example 2A]  **19.** 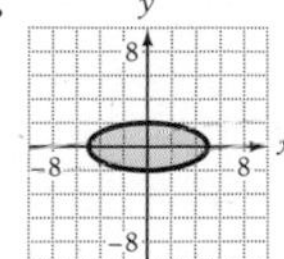 [12.5.1, Problem 1A]

20. 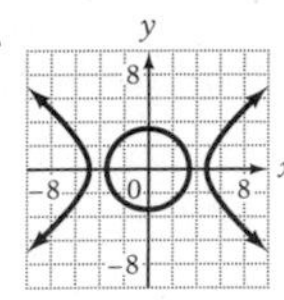 The system of inequalities has no real solution. [12.5.2, Example 2B]

FINAL EXAM

1. -31 [1.2.2] **2.** -1 [1.4.2] **3.** $33 - 10x$ [1.4.3] **4.** The solution is 8. [2.1.1] **5.** The solution is $\frac{2}{3}$. [2.1.2]
6. The solutions are 4 and $-\frac{2}{3}$. [2.5.1] **7.** The solution set is $\{x | -4 < x < -1\}$. [2.5.2]
8. The solution set is $\left\{x \middle| x > \frac{3}{2}\right\}$. [2.4.2] **9.** The equation is $y = -\frac{2}{3}x - \frac{1}{3}$. [3.6.1] **10.** $6a^3 - 5a^2 + 10a$ [5.3.1]
11. $\frac{6}{5} - \frac{3}{5}i$ [7.5.4] **12.** The equation is $2x^2 - 3x - 2 = 0$. [8.1.1] **13.** $(2 - xy)(4 + 2xy + x^2y^2)$ [5.7.2]
14. $(x - y)(1 + x)(1 - x)$ [5.7.4] **15.** $x^2 - 2x - 3 - \frac{5}{2x - 3}$ [5.4.2] **16.** $\frac{x(x - 1)}{2x - 5}$ [6.2.1]
17. $-\frac{10x}{(x + 2)(x - 3)}$ [6.2.2] **18.** $\frac{x + 3}{x + 1}$ [6.3.1] **19.** The solution is $-\frac{7}{4}$. [6.4.1] **20.** $d = \frac{a_n - a_1}{n - 1}$ [6.6.1]
21. $\frac{y^4}{162x^3}$ [5.1.2] **22.** $\frac{1}{64x^8y^5}$ [7.1.1] **23.** $-2x^2y\sqrt{2y}$ [7.2.2] **24.** $\frac{x^2\sqrt{2y}}{2y^2}$ [7.2.4]
25. The solutions are $\frac{3 + \sqrt{17}}{4}$ and $\frac{3 - \sqrt{17}}{4}$. [8.2.2] **26.** The solutions are 27 and -8. [8.3.1]
27. The equation is $y = -3x + 7$. [3.5.2] **28.** The solutions are $\frac{3}{2}$ and -2. [8.3.3] **29.** The solution is $(3, 4)$. [4.2.1]
30. 10 [4.3.1] **31.** The solution is 6. [10.4.2] **32.** $2y + 2y^2 + 2y^3 + 2y^4 + 2y^5$ [11.1.2]
33. $\frac{23}{45}$ [11.3.3] **34.** $144x^7y^2$ [11.4.1] **35.** The solution is $\left(\frac{5}{2}, -\frac{3}{2}\right)$. [12.4.1] **36.** $f^{-1}(x) = \frac{3}{2}x + 6$ [9.3.2]
37. $\log_2 \frac{a^2}{b^2}$ [10.2.2] **38.** x-intercept: $\left(\frac{9}{2}, 0\right)$ [3.3.2] **39.** [3.7.1]
y-intercept: $(0, -3)$

40. 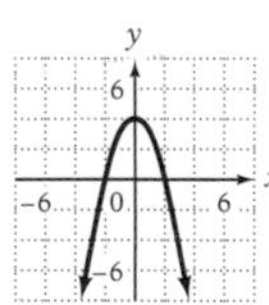 [8.5.1/12.1.1] **41.** 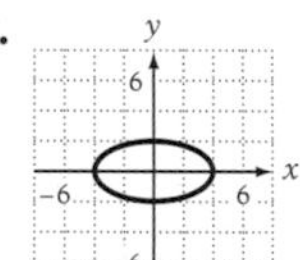 [12.3.1] **42.** 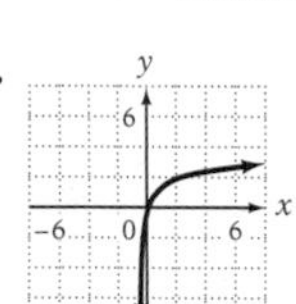 [10.3.1]

43. 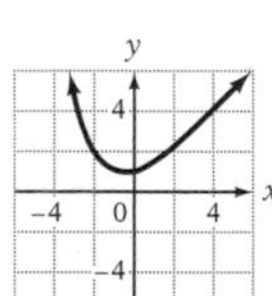 $-2, 1.7$ [10.1.2] **44.** 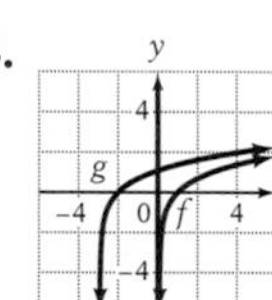 [9.1.1] **45.** The range of scores is 69 or better. [2.4.3]

46. The cyclist traveled 40 mi. [2.2.2] **47.** The amount invested at 8.5% is \$8000. The amount invested at 6.4% is \$4000. [2.3.1] **48.** The width is 7 ft and the length is 20 ft. [8.4.1] **49.** The number of additional shares is 200. [6.5.1] **50.** The rate of the plane was 420 mph. [6.4.3] **51.** The distance the object has fallen is approximately 88 ft. [7.4.2] **52.** The rate of the plane for the first 360 mi was 120 mph. [8.4.1] **53.** The intensity is 200 lumens. [6.5.2] **54.** The rate of the boat in calm water is 12.5 mph. The rate of the current is 2.5 mph. [4.4.1] **55.** The value of the investment is \$4785.65. [10.5.1]
56. The value of the house would be approximately \$577,284. [11.3.4]

Glossary

abscissa The first number in an ordered pair; it measures a horizontal distance and is also called the first coordinate of an ordered pair. (Sec. 3.1)

absolute value equation An equation that contains the absolute value symbol. (Sec. 2.5)

absolute value of a number The distance of the number from zero on the number line. (Sec. 1.1/2.5)

addition method An algebraic method of finding an exact solution of a system of linear equations. (Sec. 4.2)

additive inverse of a polynomial The polynomial with the sign of every term changed. (Sec. 5.2)

additive inverses Numbers that are the same distance from zero on the number line but lie on different sides of zero; also called opposites. (Sec. 1.1/1.4)

analytic geometry Geometry in which a coordinate system is used to study relationships between variables. (Sec. 3.1)

antilogarithm If $\log_b M = N$, then the antilogarithm, base b, of N is M. (Sec. 10.2)

arithmetic progression A sequence in which the difference between any two consecutive terms is constant; also called an arithmetic sequence. (Sec. 11.2)

arithmetic sequence A sequence in which the difference between any two consecutive terms is constant; also called an arithmetic progression. (Sec. 11.2)

arithmetic series The indicated sum of the terms of an arithmetic sequence. (Sec. 11.2)

asymptotes The two straight lines that a hyperbola "approaches." (Sec. 12.3)

average rate of change The average rate of change of y with respect to x is the change in y divided by the change in x. (Sec. 3.4)

axes The two number lines that form a rectangular coordinate system; also called coordinate axes. (Sec. 3.1)

axis of symmetry of a hyperbola A line of symmetry that passes through the vertices of a hyperbola. (Sec. 12.3)

axis of symmetry of a parabola A line of symmetry that passes through the vertex of a parabola. (Sec. 8.5/12.1)

base In an exponential expression, the number that is taken as a factor as many times as indicated by the exponent. (Sec. 1.2)

binomial A polynomial of two terms. (Sec. 5.2)

binomial factor A factor that has two terms. (Sec. 5.5)

center of a circle The central point that is equidistant from all the points that make up a circle. (Sec. 12.2)

center of an ellipse The intersection of the two axes of symmetry of the ellipse. (Sec. 12.3)

center of a hyperbola The point halfway between the two vertices of a hyperbola. (Sec. 12.3)

circle The set of all points (x, y) in the plane that are a fixed distance from a given point (h, k) called the center. (Sec. 12.2)

clear denominators Remove denominators from an equation that contains fractions by multiplying each side of the equation by the least common multiple (LCM) of the denominators. (Sec. 2.1/6.4)

closed interval In interval notation, an interval that contains its endpoints. (Sec. 1.1)

cofactor of an element of a matrix $(-1)^{i+j}$ times the minor of that element of the matrix, where i is the row number of the element and j is its column number. (Sec. 4.3)

combined variation A variation in which two or more types of variation occur at the same time. (Sec. 6.5)

combine like terms Use the Distributive Property to add the coefficients of like variable terms; add like terms of a variable expression. (Sec. 1.4)

common difference of a sequence The difference between any two consecutive terms in an arithmetic sequence. (Sec. 11.2)

common logarithms Logarithms to the base 10. (Sec. 10.2)

common monomial factor A monomial factor that is a factor of the terms in a polynomial. (Sec. 5.5)

common ratio of a sequence In a geometric sequence, the nonzero constant that is used to find each successive term of the sequence. (Sec. 11.3)

complete the square Add to a binomial the constant term that makes it a perfect-square trinomial. (Sec. 8.2)

complex fraction A fraction whose numerator or denominator contains one or more fractions. (Sec. 1.3/6.3)

complex number A number of the form $a + bi$, where a and b are real numbers and $i = \sqrt{-1}$. (Sec. 7.5)

composite number A natural number that is not a prime number. (Sec. 1.1)

composition of functions The operation on two functions f and g denoted by $f \circ g$. The value of the composition of f and g is given by $(f \circ g)(x) = f[g(x)]$. (Sec. 9.2)

compound inequality Two inequalities joined with a connective word such as "and" or "or." (Sec. 2.4)

compound interest Interest that is computed not only on the original principal but also on the interest already earned. (Sec. 10.5)

conditional equation An equation that is true if the variable it contains is replaced by the proper value. $x + 2 = 5$ is a conditional equation. (Sec. 2.1)

conic section A curve that can be constructed from the intersection of a plane and a right circular cone. The four conic sections are the parabola, hyperbola, ellipse, and circle. (Sec. 12.1)

conjugates Binomial expressions that differ only in the sign of a term. The expressions $a + b$ and $a - b$ are conjugates. (Sec. 7.2/7.5)

consecutive even integers Even integers that follow one another in order. (Sec. 2.1)

consecutive integers Integers that follow one another in order. (Sec. 2.1)

consecutive odd integers Odd integers that follow one another in order. (Sec. 2.1)

constant function A function given by $f(x) = b$, where b is a constant. (Sec. 3.3)

constant of proportionality k in a variation equation; also called the constant of variation. (Sec. 6.5)

constant of variation k in a variation equation; also called the constant of proportionality. (Sec. 6.5)

constant term A term that contains no variable part. (Sec. 1.4)

coordinate axes The two number lines that form a rectangular coordinate system; also called simply axes. (Sec. 3.1)

coordinates of a point The numbers in the ordered pair that is associated with the point. (Sec. 3.1)

cube root of a perfect cube One of the three equal factors of the perfect cube. (Sec. 5.7)

cubic function A third-degree polynomial function. (Sec. 5.2)

degree of a monomial The sum of the exponents of the variables in a monomial. (Sec. 5.1)

degree of a polynomial The greatest of the degrees of any of the terms of a polynomial. (Sec. 5.2)

dependent system of equations A system of equations whose graphs coincide. (Sec. 4.1)

dependent variable A variable whose value depends on that of another variable, which is known as the independent variable. (Sec. 3.2)

descending order The terms of a polynomial in one variable arranged so that the exponents of the variable decrease from left to right. (Sec. 5.2)

determinant The number associated with a square matrix. (Sec. 4.3)

difference of two perfect cubes A polynomial of the form $a^3 - b^3$. (Sec. 5.6)

difference of two perfect squares A polynomial of the form $a^2 - b^2$. (Sec. 5.6)

direct variation A special function that can be expressed as the equation $y = kx$, where k is a constant called the constant of variation or the constant of proportionality. (Sec. 6.5)

discriminant For an equation of the form $ax^2 + bx + c = 0$, the quantity $b^2 - 4ac$. (Sec. 8.2)

domain The set of the first coordinates of all the ordered pairs of a function. (Sec. 1.1/3.2)

double root The two equal roots of a quadratic equation. (Sec. 8.1)

element of a matrix A number in a matrix. (Sec. 4.3)

elements of a set The objects in the set. (Sec. 1.1)

ellipse An oval shape that is one of the conic sections. (Sec. 12.3)

empty set The set that contains no elements. (Sec. 1.1)

equation A statement of the equality of two mathematical expressions. (Sec. 2.1)

evaluate a function Determine $f(x)$ for a given value of x. (Sec. 3.2)

evaluate a variable expression Replace the variable in a variable expression by a numerical value and then simplify the resulting expression. (Sec. 1.4)

even integer An integer that is divisible by 2. (Sec. 2.1)

expand by cofactors A technique for finding the value of a 3×3 or larger determinant. (Sec. 4.3)

exponent In an exponential expression, the raised number that indicates how many times the factor, or base, occurs in the multiplication. (Sec. 1.2)

exponential equation An equation in which the variable occurs in the exponent. (Sec. 10.4)

exponential form The expression 2^6 is in exponential form. Compare factored form. (Sec. 1.2)

exponential function The function with base b defined by $f(x) = b^x$, where b is a positive real number not equal to 1. (Sec. 10.1)

extraneous solution When each side of an equation is raised to an even power, the resulting equation may have a solution that is not a solution of the original equation. Such a solution is called an extraneous solution. (Sec. 6.4)

factor a polynomial Write the polynomial as a product of other polynomials. (Sec. 5.5)

factor by grouping Process of grouping and factoring terms in a polynomial in such a way that a common binomial factor is found. (Sec. 5.5)

factored form The multiplication $2 \cdot 2 \cdot 2 \cdot 2 \cdot 2$ is in factored form. Compare exponential form. (Sec. 1.2)

finite sequence A sequence that contains a finite number of terms. (Sec. 11.1)

finite set A set in which all the elements can be listed. (Sec. 1.1)

first coordinate of an ordered pair The first number of the ordered pair; it measures a horizontal distance and is also called the abscissa. (Sec. 3.1)

first-degree equation in one variable An equation that contains only one variable and that variable has an exponent of 1. (Sec. 2.1)

FOIL method A method of finding the product of two binomials in which the sum of the products of the first terms, the outer terms, the inner terms, and the last terms is found. (Sec. 5.3)

formula A literal equation that states a rule about measurement. (Sec. 6.6)

function A relation in which no two ordered pairs that have the same first coordinate have different second coordinates. (Sec. 3.2)

function notation Notation used for those equations that define functions. The letter f is commonly used to name a function. (Sec. 3.2)

general term of a sequence In the sequence $a_1, a_2, a_3, \ldots, a_n, \ldots$, the term a_n. (Sec. 11.1)

geometric progression A sequence in which each successive term of the sequence is the same nonzero constant multiple of the preceding term; also called a geometric sequence. (Sec. 11.3)

geometric sequence A sequence in which each successive term of the sequence is the same nonzero constant multiple of the preceding term; also called a geometric progression. (Sec. 11.3)

geometric series The indicated sum of the terms of a geometric sequence. (Sec. 11.3)

graph of a function A graph of the ordered pairs that belong to the function. (Sec. 3.2/3.3)

graph of an ordered pair The dot drawn at the coordinates of the point in the plane. (Sec. 3.1)

graph of a real number A heavy dot placed directly above the number on the number line. (Sec. 1.1)

graph of $x = a$ A vertical line passing through the point whose coordinates are $(a, 0)$. (Sec. 3.3)

graph of $y = b$ A horizontal line passing through the point whose coordinates are $(0, b)$. (Sec. 3.3)

greatest common factor of two numbers The greatest number that is a factor of the two numbers. (Sec. 1.3)

grouping symbols Parentheses ( ), brackets [ ], braces { }, the absolute value symbol, and the fraction bar. (Sec. 1.2)

half-plane The solution set of a linear inequality in two variables. (Sec. 3.7)

horizontal shift The displacement of a graph to the right or left on the coordinate axes; also known as a horizontal translation. (Sec. 9.1)

horizontal translation The displacement of a graph to the right or left on the coordinate axes; also known as a horizontal shift. (Sec. 9.1)

hyperbola A conic section formed by the intersection of a right circular cone and a plane perpendicular to the base of the cone. (Sec. 12.3)

hypotenuse In a right triangle, the side opposite the 90° angle. (Sec. 3.1/7.4)

i The imaginary unit defined so that $i^2 = -1$. (Sec. 7.5)

identity An equation in which any replacement for the variable will result in a true equation. $x + 2 = x + 2$ is an identity. (Sec. 2.1)

imaginary number A number of the form ai, where a is a real number and $i = \sqrt{-1}$. (Sec. 7.5)

imaginary part of a complex number For the complex number $a + bi$, b is the imaginary part. (Sec. 7.5)

inconsistent system of equations A system of equations that has no solution. (Sec. 4.1)

independent system of equations A system of equations that has exactly one solution. (Sec. 4.1)

independent variable A variable whose value determines that of another variable, which is known as the dependent variable. (Sec. 3.2)

index In the expression $\sqrt[n]{a}$, n is the index of the radical. (Sec. 7.1)

index of summation The variable used in summation notation. (Sec. 11.1)

infinite geometric series The indicated sum of the terms of an infinite geometric sequence. (Sec. 11.3)

infinite sequence A sequence that contains an infinite number of terms. (Sec. 11.1)

infinite set A set in which all the elements cannot be listed. (Sec. 1.1)

integers The numbers $\ldots, -3, -2, -1, 0, 1, 2, 3, \ldots$. (Sec. 1.1)

intersection of two sets The set that contains all elements that are common to both of the sets. (Sec. 1.1)

interval notation A type of set-builder notation in which the property that distinguishes the elements of the set is their location within a specified interval. (Sec. 1.1)

inverse of a function The set of ordered pairs formed by reversing the coordinates of each ordered pair of the function. (Sec. 9.3)

inverse variation A function that can be expressed as the equation $y = k/x$, where k is a constant. (Sec. 6.5)

irrational number A number that cannot be written in the form a/b, where a and b are integers and b is not equal to zero. (Sec. 1.1/7.1)

joint variation A variation in which a variable varies directly as the product of two or more variables. A joint variation can be expressed as the equation $z = kxy$, where k is a constant. (Sec. 6.5)

leading coefficient In a polynomial, the coefficient of the variable with the largest exponent. (Sec. 5.2)

least common multiple (LCM) of two numbers The smallest number that is a multiple of each of the numbers. (Sec. 1.3)

least common multiple (LCM) of two polynomials The polynomial of least degree that contains the factors of each polynomial. (Sec. 6.2)

leg of a right triangle In a right triangle, one of the two sides that are not opposite the 90° angle. (Sec. 3.1/7.4)

like terms Terms of a variable expression that have the same variable part. Having no variable part, constant terms are like terms. (Sec. 1.4)

linear equation in three variables An equation of the form $Ax + By + Cz = D$, where A, B, and C are coefficients of the variables and D is a constant. (Sec. 4.2)

linear equation in two variables An equation of the form $y = mx + b$, where m is the coefficient of x and b is a constant, or of the form $Ax + By = C$. Also called a linear function. (Sec. 3.3)

linear function An equation of the form $y = mx + b$, where m is the coefficient of x and b is a constant; also called a linear equation in two variables. Its graph is a straight line. (Sec. 3.3/5.2)

linear inequality in two variables An inequality of the form $y > mx + b$ or $Ax + By > C$. The symbol $>$ could be replaced by $\geq$, $<$, or $\leq$. (Sec. 3.7)

literal equation An equation that contains more than one variable. (Sec. 3.3/6.6)

logarithm The mathematical operation that is the inverse of exponentiation. The logarithm of a number x in base b is the number y such that $b^y = x$, written $\log_b x = y$. (Sec. 10.2)

lower limit In a tolerance, the lowest acceptable value. (Sec. 2.5)

lowest common denominator The smallest number that is a multiple of each denominator in question. (Sec. 1.3)

main diagonal The elements $a_{11}, a_{22}, a_{33}, \ldots, a_{nm}$ of a matrix. (Sec. 4.3)

main fraction bar The fraction bar that is placed between the numerator and denominator of a complex fraction. (Sec. 1.2)

matrix A rectangular array of numbers. (Sec. 4.3)

maximum value of a function The greatest value that the function can have. (Sec. 8.6)

midpoint of a line segment The point on a line segment that is equidistant from its end points. (Sec. 3.1)

minimum value of a function The least value that the function can have. (Sec. 8.6)

minor of an element The minor of an element in a 3×3 determinant is the 2×2 determinant obtained by eliminating the row and column that contain the element. (Sec. 4.3)

monomial A number, a variable, or a product of a number and variables; a polynomial of one term. (Sec. 5.1/5.2)

multiplicative inverse The multiplicative inverse of a nonzero real number a is $1/a$; also called the reciprocal of a. (Sec. 1.3/1.4)

***n* factorial** The product of the first n natural numbers, written $n!$. (Sec. 11.4)

natural exponential function The function defined by $f(x) = e^x$, where $e \approx 2.71828$. (Sec. 10.1)

natural logarithm Logarithm to the base e (the base of the natural exponential function), abbreviated $\ln x$. (Sec. 10.2)

natural numbers The numbers 1, 2, 3, ...; also called the positive integers. (Sec. 1.1)

negative integers The numbers $\ldots, -3, -2, -1$. (Sec. 1.1)

negative slope The slope of a line that slants downward to the right. (Sec. 3.4)

nonfactorable over the integers A description of a polynomial that does not factor using only integers. (Sec. 5.6)

nonlinear system of equations A system of equations in which one or more of the equations are not linear equations. (Sec. 12.4)

n*th power of *a $a^n = a \cdot a \cdot a \cdots a$, where n is a positive integer and a is a factor n times. (Sec. 1.2)

n*th root of *a If n is a positive integer, then the nth root of a, $a^{1/n}$, is the number whose nth power is a. (Sec. 7.1)

null set A set that contains no elements; also called an empty set. (Sec. 1.1)

numerical coefficient The number part of a variable term. (Sec. 1.4)

odd integer An integer that is not divisible by 2. (Sec. 2.1)

one-to-one function A function in which, given any y, there is only one x that can be paired with the given y. (Sec. 9.3)

open interval In interval notation, an interval that does not contain its endpoints. (Sec. 1.1)

opposites Numbers that are the same distance from zero on the number line but lie on different sides of zero; also called additive inverses. (Sec. 1.1)

order $m \times n$ A matrix of m rows and n columns. (Sec. 4.3)

Order of Operations Agreement Rules that specify the order in which operations are performed in simplifying numerical expressions. (Sec. 1.2)

ordered pair A pair of numbers expressed in the form (a, b) and used to locate a point in a rectangular coordinate system. (Sec. 3.1)

ordered triple Three numbers expressed in the form (x, y, z) and used to locate a point in an xyz-coordinate system. (Sec. 4.2)

ordinate The second number of an ordered pair; it measures a vertical distance and is also called the second coordinate of an ordered pair. (Sec. 3.1)

origin The point of intersection of the two number lines that form a rectangular coordinate system. (Sec. 3.1)

parabola The name given to the graph of a quadratic function. (Sec. 8.5/12.1)

parallel lines Lines that have the same slope and thus do not intersect. (Sec. 3.6)

Pascal's Triangle A triangular array of the numbers for the coefficients of the terms of the expansion of the binomial $(a + b)^n$. (Sec. 11.4)

percent mixture problem A problem that involves combining two solutions or alloys that have different concentrations of the same substance. (Sec. 2.4)

perfect cube The product of the same three factors. (Sec. 5.7)

perfect square The product of a term and itself. (Sec. 5.7)

perfect-square trinomial The square of a binomial. (Sec. 5.7)

perpendicular lines Lines that intersect at right angles. (Sec. 3.6)

plane The infinitely extending, two-dimensional space in which a rectangular coordinate system lies; may be pictured as a large, flat piece of paper. (Sec. 3.1)

polynomial A variable expression in which the terms are monomials. (Sec. 5.2)

positive integers The numbers 1, 2, 3, ...; also called the natural numbers. (Sec. 1.1)

positive slope The slope of a line that slants upward to the right. (Sec. 3.4)

power In an exponential expression, the number of times (indicated by the exponent) that the factor, or base, occurs in the multiplication. (Sec. 1.2)

prime number A number whose only whole-number factors are 1 and itself. For example, 17 is a prime number. (Sec. 1.1)

prime polynomial A polynomial that is nonfactorable over the integers. (Sec. 5.5)

principal square root The positive square root of a number. (Sec. 7.1)

proportion An equation that states the equality of two ratios or rates. (Sec. 6.5)

quadrant One of the four regions into which a rectangular coordinate system divides the plane. (Sec. 3.1)

quadratic equation An equation that can be written in the form $ax^2 + bx + c = 0$, where a and b are coefficients, c is a constant and $a \neq 0$; also called a second-degree equation. (Sec. 5.8/8.1)

quadratic formula The solutions of the quadratic equation $ax^2 + bx + c = 0, a \neq 0$, are $x = \frac{-b \pm \sqrt{b^2 - 4ac}}{2a}$. (Sec. 8.2)

quadratic function A function that can be expressed by the equation $f(x) = ax^2 + bx + c$, where a is not equal to zero. (Sec. 5.2/8.5)

quadratic inequality in one variable An inequality that can be written in the form $ax^2 + bx + c < 0$ or $ax^2 + bx + c > 0$, where a is not equal to zero. The symbols $\leq$ and $\geq$ can also be used. (Sec. 8.7)

quadratic trinomial A trinomial of the form $ax^2 + bx + c$, where a and b are nonzero coefficients and c is a nonzero constant. (Sec. 5.6)

radical In a radical expression, the symbol $\sqrt{}$. (Sec. 7.1)

radical equation An equation that contains a variable expression in a radical. (Sec. 7.4)

radical function A function that contains a variable expression in a radical. (Sec. 6.1)

radicand In a radical expression, the expression within the radical. (Sec. 7.1)

radius of a circle The fixed distance, from the center of a circle, of all points that make up the circle. (Sec. 12.2)

range The set of the second coordinates of all the ordered pairs of a function. (Sec. 3.2)

rate The quotient of two quantities that have different units. (Sec. 6.5)

rate of work The part of a task that is completed in one unit of time. (Sec. 6.4)

ratio The quotient of two quantities that have the same unit. (Sec. 6.5)

rational expression A fraction in which the numerator and denominator are polynomials. (Sec. 6.1)

rational function A function that is written in terms of a rational expression. (Sec. 6.1)

rationalize the denominator The procedure used to remove a radical from the denominator of a fraction. (Sec. 7.2)

rational number A number of the form a/b, where a and b are integers and b is not equal to zero. (Sec. 1.1)

real numbers The rational numbers and the irrational numbers taken together. (Sec. 1.1/7.5)

real part of a complex number For the complex number $a + bi$, a is the real part. (Sec. 7.5)

reciprocal of a fraction The fraction that results when the numerator and denominator of a fraction are interchanged. (Sec. 1.3/1.4)

reciprocal of a rational expression A rational expression in which the numerator and denominator have been interchanged. (Sec. 6.2)

rectangular coordinate system A coordinate system formed by two number lines, one horizontal and one vertical, that intersect at the zero point of each line. (Sec. 3.1)

relation A set of ordered pairs. (Sec. 3.2)

repeating decimal A decimal formed when dividing the numerator of a fraction by its denominator, in which a digit or a sequence of digits in the decimal repeats infinitely. (Sec. 1.1)

right triangle A triangle that contains a 90° angle. (Sec. 3.1)

root(s) of an equation The replacement value(s) of the variable that will make the equation true; also called the solution(s) of the equation. (Sec. 2.1)

roster method A method of designating a set by enclosing a list of its elements in braces. (Sec. 1.1)

scatter diagram A graph of ordered-pair data. (Sec. 3.4)

scientific notation Notation in which each number is expressed as the product of a number between 1 and 10 and a power of 10. (Sec. 5.1)

second coordinate of an ordered pair The second number of the ordered pair; it measures a vertical distance and is also called the ordinate. (Sec. 3.1)

second-degree equation An equation of the form $ax^2 + bx + c = 0$, where a and b are coefficients, c is a constant, and $a \neq 0$; also called a quadratic equation. (Sec. 8.1)

sequence An ordered list of numbers. (Sec. 11.1)

series The indicated sum of the terms of a sequence. (Sec. 11.1)

set A collection of objects. (Sec. 1.1)

set-builder notation A method of designating a set that makes use of a variable and a certain property that only elements of that set possess. (Sec. 1.1)

sigma notation Notation used to represent a series in a compact form; also called summation notation. (Sec. 11.1)

slope A measure of the slant, or tilt, of a line. (Sec. 3.4)

slope-intercept form of a straight line The equation $y = mx + b$, where m is the slope of the line and $(0, b)$ is the y-intercept. (Sec. 3.4)

solution(s) of an equation The replacement value(s) of the variable that will make the equation true; also called the root(s) of the equation. (Sec. 2.1)

solution of an equation in two variables An ordered pair whose coordinates make the equation a true statement. (Sec. 3.1)

solution of a system of equations in three variables An ordered triple that is a solution of each equation of the system. (Sec. 4.2)

solution of a system of equations in two variables An ordered pair that is a solution of each equation of the system. (Sec. 4.1)

solution set of an inequality A set of numbers, each element of which, when substituted for the variable, results in a true inequality. (Sec. 2.4)

solution set of a system of inequalities The intersection of the solution sets of the individual inequalities. (Sec. 4.5)

solve an equation Find a root, or solution, of the equation. (Sec. 2.1)

square matrix A matrix that has the same number of rows as columns. (Sec. 4.3)

square of a binomial A polynomial that can be expressed in the form $(a + b)^2$. (Sec. 5.3)

square root of a perfect square One of the two equal factors of the perfect square. (Sec. 5.7)

standard form of a quadratic equation A quadratic equation written with the polynomial in descending order and equal to zero. $ax^2 + bx + c = 0$ is in standard form. (Sec. 5.8/8.1)

substitution method An algebraic method of finding a solution of a system of linear equations. (Sec. 4.1)

summation notation Notation used to represent a series in a compact form; also called sigma notation. (Sec. 11.1)

synthetic division A shorter method of dividing a polynomial by a binomial of the form $x - a$. This method uses only the coefficients of the variable terms. (Sec. 5.4)

system of equations Two or more equations considered together. (Sec. 4.1)

system of inequalities Two or more inequalities considered together. (Sec. 4.5)

term of a sequence A number in a sequence. (Sec. 11.1)

terminating decimal A decimal formed when dividing the numerator of a fraction by its denominator and the remainder is zero. (Sec. 1.1)

terms of a variable expression The addends of the variable expression. (Sec. 1.4)

tolerance of a component The acceptable amount by which the component may vary from a given measurement. (Sec. 2.5)

trinomial A polynomial of three terms. (Sec. 5.2)

undefined slope The slope of a vertical line is undefined. (Sec. 3.4)

uniform motion The motion of an object whose speed and direction do not change. (Sec. 2.2/6.4)

union of two sets The set that contains all elements that belong to either of the sets. (Sec. 1.1)

upper limit In a tolerance, the greatest acceptable value. (Sec. 2.5)

value mixture problem A problem that involves combining two ingredients that have different prices into a single blend. (Sec. 2.2)

value of a function The value of the dependent variable for a given value of the independent variable. (Sec. 3.2)

variable A letter of the alphabet used to stand for a number that is unknown or that can change. (Sec. 1.1)

variable expression An expression that contains one or more variables. (Sec. 1.4)

variable part In a variable term, the variable or variables and their exponents. (Sec. 1.4)

variable term A term composed of a numerical coefficient and a variable part. When the numerical coefficient is 1 or -1, the 1 is not usually written. (Sec. 1.4)

vertex of a parabola The point on the parabola with the smallest y-coordinate or the largest y-coordinate. (Sec. 8.5/12.1)

vertical shift The displacement of a graph up or down on the coordinate axes; also known as a vertical translation. (Sec. 9.1)

vertical translation The displacement of a graph up or down on the coordinate axes; also known as a vertical shift. (Sec. 9.1)

whole numbers The numbers 0, 1, 2, 3,.... The whole numbers include the natural numbers and zero. (Sec. 1.1)

***x*-coordinate** The abscissa in an ordered pair. (Sec. 3.1)

***x*-intercept** The point at which a graph crosses the x-axis. (Sec. 3.3)

***xy*-coordinate system** A rectangular coordinate system in which the horizontal axis is labeled x and the vertical axis is labeled y. (Sec. 3.1)

***xyz*-coordinate system** A three-dimensional coordinate system formed when a third coordinate axis (the z-axis) is located perpendicular to the xy-plane. (Sec. 4.2)

***y*-coordinate** The ordinate in an ordered pair. (Sec. 3.1)

***y*-intercept** The point at which a graph crosses the y-axis. (Sec. 3.3)

zero of a function A value of x for which $f(x) = 0$. (Sec. 3.3)

zero slope The slope of a horizontal line. (Sec. 3.4)

Index

Index of Applications

Area

in^2	square inc
ft^2	square fee

Area

cm^2	square centimeters
m^2	square meters

1 g)
g)
g)

g)
00 g)
00 g)

gree (for angles and temperature)
he principal square root of a
the absolute value of a

Slope-intercept Form of a Straight Line

$y = mx + b$

Quadratic Formula

$$= \frac{-b \pm \sqrt{b^2 - 4ac}}{2a}$$

ean Theorem

$a^2 + b^2 = c^2$

rea of a Square

$P = 4s$
$A = s^2$

$A = \frac{1}{2}h(b_1 + b_2)$

of a Cube

$= s^3$
$= 6s^2$

Table of Measurement Abbreviations

U.S. Customary System

Length		Capacity		Weight	
in.	inches	oz	fluid ounces	oz	ounces
ft	feet	c	cups	lb	pounds
yd	yards	qt	quarts		
mi	miles	gal	gallons		

Metric System

Length		Capacity		Weight/Mass	
mm	millimeters (0.001 m)	ml	milliliters (0.001 L)	mg	milligrams (0.00
cm	centimeters (0.01 m)	cl	centiliters (0.01 L)	cg	centigrams (0.0
dm	decimeters (0.1 m)	dl	deciliters (0.1 L)	dg	decigrams (0.1
m	meters	L	liters	g	grams
dam	decameters (10 m)	dal	decaliters (10 L)	dag	decagrams (1
hm	hectometers (100 m)	hl	hectoliters (100 L)	hg	hectograms (
km	kilometers (1000 m)	kl	kiloliters (1000 L)	kg	kilograms (1

Time

h	hours	min	minutes	s	seconds

Table of Symbols

$+$	add
$-$	subtract
$\cdot$, $(a)(b)$	multiply
$\frac{a}{b}$, $\div$	divide
( )	parentheses, a grouping symbol
[ ]	brackets, a grouping symbol
π	pi, a number approximately equal to $\frac{22}{7}$ or 3.14
$-a$	the opposite, or additive inverse, of a
$\frac{1}{a}$	the reciprocal, or multiplicative inverse, of a
$=$	is equal to
$\approx$	is approximately equal to
$\neq$	is not equal to
$<$	is less than
$\leq$	is less than or equal to
$>$	is greater than
$\geq$	is greater than or equal to
(a, b)	an ordered pair whose first component is a and whose second component is b
$^\circ$	d
$\sqrt{a}$	
$\lvert a \rvert$	

TI-30X IIS

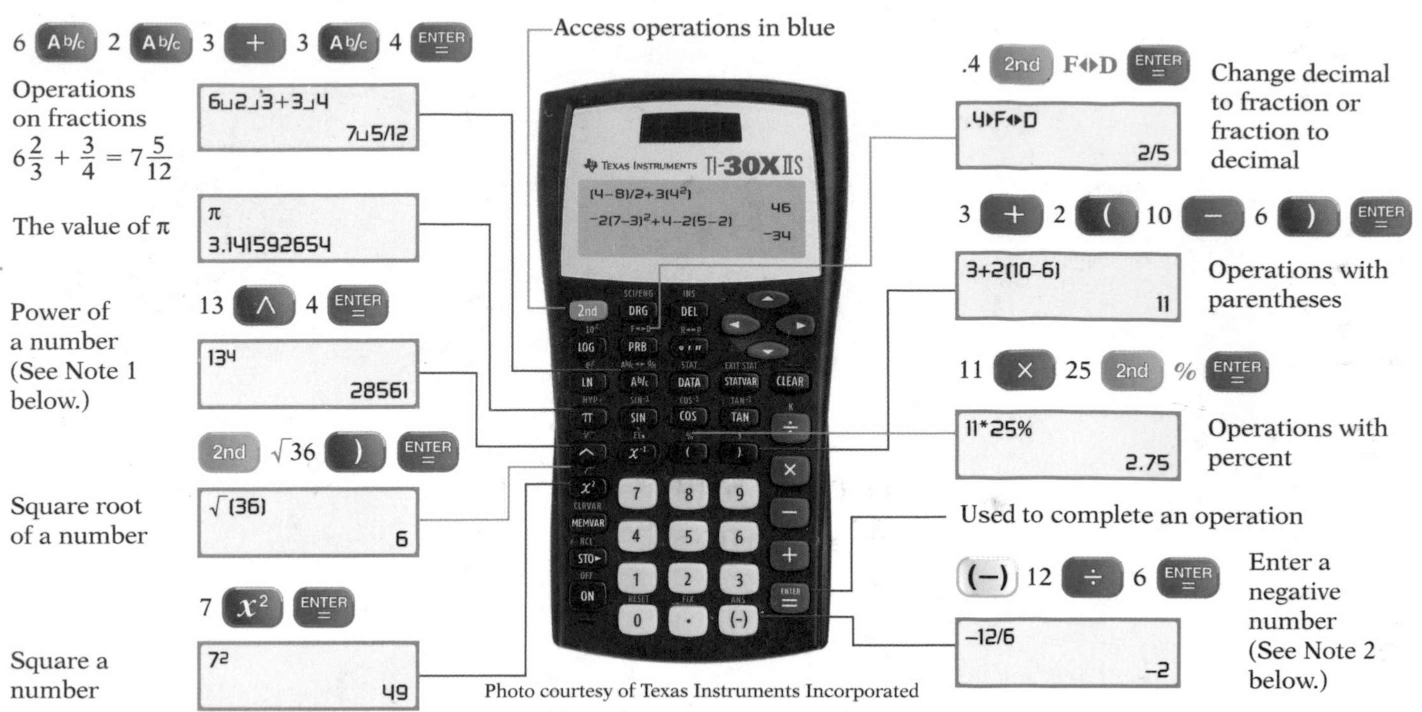

Photo courtesy of Texas Instruments Incorporated

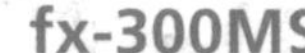

fx-300MS

Access operations in gold

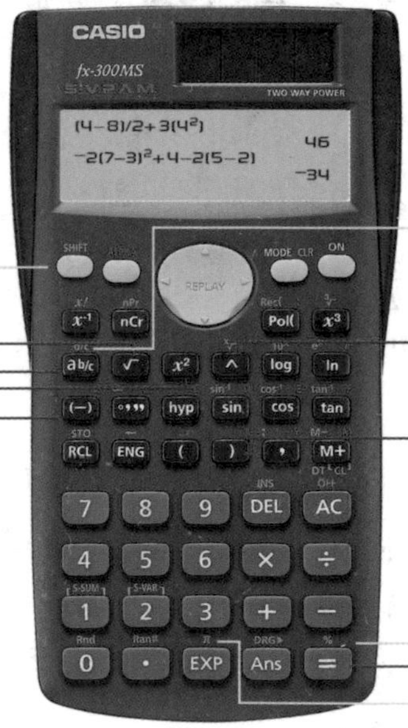

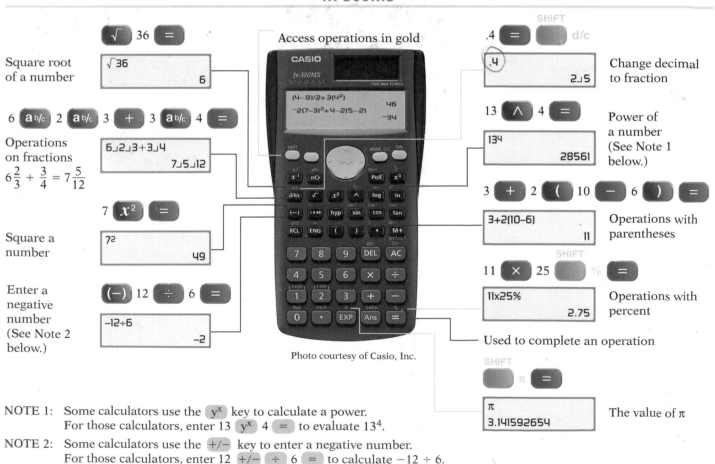

√ 36 =

Square root of a number

√36
6

6 a b/c 2 a b/c 3 + 3 a b/c 4 =

Operations on fractions $6\frac{2}{3} + \frac{3}{4} = 7\frac{5}{12}$

6⌟2⌟3+3⌟4
7⌟5⌟12

7 x^2 =

Square a number

7²
49

Enter a negative number (See Note 2 below.)

(−) 12 ÷ 6 =

−12÷6
−2

Photo courtesy of Casio, Inc.

.4 = SHIFT d/c

Change decimal to fraction

.4
2⌟5

13 ^ 4 =

Power of a number (See Note 1 below.)

13^4
28561

3 + 2 (10 − 6) =

Operations with parentheses

3+2(10−6)
11

11 × 25 SHIFT % =

Operations with percent

11x25%
2.75

Used to complete an operation

SHIFT π =

The value of π

π
3.141592654

NOTE 1: Some calculators use the y^x key to calculate a power. For those calculators, enter 13 y^x 4 = to evaluate 13^4.

NOTE 2: Some calculators use the +/− key to enter a negative number. For those calculators, enter 12 +/− ÷ 6 = to calculate $-12 \div 6$.